AF770322

Proceedings of the North Pacific Symposium on Invertebrate Stock Assessment and Management

NRC Monograph Publishing Program

Editor: R.H. Haynes, OC, FRSC (York University)

Editorial Board: W.G.E. Caldwell, FRSC (University of Western Ontario); P.B. Cavers (University of Western Ontario); G. Herzberg, CC, FRS, FRSC (NRC, Steacie Institute of Molecular Sciences); K.U. Ingold, OC, FRS, FRSC (NRC, Steacie Institute for Molecular Sciences); W. Kaufmann (Editor-in-Chief Emeritus, Annual Reviews Inc., Palo Alto, CA); M. Lecours (Université Laval); W.H. Kewis (Washington University); L.P. Milligan, FRSC (University of Guelph); G.G.E. Scudder, FRSC (University of British Columbia); E.W. Taylor, FRS (University of Chicago); B.P. Dancik, *Editor-in-Chief, NRC Research Press* (University of Alberta)

Inquiries: Monograph Publishing Program, NRC Research Press, National Research Council of Canada, Ottawa, Ontario K1A 0R6, Canada

Cover: The cover design represents a female Dungeness crab (*Cancer magister*) in the traditional Haida art form. The design was contributed by the Haida Fisheries Program of the Council of the Haida Nation and was drawn by a self-taught Haida artist, Patrick Wesley.

Correct citation for this publication: Jamieson, G.S., and Campbell, A. (*Editors*). 1998. *Proceedings of the North Pacific Symposium on Invertebrate Stock Assessment and Management.* Can. Spec. Publ. Fish. Aquat. Sci. 125. 462 p.

CANADIAN SPECIAL PUBLICATION OF FISHERIES AND AQUATIC SCIENCES 125

Proceedings of the North Pacific Symposium on Invertebrate Stock Assessment and Management

Edited by

Glen S. Jamieson
and
Alan Campbell

Department of Fisheries and Oceans
Pacific Biological Station
Nanaimo, BC V9R 5K6, Canada

NRC · CNRC

NRC Research Press
Ottawa 1998

© 1998 National Research Council of Canada

All rights reserved. No part of this publication may be reproduced, stored in a retrieval system, or transmitted by any means, electronic, mechanical, photocopying, recording or otherwise, without the written permission of the National Research Council of Canada, Ottawa, Ontario K1A 0R6, Canada.
Printed in Canada on acid-free paper ♾

ISBN 0-660-17221-6
NRC No. 41392

Canadian Cataloguing in Publication Data

(Canadian Special Publication of Fisheries and Aquatic Sciences,
ISSN 0706-6481; 125)
"A publication of the National Research Council of Canada
Monograph Publishing Program."
Includes abstracts in French.
Includes bibliographical references.
ISBN 0-660-17221-6

1. Marine invertebrates — North Pacific Ocean — Congresses. 2. Invertebrate populations — North Pacific Ocean — Congresses. 3. Shellfish fisheries — North Pacific Ocean — Congresses. 4. Fishery management — North Pacific Ocean — Congresses.
I. National Research Council Canada. II. Title. III. Series.

SH370.N67 1998 639.4 C98-900029-X

Contents

Abstract/Résumé

The international North Pacific Symposium on Invertebrate Stock Assessment and Management was held on March 6–10, 1995, in Nanaimo, British Columbia, Canada. Emphasis was on temperate, benthic species and the Symposium focused on new, innovative evaluation of the implications and needs for changing management approaches and demands in invertebrate fishery science. This Symposium was organized by the Canadian Department of Fisheries and Oceans, with the Province of British Columbia, two aboriginal First Nations, and two industry associations as co-sponsors. One hundred and fifteen participants attended from Canada, U.S.A., Mexico, Chile, Argentina, Russia, Australia, New Zealand, Indonesia, Philippines, South Africa, Norway, U.K., and Italy. Of 63 presentations (55 oral and 8 posters), 22 were on crustaceans, 11 were on gastropods, 7 were on echinoderms, 6 were on bivalves, and 17 were non-species specific (analytical or theoretical content). Presentations at the Symposium were structured around the following six sessions: (1) estimation of abundance within specified areas: design-based versus model-based surveys, kriging, absolute estimates versus indices, and simultaneous multiple biomass estimation approaches, (2) spatial distribution of abundance and implications: concentration profiles, behavior of fishers and effort allocation, fertilization success in sessile broadcast spawners, adult–juvenile interactions, and aggregation behaviors, (3) dynamics of harvested stocks and ecosystems: metapopulations, larval dispersal and gene flow, fisheries and their selective effects, and management implications of life history and behavior variability, (4) population modeling and parameter estimation: interpretation of fishery data, forecasting, modeling combined fishery and research data, and stock recruitment–environment relationships, (5) invertebrate fisheries management: rotating harvest models, community-based versus centrally dictated management, management with reduced research resources, territorial user's rights, and co-management, and (6) decision making in invertebrate fisheries management: reducing overexploitation risk, management under uncertainty, bio-socio-economic modeling, and problems and perspectives. There are 43 papers in the Proceedings, reorganized under new headings, including a Symposium Overview.

Le Symposium international du Pacifique Nord sur l'évaluation et la gestion des stocks d'invertébrés s'est tenu à Nanaimo (Colombie-Britannique), Canada, du 6 au 10 mars 1995. Avec pour champ d'étude les espèces benthiques des eaux tempérées, les participants au Symposium se sont penchés sur les moyens d'innover dans l'évaluation des répercussions et des besoins pour changer les approches et les exigences de la gestion dans le domaine de la recherche halieutique sur les invertébrés. Organisée par le ministère des Pêches et des Océans du Canada, la rencontre était parrainée par la province de Colombie-Britannique, deux Premières Nations et deux associations de l'industrie des pêches. Elle a rassemblé 115 participants représentant le Canada, les États-Unis, le Mexique, le Chili, l'Argentine, la Russie, l'Australie, la Nouvelle-Zélande, l'Indonésie, les Philippines, l'Afrique du Sud, la Norvège, le Royaume-Uni et l'Italie. Sur les 63 communications (55 exposés et 8 affichages), 22 concernaient les crustacés, 11 les gastropodes, 7 les échinodermes, 6 les bivalves, et 17 ne portaient pas sur des espèces, mais avaient plutôt un caractère analytique ou théorique. Les diverses communications étaient structurées en six sessions : (1) estimation de l'abondance dans des secteurs précis : relevés construits spécifiquement contre relevés basés sur un modèle; krigeage; estimations absolues contre indices; méthodes multiples simultanées d'estimation de la biomasse; (2) distribution spatiale de l'abondance et incidences : profils de concentration; comportement des pêcheurs et allocation de l'effort; succès de la fécondation en pleine eau chez les géniteurs sessiles; interactions adultes-juvéniles; comportements de regroupement; (3) dynamique des stocks exploités et des écosystèmes : métapopulations, dispersion des larves et flux génétique; les pêches et leurs effets sélectifs; incidences pour la gestion de la variabilité du cycle biologique et du comportement; (4) modélisation des populations et estimation des paramètres : interprétation des données sur les pêches; prévision; modélisation des pêches combinées et données de recherche; relations entre le recrutement et l'environnement; (5) gestion des pêches des invertébrés : modèles d'exploitation en rotation; gestion régie par la communauté plutôt que de façon centralisée; gestion dans un contexte de réduction des ressources allouées à la recherche; droits des usagers du territoire; cogestion; (6) prise de décision dans la gestion des pêches des invertébrés : réduction du risque de surexploitation; gestion dans l'incertitude; modélisation bio-socio-économique; enfin, problèmes et perspectives. Les Actes rassemblent 43 communications, rassemblées sous de nouvelles rubriques, ainsi qu'une vue d'ensemble du Symposium.

List of contributors

Addison, J.T., Ministry of Agriculture, Fisheries and Food, Directorate of Fisheries Research, Fisheries Laboratory, Lowestoft, Suffolk NR33 0HT, U.K.

Alvarado, J., Estación Costera de Investigaciones Marinas, Las Cruces, Facultad de Ciencias Biológicas, Pontificia Universidad Católica de Chile, Casilla 114-D, Santiago, Chile.

Andrew, N.L., NSW Fisheries Research Institute, P.O. Box 21, Cronulla, New South Wales, Australia 2230.

Arifin, Z., Division for Marine Resources R&D, Indonesian Institute of Sciences, Guru-Guru, Poka, Ambon 97233, Indonesia.

Baker, J.L., South Australian Aquatic Sciences Centre, South Australian Research and Development Institute, P.O. Box 120, Henley Beach, South Australia 5022, Australia.

Bangi, H.G.P., Marine Science Institute, University of the Philippines, Diliman, Quezon City, Philippines 1001.

Bannister, R.C.A., Ministry of Agriculture, Fisheries and Food, Directorate of Fisheries Research, Fisheries Laboratory, Lowestoft, Suffolk NR33 0HT, U.K.

Barry, S.T., Washington Department of Fish and Wildlife, 48 Devonshire Road, Montesano, WA 98563, U.S.A.

Bazhin, A.G., Kamchatka Research Institute of Fisheries and Oceanography, Naberezhnaya 18, Petropavlovsk-Kamchatsky 683002, Russia.

Botsford, L.W., Department of Wildlife, Fish and Conservation Biology, University of California, Davis, CA 95616, U.S.A.

Bradbury, A., Washington Department of Fish and Wildlife, Pt. Whitney Shellfish Laboratory, 1000 Point Whitney Road, Brinnon, WA 98320, U.S.A.

Breen, P.A., NIWA, National Institute of Water and Atmospheric Research Ltd., P.O. Box 14901, Kilbirnie, Wellington, New Zealand.

Brownie, C., Department of Statistics, North Carolina State University, Raleigh, NC 27695, U.S.A.

Caddy, J.F., Fisheries Department, F.A.O., Via delle Terme di Caracalla, Rome 00100, Italy.

Campbell, A., Fisheries and Oceans Canada, Pacific Biological Station, Nanaimo, BC V9R 5K6, Canada

Caputi, N., Fisheries Department of Western Australia, Western Australian Marine Research Laboratories, P.O. Box 20, North Beach WA 6020, Australia.

Castilla, J.C., Estación Costera de Investigaciones Marinas, Las Cruces, Facultad de Ciencias Biológicas, Pontificia Universidad Católica de Chile, Casilla 114-D, Santiago, Chile.

Chen, C.-L., Department of Fisheries and Oceans, P.O. Box 5667, St. John's, NF A1C 5X1, Canada.

Chiasson, Y., Department of Fisheries and Oceans, Gulf Fisheries Centre, Science Branch, P.O. Box 5030, Moncton, NB E1C 9B6, Canada.

Chubb, C.F., Fisheries Department of Western Australia, Western Australian Marine Research Laboratories, P.O. Box 20, North Beach WA 6020, Australia.

Collie, J.S., Graduate School of Oceanography, University of Rhode Island, Narragansett, RI 02882, U.S.A.

Dawe, E.G., Department of Fisheries and Oceans, P.O. Box 5667, St. John's, NF A1C 5X1, Canada.

Defeo, O., Estación Costera de Investigaciones Marinas, Las Cruces, Facultad de Ciencias Biológicas, Pontificia Universidad Católica de Chile, Casilla 114-D, Santiago, Chile.

Espoz, C., Estación Costera de Investigaciones Marinas, Las Cruces, Facultad de Ciencias Biológicas, Pontificia Universidad Católica de Chile, Casilla 114-D, Santiago, Chile.

Fogarty, M.J., Center for Environmental and Estuarine Studies, University of Maryland System, Chesapeake Biological Laboratory, Solomons, MD 20688, U.S.A.

Frusher, S.D., Department of Primary Industry and Fisheries, Marine Research Laboratories, Taroona, 7053, Tasmania, Australia.

Garza, D.A., Marine Advisory Program, University of Alaska Fairbanks, 700 Katlian Street, Suite D, Sitka AK 99835, U.S.A.

Gibson, I.D., Department of Primary Industry and Fisheries, Marine Research Laboratories, Taroona, 7053, Tasmania, Australia.

Hall, M.A., Inter-American Tropical Tuna Commission, c/o Scripps Institution of Oceanography, 8604 La Jolla Shores Dr., La Jolla, CA 92093, U.S.A.

Hand, C.M., Fisheries and Oceans Canada, Pacific Biological Station, Nanaimo, BC V9R 5K6, Canada.

Harbo, R., Fisheries and Oceans Canada, Fisheries Management, South Coast Division, 3225 Stephenson Point Road, Nanaimo, BC V9T 1K3, Canada.

Hastings, A., Division of Environmental Studies, University of California, Davis, CA 95616, U.S.A.

Hilborn, R., School of Fisheries, Box 357980, University of Washington, Seattle, WA 98195-7980, U.S.A.

Hoenig, J.M., Department of Fisheries and Oceans, P.O. Box 5667, St. John's, NF A1C 5X1, Canada.

Ivanov, B.G., Russian Research Institute of Fisheries and Oceanography, (VNIRO), 107140 Moscow, Russia.

Jamieson, G.S., Fisheries and Oceans Canada, Pacific Biological Station, Nanaimo, BC V9R 5K6, Canada.

Joll, L.M., Fisheries Department of Western Australia, Western Australian Marine Research Laboratories, P.O. Box 20, North Beach, WA 6020, Australia.

Jones, R.R., Haida Fisheries Program, Council of the Haida Nation, P.O. Box 98, Skidegate, Haida Gwaii, BC V0T 1S0, Canada.

Juinio-Meñez, M.A., Marine Science Institute, University of the Philippines, Diliman, Quezon City, Philippines 1001.

Keesing, J.K., South Australian Aquatic Sciences Centre, South Australian Research and Development Institute, P.O. Box 120, Henley Beach, South Australia 5022, Australia.

Kendrick, T.H., NIWA, National Institute of Water and Atmospheric Research Ltd., P.O. Box 14901, Kilbirnie, Wellington, New Zealand.

Kennedy, R.B., Department of Primary Industry and Fisheries, Marine Research Laboratories, Taroona, 7053, Tasmania, Australia.

Kruse, G.H., Alaska Department of Fish and Game, Commercial Fisheries Management and Development Division, P.O. Box 25526, Juneau, AK 99802, U.S.A.

Lai, H.-L., Northeast Fisheries Science Center, National Marine Fisheries Service, NOAA, 166 Water Street, Woods Hole, MA 02543, U.S.A.

Largier, J.L., Center for Coastal Studies, Scripps Institution of Oceanography, University of California, San Diego, La Jolla, CA 92093, U.S.A.

LaRiviere, P.E., Washington Department of Fish and Wildlife, 48 Devonshire Road, Montesano, WA 98563, U.S.A.

Lee, L., Haida Fisheries Program, Council of the Haida Nation, P.O. Box 98, Skidegate, Haida Gwaii, BC V0T 1S0, Canada.

Levitan, D.R., Department of Biological Science, Florida State University, Tallahassee, FL 32306-2043, U.S.A.

Lilly, G.R., Department of Fisheries and Oceans, Science Branch, P.O. Box 5667, St. John's, NF A1C 5X1, Canada.

McGarvey, R., South Australian Research and Development Institute, South Australian Aquatic Sciences Centre, 2 Hamra Avenue, West Beach, SA 5024, Australia.

McShane, P.E., National Institute of Water and Atmospheric Research Ltd., P.O. Box 14-901, Wellington, New Zealand.

Macawaris, N.N.D., Marine Science Institute, University of the Philippines, Diliman, Quezon City, Philippines 1001.

Manríquez, P., Estación Costera de Investigaciones Marinas, Las Cruces, Facultad de Ciencias Biológicas, Pontificia Universidad Católica de Chile, Casilla 114-D, Santiago, Chile.

Moloney, C.L., Marine Science Research Institute, Zoology Department, University of Cape Town, Rondebosch 7700, South Africa.

Momot, W.T., Department of Biology, Lakehead University, Thunder Bay, ON P7B 5E1, Canada.

Moriyasu, M., Department of Fisheries and Oceans, Gulf Fisheries Centre, Science Branch, P.O. Box 5030, Moncton, NB E1C 9B6, Canada.

Murphy, M.C., Alaska Department of Fish and Game, Commercial Fisheries Management and Development Division, P.O. Box 25526, Juneau, AK 99802, U.S.A.

Oliva, D., Estación Costera de Investigaciones Marinas, Las Cruces, Facultad de Ciencias Biológicas, Pontificia Universidad Católica de Chile, Casilla 114-D, Santiago, Chile.

Orensanz, J.M., School of Fisheries, University of Washington, Seattle, WA 98195, U.S.A.

Otto, R., National Marine Fisheries Service, Kodiak Laboratory, P.O. Box 1638, Kodiak, AK 99615-1638, U.S.A.

Parma, A.M., International Pacific Halibut Commission, P.O. Box 95009, Seattle, WA 98145-2009, U.S.A.

Parsons, D.G., Department of Fisheries and Oceans, Science Branch, P.O. Box 5667, St. John's, NF A1C 5X1, Canada.

Penn, J.W., Fisheries Department of Western Australia, Western Australian Marine Research Laboratories, P.O. Box 20, North Beach, WA 6020, Australia.

Phillips, A., Fisheries and Oceans Canada, Pacific Biological Station, Nanaimo, BC V9R 5K6, Canada.

Pino, C., Estación Costera de Investigaciones Marinas, Las Cruces, Facultad de Ciencias Biológicas, Pontificia Universidad Católica de Chile, Casilla 114-D, Santiago, Chile.

Pollock, K.H., Department of Statistics, North Carolina State University, Raleigh, NC 27695, U.S.A.

Prescott, J.H., South Australian Research and Development Institute, South Australian Aquatic Sciences Centre, 2 Hamra Avenue, West Beach, SA 5024, Australia.

Prince, J., Biospherics Pty Ltd., P.O. Box 168, South Fremantle, Western Australia 6162.

Purnomo, A.H., Assessment Institute for Agricultural Technology, Jalan chr. Soplanit Rumahtiga, Ambon 97233, Indonesia.

Purwati, P., Division for Marine Resources R&D, Indonesian

Institute of Sciences, Guru-Guru, Poka, Ambon 97233, Indonesia.

Quinn, J.F., Division of Environmental Studies, University of California, Davis, CA 95616, U.S.A.

Robert, G., Department of Fisheries and Oceans, P.O. Box 550, Halifax, NS B3J 2S7, Canada.

Rosson, A., Estación Costera de Investigaciones Marinas, Las Cruces, Facultad de Ciencias Biológicas, Pontificia Universidad Católica de Chile, Casilla 114-D, Santiago, Chile.

Ruiz-Avila, R., Biospherics Pty Ltd., P.O. Box 168, South Fremantle, Western Australia 6162

Schwarz, C., Department of Mathematics and Statistics, Simon Fraser University, Burnaby, BC V5A 1S6, Canada.

Seijo, J.C., Centro de Investigación y de Estudios Avanzados del IPN, Unidad Mérida, Apartado Postal 73 Cordemex, Mérida, México.

Sewell, M.A., Department of Larval Ecology, Harbor Branch Oceanographic Institution, 5600 Old Dixie Highway, Port Pierce, FL 34946, U.S.A.

Shepherd, S.A., South Australian Research and Development Institute, Aquatic Sciences Centre, P.O. Box 120, Henley Beach, South Australia 5022, Australia.

Sinclair, A., Department of Fisheries and Oceans, Gulf Fisheries Centre, Science Branch, P.O. Box 5030, Moncton, NB E1C 9B6, Canada.

Sluczanowski, P., South Australian Research and Development Institute, P.O. Box 1625, Adelaide, South Australia 5000.

Smith, B.D., Canadian Wildlife Service, Pacific Wildlife Research Centre, R.R. No. 1, 5421 Robertson Road, Delta, BC V4K 3N2, Canada.

Smith, S.J., Department of Fisheries and Oceans, Bedford Institute of Oceanography, P.O. Box 1006, Dartmouth, NS B2Y 4A2, Canada.

Soto, R., Estación Costera de Investigaciones Marinas, Las Cruces, Facultad de Ciencias Biológicas, Pontificia Universidad Católica de Chile, Casilla 114-D, Santiago, Chile.

Veitch, P.J., Department of Fisheries and Oceans, Science Branch, P.O. Box 5667, St. John's, NF A1C 5X1, Canada.

Wade, E., Department of Fisheries and Oceans, Gulf Fisheries Centre, Science Branch, P.O. Box 5030, Moncton, NB E1C 9B6, Canada.

Walters, C., Fisheries Centre, University of British Columbia, Vancouver, BC V6T 1Z4, Canada.

Warren, W.G., Department of Fisheries and Oceans, Science Branch, P.O. Box 5667, St. John's, NF A1C5X1, Canada.

Wing, S.R., Wildlife, Fish and Conservation Biology, University of California, Davis, CA 95616, U.S.A.

Woodby, D., Alaska Department of Fish and Game, P.O. Box 240020, Douglas, AK 99824, U.S.A.

Worthington, D.G., NSW Fisheries Research Institute, P.O. Box 21, Cronulla, New South Wales, Australia 2230.

Zheng, J., Alaska Department of Fish and Game, Commercial Fisheries Management and Development Division, P.O. Box 25526, Juneau, AK 99802-5526, U.S.A.

List of contributors (by senior author) — Poster Session

Bannister, R.C.A., Ministry of Agriculture, Fisheries and Food, Directorate of Fisheries Research, Fisheries Laboratory, Lowestoft, Suffolk NR33 0HT, U.K.

Bureau, D., Department of Biological Sciences, Simon Fraser University, Burnaby, BC V5A 1S6, Canada.

Keesing, J.K., South Australian Aquatic Sciences Centre, P.O. Box 120, Henley Beach 5022, Australia.

Kruse, G.H., Alaska Department of Fish and Game, Commercial Fisheries Management and Development Division, P.O. Box 25526, Juneau, AK 99802, U.S.A.

Morrell, M., Mike Morrell Consulting, Denman Island, BC V0R 1T0, Canada.

Nash, W.J., Marine Resources Division, Department of Primary Industry and Fisheries, Crayfish Point, Taroona, Tasmania 7053, Australia.

Richards, R.A., National Marine Fisheries Service, Northeast Fisheries Science Center, Woods Hole, MA 02543, U.S.A.

Van der Meeren, G.I., Institute of Marine Research, Austevoll Aquaculture Research Station, N-5392, Storebø, Norway.

Acknowledgments

This Symposium required the co-operation and support of a large number of people. Financial assistance was provided by Fisheries and Oceans Canada. Co-sponsors included Haida Fisheries Program; Ministry of Agriculture, Fisheries and Food, Province of British Columbia; Nuu-chal-nulth Fisheries Council; Pacific Underwater Harvesters Association; and Underwater Harvesters Association. I extend our sincere appreciation to each of these organizations for their support and, in particular, Dr. Don Noakes, Russ Jones, Don Hall, Mike Featherstone, and Jamie Austin, respectively.

Many people contributed to the local arrangements of the Symposium and we would particularly like to thank Ann Thompson for her organizational and secretarial assistance, together with Tricia Boxwell and Dr. Alan Campbell, Dr. Neil Bourne, Antan Phillips, Graham Gillespie, Bruce Clapp, Claudia Hand, Dwight Heritage, and Kelly Francis for their untiring and devoted efforts to making the Symposium a success. We acknowledge the significant contribution of Dr. J.M. (Lobo) Orensanz who provided advice on the structure of the Symposium. Drs. John Davis and Don Noakes gave opening remarks and Dr. W. G. Doubleday was the guest speaker at the Symposium banquet. Senna Charleson (Hesquiat First Nation), Russ Jones (Haida First Nation), and Chief Gerry Brown (Nanaimo First Nation) gave introductions of welcome at a reception hosted by First Nation people. We also thank Dr. A. Campbell, Dr. L. Botsford, Dr. M. Fogarty, Dr. B.D. Smith, Dr. J. Hoenig, Dr. B.G. Ivanov, Dr. J. Castilla, and Dr. G.H. Kruse for chairing Sessions. Finally, I extend the thanks of all of us to the participants.

Papers in these proceedings were reviewed by at least two referees. The reviewing, editing, and revising of manuscripts required the co-operation of reviewers and authors. As editors, we acknowledge their efforts. Finally, we sincerely appreciate the editorial assistance of Ann Thompson. Without her great effort, neither the Symposium nor the Proceedings would have been possible.

Preface

The first North Pacific Workshop, held in 1984, characterized North Pacific fishery resources and fisheries on the basis of broad taxonomic groupings (e.g., bivalve, gastropod, crustacean, etc.) and discussed then current issues relating to single species population dynamics, modeling, and multispecies interactions. At that time, many scientists and managers felt they had a good understanding of stock assessment and management issues and what was required to overcome the deficiencies. Today, most appreciate that the goal of achieving optimal management is more elusive than many initially thought and a great deal of new information has been obtained over the past 10 years.

This realization provided the impetus for Canada's decision to host another North Pacific meeting of fishery scientists, managers, and industry representatives focused mainly on temperate, benthic invertebrates. The Symposium brought together scientists and managers interested in invertebrates both to address issues associated with assessment of resouces and to discuss and determine if better management advice and process, applicable to both developed and developing countries, could be formulated. Participants from around the world were invited to attend. Over 14 countries were represented at the Symposium, including most major fishing countries in both the northern and southern hemispheres.

At the time of the Symposium, papers were grouped on the basis of their abstract content into the Session structure described in Appendix 1. For the purposes of the Proceedings, however, the manuscripts have been regrouped under the headings listed in the Table of Contents. During the Symposium, each Session was summarized by representatives (Appendix 1) from each of three sectors: (1) fisheries science, (2) fishery management, and (3) the fishing industry. These latter perspectives are not included per se but their content has been incorporated into a Symposium Overview paper, authored by J.M. Orensanz and G.S. Jamieson, included as the last paper in this Proceedings.

Nowadays, scientists and managers better appreciate the complexity of marine ecosystems, the effectiveness of modern fishing technologies to rapidly decimate populations if unconstrained, the significant role natural processes can have on species abundance, the politics and conflict around competition among harvest sectors for dwindling species' abundances, and the difficulties both scientists and managers have in maintaining programs and communcating recommendations, while being asked to do more with less resources.

Currently, fishery science and management seem to be increasingly crisis-driven, with available resources more often being directed by administrative managers towards high profile problems arising from unanticipated environmental change (natural or man-induced), resource sector conflicts, or earlier failure to have planned appropriately. In the meantime, proactive and/or long-term monitoring studies are frequently given lower priority, resulting often in the termination of continued sampling in existing data time series and/or difficulty in getting long-term resource commitment to establish new data series.

I trust that everyone at the Symposium benefited from our deliberations and, as in 1984, I hope that this Symposium will be one of a long series of opportunities for Pacific invertebrate biologists to present results and exchange views in an international forum.

Glen S. Jamieson
Symposium Chairman

Dedication

The Symposium was dedicated to two eminent Canadian scientists, now deceased, Drs. Frank R. Bernard and Dan B. Quayle, who were both based at the Pacific Biological Station in Nanaimo. Dr. Quayle was the Honorary Chairman of the previous North Pacific Workshop in 1984 and, as a molluscan expert, had considerable experience with invertebrate fisheries throughout he world. Dr. Quayle retired from the Fisheries Research Board of Canada, the precursor to the present Department of Fisheries and Oceans, in 1972, but remained active as a researcher and worked daily at the laboratory until shortly before his death in 1994.

Dr. Bernard, initially hired by Dr. Quayle as a research technician in 1962, worked his way up through the ranks, and ultimately held every job classification at the Station, including that of Acting Director. Dr. Bernard worked on virtually every invertebrate species exploited in British Columbia at one time or another and published extensively on most of them. However, he is perhaps best known as one of the region's major bivalve taxonomists and published many definitive articles in this field before his untimely death in 1989.

Assessment of Abundance and Related Parameters

Getting more out of your survey designs: an application to Georges Bank scallops (*Placopecten magellanicus*)

Stephen J. Smith and Ginette Robert

Abstract: Field surveys which use a randomized or finite population survey design (e.g., stratified random) are often used to monitor fish and invertebrate populations. Inferences from data collected in this way have been mainly restricted to the estimation of means, totals, and their respective standard errors. On the other hand, the incorporation of information on ancillary measures (e.g., depth, sediment type) or spatial structure into these kinds of estimates has often been left to model-based methods which usually ignore the survey design. In this paper we present applications of a number of methods which incorporate both additional information and the survey design into estimates of means, totals, etc., with the aim of increasing precision of these estimates. For the case of Georges Bank scallops (*Placopecten magellanicus*), bottom sediment type whether included as a poststratifying variable or used in a predictive relationship resulted in increased precision of the estimate of the mean number of scallops.

Résumé : Les relevés sur le terrain qui reposent sur un plan d'échantillonnage d'une population aléatoire ou finie (p. ex. aléatoire stratifié) sont souvent utilisés pour surveiller les populations de poissons et d'invertébrés. Les inférences établies à partir des données obtenues de cette façon ont été principalement limitées à l'estimation de moyennes, de totaux, et de leur écarts-types respectifs. Par ailleurs, l'incorporation de données relatives à des mesures accessoires (p. ex. profondeur, type de sédiments) ou à la structure spatiale dans ces sortes d'estimations a souvent été laissée à des méthodes fondées sur des modèles qui habituellement ne tiennent pas compte du plan d'échantillonnage. Dans la présente communication, nous présentons des applications d'un certain nombre de méthodes qui incorporent à la fois de l'information additionnelle et le plan d'échantillonnage pour estimer les moyennes, les totaux, etc., qui visent à accroître la précision de ces estimations. Dans le cas du pétoncle du banc Georges (*Placopecten magellanicus*), le type de sédiments de fond, qu'il soit incorporé comme variable post-stratification ou utilisé dans une relation prédictive, a fait augmenter la précision de l'estimation du nombre moyen de pétoncles.
[Traduit par la Rédaction]

Introduction

The abundance of populations or stocks of commercial fish and invertebrate species are often monitored through the use of field surveys where areas inhabited by the species are sampled in some way by commercial or specialized sampling gear. Approaches used to sample and estimate abundance from such monitoring programs generally fall into one of two categories. In the first category, finite population theory or survey sampling methods are used to define the survey design and the appropriate estimates of abundance along with associated standard errors. Examples of surveys in this category are the trawl surveys of groundfish for the eastern coasts of Canada (Doubleday 1981) and the United States (Azarovitz 1981), dredge surveys for scallops (*Placopecten magellanicus*) on Georges Bank (Mohn et al. 1987) and Northeastern United States (Serchuk and Wigley 1986), and acoustic surveys for pelagic fish off of South Africa (Jolly and Hampton 1990).

In the second general category we placed fixed-station trawl surveys (Traynor et al. 1990), or nonrandom transect style surveys common to acoustic surveys (Simmonds et al. 1992) and also used for other types of gear (Pelletier and Parma 1994) (e.g., longline survey for halibut). In cases like these, estimation methods are not uniquely determined by the survey design and instead models with implicit (Robert et al. 1994) (e.g., contouring, Delaunay triangles) or explicit spatial structures (e.g., kriging) have often been used to estimate abundance (e.g., Conan and Wade 1989; Guillard et al. 1992; Simard et al. 1992).

Spatial methods have also been applied to data from finite population-based surveys to estimate abundance and other quantities. Often such applications have reported that finite-population methods require the assumption of spatial independence for appropriateness of their estimates, especially the variance estimates (Simard et al. 1992; Foote and Stefánsson 1993; Ecker and Heltshe 1994). While finite-population methods may ignore fine-scale spatial structure in the population being sampled, it is not necessarily true that spatial independence is required as an assumption or that spatial structure need be completely ignored.

In this paper, we show that while the properties of finite population estimators are actually independent of spatial

S.J. Smith. Department of Fisheries and Oceans, P.O. Box 1006, Bedford Institute of Oceanography, Dartmouth, NS B2Y 4A2, Canada.
G. Robert. Department of Fisheries and Oceans, P.O. Box 550, Halifax, NS B3J 2S7, Canada.

Correct citation: Smith, S.J., and Robert, G. 1998. Getting more out of your survey designs: an application to Georges Bank scallops (*Placopecten magellanicus*). *In* Proceedings of the North Pacific Symposium on Invertebrate Stock Assessment and Management. *Edited by* G.S. Jamieson and A. Campbell. Can. Spec. Publ. Fish. Aquat. Sci. 125. pp. 3–13.

structure, the incorporation of additional information possibly reflective of spatial structure into such estimators may have a benefit in increasing the precision of the estimates. Data collected from stratified random surveys for scallops on Georges Bank in 1989 and 1990 were used to compare estimates of the mean number of scallops from a stratified random survey with those from incorporating bottom sediment type into the survey design or estimation process.

Material and methods

Scallop survey data

The design of the annual Georges Bank scallop survey is based on a stratification by commercial effort (Robert and Jamieson 1986). Logbooks of the commercial fleet in the preceding 9 months are analyzed to determine areas or strata of arbitrary levels of catch rates (very low, low, medium, and high). Areas of high catch rates are sampled more heavily as they represent the area most important to the fleet (and presumably the areas of greatest abundance). A regular array of data points 0.2° of latitude and longitude apart, is established from which random sampling stations are chosen for each stratum.

Surveys are carried out with a 2.44-m New Bedford scallop dredge with 75-mm rings and a 38-mm mesh liner. Survey data is standardized to a tow distance of 800 m or an area of 1951 m^2 per tow. Actual tow distances vary between 1100 and 1300 m but all tows are standardized to a distance of 800 m assuming a linear relationship between catch and tow duration. Shell height frequencies for live scallops caught are measured in 5-mm intervals; scallop age is inferred from shell height by a von Bertalanffy growth curve (Robert et al. 1994).

This survey series began in 1981 and continues to the present. We will only consider two years here, 1989 and 1990. Survey estimates for ages 3, 4, and 5 years were selected for this study because these ages are relatively abundant and hence well represented in the survey data. Also, these ages represent distinct size-classes corresponding to fishery prerecruits (age 3), young recruits (age 4) and heavily fished recruits (age 5).

Estimation methods

The finite population basis for sample survey methods such as the stratified random design assumes that we can identify N nonoverlapping sample units in a population for which some characteristic(s) y can be measured on each of them. In our case we are interested in y_{hia}, the number of scallops caught in stratum h, tow i of age a, where we have N possible locations to tow our dredge, grouped into L geographical strata ($N = \Sigma_{h-1}^{L} N_h$). Given that the total number of sample units is too large to completely enumerate we take n ($n < N$) sample units ($n = \Sigma_{h-1}^{L} n_h$) at random from the population of units and use measurements from these to infer the population values.

Our notation will follow that of (Cochran 1977) and (Thompson 1992) with upper case letters indicating population quantities and lower case letters used for sample quantities. The population stratified mean is defined as,

$$\bar{Y}_{st,a} = \sum_{h=1}^{L} \frac{N_h}{N} \sum_{i=1}^{N_h} \frac{y_{hia}}{N_h},$$

$$= \sum_{h=1}^{L} \frac{N_h}{N} \bar{Y}_{ha},$$

and is estimated by the sample stratified mean with known weights N_h/N,

$$\bar{y}_{st,a} = \sum_{h=1}^{L} \frac{N_h}{N} \sum_{i=1}^{n_h} \frac{y_{hia}}{n_h},$$

[1] $$= \sum_{h=1}^{L} \frac{N_h}{N} \bar{y}_{ha}.$$

Properties of the sample mean $\bar{y}_{ha}$ (and hence $\bar{y}_{st,a}$) are derived on the basis of distinct samples of size n_h repeatedly chosen from the N_h possible samples within each stratum. For example, the sample mean is unbiased because the average of the $\bar{y}_{ha}$ obtained from all of the possible samples of size n_h from N_h, $\binom{N_h}{n_h}$, is equal to $\bar{Y}_{ha}$. The precision of $\bar{y}_{ha}$ is the average of $(\bar{y}_{ha} - \bar{Y}_{ha})^2$ over all possible sample draws. This average can be written as,

[2] $$\mathrm{Var}(\bar{y}_{ha}) = \frac{(N_h - n_h)}{n_h N_h} \sum_{i=1}^{N_h} \frac{(y_{hia} - \bar{Y}_{ha})^2}{N_h - 1}$$

[3] $$= \frac{(N_h - n_h)}{n_h N_h} S_{ha}^2.$$

An unbiased estimate for this variance is given by,

$$\widehat{\mathrm{Var}}(\bar{y}_{ha}) = \frac{(N_h - n_h)}{n_h N_h} \sum_{i=1}^{n_h} \frac{(y_{hia} - \bar{y}_{ha})^2}{n_h - 1}$$

[4] $$= \frac{(N_h - n_h)}{n_h N_h} s_{ha}^2.$$

The variance of the stratified mean is

[5] $$\widehat{\mathrm{Var}}(\bar{y}_{st}) = \frac{1}{N^2} \sum_{h=1}^{L} N_h (N_h - n_h) \frac{s_{ha}^2}{n_h}.$$

Note that none of these derivations required assumptions of independent and identically distributed (e.g., normal distribution) random variables. The only assumption required is that samples within strata are taken in a random manner with a known probability of being chosen.

Spatial information can be built into these survey designs by defining strata according to variables (e.g., sediment type, depth) which exhibit relationships with scallop abundance.

However, this exercise would only be useful if the new strata resulted in an increase in precision. After-the-fact evaluation of potential new stratifying variables can be accommodated by *post-stratifying* the original data into new strata defined by these variables. The new strata are established within the existing strata so that a double stratification is in effect. Given that samples were originally sampled by simple random sampling within the original strata, the variance of the stratified mean from the new strata within the original strata is, (Thompson 1992),

$$\mathrm{Var}(\bar{y}_{st}) \approx \frac{N-n}{nN}\sum_{h=1}^{L}\left(\frac{N_h}{N}\right)S_{ha}^2 + \frac{1}{n^2}\left(\frac{N-n}{N-1}\right)$$

[6] $$\sum_{h=1}^{L}\frac{N-N_h}{N}S_{ha}^2,$$

with the first term on the right giving the stratified variance under proportional allocation of samples to strata and the second term reflecting the fact that the sample sizes in the new strata are now random variables. The poststratification variance in eq. 6 can be estimated by substituting sample variances $s_{ha}2$ for $S_{ha}2$ in a manner similar to previous variance estimates given here. Rao (1988) recommends using eq. 5 to estimate the variance when constructing confidence intervals for poststratified estimates. However, this estimate will be affected by the chance allocation of sets to the post strata, which will tend to minimize (or exaggerate) any advantages of the poststratification and hence cloud the issue. We use the variance in eq. 6 as it should only reflect gains in precision due to the strata themselves. Allocation schemes can be introduced if the decision to establish new strata for future surveys based on the poststratification is made.

An alternative approach of incorporating spatial information is to use *predictive* type estimates in sample surveys. When we choose n_h i's from $i = 1, \dots, N_h$ to obtain the sample, we partition the population of sample units into those that we observe ($i \in k_h$) and those that were unobserved ($i \notin k_h$). That being the case, we can write the population mean for the stratum as,

$$\bar{Y}_{ha} = \frac{1}{N_h}\left\{\sum_{i\in k_h}^{n} y_{hia} + \sum_{i\notin k_h}^{N_h-n_h} y_{hia}\right\}.$$

The standard estimate of $\bar{Y}_{ha}$ in a predictive framework (Little 1983) uses $\bar{y}_{ha}$ to estimate the y_{hia} that are not in the sample ($i \notin k_h$) to give

$$\hat{Y}_{ha} = \frac{1}{N_h}(n_h\bar{y}_{ha} + (N_h - n_h)\bar{y}_{ha})$$
$$= \bar{y}_{ha}.$$

However, we may have access to an ancillary variable x_{hi} such as depth or sediment type or some other quantity to which scallop distribution or even abundance may be related. If we know the x_{hi} for all i population units (tows) in stratum h and know the relationship between scallops and these x's, then it may be possible to predict scallop abundance at unsampled locations simply by knowing the value of x there. If we define the relationship between scallop abundance and some x_{hi} as $y_{hia} = f(x_{hi})$, then we can write the predictive mean and its estimate of the mean abundance of scallops of age a in stratum h as,

$$\bar{Y}_{ha} = \frac{1}{N_h}\left\{\sum_{i\in k_h}^{n_h} y_{hia} + \sum_{i\notin k_h}^{N_h-n_h} f(x_{hi})\right\},$$

$$\hat{y}_{ha} = \frac{1}{N_h}\left\{n_h\bar{y}_{ha} + \sum_{i\notin k_h}^{N_h-n_h}\hat{y}_{hia}\right\}.$$

The estimate of the variance of $\hat{y}_{ha}$ is (Thompson 1992),

$$\widehat{\mathrm{Var}}(\hat{Y}_{ha}) = \frac{(N_h - n_h)}{N_h n_h (n_h - p)}\sum_{i=1}^{n_h}(y_{hia} - \hat{y}_{hia})^2,$$

where p is the number of parameters used in the model for $f(x_{hi})$. Over all strata the stratified predictive mean and its associated variance are estimated as, respectively,

$$\hat{y}_{st,a} = \sum_{h=1}^{L}\frac{N_h}{N}\hat{y}_{ha}$$

and

$$\widehat{\mathrm{Var}}(\hat{y}_{st,a}) = \frac{1}{N}\sum_{h=1}^{L}\widehat{\mathrm{Var}}(\hat{y}_{ha}).$$

Results

Basic principles

Independence of the properties (bias and precision) of standard finite population estimates (the sample mean and its variance, eq. 4) from the frequency distribution of the y-values or spatial structure is illustrated by the following simulation study. Two populations of $N = 1600$ random variates were generated. For the first population we sought to imitate the main characteristics we have commonly observed in survey data, i.e., long-tailed frequency and a large number of zeroes. While discrete distributions such as the negative binomial may be a natural choice for our simulation, these distributions do not accommodate both a long tail and a large number of zeroes. Instead we used a Weibull distribution (scale = 1, shape = 30) which gave the desired amount of mass near zero and a long tail. Weibull random variates are always greater than zero and so zero observations were introduced by multiplying the Weibull variates by 1600 Bernoulli random variables, z, where $P(z = 1) = 0.75$. The resultant population had a mean of $\bar{Y} = 28.92$ and a variance of $S^2 = 816.80$. The 1600 random variates were randomly assigned to a 40×40 grid.

The second population was generated as a random field on a 40×40 grid with a Gaussian error process using the Cholesky decomposition method discussed in Cressie (1993).

Fig. 1. Quantile–quantile plot of simulated data. (*a*) Weibull-Bernoulli random variates. (*b*) Normally distributed random field on a 40 × 40 grid with a spherical covariance function.

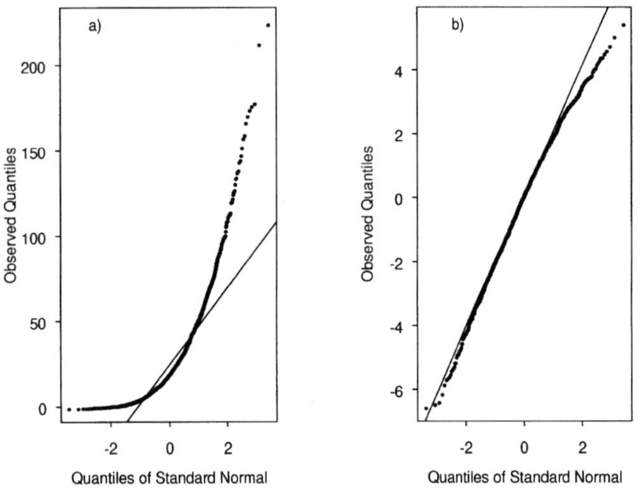

Table 1. Results of the simulation comparing estimates and population values from two finite populations of $N = 1600$ sampling units.

	Weibull Model		Spherical Model	
	n=10	n=30	n=10	n=30
Population quantities				
$\bar{Y}$	28.92	28.92	0.00	0.00
S^2/n	81.68	26.88	0.38	0.13
Sample estimates				
Mean $[\bar{y}]$	28.84	28.97	0.00	0.00
Mean $[\mathrm{Var}(\bar{y})]$	81.84	26.93	0.38	0.13

Note: The first population was generated as 1600 random numbers from a Weibull distribution (shape = 30, scale = 1) multiplied by 1600 Bernoulli random variables with $P(z = 1) = 0.75$. The second population was constructed as a normally distibuted random field on a 40 × 40 grid with a spherical covariance function (sill = 4, range = 10 units). Each finite population was resampled 5000 times (without replacement) for sample sizes of 10 and 30.

Fig. 2. Variogram plot of the two simulated data sets. The Weibull model data consists of Weibull random variates (scale = 1, shape = 30) multiplied by Bernoulli random variables where $P(z = 1) = 0.75$. The spherical model data were generated from a normally distributed random field on a 40 × 40 grid with a spherical covariance function (sill = 4, range =10 units).

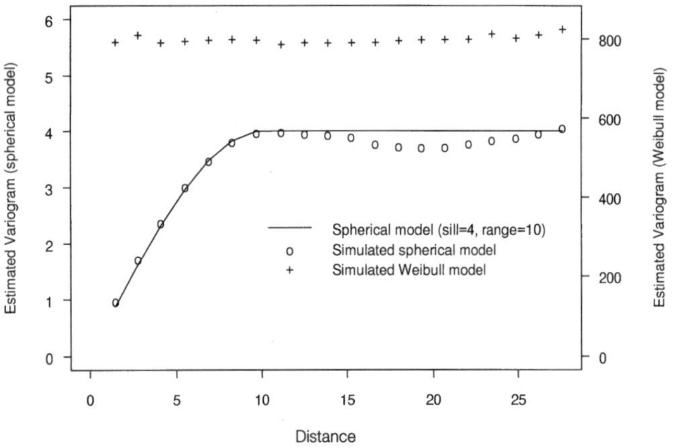

Fig. 3. Quantile–quantile plot of means from simulated data. (*a*) Weibull-Bernoulli random variates for sample size equal to 30. (*b*) Normally distibuted random field on a 40 × 40 grid with a spherical covariance function and a sample size of 10.

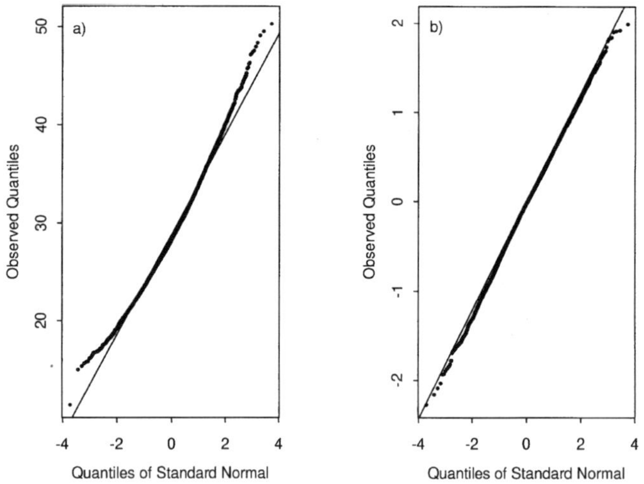

A spherical covariance function was used with the sill equal to 4 grid units and range equal to 10 grid units. The S-PLUS function **`rfsim()`** was used to generate the random variates (Mathsoft 1996). The mean for this population was 0.00 with variance equal to 3.8.

Quantile–quantile plots show the high degree of skewness in the first population with a larger mass at zero and a heavier right-hand tail than exhibited by the reference normal distribution (Fig. 1*a*). The second population which was generated from a Gaussian process is indistinguishable from the normal distribution except in the upper tail where the random variates have fewer large values than expected for a normal distribution (Fig. 1*b*).

Variograms for the two populations presented in Fig. 2 indicate that the random variates do not have any spatial structure in the first population but are highly correlated in the second population with the degree of correlation being a function of distance apart. The observed variogram for the second population matches the theoretical spherical function quite well. This kind of population is typical of the theoretical populations expected for geostatistical analysis.

Each of the finite populations were resampled 5000 times for a sample size of 10 and then again for a sample size of 30. For each sample we calculated the sample mean and sample variance of the mean. Results of the simulation study are summarized in Table 1. Sample estimates of the means in all four cases were unbiased within the resolution of the number of replications used. Similarly, estimates of the population variances of the mean were also unbiased in all four cases despite the nonnormality of the first population and the spatial structure in the second population.

In addition, the distribution of the sample means from this

Table 2. Estimates from stratified random survey for Georges Bank scallops in 1989 and 1990.

					% Efficiency		
Year	Age	$\hat{Y}_{st}$	$\bar{y}_{st}$	var[$\bar{y}_{st}$]	Allocation	Strata	Total
1989	3	263 545 224	105.26	180.10	−1.69	5.97	4.28
	4	194 929 023	77.86	188.24	−70.74	0.00	−70.74
	5	35 321 756	14.11	3.96	−36.38	1.99	−34.39
1990	3	183 368 283	79.13	349.32	−35.08	0.32	−34.76
	4	79 420 328	34.35	9.42	−14.58	5.46	−9.12
	5	33 622 651	14.54	2.10	−22.80	1.50	−21.30

Note: Efficiency of stratified design with respect to a simple random sample expressed as a percentage of simple random sampling variance.

Table 3. Mean number of scallops for ages 3, 4, and 5 for each stratum where the strata have been determined by commercial catch-per-unit-effort.

					Rank of $\bar{y}$ by sediment type					
Stratum	Year	Age	n_h	$\bar{y}_h$	Biogenic	Mixed	Sand-gravel	Sand-shell	Deep sand-shell	Sand dunes
1	89	3	31	55.32	3	2	1	5	4	6
		4		94.52	2	4	1	3	5	6
		5		16.03	3	4	1	2	5	6
	90	3	30	40.60	3	6	1	2	4	5
		4		33.03	2	5	1	3	4	6
		5		18.60	2	5	1	4	3	6
Mean rank					2.5	4.3	1	3.2	4.2	5.8
2	89	3	9	67.78	—	4	3	2	1	5
		4		73.33	—	4	3	2	1	5
		5		12.44	—	4.5	2	3	1	4.5
	90	3	10	39.40	6	1	3	4	2	5
		4		27.30	3	1	2	5	4	6
		5		9.60	4	5	1	3	2	6
Mean rank					4.3	3.2	2.3	3.2	1.8	5.2
3	89	3	23	143.17	3	4	2	1	—	—
		4		87.78	3	4	2	1	—	—
		5		22.22	1	4	2	3	—	—
	90	3	20	142.40	3	4	6	1	2	5
		4		20.90	6	4	2	3	1	5
		5		13.00	4	5	1	3	2	6
Mean rank					3.3	4.2	2.5	2	1.7	5.3
4	89	3	87	161.15	3	4	2	1	5	6
		4		57.03	1	4	2	5	3	6
		5		8.59	2	3	4	5	1	6
	90	3	90	99.22	3	2	1	4	5	6
		4		46.88	2	4	1	5	3	6
		5		15.02	2	3	4	5	1	6
Mean rank					2.2	3.3	2.3	4.2	3	6

Note: Data from Georges Bank scallop surveys for 1989 and 1990. Strata 1, 2, 3, and 4 refer to very low, low, intermediate, and high catch-per-unit-effort strata, respectively. The right-most six columns give the rank (in descending order) of the mean number of scallops by sediment type within each stratum.

simulation are quite close to normal distributions as illustrated by the quantile–quantile plots in Fig. 3. The agreement between the sample means and the normal distribution for the first population was much better for sample sizes of 30 than for those of 10 (not shown), whereas the agreement was very good for sample sizes of 10 for the second population. The need for larger sample sizes, for the distribution of the mean of the Weibull-Bernoulli random variates to approach the normal, is a direct function of the skewness present in the first population.

Evaluating the current design

We now turn our attention to the stratified design used for scallops on Georges Bank. The basic stratified estimates for

Fig. 4. Sediment type map for the Canadian zone of Georges Bank, adapted from Thouzeau et al. 1991*a*.

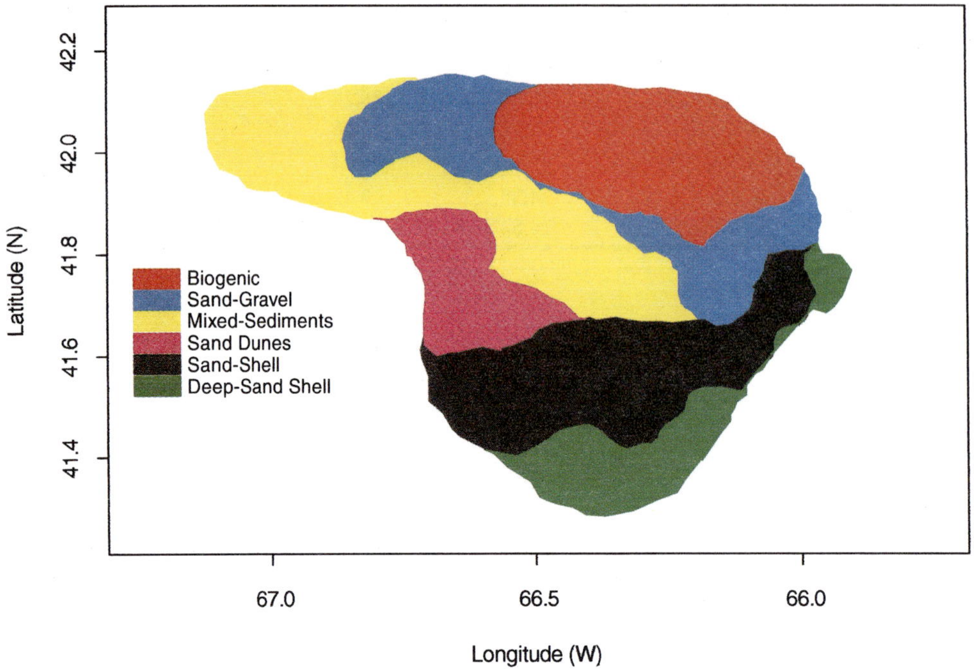

Fig. 5. Location of sets with numbers of age-3 scallops caught during the 1989 Georges Bank survey indicated by color coding.

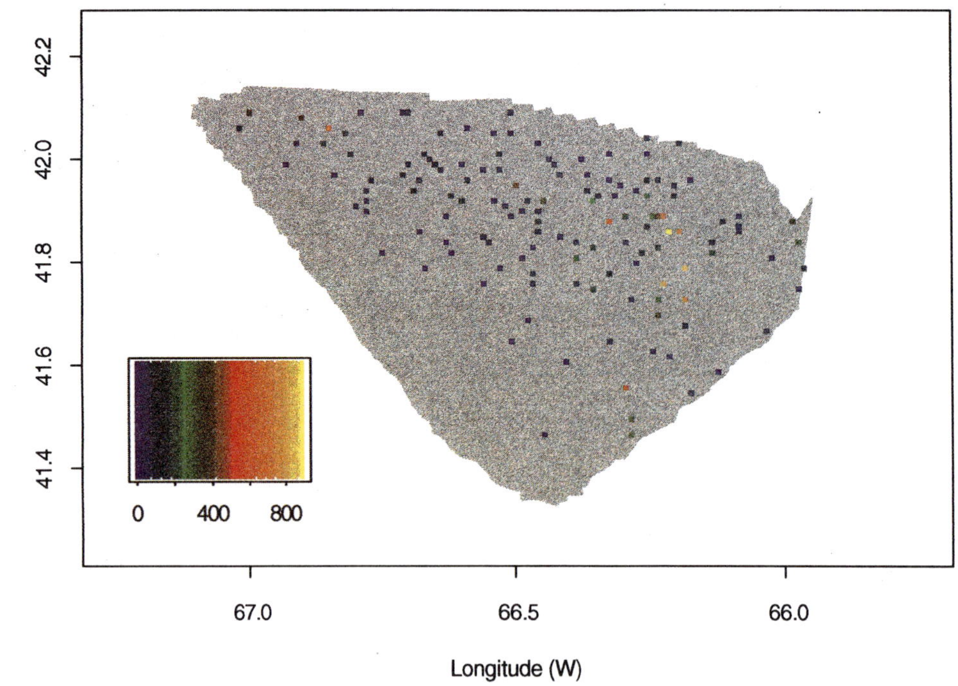

total number and mean number of scallops per tow (with associated estimate of standard error) for ages 3, 4, and 5 for 1989 and 1990 are presented in Table 2.

Currently spatial information is incorporated into the design by defining the strata boundaries according to levels of commercial catch-per-unit-effort and allocating more samples to the higher catch rate strata. The degree to which this design may increase (or decrease) precision can be evaluated by comparing the current stratified variance with that which would have been obtained if no spatial information had been used, i.e., simple random sampling. The difference between the stratified and simple random variance known as efficiency was used by Smith and Gavaris (1993) to evaluate stratified designs for Atlantic cod (*Gadus morhua*). The difference between the two variances is comprised of two components, one due to the allocation scheme and the other measures the effectiveness of

Table 4. Comparison of three types of survey estimates of scallop abundance within strata.

Year	Age	Strata	Standard		Poststratified		Predictive	
			$\bar{y}_h$	var $(\bar{y}_h)$	$\bar{y}_h$	var $(\bar{y}_h)$	$\bar{y}_h$	var $(\bar{y}_h)$
1989	3	1	55.32	324.72	46.09	277.12	46.09	238.39
		2	67.78	1154.64	48.32	1088.42	48.32	1235.52
		3	143.17	1951.87	144.66	1720.49	144.66	2022.30
		4	161.15	506.25	136.80	535.89	136.80	470.46
	4	1	94.52	1030.41	85.31	876.18	85.31	808.26
		2	73.33	1557.88	44.59	1156.13	44.59	1355.71
		3	87.78	450.29	82.45	344.30	82.45	440.16
		4	57.03	57.61	45.62	37.16	45.62	53.14
	5	1	16.03	17.47	15.73	19.62	15.73	14.52
		2	12.44	11.63	11.03	15.17	11.03	11.56
		3	22.22	55.06	14.83	36.25	14.83	53.00
		4	8.60	0.92	7.35	0.70	7.35	0.86
1990	3	1	40.60	99.20	44.51	99.63	44.51	85.19
		2	39.40	278.56	42.65	234.35	42.65	172.40
		3	142.40	9192.97	135.78	9560.47	135.78	8757.22
		4	99.22	102.41	101.48	87.46	101.48	82.08
	4	1	33.03	67.90	37.13	68.89	37.13	58.37
		2	27.30	49.42	25.40	40.73	25.40	35.64
		3	20.90	10.76	20.00	14.41	20.00	9.92
		4	46.88	17.22	47.92	14.99	47.92	14.06
	5	1	18.60	19.98	19.27	22.82	19.27	18.58
		2	9.60	4.49	9.69	5.55	9.69	5.02
		3	13.00	5.52	14.86	4.70	14.86	4.54
		4	15.02	2.04	15.56	2.14	15.56	1.93

Note: Standard refers to standard estimates of mean and variance within a stratum. Strata 1, 2, 3, and 4 refer to very low, low, intermediate, and high catch-per-unit-effort strata, respectively. Poststratified and predictive estimates make use of the sediment type data.

the stratification. These components and their total are presented in Table 2 for each age and year as percentages of the simple random sampling variance. In all cases, allocation components were less than zero indicating that sample sizes were not allocated in an optimal manner, i.e., in proportion to the strata variances. Strata contributed little to an increase in precision and hence total efficiency estimates reflected the suboptimal allocation scheme. Therefore, with the exception of the meager increase in precision for age 3 in 1989, variances of the mean were between 9 and 71% larger than they would have been, had simple random sampling been used.

Efficiency estimates indicate that commercial catch-per-unit-effort from a period prior to the survey does not appear to convey useful information with respect to the spatial variability of scallop abundance. That is, stratification was unsuccessful here in increasing the precision of the mean over that expected from a simple random sample because variances within strata were greater than between strata overall.

Alternatives to the current design

Scallop distribution on Georges Bank has been shown to be related to bottom sediment type by Thouzeau et al. (1991*a*). They reported that the abundance of scallops was higher on gravel than other types of sediment. Further, Thouzeau et al. (1991*b*) identified six faunal groups based on the relative amounts of sand, gravel, shell, and biogenic material. The three gravel type groups of biogenic gravel, mixed sediments, and sand-gravel are located in the Northeast Peak. The central and south part of Georges Bank was characterized by the three sand groups of sand-shell, deep sand-shell, and sand-dunes. At present very little sediment information is recorded at the dredge sites in the Georges Bank survey. Instead we have determined the approximate sediment type at each site by locating the sample sites on a sediment type map adapted from Thouzeau et al. (1991*b*) (Fig. 4).

The ranks in descending order of mean numbers of scallops by age for each sediment type within the existing strata are given in Table 3. The sediment type of sand-gravel stands out as being highly ranked for all four of the existing strata. Sand-shell, deep sand-shell, and biogenic also are ranked highly for some of the strata while sand-dunes are consistently rated as the least likely sediment type with which scallops were associated.

We used the sediment type information to define strata within the existing strata and as factor levels in a simple linear model. In the former case these new strata were used to poststratify the data while in the latter predictive estimates were derived. Resulting estimates are compared in Table 4. While there were a few exceptions, overall the variances for poststratified and predictive estimates were generally smaller than those for the standard method used for existing catch-per-unit-effort strata. In addition, variances for predictive estimates were usually smaller than those from poststratification.

Variances for the poststratification estimates are overestimates of what would be expected had sediment type been used to define strata beforehand. The random sample size

Table 5. A comparison of the two components in the variance of the mean from poststratification.

Year	Age	Strata	Total	Components of variance: Proportional	Components of variance: Random sample size	Percent
1989	3	1	277.12	226.18	50.94	18
		2	1088.42	836.09	252.34	23
		3	1720.49	1488.81	231.68	13
		4	535.89	511.03	24.86	5
	4	1	876.18	747.06	129.12	15
		2	1156.13	727.41	428.72	37
		3	344.30	290.62	53.68	16
		4	37.16	35.01	2.15	6
	5	1	19.62	17.31	2.31	12
		2	15.17	12.06	3.11	21
		3	36.25	31.83	4.42	12
		4	0.70	0.67	0.03	4
1990	3	1	99.63	90.91	8.72	9
		2	234.35	205.02	29.33	13
		3	9560.47	8402.39	1158.08	12
		4	87.46	84.59	2.87	3
	4	1	68.89	62.98	5.91	9
		2	40.73	31.58	9.15	15
		3	14.41	12.54	1.87	13
		4	14.99	14.51	0.48	3
	5	1	22.82	20.28	2.54	11
		2	5.55	4.09	1.46	26
		3	4.70	3.75	0.95	20
		4	2.14	2.06	0.08	4

Note: The *Proportional* component refers to the assumption of proportional allocation given simple random sampling within original commercial catch rate strata.The second component, *Random sample size*, refers to the fact that sample sizes are random and not fixed as is normally the case in sample surveys. The right-most column gives the random component as a percentage of the total.

Table 6. Comparison of different estimates of stratified mean number of scallops by age for Georges Bank scallops.

Year	Age	Stratified $\bar{y}_{st}$	Stratified var $(\bar{y}_{st})$	Poststratified $\bar{y}_{st}$	Poststratified var $(\bar{y}_{st})$	Predictive $\bar{y}_{st}$	Predictive var $(\bar{y}_{st})$
1989	3	105.26	180.10	91.29	168.74	91.29	166.15
	4	77.86	188.24	65.03	152.03	65.03	137.59
	5	14.11	3.96	12.11	3.84	12.11	3.65
1990	3	79.31	349.32	80.57	358.34	80.57	323.64
	4	34.35	9.47	35.28	9.00	35.28	7.45
	5	14.54	2.10	14.68	2.34	14.68	1.99

Note: Stratified refers to estimates from using strata defined by commercial catch-per-unit-effort only. Poststratified and predictive estimates use sediment type information within existing strata.

component of eq. 6 is the main contributor to this overestimation. Assuming that a proportional-to-area allocation scheme is used then our results in Table 5 indicate that variances could have been smaller and in some cases much smaller than those given in Table 4.

Combining variance estimates in Table 4 over strata gives stratified estimates of the mean in Table 6. Gains in precision of poststratified and predictive estimates over standard stratified estimates were fairly minor when compared to losses in Table 2 due to suboptimal allocation of samples to strata. Indeed, poststratified estimates for ages 3 and 5 in 1990 had poorer precision than the original estimates. While predictive estimates always provided a gain in precision, gains were only substantive enough to offset losses due to poor or minimal allocation effects for age 3 in 1989 and age 4 in 1990.

Spatial patterns

All estimation methods presented above employ a spatial structure of some sort. This structure will determine what values will be predicted for nonsampled locations using data from the sampled locations (e.g., Fig. 5). In the case of the standard stratified estimates, locations simply tie strata means to the location of the strata — all unsampled locations in the strata are assigned the strata mean (Fig. 6). Poststratified and

Fig. 6. Map of abundance of age-3 scallops from the 1989 Georges Bank survey estimated from strata means.

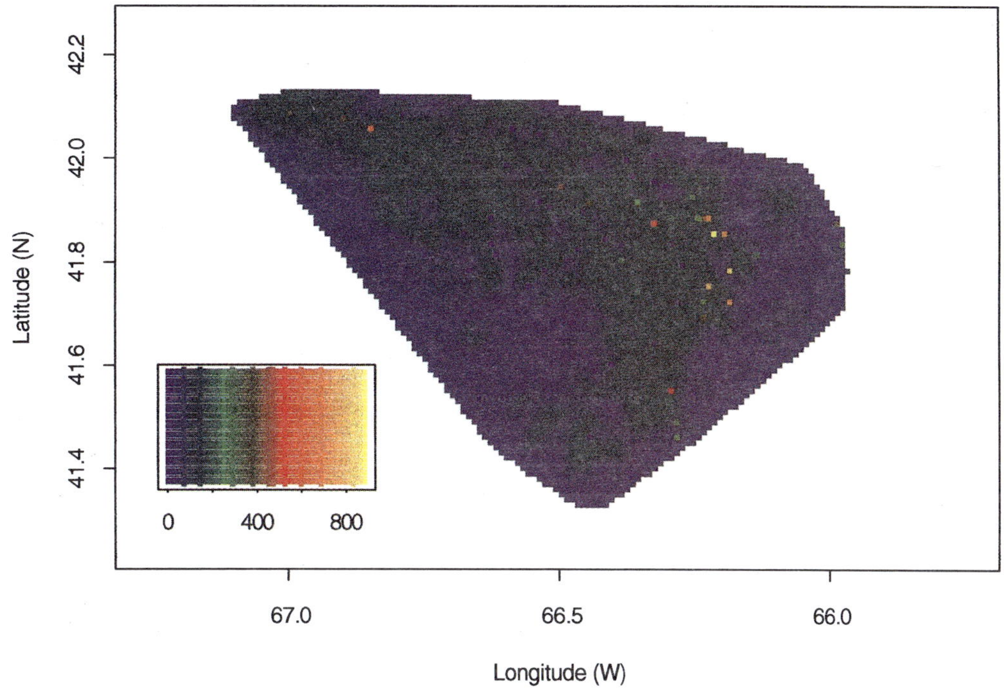

Fig. 7. Map of abundance of age-3 scallops from the 1989 Georges Bank survey as predicted by the poststratification/predictive models.

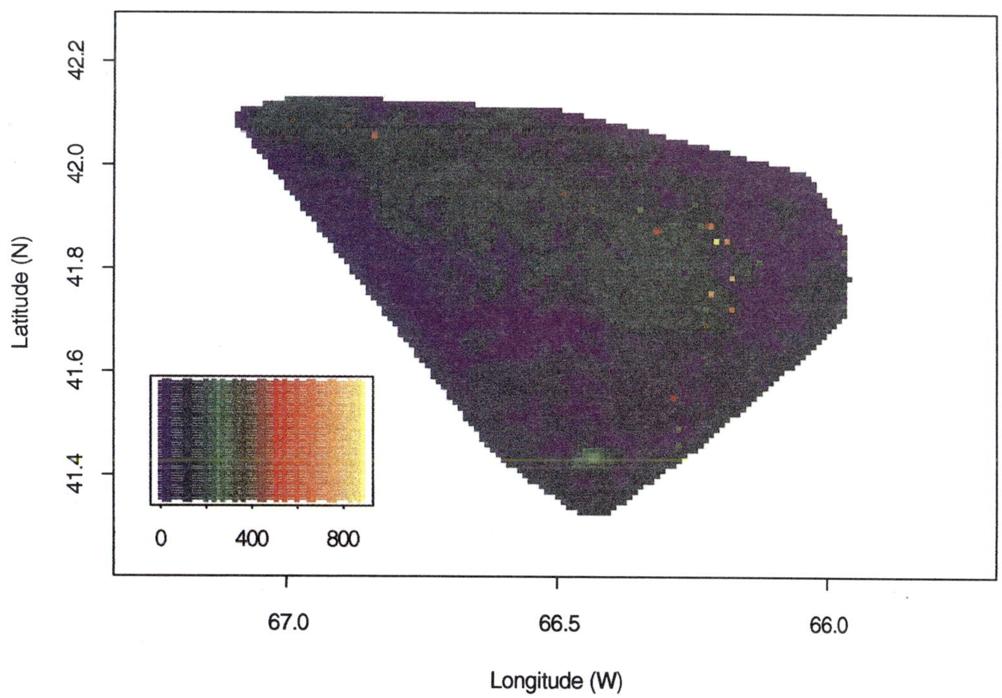

predictive estimates are located by their original strata and the sediment type. Unsampled sites are assigned predicted values based on their membership in the original stratification scheme and associated sediment type.

Note that ranges for predicted values for all of these maps differ with the strata means map (Fig. 6) having the least range and the poststratified/predictive map (Fig. 7) having the most range. The strata means map has the simplest spatial structure and the predicted value at any location is just one of four strata means, depending on what stratum the location belongs to.

Note that the strata means for age 3 in 1989 (Table 3) are very similar for strata 1 and 2 (55.32 and 67.78, respectively) and for strata 3 and 4 (143.17 and 161.15, respectively). Therefore, spatial structure for unobserved locations in Fig. 6 simply represents the contrast between the two lower catch rate strata and the two higher catch rate strata.

Poststratified and predictive estimates will have one of 24 possible values depending upon what stratum and sediment type the location belongs to. The main spatial structure in Fig. 7 reflects generally higher numbers of scallops caught on shell-gravel and sand-shell type sediments (see Fig. 4).

Discussion

Results of the simulation study illustrate the fact that spatial structure or nonnormal distributions do not affect properties of estimates from finite population based methods. Sample means and variances are still unbiased estimates of population quantities. In the case of the variance (eq. 4), we are essentially estimating a population sum of squares with a sample sum of squares and distribution type or spatial correlation do not enter into the situation. While the normal distribution may be suitable in some cases for characterizing the distribution of samples means, sample size limitations and skewness in the data may result in less than satisfactory confidence limits (e.g., negative lower bounds). Methods such as bootstrapping offer an alternative means of calculating confidence limits (Smith 1994).

Incorporating spatial structure into the design, the estimates, or both can be important when doing so increases the precision of estimates. The basic level that spatial structure can be used to increase precision is in the stratification scheme. The stratification scheme for the Georges Bank scallop surveys using commercial data has been modified several times since it was first attempted for this stock. Commercial fishing effort from 10 min × 10 min unit areas was tried first. However, Caddy and Chandler (1969) found that catch was a better index of stock abundance. They also examined depth and sediment variables; depth was found to be the next best variable. Jamieson et al. (1981) advocated a stratification design based on commercial catch rate data from several months preceding the survey. Subsequently, scallop surveys have been based on commercial catch rates.

However, our study and those by Mohn et al. (1987) and Smith and Mohn (1987) show that strata and allocation schemes based on these catch rates do not provide gains in precision over that obtained from simple random sampling for this survey. The biggest culprit in this loss of precision was the allocation scheme — areas with the highest commercial catch rates do not necessarily exhibit the highest variability in abundance. This trend is evident in Table 4 where the highest commercial catch rate stratum (stratum 4) generally had the lowest variances.

Strata defined by commercial catch rate for scallops provided little if any gains in precision. Gavaris and Smith (1987) found that suboptimal allocation schemes and minimal gains from strata boundaries were characteristic of strata used in the design for Scotian Shelf groundfish surveys when Atlantic cod was the target species. In the case of cod, strata were mainly based upon depth ranges.

The current design allows a research vessel to mimic commercial fishing patterns from the previous nine months to see what the fishing fleet saw with respect to size composition and abundance. However, interpretation of survey estimates as abundance indices should require that precision of estimates be as high as possible. To that end the present method of basing the strata on commercial catch rates can be retained but sample allocation could be made to be proportional to the area of the strata. This will have the effect of making allocation effects equal to zero (Smith and Gavaris 1993) and provide gains in precision over that presently being obtained.

Sediment type strata could then be designed within each of the commercial catch rate strata and at a minimum, samples could be allocated to these strata according to area as well. Results presented here suggest that this two-tiered approach would likely provide some gains in precision over simple random sampling. Note that because variances estimated by eq. 6 were overestimates, variances from the sediment type strata could be between 3 and 37% smaller than reported here.

Sediment type boundaries may be loosely defined as strong tidal currents and storm events can play important roles in redistributing sand and finer size particles over considerable distances at times. Actual boundaries during the survey may be determined by deploying equipment such as the ROXANNE (Marine Micro Systems Ltd., The Innovation Centre, Offshore Technology Park, Aberdeen, AB23 8GX, Scotland, United Kingdom) sounder system prior to conducting the actual scallop dredge survey.

Of the two model-based methods presented here, predictive estimates appeared to provide the most promising means for increasing precision. Other covariates such as depth could be included if they contributed to increasing precision of the estimates.

Acknowledgments

The authors gratefully acknowledge Gerry Black's (Science Branch, DFO, Halifax) assistance in extracting and processing the data. Prof. Brian Ripley (Oxford University) provided the S-PLUS functions to estimate the variograms. These functions and those in the new S+SpatialStats module from S-Plus (S-PLUS Version 3.3 for Windows (1995), Statsci Division, MathSoft Inc., Seattle, Washington) were used for the variogram and spatial simulation analyses. Comments from two anonymous referees contributed towards a much improved version of this paper. We thank these referees for their time and attention.

References

Azarovitz, T.R. 1981. A brief historical review of the Woods Hole trawl survey time series. *In* Bottom trawl surveys. *Edited by* W.G. Doubleday and D. Rivard. Can. Spec. Publ. Fish. Aquat. Sci. 58. pp. 62–67.

Caddy, J.F., and Chandler, R.A. 1969. Georges Bank scallop survey August 1966: a preliminary study of the relationship between research vessel catch, depth, and commercial effort. Fisheries Research Board of Canada Manuscript Report No. 1054. 15 p.

Cochran, W.G. 1977. Sampling techniques. John Wiley & Sons, New York, N.Y.

Conan, G.Y., and Wade, E. 1989. Geostatistical mapping and global estimation of harvestable resources in a fishery of northern shrimp (*Pandalus borealis*). ICES C.M. 1989/D: 1. 11 p.

Cressie, N.A.C. 1993. Statistics for spatial data. Revised ed. John Wiley & Sons, New York, N.Y.

Doubleday, W.G. (*Editor*). 1981. Manual on groundfish surveys in the Northwest Atlantic. NAFO Scientific Council Studies. No. 2. 55 p.

Ecker, M.D., and Heltshe, J.F. 1994. Geostatistical estimates of scallop abundance. Chapter 6. *In* Case studies in biometry. *Edited by* N. Lange, L. Ryan, L. Billard, D. Brillinger, L. Conquest, and J. Greenhouse. John Wiley & Sons Inc., New York, N.Y. pp. 125–144.

Foote, K.G., and Stefánsson, G. 1993. Definition of the problem of estimating fish abundance over an area from acoustic line-transect measurents of density. ICES. J. Mar. Sci. **50**: 369–381.

Gavaris, S., and Smith, S.J. 1987. Effect of allocation and stratification strategies on precision of survey abundance estimates for Atlantic cod (*Gadus morhua*) on the Eastern Scotian Shelf. J. Northwest Atl. Fish. Sci. **8**: 137–145.

Guillard, J., Gerdeaux, D., Brun, G., and Chappez, R. 1992. The use of geostatistics to analyze data from an echo-integration survey of fish stock in Lake Sainte-Croix. Fish. Res. **13**: 395–406.

Jamieson, G.S., Lundy, M.J., Kerr, G.L., and Witherspoon, N.B. 1981. Fishery characteristics and stock status of Georges Bank scallops. Canadian Atlantic Fisheries Scientific Advisory Committee, Research Document No. 81/70. 35 p.

Jolly, G.M., and Hampton, I. 1990. A stratified random transect design for acoustic surveys of fish stocks. Can. J. Fish. Aquat. Sci. **47**: 1282–1291.

Little, R.J.A. 1983. Estimating a finite population mean from unequal probability samples. J. Am. Stat. Assoc. **78**: 596–604.

Mathsoft 1996. S+SPATIALSTATS User's manual, version 1.0. Mathsoft, Inc.. Seattle, Wash.

Mohn, R.K., Robert, G., and Roddick, D.L. 1987. Research sampling and survey design of Georges Bank scallops (*Placopecten magellanicus*). J. Northwest Atl. Fish. Sci. **8**: 117–122.

Pelletier, D., and Parma, A.M. 1994. Spatial distribution of Pacific Halibut (*Hippoglossus stenolepis*): an application of geostatistics to longline survey data. Can. J. Fish. Aquat. Sci. **51**: 1506–1518.

Rao, J.N.K. 1988. Variance estimation in sample surveys. *In* Handbook of statistics, Vol. 6 (Sampling). *Edited by* P.R. Krishnaiah and C.R. Rao. Elsevier Science Publishers, Amsterdam, The Netherlands. pp. 427–447.

Robert, G., and Jamieson, G.S. 1986. Commercial fishing isopleths and their use in offshore sea scallop *(Placopecten magellanicus)* stock evaluations. *In* North Pacific Workshop on Stock Assessment and Management of Invertebrates. *Edited by* G.S. Jamieson and N. Bourne. Can. Spec. Publ. Fish. Aquat. Sci. 92. pp. 76–82.

Robert, G., Black, G.A.P., and Butler, M.A.E. 1994. Georges Bank scallop stock assessment — 1994. DFO Atlantic Fisheries Research Document No. 94/97. 42 p.

Serchuk, F.M., and Wigley, S.E. 1986. Evaluation of USA and Canadian research vessel surveys for sea scallops (*Placopecten magellanicus*) on Georges Bank. J. Northwest Atl. Fish. Sci. **7**: 1–13.

Simard, Y., Legendre, P., Lavoie, G., and Marcotte, D. 1992. Mapping, estimating biomass and optimizing sampling programs for spatially autocorrelated data: case study of the northern shrimp *(Pandalus borealis*). Can. J. Fish. Aquat. Sci. **48**: 32–45.

Simmonds, E.J., Williamson, N.J., Gerlotto, F., and Aglen, A. 1992. Acoustic survey design and analysis procedures: a comprehensive review of current practice. ICES Cooperative Research Report No. 187. 131 p.

Smith, S.J. 1994. Evaluating statistical properties of trawl survey estimates of mean abundance. ICES, C.M. (B+D+G+H): 3. 23 p.

Smith, S.J., and Gavaris, S. 1993. Improving the precision of fish abundance estimates of Eastern Scotian Shelf cod from bottom trawl surveys. N. Am. J. Fish. Manage. **13**: 35–47.

Smith, S.J., and Mohn, R.K. 1987. Considerations on the representation and analysis of a spatially aggregated resource: Georges Bank scallops. ICES, C.M. 1987/K: 26. 18 p.

Thompson, S.K. 1992. Sampling. John Wiley & Sons, Inc., New York, N.Y.

Thouzeau, G., Robert, G., and Smith, S.J. 1991*a*. Spatial variability in distribution and growth of juvenile and adult sea scallops *Placopecten magellanicus* (Gmelin) on eastern Georges Bank (Northwest Atlantic). Mar. Ecol. Prog. Ser. **74**: 205–218.

Thouzeau, G., Robert, G., and Ugarte, R. 1991*b*. Faunal assemblages of benthic megainvertebrates inhabiting sea scallop grounds from eastern Georges Bank, in relation to environmental factors. Mar. Ecol. Prog. Ser. **74**: 61–82.

Traynor, J.J., Karp, W.A., Furusawa, M., Sasaki, T., Teshima, K., Sample, T.M., Williamson, N.J., and Yoshimura, T. 1990. Methodology and biological results from surveys of walleye pollock (*Theragra chalcogramma*) in the eastern Bering Sea and Aleutian Basin in 1988. *In* Proceedings of the Symposium on Application of Stock Assessment Techniques to Gadids. *Edited by* L.L. Low. Int. North Pac. Fish. Comm. Bull. No. 50. pp. 69–99.

Adaptive cluster sampling: efficiency, fixed sample sizes, and an application to red sea urchins (*Strongylocentrotus franciscanus*) in southeast Alaska

Doug Woodby

Abstract: Adaptive cluster sampling designs and estimators (Thompson 1990) were evaluated for simulated populations across a gradient of aggregations indexed by their variance to mean ratio (VMR). The modified Hansen–Hurwitz estimator was inefficient relative to simple random sampling for all but the more highly aggregated populations, whereas the modified Horvitz–Thompson estimator was generally more efficient, especially for highly aggregated populations and high initial sampling fractions. A constrained grid adaptive design was implemented to overcome the problem of random sample sizes for adaptive cluster designs. This fixed sample size method was implemented for systematic adaptive cluster sampling of simulated one dimensional populations, and was found to have higher efficiency for both Hansen–Hurwitz and Horvitz–Thompson estimators relative to simple random sampling. Efficiency was highest with low initial sampling fractions. The fixed sample size design provided improved precision when applied to a red sea urchin (*Strongylocentrotus franciscanus*) population in southeast Alaska, USA.

Résumé : Les plans d'échantillonnage par grappe adaptatifs et les estimateurs (Thompson 1990) ont été évalués pour des populations simulées le long d'un gradient d'agrégation indexé par le rapport variance/moyenne (RVM) de ces populations. L'estimateur modifié de Hansen–Hurwitz était inefficace par rapport à l'échantillonnage aléatoire simple de toutes les populations, sauf celles présentant les degrés d'agrégation les plus élevés, alors que l'estimateur Horvitz–Thompson était généralement plus efficace, surtout pour les populations à forte agrégation et les fractions d'échantillonnage initiales élevées. Un plan adaptatif à contrainte de grille a été mis en oeuvre pour surmonter le problème de la taille des échantillons aléatoires pour les plans d'échantillonnage par grappe adaptatifs. Cette méthode de taille d'échantillon fixe a été appliquée à l'échantillonnage par grappe adaptatif systématique de populations unidimensionnelles simulées et on a constaté qu'elle présentait une plus grande efficacité que les deux estimateurs Hanse–Hurwitz et Horvitz–Thompson pour l'échantillonnage aléatoire simple. L'efficacité était la plus grande dans le cas des fractions d'échantillonnage initiales faibles. Le plan d'échantillonnage avec taille d'échantillon fixe a fourni une précision améliorée lorsqu'il a été appliqué à la population d'oursins rouges (*Strongylocentrotus franciscanus*) au sud-est de l'Alaska, É.-U.
[Traduit par la Rédaction]

Introduction

Aggregated populations are difficult to sample efficiently with conventional designs, such as simple random sampling (SRS). Random samples of highly aggregated populations often yield many zero counts, and researchers may feel an intuitive urge to continue sampling in areas of aggregations, even though such sampling violates prescriptions for equal probability for selection of samples and would lead to biased estimates of the mean and variance. This urge has been satisfied, in part, by the recent development of adaptive cluster sampling (ACS) methods with unbiased estimators (Thompson 1990, 1991, 1992). As yet, there have been few applications of ACS methods to marine invertebrates or to sampling biological populations in general. This study provides some numerical results on the statistical efficiency of ACS relative to SRS, and a systematic adaptive cluster sampling method (SysACS) with fixed sample size is described that overcomes the problem of random sample sizes in ACS. Also, an application of the fixed sample size method is made to estimate the abundance of red sea urchins (*Strongylocentrotus franciscanus*) in southeast Alaska.

ACS is a two-stage process beginning with a random sample. In the second stage, additional samples are taken near the initial samples having the highest counts. This approach would normally provide a biased estimate of population density if the usual formula for the mean, which assumes equal probability of selection for all samples, were used. Instead, the ACS estimators are based on unequal selection probabilities. These estimators have the desirable attribute of being design unbiased, meaning that they are unbiased due to the method for selecting samples, and that this is independent of any assumptions about the population. ACS is a special class of sequential sampling methods with a theoretical basis of development spanning the past quarter century (Thompson 1992). The few applications

D. Woodby. Alaska Department of Fish and Game, P.O. Box 240020, Douglas, AK 99824, U.S.A.

Correct citation: Woodby, D. 1998. Adaptive cluster sampling: efficiency, fixed sample sizes, and an application to red sea urchins (*Strongylocentrotus franciscanus*) in southeast Alaska. *In* Proceedings of the North Pacific Symposium on Invertebrate Stock Assessment and Management. *Edited by* G.S. Jamieson and A. Campbell. Can. Spec. Publ. Fish. Aquat. Sci. 125. pp. 15–20.

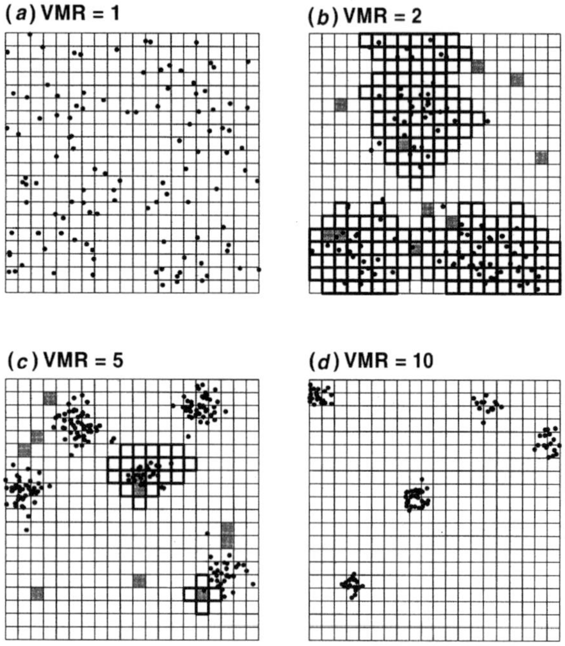

Fig. 1. Examples of populations of 200 points on 20 × 20 grids with variance to mean ratios (VMR) = 1, 2, 5, and 10. Populations with VMR = 2 and 5 include *n* initial sample units (shaded) and n_a units sampled adaptively (outlined in bold).

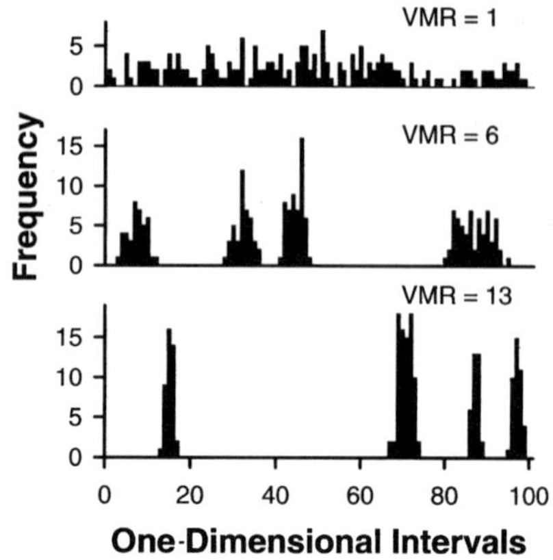

Fig. 2. Examples of populations of 100 points on 1 × 100 grids with variance to mean ratios (VMR) = 1, 6, and 13.

of ACS to aggregated populations include a modified ACS method for forest sampling (Roesch 1993), and an evaluation of the efficiency of ACS for estimating duck populations surveyed from the air (Smith et al. 1995).

Sample populations

Simulated populations — two dimensions

Adaptive cluster sampling designs were evaluated using simulated populations in a procedure similar to that of Thompson (1990, 1992). In each simulation, 200 individual points were distributed according to a Poisson cluster process (Diggle 1983) on a 20 × 20 grid as follows. Five parent locations were selected without replacement using uniform random deviates on the interval (0, 20), which when selected in pairs gave *x*, *y* coordinates on the grid. Uniform deviates were generated by the "uniform" function in SAS (1990), based on the algorithm by Fishman and Moore (1982). The expected number of daughter values for each parent was selected using Poisson random deviates ("ranpoi" function in SAS 1990) with mean = 40. Daughter points were assigned to one of the parents with probability proportional to the expected size of each parent's daughter population. Daughter points were distributed about the parent locations using a bivariate normal distribution with mean and covariance = 0 and variance in both the *x* and *y* directions = 5, 1.37, 0.94, and 0.74. The resulting population distributions ranged from complete spatial randomness to highly clumped, with variance to mean ratios averaging 1, 2, 5, and 10, respectively. Daughter points falling outside the grid were discarded. Assignment and location of daughter points was continued until the population totaled 200 points. Fixed population size was a departure from the method of Thompson (1990), and was used to simplify comparisons of results between the various sampling designs presented here. Examples of simulated populations are shown in Fig. 1 for each of the expected variance to mean ratios.

Simulated populations — one dimension

The Alaska Department of Fish and Game conducts surveys of red sea urchins and other benthic invertebrate populations with densities estimated as numbers per linear metre of shoreline. One-dimensional populations of this type were simulated for an evaluation of the relative efficiency of SysACS designs. Simulated counts were created using the methods for two dimensions, except that the grid size was collapsed and elongated to 1 × 100 cells, and a mean of 100 daughter values were distributed in just one dimension from three parent locations. Variance in displacement from the parent location in the *x* direction was limited to three levels, 5, 1.37, and 0.94. The population variance to mean ratio averaged 1, 6, and 13 for the three levels of aggregation. Examples of these one-dimensional populations are shown in Fig. 2.

Urchin counts in one dimension

Counts of red sea urchins on three shoreline areas in southern southeast Alaska (Fig. 3) were sampled systematically with random starts in August 1994. Counts were made on pairs of 1-m wide transects extending perpendicular to shore at each interval location. Transects in pairs were approximately 20 m apart. All urchins 60 mm test diameter were counted from below the mean lower low tide stage to a depth of 10 m datum. Counts were averaged for each transect pair, and there were 20 transect pairs in areas *a* and *c*, and 30 in area *b* (Fig. 4). The interval between transect pairs ranged from 2.4 to 6.8 km (Table 1). These samples were used as the populations from which systematic and adaptive samples on a constrained grid were drawn (below). Variance to mean ratios were much higher for these urchin samples than for the simulated populations, ranging from 92 to 1211 (Table 1).

Sampling designs

Adaptive cluster sampling (ACS)

Stage 1 of ACS was implemented on the two-dimensional grid as a simple random sample without replacement of *n* cells from

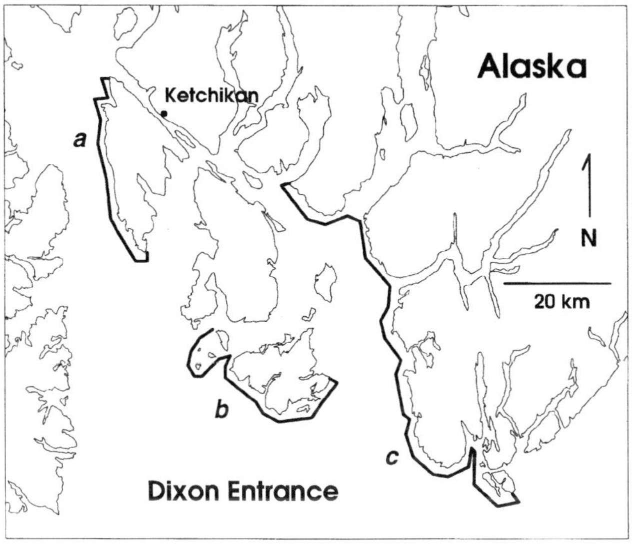

Fig. 3. Map of southern southeast Alaska, USA, showing the location of urchin surveys on shorelines of (*a*) Gravina Is., (*b*) Duke Is. and Percy Is., and (*c*) the mainland, August 1994.

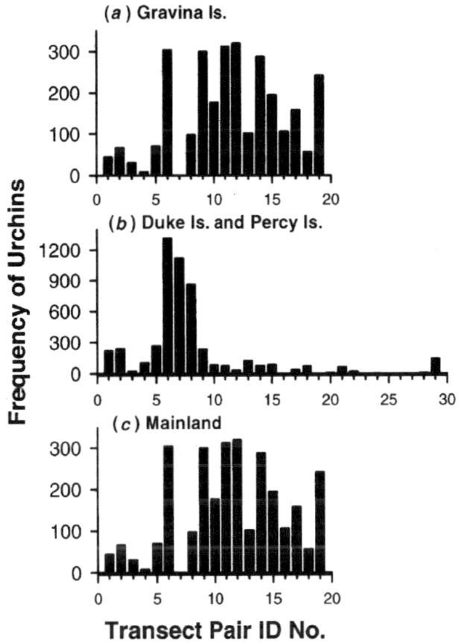

Fig. 4. Average frequencies of red sea urchins on transect pairs on shorelines of (*a*) Gravina Is., (*b*) Duke Is. and Percy Is., and (*c*) the mainland, August 1994. Variance to mean ratios were 92, 1211, and 457 in areas *a*, *b*, and *c*, respectively.

Table 1. Summary of survey data for red sea urchins in southern southeast Alaska, August 1994.

Area	N	Urchins	Density (No./m)	VMR[a]	Interval[b] (km)
(*a*) Gravina Is.	20	5744	144	92	2.4
(*b*) Duke Is., Percy Is.	30	5309	88	1211	3.6
(*c*) Mainland	20	1852	46	457	6.8

[a]Variance to mean ratio.
[b]Systematic spacing between transect pairs. Spacing is unequal between areas because fixed sample sizes for each area were distributed to shorelines of unequal lengths.

a total of $N = 400$ by selection of uniform random deviates on the interval (1, N). Stage 2 was the selection of cells in the neighborhood of the ith cell selected in stage 1, provided that $y_i \geq c$, where y_i is the number of daughter points in cell i and c is a selection condition. The condition $c = 1$ was used for designs in this study. The neighborhood was defined as those cells adjacent to the ith cell. Diagonal cells were not included in the neighborhood. If the condition was met in a neighborhood cell, then the cells in the neighborhood of that cell were sampled and so on until no new cells were added. The resulting group of cells is a cluster (Figs. 1*b*, 1*c*). Clusters may include many neighboring cells that meet the condition $y_i \geq c$ and edge cells that do not. Those cells meeting the condition share a network, whereas edge cells are each defined to be in their own network of size 1. The number of cells sampled adaptively after stage 1 is n_a. The total sample size $n_t = n + n_a$.

Three unbiased estimates of the mean population density were made. A classic SRS estimator was calculated as

$$\hat{\mu}_{\text{SRS}} = \frac{1}{n_t} \sum_{i=1}^{n_t} y_i$$

for the initial n random sample plus an additional n_a samples were also selected at random without replacement.

Two adaptive cluster sampling estimators derived by Thompson (1990, 1992) provide unbiased estimates based on unequal probabilities of sampling. His modified Hansen–Hurwitz (HH) estimator takes a simple average of the network means for those networks intersected by the initial samples, and was calculated as

$$\hat{\mu}_{\text{HH}} = \frac{1}{n} \sum_{i=1}^{n} w_i$$

where the network means were calculated as

$$w_i = \frac{1}{m_i} \sum_{j \in \Psi_i} y_j$$

where Ψ_i is the network that includes cell i, and m_i is the number of cells in Ψ_i.

Thompson's (1990) modified Horvitz–Thompson (HT) estimator is based on the estimated probabilities for including each network $k = 1$ to K for each observed value:

$$\hat{\mu}_{HT} = \frac{1}{N} \sum_{k=1}^{K} \frac{y_k z_k}{\alpha_k}$$

where y_k is the sum of values in network k, z_k is 1 if any cell in network k is in the initial sample and 0 otherwise, and α_k is the probability that any unit in network k is included in the estimator. This probability was calculated as

$$\alpha_k = 1 - \binom{N - x_k}{n} \Big/ \binom{N}{n}$$

where x_k is the number of cells in the kth network.

Systematic adaptive cluster sampling (SysACS)

Systematic adaptive cluster sampling (SysACS) designs were evaluated for simulated populations in one dimension. Thompson (1991, 1992) provides counterparts of the two ACS estimators for SysACS that are roughly similar to the standard ACS estimators. SysACS differs from ACS in having N^* total possible primary units each with M secondary units located on a systematic grid. A sample of n^* primary units are chosen at random from the N^* possible units.

The modified Hansen–Hurwitz type estimator for SysACS is:

$$\hat{\mu}^*_{HH} = \frac{1}{n^*} \sum_{i=1}^{n^*} w^*_i$$

where w_i* is the average of network values sampled by the ith primary unit. This average is calculated as

$$w^*_i = \frac{1}{M} \sum_{k=1}^{K} \frac{y_k I_{ik}}{x^*_k}$$

where y_k is the sum of observations in network k, I_{ik} is an indicator variable equal to 1 if the kth network intersects the ith primary unit, and is otherwise 0; and x^*_k is the number of primary units (of N^* possible) intersecting the kth network.

The modified Horvitz–Thompson estimator for SysACS is

$$\hat{\mu}^*_{HT} = \frac{1}{MN} \sum_{k=1}^{K} \frac{y_k z_k}{\alpha^*_k}$$

where α^*_k is the probability of including in the sample at least one of the primary units (of N^* possible) that intersect the kth network, calculated as

$$\alpha^*_k = 1 - \binom{N^* - x^*_k}{n^*} \Big/ \binom{N^*}{n^*}.$$

Constrained grid (fixed sample size) SysACS

The SysACS design was modified to allow for a fixed total sample size. For any adaptive cluster design, it is possible to specify that the neighborhood for adaptive samples be limited to a specific pattern (Thompson 1992). For example, in the rectangular layout described above for ACS, only the adjacent cells were included in the neighborhood, whereas cells on the diagonal were excluded.

A simple neighborhood restriction for the one-dimensional SysACS design is to adaptively sample only those cells midway between secondary unit cells. This restriction will provide for a limit to sample size because there are no new cells to sample adaptively after the middle cells have been sampled in any primary unit. The sample size can be limited further by setting c, the selection condition, to equal any value that excludes the desired number of observations in the initial systematic sample from serving as starting points for the adaptive sample. In this way the final sample size can be fixed to nearly any value between Mn^* and $1.5 \times Mn^*$, except in cases where c is equal to tied values in the initial sample.

This "midway" design can be extended to two dimensions if the midway location is defined such that any adaptive samples selected subsequent to the midway locations are the same as the initial samples in the primary unit.

Monte Carlo estimates

The variances of $\hat{\mu}_{SRS}$, $\hat{\mu}_{HH}$, $\hat{\mu}_{HT}$, $\hat{\mu}^*_{HH}$, and $\hat{\mu}^*_{HT}$ were estimated using Monte Carlo methods. Populations were simulated 1000 times for each set of sampling and population characteristics. To investigate the effect of various initial sample sizes on final sample sizes and efficiency, n was set to values of 5, 10, 25, 50, 100, and 200 for the ACS design on the two-dimensional grid. M was fixed at 10 for the SysACS design applied to the simulated populations in one dimension. M was fixed at 2 for urchin populations in areas a and b, and 3 for area b. The number of primary sample units, n^*, varied from 1 to 5 for all SysACS designs.

Efficiency (E) of the adaptive estimators was estimated relative to random sampling by comparing variances $E_{HH} = \sigma^2_{SRS}/\sigma^2_{HH}$ and $E_{HH} = \sigma^2_{SRS}/\sigma^2_{HT}$.

Results

Populations in two dimensions

The Horvitz–Thompson estimator was always more efficient than the Hansen–Hurwitz estimator for the two-dimensional grid populations sampled by ACS (Fig. 5). The Horvitz–Thompson estimator was particularly efficient for the more highly aggregated populations, and for larger initial sampling fractions ($f = n/N$) of n between 100 and 400 total cells. When the initial sample was 200, the most highly aggregated population was completely sampled and there was no variation in the mean across all Monte Carlo trials.

The Hansen–Hurwitz estimator was notably inefficient relative to SRS for all but the most highly aggregated populations, and then only at the higher initial sampling fractions. The HH estimator was never more than twice as efficient as the SRS estimator, and was frequently less efficient, especially for the more dispersed populations and lower sampling fractions.

The adaptive sample size (n_a) was greatest for the less highly aggregated populations, that is, those with less space between aggregations (Fig. 6*a*). For example, after an initial random sample of 10 cells, an additional 192 cells were sampled adaptively in the loosely aggregated population in Fig. 1*b* (VMR = 2). Stage 2 sample sizes increased with increasing

Fig. 5. Efficiency (ratio of variances) of (*a*) the modified Hansen–Hurwitz (HH) and (*b*) Horvitz–Thompson (HT) estimators relative to SRS for ACS designs applied to the simulated populations in two dimensions. Variance to mean ratios were 1 (solid circles), 2 (squares), 5 (triangles), and 10 (open circles).

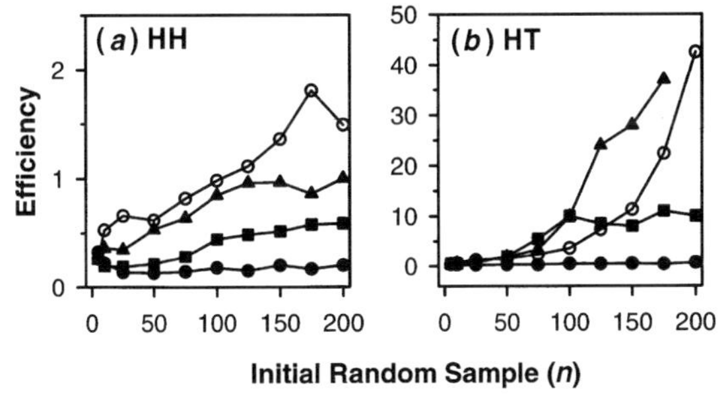

Fig. 6. The size of adaptive samples added to initial sample sizes for (*a*) ACS designs applied to simulated populations in two dimensions and (*b*) SysACS designs applied to simulated populations in one dimension. Variance to mean ratios were 1 (circles), 10 (squares in graph *a*), and 6 (squares in graph *b*). Error bars are 95% confidence intervals. *M* is the number of secondary sampling units in the systematic design and n^* is the number of primary units (random starting points for the layout of systematic secondary units).

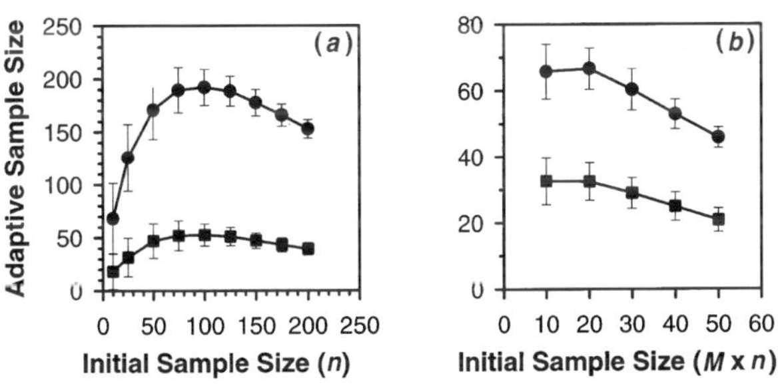

sizes of stage 1 samples, and then declined as the sampling fraction increased. In effect, there were fewer new samples to add as *n* increased. The variation in n_a was also greatest for the less aggregated populations and for smaller sampling fractions.

Populations in one dimension

The HT estimator was especially efficient for sampling the more aggregated, simulated populations in one dimension under SysACS (Fig. 7*b*). The HH estimator also showed improved efficiency for more aggregated populations, but was less efficient than the HT estimator (Fig. 7*a*). The HH estimator was highly inefficient for completely random populations (VMR = 1). The adaptive sample size decreased as the initial sampling fraction increased (Fig. 6*b*). The average deviation in n_a was in the range 3–9 for all simulations.

Efficiency of the HH and HT estimators was virtually the same for fixed sample size SysACS for all simulated populations (Fig. 8). In fact, the estimators are the same where there is only one primary unit. Efficiency was highest for the moderately aggregated population, approximately 1 for the completely random populations, and somewhat less than 1 for the most highly aggregated populations. Efficiency decreased as

Fig. 7. Efficiency (ratio of variances) of (*a*) the modified Hansen–Hurwitz (HH) and (*b*) Horvitz–Thompson (HT) estimators relative to SRS for SysACS designs applied to the simulated populations in one dimension. Variance to mean ratios were 1 (circles), 6 (squares), and 13 (triangles). *M* and n^* are defined in Fig. 6.

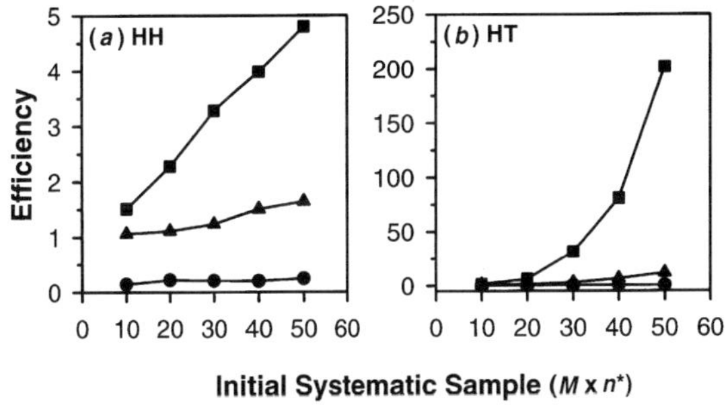

Fig. 8. Efficiency (ratio of variances) of (*a*) the modified Hansen–Hurwitz (HH) and (*b*) Horvitz–Thompson (HT) estimators relative to SRS for the constrained grid (fixed sample size) SysACS designs applied to the simulated populations in one dimension. Variance to mean ratios were 1 (circles), 6 (squares), and 13 (triangles). *M* and n^* are defined in Fig. 6.

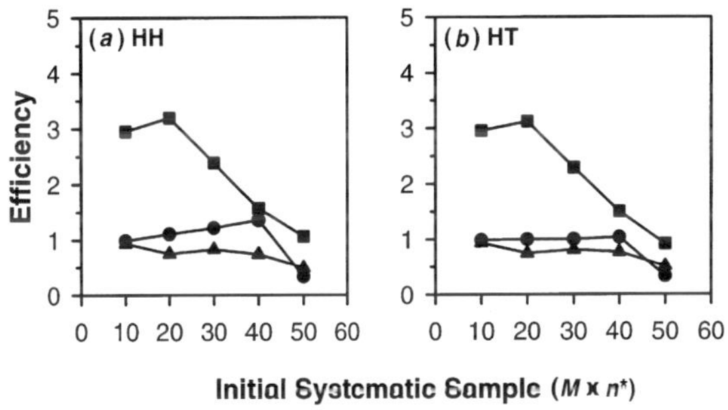

the initial sampling fraction increased. These results stand in contrast to random sample size methods, and illustrate the effect of sampling on a constrained grid where neighborhood cells are not cells immediately adjacent to the cells in the initial sample.

The two SysACS provided modest gains in relative efficiency for the urchin population samples (Fig. 9). Relative efficiency was highest for urchins sampled in area *b*, which had the highest variance to mean ratio (Table 1). The Horvitz–Thompson estimator performed only slightly better than the Hansen–Hurwitz estimator.

Discussion

Marine invertebrate populations are often highly aggregated and are likely candidates for adaptive sampling methods. When sampling additional units is costly, as with research vessels, SCUBA, remotely operated vehicles, or submersibles, adaptive sampling may be used to reduce sample sizes while still meeting precision goals.

The two-stage approach of adaptive cluster sampling is particularly applicable to populations for which only limited distribution data is available. For example, the red sea urchin

Fig. 9. Efficiency (ratio of variances) of the modified Hansen–Hurwitz (circles) and Horvitz–Thompson (squares) estimators relative to SRS for the constrained grid (fixed sample size) SysACS designs applied to the three urchin populations of Fig. 4.

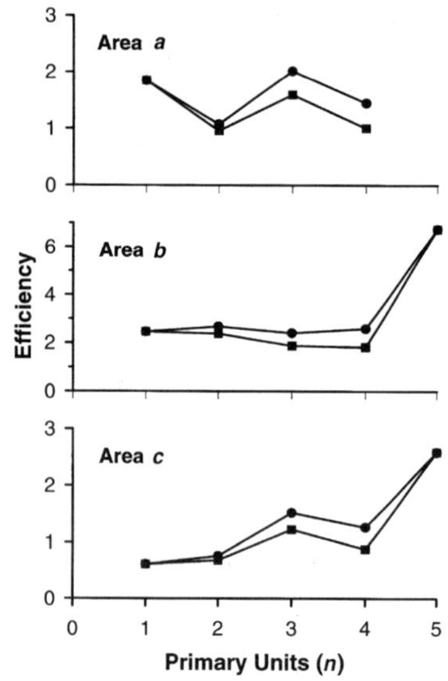

populations sampled by the Alaska Department of Fish and Game in southeast Alaska are known to be aggregated to some extent, but there is limited knowledge on locations of dense concentrations. For these populations, a simple random sample distributed across the area of interest provides information to locate future sampling efforts.

ACS designs may be inappropriate for mobile populations where the time between initial and final samples is long enough to allow significant movements away from initial sampling locations. These designs would also be inappropriate where the sampling unit is so large relative to the size of aggregations that adaptive sample units are unlikely to encounter individuals.

Few analytical guidelines are available to help in choosing design characteristics, including initial sample size and selection condition, to achieve a desired precision with ACS. Thompson (1990) and Smith et al. (1995) provide a general prescription that the Hansen–Hurwitz estimator is more efficient than SRS when the within-network variance is close to (nearly as large as) the population variance and when the initial sampling fraction is close to the final sampling fraction. This result applies to the Horvitz–Thompson estimator whenever it is at least as efficient as the Hansen–Hurwitz estimator, which was true for several examples of Thompson (1990), for the several duck populations surveyed by Smith et al. (1995), and in the simulations here. This guidance is useful once data are available on within-network variances, but these data are unlikely to be available until networks have been sampled adaptively.

The variance to mean ratio was used here as an index of aggregation and to serve as a guide to populations for which ACS may be appropriate. The ratio is simple to compute and will generally be available from pilot survey data collected randomly. More sophisticated, geostatistical methods of describing spatial distributions are available (e.g., Cressie 1991). These methods, including variograms, provide information on spatial pattern that is lacking in the variance to mean ratio. However, geostatistical methods require more frequent sampling, and may be inappropriate for invertebrate populations that must be sampled under water at high cost.

The fixed sample size method reduces some of the uncertainty in sampling design. In the simple method proposed here, the final sample size can be constrained to not exceed 1.5× the initial sample size. The trade-off for certainty in sample size is that gains in precision may be limited if the spacing between adaptive samples and the initial secondary units is larger than the size of aggregations.

More elaborate constrained grid designs may be needed to improve efficiency. A simple extension of the present method is to adaptively sample two or more equally spaced plots between initial samples. This design would lose some of the flexibility in controlling sample size, since the neighborhood of each adaptively sampled plot may include previously unsampled plots. If desired, sample size can be limited further by restricting the selection condition.

Adaptive sampling must be applied judiciously, as it may be more efficient to use equal probability methods, such as SRS, when the population is not highly aggregated. The key is to judge whether adaptive samples placed in the neighborhood of initial samples will provide a more representative final sample than a continuation of random, equal probability sampling. Continued equal probability sampling is likely to be more efficient when there are large areas of unsampled populations for which the initial samples are not representative.

Acknowledgments

I thank Steve Thompson for his advice early in the development of ideas for this paper and his later help in technical issues. I also thank two anonymous reviewers. Ocean Fresh Seafoods of Fort Bragg, CA, provided financial support for the urchin surveys. The Alaska Department of Fish and Game supported all other aspects of the work. Urchin surveys were conducted by Bill Davidson, Tim Koeneman, Robert Larson, Marc Pritchett, Randy Timothy, Scott Walker, and the author. This paper is contribution PP-114 of the Alaska Department of Fish and Game, Commercial Fisheries Management and Development Division, Juneau.

References

Cressie, N.A. 1991. Statistics for spatial data. John Wiley & Sons, Inc., New York, N.Y. 900 p.

Diggle, P.J. 1983. Statistical analysis of spatial point patterns. Academic Press, New York, N.Y. 148 p.

Fishman, G.S., and Moore, L.R. 1982. A statistical evaluation of multiplicative congruential generators with modulus (2^{31} - 1). J. Am. Stat. Assoc. **77**: 129–136.

Roesch, F.A., Jr. 1993. Adaptive cluster sampling for forest inventories. For. Sci. **39**: 655–669.

SAS Institute Inc. 1990. SAS/IML software, version 6. SAS Institute Inc., Cary, N.C.

Smith, D.R., Conroy, M.J., and Brakhage, D.H. 1995. Efficiency of adaptive cluster sampling for estimating density of wintering waterfowl. Biometrics, **51**: 777–786.

Thompson, S.K. 1990. Adaptive cluster sampling. J. Am. Stat. Assoc. **85**: 1050–1059.

Thompson, S.K. 1991. Adaptive cluster sampling: designs with primary and secondary units. Biometrics, **47**: 1103–1115.

Thompson, S.K. 1992. Sampling. John Wiley & Sons Inc., New York, N.Y. 343 p.

Spatial analysis for marine populations: factors to be considered

William G. Warren

Abstract: Spatial analysis (geostatistics, kriging) can be used not only for mapping but also for estimating the abundance or density of relatively sedentary marine populations. In this paper, basic kriging theory is outlined and the difference between this and the conventional methods of analyzing (stratified) random samples is described. What is meant by kriging being an exact interpolator is explained along with how this can be affected by taking measurement error into account. The use of covariates and kriging neighbourhoods is discussed. Problems that arise from a relatively high proportion of zero observations and nonzero observations that are positively skewed, are illustrated with data collected during a survey of scallop stocks.

Résumé : L'analyse spatiale (géostatistique et krigeage) peuvent servir non seulement à l'établissement de cartes, mais également à l'estimation de l'abondance ou de la densité de populations marines relativement sédentaires. Dans le présent article, on présente les bases théoriques du krigeage et on décrit la différence entre cette méthode et les méthodes traditionnelles fondées sur l'analyse d'échantillons aléatoires (stratifiés). On explique ce que l'on veut dire lorsqu'on dit que le krigeage est un interpolateur exact, et comment cette méthode peut être affectée par la prise en compte de l'erreur de mesure. On traite de l'utilisation de covariables et de voisinages de krigeage. Les problèmes attribuables à une proportion relativement élevée de valeurs zéro et de valeurs non-zéro qui présentent une asymétrie vers la droite sont illustrés à l'aide de données obtenues au cours d'un relevé des stocks de pétoncles.
[Traduit par la Rédaction]

Introduction

In recent years there has been a recognition that, for many marine populations, the density at a location is more likely to be similar to the density at a nearby location than at one that is further away; in other words, there is some degree of spatial correlation. There has been a feeling that stock assessment could be improved if, somehow, this spatial correlation could be incorporated into the analysis. Unfortunately, there has also been an erroneous belief that the existence of spatial correlation causes the conventional design-based (stratified random) estimators to be biased (cf. Brus and de Gruijter 1993). This has given rise to an interest in the methods of spatial analysis, especially kriging, for the assessment of marine populations, e.g., Conan et al. (1988*a*, 1988*b*), Conan and Wade (1989), Guillard et al. (1990), Anonymous (1991), Freire et al. (1991), Sullivan (1991), Simard et al. (1992, 1993), González-Gurriarán et al. (1993), Petitgas and Poulard (1989), Petitgas (1993). As with conventional design-based estimation, there is a tacit assumption that the spatial distribution, as observed, is fixed in time, at least for the period of the survey. Accordingly, most applications have dealt with relatively sedentary species, such as crab, or with acoustic surveys in which the population under study is surveyed in a relatively short time period.

The advocates of kriging claim that it should be used in preference to conventional design-based estimation because it produces estimates with smaller standard error. However, as will be shown below, there is a conceptual difference between the estimation variance obtained by kriging and the conventional variance of the estimate obtained via a (stratified) random sample. The variances then, strictly speaking, cannot be compared.

I will first give a brief outline of kriging methodology. I will then point out some fundamental differences between kriging and conventional design-based analysis. Next, I will describe some of the features of kriging, for example, what is meant by its being an exact interpolator and how this can be affected when measurement error is taken into account, and the incorporation of covariates. Some of the problems that arise will be illustrated with a typical set of survey data that would be collected to assess stocks of invertebrates or, indeed, other fish species.

W.G. Warren. Fisheries and Oceans Canada, Science Branch, P.O. Box 5667, St. John's, NF A1C 5X1, Canada.

Correct citation: Warren, W.G. 1998. Spatial analysis for marine populations: factors to be considered. *In* Proceedings of the North Pacific Symposium on Invertebrate Stock Assessment and Management. *Edited by* G.S. Jamieson and A. Campbell. Can. Spec. Publ. Fish. Aquat. Sci. 125. pp. 21–.

Basic kriging methodology

It is assumed that the catch $z(s)$ at any location s comes from a realization of an underlying stochastic process such that the expected catch at any location is constant, i.e., $E[z(s)] = c$, and that, in the isotropic case, the variance of the difference between catches at two locations depends solely on the distance, h, between the locations, i.e., $\mathrm{Var}[z(s) - z(s + h)] = 2\gamma(h)$. These, taken together, form what is called the intrinsic hypothesis or *intrinsic stationarity*. The function $\gamma(h)$ is called the (semi)variogram. More generally, h is a vector denoting direction as well as distance; this is the *anisotropic* case which, to keep the presentation reasonably simple, will not be considered in this paper.

We will add a stronger assumption, namely that the covariance of the catches at two locations is dependent solely on the distance between stations, i.e., $\mathrm{Cov}[z(s + h), z(s)] = C(h)$. This is referred to as *second order stationarity*. It follows that $\gamma(h) = C(0) - C(h)$. Further $C(0) = \mathrm{Var}[z(s)] = \sigma^2$.

The classical estimator of the variogram is

$$\gamma(h) = \frac{1}{|N(h)|} \sum_{N(h)} \frac{[z(s_i) - z(s_j)]^2}{2}$$

where $N(h)$ denotes the set of sample location pairs such that the distance between locations is h and $|N(h)|$ is the number of pairs within the set. In practice, we usually work with distance classes rather than distances *per se*; i.e., we take distances between $h - \delta_h/2$ and $h + \delta_h/2$, for δ_h sufficiently small, as being h. (We also need to make δ_h large enough to have a reasonable number of locations pairs, say at least 30, in each distance class). We will refer to this as the empirical variogram. Other estimators are possible (see e.g., Cressie 1991, Section 2.4.3) but are outside the scope of the present paper.

In practice, we would like to smooth the estimates given by the above for each value of h. Not every function is valid for a variogram. A commonly used and, in my experience, very serviceable function is that of the spherical variogram, namely

$$\gamma(h) = c_0 + c_s[3h/2a - (1/2)(h/a)^3],\ 0 < h \le a$$

$$= c_0 + c_s,\ h > a$$

with, of course, $\gamma(0) = 0$.

As h increases from 0, $\gamma(h)$ increases until $h = a$ after which it remains constant, equal to $c_0 + c_s$ which is referred to as the *sill*; a is known as the *range*. Under second-order stationarity, a is the distance at which catches cease to be correlated. It follows that the expected value of the sill is $\mathrm{Var}[z(s)] = \sigma^2$.

If we let $h \to 0$ we find that $\gamma(h) \to c_0$, which must be nonnegative but not necessarily zero, although $\gamma(0)$ is, by definition equal to 0; there is thus, in general, a discontinuity at the origin. The parameter c_0 is known as the *nugget* and reflects the presence of microscale variation, i.e., variation at distances less than the smallest attained in the data set. It may also include variation due to random measurement error. A spherical variogram is not always appropriate; other relatively common models are the linear, exponential, and Gaussian (for further detail see e.g., Cressie 1991, Section 2.3.1; Zimmerman and Zimmerman 1991). Indeed, variography, or the practice of finding a suitable model for the data at hand, including the use of diagnostics (see e.g., Cressie 1991, Section 2.6.3), is an important step in the kriging process.

I have usually found it adequate to fit the chosen variogram model to the empirical variogram by weighted least squares, with the weights being the number of location pairs in each distance class. For the larger distance classes the empirical variogram often becomes highly erratic; accordingly, for fitting, distances beyond a certain value (usually from 1/2 to 3/4 of the maximum distance) are often ignored in practice. There are, however, a variety of fitting methods available with no one method being universally superior (see e.g., Zimmerman and Zimmerman 1991).

Kriging is most commonly used for interpolation, i.e., for estimating what the value of the realization would be at some unobserved location, s_0. With observations at locations $s_1, s_2, \ldots s_n$, the estimator is taken to be of the form $\Sigma_i^n \lambda_i z(s_i) = \hat{z}(s_0)$. The objective is to have $E[\hat{z}(s_0)] = z(s_0)$ and to minimize $\mathrm{Var}[\hat{z}(s_0) - z(s_0)]$. (Note that this differs from minimizing $\mathrm{Var}[\hat{z}(s_0)]$ *per se*). Unbiasedness requires $\Sigma_i^n \lambda_i = 1$. Introduction of a Lagrange multiplier, μ, leads to the λ_i and μ being estimated through the solution of

$$\begin{vmatrix} \Gamma & \underline{1} \\ \underline{1}' & 0 \end{vmatrix} \begin{vmatrix} \underline{\lambda} \\ \mu \end{vmatrix} = \begin{vmatrix} \underline{\gamma} \\ 1 \end{vmatrix}$$

where $\Gamma = [\gamma(s_i, s_j)]$, $i, j = 1, 2, \ldots n$, $\underline{1}$ is an $n \times 1$ vector of 1's and $\underline{\gamma} = [\gamma(s_0, s_i)]$. Note that $\gamma(s_i, s_j)$ is used to denote $\gamma(\|s_i - s_j\|)$ where $\|s_i - s_j\|$ is the distance between locations s_i and s_j. It turns out that the minimized variance of $(\hat{z}(s_0) - z(s_0))$ is $\underline{\lambda}'\underline{\gamma} + \mu$; this is referred to as the *estimation variance*.

Kriging can also be used to estimate the mean. Again the estimate is of the form $\hat{c} = \Sigma_i^n \lambda_i z(s_i)$. The requirements of unbiasedness and minimization of $\mathrm{Var}[\hat{c} - c]$ lead to kriging equations

$$\begin{vmatrix} C & -\underline{1} \\ \underline{1}' & 0 \end{vmatrix} \begin{vmatrix} \underline{\lambda} \\ \mu \end{vmatrix} = \begin{vmatrix} \underline{0} \\ 1 \end{vmatrix}$$

where $C = \sigma^2 I_n - \Gamma$. The estimation variance turns out to be simply the value of μ.

The above is an outline of the basics of ordinary point kriging. It puts aside such topics as anisotropy, block kriging, universal kriging, and intrinsic functions of order k, any of which may be applicable in some applications. The objective here is, however, to focus on a few of the more fundamental factors that have to be considered when applying kriging for mapping or estimation of abundance.

Comparison of model-based and design-based methods

To illustrate the difference between model-based inferences from kriging and conventional design-based analysis of (stratified) random sampling, I will follow Brus and de Gruijter (1993). As before, let $z(s_i)$ denote the observation made at location s_i. In the design-based approach, the s_i are randomly selected from all the possible locations, $s_1, s_2, \ldots s_N$ where, in theory, N is finite but, in practice, can be regarded as indefinitely large. (Cordy (1993) has considered the problem of applying finite population theory where areal samples are drawn from a continuous region, such as a geographical area, where there is no natural, well defined sampling frame and, thus, in effect, a noncountable infinity of sampling units. The notion of a finite population, with N large but countable, has, however, been widely used in this context and is perceived by its users as a viable strategy. Accordingly, it forms the basis of what follows.) At any location, s_i, $z(s_i)$ is regarded as fixed, but unknown until measured. It is the s_i that are regarded as random; to emphasize this Brus and de Gruijter write $z(\underline{s}_i)$. In contrast, in the model-based approach, kriging, the s_i of the sample are regarded as fixed and the $z(s_i)$ are assumed to be a realization of some underlying stochastic process (or distribution). Here it is the z's that are random and, thus, Brus and de Gruijter write $\underline{z}(s_i)$. The distinction is important. In the design-based approach, expectations are taken over all possible sets of s_i of size n; in the model-based approach, the expectations are with respect to some assumed underlying stochastic process (or distribution). In both approaches we can imagine that our data are a subset of a realization of some stochastic process. In the design-based approach, we focus on this realization and look at what would be the average over all possible samplings of this realization; in the model-based approach, the set of locations is taken as fixed and we consider what would be the average of the (conceptual) realizations over this set. Accordingly, the estimators and, in particular, their variances, are conceptually different and thus, not really comparable.

In the conventional design-based situation let x_i, $i = 1, 2, \ldots N$ denote the observations at the N possible locations and define $M = \Sigma^N x_i/N$ and $S^2 = \Sigma^N (x_i - M)^2/(N-1)$. Further, let $m = \Sigma^n x_i/n$ and $s^2 = \Sigma^n (x_i - m)^2/(n-1)$ be calculated from the observations at the n sampled locations. Then, as shown in Cochran (1977), $E(m) = M$ and $E(s^2) = S^2$, where the expectations are obtained by averaging over all possible samples, i.e., over all possible combinations of the N locations chosen n at a time. Further, $\mathrm{Var}(m) = S^2/n$ so that, in this sense, s^2/n is an unbiased estimator of $\mathrm{Var}(m)$.

If the same data were used for kriging, the variogram would be estimated from components $(x_i - x_j)^2/2$, for $i = 1, 2, \ldots n-1$, $j > i$. Further

$$\sum_i \sum_{j>i} (x_i - x_j)^2/2(n(n-1)/2)$$

$$= [\sum_i^n (n-1)\, x_i^2 - 2 \sum_i \sum_{j>i} x_i x_j] / n\,(n-1)$$

$$= n\,[\sum_i^n x_i^2 - \sum_j^n \sum_j^n x_i x_j / n] / n\,(n-1)$$

$$= \underline{x}'\, [I_n - J_{n,n}/n]\, \underline{x} / (n-1)$$

$$= \sum_i^n (x_i - m)^2 / (n-1) = s^2$$

where $J_{n,n}$ denotes an $n \times n$ matrix of 1's.

In contrast to the design-based situation, the x_i here are regarded as a realization from some second-order stationary process with, for simplicity of illustration, a spherical variogram. The sill of the latter is given by $\mathrm{Var}(x) = \sigma^2$ and estimated by the average of those $(x_i - x_j)^2/2$ for which the distance between locations i and j is greater than the range of the variogram. This implies that the estimate of the sill, $\hat{\sigma}^2$, must be greater than s^2 with equality in the case of a pure nugget variogram, i.e., range = 0.

Further, in this latter case, i.e., $\gamma(h) = \hat{\sigma}^2$, $h > 0$, $\gamma(0) = 0$. The equations for obtaining the kriged mean are

$$\begin{vmatrix} \hat{\sigma}^2 I_n & -\underline{1} \\ \underline{1}' & 0 \end{vmatrix} \begin{vmatrix} \lambda \\ \mu \end{vmatrix} = \begin{vmatrix} \underline{0} \\ 1 \end{vmatrix}.$$

Hence $\lambda_i = 1/n$, for all i, the kriged mean is $\Sigma^n x_i/n$ (the same as in design-based estimation) and the estimation variance $\hat{\sigma}^2/n = s^2/n$. Thus, in this special case, by either method we get the same numerical estimates, although they have a different interpretation.

But, in general, how does the estimation variance of the kriged mean relate numerically to s^2/n ? Consider the special case of $n = 3$ and the data located such that $\mathrm{Var}(x_i) = \sigma^2$, $\mathrm{Cov}(x_1, x_2) = \mathrm{Cov}(x_2, x_3) = \rho\sigma^2$, $\mathrm{Cov}(x_1, x_3) = 0$, valid for $|\rho| < 1/\sqrt{2}$. Then the kriged mean is given by $\Sigma\, \lambda_i\, x_i$ with estimation variance μ where λ_1, λ_2, λ_3, and μ satisfy

$$\begin{vmatrix} \hat{\sigma}^2 & \rho\hat{\sigma}^2 & 0 & -1 \\ \rho\hat{\sigma}^2 & \hat{\sigma}^2 & \rho\hat{\sigma}^2 & -1 \\ 0 & \rho\hat{\sigma}^2 & \hat{\sigma}^2 & -1 \\ 1 & 1 & 1 & 0 \end{vmatrix} \begin{vmatrix} \lambda_1 \\ \lambda_2 \\ \lambda_3 \\ \mu \end{vmatrix} = \begin{vmatrix} 0 \\ 0 \\ 0 \\ 1 \end{vmatrix}.$$

Hence $\lambda_1 = \lambda_3 = (1-\rho)/(3-4\rho)$, $\lambda_2 = (1-2\rho)/(3-4\rho)$ and $\mu = \hat{\sigma}^2 (1-2\rho^2)/(3-4\rho) = \sigma_K^2$. Thus, σ_K^2 will be greater than $s^2/3$ if

$$\hat{\sigma}^2 > \frac{3-4\rho}{1-2\rho^2} \frac{s^2}{3}.$$

Since it has been established that $\hat{\sigma}^2 > s^2$, this will certainly be true if, for $\rho > 0$,

$$\frac{3-4\rho}{3(1-2\rho^2)} \leq 1$$

or $\rho < 2/3$. This demonstrates that, numerically, the estimation variance of the kriged mean is not necessarily less than the conventional design-based estimate of the variance of the sample mean. Note, however, that under the model-based approach, $\mathrm{Var}(\bar{x}) = (3 + 4\rho)\,\hat{\sigma}^2/9$ which, for $0 < \rho < 1/\sqrt{2}$ is always greater than σ_K^2. But, again, it must be emphasized that, with the model- and design-based approaches, we are dealing with two different concepts.

An exact or smoothed interpolator?

The description of kriging as an exact interpolator simply reflects the fact that the estimate at the location of a data point is the value observed at that location. With no loss of generality, we may suppose that the value at location s_1 is to be estimated. The kriging equations are then

$$\begin{vmatrix} 0 & \gamma_{12} & \ldots & \gamma_{1n} & 1 \\ \gamma_{21} & 0 & \ldots & \gamma_{2n} & 1 \\ \cdot & \cdot & \cdot & \cdot & \cdot \\ \cdot & \cdot & \cdot & \cdot & \cdot \\ \cdot & \cdot & \cdot & \cdot & \cdot \\ \gamma_{n1} & \gamma_{n2} & \ldots & 0 & 1 \\ 1 & 1 & \ldots & 1 & 0 \end{vmatrix} \begin{vmatrix} \lambda_1 \\ \lambda_2 \\ \cdot \\ \cdot \\ \cdot \\ \lambda_n \\ 0 \end{vmatrix} = \begin{vmatrix} 0 \\ \gamma_{12} \\ \cdot \\ \cdot \\ \cdot \\ \gamma_{1n} \\ 1 \end{vmatrix}.$$

Note $\gamma_{ij} = \gamma_{ji}$. Thus $\lambda_1 = 1$, $\lambda_j = 0, j \neq 1$, $\mu = 0$. That is, $\hat{z}(s_1) = z(s_1)$, $\mathrm{Var}(\hat{z}(s_1)) = 0$.

Suppose now that the observations are subject to measurement error, with variance $\sigma_e 2$. It may be thought better to allow for this error and obtain a "smoothed" estimate at a data location, i.e., one that is more consistent with the observations at neighbouring locations. Following Cressie (1991), the only change that needs to be made to the kriging equations is to substitute $\sigma_e 2$ for 0 as the first element of the right-hand side. To see what happens we consider the special case with $n = 2$. The kriging equations are then

$$\begin{vmatrix} 0 & \gamma & 1 \\ \gamma & 0 & 1 \\ 1 & 1 & 0 \end{vmatrix} \begin{vmatrix} \lambda_1 \\ \lambda_2 \\ \mu \end{vmatrix} = \begin{vmatrix} \sigma_e^2 \\ \gamma \\ 1 \end{vmatrix}.$$

Hence $\lambda_1 = 1 - \sigma_e^{2/2\gamma}, \lambda_2 = \sigma_e^2/2\gamma$, and $\mu = \sigma_e^2/2$.

Thus the estimate is shifted away from the observed value towards the average of $z(s_1)$ and $z(s_2)$, the magnitude of the shift depending on the magnitude of the error measurement variance relative to the covariance between the two locations. The effect is similar in the more general case with $n > 2$.

Use of covariates

Instead of assuming that the expectation of $z(s)$ is constant throughout the region of interest, we now suppose that it varies with some covariate(s), temperature or depth. Accordingly we replace $z(s) = c + \varepsilon(s)$ by $z(s) = f(s) + \varepsilon(s)$ where, for example, $f(s)$ might take the form $\alpha_0 + \alpha_1 s$. The assumption of second-order stationarity (or, more generally, the intrinsic hypothesis) then applies to, $\varepsilon(s_i) = z(s_i) - f(s_i) = z(s_i) - \alpha_0 - \alpha_1 s_i$. To obtain the best linear unbiased estimates of α_0 and α_1 we need to know the covariances between the $\varepsilon(s_i)$, or, equivalently, their variogram. But to determine the latter we first need α_0 and α_1. (This is the circularity argument of universal kriging — see, e.g., Cressie (1991) Section 3.4.3.) The simplest approach is to estimate the α_j by ordinary least squares. We may then compute the residuals and, from these, construct a variogram. The α_j may then be reestimated by generalized least squares, using the covariances obtained from this variogram. New residuals are then obtained and used to generate a new variogram and the process repeated until convergence. Cressie (1991) points out that this iterative procedure does not fully eliminate a bias problem but, for most practical situations, the bias should be small. Further, my experience so far is that convergence is rapid and the difference between the result from ordinary least squares and the first iteration of generalized least squares is inconsequential relative to the other uncertainties in the system.

For a geographical drift in the mean it may be possible to take $f(s)$ as a simple polynomial in latitude and/or longitude, e.g., $f(s) = \alpha_0 + \alpha_1 Latitude(s) + \alpha_2 Longitude(s)$.

Use of a kriging neighbourhood

In theory $\hat{z}(s_0)$ is a weighted average of the observations at *all* sample locations. In principle, its calculation involves the inversion of an $(n + 1) \times (n + 1)$ matrix. Large values of n may make this computationally prohibitive — although I have experienced no problems with n as great as 250. The estimate is, however, influenced predominantly by the observations at nearby locations with little weight being given to the observations at the more distant locations, particularly at distances greater than the range of the variogram. However, as Cressie (1991) observes, it is incorrect to assume that these observations receive zero weight. Accordingly, it is customary to define a kriging neighbourhood, i.e., a region about the location at which estimation is desired, with the estimate formed as a weighted average of the observations at locations within the neighbourhood. Cressie (1991) points out that the range, by itself, does not provide sufficient guidance to determine the kriging neighbourhood. Some criteria for the choice of a kriging neighbourhood have been suggested (e.g., Rivoirard 1987) but Cressie (1991) concludes that: "The compromise between kriging variance and the size of kriging neighbourhood is worthy of more research."

The use of a kriging neighbourhood would seem to work best when the data locations form a regular grid. With data locations placed randomly, the number of locations within a neighbourhood of fixed size and shape would vary. Consequently, a different matrix inversion is needed for each location at which an estimate is required. Admittedly, these will be smaller matrices but there may be a limit beyond which, unless the kriging neighbourhood is very small, it is more expedient to carry out one large matrix inversion than a large number of smaller inversions.

What are the consequences of mapping from irregularly spaced data when there are locations for which the distance to the nearest data location exceeds the range of the variogram? (This situation could also arise, but is probably less frequent, when the data locations form a regular grid.) Let me address this question with a simple example with $n = 3$. The kriging equations will be

$$\begin{vmatrix} 0 & \gamma_1 & \gamma_2 & 1 \\ \gamma_1 & 0 & \gamma_3 & 1 \\ \gamma_2 & \gamma_3 & 0 & 1 \\ 1 & 1 & 1 & 0 \end{vmatrix} \begin{vmatrix} \lambda_1 \\ \lambda_2 \\ \lambda_3 \\ \mu \end{vmatrix} = \begin{vmatrix} \hat{\sigma}^2 \\ \hat{\sigma}^2 \\ \hat{\sigma}^2 \\ 1 \end{vmatrix}$$

since $\gamma(s_0, s_i) = \hat{\sigma}^2$, $i = 1, 2, 3$. Hence

$$\lambda_1 = \gamma_3 (\gamma_1 + \gamma_2 - \gamma_3)/\Delta$$

$$\lambda_2 = \gamma_2 (\gamma_3 + \gamma_1 - \gamma_2)/\Delta$$

$$\lambda_3 = \gamma_1 (\gamma_2 + \gamma_3 - \gamma_1)/\Delta$$

where $\Delta = (\gamma_1 + \gamma_2 + \gamma_3)^2 - 2(\gamma_1^2 + \gamma_2^2 + \gamma_3^2)$. If $\gamma_1 = \gamma_2 = \gamma_3$, then $\lambda_i = 1/3$, $i = 1, 2, 3$. On the other hand, let $\gamma_2 = \gamma_3 \to \infty$ and $\gamma_1 \to 0$; then $\lambda_1 : \lambda_2 : \lambda_3 = 2 : 1 : 1$. Because we cannot have $\gamma_2 = \gamma_3 \to 0$, $\gamma_1 \to \infty$, the interpolated value lies between $[z(s_1) + z(s_2) + z(s_3)]/3$ and $[2z(s_1) + z(s_2) + z(s_3)]/4$.

The implication for more realistic situations is that the estimates at all locations that are at distances from data locations of more than the range of the variogram are the same, with a value that is not far removed from the overall mean, $(\Sigma^n z(s_i)/n)$. When a kriging neighbourhood is used, estimates at such locations will take on a value that is roughly the average of the observations at locations within the particular neighbourhood. The latter estimate may seem to be more rational than one that is close to the average over *all* data locations but, because it is based on less information, it will have a larger estimation variance (albeit, sometimes negligibly so). Likewise, when a covariate is used, since $\Sigma^n \varepsilon(s_i) = 0$, the estimate will tend to take on a value close to that predicted by the fitted regression. For such interpolation should we be using kriging neighbourhoods and/or predictions from covariates? There does not seem to be any definitive answer to this question.

An example

The data from a scallop survey are presented in the Appendix. The numbers and weights represent the catches made by a 12-foot-wide dredge in nominal 1 nm tows. There was

Fig. 1. Variogram based in the indicator variable (1 = present, 0 = absent). Distance h is measured in equivalents to degrees of latitude, i.e., 0.1 = 6 *n*.

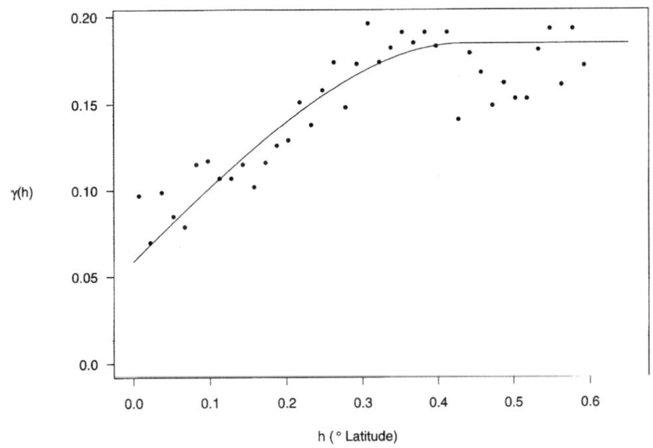

Fig. 2. Variogram of untransformed nonzero scallop catches. Distance as in Fig. 1.

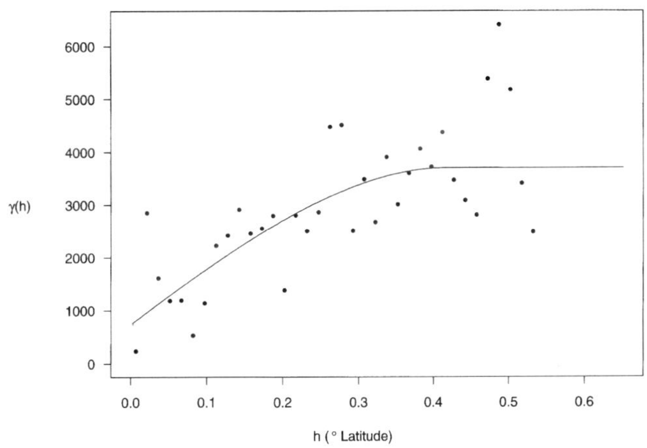

however variation in the distance towed. Accordingly, the actual distances were calculated from the recorded start and stop latitudes and longitudes, and these distances used to recalculate catch weights per nautical mile of tow. The location was then taken as the midpoint of the tow.

These data exhibit two of the problems that plague the analysis of fisheries survey data, namely a nonnegligible proportion of zero catches and a strongly positively skewed distribution of the nonzero catches. Since, with these data, the zero catches tend to be spatially aggregated, it is assumed that they represent areas with no scallops, perhaps because of unsuitable habitat. Accordingly, the analysis was split into two components, (*i*) utilizing an indicator variable (0 or 1) denoting the absence or presence of scallop to estimate the unoccupied area, and (*ii*) kriging the mean conditional on there being scallops present.

The variogram estimated from the indicator variable is given in Fig. 1. The interlocation distances were grouped by distance classes of width equivalent to 0.015° of latitude (= 0.9 nm). The first 30 distance classes were used for fitting

Fig. 3. Variogram of logarithmically transformed nonzero scallop catches. Distance as in Fig. 1.

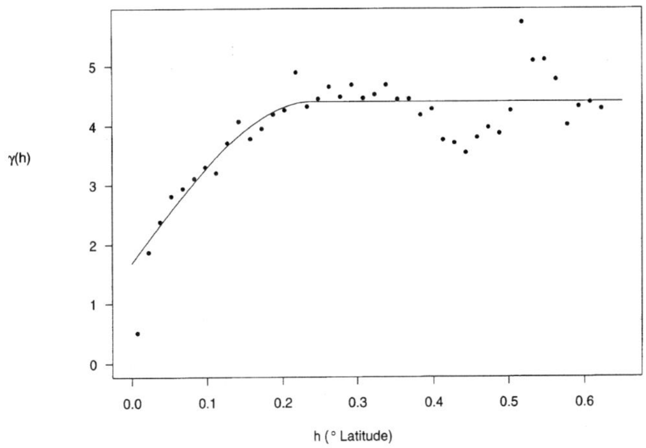

a spherical variogram by weighted least squares. The resulting kriged estimate of the mean was 0.7114, so that 0.2886 of the total surveyed area of 867 nm^2 was estimated to be unoccupied.

Three approaches were taken to estimating a kriged mean conditional on there being scallops present. Firstly, a variogram was constructed from the nonzero data *per se*, i.e., without any transformation. The resulting variogram is given in Fig. 2. The class width was again taken as 0.9 nm. It will be seen that there is a wide scatter of points and that the variogram is poorly defined; this is common with highly skewed data. Notwithstanding, a spherical variogram tracked the points reasonably well, and that plotted is based on the first 30 distance classes. This led to a kriged mean of 17.3192 which would give an overall density per nautical mile towed of $0.7114 \times 17.3192 = 12.3209$ kg/nm. This is substantially less than the overall mean of the raw data of 17.9146 kg/nm.

Kriging gives the best linear predictor under the assumption that the data are the realization of a Gaussian (normal) process. This is clearly not the case here. Normality can be approximated, if not attained, by logarithmically transforming the data. (Since we are focusing on nonzero catches, the taking of logarithms here presents no difficulty.) The resulting variogram is given in Fig. 3. Clearly this variogram is well defined for at least the first 30 or so distance classes. Here, fitting by weighted least squares and using the first 40 distance classes (the difference is negligible whether the first 30, 35, or 40 distance classes are used) we have a sill of 4.4015. This leads to a kriged mean of 0.9925. The problem with this approach is that we have to back transform to the original scale. A first-order approximation to this is $\exp(0.9925 + 4.4015/2) = 24.3680$. This converts to an overall density of $0.7114 \times 24.2680 = 17.3352$ kg/nm.

The third approach, which avoids the back transformation, comes from the fact that the variogram in the original scale is proportional to $1 - 1/e^{\gamma(h)}$ where $\gamma(h)$ denotes the variogram on the (logarithmically) transformed scale. Thus the kriging weights obtained from $1 - 1/e^{\gamma(h)}$ will be the same as for a backtransformed variogram (although this approach does not permit direct calculation of the estimation variance). This leads to a kriged mean of 23.0053 and thus an overall density of $0.7114 \times 23.0053 = 16.3658$ kg/nm.

Conversion to total weight is accomplished by multiplying by $867 \times 6080/12$. The three estimates are then 5412, 7615, and 7189 tonnes, respectively, compared with an estimate based on the arithmetic mean of the data of 7870 tonnes.

Discussion

Which of the estimators is best? I do not believe that there is an absolute answer to this question. Because of the inherent uncertainty in the variogram for the raw data, the estimate so obtained would seem to be the most questionable (not simply because it differs substantially from the others). In this example, the remaining three do not differ substantively. Variances have not been compared, partly because of the complications in back transformation, but also because of the inappropriateness of comparing design- and model-based estimators. The simple arithmetic mean (or, in the case of stratified random sampling, the stratified mean), has the merit of uniqueness. No assumptions (other than random selection of locations) are required and there are no choices that have to be made in the analysis. Perhaps one should not generalize from a single example, but it does suggest that, for estimating total abundance, there may be no particular advantage in using a kriged mean in preference to the conventional design-based estimator.

The situation is somewhat different if mapping is involved. Here kriging is a viable alternative to other methods of spatial interpolation and perhaps is the only one based on optimizing a measure of precision. But again, there are no unique answers. At each stage in the analytical process there are choices to be made that can affect the final estimates. The viability of the results would then appear to depend on the skill of the analyst; there is, or should be, no black box taking his/her place.

Finally, there have been efforts to bridge the gap between the extremes of model- and design-based inference as described above. Cordy and Thompson (1995) observe that: "Although spatial statistics is presented in a probabilistic framework, there have been some attempts to develop a deterministic view of geostatistics, which is more consistent with the design-based perspective." They go on to propose a design-based method of variance estimation which incorporates a measure of spatial continuity, a key component in their development being the reexpression of integral formulae for the variance in terms of the deterministic versions of the variogram and the covariance function (cf. Isaaks and Srivastava 1988). This new and interesting approach is, however, outside the scope of the present paper.

Acknowledgments

I am indebted to Dr. S. Naidu for making available the scallop survey data. The comments of two anonymous referees were helpful and much appreciated.

References

Anonymous. 1991. Report of the Workshop on the Applicability of Spatial Staistical Techniques to Acoustic Survey Data. ICES, Cooperative Research Report No. 195. Denmark.

Brus, D.J., and de Gruijter, J.J. 1993. Design-based versus model-based estimates of spatial means; theory and application in environmental soil science. Environmetrics, **4**: 123–152.

Cochran, W.G. 1977. Sampling techniques, 3rd ed. John Wiley & Sons, New York, N.Y.

Conan, G.Y., and Wade, E. 1989. Geostatistical analysis, mapping and global estimation of harvestable resources in a fishery of northern shrimp (*Pandalus borealis*). ICES C.M. 1989/D:1. International Council for the Exploration of the Sea, Denmark.

Conan, G.Y., Moriyasu, M., Wade, E., and Comeau, M. 1988*a*. Abundance and spatial distribution surveys of snow crab stocks by geostatistics. ICES C.M. 1988/K:10. International Council for the Exploration of the Sea, Denmark.

Conan, G.Y., Buerkle, U., Wade, E., Chadwick, M., and Comeau, M. 1988*b*. Geostatistical analysis of spatial distribution of a school of herring. ICES C.M. 1988/D:25. International Council for the Exploration of the Sea, Denmark.

Cordy, C.B. 1993. An extension of the Horvitz-Thompson theorem to point sampling from a continuous universe. Stat. Probab. Lett. **18**: 353–362.

Cordy, C.B., and Thompson, C.M. 1995. An application of the deterministic variogram to design-based variance estimation. Math. Geology, **27**: 173–205.

Cressie, N.A.C. 1991. Statistics for spatial data. John Wiley & Sons, New York, N.Y.

Freire, J., Fernández, L., and González-Gurriarán, E. 1991. Spatial distribution of decapod crustaceans in the Galician continental shelf (NW Spain) using geostatistical analysis. ICES C.M. 1991/K:4. International Council for the Exploration of the Sea, Denmark.

González-Gurriarán, E., Freire, J., and Fernández, L. 1993. Geostatistical analysis of spatial distribution of *Liocarcinus depurator*, *Macropipus tubersulatus* and *Polybius henslowii* (Crustacea: Brachyura) of the Galacian continental shelf (NW Spain). Mar. Biol. (Berlin), **115**: 453–461.

Guillard, J., Gerdeaux, D., and Chautrin, J.M. 1990. The use of geostatistics for abundance estimation by echointegration in lakes: the example of Lake Annecy. Rapp. P.-v. Réun. Cons. Int. Explor. Mer. **189**: 410–414.

Isaaks, E.H., and Srivastava, R.M. 1988. Spatial continuity measures for probabalistic and deterministic geostatistics. Math. Geology, **20**: 313–341.

Petitgas, P. 1993. Geostatistics for fish stock assessment: a review and an acoustic application. ICES J. Marine Sci. **50**: 285–298.

Petitgas, P., and Poulard, J.C. 1989. Applying stationary geostatistics to fisheries: a study on hake in the Bay of Biscay. ICES C.M. 1989/G:62. International Council for the Explorations of the Sea, Denmark.

Rivoirard, J. 1987. Two key parameters when choosing the kriging neighborhood. Math. Geology, **19**: 851–856.

Simard, Y,. Legendre, P., Lavoie, G., and Marcotte, D. 1992. Mapping, estimation biomass and optimizing sampling programs for spatially autocorrelated data: case study of the northern shrimp (*Pandalus borealis*). Can. J. Fish. Aquat. Sci. **49**: 32–45.

Simard, Y., Marcotte, D., and Bourgault, G. 1993. Exploration of geostatistical methods for mapping and estimating acoustic biomass of pelagic fish in the Gulf of St. Lawrence: size of echo-integration unit and auxiliary environmental information. Aquat. Living Resour. **6**: 185–199.

Sullivan, P.J. 1991. Stock abundance using depth-dependent trends and spatially correlated variation. Can. J. Fish. Aquat. Sci. **48**: 1691–1703.

Zimmerman, D.L., and Zimmerman, M.B. 1991. A comparison of spatial semivariogram estimators and corresponding ordinary kriging predictors. Technometrics, **33**: 77–91.

Appendix

Scallop survey data

Start				Finish					
Latitude		Longitude		Latitude		Longitude		*N*	*CW*
45	257	49	214	45	247	49	212	16	1.50
45	244	49	213	45	252	49	217	0	0
45	247	49	223	45	238	49	223	11	1.00
45	239	49	215	45	245	49	203	5	0.49
45	243	49	195	45	234	49	189	839	98.26
45	248	49	166	45	256	49	160	0	0
45	254	49	154	45	245	49	148	3	0.40
45	248	49	130	45	252	49	119	17	1.75
45	254	49	102	45	256	49	89	20	2.40
45	252	49	85	45	244	49	79	10	1.25
45	242	49	76	45	252	49	77	16	2.20
45	215	49	60	45	225	49	60	5	0.40
45	214	49	51	45	204	49	45	439	47.65
45	219	49	36	45	228	49	32	0	0
45	229	49	13	45	231	48	599	0	0
45	253	48	583	45	255	48	565	12	1.65
45	237	48	577	45	232	48	589	0	0
45	233	48	560	45	234	48	546	75	9.24
45	236	48	521	45	237	48	507	125	14.74
45	238	48	493	45	248	48	491	3	0.25
45	243	48	454	45	233	48	451	90	6.65
45	226	48	462	45	231	48	473	4	0.50
45	212	48	509	45	205	48	518	4	0.70
45	200	48	538	45	198	48	550	277	27.36
45	210	48	583	45	215	48	594	159	24.47
45	189	48	553	45	182	48	541	7	1.70
45	181	48	522	45	189	48	529	0	0
45	158	48	546	45	161	48	560	1	0.10
45	158	48	546	45	156	48	532	0	0
45	154	48	523	45	152	48	510	0	0
45	148	48	514	45	138	48	518	0	0
45	150	48	505	45	159	48	501	0	0
45	136	48	491	45	126	48	496	16	1.00
45	128	48	523	45	132	48	536	0	0
45	127	48	521	45	117	48	520	0	0
45	110	48	519	45	103	48	524	0	0
45	115	48	555	45	115	48	541	0	0
45	119	48	585	45	122	48	598	5	0.45
45	110	48	584	45	101	48	589	0	0
45	87	48	566	45	84	48	579	0	0
45	79	48	531	45	69	48	532	0	0
45	36	48	554	45	26	48	559	0	0
45	32	48	563	45	41	48	565	0	0
45	24	48	559	45	16	48	566	0	0
45	23	49	0	45	26	49	14	0	0
45	12	49	20	45	2	49	19	1	0.10
44	598	49	16	45	6	49	10	0	0
44	590	49	26	44	582	49	31	3	0.25
44	569	49	3	44	579	49	3	1434	113.89
44	566	49	7	44	575	49	6	693	46.54
44	563	49	28	44	568	49	17	6345	506.85
44	553	49	95	44	543	49	98	440	42.22
44	579	49	65	44	583	49	52	54	5.95
45	21	49	52	45	30	49	48	542	45.36
45	25	49	64	45	18	49	73	277	23.08

Appendix (*continued*).

Start				Finish					
Latitude		Longitude		Latitude		Longitude		*N*	*CW*
45	25	49	95	45	33	49	101	102	9.30
45	33	49	77	45	33	49	63	279	24.27
45	45	49	36	45	50	49	25	1277	82.70
45	59	49	26	45	67	49	33	336	24.35
45	60	49	63	45	55	49	74	44	3.43
45	48	49	97	45	50	49	109	91	7.90
45	45	49	110	45	55	49	111	219	16.35
45	55	49	107	45	51	49	119	153	13.36
45	70	49	83	45	76	49	73	970	89.49
45	86	49	45	45	76	49	44	1143	93.79
45	80	49	22	45	86	49	11	625	45.23
45	91	49	6	45	95	49	18	564	42.99
45	104	49	59	45	114	49	68	538	43.07
45	93	49	100	45	85	49	113	744	68.89
45	110	49	95	45	109	49	109	443	39.01
45	118	49	85	45	122	49	73	468	41.22
45	119	49	53	45	114	49	40	283	28.32
45	129	49	11	45	138	49	8	29	3.50
45	148	49	3	45	148	49	17	1378	114.92
45	159	49	11	45	158	49	24	2068	183.33
45	174	49	38	45	166	49	47	315	35.09
45	182	49	30	45	189	49	20	477	47.96
45	193	49	16	45	185	49	25	1092	100.15
45	174	49	69	45	175	49	83	320	30.77
45	148	49	97	45	138	49	98	138	12.66
45	167	49	110	45	177	49	112	238	18.74
45	197	49	79	45	199	49	65	223	21.46
45	207	49	103	45	198	49	106	265	20.21
45	195	49	128	45	197	49	140	613	49.73
45	201	49	154	45	204	49	165	247	20.78
45	211	49	161	45	213	49	174	414	34.15
45	223	49	180	45	226	49	193	88	7.80
45	228	49	220	45	229	49	234	19	1.40
45	228	49	225	45	223	49	213	21	1.75
45	210	49	206	45	207	49	193	81	7.20
45	197	49	197	45	200	49	210	12	1.20
45	198	49	220	45	196	49	233	0	0
45	176	49	245	45	186	49	246	5	0.55
45	177	49	227	45	167	49	227	13	1.00
45	178	49	215	45	188	49	216	110	9.08
45	186	49	191	45	186	49	176	217	18.03
45	179	49	178	45	169	49	179	67	6.45
45	171	49	177	45	171	49	163	76	6.75
45	175	49	144	45	186	49	143	466	37.89
45	173	49	154	45	164	49	154	121	9.96
45	157	49	158	45	155	49	170	39	3.60
45	156	49	178	45	155	49	165	53	4.50
45	153	49	195	45	154	49	208	1	0.10
45	149	49	235	45	148	49	248	74	7.50
45	142	49	253	45	133	49	258	10	0.75
45	133	49	252	45	140	49	242	9	0.90
45	134	49	241	45	125	49	245	17	1.50
45	122	49	246	45	113	49	239	15	1.45
45	125	49	219	45	134	49	211	0	0
45	144	49	201	45	136	49	192	24	1.35

Appendix (*continued*).

Start				Finish					
Latitude		Longitude		Latitude		Longitude		*N*	*CW*
45	143	49	200	45	153	49	200	12	1.50
45	126	49	204	45	116	49	204	20	1.50
45	122	49	193	45	132	49	193	13	0.95
45	126	49	180	45	116	49	180	6	0.50
45	127	49	179	45	137	49	178	30	2.94
45	130	49	167	45	123	49	157	6	0.15
45	134	49	156	45	141	49	164	23	1.45
45	143	49	163	45	139	49	151	21	1.40
45	133	49	137	45	132	49	123	64	5.23
45	120	49	126	45	113	49	133	263	17.76
45	112	49	126	45	113	49	113	944	81.30
45	106	49	133	45	104	49	146	35	2.95
45	104	49	155	45	98	49	162	122	7.94
45	94	49	164	45	91	49	150	215	17.91
45	99	49	169	45	100	49	183	79	4.85
45	104	49	231	45	94	49	231	12	0.85
45	89	49	245	45	79	49	245	2	0.10
45	87	49	239	45	76	49	239	4	0.40
45	66	49	226	45	58	49	216	2	0.20
45	64	49	173	45	65	49	159	22	1.50
45	64	49	172	45	65	49	184	23	1.95
45	55	49	207	45	53	49	219	12	1.25
45	49	49	245	45	48	49	258	4	0.20
45	40	49	256	45	33	49	249	0	0
45	29	49	245	45	20	49	241	10	0.50
45	36	49	188	45	38	49	174	33	1.95
45	33	49	183	45	28	49	193	36	2.50
45	20	49	192	45	10	49	190	250	19.69
45	7	49	134	45	13	49	122	7	0.35
45	5	49	117	45	9	49	128	14	0.80
45	3	49	144	45	3	49	158	28	2.00
45	5	49	163	45	2	49	176	5	0.35
45	4	49	206	45	6	49	220	695	63.51
44	596	49	249	44	585	49	253	73	7.99

Appendix (*concluded*).

Start				Finish					
Latitude		Longitude		Latitude		Longitude		*N*	*CW*
44	586	49	255	44	594	49	249	21	2.50
44	582	49	239	44	583	49	225	11	1.00
44	588	49	172	44	588	49	158	0	0
44	589	49	163	44	594	49	173	2	0.20
44	569	49	156	44	566	49	143	1	0.07
44	552	49	139	44	552	49	124	39	4.55
44	553	49	125	44	552	49	137	26	3.10
44	563	49	160	44	568	49	171	4	0.40
44	565	49	175	44	556	49	175	0	0
44	550	49	208	44	550	49	222	32	3.75
44	533	49	229	44	528	49	240	146	18.21
44	523	49	226	44	531	49	232	178	20.91
44	531	49	177	44	525	49	165	17	1.50
44	543	49	151	44	535	49	159	477	34.36
44	533	49	149	44	541	49	157	335	24.64
44	522	49	137	44	518	49	125	196	19.51
44	511	49	153	44	520	49	168	393	25.95
44	516	49	227	44	518	49	240	237	21.53
44	509	49	240	44	501	49	238	21	3.00
44	499	49	236	44	499	49	222	13	1.10
44	502	49	218	44	510	49	221	242	23.71
44	506	49	160	44	505	49	147	0	0
44	505	49	139	44	505	49	152	0	0
44	507	49	83	44	509	49	70	18	1.50
44	507	49	60	44	510	49	72	369	32.64
44	488	49	35	44	497	49	39	546	48.20
44	484	49	49	44	493	49	50	1	0.08
44	476	49	65	44	481	49	76	0	0
44	474	49	133	44	474	49	150	0	0
44	476	49	137	44	482	49	126	0	0
44	493	49	168	44	495	49	181	2	0.10
44	474	49	197	44	474	49	210	2	0.15
44	483	49	230	44	474	49	233	2	0.15
44	474	49	233	44	474	49	248	4	0.50

Note: Starting and finishing latitude and longitude given in degrees and minutes × 10. *N* denotes the number caught and *CW* the catch weight (kg) by station.

Snow crab, *Chionoecetes opilio*, stock assessment in the southwestern Gulf of St. Lawrence by bottom trawl survey

M. Moriyasu, E. Wade, A. Sinclair, and Y. Chiasson

Abstract: The exploitation of snow crab, *Chionoecetes opilio*, in the Southwestern Gulf of St. Lawrence (area 12) started in the mid-1960's and developed rapidly to a peak landing of 31 500 t in 1982. It then fluctuated around 25 000 t before dropping suddenly to 11 700 t in 1987. In 1989, the fishery closed prematurely because of a further rapid decline in catch and a high incidence of soft-shelled crab, which for snow crab is a sign of a depressed stock. Starting in 1988, a postseason trawl survey has been conducted in area 12. The survey has provided both a better understanding of stock dynamics and better management. Kriging is now used to forecast biomass and to map distributions of harvestable crab, recruitment to the fishery, and abundance of soft-shelled crab. There is a strong correlation between biomass estimates from the surveys and CPUE calculated from an 8-year data set. The current high biomass of commercial-sized crab resulted from good recruitment first observed in the 1988 trawl survey. Recruitment to the benthic stage has not been constant over time and the crab fishery must depend on waves of recruitment. Results from the current survey suggest recruitment will be low for 3–4 years starting in 1996. The postseason trawl survey is the most useful tool for predicting, one year in advance, abundance of the snow crab in the southwestern Gulf of St. Lawrence and for providing useful information on long-term trends in the stock.

Résumé : L'exploitation du crabe des neiges (*Chionoecetes opilio*) dans le sud-ouest du golfe du Saint-Laurent (zone 12) a commencé au milieu des années 60 et a rapidement atteint un pic de 31 500 t en 1982. Par la suite, les prises ont fluctué à quelque 25 000 t avant de chuter brusquement à 11 700 t en 1987. La saison de pêche a pris fin prématurément en 1989 à cause d'un déclin rapide des prises et de la capture élevée de crabes à carapace molle, phénomène qui indique un signe de dépression des stocks dans le cas du crabe des neiges. À partir de 1988, on a effectué un relevé de chalutage dans la zone 12 après la clôture de la saison de pêche; les données recueillies ont permis de mieux comprendre la dynamique du stock et d'améliorer la gestion. Le krigeage est maintenant utilisé pour prévoir la biomasse et produire des cartes illustrant la répartition des crabes exploitables, le recrutement à la pêcherie et l'abondance de crabes à carapace molle. Il y a une forte corrélation entre les estimations de biomasse obtenues grâce aux relevés et les captures par unité d'effort (CPUE) calculées à partir d'une base de données couvrant une période de huit ans. La biomasse actuelle élevée du crabe de taille commerciale résulte d'un bon recrutement observé pour la première fois pendant la campagne de chalutage de 1988. Le recrutement d'individus dans le stade benthique n'est pas régulier, et la pêche au crabe doit tenir compte des poussées de recrutement. Les résultats de la campagne menée à l'heure actuelle laissent croire que le recrutement sera faible pendant trois ou quatre ans à partir de 1996. Le relevé de chalutage effectuée après la saison de pêche est l'outil le plus utile pour prévoir un an à l'avance l'abondance de crabe des neiges dans le sud-ouest du golfe du Saint-Laurent et obtenir des renseignements utiles sur les tendances à long terme du stock. [Traduit par la Rédaction]

Introduction

Snow crab, *Chionoecetes opilio*, is the most important commercial crab species in Atlantic Canada with total landings exceeding 50 000 t in 1995 (Anonymous 1996*a*,1996*b*; Dawe et al. 1996; Tremblay and Eagles 1996). There are 30 permanent fishing areas in eastern Canada and the southwestern Gulf of St. Lawrence fishery (area 12) had the highest landings (Fig. 1), 19 944 tonnes (t) in 1995. Crab in area 12 are fished by 130 fishers from New Brunswick, Quebec, and Nova Scotia with a maximum of 150 traps allocated per license. Only males with carapace width ≥95 mm are harvested. The fishing season starts in the spring when the Gulf is free of ice and continues until the quota is reached, usually in June–July.

Exploitation of snow crab in area 12 started around the mid-1960's and rapidly developed to a peak of 31 500 t in 1982. Landings between 1966 and 1981 (Fig. 2) are probably not directly related to abundance of the resource, but rather reflect changes in fishing technology and expansion of fishing grounds (Chiasson et al. 1992). Fishers became better equipped with bigger boats, more powerful engines, larger cargo holding capacity, and better navigational and communication equipment. Catches fluctuated around 25 000 t until

M. Moriyasu[1], E. Wade, A. Sinclair, and Y. Chiasson. Department of Fisheries and Oceans, Gulf Fisheries Centre, Science Branch, P.O. Box 5030, Moncton, NB E1C 9B6, Canada.

Correct citation: Moriyasu, M., Wade, E., Sinclair, A., and Chiasson, Y. 1998. Snow crab, *Chionoecetes opilio*, stock assessment in the southwestern Gulf of St. Lawrence by bottom trawl survey. *In* Proceedings of the North Pacific Symposium on Invertebrate Stock Assessment and Management. *Edited by* G.S. Jamieson and A. Campbell. Can. Spec. Publ. Fish. Aquat. Sci. 125. pp. 29–40.

[1]Author to whom all correspondence should be addressed. e-mail: moriyasum@mar.dfo-mpo.gc.ca

Fig. 1. Southern Gulf of St. Lawrence snow crab, *Chionoecetes opilio*, management areas and distribution of fishing effort in 1995 based on the logbook data provided by 264 fishing vessels in the southern Gulf of St. Lawrence snow crab fisheries.

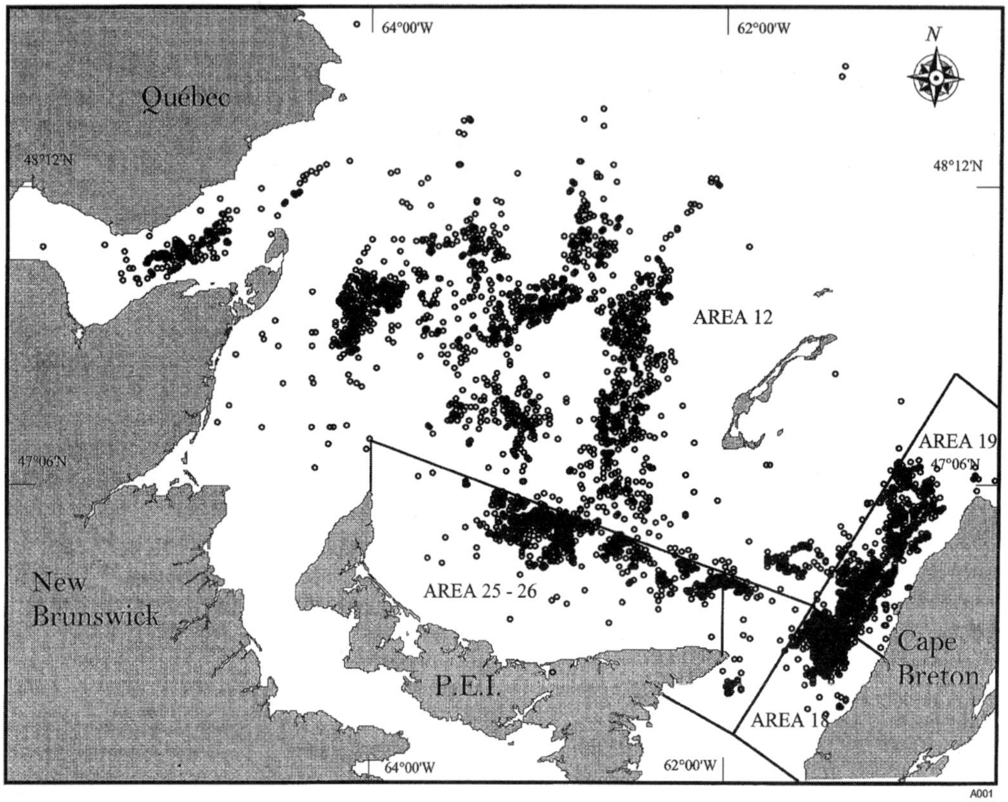

Fig. 2. Annual fluctuation of total landings of snow crab in the southwestern Gulf of St. Lawrence (area 12) between 1969 and 1995.

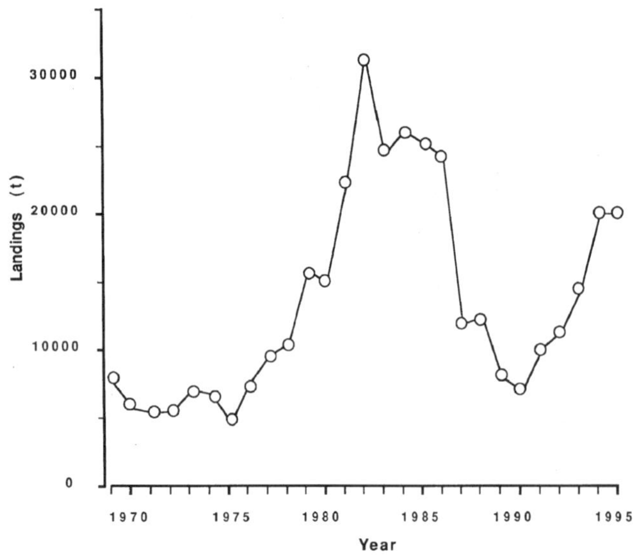

1986 before dropping to 11 700 t in 1987 and 1988 (Fig. 2). In 1989, after landing 7880 t, the fishery prematurely closed due to a high incidence of postmolt soft-shelled crab in the commercial catch.

Miller (1975) and Conan and Maynard (1987) attempted direct observation to estimate snow crab abundance and both concluded that, because of the highly aggregated distribution, standard statistical techniques were not acceptable. Based on strong spatial correlation among sampling units, Conan and Maynard (1987) suggested that geostatistical methods such as kriging would improve the accuracy of abundance estimates. In 1989, geostatistical techniques were applied to the trawl survey. New management measures were introduced in 1990 and one goal involved determining a total allowable catch (quota) based on the biomassf com adultond mal was atchesed (neab be low cvalue. These crab should have opportunity to mate and will be of higher value when they recruit the following fishing season. Soft-shelled crab has low meat content and is discarded at sea by fishers, but these crab are physically fragile and may have significant mortality in being released. Monitoring soft-shelled crab in catches is done objectively with a durometer gauge (Foyle et al. 1989) and the fishery is closed when certified on-board observers record the percentage of soft-shelled crab exceeding 20% for more than two consecutive weeks.

Based on trawl survey results, a quota system was established in 1990 in co-operation with industry, which was more conservative (exploitation rate of 32–43%) than the 50–60% exploitation rate targeted for previously. To ensure a conservative initial quota in 1990, fishery managers set the quota at 7000 t, less than the lower limit of the biomass estimate and at a level stated by the crab industry to be the threshold of economic profitability. Since 1990, this exploitation level of 32% has been the target for area 12. Catches have since increased

Table1. Classification of carapace stage based on carapace condition, durometer reading and corresponding approximative age after terminal molt (modified from Anonymous 1994 and Sainte-Marie and Dufour 1994).

Category	Stage	Durometer reading	Carapace condition	Approximative age after terminal molt
New soft	I	<68	Brightly coloured, iridescent, soft, no epibionts, chelae easily bent	0–5 mo
Clean	II	Variable	Brightly coloured, some iridescence, may have epibionts, chelae not easily bent	5 mo – 1 yr
Intermediate	III	>68	Dull brown dorsally and yellow-brown ventrally, no irridescence, shell abrasion evident, epibionts	8 mo – 3 yr
Old	IV	>68	Carapace very dirty but hard, decay may be present at leg joints, epibiont removable at processing plant	2–5 yr
Very old	V	Variable	Carapace very dirty and may be soft (durometer reading <68), progression of decay may be evident, epibiont not removable at processing plant	4–6 yr

to about 20 000 t in both 1994 and 1995 (quotas were 20 000 t for both years) with a value of around C$125 million (Loch et al. 1996).

Here, we emphasize analysis of data collected during 8 years of survey (1988–1995) of the southwestern Gulf of St. Lawrence snow crab fishery and support the usefulness of geostatistic techniques in stock assessment of this species.

Material and methods

Commercial catch rates

Since quotas were introduced in 1990, catch-per-unit-of-effort (CPUE) data have been used to provide an index of stock abundance. These data are obtained from logbooks and sales slips. Logbooks collect information on catch, effort in numbers of trap hauls, trap type, depth, position (latitude/longitude), and weather. Sales slip and logbook data are compiled by the Statistics and Informatics Branches of the Maritimes and Laurentian Regions of the Department of Fisheries and Oceans. Because there is not complete compliance in recording logbooks, sales slips are used to estimate total catch. Total effort in number of traps hauled is estimated by dividing total landings by mean CPUE. Information on fishing location is used to map the distribution of fishing effort.

Trawl survey

A post-fishing season trawl research vessel (RV) survey has been conducted since 1988 using a chartered snow crab fishing vessel. The gear is a Bigouden *Nephrops* trawl, with a 20-m head line, a 27.3-m foot rope with a 3.2-m, 8-mm galvanized chain, and mesh size of 80 mm in the wings and 50 mm in the cod-end (Conan et al. 1994*a*). A stratified-random survey is used: one–two locations are randomly chosen among nine grids within each 10-min longitude by 10-min latitude rectangle. The same stations are used every year. In 1988 there were 173 stations in area 12. In 1990, the survey was extended to inshore fisheries for all areas in the southern Gulf of St. Lawrence, with a total of 262 stations. The survey is done during daylight on a 12-h basis. Tows vary between 5 to 8 min at a speed of approximately 2 knots. The appropriate length of a tow was determined in 1987. *Nephrops* trawls dig into the bottom sediment and tows longer than 5–8 min result in a full load of mud, reduced catchability, and significant net damage. A trawl starts when the predetermined amount of warp is let out and winch drums are locked. Net behavior is monitored by a SCANMAR net-height sensor indicating net height from the sea floor. In cases of abnormal net behavior, tows are repeated. Recorded for each tow are: duration, position at start and end, horizontal opening of the trawl using SCANMAR, and water depth. The survey usually starts when the fishery has ended, late June or early July, and is completed in October. From 45 to 65 sea-days are required to cover the entire southern Gulf of St. Lawrence. Water depths range from 30 to 130 m. To minimize interannual variation in catchability, the same vessel and trawl net have been used throughout the time series.

All male crab were measured for carapace width (CW) and claw height (CH) using modified vernier calipers (Watson and Wells 1970), for carapace hardness using a hardness gauge (Foyle et al. 1989), and for carapace condition, which was categorized into five groups (Table 1). New-soft (stage I) and clean crab (stage II) with durometer reading <68 is considered as postmolt soft-shelled crab. The terminology of male maturity phase follows Sainte-Marie et al. (1995). Size–frequency histograms were standardized to the surface area swept by the trawl.

Commercially harvestable biomass was estimated for hard- and soft-shelled adult males ≥95 mm CW. Soft-shelled adult males will be recruited to the fishery the following year as hard-shelled terminal molt crab (Conan and Comeau 1986). The incidence of adolescent males in the survey is also used to indicate the likelihood of encountering soft-shelled crab in the following fishing season. Soft-shelled crab are vulnerable to traps at carapace sizes larger than approximately 70 mm (M. Moriyasu, unpublished data). This size group can also be predicted from smaller modes of adolescent crab ≥56 mm CW.

Table 2. Fishery performance in area 12 between 1986 and 1995.

Year	Catch (t)	Fishing effort (number of traps hauled)	CPUE (kg/trap haul)	Soft-shelled crab (%)
1986	24267	454657	53.4	4.9
1987	11782	449293	26.2	10.8
1988	12355	528844	23.4	7.5
1989	7882	356442	22.1	37.2
1990	6950	254578	27.3	16.8
1991	10019	326671	30.7	11.5
1992	11235	362967	31.0	8.6
1993	14336	344698	41.6	6.1
1994	19995	390833	51.2	5.6
1995	19944	416890	47.8	2.5

Mapping and estimation of abundance by kriging

Kriging (Matheron 1970; Clark 1979) is used to estimate biomass and to map density of different categories of crab (Conan 1985; Conan et al. 1988; Conan et al. 1994*b*). Kriging consists of two procedures: (*i*) analyzing and modeling the covariance between sampling units as a function of distance between their locations; and (*ii*) calculating optimal weighting to be attributed to each sampling unit. Separate variograms and density estimates are made for each sampling unit, which are summed to give a total for the southern Gulf, an area of 36 931 km^2 (Fig. 1). Numbers of crab are converted to biomass using a CW–weight relationship, $W = 2.665 \times 10^{-4}$ CW$^{3.098}$, which was calculated from all years combined (Hébert et al. 1992*b*). Natural mortality between date of survey completion and the next year's fishing season, 6–7 months later, is considered to be negligible.

Stock biomass and CPUE prediction model

Given the consecutive estimates of survey biomass as well as information on catches, it is possible to construct a model to predict the stock biomass and commercial CPUE in the year following the most recent RV survey. An approach similar to that described by Conser and Idoine (1992) was used.

Assuming no natural mortality of adult males, the biomass of hard-shelled adult males ≥95 mm CW that survive the fishery in year y and that will be available to the fishery in year $y+1$ (S_{y+1}) is given by:

$$[1] \quad S_{y+1} = S_y + R_y - C_y$$

where R_y = biomass (*t*) of soft-shelled adult males (fishery recruits) in year y and C_y = the catch biomass (*t*) in year y. If one assumes that the RV survey estimate of the survivor's biomass (s) is proportional to the true survivor's biomass (S) with a catchability of k_s, then:

$$[2] \quad s_y = k_s\, S_y$$

If the ratio of the catchabilities of the fishery recruits (k_r) and survivors (k_s) is p, then the RV survey estimate of recruitment (r) is

$$[3] \quad r_y = k\, p\, R_y$$

The model calculates survivor and fishery recruit components of the fishable population in a stepwise manner. One parameter to be estimated is the total fishable population biomass at the beginning of the initial year (B_1). The survivor component of the harvestable population in the next year is

$$[4] \quad \hat{S}_2 = B_1 - C_1$$

In subsequent years, the survivor component is calculated by rearranging and substituting eq. 3 into eq. 1.:

$$[5] \quad \hat{S}_{y+1} = \hat{S}_y + \frac{r_y}{kp} - C_y$$

The RV survey estimate of recruitment in year y may be predicted from RV results and catch of the previous year and RV results from the current year:

$$[6] \quad \hat{r}_y = s_y + \frac{r_y}{p} - (s_{y-1} + r_{y-1}) + \frac{C_{y-1}}{k}$$

The commercial catch per unit effort may be predicted from the modeled total fishable biomass from eqs. 3, 4, and 5 ($\hat{B}_y = \hat{S}_y + \hat{R}_y$). An initial examination of data suggested that a linear description was most appropriate.

$$[7] \quad \hat{U}_y = a + b\,\hat{B}_y$$

The objective of the model is to choose the set of parameters B_1, k, p, a, and b that minimizes the following set of residuals (log normal error structure assumed):

recruitment estimates: $\ln\left(\frac{r_y}{\hat{r}_y}\right)$

survivor estimates: $\ln\left(\frac{s_y}{\hat{S}_y\, k}\right)$

CPUE estimates, where u_y is the observed value:

$$\ln\left(\frac{u_y}{\hat{U}_y}\right)$$

The sums of squared residuals were minimized using the Marquardt algorithm. No residual weighting was used. The observed recruits, survivors, and catch per unit effort were assumed to be independent.

Bootstrapping was used to examine the precision and correlation of parameter estimates. The original residuals from the base fit were resampled at random, with replacement, and added to the predicted values from the base fit to obtain a new set of observations. The model was fit to the data and the resulting parameter estimates were saved. The distribution and correlations among 300 replicates of this process were examined to get an idea of the uncertainties of the predictions.

Results

Fishery performance

The quality of information provided in logbooks was extremely good and compliance rate was >95%. Although the fishing fleet covered the entire fishing ground (Fig. 1), fishers tended to move from patch to patch when either catch rate decreased or incidence of soft-shelled crab increased. The degree of movement varied from year to year, month to month, and even week to week during the fishing season in relation to the availability of snow crab. Between 1989 and 1993, the searching activity of the fleet was more evident than during the period of high biomass, 1994–1995. A continuous

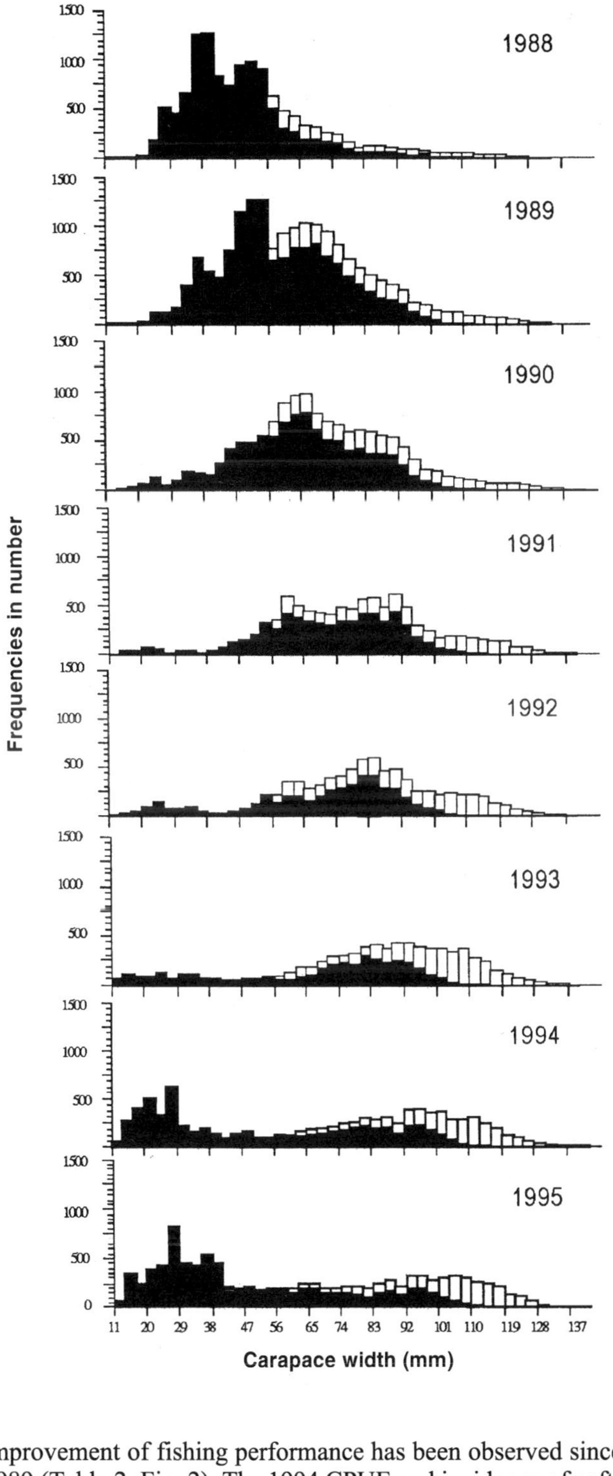

Fig. 3. Size frequency distributions of male crab caught during the trawl survey from 1988 to 1995. (closed bar, adolescent male; open bar, adult male).

improvement of fishing performance has been observed since 1989 (Table 2, Fig. 2). The 1994 CPUE and incidence of soft-shelled crab are similar to levels in 1986.

Size structure and biomass estimates

The three principal modes (27.5, 39.5, and 51.5 mm CW) of small crab that appeared in 1988 could be followed in size–frequency distributions from 1988 to 1995. These animals

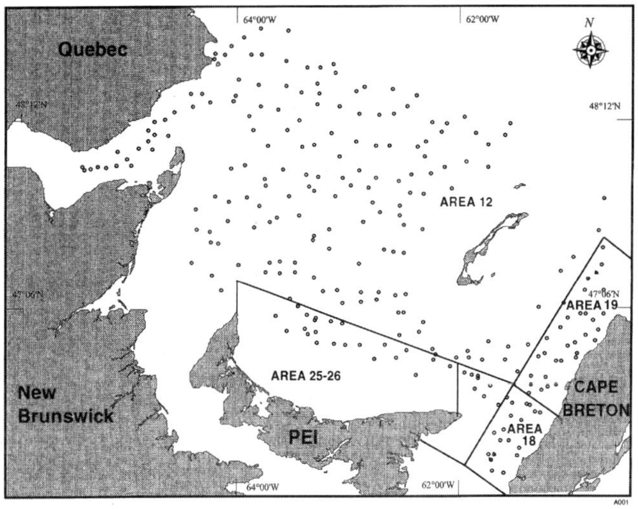

Fig. 4. Geographic distribution of the 1995 trawl survey stations (total of 262 stations were successfully trawled).

started to recruit to the fishery in 1991. Similar modes of pre-recruits were not observed until 1994 (Fig. 3).

Positions of trawl stations are shown in Fig. 4. The variograms from kriging showed a correlation or sill effect varying between 20 and 100 km depending on the survey year and category of crab assessed (Fig. 5). Biomass projections showed a continuous increase since 1989 and a slight decline for 1995 and 1996 (Table 3, Fig. 6). A decrease in annual recruitment to the fishery was observed from 1990 to 1992, followed by a rapid increase from 1992 to 1994, and a decrease during 1995 and 1996. There was an increase in the size of high-density patches of adult crab ≥95 mm CW up to 1993 followed by a slight decrease in recent years (Fig. 7). Exploitation rate was 90.9% in 1989, but decreased to a low level, varying from 32.0 to 42.7%, between 1990 and 1996.

The abundance of adolescent crab ≥56 mm CW increased from 1989 to 1991 but has generally decreased since 1992 (Table 3, Fig. 6). There was an increase in the size of high-density patches of adolescent crab ≥56 mm CW up to 1992 followed by a continuous decrease in recent years (Fig. 8). The 1996 projection for soft-shelled crab abundance was slightly lower than in 1994 and represents a 60% decrease from the peak observed in 1991.

Stock biomass and CPUE prediction model

The model fit for estimates of survivors, recruitment, and CPUE was good (Fig. 9) with close agreement between observed and predicted values. The correlation among parameter estimates was relatively low (Table 4), except for a versus b and k versus p. The latter indicated that there was little difference between catchabilities of survivors ($k = 1.250$) and recruits (p = ratio of adult and recruit catchability = 0.975) (Table 4). The fact that the estimated catchability of survivors was >1.0 could result from incorrect assumptions regarding natural mortality, errors in survey swept surface, and underreported total catch. The cumulative frequencies of the bootstrap estimates from 300 replications of the model fit indicated relatively wide 95% confidence intervals. The predicted biomass of survivors remaining after the 1995 fishery was between

Table 3. Biomass estimation of total harvestable crab, annual recruitment to the fishery, and the forecast of incidence of soft-shelled crab in the southwestern Gulf of St. Lawrence fishery (area 12) based on kriging analysis of the data from trawl surveys conducted in the previous year.

	Prediction for							
	1989 (1988)[a]	1990 (1989)	1991 (1990)	1992 (1991)	1993 (1992)	1994 (1993)	1995 (1994)	1996 (1995)
Total harvestable biomass (all adult male ≥95 mm)								
Crab density per km^2	497	1241	1489	1782	2341	3759	3613	2881
Biomass (t)	8676	21748	23444	29443	37771	61936	58682	49517
±2 SD	±3635	±11614	±12402	±14714	±14175	±8176	±6748	±7923
Recruitment to the fishery (soft-shelled adult male ≥ 95 mm)								
Crab density per km^2	—[b]	1195	913	827	1394	2067	1700	1074
Biomass (t)	—[b]	19741	15082	13659	22491	34052	27606	18458
±2 SD		±9983	±6477	±6276	±9266	±7968	±4831	±4319
Soft-shelled male of postmolt CW ≥ 70 mm[c]								
Crab density per km^2	3511	7310	9481	5659	6733	7202	4370	3910
Abundance (× 10^6)	102.7	213.8	277.4	165.5	197.0	210.7	127.8	114.4
±2 SD	±21.0	±46.6	±54.8	±46.5	±76.5	±62.9	±37.6	±25.5

[a]Years in parentheses represent survey years.

[b]Recruitment to the fishery in 1989 could not be estimated because of the absence of information on carapace condition.

[c]Projection in the following spring was made based on the abundance of adolescent males of CW ≥ 56 mm at the time of the survey in the previous year by assuming that all adolescent males molt to CW ≥70 mm in the following spring.

10 000 and 35 000 t ($p = 0.05$). Intervals for recruitment were 7000–30 000 t and the commercial CPUE interval range was 30–55 kg per trap. The relationship between CPUE (C_i) estimated from logbook data and the biomass of commercially harvestable crab estimated by kriging (B_i) showed a strong linear correlation (Fig. 10):

$$B_i = 1761.261 \quad C_i - 28801.348 \quad r^2 = 0.97$$

Discussion

Methodology of survey and data analysis

Many attempts have been made to estimate the abundance of snow crab in Atlantic Canada (Bailey and Elner 1989). Miller (1975) used a Perry PC8 submersible and bottom photography to count snow crab in different locations. Conan and Maynard (1987) used an underwater television camera towed on a sledge. However, these methods did not allow for information to be collected on size, maturity phase, and sex and provided only coarse estimates of abundance. In addition, submersible observations tend to underestimate density of large crab (Miller 1975). Beam trawl (Miller 1975) and modified otter trawl (Powles 1968) were also used but these methods also appeared to underestimate abundance (Miller 1975). However, Conan et al. (1994*a*) compared two types of otter trawls, *Nephrops* trawl and Devismes trawl, and suggested that a combination of both type of trawls would collect a wide size range of snow crab. Experimental traps combined with data on effective area fished was proposed by Miller (1975) for assessing the snow crab stock. However, efficiency of traps varies greatly as a function of the physiological state of individuals (molting history) (Moriyasu and Mallet 1986; O'Halloran and O'Dor 1988), type of bait (Perry 1969), and soak time (Bailey and Dufour 1983). In addition, trapping is size selective and it is difficult to estimate the abundance of smaller size crab (CW < 70 mm) and females (M. Moriyasu, unpublished data).

The *Nephrops* trawl was originally developed for *Nephrops norvegicus*. This gear is efficient at digging buried snow crab out of the sediment and can be fished over most substrates inhabited by this species. However, caution must be used when interpreting the catch of smaller size snow crab (CW ≤ 15–20 mm) because of size-dependent selectivity of the net.

The Leslie method (Ricker 1975) has been commonly used to estimate stock size since 1978 (Bailey 1978). This method requires abundance indices and landings at several time intervals throughout the fishing season. Because the target population and fishing effort likely change over a fishing season, important assumptions of the Leslie method are violated (Miller and Mohn 1989). Miller and Mohn (1989) suggested that a seasonal mean CPUE is a better alternative to Leslie analysis. Dawe et al. (1993) demonstrated the possibility of applying index-removal methods for assessment based on sampling a population before and after the fishery. A tag–recapture method has also been attempted for estimating stock size of the Atlantic snow crab (Bailey 1978; Elner and Robichaud 1980, 1981, 1984; Taylor and O'Keefe 1981, 1983, 1984; Bailey and Coutu 1987; Dufour 1988). This method, however, relies on complete reporting by fishers and underlying assumptions are often violated (Miller and Mohn 1989).

Bailey and Dufour (1987) demonstrated a better method using an injected ferromagnetic tag. This method requires reasonable coverage of landings to detect tags, which would not be appropriate for a large area such as the southern Gulf of St. Lawrence. Even if these methods were effective, however, they would not have any predictive capacity and would provide no information on the abundance of different categories and size-classes of snow crab.

Kriging is a collection of generalized linear regression techniques used to minimize a variance (Deutsch and Journel

Fig. 5. Example of variograms computed for adult male ≥95 mm CW and adolescent male ≥56 CW from the 1995 trawl survey data. The range of spatial covariance effects is approximately 20 and 40 km, respectively.

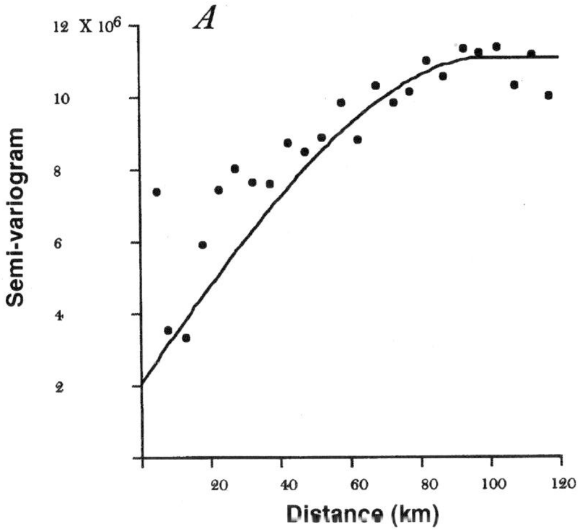

Parameters	Variable Limits	Model Information
# of pairs: 11302 Direction: 0.0 Tolerance: 90.0 Bandwidth: MAX	Min. 0.00 Max. 14818.14 Mean: 2022.21 Variance: 9014300.00	Nugget = 2.1 X 10^6 Spherical: Sill 9.0 X 10^6 Range of influence = 100 km

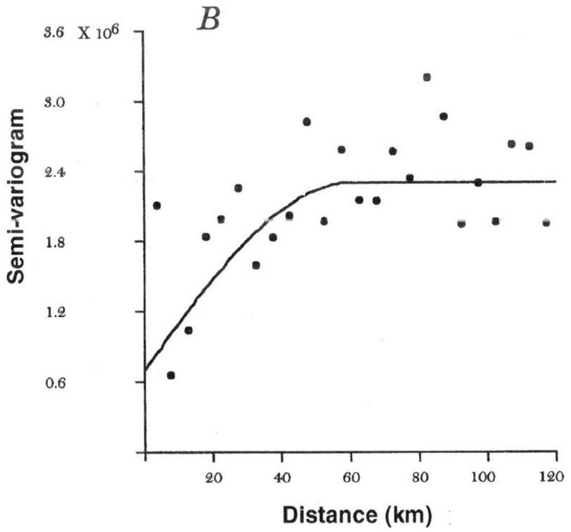

Parameters	Variable Limits	Model Information
# of pairs: 11319 Direction: 0.0 Tolerance: 90.0 Bandwidth: MAX	Min. 0.00 Max. 11831.77 Mean: 642.26 Variance: 1933569.88	Nugget = 0.7 X 10^6 Spherical: Sill 1.6 X 10^6 Range of influence = 60 km

1992). In simple kriging, the sum of the weights is equal to 1. The method assumes that the mean and variance of the estimate at a given area are independent of surrounding areas (stationarity) and that covariances are identical for all locations of the area surveyed. The method assigns lower weights to closely spaced sample points and can be used to generate high definition maps. Kriging does not, however, take into account coastal or island boundaries when fitting a variogram (Conan et al. 1994*b*). In our study, areas were subdivided into small blocks that were summed for total estimates. Recently,

Fig. 6. Annual variation of predicted biomass of commercially harvestable crab, i.e. adult male ≥95 mm CW (closed circle), annual recruitment to the fishery, i.e. soft-shelled adult male ≥95 mm CW (open circle) and the abundance of adolescent male ≥56 mm CW at the time of survey (closed square).

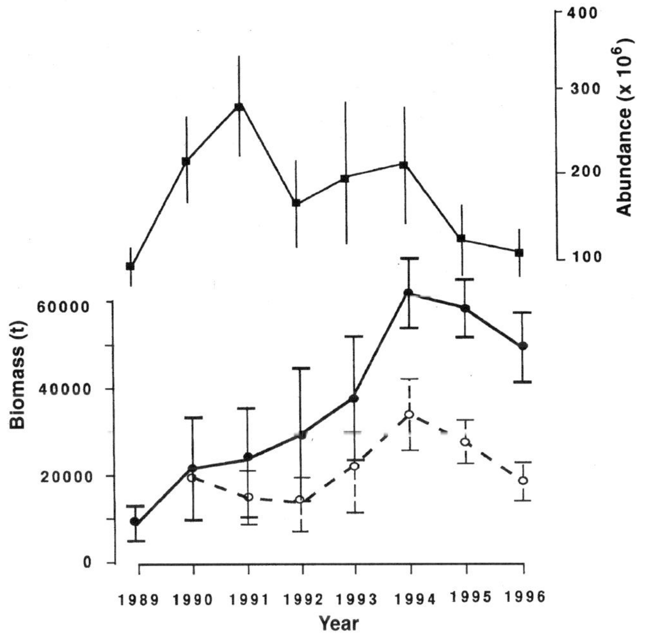

accuracy of estimates has been improved by accounting for border discontinuities. Multivariable co-kriging techniques are also being exploited to account for environmental variables such as sediment, depth, and temperature in the distribution of the resource. Estimates of harvestable biomass and soft-shelled crab by kriging have been consistent, providing a coherent trend in the abundance of southern Gulf snow crab.

Uncertainty of the forecast

Three distinct size modes (27.5, 39.5, and 51.5 mm CW) corresponding to instars VII, VIII, and IX (Sainte-Marie et al. 1995) most likely reached commercial size between 1991 and 1993 based on the succession of modes. The group of small immature crab (instars VI and VII) observed in the 1994 survey may, therefore, enter the fishery in 5–6 years, i.e., 1999–2000. Although estimation of timing of future recruitment to the fishery is reasonable compared to results obtained in the northern Gulf of St. Lawrence (Sainte-Marie et al. 1995), the long-term forecast is based on the assumption that there is no skip molting and that terminal molt does not occur at CW < 95 mm during that period. Terminal molt, however, occurs over a wide size range, from crab 38 mm CW (M. Moriyasu, personal observation) to 150 mm CW (Bailey and Elner 1989). Factors affecting skip molting are unknown, but in the southern Gulf of St. Lawrence the percentage of skip molters varies from 13% (Chiasson et al. 1991) to 28% (Hébert et al. 1992*a*). Other factors can also affect timing and abundance of recruitment. Bailey (1982) reported that recruitment of snow crab seems to be negatively dependent on cod abundance 3–6 years earlier. Waiwood and Elner (1982) disagreed with Bailey's (1982) conclusion and proposed an alternative hypothesis that the removal of large old crab would have released the snow crab

Fig. 7. Density contours of adult crab ≥95 mm CW (commercially harvestable biomass in the following season) calculated by kriging from 1989 to 1995.

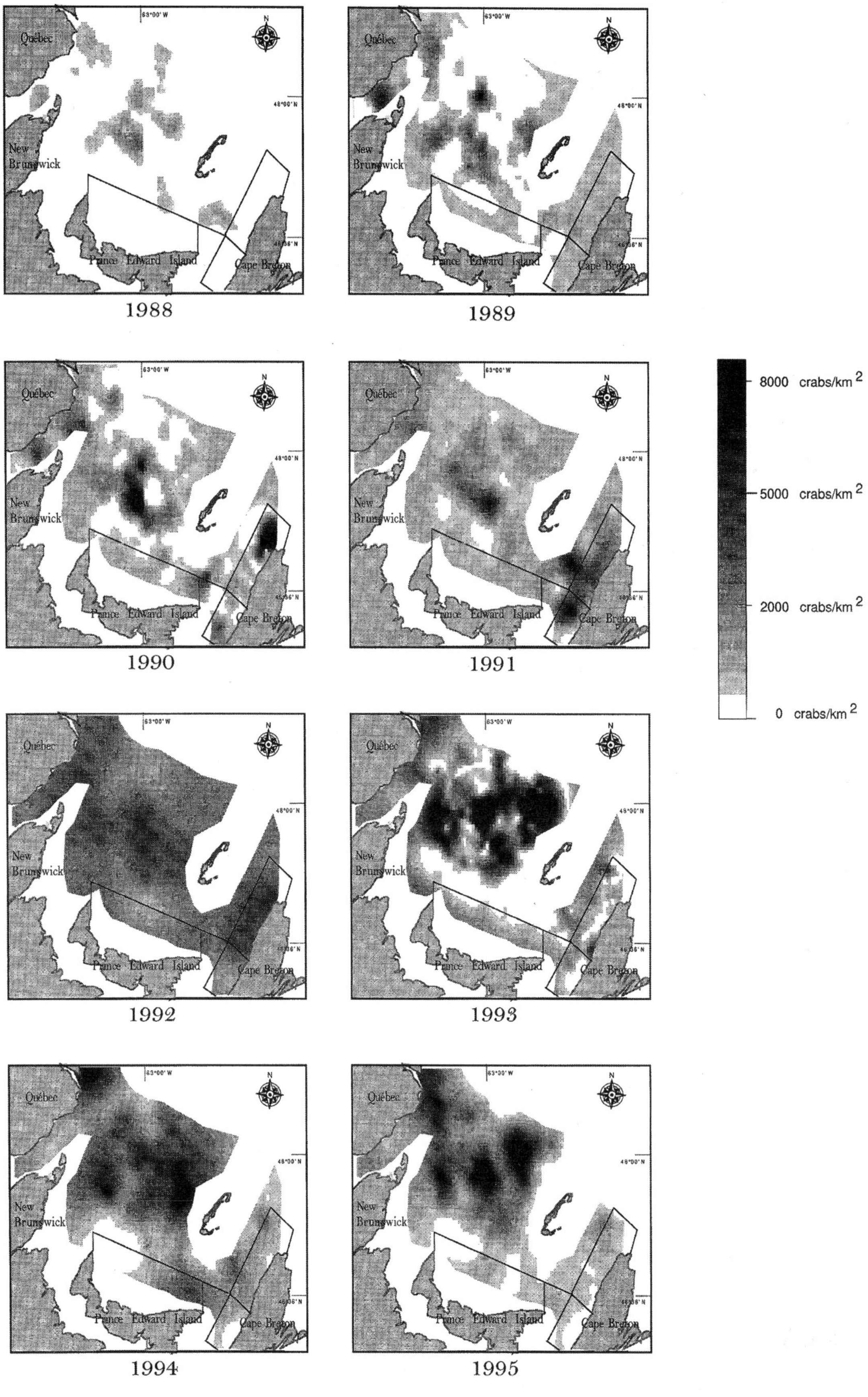

Fig. 8. Density contours of adolescent males of premolt CW ≥ 56 mm (abundance of soft-shelled crab in the following season by assuming that the molting occurs in the following spring and the minimum postmolt size is 70 mm CW) calculated by kriging from 1989 to 1995.

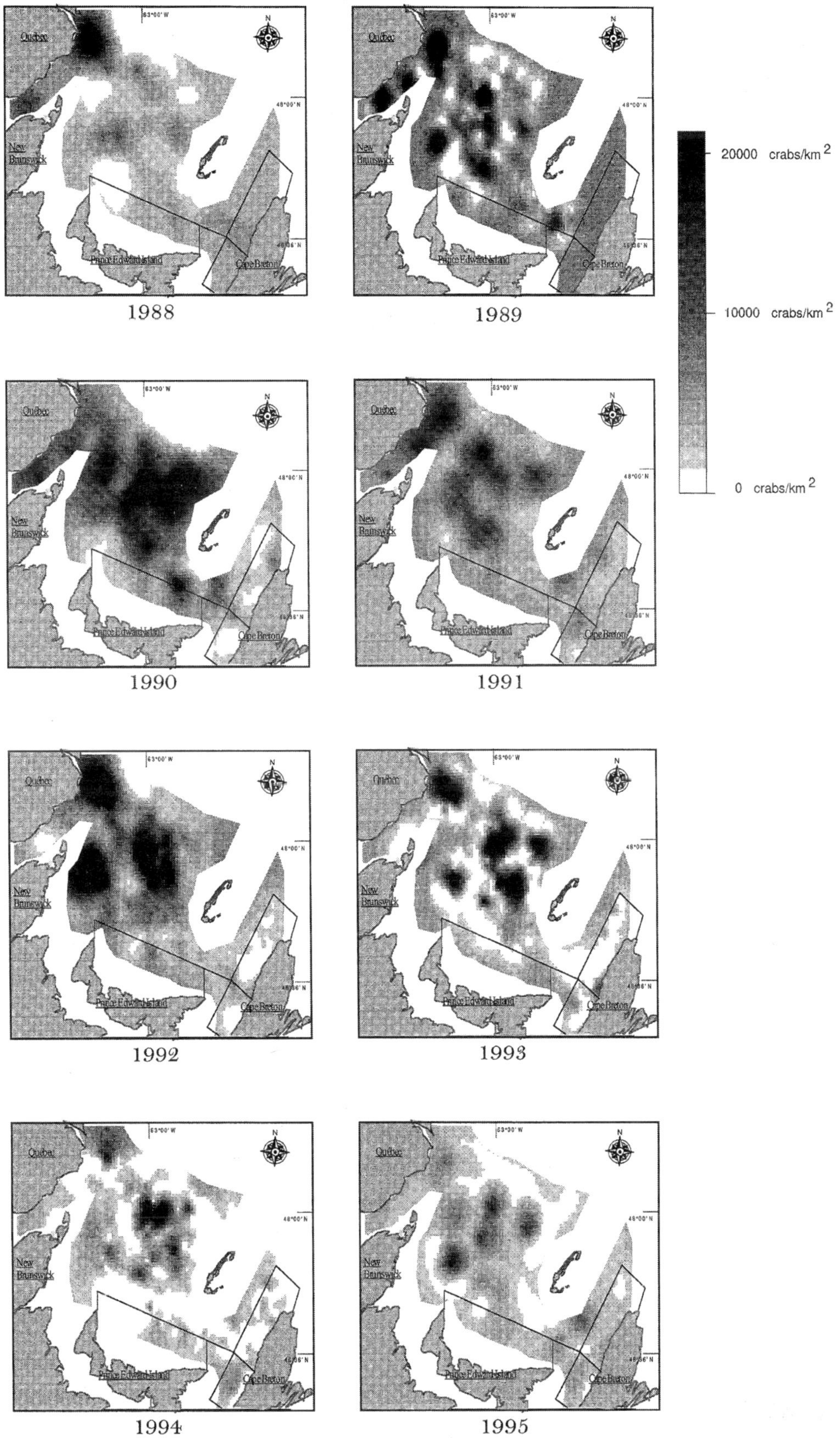

Table 4*a*. Correlation of parameters estimated with the Marquardt algorithm.

	N	*a*	*b*	*k*	*p*
N	1	—	—	—	—
a	−0.038	1	—	—	—
b	−0.0129	−0.884	1	—	—
k	−0.236	−0.0956	0.228	1	—
p	0.297	−0.00992	−0.00676	−0.712	1

Table 4*b*. Parameter estimates and associated coefficients of variation (CV).

Parameter	Estimate	CV
B_1	9281	0.036
a	16.78	0.44
b	7.96×10^{-4}	0.36
k	1.250	0.067
p	0.9750	0.046

Note: *N*, number of data set; B_1, initial fishable biomass; *a* and *b*, constant; *k*, biomass of survivors; *p*, ratio of the catchability of fishery recruits (k_r) and survivors (k_s).

Fig. 9. Observed (square) and predicted (diamond) research survey estimates of survivors (upper panel), recruitment (middle panel), and commercial CPUE (lower panel) by modified Delury model (Conser and Idoine 1992).

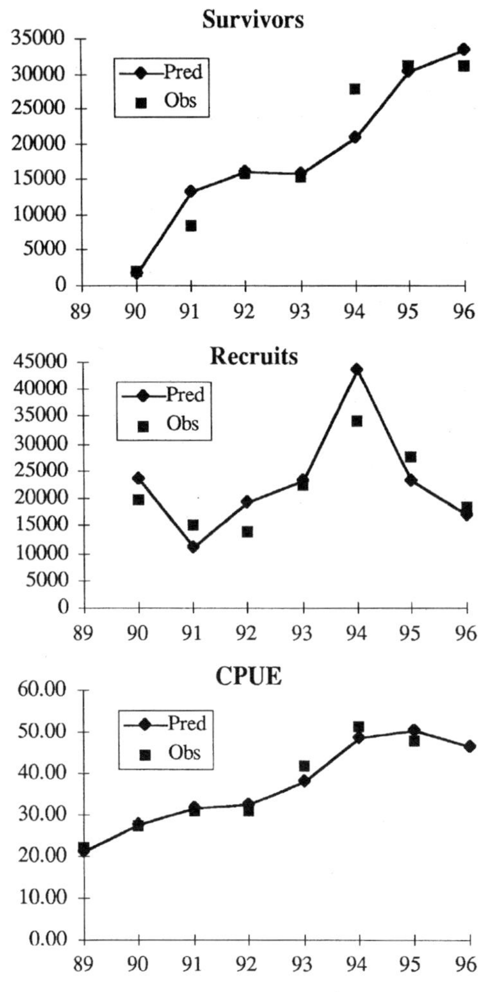

Fig. 10. Regression plot of CPUE (kg/trap haul) from fishers' logbooks versus biomass estimate of adult crab ≥95 mm CW from the trawl survey between 1989 and 1995 in the southwestern Gulf of St. Lawrence snow crab fishery (area 12).

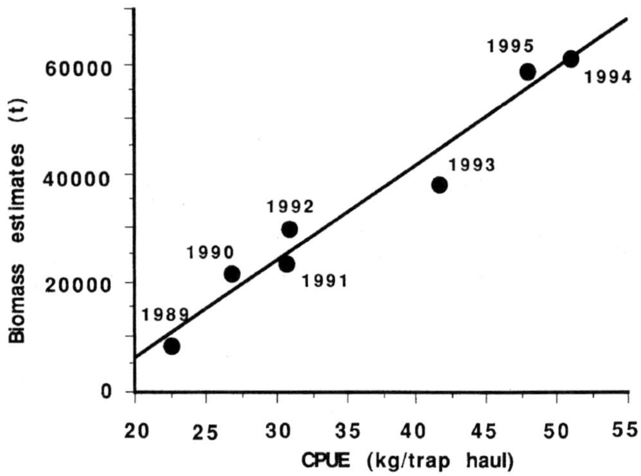

population from a "stagnant" phase to a "dynamic" high-growth phase. Comeau and Conan (1992) further hypothesized that removal by fishing of large "morphometrically mature" males from the accumulated biomass reduced the mortality of "juvenile" males. Forecasts of recruitment should thus be interpreted with caution until terminal and skip molting are better understood.

In this study, we assumed that natural mortality was negligible. During a period of strong recruitment to the fishery (1991–1994), it was expected that natural mortality might have little impact on results. However, the current population consists of a large number of older adult males and natural mortality rate is likely an important consideration.

Stock management

Since 1989, the area 12 fishery has been managed by quotas corresponding to an exploitation rate of 30–40%. The main reason for the rapid recovery of this stock since 1989 was due to strong recruitment of at least three different molt-classes. As reported by Loch et al. (1996), good communication among scientists, managers, and industry representatives throughout the period of low abundance contributed to the creation of a strong partnership in stock management. However, it should be noted that in other snow crab fisheries in Atlantic Canada, the same pattern of stock fluctuation was observed despite different stock management regimes (Taylor et al. 1994; Sainte-Marie and Dufour 1994, 1995), suggesting large-scale patterns in recruitment across eastern Canada (Sainte-Marie 1997). Fisheries depend on the arrival and strength of recruitment waves. Protective measures have been set to optimize recruitment by avoiding soft-shelled crab capture and to maximize yield per recruit (Comeau and Conan 1992; Sainte-Marie et al. 1995). Future recruitment is protected when adolescent males are avoided. In the southwestern Gulf of St. Lawrence, the majority of fishers discard adolescent males, called small- or pencil-clawed males, and can distingush them from "large-clawed males" (Sainte-Marie et al. 1995 in northern Gulf of

St. Lawrence fisheries; M. Moriyasu personal observation in the southern Gulf of St. Lawrence fisheries).

The conservative exploitation rate of around 30–40% did not apparently exceed the level of annual recruitment. This level of exploitation allowed fishing of a year-class to be spread over several years. Nevertheless, the recent accumulation of large old males with this exploitation level could negatively affect future recruitment (Comeau and Conan 1992; Waiwood and Elner 1982). In addition, commercial quality of crab decreases with age after terminal molt. Therefore, a better harvesting strategy might take the following factors into account: (*i*) optimizing yield per recruit by avoiding soft-shelled and adolescent crab; and (*ii*) maximizing value by avoiding an accumulation of old crab.

Natural mortality of old crab was not accounted for in our analysis. It is reasonable to consider that a majority of very old crab present in a given survey will not be available to a following fishery due either to natural mortality or to reduced catchability by traps (Sainte-Marie and Dufour 1994, 1995). Longevity of crab after terminal molt is most likely between 4 and 6 years (Sainte-Marie and Dufour 1994, 1995); old crab proportion will be more pronounced as the population ages.

In conclusion, trawl surveys still appear to be an appropriate method for assessing stock abundance and recruitment and geostatistics appears to be an appropriate method to describe the patchy distribution of snow crab. Precise estimation of biomass cannot be made more than one year in advance until we both understand factors determining terminal and skip molting and have better estimation of natural mortality.

Acknowledgments

We are indebted to those involved in this project over the years, particularly Dr. G.Y. Conan who established the original survey plan and first introduced geostatistical techniques to snow crab assessment in the southern Gulf of St. Lawrence. We thank P. Mallet, M. Hébert, P. DeGrâce, and R. Campbell for conducting laborious field work and for participating in data analysis. We also thank Drs. E.M.P. Chadwick, M. Mallet, B. Smith, and I. Vining, and Messrs. M. Comeau and R. Dufour for reviewing the manuscript and providing constructive criticisms. This study was financed by the Atlantic Fisheries Adjustment Program (AFAP) of the Department of Fisheries and Oceans.

References

Anonymous. 1994. General overview of Atlantic coast snow crab and report on the status of snow crab in the southern Gulf of St. Lawrence (Fishing Areas 12, 18, 19, 25 and 26). DFO Atl. Fish. Stock Status Rep. 94/1.

Anonymous. 1996*a*. Southern Gulf snow crab. DFO Atl. Stock Status Rep. 96/1.

Anonymous. 1996*b*. Le crabe des neiges de l'estuaire et du nord du golfe Saint-Laurent. DFO Atl. Stock Status Rep. 96/6.

Bailey, R. 1978. Analysis of the snow crab population in northwestern Cape Breton, 1978. Can. Atl. Fish. Sci. Adv. Comm. Res. Doc. 1978/41.

Bailey, R.F.J. 1982. Relation between catches of snow crab, *C. opilio* (O. Fabricius) and abundance of cod (*Gadus morhua* L.) in the southwestern Gulf of St. Lawrence. *In* Proceedings of the International Symposium on the Genus *Chionoecetes*. *Edited by* B. Melteff. University of Alaska, Alaska Sea Grant Rep. 82-10. pp. 485–497.

Bailey, R., and Coutu, J.-M. 1987. Crabe des neiges (*Chionoecetes opilio*) de l'estuaire et du nord du golfe Saint-Laurent: evaluation de 1986. Can. Atl. Fish. Sci. Adv. Comm. Res. Doc. 87/70.

Bailey, R., and Dufour, R. 1983. Comparaison des rendements de crabe des neiges obtenus par deux types de casiers differents. Can. Atl. Fish. Sci. Adv. Comm. Res. Doc. 83/19.

Bailey, R., and Dufour, R. 1987. Field use of an injected ferromagnetic tag on the snow crab (*Chionoecetes opilio* O. Fab.). J. Cons. Int. Explor. Mer. **43**: 237–244.

Bailey, R.F.J., and Elner, R.W. 1989. Northwest Atlantic snow crab fisheries: lessons in research and management. *In* Marine Invertebrate fisheries: their assessment and management. *Edited by* J.F. Caddy. John Wiley & Sons, New York, N.Y. pp. 261–280.

Chiasson, Y., Hébert, M., Wade, E., Gallant, C., Degrâce, P., Mallet, P., and Moriyasu, M. 1991. La pêche au crabe des neiges (*Chionoecetes opilio*) dans le sud-ouest du golfe du Saint-Laurent en 1990: état de la ressource et l'estimation de la biomasse. Can. Atl. Fish. Sci. Adv. Comm. Res. Doc. 91/27.

Chiasson, Y.J., Hébert, M., and Moriyasu, M. 1992. A retrospective look at the development and expansion of southwestern Gulf of St. Lawrence snow crab, *Chionoecetes opilio*, fishery. Can. Tech. Rep. Fish. Aquat. Sci. 1847.

Clark, I. 1979. Practical geostatistics. Elsvier Science Publications, London and New York.

Comeau, M., and Conan, G.Y. 1992. Morphometry and gonad maturity of male snow crab, *Chionoecetes opilio*. Can. J. Fish. Aquat. Sci. **49**: 2460–2468.

Conan, G.Y. 1985. Assessment of shellfish stock by geostatistical techniques. Int. Counc. Explor. Sea, C.M. 1985/K:30.

Conan, G.Y., and Comeau, M. 1986. Functional maturity of male snow crab, (*Chionoecetes opilio*). Can. J. Fish. Aquat. Sci. **43**: 1710–1719.

Conan, G.Y., and Maynard, D.R. 1987. Estimates of snow crab (*Chionoecetes opilio*) abundance by underwater television – a method for population studies on benthic fisheries resources. J. Appl. Ichtyol. **3**: 158–165.

Conan, G.Y., Moriyasu, M., Wade, E., and Comeau, M. 1988. Assessment and spatial distribution surveys of snow crab stocks by geostatistics. Int. Counc. Explor. Sea, C.M. 1988/K:10.

Conan, G.Y., Comeau, M., Gosset, C., Robichaud, G., and Garaïcoechea, C. 1994*a*. The Bigouden *Nephrops* trawl and the devismes trawl, two otter trawls efficiently catching benthic stages of snow crab (*Chionoecetes opilio*), and the American lobster (*Homarus americanus*). Can. Tech. Rep. Fish. Aquat. Sci. 1992.

Conan, G.Y., Maynou, F., Stolyarenko, D., and Mayer, L. 1994*b*. Mapping and assessment of fisheries resources with coastal and depth constraints, the case study of snow crab in the Bay of Islands Fjord (Newfoundland). Int. Counc. Explor. Sea, C.M. 1994/B+D+G+H:4.

Conser, R.J., and Idoine, J. 1992. A modified DeLury model for estimating mortality rates and stock sizes of American lobster populations. Res. Doc. SAW, 14/7.

Dawe, E.G., Hoenig, J.M., and Xu, X. 1993. Change-in-ratio and index-removal methods for population assessment and their application to snow crab (*Chionoecetes opilio*). Can. J. Fish. Aquat. Sci. **50**: 1467–1476.

Dawe, E.G., Taylor, D.M., Veitch, P.J., Drew, H.J., Beck, P.C., and O'Keefe, P.G. 1996. Status of Newfoundland and Labrador snow crab. DFO Atl. Fish. Res. Doc. 96/06.

Deutsch, C.G., and Journel, A.G. 1992. GSLIB Geostatistical software library and user's guide. Oxford University Press, New York, N.Y.

Dufour, R. 1988. Marquage des crabes des neiges du nord-est du golfe du Saint-Laurent en 1987. Can. Atl. Fish. Sci. Adv. Comm. Res. Doc. 88/39.

Elner, R.K., and Robichaud, D.A. 1980. Analysis of the Cape Breton inshore fishery for snow crab, 1979. Can. Atl. Fish. Sci. Adv. Comm. Res. Doc. 80/55.

Elner, R.K., and Robichaud, D.A. 1981. Assessment of the Cape Breton inshore fishery for snow crab, 1980. Can. Atl. Fish. Sci. Adv. Comm. Res. Doc. 81/40.

Elner, R.K., and Robichaud, D.A. 1984. The fishery for snow crab off the Atlantic coast of Cape Breton Island: the 1993 assessment. Can. Atl. Fish. Sci. Adv. Comm. Res. Doc. 84/15.

Foyle, T.P., Hurley, G.V., and Taylor, D.M. 1989. Field testing shell hardness guages for the snow crab fishery. Can. Ind. Rep. Fish. Aquat. Sci. 193.

Hébert M., Chiasson, Y., Mallet, P., Degrâce, P., Gallant, C., and Moriyasu, M. 1992*a*. Evaluation de l'état du stock de crabe des neiges (*Chionoecetes opilio*) dans le sud-ouest du golfe du Saint-Laurent (zone 12) en 1992. Can. Atl. Fish. Sci. Adv. Comm. Res. Doc. 92/109.

Hébert, M., Gallant, C., Chiasson, Y., Mallet, P., Degrâce, P., and Moriyasu, M. 1992*b*. Monitoring of occurrence of soft-shell crabs in commercial catches of snow crab (*Chionoecetes opilio*) in the Southwestern Gulf of St. Lawrence (zone 12) in 1990 and 1991. Can. Tech. Rep. Fish. Aquat. Sci. 1886E.

Loch, J.S., Moriyasu, M., and Jones, J.B. 1996. An improved link between industry, management and science: review of case history of the Southwestern Gulf of St. Lawrence snow crab fishery. Aquat. Living Resour. **8**: 253–265.

Matheron, G. 1970. La théorie des variables régionalisées et ses applications. Les cahiers du Centre de Morphologie Mathématique de Fontainebleau. Fasc. 5.

Miller, R.J. 1975. Density of the commercial spider crab, *Chionoecetes opilio*, and calibration of effective area fished per trap using bottom photography. J. Fish. Res. Board Can. **32**: 761–768.

Miller, R.J., and Mohn, R.K. 1989. Less Leslie please. Can. Atl. Fish. Sci. Adv. Comm. Res. Doc. 89/22.

Moriyasu, M., and Mallet, P. 1986. Molt stages of the snow crab *Chionoecetes opilio* by observation of morphogenesis of setae on the maxilla. J. Crust. Biol. **6**: 709–718.

O'Halloran, M.J., and O'Dor, R.K. 1988. Molt cycle of male snow crabs, *Chionoecetes opilio*, from observations of external features, setal changes and feeding behavior. J. Crust. Biol. **82**: 164–176.

Perry, L.A. 1969. Nova Scotia deep sea crab explorations, 1968. Nova Scotia Department of Fisheries, Pictou, N.S. (Mimeograph.)

Powles, H.W. 1968. Distribution and biology of the spider crab, *Chionoecetes opilio*, in the Magdalen Shallows, Gulf of St. Lawrence. Fish. Res. Board Can. MS Rep. No. 997.

Ricker, W.E., 1975. Computations and interpretation of biological statistics of fish populations. Bull. Fish. Res. Board Can. 191.

Sainte-Marie, B. 1997. Comment – an improved link between industry, management and science: review of case history of the Southwestern Gulf of St. Lawrence snow crab fishery. Can. J. Fish. Aquat. Sci. **54**: 496–500.

Sainte-Marie, B., and Dufour, R. 1994. Crabe des neiges de l'estuaire et du nord du golfe Saint-Laurent (Zones de pêche 13, 14, 15, 16 et 17). *In* Rapport sur l'état des inverébrés en 1993: Crustacés et mollusques des côtes du Québec et crevette nordique de l'estuaire et du golfe du Saint-Laurent. *Edited by* L. Savard. Can. Fish. Aquat. Sci. MS Rep. No. 2257. pp. 2–23.

Sainte-Marie, B., and Dufour, R. 1995. Crabe des neiges de líestuaire et du nord du golfe Saint-Laurent (Zones de pêche 13, 14, 15, 16 et 17). *In* Rapport sur l'état des invertèbrès en 1994: Crustacés et mollusques des côtes du Québec et crevette nordique et zooplancton de l'estuaire et du golfe du Saint-Laurent. *Edited by* L. Savard. Can. MS Rep. Fish. Aquat. Sci. No. 2323. pp. 2–27.

Sainte-Marie, B., Raymond, S., and Brêthes, J.-C. 1995. Growth and maturation of the benthic stages of male snow crab, *Chionoecetes opilio* (Brachyura: Majidae). Can. J. Fish. Aquat. Sci. **52**: 903–924.

Taylor, D.M., and O'Keefe, P.G. 1981. Assessment of snow crab (*Chionoecetes opilio*) stocks in Newfoundland in 1979. Can. Atl. Fish. Sci. Adv. Comm. Res. Doc. 81/57.

Taylor, D.M., and O'Keefe, P.G. 1983. Assessment of snow crab (*Chionoecetes opilio*) stocks in Newfoundland in 1980. Can. Atl. Fish. Sci. Adv. Comm. Res. Doc. 83/3.

Taylor, D.M., and O'Keefe, P.G. 1984. Assessment of snow crab (*Chionoecetes opilio*) stocks in Newfoundland in 1981. Can. Atl. Fish. Sci. Adv. Comm. Res. Doc. 84/3.

Taylor, D.M., O'Keefe, P.G., and Fitzpatrick, C. 1994. A snow crab, *Chionoecetes opilio* (Decapoda, Majidae), fishery collapse in Newfoundland. Fish. Bull. **92**: 412–419.

Tremblay, M.J., and Eagles, M.D. 1996. Assessment of the 1995 snow crab fishery off eastern Nova Scotia. DFO Atl. Fish. Res. Doc. 96/35.

Waiwood, K.G., and Elner, R.W. 1982. Cod predation of snow crab (*Chionoecetes opilio*) in the Gulf of St. Lawrence. *In* Proceedings of the International Symposium on the Genus *Chionoecetes*. *Edited by* B. Melteff, University of Alaska, Alaska Sea Grant Report No. 82-10. pp. 499–520.

Watson, J., and Wells, P.G. 1970. A gauge for carapace measurements of crabs. J. Fish. Res. Board Can. **27**: 1158–1161.

Assessing stocks of abalone (*Haliotis* spp.): methods and constraints

Paul E. McShane

Abstract: Stock assessments of abalone (*Haliotis* spp.) have seldom been successful because of the difficulty of deriving precise indices of abundance. This difficulty arises from the contagious distribution of individuals within populations of abalone, and the patchy distribution of populations within stocks. Stocks are usually arbitrarily defined because boundaries separating discrete stocks of abalone are poorly understood. These characteristics of abalone and the spatial distribution of fishing effort over many populations mean that catch rates provide little information on the stock abundance of abalone. Attempts have been made to derive other estimates of abundance or stock size (biomass) including tag–recapture, change-in-ratio techniques, and direct survey. These studies and methods are summarized here. Stock assessment of the New Zealand abalone (*H. iris*) is dependent on the derivation of relative abundance estimates derived from field surveys. Such surveys are more practical than the traditional quadrat or line transect methods because they quantify and describe aggregations (patches) of abalone rather than individuals. Estimates of patch frequencies derived from such surveys may provide greater precision over more realistic spatial scales than estimates of abundance produced by other survey methods.

Résumé : Les évaluations des stocks d'ormeau (*Haliotis* spp.) donnent rarement de bons résultats à cause de la difficulté que pose l'établissement d'indices précis de l'abondance. Cette difficulté vient de la distribution contagieuse des individus dans les populations d'ormeau et de la distribution irrégulière des populations dans les stocks. Les stocks eux-mêmes sont généralement définis de façon arbitraire car on connaît mal les limites qui séparent les différents stocks. À cause de ces caractéristiques des ormeaux, et de la distribution spatiale de l'effort de pêche sur de nombreuses populations, les taux de capture donnent peu d'information sur l'abondance des stocks. Diverses tentatives ont été faites pour calculer d'autres estimations de l'abondance ou de la taille des stocks (biomasse), notamment le marquage–recapture, les changements dans les méthodes de calcul des rapports et les relevés directs. L'évaluation du stock de paua (*H. iris*) est tributaire de l'établissement d'estimations de l'abondance relative à partir des relevés sur le terrain. Ces relevés sont plus pratiques que les méthodes classiques sur quadrat ou sur transect linéaire car elles quantifient et décrivent les agrégations d'ormeaux plutôt que les individus. Les estimations de la fréquence des agrégations à partir de ces relevés peut fournir une plus grande précision à des échelles spatiales plus réalistes que les estimations de l'abondance fournies par d'autres méthodes d'étude.
[Traduit par la Rédaction]

Introduction

Stock assessment relates to the provision and interpretation of biological information for decision making in fisheries management. Assessment often involves some estimate of stock size or biomass to which estimates of recruitment, growth, and survival are used to derive an estimate of productivity (Megrey and Wespestad 1988). Such estimates are used to guide managers in providing for sustainable harvests of exploitable stocks. Stock assessment might also evaluate existing or alternative management strategies. For example, is a particular catch limit, effort restriction, or size limit achieving the usual goals of conserving the stock and providing reasonable yields to the fishery?

From the above definition, few assessments of abalone (*Haliotis* spp.) fisheries have been successful. Five reasons for this are examined in this paper: (*i*) the stock structure of abalone is equivocal; (*ii*) traditional methods of stock assessment are generally inapplicable to abalone fisheries; (*iii*) direct surveys, which can provide information on the abundance, size composition, and biomass of abalone stocks are costly and usually imprecise; (*iv*) there is considerable intraspecific variation in growth, survival, and recruitment which can confound any general estimate of the potential productivity of abalone fisheries; and (*v*) abalone fisheries are usually managed over larger spatial scales than those representing discrete stocks.

Breen (1992) has reviewed application of stock-assessment methods to abalone. Methods involving catch per unit effort (CPUE), such as surplus-production analysis and Leslie-DeLury depletion for estimating biomass involve assumptions likely to be violated in assessing abalone stocks. Breen (1992) summarized the application of stock-reduction analyses (Kimura et al. 1984) for estimating the biomass, fishing mortality, and recruitment in abalone fisheries. He concluded that previously neglected techniques such as experimental fishing could be applied to the assessment of abalone stocks. In this review, survey-based methods for stock assessment and their application for estimating the biomass of populations of abalone and other spatially aggregated sedentary animals are examined. Using examples from one fishery for *H. iris*,

P.E. McShane.[1] National Institute of Water and Atmospheric Research Ltd., P.O. Box 14-901, Wellington, New Zealand.

Correct citation: McShane, P.E. 1998. Assessing stocks of abalone (*Haliotis spp.*): methods and constraints. *In* Proceedings of the North Pacific Symposium on Invertebrate Stock Assessment and Management. *Edited by* G.S. Jamieson and A. Campbell. Can. Spec. Publ. Fish. Aquat. Sci. 125. pp. 41–48.

[1] Present address: SARDI Aquatic Sciences, P.O. Box 120, Henley Beach, SA 5022, Australia.

alternative survey methods and management strategies which overcome many of the existing constraints to stock assessment of abalone are discussed.

Stock structure

Abalone may be considered as sedentary individuals spatially dispersed over their subtidal reef habitat (Shepherd 1973). The larval dispersal potential of abalone means that populations are genetically connected over large distances (hundreds of kilometres) (Brown and Murray 1992), but for at least one species of abalone most of the settlement of larvae occurs near the natal site (Prince et al. 1987; McShane et al. 1988*a*). Although abalone are capable of short-term movement over hundreds of metres (Newman 1966; Ault and DeMartini 1987; McShane 1993), migration between populations may be considered negligible. Genetic studies such as Brown and Murray's are informative in defining the spatial limits of genetically connected populations, but even very low rates of genetic exchange can maintain genetic homogeneity over large spatial areas (e.g., Solé Cava et al. 1994).

In considering the stock structure of abalone, a fishery may be comprised of many populations (Sluczanowski 1986) that inhabit coastal reefs over scales of tens of kilometres or less (Shepherd and Brown 1993). The consideration of smaller spatial scales for defining abalone populations is consistent with the biological characteristics of growth, survival, and recruitment that can vary over small distances (McShane et al. 1988*b*; McShane and Smith 1991; Day and Fleming 1992). In this paper the term population is used to describe assemblages of abalone which have similar traits of growth, survival, and recruitment that represent discrete geographical units or habitats and which are self-sustaining reproductive units.

Traditional methods of stock assessment are inappropriate for abalone fisheries

Catch-per-unit-effort data, the mainstay of the assessment of many exploitable stocks, can be unreliable for abalone fisheries because fishers can maintain catch rates against declining abundance by serially depleting populations (Hilborn and Walters 1987; Breen 1992). Age-structured models cannot be used reliably for abalone fisheries because of the difficulty of aging abalone. However, age determination of abalone remains an active field of research (Erasmus et al. 1994; Shepherd et al. 1995). Thus, stock assessment of abalone has, until recently, tended to be more qualitative than for other fisheries.

In general, the size of exploited stocks, levels of fishing, and the effects of fishing on stock abundance, have eluded abalone scientists. Assessments have focused on yield-per-recruit (YPR) or egg-per-recruit (EPR) modeling that have been used principally in the examination of the effectiveness of size limits (Sloan and Breen 1988; Tegner et al. 1989; Shepherd et al. 1991; McShane 1992; Nash 1992).

The recent attention given to EPR modeling reflects the uncertainty of the relationship between population abundance and recruitment (Prince et al. 1988*a*). Managers are reluctant to discard the notion that recruitment depends in some way on stock size despite contrary empirical evidence (McShane 1995*a*).

Population surveys

Considerable attention is given to capture–recapture and mark–recapture methods of determining population size (e.g., texts by Seber 1973 and Krebs 1989). However, such methods have little applicability to abalone because they do not disperse randomly (Shepherd 1986*a*, 1986*b*), and the probability of capture differs among marked individuals in open and cryptic habitat (Prince et al. 1988*b*) in violation of the main assumptions underpinning mark–recapture methods (Seber 1973).

Population surveys usually involve a fixed sampling unit such as a quadrat or transect which is deployed randomly within the area of interest. Such surveys for abalone have been described by Breen (1980, 1986). Long, thin sample units (e.g., transects) are more likely to encounter an aggregation than one of the same area that is circular or square (Krebs 1989). Thus, when sample size is held constant, means derived from transects will usually have a lower variance than means derived from quadrats or circles.

Surveys can provide an absolute measure of abundance such as the density of individuals, but dive surveys can be very costly to conduct in the wave-exposed shallow subtidal habitats of abalone (Sloan and Breen 1988; McShane 1994). A further disadvantage is that, in providing areal density estimates, the surveys provide little information on the frequency of aggregations in the population. A "dense" population might comprise one large aggregation in a large area devoid of other individuals. However, quadrat surveys do provide indices of dispersion such as variance-to-mean ratios of density.

The clumping of individuals in aggregations (patches), characteristic of abalone, contributes to the typically high coefficients of variation (CV) associated with density values for abalone. The frequency of occurrence of individuals in a sample unit (e.g., 1 m^2) often follows a negative binomial distribution, i.e., most of the sample units have no abalone, a few have one or two individuals and occasionally many are recorded if the sample unit falls on an aggregation.

Within each population there is spatial heterogeneity reflecting the aggregative behavior of many species of abalone. The aggregative behavior depends on species (e.g., Shepherd 1986*a*; McShane 1995*b*), and the habitat occupied by individual species (Breen 1980; McShane et al. 1994). Shepherd (1986*a*) provided evidence of aggregative behavior of *H. laevigata* near spawning, but other species such as *H. iris* (Sainsbury 1982*a*) and *H. cracherodii* (Douros 1987) form aggregations unrelated to reproductive behavior. Relatively little is known of the dynamics of aggregation in relation to the effects of harvesting.

Surveys done on abalone show high variances about estimated mean values for density (Table 1). Coefficients of variation greater than one indicate a nonuniform or patchy population structure (Krebs 1989). On this basis there is apparent interspecific variation in patchiness. For instance, *H. iris, H. laevigata*, and *H. rufescens* may be considered as being more patchily distributed than the other species of abalone. For *H. laevigata*, the high estimated variance may be because of a small sample unit (1 m^2) which is less than the area occupied by patches of this species. Table 1 shows a high CV for a sampling unit of 1 m^2 relative to larger sample units. Breen and Adkins (1979) developed a sample unit which was based on the scale of aggregation of the abalone (*H. kamtschatkana*).

Table 1. Coefficients of variation (CV) associated with the area of the sample unit employed in population surveys of abalone.

Species	Sample unit (m^2)	CV (range of values)	Source
H. corrugata	10	0.6–1.0	Tutschulte 1976
H. fulgens	10	0.8–1.7	Tutschulte 1976
H. rufescens	15	1.4–3.0	Tegner et al. 1989
H. kamtschatkana	16	0.3–1.0	Breen 1986
H. laevigata	1	1.2–4.3	Shepherd 1986*a*
H. rubra	100	0.2–0.3	McShane et al. 1988*a*
	10	0.3-4.5	Andrew and Underwood 1992
	1	0.5–0.7	McShane et al. 1988*a*
H. iris	25	0.2–2.9	McShane 1995*b*

Andrew and Underwood (1992) in surveys of *H. rubra* noted that the largest component of variance in density estimates was at the smallest among-transect scale.

Relative measures based on the collection rate of abalone by research divers (timed swims) have been evaluated (Shepherd 1985; Shepherd and Partington 1995) and applied to abalone by McShane (1994). Timed swims differ from areal counts in that search time, but not the area surveyed, is fixed. The number of individuals collected in a given search time is used as an index of abundance for comparison among populations and for comparison of abundance over time for a population (McShane 1994, 1995*b*).

Comparison of transect surveys with surveys based on timed swims revealed a four-fold difference in cost for similar precision (McShane 1995*b*). Timed swims offer high replication over large spatial scales at reasonable cost, but the method suffers from sources of systematic error such as variation in collection efficiency among research divers (Shepherd 1985; McShane 1994) or differences in catch rates due to weather or sea conditions. These additional sources of variation can be constrained (McShane 1995*b*), but they must be evaluated for survey data from timed swims to be reliable.

A further problem with abundance estimates based on timed swims, is that handling can constrain searching. When abalone are abundant, relatively little habitat is searched compared with habitat in which abalone are sparse (Beinssen 1979*a*). This can cause apparent stability in estimates of abundance while the true abundance is declining (McShane 1994). For example, an area with frequent large patches of abalone may show similar collection rates in successive surveys even though most of the patches may have been removed by fishing.

The two main limitations of timed swims (systematic error and no information on patch structure) prompted a modification described by McShane (1995*b*) to survey patches rather than individuals in a population. In contrast to the frequency of individuals, which follows a negative binomial distribution, the frequency of patches approximates a normal distribution (P.E. McShane, unpublished data). Sample means from populations with normal distributions generally have lower variances than those from skewed distributions such as the negative binomial (Krebs 1989).

Relative measures such as collection rates can be expressed in terms of absolute measures such as density by assuming a proportionality constant. The proportionality constant or "catchability" coefficient may be derived empirically such that relative index data can be used to derive absolute measures of abundance or density (Eberhardt 1978).

The relationship of collection rates of divers and density of abalone has been shown to be linear (Shepherd 1985; Prince 1992; McShane 1994, 1995*b*; Shepherd and Partington 1995). Density can be estimated from information on searching time and distance covered in recording abundance and patch frequency as described by McShane (1995*b*). Such a procedure corrects for the negative effects of handling time so that the abundance of individuals or patches is expressed in terms of searching time or distance covered in the survey (McShane 1995*b*).

Change-in-ratio methods

The change-in-ratio (CIR) method is based on the following assumptions (Seber 1973): (*i*) the population is composed of two types of organisms, e.g., recruits and prerecruits; (*ii*) a differential change in the numbers of the two types of organisms occurs during the observation period; (*iii*) the population under consideration is closed; and (*iv*) all individuals have the same chance of being sampled in both the first and second samples.

Seber (1973, chapter 9) and Krebs (1989, chapter 4) provide a theoretical and methodological discussion on the application of CIR methods to wildlife surveys including calculation of confidence limits for the estimate of population size. Seber points out that the expected change in the proportion has a large effect on the sample size necessary to estimate population size with acceptable error. For small changes in the proportion of types (<0.10), large sample sizes are necessary.

Potential biases in the CIR method discussed by Dawe et al. (1993) include differential catchability of the two types. This source of bias does not necessarily affect abalone fisheries in which only individuals above a certain size limit are harvested. Abalone populations generally conform with most of the assumptions governing application of the method. Breen (1980) showed large changes in the proportion of harvestable abalone (>60%) associated with fisheries for *H. kamtschatkana*. McShane (1994) provided a crude estimate of the exploited biomass accompanying a seasonal change in the relative abundance of *H. rubra* in the Victorian fishery. A more robust application resulted in the estimation of the abundance of some Tasmanian stocks of *H. rubra* (W.J. Nash, Marine Research Laboratories, Taroona, Dept. of Primary Industry, Tasmania, Australia, unpublished data). Spatial variation in the size composition of populations of abalone can confound assessment of stocks by CIR if fishers are serially depleting populations. However, when extraction rates are high (>40% of the population) CIR methods can provide precise estimates of abundance (SE < 10% of the mean) (W.J. Nash, Marine Research Laboratories, Taroona, Dept. of Primary Industry, Tasmania, Australia, unpublished data).

Precision and detecting fishing-induced change in populations

Power analysis is routinely considered in mainstream statistical and biometric texts (e.g., Snedecor and Cochran 1967;

Winer 1971; Sokal and Rohlf 1981; Zar 1984) but is rarely considered in assessing the response of exploited populations to fishing (Peterman 1990). The prevalence of Type II error (falsely retaining the null hypothesis) may be high when comparing imprecise estimates of abundance as is often the case in the assessment of fish stocks. Surveys of abalone populations yielding demonstrably high variances associated with the density estimates may be expected to have low power to detect a change in abundance. Sufficient power to detect significant differences in the abundance of abalone populations only occurs for large effect sizes or with unrealisticaly high levels of replication (so that means can be estimated precisely).

Sloan and Breen (1988) briefly discussed power of detecting a change in sequential surveys of *H. kamtschatkana*. They suggested that a 25% decline from 1 m^2 quadrats would require surveys of 100 quadrats at one site. McShane (1994) specified power in assessing stocks of abalone with timed swims. For power of 0.8 or greater, changes of 15–63% of mean values would be detected for populations of *H. rubra* and *H. iris*. Not surprisingly, only large changes in abundance have been recorded with abalone surveys. Breen (1986) and Prince and Guzman del Próo (1993) report large decreases (>70%) in the density of individuals within populations of *H. kamtschatkana* and *H. fulgens – H. corrugata*, respectively.

Fixed versus random sites

Breen and Adkins (1981) suggested that a series of permanently identifiable benchmark sites be established to overcome imprecision associated with small-scale variation in abundance. Sampling of a number of fixed sites corresponding to known aggregations is being followed for the fishery for *H. rubra* (N.L. Andrew, Fisheries Research Institute, Cronulla, N.S.W., Australia, personal communication). Hilborn and Walters (1992), in assessing the relative merits of fixed versus random surveys of fish stocks, advocate fixed sites for simplicity of operation. However, there are some disadvantages and limitations with the use of fixed sites in assessing the response of abalone stocks to fishing. First, the use of fixed sites can violate the assumption of nonindependence of treatments (Underwood 1981; Sokal and Rohlf 1981). A nonfishing-induced change at a fixed site could confound any assessment based on sequential censuses at that fixed site. For instance, an aggregation of abalone might periodically occur at a fixed site through small-scale movement patterns (Shepherd 1986*b*).

A series of fixed sites does not necessarily represent the harvestable stock and a change in the status of fixed sites cannot be assumed to apply to that stock. In comparing a known harvest with a measurable change in abundance or patch frequency, fixed sites cannot be used to derive estimates of harvestable biomass whereas they can be for random samples.

An advantage of fixed sites is that they can be chosen to represent areas likely to receive fishing-induced impacts. Thus, large aggregations can be monitored over time to demonstrate effects of fishing. If there is a sufficiently large number of fixed sites (or known aggregations) then the problem of nonindependence may be overcome by subsampling. The problem of relating the measured impact to the actual impact on the harvestable stock remains.

Estimating the productivity of abalone fisheries

Productivity is defined here as the surplus of growth and recruitment over natural mortality. Surplus production theory is based on the premise that a reduction in density associated with harvesting causes increases in growth and decreases in natural mortality (Ricker 1975). Examination of processes of growth, survival, and recruitment may help explain the interspecific differences in the productivity of abalone fisheries.

Case studies

There are few estimates of productivity available for abalone fisheries. Breen (1986) concluded that recruitment of *H. kamtschatkana* was insufficient to maintain the stock in the British Columbian fishery. Prince and Guzman del Próo (1993) described a stock reduction analysis of *H. fulgens* and *H. corrugata* suggesting these species had low productivity. However, the latter assessment contained little information about the potential productivity of abalone stocks under more optimal management regimes. Smith (1972) suggested from length–frequency samples of large animals that productivity was about 0.05 (annual turnover rate). Estimates provided by McShane (1994) suggest an annual productivity of about 10% for the Victorian *H. rubra* fishery. Hines and Pearse (1982) postulated that populations of *H. rufescens* and *H. wallensis* were highly productive (30–100% per annum) in order to maintain populations in crevice refugia under very high predation pressure from sea otters. Similarly, Kojima et al. (1977) describe a Japanese artisinal fishery sustainably harvesting populations of young individuals (1–3 years of age) at high rates of fishing.

Prince (1989) crudely expressed productivity of abalone fisheries in terms of catch per linear distance of coastline and showed considerable interspecific variation in production. For example, the eastern zone (Victoria) fishery for *H. rubra* produces 500 $t \cdot yr^{-1}$ over 100 km of coastline (McShane et al. 1986) compared with the southern fishery for *H. iris* which produces a similar harvest over 1500 km of coastline. These comparisons are admittedly coarse but differences of an order of magnitude suggest real differences in productivity among species of abalone.

Growth, survival, and recruitment

Growth and survival (or natural mortality, *M*) are correlated processes (Pauly 1980). Abalone are often growth dominated with rates of growth in excess of rates of mortality (Day and Fleming 1992). Assuming constant recruitment, abalone populations will be represented by an accumulation of large, old individuals (Sainsbury 1982*a*). Estimates of natural mortality show intra- and inter-specific variation (Shepherd and Breen 1992) suggesting similar variation in productivity. Estimates for *H. iris* indicate low instantaneous rates of natural mortality ($M < 0.1$, Sainsbury 1982*a*; McShane and Naylor 1997) compared with rates for *H. rubra* ($M > 0.2$, Shepherd et al. 1982; Prince et al. 1988*b*; McShane 1992). In contrast, the range of rates of growth are similar for the two species (Sainsbury 1982*a*; Prince et al. 1988*b*; McShane et al. 1988*b*; McShane and Naylor 1995).

Recruitment is perhaps the most important parameter affecting variation in the productivity of abalone populations.

Recruitment is usually measured as the abundance of individuals entering the fishery. In the case of abalone, this definition of recruitment is somewhat arbitrary because legal minimum sizes are imposed in most fisheries. More recently, recruitment of abalone has been assessed as the frequency of newly settled individuals, which can vary over several orders of magnitude for *H. rubra* (McShane and Smith 1991). Such estimates can provide clues as to sources of variation in recruitment such as the abundance of spawners and habitat-related factors. Recruitment of *H. rubra* is consistently high compared with *H. laevigata* or *H. iris* (McShane and Smith 1991; Shepherd et al. 1992; McShane and Naylor 1996). One possible explanation of the comparatively high levels of recruitment in *H. rubra* is their association with stands of fucoid macroalgae that promote retention of larvae near the natal site (McShane et al. 1988*a*). *Haliotis iris* and *H. laevigata* occupying open habitat have generally lower rates of recruitment than *H. rubra* (Shepherd 1990; Shepherd et al. 1992; McShane and Naylor 1996).

Information from length–frequency distributions

Length–frequency distributions from samples of populations will reflect growth, survival, and recruitment processes. Breen (1986) surveyed many populations of *H. kamtschatkana* and found juveniles absent or rare in some populations. McShane (1993) and McShane et al. (1994) reported consistently low numbers of juvenile *H. iris*. Sainsbury (1982*b*) and Fournier and Breen (1983) showed how stochastic or episodic recruitment introduced heterogeneity into length–frequency distributions of abalone populations. Skewed length–frequency distributions shown for *H. iris* were interpreted by Sainsbury (1982*b*) as being caused by low or episodic recruitment. The stability of length–frequency distributions of some exploited populations over time (McShane et al. 1994) suggests that recruitment of *H. iris* is uniformly low rather than episodic.

The normal distribution of lengths of individuals in *H. iris* populations (McShane 1996) is consistent with Sainsbury's (1982*b*) hypothesis of low recruitment and growth rates in excess of mortality rates. However, even heavily exploited populations of *H. iris* have normally distributed length frequencies with little evidence of size-selective harvesting (McShane 1996). For *H. iris*, the length–frequency distribution in a population may be approximated by the variance associated with asymptotic length (P.E. McShane, unpublished data). This suggests that recruitment is low and that exploitation rates are low in relation to the accumulated individuals in the population. McShane (1993) suggests that recruitment to exposed populations of *H. iris* is by migration rather than larval dispersal. In such populations, there are few juveniles suggesting that migrants are mature individuals and that recovery from the effects of harvesting is a slow process.

Yield-based modeling and biological reference points

Yield-per-recruit models feature prominently in assessment of abalone fisheries (review by Breen 1992). Abalone fisheries tend to produce maximum YPR under high fishing intensities and sizes less than the minimum legal sizes imposed by managers (Breen 1986). However, the variation in growth rate means that YPR can vary substantially among populations (McShane 1992). For abalone, growth rates slow at the onset of reproductive maturity (Tutschulte and Connell 1988). YPR modeling of abalone populations often show a yield curve unresponsive to high rates of fishing. This is because abalone typically grow rapidly to nearly maximum size and the increase in yield with age is relatively small thereafter (Breen 1986; Sluczanowski 1986). A weakness of YPR models is that they provide no information on future recruitment levels; yields are determined primarily by growth and survival (cf., Alverson and Carney 1975).

More recently, YPR modeling has been used to provide biological reference points for the estimation of yields for use in the management of fisheries (Mace 1988*a*, 1988*b*). Such reference points which usually relate to a target fishing mortality, provide yardsticks for use by fisheries research institutions to account for differences in productivity among fish stocks. They are most commonly expressed as constant yield strategies (e.g., maximum sustainable yield, MSY) or constant fishing mortality strategies (e.g, F_{MSY} or F_{max}). F_{max} is the fishing mortality which produces the maximum YPR.

The concern that MSY strategies could result in recruitment overfishing, particularly for stocks subject to stochastic variation in recruitment, has prompted acceptance of $F_{0.1}$, the fishing mortality corresponding to the point on a YPR curve where the slope is 10% of that at the origin (Mace 1988*a*). This reference point is popular because it results in yields only slightly smaller than F_{max}, maintains higher levels of spawning stock biomass at lower levels of effort, and may be closer to the economic optimum (Mace 1988*b*). Because abalone tend to produce asymptotic YPR relationships with fishing mortality, F_{max} is often difficult to estimate.

The adoption of $F_{0.1}$ in favour of other reference points such as F_{MSY} and F_{max} reflects the concern that high levels of fishing mortality may cause a deleterious decline in spawning stock abundance (Mace 1988*b*). There is also concern that management under MSY that imposes constant annual yields introduces unacceptably high levels of risk for fish stocks subject to stochastic environmental variation (Larkin 1977).

In the absence of a stock-recruitment relationship, abalone populations could be managed under F_{max} strategies (Breen 1986, 1992). However, there is a general reluctance for managers to discard the notion of recruitment overfishing for abalone (McShane 1995*a*). The more recent focus on egg-per-recruit models reflects this concern and the need to conserve "appropriate" levels of egg production (Breen 1992; McShane 1995*a*).

The concerns over potential recruitment overfishing might lead to acceptance of $F_{0.1}$ as a safer reference point than F_{max}. However, until reliable estimates of biomass are available, reference fishing mortalities have little meaning for abalone fisheries. Furthermore, for some abalone fisheries such as those for *H. iris* in which recruitment is evidently low (Sainsbury 1982*a*; McShane et al. 1994), even conservative reference points such as $F_{0.1}$ might overestimate production. Application of biological reference points to abalone populations requires a strict definition of the spatial distribution of stocks, a reliable estimate of exploitable biomass, and an understanding of recruitment dynamics. Most of these requirements are unavailable for existing abalone fisheries.

Fleet dynamics of abalone fisheries

Allen and McGlade (1986) included an assessment of the behavior of harvesters in evaluating the response of Nova Scotian groundfish fisheries in terms of discovery and exploitation. They identified two types of behavior: directed exploitation based on knowledge of distribution and abundance; and random fishing behavior involving risk taking. In modeling the dynamics of the groundfish fishery based on these disparate fishing behaviors, the risk takers survived at the expense of the focused harvesters. Similar behavioral dynamics were considered for abalone fisheries by Hilborn and Walters (1987). In fishing, the expected return from catch and anticipated catch rate is weighed against the cost of transferring to another area and the distance of that area from home port. For the New Zealand fishery for *H. iris*, there is an uneven spatial distribution of effort reflecting the preference of fishers to stay close to a home port (McShane et al. 1994). In this latter fishery, remote areas with abundant populations of abalone are forsaken for local populations. The local populations have a low abundance of abalone and a low frequency of large aggregations, but the remote areas have few safe anchorages and the conditions may not favor diving. This uneven concentration of effort in the *H. iris* fishery has led to spatial depletion of some populations (McShane et al. 1994).

Another aspect of fishing behavior pertinent to abalone fisheries is the targetting of large aggregations. Large aggregations are relatively infrequent in *H. iris* populations; most of the aggregations are small (<5 individuals) (McShane et al. 1994) but large aggregations determine the overall abundance of the population (McShane 1995*b*). In harvesting only large aggregations, fishing can have a relatively low impact on the total number of aggregations but a large impact on the number of individuals (McShane 1996). The additional searching time necessary for the harvesting of small aggregations means that it is more cost effective to harvest large aggregations (McShane 1995*b*).

If habitat which supports large aggregations of abalone is known to fishers then there will be little exploratory fishing (the risk takers of Allen and McGlade 1986). This behavior could differ among fisheries because of differences in stock structure or abundance. For example, in the Victorian fishery for *H. rubra*, fishers expect catch rates to be in excess of $100\ \mathrm{kg \cdot h^{-1}}$ and fish known aggregations accordingly (McShane and Smith 1989). If one population has insufficient aggregations to support this latter catch rate, the fisher may harvest another population. In contrast, the NSW fishery for *H. rubra* has harvestable populations of low abundance and a low frequency of large aggregations. Consequently, the expected catch rate of divers is much lower and the diving behavior less focused on large aggregations (N.J. Andrew, Fisheries Research Institute, NSW, Australia, personal communication).

Spatial management of abalone fisheries

Spatial management of abalone fisheries was considered in a simulation of stock and fleet dynamics by Hilborn and Walters (1987). They showed that catch rates could be maintained by serial depletion of substocks. However, the assumption that abalone fishers can extract most of the harvestable population in fishing (e.g., Beinssen 1979*b*) was challenged by McShane and Smith (1989) who demonstrated a relatively low impact of intensive fishing on a population of *H. rubra*.

The spatial management of abalone fisheries in which spatial units are defined and the biological characteristics of the populations within each unit incorporated into management strategies have not been adequately resolved. Most abalone fisheries are managed over broad spatial areas despite the fact that the biological characteristics influencing the productivity of populations vary over much smaller spatial scales.

Shepherd (1992) summarizes the application of annual catch quotas to abalone fisheries. Characteristics of abalone identified by Shepherd include long life and slow growth. Shepherd points out that the empirical basis of quota setting and quota adjustment is poor. Ideally assessment should involve identification of the principal stocks, measurement of their spatial extent, and estimation of the size of the population.

Management of abalone fisheries, taking into account spatial differences in productivity, compares with forestry and subsequent considerations for optimal management. Nonproductive abalone populations characterized by low or zero recruitment may be considered as "old growth" forests and harvesting involves removal without replacement (mining) (Conrad and Ludwig 1994). For abalone it may be economic to mine unproductive populations because in the short run, high yields are produced from an accumulation of large individuals. Productive abalone populations often have fewer large aggregations and the size composition is skewed towards smaller individuals. Research to identify and quantify productive populations is necessary to implement more effective spatial management of abalone fisheries.

Acknowledgments

I thank Paul Breen, Jeremy Prince, and an anonymous reviewer for helpful comments on a previous draft.

References

Allen, P.M., and McGlade, J.M. 1986. Dynamics of discovery and exploitation: the case of the Scotian Shelf groundfish fisheries. Can. J. Fish. Aquat. Sci. **43**: 1187–1200.

Alverson, D.L., and Carney, M.J. 1975. A graphic review of the growth and decay of population cohorts. J. Cons. Int. Explor. Mer. **36**: 133–143.

Andrew, N.L., and Underwood, A.J. 1992. Associations and abundance of sea urchins and abalone on shallow subtidal reefs in southern New South Wales. Aust. J. Mar. Freshwater Res. **43**: 1547–1559.

Ault, J.S., and DeMartini, J.D. 1987. Movement and dispersion of red abalone, *Haliotis rufescens*, in northern California. Calif. Fish Game, **73**: 196–213.

Beinssen, K. 1979*a*. Fishing power of divers in the abalone fishery of Victoria, Australia. Rapp. P.-V. Reun, Cons. Int. Explor. Mer. **175**: 20–22.

Beinssen, K. 1979*b*. The estimation of fishing mortality of abalone based on experiments to measure the rate of reef coverage by the fishing fleet. Rapp. P.-V. Reun., Cons. Int. Explor. Mer. **175**: 27–29.

Breen, P.A. 1980. Measuring fishing intensity and annual production in the abalone fishery of British Columbia. Tech. Rep. Can. Fish. Res. Board. No. 947. 49 p.

Breen, P.A. 1986. Management of the British Columbia fishery for northern abalone (*Haliotis kamtschatkana*). *In* North Pacific Workshop on Stock Assessment and Management of Invertebrates. *Edited by* G.S. Jamieson and N. Bourne. Can. Spec. Publ. Fish. Aquat. Sci. 92. pp. 300–312.

Breen, P.A. 1992. A review of models used for stock assessment in abalone fisheries. *In* Abalone of the world: biology, fisheries and culture. *Edited by* S.A. Shepherd, M.J. Tegner, and S.A. Guzmán del Próo. Fishing News Books, Oxford, U.K. pp. 253–275.

Breen, P.A., and Adkins, B.E. 1979. A survey of abalone populations on the east coast of the Queen Charlotte Islands, August 1978. Fish. Mar. Serv. Tech. Rep. 1490. 125 p.

Breen, P.A., and Adkins, B.E. 1981. Abalone surveys and tagging conducted during 1979. Can. Tech. Rep. Fish. Aquat. Sci. 1623. 88 p.

Brown, L.D., and Murray, N.D. 1992. Population genetics, gene flow, and stock structure in *Haliotis rubra* and *Haliotis laevigata*. *In* Abalone of the world: biology, fisheries and culture. *Edited by* S.A. Shepherd, M.J. Tegner, and S.A. Guzmán del Próo. Fishing News Books, Oxford, U.K. pp. 24–33.

Conrad, J., and Ludwig, D. 1994. Forest land policy: the optimal stock of old-growth forest. Nat. Resour. Model. **8**: 27–45.

Dawe, E.G., Hoenig, J.M., and Xu, X. 1993. Change-in-ratio and index-removal methods for population assessment and their application to snow crab (*Chionoecetes opilio*). Can. J. Fish. Aquat. Sci. **50**: 1467–1476.

Day, R.W., and Fleming, A.E. 1992. The determinants and measurement of abalone growth. *In* Abalone of the world: biology, fisheries and culture. *Edited by* S.A. Shepherd, M.J. Tegner, and S.A. Guzmán del Próo. Fishing News Books, Oxford, U.K. pp. 141–168.

Douros, W.J. 1987. Stacking behavior of an intertidal abalone: an adaptive response or a consequence of space limitation? J. Exp. Mar. Biol. Ecol. **108**: 1–14.

Eberhardt, L.L. 1978. Appraising variability in population studies. J. Wildl. Manage. **42**: 207–238.

Erasmus, J., Cook, P.A., and Sweijd, N. 1994. Internal shell structure and growth lines in the shell of the abalone *Haliotis midae*. J. Shellfish Res. **13**: 493–501.

Fournier, D.A., and Breen, P.A. 1983. Estimation of abalone mortality rates with growth analysis. Trans. Am. Fish. Soc. **112**: 403–411.

Hilborn, R. and Walters, C.J. 1987. A general model for simulation of stock and fleet dynamics in spatially heterogeneous fisheries. Can. J. Fish. Aquat. Sci. **44**: 1366–1369.

Hilborn, R., and Walters, C.J. 1992. Quantitative fisheries stock assessment. Chapman and Hall Ltd., New York, N.Y.

Hines, A.H., and Pearse, J.S. 1982. Abalones, shells, and sea otters: dynamics of prey populations in central California. Ecology, **63**: 1547–1560.

Kimura, D.K., Balsinger, J.W., and Ito, D.H. 1984. Generalized stock reduction analysis. Can. J. Fish. Aquat. Sci. **41**: 1325–1333.

Kojima, H., Nakahisa, Y., and Tanimoto, H. 1977. A study on the stock of Japanese black abalone, *Haliotis discus discus* in Tokushima Prefecture – 1. Growth of shells. Bull. Tokai Reg. Fish. Res. Lab. **90**: 25–37.

Krebs, C.J. 1989. Ecological methodology. Harper Row. New York, N.Y. 654 p.

Larkin, P.A. 1977. An epitaph for the concept of maximum sustainable yield. Trans. Am. Fish. Soc. **106**: 1–11.

Mace, P.M. 1988*a*. A survey of stock assessment methods and results. N.Z. Ministry of Agriculture and Fisheries. N.Z. Fisheries Assessment Research Document 88/6. 35 p.

Mace, P.M. 1988*b*. The relevance of MSY and other biological reference points to stock assessment in New Zealand. N.Z. Ministry of Agriculture and Fisheries. N.Z. Fisheries Assessment Research Document 88/30. 31 p.

McShane, P.E. 1992. Exploitation models and catch statistics of the Victorian fishery for abalone *Haliotis rubra*. Fish. Bull. **90**: 139–146.

McShane, P.E. 1993. Evidence for localised recruitment failure in the abalone *Haliotis iris* (Mollusca:Gastropoda). *In* Proceedings of the Second International Temperate Reefs Symposium. *Edited by* C.N. Battershill, D.R. Schiel, G.P. Jones, R.G. Creese, and A.B. MacDiarmid. 7–10 January 1992, Auckland, New Zealand, NIWA Marine, Wellington, N.Z. 252 p.

McShane, P.E., 1994. Estimating the abundance of abalone (*Haliotis* spp.) stocks — examples from Victoria and southern New Zealand. Fish. Res. **19**: 379–394.

McShane, P.E. 1995*a*. Recruitment variation in abalone (*Haliotis* spp.): its importance to fisheries management. Mar. Freshwater Res. **46**: 555–570.

McShane, P.E. 1995*b*. Estimating the abundance of abalone: the importance of patch size. Mar. Freshwater Res. **46**: 657–662.

McShane, P.E. 1996. Patch dynamics and exploitation of abalone (*Haliotis iris*) populations. Fish. Res. **25**: 191–199.

McShane, P.E., and Naylor, J.R. 1995. Density independent growth in *Haliotis iris* Martyn (Mollusca: Gastropoda). J. Exp. Mar. Biol. Ecol. **190**: 51–60.

McShane, P.E., and Naylor, J.R. 1996. Variation in spawning and recruitment of *Haliotis iris* (Mollusca: Gastropoda). N.Z. J. Mar. Freshwater Res. **30**: 325–332.

McShane, P.E., and Naylor, J.R. 1997. Direct estimation of natural mortality of the New Zealand abalone *Haliotis iris*. N.Z. J. Mar. Freshwater Res. **31**: 143–145.

McShane, P.E., and Smith, M.G. 1989. Direct measurement of fishing mortality in abalone (*Haliotis rubra* Leach) off southeastern Australia. Fish. Res. **8**: 93–102.

McShane, P.E., and Smith, M.G. 1991. Recruitment variation in sympatric populations of the abalone *Haliotis rubra* (Mollusca: Gastropoda). Mar. Ecol. Prog. Ser. **73**: 203–210.

McShane, P.E., Beinssen, K.H.H., and Foley, S. 1986. Abalone reefs in Victoria — a resource atlas. Vic. Min. Cons. For. Lands M.S.L. Tech. Rep. No. 47. 50 p.

McShane, P.E., Black, K.P., and Smith, M.G. 1988*a*. Recruitment processes in *Haliotis rubra* (Mollusca: Gastropoda) and regional hydrodynamics in southeast Australia imply localised dispersal of larvae. J. Exp. Mar. Biol. Ecol. **124**: 175–203.

McShane, P.E., Smith, M.G., and Beinssen, K.H.H. 1988*b*. Growth and morphometry in abalone (*Haliotis rubra*) from Victoria. Aust. J. Mar. Freshwater Res. **39**: 161–166.

McShane, P.E., Mercer, S.F., and Naylor, J.R. 1994. Spatial variation and commercial fishing of New Zealand populations of abalone (*Haliotis iris* and *H. australis*). N.Z. J. Mar. Freshwater Res. **28**: 345–355.

Megrey, B.A., and Wespestad, V.G. 1988. A review of biological assumptions underlying fishery assessment models. *In* Fishery science and management — objectives and limitations. *Edited by* W.S. Wooster. Lect. Notes Coastal Estuarine Stud. **28**: 31–74.

Nash, W.J. 1992. An evaluation of egg-per-recruit analysis as a means of assessing size limits for blacklip abalone (*Haliotis rubra*) in Tasmania. *In* Abalone of the world: biology, fisheries and culture. *Edited by* S.A. Shepherd, M.J. Tegner, and S.A. Guzmán del Próo. Fishing News Books, Oxford, U.K. pp. 318–338.

Newman, G.G. 1968. Growth of the South African abalone *Haliotis midae*. S. Afr. Div. Sea Fish. Invest. Rep. 67. pp. 1–24.

Pauly, D. 1980. On the interrelationships between natural mortality growth parameters, and mean environmental temperature in 175 fish stocks. J. Cons. Int. Explor. Mer. **39**: 175–192.

Peterman, R.M. 1990. Statistical power analysis can improve fisheries research and management. Can. J. Fish. Aquat. Sci. **47**: 2–15.

Prince J.D. 1989. World abalone fisheries offer a grim warning to Tasmanians. Fishing Today, **23**: 59–66.

Prince, J.D. 1992. Using a spatial model to explore the dynamics of an exploited stock of the abalone *Haliotis rubra*. *In* Abalone of the world: biology, fisheries and culture. *Edited by* S.A. Shepherd, M.J. Tegner, and S.A. Guzmán del Próo. Fishing News Books, Oxford, U.K. pp. 305–317.

Prince, J.D., and Guzmán del Próo, S.A. 1993. A stock reduction analysis of the Mexican abalone (Haliotid) fishery. Fish. Res. **16**: 25–49.

Prince, J.D., Sellers, T.L., Ford, W.B., and Talbot, S.R 1987. Experimental evidence for limited dispersal of haliotid larvae (genus *Haliotis*; Mollusca: Gastropoda). J. Exp. Mar. Biol. Ecol. **106**: 243–263.

Prince, J.D., Sellers, T.L., Ford, W.B., and Talbot, S.R. 1988*a*. Confirmation of a relationship between the localized abundance of breeding stock and recruitment for *Haliotis rubra* Leach (Mollusca: Gastropoda). J. Exp. Mar. Biol. Ecol. **122**: 91–104.

Prince, J.D., Sellers, T.L., Ford, W.B., and Talbot, S.R. 1988*b*. Recruitment, growth, mortality and population structure in a southern Australian population of *Haliotis rubra* (Mollusca: Gastropoda). Mar. Biol. (Berlin), **100**: 75–82.

Ricker, W.E. 1975. Computation and interpretation of biological statistics of fish populations. Bull. Fish. Res. Board Can. 191. 382 p.

Sainsbury, K.J. 1982*a*. Population dynamics and fishery management of the paua, *Haliotis iris*. 1. Population structure, growth, reproduction and mortality. N.Z. J. Mar. Freshwater Res. **16**: 147–161.

Sainsbury, K.J. 1982*b*. Population dynamics and fishery management of paua, *Haliotis iris*. 2. Dynamics and management as examined using a size class model. N.Z. J. Mar. Freshwater Res. **16**: 163–173.

Seber, G.A.F. 1973. The estimation of animal abundance and related parameters. Griffin, London, U.K. 506 p.

Shepherd, S.A. 1973. Studies on southern Australian abalone (genus *Haliotis*) I. Ecology of five sympatric species. Aust. J. Mar. Freshwater Res. **24**: 217–257.

Shepherd, S.A. 1985. Power and efficiency of a research diver, with a description of a rapid underwater measuring guage: their use in measuring recruitment and density of an abalone population. *In* Diving for science. *Edited by* C.T. Mitchell. American Academy of Underwater Science, La Jolla, Calif. pp. 263–272.

Shepherd, S.A. 1986*a*. Studies on southern Australian abalone (genus *Haliotis*) VII. Aggregative behavior of *H. laevigata* in relation to spawning. Mar. Biol. (Berlin), **90**: 231–236.

Shepherd, S.A. 1986*b*. Movement of the southern Australian abalone *Haliotis laevigata* in relation to crevice abundance. Aust. J. Ecol. **11**: 295–302.

Shepherd, S.A. 1990. Studies on southern Australian abalone (genus *Haliotis*) XII. Long-term recruitment and mortality dynamics of an unfished population. Aust. J. Mar. Freshwater Res. **41**: 475–492.

Shepherd, S.A. 1992. Managing abalone fisheries by quota in Australia, South Africa and New Zealand: a review. *In* The management and enhancement of sea-urchins and other kelp bed resources. Report No. 1-CSGCP-028. Sea Grant College, University of California, La Jolla, Calif.

Shepherd, S.A., and Breen, P.A. 1992. Mortality in abalone: its estimation, variability and causes. *In* Abalone of the world: biology, fisheries and culture. *Edited by* S.A. Shepherd, M.J. Tegner, and S.A. Guzmán del Próo. Fishing News Books, Oxford, U.K. pp. 276–304.

Shepherd, S.A., and Brown, L. 1993. What is an abalone stock? Implications for the role of refugia in conservation. Can. J. Fish. Aquat. Sci. **50**: 2001–2009.

Shepherd, S.A., and Partington, D. 1995. Studies on southern Australian abalone (genus *Haliotis*) XVI. Recruitment, habitat and stock relations. Aust. J. Mar. Freshwater Res. **46**: 669–680.

Shepherd, S.A., Kirkwood, G.R., and Sandland, R.L. 1982. Studies on Southern Australian abalone (genus *Haliotis*). III. Mortality of two exploited species. Aust. J. Mar. Freshwater Res. **33**: 265–272.

Shepherd, S.A., Guzmán del Próo, S.A., Turrubiates, J., Belmar, J., Baker, J., and Sluczanowski, P.R. 1991. Growth, size at sexual maturity, and egg-per-recruit analysis of the abalone *Haliotis fulgens* in Baja California. The Veliger, **34**: 324–330.

Shepherd, S.A., Lowe, D., and Partington, D. 1992. Studies on southern Australian abalone (genus *Haliotis*) XIII: larval dispersal and recruitment. J. Exp. Mar. Biol. Ecol. **164**: 247–260.

Shepherd, S.A., Avalos-Borja, M., and Ortiz, M. 1995. Toward a chronology of *Haliotis fulgens* with a review of abalone shell microstructure. Mar. Freshwater Res. **46**: 607–615.

Sloan, N.A., and Breen, P.A. 1988. Northern abalone, *Haliotis kamtschatkana*, in British Columbia: Fisheries and synopsis of life history information. Can. Spec. Publ. Fish. Aquat. Sci. 103. 46 p.

Sluczanowski, P.R. 1986. A disaggregate model for sedentary stocks: The case of the South Australian abalone. *In* North Pacific Workshop on Stock Assessment and Management of Invertebrates. *Edited by* G.S. Jamieson and N.F. Bourne. Can. Spec. Publ. Fish. Aquat. Sci. 92. pp. 393–401.

Smith, S.V. 1972. Production of calcium carbonate on the mainland shelf of southern California. Limnol. Oceanogr. **17**: 28–41.

Snedecor, G.W., and Cochran, W.G. 1967. Statistical methods. Iowa State University Press. Iowa. 593 p.

Sokal, R.R., and Rohlf, F.J. 1981. Biometry. 2nd ed. W.H. Freeman, San Francisco, Calif.

Solé-Cava, A.M., Thorpe, J.P., and Todd, C.D. 1994. High genetic similarity between geographically distant populations in a sea anemone with low dispersal capabilities. J. Mar. Biol. Ass. U.K. **74**: 895–902.

Tegner, M.J., Breen, P.A., and Lennert, C.J. 1989. Population biology of red abalones (*Haliotis rufescens*) in southern California and management of the red and pink (*H. corrugata*) abalone fisheries. Fish. Bull. **87**: 313–339.

Tutschulte, T.C. 1976. The comparative ecology of three sympatric abalones. Ph.D. thesis, University of California, San Diego, Calif. 355 p.

Tutschulte, T.C., and Connell, J.H. 1988. Growth of three species of abalone (*Haliotis*) in southern California. Veliger, **31**: 204–213.

Underwood, A.J. 1981. Techniques of analysis of variance in experimental marine biology and ecology. Oceanogr. Mar. Biol. Ann. Rev. **19**: 513–605.

Winer, B.J. 1971. Statistical principles in experimental design. McGraw-Hill, New York, N.Y. 907 p.

Zar, J.H. 1984. Biostatistical analysis. Prentice Hall, N.J. 718 p.

New developments in change-in-ratio and index-removal methods, with application to snow crab (*Chionoecetes opilio*)

Chiu-Lan Chen, John M. Hoenig, Earl G. Dawe, Cavell Brownie, and Kenneth H. Pollock

Abstract: Change-in-ratio and index-removal estimates of population size and related parameters can be obtained when a survey is conducted before and after a fishery and the magnitude and composition of the landings are determined. The former looks at how a known, selective removal affects the composition of the population. The latter looks at how a known removal affects catch rate. If spatial pattern remains relatively stable between the two surveys, increased precision can be obtained by occupying the same sampling stations in the two surveys rather than rerandomizing locations for the second survey. We derive expressions for the biases and variances of the estimates for the case where stations are reoccupied. For snow crab (*Chionoecetes opilio*) in Newfoundland, reuse of stations led to reductions in standard errors of 33–47% from the usual rerandomized-station design. An assumption of the index-removal method is that all animals have the same probability of capture in the surveys. This is not true when survey sampling gear is size selective. The bias can be minimized by making separate estimates for each size-class. However, when large animals have greater catchability than smaller animals, we can achieve greater precision by utilizing information on relative catchability. We propose that separate estimates for each size-class first be computed. If the estimated catchability coefficients show an increasing trend with size, these may then be constrained to increase monotonically with size. It could also be assumed that the catchability coefficients must be a specified function of size. We conclude that these methods may have considerable value for assessing invertebrate stocks.

Résumé : On peut obtenir des estimations dla tailationres congemenprélèvur lorfectue un relevé avant et après une pêche, et que l'on connaît l'importance et la composition des prises. Dans le premier cas, on examine de quelle manière un prélèvement sélectif connu influe sur la composition de la population. Dans le second, on examine de quelle manière un prélèvement connu influe sur le taux de capture. Si la distribution spatiale reste relativement stable entre les deux relevés, on peut avoir une meilleure précision en occupant les deux fois les mêmes stations d'échantillonnage plutôt qu'en choisissant de nouveau les sites de manière aléatoire pour le second relevé. Nous tirons des expressions des biais et variances des estimations pour les cas où l'on a occupé les mêmes stations les deux fois. Pour les crabes des neiges (*Chionoecetes opilio*) de Terre-Neuve, l'utilisation des mêmes stations a permis de réduire l'erreur-type de 33 à 47 % par comparaison avec les techniques habituelles de randomisation du choix des stations. Une des hypothèses de la méthode du prélèvement indicateur est que tous les individus ont la même probabilité d'être capturés au cours des relevés. Ce n'est pas le cas lorsque les engins d'échantillonnage sont sélectifs sur le plan de la taille. On peut réduire le biais à un minimum en effectuant des estimations séparées pour chaque classe de taille. Cependant, quand les gros spécimens ont une plus grande capturabilité que les petits, on peut arriver à une meilleure précision en utilisant l'information sur la capturabilité relative. Nous proposons de calculer en premier des estimations séparées pour chaque classe de taille. Si les coefficients de capturabilité estimatifs montrent que la tendance croît avec la taille, on peut les contraindre pour obtenir une augmentation monotone avec la taille. On peut aussi postuler que les coefficients de capturabilité sont une fonction donnée de la taille. Nous en concluons que ces méthodes peuvent être précieuses pour évaluer les stocks d'invertébrés. [Traduit par la Rédaction]

Introduction

In this paper, we consider how population size, exploitation rate, and related parameters can be estimated from survey data collected just before and just after a known (or estimated) harvest. It is assumed that the time between the start of the first and the end of the second survey is sufficiently short that the population can be considered closed except for the removals. There are three types of estimators that can be used with these

C.-L. Chen. Biomathematics Program, North Carolina State University, Raleigh, NC 27695, U.S.A.

J.M. Hoenig[1] and Earl G. Dawe. Department of Fisheries and Oceans, P.O. Box 5667, St. John's, NF A1C 5X1, Canada.

C. Brownie and K.H. Pollock. Department of Statistics, North Carolina State University, Raleigh, NC 27695, U.S.A.

Correct citation: Chen, C.-L., Hoenig, J.M., Dawe, E.G., Brownie, C., and Pollock, K.H. 1998. New developments in change-in-ratio and index-removal methods, with application to snow crab (*Chionoecetes opilio*). *In* Proceedings of the North Pacific Symposium on Invertebrate Stock Assessment and Management. *Edited by* G.S. Jamieson and A. Campbell. Can. Spec. Publ. Fish. Aquat. Sci. 125. pp. 49–61.

[1] Author to whom all correspondence should be sent at the following address: Virginia Institute of Marine Science, P.O. Box 1346, Gloucester Point, VA 23062, U.S.A.

data: change-in-ratio, index-removal, and removal methods. The methods are simple and have been known for at least 40 years but, surprisingly, they are not well known to fishery scientists and much work remains to be done to elucidate their statistical properties.

We begin by reviewing the available methods. We then show that increased statistical efficiency can generally be achieved by using the same sampling locations for both surveys rather than rerandomizing the stations for the second survey. The improvements in precision are studied by simulation and in an example involving snow crab (*Chionoecetes opilio*) from St. Mary's Bay, Newfoundland, Canada. We also show that, when information exists on the relative catchability of different groups in the population (e.g., size groups), this information can be incorporated in the index-removal estimation scheme.

Review of models

Change-in-ratio estimation

Change-in-ratio methods make use of the change in population composition caused by a known, selective removal. For example, if a known removal of males from a population causes the proportion that is male to change a great deal, then the population must be relatively small whereas, if the removal only changes the sex ratio by a small amount, then the population must be relatively large. For the change-in-ratio method, the population must be divided into two classes (say, x- and y-type) on the basis of size, sex, or another distinguishing feature and the removal must be selective for one of the classes. Formally, we have

$$[1] \quad \hat{N}_{\text{cir}} = \frac{\hat{R}_x - \hat{P}_2 \hat{R}}{\hat{P}_1 - \hat{P}_2}$$

$$[2] \quad \hat{X}_{\text{cir}} = \hat{P}_1 \hat{N}_{\text{cir}}$$

where $\hat{N}_{\text{cir}}$ is the change-in-ratio estimate of initial population size ($x + y$ types combined), $\hat{X}_{\text{cir}}$ is the estimate of the initial number of x-type animals, $\hat{R}$ is the estimated number of animals removed, $\hat{R}_x$ is the estimated number of x-type animals (e.g., males) removed, $\hat{P}_1$ is the estimated proportion that is x-type during the survey prior to the selective removal, and $\hat{P}_2$ is the estimated proportion that is x-type in the post-removal survey. Equations 1 and 2 are the maximum likelihood estimators when the estimates of the proportions and removals are maximum likelihood estimates (Dawe et al. 1993). Change-in-ratio estimation has been known at least since Kelker (1940).

It is possible to generalize change-in-ratio estimators to allow for more than two classes in the population and for additional surveys and removals. It is also possible to relax the assumptions. Recent developments are presented by Udevitz and Pollock (1991, 1995) and reviewed by Pollock and Hoenig (1998).

In order to obtain (asymptotically) unbiased estimates of the sizes of the x- and y-type populations, the probability of capture in a survey must be equal for the two classes. However, if only one class is harvested then the estimate of population size for that class is unbiased even if the proportions estimated in the surveys are biased due to differential catchability of the two types of animals (see Seber 1982 and Pollock et al. 1985 for discussions).

Index-removal estimators

The index-removal method makes use of the decline in relative abundance due to a known removal. If catch rate (catch per unit of sampling effort) is proportional to animal abundance, and if a known removal causes the catch rate to decline by a specified proportion P, then the removal is equal to 100 P % of the population. For example, if the catch rate is 10 before the removal of 300 animals and is 7 after the removal is made, then we calculate that the removal of 300 animals resulted in a loss of $(10 - 7)/10 = 3/10$ of the population. Thus, the population size (before the removal) must have been 1000 animals. More formally, if $E(c_{x1})$ and $E(c_{x2})$ are the expected values of the observed catch rates of a component, x, of the population before and after the removal, respectively, and if R_x is the number of x-type animals removed, then the initial size of the population of x-type animals is given by

$$X_{\text{ir}} = \frac{E(c_{x1})\, R_x}{E(c_{x1}) - E(c_{x2})}$$

and an estimate can be obtained as

$$[3] \quad \hat{X}_{\text{ir}} = \frac{\hat{c}_{x1}\, \hat{R}_x}{\hat{c}_{x1} - \hat{c}_{x2}}$$

where $\hat{c}_{x1}$ and $\hat{c}_{x2}$ are the estimated mean catch rates in the first and second surveys, respectively, and $\hat{R}_x$ is the removal or an estimate of the removal.

This approach is well known in the wildlife literature (Petrides 1949; Eberhardt 1982; Seber 1982; Roseberry and Woolfe 1991) but has received little attention in the fisheries literature (Dawe et al. 1993). Seber (1982) and Routledge (1989) discuss the statistical theory in detail. In particular, Routledge (1989) generalized the approach to include J removal periods and $J + 1$ surveys. As developed above, eq. 3 is a moment estimator. Eberhardt (1982) showed it is also the maximum likelihood estimator when the catches in the two surveys are independent Poisson random variables. Dawe et al. (1993) pointed out that eq. 3 is the maximum likelihood estimator for population size whenever the estimates of the catch rates and removal are maximum likelihood. See also Hoenig and Pollock (1998) for a review.

For the simplest case described above, the assumptions of the method are as follows: (*i*) the population is closed except for the removals which are known exactly, and (*ii*) all animals have the same probability of capture which does not change from survey to survey. It is easily verified that heterogeneity of capture probabilities can introduce bias. Suppose, for example, that the population is composed of 500 males and 500 females, that males have a catchability coefficient of 0.01 whereas females have a catchability coefficient of 0.005 (i.e., half that of the males), and that 300 males and 100 females are removed from the population between the time of the two surveys. In the first survey we would expect to catch $0.01 \times 500 = 5$ males if one randomly placed unit of sampling effort is expended. In the second survey we would expect to catch $0.01 \times (500 - 300) = 2$ males. Thus, the calculated size of the initial population of males would be

$$X_{\text{males}} = \frac{5}{5-2} \times 300 = 500$$

which is what we want. Similarly, the size of the female population would be calculated to be 500, as desired. However, suppose it was not known that males have a different catchability coefficient than females and the size of the total population was calculated from combined data on males and females. In the first survey, one would expect that 7.5 animals (5 males + 2.5 females) would be caught with one randomly placed unit of sampling effort. In the second survey, the expectation would be 2 + 2 = 4 animals. Consequently, the calculated population size would be

$$N = X_{\text{males}} + X_{\text{females}} = \frac{7.5}{7.5-4} \times 400 = 857$$

instead of the actual value of 1000. Note that the heterogeneity of capture probabilities is a problem because the removal was selective with respect to capture probabilities (i.e., proportionately more of the males were removed than of the females). The problem of heterogeneity can be avoided by making separate estimates for each component of the population provided that capture probabilities within a population component are homogeneous (and do not change over time) and provided that the removals are known by component.

Removal estimators

The simplest removal estimator is obtained by dividing the fishing season in two parts with the fishing effort in the first part equal to that in the second part. The population is closed except for the removals and catch rate is assumed proportional to abundance. Thus, we would expect the catch in the second part of the season to be smaller than that in the first part because abundance was reduced in the first part by harvest. The maximum likelihood estimator is

$$\hat{N}_r = \frac{\hat{R}_1 \hat{R}_1}{(\hat{R}_1 - \hat{R}_2)} = \frac{\hat{R}_1^{\,2}}{(\hat{R}_1 - \hat{R}_2)}$$

where $\hat{R}_1$ and $\hat{R}_2$ are the catches (or, more generally, estimated catches) in the first and second parts of the season, respectively. This estimator is very similar in structure to the index-removal estimator. However, in the index-removal estimator the relative abundance is measured by the pre- and post-season surveys and the removal is done separately by the fishery. In contrast, in the removal estimator, the fishery information is used both to measure relative abundance and to deplete the population. Pre- and post-season surveys don't have to be conducted to use the removal method but catch and effort data for each subdivision of the fishing season must be obtained.

The removal method dates back to the work of Zippin (1956), Leslie and Davis (1939), and DeLury (1947). Modern generalizations are reviewed by Seber (1982).

Combining methods

It is possible to combine methods. Routledge (1989) combined the removal and index-removal methods using a likelihood framework. Dawe et al. (1993) combined the change-in-ratio and index-removal methods by computing a weighted mean. Recently, Chen (1995) combined change-in-ratio, index-removal, and removal estimators in a single likelihood function.

When more than one method is appropriate, increased efficiency can be achieved by combining methods. However, use of an estimator when the assumptions are not met will generally lead to biased results. Assessment of the assumptions is thus important. The index-removal method requires a stronger assumption (constant gear efficiency over time) than the change-in-ratio estimator (equal gear efficiencies not necessarily constant over time or constant ratio of gear efficiencies over time if only one group is exploited).

Estimation of exploitation rate, catchability coefficient, and fishing power of the gear

In addition to estimating population size (for various subsets of the population), it is also possible to estimate exploitation rate, catchability coefficient, and fishing power of the sampling gear. The exploitation rate, u, is the fraction of the population present at the beginning of the study that is harvested. It can be estimated by dividing the harvest or estimated harvest by the estimated initial population, $\hat{X}_{\text{cir}}$ or $\hat{X}_{\text{ir}}$. Thus, the estimated exploitation rate by the index-removal method is (Eberhardt 1982)

$$\hat{u}_{\text{xir}} = \frac{\hat{R}_x}{\hat{X}_{\text{ir}}} = \frac{\hat{c}_{x1} - \hat{c}_{x2}}{\hat{c}_{x1}}$$

which does not require that the total removal be known. The estimated exploitation rate by the change-in-ratio method is

$$\hat{u}_{\text{xcir}} = \frac{\hat{R}_x}{\hat{X}_{\text{cir}}} = \frac{\hat{f}(\hat{P}_1 - \hat{P}_2)}{\hat{f}\hat{P}_1 - \hat{P}_1 \hat{P}_2}$$

for which it is necessary to know the fraction of the harvest which is x-type (f) but not the magnitude of the harvest.

The catchability coefficient, q, is the fraction of the population taken by one randomly placed unit of sampling effort when the fraction taken is small (e.g., less than 2%; see Ricker 1975). It can be estimated by dividing the initial catch rate by the estimated initial population size. The catchability coefficient can be expressed as

$$[4] \quad q = \frac{\rho \phi a}{\phi A} = \frac{\rho a}{A}$$

where ϕ is the average density of animals (per unit area), A is the area inhabited by the stock, a is the area fished by the sampling gear (per unit of sampling effort), and ρ is the proportion of the animals encountering the sampling gear that is retained. The denominator is simply the total population size while the numerator is the expected catch from one unit of sampling effort. The catchability coefficient is population-specific because its definition includes the area inhabited by the population. If this area is known and if the estimate of q is multiplied by A, then we obtain an estimate of the fishing power of the gear, ρa, which is a property of the sampling gear. Consequently, we can obtain an estimate of the catchability coefficient q^* for a new population inhabiting an area A^* by dividing the estimated fishing power by A^*.

Efficient sampling design

An important feature of the work of Eberhardt (1982) and Dawe et al. (1993) was the recognition that, in practice, catch rates are usually estimated with replicate samples so that the variance can be estimated empirically rather than by resorting to theoretical variances under the assumption of a Poisson, binomial, or hypergeometric distribution of catches as described by Seber (1982). Also, the population proportions are usually estimated by cluster sampling rather than by simple random sampling (Dawe et al. 1993).

Thus, if a standard method of sampling is employed at each of n randomly selected locations at time j (j = 1, 2), then the proportion of animals which is type "x" is estimated by

$$[5] \quad \hat{P}_j = \frac{\sum_{i=1}^{n} x_{ij}}{\sum_{i=1}^{n} z_{ij}}$$

where x_{ij} and z_{ij} are the number of x-type animals and the total number of animals caught, respectively, at the ith location (i = 1, ... n). The variance of this ratio estimator can be estimated by (Cochran 1977)

$$[6] \quad \hat{V}(\hat{P}_j) = \frac{\sum_{i=1}^{n} (x_{ij} - \hat{P}_j z_{ij})^2}{n\,(n-1)\,\bar{z}_j^2}$$

where $\bar{z}_j$ is the mean of the n observations of z_{ij}.

If the same stations are occupied in the pre- and post-removal surveys, the estimated proportions will not be independent. The covariance of $\hat{P}_1$ and $\hat{P}_2$ can be estimated by

$$[7] \quad \widehat{\text{Cov}}(\hat{P}_1, \hat{P}_2) = \frac{\sum_{i=1}^{n} (x_{i1} - \hat{P}_1 z_{i1})\,(x_{i2} - \hat{P}_2 z_{i2})}{n\,(n-1)\,\bar{z}_1\,\bar{z}_2} \; .$$

If the spatial pattern is persistent (areas with a high proportion of x-type animals before the removal tend to have a high proportion after the removal and areas with a low proportion before tend to have a low proportion after the removal) then we would expect the estimated covariance to be positive.

The catch rate for x-type animals in survey j can be estimated by

$$[8] \quad \hat{c}_{xj} = \frac{1}{n} \sum_{i=1}^{n} x_{ij}$$

with variance estimated by the usual formula

$$[9] \quad \hat{V}(\hat{c}_{xj}) = \frac{\sum_{i=1}^{n} (x_{ij} - \hat{c}_{xj})^2}{n\,(n-1)} \; .$$

The covariance of the estimated catch rates in the two surveys can be estimated by

$$[10] \quad \widehat{\text{Cov}}(\hat{c}_{x1}, \hat{c}_{x2}) = \frac{\sum_{i=1}^{n} (x_{i1} - \hat{c}_{x1})\,(x_{i2} - \hat{c}_{x2})}{n\,(n-1)} \; .$$

Again, we would expect the estimated covariance to be positive if the spatial pattern is persistent over time (high abundance areas tend to remain high and low abundance areas tend to remain low).

We may also need estimates of $\text{Cov}(\hat{P}_j, \hat{c}_{xk})$ for j = 1, 2 and k = 1, 2. These can be estimated by

$$[11] \quad \widehat{\text{Cov}}(\hat{P}_j, \hat{c}_{xk}) = \frac{\sum_{i=1}^{n} (x_{ij} - \hat{P}_j z_{ij})\,(x_{ik} - \hat{c}_{xk})}{n\,(n-1)\,\bar{z}_j} \; .$$

The total removal from the population, by type of animal, is assumed to be known or estimated. It is convenient to parameterize the removals in terms of the total removals, R, and the fraction, f, that is x-type, such that the removal of x-type animals, R_x, is $R_x = fR$. The sampling design and the resulting variance of the estimated removals will depend on the particular circumstances and thus are not considered here. We assume that the variances have been estimated and that the estimates are independent of the estimates of the proportions and catch rates in the surveys.

All of the estimators are functions of the random variables $\{\hat{P}_1, \hat{P}_2, \hat{c}_{x1}, \hat{c}_{x2}, \hat{f}, \hat{R}\}$ and we can estimate the variances and covariances of these random variables as discussed above. Estimates of the variances of the estimators of population parameters can be obtained using the Taylor's series or delta method (see Seber 1982, pp. 7–8). In general terms, the variance of a function, g, of a set of random variables $\boldsymbol{Y} = \{Y_1, Y_2, ..., Y_p\}$ can be estimated by

$$[12] \quad \hat{V}(g(\boldsymbol{Y})) = \sum_{r=1}^{p} \sum_{s=1}^{p} \frac{\partial g}{\partial Y_r} \frac{\partial g}{\partial Y_s} \widehat{\text{Cov}}(Y_r, Y_s)$$

where the derivatives are evaluated at the parameter estimates.

The variances of the estimates of the initial number of x-type animals are thus found to be

$$[13] \quad \hat{V}(\hat{X}_{\text{ir}}) = \frac{\hat{X}_{\text{ir}}^2\,\hat{V}(\hat{c}_{x2}) + \hat{X}_2^2\,\hat{V}(\hat{c}_{x1})}{(\hat{c}_{x1} - \hat{c}_{x2})^2} + \frac{\hat{c}_{x1}^2\,\hat{V}(\hat{R}_x)}{(\hat{c}_{x1} - \hat{c}_{x2})^2} - \frac{2\hat{X}_{ir}\hat{X}_2\,\widehat{\text{Cov}}(\hat{c}_{x1}, \hat{c}_{x2})}{(\hat{c}_{x1} - \hat{c}_{x2})^2}$$

where $\hat{X}_2 = \hat{X}_{\text{ir}} - \hat{R}_x$, and

$$[14] \quad \hat{V}(\hat{X}_{cir}) = \frac{\hat{N}_1^2\,\hat{P}_2^2\,\hat{V}(\hat{P}_1) + \hat{N}_2^2\,\hat{P}_1^2\,\hat{V}(\hat{P}_2)}{(\hat{P}_1 - \hat{P}_2)^2} + \frac{\hat{P}_1^2\,\hat{R}^2\,\hat{V}(\hat{f}) + \hat{P}_1^2(\hat{f} - \hat{P}_2)^2\,\hat{V}(\hat{R}) + 2\hat{R}\,\hat{P}_1^2\,(\hat{f} - \hat{P}_2)\,\widehat{\text{Cov}}(\hat{f}, \hat{R})}{(\hat{P}_1 - \hat{P}_2)^2} - \frac{2\hat{P}_1\,\hat{P}_2\,\hat{N}_1\,\hat{N}_2\,\widehat{\text{Cov}}(\hat{P}_1, \hat{P}_2)}{(\hat{P}_1 - \hat{P}_2)^2}$$

where $\hat{N}_2 = \hat{N}_{\text{cir}} - \hat{R}$. In eqs. 13 and 14, the first term is the uncertainty in the survey results, the middle term is the uncertainty in the removals, and the third term is the reduction in variance due to the reoccupying of the same stations in survey 2 as in survey 1.

If the catchability of the two types of animals in the surveys is equal (so that the change-in-ratio estimate of total population, i.e., x- and y-types, is valid), then the variance of the estimated total can be estimated as

$$[15]\quad \hat{V}(\hat{N}_{\text{cir}}) = \frac{\hat{N}_{\text{cir}}^2\,\hat{V}(\hat{P}_1) + \hat{N}_2^2\,\hat{V}(\hat{P}_2)}{(\hat{P}_1-\hat{P}_2)^2} + \frac{\hat{R}^2\,\hat{V}(\hat{f}) + (\hat{f}-\hat{P}_2)^2\,\hat{V}(\hat{R}) + 2\hat{R}\,(\hat{f}-\hat{P}_2)\,\widehat{\text{Cov}}\,(\hat{f},\hat{R})}{(\hat{P}_1-\hat{P}_2)^2} - \frac{2\hat{N}_{\text{cir}}\,\hat{N}_2\,\widehat{\text{Cov}}\,(\hat{P}_1,\hat{P}_2)}{(\hat{P}_1-\hat{P}_2)^2}\ .$$

Also, if the catchability of the two types of animals in the surveys is equal then the total population of the two types can be estimated by $\hat{N}_{\text{ir}} = \hat{X}_{\text{ir}}/\hat{P}_1$, with variance estimated by

$$[16]\quad \hat{V}(\hat{N}_{\text{ir}}) = \frac{\hat{N}_{\text{ir}}^2\,(\hat{V}(\hat{c}_{x1}) + \hat{V}(\hat{c}_{x2})) + \hat{c}_{x1}^2\hat{R}_x^2\,\hat{P}_1^{-4}\,\hat{V}(\hat{P}_1)}{(\hat{c}_{x1}-\hat{c}_{x2})^2} + \frac{\hat{c}_{x1}^2\,\hat{V}(\hat{R}_x)}{\hat{P}_1^2\,(\hat{c}_{x1}-\hat{c}_{x2})^2} - \frac{2\hat{N}_{\text{ir}}^2\,\widehat{\text{Cov}}\,(\hat{c}_{x1},\hat{c}_{x2})}{(\hat{c}_{x1}-\hat{c}_{x2})^2}\ .$$

We estimate the total population size this way, rather than as the sum of two index-removal estimates, because one of the classes is unexploited (or lightly exploited) thus giving rise to no estimate (or a very imprecise estimate) for the class.

Variance formulae for estimates of exploitation rate and catchability coefficient are given in Appendix 1.

It is also possible to use the Taylor's series approach to estimate the bias of the estimated population parameters (Appendix 2).

Evaluation of the efficient design

Modeling catches

We conducted a simulation study to see how the change-in-ratio and index-removal methods perform when animals have a patchy distribution with a degree of persistence in spatial pattern over time. This is a common pattern in nature and has been observed for snow crab as shown in the example in the next section.

When animals are independently and randomly distributed over space, the catch obtained from a standard unit of sampling effort has a Poisson distribution and the variance of catches is equal to the mean catch. Also, the sum of the catches from J units of sampling effort (i.e., from J locations) has a Poisson distribution with parameter equal to J times the mean catch per unit effort. Thus, following Chapman and Murphy (1965), we might assume that the catches of x- and y-type animals in the surveys have the following distributions:

$$x_{1j} \sim \text{Poisson}(q_1 f_1 X_1), j = 1, 2, \ldots J$$
$$y_{1j} \sim \text{Poisson}(q_1 f_1 Y_1), j = 1, 2, \ldots J$$
$$x_{2j} \sim \text{Poisson}(q_2 f_2 (X_1 - R_x)), j = 1, 2, \ldots J$$
$$y_{2j} \sim \text{Poisson}(q_2 f_2 (Y_1 - R_y)), j = 1, 2, \ldots J$$

where q_i is the catchability coefficient in the ith survey, f_i is the sampling effort expended at each location in the ith survey, and the other symbols are as defined previously. For index-removal estimation, we assume further that $q_1 = q_2$. We refer to this as the common Poisson model because the catches for each location follow a common Poisson distribution (there is no location effect). It is clear that, because there is no location effect, there is no advantage to reoccupying the same stations in the second survey as in the first.

In the common Poisson model, the variances of the catches equal the means. However, when animals have a patchy distribution the variance is greater than the mean. We therefore modify the common Poisson model to allow for this overdispersion by introducing random location effects. (There may be a spatial correlation in animal distribution but, because the sampling station locations are chosen randomly, the station effects are independent, identically distributed random variables.) For convenience, we let the location effects have a gamma distribution. The gamma distribution is extremely flexible and, when a variable Z follows a Poisson distribution whose parameter, λ, is drawn from a gamma distribution, the unconditional distribution of Z is negative binomial.

We used two models to simulate sampling from a population. In model I, the random location effects for x-type animals are independent of those for y-type animals. The sample catches then have the following distributions:

$$x_{1j} \sim \text{Poisson}(q_1 f_1 X_1\, \delta x_{1j}), j = 1, 2, \ldots J$$
$$y_{1j} \sim \text{Poisson}(q_1 f_1 Y_1\, \delta y_{1j}), j = 1, 2, \ldots J$$
$$x_{2j} \sim \text{Poisson}(q_2 f_2 (X_1 - R_x)\, \delta x_{2j}), j = 1, 2, \ldots J$$
$$y_{2j} \sim \text{Poisson}(q_2 f_2 (Y_1 - R_y)\, \delta y_{2j}), j = 1, 2, \ldots J$$

where

$$\delta x_{ij} \sim \text{Gamma}(\alpha_1, \beta_1),\ i = 1, 2; j = 1, 2, \ldots J$$
$$\delta y_{ij} \sim \text{Gamma}(\alpha_2, \beta_2),\ i = 1, 2; j = 1, 2, \ldots J.$$

The expected value of the catch of x-type animals (at any location) is

$$E(x_{ij}) = q_i f_i X_i\, \alpha_1\beta_1$$

while for y-type animals it is

$$E(y_{ij}) = q_i f_i Y_i\, \alpha_2\beta_2.$$

In order to keep the catchabilities for the two types of animals equal, we introduce the constraint that $\alpha_1\beta_1 = \alpha_2\beta_2$. Note that for the paired design, $\delta x_{1j} = \delta x_{2j}$ and $\delta y_{1j} = \delta y_{2j}$ because the same stations are used for the two surveys. For index-removal estimation, q_1 is set equal to q_2.

In general, the catches of the two subclasses at a given time are correlated and the catches of a given subclass (at a set of fixed locations) are correlated over time. We use model II, a generalization of model I, to mimic this. The catches are distributed as follows:

$$x_{1j} \sim \text{Poisson}(q_1 f_1 X_1\, (\delta_{1j} + \delta x_{1j})), j = 1, 2, \ldots J$$
$$y_{1j} \sim \text{Poisson}(q_1 f_1 Y_1\, (\delta_{1j} + \delta y_{1j})), j = 1, 2, \ldots J$$
$$x_{2j} \sim \text{Poisson}(q_2 f_2 (X_1 - R_x)\, (\delta_{2j} + \delta x_{2j})), j = 1, 2, \ldots J$$
$$y_{2j} \sim \text{Poisson}(q_2 f_2 (Y_1 - R_y)\, (\delta_{2j} + \delta y_{2j})), j = 1, 2, \ldots J$$

where

$$\delta_{ij} \sim \text{Gamma}(\alpha, \beta),\ i = 1, 2; j = 1, 2, \ldots J$$
$$\delta x_{ij} \sim \text{Gamma}(\alpha_1, \beta_1),\ i = 1, 2; j = 1, 2, \ldots J$$
$$\delta y_{ij} \sim \text{Gamma}(\alpha_2, \beta_2),\ i = 1, 2; j = 1, 2, \ldots J.$$

Again, we introduce the constraint that $\alpha_1\beta_1 = \alpha_2\beta_2$ so that the

Table 1. Comparison of change-in-ratio estimates for the paired and unpaired designs when data are generated according to overdispersion model I.

	Bias($\hat{X}_{cir}$)		SE($\hat{X}_{cir}$)	
n_i/N_i	Paired	Unpaired	Paired	Unpaired
0.9	2.0	49.6	42.3	2 079.6
0.7	2.8	59.3	48.1	692.9
0.5	4.4	70.4	57.3	2 245.3
0.3	7.8	77.0	75.5	2 323.1
0.1	31.6	78.5	173.2	8 659.3
	Bias($\hat{Y}_{cir}$)		SE($\hat{Y}_{cir}$)	
0.9	26.5	98.1	235.2	3 508.1
0.7	27.9	115.5	239.2	1 141.9
0.5	29.7	135.0	248.5	3 828.0
0.3	38.1	146.1	269.2	4 041.4
0.1	81.7	158.9	405.1	11 089.9

Note: Estimates are based on 40 000 simulations of the paired design and 100 000 simulations for the unpaired design. R_x/X_1 (fraction of the *x*-type population removed) was fixed at 0.7, R_y/Y_1 was fixed at 10%. n_i/N_i is the expected fraction of the population seen in each survey. Other parameters are given in the text. The sizes of the *x*- and *y*-populations being estimated are 1000.

Table 2. Comparison of change-in-ratio estimates for the paired and unpaired designs when data are generated according to overdispersion model II.

	Bias($\hat{X}_{cir}$)		SE($\hat{X}_{cir}$)	
n_i/N_i	Paired	Unpaired	Paired	Unpaired
0.9	2.8	26.1	41.7	132.5
0.7	2.9	27.0	47.4	165.3
0.5	4.7	30.8	56.6	126.8
0.3	8.1	34.8	75.6	288.2
0.1	30.8	64.4	186.2	1014.6
	Bias($\hat{Y}_{cir}$)		SE($\hat{Y}_{cir}$)	
0.9	15.7	52.0	172.2	285.9
0.7	16.0	52.4	180.1	325.6
0.5	19.6	58.3	188.6	372.1
0.3	25.8	66.0	214.1	462.0
0.1	67.5	119.5	370.8	1693.7

Note: Estimates are based on 40 000 simulations. R_x/X_1 (fraction of the *x*-type population removed) was fixed at 0.7, R_y/Y_1 was fixed at 10%. n_i/N_i is the expected fraction of the population seen in each survey. Other parameters are given in the text. The sizes of the *x*- and *y*-populations being estimated are 1000.

catchabilities for the two types are equal. Also, note that for the paired design $\delta_{1j} = \delta_{2j}$, $\delta x_{1j} = \delta x_{2j}$, and $\delta y_{1j} = \delta y_{2j}$.

Design of the simulation study

We compared the bias and standard error of the estimates for the paired and unpaired design. The simulation data were generated from the two overdispersion models under the following conditions: (*i*) population size $N_1 = 2000$ with $X_1 = 1000$ and $Y_1 = 1000$; (*ii*) $q_1 = q_2 = 0.001$ with effort chosen so that the expected fraction of the population seen in each survey (n_i/N_i) ranged from 0.1 to 0.9 by increments of 0.2; (*iii*) 10% of *y*-type animals removed (assumed known exactly); (*iv*) 70% of *x*-type animals removed; (*v*) overdispersion parameters $\alpha_1 = \alpha_2 = 1.0$, $\beta_1 = \beta_2 = 1.0$ for model I and $\alpha = \alpha_1 = \alpha_2 = 0.5$, $\beta = \beta_1 = \beta_2 = 1.0$ for model II (thus, expected catches in the surveys are the same for the two models); (*vi*) at least 10 000 data sets were generated for each scenario; and (*vii*) new values of the δ_{ij}, δx_{ij}, and δy_{ij} were generated for each simulated data set. These are conditions in which the methods can be expected to perform well. We also conducted an additional simulation which may be more typical of a commercial fishery. The initial population size of *x*-type animals was 3.2×10^6 animals while for *y*-type animals it was 3.5×10^6. The exploitation rates on the *x*- and *y*-type animals were 18 and 5%, respectively. The expected fraction of the population seen in each survey was 1%. The overdispersion parameters were $\alpha = \alpha_1 = \alpha_2 = 0.5$ and $\beta = \beta_1 = \beta_2 = 1.0$. The simulation of additional scenarios is discussed by Chen (1995).

Whenever an estimate was infeasible (e.g., negative) we replaced it with the minimum feasible estimate which is equal to the removal plus the number of animals seen in the second survey.

Table 3. Comparison of unpaired and paired designs for change-in-ratio and index-removal estimation for a simulated population with exploitation rates of 18 and 5% for *x*- and *y*-type animals, respectively.

Type	$E(\hat{X}_1)$	SE($\hat{X}_1$)	$E(\hat{Y}_1)$	SE($\hat{Y}_1$)
Change-in-ratio estimator				
Unpaired, no lower bound	1.31	2080	1.66	2380
Unpaired, lower bound	10.7	1380	12.4	1710
Paired	3.21	0.273	3.57	0.728
Index-removal estimator				
Paired[a]	3.21	0.123	—	—

Note: True population sizes were $X_1 = 3.2 \times 10^6$ and $Y_1 = 3.5 \times 10^6$. Number of simulated data sets was 5 000 000 for the change-in-ratio estimators and 10 000 for the index-removal methods. See text for other parameters. When estimated population size was infeasible (e.g., negative) the estimate was replaced by the minimal estimate (lower bound) equal to the removal plus the number of animals observed in the second survey. $E(\cdot)$ = expected value; SE($\cdot$) = standard error. Estimates should be multiplied by 10^6.

[a]Standard errors for the unpaired cases (with and without a lower bound) are in excess of 100×10^6 but estimates are not precise and are not reported here.

Simulation results

Change-in-ratio estimation

In all of the simulations the paired design worked far better than the unpaired design (Tables 1–3). In fact, the variability in the results with the unpaired design was so large that 10 000 simulations were not enough to enable us to determine precisely the bias and variance. Nonetheless, it is clear that even when 70% of the *x*-type population is removed, compared to only 10% of the *y*-type population, the bias is substantial and the standard errors are so large under the unpaired design as to call the utility of the estimator into question.

The simulation of conditions representing a fishery with an

Table 4. Comparison of index-removal estimates for the paired and unpaired designs when data are generated according to overdispersion model I.

	Bias($\hat{X}_{ir}$)		SE($\hat{X}_{ir}$)	
x_i/X_i	Paired	Unpaired	Paired	Unpaired
0.9	1.7	19.8	30.3	107.8
0.7	1.9	21.4	34.4	114.6
0.5	2.6	21.9	41.0	115.6
0.3	4.9	24.9	54.2	127.6
0.1	14.3	35.2	101.0	228.2

Note: Estimates are based on 10 000 simulations. R_x/X_1 is the fraction of the x-type population removed and x_i/X_i is the fraction of the x-type population seen in each survey. Other parameters are given in the text. The size of the x-population being estimated is 1000.

Table 5. Results from the pre- and post-season surveys of snow crab in St. Mary's Bay using small mesh traps.

Catch		Catch rate	Covariance	Correlation, r
X before	X before	44.0727	11.3911	1.0
X before	Y before		14.0993	0.42776
Y before	Y before	100.3455	95.3727	1.0
X before	X after		5.3555	0.58060
X after	Y before		6.8309	0.25593
X after	X after	33.0727	7.4693	1.0
X before	Y after		14.0305	0.44432
Y before	Y after		66.9709	0.73295
X after	Y after		9.8238	0.38419
Y after	Y after	88.1091	87.5364	1.0

$\hat{P}_1 = 0.30517$; $\hat{V}(\hat{P}_1) = 0.40287 \times 10^{-3}$; $\widehat{\mathrm{Cov}}(\hat{P}_1, \hat{P}_2) = 0.23468 \times 10^{-3}$; $\hat{P}_2 = 0.27292$; and $\hat{V}(\hat{P}_2) = 0.44739 \times 10^{-3}$.

Note: Covariances are covariances of the means, i.e., covariance of the paired observations divided by the number of sampling stations (= 55). (X-type = legal size, Y-type = sublegal size.)

exploitation rate of 18% on x-type animals and 5% on y-type animals was very informative. The results for the unpaired design were so enormously variable that it took 1 880 000 simulated data sets to get the estimates of expected value and standard error to stabilize somewhat but even after this many simulations the results were not very precise. For this scenario, the change-in-ratio estimator is badly biased and has enormous variance when the unpaired design is used (Table 3). Also, infeasible (negative) estimates arise frequently. In contrast, the estimator worked much better with the paired design: none of the estimates were infeasible, the bias appeared small, but the standard errors were very large.

Index-removal estimation

Under model I, the paired design provided nearly unbiased estimates and small standard errors for all scenarios considered (Table 4). The unpaired results were uniformly inferior to the paired results.

For the exploited fishery scenario (Table 3), the results are similar to those found for the change-in-ratio estimator: the unpaired design performed very poorly but the paired design

Table 6. Results from the pre- and post-season surveys of snow crab in St. Mary's Bay using large mesh traps.

Catch		Catch rate	Covariance	Correlation, r
X before	X before	61.3454	17.5173	1.0
X before	Y before		15.0271	0.53503
Y before	Y before	67.4546	45.0327	1.0
X before	X after		10.7164	0.72335
X after	Y before		9.1520	0.38530
X after	X after	44.4546	12.5291	1.0
X before	Y after		12.5767	0.46584
Y before	Y after		31.4836	0.72732
X after	Y after		13.4195	0.58773
Y after	Y after	58.8000	41.6091	1.0

$\hat{P}_1 = 0.47628$; $\hat{V}(\hat{P}_1) = 0.45351 \times 10^{-3}$, $\widehat{\mathrm{Cov}}(\hat{P}_1, \hat{P}_2) = 0.32587 \times 10^{-3}$; $\hat{P}_2 = 0.43053$; and $\hat{V}(\hat{P}_2) - 0.48732 \times 10^{-3}$.

Note: Covariances are covariances of the means, i.e., covariance of the paired observations divided by the number of sampling stations (= 55). (X-type = legal size, Y-type = sublegal size.)

always produced feasible estimates, appeared to have minimal bias, and had a very small standard error.

Example: snow crab fishery in St. Mary's Bay, Newfoundland

The fisheries for snow crab in Atlantic Canada are exclusively trap fisheries and only males above the minimum legal size of 95 mm carapace width are harvested. There is a mesh size restriction to reduce the catch of sublegal-sized crab and fishermen have been taught to return sublegal-sized crab to the water as quickly as possible to protect the prerecruits. We define x-type animals to be males ≥95 mm in width and y-type animals to be males with width ≥78 but <95 mm. In 1992, the fishery in St. Mary's Bay occurred from September 1 to September 10. From sales slip data, it is known that 306 t of snow crab were harvested. Extensive sampling of crab in the fish plants revealed that the mean weight of individual crab was 480.49 g and that 90.826% of the harvested crab were of legal size (x-type). Thus, we calculate that 578 425 legal-sized and 58 425 sublegal-sized crab were harvested. For purposes of this analysis we assume these figures are exact.

The preseason research survey was conducted from August 20 to August 28 and the postseason survey was from September 14 to September 28. Fifty-five fleets of traps were set at randomly selected locations in that part of the bay deeper than 40 m (approximately 660 km^2). The same locations were occupied in both surveys. Each fleet of traps contained 2 large-meshed and 2 small-meshed traps. The large-meshed traps were the same type as used in the commercial fishery. The primary sampling unit was the combined catch from all of the traps of a given type in a fleet of traps. However, catches were expressed on a per-trap basis to account for the fact that occasionally a trap in a fleet didn't fish properly. The survey methods are described in Dawe et al. (1993).

For both small- and large-meshed traps, the proportion of legal-sized crab in the second survey was slightly lower than in the first survey (Tables 5 and 6). For small-meshed traps, the proportion declined from 0.31 to 0.27 between the two

Fig. 1. Scatterplots of catches of crab of two size-classes in survey *i* versus catches of crab at the same location in survey *j* for $i, j = 1, 2$. Note that the catches of legal-sized crab in survey 1 are highly correlated with catches of legal-sized crab in survey 2, and similarly for sublegal-sized crab. In contrast, catches of legal-sized crab are only weakly correlated with catches of sublegal-sized crab both within a survey and between surveys. (*a*) Catches from small-mesh traps. (*b*) Catches from large-mesh traps.

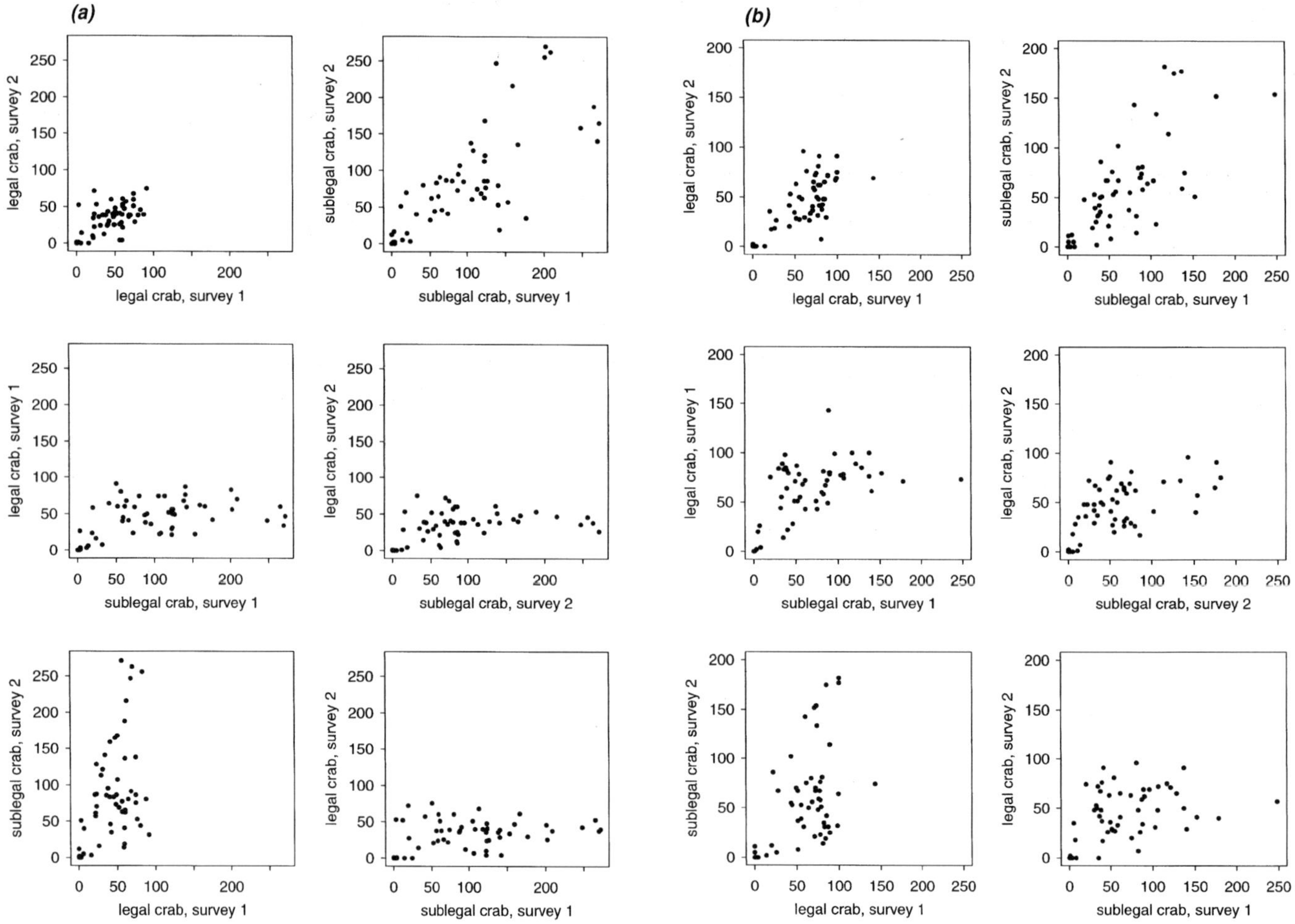

surveys while for large-meshed traps the proportion declined from 0.48 to 0.43. The catch rates also declined slightly from the first survey to the second (Tables 5 and 6). For small-meshed traps, the catch rate declined from 44 to 33 legal-sized crab per trap and from 100 to 88 sublegal-sized crab per trap. For large mesh traps, the catch rate declined from 61 to 44 legal-sized crab per trap and from 67 to 59 sublegal-sized crab per trap.

The correlation, *r*, between catch per trap in the preseason survey and the catch at the corresponding location in the post-season survey was large and positive (Tables 5 and 6, Fig. 1) thus indicating that increased efficiency can indeed be attained by reusing rather than rerandomizing the sampling locations. For small-mesh traps, the correlation between catches of legal-sized crab in the two surveys was 0.58 and for sublegal-sized crab it was 0.73. For large-mesh traps, the correlation was 0.72 for legal-sized crab and 0.73 for sublegal-size crab.

The estimates of the number of legal-sized crab at the beginning of the season ranged from 2.1 to 3.8 million animals with standard errors for the paired design ranging from 0.3 million (for the smallest estimate) to 2.1 million (for the largest estimate) (Table 7). The standard errors were also calculated without including the covariance term. The reduction in estimated standard error achieved by including the covariance term was 33–47%.

The exploitation rate was estimated to be from 15 to 28% (Table 8) with standard errors ranging from 4 to 8% for the paired design. As with the estimates of population size, these values were 33–47% less than those obtained when the covariance term was ignored.

Incorporating information about relative catchability in index-removal estimators

The problem of heterogeneity of capture probabilities can be minimized by making separate estimates for various subsets of the population. For example, separate estimates could be made for males, and for females, or for different size groups of animals. However, when information is available on the relative catchability of different groups, this information can be incorporated in the estimation procedure to increase the statistical efficiency of the estimator. For example, one may have good

Table 7. Estimates of initial population of legal-sized snow crab ($\times 10^6$).

Method	Estimate	SE with covariance	SE without covariance	% reduction in SE
Estimates from large-mesh traps				
cir	3.1673	1.0642	1.9199	45
ir	2.1008	0.3096	0.5797	47
Estimates from small-mesh traps				
cir	3.8287	2.1465	3.1974	33
ir	2.3175	0.5094	0.7850	35

Note: To estimate how reoccupying stations improves efficiency, the standard error (SE) was calculated with and without the covariance term.

Table 8. Estimates of exploitation rate for legal-sized snow crab in St. Mary's Bay.

Method	Estimate	SE with covariance	SE without covariance	% reduction in SE
Estimates from large-mesh traps				
cir	0.183	0.0614	0.1107	45
ir	0.275	0.0406	0.0760	47
Estimates from small-mesh traps				
cir	0.151	0.0847	0.1262	33
ir	0.250	0.0549	0.0845	35

reason to believe that male crab are more catchable than females in a trap survey because of differential behavior or differential body size among the sexes. Similarly, one may believe that large crab may be more catchable by traps than small crab. In these cases, one may wish to introduce order restrictions in the estimation procedure to ensure that the estimated catchabilities are consistent with the available information on relative catchabilities of the various groups.

We consider a suite of four models that vary in the amount of information assumed about the relative catchabilities of the different groups. The simplest approach is to make separate, independent estimates for each group. If qualitative information is available about the relative catchabilities of the groups, then one can introduce order restrictions for the catchability coefficients. One might also assume a functional relationship for the way catchability coefficients vary with a covariate. In particular, the catchability coefficient might be a logistic function of body size. In this case, we would estimate the parameters of the functional relationship rather than the catchability coefficients for each size group. Finally, if a sampling gear selectivity curve is available from some other study, then the parameter estimates of the selectivity curve can be incorporated directly in the population estimation procedure.

We illustrate the approach by assuming that the catches per unit of sampling effort in the surveys follow Poisson or multinomial distributions. This is consistent with previous treatments of the subject in the literature. It is possible to assume catch rate follows a different distribution such as a normal distribution as suggested by Routledge (1989). The normal distribution appears to be a more reasonable model for many fishery applications and the modifications to handle this are straightforward.

Four models for catches that follow Poisson distributions

We assume that the expected value of the total number of animals caught when one unit of sampling effort is expended at each of n randomly selected locations is given by $E(C) = qnX = \lambda$ (say), where $E(\cdot)$ denotes expected value of the quantity in parentheses, q is the catchability coefficient, n is the number of units of sampling effort (locations), and X is the population size. Thus, catch is assumed proportional to sampling effort and to abundance. This assumption is justified if the sampling is with replacement (animals are released unharmed after being caught) or the fraction of the population caught is negligible so that the population size, X, does not change due to the random sampling. Furthermore, we assume that the total number of animals caught during the survey, C, follows a Poisson distribution with parameter λ, i.e., $C \sim P(\lambda)$. Thus, the probability density function for the number of animals caught is

$$f(C) = \frac{\lambda^C e^{-\lambda}}{C!} = \frac{(qnX)^C e^{-qnX}}{C!} .$$

If R animals are removed from the population, the abundance becomes $X - R$ and, under the assumption that expected value of the catch is proportional to abundance, the expected value of the catch becomes $E(C) = qn(X - R)$. If we assume that the distribution of catch remains Poisson then we can write the likelihood, Λ, for obtaining a series of catches $\{C_1, C_2, \ldots C_J\}$ from J surveys having respective sampling efforts $\{n_1, n_2, \ldots n_J\}$ as

$$[18] \quad \Lambda = \prod_{j=1}^{J} \frac{\lambda_j^{C_j} e^{-\lambda_j}}{C_j!} = \prod_{j=1}^{J} \frac{(qn_j X_j)^{C_j} e^{-qn_j X_j}}{C_j!} .$$

where λ_j is the Poisson parameter for the jth random survey and X_j is the number of animals in the population just before the jth survey. Thus, X_1 is equal to the original population X, and

$$[19] \quad X_j = X - \sum_{k=1}^{j-1} R_j \quad \text{for } j \geq 2$$

where R_k is the number of animals removed from the population after the kth survey. Here, we have treated the removals R_k as known, fixed values.

There are two unknowns, q and X_1, in eq. 18. When the data consist of two surveys and one removal, the estimates which maximize the likelihood are given by

$$\hat{X}_1 = \frac{\hat{c}_1 \hat{R}}{\hat{c}_1 - \hat{c}_2}$$

$$\hat{q} = \frac{\hat{c}_1}{\hat{X}_1} = \frac{\hat{c}_1 - \hat{c}_2}{\hat{R}} .$$

Here, $\hat{c}_j$ is the catch per unit of sampling effort in the jth survey for $j = 1, 2$.

Method 1: independent estimates by size group

Suppose we have reason to believe that the I subgroups in the population have different catchabilities, q_i. Suppose, further, that we believe the catches of the various subgroups are independent Poisson random variables. This assumption is made explicitly for some change-in-ratio estimation models (see Seber 1982). Then the likelihood (eq. 18) can be generalized by the introduction of an index i denoting subgroup-specific population sizes and catchabilities. Thus, the likelihood becomes

$$[20] \quad \Lambda = \prod_{i=1}^{I} \prod_{j=1}^{J} \frac{\lambda_{ij}^{\,C_{ij}} e^{-\lambda_{ij}}}{C_{ij}\,!} = \frac{(q_i\, n_j\, X_{ij})^{C_{ij}} e^{-q_i n_j X_{ij}}}{C_{ij}\,!} \,.$$

Here, the X_{ij} are defined in a manner analogous to eq. 19. That is, X_{i1} is the original number of animals in the population in subgroup i and

$$[21] \quad X_{ij} = X_{i1} - \sum_{k=1}^{j-1} R_{ik} \quad \text{for } j \geq 2$$

where R_{ik} is the number of animals removed from the ith subgroup of the population after the kth survey. Here, we have again treated the removals R_{ik} as known, fixed values.

Equation 20 has $2I$ unknowns: I initial abundances and I catchability coefficients. It is easily verified that maximizing eq. 20 with respect to the $2I$ unknowns is equivalent to maximizing eq. 18 separately for each subgroup in the population.

Method 2: introducing order restrictions for the catchabilities

Suppose we have good reason to believe that $q_1 < q_2 < ... q_I$ and we wish to introduce these order restrictions into the estimation procedure. Let $q_0 = 0$ and let $q_i = q_{i-1} + \delta_i^2$, for $i = 1, 2, .. I$, and substitute these definitions into the likelihood (eq. 20). We now have I initial abundances and I values of δ_i^2 to estimate. Note that, regardless of the value of the estimate $\hat{\delta}_i$, the value of $\hat{\delta}_i^2$ must be nonnegative and, thus, the estimate of q_i must be greater than or equal to the estimate of q_{i-1}.

Method 3: introducing a functional relationship for catchability

Often, the catchability of an animal will vary with the animal's body size. For example, the chances of a fish escaping through the meshes of a trawl generally decrease as the size of the fish increases. In contrast, the ability of some animals to avoid sampling gear may increase as the animals increase in age or size. In fisheries work, it is common to model the selectivity of fishing gear as a logistic function of body size. Thus, the proportion of the animals of length l that is retained by the gear, $p(l)$, can be described by

$$[22] \quad p(l) = \frac{1}{1 + \exp(-\alpha\,(1 - l_0))}$$

where α is a shape parameter and l_0 is a location parameter. The catchability of animals in the ith size group would then be proportional to the selectivity for that group:

$$[23] \quad q_i = \beta\, p(l)$$

where β is an additional parameter relating the catchability to the selectivity of the gear. Equation 23 can be substituted for the q_i in eq. 20. In this case, we estimate the parameters α and l_0 of the logistic curve and the scaling parameter β, instead of the I catchabilities, q_i. When only a limited range of sizes of animals is caught, it might be better to model catchability as an asymptotic or linear function of size. The basic idea remains the same.

Method 4: using independent estimates of gear selectivity

It is often the case that the gear selectivity can be estimated by comparing the catches from two sampling gears with different mesh sizes. If gear selectivity parameter estimates are available from an independent study then one need only estimate the initial population sizes, X_{i1}, and the scaling parameter, β, of the logistic curve.

Discussion

Efficient design

The present study appears to be the first to point out the potential gains in efficiency obtainable by the reuse of the same stations in the two surveys. These gains can be achieved at no additional cost of sampling, i.e., without increasing the sampling effort. In the snow crab example, the increase in precision was substantial: standard errors of estimated population size, calculated with the covariance term, were 33–47% less than those calculated without the covariance term. In the simulation studies, the variances of the catch rates were considerably smaller than those observed in St. Mary's Bay. However, the correlations (between types of animals in a survey and between surveys) were grossly similar to what was observed for St. Mary's Bay (see Chen 1995, Table 11). The simulations clearly demonstrate the reduction in bias and increase in efficiency attainable through use of the efficient design.

Gains in efficiency can be expected when the spatial pattern persists from one survey to the next and when the scale of the patchiness is large enough that the researcher is able to place the sampling gear in the same patches in the subsequent survey. Short-term movements and migrations and imprecision in setting fishing gear will lessen the effectiveness of reusing sampling locations.

Use of the paired design also results in a substantial reduction in the small-sample bias and in the incidence of infeasible estimates when sample sizes are small. Bias of the change-in-ratio and index-removal estimators has not previously been studied although it is known that the estimators are asymptotically (as sample sizes approach infinity) unbiased. Bias of the estimates of the total and x-type population sizes, when the covariance (between surveys) is zero, is positive for both the change-in-ratio and index-removal methods. Magnitude of bias increases as the change in catch rates approaches zero and as the change in proportions approaches zero. Positive bias in estimates of population size implies that the estimates of exploitation rate have a negative bias.

Change-in-ratio estimators can also be used to estimate the relative survival rate of two groups (see Seber 1982). For example, Frusher et al. (1998) estimate the ratio of survival of legal lobsters to that of sublegal-sized lobsters of the same sex. By similar methods to those used here, it can be shown that gains in efficiency can be achieved for this estimator when the sampling stations in the second survey are the same as in the first survey. Gains in efficiency with the paired design can also be expected for the change-in-ratio estimator of relative habitat usage and relative catchability described by Heimbuch and Hoenig (1989).

Incorporating information on catchability

The methods described for incorporating information on relative catchability of different population components can be used as part of a general model building strategy. One can start by looking at separate estimates of catchability by size group. If these show a general trend or pattern that is consistent with expectation based on knowledge of the biology of the species and the characteristics of the sampling gear, then the estimates might be smoothed somewhat by imposing order restrictions. One might also estimate the parameters of a selectivity-with-size model. However, this would require sufficient contrast in the data, i.e., a sufficient range of sizes in the data. One could also use assumed selectivity parameters if these are available from an external study. A likelihood ratio test could be used to test if selectivity parameters estimated by the index-removal method are significantly different from assumed values from an external study.

The problem of heterogeneity of capture probabilities also occurs with removal estimators. One way to handle this is to make separate estimates for each component of the population. In this case, one may wish to constrain or model the catchability coefficients as functions of the sizes of the animals as was done for the index-removal estimator.

Conclusions about snow crab in St. Mary's Bay

The estimates of population size are rather variable and the standard errors are quite large. This is what would be expected if the removal accounted for only a small fraction of the legal-sized population. Judging from the small change in catch rate and the small change in composition, this appears to be the case. Estimates of the fraction of the legal-sized population removed vary from 15 to 28% (Table 8). Thus, no matter which portion of the data is considered, and no matter which analytical method is used, the results are that the population in St. Mary's Bay is lightly exploited relative to other areas where target exploitation rates of 50–60% have been achieved. These results are consistent with estimates for St. Mary's Bay for the previous year (1991) of 9–29% exploitation rate (Dawe et al. 1993).

We computed index-removal estimates of population size of legal-sized crab by 5-mm length-class. We expected estimates of the catchability coefficient to show an increasing trend with size but, instead, the estimates decreased with size. We have no explanation for this.

The mean catch rate for sublegal-sized crab declined from the first survey to the second one. Although the decline is not statistically significant, it is interesting to consider the implications of such a decline. One possibility is that the catchability of all crab declined over the course of the study. In this case, the estimated decline in catch rate for legal-sized crab would be too large and the index-removal estimate of population size of legal-sized crab would be too low. The change-in-ratio estimator would be unaffected by a change in catchability provided that both size groups were equally affected (same percentage decline in catchability). Another possibility is that the population of sublegal-sized crab declined not only because of the known harvest but also because of mortality experienced by discarded sublegal-sized crab. In this case, the index-removal estimator of the legal-sized crab population is unaffected but the change-in-ratio estimator is biased low (see Seber 1982, page 361). The fact that the interval of time between the start of the first survey and the end of the second was short (40 days) suggests that any change in catchability (as well as failure of the assumption of closure) should be minimal. Studies have shown that if snow crab are returned to the water quickly the probability of survival can be very high and these studies have been shown to crab fishermen who now appear to recognize that discarded small crab are their future income. Therefore, it is suspected that discard mortality is relatively small.

Estimates of population size from large-mesh trap data are smaller than the corresponding estimate from small-mesh trap data for both the change-in-ratio and index-removal methods. However, the reverse was observed in the previous year's study (Dawe et al. 1993) so this does not appear to be significant.

It appears that change-in-ratio and index-removal estimators may be quite useful for assessing invertebrate stocks and estimating a variety of parameters. However, up until now it appeared to be difficult to obtain precise estimates. The current work suggests that through careful study design it may be possible to achieve acceptable precision.

Acknowlegments

We thank Paul Beck and Joe Drew for help with all aspects of the field work. Robert Kennedy and Mark Udevitz provided helpful comments.

References

Chapman, D.G., and Murphy, G.I. 1965. Estimates of mortality and population from survey-removal records. Biometrics, **21**: 921–935.

Chen, C.L. 1995. Change-in-ratio methods. Ph.D. thesis. Department of Statistics, North Carolina State University, Raleigh, N.C.

Cochran, W.G. 1977. Sampling techniques. Third edition. John Wiley & Sons, Inc., New York, N.Y.

Dawe, E.G., Hoenig, J.M., and Xu, X. 1993. Change-in-ratio and index-removal methods for population assessment and their application to snow crab (*Chionoecetes opilio*). Can. J. Fish. Aquat. Sci. **50**: 1467–1476.

DeLury, D.B. 1947. On the estimation of biological populations. Biometrics, **3**: 145–167.

Eberhardt, L.L. 1982. Calibrating an index using removal data. J. Wildl. Manage. **46**: 734–740.

Frusher, S.D., Kennedy, R.B., and Gibson, I.D. 1998. Preliminary estimates of exploitation rates in the Tasmanian rock lobster (*Jasus edwardsii*) fishery using the change-in-ratio and index-removal techniques with tag–recapture data. *In* Proceedings of the North Pacific Symposium on Invertebrate Stock Assessment and Management. *Edited by* G.S. Jamieson and A. Campbell. Can. Spec. Publ. Fish. Aquat. Sci. 125. pp. 63–71.

Heimbuch, D.G., and Hoenig, J.M. 1989. Change-in-ratio estimators for habitat usage and relative population size. Biometrics, **45**: 439–451.

Hoenig, J.M., and Pollock, K.H. 1998. Index-removal methods. *In* Encylopedia of statistical sciences. Update volume. *Edited by* S. Kotz, C.B. Read, and D.L. Banks. John Wiley & Sons, Inc., New York, N.Y. pp. 341–346.

Kelker, G.H. 1940. Estimating deer populations by a differential hunting loss in the sexes. Proc. Utah Acad. Sci. Arts Lett. **17**: 6–69.

Leslie, P.H., and Davis, D.H.S. 1939. An attempt to determine the absolute number of rats on a given area. J. Anim. Ecol. **8**: 94–113.

Petrides, G.A. 1949. View points on the analysis of open season sex

and age ratios. Trans. N. Am. Wildl. Nat. Resour. Conf. **14**: 391–410.
Pollock, K.H., and Hoenig, J.M. 1998. Change in ratio estimators. *In* Encylopedia of statistical sciences. Update volume. *Edited by* S. Kotz, C.B. Read, and D.L. Banks. John Wiley & Sons, Inc., New York, N.Y. pp. 109–112.
Pollock, K.H., Lancia, R.A., Conner, M.C., and Wood, B.L. 1985. A new change-in-ratio procedure robust to unequal catchability of types of animal. Biometrics, **41**: 653–662.
Ricker, W.E. 1975. Computation and interpretation of biological statistics of fish populations. Bull. Fish. Res. Board Can. 191.
Roseberry, J.L., and Woolfe, A. 1991. A comparative evaluation of techniques for analyzing white-tailed deer harvest. Wildl. Monogr. **117**: 1–59.
Routledge, R. 1989. The removal method for estimating natural populations: incorporating auxiliary information. Biometrics, **45**: 111–121.
Seber, G.A.F. 1982. The estimation of animal abundance and related parameters. 2nd edition. Macmillan Publishing Co., New York, N.Y.
Udevitz, M.S., and Pollock, K.H. 1991. Change-in-ratio estimates for populations with more than two subclasses. Biometrics, **47**: 1531–1546.
Udevitz, M.S., and Pollock, K.H. 1995. Using effort information with change-in-ratio data for population estimation. Biometrics, **51**: 471–481.
Zippin, C. 1956. An evaluation of the removal method of estimating animal populations. Biometrics, **12**: 163–169.

Appendix 1. *Variance formulae for estimates of exploitation rate and catchability coefficient*

Here, we present formulae for estimating the variance of estimates of exploitation rate and catchability coefficient obtained from index-removal and change-in-ratio methods. As in the main text, it is assumed that the estimates of removals by type are independent of the pre- and post-season survey results. The exploitation rate on x-type animals can be estimated from index-removal analysis as

$$\hat{u}_{\mathrm{xir}} = \frac{\hat{c}_{x1} - \hat{c}_{x2}}{\hat{c}_{x1}} \ .$$

The variance of this can be estimated as

$$\hat{V}(\hat{u}_{\mathrm{xir}}) = \frac{\hat{c}_{x2}^{\,2}}{\hat{c}_{x1}^{\,4}} \hat{V}(\hat{c}_{x1}) + \frac{1}{\hat{c}_{x1}^{\,2}} \hat{V}(\hat{c}_{x2}) - 2\frac{\hat{c}_{x2}^{\,2}}{\hat{c}_{x1}^{\,3}} \widehat{\mathrm{Cov}}(\hat{c}_{x1}, \hat{c}_{x2}).$$

Exploitation rate can be estimated from change-in-ratio analysis as

$$\hat{u}_{\mathrm{xcir}} = \frac{\hat{f}(\hat{P}_1 - \hat{P}_2)}{\hat{f}\hat{P}_1 - \hat{P}_1\hat{P}_2} \ .$$

The variance can be estimated as

$$\hat{V}(\hat{u}_{\mathrm{xcir}}) = \frac{\hat{f}^2\hat{P}_2^{\,2}}{\hat{P}_1^{\,4}(\hat{f} - \hat{P}_2)^2} \hat{V}(\hat{P}_1) + \frac{(\hat{P}_1 - \hat{f})^2\hat{f}^2}{\hat{P}_1^{\,2}(\hat{f} - \hat{P}_2)^4} \hat{V}(\hat{P}_2)$$

$$+ \frac{(\hat{P}_2 - \hat{P}_1)^2\hat{P}_2^{\,2}}{\hat{P}_1^{\,2}(\hat{f} - \hat{P}_2)^4} \hat{V}(\hat{f}) + 2\frac{\hat{f}^2\hat{P}_2(\hat{P}_1 - \hat{f})}{\hat{P}_1^{\,3}(\hat{f} - \hat{P}_2)^3} \widehat{\mathrm{Cov}}(\hat{P}_1, \hat{P}_2) \ .$$

The catchability coefficient for x-type animals can be estimated from index-removal analysis as

$$\hat{q}_{\mathrm{ir}} = \frac{\hat{c}_{x1}}{\hat{X}_{\mathrm{ir}}} = \frac{\hat{c}_{x1} - \hat{c}_{x2}}{\hat{R}_x}$$

and the variance can be estimated as

$$\hat{V}(\hat{q}_{\mathrm{ir}}) = \frac{1}{\hat{R}_x^{\,2}} (\hat{V}(\hat{c}_{x1}) + \hat{V}(\hat{c}_{x2})) \ .$$

$$+ \frac{(\hat{c}_{x1} - \hat{c}_{x2})^2}{\hat{R}_x^{\,4}} \hat{V}(\hat{R}_x) - \frac{2}{\hat{R}_x^{\,2}} \widehat{\mathrm{Cov}}(\hat{c}_{x1}, \hat{c}_{x2}) \ .$$

Finally, the catchability coefficient for x-type animals is obtained from change-in-ratio analysis as

$$\hat{q}_{\mathrm{cir}} = \frac{\hat{c}_{x1}}{\hat{X}_{\mathrm{cir}}} = \frac{\hat{c}_{x1}(\hat{P}_1 - \hat{P}_2)}{\hat{P}_1 \hat{R}(\hat{f} - \hat{P}_2)}$$

with variance estimated as

$$\hat{V}(\hat{q}_{\mathrm{cir}}) = \frac{\hat{q}_{\mathrm{cir}}^{\,2} \hat{V}(\hat{c}_{x1})}{\hat{c}_{x1}^{\,2}} + \frac{\hat{q}_{\mathrm{cir}}^{\,2} \hat{P}_2^{\,2} \hat{V}(\hat{P}_1)}{\hat{P}_1^{\,2}(\hat{P}_1 - \hat{P}_2)^2} + \frac{\hat{c}_{x1}^{\,2}(\hat{P}_1 - \hat{f})^2 \hat{V}(\hat{P}_2)}{\hat{P}_1^{\,2} \hat{R}^2 (\hat{f} - \hat{P}_2)^2}$$

$$+ \frac{\hat{q}_{\mathrm{cir}}^{\,2} \hat{V}(\hat{R})}{\hat{R}^2} + \frac{\hat{q}_{\mathrm{cir}}^{\,2} \hat{V}(\hat{f})}{(\hat{f} - \hat{P}_2)^2} + \frac{2\hat{c}_{x1}^{\,2} \hat{P}_2(\hat{P}_1 - \hat{f}) \widehat{\mathrm{Cov}}(\hat{P}_1, \hat{P}_2)}{\hat{P}_1^{\,3} \hat{R}^2 (\hat{f} - \hat{P}_2)^3}$$

$$+ \frac{2\hat{q}_{\mathrm{cir}} \hat{P}_2 \widehat{\mathrm{Cov}}(\hat{c}_{x1}, \hat{P}_1)}{\hat{P}_1^{\,2} \hat{R}(\hat{f} - \hat{P}_2)} + \frac{2\hat{q}_{\mathrm{cir}} (\hat{P}_1 - \hat{f}) \widehat{\mathrm{Cov}}(\hat{c}_{x1}, \hat{P}_2)}{\hat{P}_1 \hat{R}(\hat{f} - \hat{P}_2)^2}$$

$$+ \frac{2\hat{q}_{\mathrm{cir}}^{\,2} \widehat{\mathrm{Cov}}(\hat{f}, \hat{R})}{\hat{R}(\hat{f} - \hat{P}_2)} \ .$$

Appendix 2. *Biases of estimators of population parameters*

By the Taylor's series approach, the asymptotic bias of a function of estimators, $\boldsymbol{Y}$, is

$$\mathrm{Bias}\,(g(\boldsymbol{Y})) = \sum_{i=1}^{n} \frac{1}{2} \frac{\partial^2 g}{\partial y_i^2} V(y_i) + \sum_{i<j}^{n} \frac{\partial^2 g}{\partial y_i \partial y_j} \mathrm{Cov}(y_i, y_j).$$

This relationship can be used to derive the following expressions.

$$\mathrm{Bias}\,(\hat{X}_{\mathrm{cir}}) = \frac{N_1 P_2}{(P_1 - P_2)^2} V(\hat{P}_1) + \frac{N_2 P_1}{(P_1 - P_2)^2} V(\hat{P}_2)$$

$$- \frac{N_1 P_1 + N_2 P_2}{(P_1 - P_2)^2} \mathrm{Cov}(\hat{P}_1, \hat{P}_2) + \frac{P_1}{P_1 - P_2} \mathrm{Cov}(\hat{R}, \hat{f})$$

$$\mathrm{Bias}\,(\hat{X}_{\mathrm{ir}}) = \frac{X_2}{(c_{x1} - c_{x2})^2} V(\hat{c}_{x1}) + \frac{X_1}{(c_{x1} - c_{x2})^2} V(\hat{c}_{x2})$$

$$- \frac{X_1 + X_2}{(c_{x1} - c_{x2})^2} \mathrm{Cov}(\hat{c}_{x1}, \hat{c}_{x2})$$

$$\text{Bias}(\hat{N}_{\text{cir}}) = \frac{N_1}{(P_1 - P_2)^2} V(\hat{P}_1) + \frac{N_2}{(P_1 - P_2)^2} V(\hat{P}_2) - \frac{N_1 + N_2}{(P_1 - P_2)^2} \text{Cov}(\hat{P}_1, \hat{P}_2) + \frac{1}{P_1 - P_2} \text{Cov}(\hat{R}, \hat{f})$$

$$\text{Bias}(\hat{N}_{\text{ir}}) = \frac{X_2}{P_1(c_{x1} - c_{x2})^2} V(\hat{c}_{x1}) + \frac{X_1}{P_1(c_{x1} - c_{x2})^2} V(\hat{c}_{x2}) + \frac{X_1}{P_1^3} V(\hat{P}_1) - \frac{X_1 + X_2}{P_1(c_{x1} - c_{x2})^2} \text{Cov}(\hat{c}_{x1}, \hat{c}_{x2}) + \frac{X_2}{P_1^2(c_{x1} - c_{x2})} \text{Cov}(\hat{P}_1, \hat{c}_{x1}) - \frac{X_1}{P_1^2(c_{x1} - c_{x2})} \text{Cov}(\hat{P}_1, \hat{c}_{x2})$$

where here c_{xi} is the expected value of the catch rate of x-type animals in survey i. Estimates of the biases can be obtained by substituting estimates for the parameters in the above expressions.

Preliminary estimates of exploitation rates in the Tasmanian rock lobster (*Jasus edwardsii*) fishery using the change-in-ratio and index-removal techniques with tag–recapture data

S.D. Frusher, R.B. Kennedy, and I.D. Gibson

Abstract: Estimation of exploitation rates in the Tasmanian rock lobster (*Jasus edwardsii*) fishery using traditional methods has been of limited success. Lesser-known techniques called the change-in-ratio (CIR) technique based on comparisons of relative abundance of two discrete sections of a population over time, and the index-removal (IR) technique comparing the decline in catch rates over time, appeared appropriate for this fishery and were evaluated. This paper presents our preliminary results for the east and south coast regions of the fishery using these techniques. At sites on the east coast male lobsters molt after the 'preseason' sampling and thus invalidate the assumption of closed discrete units within the population. An adjustment to both techniques using tag–recapture data is presented to address this issue. The precision of derived estimates of exploitation rates, and the appropriate sample sizes required to detect significant differences in the levels of exploitation between seasons, were examined using a Monte Carlo bootstrap procedure.

Résumé : L'estimation des taux d'exploitation de la pêcherie de langouste de Tasmanie (*Jasus edwardsii*) à l'aide des méthodes traditionnelles n'a connu qu'un succès limité. Deux techniques moins connues, une appelée technique de changement dans le rapport (*change-in-ratio technique*), qui est fondée sur des comparaisons de l'abondance relative de deux segments discrets d'une population dans le temps, et l'autre appelée technique de l'indice de retrait (*index-removal technique*), qui compare la diminution du taux de prises dans le temps, semblent appropriées pour cette pêcherie et ont été évaluées. Dans la présente communication, nous présentons les résultats préliminaires obtenus avec ces techniques pour les régions des côtes est et sud de la pêcherie. Dans des sites de la côte est, les langoustes mâles muent après l'échantillonnage « pré-saison » et, ainsi, invalident l'hypothèse d'unités discrètes fermées dans la population. Un ajustement aux deux techniques à l'aide des données de marquage-recapture est présenté pour résoudre cette difficulté. La fidélité des estimations dérivées des taux d'exploitation et les tailles d'échantillon appropriées pour déceler les différences statistiquement significatives dans les niveaux d'exploitation entre les saisons, ont été examinés à l'aide de la méthode d'auto-amorçage de Monte Carlo.
[Traduit par la Rédaction]

Introduction

The Tasmanian rock lobster (*Jasus edwardsii*) fishery is over-exploited and recent management decisions have tried to both cap effort and lower exploitation. To assess the impact of such management decisions it is considered important that significant declines in exploitation rates could be measured. To address this requirement, an evaluation of techniques for obtaining accurate estimates of exploitation rate is an important component of a newly established long-term monitoring project for the Tasmanian rock lobster fishery.

S. D. Frusher, R.B. Kennedy, and I. D. Gibson. Department of Primary Industry and Fisheries, Marine Research Laboratories, Taroona 7053, Tasmania, Australia.

Correct citation: Frusher, S.D., Kennedy, R.B., and Gibson, I.D. 1998. Preliminary estimates of exploitation rates in the Tasmanian rock lobster (*Jasus edwardsii*) fishery using the change-in-ratio and index-removal techniques with tag–recapture data. *In* Proceedings of the North Pacific Symposium on Invertebrate Stock Assessment and Management. *Edited by* G.S. Jamieson and A. Campbell. Can. Spec. Publ. Fish. Aquat. Sci. 125. pp. 63–71.

Trials using catch curve and depletion analyses were of limited success in determining exploitation rate due to significant variation in fishers' catch rates throughout the year, inability to age lobsters, and lack of discrete age-classes due to high level of exploitation. Exploitation rate estimation using change-in-ratio (CIR; Paulik and Robson 1969) and index-removal (IR; Eberhart 1982) techniques appeared suitable for this fishery and were tested.

The CIR technique looks at the change in ratio of the relative abundance of two discrete components of a population between two sampling times. It assumes (*i*) that all factors, such as emigration, natural mortality, etc., that affect the population, affect the two components of the population under study equally; (*ii*) the population can be divided into two distinct and non-overlapping components; and (*iii*) that the probability of capturing animals does not vary by component for each sampling period. However, the probability of capturing animals can change from one sampling period to the next.

The IR technique looks at the change in catch rate between sampling periods and attributes this change to removal from the fishery (i.e., harvested lobsters). Hoenig and Pollock (1995) outline the assumptions of this method as follows:

Fig. 1. Sampling strategy used to determine exploitation rates with hypothetical ratios of undersized (us_1) and legal-sized (s_1) lobsters to illustrate the change-in-ratio technique.

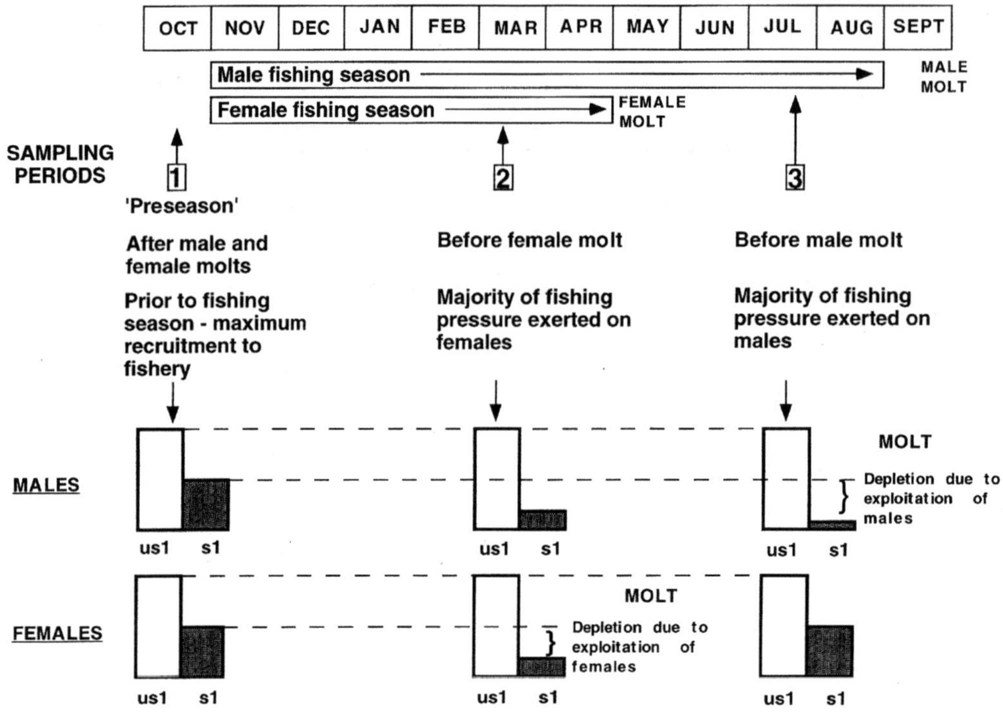

(*i*) the population is closed except for removals; (*ii*) all animals have the same probability of capture in surveys and this probability does not vary from survey to survey; and (*iii*) sampling is with replacement or the fraction of the population taken in the survey is negligible.

The IR technique was initially thought to be of little use for estimating exploitation rate as catch rate declines over the course of the fishery were improbably low. This was considered to be due to the large area in which fishers record their catches in logbooks as catch rates could have been maintained by serial stock depletion within the reporting areas. The current study overcomes this by using identical pots fished in the same region in both pre- and post-fishing surveys. However, it was considered an advantage of the CIR technique over the IR technique that the probability of capturing animals could change from one sampling period to the next.

Both the CIR technique (Kelker 1940; Pollock 1991) and the IR technique (Petrides 1949; Eberhardt 1982) have been predominantly used in wildlife studies. Although Chapman and Murphy (1965) applied the CIR method to a fisheries scenario in 1965, it has not been until recently that these methods have been used in a fisheries situation. Dawe et al. (1993), Chen et al. (1998), and Chen (1995) apply these methods to the Canadian snow crab fishery in Newfoundland. The CIR and IR methods have been recently reviewed by Pollock and Hoenig (1995) and Hoenig and Pollock (1995), respectively.

The rock lobster fishery has a mandatory size limit for each sex and thus the catch can be split into 'legal size and above' (legal-sized lobsters) and 'below legal size' (undersized lobsters) for each sex. The ratio of legal-sized and undersized components of each sex recorded during research surveys are used for estimation of exploitation rates using the CIR technique. Catch rates (number per pot) of the legal-sized component of each sex are used in the IR technique.

In Tasmania, lobsters are trapped in baited pots and the fishing season commences in November of each year and closes in the following May for females and September for males. Males molt around October in southern waters and slightly later in northern waters. Females molt in May–June. Thus most molting occurs outside the fishing season and there is little recruitment to the fishery during the fishing season. Natural mortality estimates are considered to be low (<0.15; R.B. Kennedy, personal communication) and are considered to affect both components of each sex equally. Both the decrease in the ratio of legal-sized to undersized lobsters and the decline in catch rates between the start and end of the fishing season are considered to be due to exploitation.

This paper outlines initial trials in using the CIR and IR techniques to determine exploitation rates of the rock lobster fishery in southern and eastern Tasmanian waters. We present an adjustment to both techniques incorporating tag–recapture data which overcomes problems encountered when molting occurs after the initial preexploitation sampling. The precision of exploitation rate estimates, as a function of sample size, were examined using a Monte Carlo bootstrap procedure.

Materials and methods

Sampling design

The first sample (preseason) was taken in late October of each year, which is after the male and female molts and before the start of the fishing season (Fig. 1). The second sample was taken each March prior to the female molt to determine female exploitation rates and partial male exploitation rates. A final

Table 1. Size limits (mm carapace length) used in determining exploitation rates with the CIR technique.

Region	Sex	us_2	us_1	s_1
East coast	M	93.5–101.4	101.5–109.9	110.0–118.5
East coast	F	98.5–101.4	101.5–104.9	105.0–108.5
South coast	M	102.5–106.4	106.5–109.9	110.0–113.5

Note: us_1 and s_1 size ranges represent one molt increment below and above the legal size limits of 110 and 105 mm CL for males and females, respectively. The us_2 size range corresponds to one molt increment immediately below the us_1 size range.

sample was taken in July–August prior to the male molt to determine the male exploitation rate.

Lobsters were caught in trapezoid-shaped pots with a square base of 0.6 m^2, a top of 0.5 m^2, and a height of 0.4 m. These measurements are similar to those used in the commercial fishery. For the CIR technique, lobsters were divided into three categories to minimize any effects due to size-related variation in catchability. The first undersized category (us_1) included lobsters in the size-class immediately below the legal size limit which corresponds to growth obtained from one molt increment. The second undersize category (us_2) is the size range immediately below the us_1 size-class which corresponds to the growth obtained from molting at this size. For example, the average molt increment for male lobsters just below the legal size limit of 110 mm carapace length (CL) is 8.5 mm CL. The us_1 size-class is therefore 8.5 mm below 110 mm and includes lobsters ranging in size from 101.5 mm CL to 109.9 mm CL. Legal-sized lobsters (s_1) are in the size-class immediately above the legal size limit which corresponds to growth obtained from one molt increment (Table 1). Molt increments were determined from tag–recapture data (see tagging). All legal-sized lobsters caught, including lobsters greater than the s_1 size-class, were used in the IR technique.

Two sites, replicates of each other, were chosen on both the east and south coasts (Fig. 2). Size of the sites varied from 0.2 to 0.5 km^2 depending on available lobster habitat. Replicate sites were <1 km apart. The number of pots used on each coast varied with the number of lobster caught. On the south coast 25 pots were used in each site with an average of 25–30 lobsters caught per pot. In contrast, 50–100 pots were used on the east coast with approximately 8–15 lobsters caught per pot.

CIR technique

The proportion (p) of legal-sized lobsters caught at each sampling period within each site is given by:

$$[1] \quad p = \frac{N_{s_1}}{N_{s_1} + N_{us_1}}$$

where N is the number of lobsters in the appropriate size category.

The estimated proportion of legal-sized lobsters at the beginning of the season (October in Fig. 1), midseason (March), and at the end of the season (July–August) were called $\hat{p}_1$, $\hat{p}_2$, and $\hat{p}_3$, respectively. Unfortunately, inclement weather in July–August prevented sampling at several sites and this analysis has therefore been restricted to the October ($\hat{p}_1$) to March ($\hat{p}_2$) period. Fishing pressure is highest during the November–March period with commercial catch returns

Fig. 2. Map of Tasmania showing the regions sampled in the study.

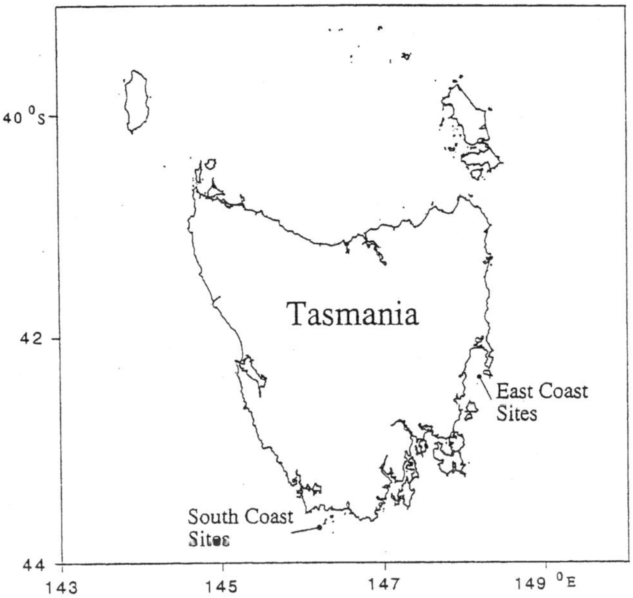

showing 86.3 and 88.5% of the total catch being taken during this period in the 1992–1993 and 1993–1994 seasons, respectively. This trend in fishing effort is reflected in all regions of the fishery. Thus, November–March exploitation rate estimates should reflect the bulk of the catch for each season.

Exploitation rates ($\hat{u}$) were determined using Paulik and Robson's equation

$$[2] \quad \hat{u} = \frac{\hat{p}_1 - \hat{p}_2}{\hat{p}_1\,(1 - \hat{p}_2)}$$

An adjustment to the CIR technique was used to compensate for molting which was found to occur for males after the 'preseason' sampling on the east coast. To undertake the adjustment, a third category, N_{us_2}, was required to account for lobsters molting into the N_{us_1} category. The us_2 size category is described above and size-classes are given in Table 1.

Proportions for use in determining exploitation rates with the modified technique were obtained by:

$$[3] \quad \hat{p}_{1\_\mathrm{adj}} = \frac{\hat{N}_{s\_\mathrm{adj}}}{\hat{N}_{s\_\mathrm{adj}} + \hat{N}_{us\_\mathrm{adj}}}$$

where

$$\hat{N}_{s\_\mathrm{adj}} = \hat{N}_{s_1} - \hat{g}_{s_1}\hat{N}_{s_1} + \hat{g}_{us_1}\hat{N}_{us_1}$$

$$\hat{N}_{us\_\mathrm{adj}} = \hat{N}_{us_1} - \hat{g}_{us_1}\hat{N}_{us_1} + \hat{g}_{us_2}\hat{N}_{us_2}$$

and $\hat{g}$ is the proportion molting after the sample was collected (obtained from tag–recapture data). The adjusted exploitation

Fig. 3. Percentage of east coast male lobsters in each of the size categories used in the adjusted CIR technique which molted after the preseason (October) sampling for the 1992–1993 (stippled) and 1993–1994 (hatched) seasons.

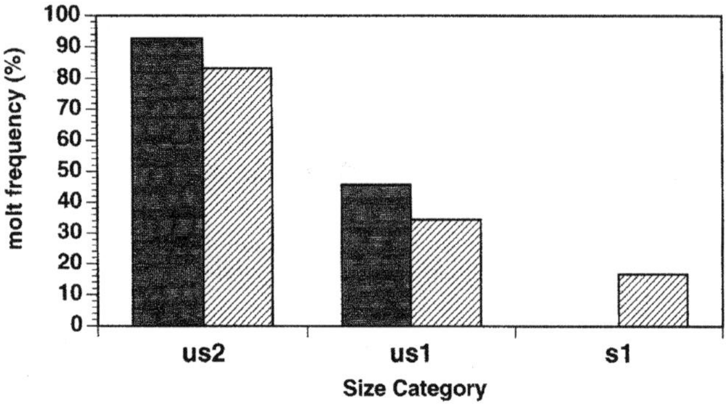

rate estimate is then obtained by substituting $\hat{p}_{1\_adj}$ for $\hat{p}_1$ in eq. 2.

IR technique

Catch rate (c) of legal-sized lobsters caught at each sampling period within each site is given by:

$$[4] \quad c = \frac{N_s}{P}$$

where N_s is the total number of legal-sized lobsters and P is the number of pots used to obtain the sample.

Exploitation rates ($\hat{u}$) were determined using Eberhardt's equation

$$[5] \quad \hat{u} = \frac{\hat{c}_1 - \hat{c}_2}{\hat{c}_1}$$

where $\hat{c}_1$ and $\hat{c}_2$ refer to catch rates obtained in the October and March sampling periods, respectively.

To adjust for molting the $\hat{c}_1$ value was adjusted accordingly

$$[6] \quad \hat{c}_{1\_adj} = \frac{(\hat{N}_s + \hat{g}_{us_1}\hat{N}_{us_1})}{P}$$

The adjusted exploitation rate estimate is then obtained by substituting $\hat{c}_{1\_adj}$ for $\hat{c}_1$ in eq. 5.

Tagging

Lobsters were tagged by inserting a tag (Hallprint T-bar anchor tag; TBA1, Hallprint Pty. Ltd., 27 Jacobsen Crescent, Holden Hill, South Australia 5088, Australia) ventrally in the first abdominal segment. Recaptured lobsters which had molted were identified by partial regeneration of a pleopod which was clipped at tagging. Molt increments used for determining size categories were determined from the average change in size of tagged lobsters which had molted between sampling periods. Tag–recapture data was also used to determine the proportion that had molted in each size category.

Table 2. Comparison of adjusted and unadjusted exploitation rates for the 1992–1993 and 1993–1994 seasons for both sites on the east coast using the CIR and IR techniques.

Site	Season	Unadjusted	Adjusted	% Change
		CIR technique		
1	1992–1993	0.82	0.87	6
	1993–1994	0.75	0.80	7
2	1992–1993	0.72	0.82	14
	1993–1994	0.61	0.70	15
1 and 2	1992–1993	0.76	0.84	11
	1993–1994	0.67	0.74	9
		IR technique		
1	1992–1993	0.58	0.73	26
	1993–1994	0.78	0.84	8
2	1992–1993	0.63	0.80	27
	1993–1994	0.75	0.83	11
1 and 2	1992–1993	0.61	0.77	26
	1993–1994	0.76	0.83	9

Bootstrapping

To examine both the CIR and IR techniques in relation to sample size, and to determine the precision of exploitation rate estimates, a Monte Carlo bootstrap procedure was used. For each actual exploitation rate estimate, 10 000 bootstrap simulations were conducted. Number of pots used in each simulation was the same as that of the original sample for determining the precision of existing estimates and was set at a variety of larger values for estimating effects of sample size. For each simulation, the appropriate number of pots (with associated data) was randomly selected (with replacement) from each sample period. Uncertainty of the growth increment was incorporated in the analysis by randomly selecting an increment from a normal distribution of molt increments (with mean and standard deviations being obtained from tag–recapture data) for each of the 10 000 simulations. No matching of pots between samples (Chen et al. 1998) was conducted due to the random placement of pots within each site during each trip. This provided simulated $\hat{p}_1$ and $\hat{p}_2$ values (CIR technique) and $\hat{c}_1$ and $\hat{c}_2$ values (IR technique) which were then combined to obtain a distribution of exploitation rates. Confidence limits of 95% were obtained using the percentile method described by Efron and Tibshirani (1986). Confidence limits take into account the asymmetric nature of derived exploitation rates. Differences in exploitation rate estimates were considered significant when there was no overlap of 95% confidence limits. The distribution of exploitation rates can contain negative values where there is an overlap between distributions of $\hat{p}_1$ and $\hat{p}_2$ or $\hat{c}_1$ and $\hat{c}_2$ values. This occurred in some of our results where the distribution of simulated exploitation rates was broad and is accounted for in the lower 95% confidence limit.

Results

Molting

Analysis of tag–recapture data indicated that some molting of male lobsters occurred after the 'preseason molt' on the east coast. Further analysis showed that the proportion molting

Table 3. Exploitation rates for the 1992–1993 and 1993–1994 rock lobster fishing seasons on the south and east coasts, using the CIR and IR techniques.

		CIR	IR	CIR	IR
Sex	Site	1992–1993		1993–1994	
		South coast			
Male	1	0.88	0.85	0.60	0.58
		(0.70–0.99)	(0.63–0.95)	(0.17–0.84)	(0.15–0.84)
	2	0.74	0.90	0.41	0.56
		(0.38–0.93)	(0.80–0.96)	(0.05–0.80)	(0.15–0.80)
	1 and 2	0.82	0.87[a]	0.53	0.58[a]
		(0.65–0.93)	(0.78–0.94)	(0.15–0.76)	(0.27–0.77)
		East coast			
Male	1	0.87	0.73	0.80	0.83
		(0.67–0.95)	(0.41–0.91)	(0.49–0.93)	(0.65–0.93)
	2	0.82	0.80	0.70	0.83
		(0.56–0.93)	(0.62–0.91)	(0.36–0.87)	(0.69–0.91)
	1 and 2	0.84	0.77	0.74	0.83
		(0.65–0.93)	(0.61–0.87)	(0.50–0.87)	(0.73–0.89)
Female	1	0.86	0.78	0.62	0.72
		(0.70–0.96)	(0.52–0.94)	(0.31–0.82)	(0.47–0.87)
	2	0.43	0.80	0.63	0.78
		(0.08–0.75)	(0.63–0.90)	(0.28–0.83)	(0.61–0.89)
	1 and 2	0.71	0.79	0.62	0.75
		(0.50–0.84)	(0.66–0.88)	(0.39–0.77)	(0.62–0.85)

Note: 95% confidence limits are presented in brackets below each estimate.
[a]Significant differences in exploitation rates found between years.

varied according to size (Fig. 3) such that smaller males molted later than larger males. The low proportion of legal-sized males molting after the survey is also affected by level of exploitation. As exploitation rate increases, the fishery becomes principally based on recruits and there will be fewer lobsters in the legal-sized category at the end of each season. As molting occurs after the end of the season (after the majority of s_1 lobsters have been caught) and has not been completed prior to the preseason sampling, the s_1 category from the preseason sample will consist mainly of lobsters which have recently molted.

As only the legal-sized component (N_s) of the fishery is measured, molting from undersized to legal size which may occur after the initial sample will always increase the adjusted exploitation rate determined by the IR method. The CIR method adjusts both undersized and legal-sized components and also allows for legal-sized lobsters molting out of the s_1 category. Adjusted exploitation rates using the CIR technique can therefore be either larger or smaller.

All adjusted exploitation rates using both methods, however, showed an increase (Table 2). Molting had the largest effect on exploitation rates determined by the IR method, with increases of up to 27%. In contrast, the largest increase using the CIR method was 15%. As molting has an effect on exploitation rates, all exploitation rates presented in the remainder of this document are adjusted exploitation rates.

Exploitation rates

On both the south and east coasts, no significant difference was found in exploitation rates between techniques, between sites, sexes (east coast only), or seasons (Table 3). No estimates are presented for south coast females as female lobsters do not grow to legal size in that region.

Although not significant, there was a consistent decline in exploitation rates at both south coast sites using both techniques from the 1992–1993 to the 1993–1994 fishing season. Only in females using the CIR technique was there a decline between seasons on the east coast, and this was considerably less than for the south coast.

The IR method produced more consistent results between sites for both regions and when sites were combined, gave a significant decline in exploitation rate between seasons for the south coast (Table 3).

The low exploitation rate estimate for site 2 females in the 1992–1993 season using the CIR technique is odd given the consistency between site 1 using the CIR technique and sites 1 and 2 using the IR technique. The reason for this is unknown; however, it may be associated with the precision of our estimates and this is addressed in the bootstrap analysis.

Bootstrapping

Results of simulated proportions of legal-sized lobsters obtained for the October 1992 and March 1993 sampling surveys at site 1 on the south coast (Figs. 4*a* and 4*b*) demonstrate that considerable variation exists in proportions that could have been obtained. In October 1992, the p_1 value could have realistically ranged from 0.4 to 0.6, and the March 1993, p_2 value from 0.0 to 0.2. Despite this, it is clear there was a decrease in

Fig. 4. Bootstrapped simulations of the proportion of male legal-sized lobsters (s_1 only) caught at site 1 on the south coast in (*a*) October 1992 and (*b*) March 1993, and (*c*) the resultant exploitation rate. p_1, p_2, and u are the actual proportions and exploitation rate, respectively, derived from the CIR formulae.

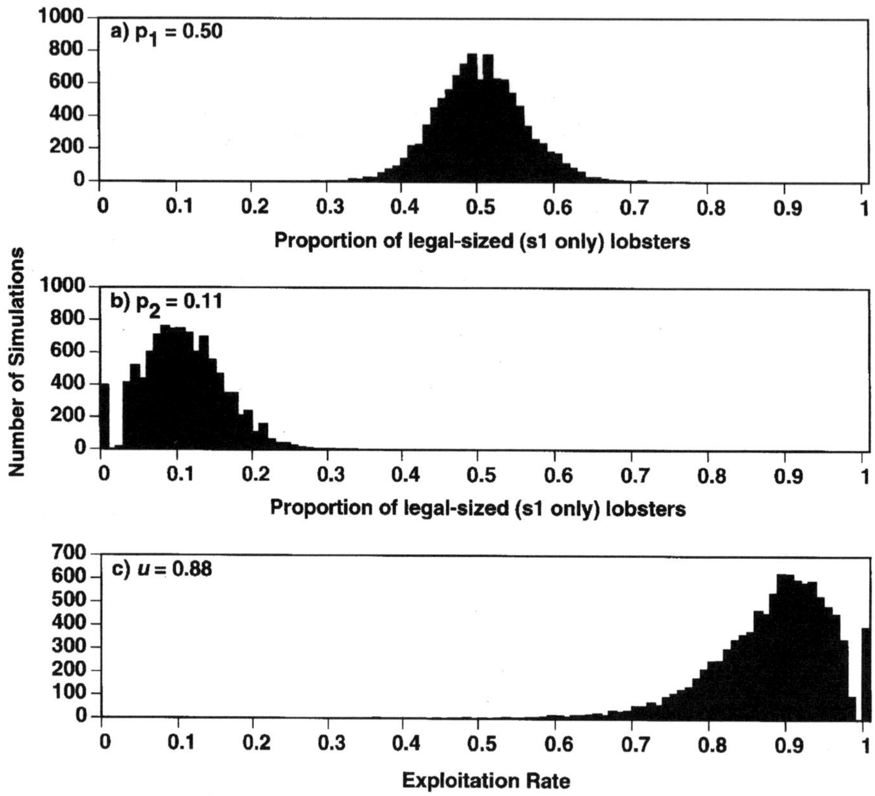

the proportion of legal-sized lobsters between the two sampling periods. These proportions were combined using the CIR technique to produce a range of exploitation rates (Fig. 4*c*). As expected, variation in exploitation rate is greater than in estimates of individual proportions with probable exploitation rates varying from 0.7 to full exploitation. Similarly, catch rates in October 1992, could have realistically varied from 1 to 2.7 and from 0 to 0.6 in March 1993 (Figs. 5*a* and 5*b*). Again, a decrease occurred between samples and exploitation rate could have varied from 0.65 to 0.98 (Fig. 5*c*).

Simulated exploitation rates, using the actual number of pots sampled, show a decline in exploitation from 1992–1993 to 1993–1994 for site 1 using both techniques on the south coast (Fig. 6*a*). However, the degree of overlap clearly indicates why no significant difference between seasons was found at this site.

To indicate the sample size required for detecting a significant change in exploitation rates, simulations were undertaken using 100 pots. Variation in exploitation rate estimates and the amount of overlap between seasons substantially decreased at both sites (Fig. 6*b*) and the difference would likely have been significant if the study had been based on 100 pots.

Discussion

Exploitation rates

As suggested by both the distribution of the fishing fleet and the intensity of effort, exploitation rates were high. The lower exploitation rates of males at all sites for the 1993–1994 season compared to the previous season can be explained by changes in the fishing season and fishing pattern between years. The opening of the 1993–1994 season in November was delayed by 17 days. November is the peak fishing month and accounted for over 28% of the catch in the 1992–1993 season. By contrast, in the 1993–1994 season with the 17-day closure, less than 20% of the total catch was taken in November. This delayed opening was followed by periods of strong westerly winds which further decreased fishing effort at the more exposed sites in the south. Fishing effort on the south coast declined by over 30% between the 1992–1993 and 1993–1994 fishing seasons.

Although winds had very limited impact on the east coast fishery, there was a 20% decline in fishing effort due to the seasonal closure. This decreased effort appeared to have no impact on the estimated exploitation rates between years and this may be related to a lack of recruits which was observed for this region (S. Frusher, personal communication).

Adjustment of the CIR and IR techniques to compensate for the effect of molting had a larger impact on IR values and thus the CIR technique appears less sensitive to molting than the IR technique. Adjustment did decrease differences between all replicates using the CIR technique. Differences in exploitation rate between replicates using the IR technique were small for both adjusted and unadjusted estimates.

We advocate that it is important to incorporate molting and to include molt frequencies for each size category. It must be stressed that incorporation of the molt will only be satisfactory when it occurs during or shortly after the first sample. Molting

Fig. 5. Bootstrapped simulations of the proportion of all male legal-sized lobsters caught per pot at site 1 on the south coast in (*a*) October 1992 and (*b*) March 1993, and (*c*) the resultant exploitation rate. c_1, c_2, and u are the actual catch rates and exploitation rate, respectively, derived from the IR formulae.

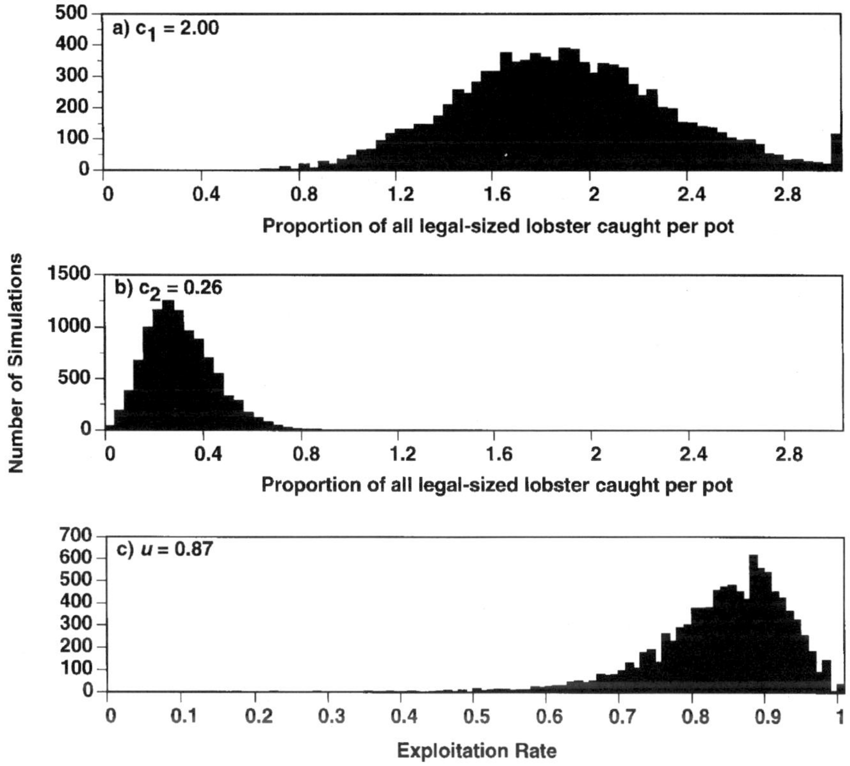

that occurs halfway through or towards the latter part of the between-survey period can not be accounted for by this technique as new legal-sized lobsters will not have had the same probability of being exploited as those lobsters which were legal size at the time of, or shortly after, the first survey period. Sampling strategy needs to account for molting by planning sampling times to occur after or very close to molt periods.

Bootstrapping

Bootstrap results showed considerable variation around the derived exploitation rate estimates using both the CIR and IR techniques. This is not all that surprising given the gregarious nature of lobsters and their nonrandom distribution in reef habitats (Chittleborough 1975).

To be able to accurately assess the impact of management decisions which aim at both limiting effort and reducing exploitation rates, it is necessary that declines in exploitation rates be both measurable and significant. With the broad confidence limits found using the small number of pots in this study (25), it is unlikely that a 25–30% reduction in exploitation rate, similar to what occurred between the 1992–1993 and 1993–1994 seasons, could be confidently detected. In contrast, the use of approximately 100 pots would be expected to enable such a decline to be significantly detected.

Both the CIR and IR techniques gave similar estimates of exploitation rate and both lacked precision with the sample size used. Although the data set is limited, the IR technique gave more similar results between sites, and when both sites were combined on the south coast, demonstrated a significant difference between fishing seasons. The IR technique also appeared more robust than the CIR technique in that it did not estimate an unusually low and inconsistent exploitation rate for females at site 2 on the east coast. Despite these improved estimates using the IR technique, the authors advocate, where possible, the use of both methods to improve the reliability of estimates derived from either of the techniques used individually.

As monitoring of exploitation rates is part of a long-term project for the rock lobster fishery, routine revision of this analysis for ongoing data will be required to evaluate the accuracy of future estimates.

Summary

Our initial trials with CIR and IR techniques have demonstrated them to be a useful means of determining exploitation rates in the Tasmanian rock lobster fishery. Similarities in exploitation rates between replicate sites for males on both the east and south coasts and decreases between years matching observed effort restrictions were encouraging results in evaluating both techniques. However, our results indicate that to obtain precise estimates of exploitation rate, precise estimates of proportions and catch rates are required and the distributions of expected proportions and catch rates from the two sampling periods should have minimal overlap. Thus, for most accurate results, the technique should be restricted to fisheries where exploitation rate is high and size composition is relatively uniform throughout the survey area. However, matching pot locations between samples, as suggested by Chen et al. (1998), may improve the precision of results. By bootstrapping

Fig. 6. Bootstrapped exploitation rates derived from the CIR and IR techniques for 1992–1993 (open bars) and 1993–1994 (closed bars) for male lobsters sampled at south coast site 1 using (*a*) the actual number of pots used and (*b*) 100 pots (i and ii refer to the number of pots sampled/total number of lobsters used in analysis, for 1992–1993 and 1993–1994, respectively).

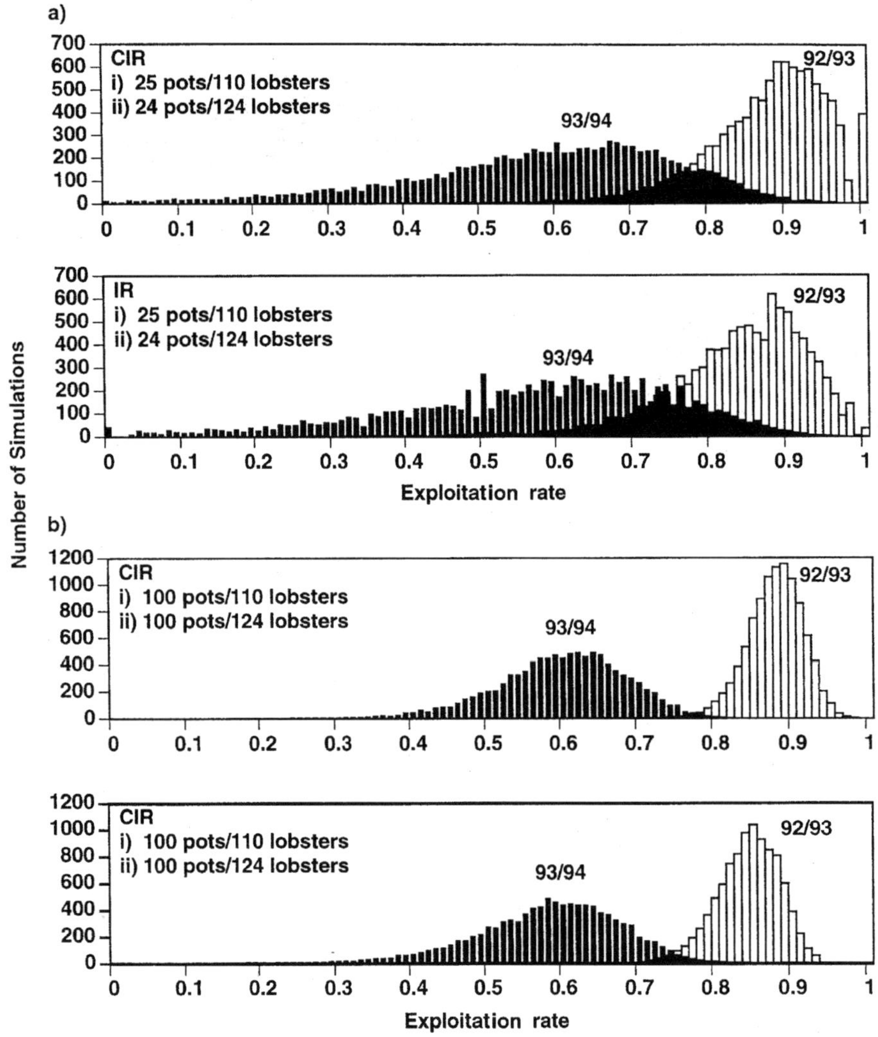

preliminary data, information pertaining to the accuracy of derived exploitation rate estimates and appropriate sample sizes for improved estimates can be achieved.

Acknowledgments

We would like to thank John Hoenig for advising on use of the IR method and constructive comments on the initial review, and Pascale Baelde, Warwick Nash, Christine Crawford, Alan Jordan, and Jeremy Lyle for helpful comments on the manuscript. The crew of the FRV Challenger were invaluable for the encouragement and support they gave while sampling in remote regions and often during inhospitable weather conditions.

References

Chapman, D.G., and Murphy, G.I. 1965. Estimates of mortality and population from survey-removal records. Biometrics, **21**: 921–935.

Chen, C. 1995. New developments in change-in-ratio estimation of population size. Ph.D. dissertation, North Carolina State University, Raleigh, N.C. 140 p.

Chen, C., Hoenig, J.M., Dawe, E.G., Brownie, C., and Pollock, K.H. 1998. New developments in change-in-ratio and index-removal methods, with application to snow crab (*Chinoecetes opilio*). *In* Proceedings of the North Pacific Symposium on Invertebrate Stock Assessment and Management. *Edited by* G.S. Jamieson and A. Campbell. Can. Spec. Publ. Fish. Aquat. Sci. 125. pp. 49–61.

Chittleborough, R.G. 1975. Environmental factors affecting growth and survival of juvenile rock lobster. Aust. J. Mar. Freshwater Res. **26**: 177–196.

Dawe, E.G., Hoenig, J.M., and Xu, X. 1993. Change-in-ratio and index-removal methods for population assessment and their application to snow crab (*Chionoecetes opilio*). Can. J. Fish. Aquat. Sci. **50**: 1467–1476.

Eberhart, L.L. 1982. Calibrating an index using removal data. J. Wildl. Manage. **46**: 734–740.

Efron, B., and Tibshirani, R. 1986. Bootstrap methods for standard errors, confidence intervals, and other measures of statistical accuracy. Stat. Sci. **1**: 54–77.

Hoenig, J.M., and Pollock, K.H. 1995. Index-removal methods. *In* Encylopedia of statistical sciences update volume. *Edited by* S. Kotz, C.B. Read, and D.L. Banks. John Wiley & Sons, Inc., New York, N.Y.

Kelker, G.H. 1940. Estimating deer populations by a differential hunting loss in the sexes. Proc. Utah Acad. Sci. Arts Lett. **17**: 6–69.

Paulik, G.J., and Robson, D.S. 1969. Statistical calculations for change-in-ratio estimators of population parameters. J. Wildl. Manage. **33**: 1–27.

Petrides, G.A. 1949. View points on the analysis of open season sex and age ratios. Trans. N. Am. Wildl. Nat. Resour. Conf. **14**: 391–410.

Pollock, K.H. 1991. Modelling capture, recapture, and removal statistics for estimation of demographic parameters for fish and wildlife populations: past, present and future. J. Am. Stat. Assoc. **86**: 225–238.

Pollock, K.H., and Hoenig, J.M. 1995. Change-in-ratio estimators. *In* Encylopedia of statistical sciences update volume. *Edited by* S. Kotz, C.B. Read, and D.L. Banks. John Wiley & Sons, Inc., New York, N.Y.

Estimating king crab (*Paralithodes camtschaticus*) abundance from commercial catch and research survey data

Jeremy S. Collie and Gordon H. Kruse

Abstract: Reliable estimates of male abundance are required to set harvest guidelines for red king crab (*Paralithodes camtschaticus*). Data available for population estimation include commercial catch and relative abundance estimates from research surveys. Although absolute age of king crabs cannot be determined, shell age and size can be used to classify legal-sized crabs into recruits and post-recruits. We modified a method of catch-survey analysis, originally developed by Collie and Sissenwine, for estimating legal male king crab abundance. The method scales the relative abundance index to the commercial catch data and smoothes the abundance estimates by accounting for errors in measuring relative abundance. We conducted Monte Carlo simulation trials with known data sets to test the precision and accuracy of this estimation method and its sensitivity to violations of the underlying assumptions. Bias in the abundance estimates was less than 5% for realistic levels of measurement and process errors. The simulation framework was used to calculate bootstrap estimates of population abundances. We applied the method to estimate the abundance of Kodiak and Bristol Bay red king crabs. In the early 1980's, dramatic declines in the abundances of both stocks coincided with decreasing recruitment and high harvest rates. The decline in Bristol Bay was compounded by a sharp, temporary increase in natural mortality.

Résumé : Une estimation fiable de l'abondance des mâles est nécessaire pour établir les règles régissant la récolte du crabe royal (*Paralithodes camtschatica*). Les données disponibles pour estimer la population comprennent les valeurs estimées des prises commerciales et de l'abondance relative provenant des relevés de recherche. Bien qu'on ne puisse déterminer l'âge absolu du crabe royal, l'âge et la taille de la carapace peuvent être utilisés pour classer les crabes de taille légale en recrues et post-recrues. Nous avons modifié une méthode d'analyse prises-relevé, élaborée initialement par Collie et Sissenwine, pour estimer l'abondance du crabe royal mâle de taille légale. La méthode établit une correspondance entre l'indice d'abondance relative et les données sur les prises commerciales et ajuste les valeurs estimées de l'abondance en tenant compte des erreurs liées à la mesure de l'abondance relative. Nous avons réalisé des essais de simulation de Monte Carlo avec des ensembles de données connues pour vérifier la fidélité et la justesse de la méthode d'estimation et sa sensibilité aux violations des hypothèses sous-jacentes. L'erreur touchant les valeurs estimées de l'abondance était inférieure à 5 % pour des degrés réalistes d'erreur de mesure et de méthode. Le cadre de simulation a été utilisé pour calculer des valeurs estimées par auto-amorçage des abondances de population. Nous avons appliqué la méthode pour estimer l'abondance du crabe royal de l'île Kodiak et de la baie de Bristol. Au début des années 80, un déclin marqué de l'abondance des deux stocks a coïncidé avec une diminution du recrutement et une augmentation des taux de récolte. Le déclin dans la baie de Bristol a été aggravé par une augmentation brusque et temporaire de la mortalité naturelle. [Traduit par la Rédaction]

Introduction

Most red king crab *(Paralithodes camtschaticus)* fisheries in Alaska are currently managed with an exploitation rate strategy in which annual catches of legal-sized males are prescribed by applying a harvest rate toward estimates of spawning male biomass (Kruse 1993). Abundances of red king crab populations have been estimated by annual stock assessment surveys.

J.S. Collie. Graduate School of Oceanography, University of Rhode Island, Narragansett, RI 02882-1197, U.S.A.

G.H. Kruse. Commercial Fisheries Management and Development Division, Alaska Department of Fish and Game, Juneau, AK 99802, U.S.A.

Correct citation: Collie, J.S., and Kruse, G.H. 1998. Estimating king crab (*Paralithodes camtschaticus*) abundance from commercial catch and research survey data. *In* Proceedings of the North Pacific Symposium on Invertebrate Stock Assessment and Management. *Edited by* G.S. Jamieson and A. Campbell. Can. Spec. Publ. Fish. Aquat. Sci. 125. pp. 73–83.

Since 1968, the National Marine Fisheries Service (NMFS) has used area-swept methods to calculate trawl-survey indices of absolute abundance of Bering Sea stocks (Otto 1986). During the 1970's through late 1980's, relative abundances of Gulf of Alaska stocks were estimated with pot surveys conducted by the Alaska Department of Fish and Game (ADF&G). For a few stocks, absolute abundances of legal males were estimated by Petersen estimators applied to tag-recapture data (Blau 1985). With the exception of stocks in Southeast Alaska, where topography prevents trawling in most king crab habitats, ADF&G switched to trawl area-swept methods in the late 1980's. Occasionally, alternative methods based on inseason changes in fishery catch per unit effort (Leslie and Davis 1939) have been attempted (Otto 1986), but these have not been used routinely for assessments or for fishery management.

More recently, alternative methods incorporating an underlying population model have been applied. These methods are catch-survey analysis (CSA, Kruse and Collie 1991) and length-based analysis (LBA, Zheng et al. 1995). The advantages of the LBA over CSA are that it can be applied to females as well as males and it more fully utilizes data on size and shell condition of crabs in the survey and commercial catches. The population dynamics model underlying CSA is more parsimonious, which makes it less data intensive, and therefore it can be applied to a wider range of sampling situations. We believe it prudent to apply multiple approaches in any one case to develop confidence in the various alternative estimation methods.

In this paper, two red king crab stocks are analyzed: one residing around Kodiak Island in the Gulf of Alaska and one in Bristol Bay in the eastern Bering Sea. These two stocks once supported the two largest king crab fisheries in the world. Small catches of red king crabs were first delivered to Kodiak in 1936, but landings were not recorded until 1950 when 27 t were landed (Spalinger and Jackson 1994). Annual landings peaked at 43 000 t in 1965, stabilized at 5 000 – 11 000 t in the late 1960's until the early 1980's when landings declined sharply. The fishery has been closed since 1983 due to extremely low stock abundance. Red king crabs in the eastern Bering Sea were exploited by Japanese fisheries from 1930 to 1940 and 1953 to 1974. Russian fisheries operated during 1959–1971 (Morrison and Gish 1994). Early U.S. red king crab explorations began in the early 1940's (Anonymous 1942). A small domestic fishery began in 1947, but domestic landings remained low until the 1970's when the foreign fishery was phased out (Morrison and Gish 1994). Total annual landings from Bristol Bay were <6100 t in the 1950's, 11 100 – 29 000 t in the 1960's, and 8 700 – 49 000 t in the 1970's. Landings peaked at 59 000 t in 1980 and declined sharply in the early 1980's. The fishery was closed in 1983 due to low abundance. During 1984–1993 annual landings were 1900–9200 t. The Bristol Bay fishery was again closed in 1994 due to low abundance; a small harvest (600 t) was taken from the Pribilof Islands area of the eastern Bering Sea.

The population estimation method described in this paper is called catch-survey analysis (CSA) because it calibrates relative abundance to absolute abundance and smoothes variability in the relative abundance data. The method is derived from the classic depletion estimators of Leslie and Davis (1939), which have long been used to estimate the abundance of invertebrate populations such as lobster (DeLury 1947). For a closed population, the catchability is estimated from the decline in catch rate and then used to estimate initial abundance. In applying depletion estimators to whale populations, Allen (1966) and others modified the methods for open populations with recruitment and natural mortality. Collie and Sissenwine (1983) developed a two-stage model for stocks that can be classified into recruits and post-recruits and applied it to east coast groundfish stocks for which the full age structure was unknown. Kruse and Collie (1991) adapted CSA to king crabs (for which age cannot be determined); the method has since been used to estimate the abundance of red king crabs in Southeast Alaska (Woodby 1994). With further refinements, CSA has been used to assess the abundance of Atlantic sea scallops, *Placopecten magellanicus* (Conser 1991), and American lobster, *Homarus americanus* (Conser and Idoine 1992).

Collie and Sissenwine (1983) recommended testing the reliability of CSA on simulated data sets for which true abundance is known. Although the method has become more widely used, its performance has not yet been tested with Monte Carlo simulations. In this paper we test the sensitivity of CSA to violations of the input assumptions and to the magnitude and form of residual errors. Finally, the simulation framework is used to calculate bootstrap estimates and confidence intervals.

Methods

Abundance and landings data

Time series of pot surveys (Kodiak) and trawl surveys (eastern Bering Sea) afford contrast in application of our approach. The Kodiak pot survey was described in detail by Peterson et al. (1986), and historical commercial landings were reported by Spalinger and Jackson (1994). We gathered pot survey and landings data for 1973 to 1986. Since 1986, surveys have been conducted by trawl rather than pot gear. We confined our Kodiak analysis to the period spanning the pot-survey data series because pots and trawls have different catchabilities and because the lack of a fishery since 1983 prevented use of commercial catch data to estimate the trawl catchability coefficient. We regarded Kodiak red king crabs as one stock because (*i*) they are managed as a single unit (Spalinger and Jackson 1994), (*ii*) crab habitats around Kodiak Island are distinct from adjacent areas on the mainland, and (*iii*) isozyme analyses indicate genetic differentiation among broad provinces (e.g., Gulf of Alaska, Aleutian Islands, Bering Sea) rather than within smaller geographic areas (Seeb et al. 1990).

The eastern Bering Sea trawl survey was described in detail by Otto (1986), and historical landings were reported by Morrison and Gish (1994). We obtained trawl survey and landings data for 1975 to 1994. Red king crabs from the eastern Bering Sea occupy two semi-discrete areas: Bristol Bay and the Pribilof Islands. The Pribilof Islands have accounted for an extremely small fraction of overall historical red king crab abundance in the eastern Bering Sea; no directed fishery occurred except in 1993 and 1994 when small catches (1200 t and 600 t, respectively) were taken. This being the case, combined abundances of red king crabs from the two areas are routinely reported by NMFS in annual stock assessment documents (e.g., Stevens et al. 1994). To compare our results to published abundance estimates, for purposes of this analysis we treated red king crabs from the eastern Bering Sea as a single stock and hereafter refer to them as Bristol Bay.

Crabs caught during surveys were classified based on commonly used measurements of size and shell condition. "Carapace width" (CW) is the greatest straight-line distance across the width of the carapace including spines, whereas "carapace length" (CL) is the distance from the posterior margin of the right eye orbit to the posterior carapace margin. Legal crabs are males ≥178 mm CW (≈145 mm CL) for Kodiak and ≥165 mm CW (≈135 mm CL) for Bristol Bay (ADF&G 1994). Shell condition (e.g., discoloration, wear, presence of barnacles) allows biologists to classify crabs as "newshell," those that molted within the past year, and "oldshell," those that have molted more than a year previously. Shell size and condition are used to categorize legal crabs as "recruits," those that have been legal for only one year, and "post-recruits," those that have been legal for more than one year. Recruits are newshell crabs between legal size and one mean growth increment larger than legal size. Post-recruits are all remaining legal crabs. For Kodiak, recruits are newshell crabs ≥145 mm CL and <165 mm CL, whereas for Bristol Bay, recruits are newshell crabs ≥135 mm CL and <150 mm CL. The use of shell age and lack of true age makes king crabs ideal candidates for CSA.

Population model

According to the population dynamics model, post-recruits in year $t + 1$ are the sum of recruits (R_t) and post-recruits (P_t) in year t discounted by natural mortality (M_t), minus the commercial catch (C_t) adjusted for natural mortality between the time of the commercial fishery in year t and the survey in year $t + 1$:

$$[1] \quad P_{t+1} = (P_t + R_t)e^{-M_t} - C_t e^{-M_t(1-T_t)}$$

where T_t is the proportion of the year between the survey and the fishery, and annual estimates of M_t and T_t are used where known.

Relative abundance and absolute abundance are related with an observation model. Relative post-recruit (p_t) and recruit (r_t) abundances are assumed proportional to absolute abundance with a survey catchability coefficient, q:

$$[2] \quad p_t = qP_t$$
$$r_t = q\phi R_t$$

where ϕ denotes recruit catchability relative to post-recruits. Substituting eq. 2 into eq. 1 and multiplying through by q, the population model is rewritten in units of relative abundance:

$$[3] \quad p_{t+1} = (p_t + r_t/\phi)e^{-M_t} - qC_t e^{-M_t(1-T_t)}$$

In this application ϕ and variables M_t, T_t, and C_t are inputs; q is estimated along with the sequence of relative abundance (r and p) that best matches the observed relative abundance.

In this study we assumed an all-measurement-error structure. Observed relative abundances of post-recruits ($\tilde{p}_t$) and recruits ($\tilde{r}_t$) are assumed equal to the true but unknown relative abundances times lognormally distributed measurement errors:

$$[4] \quad \tilde{p}_t = p_t e^{\eta_t}$$
$$\tilde{r}_t = r_t e^{\delta_t}$$

The residual errors (η and δ) are minimized by weighted nonlinear least squares:

$$[5] \quad \mathrm{MIN}\left[\sum_{t=1}^{n}(\eta_t)^2 + \sum_{t=1}^{n-1}(w\delta_t)^2\right]$$

where w is the weight of recruit errors relative to post-recruit errors. The set of parameters to estimate includes q, p in the initial year, and r for each year but the last. With n years of data, there are $2n - 1$ residual errors, $n + 1$ parameters to estimate, and $n - 2$ degrees of freedom. We used the modified Levenberg–Marquardt algorithm in subroutine BCLSF of the IMSL Math Library (IMSL 1991) to minimize eq. 5.

Unlike the original Collie–Sissenwine (1983) formulation, we chose not to include process errors along with measurement errors in the parameter estimation. There were three reasons for omitting process errors. Although the mixed-error model has the same degrees of freedom as the measurement-error model, it has more parameters ($2n$), and may become overparameterized (Deriso et al. 1985). In previous trials, we found data sets for which the mixed-error model would not converge to a unique solution. For the two king crab data sets, we found that measurement-error and mixed-error models produced similar estimates of absolute abundance (Kruse and Collie 1991). When fitting with a single error structure, the all measurement-error assumption is generally preferable to all process error (Walters 1986) even though the latter reduces to a simple linear regression.

Model parameters

In this application, recruits and post-recruits were assumed to have equal catchability ($\phi = 1$) because both stages are legal size. To apply natural mortality over an appropriate period, we estimated the time lag, T_t, between the midpoint of the assessment survey and the midpoint of the commercial fishery. In some years a small amount of harvest was taken during a second fishing season. In these cases, we computed the midpoint of the fishery as the weighted (by catch) average of the midpoints of the first and second seasons.

Estimates of natural mortality, M_t, are required inputs. In our trials, we investigated three mortality schedules: (*i*) fixed, (*ii*) time-varying based on published sources (Bristol Bay only), and (*iii*) a proxy series related to annual temperature variations. Published estimates of M tend to range between 0.1 and 0.5 (see review by Zheng et al. 1995). The North Pacific Fishery Management Council (NPFMC 1990) used $M = 0.3$ for Bristol Bay red king crabs. We set $M = 0.36$ for fixed M trials for both stocks to facilitate comparison with the results of Zheng et al. (1995). In a recent analysis of the Bristol Bay stock, Zheng et al. (1995) estimated M to be 0.19 for 1972–1979, 0.71 for 1980 and 1983–1984, 1.26 for 1981–1982, and 0.18 for 1985–1992. Accordingly, in another set of trials, we adopted this time-varying mortality schedule for Bristol Bay only, except that we used single value, $M = 0.93$, from 1981 to 1984, and we assumed that $M = 0.18$ also applied to 1993 and 1994.

A reviewer (B. Stevens, NMFS, Kodiak, Alaska) of an early draft of this paper suggested to us that crab mortality could be related to water temperature: warmer temperatures may lead to increased stress and mortality either from more rapid growth or higher susceptibility to disease. To explore this hypothesis, we generated a proxy mortality schedule with temperature data. We hypothesized that temperatures during molting were most critical. In late winter and spring, red king crabs tend to migrate to shallow coastal areas prior to molting and mating (Powell et al. 1973; Powell and Nickerson 1965). Males tend to molt first, in February and March. After the males' shells harden, they grasp molting females just prior to mating, which generally occurs during April and May.

To generate a proxy mortality schedule for Kodiak, we obtained subsurface (50 m) temperature data from a location (59°50.7′N, 149°28.0′W) over the 263-m isobath at the mouth of Resurrection Bay near Seward, Alaska, in the northern Gulf of Alaska. This temperature series was described by Xiong and Royer (1984) and others. The sampling site is well connected to continental shelf circulation because of its position within the Alaska Coastal Current, which flows counterclockwise around the Gulf of Alaska. It is generally accepted that temperature variations at this site are broadly representative of the gulf (T. Royer, University of Alaska, Fairbanks, personal communication). Because red king crabs generally molt in shallow water in February–May, we computed annual anomalies of 50-m temperatures in the following way. First we computed long-term monthly mean temperatures for the period 1971–1994, and then computed monthly anomalies by subtracting the long-term monthly means from the monthly mean temperatures for each year. The annual temperature anomaly was computed as the mean of the February–May monthly anomalies. There were three years in which no February–May temperature observations were made: 1973, 1980, and 1981. For these years, we computed temperature anomalies by linear interpolation from adjacent years.

Unfortunately, no routine subsurface temperatures are available for Bristol Bay from February to May. However, bottom-temperature data are collected during the annual NMFS trawl survey, typically conducted during June and July. Stevens et al. (1994) calculated a bottom-temperature index by computing the mean bottom temperature from 36 shallow-water (35–75 m) stations located just north of the Alaska Peninsula. From these data we computed annual bottom-temperature anomalies by subtracting the long-term (1971–1994) mean from Stevens' annual bottom-temperature index. The temperature anomalies were scaled to a mean M of 0.36 and a standard deviation of 0.15 to give maximum variability while ensuring that none of the annual M_t values would be negative.

CSA was performed with the three different mortality schedules, and in each case, the same number of parameters was estimated. On output, the relative abundance estimates (r_t and p_t) were divided by $\hat{q}$ to obtain the absolute abundance of legal crabs, $\hat{R}_t + \hat{P}_t$ (from eq. 2). Annual harvest rates (h_t) were calculated as the catch divided by the number of legal crabs at the time of the fishery:

$$[6] \quad h_t = \frac{C_t}{(\hat{R}_t + \hat{P}_t)e^{-M_t T_t}}$$

Sensitivity of CSA to input data and assumptions

We tested how well the model performs with a known data set, without error, based on the 20 years of data estimated for Bristol Bay. First, we tested how errors in the input parameters affect the estimation of the output parameters, especially q. We focus on q because division of relative abundance by the wrong q will result in biased absolute abundance estimates. We measured the deviation in estimated q from its known value of 1.0, with errors in the input values of T, M, and ϕ.

Next we tested the sensitivity of CSA to errors in classifying recruits and post-recruits. The magnitude of these errors can be estimated from crabs tagged by ADF&G biologists during summer surveys and recovered by ADF&G technicians during the fall/winter fishery. Molting did not occur during this period, so the tagged crabs should have the same size and shell age as they did when tagged. The size measurements were accurate to ±1 mm, but errors in shell aging apparently did occur. If we assume that the original survey classification was correct, 3% of the recruits were later misclassified as post-recruits and 3% of post-recruits were misclassified as recruits (Blau 1988). These misclassification rates were used to calculate observed relative abundance of recruits (r_{obs}) and post-recruits (p_{obs}) with aging errors in each year.

$$[7] \quad r_{obs} = 0.97r + 0.03p$$

$$p_{obs} = 0.03r + 0.97p$$

We then tested the ability of CSA to estimate the original correct values. We consider these to be worst-case estimates of the misclassification rates in the survey because the resampling was done by technicians with less experience than the ADF&G biologists who conducted the survey. Also, the resampling occurred during the fishery which is later in the molting cycle when newshell crabs become progressively darker, scratched, and more difficult to distinguish from oldshell crabs.

The next step was to test how the magnitude and structure of the residual errors affect parameter estimation. The Monte Carlo simulations start with the same known data set with no error. The known data were modified with lognormally distributed random errors of the form $\exp(\varepsilon - \sigma^2/2)$, where ε is a random normal deviate with mean 0 and standard deviation σ. The factor $\sigma^2/2$ ensures that the lognormal errors have mean 1. We tested three types of error structure (*a*) measurement errors as assumed in fitting the model with eq. 4, with $\sigma = 0.1$, 0.3, 0.5, 0.7, and 0.9; (*b*) correlated measurement errors as might occur because the observed $\tilde{r}$ and $\tilde{p}$ each year come from the same survey (the input correlation between η_t and δ_t was 0.5); and (*c*) process errors, which could result from year-to-year variations in M or errors in estimating C (in this case the population dynamics eq. 3 was multiplied by lognormal errors with $\sigma = 0.1$, 0.3, and 0.5). In addition to trials with single sources of error, trials were run with mixed levels of measurement and process errors and also with an assumed 10% error in classifying recruits and post-recruits (as in eq. 7, but with larger errors). Regardless of the structure of the input error, the same measurement-error model was used for parameter estimation (eq. 5). The randomization process was repeated 200 times and the results summarized according to the bias and dispersion of the parameter estimates.

Finally, the Monte Carlo simulation structure was modified to provide bootstrap estimates of the bias and dispersion of abundance estimates in each year. The first pass through the CSA used the observed data set. The result of this fit was the input data for the parametric bootstrap. These data were modified with measurement errors that had the same standard deviation as the observed residuals. The results of 1000 random trials were summarized by the mean and the empirical 95% confidence interval.

Results

Observed and estimated recruits and post-recruits for Bristol Bay red king crabs are in general agreement (Fig. 1). Changes in post-recruit abundance are offset by one year from the recruits as predicted by the population dynamics model (eq. 3). There is a temporal pattern in the residuals of the constant-M fit. In early and later years the model overpredicts recruits and underpredicts post-recruits; in middle years the opposite is true. This pattern suggests that natural mortality was above average from 1981 to 1985 and below average in the early and later years. The variable-M fit accounts for this shift and fits the observed data much better, with $q = 0.95$ and a reduction in the sum of squared residuals (SSQ) to 1.293 (Table 1).

The population model smoothes year-to-year variability in

Fig. 1. Abundances of Bristol Bay red king crabs measured with a trawl survey and estimated with catch-survey analysis. The constant *M* was 0.36 and the variable *M*s are from Zheng et al. (1995).

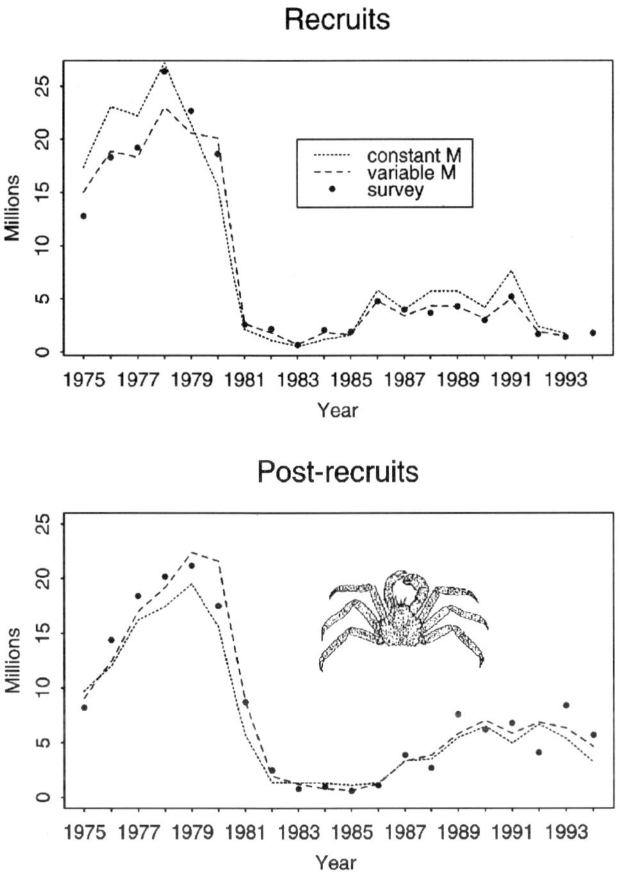

Fig. 2. Legal male abundance, harvest rate, and natural mortality of Bristol Bay red king crabs.

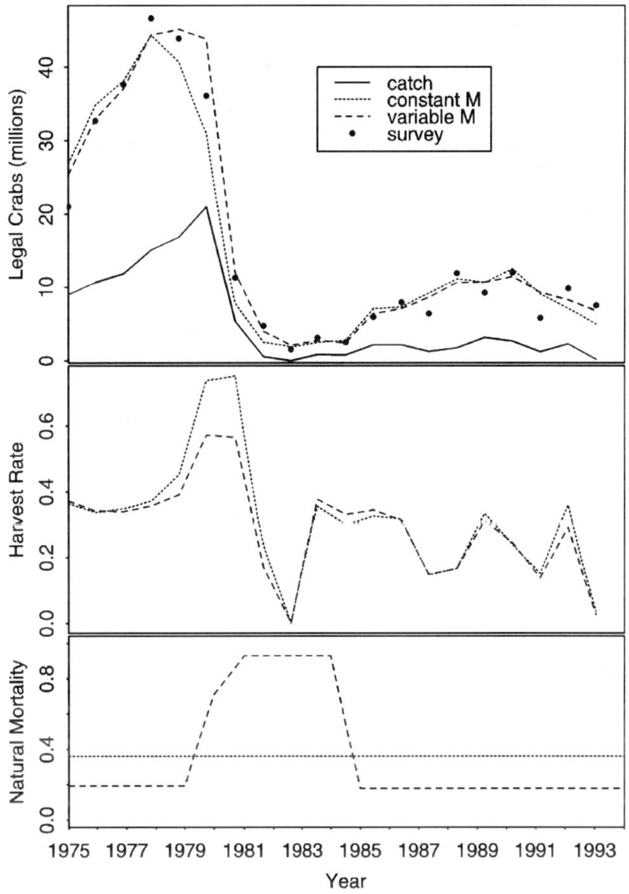

Table 1. Residual analysis of catch-survey analysis fits to Bristol Bay and Kodiak data with constant natural mortality.

Statistic	Bristol Bay	Kodiak
Number of years	20	13
Catchability coefficient (q)	1.01	0.97
Standard error of q		
linear approximation	0.58	0.31
bootstrap	0.15	0.10
Sum of squared residuals (SSQ)	4.313	1.009
Standard deviation of residuals	0.49	0.30
Kendall's rank correlation (τ) between recruit and post-recruit residuals	0.041	0.091
probability $\tau = 0$	0.807	0.681
First-order autocorrelation of residuals		
recruits	0.70	0.22
post-recruits	0.56	−0.29

the observed abundance estimates. A small increase or decrease in abundance can make the difference of the fishery being open or closed because thresholds are used to protect depressed stocks. If the increase or decrease were actually a measurement error, a management decision could also be made in error. Recruit abundance is not predicted in the final year because of the recursive nature of the population equation (eq. 3). To estimate legal abundance we must rely on observed recruits in the final year.

Estimates of legal crab abundance are superimposed on the survey data along with the commercial catch (Fig. 2). Large catches were extracted from the large stock that existed from 1975 to 1980, prior to the crash in 1981. The fishery was closed in 1983 and continued at low levels until 1994, when only a small catch was taken around the Pribilof Islands. The collapse was caused by a combination of high harvest rates and an apparent increase in natural mortality. With the variable-*M* fit, harvest rates were still high leading up to the crash, but more of the deaths were attributed to natural causes. Recruits which were there one year disappeared by the next. The instantaneous natural mortality rate apparently increased to almost one. The trial with *M* scaled to sea-temperature anomalies did not improve the fit ($q = 0.87$, SSQ = 4.478) because the increase in temperature preceded the apparent increase in *M*.

Kodiak can be considered a test data set because there was no calibrated trawl survey for this stock to give a prior estimate of q. We performed two runs, one with constant natural mortality and one in which *M* was scaled to anomalies in subsurface temperature in the Gulf of Alaska ($q = 1.00$, SSQ = 0.617). The fit with variable *M* was improved in earlier years (Fig. 3), because the decline in abundance coincided with the increase in ocean temperature in the Gulf of Alaska. Observed high recruitment in 1979 could be considered a

Fig. 3. Relative abundances of Kodiak red king crabs measured with a pot survey and estimated with catch-survey analysis. The constant M was 0.36 and the variable Ms were scaled to ocean temperature anomalies.

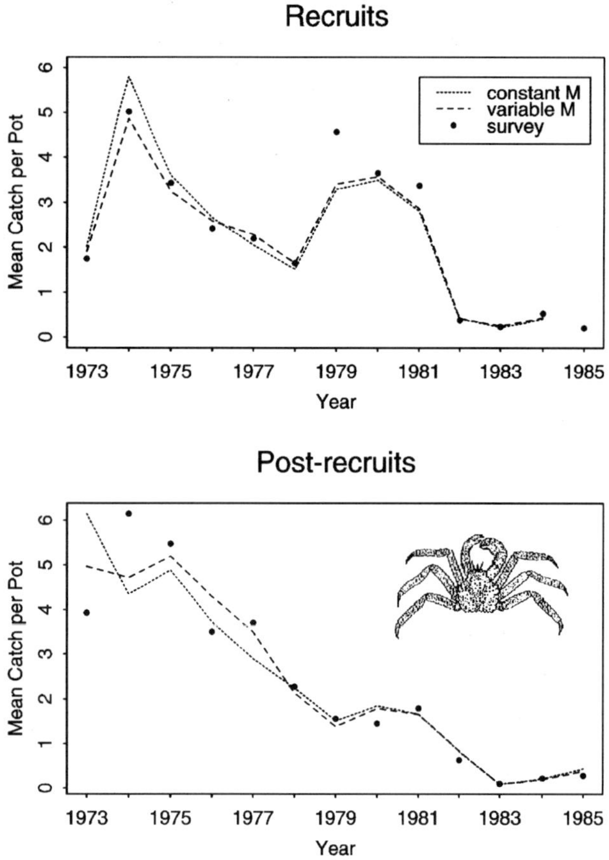

Fig. 4. Legal male abundance, harvest rate, and natural mortality of Kodiak red king crabs.

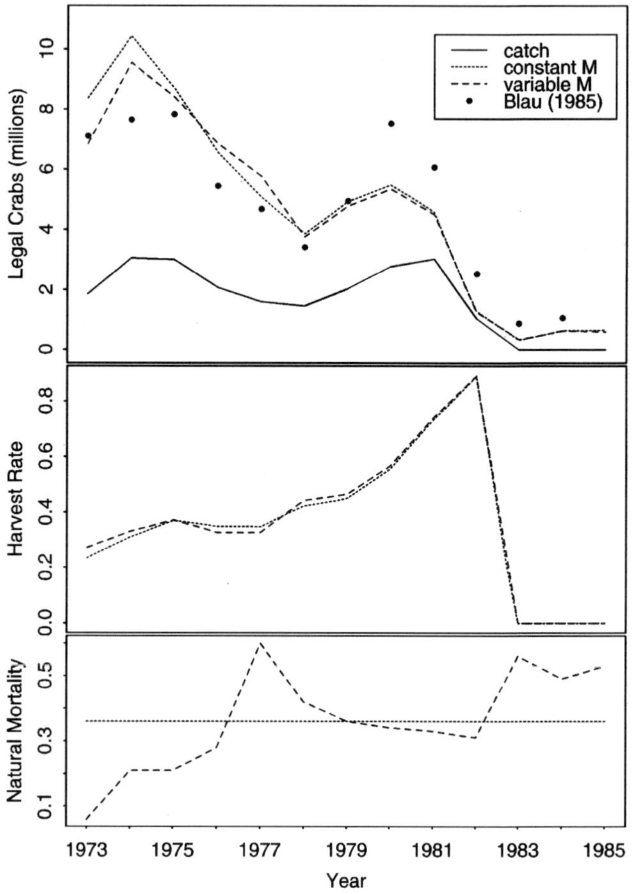

measurement error because post-recruit abundance did not increase correspondingly the following year.

The estimates of legal crab abundance for Kodiak are compared with the mark-recapture estimates of Blau (1985) in Fig. 4. Blau's mark-recapture estimates tend to be somewhat lower in 1973–1978 and higher in 1980–1984. Higher mark-recapture estimates during the latter period correspond to years of reduced tag-recovery programs. Declining stocks also provided added incentives to some fishers who retained large numbers of tags to increase apparent stock size from Petersen estimation methods. The lower abundance estimates of CSA are associated with higher estimates of harvest rates than were estimated at the time of the fishery. Harvest rates increased until 1982 when the fishery collapsed; it has been closed since 1983 due to extremely depressed stock size.

The residuals of the CSA had a larger standard deviation for Bristol Bay than for Kodiak (Table 1). The recruit and post-recruit residuals for a given year tended to be positively correlated, as might be expected because they came from the same survey and the total survey catch was subject to a common measurement error. However, neither correlation was significant. The first-order autocorrelation of each residual time series was significant ($p = 0.05$) only for the Bristol Bay recruit residuals. This fit had a temporal pattern in the residuals attributable to the apparent shift in M (Fig. 1). Otherwise there were no apparent patterns in the residuals that would indicate a violation of the model assumptions.

Sensitivity to errors in input data

Results are expressed as the deviation of q from its known value of 1.0 with errors in the input parameters (Fig. 5). The goodness-of-fit is indicated by the circles, the area of which is proportional to SSQ. With correct values of the input parameters and no errors in the relative abundance data, SSQ = 0. The estimate of q is insensitive to errors in specifying the timing of the commercial catch. Even though we specify an accurate annual value, the consequence of not doing so is minor. Likewise, discretizing a continuous catch would not be a problem. In this example T was constant at 0.25; increasing or decreasing the input T did not affect the sum of squares because e^{-MT} is a constant multiplier of q in eq. 3.

The q estimate is negatively correlated with M because the decline in numbers can be attributed to the commercial catch or natural mortality. Likewise, it can be seen from (eq. 3) that if we assume recruit mortality equals post-recruit mortality, when in fact the former is higher, q would be overestimated. The estimated q is also negatively correlated with recruit catchability (ϕ). For M and especially for ϕ, we must substantially deviate from the known input value to appreciably increase the sum of squares. The sum of squares of the fit to the actual data

Fig. 5. Sensitivity of the estimated catchability coefficient to errors in specifying the input values of T, M, and ϕ in eq. 3. The correct values were $q = 1$, $T = 0.25$, $M = 0.3$, and $\phi = 1$. The area of each circle is proportional to the sum of squared residuals of the fit.

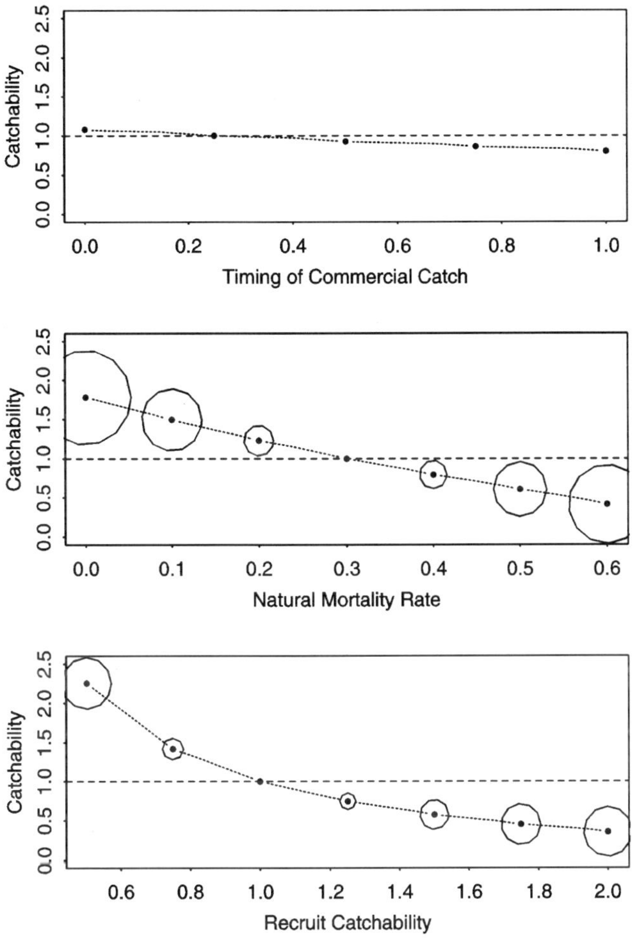

Fig. 6. Distributions of the catchability coefficients estimated in Monte Carlo simulations of catch-survey analysis. Each box plot summarizes the distribution of 200 q values. The white bar in the interior of the box is the median. The height of the box is equal to the interquartile distance (IQD). The whiskers extend 1.5× the IQD, or to the extreme range of the data, whichever is less. Individual lines beyond the whiskers are outliers. The three numbers under each box indicate the standard deviation of measurement errors, the correlation of measurement errors, and process error standard deviation (from left to right).

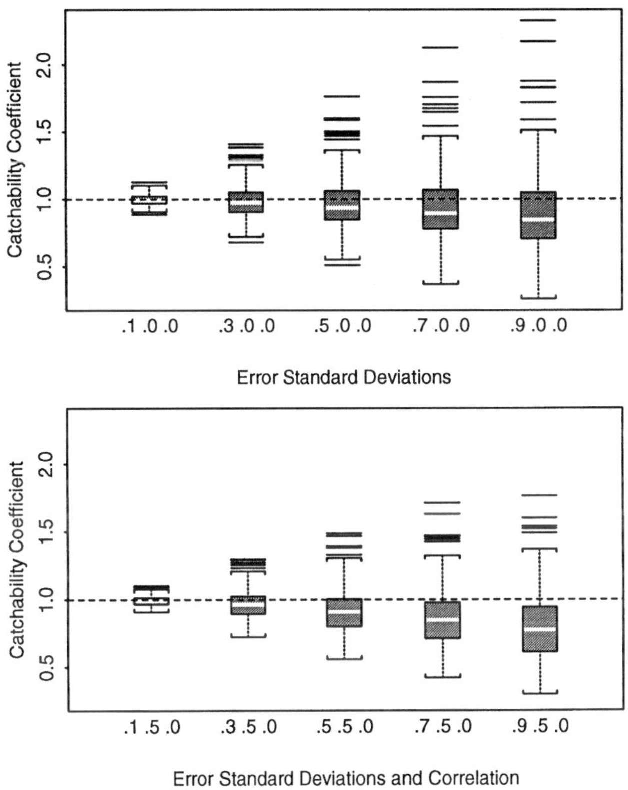

was more than twice as big as the largest circle in Fig. 5. It is therefore unlikely that we can estimate M or ϕ from real data sets in addition to q. These parameters are not structurally confounded in eq. 3, but their estimates will be strongly negatively correlated.

The simulated shell aging errors caused q to be slightly overestimated ($q = 1.01$, SSQ = 0.007). The sum of squared residuals between the CSA estimates and the original correct data was <0.001, which indicates that CSA tends to correct for the aging errors. The simulated errors had little effect because the misclassification rates were small and symmetric. The bias in q would be larger if the aging errors were larger and there was a difference in magnitude between recruits and post-recruits, or if the aging errors were asymmetric.

Monte Carlo simulations

Box plots summarize the distributions of the q estimates relative to the true value of 1.0 (Figs. 6 and 7). A quantile plot of the log-transformed q estimates against the quantiles of the normal distribution (not shown) indicated that the distribution of estimated q's was lognormal. Increasing the measurement-error standard deviation caused the distribution to be skewed downward and to have greater dispersion. The boxes in the bottom panel of Fig. 6 correspond to the ones in the top panel, with the addition of correlated measurement errors. With correlated measurement error the bias was greater and the dispersion less. The two real data sets had error standard deviations of 0.3 and 0.5; observed correlation in the residuals was low (Table 1).

Measurement and process errors caused different patterns in the simulated abundance data. With measurement errors, the distribution of post-recruit abundance fluctuated around the true values. With process errors, simulated abundance remained above or below the true abundance for extended periods, and was only tied to the true values because the correct recruitment was input each year. We therefore expected CSA to have more trouble reconstructing the true abundance with process errors, especially with an all measurement-error assumption. As the standard deviation of process errors was increased, the distribution of q was skewed upward (Fig. 7). At a standard deviation of 0.5, the dispersion was such that convergence difficulties were encountered. With mixed errors the negative bias from the measurement errors to some extent

Fig. 7. Distribution of the catchability coefficients estimated in Monte Carlo simulations of catch-survey analysis with mixed errors. The explanation of box plots is the same as for Fig. 6.

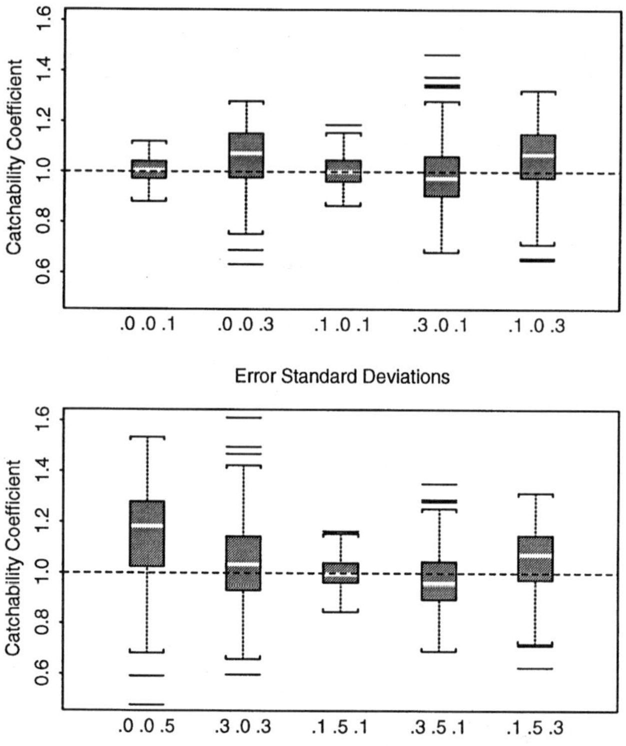

counteracted the positive bias from the process errors. Adding correlated measurement errors shifted the medians downward as in Fig. 6. The real data probably corresponded most closely to the cases in the upper right of Fig. 7.

Although it is convenient to express the simulation results with distributions of q, we are ultimately more interested in the distribution of absolute abundance estimates. Table 2 lists the average percent bias and coefficient of variation of legal abundance for combinations of measurement and process errors. With only measurement errors, the bias was low and the CV increased with the standard deviation. Process errors resulted in an increasingly negative bias, because q was overestimated. With mixed errors the bias was less than 5%. Simulated aging errors increased the bias and decreased the dispersion of the parameter estimates. The real data probably correspond most closely to the mixed-error simulations.

A parametric bootstrap was used to calculate the bias and dispersion of abundance estimates in each year for Bristol Bay and Kodiak red king crabs. The bootstrap standard errors of q were lower than those based on the asymptotic normality of the least-squares estimator (Table 1). We consider the bootstrap estimates to be more reliable because they are consistent with the Monte Carlo simulations with σ in the range 0.3–0.5. The bootstrap mean abundances were very close to the initial CSA estimates (Fig. 8), which indicates very small bias in the estimates. The empirical 95% confidence limits were quite symmetric and corresponded closely to ± 2 times the bootstrap standard deviation of q. The parametric confidence interval, which is slightly simpler to calculate, may therefore be adequate, although the empirical confidence limit is generally preferred. The confidence intervals were narrowest from 1980 to 1982 because of the high harvest rates. Depletion estimators perform better when a stock is heavily exploited.

Discussion

Catch survey analysis is useful for estimating the abundance of populations that can be classified into recruits and post-recruits

Table 2. Results of Monte Carlo simulations of catch-survey analysis.

Standard deviation of measurement errors	Correlation of measurement errors	Standard deviation of process errors	Proportion misaged	Bias of q estimates	Standard error of q estimates	Average % bias of abundance estimates	Average CV of abundance estimates	Number of replicates that converged
0.1	0.0	0.0	0.0	–0.00	0.04	0.2	5.1	200
0.5	0.0	0.0	0.0	–0.04	0.19	0.7	25.4	200
0.9	0.0	0.0	0.0	–0.10	0.31	–1.1	47.1	200
0.1	0.5	0.0	0.0	–0.01	0.04	0.30	4.8	200
0.5	0.5	0.0	0.0	–0.08	0.18	3.72	24.8	200
0.9	0.5	0.0	0.0	–0.20	0.27	7.16	48.3	196
0.0	0.0	0.1	0.0	0.01	0.05	0.0	12.0	200
0.0	0.0	0.3	0.0	0.06	0.13	–3.6	34.4	200
0.0	0.0	0.5	0.0	0.14	0.19	–11.6	54.3	192
0.1	0.0	0.1	0.0	0.00	0.06	0.1	12.9	200
0.3	0.0	0.1	0.0	–0.01	0.13	0.4	19.3	200
0.1	0.0	0.3	0.0	0.06	0.13	–3.0	34.5	194
0.3	0.0	0.3	0.0	0.04	0.17	–3.4	37.0	198
0.3	0.0	0.3	0.1	0.06	0.13	–5.7	31.3	200
0.3	0.5	0.3	0.1	0.04	0.14	–4.4	32.4	200

Note: The known data set was based on the Bristol Bay data set with $q = 1$. The averages are over the 20 years of the data set.

Fig. 8. Bootstrap abundance estimates of Bristol Bay and Kodiak red king crabs. The 95% confidence intervals (CI) are the 2.5 and 97.5 percentiles of 1000 bootstrap replicates.

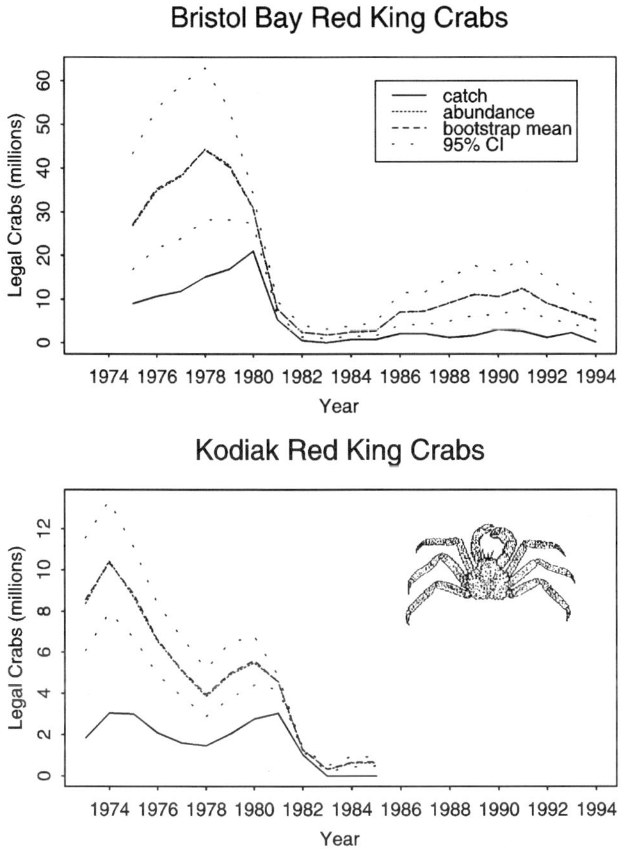

and for which the commercial catch in numbers is recorded. It is particularly useful for stocks with limited age data, either because they cannot be aged or because of the expense of age determination. King crabs are excellent candidates for CSA because shell age can be used to classify them into recruits and post-recruits, even though the absolute age is unknown. For many other exploited fish and shellfish stocks a recruitment index can be obtained from the size–frequency distribution of survey data. Additional auxiliary information (e.g., commercial catch per unit effort) can be included in the estimation. The CSA framework can readily be customized for application to a wide range of populations (Conser 1991; Conser and Idoine 1992).

An advantage over direct area-swept estimates from trawl surveys is that CSA attempts to separate measurement errors from true changes in stock abundance, whereas area-swept estimation methods ignore information about the stock from prior years by implicitly assuming that each year's observation is completely independent. Another advantage is that CSA provides a method to estimate catchability of survey gear. This calibration is important for trawl surveys in which catchability is uncertain, and it is critical for pot surveys which would otherwise provide only an estimate of relative abundance.

In this study we extended the original CSA formulation (Collie and Sissenwine 1983) by allowing for annual estimates of the timing of the commercial catch and natural mortality. More importantly, we rigorously tested the performance of the estimation method with known data sets and Monte Carlo simulation. The measurement-error model provided robust estimates of q and abundance. Of 3000 Monte Carlo trials reported in Table 2, only 20 (0.67%) failed to converge to a solution, and these failures were for high error levels unlikely to be encountered with real data sets. A mixed measurement- and process-error model may still be appropriate for larger data sets with relatively small measurement errors; the measurement-error model is a robust alternative when these conditions are not met.

The deviations in the median q values from the true value (Figs. 6 and 7) can be explained by the fact that the multiplicative lognormal errors, $\exp(\varepsilon - \sigma^2/2)$, were scaled to have mean 1. The median of the errors is less than one and therefore the median q was either less than or greater than one, depending on the assumed error structure. The bias in the mean q values (Table 2) probably results from the propagation of the errors over time. With simulated measurement errors there was a negative bias in q, that increased with increasing error standard deviation (σ). Interestingly, the corresponding estimates of abundance were less biased than q (Table 2), perhaps because of covariance between the estimates of q and relative abundance.

Simulated process errors caused a positive bias in q (Table 2) which can be explained by the observation that, in many of the random data sets, the simulated post-recruits remained below the true abundance for most of the years. To account for the apparent greater mortality of post-recruits, the effect of the catch was increased and q overestimated. If real data sets contain high levels of process error ($\sigma \geq 0.5$), the positive bias in q and corresponding negative bias in abundance (Table 2) would be of concern. Whether an alternative error structure is desirable in this situation could only be determined with Monte Carlo simulation of a mixed-error estimation model with mixed-error data sets.

In simulation trials with mixed errors, the negative bias in q caused by the measurement errors tended to cancel the positive bias caused by process errors (Table 2). Correlated measurement errors and aging errors increased the bias of q and decreased the dispersion (Figs. 6 and 7). Both types of error accentuate deviations of the simulated data from the correct data. If recruit and post-recruit abundances are of the same magnitude and aging errors are symmetric, the 10% of recruits misclassified as post-recruits tends to be replaced by the 10% of post-recruits misclassified as recruits, with little net effect on the relative abundance estimates.

Turning to the real data sets for Bristol Bay and Kodiak, the error standard deviations (σ) of 0.3 and 0.5 are probably partitioned between measurement and process errors, though the exact ratio is unknown. The various error combinations with $\sigma = 0.1$, 0.3, or 0.5 probably bracket the true error variances. Correlation of the residuals from the observed data sets was low (Table 1), and even a correlation of 0.5 had a minor effect on parameter estimation. We consider the aging errors represented by eq. 7 to be conservative estimates because the reclassification was done by different people with less training at a later date. Shell age misclassification appears to be a minor source of error in estimating the abundance of these king crab stocks. Considering the most likely magnitudes of all the potential sources of error, we conclude that the biases will in most cases be acceptable (<5%) and in many cases negligible. Correcting the abundance estimates for bias is inadvisable because the direction of the bias depends on the source of the errors, which is unknown in real data sets. Even with the shorter Kodiak data set (13 yr), the precision of the abundance estimates was acceptable (Fig. 8). The CSA performed well because both stocks were depleted in the early 1980's. Lower precision is expected for shorter time series and stocks that have been less heavily exploited.

Both red king crab stocks underwent major changes. In Bristol Bay, abundance of legal males increased during the late 1970's, declined sharply in the early 1980's, increased slightly in the late 1980's, and declined since then (Fig. 2). For Kodiak, legal abundance generally declined by 1 million crabs per year in 1973–1985 with the exception of a slight increase during the late 1970's (Fig. 4). These absolute abundance estimates are consistent with estimates obtained with mark-recapture studies (Blau 1985), area-swept calculations (Otto 1986), and length-based analysis (Zheng et al. 1995). This correspondence is partly due to our choice of $M = 0.36$, but even with the M value commonly used by the North Pacific Fishery Management Council (0.3) our absolute abundance estimates would be only slightly lower. This cross-validation of abundance estimators supports the use of CSA for stocks, such as those in Southeast Alaska, where no other absolute abundance estimates are available (Woodby 1994).

What caused the decline of these red king crab populations? Although the primary purpose of our study was to test the sensitivity of CSA to various errors and model assumptions, our results provide some insight into this difficult question. Comparison of abundance and catch partly implicates the fishery in the declines (Figs. 2 and 4). In both cases, harvest rates <40% generally corresponded to periods of relatively stable or increasing abundance, whereas harvests >40% were associated with periods of stock decline. Moreover, the highest harvest rates (>50% for Bristol Bay and >80% for Kodiak) coincided with the sharpest declines in the early 1980's. However, fishing does not fully explain stock changes. A sharp increase in apparent natural mortality during the early 1980's contributed to the collapse in Bristol Bay (Fig. 2).

Our analysis of the potential effects of temperature on M produced mixed results for Kodiak and Bristol Bay. The potential role of temperature deserves further study, because the effects could be more complex (e.g., nonlinear relationships and time lags) than we modeled. Otto (1986) speculated that Pacific cod (*Gadus macrocephalus*), a major predator of red king crabs, could have a substantial effect on crab mortality. To explore this, we ran additional CSA trials using age 3–12 cod abundances for the Gulf of Alaska (Thompson and Zenger 1994) and Bering Sea (Thompson 1994). For Bristol Bay, the fit was improved ($q = 1.05$, SSQ = 3.326) but for Kodiak the fit was worse than with constant M ($q = 1.09$, SSQ = 1.175). Livingston (1989) estimated that cod ate only a small percentage (1.4–3.8%) of the red king crab stock in the Eastern Bering Sea during sampling years 1981, 1984, and 1985; she concluded that cod were not a major cause of the red king crab decline in the early 1980's.

Low recruitment in the early 1980's also contributed to the collapse of both king crab stocks. The close relations between recruit (Figs. 1 and 3) and legal male abundances (Figs. 2 and 4) show that both populations are largely driven by recruitment variations. The lack of recovery of the Bristol Bay stock since the early 1980's has been attributed to low spawning biomass (Zheng et al. 1995). These authors estimated two stock-recruitment relationships for Bristol Bay that provide competing interpretations of recruitment trends: a dome-shaped curve stresses the importance of density-dependent effects, whereas a curve that is flat over intermediate to high levels of spawning biomass emphasizes autocorrelated environmental effects. The fact that these two relationships fit the data equally well illustrates the classic problem of separating stock and environmental effects on recruitment. Armstrong (1983) speculated that year-class strength of red king crabs was determined by a suite of biotic and abiotic factors affecting reproductive success, larval survival, and survival of young juveniles and that the fishery could exacerbate these effects. Attempts to better understand this recruitment process are ongoing (Tyler and Kruse 1995).

New developments for application to red king crabs could include a three-stage model. Because molting probability is nearly unity until legal size is attained, the extension of our model to include a third stage, pre-recruits, is possible. This development would incorporate the mature but sublegal male component of the population which is part of the harvest strategy of Bristol Bay red king crabs. A three-stage model would be intermediate between the two-stage catch-survey analysis and a full length-based analysis.

Acknowledgments

We thank the following scientists for providing data: B.A. Johnson for Kodiak pot surveys, J. Reeves and B. Stevens for Bering Sea trawl surveys, B. Stevens for Bering Sea temperatures, and T. Royer for Seward temperatures. Terry Quinn provided statistical advice and Jie Zheng performed a constructive review. This study was funded, in part, by the Commercial Fisheries Management and Development Division of the Alaska Department of Fish and Game.

References

Alaska Department of Fish and Game (ADF&G). 1994. Commercial shellfish fishing regulations, 1994–95 edition. Alaska Department of Fish and Game, Division of Commercial Fisheries Management and Development, Juneau, Alaska.

Allen, K.R. 1966. Some methods for estimating exploited populations. J. Fish. Res. Board Can. **23**: 1553–1574.

Anonymous. 1942. Report of the Alaska crab investigation. U.S. Fish and Wildlife Service, Washington, D.C. Fishery Market News **4**(5a).

Armstrong, D.A. 1983. Cyclic crab abundances and relationship to environmental causes. *In* From year to year: interannual variability of the environment and fisheries of the Gulf of Alaska and Eastern Bering Sea. *Edited by* W.S. Wooster. Washington Sea Grant Publication No. 83-3. pp. 102–110.

Blau, S.F. 1985. Overview and comparison of the major red king crab *(Paralithodes camtschatica)* surveys and fisheries in western Alaska 1969–1984. *In* Proceedings of the International King Crab Symposium. Alaska Sea Grant College Program Report No. 85-12. pp. 23–47.

Blau, S.F. 1988. Commercial catch sampling and estimated harvest by sizes and exoskeletal ages of red king crabs 1960–86, Kodiak, Alaska. Alaska Department of Fish and Game, Division of Commercial Fisheries, Juneau, Alaska. Fisheries Research Bulletin No. 88-02.

Collie, J.S., and Sissenwine, M.P. 1983. Estimating population size from relative abundance data measured with error. Can. J. Fish. Aquat. Sci. **40**: 1871–1879.

Conser, R.J. 1991. A DeLury model for scallops incorporating length-based selectivity of the recruiting year-class to the survey gear and partial recruitment to the commercial fishery. Northeast Regional Stock Assessment Workshop Report. Woods Hole, MA. Res. Doc. SAW12/2. Appendix to CRD-91-03. 18 p.

Conser, R.J., and Idoine, J.S. 1992. A modified DeLury model for estimating mortality rates and stock sizes of American lobster populations. Northeast Regional Stock Assessment Workshop Report. Woods Hole, Mass. Res. Doc. SAW14/7. Appendix to CRD-92-07. 28 p.

DeLury, D.B. 1947. On the estimation of biological populations. Biometrics, **3**: 145–167.

Deriso, R.B., Quinn, T.J. II, and Neal, P.R. 1985. Catch-age analysis with auxiliary information. Can. J. Fish. Aquat. Sci. **42**: 815–824.

International Mathematics and Statistics Library (IMSL). 1991. International Mathematics and Statistics Library. IMSL, Inc. Houston, Tex.

Kruse, G.H. 1993. Biological perspectives on crab management in Alaska. *In* Proceedings of the International Symposium on Management Strategies for Exploited Fish Populations. *Edited by* G.H. Kruse, D.M. Eggers, R.J. Marasco, C. Pautzke, and T.J. Quinn. Alaska Sea Grant College Program Report No. 93-02. pp. 355–384.

Kruse, G.H., and Collie, J.S. 1991. Preliminary application of a population size estimation model to the Bristol Bay stock of red king crabs. Alaska Department of Fish and Game, Division of Commercial Fisheries, Juneau, Alaska. Regional Information Report No. 5J91-09. 25 p.

Leslie, P.H., and Davis, D.H.S. 1939. An attempt to determine the absolute number of rats on a given area. J. Anim. Ecol. **8**: 94–113.

Livingston, P.A. 1989. Interannual trends in Pacific cod, *Gadus macrocephalus*, predation on three commercially important crab species in the eastern Bering Sea. Fish. Bull. (U.S.), **87**: 807–827.

Morrison, R., and Gish, R.K. 1994. Annual management report for the shellfish fisheries of the Bering Sea area, 1993. *In* Annual management report for the shellfish fisheries of the westward region, 1993. Alaska Department of Fish and Game, Commercial Fisheries Management and Development Division, Kodiak, Alaska. Regional Information Report No. 4K94-29, pp. 151–219.

North Pacific Fishery Management Council (NPFMC). 1990. Environmental assessment for amendment 1 to the fishery management plan for the commercial king and Tanner crab fisheries in the Bering Sea/Aleutian Islands. North Pacific Fishery Management Council, Anchorage, Alaska.

Otto, R.S. 1986. Management and assessment of eastern Bering Sea king crab stocks. *In* North Pacific Workshop on Stock Assessment and Management of Invertebrates. *Edited by* G.S. Jamieson and N. Bourne. Can. Spec. Publ. Fish. Aquat. Sci. 92. pp. 83–106.

Peterson, R., Donaldson, B., Hornig, L., and Harris, S. 1986. Red king crab, *Paralithodes camtschatica*, explorations around the Kodiak Archipelago: the history and results, 1971–1985. Alaska Department of Fish and Game, Division of Commercial Fisheries, Juneau, Alaska. Technical Data Report No. 189.

Powell, G.C., and Nickerson, R.B. 1965. Reproduction of king crabs, *Paralithodes camtschatica*. J. Fish. Res. Board Can. **22**: 101–111.

Powell, G.C., Shafford, B., and Jones, M. 1973. Reproductive biology of young adult king crabs, *Paralithodes camtschatica* (Tilesius) at Kodiak, Alaska. Proceedings of the National Shellfisheries Association **63**: 77–87.

Seeb, J.E., Kruse, G.H., Seeb, L.W., and Weck, R.G. 1990. Genetic structure of red king crab populations in Alaska facilitates enforcement of fishing regulations. *In* Proceedings of the International Symposium on King and Tanner Crabs. Alaska Sea Grant College Program Report No. 90-04. pp. 491–502.

Spalinger, J.A., and Jackson, D.R. 1994. Annual management report for the shellfish fisheries of the Kodiak area, 1993. *In* Annual management report for the shellfish fisheries of the westward region, 1993. Alaska Department of Fish and Game, Commercial Fisheries Management and Development Division, Kodiak, Alaska. Regional Information Report No. 4K94-29. pp. 13–69.

Stevens, B.G., Haaga, J.A., and MacIntosh, R.A. 1994. Report to industry on the 1994 eastern Bering Sea crab survey. National Marine Fisheries Service, Alaska Fisheries Science Center, Kodiak, Alaska. Processed Report No. 94-07.

Thompson, G.G. 1994. Pacific cod. *In* Stock assessment and fishery evaluation report for the groundfish resources of the Bering Sea/Aleutian Islands regions as projected for 1995. North Pacific Fishery Management Council, Anchorage, Alaska. pp. 2-1 to 2-31.

Thompson, G.G., and Zenger, H.H. 1994. Pacific cod. *In* Stock assessment and fishery evaluation report for the groundfish resources of the Gulf of Alaska as projected for 1995. North Pacific Fishery Management Council, Anchorage, Alaska. pp. 2-1 to 2-39.

Tyler, A.V., and Kruse, G.H. 1995. Report on the modeling workshop on year-class formation of red king crab. Alaska Department of Fish and Game, Commercial Fisheries Management and Development Division, Juneau, Alaska. Regional Information Report 5J95-11.

Walters, C.J. 1986. Adaptive management of renewable resources. Macmillan Publishing Co., New York, N.Y. 374 p.

Woodby, D. 1994. Population sizes and harvest rates of red king crabs in Southeast Alaska, 1993. Alaska Department of Fish and Game, Commercial Fisheries Management and Development Division, Juneau, Alaska. Regional Information Report No. 1J94-20.

Xiong, Q., and Royer, T.C. 1984. Coastal temperature and salinity in the northern Gulf of Alaska, 1970–1983. J. Geophys. Res **89**(C5): 8061–8068.

Zheng, J., Murphy, M.C., and Kruse, G.H. 1995. A length-based population model and stock-recruitment relationships for red king crab, *Paralithodes camtschaticus*, in Bristol Bay, Alaska. Can. J. Fish. Aquat. Sci. **52**: 1229–1246.

A modified catch-at-size analysis model for a red sea urchin (*Strongylocentrotus franciscanus*) population

Han-Lin Lai and Alex Bradbury

Abstract: A modified catch-at-size analysis model was developed to estimate the life-history parameters and to evaluate management strategies for a red sea urchin (*Strongylocentrotus franciscanus*) population in Washington, U.S.A., using the size–frequency distributions collected from surveys as input. The population underwent heavy exploitation in 1988 but the abundance during 1989–1994 remained at 48% of that in 1988. Although the stock condition was stable, weak recruitment in 1992–1994 posed great uncertainty for the future. Simulation results indicate that a maximum size limit such as the current 140 mm is necessary for a constant harvest rate policy in order to reduce variability of yield and risk of stock collapse, defined as the probability for which spawning stock number (SSN) of the population declines to less than 9×10^6. Conversely, an upper limit seems not very critical for a constant catch policy. A periodic harvest schedule does not increase cumulative yield but is helpful in reducing variability of yield, risk, and probably the costs of management and enforcement. The harvest rate corresponding to maximum equilibrium yield (u_{MYEQ}) is derived from simulation and used as a biological reference point. Because recruitments in 1992–1994 were weak and the stock has yet to rebuild from the heavy exploitation in 1988, the future target harvest rate or quota should be substantially less than the maximum equilibrium yield (MYEQ) depending on the level of SSN.

Résumé : Un modèle modifié d'analyses des prises selon la taille a été élaboré pour estimer les paramètres du cycle vital et évaluer les stratégies de gestion d'une population d'oursins rouges (*Strongylocentrotus fransciscanus*) de l'État de Washington, É.-U., en utilisant comme données d'entrée, les distributions de fréquences de taille obtenues à partir des relevés. La population a subi une forte exploitation en 1988, mais l'abondance au cours de la période 1989–1994 est demeurée à 48 % de celle de 1988. Bien que l'état du stock ait été stable, un faible recrutement au cours de la période 1992–1994 a causé une vive incertitude pour l'avenir de ce stock. Les résultats de simulations indiquent qu'une limite de taille maximale, comme la valeur actuelle de 140 mm, est nécessaire pour une politique de taux de récolte constant, si l'on veut réduire la variabilité du rendement et le risque d'effondrement du stock, défini par la probabilité à laquelle l'effectif du stock de reproduction (ESR) de la population tombe à moins de 9×10^6. À l'opposé, une limite supérieure ne semble pas très déterminante pour une politique des prises constantes. Un calendrier de récolte périodique n'accroît pas le rendement cumulé, mais est utile pour réduire la variabilité du rendement, le risque et, probablement, les coûts de gestion et de mise en application de la loi. Le taux de récolte correspondant au rendement maximum à l'équilibre est dérivé par simulation et utilisé comme point de référence biologique. Parce que le recrutement au cours de la période 1992–1994 a été faible et que le stock ne s'était pas encore reconstitué à la suite de la forte exploitation de 1988, les taux de récolte ou les quotas cibles pour l'avenir devraient être nettement inférieurs au rendement maximum à l'équilibre, selon l'importance de l'ESR. [Traduit par la Rédaction]

Introduction

The red sea urchin (*Strongylocentrotus franciscanus*), which is distributed from Northern Japan, east to Alaska and south to Baja California, has been harvested by commercial divers since 1971 in Washington. The red sea urchin are harvested for their gonad and exported to Japan where urchin "roe" is a delicacy. The biology of the red sea urchin has been reviewed by Kato and Schroeter (1985) and Tegner (1989). Botsford et al. (1993) reported that the growth and adult natural mortality rates are low and that recruitment is generally low but variable over spatial and temporal scales. These parameters, together with the high value of the red sea urchin and the efficiency of divers, suggest that the species is highly susceptible to overfishing.

In the past decade, several regulations including size limits, 3-year areal rotation, limited entry, and restricted season length, have been imposed to manage the red sea urchin fisheries in Washington. Most of these regulations are based on the logistics of enforcement and scientific studies from California and British Columbia. Nonetheless, research surveys reveal that the abundance of red sea urchin populations in Washington declined substantially over the past decade

H.L. Lai. Northeast Fisheries Science Center, National Marine Fisheries Service, NOAA, 166 Water Street, Woods Hole, MA 02543, U.S.A.

A. Bradbury. Washington Department of Fish and Wildlife, Pt. Whitney Shellfish Laboratory, 1000 Point Whitney Road, Brinnon, WA 98320, U.S.A.

Correct citation: Lai, H.L., and Bradbury, A. 1998. A modified catch-at-size analysis model for a red sea urchin (*Strongylocentrotus franciscanus*) population. *In* Proceedings of the North Pacific Symposium on Invertebrate Stock Assessment and Management. *Edited by* G.S. Jamieson and A. Campbell. Can. Spec. Publ. Fish. Aquat. Sci. 125. pp. 85–96.

Fig. 1. Red sea urchin fishing ground and fishing districts in Washington State.

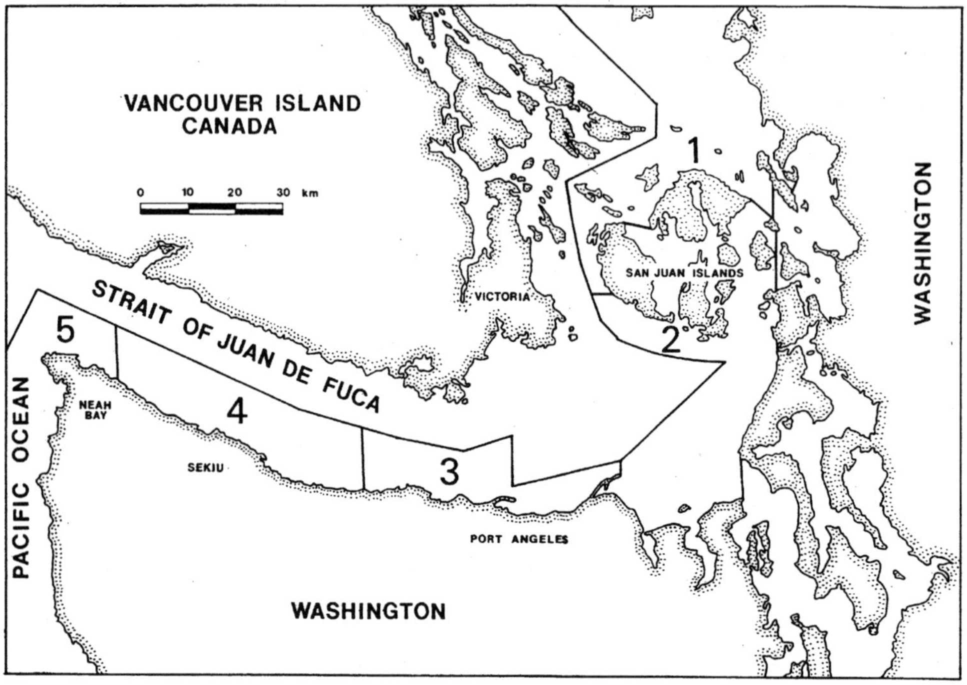

(Pfister and Bradbury, 1996). In particular, the populations in the San Juan Islands declined almost 50% between 1988 and 1989 after heavy exploitation in 1988. The effectiveness of past management strategies is thus questionable. Although there is a consensus on the desirability of limited entry and restricted season length among industry and management agencies, regulations pertaining to the 3-year areal rotation and minimum and maximum size limits have been the target of industry criticism.

Management has been hampered by insufficient knowledge of life-history parameters and the lack of total biomass estimates, so that stock assessments could not be properly carried out for the species. For the red sea urchin, the parameters need to be estimated from size–frequency data because a valid technique of age determination is not available. Also, size-dependent recruitment and natural and fishing mortalities are likely to occur and vary over time. Because of these stochastic and non-steady state conditions, the catch-at-size analysis (CASA) model (Sullivan et al. 1990) was applied; however, the CASA model was originally developed to use catch-at-size data of finfish. Therefore, extensive modifications of the CASA model are required to accommodate the special characteristics of the life history of the red sea urchin and the uses of size–frequency data obtained from surveys.

This paper was motivated by the urgency to develop a size-based method for estimation of life-history parameters and evaluation of various management regulations for a red sea urchin population using survey size–frequency data as the input. A Monte Carlo simulation was developed based on this model to explore alternative management strategies, mainly constant catch and constant harvest rate policies, and to evaluate the effectiveness of periodic harvest schedules and size limits. This paper, as well as that of Zheng et al. (1998), demonstrate that the framework of CASA can be easily modified depending on the suitable theoretical or empirical bases of the applications.

Materials and methods

The fishery

Commercial fishing areas for the red sea urchin in Washington extend from the northern end of the San Juan Islands through the Strait of Juan de Fuca to Cape Flattery on the Pacific Ocean (Fig. 1). Annual landings increased in the mid-1980's reaching a peak of 3658 t (metric ton) by 1988 (Fig. 2). After 1988, earlier closures, shortened seasons, and quotas were imposed. These measures reduced landings to 552 and 616 t in 1993 and 1994, respectively, but the price increased substantially. Current regulations are discussed below.

Size limits

Current size limits are 102–140 mm in the San Juan Islands (Districts 1 and 2) and 82–133 mm in the Strait of Juan de Fuca (Districts 3, 4, and 5). A minimum size limit was based on the first size of sexual maturity at 50 mm (Bernard and Miller 1973) and allowed urchins to spawn several times before entering the fishery. A maximum size limit was introduced based on studies suggesting that adults provide spine shelter and possibly augment the food supply for juveniles (Tegner and Dayton 1977; Breen et al. 1985; Sloan et al. 1987). Kato and Schroeter (1985) recommended a maximum size limit as a management device. The size limits imposed in Washington were intended to protect the smallest and largest 20% of the population, an arbitrary threshold that managers felt was satisfactory.

Periodic harvest

Starting in 1977, a 3-year periodic rotation of the fishing

Fig. 2. The (*a*) catches and fleet size value and (*b*) price of the red sea urchin in Washington.

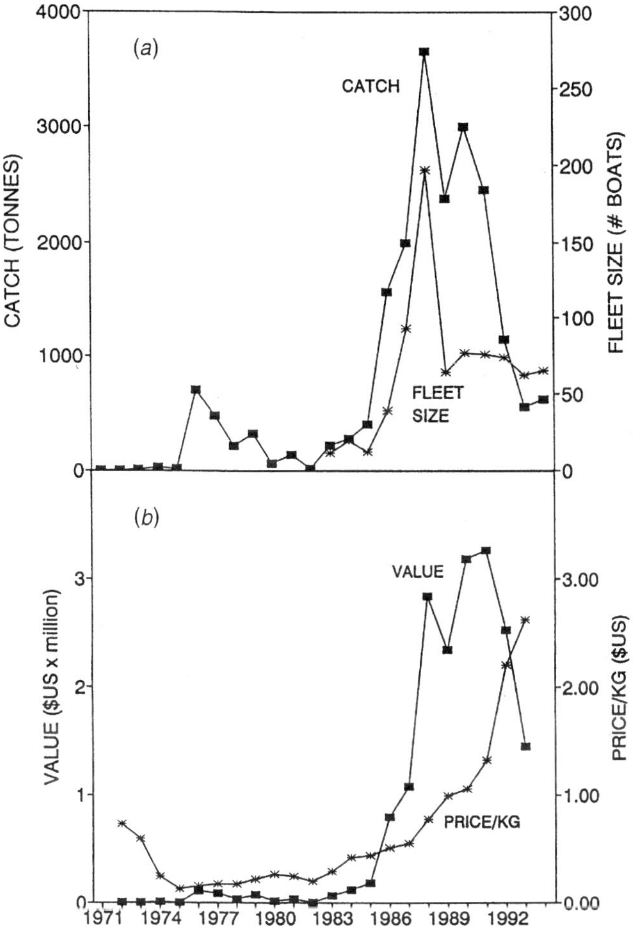

grounds was imposed for easier fleet monitoring and enforcement of fishing regulations, and the presumed recovery of harvestable stocks through recruitment and redistribution of stock. It was believed then that a closed period would allow some urchins to grow through the "window of vulnerable size" between the minimum and maximum size limits, and thus maintain sufficient stock size of adults.

Length of season

After 1977, the fishing season was restricted to the winter prespawning period when gonad yield was high. The length of the season was determined *ad hoc*, based on annual trends in survey density at established sites, predicted fishing effort, and landings in the previous season. Recently, the length of the season was reduced to less than a month.

Limited entry

Following the dramatic increase in fleet size (Fig. 2) and numerous poaching violations in 1988, limited-entry licensing was jointly proposed by the management agency and industry. A retroactive limited-entry law reduced the fleet from 197 boats in 1988 to 65 boats in 1994. The long-term goal is to reduce the fleet to 45 boats. Starting in 1988, only two divers per boat were allowed in the water at any one time. However, with the reduction of stock and an increase in fishing efficiency, only one diver per boat was allowed after 1990.

Data

Among the five districts, only District 1 has a sufficient history of survey data, and therefore the analyses are based on this district. Eight sites in District 1 were surveyed annually from 1988 to 1994 for density and size–frequency distributions (SFD), except 1989 when only six stations were surveyed and 1992 when the survey was suspended. These sites were originally selected based on a 1984 pilot survey within known productive commercial fishing grounds. At each site, two SCUBA divers counted red sea urchins within a 150 ft × 6 ft transect (total area = 83.6 m^2) and measured the test diameters of the first 20 individuals encountered at each end of the transect. Each of the transects roughly followed a depth contour, with the shallowest at 4.3 m and the deepest at 13.1 m.

Size selectivity (v_i) of urchins harvested by commercial divers in District 1 was estimated using a random SFD sample of 518 urchins from all boats unloading their landings in a processing plant in 1991. The size–weight relationship and proportion of maturity at size were based on 287 urchins collected in October 1986 – January 1987, at Point Wilson and the Southern San Juan Islands. The peristomal membranes were slit and the urchins were drained for 10 min prior to being weighed.

The modified CASA model

The catch-at-size analysis (CASA, Sullivan et al. 1990) model was modifed so that SFD collected from surveys could be used as the input data to estimate life-history parameters of the red sea urchin population. For convenience, the beginning of a year was assumed to be August 1 because the surveys were carried out in early August. The modified CASA (m-CASA) assumes that the urchin population will be exposed to a short fishing season immediately if it is on the periodic harvest schedule, and that the survivors from fishing will be continuously exposed to size-specific natural mortality during the year. At the end of the year (assumed to be July 31 of the next calendar year), the surviving individuals in any size category will grow into subsequent size categories in a way dependent upon a probability of growth from one size category into another in one unit of time interval. The resultant SFD, including the size-specific recruitment, is the SFD at the beginning of the next year.

Let $N_{i,t}$ be the population abundance in the ith size category in year t; $n_{i,t}$ be the population abundance index in the ith size category measured from survey in year t; R_t be the abundance of recruits in year t; p_{ij} be the probability that an individual grows from the ith size category to the jth one; S_i be the natural survival rate of the ith size category; u_t be the harvest rate of the fully vulunerable size category in year t; q_i be the catchability, i.e., the probability of sighting an urchin in the ith size category by a diver; v_i be selectivity, i.e., the probability of harvesting if a urchin in the ith size category is sighted and judged to be harvestable based on the size-limit regulation; and d_i be the proportion of recruits in the ith size category. The m-CASA is written mathematically by a system equation which describes the relationship of the states of the population between two consecutive years:

Fig. 3. Some graphic representations of life-history parameters. (A) Selectivity curves: Solid line indicates the estimated curve for the size limit [100, 140] based on 1991 samples; × indicates the lower leg of the hypothetical curve with minimum size limit of 50 mm; + indicates the upper leg of the hypothetical curve without maximum size limit. (B) Observed and estimated abundance indices. (C) The SR relationship and its estimated parameters. (D) Survival rates. (E) Y/R curve and the biological reference points, F_{max} and $F_{0.1}$. (F) Catchability and maturity curves.

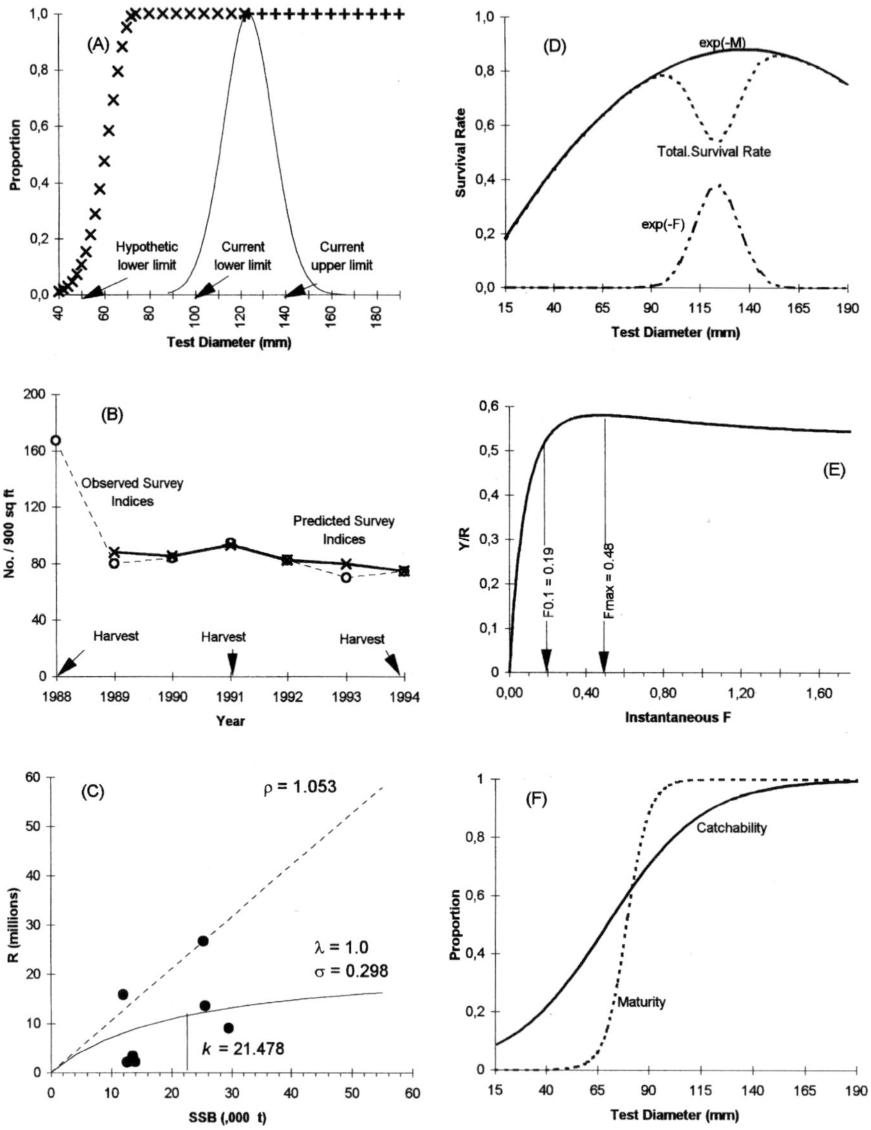

$$[1]\quad \begin{bmatrix} N_{1,t+1} \\ N_{2,t+1} \\ \vdots \\ \vdots \\ N_{L,t+1} \end{bmatrix} = \begin{bmatrix} S_1(1-v_1u_t)p_{11} & 0 & \cdots & \cdots & 0 \\ S_1(1-v_1u_t)p_{12} & S_2(1-v_2u_t)p_{22} & \ddots & & \vdots \\ \vdots & \vdots & & \ddots & 0 \\ S_1(1-v_1u_t)p_{1L} & S_2(1-v_2u_t)p_{2L} & \cdots & \cdots & S_L(1-v_Lu_t)p_{LL} \end{bmatrix} \times \begin{bmatrix} N_{1,t} \\ N_{2,t} \\ \vdots \\ N_{L,t} \end{bmatrix} + R_{t+1} \begin{bmatrix} d_1 \\ d_2 \\ \vdots \\ d_L \end{bmatrix}$$

and by an observation equation that links together the state of the population and the observed survey abundance indices

$$[2]\quad \begin{bmatrix} n_{1,t} \\ n_{2,t} \\ \vdots \\ \vdots \\ n_{L,t} \end{bmatrix} = \begin{bmatrix} q_1 & 0 & \cdots & \cdots & 0 \\ 0 & q_2 & \ddots & & \vdots \\ \vdots & \ddots & \ddots & \ddots & \vdots \\ \vdots & & \ddots & \ddots & 0 \\ 0 & \cdots & \cdots & 0 & q_L \end{bmatrix} \times \begin{bmatrix} N_{1,t} \\ N_{2,t} \\ \vdots \\ N_{L,t} \end{bmatrix} + \begin{bmatrix} e_{1,t} \\ e_{2,t} \\ \vdots \\ e_{L,t} \end{bmatrix}$$

where $e_{i,t}$ is the random error and is assumed to be normally distributed with mean 0 and variance $\sigma_e 2$. Although the additive random error is made for convenience, a multiplicative random error is also possible. However, residual plots obtained from the two assumptions do not reveal a substantial difference.

The probability that an individual grows from the ith size category to the jth one is assumed dependent on a von Bertalanffy growth equation and a gamma distribution. Using a gamma distribution with parameters α and β, the probability is calculated by

$$[3] \quad p_{ij} = \int_{l_i-\delta/2}^{l_i+\delta/2} \frac{x^{\alpha-1}e^{-x/\beta}}{\Gamma(\alpha)\,\beta^{\alpha}}\,dx$$

where l_i and δ are the midsize and the width of the ith size category, respectively. The mean (μ_i) and variance ($\sigma_i 2$) of this gamma distribution can be reparameterized by the size increment, via a von Bertalanffy growth equation, in a 1-year period

$$[4] \quad \mu_i = \alpha\beta = \Delta l_i = (L_\infty - l_i)(1 - e^{-K}) \text{ and}$$

$$\sigma_i 2 = \alpha\beta^2 = \beta\Delta l_i$$

Then, the parameters β, L_∞, and K can be estimated. The gamma distribution is used because of the properties given in eq. 4 and because it can mimic various distributions (Sullivan et al. 1990). Other distributions, such as beta, normal, and lognormal, can be used depending on the applications.

Juvenile urchins have a low survival rate due to predation (Tegner and Dayton 1977; Sloan et al. 1987). In California waters, sea otters, spiny lobsters, and other predators are normally abundant and prey on adolescent urchins, whereas similarly heavy predation has not been reported in Washington. Also, we found that the survival rate for adolescents is probably higher than that for juveniles and elders when exploratory analyses were carried out in the early stage of model development. Therefore, the natural survival rate is assumed to be a quadratic curve

$$[5] \quad S_i = c_0 + c_1 l_i + c_2 l_i 2$$

Other forms of natural mortality can be used if they are valid.

The size-dependent catchability of the dive survey is assumed to be a logistic curve and is equal to 1.0 for at least one fully observable size category

$$[6] \quad q_i = [1 + \exp(\alpha_q + \beta_q l_i)]^{-1}/\max(q_i)$$

The catchability q_i is the probability of an urchin in the ith size category being found by a diver. We assumed that the catchability (q_i) of survey and commercial divers are equal. The selectivity (v_i) conditioned on q_i is then estimated from a random SFD sample of landings in 1991 and the survey SFD (Fig. 3A). The probability of harvesting an urchin from the population (N_{it}) in the ith size category is the product of q_i and v_i.

After 1989, the fishing season in District 1 was less than a month. Because of the short season, we assumed that harvest was the instantaneous removal of a $v_i u_t q_i$-portion of individuals from the population so that the catch-at-size in year t can be calculated by

$$[7] \quad C_{i,t} = v_i u_{i,t} q_i N_{i,t}$$

However, the Baranov catch equation should be used if the fishery is year-round. Because harvest occurred only in 1991 during the period 1989–1994, only u_{91} was estimated.

By analogy with Sullivan et al. (1990), the following relationship is derived

$$[8] \quad n_{i,89} = n_{89}\frac{f_i/q_i}{\Sigma f_i/q_i}$$

where n_{89} (= $\Sigma n_{i,89}$) and $f_i = n_{i,89}/n_{89}$. Therefore, only n_{89} was estimated. The estimated recruitment (R_t) is distributed into different size categories by a set proportion $\{d_i\}$. We examined the SFD of recruits from different years and areas and found that the SFD's generally fit well to normal distributions. Therefore, d_i is calculated by

$$[9] \quad d_i = \int_{l_i-\delta/2}^{l_i+\delta/2} \Phi(x|\mu_r, \sigma_r^2)dx$$

where Φ is a normal distribution with mean size μ_r and variance $\sigma_r 2$ to be estimated.

In summary, the parameters to be estimated are n_{89}, R_{90}, R_{91}, R_{92}, R_{93}, R_{94}, μ_r, $\sigma_r 2$, L_∞, K, β, c_0, c_1, c_2, u_{91}, α_q, and β_q. Using the SFD's collected from 1989 to 1994 (Fig. 4) and the selectivity in 1991 as the input data, a least-squares technique is applied to minimize the sum of squares

$$[10] \quad SSQ = \Sigma_t\Sigma_i\,(n_{i,t} - \hat{n}_{i,t})^2$$

for estimating parameters and where $\hat{n}_{i,t}$ is the estimate of $n_{i,t}$ from the model (eqs. 1 and 2). Using eq. 7 and knowing $C_{i,91}$, the values of $N_{i,91}$ for all i's are calculated. Then, the estimates of life-history parameters (Table 1) and eq. 1 are applied to calculate the values of $N_{i,94}$, which are the essential input for the simulation model described in the following.

Simulation model for management strategies

In addition to the parameters estimated from m-CASA, parameters for the stock-recruitment (SR) model also need to be estimated for the simulation model. The recruitment is estimated by inputting estimated recruits from m-CASA into the Shepherd SR model (Shepherd 1982) with lognormally distributed variance

$$[11] \quad R_t = \frac{\rho\, SSB_{t-2}}{1 + (SSB_{t-2}/\kappa)^{\lambda}}\, e^{\varepsilon_t}$$

where $\varepsilon_t \sim N(0, \sigma^2)$, SSB is spawning stock biomass, and ρ, κ, and λ are parameters of the model. The SSB is calculated by

$$[12] \quad SSB_t = \Sigma_i w_i m_i N_{i,t}$$

where m_i is the proportion of sexually mature individuals in the ith size category (Fig. 3F) and w_i is the weight in grams of an urchin in the ith size category, calculated by the size–weight relationship, $w_i = 0.00116\, l_i 2.69$ which is derived from the data mentioned above. The advantages of using the Shepherd SR model are discussed later. Because of the short time series of SFD, the recruitments in 1984, 1988, and 1989 are approximately estimated from the respective SFD using the estimates of μ_r and $\sigma_r 2$ from the m-CASA. The SSB in 1982 is assumed to be equal to that in 1984, and the SSB in 1986 and 1987 equal to that in 1988. The parameters in eq. 11 are estimated by the procedure described in Shepherd (1982) with $\lambda = 1$, which is analogous to the Beverton-Holt SR relationship.

Fig. 4. Observed and estimated size–frequency distributions from 1989 to 1994. SDF was not collected in 1992. (×, observations; —, estimates).

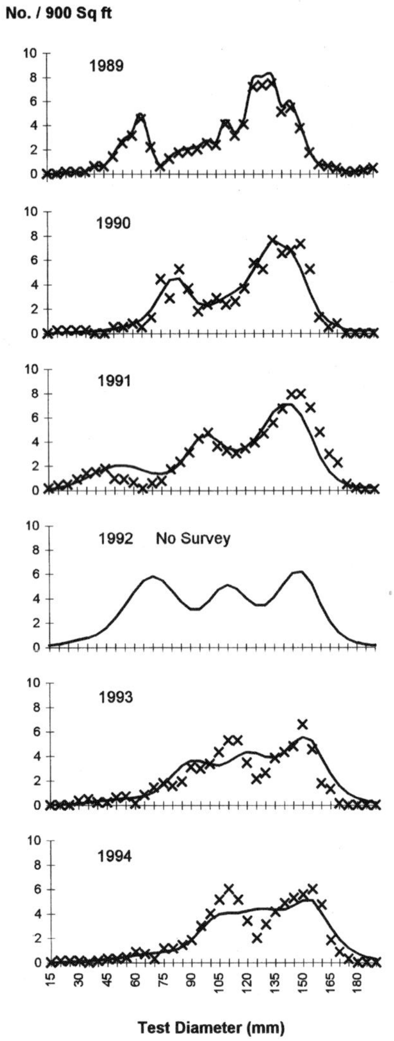

Using the framework of m-CASA described in eq. 1 and the SR relationship in eq. 11, a simulation model was developed to compare the effects of constant-catch (Const-C) and constant-harvest-rate (Const-F) policies, with combinations of four different periodic harvest schedules and four different size-limits, on the abundance and yield of the population. The four periodic harvest schedules are: (*i*) harvest occurs every year, 1-year periodic harvest schedule, (*ii*) harvest occurs once every other year, 2-year periodic harvest schedule, (*iii*) harvest occurs once every three years, 3-year periodic harvest schedule, and (*iv*) harvest occurs once every five years, 5-year periodic harvest schedule. The 3-year periodic harvest schedule is currently employed.

The four size-limit regulations (in millimetre test diameter) are: (*i*) the currently employed minimum and upper limits, size limit [100, 140], (*ii*) the currently employed minimum size limit but without the maximum size limit, size limit [100, ∞], (*iii*) size limit [50, 140], and (*iv*) size limit [50, ∞]. Size limit (*ii*) is used to address whether a maximum size limit is necessary under the current management regime. Size limit (*iii*) is used to address whether a minimum size limit is necessary and size limit (*iv*) is used to address the impacts on population if there is no size limit regulation. The minimum size limit of 50 mm is used because the red sea urchin becomes sexually mature at 50 mm (Bernard and Miller 1973) and because urchins smaller than 50 mm are not marketable. Furthermore, the selectivity curve around 50 mm is assumed to be the same as that around the current minimum size limit of 100 mm. The maximum size limit in (*iii*) is assumed to be the same as that around the current maximum size limit of 140 mm. The selectivity curves of the four size-limit regulations are depicted in Fig. 3A.

Simulation started in August 1994 and was projected into a 200-year time horizon with 500 Monte Carlo simulations for all combinations of harvest policies and size-limit regulations. A threshold level of 30 urchins per 83.6 m^2 was subjectively selected. This is around 20% of the spawning stock in August 1988 when the stock size was still high before the heavy harvest. This threshold level was converted to an equivalent population of 9×10^6 urchins in District 1. Subsequently, the probability of spawning stock number (SSN) declining to $<9 \times 10^6$ urchins, $P(\mathrm{SSN} \le 9 \times 10^6)$, is called the risk of stock collapse in the following text. The yield (Y), variability (coefficient of variation, CV) of yield, spawning stock number, and probability of SSN becoming less than nine million urchins, $P(\mathrm{SSN} \le 9 \times 10^6)$, are used as criteria to compare various combinations of harvest policies and size-limit regulations. The yield and SSN are calculated as

[13] $Y_t = \Sigma q_i v_i u_i w_i N_{i,t}$ and

[14] $\mathrm{SSN}_t = \Sigma_i m_i N_{i,t}$

where m_i is defined as in eq. 12.

Results

Parameter estimates and stock conditions

The SFD collected in District 1 from 1989 to 1994 (Fig. 4) were input into m-CASA to estimate the life-history parameters (Table 1). The estimated survey abundance indices decreased slightly from 1989 to 1994 with an average around 48% of that in 1988 (Fig. 3B). In 1991, the harvest season in District 1 was shortened. The 1991 catch was 1.3×10^3 t and the estimated harvest rate in 1991 was 0.39 for the fully selected size category (i.e., 123 mm). This restricted harvest in 1991, which was intended to stop further decline of the stock, has likely stabilized the abundance since 1991 (Fig. 3B).

The SR relationship and its estimated parameters are shown in Fig. 3C. The year-to-year recruitments are highly variable. Since 1989, the SSB has been less than κ (the level of SSB above which density-dependent effects take over), and the recruitments in 1992–1994 were at a record low and are all smaller than the model predicted value. Because of these circumstances, stock status of the red sea urchin should be assessed more frequently.

The estimated size-dependent natural, fishing, and total survival rates are shown in Fig. 3D. Weighted by the SFD over years, the estimated average natural mortality rate is 0.16. Using the estimates of L_∞ and K and assuming knife-edged recruitment at age 4 (approximately 100 mm in test diameter), biological reference points based on the *Y/R* model (Beverton

Table 1. List of life-history parameters for red sea urchins in District 1, estimated from the m-CASA.

Category	Parameters	Value
Growth	L_∞ (mm)	152.500
	K	0.249
	β	0.391
1989 abundance index	n_{89}	87.921
Survey catchability	α_q	3.023
	β_q	−0.043
Natural survival rate	c_0	9.980×10^{-2}
	c_1	1.278×10^{-2}
	c_2	-4.651×10^{-4}
1991 harvest rate	u_{91}	0.378
Recruitment SDF	μ_r	15.472
	σ_r	0.378
Recruitment	$R_{90}\ (\times 10^6)$	2.211
	$R_{91}\ (\times 10^6)$	15.788
	$R_{92}\ (\times 10^6)$	2.159
	$R_{93}\ (\times 10^6)$	3.376
	$R_{94}\ (\times 10^6)$	2.123
1994 population SDF	$[N_{i,94}]\ (\times 10^6)$	[0.045, 0.078, 0.124, 0.177, 0.229, 0.273, 0.318, 0.359, 0.394, 0.427, 0.451, 0.461, 0.463, 0.488, 0.590, 0.804, 1.089, 1.333, 1.440, 1.423, 1.385, 1.397, 1.428, 1.414, 1.363, 1.356, 1.445, 1.558, 1.545, 1.211, 0.863, 0.570, 0.355, 0.212, 0.123, 0.072]

and Holt 1957) are estimated by the method of Thompson and Bell (Ricker 1975). The estimate of F_{max} (fishing mortality at which Y/R is maximized) is 0.48 and $F_{0.1}$ (at which the slope of the Y/R curve is 10% of the slope at origin) is 0.19 (Fig. 3E). Because the calculations of F_{max} and $F_{0.1}$ are based on knife-edge recruitment and annual harvest, their uses for a fishery with a maximum size limit and periodic harvest schedule need more investigation.

Const-F versus Const-C policy for the 30-year time horizon

Yield and its variability

Under a Const-F policy, the average annual yields of the next 30 year will increase gradually to a peak and then decrease to asymptotes as the target harvest rate increases, except that of the 3-year and 5-year periodic harvest schedules with minimum and upper limits (Fig 5*a*). Under a Const-C policy, the average yields increase linearly with the target quota initially when the population can provide sufficient harvestable biomass to meet the target quota. However, the average yield starts to depart from linearity when the population does not have enough harvestable biomass to meet the target quota. After reaching a peak, the average yields decrease and approach asymptotes (Fig. 5B). It is no surprise that a greater average yield is associated with the size limit [100, ∞] than with [100, 140] because there is more harvestable biomass. When excessive harvest pressures, either in the form of target harvest rate or target quota, are applied, however, the average yield converges. Also, the results show that the periodic harvest schedule produces only a large one-time catch. When the yields of 2-year and greater periodic harvest schedules are divided by the length of the period, the annual yields are not substantially higher than that of the 1-year periodic harvest schedule (Figs. 5A and 5B).

In general, the variability of average annual yield is projected to decrease as the frequency of periodic harvests decreases, but increase with increasing target harvest rate or target quota (Figs. 5C and 5D). Under a Const-F policy (Fig. 5C), the CV of yield would be smaller for the size limit [100, 140] than [100, ∞]. Under a Const-C policy (Fig. 5D), there is no variability (CV = 0) as long as the population can provide the target quota. Thereafter, the CV of yield is greater for the size limit [100, 140] than [100, ∞] until the target quota is reached. Because the target quota produces the peak yield, the population is likely to provide enough harvestable biomass to meet the quota. Under the size limit [100, 140], the CV of yield under a Const-F policy is greater than or equal to that under a Const-C policy (at $u = 1.0$). Under the size limit [100, ∞], the Const-C policy has a smaller CV than the Const-F policy.

Spawning stock number and probability of SSN less than the threshold

In general, the average SSN over the next 30 year increases (Figs. 6A and 6B) and the $P(\text{SSN} \leq 9 \times 10^6)$ decreases with increasing years of periodic harvest schedule (Figs. 6C and 6D). Under the Const-F policy, the SSN is greater (Fig. 6A) and the $P(\text{SSN} \leq 9 \times 10^6)$ is smaller (Fig. 6C) for the size limit [100, 140] than for [100, ∞]. Under the Const-C policy, the SSN is smaller (Fig. 6B) and the $P(\text{SSN} \leq 9 \times 10^6)$ is greater (Fig. 6D) for the size limit [100, 140] than for [100, ∞] up to the target quota that produces the peak yield. Thus, with the same amount of target quota, the size limit [100, 140] poses

Fig. 5. Comparison of mean annual yield and CV of yield for Const-F (A and C), Const-C (B and D) policies with size limit [100, 140] (broken lines), and size limit [100, ∞] (solid lines) for 1-year, 2-year, 3-year, and 5-year periodic harvest schedules. Numbers in each panel indicate four periodic harvest schedules.

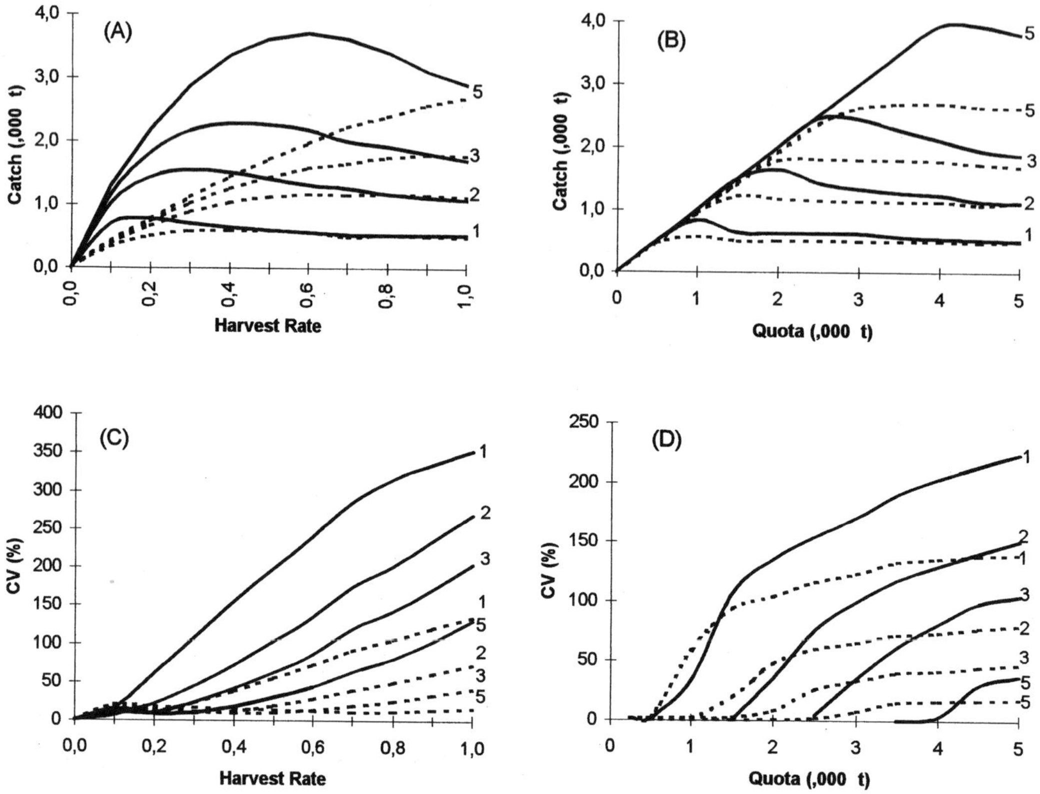

Fig. 6. Comparison of mean SSN and probability of SSN less than the threshold of nine million urchins for Const-F (A and C) and Const-C (B and D) with size limit [100, 140] (broken lines) and size limit [100, ∞] (solid lines) for 1-year, 2-year, 3-year, and 5-year periodic harvest schedules. Numbers in each panel indicate four periodic harvest schedules.

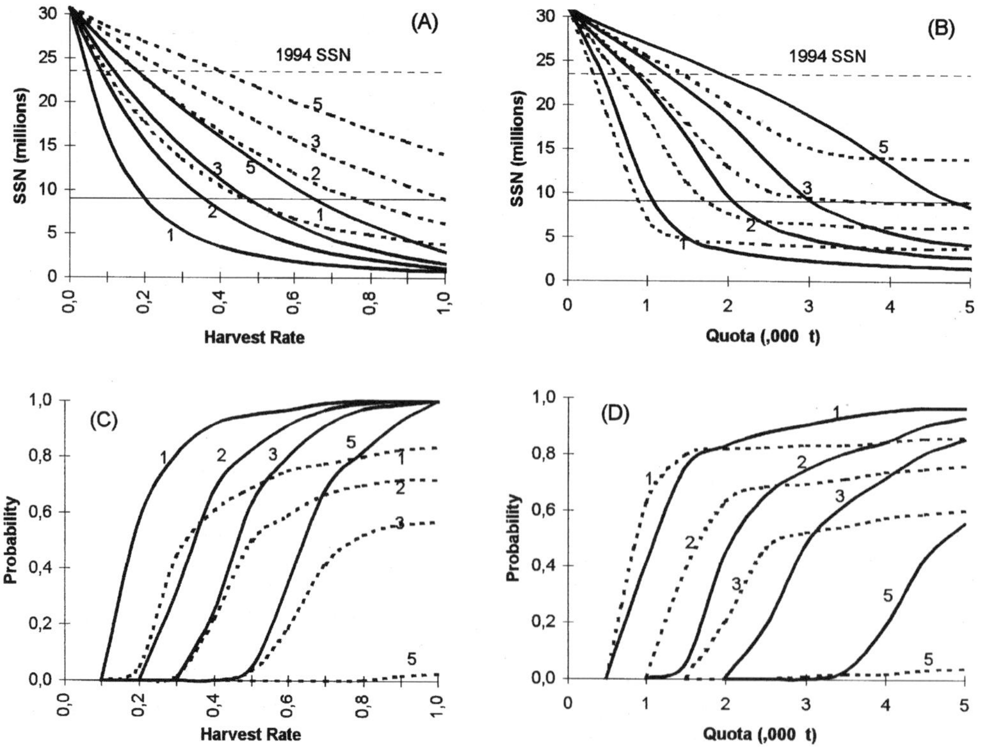

Fig. 7. Comparison of mean SSN and probability of SSN less than the threshold of nine million urchins for Const-F (A and C) and Const-C (B and D) with size limit [50, 140] (broken lines) and size limit [50, ∞] (solid lines) for 1-year, 2-year, 3-year, and 5-year periodic harvest schedules. Numbers in each panel indicate four periodic harvest schedules.

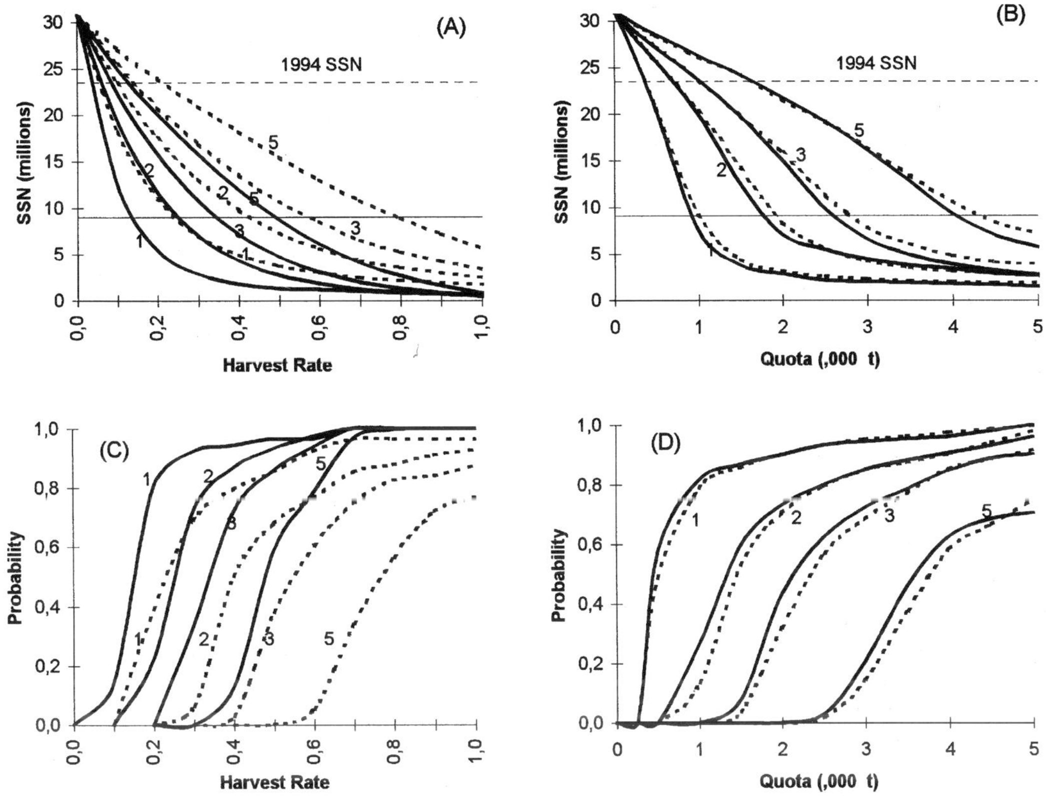

more danger to the population than [100, ∞] because most of the urchins within [100, 140] are harvested and unable to contribute to spawning stock. Also, under a Const-C policy, $P(\text{SSN} \leq 9 \times 10^6) = 0$ when the populations can provide enough harvestable biomass to meet the target quota.

Under the size limit [100, 140], the $P(\text{SSN} \leq 9 \times 10^6)$ is smaller for the Const-F policy than the Const-C policy. In contrast, under the size limit [100, ∞], the $P(\text{SSN} \leq 9 \times 10^6)$ is greater for the Const-F policy than the Const-C policy. From the results shown in Figs. 5 and 6, the $P(\text{SSN} \leq 9 \times 10^6)$ will be small in the next 30 year if the harvest level is established at a harvest rate of 0.39 (for the fully vulnerable size at 123 mm) or a quota of 1.3×10^3 t under the current management regime in District 1, i.e., a 3-year periodic harvest schedule with the size limit [100, 140], providing that the biological and fishery conditions of the stock remain unchanged.

The need for a minimum size limit

Simulations were carried out for the size limits [50, 140] and [50, ∞] under both Const-F and Const-C policies to address the effects of the no minimum size limit regulations on the population. The average SSN under the no minimum size limit regulations (Figs. 7A and 7B) decreases faster than that under the minimum size limit regulations (Figs. 6A and 6B). The risk without the minimum size limit is twice that with the minimum size limit (Fig. 7C versus Fig. 6C and Fig. 7D versus Fig. 6D). The sustainability of the population under harvest is in question and a more restricted control of effort or catch is required.

Equilibrium analysis and the biological reference point

In searching for a useful biological reference point, the simulations were allowed to run into a 200-year time horizon. The results reveal that the population converges to an almost stationary state after 100 year. For convenience, let the average yield and the average parameters of the stock from years 171 to 200 be in equilibrium states. Applying the method of Lai et al. (1993), the relationship of equilibrium yield versus harvest rate obtained from the results of a Const-F policy mimics the surplus production model (Fig. 8A). At the maximum equilibrium yield (MYEQ), which is considered to be the MSY of the surplus production model, the corresponding harvest rate is u_{MYEQ}. The intersections of the yield curves and the SSN curves with the x-axis are the harvest rates at which the stock eventually becomes extinct (Figs. 8A and 8B) and thus this harvest rate is defined as u_{ext}.

The MYEQ is greater and the SSN_{eq} and u_{ext} are smaller for the size limit [100, ∞] than for [100, 140] (Figs. 8A and 8B). The u_{ext} under a 1-year periodic harvest schedule with the size limit [100, ∞] is 0.2, the smallest value for all regulations (Fig. 8C). Comparing $u_{\text{ext}} = 0.2$ with $M \approx 0.16$ and $F_{0.1} = 0.19$ for this stock, the regulation of a 1-year periodic harvest schedule with the size limit [100, ∞] is very risky. The $P(\text{SSN} \leq 9 \times 10^6)$ increases exponentially by increasing the harvest rate by 0.1 above u_{MYEQ}, although there is no risk at u_{MYEQ} (Fig. 8C). This result suggests that managers should avoid setting target quotas too close to MYEQ. As shown also in Figs 5, 6, and 7, a regulation without a maximum size limit may

Fig. 8. The mean equilibrium yield versus harvest rate for the four periodic harvest schedules with the size limits [100, 140] (broken lines) and [100, ∞] (solid lines).

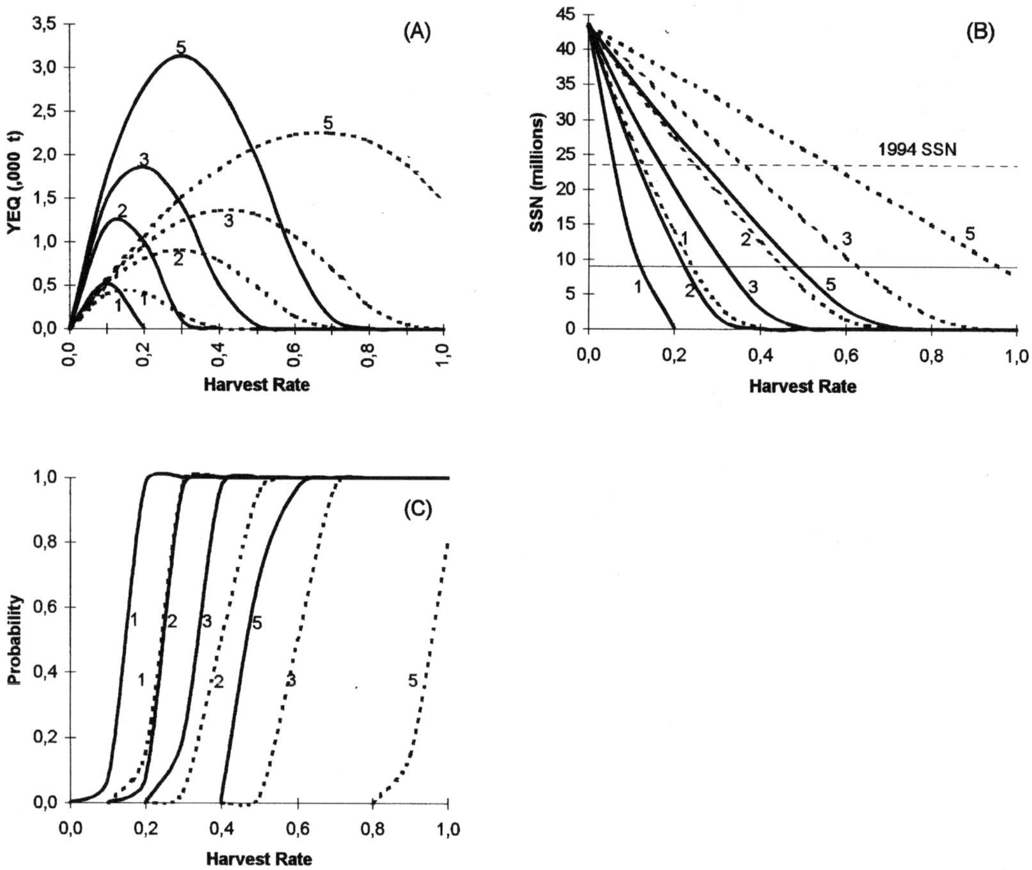

slightly increase annual yield but it also increases the fluctuation of yield and the risk of stock collapse.

Discussion

The general framework of the CASA model (Sullivan et al. 1990) is flexible, so that it can be modified in many ways to fit the empirical and theoretical bases of the interested fishery resource. The model's independence of age data is particularly well suited to many invertebrates and finfish for which the aging technique is not available or reliable. Sea urchins fit into this category. Zheng et al. (1998), for example, modified the CASA model for Tanner crab in the eastern Bering Sea. Unfortunately, SFD's from harvested sea urchin are not available; therefore, the analysis is solely dependent on survey data. The precision and accuracy of estimates will improve if both catch and survey SFD's are available. A short history of data and a missing critical survey before the 1992 harvest season may be the causes of discrepancies between observed and predicted SFD's in 1993 and 1994. The m-CASA model, nonetheless, captures the trend of abundance reasonably well under these constraints. It should be pointed out that if the fitted values are plotted with the moving averages of five consecutive size categories, the discrepancies in 1993 and 1994 disappear.

Either a Const-F or a Const-C policy can be used for the management of the red sea urchin population in Washington. The choice is dependent on whether managers intend to minimize the year-to-year variability of yield or the risk of stock collapse, e.g., $P(\mathrm{SSN} \le 9 \times 10^6)$. If a Const-F policy is recommended, a maximum size limit such as that currently employed, 140 mm, will reduce both year-to-year variability of yield and $P(\mathrm{SSN} \le 9 \times 10^6)$; however, if a Const-C policy is selected, a maximum size limit seems not as critical.

Botsford et al. (1993) has assumed that a minimum of 0.7 urchins·m^{-2} is necessary for a successful broadcast spawning based on the studies of Levitan (1991) and Levitan et al. (1989). In this paper, however, we found that a threshold density of 0.7 urchins·m^{-2} (equivalently, SSN $\approx 17 \times 10^6$ in 1994) would be too low because the probability of the population declining to less than this level is >90% in all cases. Although survey results show that recruitment fluctuated widely in past years, there has been no apparent recruitment failure. We realize that the risk defined as $P(\mathrm{SSN} \le 9 \times 10^6)$ is a subjective selection although it is around 20% of the 1988 SSN. Further studies are essential to define this threshold SSN level.

Because weak recruitments are likely associated with low SSB (Fig. 3C), Shepherd's SR model is used in this paper. In this SR model, recruitment becomes more density-independent when SSN < κ, indicating that the population is subject to a greater danger of recruitment failure. The estimated SSN is always smaller than the estimated value of κ during the period of 1992–1994 (Fig. 3C). If SSN and recruitment remained at these average levels after 1995, our projected annual yield and stock condition are too optimistic. Using the other simulation, in which the recruitment after 1994 was assumed to be normally distributed with the mean and variance estimated from the recruitments from 1992 to 1994, it was shown that there are only five harvest seasons under a target quota of 1×10^3 t and under the current management regime until the SSN < 9×10^6. Simulations with a larger value of κ or with a smaller value of λ were also run. The results of these simulations showed that there is a greater variability of yield and SSN and a higher risk of stock collapse.

One reviewer pointed out that the projections of SSN, risk, and yield over 200 year have little meaning because of lack of a time-dependent term in the SR model and potential nonstationary recruitment. We emphasize that the simulation technique was only used to approximate an equilibrium condition, to obtain biological reference points, and to infer the potential crises due to harvest. The biological reference points (e.g., F_{max} and $F_{0.1}$ from the yield per recruit model and F_{MSY} from the production model) are traditionally calculated by equilibrium analysis and serve as a threshold harvest level. They should not be used as the target harvest rate.

The uncertainties surrounding the SR relationship have not been completely modeled in the simulation partially due to a short history of data. The crucial uncertainties include autocorrelation and nonstationary recruitment series (for analytical methods, see Sigler and Fujioka 1993), spine canopy shelter (Tegner and Dayton 1977; Breen et al. 1985; Sloan et al. 1987), successful broadcast spawning (Levitan 1991; Levitan et al. 1989), and spatial exchange of larvae (Botsford et al. 1993). The SR relationship plays the critical role in harvest management. The long-term and timely collection of data should be a priority of sea urchin management.

The use of u_{MYEQ} as a biological reference point under the current management regime is more appropriate than the uses of traditional F_{max} and $F_{0.1}$. This is because F_{max} and $F_{0.1}$ are based on knife-edged recruitment to the fishery occurring annually. In Washington, minimum and maximum size limits and periodic harvest are used in the regulation. Gulland and Boerema (1973) and many others have recommended that the harvest level should be set below u_{MYEQ} for the conservation of stock. For a stock with SSN < SSN_{MYEQ}, the harvest level should be reduced further to allow the stock to rebuild. Because the red sea urchin population in District 1 is close to SSN_{MYEQ}, a more conservative harvest level should be used.

As shown in Figs. 5A and 5B, the periodic harvest schedule produces a large one-time catch and the average annual yield is only slightly greater than that of a 1-year periodic harvest schedule. This finding is in agreement with that found in Botsford et al. (1993). The major advantages of a periodic harvest schedule are to reduce year-to-year variability of yield and the risk of stock collapse. Another advantage might be to reduce the costs of management and enforcement because the distribution of the fleet becomes more concentrated.

The Const-C policy will produce minimum average catches and a higher probability of stock collapse than the Const-F policy (Murawski and Idoine 1989; Overholtz 1993). Conversely, the Const-F policy will produce high variability in yield (Getz et al. 1987; Overholtz 1993). From our results, these generalizations are only true with the size limit [100, ∞]. The probability of stock collapse is higher under the Const-C policy with the size limit [100, 140] (Fig. 6). Managers should recognize that regardless of the selection of policy and regulations there is always a ceiling of harvest level (in terms of harvest rate or quota), a risk of stock collapse, and a fluctuation of yield over the years. Managers can select either a risk-prone policy with maximal yield (or revenue), or a risk-averse policy with lower yield (or revenue).

Acknowledgments

Part of this study was conducted when the senior author was an employee of the Washington Department of Fish and Wildlife (WDFW). The senior author thanks Northeast Fisheries Science Center (NEFSC), NMFS, NOAA, for the support to present this article in the Symposium and valuable discussions and comments from the NEFSC scientists on the contents of this paper. We thank the constructive comments from two reviewers. Thanks to L. Goodwin, D. Rothaus, and B. Sizemore of WDFW for participating in dive surveys and M. Barker of WDFW for his courageous support of this study.

References

Bernard, F.R, and Miller, D.C. 1973. Preliminary investigation of the red sea urchin (*Strongylocentrotus franciscanus*, Agassiz) resources of British Columbia. Fish. Res. Board Can. Tech. Rep. 400. 37 p.

Beverton, R.J.H., and Holt, S.J. 1957. On the dynamics of exploited fish populations. Chapman and Hall Ltd., London, U.K. Reprinted 1993. 533 p.

Botsford, L.W., Quinn, J.F., Wing, S.R., and Brittnacher, J.G. 1993. Rotating spatial harvest of a benthic invertebrate, the red sea urchin, *Strongylocentrotus franciscanus*. *In* Proceedings of the International Symposium on Management Strategies for Exploited Fish Populations. *Edited by* G.H. Kruse, D.M. Eggers, R.J. Marasco, C.C Pautzke, and T.J. Quinn, II. Alaska Sea Grant, College Program Report, AK-SG-93-02. University of Alaska, Fairbanks, AK. pp. 409–428.

Breen, P.A., Carolsfeld, W., and Yamanaka, K.L. 1985. Social behavior of juvenile red sea urchins, *Strongylocentrotus franciscanus* (Agassiz). J. Exp. Mar. Biol. Ecol. **92**: 45–61.

Getz, W.M., Francis, R.C., and Swartzman, G.L. 1987. On managing variable marine fisheries. Can. J. Fish. Aquat. Sci. **44**: 1370–1375.

Gulland, J.A., and Boerema, L.K. 1973. Scientific advice on catch level. Fish. Bull. (U.S.), **71**: 325–335.

Kato, S., and Schroeter, S.C. 1985. Biology of the red sea urchin, *Strongylocentrotus franciscanus*, and its fishery in California. Mar. Fish. Rev. **47**: 1–20.

Lai, H.L., Mug, M., and Gallucci, V.F. 1993. Management strategies for the tropical corvina reina, *Cynoscion albus*, in a multi-mesh size gillnet artisanal fishery. *In* Proceedings of the International Symposium on Management Strategies for Exploited Fish Populations. *Edited by* G.H. Kruse, D.M. Eggers, R.J. Marasco, C.C. Pautzke, and T.J. Quinn, II. Alaska Sea Grant, College Program Report, AK-SG-93-02. University of Alaska, Fairbanks, AK. pp. 21–37.

Levitan, D.R. 1991. Influence of body size and population density on fertilization success and reproductive output in a free spawning invertebrate. Biol. Bull. **181**: 261–268.

Levitan, D.R., Sewell, M.A., and Chia, F.-S. 1989. How distribution and abundance influence fertilization success in the sea urchin *Strongylocentrotus franciscanus*. Ecology, **73**: 248–254.

Murawski, S.A., and Idoine, J.S. 1989. Yield sustainability under constant-catch policy and stochastic recruitment. Trans. Am. Fish. Soc. **118**: 349–367.

Overholtz, W.J. 1993. Harvesting strategies and fishing mortality reference point comparisons for the northwest Atlantic stock of Atlantic mackerel (*Scomber scomber*). Can. J. Fish. Aquat. Sci. **50**: 1749–1756.

Pfister, C.A., and Bradbury, A. 1996. Harvesting red sea urchins: recent effects and future predictions. Ecol. Appl. **6**: 298–310.

Ricker, W.E. 1975. Computation and interpretation of biological statistics of fish populations. Bull. Fish. Res. Board. Can. No. 191. 382 p.

Shepherd, J.G. 1982. A versatile new stock-recruitment relationship for fisheries and the construction of sustainable yield curves. J. Cons. Int. Explor. Mer. **40**: 67–75.

Sigler, M.F., and Fujioka, J.T. 1993. Comparison of policies for harvesting sablefish, *Anaplopoma fimbria*, in the Gulf of Alaska. *In* Proceedings of the International Symposium on Management Strategies for Exploited Fish Populations. *Edited by* G.H. Kruse, D.M. Eggers, R.J. Marasco, C.C Pautzke, and T.J. Quinn, II. Alaska Sea Grant, College Program Report, AK-SG-93-02. University of Alaska, Fairbanks, Alaska. pp. 7–19.

Sloan, N.A., Lauridsen, C.P., and Harbo, R.M. 1987. Recruitment characteristics of the commercially harvested red sea urchin, *Strongylocentrotus franciscanus*, in southern British Columbia. Fish. Res. (Amsterdam), **5**: 55–69.

Sullivan, P.J., Lai, H.L., and Gallucci, V.F. 1990. A catch-at-length analysis that incorporates a stochastic model of growth. Can. J. Fish. Aquat. Sci. **47**: 184–198.

Tegner, M.J. 1989. The feasibility of enhancing red sea urchin, *Strongylocentrotus franciscanus*, stocks in California: an analysis of options. Mar. Fish. Rev. **51**: 1–22.

Tegner, M.J., and Dayton, P.K. 1977. Sea urchin recruitment patterns and implications of commercial fishing. Science (Washington, D.C.), **196**: 324–326.

Zheng, J., Kruse, G.H., and Murphy, M.C. 1998. A length-based approach to estimate population abundance of Tanner crab, *Chionoecetes bairdi*, in the eastern Bering Sea. *In* Proceedings of the North Pacific Symposium on Invertebrate Stock Assessment and Management. *Edited by* G.S. Jamieson and A. Campbell. Can. Spec. Publ. Fish. Aquat. Sci. 125. pp. 97–105.

A length-based approach to estimate population abundance of Tanner crab, *Chionoecetes bairdi*, in Bristol Bay, Alaska[1]

Jie Zheng, Gordon H. Kruse, and Margaret C. Murphy

Abstract: We constructed a length-based population model to estimate population abundance of Tanner crab, *Chionoecetes bairdi*, in Bristol Bay, Alaska. The model incorporates variable growth and gradual recruitment over length. A nonlinear least-squares approach was used to estimate abundance, recruitment, and natural mortality. The model was applied to abundance and catch data from 1975 to 1994 and fit observed abundances well. The model-reduced survey measurement error and estimated year-to-year abundances were less variable than those estimated directly by trawl surveys. Recruitment was periodic, strong during the mid-1970's, weak during the early and mid-1980's, and good during the late 1980's and the early 1990's. Recruitment declined dramatically after 1992. Over the past 20 years, periodic recruitment, coupled with high natural mortality and harvest rate, resulted in a 20-fold difference between peak and low legal male crab abundances.

Résumé : Nous avons élaboré un modèle de population fondé sur la longueur des individus pour estimer l abondance de la population de crabes des neiges du Pacifique (*Chionoecetes bairdi)*, dans la baie Bristol, en Alaska. Les paramètres de croissance variable et de recrutement graduel en fonction de la longueur sont intégrés à ce modèle. Une analyse non linéaire par la méthode des moindres carrés a permis d évaluer l abondance, le recrutement et la mortalité naturelle. Le modèle a été appliqué aux données recueillies concernant l abondance et les prises entre 1975 et 1994, et l on a obtenu des résultats similaires aux données enregistrées sur l abondance. Grâce à ce modèle, on a réduit l erreur liée aux mesures effectuées au cours des relevés, et les variations entre les estimations annuelles de l abondance sont moins grandes que celles qui sont obtenues directement au moyen de campagnes de chalutage. Le recrutement est un phénomène périodique; important au milieu des années 70, il était faible au début et au milieu des années 80, et bon à la fin des années 80 ainsi qu au début des années 90. On a ensuite constaté une baisse importante du recrutement à partir de 1992. Pendant les 20 dernières années, le recrutement périodique, conjugué à des taux élevés de mortalité naturelle et de récolte, s est traduit par une différence entre le sommet et le plancher d abondance des crabes mâles de taille légale telle que le nombre maximal d individus égale 20 fois le nombre minimal.
[Traduit par la Rédaction]

Introduction

Tanner crab, *Chionoecetes bairdi*, in the eastern Bering Sea support an important fishery in the United States. These crab were first targeted by Japanese and Russian fleets in 1965. The fishery expanded quickly in the late 1960's, and the catch reached 28 700 t in 1968 (Otto 1990). Foreign fishing for Tanner crab has been prohibited under the Magnuson Fisheries Conservation and Management Act since 1980. Directed fisheries for Tanner crab by the U.S. fleet began in 1974. Catch peaked in 1978 at 31 700 t (Otto 1990). The population collapsed in the mid-1980's and no fishing was allowed in 1986 and 1987 (Griffin and Ward 1994). The catch has been maintained at a moderate level in recent years and annual ex-vessel values averaged US$46 million during 1990–1992 (Griffin and Ward 1994).

In the eastern Bering Sea, Tanner crab are primarily found in Bristol Bay and the vicinity of the Pribilof Islands. For the purpose of this study, we used Stevens' et al. (1994) definition of Bristol Bay: the eastern Bering Sea south of 58°39′N and east of 168°W. Somerton (1981) demonstrated that at 50% maturity both male and female Tanner crab in Bristol Bay were slightly more than 10 mm larger in carapace width (CW) than those of the Pribilof Islands. From 1972 to 1993 about 75% of Tanner crab catch in the eastern Bering Sea originated from the Bristol Bay area. Our study focused on the Bristol Bay Tanner crab population.

More than 20 years of catch and survey data are available, yet no comprehensive population models have been developed for this population. Population abundances in Bristol Bay are estimated from trawl survey data collected by the National Marine Fisheries Service (NMFS) and used to set harvest quotas each year. Estimated abundances fluctuate greatly from year to year because of changes in stock size and survey measurement error. To help distinguish true changes in stock size

J. Zheng, G.H. Kruse, and M.C. Murphy. Alaska Department of Fish and Game, Commercial Fisheries Management and Development Division, P.O. Box 25526, Juneau, AK 99802–5526, U.S.A.

[1]Contribution PP-112 of the Alaska Department of Fish and Game, Commercial Fisheries Management and Development Division, Juneau, Alaska.

Correct citation: Zheng, J., Kruse, G.H., and Murphy, M.C. 1998. A length-based approach to estimate population abundance of Tanner crab, *Chionoecetes bairdi*, in Bristol Bay, Alaska. *In* Proceedings of the North Pacific Symposium on Invertebrate Stock Assessment and Management. *Edited by* G.S. Jamieson and A. Campbell. Can. Spec. Publ. Fish. Aquat. Sci. 125. pp. 97–105.

from survey measurement error, we constructed a population model that links population abundances by length-classes in adjacent years. The model is similar to length-based models developed by Sullivan et al. (1990) and Zheng et al. (1995). It incorporates variable growth and gradual recruitment over length. A nonlinear least-squares approach was used to estimate abundance, recruitment, and natural mortality.

Methods

Data

Trawl survey data for Bristol Bay Tanner crab from 1975 to 1994 were obtained from NMFS. To obtain a better length composition in the initial year (1975), we removed an extreme outlier from the survey data in that year by replacing the abundance in the outlier station (E05) with the average of abundances in the four adjacent survey stations. The crab caught in this station were primarily large males. Tanner crab sizes are measured by carapace width, but for convenience and to prevent confusion with previous modeling efforts (Zheng et al. 1995), we use the term "length" in this study to represent carapace width. Population abundances by 5-mm length-classes, sex, and shell condition were estimated using the "area-swept" method (Alverson and Pereyra 1969). If multiple hauls were conducted for a single station during a survey, the average of abundances from all hauls within the station was used.

Catches of Bristol Bay Tanner crab by length were obtained from the Alaska Department of Fish and Game for the U.S. fleet from 1975 to 1994 and from Otto (1982) for the catch sampling data for the Japanese fleet from 1975 to 1977. The Japanese fleet has been prohibited from this fishery since 1978.

Male population model

Our approach to model development was to restrict the number of parameters to be estimated, while at the same time we wanted to include structural relationships for which there was good evidence. We assumed that most Tanner crab in Bristol Bay complete molting before June each year when NMFS conducts its trawl survey of the region. Thus, we modeled population abundance at the end of June. Because small juvenile crab do not completely recruit to the survey gear and survey catchability of small juveniles is highly variable from year to year, we modeled abundances of only mature, or close to mature, crab with lengths ≥93 mm. The male crab model is similar to the one developed for red king crab (*Paralithodes camtschaticus*) in Bristol Bay (Zheng et al. 1995), except for different molting probabilities for old-shell crab.

Growth is an essential structural component of our length-based model. Growth is divided into two components which are highly dependent on size: molting probability and growth increment. Mean growth increment per molt, G_l, is a linear function of premolt mean length ι of the length-class l (Weber and Miyahara 1962; Donaldson et al. 1981):

$$[1] \quad G_l = a + b\iota$$

Growth increment varies among individual crab of the same premolt size. A gamma distribution was used to describe variation in growth increment per molt for flexibility:

$$[2] \quad g(x|\alpha_l, \beta) = x^{\alpha_l - 1} e^{-x/\beta} / (\beta^{\alpha_l}\, \Gamma(\alpha_l))$$

where x is growth increment per molt for a given length and α_l and β are parameters. For a given length-class l, $\alpha_l = G_l/\beta$, and the growth is represented by two parameters, G_l and β. The expected proportion of crab molting from length-class l to length-class l' is equal to the integral of the gamma function over length interval (l_1, l_2) of the receiving length-class l :

$$[3] \quad P_{l,l'} = \int_{l_1-\iota}^{l_2-\iota} g(x|\alpha_l, \beta)\, dx$$

where ι is the midlength of length-class l. For the last length-class L, $P_{L,L} = 1$.

Observations on shell condition are commonly used to classify crab into two categories: new-shell crab, those that molted within the past year, and old-shell crab, those that did not (Stevens et al. 1994). Following Balsiger's (1974) findings for red king crab, we used a reverse logistic function to fit molting probabilities for new-shell crab as a function of length and time:

$$[4] \quad mn_{l,t} = 1 - \frac{1}{1 + \phi_\tau e^{-\omega_t \iota}}$$

where ϕ_t and ω_t are parameters and ι is the mean length of length-class l. Tag–recapture data showed major shifts in molting probabilities over time for red king crab in Bristol Bay (Balsiger 1974). Based on changes in ratios of new-shell to old-shell crab abundances by length over time, we suspected that similar shifts were likely for Tanner crab. Accordingly, three logistic functions were used to describe molting probabilities for new-shell crab during different periods: a high molting period (1975–1979), low molting periods (1980–1986 and 1990–1994), and a good molting period (1987–1989), with parameters ϕ_1 and ω_1, ϕ_2 and ω_2, and ϕ_3 and ω_3, respectively. Molting probabilities for old-shell crab were assumed constant over length and were described for the same three periods as new-shell crab by three parameters, ϕ_4, ϕ_5, and ϕ_6. A special case of the model is a terminal molt at maturity (Conan and Comeau 1986) with parameters ϕ_4, ϕ_5, and ϕ_6 close or equal to zero. Molting probabilities for different periods were determined by the best fit of the data. An *F*-test described by Schnute (1981) was used to test hypotheses of constant molting probability versus variable molting probabilities over time for both new- and old-shell crab.

Number of recruits, R_t, and proportion of recruits belonging to each length-class, U_l, determine recruitment into the modeled population:

$$[5] \quad R_{l,t} = R_t\, U_l$$

where U_l is described by a gamma distribution, such as the one in eq. 2, with parameters α_r and β_r. Male recruits to the modeled population were defined as a portion of new-shell crab of size 93–132 mm CW. These are not to be confused with crab that are recruited to the fishery at a legal size ≥138 mm CW.

We modeled new- and old-shell male crab separately because observed abundance data by shell condition suggested that they may have different probabilities of molting. Annual abundance of new-shell crab is the combined result of growth, molting probability, mortality, and recruitment such that

$$[6] \quad N_{l+1,t+1} = \sum_{l'=1}^{l'=l+1} P_{l',l+1} \left[\left(N_{l',t} e^{-M} - \frac{C_{l',t} e^{(y_t-1)M} N_{l',t}}{N_{l',t} + O_{l',t}} \right) mn_{l',t} + \left(O_{l',t} e^{-M} - \frac{C_{l',t} e^{(y_t-1)M} O_{l',t}}{N_{l',t} + O_{l',t}} \right) mo_t \right] + R_{l+1,t+1}$$

where $N_{l,t}$ and $O_{l,t}$ are abundances of new-shell and old-shell crab in length-class l in year t, $C_{l,t}$ is catch in length-class l in year t, y_t is the time between the survey and the fishery, $mn_{l',t}$ and mo_t are molting probabilities for new- and old-shell crab, and M is instantaneous natural mortality which we assumed constant and independent of size and shell condition for simplicity. Annual abundance of old-shell crab, as the nonmolting portion of survivors of crab from the previous year, is

$$[7] \quad O_{l+1,t+1} = \left(N_{l+1,t} e^{-M} - \frac{C_{l+1,t} e^{(y_t-1)M} N_{l+1,t}}{N_{l+1,t} + O_{l+1,t}} \right) (1 - mn_{l+1,t}) + \left(O_{l+1,t} e^{-M} - \frac{C_{l+1,t} e^{(y_t-1)M} O_{l+1,t}}{N_{l+1,t} + O_{l+1,t}} \right) (1 - mo_t)$$

Males >163 mm CW were grouped together to form the last length-class. Sublegal males (<138 mm CW) were not legally retained in the catch but were sorted, discarded, and subjected to handling mortality. Due to complexity and lack of data, we did not model handling mortality. Instead we considered handling mortality to be included in the M estimates.

Female population model

Major differences between the male and female population models are molting probability and fishing mortality. Before maturity, we assumed females molt once per year (Donaldson et al. 1981). Upon maturity, female crab stop molting and growing (Donaldson et al. 1981). Female crab are not allowed to be retained in catch and are returned to the sea. As with sublegal males, we considered handling mortality as a part of M. Natural mortality was assumed constant over time for all lengths and shell conditions. The growth for immature females was modeled by growth functions similar to males (eqs. 1–3). Similar to male recruitment, a gamma function was used to describe female recruitment into the female population. Because females mature at smaller sizes than males (Donaldson et al. 1981), we modeled female crab with length ≥70 mm and recruits to the modeled population were defined as a portion of new-shell length-classes with lengths ranging from 70 to 109 mm CW.

Because of different molting probabilities for mature and immature females, a maturity probability function is needed to separate new-shell females as immature or mature females. A logistic function similar to eq. 4 was used to model maturity probability as a function of length. Two logistic functions were used to describe maturity probabilities during different periods: periods when crab matured at larger size (1975–1979 and 1984–1988) and periods when crab matured at smaller size (1980–1983 and 1989–1994), with parameters ϕ_1 and ω_1, and ϕ_2 and ω_2, respectively. Maturity probability functions for different periods were determined by the best fit of the data. The F-test (Schnute 1981) was used to test the hypothesis of constant maturity versus variable maturity over time.

New-shell females are either immature or primiparous crab and their abundance results from a combination of recruitment, growth, and natural mortality:

$$[8] \quad N_{l+1,t+1} = \sum_{l'=1}^{l'=l+1} [P_{l',l+1} N_{l',t} (1 - m_{l',t}) e^{-M}] + R_{l+1,t+1}$$

where $m_{l',t}$ is maturity probability for length-class l' and year t. Old-shell mature females are survivors of the mature females from the previous year,

$$[9] \quad O_{l+1,t+1} = (N_{l+1,t}\, m_{l+1,t} + O_{l+1,t}) e^{-M}$$

Parameter estimation

Mean growth increment per molt was estimated for Kodiak Tanner crab by Donaldson et al. (1981). The intercept and slope parameters for the linear relationships between mean increment per molt and premolt length in mm are 15.75 and 0.07 for male crab and 25.60 and −0.1337 for female crab. We assumed that males >130 mm have the same growth increment per molt as 130-mm males and females >95 mm have the same growth increment as 95-mm females. The growth increment per molt for Bristol Bay Tanner crab was assumed to be the same as Kodiak Tanner crab.

A measurement error estimation procedure was used to estimate parameters. Measurement errors were assumed to be log-normally distributed such that:

$$[10] \quad \tilde{N}_{l,t} = N_{l,t} e^{\eta_{l,t}} \text{ and } \tilde{O}_{l,t} = O_{l,t} e^{\delta_{l,t}}$$

where $\tilde{N}_{l,t}$ and $\tilde{O}_{l,t}$ are estimated abundances of new- and old-shell crab in length-class l and year t from trawl survey data, and $\eta_{l,t}$ and $\delta_{l,t}$ are normally distributed measurement errors of abundances. Observed catches by length were assumed to be free of measurement errors because of accurate measurement of length, a small number of length-classes, and large sample sizes (normally >5000 crab·yr^{-1}). Parameters of the model were estimated using a nonlinear least squares approach to minimize the residual sum of squares (RSS):

$$[11] \quad \text{RSS} = \sum_{l,t} [(\ln(N_{l,t} + \kappa) - \ln(\tilde{N}_{l,t} + \kappa))^2 + (\ln(O_{l,t} + \kappa) - \ln(\tilde{O}_{l,t} + \kappa))^2]$$

where κ is a constant set equal to 100 000 crab (about 0.5% of the largest observed abundance by length) to prevent taking the logarithm of zero and to reduce the effect of length-classes with zero or very low abundances on parameter estimation. A larger κ weights high abundances more, and vice versa. Residuals between adjacent length-classes and between years were examined to check the assumption of independence.

Model parameters were estimated separately for male and female crab and consisted of recruits for each year (R_t for t = 76–94), total abundance in the first year (1975), parameters α_r and β_r, instantaneous natural mortality M, growth parameter β, molting probability parameters (ϕ_1, ϕ_2, ϕ_3, ϕ_4, ϕ_5, ϕ_6, ω_1, ω_2, and ω_3) for male crab, and maturity parameters (ϕ_1, ϕ_2, ω_1, and ω_2) for female crab. To reduce parameter numbers, we assumed that length and shell compositions from the first year (1975) survey data approximate true compositions within sex. Starting in the second year, abundances by length, sex, and

Table 1. Summary of parameter estimates for a length-based population model of *Chionoecetes bairdi* crab in Bristol Bay, Alaska.

	Male crab		Female crab	
Parameter	Scenario 1	Scenario 2	Scenario 1	Scenario 2
N_{75}	181.811	211.865	164.328	151.860
β	0.995	0.580	1.000	1.000
α_r	64.348	59.181	80.158	49.620
β_r	1.574	1.723	0.966	1.577
ϕ_1	120864.6	7274.8	4617109.0	37110.4
ω_1	0.085	0.066	0.172	0.126
ϕ_2	42690.2	NA	24190.8	NA
ω_2	0.090	NA	0.129	NA
ϕ_3	108632.7	NA	NA	NA
ω_3	0.085	NA	NA	NA
ϕ_4	0.185	0.000	NA	NA
ϕ_5	0.091	NA	NA	NA
ϕ_6	0.000	NA	NA	NA
M	0.489	0.495	0.5231	0.551
R_{76}	90.898	97.339	67.150	83.573
R_{77}	55.106	79.591	26.616	34.384
R_{78}	38.252	36.443	16.110	15.006
R_{79}	21.598	22.290	25.361	22.638
R_{80}	21.479	14.254	18.975	23.460
R_{81}	9.220	7.275	17.356	16.951
R_{82}	14.020	12.027	3.695	4.129
R_{83}	8.300	8.342	2.913	3.149
R_{84}	3.805	3.659	5.535	4.006
R_{85}	3.908	3.467	4.659	5.582
R_{86}	17.020	15.361	4.697	5.034
R_{87}	30.079	31.623	98.322	59.465
R_{88}	127.672	127.622	36.520	76.161
R_{89}	68.069	51.732	68.567	57.893
R_{90}	84.093	55.828	81.403	64.293
R_{91}	00.020	80.738	31.899	30.037
R_{92}	49.403	51.403	12.031	10.812
R_{93}	15.039	16.100	3.843	3.852
R_{94}	4.292	4.695	0.859	0.894
RSS	117.825	164.307	109.937	140.312
df	538	544	353	355

Note: For male crab, scenario 1 is three levels of molting probability and scenario 2 is one level of molting probability. For female crab, scenario 1 is two levels of maturity and scenario 2 is one level of maturity. Recruits *R* are in millions of crab.

shell condition were computed recursively from abundances by length, sex, and shell condition in the first year and by annual recruitment, catch, and model parameters. There are a total of 571 data points and 33 or 27 parameters (depending on molting probabilities) for males and 381 data points and 28 or 26 parameters (depending on maturity probabilities) for females.

IMSL math library FORTRAN subroutine DBCLSF (IMSL 1991) was used to perform nonlinear least-squares parameter estimation on the models through a modified Levenberg–Marquardt algorithm and a finite-difference Jacobian. All parameters were bounded to be nonnegative. Initial estimates of total abundance in the first year and annual recruits were derived from the survey data. Initial estimates of natural mortalities were set to 0.3 for males (NPFMC 1990) and 0.45 for females. All initial ϕ and ω parameter values for molting probabilities and maturities were 290974.7 and 0.091, based on results for Bristol Bay red king crab (Zheng et al. 1995). Initial values of parameters α_r and β_r were 68.50 and 1.5 for males and 55.13 and 1.5 for females. Initial values of all parameters except growth parameter β were updated and changed interactively to minimize RSS. Due to confounding with recruitment parameters, growth parameter β was fixed initially from 0.5 to 2.0 with a 0.25 interval for both males and females. After all other parameters were stable, β was estimated

Fig. 1. Comparison of observed and estimated male Tanner crab abundances (>92 mm CW) in Bristol Bay using two scenarios of molting probability.

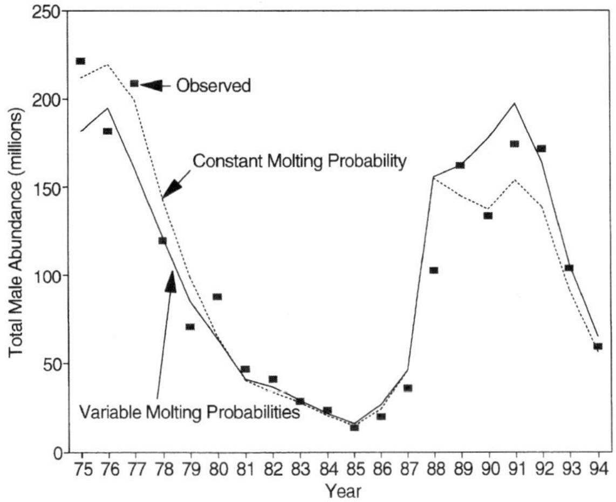

Fig. 2. Comparison of the logarithms of observed and estimated length frequencies of Bristol Bay male Tanner crab by year for the variable molting probability scenario.

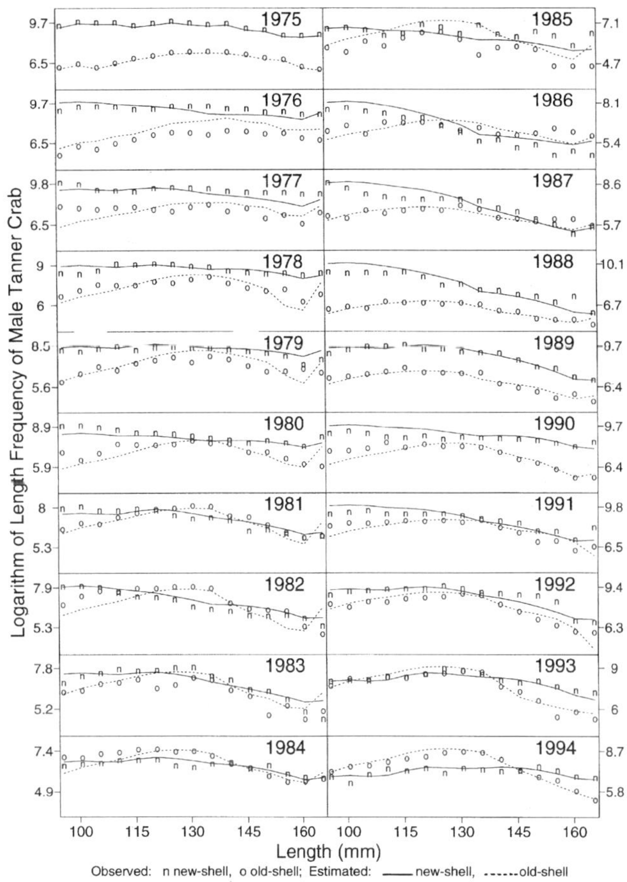

simultaneously with all parameters. The final parameter values were those that minimized RSS.

Results

Male abundance

Scenario 1, with three levels of molting probability over time, fit observed abundances much better than scenario 2, which had a constant level of molting probability (Table 1). This was confirmed by the F-test with $P < 0.00001$. Parameter estimates for both scenarios were quite different. Natural mortalities for both scenarios were similar and above 0.4 (Table 1). Variation in growth increment per molt was higher for scenario 1 than for scenario 2. Molting probability for old-shell crab was 0 for scenario 2 and >0 for most years for scenario 1.

Trends in recruitment to the modeled population were similar for both scenarios. Recruitment was high during the mid-1970's, declined in the late 1970's, was low during the early and mid-1980's, and increased to high levels during the late 1980's (Table 1). Poor recruitment occurred during 1993–1994. Recruitment was generally lower for scenario 2 than for scenario 1 during the early 1980's (Table 1).

Total male abundance (>92 mm CW) was high during the mid-1970's, late 1980's, and early 1990's, and was low during the other periods (Fig. 1). Abundances estimated by both scenarios had a similar trend over time. Abundance estimated by scenario 2 was higher during the 1970's and lower during the late 1980's and early 1990's than that by scenario 1 (Fig. 1). Abundances estimated by both scenarios were much higher than the observed abundance in 1988.

Logarithms of observed and estimated abundances by length by shell condition are illustrated for scenario 1 (Fig. 2). Overall, the model closely fit observed size frequencies. Largest discrepancies occurred in 1976 and 1985, when estimated old-shell crab abundances were higher than those observed, and in 1986 when estimated new-shell crab abundances were higher (Fig. 2).

For scenario 1, residuals for adjacent length-classes within a shell condition and year were significantly correlated ($r = 0.635$ and df = 558), but annual means of residuals were not significantly associated between years ($r = -0.384$ and df = 17). Annual means of residuals from new-shell crab were also not significantly correlated with those from old-shell crab with a one-year time lag ($r = 0.039$ and df = 17). Correlations of residuals for scenario 2 were similar to those for scenario 1.

Variation in growth increment was high and crab within a length-class usually molted to one of three larger length-classes, with a mean growth increment per molt slightly larger than 20 mm (Fig. 3). Molting probabilities for new-shell crab with scenario 1 were high during 1975–1979 and 1987–1989 and low during 1980–1986 and 1990–1994 (Fig. 4). As a contrast, molting probabilities for old-shell crab were zero during 1987–1989, 0.091 during 1980–1986 and 1990–1994, and 0.185 during 1975–1979 (Table 1).

Recruitment occurred in the first eight length-classes and 76.5% was confined to the first four length-classes (Fig. 4). The gamma distribution of proportions of recruits over length was truncated because crab with sizes smaller than the first length-class were not modeled. Thus, recruitment estimated here was an approximation of absolute recruitment. Mean length at recruitment was 106.6 mm.

Trend in catch is similar to that in legal male abundance because a harvest rate strategy has been used to manage this fishery (Fig. 5). We estimated two harvest rates for this

Fig. 3. Proportions of growth increments from a given premolt length-class to postmolt length-classes for Bristol Bay male Tanner crab with the variable molting probability scenario. Midlengths of premolt length-classes are indicated inside the plot.

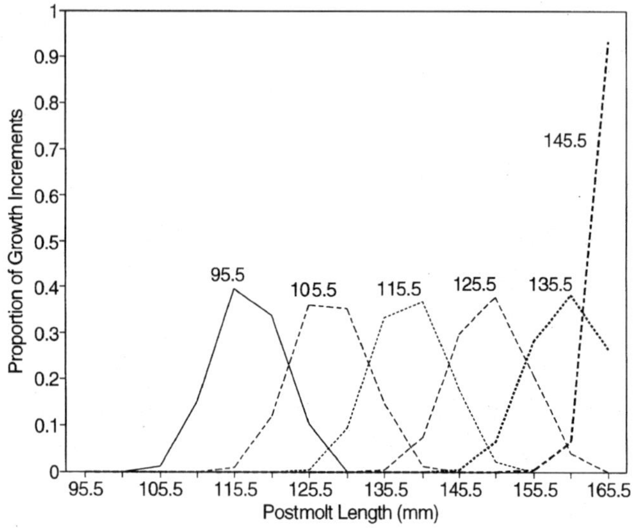

Fig. 4. Molting probabilities for new-shell crab and proportion of recruits by length-class for Bristol Bay male Tanner crab with the variable molting probability scenario.

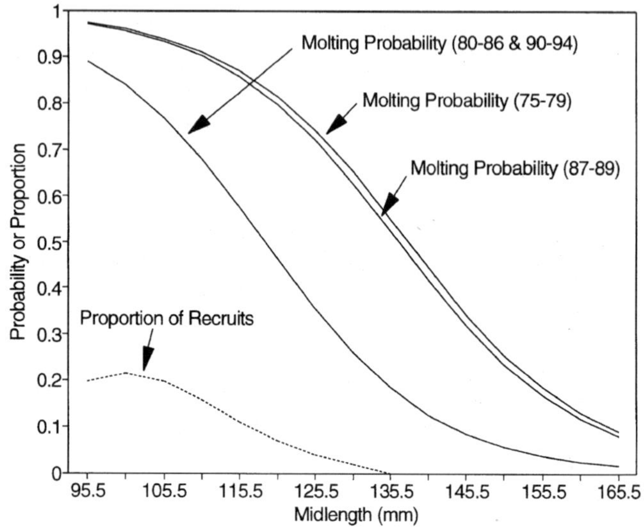

population: (*i*) total annual catch divided by total legal male abundance at the end of June, and (*ii*) total annual catch divided by total legal male abundance adjusted by natural mortality to the midpoint of the fishing season. The first harvest rate is currently used to set harvest quotas, whereas the second harvest rate more accurately represents the impact of harvest on population abundances because time from the end of June to the midpoint of the fishing season was lagged in most years by over half a year. Low harvest rates were applied during 1982–1988 and no fishery occurred from July 1985 to June 1987 (Fig. 5). The mean of the 1989–1993 harvest rates adjusted for natural mortality was 0.507 (Fig. 5). Strongly autocorrelated recruitment makes it difficult for the population to remain abundant while experiencing such high harvest rates.

Female abundance

Scenario 1 (two levels of maturity over time) fit observed data better than scenario 2 (constant maturity over time) (Table 1). The *F*-test produced a *P*-value <0.00001. RSS was insensitive to change in variation of growth increment per molt for immature females, so β was fit at a biologically reasonable value of 1.0 for both scenarios. Estimated natural mortality was slightly higher for scenario 2 than for scenario 1 and was higher for both scenarios than that for male crab (Table 1). Variation in recruitment over length was less for scenario 1 than for scenario 2. Trends in recruitment were similar for both scenarios, except in 1987 and 1988 when trends contrasted between the two scenarios (Table 1).

Trends in female abundance (Fig. 6) were similar to those in males (Fig. 1). Total female abundance (>69 mm CW) generally declined from the mid-1970's to the mid-1980's, increased sharply during the late 1980's, and decreased again in the early 1990's (Fig. 6). Scenario 1 fit total observed abundance better than scenario 2, especially after 1986 (Fig. 6). Estimated abundances in 1977 and 1991 were much lower than observed abundances, which were dominated by an extremely high catch at one survey station.

Observed abundances by length and shell condition fit well with scenario 1 (Fig. 7). Estimated new-shell crab abundances were lower than observed abundances in 1977 and higher in 1978 and 1990. Estimated old-shell crab abundances exceeded observed abundances in 1976, 1979, and 1988, and were lower in 1991. Abundance of old-shell females was much higher than new-shell female abundance during 1981–1984 and 1991–1994, indicating relatively poor recruitment during these periods.

Similar to male crab, residuals for adjacent length-classes within a shell condition and year were significantly correlated ($r = 0.765$ and 0.787 for scenarios 1 and 2, respectively, and df = 178). Correlation of annual means of residuals was not significant ($r = -0.109$ and -0.137 for scenarios 1 and 2, respectively, and df = 17).

For scenario 1, maturity probability over length was higher during 1980–1983 and 1989–1994 than during 1975–1979 and 1984–1988 (Fig. 8). The 50% maturity probability was about 80 mm CW for the former two periods and about 90 mm CW for the latter two periods. Ninety percent of recruitment occurred within the first four length-classes with a mean recruit length of 80.7 mm (Fig. 8). As with males, the gamma distribution of female recruits over length was truncated at the first length-class.

Discussion

The advantage of a length-based approach to estimate population abundance is that it incorporates survey, fisheries, and tagging data, and links abundances from multiple years together. Thus, the random component of measurement errors in abundances estimated directly from survey data can be filtered out. The length-based model fit survey data well for Bristol Bay Tanner crab and population abundance estimates from the

Fig. 5. Total catch, legal male abundance, and harvest rates of Bristol Bay Tanner crab for the variable molting probability scenario from July 1, 1975 to June 30, 1995. (*a*) harvest rate defined as total annual catch divided by total legal male abundance adjusted by natural mortality to the midpoint of each fishing season; (*b*) harvest rate defined as total annual catch divided by total legal abundance at the end of June; (*c*) total legal male abundance adjusted by natural mortality to the midpoint of each fishing season; and (*d*) total annual catch from July 1 to June 30.

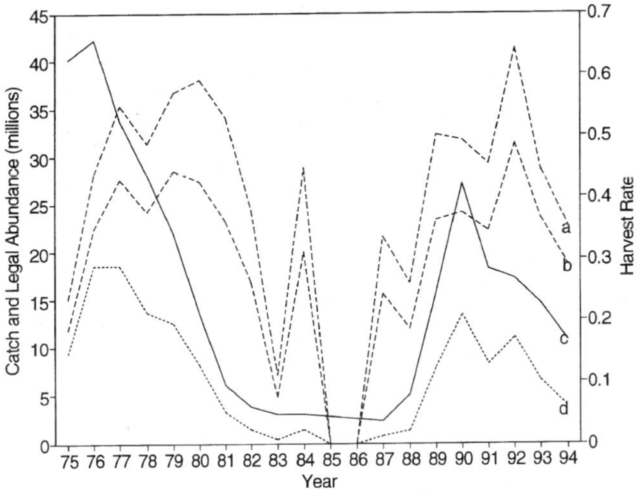

Fig. 6. Comparison of observed and estimated female Tanner crab abundances (>69 mm CW) in Bristol Bay using two scenarios of maturity.

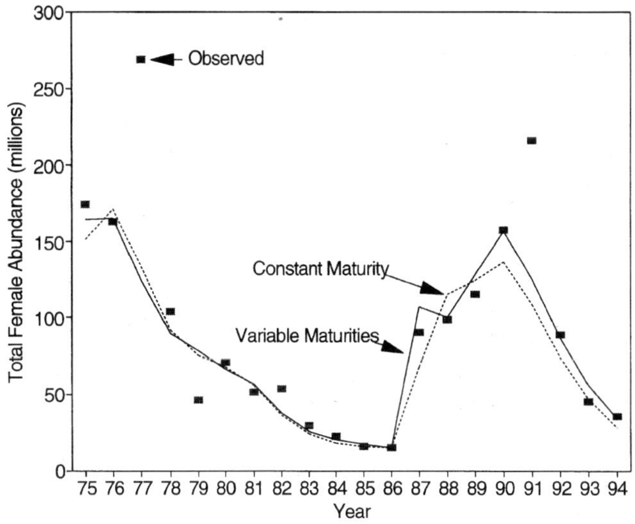

model were much smoother than those estimated by survey data directly. We believe that our length-based approach leads to more accurate estimates of annual abundance than the "area-swept" approach in which yearly estimates are treated independently and measurement errors are not removed.

Natural mortality is one of the most important parameters affecting population abundance estimates by a length-based model. Using survey data and Japanese catch and effort data from 1969 to 1979, Somerton (1981) combined a simple production model and the Leslie method to obtain natural mortality estimates of 0.35 for prerecruit male Tanner crab (less than commercial size) and from 0.22 to 0.28 for commercial-size male Tanner crab in the eastern Bering Sea. To our knowledge, these are the only published estimates of natural mortality for this population and they are much lower than natural mortalities derived from our study. Somerton's low estimates are primarily caused by rapid increase in estimates of surveyed abundance from 1972 to 1975, during which time estimated natural mortality was negative. On the contrary, time series of abundance by length for Bristol Bay Tanner crab since 1975 suggests that high natural mortality occurs. The natural mortality we estimated also includes mortality due to bycatches from directed crab fisheries and groundfish trawl fisheries. Taken together, annual bycatches of all components of the population from 1990 to 1994 are higher than legal male catch from the commercial Tanner crab fishery (North Pacific Fishery Management Council 1994; Tracy 1995). However, mortality associated with this bycatch is difficult to estimate and therefore was not separated from the natural mortality rate.

Because bycatch rate depends on crab size and bycatch mortality rate may vary over time, natural mortality may be size- and time-specific. We chose not to attempt to estimate size- and time-dependent natural mortality because our model would become overparameterized in the absence of additional independent data on mortality. Bycatches from crab fisheries have been collected since 1990 (Tracy 1995). When a long time series of bycatch data is accumulated, it will be possible to further refine the model to include this factor.

Because molting probability declines with increasing size (Fig. 4), old-shell crab tend to be older than new-shell crab and may suffer a higher natural mortality. We modified the model and estimated natural mortalities for new- and old-shell crab separately. Natural mortality for old-shell crab was bounded to be equal to or greater than that for new-shell crab. This resulted in natural mortality for old-shell females always converging to that for new-shell females, whereas natural mortality for old-shell males depended on assumptions for molting probability. Without mark–recapture or laboratory estimates of molting probabilities, it is difficult to estimate natural mortality for old-shell and new-shell male crab separately from survey and fishery data alone. Therefore, we assumed that natural mortality is identical for both new- and old-shell crab.

We assumed that catchability of survey gear has been equal to unity for the modeled population since 1975 to allow us to estimate natural mortality. This assumption is approximately valid for mature, or close to mature, crab because the survey gear is a standardized bottom trawl and the "area-swept" method was used to estimate population abundances. If survey catchability was significantly less than one, natural mortality was underestimated, and vice versa.

Our results suggested that molting and maturity probabilities were variable over time, which makes it difficult to project population abundances. Old-shell males molted in most years but with an overall lower molting probability than new-shell males. We believe that the zero molting probability for old-shell males during 1987–1989 was an artifact primarily caused by a relatively high abundance of small size, old-shell males. Our result of low molting probability of old-shell males may be partially caused by survey timing. Survey timing in the Bering Sea may precede molting of some male Tanner crab in

Fig. 7. Comparison of the logarithms of observed and estimated length frequencies of Bristol Bay female Tanner crab by year for the variable maturity probability scenario.

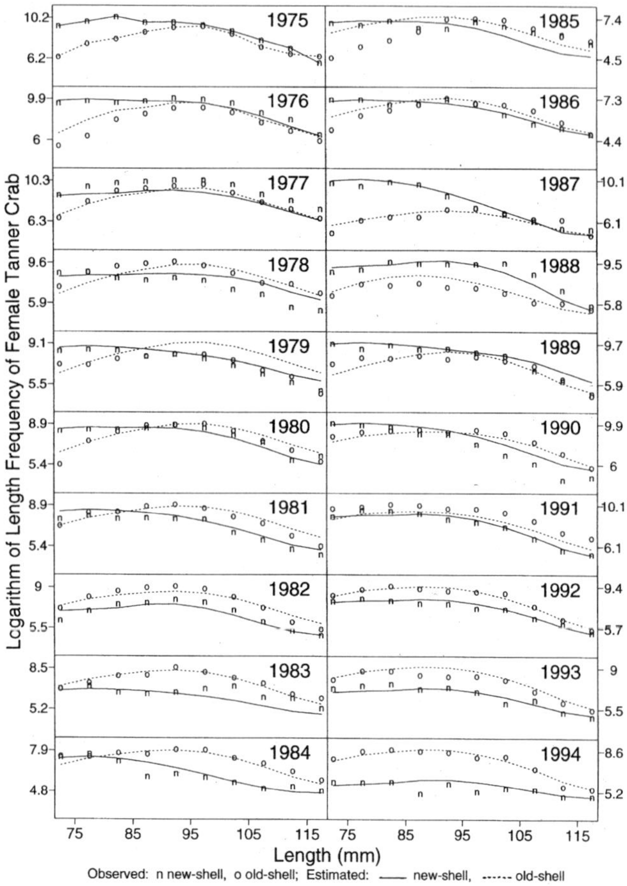

Fig. 8. Maturity probabilities and proportion of recruits by length-class for Bristol Bay female Tanner crab with the variable maturity probability scenario.

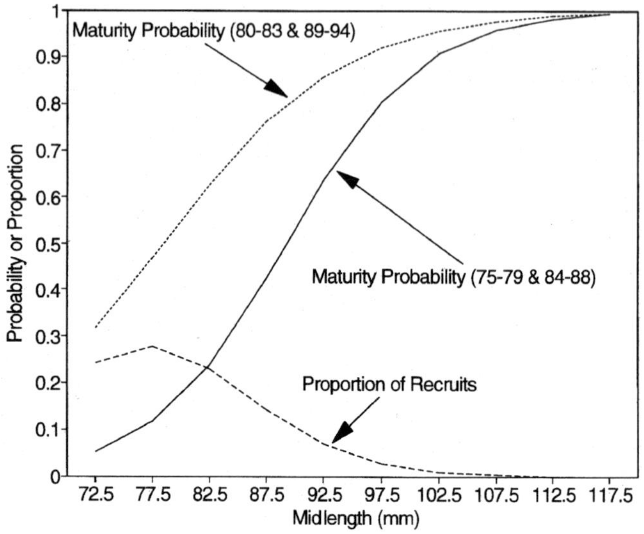

some years (Somerton 1981), resulting in underestimation of molting probability. However, the commercial Tanner crab fishery in Bristol Bay occurs well after the molting season and proportions of new-shell male crab from commercial and survey catches were similar, except during the early 1990's when the survey catch had lower proportions of new-shell males. This observation suggests that molting probabilities may not be greatly underestimated for Bristol Bay Tanner crab.

Low molting probability of old-shell males may also result from skipping molt and terminal molt at maturity. Large male Tanner crab off Kodiak Island and in Cook Inlet have intermolt periods of more than two years (Donaldson et al. 1981; Paul and Paul 1995). Intermolt periods for Bristol Bay male Tanner crab may be longer because of colder temperatures. Conan and Comeau (1986) suggested that male crabs of genus *Chionoecetes* undergo a terminal molt at maturity. On the contrary, laboratory (Paul and Paul 1995) and field (Donaldson et al. 1981) observations off Kodiak Island and in Cook Inlet show that functionally mature male Tanner crab are capable of molting. Based on low molting probabilities of old-shell males that we estimated, it is likely that a portion of old-shell, mature males in Bristol Bay have stopped molting. However, our size and shell abundance and fishery data do not allow us to distinguish skip-molt males from terminal-molt males. Claw size and maturity data are needed to address the terminal molt hypothesis.

Mean growth increment per molt is relatively constant over time. Applying simple length–frequency analysis to survey data, Somerton (1981) determined that mean growth increment for Bering Sea Tanner crab decreased as a function of length. In contrast, mean growth increment estimated by Donaldson et al. (1981) using tagging data for Kodiak Tanner crab increased slightly over length. Mean growth increment estimated by Somerton (1981) was also much smaller than that by Donaldson et al. (1981). Because punctuated growth and variable molting probabilities over time were not incorporated into Somerton's simple length–frequency analysis, his estimates of growth increment are questionable. Unfortunately, tagging data are lacking for Bristol Bay; therefore, we assumed that both Kodiak and Bristol Bay Tanner crab populations have similar growth increments per molt and adopted the results from Donaldson et al. (1981). This assumption was partially relaxed by estimating variation in growth increment per molt in the model. Generally, lower mean growth increment per molt would increase estimated variation in growth increment per molt and decrease estimated recruitment and natural mortality.

Our results indicate that measurement errors are likely to be lognormally distributed and independent between years, but not independent between length-classes within a year. The correlation of residuals between adjacent length-classes in a year reduces the effective degrees of freedom below those presented in Table 1 and makes our estimator less efficient than that presented by Myers and Cadigan (1995) for catch-at-age data. However, it is unlikely that this substantively affects the results of our *F*-test. The difference between the two scenarios is so significant that our conclusions will hold true even with a 90% reduction of the degrees of freedom. Further, abundance estimates in our model may not greatly be affected by this correlation because our objectives are still to minimize measurement errors of observed abundances. Effects of different structures of measurement errors on parameter and abundance estimates under a length-based model should be evaluated in future studies.

Recruitment was highly autocorrelated for Bristol Bay Tanner crab. A combination of high natural mortality, high harvest rate, and strongly autocorrelated recruitment produced a highly fluctuating time series of population abundance. Because of autocorrelation, effects of spawning stock and environmental conditions on recruitment success are probably confounded. Work is needed to determine what environmental conditions are favorable to strong recruitment and whether a stock–recruitment relationship exists. Understanding recruitment dynamics is essential for optimal harvest of the resource (Zheng 1994). With this in mind, a study of oceanographic conditions and Tanner crab year-class success has been initiated (A.V. Tyler, University of Alaska, Fairbanks, personal communication).

Acknowledgments

Survey data of Bristol Bay Tanner crab were provided by Drs. B.G. Stevens and J.E. Reeves of NMFS. This paper is funded in part by co-operative agreement NA37FL0333 from the National Oceanic and Atmospheric Administration. The views expressed herein are those of the authors and do not necessarily reflect the views of NOAA or any of its subagencies.

References

Alverson, D.L., and Pereyra, W.T. 1969. Demersal fish in the Northeastern Pacific Ocean – an evaluation of exploratory fishing methods and analytical approaches to stock size and yield forecasts. J. Fish. Res. Board Can. **26**: 1985–2001.

Balsiger, J.W. 1974. A computer simulation model for the eastern Bering Sea king crab. Ph.D. dissertation, University of Washington, Seattle, Wash.

Conan, G.Y., and Comeau, M. 1986. Functional maturity and terminal molt of male snow crabs, *Chionoecetes opilio*. Can. J. Fish. Aquat. Sci. **43**: 1710–1719.

Donaldson, W.E., Cooney, R.T., and Hilsinger, J.R. 1981. Growth, age, and size at maturity of Tanner crab, *Chionoecetes bairdi* M.J. Rathbun, in the northern Gulf of Alaska. Crustaceana, **40**: 286–302.

Griffin, K.L., and Ward, M.L. 1994. Annual management report for the shellfish fisheries of the Bering Sea area, 1992. *In* Annual management report for the shellfish fisheries of the westward region, 1992. Alaska Department of Fish and Game, Division of Commercial Fisheries Management and Development, Kodiak. Reg. Inf. Rep. 4K94-9. pp. 131–192.

International Mathematics and Statistics Library (IMSL). 1991. Math/library users manual, version 2.0. International Mathematics and Statistics Library, Houston, Tex.

Myers, R.A., and Cadigan, N.G. 1995. Statistical analysis of catch-at-age data with correlated errors. Can. J. Fish. Aquat. Sci. **52**: 1265–1273.

North Pacific Fishery Management Council (NPFMC). 1990. Environmental assessment for Amendment 1 to the fishery management plan for the commercial king and Tanner crab fisheries in the Bering Sea/Aleutian Islands. North Pacific Fishery Management Council, Anchorage, Alaska.

North Pacific Fishery Management Council (NPFMC). 1994. Stock assessment and fishery evaluation for the groundfish resources of the Bering Sea/Aleutian Islands regions as projected for 1995. North Pacific Fishery Management Council, Anchorage, Alaska.

Otto, R.S. 1982. An overview of the eastern Bering Sea Tanner crab fisheries. *In* Proceedings of the International Symposium on Genus *Chionoecetes*. Alaska Sea Grant Rep. 82-10. pp. 83–115.

Otto, R.S. 1990. An overview of eastern Bering Sea king and Tanner crab fisheries. *In* Proceedings of the International Symposium on King and Tanner Crabs. Alaska Sea Grant Rep. 90-04. pp. 9–26.

Paul, A.J., and Paul, J.M. 1995. Molting of functionally mature male *Chionoecetes bairdi* Rathbun (Decapoda: Majidae) and changes in carapace and chela measurements. J. Crustac. Biol. **15**: 686–692.

Schnute, J. 1981. A versatile growth model with statistically stable parameters. Can. J. Fish. Aquat. Sci. **38**: 1128–1140.

Somerton, D.A. 1981. Life history and population dynamics of two species of Tanner crab, *Chionoecetes bairdi* and *C. opilio*, in the eastern Bering Sea with implications for the management of the commercial harvest. Ph.D. dissertation, University of Washington, Seattle, Wash.

Stevens, B.G., Haaga, J.A., and MacIntosh, R.A. 1994. Report to industry on the 1994 eastern Bering Sea crab survey. Alaska Fisheries Science Center, Processed Rep. 94-07. National Marine Fisheries Service, National Oceanic and Atmospheric Administration, 7600 Sand Point Way NE, Seattle, WA 98115, U.S.A.

Sullivan, P.J., Lai, H.L., and Gallucci, V.F. 1990. A catch-at-length analysis that incorporates a stochastic model of growth. Can. J. Fish. Aquat. Sci. **47**: 184–198.

Tracy, D.A. 1995. Alaska Department of Fish and Game summary of the 1994 mandatory shellfish observer program database. Alaska Department of Fish and Game, Division of Commercial Fisheries Management and Development, Kodiak. Reg. Inf. Rep. 4K95-32.

Weber, D.D., and Miyahara, T. 1962. Growth of the adult male king crab, *Paralithodes camtschatica* (Tilesius). Fish. Bull. (U.S.), **62**: 53–75.

Zheng, J. 1994. Threshold management strategies for exploited fish populations. Ph.D. dissertation, University of Alaska Fairbanks, Fairbanks, Alaska.

Zheng, J., Murphy, M.C., and Kruse, G.H. 1995. A length-based population model and stock-recruitment relationships for red king crab, *Paralithodes camtschaticus*, in Bristol Bay, Alaska. Can. J. Fish. Aquat. Sci. **52**: 1229–1246.

Growth, Mortality, and Yield-per-Recruit

Assessment of the eastern Bering Sea snow crab, *Chionoecetes opilio*, stock under the terminal molting hypothesis

Robert S. Otto

Abstract: Strong recruitment to the eastern Bering Sea (EBS) snow crab (*Chionoecetes opilio*) stock during the late 1980's resulted in high abundance and record landings from 1987 to 1991. Abundance and landings declined sharply thereafter. Paired carapace width (CW) – chela height (CH) measurements were recorded from 29 641 male snow crab during annual trawl surveys of the EBS from 1989 to 1994. Changes in the allometric relationship between CW and CH were used to estimate the proportion of morphometrically mature (MM) males and plot changes in their proportion, P_{MM}, with size. These data were interpreted under the terminal molt hypothesis in which the molt to morphometric maturity is the last molt in the crab's life span. The population frequency of MM males with progressively older relative shell ages was used to estimate size-specific total mortality rates. Published data on growth and fishery landings were modeled along with size-specific mortality rates and the size frequency of MM males to elucidate recruitment patterns and better predict the growth and decay of cohorts. The relationship between P_{MM} and CW was unstable in time and plotted data indicated an inverse relationship between cohort abundance and size at maturity. Different CW–P_{MM} relationships characterized various geographic segments of the stock and appeared to be influenced by an ontogenetic migration from shallower to deeper waters and consequent changes in the geographic distribution of dominant cohorts.

Résumé : Les forts taux de recrutement observés à la fin des années 80 chez le stock de crabe des neiges (*Chionoecetes opilio*) de la mer de Béring orientale ont été suivis d'une augmentation substantielle des effectifs et ont conduit à l'enregistrement de débarquements records entre 1987 et 1991. On a cependant noté par la suite un brusque déclin des effectifs et du nombre de débarquements. Des mesures appariées de la largeur de la carapace (CW) et de la hauteur des chélipèdes (CH) ont été prises sur 29 641 crabes mâles durant des relevés annuels effectués au chalut dans la mer de Béring orientale, de 1989 à 1994. On a utilisé les variations de la relation allométrique établie entre CW et CH pour estimer la proportion de mâles morphométriquement matures (MM) et représenter graphiquement les changements de leur proportion, P_{MM}, en fonction de la taille. On a ensuite interprété ces données en tenant pour avérée l'hypothèse de la mue terminale, selon laquelle la mue qui mène à la maturité morphométrique est la dernière que connaît le crabe durant sa vie. On s'est fondé sur la fréquence dans la population de mâles MM présentant des âges relatifs de carapace de plus en plus avancés pour estimer les taux de mortalité totale par taille. On a intégré dans des modèles les données publiées sur la croissance et les débarquements, ainsi que les taux de mortalité par taille et la distribution de la taille des mâles MM, afin d'expliquer les tendances de recrutement et de mieux prévoir la croissance et le déclin des cohortes. La relation entre P_{MM} et CW s'est révélée instable dans le temps, et les données représentées graphiquement reflétaient l'existence d'une relation inverse entre l'abondance des cohortes et la taille à maturité. Différentes relations entre CW et P_{MM} ont été observées chez les différents segments géographiques du stock. Ces variations semblaient liées à la migration qu'effectue le crabe des neiges durant son ontogénèse, des eaux peu profondes vers les eaux profondes, et à l'incidence de ces déplacements sur la répartition géographique des cohortes dominantes. [Traduit par la Rédaction]

Introduction

The eastern Bering Sea (EBS) snow crab (*Chionoecetes opilio*) fishery is among the world's largest single species crab fisheries. It began as an outgrowth of Japanese crab fisheries during the 1970's (Otto 1981) and was further developed by U.S. fishermen in the 1980's. Indices of abundance from the annual National Marine Fisheries Service (NMFS) trawl survey were declining in the early 1980's, reached their lowest level in 1985, and then rebounded rapidly (Otto 1990; Stevens et al. 1994). The survey abundance index of commercial-sized males (≥102 mm CW) peaked in 1991 and then declined rapidly as a large cohort passed through the fishery. Landings followed similar trends (Table 1, Morrison and Gish 1994). Since the dynamics of cohort growth and decay could be radically affected by the timing of sexual maturity, I initiated a study of maturity in 1989 as a means of improving predictions of stock abundance.

Under the terminal molt hypothesis, males having the large chela characteristic of morphometrically mature (MM) crab do not molt again in their life spans. Before Conan and Comeau's (1986) research, it was generally believed that growth of male snow crab continued after the maturity molt although

R.S. Otto. National Marine Fisheries Service, Kodiak Laboratory, P.O. Box 1638, Kodiak, AK 99615-1638, U.S.A.

Correct citation: Otto, R.S. 1998. Assessment of the eastern Bering Sea snow crab, *Chionoecetes opilio,* stock under the terminal molting hypothesis. *In* Proceedings of the North Pacific Symposium on Invertebrate Stock Assessment and Management. *Edited by* G.S. Jamieson and A. Campbell. Can. Spec. Publ. Fish. Aquat. Sci. 125. pp. 109–124.

Table 1. Trawl survey abundance indices of commercial size (≥102 mm CW) male snow crab in comparison to fishery landings for the eastern Bering Sea.

		Landings in following year ($Y + 1$)		
Survey year (Y)	Survey index (millions of crab)	Crab (millions)	Tons (thousands)	U.S. dollars (millions)
1985	75	76.5	44.4	60.0
1986	83	81.3	46.2	75.7
1987	151	105.7	60.8	100.7
1988	171	112.6	67.8	110.7
1989	187	129.0	73.4	102.3
1990	420	265.1	149.1	162.6
1991	484	227.4	143.0	156.5
1992	256	169.6	104.7	171.9
1993	135	114.8	67.9	192.4
1994	72	60.6	34.1	180.0

that of females did not. The genus *Chionoecetes* was thought anomalous among the spider crab (Brachyura, Majidae). Evidence for the universality of a terminal molt at morphometric maturity in Majidae and for snow crab specifically was reviewed by Conan et al. (1990) along with the historical development of the terminal molt hypothesis. Elner and Beninger's (1992) review of reproductive biology places terminal molting in a larger context. Although the terminal molt hypothesis for *Chionoecetes* males generated considerable controversy (Donaldson and Johnson 1988; Conan et al. 1988; Jamieson and McKone 1988), it is generally accepted that a terminal molt occurs in both sexes of snow crab. There is good evidence that it occurs in the Tanner crab (*C. bairdi*) as well (Stevens et al. 1993). I refer to the "terminal molt hypothesis" as there are documented occurrences of molting among MM *Chionoecetes* males (Donaldson and Johnson 1988). In this paper, I assume that molting frequency among MM males is negligibly small.

The existence of a terminal molt in male snow crab complicates interpretation of size–frequency distributions and recruitment patterns. Interpretation requires annual estimation of the proportion (P) of nonmolting MM males and how P_{MM} is related to CW. Terminal molting offers a unique opportunity to quantify survival since abundance of crab with progressively older shells over a period of years should be related by a simple ratio. The relationship of mortality or survival to size is hence obtainable if the proportion of MM crab for progressively older shell ages is known and a consistent index of abundance is available. Here, I examine the utility of using the terminal molt hypothesis to interpret fishery data and abundance indices from the annual NMFS trawl survey of the EBS as an aid in predicting abundance and hence improving stock assessments.

Materials and methods

The EBS survey

The NMFS conducts an annual trawl survey of the EBS to collect data on the abundance, distribution, and biology of 5 species of crab and over 10 species of groundfish. The survey area (Fig. 1) encompasses the adult distribution of most commercial species but frequently does not cover the entire species range. For example, large numbers of immature snow crab are found in areas to the north of the annual survey area (Wolotira et al. 1977; Otto et al. 1980; Jewett 1981; Stevens and MacIntosh 1986) and some immature crab are thought to immigrate to the survey area as they grow and mature.

The EBS survey takes place during June, July, and August to avoid the spring crab molting periods. Snow crab that were in the process of molting (had well-formed second shells) were recorded in only 65 cases (0.03%) among 205 341 snow crab that were measured during the 1989–1994 study period. The survey area is divided into a grid of 20 nautical mile (37.1 km) squares and stations are usually towed at the center of each grid square. Over the period from 1989 to 1994, the area covered, sampling density, and timing (Table 2) of the survey were similar.

The survey employed an 83–112 otter trawl (Wathne 1977) throughout this investigation. This trawl has been modified to maintain strong contact with the bottom (Bakkala 1993). The headrope is 35% longer than the cable-chain footrope and is not rigged with bobbins or rollers. A trawl mensuration system provided data on trawl width and headrope height. This insured that the trawl was fishing normally as well as allowing for computation of the area swept by the trawl over the distance towed. Trawling was conducted for 30 min at each station and the distance towed was determined with either global positioning or LORAN C systems. Bottom water temperatures were recorded at almost all stations with expendable bathythermographs or temperature probes attached to the net. All tows were made during daylight hours. Tows were usually made every 2 h and 4–6 catches were usually processed each day.

Sampling survey catches

All crab were removed from the catch following each tow. Due to large catches and limited time between tows, it was not always possible to measure all crab caught or take chela height (CH) measurements from all males. Crab were separated by species and sex and a stratified sample was taken if more crab were captured than could be measured. Sampling factors were recorded for the extrapolation of crab measured to an estimate of those caught. On an annual basis, 11–23% of captured snow crab were measured (Table 2).

Crab were measured (nearest 1.0 mm CW or CH) with vernier calipers and scored by progressively older shell age (SA). SA1 (new, soft) and SA2 (new, clean, hard) crab are believed to have molted during the spring or early summer of the current year. Later, I refer to SA1 and SA2 males as "new shell" (NS). Old shell (OS) categories are SA3 (old, moderately stained, epifauna usually attached), SA4 (very old, heavily stained, and often encrusted), and SA5 (very, very old, usually stained dark brown or black, frequent rotting or soft portions). The process of scoring is subjective and correspondence between shell age scores and time since last molting becomes less certain as shell age increases.

A representative subsample of males was chosen for CH measurements (Wallace et al. 1949; measurement 4 of Conan and Comeau 1986). Instructions called for measuring: (*i*) the right chela, (*ii*) a higher percentage of OS crab due to their less common occurrence, (*iii*) a representative size range of crab, and (*iv*) not measuring crab that are missing one cheliped or for which left and right chelipeds differed in size. Annual sample sizes exceeded 3000 CW–CH pairs in all years (Table 2).

Fig. 1. Plots of the CPUE (numbers per square nautical mile towed) of male *C. opilio* in the eastern Bering Sea showing geographic distributions of the 1989–1994 means for (*a*) >99 mm carapace width, (*b*) 50–99 mm carapace width, (*c*) <50 mm carapace width, along with (*d*) total numbers of carapace width – chela height measurements. The smallest circles indicate that one or more (usually 6) tows were made but specified crab did not occur. Depth increases from northeast to southwest.

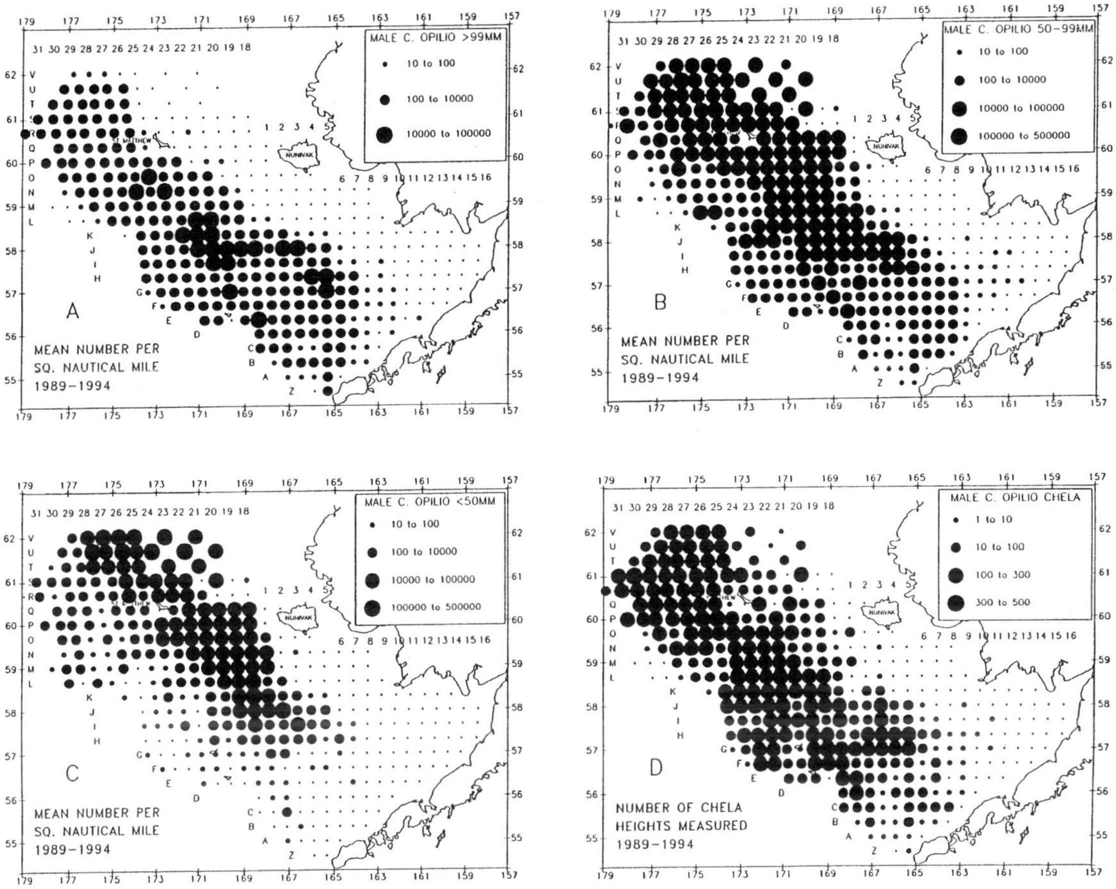

Table 2. Catch sampling of snow crab in NMFS trawl surveys of the eastern Bering Sea.

		Males			Females	
Year	Sampling dates	Number caught	Number measured	Chela measured	Number caught	Number measured
1989	6/3–8/14	117 749	25 288	3 851	173 900	9 916
1990	6/1–8/06	108 292	24 834	4 990	129 048	9 487
1991	6/1–8/01	154 480	27 411	6 615	217 449	12 740
1992	6/5–8/03	99 389	18 832	5 445	132 872	9 437
1993	6/4–7/26	175 668	20 721	3 394	190 438	9 312
1994	6/3–7/24	198 193	22 478	5 346	196 851	14 374

I used Somerton's (1980) technique to classify morphologically immature (MI) versus MM males and hence calculate P_{MM} in each 5 mm CW-shell age group. In this method CW–CH pairs are transformed to natural logarithms and straight lines characterizing MM and MI males are iteratively fitted. Resultant lines are nearly parallel and I computed coefficients of a dividing line that falls exactly one half the distance between them. Pairs falling above the dividing line were classed MM males while those below it were classed as MI males. This is virtually the same method used by Stevens et al. (1993). Although this procedure was applied annually (Table 3), a single line was later used for the entire study period. After pairs were scored as MM or MI, they were treated as a priori classes and subjected to discriminant analysis (SAS Institute 1988) as a means of comparing my classification with that of others (e.g., Conan and Comeau 1986). Misclassification errors ranged from 0.4 to 2.3% over the 6 years of data (Table 4).

Females were scored for the presence or absence of brooded eggs or egg remnants. Females that were brooding eggs or had

Table 3. Parameters of lines describing the linear relationship between the natural logarithms of carapace width and chela height for morphometrically mature and immature male *C. opilio*.

Year	*n*	*a*	*b*	*r*
(*a*) Morphometrically immature males				
1989	1164	–3.0552	1.3030	0.974
1990	1946	–2.5466	1.1882	0.982
1991	2487	–2.9094	1.2715	0.979
1992	1482	–2.8189	1.2482	0.983
1993	842	–3.1664	1.3384	0.983
1994	1948	–2.8847	1.2683	0.975
All	9869	–2.8969	1.2696	0.981
Watson	—	–2.481	1.166	0.997
Year	***n***	***a***	***b***	***r***
(*b*) Morphometrically mature males				
1989	2687	–2.7875	1.2981	0.985
1990	3044	–2.7495	1.2910	0.986
1991	4128	–2.8056	1.3034	0.983
1992	3963	–2.8131	1.3079	0.983
1993	2552	–2.7325	1.2910	0.982
1994	3398	–2.7692	1.2953	0.985
All	19772	–2.7762	1.2978	0.983
Watson	—	–2.481	1.332	0.991
Year	***n***	***a***	***b***	**% mature**
(*c*) Dividing lines				
1989	3851	–2.9214	1.3005	69.8
1990	4990	–2.6480	1.2396	62.0
1991	6615	–2.8575	1.2874	62.4
1992	5445	–2.8160	1.2780	72.8
1993	3394	–2.9494	1.3147	75.2
1994	5346	–2.8269	1.2818	63.6
All	29641	–2.8365	1.2837	66.7

Note: Data from Watson (1970) are provided for comparison. Dividing lines are those which best separate mature from immature males.

Table 4. Discriminant functions by year for carapace width – chela height pairs on natural logarithmic axis for eastern Bering Sea male *C. opilio*.

			Discriminant score coefficients		
Year	*n*	% misclassed	Constant	Width	Chela
1989	3851	1.56	–161	73.997	–59.615
1990	4990	0.88	–160	74.760	–60.412
1991	6615	1.00	–160	75.616	–60.709
1992	5445	2.13	–159	73.608	–59.811
1993	3394	1.15	–166	78.000	–62.628
1994	5346	0.49	–174	80.000	–64.121
All	29641	0.90	–162	74.933	–60.530
Scaled by 0.01			1.62	–0.749	0.605
Conan and Comeau (1986)			1.76	–0.789	0.614

Note: The a priori classification for all years was by the dividing line $\log_e CH = -2.8365 + 1.2837 \log_e CW$. Males with chela heights above the line were classified as morphometrically mature and those with chela below the line were classified as morphometrically immature.

egg remnants attached to their pleopods were considered mature. Data on mature females are of interest as a means of corroborating minimum size at male maturity and geographic trends in size.

Abundance

Abundance indices were calculated by the area-swept technique (Alverson and Pereyra 1969; Hoopes and Greenough 1970) by assuming a catchability of 1.0. Abundance indices are relative rather than absolute because there were no data on catchability of snow crab in the trawl. For management purposes, it is usually assumed that virtually all commercial-sized (≥102 mm CW) male snow crab in the trawl's path are captured. Smaller males and females are

Fig. 2. Logarithmic plots of the chela height – carapace width relationship for male *C. opilio* showing the dividing line used to separate morpometrically mature (large chela) from morphometrically immature crab (small chela). The dividing line ln(CH)= –2.8365 + 1.2837 ln(CW) was used in all years and is derived from 29 641 paired measurements. Data points have been "jittered" to enhance visibility.

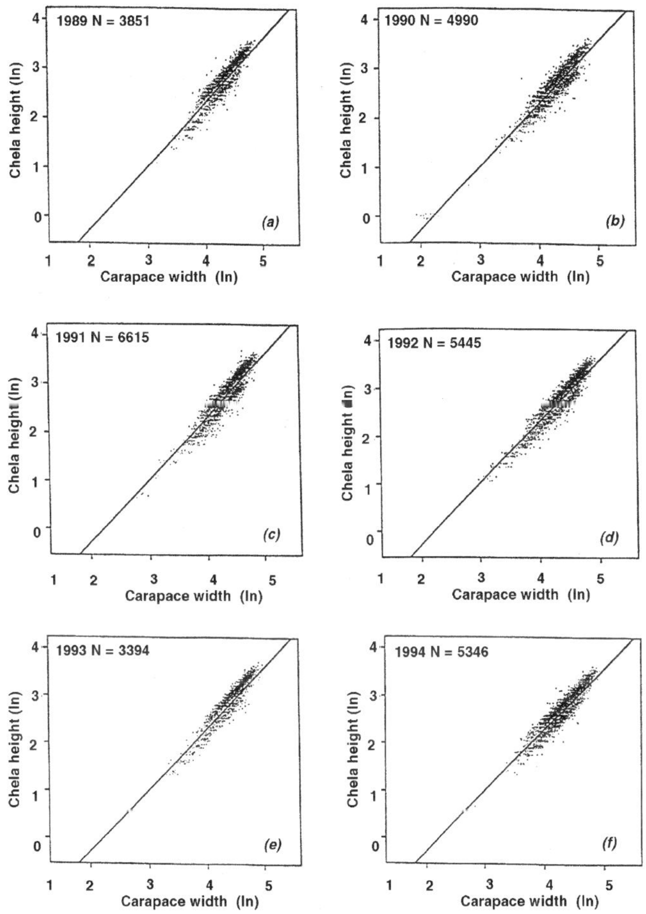

probably less vulnerable than commercial-sized males. Biases in calculating natural mortality rates are not incurred by use of abundance indices that are unadjusted for catchability if it is the same for a given size group from year to year. NMFS estimates of abundance are usually made by 5-mm CW group and I use 5-mm groupings in all calculations below.

Abundance for each 5-mm CW group by shell age was calculated by multiplying the proportion of crab of a given shell age by the total abundance index for that CW group. Abundance for MM males was taken as the product of P_{MM} and abundance per CW shell age group. I did not attempt to compute confidence intervals for abundance indices.

Mortality estimates

Survival calculations for MM males for each 5-mm CW group were made by computing an abundance index for each SA and then following the index and those for successively older SA's over several years. I used simple ratios of abundance, catch curve analysis, Robson–Chapman estimators, and Heincke estimators (Ricker 1975) to characterize survival of MM males. Heincke's survival estimate was used for predictive purposes because it is less subject to errors associated with the subjectivity of categorizing crab by shell age. Total survival (S) and hence mortality ($A = 1 - S$) for each 5-mm CW group was partitioned into natural and fishing components by equating the rate of exploitation to the conditional rate of fishing mortality (m) in a Type 1 fishery and calculating the conditional rate of natural mortality (n) from $A = m + n - mn$ (Ricker 1975). The relationship between "natural survival," $S' = (1 - n)$, and CW was characterized by a three parameter logistic curve and fitted using a Levenberg–Marquardt algorithm (IMSL 1989, subroutine BCLSF).

Mortality rates for MI males were computed iteratively for crab summed over all shell ages. I used a constant mortality rate for all MI males because I had no means of characterizing the size – mortality rate relationship. In predicting future commercial-sized male abundance of MI males, natural mortality was varied until the residual sum of squares between observed (survey) and predicted abundance was minimized.

Data on the CW frequency in snow crab landings were obtained from the Alaska Department of Fish and Game through their observer and port sampling programs. These data were grouped by 5 mm CW and proportions of the total sample calculated. The total catch of crab in each year (Morrison and Gish 1994) was prorated by these proportions to obtain estimates of total catch by 5-mm CW group. The ratio of estimated total commercial catch to the survey index for each 5-mm CW group was considered as the rate of exploitation although it is more properly a rate of utilization (Ricker 1975).

Predicting abundance

Growth per molt data taken from McBride (1982), Taylor and Hoenig (1990), and Hoenig et al. (1994) were used to extrapolate the proportion of surviving MI males in one 5-mm CW group that would grow and enter larger 5-mm CW groups the following year. This was done by regressing the increment of CW growth per molt against premolt CW and then using the deviations from the regression line to construct a growth transition matrix. The regression equation was used to calculate the expected transfer and transfers to 5-mm CW intervals above and below the expected interval were allocated according to the distribution of deviations about the regression line. This incremental growth model was simply implemented in a spread sheet (Lotus Development Corporation 1989).

Extrapolation of commercial-sized male abundance was limited to 1 future year. Calculations were essentially the same as those used to extrapolate king crab (*Paralithodes*) abundance (Otto 1986), except that they were made separately for MM and MI males. Recruitment was calculated by "growing" MI males (as discussed earlier) and then subtracting natural mortality. I assumed that MI males were an insignificant portion of the fishery. I had no data on the frequency of MI males in the EBS fishery, but MI males are apparently less vulnerable to pot fishing than are MM males (Hoenig and Dawe 1991; Comeau et al. 1991). Escapement was computed by subtracting the estimated commercial catch by 5-mm CW group from the survey abundance estimates. Escapees were allowed to "survive" according to the logistic relationship described above and then summed over all 5-mm CW groups above 100 mm CW. The extrapolated population is the sum of recruitment and surviving escapement.

Fig. 3. Proportions classified as morphometrically mature for combined 1989–1994 eastern Bering Sea male snow crab by 5 mm carapace width group, (*a*) proportions by shell age and (*b*) proportions for all shell ages weighted for shell age frequencies (population) and the unweighted combined sample. Sample sizes range from 419 to 973 at widths <50 mm, from 1103 to 2462 for widths 50–120, and from 90 to 715 above 120 mm. Data for carapace widths <40 mm are omitted.

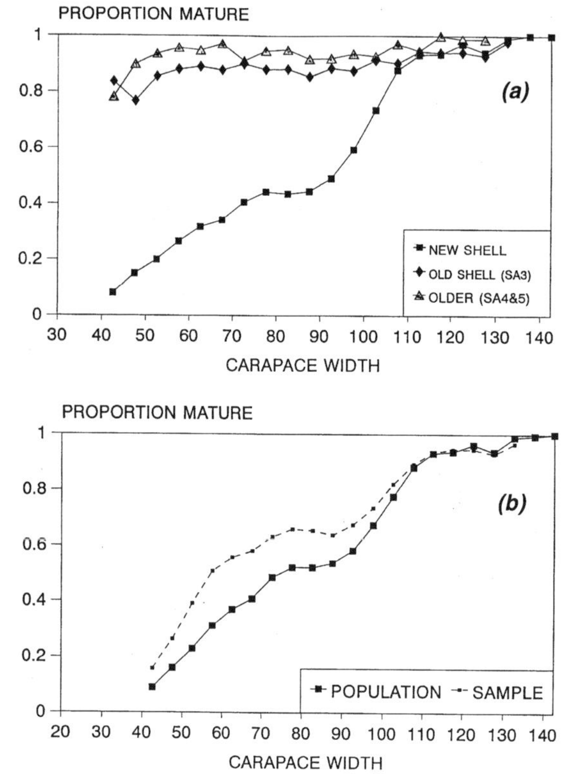

Results and discussion

Distribution of snow crab

Male snow crab in the EBS appear to make an ontogenetic migration from shallow waters to the north and northeast to deeper waters of the outer continental shelf. Large males (>99 mm CW) tend to be found from the middle to outer shelf portion of the distribution (Fig. 1*a*). The outer edge of the large male's distribution is about 200 m. Intermediate sizes of males (50–99 mm CW) are found broadly across the shelf but are concentrated in shallower portions (Fig. 1*b*). Small males (<50 mm CW) are concentrated in the shallow waters to the north and northeast (Fig. 1*c*), but south of 61°N latitude, small males are almost always found at depths >50 m. Mature female distribution resembles that of intermediate males.

Snow crab are strongly associated with the EBS middle-shelf region that is characterized by low circulation and cold, summer, bottom water temperatures. Of 175 625 snow crab that were measured during the study period at stations with known bottom temperatures, 97.8% came from temperatures between –2.0 and 4.0°C. This is similar to Bailey and Elner's (1989) –1 to 4°C range for commercial snow crab grounds. The middle-shelf hydrographic domain is separated from the coastal domain at approximately the 50-m isobath (Kinder and Schumacher 1981*a,* 1981*b*). Coastal domain bottom waters may reach 8–12°C in summer. Foyle et al. (1989) showed that snow crab are energetically restricted to cold waters since metabolic costs exceed caloric intake at about 7°C and feeding rate increases up to 6°C but decreases at higher temperatures. Snow crab also tend to be excluded from shelf edge waters by warm upwelling waters that may exceed 5.0°C. The highest temperature where crab were taken during 1989–1994 surveys was 7.9°C but only 432 (0.25%) of 175 625 measured snow crab were taken at temperatures greater than 5°C. Migration of snow crab into the survey area is necessarily from the north and confined to the middle shelf. Because annual surveys do not extend into northern portions of the middle shelf, immigration may provide recruitment that is not accounted for in the following sections.

Chela allometry and morphometric maturity

Large numbers of CW–CH measurements were collected over the range of snow crab within the survey area (Table 2, Fig. 1*d*). A single dividing line on logarithmic axes was used to separate MM males from MI males (Fig. 2). Parameters of regression lines characterizing the allometric relationship between CH and CW for both groups as well as the dividing lines are shown in Table 3 for all years and in summation. Interannual variations in parameters are small relative to dividing lines. Annual deviations in parameters of the dividing lines relative to the line derived from all data ranged from –6.6% (1990) to +4.0% (1993) for the intercept and from –3.4 to +2.4% for the slope. I chose to use the dividing line derived from the combined data set because it represents more data. Parameters for the upper (MM) lines appear more stable than those for the lower (MI) lines reflecting the fact that the majority of crab were classified as MM (Table 3).

The dividing line $\ln(CH) = -2.8365 + 1.2837 \ln(CW)$ was used to separate MM from MI males in all years and these were used as a priori groups in discriminant analysis (Table 4). Discriminant analysis indicates that a maximum of 2.1% of CW–CH pairs would be misclassified in any year and that this percentage is more typically near 1%. When scaled to the same decimal places, the discriminant function (Table 4) was very similar to that of Conan and Comeau (1986). Conan and Comeau's dividing line, converted to logarithmic scale, was $\ln(CH) = -4.1634 + 1.5787 \ln(CW)$. At sizes above $\ln(CW) = 4.5$ (90 mm CW) their line would be above that shown in Fig. 2 and at smaller sizes it would fall below. Differences between dividing lines and discriminant functions in this study and those of Conan and Comeau (1986) perhaps reflect differing morphology between snow crab inhabiting the southwestern Gulf of St. Lawrence and the EBS. Watson's (1970) data (Table 3) also suggest such differences.

Relationship of morphometric maturity to carapace width

Partitioning CW–CH pairs by shell condition shows the overwhelming tendency for OS (SA3–SA5) male crab to be MM (Fig. 3*a*). This was true in every year in various geographic subdivisions of the survey area and for all sizes of crab. The proportion of MM males among old shell (SA3) is only slightly less than for older shell condition categories (SA4 and SA5 combined). When these sample data are weighted for the proportion of each shell age observed (Fig. 3*b*), it is apparent that

Fig. 4. Proportions of new shell male snow crab that were classified as morphometrically mature plotted with population abundance indices (billions) by 5 mm carapace width group. Sample sizes range from 32 to 346 and 95% confidence interval half widths range from 0.0275 to 0.141. Data for carapace widths <40 mm are omitted.

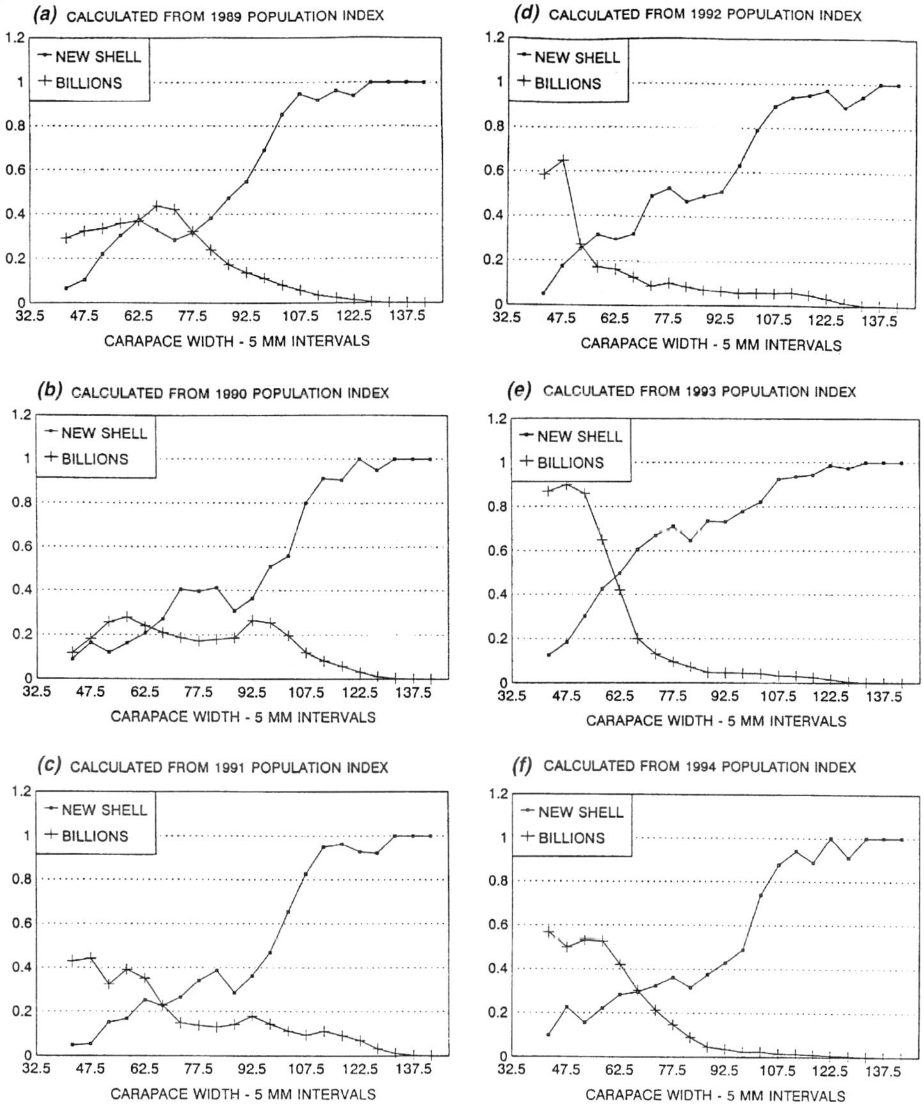

the P_{MM} among new shell crab are most important relative to the population as a whole. Use of the unweighted curve (Fig. 3*b*) would seriously underestimate size at maturity for any stated percentile. The strong tendency of SA3 SA5 males to be morphometrically mature is entirely consistent with the terminal molt hypothesis.

The CW–P_{MM} relationship was not a smooth ogive even when data were aggregated over the 6 years of study (Fig. 3*b*). This was unexpected given previous descriptions of male size at maturity in the EBS (Somerton 1981*a*, 1981*b*) and the Gulf of St. Lawrence (Watson 1970). Data from the Japan Sea (Yamasaki and Kuwahara 1991), however, are more similar to the situation described below. For example, data from 1989 (Fig. 4*a*) indicated that proportions of MM males were 6.6% (±6.4% 95% C.I., n = 60) at CW's 40–44 mm, increasing smoothly to 37% (±9.2%, n = 110) at CW's 60–64 mm, decreasing slightly to 28.3% (± 6.5%, n = 191) at 70–74 mm, and then increasing to approach 100% at the largest sizes. There was hence a flat portion or a trough in the maturity curve that was associated with a strong mode in abundance centered at ~70 mm CW in 1989. A flat portion of the curve was associated with this same dominant mode in abundance as it was displaced 20–25 mm to the right in 1990 (Fig. 4*b*) but began to disappear in 1991 (Fig. 4*c*) and 1992 (Fig. 4*d*) as the dominant cohort decayed. In 1993 the maturity curve is convex upward (Fig. 4*e*) but by 1994 it shows a flat portion associated with the leading edge of another apparently dominant mode (Fig. 4*f*). Comeau et al. (1991) also found evidence of density dependence in the molt to the MM phase. Tremblay et al. (1994) found that the percentage of male snow crab in the 78–94 mm CW group was positively correlated with the percentage of these crab that were MI, which would be similar to the situation described above. Sainte-Marie et al. (1995) show a plateau in the maturity curve that is associated with a dominant size–frequency mode in their samples, which is also similar to the situation for the EBS.

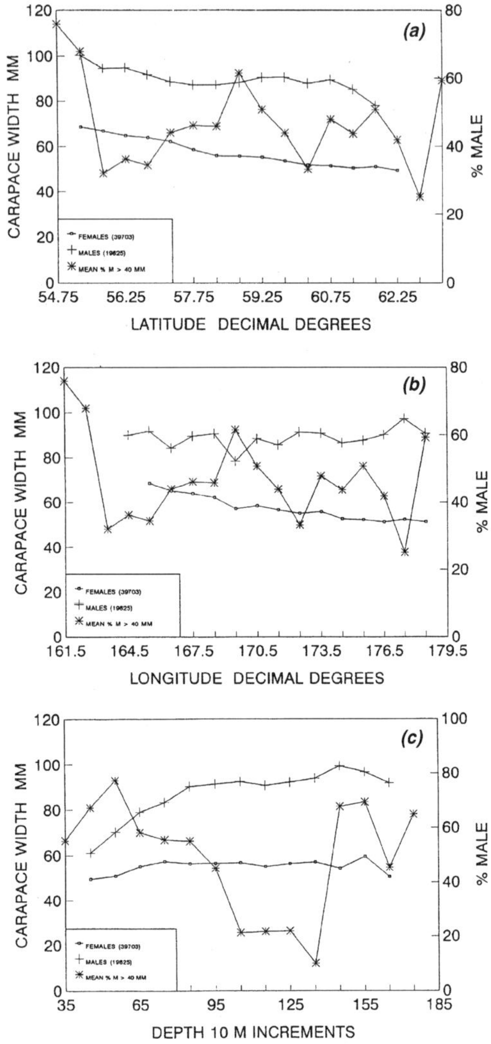

Fig. 5. Geographic trends in mean carapace widths of morphometrically mature male and ovigerous female *C. opilio* plotted with the 6 year mean percentage of males for (*a*) 0.5° increments of latitude and (*b*) 1.0° increments of longitude and (*c*) 10 m increments of depth. Mean carapace widths based on less than 100 observations are omitted.

Size at maturity

The grand mean CW for mature females was 56.1 ($n = 39\,659$). The smallest mature females encountered were 26 mm CW but they were rare at sizes below 35 mm. The largest mature female was 89 mm but very few females reach sizes greater than 80 mm. To the north in the Chuckchi Sea, Paul et al.'s (1997) data showed 52% of 29 females of 35–39 mm CW were ovigerous but they did not observe mature females <30 mm CW or >65 mm CW. In the EBS the size at 50% maturity (SM50) for females is near 50 mm (Otto et al. 1980; Somerton 1981*a*, 1981*b*) as was true for the Gaspé area (Watson 1970).

The grand mean CW for MM males was 88.5 mm ($n = 19\,848$). Males classed as MM were rarely encountered below 30 mm CW and there were few occurrences below 40 mm CW. Due to small sample size, I did not plot proportions of MM males for CW groups smaller than 40–44 mm CW (Figs. 3, 4, and 6). SM50 was 75 mm (Fig. 3) which is 15% smaller than the grand mean but larger than SM50 = 65 mm CW found by Somerton (1981*a,* 1981*b*) or SM50 = 57 mm CW found by Watson (1970). Somerton's and Watson's values were unweighted for population shell-age frequencies which could account for a smaller size (Fig. 3*b*). The largest MM male was 146 mm CW while MI males are uncommon above 110 mm (Figs. 3 and 4). Sainte-Marie et al. (1995) show that 50% of males have spermatophores at 38.5 mm CW and that none had spermatophores below 36.9 mm CW. Paul et al. (1997) found 19% of 37 males of 20–29 mm CW had spermatophores and 100% of 25 males of 30–35 mm CW had them. The minimum size for male morphometric maturity appears to be near 30 mm CW in the EBS. The minimum size at maturity for females and males is very similar and probably occurs during the second year (Somerton 1981*a*) after hatching. Potential effects of terminal molting on cohort growth patterns can hence occur over the majority of a cohort's life span.

Because the male maturity curve was not a smooth ogive in most years, I considered whether the shape of the maturity curve could be related to geographic differences in size at maturity. I plotted mean sizes of MM males and mature females along with the sex ratio by latitude, longitude, and depth (Fig. 5) in order to see if there were areas characterized by sharp differences or transitions in the size at maturity. There were trends in mean CW of MM males and ovigerous females with latitude, longitude, and depth, but these were gradual and apparently unrelated to sex ratio (percent males for crab >40 mm CW) except with respect to depth.

Size-related distributional differences (Fig. 1) were related to depth with increasing mean sizes of MM males up to about 80 m (Fig. 5*c*). I partitioned the data at 80 m in order to further examine patterns of interaction between cohort size and the maturity curve. I present maturity curves and size–frequency data (Fig. 6) for the years 1989–1991 since differences between shallow and deep water are small in later years and both areas resemble the maturity curves of Fig. 4*d*–4*f*. In both, deep and shallow water troughs or plateaus in maturity curves were associated with strong modes in CW–frequency distributions. This effect was pronounced in shallow water in 1990 (Fig. 6*b*). Because the dominant CW mode disappears from waters <80 m after 1990 (Fig. 6*b*–6*c*), presence of this mode in deeper water in 1991 (Fig. 6*f*) was explained by migration. Interaction between ontogenetic migration and a dominant cohort's effect on the CW–P_{MM} curve probably explains recruitment patterns to the commercial fishery over most of the EBS.

Estimating survival of morphometrically mature males

A ratio of the abundance in year $t + 1$ for an older shell age to that in year t for a younger shell age is a natural estimate of survival within a CW group. For example, the ratio of abundance of old shell (SA3) MM crab in 1990 to the abundance of NS (SA1 + SA2) MM crab in 1989 is an estimate of survival for males maturing in 1989. Grouping SA1 and SA2 as new shells, there are three ratios that may be used: old (SA3)/NS, very old (SA4)/old, and very, very old (SA5)/very old. Because 2 years of abundance indices must be used in each ratio, there are five estimates of survival for each ratio within each CW group from the 6 years of data. I averaged the five estimates for each ratio (Fig. 7*a*) as a means of comparison. Survival decreases with increasing shell age. The pattern of

Fig. 6. Maturity curves for male *C. opilio* of all shell ages by 5-mm carapace width group, P_{MM}, plotted with the population proportion of all males >40 mm carapace width caught. The data are for 1989, 1990, and 1991 in shallow waters (<80 m, *a–c*) and deep waters (>80 m, *d–f*). Points with <20 observations are excluded.

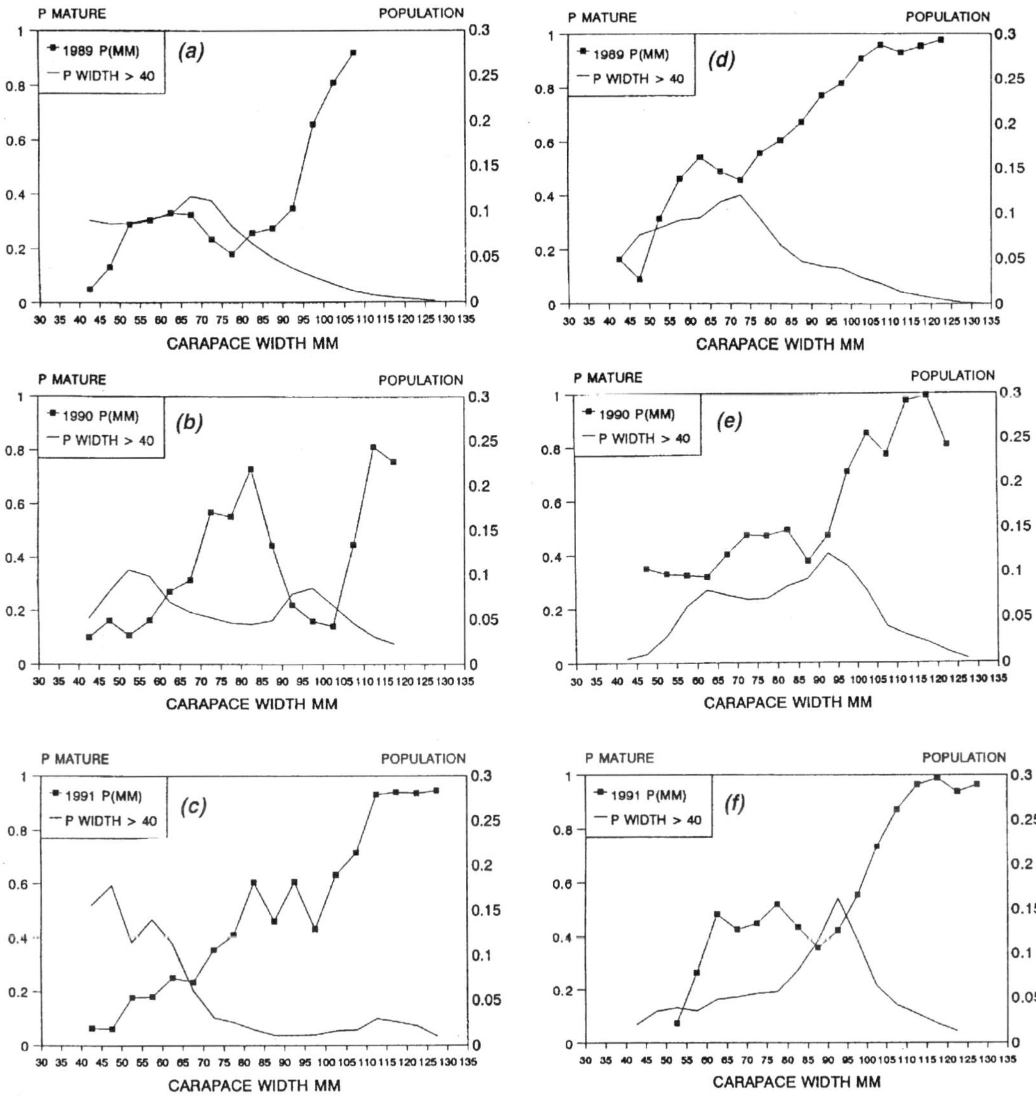

survival relative to CW, however, is similar for the three ratios, showing low survival at small CW's, increasing to a broad peak at intermediate CW's, and decreasing rapidly above 100 mm CW as MM males enter the fishery. There was considerable interannual variability in all ratios but the above patterns held true in each year.

Using abundance estimates for four shell categories as described earlier, there are three estimates of survival from the 6 years of data that can be applied to each CW group from catch curve analysis, Heincke's and Robson–Chapman's estimators. Averaged values for these estimators (Fig. 7*b*) show size-related patterns of survival that are the same as those for ratios (Fig. 7*a*). The Robson–Chapman and Heincke's estimates were similar. Catch curve analysis indicated lower survival rates that were more stable with size. Catch curve estimates would tend to be biased downward by a reduction in survival with increasing shell age (Fig. 7*a*). Annual variability in Heincke's estimate (Fig. 7*c*) was considerable but approximately the same size-related pattern for the three possible starting years was evident. The average of the three Heincke estimates by CW group shows a steeper assent at smaller sizes than is evident in other methods. The steepness of assent was derived largely from the 1991 data set (Fig. 7*c*). Survival at intermediate and larger sizes was reasonably stable. Since assigning shell ages is a subjective process, use of progressively older shell ages as proxies for time is questionable. Heincke's estimate is least restrictive since it relies on only the youngest group and the sum of older groups.

Estimates of survival exclusive of fishing mortality were necessary to compute natural mortality for commercial-sized crab. Annual exploitation rates from the fishery (Fig. 8*c*) were used as conditional fishing mortality (m) to estimate conditional annual rates of natural mortality (n) from total mortality rates ($A = m + n - mn$) specified by Heincke estimates (Fig. 7*c*). These were computed for each CW group by year. Conditional rates of natural mortality were converted to

Fig. 7. Estimates of survival for morphometrically mature male *C. opilio*. Mean ratios of abundance by shell age (*a*) are for progressively older shell ages lagged by one year, for example new-shelled crab (SA1 and 2, year *t*) to old shelled (SA3, year *t* + 1); the mean of Heincke estimates shown for comparison; (*b*) means of survival estimates treating shell ages as proxies for years since molting; (*c*) annual variation in Heincke survival estimates; and (*d*) logistic curve fitted to Heincke estimates adjusted for fishery removals.

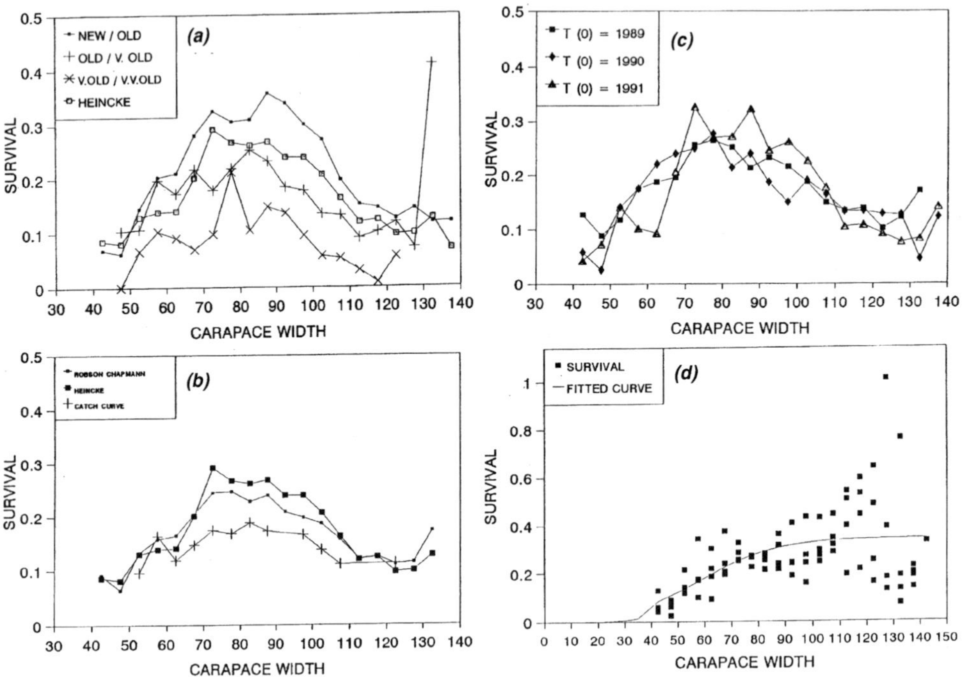

survival and plotted along with survival rates from prerecruit crab and the relationship of survival to size was fit by a three-parameter logistic function (Fig. 7*d*). Asymptotically, "natural survival" ($S' = 0.342$) corresponded to a conditional rate of natural mortality ($n = 0.658$) or instantaneous rate $M = 1.07$. Survival rates were interpolated from this curve [$S' = 0.3421/(1 + 61.0747 \cdot \exp(-0.0703 \cdot \mathrm{CW}))$] to predict future abundance of MM males from a given year's standing stock. Variation about the line is heavily influenced by apparent variation in exploitation rates (compare Figs. 7*c* and 8*c*) and is largest for males >125 mm CW for which abundance and hence sample sizes were low.

Comeau et al. (1991) cited radio element dating that showed a maximum carapace age for mature snow crab of both sexes of 4 years as evidence for high mortality among MM males. Recently the NMFS contracted with the University of Washington to perform radio element dating of EBS snow crab carapaces. Of three carapaces analyzed: an age of 40 days for a SA1 (soft) male, 4.4 years for one SA4 (very old) male, and 4.9 years for a second SA4 male (J.M. Orensanz, University of Washington, Seattle, Wash., personal communication). If SA were a proxy for time in the MM phase, SA4 should correspond to only 2 years, which questions the use of the SA as a proxy and also indicates that MM male snow crab may live longer than 4 years in the EBS because SA5 individuals are regularly encountered. Estimates of survival described earlier indicate that survival is about 30% per year at most sizes. At this rate about 7 of 10 000 crab would survive for 6 years. High mortality of MM males may explain Donaldson and Johnson's (1988) observation that old shell (SA > 2) male snow crab do not seem to accumulate in the EBS.

Low survival for large crab is to be expected since males enter the fishery at about 100 mm CW (Fig. 8) and the target exploitation rate for the fishery is 58% of the survey index under current management practice. Males are fully recruited at ~110 mm and survival at sizes above this are ~10% (Fig. 7*c*). Survival rates exclusive of fishing would be 34% for CW's ≥ 110 mm or a conditional natural mortality (*n*) of 66%. If the conditional rate of fishing (*m*) is near the target of 58%, the actual total mortality rate ($A = m + n - mn$) would be 80%, rather than near 90% as determined earlier. This may indicate that catchability of commercial-sized male snow crab is considerably less than 1.0.

Livingston (1991) and Livingston et al. (1993) estimated total consumption of snow crab by groundfish predators for the years 1984–1989 ranged from 10.6 to 30.9 billion crab per year with an average of 16.5 billion. Pacific cod (*Gadus macrocephalus*) is a major predator on snow crab and accounted for 64% of the aggregate total finfish predation on snow crab. Annual average consumption by cod was 10.5 billion crab, almost all <70 mm CW and most of which were 20–60 mm CW. Yellowfin sole (*Pleuronectes asper*), rock sole (*Pleuronectes bilineatus*), and flathead sole (*Hippoglossoides elassodon*) accounted for most additional consumption consisting almost entirely of crab <40 mm CW. It appears that snow crab tend to escape their major groundfish predators at sizes above 70 mm and perhaps this accounts for the increase in survival (Fig. 7) at intermediate sizes.

Growth

Data for Hiatt plots and incremental growth plots (Fig. 9) came mostly from Canadian studies (62 points from Hoenig et al. 1994, 20 from first year recoveries in Taylor and Hoenig 1990, and 6 EBS records from McBride 1982). A Hiatt plot was fitted

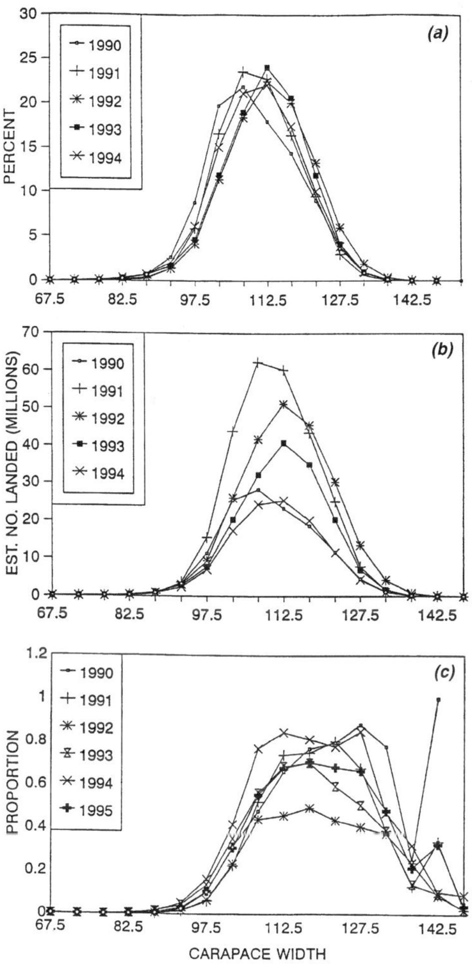

Fig. 8. Fishery removals of male *C. opilio*, showing (*a*) size frequency of removals expressed as a percentage of all removals; (*b*) estimated numbers of crab landed; and (*c*) exploitation rates calculated relative to the previous year's survey abundance index (*c*). Sample sizes by the Alaska Department of Fish and Game were: 32 110 (1990), 40 400 (1991), 168 325 (1992), 205 592 (1993), and 173 293 (1994).

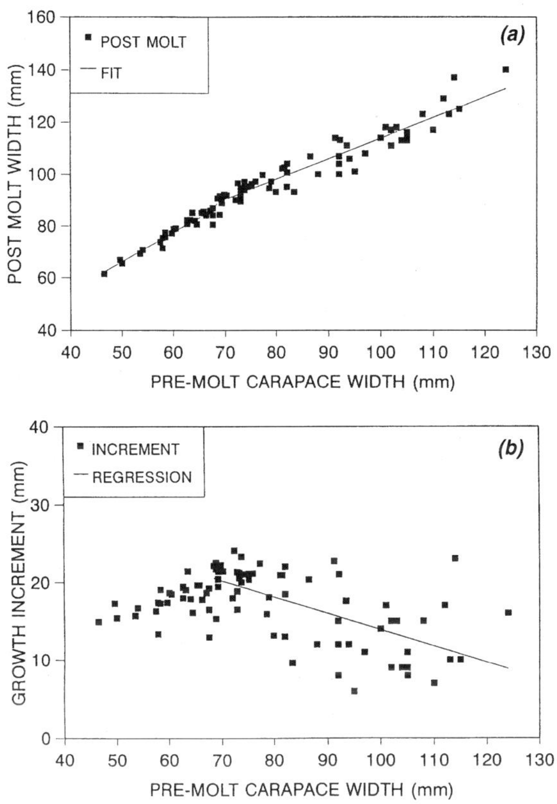

Fig. 9. Growth of male *C. opilio* expressed as: (*a*) a two-phase Hiatt diagram with a change in slope at 67 mm carapace width, and (*b*) an incremental growth plot with linear regression line ($Y = 34.945 - 0.2104\,X$, $r = 0.389$, $n = 66$). All but six data points are from eastern Canada.

with two straight lines according to a "broken stick" model. The sum of squares was minimized with a break point at 67 mm CW (Fig. 9*a*). The line $Y = 9.663 + 1.130X$ ($r = 0.964$, $n = 27$) characterized smaller crab and $Y = 34.945 \pm 0.780X$ ($r = 0.949$, $n = 61$) fit larger crab. I used these lines to predict when a change in growth occurred but used molting increments to characterize growth.

The line $Y = 34.945 - 0.21048X$ ($r = 0.389$, $n = 61$) was fit to growth per molt data for crab >67 mm CW (Fig. 9*b*) and used to predict transition from one 5-mm CW group to another upon molting. Plotting deviations from this line showed that 43% fell within ±2 mm of the predicted growth increment and this was taken as the proportion that would transfer to the predicted 5-mm CW group upon molting. Similarly, 25% of the deviations were 5–9 mm greater and 28% were 5–9 mm smaller than the predicted increment and these were used as proportions that would fall ±1 5-mm group from the predicted group. The remaining 3% fell ±2 groups away from the predicted group. These proportions were used to construct a molting transition matrix that introduced realistic variability around predicted transitions from the regression line.

Predicting abundance

Indices of abundance for MM and MI males (Fig. 10) were computed for each 5-mm CW group from the data depicted in Fig. 4. Estimated catch (Fig. 8*b*) in the following year (fishing begins January 15) was subtracted from MM male abundance and the remainder assumed to survive to the survey season according to the fitted equation in Fig. 7*d*. I assumed that only MM males are landed because MI males are rare for CW > 100 mm (Fig. 4) and because MI males may be less vulnerable to pot fishing than to trawling (Comeau et al. 1991). This procedure provided an estimate of the MM males in each CW group surviving from year t to year $t + 1$. Growth of MI males was according to the transition process described in the previous paragraph. Mortality of MI males was assumed constant over all size groups. The smallest size group that would contribute to the pool of males above 100 mm CW was 70–74 mm CW and so, assuming constant mortality, is similar to what appears to be the case for MM males. Apparent natural mortality for MI males is actually confounded with vulnerability to the survey.

Because M was unknown for MI crab, I used trial values ranging from 0.0 to 0.4, iteratively. MI males were "killed" according to a trial value, grown according to the transition matrix, and the sum of all MI and MM males above 100 mm

Fig. 10. Frequency distributions by 5 mm carapace width group for abundance of (*a*) morphometrically immature (MI) males (grey), (*b*) morphometrically mature (MM) males (clear), and (*c*) fishery removals (black).

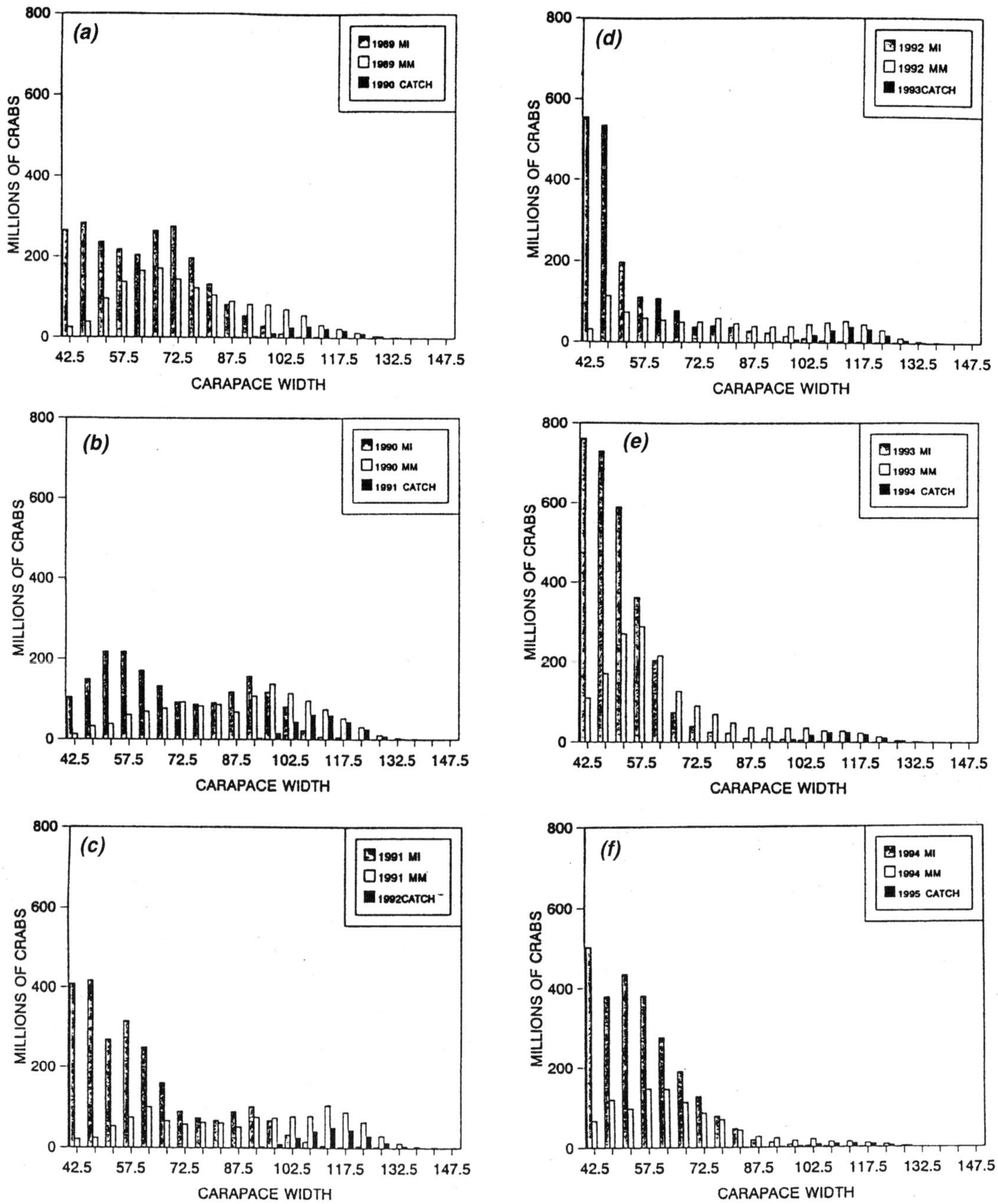

CW was compared to the survey index of abundance in the following year (Fig. 11). Predictions for 1995 were made assuming average values of exploitation rates (Fig. 8*c*). Trial values of M were evaluated according to the sum of squares (SS) of predicted values about the survey values for 1990–1994 and 1991–1994. The SS for 1990–1994 was minimized at $M = 0.0$ and the SS for 1991–1994 was minimized at $M = 0.1$. This could imply that almost all MI males survive or, more likely, that prerecruit MI males are less vulnerable to the survey than commercial-sized males. The response of SS to M, however, is not great over reasonable values of M and $M = 0.3$ fits the data nearly as well as $M = 0.1$ (Fig. 11). Predictions for the years 1991–1994 appear useful and both levels of M show the 1995 commercial stock to be similar to that of 1994. Subsequent to this Symposium, the 1995 survey index was 81.9 million males ≥ 100 mm, which lies between the 53.8 predicted for $M = 0.3$ and 100 for $M = 0.1$.

If $M = 0.3$, $S = 0.74$ for MI males, which is about twice the asymptotic rate for MM males. This is countered by the idea that MI males would tend to suffer higher mortalities related to the molting and soft-shell periods. Alternatively, Somerton's (1981*a*) model of snow crab growth indicates that males would reach 136 mm CW at about 20 years of age and Hoenig's (1983) analysis of the relationship between mortality

Fig. 11. Observed and predicted survey abundance indices for male *C. opilio* ≥ 100 mm carapace width.

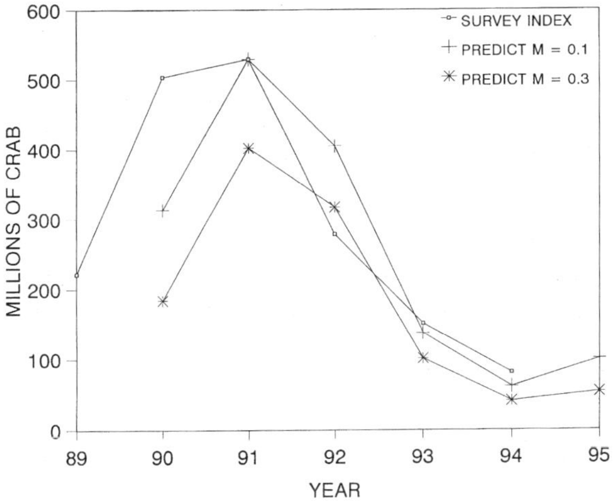

and age shows that fish reaching a maximum age of 20 years would have a total mortality rate of 0.21 while molluscs would have a rate of 0.28. Unfortunately, there are no such analyses for crustaceans. Since very few snow crab reach sizes >136 mm (Figs. 8 and 10), a mortality rate in the vicinity of 0.3 may be reasonable. If so, then vulnerability of snow crab at 70–100 mm CW to the survey trawl would need to be near 0.5 to account for the fits at $M = 0.0$ or $M = 0.1$.

Predictions from 1989 survey and 1990 fishery data underestimate 1990 survey abundance at any reasonable value of M. MI males may be less vulnerable to trawling than MM males and perhaps their prevalence in 1989 was underestimated. Conan and Maynard (1987) observed snow crab using a dredge-mounted television camera and found that 37% of them were buried in mud. Of those buried, 79% were smaller than 100 mm CW and "either females or juveniles." Miller (1975) found few buried crab using underwater photography but cited Watson and Simpson's (1969) observations from a submersible that indicated burying was common. Perhaps the dominant mode that lead to sharp increases in abundance in 1990 was underrepresented in 1989 but became more vulnerable to trawling in subsequent years as CW increased. There is also indication that many of the crab which composed this mode were in shallow water (<80 m) in 1989 (Fig. 6) and perhaps immigration was occurring. Finally the mode that was present at 65–74 mm in 1989 moved to 90–99 mm CW in 1990, implying growth of about 25 mm and incremental growth data (Fig. 9*b*) indicates that crab of 65–74 mm should grow about 17–19 mm. The effect of underestimating incremental growth would be less important because, as the dominant mode moved closer to 100 mm, it would be entirely captured by the growth transition process even at smaller average increments.

Management, assessment, and future research concerns

Management of the EBS snow crab fishery is similar to that in eastern Canada as described by Bailey and Elner (1989) and most rationale follows that for other North American crab fisheries (Miller 1976). Kruse (1993) reviews the current biological rationale for Alaskan management measures. Biologically oriented management measures include a guideline harvest level (GHL, similar to a quota) that is apportioned by sub-districts east and west of 173°W longitude, a prohibition on the retention of females, a fishing season which begins January 15, and a size limit of 78 mm CW. The EBS snow crab fishery is managed under a fishery management plan (FMP) developed by the North Pacific Fishery Management Council (NPFMC 1989) that specifies factors that must be considered in management measures but defers setting of regulations to the State of Alaska.

Biological factors required by the FMP to be considered in setting GHL's include estimates of exploitable biomass, recruitment, threshold (minimum viable reproductive stock below which fishing is prohibited), and acceptable biological catch (ABC, exploitable biomass modified by other concerns). The NMFS trawl survey of the EBS has been the principal tool in setting GHL's. Actually a GHL is not a quota in that the fishery may be allowed to exceed the GHL if in-season fishery monitoring indicates that survey abundance indices appear to underestimate commercial stock. Analysis described earlier indicates that this practice is warranted since apparent catchabilityin the survey may well be <1.0. Further research on catchability is badly needed.

A better understanding of stock structure is possible by applying the terminal molt hypothesis and partitioning males between MM and MI life phases. This process is essential in assessing recruitment relative to GHL's as well as longer term analysis of stock-recruitment relationships that may be used in determining thresholds. While predictive capability is uncertain at present, it has the potential to allow validation of a current year's abundance estimate based on the previous year's data and hence apply multiple years of data to the assessment process. Additionally, budgetary limitations may result in conducting biennial rather than annual surveys. A predictive capability would then become essential in setting a GHL.

The FMP also espouses goals relative to crab habitat. The distribution of snow crab in the EBS and the tendency of large numbers of MI crab to occur in shallow (<80 m) water suggest that some shallow water portions of the EBS are nursery areas. These nursery areas are all confined to the midshelf hydrographic regime due to the snow crab's thermal limits. However, the snow crab fishery tends to occur in deeper waters of the shelf or shelf edge. The distribution of fishing effort and catch is an inadequate measure of the importance of various habitats relative to potential problems associated with exploration and development of hydrocarbon resources. Similarly, bycatch of snow crab in other fisheries needs to be evaluated relative to its distribution as well as the prevalence of MI and MM males in order to assess potential effects on the snow crab stock.

Geographic clines in size at maturity suggest no substocks that require separate management measures. Dividing the GHL between areas east and west of 173°W longitude is not justifiable on this basis. Concentrations of commercial-sized males were often conveniently separated by this line and continuance of this practice may be warranted on the basis of spreading out harvests over the distribution of commercial-sized males. At present the EBS snow crab population is considered a single stock for management purposes and there seems little reason to change this practice. Differences in the CW–P_{MM} relationship were apparently related to an interaction between cohort size and depth. Such geographic differences dictate that extensive sampling in time and space is necessary to characterize male snow crab maturity.

Biological factors that the FMP requires in setting minimum size limits include "biological and functional" size at maturity, protecting reproductive capability, growth rates, and yield-per-recruit considerations. Size limits strongly affect GHL's because these are set by applying a target exploitation to the biomass of crab above the size limit. Snow crab reproductive biology is complex because males as small as 30 mm CW contain mature spermatophores (Paul et al. 1997), MI males may mate under some circumstances, and females are known to be capable of storing sufficient spermatophores to fertilize successive broods of eggs (Elner and Beninger 1992). There is still considerable debate as to whether morphometric maturity equates to functional maturity and there is no information on mating pairs in the EBS. The current target exploitation rate (0.58) was developed on the basis of yield-per-recruit analyses that predate acceptance of terminal molting as the usual male life history pattern (Somerton and Low 1977). Target exploitation rates will need reevaluation when better growth models become available through continued examination of the CW–P_{MM} relationship and its dynamics relative to cohort strength.

The current 78 mm CW size limit is intended to allow males, on the average, to mate at least once before being available to the fishery (NPFMC 1989). This is considerably less than the mean size of MM males (88.5 mm CW) and near the SM50 (75 mm CW, Fig. 4). A 75 mm CW male would be expected to reach 94 mm CW after one molt at which point the vast majority would be MM, and either recruited to the fishery or about to recruit to the fishery. Historically, the size limit has not generally been at issue since processors have usually been unwilling to purchase crab smaller than 102 mm CW and most landed crab were considerably larger than this (Fig. 8). As abundance of commercial males decreased, exploitation rates of smaller males in the fishery increased slightly (Fig. 8*c*, 1994) and there has been discussion of raising the size limit to match current market practice. The present analysis indicates that the 95 mm size limit applied in eastern Canada may be appropriate for the EBS as well.

The CW–P_{MM} relationships described above indicate that almost all snow crab harvested in the EBS fishery are MM and also show that only those crab that mature late in their life spans recruit to the fishery. Since exploitation rates in the fishery commonly exceed 0.6 over much of the exploited size range (Fig. 8*c*) and natural mortality rates appear to be high, opportunities for mating by commercial-sized males are few relative to males that reach MM at smaller sizes. If MM is equatable with functional maturity as Conan and Comeau (1986) suggested, then Kruse's (1993) concern that individuals with genotypes responsible for late maturity or fast growth may be lost to the population may need closer examination. The current restriction against harvesting females implies that about half of the population genome is totally protected from fishery-induced selection, which argues against fishery-induced genetic selection. Males, however, mature over a much larger size range than females and it is possible that growth rates and timing of maturity are sex-linked characteristics. Management measures that foster long-term reproductive viability may be inadequate to protect against fishery-induced genetic selection favoring slow-growing or early-maturing males. Alternatively, the tendency to mature later at larger sizes may depend largely on the population structure encountered by each cohort, implying a great deal of phenotypic plasticity and relatively weak genotypic control. A great deal of research will be necessary to examine whether genotypic control or phenotypic plasticity is more important in determining growth rates and size at maturity in snow crab and other exploited majid crab populations.

Acknowledgments

I wish to thank D.A. Somerton for providing the program MATURE and for fruitful conversations concerning the reproductive biology of snow crab; B.A. Johnson for statistical advice, help with discriminate analysis routines, and for jittered plots of chela allometry; and J.A. Haaga for producing geographic distribution charts. Dave Taylor, Canadian Division of Fisheries and Oceans, St. John's, Newfoundland, kindly provided his data on incremental growth per molt. Numerous NMFS personnel and visiting scientists have participated in the EBS survey and are thanked for their help and support. A memorable EBS cruise with R.W. Elner provided opportunity for direct comparison of methods and stimulating discussion.

References

Alverson, D.L., and Pereyra, W.T. 1969. Demersal fish explorations in the northeast Pacific ocean — an evaluation of exploratory fishing methods and analytical approaches to stock size and yield forecasts. J. Fish. Res. Board Can. **26**: 1985–2001.

Bailey, R.F., and Elner, R.W. 1989. Northwest Atlantic snow crab fisheries: lessons in research and management. *In* Marine invertebrate fisheries. Their assessment and management. *Edited by* J.F. Caddy. John Wiley & Sons, New York, N.Y. pp. 261–280.

Bakkala, R.G. 1993. Structure and historic changes in the groundfish complex of the eastern Bering Sea. NOAA Tech. Rep. NMFS 114. 91 p.

Comeau, M., Conan, G.Y., Robichaud, G., and Jones, A. 1991. Life history and population fluctuations of snow crab (*Chionoecetes opilio*) in the fjord of Bonne Bay on the west coast of Newfoundland, Canada — from 1983 to 1990. Can. Tech. Rep. Fish. Aquat. Sci. 1817.

Conan, G.Y., and Comeau, M. 1986. Functional maturity and terminal molt of male snow crab, *Chionoecetes opilio*. Can. J. Fish. Aquat. Sci. **43**: 1710–1719.

Conan, G.Y., and Maynard, D.R. 1987. Estimates of snow crab (*Chionoecetes opilio*) abundance by underwater television — a method for population studies on benthic fisheries resources. J. Appl. Ichthyol. **3**: 158–165.

Conan, G.Y., Moriyasu, M., and Cormier, R. 1988. Reply to Donaldson and Johnson. Can. J. Fish. Aquat. Sci. **45**: 1501–1503.

Conan, G.Y., Elner, R.W., and Moriyasu, M. 1990. Review of literature on life histories in the genus *Chionoecetes* in light of recent findings on growth and maturity of *C. opilio* in eastern Canada. *In* Proceedings of the International Symposium on King and Tanner Crabs. Lowell Wakefield Fish. Symp. Ser., Alaska Sea Grant Rep. 90-04, University of Alaska, Fairbanks, Alaska. pp. 163–180.

Donaldson W.E., and Johnson, B.A. 1988. Some remarks on "functional maturity and terminal molt of male snow crab, *Chionoecetes opilio*" by Conan and Comeau. Can. J. Fish. Aquat. Sci. **45**: 1499–1501.

Elner, R.W., and Beninger, P.G. 1992. The reproductive biology of snow crab, *Chionoecetes opilio*: a synthesis of recent contributions. Am. Zool. **32**: 524–533.

Foyle, T.P., O'Dor, R.K., and Elner, R.W. 1989. Energetically defining the thermal limits of the snow crab. J. Exp. Biol. **145**: 371–393.

Hoenig, J.M. 1983. Empirical use of longevity to estimate mortality rates. Fish. Bull. U.S. **81**: 898–903.

Hoenig, J.M., and Dawe, E.G. 1991. Relative selectivity of four sampling methods using traps and trawls for male snow crabs. CAFSAC Res. Doc. 91/32. 17 p. (Cited with permission.)

Hoenig, J.M., Dawe, E.G., and O'Keefe, P.G. 1994. Molt indicators and growth per molt for male snow crabs (*Chionoecetes opilio*). J. Crustac. Biol. **14**: 273–279.

Hoopes, D.T., and Greenough, J.W. 1970. King crab research. Int. North Pac. Fish. Comm. Annu. Rep. **1968**: 116–124.

IMSL Inc. 1989. Math/Library. FORTRAN subroutines for mathematical applications. Version 1.1. IMSL Inc., Houston, Tex.

Jamieson, G.S., and McKone, W.D. (*Editors*). 1988. Proceedings of the International Workshop on Snow Crab Biology, December 8–10, 1987, Montrèal, Quebec. Can. MS Rep. Fish. Aquat. Sci. 2005. 163 p.

Jewett, S.C. 1981. Variations in some reproductive aspects of female snow crabs *Chionoecetes opilio.* J. Shellfish Res. **1**: 95–99.

Kinder, T.H., and Schumacher, J.D. 1981*a*. Hydrographic structure over the continental shelf of the southeastern Bering Sea. *In* The eastern Bering Sea shelf: oceanography and resources. (Vol. 1). *Edited by* D.W. Hood and J.A. Calder. University of Washington Press, Seattle, Wash. pp. 31–52.

Kinder, T.H., and Schumacher, J.D. 1981*b*. Circulation over the continental shelf of the southeastern Bering Sea. *In* The eastern Bering Sea shelf: oceanography and resources. (Vol. 1). *Edited by* D.W. Hood and J.A. Calder. University of Washington Press, Seattle, Wash. pp. 31–52.

Kruse, G.H. 1993. Biological perspectives on crab management in Alaska. *In* Proceedings of the International Symposium on Management Strategies for Exploited Fish Populations. *Edited by* G. Kruse, D.M. Eggers, R.J. Marasco, C. Pautzke, and T.J. Quinn. University of Alaska Sea Grant College Program Report No. 93-02. pp. 355–384.

Livingston, P.A. (*Editor*). 1991. Groundfish food habits and predation on commercially important prey species in the eastern Bering Sea from 1984 to 1986. NOAA Tech. Memo. NMFS/NWC-207. 240 p.

Livingston, P.A., Ward, A., Lang, G.M., and Yang, M. 1993. Groundfish food habits and predation on commercially important prey species in the eastern Bering Sea from 1987 to 1989. NOAA Tech. Memo. NMFS-AFSC-11. 192 p.

Lotus Development Corporation. 1989. Lotus 1-2-3 release 2.2. Lotus Development Corporation, Cambridge, Mass.

McBride, J. 1982. Tanner crab tag development and tagging experiments 1978–1982. *In* Proceedings of the International Symposium on the Genus *Chionoecetes*. Lowell Wakefield Fish. Symp. Ser., Alaska Sea Grant Rep. 82-10. University of Alaska, Fairbanks, Alaska. pp. 383–403.

Miller, R.W. 1975. Density of the commercial spider crab, *Chionoecetes opilio*, and the calibration of effective area fisher per trap using bottom topography. J. Fish. Res. Board Can. **23**: 761–768.

Miller, R.J. 1976. North American crab fisheries: regulations and their rationales. Fish. Bull. U.S. **74**: 623–633.

Morrison, R., and Gish, R.K. 1994. Annual management report for the shellfish fisheries of the Bering Sea area, 1993. *In* Annual management report for the shellfish fisheries of the Westward Region, 1993. Regional Information Report No. 4K94-29. Alaska Department of Fish and Game, Commercial Fisheries Management and Development Division, 211 Mission Road, Kodiak, Alaska 99615. pp. 151–219.

North Pacific Fishery Management Council (NPFMC). 1989. Fishery management plan for the king and Tanner crab fisheries of the Bering Sea/Aleutian Islands. North Pacific Fishery Management Council, P.O. Box 103136, Anchorage, Alaska 99510.

Otto, R.S. 1981. Eastern Bering Sea crab fisheries. *In* The eastern Bering Sea shelf: oceanography and resources. (Vol. 2). *Edited by* D.W. Hood and J.A. Calder. University of Washington Press, Seattle, Wash. pp. 1037–1066.

Otto, R.S. 1986. Management and assessment of eastern Bering Sea king crab stocks. *In* North Pacific Workshop on Stock Assessment and Management of Invertebrates. *Edited by* G.S. Jamieson and N. Bourne. Can. Spec. Publ. Fish. Aquat. Sci. 92. pp. 83–106.

Otto, R.S. 1990. An overview of eastern Bering Sea king and Tanner crab fisheries. *In* Proceedings of the International Symposium on King and Tanner Crabs. Lowell Wakefield Fish. Symp. Ser., Alaska Sea Grant Rep. 90-04, University of Alaska, Fairbanks, Alaska. pp. 9–26.

Otto, R.S., Fukuyama, A.K, Armetta, T.M., MacIntosh, R.A., and McBride, J. 1980. King and tanner crab research in the eastern Bering Sea, 1979. Int. North Pac. Fish. Comm. Annu. Rep. **1979**: 78–95.

Paul, J.M., Paul, A.J., and Barber, W.E. 1997. Reproductive biology and distribution of snow crab from the Northeastern Chuckchi Sea. *In* Fish ecology in Arctic North America. *Edited by* J. Reynolds. Am. Fish. Soc. Symp. **19**: 287–294.

Ricker, W.E. 1975. Computation and interpretation on biological statistics of fish populations. Fish. Res. Board Can. Bull. 191. 382 p.

Sainte-Marie, B., Raymond, S., and Brêthes, J. 1995. Growth and maturation of the male snow crab, *Chionoecetes opilio* (Brachyura: Majidae). Can. J. Fish. Aquat. Sci. **52**: 903–924.

SAS Institute Inc. 1988. SAS/STAT users guide. Release 6.03 edition. SAS Institute. Cary, N.C. 1028 p.

Somerton, D.A. 1980. A computer technique for estimating the size at sexual maturity in crabs. Can. J. Fish Aquat. Sci. **37**: 1488–1494.

Somerton, D.A. 1981*a*. Life history and population dynamics of two species of Tanner crab, *Chionoecetes bairdi* and *C. opilio*, in the eastern Bering Sea with implications for the management of the commercial harvest. Ph.D. dissertation. University of Washington, Seattle, Wash. 220 p.

Somerton, D.A. 1981*b*. Regional variation in the size of maturity of two species of Tanner crab (*Chionoecetes bairdi* and *C. opilio*) in the eastern Bering Sea and its use in defining management subareas. Can. J. Fish. Aquat. Sci. **38**: 163–174.

Somerton, D.A., and Low, L.L. 1977. Determination of minimum size and yield limitations for Tanner crabs in the eastern Bering Sea. Northwest and Alaska Fisheries Center Processed Report. 49 p.

Stevens, B.G., and MacIntosh, R.A. 1986. Analysis of crab data from the 1985 NMFS survey of the northeast Bering Sea and Norton Sound. AFSC Processed Report 86-16. Alaska Fisheries Science Center, 7600 Sand Point Way NE, Seattle, Wash. 35 p.

Stevens, B.G., Donaldson, W.E., Haaga, J.A., and Munk, J.E. 1993. Morphometry and maturity of paired Tanner crabs, *Chionoecetes bairdi,* from shallow and deepwater environments. Can. J. Fish. Aquat. Sci. **50**: 1504–1516.

Stevens, B.G., Haaga, J.A., and MacIntosh, R.A. 1994. Report to industry on the 1994 eastern Bering Sea crab survey. AFSC Processed Report 94-07. Alaska Fisheries Science Center, 7600 Sand Point Way NE, Seattle, Wash. 52 p.

Taylor, D.M., and Hoenig, J.M. 1990. Growth per molt of male snow crab *Chionoecetes opilio* from Conception and Bonavista Bays, Newfoundland. Fish. Bull. U.S. **88**: 753–760.

Tremblay, M.J., Eagles, M.D., and Elner, R.W. 1994. Catch, effort and population structure in the snow crab fishery off eastern Cape Breton, 1978–1993: a retrospective. Can. Tech. Rep. Fish. Aquat. Sci. 2021. 47 p.

Wallace, M.M., Pertuit, C.J., and Hvatum, A.R. 1949. Contribution to the biology of the king crab *Paralithodes camtschatica* (Tilesius). U.S. Fish. Wildl. Serv., Fish. Leaflet 340. 50 p.

Wathne, F. 1977. Performance of trawls used in resource assessment. Mar. Fish. Rev. **39**: 16–23.

Watson, J. 1970. Maturity, mating and egg laying in the spider crab, *Chionoecetes opilio*. J. Fish. Res. Board Can. **27**: 1607–1616.

Watson, J., and Simpson, S.L. 1969. Submarine operations in queen crab research. *In* Observations made by St. Andrews staff during submersible operations. *Edited by* A.V. Tyler. Fish. Res. Board. Can. Tech. Rep. 102. pp. 26–35.

Wolotira, R.J., Sample, T.M., and Morin, M. 1977. Demersal fish and shellfish resources of Norton Sound, the southeastern Chukchi Sea and adjacent waters in the baseline year 1976. Northwest and Alaska Fisheries Center, 7600 Sand Point Way NE, Seattle, Wash. 292 p.

Yamasaki, A., and Kuwahara, A. 1991. The terminal molt of the male snow crab in the Japan Sea. Nippon Suisan Gakkai-shi, **57**: 1839–1884.

Interpretation of growth, mortality, and recruitment patterns in size-at-age, growth increment, and size frequency data

Barry D. Smith and Louis W. Botsford

Abstract: The determination of age structure for many invertebrate populations is often difficult because individuals lack the hard body parts that record growth (e.g., otoliths, scales), and population size distributions often lack patterns expressing an age structure. Consequently, assessments of invertebrate fisheries are often of necessity length-based. Here we review the quantitative basis for inferring life-history characteristics such as growth and mortality rates from size frequency distributions that do not necessarily exhibit age pulses. We use an equilibrium solution to the von Foerster equation to describe the dependence of size distributions and optimal lower size limits for harvest on life-history characteristics. We then describe various forms of the von Bertalanffy growth model that include stochasticity in different ways, and the patterns of size versus age, growth increment versus size and time-at-large, and numbers versus size to which the different models lead. We show how our portrayals can help an analyst develop a general intuitive basis for visual interpretation of size-related patterns of growth and mortality in their data by formally analysing size frequency and growth data from a few typical invertebrate populations.

Résumé : La détermination de la structure d'âge de nombreuses populations d'invertébrés est souvent difficile parce que ces organismes ne possèdent pas de structures corporelles dures qui permettent d'enregistrer la croissance (p. ex. otolithes ou écailles) et parce que, souvent, les distributions de tailles de la population ne montrent pas de caractéristiques permettant d'exprimer la structure d'âge. Par conséquent, les évaluations des pêcheries d'invertébrés sont souvent, par nécessité, fondées sur la longueur des organismes. Dans la présente communication, nous passons en revue les fondements quantitatifs permettant de déduire des caractéristiques du cycle vital comme les taux de croissance et de mortalité à partir des distributions de fréquences de tailles qui ne présentent pas nécessairement de pointes liées à l'âge. Nous utilisons une solution d'équilibre à l'équation de von Foerster pour décrire la dépendance, à l'égard des caractéristiques du cycle vital, des distributions de taille et des limites de taille inférieure optimales pour la récolte. Nous décrivons ensuite diverses formes du modèle de croissance de von Bertallanffy qui tiennent compte de la stochasticité de différentes façons; nous décrivons également les caractéristiques taille selon l'âge, augmentation de croissance selon la taille et la période en liberté et effectifs selon la taille, auxquelles mènent les différents modèles. Nous montrons comment nos caractérisations peuvent aider un analyste à développer une base intuitive générale pour l'interprétation visuelle des caractéristiques de croissance et de mortalité liées à la taille dans leurs données en analysant formellement les données sur la croissance et la fréquence de tailles tirées de quelques populations d'invertébrés typiques. [Traduit par la Rédaction]

Introduction

Even the most basic of population assessments require information on the growth and mortality rates of individuals in a population (Ricker 1975; Gulland 1983). Size-at-age data have typically been used to provide growth information since growth is a measure of change in size over time. Since high quality size-at-age and numbers-at-age data are often difficult to acquire for invertebrates, invertebrate fisheries usually cannot be assessed using techniques such as virtual population analysis (VPA, Pope 1972) and its descendants (see Hilborn and Walters 1992) which are often applied to fish. Size frequency analysis is often the best alternate methodology for estimating growth and mortality when individuals do not have permanent anatomical structures which record the passage of time.

Recruitment pulses in a size frequency distribution can aid visual and analytic interpretation of year-class patterns for temperate species undergoing continuous growth (Schnute and Fournier 1980; Smith and McFarlane 1990; Botsford et al. 1994). The growth and mortality processes can be visualized as the dissipation of recruitment pulses through time (Fournier and Breen 1983). A conceptually and analytically more challenging problem is the interpretation of growth and mortality rates from size frequency distributions which lack multiple age pulses. Such distributions would be typical of species which tend not to have annual recruitment pulses, but which have

B.D. Smith.[1] Department of Fisheries and Oceans, Maurice Lamontagne Institute, P.O. Box 1000, Mont-Joli, QC G5H 3Z4, Canada.

L.W. Botsford. Department of Wildlife, Fish and Conservation Biology, University of California, Davis, CA 95616, U.S.A.

Correct citation: Smith, B.D., and Botsford, L.W. 1998. Interpretation of growth, mortality, and recruitment patterns in size-at-age, growth, increment, and size frequency data. *In* Proceedings of the North Pacific Symposium on Invertebrate Stock Assessment and Management. *Edited by* G.S. Jamieson and A. Campbell. Can. Spec. Publ. Fish. Aquat. Sci. 125. pp. 125–139.

[1]Present address: Environment Canada - Canadian Wildlife Service, Pacific Wildlife Research Centre, 5421 Robertson Road, Delta, BC V4K 3N2, Canada.

individuals recruiting at a constant rate into the population of interest. In temperate climates recruitment pulses are generally seasonal and are apparent for the smaller individuals in size distributions. When growth is asymptotic the pulses tend to become smeared at larger sizes due to growth variability (Botsford et al. 1994). In contrast, tropical and subtropical species are more likely to be characterized by size frequency distributions lacking age pulses.

A size frequency distribution provides static information on the size composition of a population. If there are no age pulses, then the distribution will be only qualitatively informative about growth, mortality, and recruitment. However, size-at-age and growth increment information, i.e., that obtained from a laboratory growth study, or a field mark–recapture study, can be quantitatively informative about growth rates. When such growth information (i.e., from individuals) is combined with size frequency data (population aggregate information) together they can be quantitatively informative about growth and mortality rates for a population. Our intention with this paper is to inform readers (*i*) of what visual inspection of size frequency distributions qualitatively can tell us about the relationship between growth and mortality for a population, and (*ii*) what visual inspection of size-at-age and growth increment plots qualitatively can tell us about growth and growth variability for a population.

The mathematical framework upon which we develop our concepts of growth and mortality also provides the basis for a general methodology for analyzing size frequency data and growth increment data in combination to measure growth and mortality, and for calculating an optimal minimum size limit. The methodologies for analyzing growth increment and size frequency data are separable thereby allowing the growth increment analysis to be linked to an analysis of size frequency data characterized by either constant or pulsed (i.e., age-structured) recruitment. The details (including precision, bias, and robustness to assumptions) of formal parameter estimation appear in other papers such as Smith and McFarlane (1990) and Smith et al. (1998) where we apply our models to the particular problems of assessing the population dynamics of the lingcod (*Ophiodon elongatus*) in British Columbia, and the red sea urchin (*Strongylocentrotus franciscanus*) in California.

Size frequency distributions

We need a formal basis upon which to build a model for size frequency distributions based on growth and mortality functions of size. The von Foerster size-structured equation (von Foerster 1959; Van Sickle 1977; DeAngelis and Mattice 1979; Huston and DeAngelis 1987) provides the basis for describing how the density of individuals in a population changes over time. We present it here as

$$[1] \quad \frac{\partial n(l,t)}{\partial t} = -\frac{\partial}{\partial l}[n(l,t)g(l)] - D(l)n(l,t)$$

where $n(l, t)$ is the density of individuals of size l at time t, $g(l)$ is the growth rate of an individual of size l, and $D(l)$ is the mortality rate of a individual of size l (see Table 1 for symbol definitions). As shown in Botsford et al. (1994) we can obtain a density expression for a size frequency distribution under steady-state and constant recruitment conditions by setting $\partial n(l, t)/\partial t = 0$ which yields

$$[2] \quad \frac{\partial n(l)}{\partial l} = -\frac{n(l)}{g(l)}\left[\frac{\partial g(l)}{\partial l} + D(l)\right]$$

Equation 2 thus describes the size-based steady-state density of numbers-at-size for any size frequency distribution. A representation such as eq. 2 would be valid for interpreting size frequencies which appear not to change appreciably in form over time. The convenience of eq. 2 is that it allows an analyst to explore the consequences of various size-dependent functions of growth, $g(l)$, and mortality, $D(l)$. If we assume von Bertalanffy growth then

$$[3] \quad g(l) = K'(L - 1)$$

As an extension of eq. 3 we let K' represent $K + bl$ below (and in Figs. 1 and 2) and choose a form of von Bertalanffy growth that allows K' to change with size l,

$$[4] \quad g(l) = (K + bl)(L - l)$$

with

$$[5] \quad \frac{\partial g(l)}{\partial l} = -K + b(L - 2l)$$

(Note that the equivalent of eq. 5 in Table 1 of Botsford et al. (1994) contains a sign error.) The advantage of eq. 4 is its potential to generate a sigmoidal growth curve when $b > 0$. (We recognize, however, that other functional forms, e.g., Richards, logistic, have a similar ability.)

We therefore have

$$[6] \quad \frac{\partial n(l)}{\partial l} = -\left[\frac{-K + b(L - 2l) + D(l)}{(K + bl)(L - l)}\right] n(l)$$

if we define $D(l)$ as positive. Equation 6 can be readily integrated numerically over size l to generate a size frequency distribution.

The general (non-steady-state) solution to eq. 2 is

$$[7] \quad n(l,t) = R[t - A(l)]\,\frac{g(l_*)}{g(l)}\,e^{-\int_{l_*}^{\infty} [D(l')/g(l')]\,\partial l'}$$

(Banks et al. 1991). $R[t - A(l)]$ represents the time varying recruitment flux, $n(l_*, t - A(l))g(l_*)$, across the recruitment boundary l_* for an individual of length l and age $A(l)$. If, for example, we retain the mortality rate constant over size (i.e., $D(l) = Z$) and assume growth according to eq. 4, then the following analytical solution exists:

$$[8] \quad n(l,t) = R[t - A(l)]\left(\frac{L - l}{L - l_*}\right)^{[Z/(K+bL)]-1}\left(\frac{K + bl_*}{K + bl}\right)^{[Z/(K+bL)]+1}$$

If growth is simplified to be strictly von Bertalanffy, i.e., $b = 0$, then

$$[9] \quad n(l,t) = R[t - A(l)]\left(\frac{L - l}{L - l_*}\right)^{(Z/K)-1}$$

If recruitment is assumed to be a constant value R, then the dependencies on time t and age-at-size $A(l)$ can be dropped. Analytical solutions are rare for models where mortality is a function of size, e.g.,

$$[10] \quad D(l) = Z\,e^{-cl}$$

Table 1. Definitions for variables appearing in this paper.

Symbol	Definition
t	time
l	size
l_*	the size in a size frequency distribution to which individuals first recruit
l_x	size when fully exploited yield-per-recruit is maximized
L	deterministic asymptotic size for an individual
K'	represents $K + bl$
K	deterministic instantaneous rate of change in growth rate for an individual
b	shape parameter for a modified von Bertalanffy or Richards growth function
$n(\bullet)$	density of individuals in a size distribution
$R[\bullet]$	recruitment rate
$g(l)$	growth rate at size l
$D(l)$	mortality rate at size l
Z	natural mortality rate at size $l = 0$
c	l-dependent mortality coefficient
μ_L	mean of L for a population
σ_L^2	variance of L for a population
μ_K	mean of K for a population
σ_K^2	variance of K for a population
A	age
A_0	age at $l = 0$ in size-at-age data
A_0'	age at $l = 0$ in size frequency data
μ_A	expected size-at-age A
σ_A^2	expected variance in sizes-at-age A
ϕ_A	in size frequency data, the deviation of the mean of an age pulse from that predicted by a growth function
$V_f(l)$	variance of measurement error at size l
T	time period over which growth increments are measured
μ_I	expected growth increment at size l during time T
σ_I^2	variance in growth increments at size l during time T
$w(l)$	weight-at-size l
φ	scaling parameter of weight-at-size function
ν	power parameter of weight-at-size function
$E[\bullet]$	expected value of the bracketed quantity
$V[\bullet]$	variance of the bracketed quantity

with $c > 0$. For such models we rely on numerical solutions to the differential equations to generate a corresponding size frequency distribution.

An analogy of eq. 9 has previously appeared in the literature accompanied with a discussion of its implications for the shape of a size frequency distribution (Barry and Tegner 1990). Other authors have exploited the benefit of knowing the ratio Z/K for interpreting fish population dynamics (see Pauly 1984; Pauly and Morgan 1987) or using size data to interpret mortality rates (Ebert 1973; Van Sickle 1977; Fournier and Breen 1983). Here we extend these interpretations by showing that eqs. 2 and 7 coupled with independent stochastic variability in von Bertalanffy's K and L parameters (Sainsbury 1980) have a particular ability to be informative of the underlying growth and mortality functions that give observed size frequency distributions their characteristic equilibrium shapes. These equations also form the basis for analytical models which facilitate estimation of stochastic growth and deterministic mortality parameters from size frequency, size-at-age and growth increment data (Smith and McFarlane 1990; Smith et al. 1998).

To appreciate the utility of these models observe that eq. 2 has special diagnostic value in that for deterministic growth and mortality $\partial n(l)/\partial l$ will equal zero when $\partial g(l)/\partial l = -D(l)$. Immediately this gives us a tool for visually assessing size frequency distributions then interpreting the underlying population dynamics, i.e., a plateau in the size distribution will occur when $\partial g(l)/\partial l + D(l) = 0$. Deterministic and stochastic portrayals of distributions arising from these models appear in Fig. 1 with the corresponding values for the derivative of the growth rate, $-\partial g(l)/\partial l$, and the mortality rate, $D(l)$, appearing in Fig. 2. For example (Figs. 1*a* Deterministic, 2*a*), von Bertalanffy growth and a constant mortality rate with $D(l) = Z = K$, generates a plateau that occurs over the entire domain of l.

In Fig. 1 we have admitted variability in von Bertalanffy's asymptotic size L by allowing L to be a normally distributed random variable with a mean of μ_L and variance $\sigma_L 2$ (Sainsbury 1980). We have not admitted variability in von Bertalanffy's K because it has little discernible effect on an equilibrium size distribution (Botsford et al. 1994). Although no single model should be considered a unique or even biologically precise descriptor of growth, we found Sainsbury's

Fig. 1. Characteristic size frequency distributions under the constant recruitment assumption. For all examples $\sigma_K = 0$ because they are indistinguishable from otherwise similar size frequency plots with $\sigma_K > 0$ (Botsford et al. 1994). Histograms represent a random sample (n = 1000) of the analytical stochastic distribution organized into 5 unit cells. The parameter values for each plot are as follows: (*a*) $\mu_L = 100$, $\sigma_L = 0$, $\mu_K = 0.3$, $b = 0$, $D(l) = 0.3$; (*b*) $\mu_L = 100$, $\sigma_L = 10$, $\mu_K = 0.3$, $b = 0$, $D(l) = 0.1$; (*c*) $\mu_L = 100$, $\sigma_L = 10$, $\mu_K = 0.1$, $b = 0$, $D(l) = 0.3$; (*d*) $\mu_L = 100$, $\sigma_L = 10$, $\mu_K = 0.1$, $b = 0$, $D(l) = 4.0e^{-0.05l}$; and (*e*) $\mu_L = 100$, $\sigma_L = 10$, $\mu_K = 0.3$, $b = 0.005$, $D(l) = 0.1$.

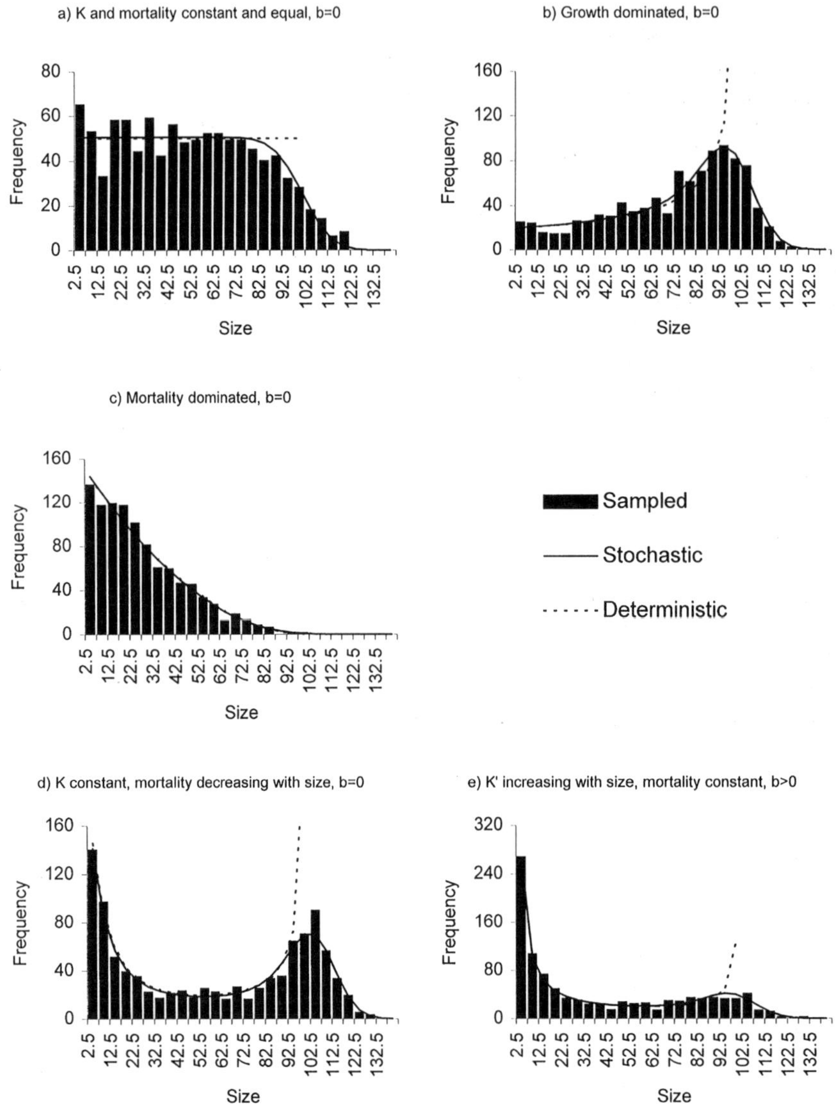

(1980) model to be particularly useful for describing stochastic growth as it is expressed in size frequency distributions, size-at-age data, and growth increment data that we have seen. This stochastic parameterization of growth overcomes the incompatibility of deterministic von Bertalanffy functions representing size-at-age data and growth increment data that was highlighted by Francis (1988).

We refer to Figs. 1*b* (with $\mu_k > Z$) and 1*c* (with $\mu_k < Z$) as growth-dominated and mortality-dominated distributions, respectively. If either $\partial g(l)/\partial l$ or $D(l)$ vary with length then there is potential for distributional forms more interesting than these. Distributions with a pronounced mode near the maximum size of individuals in the distribution appear frequently in nature as either a growth-dominated or bimodal distributional form. Figures 1*d* and 1*e* show that both

$$\partial g(l)/\partial l = -K, \text{ with } D(l) = Z\,e^{-cl} \text{ (Figs. 1}d\text{ and 2}d\text{)}$$

and $\partial g(l)/\partial l = -K + b(L - 2l)$, with $D(l) = Z$ (Figs. 1*e* and 2*e*) can produce a bimodal distribution. In both cases the peak of the right-hand mode tends to occur near the value for μ_L with the spread around μ_L giving an indication of the value of σ_L. Note that the portrayals in Fig. 1 are ideal and natural distributions are not likely to perfectly conform to those distributions. In particular, size selectivity during data collection will result in the smallest individuals being under-represented in the distribution.

At least two recruitment processes can complicate the size distributions portrayed in Fig. 1. Periodic pulsed recruitment can introduce age pulses into a distribution thus disguising the underlying growth and mortality patterns that are more evident

Fig. 2. Functions of size for $-\partial g(l)/\partial l$ and $D(l)$ for the characteristic size frequency distributions portrayed in Fig. 1. In the deterministic case a plateau in a size frequency distribution will occur when $-\partial g(l)/\partial l = D(l)$.

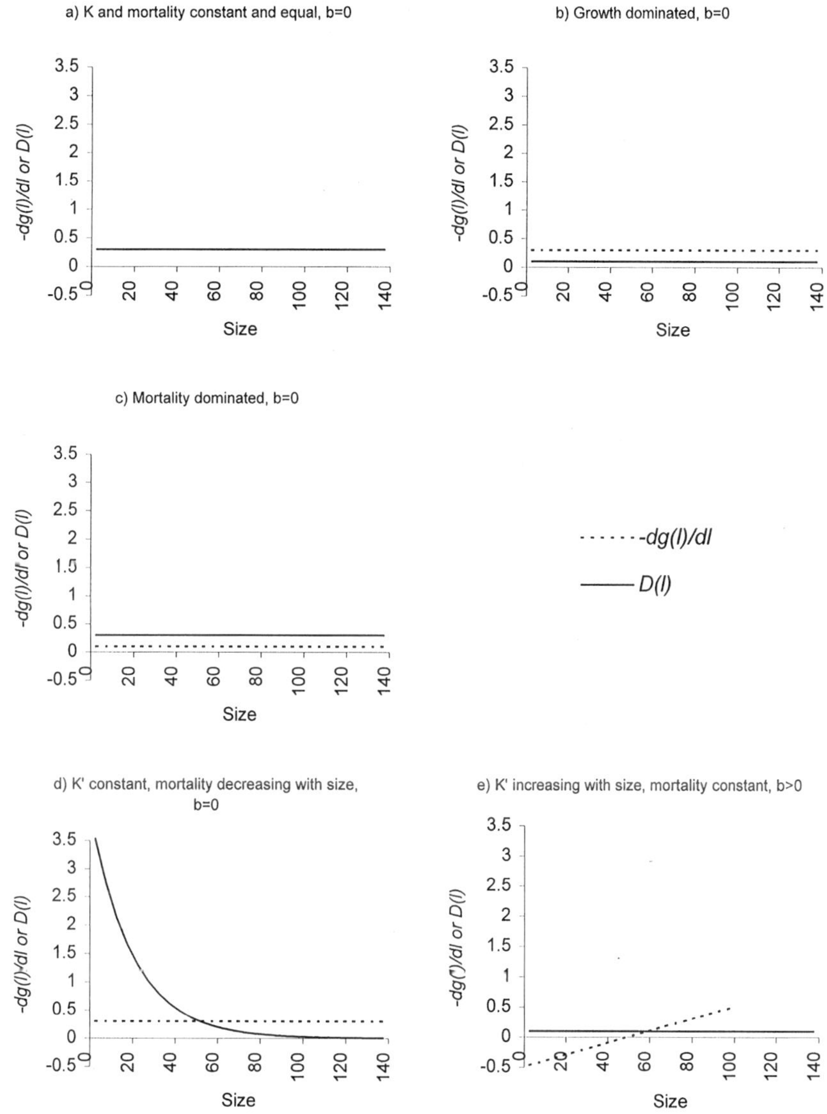

in the constant recruitment situation. Periodic pulses will introduce age pulses into a size distribution which will shift toward the right over time, tending to result in a repetition of the size distribution pattern annually if annual recruitment is somewhat constant (Fig. 3). Random recruitment pulses occurring in a population where recruitment is typically low, but which occasionally experiences a strong recruitment event, can lead an analyst to suspect regular recruitment pulses as in Fig. 3 unless a time series of collected distributions indicates a pattern of random recruitment (Fig. 4).

Only repeated sampling over time can resolve whether the nature of recruitment into size frequency distributions is regular or random. In principle, random recruitment such as that portrayed in Fig. 4 would be describable by a probability function. As yet we know of no analytical models for estimating growth and mortality parameters from size frequency distributions characterized by random recruitment. However, simulation studies have shown (Smith et al. 1998) that if the magnitude of recruitment variability is low, say a coefficient of variation (RV) of two or less, then growth and mortality parameters can be confidently measured from such distributions.

Maximizing yield-per-recruit

Having a simple method to calculate the maximum yield-per-recruit from size frequency data is valuable since, due to a paucity of fishery data, many invertebrate fisheries are passively managed by minimum size limits, or area and seasonal closures (see Jamieson 1986). Our model facilitates a simple calculation for determining a minimum size limit (l_x) that would maximize yield-per-recruit if it is assumed that virtually all individuals above l_x are rapidly exploited by a fishery.

For species that are candidates to be managed by a minimum size limit, and for which the instantaneous fishing mortality (Ricker 1975) can be assumed to approach infinity, i.e., a 100% exploitation rate above the minimum size l_x, maximum yield-per-recruit occurs when the value of l_x maximizes the function $n(l)g(l)w(l)$. The per-recruit flux, $n(l_x)g(l_x)$, across the

Fig. 3. Size frequency distributions characterizing pulsed recruitment for the mortality-dominated and growth-dominated examples of Fig. 1. The winter and summer portrayals depict the shift to the right of pulses for a species whose peak in annual recruitment occurs in summer. For all examples $\sigma_K = 0$, $b = 0$, $c = 0$. Histograms represent a random sample ($n = 1000$) of the analytical stochastic distribution organized into 5 unit cells. The parameter values for each plot are as follows: (*a* and *b*) $\mu_L = 100$, $\sigma_L = 10$, $\mu_K = 0.1$, $D(l) = 0.3$; and (*c* and *d*) $\mu_L = 100$, $\sigma_L = 10$, $\mu_K = 0.3$, $D(l) = 0.1$.

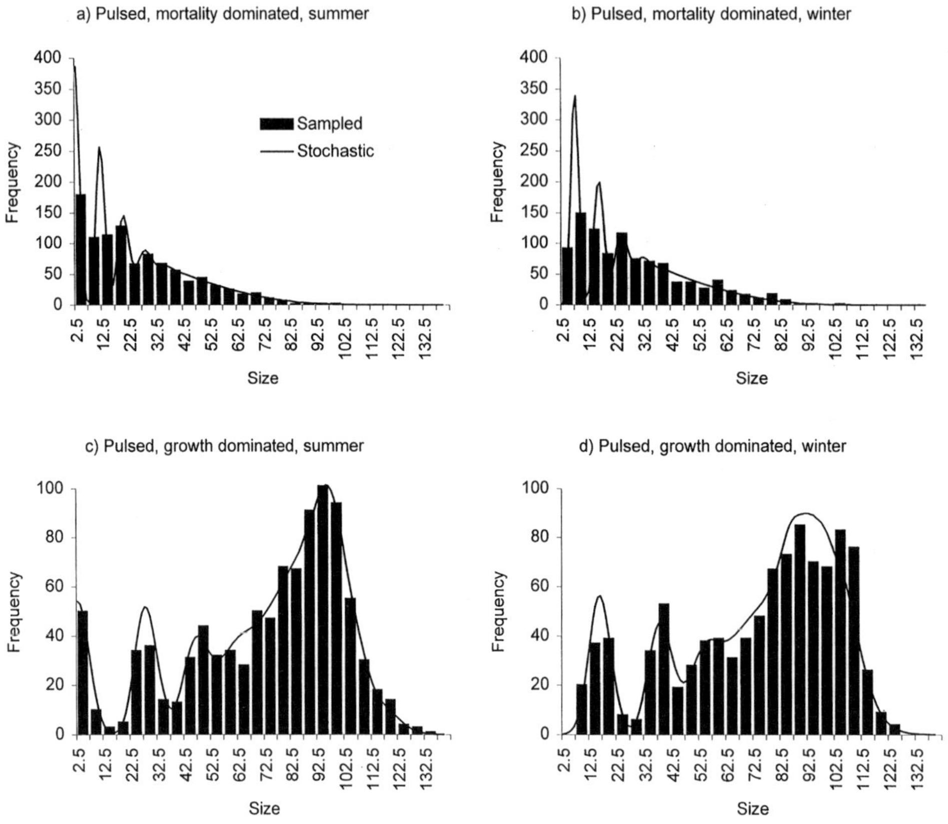

minimum size boundary l_x is converted to biomass from numbers by $w(l_x)$. The function $w(l)$ thus relates yield in weight to a linear measure of size by, for example, a power model as in $w(l) = \varphi l^{\nu}$. Typically $w(l)$ would represent total weight, but in some cases, such as that of the red sea urchin, it could represent gonad weight since gonad flesh is the commercial product.

The estimate for the size l_x ($l_x > 0$) when fully exploited yield-per-recruit is maximized is obtained by solving

$$[11]\quad \frac{\partial[n(l)\,g(l)\,w(l)]}{\partial l} = 0$$

which simplifies to

$$[12]\quad 0 = g(l_x)\left.\frac{\partial w(l)}{\partial l}\right|_{l=l_x} - w(l_x)\,D(l_x)$$

Equation 12 depends neither on population density, $n(l)$, nor recruitment history $R[t - A(l)]$ and therefore can be applied to populations without having concern about past recruitment patterns. It is noteworthy that l_x is inherently conservative. That is, if the population of interest actually experienced an exploitation rate less than 100%, when managed by a minimum size limit of l_x, then yield-per-recruit would be maximized at a size smaller than l_x.

For von Bertalanffy growth and constant mortality, Z, maximum fully exploited yield-per-recruit occurs when

$$[13]\quad l_x = \frac{L}{\left(1 + \frac{Z}{K\nu}\right)}$$

if yield in weight is related to a linear measure of size by the power parameter ν. Equation 13 bears resemblance to Hoenig's (1987) function to estimate the size when the surplus yield in numbers (i.e., $\nu = 1$) of a population is optimized. When growth is considered to be stochastic in L and K as above, then the necessary condition for the l_x that produces maximum yield is

$$[14]\quad 0 = E\,[g(l_x)]\left.\frac{\partial w(l)}{\partial l}\right|_{l=l_x} - w(l_x)\,D(l_x)$$

Continuing with the previous example, the independence of L and K leads to

$$[15]\quad l_x = \frac{\mu_L}{\left(1 + \frac{Z}{\mu_K \nu}\right)}$$

Solutions to Eq. 14 are less tractable if $g(l)$ includes power

Fig. 4. Characteristic growth-dominated size frequency distributions with random recruitment. Each figure (*a–d*) portrays 30 realizations (thin lines) of the analytical deterministic distribution (thick line) when random variation in recruitment is added. Fig. 4*a* portrays 30 realizations of a random sample (n = 1000) of the analytical stochastic distribution organized into 5 unit cells when the coefficient of variation of recruitment (RV) is zero. The other examples (*b–d*) use a value of RV = 5. The value for RV is the coefficient of variance for a lognormal probability distribution of recruitment over monthly time cells. For all examples $\mu_L = 100$, $\sigma_L = 10$, and $\sigma_K = 0$. Note that the ratio μ_K/Z determines the analytical shape of the size distribution.

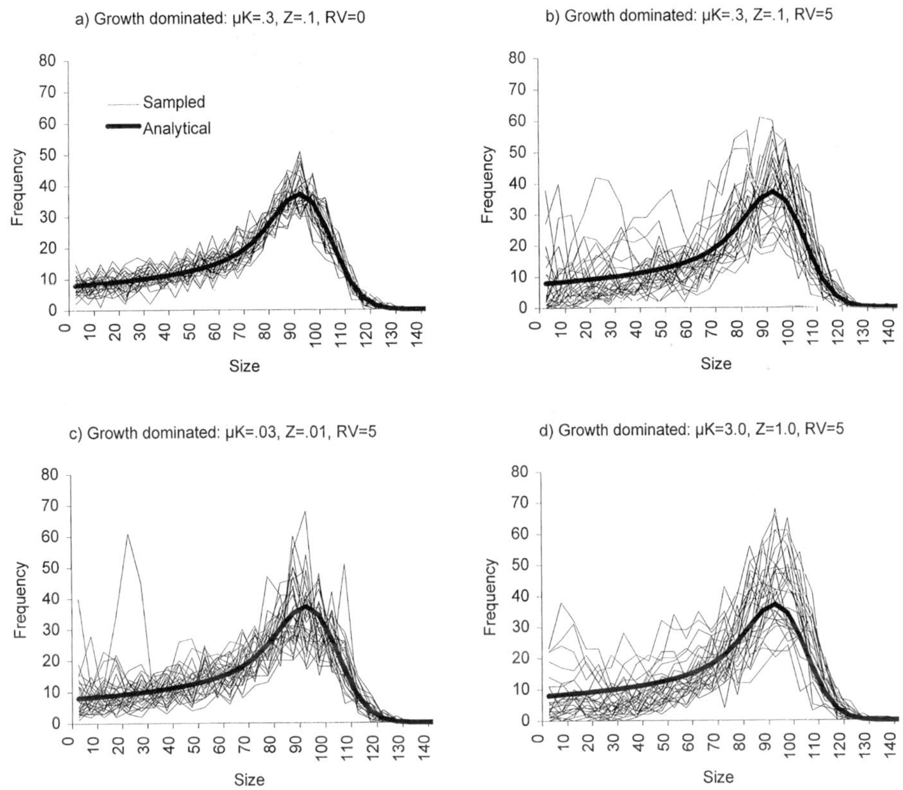

parameters such as in the Richards function (see Schnute 1981),

$$[16]\quad E\,[g(l)] = E\left[\frac{K\,l}{b}\left(\left(\frac{L}{l}\right)^{b} - 1\right)\right]$$

however, it would probably be acceptable to assume $E[L^b] \approx E[L]^b$ in the above case knowing that the coefficient of variance of L is likely to be quite small, say about 10%.

It is noteworthy that the value of l_x depends only on the ratio Z/K, but the value for the maximum harvest depends on $g(l)$. For example, the expected size distributions for the populations portrayed in Fig. 4 are identical because all four distributions are characterized by the same ratio Z/K, and will thus have the same value for l_x. However the individuals in the population represented by Fig. 4*d* grow 100× faster, i.e., individuals move much more quickly from the left to the right of the size frequency distribution, than those in Fig. 4*c*. Although the calculated value for l_x would be the same in all cases, the expected yield, $\int n(l)g(l)w(l)\partial t$, from the population in Fig. 4*c* would be only 1% of that for the population represented in Fig. 4*d*, given the same constant recruitment rate, R, and value of ν.

Size-at-age data

The stochastic von Bertalanffy growth (eq. 4) model already discussed can generate graphical representations for size-at-age which allow analysts to interpret the growth model generating their observed data. By using μ_A and σ_A2 to represent the mean size-at-age and variance-at-age, respectively, the distributions in Figs. 5*a*–5*d*, when *b = 0*, can be generated by

$$[17]\quad \mu_A = \mu_L(1 - E\,[e^{-\mu_K(A-A_0)}])$$

$$[18]\quad \sigma_A^2 = \eta\,\sigma_L^2 + \mu_L^2\,V[e^{-\mu_K(A-A_0)}] + V_I(l)$$

where

$$[19]\quad E\,[e^{-\mu_K(A-A_0)}] = \left(1 + \frac{(A - A_0)\,\sigma_K^2}{\mu_K}\right)^{-\frac{\mu_K^2}{\sigma_K^2}}$$

$$[20]\quad V\,[e^{-\mu_K(A-A_0)}] = \left(1 + \frac{2(A - A_0)\,\sigma_K^2}{\mu_K}\right)^{-\frac{\mu_K^2}{\sigma_K^2}} - \left(1 + \frac{(A - A_0)\,\sigma_K^2}{\mu_K}\right)^{-\frac{2\mu_K^2}{\sigma_K^2}}$$

and

Fig. 5. Characteristic size-at-age plots. The corresponding growth increment plots for a time-at-large of one age unit appear in Fig. 6. The parameter values for each plot are as follows: (*a*) μ_L=100, σ_L = 10, μ_K = 0.3, σ_K = 0, b = 0; (*b*) μ_L = 100, σ_L = 0, μ_K = 0.3, σ_K = 0.1, b = 0; (*c* and *d*) μ_L = 100, σ_L = 10, μ_K = 0.3, σ_K = 0.1, b = 0; and (*e*) μ_L = 100, σ_L = 10, μ_K = 0.01, σ_K = 0.0033, b = 0.005.

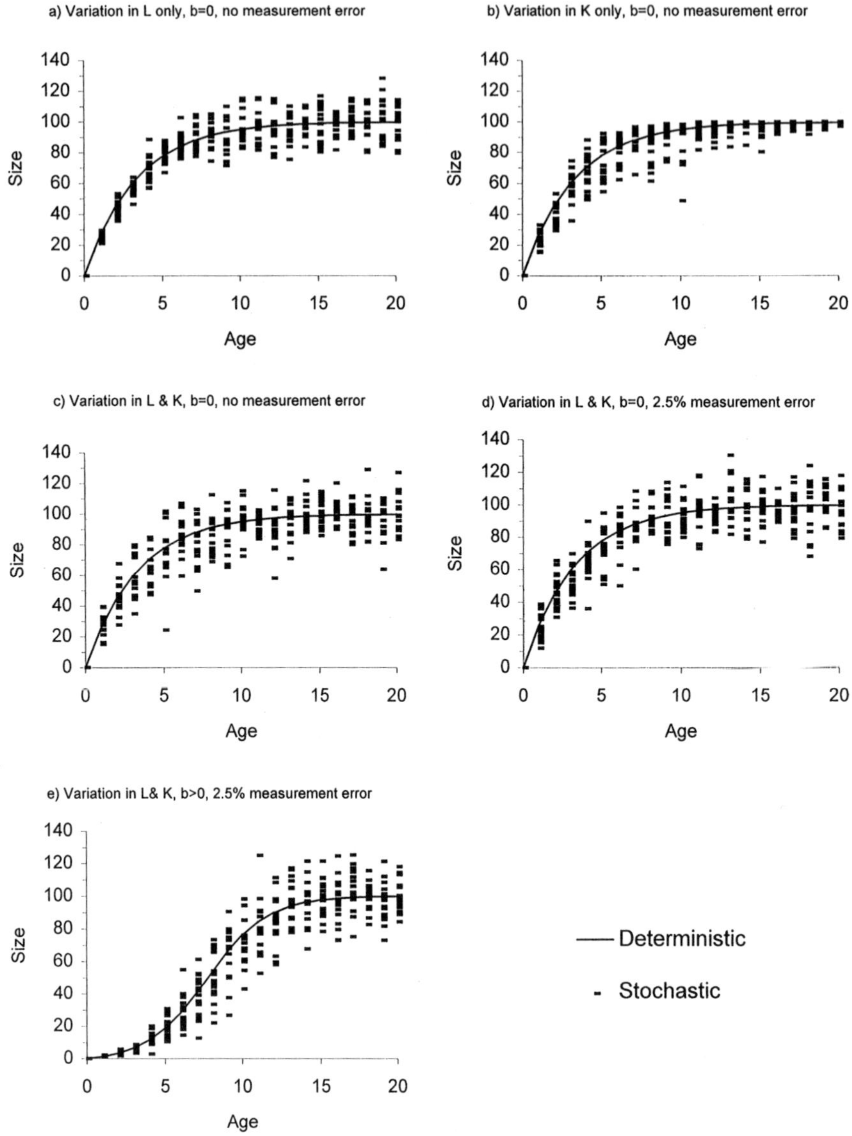

$$[21]\quad \eta = 1 - 2\left(1 + \frac{(A - A_0)\,\sigma_K^2}{\mu_K}\right)^{-\frac{\mu_K^2}{\sigma_K^2}} + \left(1 + \frac{2(A - A_0)\,\sigma_K^2}{\mu_K}\right)^{-\frac{\mu_K^2}{\sigma_K^2}}$$

(Sainsbury 1980; Smith and McFarlane 1990) where A is the age of an individual and A_0 is the age when the expected size of an individual is zero. The term $V_l(l)$ represents the variance of error in the measurement of size l as a function of size.

Note that the deterministic solution to the growth model of eq. 4 ($b > 0$) as a function of age is

$$[22]\quad l = \frac{L\,K(1 - e^{-(K+bL)(A-A_0)})}{L\,b\,e^{-(K+bL)(A-A_0)} + K}$$

Owing to its complexity there exist no simple analytical stochastic expressions for expected sizes-at-age, growth increments, and their variances for this growth model; rather we calculate these values numerically.

Figures 5*a*, 5*b*, and 5*c* (b = 0) distinctly portray the variation in L, K, or both in size-at-age data. Species expressing variance in L only will produce size-at-age distributions that will tend to increase in variance with increasing size. Species expressing variance in K only will tend to grow toward a singular value for L. Variance in size-at-age will be maximum at intermediate ages. Species expressing variance in both variables will display a maximum variance in size-at-age at intermediate sizes. Error in the measurement of size is also not discernible in size-at-age plots (e.g., compare Fig. 5*c* with 5*d*). If $b > 0$, then the same arguments with regard to variance in L and K apply as with b = 0, however, the sigmoidal form of the size-at-age curve is readily apparent (Fig. 5*e*).

Fig. 6. Characteristic growth increment plots. The corresponding size-at-age plots appear in Fig. 5. The parameter values for each plot are as follows: (*a*) $\mu_L = 100$, $\sigma_L = 10$, $\mu_K = 0.3$, $\sigma_K = 0$, $b = 0$; (*b*) $\mu_L = 100$, $\sigma_L = 0$, $\mu_K = 0.3$, $\sigma_K = 0.1$, $b = 0$; (*c* and *d*) $\mu_L = 100$, $\sigma_L = 10$, $\mu_K = 0.3$, $\sigma_K = 0.1$, $b = 0$; and (*e*) $\mu_L = 100$, $\sigma_L = 10$, $\mu_K = 0.01$, $\sigma_K = 0.0033$, $b = 0.005$.

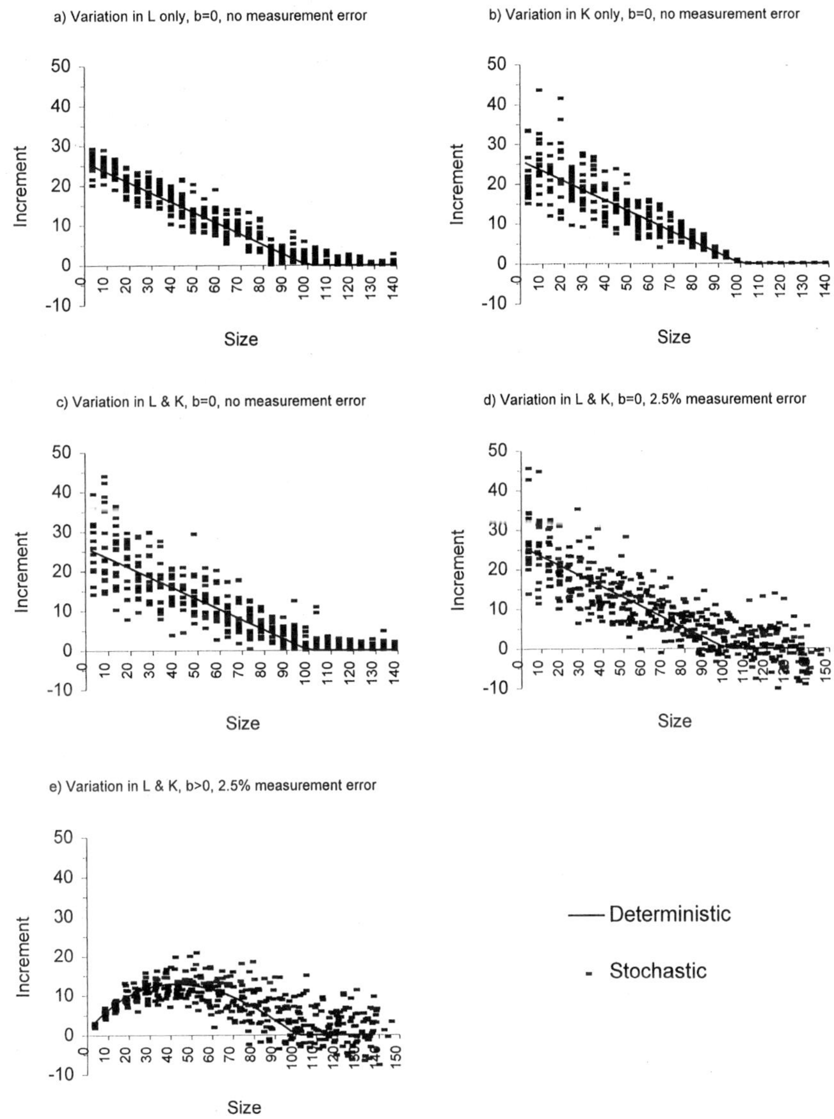

Growth increment data

Analogous growth increment plots (Fig. 6) exist for each of the five size-at-age distributional forms of Fig. 5. The patterns in Fig. 6 are generated by the following growth increment models (eqs. 23 and 24), with $b = 0$, (Sainsbury 1980; Smith and McFarlane 1990) where T is the elapsed time between an initial, then subsequent, size measurement to determine growth. We use μ_I and σ_I2 to represent the mean growth increment and the variance of growth increments, respectively, conditional on the initial size l of an individual and the time period T.

[23] $\mu_I = (\mu_L - l)(1 - E\,[e^{-\mu_K T}])$

[24] $\sigma_I^2 = \eta\,\sigma_L^2 + (\mu_L - l)^2\,V\,[e^{-\mu_K T}] + V_I(l)$

In eqs. 23 and 24 $E[e^{-\mu_K T}]$, $V[e^{-\mu_K T}]$, and η are as in eqs. 19–21 but with $A - A_0$ replaced by T.

Growth increments for a species with growth variance only in L (i.e., $\sigma_L > 0$, $\sigma_K = 0$) will generate a growth increment plot that tends to have constant variance-at-size over sizes considerably smaller than μ_L (Fig. 6*a*). As size approaches the domain of the distribution of L's, the lower growth rate limit of zero begins to affect the form of the distribution of increments-at-size causing the variance of increments-at-size to decrease. If there is variance only in K (i.e., $\sigma_L = 0$, $\sigma_K > 0$) then a growth increment plot will have a conical shape with a vertex at μ_L (Fig. 6*b*). Growth increments plots expressing variance in both L and K will display decreasing variance with size but not to a vertex as in the case with $\sigma_K = 0$ (Fig. 6*c*). Additionally, the occurrence of negative growth increments in such a plot is direct evidence of measurement error $V_I(l)$ when it is known that individuals of the species in question cannot undergo negative growth (Fig. 6*d*). As previously mentioned, such direct evidence will not be distinguishable in a plot of size-at-age (Fig. 5). A sigmoidal growth form ($b > 0$) would introduce

Fig. 7. Size-at-age (SA, $n = 1108$) and size frequency (SF, $n = 2547$) data for the waved whelk (*Buccinum undatum*) with the corresponding predicted fits using each data set independently and in combination. The parameter values corresponding to the fits to these data are in Table 2.

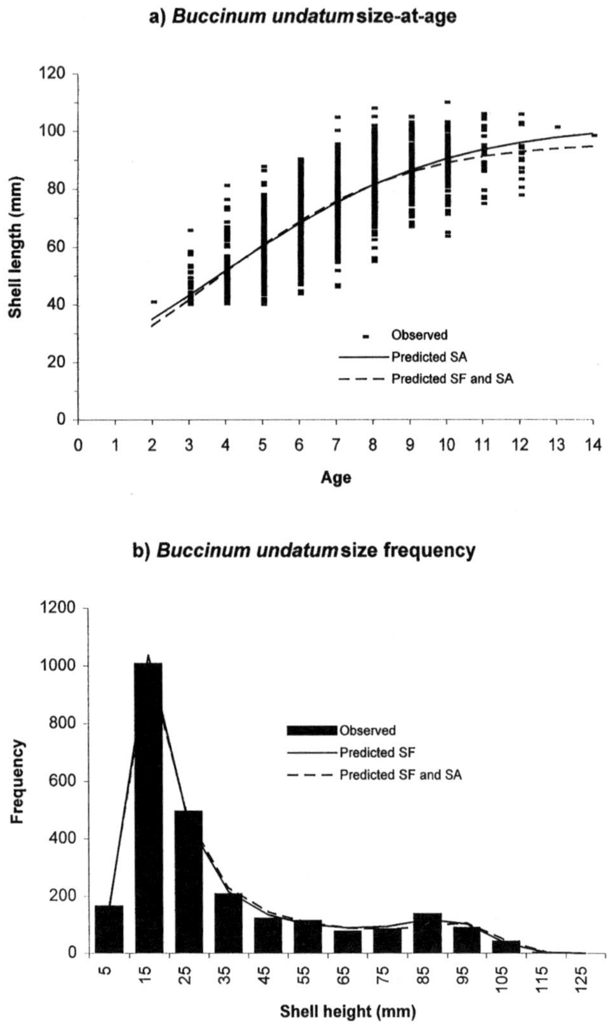

distinct convex curvature into the growth increment plot as in Fig. 6*e*.

Constant recruitment examples

Buccinum undatum

For our first example using real data we present a size-at-age (SA) plot (Fig. 7*a*) for the waved whelk (*Buccinum undatum*) from the Gulf of St. Lawrence (Gendron 1992 and Louise Gendron, Maurice Lamontagne Institute (MLI), unpublished data). Age was determined by interpreting growth annuli on an individual whelk's operculum. The absence of the smallest individuals in Fig. 7*b* is due to SCUBA divers choosing not to sample such small individuals for age analysis. Notwithstanding that individuals less than 40 mm shell length were not sampled, we can interpret from this plot that a stochastic sigmoidal growth form could adequately describe growth for this species. A corresponding size frequency (SF) plot of shell length constructed from measurements of individuals exhaustively sampled from quadrats placed on a sand-mud bottom (Jalbert et al. 1989) indicates a possible bimodal form (Fig. 7*b*), with the absence of the smallest individuals in the distribution resulting from the inability to quantitatively sample the smallest individuals.

If we assume that the left-hand mode in Fig. 7*b* is composed of several age-classes which show no apparent pattern of interannual recruitment variability, then we might interpret that recruitment in the waved whelk occurs at a relatively constant rate. Thus growth and mortality for the waved whelk can be described by growth and mortality processes leading to bimodality in an equilibrium size distribution as depicted in Figs. 1*e*, 2*e*, and 5*e*. However, the assumption of constant recruitment could later be falsified if size frequency plots collected sequentially in time portray a progression of year-classes.

The size-at-age data and size frequency data (Figs. 7*a* and 7*b*) appear to be in accordance with each other (Table 2). Note particularly that formal statistical analyses of these size-at-age and size frequency data seem to indicate a value for a μ_L near 90–103 mm shell length with a value for the standard deviation (SD) of L (σ_L) of 6–8 mm shell length. We remind you that size frequency information is uninformative of the SD of K (σ_K) (Fig. 1 and Botsford et al. 1994). Our combined analysis of these size-at-age and size frequency data estimated mortality, $D(l)$, to vary with shell length according to eq. 10. We used eq. 14 and our estimates of growth, $g(l)$, and $D(l)$ to calculate l_x, the shell length when fully exploited meat yield-per-recruit is maximized. Accepting the value of 3.1 for the power parameter ν relating meat yield to shell length (L. Gendron, MLI, unpublished data), we estimated l_x to be near 79–82 mm (Table 2).

Strongylocentrotus franciscanus

In our second example, the red sea urchin (*Strongylocentrotus franciscanus*), scrutiny of the size frequency data (SF) in Fig. 8*b* seems to suggest a value for μ_L around 100 mm test diameter. However statistical analysis of the growth increment data alone (GI, Fig. 8*a*) yielded an estimate for μ_L of 71 mm test diameter (Table 2). This seems low in comparison with the size frequency data, but the standard error (SE) of this estimate was 45 mm test diameter. The slope of the relationship between growth increment and initial size, where it is not modified by encroachment upon the domain of L, represents a value of 0.71 yr^{-1} for μ_K (SE: 0.66 yr^{-1}). There is no pattern in the data to indicate the presence of variance in K, (i.e., $\sigma_K = 0\ yr^{-1}$). The large standard errors for the growth parameter estimates obtained from these growth increment data indicate they are not particularly informative about growth on their own.

By accepting that negative growth is unlikely since these urchins were well-fed, the negative values for increments are direct evidence of measurement error. Thus the uncertainty in the estimate of μ_L can be explained in part by the measurement error that enters this analysis (about 2.5% of test diameter). The presence of measurement error signifies that individuals repeatedly measured will yield different test diameters. In the case of the red sea urchin this nonrepeatability can arise from irregularities in test diameter, asymmetry of the test and/or imprecise measurements.

Table 2. Formally estimated growth and mortality parameter values associated with the constant recruitment examples.

	Parameter									
	μ_L	σ_L	μ_K	σ_K	b	A_0	Z	c	υ	l_x
Buccinum undatum **(waved whelk)**										
Size-at-age data (SA)	102.8	6.1	1.9×10^{-3}	0*	3.3×10^{-3}	−11.4	—	—	—	—
Size frequency data (SF)	90.1	8.2	2.3×10^{-2}	0*	3.6×10^{-3}	—	1*	2.6×10^{-2}	3.1	79
Both data sets (SA and SF)	96.5	6.4	2.9×10^{-3}	0*	4.1×10^{-3}	−8.7	0.62	1.5×10^{-2}	3.1	82
Strongylocentrotus franciscanus **(red sea urchin)**										
Growth increment data (GI)	71.3	37.6	0.71	0*	0*	—	—	—	—	—
Size frequency data (SF)	111.3	16.2	3.43	0*	0*	—	1*	0*	3.6	103
Both data sets (GI and SF)	112.2	15.9	0.46	0*	0*	—	0.13	0*	3.6	104
Mactromeris polynyma **(Stimpson's surf clam)**										
Growth increment data	70.8	2.8	0.44	0.15	0*	—	—	—	—	—
Size-at-age data (SA)	115.1	5.3	1.2×10^{-2}	0*	1.3×10^{-3}	−6.8	—	—	—	—
Size frequency data (SF)	109.0	7.5	0.18	0*	1.6×10^{-2}	—	1*	0*	2.95	90
SA and SF data sets	111.4	6.2	1.9×10^{-2}	0*	1.2×10^{-3}	−4.8	0.09	0*	2.95	90

Note: These values are used to calculate the tabled values for l_x. For size frequency (SF) analyses the reported parameter values are accompanied by unreported parameter values for selectivity curves. All size dimensions are millimeters and all time dimensions are years. An asterisk (*) beside a number indicates the number was required to be fixed at that value (e.g., Z) or the number tended toward that boundary value and was subsequently fixed at that value to increase model parsimony. When Z is fixed at one, as it must be for an analysis of size frequency data alone, then μ_K is redefined to represent the ratio μ_K/Z, and b is redefined to represent the ratio b/Z. A dash (—) indicates that the parameter does not play a role in that particular analysis. Plots corresponding to these data are in Figs. 7–9.

The uncertainty in the growth estimates provided by the analysis of the growth increment data alone can be overcome by an analysis of the size frequency and growth increment data in combination. Performing this joint analysis shows that these growth increment data are reasonably compatible with field collected size frequency data (Kalvass et al. 1991; Fig. 8*b*). The analysis of size frequency and growth increment data in combination yielded parameter values ($\mu_L \approx$ 112 mm, $\sigma_L \approx$ 16 mm, $\mu_K = 0.46$ yr^{-1}) which are consistent with the range of possible values yielded by the growth increment data alone.

Since the independent growth increment data provide us with an estimate for the growth rate, $g(l)$, then a value for the mortality rate, $D(l)$, can be obtained by size frequency analysis. For the example portrayed, a constant value for $D(l)$ of 0.13 yr^{-1} was adequate to explain the observed size frequency distribution. If we assume gonad weight is related to test diameter by the power parameter $\nu = 3.6$, then the use of eq. 15 results in an estimate for l_x, the test diameter when fully exploited gonad yield-per-recruit is maximized, of 103–104 mm (Table 2). The value of $\nu = 3.6$ was interpreted from gonad weight versus test diameter data of Tegner and Levin (1983) that had been collected in December at a time when gonads are ripe and the fishery for gonads in southern California was most active (Kato and Schroeter 1985).

Mactromeris polynyma

Laboratory-obtained growth increment data for Stimpson's surf clam (*Mactromeris polynyma*) from the Gulf of St. Lawrence (Fig. 9*b*; Michel Giguère and Jean Lambert, MLI, preliminary and unpublished data) indicate von Bertalanffy growth with a μ_L near 71 mm shell length, a value for σ_L of 3 mm shell length, a value of μ_K near 0.44 yr^{-1}, and significant variance in K (i.e., $\sigma_K = 0.15$ yr^{-1}) (Table 2). In contrast, the size-at-age data (Jean Lambert, MLI, preliminary and unpublished data; Fig. 9*a*) indicate sigmoidal growth with quite different parameter values than for the growth increment data, and no pattern to suggest a value for $\sigma_K > 0$ yr^{-1}. The two forms of growth data are therefore both visually and statistically incompatible. More satisfyingly, the size-at-age and size frequency data appear compatible since they indicate a value for μ_L of about 109–115 mm shell length with a value for σ_L of about 5–8 mm shell length (Table 2). If we accept the parameter values from the joint analysis of the latter two data sets, and use a value for meat yield in relation to shell length of $\upsilon = 2.95$ (Jean Lambert, MLI, unpublished data), then we calculate a value for l_x of 90 mm (Table 2).

The value of finding such discrepancies among data sets is our ability to judge the quality of the data. In this case the conflicting interpretations of the data led the biologist studying

Fig. 8. Growth increment (GI, open circles: $n = 28$, 23 days-at-large; diamonds: $n = 32$, 27 days-at-large) and size frequency (SF, $n = 375$) data for the red sea urchin (*Strongylocentrotus franciscanus*) with the corresponding predicted fits using each data set independently and in combination. The parameter values corresponding to the fits to these data are in Table 2. Note that each series of growth increment data corresponds to a different time-at-large and thus produces its own fit using the same model parameters.

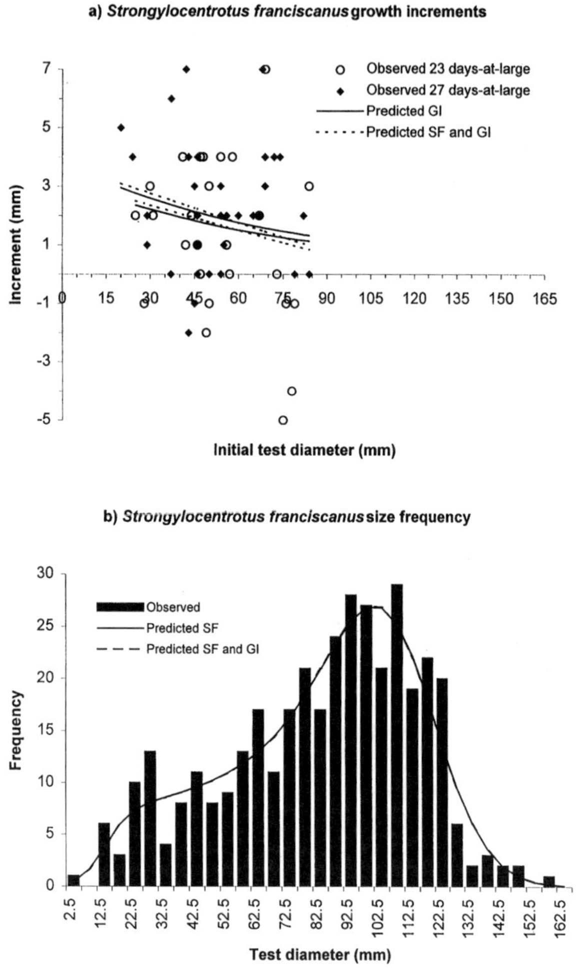

Fig. 9. Size-at-age (SA, $n = 255$), growth increment (GI, $n = 357$, time-at-large is 467 and 474 days) and size frequency (SF, $n = 1770$) data for Stimpson s surf clam (*Mactromeris polynyma*) with the corresponding predicted fits using each data set independently and using the size-at-age and size frequency data in combination. The parameter values corresponding to the fits to these data are in Table 2.

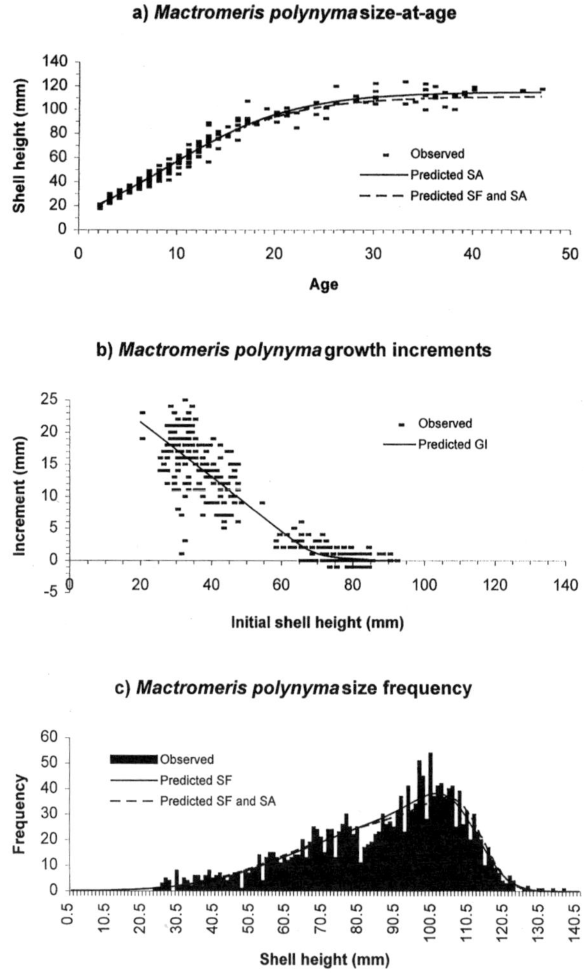

this species to conclude that the laboratory growth experiments were yielding seriously inaccurate growth increment data. His conclusion was partly based on the premise that we should have the most confidence in the size frequency data. These data tend to be least vulnerable to bias since they are collected in the field and are not subject to the unnatural conditions that accompany laboratory experiments, or the uncertainty in accepting growth checks as annuli. His next step was to execute field growth studies that would be more likely to yield growth rates compatible with the field-collected size frequency data.

Pulsed recruitment example

Placopecten magellanicus

We chose size-at-age data (Fig. 10*a*) and size frequency data (Fig. 10*b*) obtained for the giant sea scallop (*Placopecten magellanicus*) near the Îles-de-la Madeleine, Québec (Michel Giguère, unpublished data; Giguère and Miller 1993) to illustrate the benefit of using growth data in combination with size frequency data to interpret the age composition of the frequency distributions. Estimating a growth function solely from the size frequency data in Fig. 10*b* would be frustrated by the overlap of the older age pulses and no independent information on growth or mortality. The combined analysis of size-at-age and size frequency data helps relieve this frustration (Table 3).

Despite the explicit incorporation of independent growth data into an analysis, the parameter estimates obtained are still somewhat conditional upon the analyst's choice for the largest age represented in the size frequency data. Also, annual variation in growth rate due to random climatic effects or other so-called year-effects, or Lee's phenomenon (Jones 1958), can cause an age pulse to depart (shift) from that predicted by the growth curve. This departure might appear random around an expected shift of zero, or if caused by Lee's phenomenon, might tend to be progressive in one direction as size-dependent

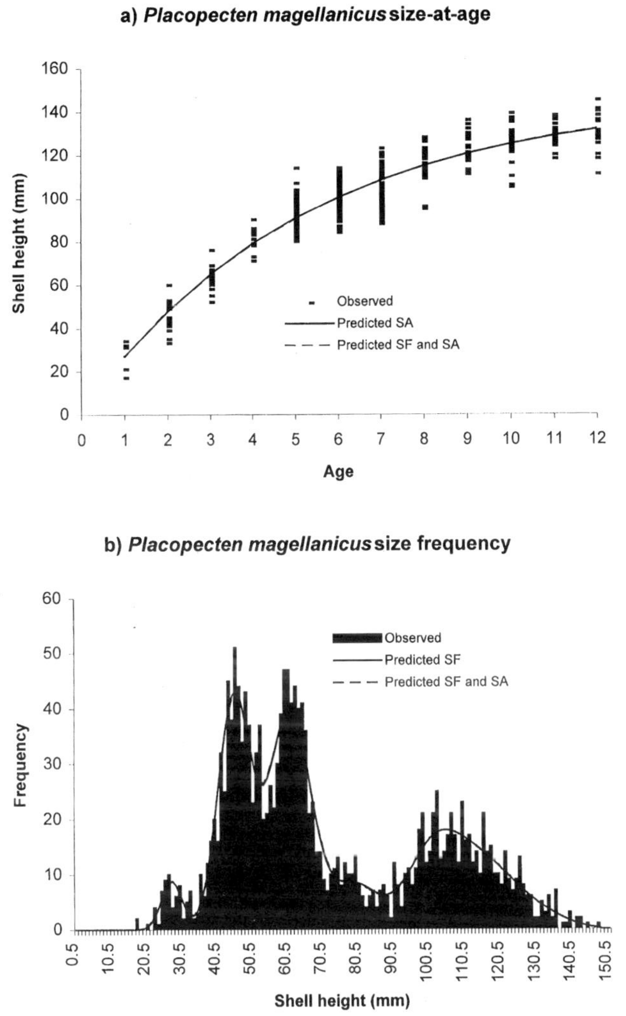

Fig. 10. Size-at-age (SA, $n = 640$) and size frequency (SF, $n = 1828$) data for the giant sea scallop (*Placopecten magellanicus*) with the corresponding predicted fits using each data set independently and in combination. The parameter values corresponding to the fits to these data are in Table 3.

mortality persistently selects against either large or small individuals. We deal with this problem by adding a vector of parameters (ϕ_A), where A represents age. Each value of ϕ_A represents the annual shift in the mean of a discernible pulse for age A from that predicted from the pulse for the previous year ($A - 1$) and the growth curve.

If an analyst has confidence in the growth parameter estimates obtained from size-at-age or growth increment data, then a size frequency analysis can be done assisted by the independent growth data. The proportions-at-age obtained from such an analysis might then be considered as representative of the recruitment and mortality history of the age-classes recognized in the distribution although typically the confidence limits around proportions-at-age are broad. If a confident estimate of mortality, $D(l)$, can be interpreted from this recruitment and mortality history, or from another independent source, then eq. 15 can be used to calculate a value for l_x. In this example we calculated a constant value for $D(l)$ of 0.43 yr^{-1} by the method of Hoenig (1983) using our knowledge that the oldest age observed in the size-at-age data was 12 yr. We then used the value of $\nu = 3.2$ from the power relationship between muscle weight, the commercial product, and shell height (Giguère and Légaré 1989) to obtain an estimate for l_x of 86 mm (Table 3).

Conclusions

Scrutiny of size-at-age, growth increment, and size frequency data can be useful for developing a conceptual understanding of the interrelationships between the underlying population processes of growth and mortality in size-structured data. We therefore recommend that a careful visual interpretation of the available data be done before an analyst undertakes a formal statistical analyses of such data in order that the analyst be more prepared to judge the likelihood of success, and the conclusions, of the analyses. Further, the relative inexpense of collecting size-based data, in conjunction with a sound conceptual basis upon which to make preliminary judgments about the dynamics of growth and mortality for a fished population, can lead to the rapid implementation of simple conservation measures founded upon a reliable first impression of the population dynamics for a particular species.

We finish by cautioning readers that the data portrayed in this paper were selected and used solely for the purpose of illustrating growth and mortality patterns. Thus they, or any parameter values reported, are not necessarily those that best or completely represent the populations from which they were collected. Readers should not consider this work as a source for definitive information on the growth or mortality dynamics of those species used as examples.

Acknowledgments

This paper is a synthesis of work performed by many biologists whose contribution in data, information, and understanding is deeply appreciated. In particular we wish to thank Louise Gendron, Michel Giguère, and Jean Lambert of the Maurice Lamontagne Institute, as well as Philippe Jalbert for his original size frequency data for the waved whelk. We also wish to thank two referees of this paper, Dr. Victor Restrepo and Dr. Thomas Wainwright, and others who critically read earlier versions of this manuscript, provided advice for improvement of the manuscript, and detected and corrected errors.

References

Banks, H.T., Botsford, L.W., Kappel, F., and Wang, C. 1991. Estimation of growth and survival in size-structured cohort data: an application to larval striped bass (*Morone saxatilis*). J. Math. Biol. **30**: 125–150.

Barry, J.P., and Tegner, M.J. 1990. Inferring demographic processes from size-frequency distributions: simple models indicate specific patterns of growth and mortality. Fish. Bull. (U.S.), **88**: 13–19.

Botsford, L.W., Smith, B.D., Wing, S.R., and Quinn, J.F. 1994. Bimodality in size distributions: the red sea urchin *Strongylocentrotus franciscanus* as an example. Ecol. Appl. **4**: 42–50.

DeAngelis, D.L., and Mattice, J.S. 1979. Implications of a partial-differential-equation cohort model. Math. Biosci. **47**: 271–285.

Ebert, T.A. 1973. Estimating growth and mortality rates from size data. Oecologia, **11**: 281–298.

Fournier, D.A., and Breen, P.A. 1983. Estimation of abalone mortality rates with growth analysis. Trans. Am. Fish. Soc. **112**: 403–411.

Table 3. Formally estimated parameter values associated with the pulsed recruitment example, the giant sea scallop (*Placopecten magellanicus*).

	μ_L	σ_L	μ_K	σ_K	b	A_0	A_0'	l_x
Growth rate parameters (size-at-age data only)	147.0	7.1	0.19	2.0×10^{-2}	—	–0.07	—	86
Growth rate parameters (size-at-age and size frequency data combined)	147.6	6.8	0.19	2.0×10^{-2}	—	–0.08	–0.08	86

Age-class descriptions from size frequency analysis

Age	Mean	SD	Annual mean shift (ϕ_A)	Proportions
1	27.2	2.9	0*	0.034
2	45.2	4.7	–2.6	0.268
3	60.9	5.9	–3.9	0.320
4	79.0	6.6	0*	0.073
5	90.7	7.1	0*	0*
6	100.4	7.3	0*	0.120
7	108.4	7.4	0*	0.068
8	115.0	7.4	0*	0.041
9	120.5	7.4	0*	0.053
10	125.1	7.3	0*	0*
11	128.9	7.2	0*	0*
12	132.0	7.1	0*	0.023

Note: All size dimensions are millimeters. The "Annual mean shift" refers to the difference in growth during one year as predicted by the growth curve and the growth represented by the size frequency data. The values for l_x were calculated using $D(l) = 0.43$ yr^{-1} (see text) and $\upsilon = 3.1$. An asterisk (*) beside a number indicates the number tended toward that boundary value and was subsequently fixed at that value to increase model parsimony. A dash (—) indicates that the parameter does not play a role in that particular analysis. Plots corresponding to these data are in Fig. 10.

Francis, R.I.C.C. 1988. Are growth parameters from tagging and age-length data comparable? Can. J. Fish. Aquat. Sci. **45**: 936-942.

Gendron, L. 1992. Determination of the size at sexual maturity of the waved whelk *Buccinum undatum* Linnaeus, 1758, in the Gulf of St. Lawrence, as a basis for the establishment of a minimum catchable size. J. Shellfish Res. **11**: 1–7.

Giguère, M., and Légaré, B. 1989. Exploitation du pétoncle aux Îles-de-la Madeleine en 1988. Canadian Atlantic Fisheries Scientific Advisory Committee, Dartmouth, Nova Scotia. Res. Doc. 89/14. 32 p.

Giguère, M., and Miller, R. 1993. Review of scallop fisheries in Québec. Can. Ind. Rep. Fish. Aquat. Sci. 217. 23 p.

Gulland, J.A. 1983. Fish stock assessment. A manual of basic methods. FAO/Wiley Series on Food and Agriculture. John Wiley & Sons, Toronto, Ont. 223 p.

Hilborn, R., and Walters, C.J. 1992. Quantitative fish stock assessment: choice, dynamics and uncertainty. Routledge, Chapman and Hall, Inc., New York, N.Y. 570 p.

Hoenig, J.M. 1983. Empirical use of longevity data to estimate mortality rates. Fish. Bull. (U.S.), **82**: 898–903.

Hoenig, J.M. 1987. Estimation of growth and mortality parameters for use in length-structured stock production models. *In* Length-based methods in fisheries research. ICLARM Conference Proceedings 13. *Edited by* D. Pauly and G.R. Morgan. Int. Cent. Living Aquat. Resourc. Manage., Manila, Philippines, and Kuwait Inst. Sci. Res., Safat, Kuwait. pp. 121–128.

Huston, M.A., and DeAngelis, D.L. 1987. Size bimodality in monospecific populations: a critical review of potential mechanisms. Am. Nat. **129**: 678–707.

Jalbert, P., Himmelman, J.H., Béland, P., and Thomas, B. 1989. Whelks (*Buccinum undatum*) and other subtidal invertebrate predators in the northern Gulf of St. Lawrence. Nat. Can. (Ottawa), **116**: 1–15.

Jamieson, G.S. 1986. A perspective on invertebrate fisheries management — the British Columbia experience. *In* North Pacific Workshop on stock assessment and management of invertebrates. *Edited by* G.S. Jamieson and N. Bourne. Can. Spec. Publ. Fish. Aquat. Sci. 92. pp. 57–74.

Jones, R., 1958. Lee's phenomenon of "apparent change in growth rate," with particular reference to haddock and plaice. Int. Comm. Northwest Atl. Fish. Spec. Publ. 1. pp. 229–242.

Kalvass, P., Taniguchi, I., Buttolph, P., and DeMartini, J. 1991. Relative abundance and size composition of red sea urchin, *Strongylocentrotus franciscanus*, populations along the Mendocino and Sonoma County coasts, 1989. Cal. Fish Game Administrative Rep. 91-3. 103 p.

Kato, S., and Schroeter, S.C. 1985. Biology of the red sea urchin, *Strongylocentrotus franciscanus*, and its fishery in California. Mar. Fish. Rev. **47**(3): 1–20.

Pauly, D., 1984. Fish population dynamics in tropical waters: a manual for use with programmable calculators. ICLARM Studies and Reviews 8. Int. Cent. Living Aquat. Resourc. Manage., Manila, Philippines. 325 p.

Pauly, D., and Morgan, G.R. (*Editors*). 1987. Length-based methods in fisheries research. ICLARM Conference Proceedings 13. Int. Cent. Living Aquat. Resourc. Manage., Manila, Philippines, and Kuwait Inst. Sci. Res., Safat, Kuwait. 486 p.

Pope, J.G. 1972. An investigation of the accuracy of virtual population analysis using cohort analysis. Int. Comm. Northwest Atl. Fish. Res. Bull. **9**: 65–74.

Ricker, W.E. 1975. Computation and interpretation of biological statistics of fish populations. Bull. Fish. Res. Board Can. 191. 382 p.

Sainsbury, K.J. 1980. Effect of individual variability on the von Bertalanffy growth equation. Can. J. Fish. Aquat. Sci. **37**: 241–247.

Schnute. J. 1981. A versatile growth model with statistically stable parameters. Can. J. Fish. Aquat. Sci. **38**: 1128–1140.

Schnute, J., and Fournier, D.A. 1980. A new approach to length frequency analysis: growth structure. Can. J. Fish. Aquat. Sci. **37**: 1337–1351.

Smith, B.D., and McFarlane, G.A. 1990. Growth analysis of Strait of Georgia lingcod by use of length-frequency and length-increment data in combination. Trans. Am. Fish. Soc. **119**: 802–812.

Smith, B.D., Botsford, L.W., Wing, S.R., and Quinn, J.F. 1998. Estimation of growth and mortality parameters from size frequency distributions lacking age patterns: an application to the red sea urchin (*Strongylocentrotus franciscanus*). *In* Proceedings of the North Pacific Symposium on Invertebrate Stock Assessment and Management. *Edited by* G.S. Jamieson and A. Campbell. Can. Spec. Publ. Fish. Aquat. Sci. 125. (Unpublished.)

Tegner M.J., and Levin, L.A. 1983. Spiny lobsters and sea urchins: analysis of a predator–prey interaction. J. Exp. Mar. Biol. Ecol. **73**: 125–150.

Van Sickle, J. 1977. Mortality rates from size distributions. Oecologia, **27**: 311–318.

von Foerster, H. 1959. Some remarks on changing populations. *In* The kinetics of cell proliferation. *Edited by* F. Stahlman. Grune and Stratton, New York, N.Y. pp. 382–407.

Spatial Pattern and its Implications

The analysis of concentration and crowding in shellfish research

J.M. (Lobo) Orensanz, Ana M. Parma, and Martín A. Hall

Abstract: The notions of concentration and crowding, density experienced by an individual within a certain neighborhood (as opposed to density measured by an observer at arbitrary locations), are central to the study both of density-dependent interactions in benthic species and of ecological effects of nonrandom distribution of fishing effort. The analysis of concentration requires meaningful conceptual models of neighborhoods and formal constructs to render these notions tractable. The theory of stochastic spatial point processes provides such a framework. Observed patterns of distribution of individual animals in space are modeled as realizations of spatial point processes; in the simplest case neighborhoods are modeled as circular isotropic regions centered around the individuals. Selected examples of studies that have addressed spatial patterns of populations of benthic invertebrates are reanalyzed for illustration. A detailed data set consisting of the location and age of all clams (*Tapes philippinarum*, $N = 6806$) found in a 59-m^2 plot, from a beach in southern Puget Sound, is used as the primary study case to analyze the first and second order properties of a pattern, as well as the distribution of concentration. Prospective applications in shellfish research and avenues for further enquiry are discussed, with emphasis on the fishing process and density-dependent response.

Résumé : Les notions de concentration et d'encombrement, la densité que subit un individu dans un certain entourage (par opposition à la densité mesurée par un observateur à des endroits choisis arbitrairement), sont au coeur de l'étude des interactions dépendantes de la densité chez les espèces benthiques et des effets écologiques d'une répartition non aléatoire de l'effort de pêche. L'analyse de la concentration exige des modèles conceptuels significatifs des voisinages, et des structures formelles pour que ces notions puissent être traitées. La théorie des processus aléatoires des points dans l'espace offre ce cadre. Les profils observés de répartition de chaque animal dans l'espace sont modélisés sous forme de réalisations de processus de points dans l'espace; dans le plus simple des cas, les voisinages sont modélisés sous forme de régions isotropiques circulaires centrées sur l'individu. Certains exemples d'études qui ont porté sur les profils spatiaux de populations d'invertébrés benthiques sont analysés de nouveau à des fins d'illustration. Un ensemble de données détaillées sur l'emplacement et l'âge de toutes les palourdes japonaises (*Tapes philippinarum*, $N = 6806$) d'un gisement de 59 m^2, sur une plage dans le sud du Puget Sound, est utilisé comme cas primaire d'étude pour analyser les propriétés de premier et de deuxième ordre d'un profil, ainsi que la répartition de la concentration. Nous analysons les perspectives d'application aux recherches sur les mollusques et crustacés, et d'autres pistes de recherche, l'accent étant mis sur la méthode de pêche et la réaction dépendante de la densité.
[Traduit par la Rédaction]

Introduction

There is perhaps no notion better established in population biology, ecology, and management than the concept of density, the number of individuals found per unit of area (or volume). Its analysis is central to the assessment of stocks of benthic sedentary organisms. In a typical study, "quadrats" are sampled at random locations (eventually stratified), density is estimated, and departure from spatial randomness is investigated using the frequency distribution of quadrat counts (e.g., Elliott 1977).

We argue that emphasis on the observation of density at arbitrary locations, together with reliance on quadrats for its measurement, have obscured the significance of a related but distinct concept: the notion of concentration (Iwao 1976; Clark 1982) or crowding (Lloyd 1967), which is density as experienced by individual organisms within some specified neighborhood. Because it refers to density experienced by individuals (as opposed to density at arbitrary locations), concentration is a relevant quantity to consider in studies involving interactions, including density dependence and the ecological effects of nonrandomness in the distribution of predation intensity or fishing effort.

The notion of concentration in ecological research has a long and convoluted history, as the concept was discovered many times independently, has received various names, and has been discussed in apparently disparate contexts. The analysis of concentration requires the explicit development of a conceptual model of neighborhoods (Orensanz 1986; Addicott

J.M. Orensanz. School of Fisheries, University of Washington, Seattle, WA 98195, U.S.A.

A.M. Parma. International Pacific Halibut Commission, PO Box 95009, Seattle, WA 98145-2009, U.S.A.

M.A. Hall. Inter-American Tropical Tuna Commission, c/o Scripps Institution of Oceanography, 8604 La Jolla Shores Dr., La Jolla, CA 92093, U.S.A.

Correct citation: Orensanz, J.M., Parma, A.M., and Hall, M.A. 1998. The analysis of concentration and crowding in shellfish research. *In* Proceedings of the North Pacific Symposium on Invertebrate Stock Assessment and Management. *Edited by* G.S. Jamieson and A. Campbell. Can. Spec. Publ. Fish. Aquat. Sci. 125. pp. 143–157.

Fig. 1. Distribution of Manila clams in a 59-m² intertidal plot, southern Puget Sound (Washington State, U.S.A.) Each dot represents the location of an individual clam. Old clams (ages 2+ and older) in black, young clams (1+ age) in red.

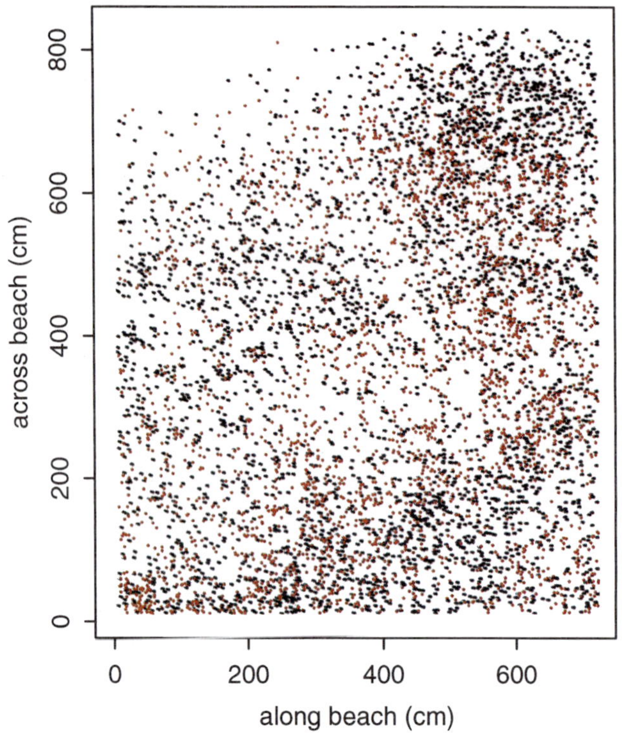

et al. 1987) and of an appropriate formal framework. We believe that the latter is provided by the theory of stochastic spatial point processes (Ripley 1977, 1981; Diggle 1983; Cressie 1991).

Below we briefly review the historical development of the concept of concentration and discuss applications that seem promising in the assessment of benthic stocks. Several methods are illustrated with data on the distribution of Manila clams (*Tapes philippinarum*) on a beach from southern Puget Sound, Washington, U.S.A.

A study case

During the summer of 1980, one of us (MAH) conducted a detailed census survey of all the clams found in an intertidal plot located at Glencove, Henderson Bay, southern Puget Sound (Washington, U.S.A.) A complete description of the study site can be found in Hall (1983). The dimensions of the rectangular plot were 7.2 m along and 8.2 m across the beach (59 m²); the upper and lower sides corresponded, respectively, to tidal shore levels +2.4 m and +1.5 m; the slope of the plot was 7°. Glencove is a sheltered, shallow environment. At the time the sample was obtained, it had been free of significant pollution, disturbance, or harvesting for at least 12 years and probably longer. Variables describing sediment characteristics were strongly correlated with shore level: organic matter content, and particle size and deviation (ϕ units) decreased, while skewness, kurtosis, and percent gravel content increased with tidal shore level.

The entire plot was carefully dug, piecewise, with a hand trowel. The location of each individual clam was recorded (Fig. 1) and each specimen was saved individually and later measured and aged by counting the number of annual growth rings. Almost all the clams were manilas or Japanese littlenecks ($N = 6806$), whose maximum lifespan in this area is 9 years. The efficiency of the sampling technique (estimated by digging quadrats after sampling was completed and sieving the sediment) was 96%. A decrease in overall density shown towards the upper-left corner of Fig. 1 is negatively correlated with total gravel content of the sediment.

It is not our intention here to make conclusive statements about the processes governing the pattern observed. Rather, we use the pattern to illustrate the application of some methods to a realistic data set. For interpretation and discussion of processes possibly underlaying the pattern we refer to Hall (1983).

Definitions

Neighborhoods

The ambit (Lloyd 1967), ecologically effective distance (Antonovis and Levin 1980), area of influence (Hall 1983; Kenkel 1988), or neighborhood (Orensanz 1986; Addicott et al. 1987; Folt et al. 1993) of an individual is a spatial region (two- or three-dimensional) within which a certain interaction between the individual and its environment (including other individuals of its own kind) is effective (or, from the viewpoint of the observer, within which its effectiveness is evaluated). Various neighborhoods can be defined, each relative to a specific process, including: (*i*) trophic, relative to resource utilization, eventually leading to competitive interactions; (*ii*) reproductive, related, in the case of broadcast spawners, to gamete dilution away from the individual; (*iii*) sensory, relative to perception of environmental conditions, including the presence of conspecific, predators, etc; and (*iv*) vulnerability to predation or fishing, relative to the area over which predators or fishermen assess density in the course of nonrandom effort allocation; etc. Territories of various types are related to the notion of a neighborhood, although in general a territory (the area effectively utilized by an individual) is contained within a respective neighborhood; the ambit of the interaction defining the boundary of the territory extends beyond the latter.

The geometry of neighborhoods can be intricate and permanently changing, particularly in relation to (*i*) advection and turbulence in a fluid environment, as in the case of chemosensory neighborhoods of predators and prey, or the trophic neighborhoods of suspension feeders; and (*ii*) individual displacements in the case of motile organisms. In most cases neighborhoods lack well-defined boundaries; the intensity of most interactions tends to gradually vanish away from the individual. Neighborhoods with well-defined boundaries are rare, perhaps best exemplified by the foraging areas of some grazing limpets and surface deposit feeders (e.g., some *Macoma* species). The neighborhoods of different individuals can (and generally do) overlap widely, an extreme case being the trophic neighborhoods of suspension feeders.

Although neighborhoods can be very difficult to map and measure in practice, it is important to be explicit, at least, about the approximate operational spatial scale relevant to the interaction being considered. Throughout this paper we deal with simple neighborhood regions: circular, centered around individuals, and isotropic (pattern insensitive to rotation). The models discussed, however, could be extended to noncircular and/or eccentric neighborhoods as well. We define the location of an individual with reference to an ideal point at its geometric center; the extension of neighborhoods can be made a function of the individual size. Some authors have proposed distance methods incorporating the area physically occupied by each individual or some part of it (e.g., a tube protruding from the sediment's surface), which is just one among other definable neighborhoods and often not the most relevant one.

Concentration and crowding

Consider first the case of individuals in an unbound region and neighborhoods centered around each of them. Location of individual i in space is denoted as $\mathbf{s}_i$, a vector of co-ordinates. Concentration (C) around reference individual i is defined as the number of individuals found within a circular neighborhood of radius h centered at $\mathbf{s}_i$ (including individual i itself),

$$C_i(h) = \sum_{j=1}^{N} I(\|\mathbf{s}_i - \mathbf{s}_j\| \le h)$$

where N is the total number of individuals in the population, $\|\mathbf{s}_i - \mathbf{s}_j\|$ is the distance between the locations of individuals i and j, and $I(\|\mathbf{s}_i - \mathbf{s}_j\| \le h)$ is an indicator variable taking the value of 1if $\|\mathbf{s}_i - \mathbf{s}_j\| \le h$ and of 0 if $\|\mathbf{s}_i - \mathbf{s}_j\| > h$. Crowding is defined analogously, with the difference that reference individual i is excluded,

$$[1] \quad C_i^*(h) = \sum_{j=1}^{N} I(\|\mathbf{s}_i - \mathbf{s}_j\| \le h) \; ; \; i \ne j$$

$$C_i^*(h) = C_i(h) - 1$$

In practice one is always interested in the assessment of the average concentration or crowding experienced by the individuals located within some bound region A (usually rectangular) of area $|A|$. In this case, some assumptions have to be made about the concentration (or crowding) experienced by individuals whose distance to the boundary is less than h (as information about the location of possible neighbors outside A is not available.) As $h/\sqrt{|A|} \to 0$, border effects could be ignored, but in reality they are always an important consideration. Corrections for border effects are discussed in the Appendix.

Spatial data

Spatial data collected to investigate the spatial distribution and the abundance of benthic sedentary organisms belong to three basic types: type 1, counts of individuals within quadrats; type 2, distances between individuals and their nearest neighbors, or between points in space and the nearest individuals; and type 3, complete maps with the coordinates of all the individuals in the study area.

In type 1, quadrat placement in the region of interest can follow a variety of designs, including complete coverage of the study area (all quadrats sampled), one-dimensional systematic samples, two-dimensional systematic samples, and random or haphazard samples (simple or stratified). Size and spacing of quadrats vary according to the purpose of the study. Spatial information about the individual organisms within quadrats is lost, so nothing can be learned for scales below quadrat size. Quadrats, however, are often easy to obtain and for that reason are the most popular technique in the study of the spatial distribution of benthic invertebrates. The use of quadrat data in benthic studies is so deeply rooted that mapped data (introduced later) are often reduced to quadrat counts before analysis, with a consequent loss of costly information.

For type 2, distances may be measured between an individual and its nearest neighbors, between sample points and the nearest individuals, or between an individual nearest a sample point and the individual's nearest neighbor. Since distances are measured only to the closest individuals, nearest-neighbor data contain information only about the smallest scales of pattern. Compared to quadrat data, nearest-neighbor data have the advantage of not depending on an (often arbitrary) choice of quadrat size or shape. They do however depend on the number of nearest neighbors considered around each individual or point selected, which also restricts the scale of the analysis. Methods based on a random sample of n individuals effectively require a complete map with the location of all the individuals (N) in the study area, in which case more appropriate methods (discussed later) are applicable. The distribution theory for nearest-neighbor tests is based on the assumption of independence of n nearest-neighbor measurements. Studies of the spatial distribution of shellfish species are usually conducted over relatively small, intensively sampled study areas, in which case nearest-neighbor measurements are not likely to be independent. For further discussion of these methods the reader is referred to Upton and Fingleton (1985, particularly table 1.10) and Cressie (1991, particularly table 8.6).

In type 3, a vector of data (size, age, sex, phenotype, etc.) can be associated with each individual. Maps are difficult to produce in the cases of highly clustered populations (e.g., multilayered mussel druses, oyster reefs) and of infaunal species (our study case) and relatively easy to obtain (at least for small study areas) in the case of epibenthic invertebrates (Ikenouye 1968; Kooijman 1979*a*, 1979*b*). Underwater photography and diving were used in two studies of giant scallops (*Placopecten magellanicus*) from eastern Canada, in study areas ranging from 9 to 256 m^2 (MacDonald and Bajdik 1992; Stokesbury and Himmelman 1993). A few studies have examined changes in spatial pattern under experimental conditions (Holme 1950; Levinton 1972). Analyses illustrated below highlight the use of mapped patterns.

Spatial pattern

Three basic types of spatial pattern (random, regular, and clustered) are generally recognized in ecological research. Regular patterns tend to occur in organisms that move away from each other, as is often the case in territorial animals; clustered patterns (most common in nature) can reflect environmental heterogeneity (e.g., a mosaic of bottom types) or interactions among individuals (e.g., reproductive aggregations). Pattern can change with scale. Consider, for example, a hypothetical population consisting of dense clusters distributed randomly over a background of low density, with individuals distributed regularly within the clusters. In this scenario (not unrealistic for many colonial benthic tubeworms and sedentary mollusks) pattern cannot be characterized simply, without reference to the multiplicity of scales. Detection and description of spatial patterns (including their change with scale) is of significance in the estimation of abundance, the assessment of fishing strategies, and inference in ecological studies.

Spatial point patterns are often compared to a realization of a completely spatially random point process. Strictly defined, complete spatial randomness (the white noise of spatial point processes) is equivalent to a homogeneous Poisson process; given an area A where N individuals are found, the N individuals are distributed independently (and with uniform probability) over A. Intuitively, this means that individuals are equally likely to occur anywhere within A and that they do not interact with each other, either repulsively or attractively. Complete spatial randomness characterizes the absence of structure (or signal) in the data. As such, it is often the null hypothesis in a statistical test to determine whether there is spatial structure in a given point pattern.

Models of spatial distribution

The spatial distribution of shellfish stocks can be analyzed through a variety of models, depending on the purpose and scale of the study and on the nature of the data. The focus in our case is on the spatial position of individual organisms (events), the ideal type of data consisting of maps with the location of every event within the study area.

Cressie (1991) introduced a very general stochastic model for spatial data. Let $\boldsymbol{s}$ be a generic data location in space (two-dimensional in our context) and suppose the potential datum $\boldsymbol{Z}(\boldsymbol{s})$ at spatial location $\boldsymbol{s}$ is a random quantity (or vector). Now let $\boldsymbol{s}$ vary over index set S (a subset of the two-dimensional space) to generate the multivariate random process (or random field)

[2] $\{\boldsymbol{Z}(\boldsymbol{s}) : \boldsymbol{s} \in S\}$

A realization of eq. 2 is denoted $\{z(\boldsymbol{s}) : \boldsymbol{s} \in S\}$. Spatial point patterns (considered in this paper) and spatial data suitable for geostatistical analysis (see Warren 1998) can be considered as special cases of this general model (Cressie 1991). In the case of geostatistical data, $\boldsymbol{Z}(\boldsymbol{s})$ is a random vector at location $\boldsymbol{s} \in S$, where S is a fixed subset of the two-dimensional space. A distinctive feature of geostatistical-type problems is that the spatial index $\boldsymbol{s}$ can vary continuously over a subset of the two-dimensional space. In the case of point patterns, S (a random set) is instead a spatial point process, a collection of random events whose realization is called a spatial point pattern. If there are measurements or observations $\{\boldsymbol{Z}(\boldsymbol{s}_i)\}$, called marks, associated with each event, then the corresponding point process is known as a marked spatial point process. Suppose that information is collected on 500 scallops from a general study region A; the locations of the observed pattern are $S = \{\boldsymbol{s}_1, ..., \boldsymbol{s}_{500}\}$. The associated $\{z(\boldsymbol{s}_1), ..., z(\boldsymbol{s}_{500})\}$ could be vectors of information, one for each scallop, including age, shell length, and total weight. If only the presence of a scallop were recorded at each location, then no $\boldsymbol{Z}(\boldsymbol{s})$ would be specified, or one could think of scalar $\boldsymbol{Z}(\boldsymbol{s}) \equiv 1$, for all $\boldsymbol{s} \in S$ (the usual spatial point process).

First-order properties of the process

The first-order properties of a spatial point process, analogous to the familiar statistical notion of mean, are summarized by the intensity function. Define the function

$$\lambda_a(\boldsymbol{s}) \equiv \frac{E[N(\boldsymbol{s}, a)]}{|a|}$$

where $N(\boldsymbol{s}, a)$ denotes the number of individuals in the region a, a circular quadrat of area $|a|$ centered around location $\boldsymbol{s}$, and $E[\cdot]$ is the expectation of $(\cdot)$. If $\lambda_a(\boldsymbol{s}) \rightarrow \lambda(\boldsymbol{s})$ as the area of the region is made smaller and smaller (i.e., as $|a| \rightarrow 0$) for all locations (all $\boldsymbol{s} \in A$), then $\lambda(\boldsymbol{s})$ is called the intensity function (Diggle 1983). The point process is known as orderly when in the limit, as $|a| \rightarrow 0$, $\Pr(N(\mathbf{s}, a) > 1)/|a| = 0$ (i.e., in a point in space there can be at most one individual). In this case the intensity function can be equivalently defined as

$$\lambda(\boldsymbol{s}) \equiv \lim_{|a| \rightarrow 0} \frac{\Pr(N(\boldsymbol{s}, a) = 1)}{|a|}$$

For a stationary process, $\lambda(\boldsymbol{s})$ takes a constant value equal to λ.

The point process N is a homogeneous Poisson process if

$$\lim_{|a| \rightarrow 0} \frac{1 - \Pr(N(\boldsymbol{s}, a) = 0)}{|a|} = \lim_{|a| \rightarrow 0} \frac{\Pr(N(\boldsymbol{s}, a) = 1)}{|a|} = \lambda$$

and $N(\boldsymbol{s}_1, a)$, $N(\boldsymbol{s}_2, a)$, ... are statistically independent for any disjoint sequence of regions of area $|a|$ centered at $\boldsymbol{s}_1, \boldsymbol{s}_2, ...$. The equality above implies again that the process is assumed to be orderly. The constant λ is the intensity of the homogeneous Poisson process. If λ varies from location to location over A, i.e., if λ is replaced by $\lambda(\boldsymbol{s})$, then the Poisson process is said to be inhomogeneous.

Suppose that we want to estimate $\lambda(\cdot)$ from a map with the location of every individual in study area A. A coarse approach would be to partition A into small quadrats of area $|a|$ using a regular grid and count the number of individuals in each cell, say $N(\boldsymbol{s}, a)$, where $\boldsymbol{s}$ is the location of the geometric center of the cell. When $\lambda(\cdot)$ does not vary much over the area of the quadrat, $N(\boldsymbol{s}, a)/|a|$ estimates $\lambda(\boldsymbol{s})$ (Cressie 1991, page 599). For $|a|$ sufficiently small, this would result in a mosaic of quadrat densities with many zeros and ones. A smoother map of the intensity surface over A can be obtained using other, more sophisticated nonparametric estimators of the intensity function $\lambda(\cdot)$; these fall under two types: kernel and nearest-neighbor (Cressie 1991; Silverman 1986). We only discuss kernel estimators here. Kernel estimators are of the form

$$\hat{\lambda}_h(\boldsymbol{s}) \equiv \frac{\sum_{i=1}^{N} k_h\left(\frac{\|\boldsymbol{s} - \boldsymbol{s}_i\|}{h}\right)}{h^2 p_h(\boldsymbol{s})}$$

where the sum is over all N individuals present in the study area. The bandwidth h determines the amount of smoothing, the kernel function $k_h(\cdot)$ is a probability density function (pdf) symmetric about the origin, and $p_h(\boldsymbol{s})$ is an edge correction (see Cressie 1991). Many kernel functions can be utilized, one of them being the normal kernel,

Fig. 2. Nonparametric kernel estimate of the intensity function for the data shown in Fig. 1 (no distinction being made between young and old clams). Grid size is 6.7 cm. Color scale ranges from black (highest density) to green (lowest). Bandwidth h is 30 cm.

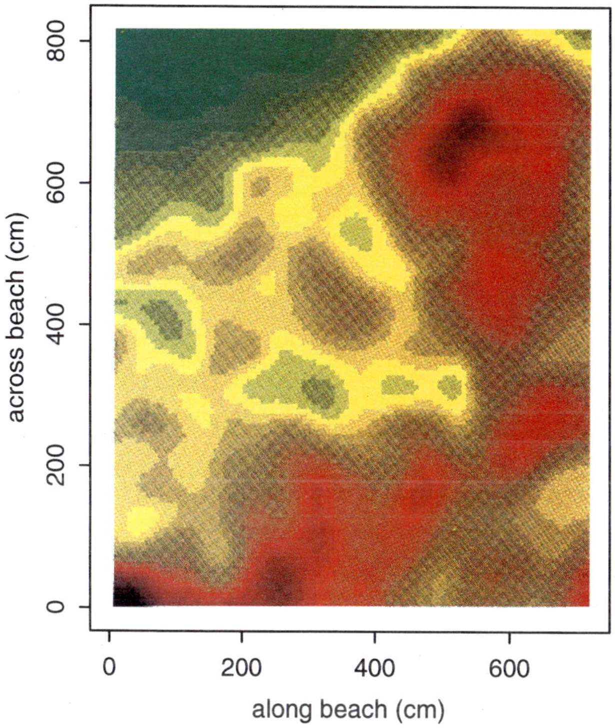

$$k_h(d) = \begin{cases} \frac{1}{2\pi} \exp\left\{\frac{-d^2}{2}\right\} & \text{for } -h \le d < h \\ 0 & \text{otherwise} \end{cases}$$

where

$$d = \frac{|s - s_i|}{h}$$

is the distance between s and s_i, in h units.

Figure 2 illustrates the estimation of the intensity function with the data shown in Fig. 1. The bandwidth h used is 30 cm and the grid used in the representation has nodes spaced at 6.7 cm. Density of the cells varied from 0 to 330 clams·m^{-2}. Notice that the degree of smoothing of the continuous surface is controlled by the bandwidth; once h has been chosen, renditions from different kernel functions differ little from each other.

Higher-order properties of the process

The probability structure of spatial point processes is generally described in terms of its first- and second-order properties. First-order properties were discussed earlier. Second-order properties, analogous to the variance–covariance structure in statistics, depend on the distance between pairs of events, rather than on counts of events. They can be characterized by the second-order intensity function, $\lambda^{[2]}(s_1, s_2)$, where s_1 and s_2 are locations in A. Just like the intensity function $\lambda(s)$ of an orderly process is the pdf of an individual occurring at location s, $\lambda^{[2]}(s_1, s_2)$ is the pdf of two individuals, one occurring at s_1 and the other at s_2, that is

$$\lambda^{[2]}(s_1, s_2) = \lim_{|a_1|, |a_2| \to 0} \frac{\Pr\,[N(s_1, a_1) = N(s_2, a_2) = 1]}{|a_1|\,|a_2|}$$

Usual models assume that the process is stationary (invariant under translation) and isotropic (invariant under rotation), which may be reasonable at least within restricted regions. Under these assumptions, $\lambda^{[2]}(s_1, s_2)$ reduces to $\lambda^{[2]}(d)$, where d is the distance between s_1 and s_2 (Diggle 1983, page 47).

Second-order properties can be characterized by the expected crowding, $E[C^*(h)]$, defined as the expected number of further individuals within distance h of an arbitrary individual. Consider first $C_i^*(h)$, the observed crowding around reference individual i, as defined by eq. 1. By assuming that the process is orderly (multiple coincident events are ruled out), the expected value of $C_i^*(h)$ can be expressed as a function of the second-order intensity:

$$E\,[C_i^*(h)] = \int_{r \le h} \frac{\lambda^{[2]}(s, s_i)}{\lambda(s_i)}\, ds$$

where $r = |s - s_i|$. This equation corresponds to the conditional density that an individual exists at s given that one exists at s_i,

$$\frac{\lambda^{[2]}(s, s_i)}{\lambda(s_i)}$$

integrated over a circular area of radius h centered at s_i. Assuming that the process is stationary, the expected crowding $E[C^*(h)]$ around an arbitrary individual located at s_0 is given by:

$$E\,[C^*(h)] - 2\pi \int_0^h \frac{\lambda^{[2]}(r)}{\lambda}\, dr$$

where $r = |s - s_0|$. The expected crowding scaled by $1/\lambda$ is known as the K-function (Ripley 1977; Diggle 1983),

$$[3] \quad K(h) = \frac{E[C^*(h)]}{\lambda}$$

The second-order intensity is related to the variance of $N(s, a)$ by

$$\mathrm{Var}\,(N(s, a)) = \int_a \int_a \lambda^{[2]}(s_1, s_2)\, ds_1\, ds_2 + \lambda|a|\,(1 - \lambda|a|)$$

(Diggle 1983, page 49) so that the variance/density ratio is

$$[4] \quad \frac{\mathrm{Var}\,(N(s, a))}{\lambda|a|} = \frac{\int_a \int_a \lambda^{[2]}(s_1, s_2)\, ds_1\, ds_2}{\lambda|a|} - \lambda|a| + 1$$

The first term on the right-hand side of the equation is a parameter related to $E[C^*(h)]$, although not quite the same: it is the expected number of further individuals occurring in a (as opposed to in the neighborhood of an arbitrary individual) in addition to any given arbitrary individual in a. Under complete spatial randomness, this quantity equals $\lambda|a|$, as $\lambda^{[2]}(d) = \lambda^2$, so that the variance/density ratio equals one.

Neither of these statistics provide a complete description of the probability structure of a spatial point process, which would require consideration of the expected numbers of specific configurations of three, four, etc. events (Diggle 1983; Andersen 1992). Realizations of a spatial point process cannot be simulated from descriptions of its first- and second-order properties alone. Higher-order properties, although definable, would be difficult to interpret.

Assessment of concentration and crowding

Quadrat data: pattern at a single scale

Most analyses of concentration or crowding in ecological research have been based on quadrat counts. Calculation of mean crowding from quadrat data requires the assumption that all the individuals found within a quadrat experience the same crowding, which would be the case only if they were regularly distributed. If the area occupied by a certain shellfish population is partitioned into Q quadrats of equal area ($|a|$), and if n_i is the number (count) of individuals within the ith quadrat, then, under that assumption,

$$[5] \quad \text{mean number of individuals per quadrat} = \bar{n} = \frac{\sum_{i=1}^{Q} n_i}{Q}$$

$$[6] \quad \text{mean concentration} = \bar{C}_a = \frac{\sum_{i=1}^{Q} n_i^2}{\sum_{i=1}^{Q} n_i} = \frac{\sum_{i=1}^{Q} n_i^2}{N}$$

where the subscript a has been added to indicate that eq. 6 corresponds to mean concentration estimated from quadrat data. Notice that $\bar{n}$ (or, equivalently, the mean density $\bar{D} = \bar{n} / |a|$) is averaged across quadrats, while concentration is averaged across individuals, all the individuals within a quadrat sharing the same concentration value. The different ecological meaning of density and concentration is evident. If the study region were expanded to include territory where no individual organisms are found, mean density would go down, but mean concentration would remain unchanged.

Under complete spatial randomness the number of organisms per quadrat has a Poisson distribution with mean $\lambda|a|$, where λ is the intensity of the Poisson process and $|a|$ is the area of a quadrat. Complete spatial randomness can be tested with a goodness-of-fit test (Elliott 1977). If it is rejected, the next step is to measure the degree of departure from it. The intuitive notion of aggregation entails a relation between concentration and density. Dividing eq. 6 by eq. 5 and remembering that the variance of n, the number of individuals in a quadrat of area $|a|$, is estimated by

$$\widehat{\text{Var}}(n) = \frac{\sum_{i=1}^{Q} n_i^2 - \frac{\left(\sum_{i=1}^{Q} n_i\right)^2}{Q}}{Q}$$

then

$$[7] \quad \frac{\bar{C}_a}{\bar{n}} = 1 + C\,V_n^2$$

which means that this ratio is a simple measure of variation among quadrats (Lewontin and Levins 1989, page 516). Notice that when $\widehat{\text{Var}}(n)$ is 0 (i.e., n_i is the same for all quadrats), $\bar{C}_a^*/\bar{n} = 1$. Also,

$$\bar{C}_a^* = \bar{n} + \left(\frac{\widehat{\text{Var}}(n)}{\bar{n}} - 1\right)$$

signifying that mean crowding (estimated from quadrat data) is the sum of the mean number of individuals per quadrat plus the amount by which the variance/mean ratio exceeds unity (Lloyd 1967, page 4). This relationship between estimators preserves the relationship between the estimated parameters in eq. 4; $\bar{C}_a^*$ is a moment estimator of the first term in the right-hand side of eq. 4.

Two ratios very similar to eq. 7,

$$I_L = \frac{\bar{C}_a^*}{\bar{n}} \quad \text{and} \quad I_\delta = \frac{N}{N-1} I_L$$

are respectively known as Lloyd's (patchiness; Lloyd 1967) and Morisita's (index of dispersion; Morisita 1959, 1962, 1971; Hurlbert 1990) indices of aggregation. Other indices of aggregation have been proposed for quadrat count data (see Cressie 1991, pp. 590–591 for a brief overview and references), all involving the variance/mean ratio in one way or another. In general, one-dimensional indices can not capture the complexity and multiplicity of scales of most spatial patterns. Location of quadrats within the study area and of individuals within quadrats are not considered and consequently most of the spatial information is effectively lost.

Mapped data: scale dependency of pattern and the *K*-function

When the location of all individuals in a given area is known, crowding can be assessed for different spatial scales by considering neighborhoods of different size around each individual. The analog of Lloyd's patchiness index (I_L, discussed above for quadrat data) is the K-function, defined earlier (eq. 3) as the ratio between expected crowding, $E[C^*(h)]$, and λ, where h determines the size of the neighborhood. The expected crowding is estimated by an empirical average,

$$\bar{C}^*(h) = \frac{\sum_{i=1}^{N}\sum_{j=1}^{N} I(\|\mathbf{s}_i - \mathbf{s}_j\| \le h)}{N}, \quad (i \neq j),$$

so that

$$\hat{K}(h) = \frac{\bar{C}^*(h)}{\hat{\lambda}}$$

where $\hat{\lambda} = N/|A|$. Because the locations of all events are now available, $\hat{K}(h)$ can be computed for any arbitrary h. The

Fig. 3. *L*-function plots of data shown in Fig. 1 (no distinction being made between young and old clams) together with envelopes (dashed lines) constructed from 100 realizations of a completely spatially random process. A: location of the clams in the plot and windows defined for close-ups on pattern. B: Plot of the *L*-function for the entire plot. C–F: Plots of the *L*-function for upper-left, upper-right, lower-left, and lower-right windows, respectively. All measurements are in cm.

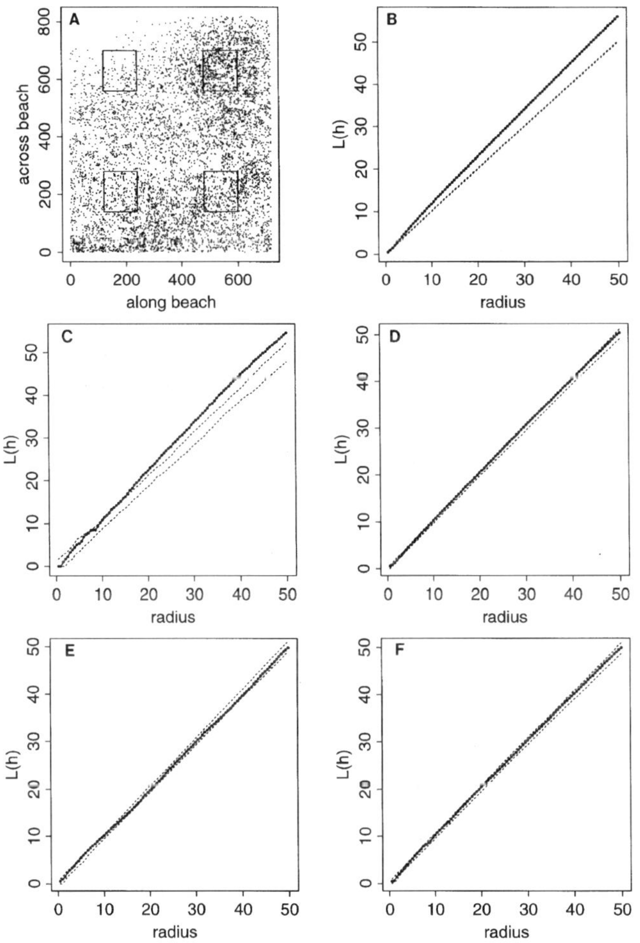

K-function thus captures the spatial dependence between different regions and is a powerful method of data summary. It presents spatial information at all scales of pattern and the location of the individuals is utilized in the estimation.

In a two-dimensional space (always our case): $K(h) = \pi h^2$ under the complete spatial randomness assumption; $K(h) < \pi h^2$ under regularity; and $K(h) > \pi h^2$ under clustering. Simulation envelopes can be computed based on realizations (conventionally 100) of $s_1, ..., s_N$ from a uniform distribution on *A*. Results can be conveniently represented with a plot of the *L*-function,

$$\hat{L}(h) = \sqrt{\frac{\hat{K}(h)}{\pi}}$$

versus *h* (Ripley 1981; Venables and Ripley 1994).

Figure 3 shows explorations of our primary study case using the *L*-function, with neighborhood radii (*h*) in the range 0–50 cm. The latter seems reasonable in our case because we were more interested in investigating local (small-scale) interactions among clams than larger-scale responses to environmental gradients. Figure 3A (top left) shows the location of all clams in the plot and the position of 4 windows that we used to explore pattern within restricted regions. The *L*-function calculated for the entire plot (Fig. 3B) suggests a random pattern at very small scales ($h < 3$ cm); at larger scales the curve gradually diverges from the envelopes towards aggregation, reflecting the underlying inhomogeneity of the process (discussed earlier). Notice that the Monte Carlo envelopes are very close to each other because of the large number of points ($N = 6806$) in the pattern. Three of the windows (Figs. 3D–3F) focus on regions within which the influence of the environmental gradient seems negligible. In these three cases, the *L*-function lies within the Monte Carlo envelopes over the entire range of interest, suggesting complete spatial randomness. One of the windows (Fig. 3C) overlaps what seems to be the steepest part of the density gradient; interestingly, the *L*-function gradually drifts away from randomness (in the direction of aggregation) for $h > 15$ cm. The pattern appears random at smaller scales.

The following three examples from the literature illustrate other patterns. (*i*) MacDonald and Bajdik (1992) collected data on the location of giant scallops (*Placopecten magellanicus*) in two 15 m × 15 m plots at two unfished sheltered sites (Colinet and Sunnyside) off Newfoundland (eastern Canada). Figure 4A (from MacDonald and Bajdik 1992, fig. 2) shows the position of scallops larger than 110 mm ($N = 198$) within one of the plots (Colinet). The *L*-function (Fig. 4D; $0 \text{ m} < h < 5 \text{ m}$) shows no discernible departure from complete spatial randomness at any scale (the same was found for the Sunnyside plot, not shown in our figure). (*ii*) Ikenouye (1968) analyzed the spatial distribution of a barnacle (*Tetraclita squamosa japonica*) at an intertidal rocky site in Chiba Prefecture (Japan). Figure 4B (digitized data from Ikenouye's fig. 2) shows the location of all barnacles ($N = 896$) within a 3 m × 3 m plot. The author partitioned the plot into quadrats of various sizes and calculated Morisita's index of dispersion for each partition. The *L*-function (Fig. 4E; $0 \text{ m} < h < 0.2 \text{ m}$) shows a clear trend towards regularity at small scales and drifts away from complete spatial randomness at larger scales, with a breakpoint at ~5 cm. (*iii*) Levin (1981) investigated the spatial distribution of a spionid polychaete, *Pseudopolydora paucibranchiata*, on an intertidal mudflat at Mission Bay (California). Figure 4C (digitized data from Levin's fig. 2) shows the location of all the tubes ($N = 337$) within a 9 cm × 9 cm quadrat. A distance method (Clark–Evans index) was used by the author to assess pattern, which was found to be uniform due to territorial behavior. The *L*-function (Fig. 4F; $0 \text{ cm} < h < 1 \text{ cm}$) shows regularity at small scales and complete spatial randomness for neighborhoods with a radius $h > 0.5$ cm.

Multivariate patterns

Individuals in a population of interest often are of more than one kind, for example young (age group 1+) and older (age groups 2+ and above) clams in our primary study case (Fig. 1). Ultimately, we want to learn about possible interactions between members of the two groups. Cross *K*-functions can be constructed to examine such bivariate patterns, analogous to the derivation in the univariate case,

Fig. 4. Three mapped point patterns of benthic invertebrates and plots of the corresponding L-functions. A and D: giant scallops in a 15 m × 15 m region (from MacDonald and Bajdik 1992). B and E: barnacles in a 3 m × 3 m region (from Ikenouye 1968). C and F: *Pseudopolydora* (a spionid polychaete) in a 9 cm × 9 cm region (from Levin 1981).

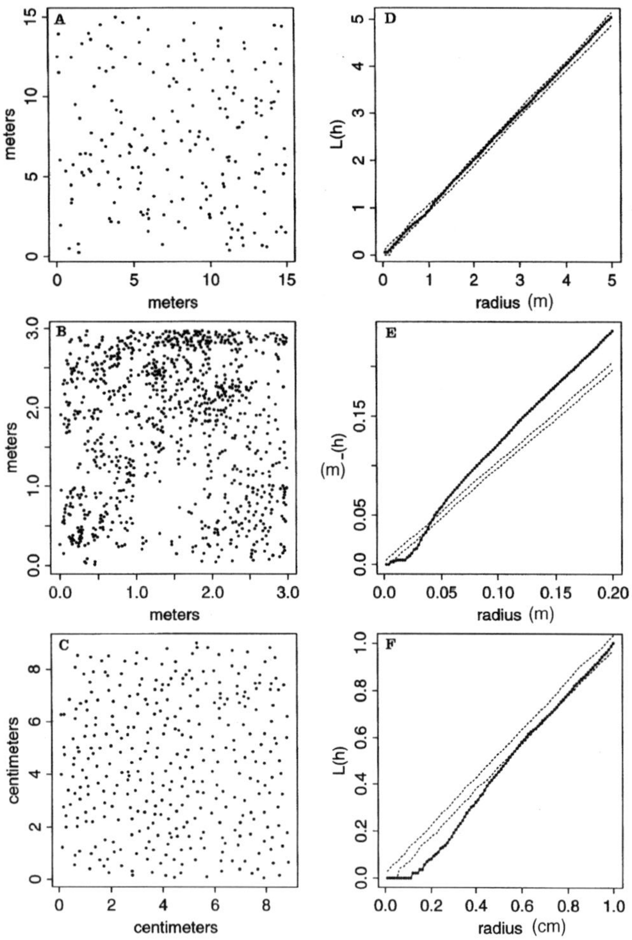

$$C_i^{*(12)}(h) = \sum_{j=1}^{N_2} I(\|\boldsymbol{s}_i^{(1)} - \boldsymbol{s}_j^{(2)}\| \le h)$$

$$\bar{C}^{*(12)}(h) = \sum_{i=1}^{N_1} \frac{C_i^{*(12)}(h)}{N_1}$$

$$\hat{K}^{(12)}(h) = \frac{\bar{C}^{*(12)}(h)}{\hat{\lambda}_2}$$

$$\hat{K}^{(12)}(h) = \frac{\sum_{i=1}^{N_1} \sum_{j=1}^{N_2} I(\|\boldsymbol{s}_i^{(1)} - \boldsymbol{s}_j^{(2)}\| \le h)}{\hat{\lambda}_1 \hat{\lambda}_2 |A|}$$

where i and j indicate individuals of types 1 and 2, respectively; $C_i{*}^{(12)}$ means crowding of organisms of type 2 around

Fig. 5. Bivariate L-function plots for data shown in Fig. 1 (young and old clams). A: based on concentration of young clams around old ones; plot B is the reverse. The analysis focused on a 360 cm × 410 cm area located at the center of the study region.

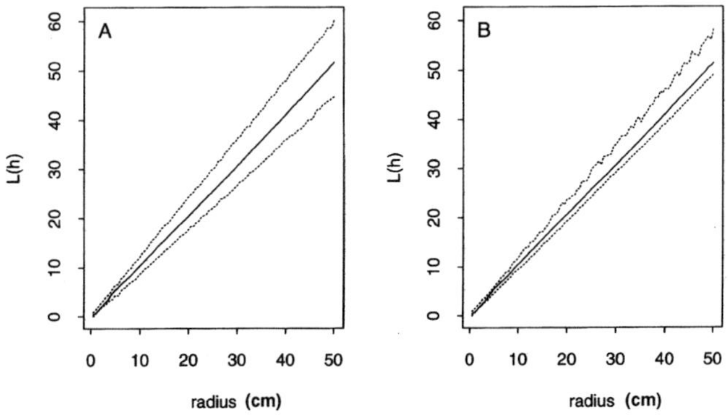

individual i of type 1 and all other conventions are as before; border corrections are omitted for simplicity of notation. If the process is stationary, then $K^{(12)}(h) = K^{(21)}(h)$ and the estimator

$$\bar{K}^{(12)}(h) \equiv \frac{N_1 \hat{K}^{(12)}(h) + N_2 \hat{K}^{(21)}(h)}{N_1 + N_2}, \quad h > 0$$

is more efficient when the component processes for types 1 and 2 are independent Poisson processes (Lotwick and Silverman 1982). Lotwick and Silverman (1982) introduced a Monte Carlo test for independence, conditional on the observed marginal patterns. The region studied is first wrapped onto a torus. Then the pattern formed by type-2 individuals is repeatedly shifted around the torus by a random amount (a vector $\boldsymbol{u}$, uniformly distributed), while the pattern formed by type-1 individuals is held fixed. For each realization of $\boldsymbol{u}$, the cross K-function is calculated; simulation envelopes are obtained as in the univariate case (notice that edge corrections are not required in the toroidal computation).

Instead of using toroidal shifts we defined a rectangular window at the center of our study plot, leaving a peripheral guarding area of width 180 cm and 205 cm on the x and y directions, respectively. The pattern formed by 1-year-old clams was then shifted with respect to the locations of the older clams, distances of up to 100 cm (uniformly distributed) in each direction, and simulation envelopes were constructed as described above for radii (neighborhood sizes) in the range (0 cm < h < 50 cm). The same was then repeated reversing the role of old and young clams. Figure 5 shows the corresponding cross-L-functions and envelopes (100 realizations). As we are interested in possible interactions at relatively small scales, we restricted the distances the patterns were shifted to a maximum of 100 cm in each direction. At large scales both types of clams are expected to respond in approximately the same way to the environmental gradient known to be present in the plot and hence to be nonindependent: both types would tend to co-occur, to depart from complete spatial randomness towards aggregation. It is towards the smaller scales, where the effect of the gradient is negligible, that interaction between both types of clams could be possibly captured in the mapped point patterns.

Fig. 6. Concentration profiles for data shown in Fig. 1 (a guarding area defined around the perimeter of the region). A: Distribution of concentration in 400-cm^2 circular neighborhoods ($N = 6806$). B: Frequency distribution of density observed in a complete partition of the region (guarding area excluded; $N = 1326$ contiguous quadrats). C: Concentration profile derived from B.

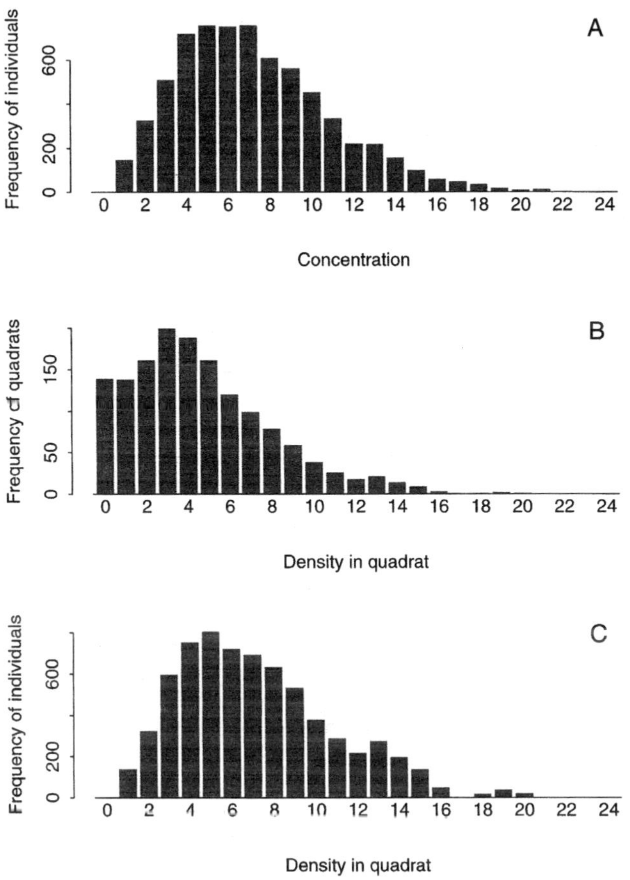

Distribution of concentration at a single scale: concentration profiles

Descriptive ecologists have traditionally focused on mean density and mean concentration, either at a single scale or over a range of scales, generally with the purpose of investigating pattern. When concentration-dependent processes are considered, the distribution of concentration among the members of the population (and not just its mean value) should be the focus of interest. We calculated the distribution of concentration for the clam plot directly, using individual concentration values, and also in the form of a concentration profile derived from a distribution of density measured in nonoverlapping quadrats. Neighborhoods and quadrats were chosen to have the same area, matching the area of influence considered by Hall (1983), 400 cm^2. The plot, containing 6806 clams, was partitioned into 1326 contiguous quadrats. Figure 6A shows the frequency distribution (in number of individuals) of individual concentration values, computed for circular neighborhoods (h = 11.284 cm) using Ripley's (1976) edge-corrected estimator (see Appendix); mean concentration was 7.25 clams per 400-cm^2 circular neighborhood (SD = 3.68). The frequency distribution (in number of quadrats) of quadrat densities in the

Fig. 7. Percentage of clams experiencing concentrations above density levels along the x-axis (A), and percentage of the total area where concentration above these levels occur (B), compared to the expected pattern under complete spatial randomness (solid lines in all graphs). Panel C shows the percentages of individuals packed in different fractions of the total area (i.e., the values along the y-axes of (A) and (B) plotted against each other); point labels indicate concentration levels which increase towards the origin; the dashed line corresponds to a regular distribution.

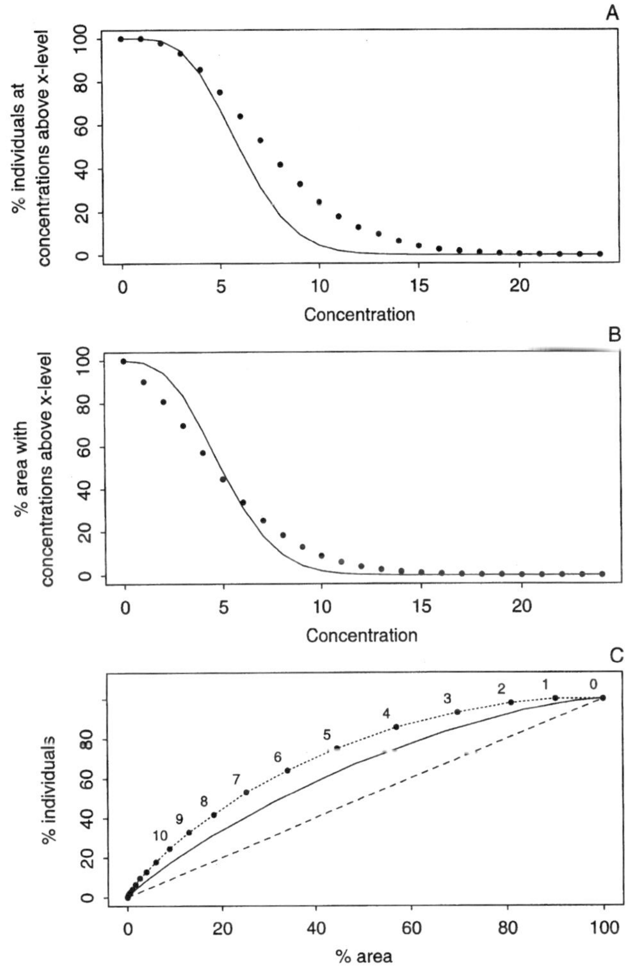

partition is shown in Fig. 6B (mean density was 5.13 per 400-cm^2 quadrat). Finally, Fig. 6C shows the concentration profile derived from the quadrat density distribution (Fig. 5B); estimated mean concentration was 7.16 per 400-cm^2 neighborhood (SD = 3.68), assuming that all individuals within a quadrat share the same concentration. Notice that the x-axis is the same in Figs. 6A–6C; the y-axis of Fig. 6C is obtained by multiplying the x- and y- axes of Fig. 6B. Since the neighborhood of interest and quadrat size have the same area, the concentration profile derived from the quadrat density distribution (Fig. 6C) is a very good approximation of the distribution of concentration calculated directly (Fig. 6A); mean concentration is virtually the same in the two.

Given a population of individuals inhabiting a certain region, the more crowded its members are, the smaller is the area that they effectively claim. Figure 7 shows the percentage of

Fig. 8. Five general types of concentration profile. First column: concentration profiles. Second column: corresponding distributions of density, *D*. Third column: accumulated concentration profiles, cumulative stock size (*S*) as a function of *D*, $S_D = \sum_0^D C_D$. Fourth column: relationship between CPUE and stock level for a sedentary species under the simplest form of sequential effort allocation in which patches of highest densities are depleted first. Depletion trajectories, to be tracked from right to left, were obtained by flipping the axes of the figures in the third column and relabeling the density axis as CPUE; solid, dashed, and dotted lines correspond to 100, 50, and 10% efficiency, respectively.

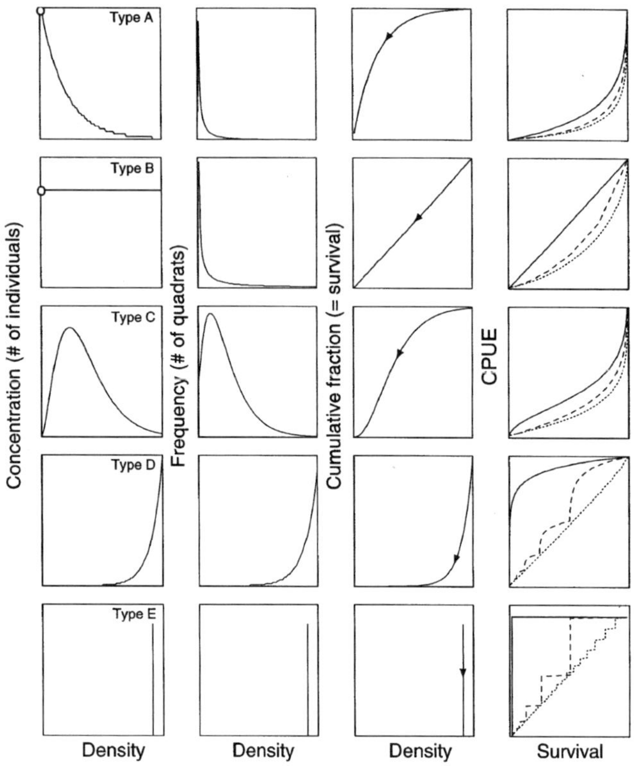

individual clams experiencing concentrations (h = 11.284 cm, as before) above a given density level (Fig. 7A) and the percentage of the total area where these concentrations occur (Fig. 7B), compared to the expected pattern under complete spatial randomness (solid line). The bottom panel shows the percentages of individuals packed in different fractions of the total area (i.e., the values along the y-axes of panels A and B plotted against each other). The dashed line corresponds to a regular distribution; the more aggregated the pattern, the steeper the curve.

Concentration profiles, as illustrated below through a simple exercise, are useful in the investigation of the fishing process.

Concentration profiles and sequential depletion

While the allocation of fishing effort in space can be best analyzed in connection to a map of density of the exploited population, the resulting level of depletion is a function of the concentration profile. Concentration profiles are commonly classified into types I–IV (Clark 1982, 1985; Hilborn and Walters 1992, pp. 184–188), but the definition of types is not consistent in the literature. To avoid confusion we denote five extreme types with letters A–E (Fig. 8). Most small-scale spatial patterns described in the literature (our primary study case included) fall within type C, in which case the corresponding distribution of density is usually described with a member of the binomial family.

We examine first the simplified case of a closed population in which (*i*) harvesters have complete knowledge of the spatial distribution of the stock, (*ii*) harvested patches are all equal in extension and do not overlap, (*iii*) fishermen remove all the individuals present in a patch during their first visit (gear efficiency equals 1), and (*iv*) the species harvested is strictly sedentary. Clark (1982) exemplified this scenario with the exploitation of a nondiffusive mineral stock. The closest case in fisheries is probably that of divers harvesting a relatively small fishing bed of an epibenthic species during a short fishing season. In the simplest form of sequential effort allocation, the sequence in which patches of the ground are harvested matches their density ranking. The cumulative concentration profile (Fig. 8, third column), tracked from right to left (as indicated by the arrows), corresponds to survival. Under the simplest form of sequential effort allocation, the relationship between CPUE and stock level (Fig. 8, fourth column) departs from proportionality in all types of concentration profile, except for type B. In the case of type A there is hyperdepletion: CPUE declines rapidly at the beginning of the season as the few high-density quadrats are depleted. Type D shows hyperstability: CPUE declines slowly as the many quadrats of high density are harvested. Type E (constant concentration, strictly regular spatial pattern) is the limit case of type D; no change in CPUE is observed until the whole stock has been completely depleted.

If the spatial allocation of effort were random, the depletion trajectory would be linear (proportionality), with some residual variance in the CPUE data; overlap between the patches harvested over time contributes to the same effect. A reduction in gear efficiency has different effects for the different types of profiles, as shown in the fourth column of Fig. 8 (dashed and dotted lines): type A, low efficiency enhances hyperdepletion; type B, pattern shifts from proportionality to hyperdepletion; type C, low efficiency enhances hyperdepletion; and types D and E, pattern shifts from hyperstability to proportionality.

Discussion

Density

We have considered nonparametric estimators of the intensity function only. Alternatively, the latter could be parameterized so that $\lambda_\Theta(s)$ is made a function of some concomitant spatial variable (e.g., some environmental gradient), of spatial location s alone, or of both. There are few examples where inhomogeneous Poisson processes have been fitted to data in two-dimensional space, although one of them (Kooijman 1979*a*, 1979*b*) refers to benthic invertebrates (sea anemones, barnacles). Environmental gradients (shore level, gravel content) could be possibly modeled in our primary study case. An interaction of most interest involves the behavior of juveniles (age 1+), which can move and possibly select microhabitats long after settlement, relative to adult clams (ages 2+ and older); such behavior could involve attraction or avoidance. Density of the adults could be mapped to incorporate the size (mass) of the individuals and then the distribution of juveniles could be modeled as a stochastic point process superimposed onto this adult biomass surface (an environmental gradient). This approach, essentially different from the multivariate analysis presented before, will be explored in a forthcoming contribution.

Besides individual organisms, shellfish scientists are interested in the spatial location of another important type of event: units of fishing effort. It is often the case that snapshot maps of the spatial distribution of fishing units can be easily obtained. Typical examples are aerial surveys of small scale or recreational fisheries over large areas. Methods discussed earlier to estimate the intensity function could be applied to smooth and map the fishing intensity function, central to the analysis of spatial effort allocation. In the case of kernel estimators (illustrated by us with the clam study case), estimated area swept could be accounted for by modulation of the bandwidth parameter. In an essentially identical situation, Worton (1989) proposed the use of kernel methods to estimate the utilization distribution in animal home-range studies.

Concentration and crowding

The notions of crowding and concentration were first made explicit in the context of ecological research by Lloyd (1967), who found it "lamentable that in most cases conclusions [have been] based on so crude a measure [of crowding] as 'mean density'" (Lloyd 1967, page 1). Lloyd referred to concentration as demand (Lloyd 1967, page 10) but this term has not been used since. The term concentration, as used here, was introduced independently by Iwao (1976; see also Watanabe 1988) and Clark (1982), albeit in disparate contexts. Discrepancy between mean crowding and mean density (which Lewontin and Levins (1989) convincingly illustrated using data from the US census) is an obvious result of the nonrandom spatial distribution of the organisms; "individuals tend to find more others of their own kind right around them than would be the case in a random distribution" (Lloyd 1967, page 28). The fact that most individuals in populations with an aggregated spatial distribution experience a high concentration, even if high density patches are scarce and small (Orensanz 1986), has important and self-evident implications in the case of density-dependent processes (resource depletion, fertilization rate in broadcast spawners, predator–prey interactions, etc.), which may affect a large fraction of the population even if they operate only over comparatively small areas. Needless to say, definition of neighborhoods (scale, shape, orientation, etc.) is a subject of utmost importance, certainly not one to be guided by the aesthetics of depicted smooth density surfaces or by the logistic convenience of a certain quadrat size.

While ecologists have focused mostly on mean concentration or crowding for assessing spatial pattern (e.g., Pielou 1974, pp. 150–155; Elliott 1977; Hurlbert 1990), interest in the distribution of concentration in a population arose in the 1980's among natural resource scientists. First, Clark (1982, 1985) introduced such distributions (as concentration profiles) as a way of describing the amount of resource found at different densities and to investigate the response of harvesters to such patterns. The observational scale of interest for the analysis of these profiles is necessarily large, commensurate with the spatial scale of the fishing process (i.e., the observational scale upon which harvesters make tactic decisions). Cumulative concentration profiles (with concentration accumulated "to the left," i.e., from high to low density; e.g., Fig. 8A), combined with size frequency distributions, were introduced by Clavier and Richard (1986, see also Baron and Clavier 1992) under the name of isobiomass curves. Profiling of concentration at smaller scales was independently proposed by Orensanz (1986, distribution of neighborhood density, his fig. 3e) and Folt and Schulze (1993, their fig. 1B). In the common case in which the data consist of a sample of quadrat counts obtained at arbitrary locations (selected randomly, haphazardly, systematically, contiguously, etc), the distribution of concentration can be calculated at a single scale, dictated by quadrat size (e.g., Fig. 6C). Measurements of concentration may be approximately meaningful if quadrat size is commensurate with the neighborhood of interest, as shown earlier with the concentration profiles calculated for our primary study case (Figs. 6A and 6C). Since information about the spatial location of individuals within elemental quadrats is not available, the assumption has to be made that all the individuals within a given quadrat experience the same concentration (the count for that quadrat).

When mapped patterns are available, concentration can be assessed for neighborhoods of varying size. It was in part motivated by mapped data in plant ecology that, starting in the late seventies, statisticians developed methods rooted in the theory of stochastic spatial point processes (Bartlett 1975; Ripley 1977, 1981; Diggle 1983; Upton and Fingleton 1985; Cressie 1991). Most important among these has been Ripley's *K*-function. As illustrated by our primary study case, the *K*-function may indicate departures from complete spatial randomness at large scales which may in fact reflect environmental gradients (nonstationarity). Evidence of nonstationarity was assessed by contrasting the *K*-function obtained for the whole area under study with those obtained by zooming into regions which can be assumed to be stationary. The analysis of the clam data was then extended to a bivariate pattern, the two types of events consisting of new recruits and older clams. The two patterns appeared to be independent of each other at small scales. Besides age groups, multivariate patterns of interest to shellfish biologists include assemblages of different species (e.g., Connell 1955, for an example involving two clam species.) Methods for the analysis of bivariate (or multivariate) patterns (Lotwick and Silverman 1982) were previously applied to some problems in terrestrial ecology (Harkness and Isham 1983; Andersen 1992).

Analysis of data gathered from the literature illustrates the merit of investigating mapped point patterns. In the scallop example (MacDonald and Bajdik 1992) the authors partitioned the study region into quadrats, using various quadrat sizes, and calculated the mean/variance ratio and a transformation of mean crowding for each partition size. They concluded that pattern was haphazard at Colinet (Fig. 4A) and aggregated at Sunnyside and speculated about the meaning of this result. The *K*-function analysis shows that both patterns fall within the range of realizations of a completely spatially random process, for all the scales about which the data are informative. Regularity at small scales in the barnacle example (Ikenouye 1986) is intriguing. The author indicates the use of a 10 cm × 10 cm grid to collect the data in the field. If the location of individual barnacles within each cell of the grid was not precisely measured, but instead was guessed in the graphical presentation, then the observed small-scale regularity could be due to a recording artifact. Interestingly, the match between the breakpoint in the pattern and the size of the grid used for recording certainly makes such a scenario plausible. Regular patterns commonly arise from such protocols; examples are easily spotted in the literature (e.g., Rawson 1980, fig. 16). Levin (1981), using distance methods, investigated several patterns corresponding to two polychaetes differing in their natural history. Our analysis of two patterns shown in her article (one from each set) confirms her main result: regularity in the species showing territoriality, random distribution in the other. The *K*-analysis expands the author's results by giving a more comprehensive depiction of the scale dependency of pattern. Regular patterns have attracted much attention in forest ecology following pioneering work by Matérn (1960). Such models are of potential interest in the case of organisms that have feeding territories, such as surface deposit feeders (Fig. 4C) and grazers.

Mapped data of benthic populations have been generally reduced before analysis, either to counts in quadrat partitions (e.g., Ikenouye 1968; Levinton 1972; Rawson 1980; MacDonald and Bajdik 1992; Stokesbury and Himmelman 1993) or to nearest-neighbor distances (e.g., Branch 1975; Underwood 1976; Levin 1981; Anderson and Kendziorek 1982; Wethey 1984), with a consequent loss of information. In a remarkable study of barnacle settlement that was well ahead of its time, Crisp (1961) captured in graphic form (his figs. 2, 4, and 5) the essence of spatial point pattern analysis. Andrew and Mapstone (1987, page 83) explicitly suggested the possible merit of point pattern analysis in benthic ecology. Ripley (1987, page 426), ten years after the publication of his seminal papers and while reviewing applications of spatial point pattern analysis in ecology, concluded that "many weak, inappropriate or even misleading methods continue to be used and can be seen in almost any issue of an ecological journal," even though "the time involved [in a full statistical analysis of an ecological data set] is unlikely to be significant compared to the fieldwork involved in collecting the data." This statement still applies to marine benthic ecology and fisheries, where appropriate methods have rarely been used to analyze mapped point patterns (but see Kooijman 1979*a*, 1979*b*, for a notable exception), although such data are often collected. Comprehensible presentations of the methods are widely available (e.g., Ripley 1981; Diggle 1983; Upton and Fingleton 1985; Cressie 1991) and they have been implemented in modern software packages (e.g., Venables and Ripley 1994).

Much of the preceding discussion dealt with static patterns of concentration. Changes of patterns over time, observed at different scales, may prove even more interesting. Kenkel (1988) and Rebertus et al. (1989), for example, used the *K*-function to explore changes in plant populations in response to competition and disturbance. Many such problems are of potential interest in shellfish research, among them the small-scale (commensurate with fertilization neighborhoods) changes in concentration profiles following density-dependent harvesting, which responds to concentration as perceived at much larger scales. This addresses an important consideration in the management of sedentary broadcast spawners, viz. the selective removal by the fishery of the densest patches of a stock, which are expected to contribute disproportionately to the population's progeny because of the high fertilization rate of individuals occurring within aggregations. The importance of this "Allee effect" (Botsford et al. 1993) was clearly recognized by early shellfish biologists (Belding 1910; Spärck 1927; Thorson 1946) and was involved in a hypothesis advanced by Gross and Smyth (1946) to explain the demise of natural oyster beds in Great Britain: "[following density-dependent depletion] the position of the remaining population probably deteriorated as a result of the scattering of the breeding oysters due to the thinning out of the stock, for this must have greatly reduced the chances of successful fertilization."

The related notions of concentration and crowding evolved independently in a variety of contexts and have received different names in ecology, resource science, spatial statistics, demography, and epidemiology. However, they can be construed into a unified research program for the study of density-dependent processes from the perspective of the individuals or phenomena under observation, as opposed to the perspective of the observer (epitomized by measurement of density with quadrats of arbitrary size placed at arbitrary locations).

Problems: focus on scale

Ultimately, sensible decisions in the analysis and modeling of concentration follow from an appropriate selection of scales for observation and representation. Concentration analysis both allows and forces focus on scale, the latter being a subject much talked about in current ecological and managerial literature, yet often relegated to the limbo of soft commentary. Examples of some problematic situations are discussed below.

Poor low-scale resolution

Abundance data routinely collected during surveys of benthic shellfish stocks (trawl and dredge surveys, quadrat counts) are generally not informative about concentration at low scales. However, it is at these small scales that many processes governing stock dynamics are operationally effective. Examples of the latter include competition for food, predator–prey interactions, fertilization, etc. While low-scale resolution is easily understood as a problem, the assessment of concentration at low scales for real life stocks is a difficult proposition. One possibility would be disjunct windowing, i.e., mapping the position of individuals within a collection of nonadjacent quadrats, followed by analysis of concentration (e.g., concentration profiles) at scales smaller than those defined by quadrat size. A very informative type of survey could consist of a combination of a systematic survey sample of the study region, each sampling unit consisting of a map of the location of the individuals within a quadrat or stripe (e.g., as derived from a photograph or TV recording in the case of epibenthic resources). Quadrat counts across the region could be analyzed with geostatistical methods (see Warren 1998), while mapped patterns within each quadrat could be used in the analysis of concentration.

Poor low-scale resolution for pattern across the entire region of interest

Stockesbury and Himmelman (1993) observed small-scale pattern over an entire shellfish ground through several disjoint windows. The authors collected mapped data for giant scallops in two unharvested beds from the Baie des Chaleurs, Gulf of St. Lawrence (eastern Canada). A total of thirty-seven 9-m^2 square plots (disjoint windows) were marked at haphazardly selected locations; location of all scallops within each plot was recorded. Pattern could be assessed within each of the windows using some of the methods outlined earlier (e.g., the *K*-function). Analyses of pattern within the disjoint windows, however, could not be simply combined to provide a consolidated description of pattern across the entire region of interest. If, for example, quadrats were partitioned and partitions were subsequently combined across quadrats into some form of agglomerative analysis (e.g., Iwao 1972; Stockesbury and Himmelman 1993), assumptions would be introduced about stationarity of the underlaying stochastic point process across the whole region under study. Such assumptions would be hardly acceptable if made explicit (notice, however, that concentration profiles are free of stationarity assumptions). Mapped pattern data observed through relatively small disjunct windows should not be combined, prima facie, to make statements about small-scale pattern across the entire region surveyed.

Poor large-scale resolution

Problems of large-scale resolution, even though pervasive, are often less apparent. Consider, for example, the exercise on concentration profiles and sequential depletion of a sedentary resource introduced earlier. Frequency distributions of density reported for benthic invertebrates (usually estimated with a sample of quadrats) are most often well described by the negative binomial distribution (type C in Fig. 8). Under sequential depletion, such distributions lead to hyperdepletion (at least at the beginning of the harvest), as initial effort goes to the small areas where much of the resource is concentrated. This entails an apparent paradox: while the overwhelming majority of investigated spatial patterns would suggest hyperdepletion as the most frequent type of CPUE trajectory, staged hyperstability seems to prevail in shellfish fisheries. Under hyperstable depletion (types D and E in Fig. 8), CPUE drops at a much slower rate than abundance as fishermen move from dense patch to dense patch, so that the resource may become seriously depleted well before it is apparent in the CPUE trajectory; in staged hyperstability beds are depleted one after another, in each case following a hyperstable pattern. Intriguingly, patterns of spatial distribution associated with hyperdepletion under strict sequential allocation (e.g., regular or quasi-regular patterns) are rarely (if ever) reported by shellfish biologists. The reason may be a mismatch between the spatial scale used in the assessment of pattern and the operational scales of the fishing process. Patterns of spatial distribution of benthic shellfish are generally measured, described, and modeled using small units (e.g., 1-m^2 quadrats). The fishing process (including perception and harvesting) operates at larger scales, even in the case of individual hand harvesters. A hypothetical diver fishing an epibenthic species will never harvest small (say, 1-m^2) patches in a perfect sequence of descending density, even in the unlikely case that he had a complete map of the fine-scale distribution of the resource. Pairs of small patches that rank next to each other in density will be variably (sometimes widely) separated in space. Since moving from patch to patch is costly, harvesting will depart from a simple response to small-scale pattern. If entire beds (rather than small patches) are the subject of a sequential harvest and each bed is harvested systematically (as is often the case in diving shellfisheries), the result is staged hyperstability, no matter what the internal spatial pattern within each bed looks like. Surveys of spatial pattern at a small scale may be of little significance to the understanding of processes with a larger operational scale.

The distribution and pattern of concentration has generally been investigated at small scales by ecologists and at large scales by fishery scientists. Yet, both are meaningful in the assessment of stocks of sedentary invertebrates when the impact of harvesting tactics on resource renewal is a matter of serious concern (i.e., when the shellfishery is not managed as a mining operation).

Acknowledgments

Kevin Stokesbury, Bruce MacDonald, and Chris Badjik kindly made available mapped data of giant scallops from eastern Canada. Glen Jamieson created the opportunity for the communication and discussion of our ideas. We acknowledge comments by an anonymous referee and Jon Schnute for a particularly thorough review. The three of us acknowledge each other for being patient and willing to compromise.

References

Addicott, J.F., Aho, J.M., Antolin, M.F., Padilla, D.K., Richardson, J.S., and Soluk, D.A. 1987. Ecological neighborhoods: scaling environmental patterns. Oikos, **49**: 340–346.

Andersen, M. 1992. Spatial analysis of two species interactions. Oecologia, **91**: 134–140.

Anderson, D.J., and Kendziorek, M. 1982. Spacing patterns in terebellid polychaetes. J. Exp. Mar. Biol. Ecol. **58**: 193–205.

Andrew, N.L., and Mapstone, B.D. 1987. Sampling and the description of spatial pattern in marine ecology. Oceanogr. Mar. Biol. Ann. Rev. **25**: 39–90.

Antonovics, J., and Levin, D.A. 1980. The ecological and genetic consequences of density-dependent regulation in plants. Ann. Rev. Ecol. Syst. **11**: 411–452.

Baron, J., and Clavier, J. 1992. Estimation of soft bottom intertidal bivalve stocks on the south-west coast of New Caledonia. Aquat. Living Resour. **5**: 99–105.

Bartlett, M.S. 1975. The statistical analysis of spatial pattern. Chapman and Hall, London, U.K. 90 p.

Belding, D.L. 1910. A report upon the scallop fishery of Massachusetts; including the habits, life history of *Pecten irradians*, its rate of growth and other facts of economic value. The Commonwealth of Massachusetts. 150 p.

Botsford, L.W., Quinn, J.F., Wing, S.R., and Brittnacher, J.G. 1993. Rotating spatial harvest of a benthic invertebrate, the red sea urchin, *Strongylocentrtus franciscanus*. *In* Management strategies for exploited fish populations. *Edited by* G. Kruse, D.M. Eggers, R.J. Marasco, C. Pautzke, and T.J. Quinn II. Alaska Sea Grant College Program, Fairbanks, Alaska. pp. 409–428.

Branch, G.M. 1975. Mechanisms reducing intraspecific competition in *Patella* spp.: migration, differentiation and territorial behaviour. J. Anim. Ecol. **44**: 575–600.

Clark, C.W. 1982. Concentration profiles and the production and management of marine fisheries. *In* Economic theory of natural resources. *Edited by* W. Eichlorn. Physica-Verlag, Wurzburg-Wien. pp. 97–112.

Clark, C.W. 1985. Bioeconomics modelling and fisheries management. John Wiley & Sons, Inc., New York, N.Y. 291 p.

Clavier, J., and Richard, O. 1986. Estimation du stock naturel d'ormeaux (*Haliotis tuberculata*) dans la région de Saint-Malo. Rev. Trav. Inst. Pêches Marit. **48**: 95–105.

Connell, J.H. 1955. Spatial distribution of two species of clams, *Mya arenaria* L. and *Petricola pholadiformis* Lamark, in an intertidal area. Rep. Fish. Inv. Mass. **8**: 15–25.

Cressie, N. 1991. Statistics for spatial data. John Wiley & Sons, Inc., New York, N.Y. 900 p.

Crisp, D.J. 1961. Territorial behavior in barnacle settlement. J. Exp. Biol. **38**: 429–446.

Diggle, P.J. 1983. Statistical analyses of spatial point patterns. Academic Press, London, U.K. 148 p.

Elliott, J.M. 1977. Some methods for the statistical analysis of samples of benthic invertebrates. 2nd. edition. Freshwater Biological Association, Ambleside, Westmorland, U.K. Scientific Publication 25. 160 p.

Folt, C.L., and Schulze, P.C. 1993. Spatial patchiness, individual performance and predator impacts. Oikos, **68**: 560–566.

Folt, C.L., Schulze, P.C., and Baumgartner, K. 1993. Characterizing a zooplankton neighborhood: small-scale patterns of association and abundance. Freshwater Biol. **30**: 289–300.

Gross, F., and Smyth, J.C. 1946. The decline of oyster populations. Nature (London), **157**: 540–542.

Hall, M.A. 1983. A spatial approach to the population dynamics of the Manila clam (*Tapes philippinarum*). Ph.D disseration, University of Washington, Seattle, Wash. 241 p.

Harkness, R.D., and Isham, V. 1983. A bivariate spatial point pattern of ant's nests. Appl. Statist. **32**: 293–303.

Hilborn, R., and Walters, C.J. 1992. Quantitative fisheries stock assessment. Chapman & Hall, New York, N.Y. 570 p.

Holme, N.A. 1950. Population-dispersion in *Tellina tenuis* Da Costa. J. Mar. Biol. Ass. U.K. **29**: 267–280.

Hurlbert, S.H. 1990. Spatial distribution of the montane unicorn. Oikos, **58**: 257–271.

Ikenouye, H. 1968. An analysis of the spatial distribution of a barnacle, *Tetraclita squamosa japonica* Pilsbry. J. Tokyo Univ. Fish. **54**: 99–105.

Iwao, S. 1972. Application of the *m*-m* method to the analysis of spatial patterns by changing the quadrat size. Res. Popul. Ecol. **14**: 97–128.

Iwao, S. 1976. A note on the related concepts 'mean crowding' and 'mean concentration.' Res. Popul. Ecol. **17**: 240–242.

Kenkel, N.C. 1988. Pattern of self-thinning in jack pine: testing the random mortality hypothesis. Ecology, **69**: 1017–1024.

Kooijman, S.A.L.M. 1979*a*. Inference about dispersal patterns. Acta Biotheor. **28**: 149–189.

Kooijman, S.A.L.M. 1979*b*. The description of point patterns. *In* Spatial and temporal analysis in ecology. *Edited by* R.M. Cormack and J.K. Ord. Internat. Co-op. Publ. House, Fairland, Maryland. pp. 305–331.

Levin, L.A. 1981. Dispersion, feeding behavior and competition in two spionid polychaetes. J. Mar. Res. **39**: 99–117.

Levinton, J. 1972. Spatial distribution of *Nucula proxima* Say (PROTOBRANCHIA): an experimental approach. Biol. Bull. **143**: 175–183.

Lewontin, R.C., and Levins, R. 1989. On the characterization of density and resource availability. Am. Nat. **134**: 513–524.

Lloyd, M. 1967. 'Mean crowding'. J. Anim. Ecol. **36**: 1–30.

Lotwick, H.W., and Silverman, B.W. 1982. Methods of analyzing spatial processes of several types of points. J. R. Statist. Soc. (B) **44**: 406–413.

MacDonald, B.A., and Bajdik, C.D. 1992. Orientation and distribution of individual *Placopecten magellanicus* (Gmelin) in two natural populations with differing production. Can. J. Fish. Aquat. Sci. **49**: 2086–2092.

Matérn, B. 1960. Spatial variation. Meddelanden fran Statens Skogsforskningsinstitut, 49 (5) [Second edition published 1986, Lecture Notes in Statistics 36, Springer, New York].

Morisita, M. 1959. Measuring of the dispersion of individuals and analysis of the distributional patterns. Mem. Faculty Sci., Kyushu Univ. (F: Biol.) **2**: 215–235.

Morisita, M. 1962. *I*-index, a measure of dispersion of individuals. Res. Popul. Ecol. **4**: 1–7.

Morisita, M. 1971. Composition of the *I*-index. Res. Popul. Ecol. **13**: 1–27.

Orensanz, J.M. 1986. Size, environment, and density: the regulation of a scallop stock and its management implications. Can. Spec. Publ. Fish. Aquat. Sci. 92. pp. 195–227.

Pielou, E.C. 1974. Population and community ecology. Principles and methods. Gordon & Breach Science Publ., New York, N.Y. 420 p.

Rawson, C.B. 1980. Scales of the spatial patterns of the clams *Protothaca staminea* (VENERIDAE) and *Macoma nasuta* (TELLINIDAE) in Garrison Bay. M.Sc. thesis, University of Washington, Seattle, Wash.

Rebertus, A.J., Williamson, G.B., and Moser, E.B. 1989. Fire-induced changes in *Quercus laevis* spatial pattern in Florida sandhills. J. Ecol. **77**: 638–650.

Ripley, B. 1976. The second-order analysis of stationary point processes. J. Appl. Probab. **13**: 255–266.

Ripley, B.D. 1977. Modelling spatial patterns. J. Roy. Stat. Soc. (B) **39**: 172–212. (Includes a multi-authored Discussion section.)

Ripley, B.D. 1981. Spatial statistics. John Wiley & Sons, Inc., New York, N.Y. 252 p.

Ripley, B.D. 1987. Spatial point pattern analysis in ecology. *In* Developments in numerical ecology. *Edited by* P. and L. Legendre. NATO ASI Ser. G14, Springer-Verlag, Berlin and Heidelberg. pp. 407–429.

Silverman, B.W. 1986. Density estimation. Chapman & Hall, New York, N.Y. 175 p.

Spärck, R. 1927. Studies on the biology of the oyster (*Ostrea edulis*). Rep. Danish Biol. Stn. 23.

Stokesbury, K.D.E., and Himmelman, J.H. 1993. Spatial distribution of the giant scallop *Placopecten magellanicus* in unharvested beds in the Baie des Chaleurs, Québec. Mar. Ecol. Prog. Ser. **96**: 159–168.

Thorson, G. 1946. Reproduction and larval development of Danish marine bottom invertebrates, with special reference to the planktonic larvae in the Sound (Öresund). Medd. Komm. Danmarks Fiskeri- og Havunders. (Plankton) 4 (1).

Underwood, A.J. 1976. Nearest neighbour analysis of spatial dispersion of intertidal prosobranch gastropods within two substrata. Oecologia, **26**: 257–266.

Upton, G.J.G., and Fingleton, B. 1985. Spatial data analysis by example. Volume 1. Point pattern and quantitative data. John Wiley & Sons, New York, N.Y. 410 p.

Venables, W.N., and Ripley, B.D. 1994. Modern applied statistics with S-Plus. Springer-Verlag, New York, N.Y. 462 p.

Warren, W.G. 1998. Spatial analysis for marine populations: factors to be considered. *In* Proceedings of the North Pacific Symposium on Invertebrate Stock Assessment and Management. *Edited by* G.S. Jamieson and A. Campbell. Can. Spec. Publ. Fish. Aquat. Sci. 125. pp. 21–28.

Watanabe, N. 1988. A new proposal for measurement of the adjusted mean crowding through consideration of size variability in habitat units. Res. Popul. Ecol. **30**: 215–225.

Wethey, D.S. 1984. Spatial pattern in barnacle settlement: day to day changes during the settlement season. J. Mar. Biol. Ass. U.K. **64**: 687–698.

Worton, B.J. 1989. Kernel methods for estimating the utilization distribution in home-range studies. Ecology, **70**: 164–168.

Appendix. *Corrections for border effects in the computation of crowding*

Border effects can be corrected in a number of ways, among others:

(1) Computing the distances $\|s_i - s_j\|$ toroidally. The pattern observed in the (rectangular) study area is copied onto each of the eight adjacent rectangular areas; distances are then measured from each of the individuals contained in the original study area to individuals within the study area itself and its replicas.

(2) Defining a guard area around the study area, in which case

$$C_i(h) \equiv \sum_{j=1}^{N^+} I(\|s_i - s_j)\| \leq h)$$

where N^+ is the number of individuals in the study area A and a surrounding guard area of width $\geq h$.

(3) Ripley's (1976) edge-corrected estimator,

$$C_i(h) \equiv \sum_{j=1}^{N} \frac{I(\|s_i - s_j)\| < h)}{w(s_i, s_j)}$$

where $w(s_i, s_j)$ is the proportion of the circumference centered at s_i and passing through s_j that is inside the study region A; it is calculated as follows: (*i*) let $d = \|s_i - s_j\|$; (*ii*) if $d^2 \leq d_1 2 + d_2 2$, then

$$w(s_i, s_j) = 1 - \frac{\cos^{-1}\left(\frac{\min(d_1, d)}{d}\right) + \cos^{-1}\left(\frac{\min(d_2, d)}{d}\right)}{\pi}$$

(*iii*) If $d^2 > d_1 2 + d_2 2$, then

$$w(s_i, s_j) = \frac{3}{4} - \frac{1}{2\pi}\left[\cos^{-1}\left(\frac{d_1}{d}\right) + \cos^{-1}\left(\frac{d_2}{d}\right)\right]$$

where d_1 and d_2 are the distances to the nearest and second nearest sides of the rectangular boundary, respectively. This correction is based on the assumption that the region outside the boundary in the vicinity of the distance measurement has a spatial pattern similar to the nearby areas within the boundary (Diggle 1983; Cressie 1991). Diggle (1983) gives an analogous border correction for circular quadrats. We used Ripley's estimator in all calculations where a border correction was required.

Fertilization success in free-spawning marine invertebrates: review of the evidence and fisheries implications

Don R. Levitan and Mary A. Sewell

Abstract: We review the evidence of how the distribution, abundance, and behavior of spawning individuals influence fertilization in free-spawning marine invertebrates. Female fertilization success is highly variable, and a considerable part of this variation is attributable to the proximity and number of spawning males. Managers of harvested populations should therefore consider aspects of fertilization when making decisions regarding refuge formation, body size restrictions, and catch limits.

Résumé : Nous avons passé en revue les données sur la façon dont la distribution, l'abondance et le comportement des sujets en état de reproduction influent sur la fécondation chez les invertébrés marins à reproduction libre. Le succès de fécondation des femelles est hautement variable et une partie considérable de cette variation est attribuable à la proximité et au nombre de mâles en reproduction. Par conséquent, les gestionnaires des populations exploitées devraient prendre en considération les aspects liés à la fécondation lorsqu'ils prennent des décisions relatives à la création de refuges et à l'établissement des limites de taille corporelle et des limites de prises. [Traduit par la Rédaction]

Introduction

When population size changes as a result of human impact or natural causes, at least two reproductive consequences are possible. The most obvious is a change in offspring production at the population level due to variation in the number of sexually mature individuals contributing to the recruitment pool. A less obvious, but potentially important, effect is a change in per capita offspring production due to variation in the reproductive efficiency of individuals. This review focuses on the latter effect, and specifically on how variation in population parameters can influence fertilization success and reproductive efficiency in free-spawning marine invertebrates.

Reduction in population size is generally associated with a reduction in intraspecific competition (e.g., Begon and Mortimer 1981). Reduced intraspecific competition, in turn, is often associated with increased survivorship, growth rate, and gamete production (Sutherland 1970; Sebens 1982; Levitan 1989). These factors, all else being equal, result in increased reproductive efficiency and increased per capita offspring production. But all else is rarely equal, and at least one factor, fertilization of eggs, appears to contribute to reproductive efficiency counter to these other density-dependent processes (Levitan 1991, 1995; Levitan and Petersen 1995).

D.R. Levitan. Department of Biological Science, Florida State University, Tallahassee, FL 32306-2043, U.S.A.
e-mail: levitan@bio.fsu.edu

M.A. Sewell. Department of Larval Ecology, Harbor Branch Oceanographic Institution, 5600 Old Dixie Highway, Port Pierce, FL 34946, U.S.A.

Correct citation: Levitan, D.R., and Sewell, M.A. 1998. Fertilization success in free-spawning marine invertebrates: review of the evidence and fisheries implications. *In* Proceedings of the North Pacific Symposium on Invertebrate Stock Assessment and Management. *Edited by* G.S. Jamieson and A. Campbell. Can. Spec. Publ. Fish. Aquat. Sci. 125. pp. 159–164.

Free-spawning, or the release of sperm and often eggs into the environment, is a common mechanism of reproduction in marine organisms (Giese and Kanatani 1987; Strathmann 1990). The probability of successful fertilization in this mating strategy depends on many factors, including the number and distribution of spawners (Pennington 1985; Yund 1990; Levitan 1991; Brazeau and Lasker 1992; Levitan et al. 1992; Babcock et al. 1994; Levitan and Young 1995), the timing of gamete release (Giese and Kanatani 1987; Levitan 1988*a*; Pearse et al. 1988; Pearse 1990; Babcock et al. 1992; Sewell and Levitan 1992), the ways in which released gametes are dispersed and interact with water currents and turbulence (Pennington 1985; Denny 1988; Denny and Shibata 1989; Epel 1991; Grosberg 1991; Havenhand 1991; Denny et al. 1992; Levitan et al. 1992; Young et al. 1992; Babcock et al. 1994; Benzie et al. 1994; Thomas 1994*a*, 1994*b*; Levitan 1995; Levitan and Young 1995; Mead and Denny 1995), and finally properties of the gametes themselves (Vogel et al. 1982; Levitan et al. 1991; Levitan 1993, 1995; Benzie and Dixon 1994; Thomas 1994*a*, 1994*b*). In this paper we will restrict our discussion to how population parameters such as distribution, abundance, and synchrony of spawning animals can influence rates of fertilization, with the idea that considering these parameters may provide information for guiding management decisions for invertebrate fisheries. A more general review of the ecological and evolutionary consequences of variation in sperm limitation and female fertilization success can be found in Levitan (1995) and Levitan and Petersen (1995).

In keeping with the tradition of viewing demographic processes from a female perspective (e.g., Deevey 1947), we consider female fertilization success, the proportion of a female's eggs that are fertilized. Although considering male fertilization success is important from a genetic and evolutionary perspective (e.g., Grosberg 1991; Yund and McCartney 1994; Levitan and Petersen 1995; Yund 1995), we address the straightforward consequences of female fertilization success for per-female zygote production, because this is more likely to be of immediate use for fisheries considerations.

It is worth noting that some of the early discussion of the importance of population parameters to successful fertilization was by scientists working with exploited species. Belding (1910), Sparck (1927), and Gross and Smyth (1946) all suggested that fertilization may be unlikely in sparse bivalve populations. More recently several fisheries reviews (Orensanz 1986; Orensanz et al. 1991; Jamieson 1993; Quinn et al. 1993) have considered fertilization in their discussions of fisheries management, but none of these studies examine in detail the evidence concerning whether, and to what degree, population parameters influence fertilization success.

Evidence that fertilization is sensitive to population parameters

Although the idea that sperm limitation might influence reproductive success and population dynamics has long been recognized (Belding 1910; Sparck 1927; Allee 1931; Mortensen 1938; Gross and Smyth 1946), none of these early papers provided field data to support these notions. All the inferences were based on laboratory work that indicated the sensitivity of fertilization to sperm concentration (e.g., Lillie 1915).

Over the past decade, field studies have been conducted, providing quantitative measurements of variation in female fertilization success. These studies, on free-spawning invertebrates and fish, have indicated that female fertilization success is variable and at times can be quite low (reviewed in Levitan 1995). This evidence comes from field experiments, field surveys, and theoretical models. Each approach has its advantages and disadvantages.

Experimental manipulations have the benefit of a high degree of control, allowing investigators to tease apart various factors influencing experimental outcomes. However, this same degree of control restricts the range of variables, which may or may not reflect natural conditions. In addition, experiments may introduce artifacts (such as the way in which eggs are manipulated, Levitan 1995) that interfere with the process of fertilization.

Natural surveys reflect natural conditions but lack controls, making it difficult to isolate factors responsible for variation in fertilization success. Natural spawns are also generally unpredictable, making data collection difficult. Those cases where spawning is more predictable (e.g., reef fish, Petersen 1991, Petersen et al. 1992; mass coral spawnings, Harrison et al. 1984, Babcock et al. 1986, 1994, Oliver and Babcock 1992; and gorgonians, Brazeau and Lasker 1992, Lasker et al. 1996), and synchrony is high, may provide a biased estimate for more unpredictable and less synchronous taxa.

Theoretical models provide an important tool for exploring how various parameters, including the level of water flow and turbulence (Denny 1988; Denny and Shibata 1989; Denny et al. 1992; Benzie et al. 1994; Levitan and Young 1995), the distribution and abundance of spawning animals (Denny and Shibata 1989; Denny et al. 1992; Young et al. 1992; Babcock et al. 1994; Benzie et al. 1994; Levitan and Young 1995), the rate of gamete release (Denny and Shibata 1989; Babcock et al. 1994; Levitan and Young 1995), and the attributes of gametes (Vogel et al. 1982; Levitan et al. 1991; Levitan 1993, 1996; Podolsky and Strathmann 1996), may influence rates of fertilization. These models also provide a mechanism for investigating how population parameters may influence levels of fertilization beyond the range possible in field experiments or noted in surveys (Levitan and Young 1995), but for these models to have true predictive ability, the assumptions must be properly tested (Levitan 1995), a process still in its infancy.

Below we discuss a variety of case studies that employ one or a combination of experiments, surveys, or theoretical modelling to address how population parameters can influence fertilization success.

Effect of population size and density on female fertilization success

In 1985, Pennington conducted the first field experiment on female fertilization using the sea urchin *Strongylocentrotus droebachiensis.* He demonstrated that fertilization decreased dramatically with distance from a single spawning male (<10% at distances >1 m). Percent fertilization was higher with three spawning males but was still negligible at a distance of 5 m. Yund (1990) did a similar distance experiment using the colonial hydrozoan *Hydractinia echinata,* which lives on gastropod shells occupied by hermit crabs. He found a similar result, negligible fertilization at distances greater than 5 m. Babcock et al. (1994) found a very different result in the sea star *Acanthaster planci.* They found a slower decline in fertilization with distance, 5% at 100 m from a spawning male. They attributed the higher fertilization to the higher level of sperm production of this large sea star (however see Benzie and Dixon 1994 for an explanation based on gamete differences). Levitan (1991), in addition to a distance experiment which again demonstrated a rapid decline in fertilization with distance, manipulated population density in the sea urchin *Diadema antillarum.* He found that increased density resulted in increased female fertilization success, from 7 to 40% over a density change of 1–16 males·m^{-2}. These studies indicate that manipulating male distance, number, or density can influence levels of female fertilization success.

Levitan et al. (1992) conducted an experiment that looked at levels of aggregation and population size simultaneously in the sea urchin *Strongylocentrotus franciscanus.* These are the two factors that determine population density (a measure that changes with quadrat size). Males were simulated by syringes from which sperm was released, and females were simulated by eggs placed in sperm-permeable containers. "Males" and "females," either 2 or 8 of each, were placed at one of two distances from each other (0.5 or 2 m) in a grid. The investigators found that fertilization was sensitive to both nearest-neighbor distances and population size. Decreasing the distance between individuals resulted in a mean additional 15% of eggs fertilized. Similarly, increasing the size of the population resulted in a mean additional 12% of eggs fertilized. These results suggest that either low abundance or a dispersed distribution can lead to decreased reproductive success, but because population size was shown to be important, and only relatively few individuals were manipulated, relative to natural populations, the question of whether sperm would be limiting in large spawning populations remains open.

To address this issue, Levitan and Young (1995) first tested and then used models developed by Denny (Denny 1988; Denny and Shibata 1989) and Vogel et al. (1982). Denny's model predicts gamete concentration at any location downstream from a sperm source, whereas Vogel et al.'s model predicts the proportion of eggs fertilized given gamete concentrations, the length of time during which eggs and sperm are in the same water mass, and species-specific gamete traits such as egg size, egg receptiveness to sperm, sperm velocity, and sperm longevity. Levitan and Young (1995) measured the appropriate biological and physical parameters for the sea biscuit *Clypeaster rosaceus* spawning in shallow tropical seas and tested the predictions of these models using small-scale experiments on in situ dye diffusion and egg fertilization. They then applied these models to predict female fertilization success when population size varied from pair spawns to mass spawnings involving hundreds of thousands of individuals over hundreds of square metres. The prediction of these simulations, based on the empirical observations of spawning rates and flow conditions, was that complete (100%) fertilization success would not be attained unless hundreds of thousands of animals spawned simultaneously.

The problem with these experiments and simulations is that they may or may not accurately reflect natural spawns. Although data from natural spawns may not be tightly controlled, they can be valuable in confirming the results and predictions of the above studies.

Natural spawns at high and low population density

Empirical data on fertilization success from natural spawns have rarely been documented, partly because of the unpredictable nature of spawning in most species. Casual observations of spawnings are usually qualitative and do not include estimates of fertilization success.

Brazeau and Lasker (1992) were able to circumvent this problem by working with the sessile brooding gorgonian *Briareum asbestinum*, because the developing embryos remain attached to the female after fertilization. They compared numbers of mature eggs with numbers of emerging larvae as an estimate of fertilization efficiency. This soft coral lives at low population density, and their estimates of fertilization success ranged from <1 to 6% fertilized over a three-year study period. They also noted a weak correlation between the density of nearby males and fertilization success. This correlation was supported by experimental manipulation of male colonies; females placed closer to male colonies produced more larvae than did more distant females.

Lasker and his colleagues (Lasker et al. 1996) measured fertilization in the Caribbean gorgonians *Plexaura kuna* and *Pseudoplexaura porosa* by collecting broadcast spawned eggs in syringes within minutes of release from female colonies. They found female fertilization success to be highly variable, ranging from 0 to 100%. Fertilization varied month to month (peaking in the middle of the three months of spawning within each year), from day to day (peaking in the middle days of spawning within each month), and even among closely spaced and timed collections in the same evening (correlated with unpredictable patterns of sperm concentration). They demonstrated that most of this enormous variation was attributable to sperm limitation, as opposed to egg quality, by placing field-collected eggs in sperm-enriched seawater and documenting an average 72% increase in the percentage of eggs fertilized.

An example of spawning at high density in a sedentary sea cucumber *Cucumaria miniata* was described by Sewell and Levitan (1992). Average density in the studied population was $46 \cdot m^{-2}$, and population size was in the thousands. Spawning was synchronous and occurred primarily at low slack tide. Fertilization of released egg pellets ranged from 1 to 100% with a mean of 92%. In this large, high-density, and aggregated (mean nearest neighbor distance 8 cm) population, fertilization approached 100%.

Reports of natural spawnings in a variety of species from the Great Barrier Reef also suggest that spawning synchrony, population size, and distance from spawning males influence levels of fertilization (Babcock and Mundy 1992; Babcock et al. 1992). Babcock and Mundy (1992) found relatively high fertilization (83%) during the peak of a major spawning event in *Acanthaster planci*, but lower levels (23%) at the end of that spawning event and also during the peak of another minor spawning event. Oliver and Babcock (1992) reported a higher level of fertilization when many colonies of the coral *Montipora digita* were observed to spawn (mean of 53%) than during a smaller spawning event when no colonies were directly observed to spawn (36%). Babcock et al. (1992) noted high fertilization (73–96%) in female sea cucumbers spawning within 1 m of one or more males. In their observation of a female 20–40 metres from several spawning males, fertilization was near zero (0–2%).

These natural observations, in at least a qualitative sense, confirm the findings of experimental manipulations and theoretical models. When organisms are at high density, large population size, and close proximity and spawn synchronously, external fertilization is a highly efficient mechanism of producing a large number of offspring. If one or more of these elements are less than ideal, fertilization efficiency declines accordingly.

Case study of how reduced population size may influence fertilization dynamics and reproductive output

In the winter of 1983–1984, a mass mortality of the sea urchin *Diadema antillarum* reduced population density by 95–99% throughout the Caribbean (Lessios 1988). Before this mortality event, *Diadema* was abundant (up to $100 \cdot m^{-2}$, Hay 1984) in some areas of the Caribbean. Levitan (1988*a*, 1989, 1991) looked at the consequences of this population reduction on per capita reproduction off St. John, U.S. Virgin Islands. At his study sites, population density declined from an average of $15–0.01 \cdot m^{-2}$ and remained low for at least 10 years following this event (Levitan 1988*b*; Karlson and Levitan 1990; D.R. Levitan, unpublished data). After the mass mortality, algal abundance increased (3000%) and concomitantly individual urchin body size increased (mean 30–70 mm test diameter, Levitan 1988*b*). This increase in test diameter was associated with a 21-fold increase in per capita egg production (Levitan 1988*a*). It was predicted that *Diadema* would recover quickly from this mortality event, because of its high fecundity (Hughes et al. 1986).

This species spawns throughout the year in the more

tropical parts of the Caribbean (Randall et al. 1964). Before the mass mortality, small group spawnings were observed (Randall et al. 1964). Spawning synchrony and patterns of aggregation were measured after the mass mortality. Spawning synchrony was weakly related to the lunar cycle (Levitan 1988*a*). The peak in spawning was near the new moon, but individuals were seen to spawn, and could be induced to spawn, for three out of the four weeks of the lunar cycle. Although individuals were always slightly aggregated, no higher levels of aggregation were associated with the lunar cycle or an individual's ability to spawn (Levitan 1988*a*). Observations of spawning during population censuses indicated that an average of only 5% of individuals spawned simultaneously. At postmortality population densities, the density of spawning individuals would therefore be very low.

Fertilization success was measured as a function of both male body size and male population density (Levitan 1991). Data were also collected on the relationship between body size and population density (Levitan 1988*c*) and on that between gamete production and body size (Levitan 1988*a*). From these data, average per capita zygote production for average-size females was calculated as a function of population density (Levitan 1991). The result indicated that although individual body size and gamete production increased dramatically with decrease in population density, per capita zygote production did not increase because female fertilization success was much reduced. Female egg production and fertilization success were inversely related as a function of population density. This result demonstrated how estimates of offspring production based on body size or gamete production can be misleading when fertilization success is not considered.

The above calculation did not include aspects of spawning synchrony, and some evidence suggests that spawning synchrony may have been higher before the mass mortality (e.g., Randall et al. 1964). This previously abundant species may have relied on localized spawning cues from conspecifics (e.g., Starr et al. 1990; Unger and Lott 1994), and without conspecifics, these cues are absent or below some critical threshold (Levitan 1988, 1991; Levitan and Petersen 1995).

The low rate, or lack, of recovery of the *Diadema* population after the mass mortality is likely to be due to the combination of low population size contributing to the recruitment pool and a drastic reduction in fertilization success, which has negated the tremendous increase in per capita egg production (Levitan 1991).

Fisheries implications

The above discussion makes clear that reducing populations of free-spawning marine invertebrates reduces not only the number of females contributing to a recruitment pool but also the fertilization efficiency of those females. The reviewed studies indicate, first, that detailed information on spawning behavior, demography, and environmental conditions are needed before we can make predictions concerning fertilization efficiency. Second, because all the studies indicated some variance in fertilization success, and thus some degree of sperm limitation, any reduction in the number of males should have some influence on fertilization efficiency and per-female offspring production. The issue for fisheries biologists will be to determine what are acceptable losses in fertilization efficiency, in light of other management considerations.

The effects of fishing pressure on fertilization may be most dramatic in situations where males are preferentially selected either because of restrictions on females or because males are larger, as in the case of protogyny. The red king crab, an example of the former, supported a male-only fishery in Bristol Bay, in the Bering Sea, until the fishery collapsed (Reeves 1982; Jamieson 1993). An example of the latter is found in finfish fisheries, where the protogynous gag grouper have been fished to the point where sex ratios have been skewed from 17% male to 1% male over the last 20 years (Coleman et al. 1996). Although the population dynamics in these two species have changed, it has yet to be demonstrated that per capita offspring production has been affected by male or sperm limitation.

Levitan et al. (1992) argued that because sperm limitation operates over a large range of population sizes, the Allee effect, i.e., the effect of reduced population growth at low population size (Allee 1931), could influence population growth in free-spawning invertebrates at a higher population threshold than previously considered. Quinn et al. (1993) modelled the Allee effect in an exploited sea urchin population and suggested that establishing high-density refuges for free-spawning invertebrates is one way to maintain sustainable catches at high levels of harvesting effort. Such measures are particularly important when the threshold of the Allee effect is unknown or difficult to estimate or restrictions on catch are hard to control (Quinn et al. 1993).

A difficult, but important, task is establishing which populations are important sources of offspring and which are likely sinks for recruitment. Obviously, if possible, source populations should be maintained as refuges while sinks are opened to harvesting (Levitan 1995). As data on oceanographic current transport of larvae become available (e.g., Ebert and Russell 1988; Roughgarden et al. 1988; Wing et al. 1995), and as molecular techniques allow for identification of the origin of recruits (Grosberg and Levitan 1993), this task will become more tractable.

In conclusion, sperm is often limiting in free spawning invertebrates, and population parameters are one important source of variation in female fertilization success. Managers of harvested populations should consider aspects of fertilization and the ways population size, distribution, and structure can influence per-female offspring production. Finally, although females are clearly important in their role as egg producers, males and their impact on the fertilization of eggs should not be ignored in management decisions.

Acknowledgments

We thank G. Jamieson for his kind invitation to this symposium. This work was supported by a grant from the National Science Foundation to DRL.

References

Allee, W.C. 1931. Animal aggregations. A study in general sociology. University of Chicago Press, Chicago, Ill.

Babcock, R.C., and Mundy, C.N. 1992. Reproductive biology, spawning and field fertilization rates of *Acanthaster planci.* Aust. J. Mar. Freshwater Res. **43**: 525–534.

Babcock, R.C., Bull, G.D., Harrison, P.L., Heyward, A.J., Oliver, J.K., Wallace, C.C., and Willis., B.L. 1986. Synchronous spawning of 105 scleractinian coral species on the Great Barrier Reef. Mar. Biol. (Berlin), **90**: 379–394.

Babcock, R., Mundy, C., Kessing, J., and Oliver, J. 1992. Predictable and unpredictable spawning events: *in situ* behavioural data from free-spawning coral reef invertebrates. Invertebr. Reprod. Dev. **22**: 213–228.

Babcock, R.C., Mundy, C.N., and Whitehead, D. 1994. Sperm diffusion models and in situ confirmation of long distance fertilization in the free-spawning asteroid *Acanthaster planci.* Biol. Bull. (Woods Hole), **186**: 17–28.

Begon, M., and Mortimer, M. 1981. Population ecology. A unified study of animals and plants. Sinauer Associates, Inc., Sunderland, Mass.

Belding, D.L. 1910. A report upon the scallop fishery of Massachusetts; including the habits, life history of *Pecten irradians,* its rate of growth and other facts of economic value. The Commonwealth of Massachusetts.

Benzie, J.A.H., and Dixon, P. 1994. The effects of sperm concentration, sperm:egg ratio, and gamete age on fertilization success in the crown-of-thorns starfish (*Acanthaster planci*) in the laboratory. Biol. Bull. (Woods Hole), **186**: 139–152.

Benzie, J.A.H., Black, K.P., Moran, P.J., and Dixon, P. 1994. Small scale dispersion of eggs and sperm of the crown-of-thorns starfish (*Acanthaster planci*) in a shallow reef habitat. Biol. Bull. (Woods Hole), **186**: 53–167.

Brazeau, D.A., and Lasker, H.R. 1992. Reproductive success in the Caribbean octocoral *Briareum asbestinum.* Mar. Biol. (Berlin), **114**: 157–163.

Coleman, F.C., Koenig, C.C., and Collins, L.A. 1996. Reproductive styles of shallow-water groupers (Pisces: Serranidae) in the eastern Gulf of Mexico and the consequences of fishing spawning aggregations. Environ. Biol. Fish. **47**: 129–141.

Denny, M.W. 1988. Biology and mechanics of the wave-swept environment. Princeton University Press, Princeton, N. J.

Denny, M.W., and Shibata, M.F. 1989. Consequences of surf-zone turbulence for settlement and external fertilization. Am. Nat. **134**: 859–889.

Denny, M.W., Dairiki, J., and Distefano, S. 1992. Biological consequences of topography on wave-swept rocky shores: I. Enhancement of external fertilization. Biol. Bull. (Woods Hole), **183**: 220–232.

Deevey, E.S., Jr. 1947. Life tables for natural populations of animals. Q. Rev. Biol. **22**: 283–314.

Ebert, T.A., and Russell, M.P. 1988. Latitudinal variation in size structure of the west coast purple sea urchin: a correlation with headlands. Limnol. Oceanogr. **33**: 286–294.

Epel, D. 1991. How successful is the fertilization process in the sea urchin egg. *In* Proceedings of the 7th International Echinoderm Conference, Atami 1990. *Edited by* T. Yanagisawa, I. Yasumasu, C. Oguro, N. Suzuki, and T. Motokawa. A.A. Balkema, Rotterdam, The Netherlands. pp. 51–54.

Giese, A.C., and Kanatani, H. 1987. Maturation and spawning. *In* Reproduction of marine invertebrates, vol. IX. Seeking unity in diversity. Blackwell Scientific/Boxwood Press, Palo Alto/Pacific Grove, Calif. pp. 252–329.

Grosberg, R.K. 1991. Sperm-mediated gene flow and the genetic structure of a population of the colonial ascidian *Botryllus schlosseri.* Evolution, **45**: 130–142.

Grosberg, R.K., and Levitan, D.R. 1993. For adults only? Supply side ecology and the history of larval ecology. Trends Ecol. Evol. **7**: 130–133.

Gross, F., and Smyth, J.C. 1946. The decline of oyster populations. Nature (London), **157**: 540–542.

Harrison, P.L., Babcock, R.C., Bull, G.D., Oliver, J.K., Wallace, C.C., and Willis, B.L. 1984. Mass spawning in tropical reef corals. Science (Washington, D.C.), **223:** 1186–1189.

Havenhand, J.N. 1991. Fertilisation and the potential for dispersal of gametes and larvae in the solitary ascidian *Ascidia mentula* Muller. Ophelia, **33**: 1–15.

Hay, M.E. 1984. Patterns of fish and urchin grazing on Caribbean coral reefs: are previous results typical? Ecology, **65**: 446–454.

Hughes, T.P., Reed, D.C., and Boyle, M.J. 1986. Herbivory on coral reefs: community structure following mass mortalities of sea urchins. J. Exp. Mar. Biol. Ecol. **113**: 39–59.

Jamieson, G.S. 1993. Marine invertebrate conservation: evaluation of fisheries over-exploitation concerns. Am. Zool. **33**: 551–567.

Karlson, R.H., and Levitan, D.R. 1990. Recruitment-limitation in open populations of *Diadema antillarum:* an evaluation. Oecologia, **82**: 40–44.

Lasker, H.R., Brazeau, D.A., Calderon, J., Coffroth, M.A., Coma, R., and Kim, K. 1996. In situ rates of fertilization among broadcast spawning gorgonian corals. Biol. Bull. (Woods Hole), **190**: 45–55.

Lessios, H.A. 1988. Mass mortality of *Diadema antillarum* in the Caribbean: what have we learned? Annu. Rev. Ecol. Syst. **19**: 371–393.

Levitan, D.R. 1988*a*. Asynchronous spawning and aggregative behavior in the sea urchin *Diadema antillarum* Philippi. *In* Proceedings of the Sixth International Echinoderm Conference. Echinoderm biology. *Edited by* R. Burke. A.A. Balkema Press, Rotterdam, The Netherlands. pp. 181–186.

Levitan, D.R. 1988*b*. Algal-urchin biomass responses following the mass mortality of the sea urchin *Diadema antillarum* Philippi at Saint John, U.S. Virgin Islands. J. Exp. Mar. Biol. Ecol. **119**: 167–178.

Levitan, D.R. 1988*c*. Density-dependent size regulation and negative growth in the sea urchin *Diadema antillarum* Philippi. Oecologia, **76**: 627–629.

Levitan, D.R. 1989. Density-dependent size regulation in *Diadema antillarum:* effects on fecundity and survivorship. Ecology, **70**: 1414–1424.

Levitan, D.R. 1991. Influence of body size and population density on fertilization success and reproductive output in a free-spawning invertebrate. Biol. Bull. (Woods Hole), **181**: 261–268.

Levitan, D.R. 1993. The importance of sperm limitation to the evolution of egg size in marine invertebrates. Am. Nat. **141**: 517–536.

Levitan, D.R. 1995. The ecology of fertilization in free-spawning invertebrates. *In* Ecology of marine invertebrate larvae. *Edited by* L. McEdward. CRC Press, Boca Raton, Fla. pp. 123–156.

Levitan, D.R. 1996. Predicting unique and optimal egg size in free-spawning marine invertebrates. Am. Nat. **148**: 174–188.

Levitan, D.R., and Petersen, C. 1995. Sperm limitation in the sea. Trends Ecol. Evol. **10**: 228–231.

Levitan, D.R., and Young, C.M. 1995. Fertilization success in large populations: empirical measures and theoretical predictions of fertilization in the sea biscuit *Clypeaster rosaceus.* J. Exp. Mar. Biol. Ecol. **190**: 221–241.

Levitan, D.R., Sewell, M.A., and Chia, F.S. 1991. Kinetics of fertilization in the sea urchin *Strongylocentrotus franciscanus:* interaction of gamete dilution, age, and contact time. Biol. Bull. (Woods Hole), **181**: 371–378.

Levitan, D.R., Sewell, M.A., and Chia, F.S. 1992. How distribution and abundance influences fertilization success in the sea urchin *Strongylocentrotus franciscanus.* Ecology, **73**: 248–254.

Lillie, F.R. 1915. Studies of fertilization. VII. Analysis of variations in the fertilization power of sperm suspensions of *Arbacia.* Biol. Bull. (Woods Hole), **28**: 229–251.

Mead, K.S., and Denny, M.W. 1995. The effects of hydrodynamic shear stress on fertilization and early development of the purple sea urchin *Strongylocentrotus purpuratus.* Biol. Bull. (Woods Hole), **188**: 46–56.

Mortensen, T. 1938. Contributions to the study of the development and larval form of echinoids. K. Dan. Vidensk. Selsk. Nat. Math. Afd. 9 raekke, **4**: 1.

Oliver, J., and Babcock, R. 1992. Aspects of the fertilization ecology of broadcast spawning corals: sperm dilution effects and in situ measurements of fertilization. Biol. Bull. (Woods Hole), **183**: 409–417.

Orensanz, J.M. 1986. Size, environment, and density: the regulation of a scallop stock and its management implications. *In* North Pacific Workshop on Stock Assessment and Management of Invertebrates. *Edited by* G.S. Jamieson and N. Bourne. Can. Spec. Publ. Fish. Aquat. Sci. 92. pp. 195–227.

Orensanz, J.M., Palmer, A.M., and Iribarne, O. 1991. Population dynamics and management of natural stocks. *In* Scallops: biology, ecology, and aquaculture. *Edited by* S.E. Shumway. Elsevier Science Publishers, Amsterdam, The Netherlands. pp. 625–713.

Pearse, J.S. 1990. Lunar reproductive rhythms in marine invertebrates: maximizing fertilization? *In* Advances in invertebrate reproduction 5. *Edited by* M. Hoshi and O. Yamashita. Elsevier Science Publishers, Amsterdam, The Netherlands. pp. 311–316.

Pearse, J.S., McClary, D.J., Sewell, M.A., Austin, W.C., Perez-Ruzafa, A., and Byrne, M. 1988. Simultaneous spawning of six species of echinoderms in Barkley Sound, British Columbia. Invertebr. Reprod. Dev. **14**: 279–288.

Pennington, J.T. 1985. The ecology of fertilization of echinoid eggs: the consequence of sperm dilution, adult aggregation, and synchronous spawning. Biol. Bull. (Woods Hole), **169**: 417–430.

Petersen, C.W. 1991. Variation in fertilization rate in the tropical reef fish, *Halichoeres bivattatus:* correlates and implications. Biol. Bull. (Woods Hole), **181**: 232–237.

Petersen, C.W., Warner, R.R., Cohen, S., Hess, H.C., and Sewell, A.T. 1992. Variation in pelagic fertilization rates: implications for production estimates, mate choice, and the spatial and temporal distribution of mating. Ecology, **73**: 391–401.

Podolsky, R.D., and Strathmann, R.R. 1996. Evolution of egg size in free spawners: consequences of the fertilization-fecundity trade-off. Am. Nat. **148**: 160–173.

Quinn, J.F., Wing, S.R., and Botsford, L.W. 1993. Harvest refugia in marine invertebrate fisheries: models and applications to the red sea urchin, *Strongylocentrotus franciscanus.* Am. Zool. **33**: 537–550.

Randall, J.E., Schroeder, R.E., and Stark, W.A., II. 1964. Notes on the biology of the echinoid *Diadema antillarum.* Caribb. J. Sci. **4**: 421–433.

Reeves, J.E. 1982. Data requirements for tanner crab management, as suggested by the Bristol Bay king crab management model. *In* Proceedings of the International Symposium on Chionoecetes, University of Alaska, December 1982. Alaska Sea Grant Report No. 85-3. pp. 679–716.

Roughgarden, J., Gaines, S., and Possingham, H. 1988. Recruitment dynamics in complex life cycles. Science (Washington, D.C.), **241**: 1460–1466.

Sebens, K.P. 1982. The limits to indeterminate growth: an optimal size model applied to passive suspension feeders. Ecology, **63**: 209–222.

Sewell, M.A., and Levitan, D.R. 1992. Fertilization success in a natural spawning of the dendrochirote sea cucumber *Cucumaria miniata.* Bull. Mar. Sci. **51**: 161–166.

Sparck, R. 1927. Studies on the biology of the oyster (*Ostrea edulis*) II-IV. Report of the Danish Biological Station No. 33. pp. 43–65.

Starr, M., Himmelman, J.H., and Therriault, J.-C. 1990. Direct coupling of marine invertebrate spawning with phytoplankton blooms. Science (Washington, D.C.), **247**: 1071–1074.

Strathmann, R.R. 1990. Why life histories evolve differently in the sea. Am. Zool. **30**: 197–207.

Sutherland, J. P. 1970. Dynamics of high and low populations of the limpet *Acmea scabra* (Gould). Ecol. Monogr. 40: 169–188.

Thomas, F.I.M. 1994*a*. Transport and mixing of gametes in three free-spawning polychaete annelids, *Phragmatopoma californica* (Fewkes), *Sabillaria cementarium* (Moore), and *Schizobranchia insignis* (Bush). J. Exp. Mar. Biol. Ecol. **179**: 11–28.

Thomas, F.I.M. 1994*b*. Physical properties of gametes in three sea urchin species. J. Exp. Biol. **194**: 263–284.

Unger, B., and Lott, C. 1994. In-situ studies of the aggregation behavior of the sea urchin *Spaerechinus granularis* Lam. (Echinodermata: Echinoidea). *In* Echinoderms through time (echinoderms Dijon). *Edited by* B. David, A. Guille, J.P. Feral, and M. Roux. A.A. Balkema, Rotterdam, The Netherlands. pp. 913–919.

Vogel, H., Czihak, G., Chang, P., and Wolf, W. 1982. Fertilization kinetics of sea urchin eggs. Math. Biosci. **58**: 189–216.

Wing, S.R., Largier, J.L., Botsford, L.W., and Quinn, J.F. 1995. Settlement and transport of benthic invertebrates in an intermittent upwelling region. Limnol. Oceanogr. **40**: 316–329.

Young, C.M., Tyler, P.A., Cameron, J.L., and Rumrill, S.G. 1992. Seasonal breeding aggregations in low-density populations of a bathyal echinoid, *Styocidaris lineata.* Mar. Biol. (Berlin), **113**: 603–612.

Yund, P. 1990. An in situ measurement of sperm dispersal in a colonial marine hydroid. J. Exp. Zool. **253**: 102–106.

Yund, P.O. 1995. Gene flow via the dispersal of fertilizing sperm in a colonial ascidian (*Botryllus schlosseri*): the effect of male density. Mar. Biol. (Berlin), **122**: 649–654.

Yund, P.O., and McCartney, M.A. 1994. Male reproductive success in sessile invertebrates: competition for fertilizations. Ecology, **75**: 2151–2167.

The Fishing Process

Quantifying potential impacts of behavioral factors on crustacean stock monitoring and assessment: modeling and experimental approaches

J.T. Addison and R.C.A. Bannister

Abstract: Catchability of crustaceans in trap fisheries is influenced by behavioral factors that may affect crustacean stock assessments. The likely impact of behavioral interactions is studied from three viewpoints. Firstly, from a theoretical viewpoint, standard assessment models are modified to include nonlinear relationships between fishing mortality and fishing effort and between catchability and abundance. The models predict that different conclusions would be reached about the level of nominal effort at which yield per recruit reaches a maximum (E_{max}) and hence about the degree of change resulting from changes in effort. When an asymptotic stock-recruitment curve is combined with a nonlinear fishing mortality – fishing effort relationship, the likelihood of stock collapse at high effort levels is changed. Secondly, experiments are described illustrating real behavioral effects. In these experiments, prestocking traps with one or more lobster or crab shows that lobster inhibit the entry of other lobster and crab and that crab inhibit other crab, but not lobster. These intra- and inter-specific effects reduced the catch rate of individual strings of traps by more than 50% in some cases. Finally, as a step towards aggregating the effect of behavioral interactions from the individual trap to the fishery level, we review recently developed modeling approaches explicitly incorporating entry and escape processes. The models provide a promising basis for describing the biological processes underlying the capture of lobster.

Résumé : Dans la pêche aux casiers, la capturabilité des crustacés dépend de facteurs comportementaux qui peuvent avoir une incidence sur les évaluations des stocks de crustacés. Les effets probables des interactions sur le comportement sont analysés sous trois aspects. Tout d'abord, du point de vue théorique, on modifie des modèles standard d'évaluation afin d'y intégrer des relations non linéaires entre la mortalité par pêche et l'effort de pêche, et entre la capturabilité et l'abondance. D'après ces modèles, on devrait obtenir des résultats différents relativement au degré d'effort nominal permettant d'atteindre un rendement maximal (E_{max}) par recrue, et donc à l'importance des variations associées aux modifications de l'effort. Lorsqu'une courbe asymptotique stock-recrutement est combinée à une relation non linéaire entre la mortalité par pêche et l'effort de pêche, la probabilité d'un effondrement des stocks varie dans le cas de coefficients d'effort élevés. Ensuite, on décrit des expériences illustrant les effets réels sur le comportement. D'après les résultats de ces expériences où l'on a fait entrer préalablement un ou plusieurs homards ou crabes des neiges dans les casiers, la présence de homards empêcherait la capture d'autres homards ou crabes, et la présence de crabes gênerait la capture d'autres crabes, mais pas celle de homards. Dans certains cas, cette influence intraspécifique et interspécifique diminue de plus de 50 % le taux de capture enregistré par chapelet de casiers. Finalement, en vue de reporter à l'ensemble de la pêche les effets des interactions observées sur le comportement dans chaque casier, on a examiné des méthodes de modélisation récemment mises au point et incorporant de façon explicite des mécanismes d'entrée et d'évasion. Ces modèles constituent un outil de base prometteur pour la description des processus biologiques qui sous-tendent la capture du homard. [Traduit par la Rédaction]

Introduction

Catchability of crustaceans in trap fisheries is influenced by behavioral factors that need to be considered when assessing crustacean stocks and developing management strategies. In particular, most assessment methods assume that the relationship between fishing effort and fishing mortality is linear and that size distribution measured by traps is unbiased. Similarly, stock-monitoring approaches using traps assume that catch per effort is proportional to abundance. It is essential, therefore, to consider the possible effect on assessment and stock-monitoring results if these assumptions are unfounded. To illustrate these effects we use the fishery for European lobster (*Homarus gammarus* (L.)) in the coastal waters of England and Wales as an example.

Lobster stocks in England and Wales have been assessed using length cohort analysis (Jones 1981), which estimates lobster mortality from size-distribution data and models how changes in minimum landing size and in fishing effort could affect equilibrium yield and biomass per recruit (Bannister and Addison 1984; Bannister 1986). The potential effect on these relationships of different types of stock and recruitment curves

J.T. Addison and R.C.A. Bannister. Centre for Environment, Fisheries and Aquaculture Science, Lowestoft Laboratory, Pakefield Road, Lowestoft, Suffolk NR33 0HT, U.K.

Correct citation: Addison, J.T., and Bannister, R.C.A. 1998. Quantifying potential impacts of behavioral factors on crustacean stock monitoring and assessment: modeling and experimental approaches. *In* Proceedings of the North Pacific Symposium on Invertebrate Stock Assessment and Management. *Edited by* G.S. Jamieson and A. Campbell. Can. Spec. Publ. Fish. Aquat. Sci. 125. pp. 167–177.

Fig. 1. Theoretical nonlinear relationships between fishing mortality and nominal fishing effort: (*a*) asymptotic; (*b*) power law; and (*c*) sigmoid. The solid line represents the proposed nonlinear relationship. The dotted line represents the standard linear model.

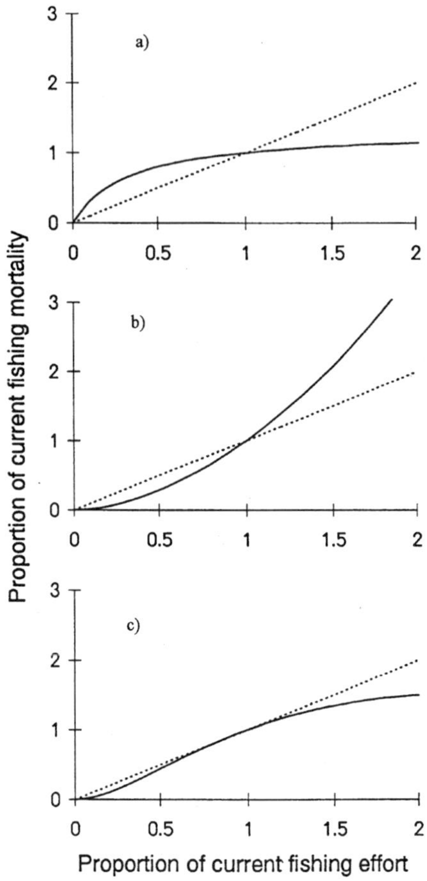

has also been simulated (Bannister and Addison 1986), thus introducing the theoretical possibility of stock collapse at high mortality levels. However, because there are no past stock and recruitment data, it is difficult to identify the threshold of stock below which recruitment overfishing is likely to occur. Consequently, to study how lobster stocks respond to fishing in the long term, stock abundance is now being monitored directly by collecting catch-per-effort data from logbooks kept by selected fishers in each main fishery, as well as spatially explicit catch-per-effort and size-distribution data from one coastal study site (Bannister and Addison 1995).

Some factors affecting size distribution have already been examined in a preliminary way. For example, Addison (1986) modeled what would happen if the abundance of large lobster is limited in a density-dependent way by habitat limitation, thus distorting the estimation of mortality rate and how it changes with time. Similarly, fishing mortality may be underestimated because size distributions of individuals in traps may be unrepresentative of the size distribution on the ground due to behavioral interactions between individuals in and around traps. In general, larger individuals are more catchable than smaller ones either due to their greater movement or because entry of smaller animals is inhibited by larger ones (Miller 1989, 1995; Miller and Addison 1995).

In this paper we model how predictions from length cohort analysis would be affected by different nonlinear relationships between fishing effort and fishing mortality and we summarize experimental results demonstrating the magnitude of intra- and inter-specific behavioral interactions likely to affect the validity of trap catch-per-effort data. We also review some recent modeling work that explicitly describes behavioral interactions underlying the capture process.

Predictions from cohort analysis

The relationship between fishing mortality and fishing effort

This section describes how predictions by the Jones length cohort analysis model (Jones 1981) are affected by departures from the basic assumption that, as nominal fishing effort increases, fishing mortality increases proportionately. Our aim is not to provide a definitive sensitivity analysis bracketing all possible assumptions, but to describe a simple range of effects illustrating the need to examine these aspects more fully in the future. To achieve this we have selected three possible variants of the mortality–effort relationship compared with the standard model, viz., an asymptotic relationship, a power law, and a sigmoid relationship (Fig. 1), and we apply these first on a yield-per-recruit basis and then with one particular choice of stock-recruitment curve. An asymptotic relationship between nominal fishing effort and fishing mortality could occur if a high density of gear saturates the grounds (e.g., Waltz 1989) or if high effort reduces lobster density to such a low level that the chance of encounter with the gear becomes disproportionately low. A power law relationship could occur if behavioral interactions inhibiting entry to the trap are pronounced when effort is low (i.e., at high lobster density), causing reduced entry to the trap and hence disproportionately low mortality, but are less pronounced when effort is high and density is reduced. The sigmoid relationship was chosen as one particular combination of the other two.

Curves of yield per recruit (Y/R) and biomass per recruit (B/R) against nominal fishing effort for each nonlinear model were compared with those for the standard linear model, using a typical size distribution for the inshore lobster stocks in England where, according to the standard model, fishing mortality F is approximately 0.85 (Figs. 2*a*–2*c*). The simulation shows that different conclusions would be reached about the level of nominal effort at which Y/R reaches a maximum (E_{max}) and hence about the degree of change resulting from changes in effort. For the asymptotic relationship, E_{max} is lower than for the linear model (Fig. 2*a*). As nominal fishing effort increases, Y/R decreases quickly but then stabilizes at a level higher than that predicted by the linear model. With this model, changes in nominal effort above or below the current value would produce smaller changes in Y/R compared with the linear model. Changes in B/R are also smaller relative to those predicted by the linear model (Fig. 2*a*). E_{max} is higher using a power law relationship model than using the linear model, but high effort will reduce Y/R more using the power law relationship than using the linear model (Fig. 2*b*). The sigmoid relationship predicts that E_{max} is higher than for the linear model, but the reduction in Y/R following an increase in nominal fishing effort is slightly less than that predicted for the linear model (Fig. 2*c*).

The scale of these effects varies according to the current

Fig. 2. Predictions from a Jones length cohort analysis model incorporating a nonlinear relationship between fishing mortality and nominal fishing effort, for typical inshore lobster fishery ($F = 0.85$): (*a*) asymptotic fishing mortality vs. fishing effort relationship; (*b*) power law relationship; and (*c*) sigmoid relationship. The solid line represents the proposed nonlinear relationship. The dotted line represents the standard linear model.

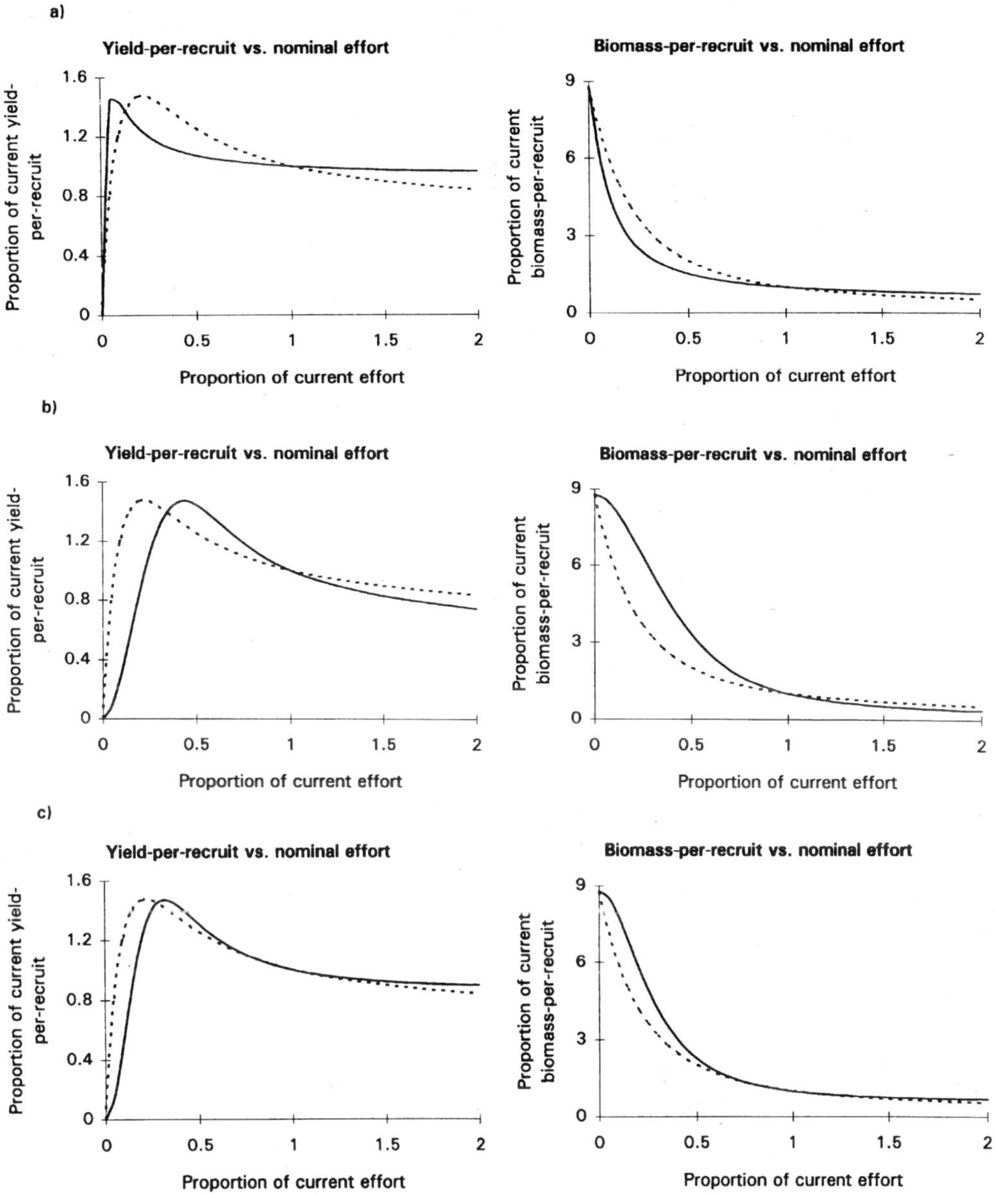

level of fishing mortality. Thus if the simulation starts with size-distribution data typical of offshore lobster stocks, where F estimated by the standard model is much lower at 0.22, the effect of introducing a nonlinear relationship (asymptotic model) is even greater (Fig. 3), E_{max} being substantially lower for the nonlinear model than the linear model and the Y/R curve being essentially flat-topped. B/R declines more rapidly for the nonlinear model than for the linear model but is then stable over a wide range of effort.

The effects are emphasized if a theoretical stock-recruitment relationship is incorporated using the technique described by Bannister and Addison (1986). Assuming a asymptotic stock-recruitment relationship for the typical inshore fishery (Beverton and Holt 1957) (Fig. 4*a*), plus a fishing effort – fishing mortality relationship that is also asymptotic, a larger reduction in nominal effort would be needed to obtain maximum yield than for the linear model (Fig. 4*b*); however, when effort is increased above its current level, stock collapse is less likely. For the power law model, optimum yield can be achieved by much smaller reductions in nominal fishing effort compared with the linear model, but when effort is increased above its current level, there is a much greater likelihood of a recruitment failure (Fig. 4*c*). For the sigmoid relationship, predicted yield is similar to that for the linear model, but there is less chance of recruitment failure (Fig. 4*d*). With this example the possibility of stock collapse at high levels of effort is therefore strongly influenced by the nature of the relationship between fishing mortality and fishing effort.

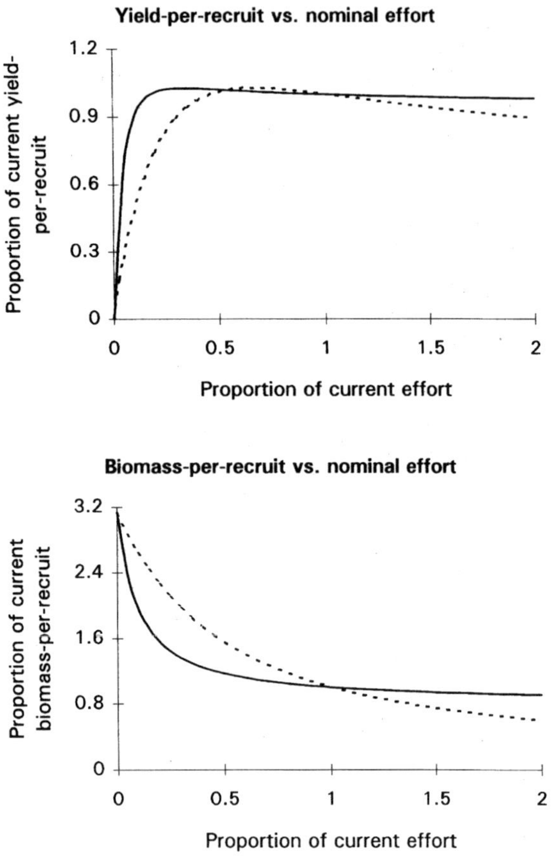

Fig. 3. Predictions from a Jones length cohort analysis model incorporating a nonlinear (asymptotic) relationship between fishing mortality and nominal fishing effort for typical offshore lobster fishery ($F = 0.22$). The solid line represents the proposed nonlinear relationship. The dotted line represents the standard linear model.

The relationship between catchability and abundance

An alternative way of expressing the nonlinearity between fishing mortality and fishing effort is to consider that catchability is not linearly related to abundance. Standard fishery models assume that catch per unit effort is proportional to abundance

[1] $$\frac{C}{f} = qN$$

where C is catch, f is effort, N is abundance, and q is a constant termed catchability. In reality however this relationship may not be linear and this would have particular consequences when using fishery logbook data to monitor trends in lobster stock abundance. Processes underlying nonlinearity include gear saturation, schooling or clumping of fish, limited search area and time, and nonrandom searching by fishers (Gulland 1964; Paloheimo and Dickie 1964; Clark 1974; Rothschild 1977; Arreguín-Sánchez 1996).

Empirical observations of catchability declining with increasing abundance come from a wide range of fisheries (e.g., Rothschild et al. 1970; Ulltang 1976; Peterman and Steer 1981; Bannerot and Austin 1983; Crecco and Savoy 1985; Winters and Wheeler 1985; Crecco and Overholtz 1990). For this exercise, we hypothesized, therefore, that catchability may decline nonlinearly with abundance in the form

Fig. 4. Incorporating a theoretical stock-recruitment curve into a Jones length cohort analysis model for the typical inshore lobster fishery. Assuming (*a*) an asymptotic stock-recruitment curve, predictions from the model for (*b*) an asymptotic, (*c*) power law, or (*d*) sigmoid relationship between fishing mortality and fishing effort. The solid lines represent the proposed nonlinear relationship. The dotted line represents the standard linear model.

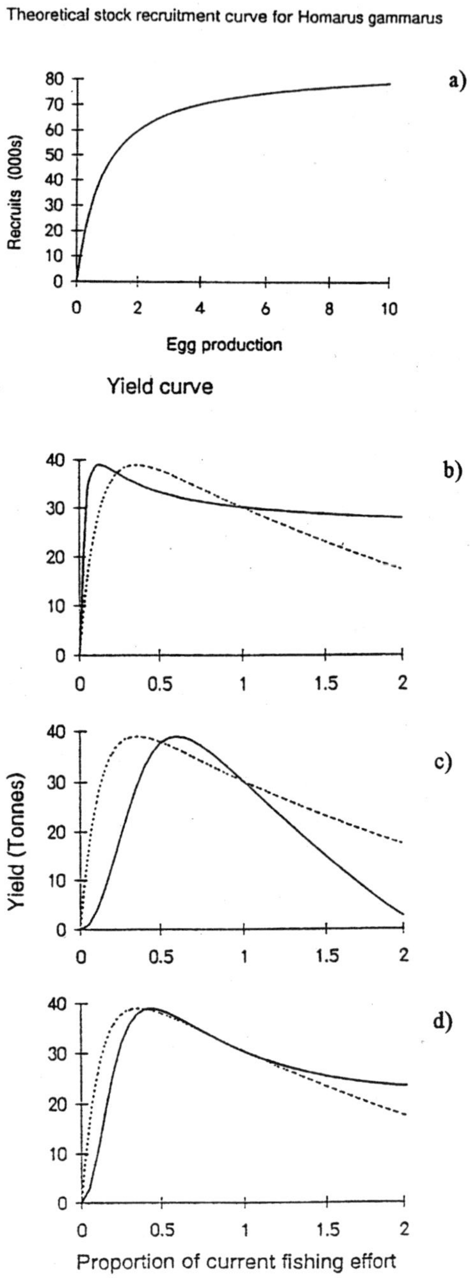

[2] $$q = \alpha N^{\beta}$$

where q is catchability and α and β are parameters. β determines the degree of nonlinearity, with high negative values of β describing strong nonlinearity. We have included this relationship (Fig. 5*a*) in a standard dynamic pool model. In this model, catchability thus increases as population numbers decline when effort increases. Catchability, and therefore fishing

Fig. 5. (*a*) Incorporating a theoretical nonlinear relationship between catchability and abundance into a dynamic pool model. Predictions from the model (*b*) for a low nonlinearity and (*c*) for a high nonlinearity. The solid line represents the proposed nonlinear relationship. The dotted line represents the standard linear model.

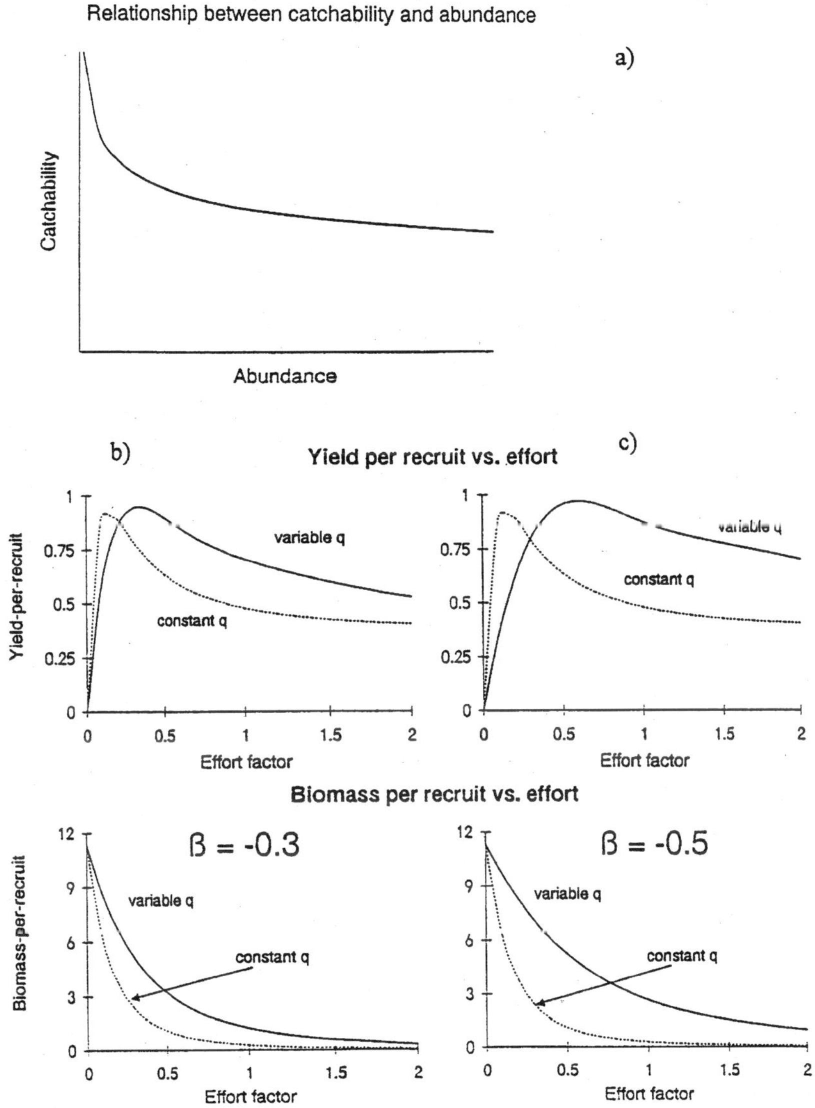

mortality-at-age (or size), also increases as population numbers within each cohort decrease with age (or size). In other words, the model assumes that catchability of larger individuals is higher than that for smaller individuals. Including such nonlinearity in the model has the effect of shifting the *Y*/*R* curve upwards and to the right. i.e., increasing *Y*/*R* and increasing E_{max} relative to the standard model (Fig. 5*b*). It also predicts higher levels of *B*/*R* at all levels of effort. The higher the degree of nonlinearity (larger negative values of β) the greater the effect (Fig. 5*c*).

Field experiments and observations

Simulation provides a generalized view of how hypothetical departures from basic assumptions might affect the results from a traditional assessment model; however, the difficulty is to measure the likely degree of nonlinearity in real-life fisheries. Long time series of reliable standardized data on fishing effort and fishing mortality are not available for crustacean fisheries in England and Wales so field and experimental investigations have been undertaken which demonstrate the actual occurrence of biological interactions and their likely magnitude. These experiments investigate the strength of intra- and inter-specific interactions between lobster (*Homarus gammarus* (L.)) and crab (*Cancer pagurus* L.) in the UK fishery. Preliminary experiments show that these can be substantial. Thus in field experiments using commercial fishers, it was shown that prestocking traps (pots) with a single lobster reduced catch rates of lobster by 54% and catch rates of crab by 60% (Table 1) (Addison 1995). Baiting traps with soft recently molted crab also significantly increased the catch of lobster, but reduced the catch of crab to negligible proportions (Addison 1995). This avoidance response was documented previously for *C. pagurus* (Chapman and Smith 1978) but it was not clear whether the increase in lobster catch is due to the exclusion of crab (the fishers' stated aim of the baiting

Table 1. Number of lobster and crab caught in traps prestocked with a lobster in comparison to traps with standard bait (110 pot-hauls of each bait type) (from Addison 1995).

	Standard bait	Bait + lobster
Lobsters	67	31
Crabs	290	115

Table 2. Numbers and mean size of lobster and crab caught in traps prestocked with large or small lobster in comparison with traps with standard bait (controls).

	Standard bait	Bait + large lobster	Bait + small lobster
Lobster			
Nos. caught	422	295	279
Mean size (mm CL)	78.6	79.5	79.4
Crab			
Nos. caught	140	55	70
Mean size (mm CW)	108.0	108.2	109.7

Note: Fished in strings of 24 parlor traps; 8 traps with a large lobster alternating with 8 traps with a small lobster and 8 control traps. Fished on 18 occasions giving a total of 144 trap hauls for each treatment and control. Lobsters were placed in the parlor end of the trap, so that they could not escape from the trap, consume the bait, or physically block the entrances to the trap.

practice) or whether damaged soft crab is simply a better attractant than standard fish bait.

Since publication of the above study, two additional experiments have been carried out prestocking traps with lobster in order to test whether the inhibitory effect on the catch of lobster and crab depends on lobster size or number. Prestocking traps with either a large (87–99 mm carapace length (CL)) or a small immature (73–83 mm CL) lobster significantly reduced the catch of both lobster (for large lobster, $\chi^2_{(1)} = 22.50, p < 0.001$ and for small lobster $\chi^2_{(1)} = 29.17, p < 0.001$) and crab (for large lobster, $\chi^2_{(1)} = 37.05, p < 0.001$ and for small lobster $\chi^2_{(1)} = 23.33, p < 0.001$) (Table 2). Size was not a factor since small lobster were just as effective as large lobster in deterring entry (Table 2) and there was no significant difference in the size composition of the lobster and crab catch in both treatments and in the control (One-way Anova, lobster $F = 1.70$, df = 2, 993, $p > 0.05$; crab $F = 0.16$, df = 2, 262, $p > 0.05$). The response of catch rates to the number of lobster prestocked in the trap produced inconsistent results however. In an initial experiment, prestocking traps with three small lobster produced a significantly lower catch of both lobster and crab than control traps ($\chi^2_{(1)} = 72.54, p < 0.001$ for lobster; $\chi^2_{(1)} = 31.80, p < 0.001$ for crab) and those prestocked with only one lobster ($\chi^2_{(1)} = 64.86, p < 0.001$ for lobster; $\chi^2_{(1)} = 19.36, p < 0.001$ for crab), but surprisingly the catch from the latter was not significantly different from the control traps ($\chi^2_{(1)} = 0.24, p > 0.05$ for lobster; $\chi^2_{(1)} = 1.80, p > 0.05$ for crab) (Table 3*a*). In this experiment there was no significant difference in the size composition of the crab catch in the treatments and the control ($F = 2.16$, df = 2, 186, $p > 0.05$), but larger lobster were caught in the traps prestocked with three lobster ($F = 11.42$, df = 2, 885, $p < 0.001$). This experiment was later repeated and on this occasion the catch rate of lobster was greater in the controls than in traps prestocked with only one lobster ($\chi^2_{(1)} = 23.32, p < 0.001$) which in turn was greater than that in traps prestocked with three lobster ($\chi^2_{(1)} = 5.46, p < 0.05$) (Table 3*b*). Similarly, in contrast with the previous experiment, there was no significant difference in the size composition of the lobster catch in the treatments and the control ($F = 1.78$, df = 2, 771, $p > 0.05$). For crab the experimental results mirrored those previously observed, with traps prestocked with three small lobster catching fewer crab than the controls ($\chi^2_{(1)} = 8.45, p < 0.01$), but catches from traps prestocked with one lobster were not significantly different from the controls ($\chi^2_{(1)} = 0.143, p > 0.05$). Again there was no significant difference in the size composition of the crab catch in both treatments and the control ($F = 0.72$, df = 2, 227, $p > 0.05$).

Prestocking traps with three large crab (129–175 mm carapace width (CW), mean width 142 mm) resulted in a small but not significant effect on the lobster catch ($\chi^2_{(1)} = 2.95, p > 0.05$), but did result in a significant reduction in the crab catch

Table 3. Numbers and mean size of lobster and crab caught in traps prestocked with one or three lobster in comparison with traps with standard bait (controls).

A. Fished on 18 occasions giving a total of 144 trap hauls for each treatment and control.

	Standard bait	Bait + 1 lobster	Bait + 3 lobsters
Lobster			
Nos. caught	366	353	169
Mean size (mm CL)	80.4	78.9	81.9
Crab			
Nos. caught	89	72	28
Mean size (mm CW)	111.9	104.5	108.2

B. Fished on 16 occasions giving a total of 128 trap hauls for each treatment and control.

	Standard bait	Bait + 1 lobster	Bait + 3 lobsters
Lobster			
Nos. caught	352	235	187
Mean size (mm CL)	79.1	79.6	80.2
Crab			
Nos. caught	90	85	55
Mean size (mm CW)	98.2	100.3	97.3

Note: Fished in strings of 24 parlor traps; 8 traps prestocked with three small lobsters alternating with 8 traps with one small lobster and 8 control traps. Lobsters placed in the parlor end of the trap.

Table 4. Numbers and mean size of lobster and crab caught in traps prestocked with three crab in comparison with traps with standard bait (controls).

	Standard bait	Bait + 3 large crabs
Lobster		
Nos. caught	225	190
Mean size (mm CL)	76.5	77.2
Crab		
Nos. caught	1208	834
Mean size (mm CW)	108.5	107.2

Note: Fished in four strings of 12 traps prestocked with 3 large crabs alternately spaced with 12 control traps. Crabs placed in parlor end of trap. Traps were fished on 6 occasions, giving 288 trap hauls of both treatment and control.

($\chi^2_{(1)} = 68.50$, $p < 0.001$) (Table 4). There was no significant difference in the size composition of the lobster and crab catch in the treatment and in the control (lobster, $d = 0.82$, $p > 0.05$; crab, $d = 1.71$, $p > 0.05$).

Linking experimental observations with assessment models

In the section on Predictions from cohort analysis, we modified traditional assessment models to illustrate how their predictions would change as a result of different assumptions about catchability. In the Field experiments and observations section, we collated results from recent experiments which quantify the effects of behavioral interactions on the catches of individual traps. An emergent problem, however, is how to integrate the conceptual approach with the field data. At the very least this means considering how to scale-up interactions with individual traps and strings to the total fishery and in particular to reflect accurately the reality of commercial fishing operations comprising many strings of traps, variable soak times, varying day-to-day weather, and possible spatial differences in abundance and trap density. It may be that variation due to behavioral interactions is relatively unimportant when such aggregation of data has occurred. For example, in an experimental and statistical approach to investigating the relationship between catch per unit effort (CPUE) and abundance, Richards and Schnute (1986) concluded that CPUE may be proportional to abundance for single species data collected over uniform environmental conditions, but that CPUE has less value as an abundance index for commercial fishery data that are likely to be combined over a range of weather conditions.

There are many possible factors to consider, but an example of how single trap effects can be masked when considering data from commercial fishing operations is demonstrated by the effect of soak time on the catch of individuals in a trap. Most experimental data show that catch numbers either vary with soak time asymptotically or show a decline after long soak times (see review by Miller (1990)), but for our study area commercial lobster data show little relationship between catch numbers and soak times (Fig. 6). We suspect that this is because the fisher will vary his soak times depending on previous catch rates in an area. The standard soak time is 2 days, but

Fig. 6. Relationship between square root of total catch of lobster (*Homarus gammarus* (L.)) and soak time (days) for commercial traps fished throughout 1989 by the same fisher at Bridlington on the east coast of England ($N = 482$; some points hidden).

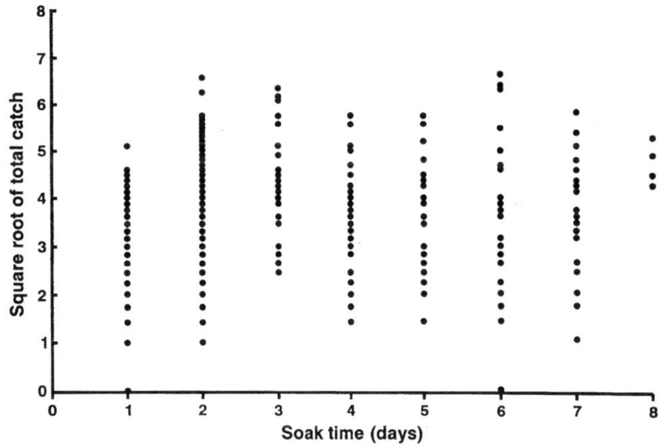

some strings of traps are occasionally hauled after 1 day if the previous few days catches in that localized area have been particularly good. Conversely, some strings of traps are left to soak for 3, 4, or more days when previous hauls produced low catches. In this way the fishing pattern of the fisher has been adjusted as a function of abundance.

New modeling approaches based on individual trap catches

Simulation model using parameters estimated from experimental fishing

The possible importance of behavioral interactions can be seen in catch data. In our study area, we have observed not only that the catch of lobster versus the catch of crab for individual traps is negatively correlated (Addison 1995), but also the monthly CPUE for lobster and crab in that fishing area is negatively correlated (Fig. 7). Secondly, the variance–mean ratio (an index of dispersion) for the catch of lobster in individual traps was less than or close to unity, suggesting a random or underdispersed distribution of lobster among traps, except at low densities, when the distribution among traps was aggregated (Addison 1995). Since underdispersion is rare in natural populations (Southwood 1978), we hypothesize that the behavioral interactions in and around the trap are reducing the incidence of high numbers of lobster per trap and hence turning an overdispersed distribution of lobster on the ground into a random or underdispersed distribution among traps. Conversely, for crab, Addison (1995) found an overdispersed distribution among traps.

In order to quantify the effect of these behavioral interactions, we are therefore developing a simulation model to investigate the time between entry of individual lobster into traps and hence the cumulative catch with varying soak time and varying local abundance of lobster. With a very simple assumption about the distribution of animals on the sea bed, we can depress the variance–mean ratio by including an interaction process, parameterized from our own experimental data. Until we have resolved statistical problems involved in

Fig. 7. Relationship between the catch per unit effort (CPUE) of lobster (*Homarus gammarus* (L.)) and of crab (*Cancer pagurus* L.) in commercial traps fished at the Bridlington study site on the east coast of England in 1989.

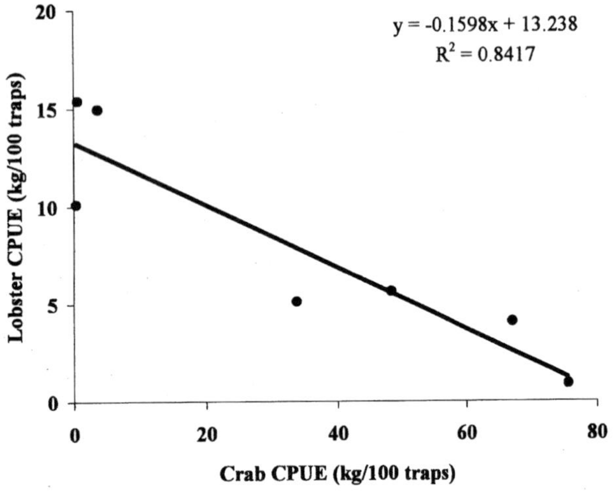

representing the encounter rate between traps and animals in a patchily distributed habitat, however, we cannot progress to scaling-up from the individual to the population level.

Fogarty and Addison model

In a rather different approach, Fogarty and Addison (1997) developed differential equation models which explicitly incorporate elements of the capture process in trap fisheries. As catch is not typically a linear function of soak time, they modeled the change in catch with soak time as follows:

$$[3] \quad \frac{dC}{dt} = f(C) - bC$$

where f(C) is a function representing entry of individuals to the traps and it is assumed that escape from the traps occurs at a constant rate b. Fogarty and Addison (1997) examined the behavior of two particular models. Entry rate was assumed either to be constant (f(C) = a), leading to the solution

$$[4] \quad C = \frac{a}{b}(1 - e^{-bt})$$

or the nonlinear form

$$[5] \quad f(C) = a\,C^{m}$$

where the parameter m generally depends on species behavior (for homarid lobster, the observed behavioral interactions suggest that m would be negative). This nonlinear entry function leads to the solution

$$[6] \quad C = \left[\frac{a}{b}(1 - e^{-(1-m)bt})\right]^{1/(1-m)}$$

These two models were then fitted to empirical data collected as scuba diver observations on the daily entry and the cumulative catch of American lobster with time (Auster 1985). Two sets of data were examined where cumulative catch was asymptotic with time. One set showed constant daily entry of lobster, but in the other, cumulative entry of individuals was markedly nonlinear. The most suitable model was then fitted depending on empirical information concerning the nature of the entry process. Fogarty and Addison (1997) thus highlight the need to have information about the entry process in order to fit the most biologically realistic model, ensuring reasonable estimates of parameters which might be used for estimation of abundance. Fogarty and Addison (1997) also developed a stochastic analogue of the constant entry model using the theory of birth–death stochastic processes. They showed that the solution of this stochastic analogue, a Poisson distribution, provided significant fits to one set of Auster's data for individual trap catches for seven days of observation. Availability of information on catch at the individual trap level, its mean and variance, plus the solution to the stochastic analogue of the model, thus provides a further diagnostic tool to determine whether the fitted model adequately describes biological processes underlying the capture of lobster.

Discussion

This paper stems from our concern to investigate whether behavioral interactions are likely to have serious implications for the assessment and monitoring of lobster and crab stocks. While our approach is not yet comprehensive, it does illustrate possible avenues providing worthwhile insights and results and scope for future work. Modeling the effect on yield-per-recruit and true-yield curves of possible nonlinearity in the relationship between fishing mortality and nominal fishing effort and between catchability and abundance certainly changes the value of E_{max} and the predicted response to changes in effort, compared to the linear model. For example, it shows that if a power law relationship between fishing effort and fishing mortality is combined with an asymptotic stock-recruitment curve, the possibility of stock collapse at high levels of fishing effort could be increased (Fig. 5) with obvious implications for assessment and management. In contrast, an asymptotic relationship between fishing effort and fishing mortality may reduce the likelihood of stock collapse and such a combination might be consistent with the seeming resilience of some lobster populations to high levels of exploitation. It should be stressed that model predictions depend on the particular choice of parameters used and we have tested only a limited range of options and assumptions. However, effects are sufficiently interesting to suggest the importance of collecting data that would help to define which fishing effort – fishing mortality relationship occurs in specific fisheries. At the very least we need to know the likely direction of the bias. Experimental studies suggest a power law relationship may be the most likely in the *H. gammarus* fishery.

Behavioral interactions between lobster and crab can have substantial effects on the catch of both species in individual traps and these are likely to operate in a density-dependent manner. Since we wish to use logbook data to monitor trends in stock abundance, implications of these effects need to be explored further. The problem of investigating behavioral interactions in this way is illustrated by the inconsistent results of experiments prestocking traps with either one or three lobster. In one of these experiments, the catch in traps prestocked with one lobster was not significantly different from control traps. An explanation is that lobster which are inhibited from entering a trap prestocked with three lobster will search for another trap, although the next one was some distance away because traps were spaced 15 fathoms (27.4 m) apart. When

Table 5. Summary of behavioral interactions in *Homarus gammarus*/*Cancer pagurus* trap fishery in the U.K.

Intraspecific interactions	Lobsters inhibit entry to traps of other lobsters
	Little or no interactions between lobsters of different sizes
	Males and females have same effect
	Some lobsters are not vulnerable to trapping
	Crabs deter entry to traps of other crabs
	Baiting with dead crabs deters entry to traps of other crabs
Interspecific interactions	Lobsters deter entry to traps of crabs
	Crabs have little or no effect on entry to traps of lobsters

density is high, lobster soon enter control traps, rendering control traps indistinguishable from those prestocked with one lobster. Both then become equally attractive as an alternative to the three-lobster pot, whereas the latter is likely to remain distinguishable from the others for the whole experimental period of 24 h. This may be difficult to resolve operationally. In principle the problem could be dealt with by checking and emptying control traps regularly throughout the day, as was done in experiments carried out close to shore by Miller (1995) and Miller and Rodger (1996), but this would be much more difficult to do under the larger scale commercial conditions of our experiments, where several strings of 24 traps were fished in 20-m depth several kilometres offshore. Preliminary experiments in the UK lobster and crab fishery show that emptying traps regularly can increase catch rates over time (J.T. Addison and S.R.J. Lovewell, unpublished). While this aspect of experimental design in our prestocking experiments needs further consideration, it is clear that the differences we obtained underestimate those that may occur if control traps could be emptied regularly. Prestocking traps with lobster had an effect on the size distribution of the lobster and crab catch in only one experiment (Table 3*a*), but clearly further field experiments are required to examine variation in catchability at size and other size-dependent effects in the fishery for these two species.

Based on the experiments described here and on other anecdotal observations, Table 5 summarizes interactions in the *H. gammarus* and *C. pagurus* fishery. Similar interactions have been observed for other crustacean fisheries. In field experiments in the fishery for *H. americanus*, prestocking traps with lobster significantly reduced the lobster catch (Richards et al. 1983). Similarly, catch rates of *H. americanus* and *Carcinus maenas* were reduced by intraspecific behavioral interactions in tank experiments (Miller and Addison 1995). Miller (1990) provided a full review of such intraspecific interactions in crustacean fisheries. In field and tank experiments, lobster deter *Cancer irroratus* and *C. borealis* from entering traps (Richards et al. 1983; Miller and Addison 1995). Size-related effects have been observed in *C. magister* (Smith and Jamieson 1989), *C. irroratus* (Miller and Addison 1995) and generally large individuals are often more catchable than small (e.g., Smith 1944; Ennis 1978; Williams and Hill 1982; Miller 1989, 1990). Male American lobster have a higher catchability than females (Miller 1995) and some individuals of *H. americanus* are 'trap-happy' while others are invulnerable to trapping over a long period of time (Karnofsky and Price 1989). The avoidance of dead conspecifics as demonstrated by *C. pagurus* has also been observed in other species (e.g., Hancock 1974; Richards and Cobb 1987). We can conclude therefore that such behavioral interactions occur in most crustacean fisheries and thus deserve serious attention in relation to their effect on catch rates.

At present we are still exploring how we might make the link between our experimental results and population models by scaling-up results obtained for individual traps on a string to the aggregate of operations by a fishery. As a start, we have concentrated on modeling intraspecific entry–escape processes in the context of varying soak time, without taking into account either the inherent patchiness of species abundance caused by habitat limitation or the size-dependent effects and interspecific interactions shown to occur experimentally. Furthermore, despite our experiments, our knowledge of individual behavior is still very rudimentary. This has already been acknowledged by other authors. For example, Smith and Jamieson (1989) developed a model to standardize the catch rate and size distribution of Dungeness crab (*C. magister*) among traps that experienced different soak times. They included agonistic interactions in their model and used the term 'virtual catch rate' to define the rate at which crab would enter a trap if catch rates did not change with increasing soak time. They noted, however, that little is known about the dynamics of agonistic interactions in *C. magister*.

Our field studies and soak time models concentrate on the likely behavior of animals in and around a trap. They take no account of the wider issue of what proportion of the true population is likely to be vulnerable to trapping at any one time. If the movement of individuals is only intermittent, because of feeding rhythms or reaction to variations in tidal flow, the proportion of the population actually exposed to fishing at any one time will be reduced, with a corresponding effect on true fishing mortality.

So far, we have yet to use our lobster and crab logbook data for management purposes or to use yield models to determine effort reduction targets. Consequently our inability to make specific corrections for behavioral interactions at the population level has not yet influenced any management decisions. We have, however, begun to investigate the possible use of study area data as a recruitment index that may be influenced by the factors discussed here. Recruitment indices have been very successfully developed for spiny lobster (e.g., Caputi and Brown 1986; Caputi et al. 1995; Cruz et al. 1995) and we feel that they could also be used to predict future catches of homarid lobster if required (Campbell 1990; Pezzack 1992; Addison et al. 1995). A good correlation was found between the catch rate of summer recruits and prerecruits one year previously for our east coast lobster study site (Fig. 8). We hypothesize two likely effects. Firstly, since prerecruit abundance is higher than recruit abundance the following year, behavioral interactions are more likely to depress catch rates in the prerecruit phase. The prerecruit catch rate is therefore a greater underestimate of abundance than is the catch rate of recruits. This would reduce the slope of the regression line, although it may not affect the prediction itself. Secondly, within years, a

Fig. 8. Relationship between summer recruit and summer prerecruit catch rates of lobster (*Homarus gammarus* (L.)) one year apart in standard traps fished commercially in a study site on the east coast of England, 1986–1992 (from Addison et al. 1995).

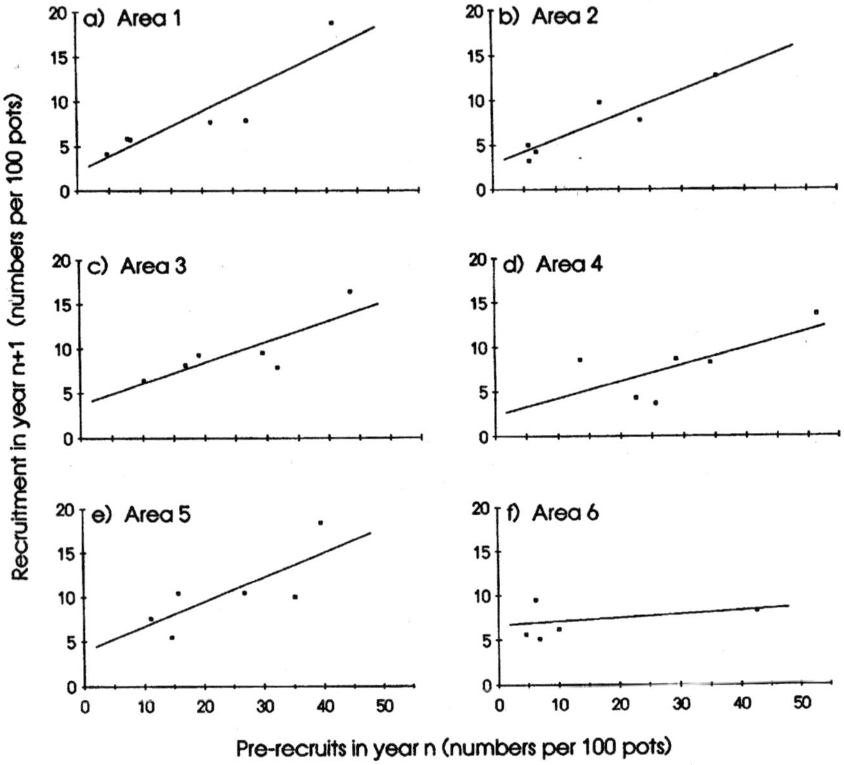

higher catch rate of recruits may have a stronger inhibitory effect on the catch rate of prerecruits and vice versa. This could increase variation about the regression line and therefore influence precision of the prediction.

Acknowledgments

This work was funded by the Ministry of Agriculture, Fisheries and Food, U.K. Steve Lovewell and Andy Lawler provided valuable assistance with field experiments.

References

Addison, J.T. 1986. Density-dependent mortality and the relationship between size composition and fishing effort in lobster populations. Can. J. Fish. Aquat. Sci. **43**: 2360–2367.

Addison, J.T. 1995. Influence of behavioural interactions on lobster distribution and abundance as inferred from pot-caught samples. ICES Mar. Sci. Symp. **199**: 294–300.

Addison, J.T., Bannister, R.C.A., and Lovewell, S.R.J. 1995. Variation and persistence of pre-recruit catch rates of *Homarus gammarus* in Bridlington Bay, England. Crustaceana, **68**: 267–279.

Arreguín-Sánchez, F. 1996. Catchability: a key parameter for fish stock assessment. Rev. Fish. Biol. Fish. **6**: 221–242.

Auster, P.J. 1985. Aspects of American lobster, *Homarus americanus*, catch in baited traps. M.S. thesis, University of Connecticut, Avery Point, Conn. 79 p.

Bannerot, S.P., and Austin, C.B. 1983. Using frequency distributions of catch per unit effort to measure fish stock abundance. Trans. Am. Fish. Soc. **112**: 608–617.

Bannister, R.C.A. 1986. Assessment and population dynamics of commercially exploited shellfish in England and Wales. *In* North Pacific Workshop on Stock Assessment and Management of Invertebrates. *Edited by* G.S. Jamieson and N. Bourne. Can. Spec. Publ. Fish. Aquat. Sci. 92. pp. 182–194.

Bannister, R.C.A., and Addison, J.T. 1984. The effect on yield and biomass per recruit of changes in fishing mortality and size at first capture in the fisheries for lobster, *Homarus gammarus* L., in England and Wales. ICES C.M. 1984/K:10. (Mimeograph.)

Bannister, R.C.A., and Addison, J.T. 1986. Effect of assumptions about the stock-recruitment relationship on a lobster (*Homarus gammarus*) stock assessment. Can. J. Fish. Aquat. Sci. **43**: 2353–2359.

Bannister, R.C.A., and Addison, J.T. 1995. Analysis of time and space variation in a lobster population on the east coast of England. ICES Mar. Sci. Symp. **199**: 334–348.

Beverton, R.J.H., and Holt, S.J. 1957. On the dynamics of exploited fish populations. Fish. Invest. Lond. Ser. II, **19**. 533 p.

Campbell, A. 1990. The lobster (*Homarus americanus*) fishery off Lower Argyle, southwestern Nova Scotia. Can. J. Fish Aquat. Sci. **47**: 1177–1184.

Caputi, N., and Brown, R.S. 1986. Relationship between indices of juvenile abundance and recruitment in the western rock lobster (*Panulirus cygnus*) fishery. Can J. Fish. Aquat. Sci. **43**: 2131–2139.

Caputi, N., Brown, R.S., and Chubb, C.F. 1995. Regional prediction of the western rock lobster, *Palinurus cygnus*, commercial catch in Western Australia. Crustaceana, **68**: 245–256.

Chapman, C.J., and. Smith, G.L. 1978. Creel catches of crab, *Cancer pagurus* L. using different bait. J. Cons. Int. Explor. Mer. **38**: 226–229.

Clark, C.W. 1974. Possible effects of schooling on the dynamics of exploited fish populations. J. Cons. Int. Explor. Mer, **36**: 7–14.

Crecco, V., and Overholtz, W.J. 1990. Causes of density-dependent catchability for Georges Bank haddock *Melanogrammus aeglefinus*. Can. J. Fish. Aquat. Sci. **47**: 385–394.

Crecco, V., and Savoy, T.F. 1985. Density-dependent catchability and its potential causes and consequences on Connecticut River American shad, *Alosa sapidissima*. Can. J. Fish. Aquat. Sci. **42**: 1649–1657.

Cruz, R., De Leon, M.E., and Puga, R. 1995. Prediction of commercial catches of the spiny lobster *Panulirus argus* in the Gulf of Batabano, Cuba. Crustaceana, **68**: 238–244.

Ennis, G.P. 1978. Growth curves for Newfoundland lobsters from data on molt increment and proportion molting. Can. Atl. Fish. Sci. Advisory Comm. Res. Doc. 78/29. 11 p.

Fogarty, M.J., and Addison, J.T. 1997. Modelling capture processes in individual traps: entry, escapement and soak time. ICES J. Mar. Sci. **54**: 193–205.

Gulland, J.A. 1964. The reliability of catch per unit of effort as a measure of abundance in North Sea trawl fisheries. Rapp. P.-V. Réun. Cons. Perm. Int. Explor. Mer, **155**: 99–102.

Hancock, D.A. 1974. Attraction and avoidance in marine invertebrates – their possible role in developing an artificial bait. J. Cons. Int. Explor. Mer, **35**: 328–331.

Jones, R. 1981. The use of length composition data in fish stock assessments (with notes on VPA and cohort analysis). FAO Fish. Circ. 734. 55 p.

Karnofsky, E.B., and Price, H.J. 1989. Behavioral response of the lobster *Homarus americanus* to traps. Can. J. Fish. Aquat. Sci. **46**: 1625–1632.

Miller, R.J. 1989. Catchability of American lobsters and *Cancer* crabs by traps. Can. J. Fish. Aquat. Sci. **46**: 1652–1657.

Miller, R.J. 1990. Effectiveness of crab and lobster traps. Can. J. Fish. Aquat. Sci. **47**: 1228–1251.

Miller, R.J. 1995. Catchability coefficients for American lobsters (*Homarus americanus*). ICES Mar. Sci. Symp. **199**: 349–356.

Miller, R.J., and Addison, J.T. 1995. Trapping interactions of crabs and American lobsters in laboratory tanks. Can. J. Fish. Aquat. Sci. **52**: 315–324.

Miller, R.J., and Rodger, R.S. 1996. Soak times and fishing strategy for American lobster. Fish. Res. **26**: 199–205.

Paloheimo, J.E., and Dickie, L.M. 1964. Abundance and fishing success. Rapp. P.-V. Réun. Cons. Perm. Int. Explor. Mer. **155**: 152–163.

Peterman, R.M., and Steer, G.J. 1981. Relation between sport-fishing catchability coefficients and salmon abundance. Trans. Am. Fish. Soc. **110**: 585–593.

Pezzack, D.S. 1992. A review of lobster (*Homarus americanus*) landing trends in the Northwest Atlantic, 1947–1986. J. Northwest Atl. Fish. Sci. **14**: 115–127.

Richards, L.J., and Schnute, J.T. 1986. An experimental and statistical approach to the question: Is CPUE an index of abundance. Can. J. Fish. Aquat. Sci. **43**: 1214–1227.

Richards, R.A., Cobb, J.S., and Fogarty, M.J. 1983. Effects of behavioral interactions on the catchability of American lobster (*Homarus americanus*) and two species of *Cancer* crabs. Fish. Bull. (U.S.), **81**: 51–60.

Richards, R.A., and Cobb, J.S. 1987. Use of avoidance response to keep spider crabs out of traps for American lobsters. Trans. Am. Fish. Soc. **116**: 282–285.

Rothschild, B.J. 1977. Fishing effort. *In* Fish population dynamics. *Edited by* J.A. Gulland. John Wiley & Sons, Inc., New York, N.Y. pp. 96–115.

Rothschild, B.J., Powell, G., Joseph, J., Abramson, N.J., Buss, J., and Eldridge, P. 1970. A survey of the population dynamics of king crab in Alaska with particular reference to the Kodiak area. Alaska Department of Fish and Game Information Leaflet 147, Juneau, Alaska.

Smith, G.F.M. 1944. The catchability of lobsters. J. Fish. Res. Board Can. **6**: 291–301.

Smith, B.D., and Jamieson, G.S. 1989. A model for standardizing Dungeness crab (*Cancer magister*) catch rates among traps which experienced different soak times. Can. J. Fish. Aquat. Sci. **46**: 1600–1608.

Southwood, T.R.E. 1978. Ecological methods with particular reference to the study of insect populations. 2nd ed. Chapman and Hall, London, U.K.

Ulltang, Ø. 1976. Catch per unit of effort in the Norwegian purse seine fishery for Atlanto-Scandian (Norwegian spring spawning) herring. FAO Fish. Tech. Pap. **155**: 91–101.

Waltz, C. 1989. The effects of temperature and trap density on the catchability of the American lobster. Proceedings of the Workshop on the Life History of the American Lobster. Nov. 29–30, 1989, University of Maine, Orono, Maine.

Williams, M.J., and Hill, B.J. 1982. Factors influencing pot catches and population estimates of the portunid crab *Scylla serrata*. Mar. Biol. **71**: 187–192.

Winters, G.H., and Wheeler, J.P. 1985. Interaction between stock area, abundance, and catchability coefficient. Can. J. Fish. Aquat. Sci. **42**: 989–998.

The benefits of catch and effort data at a fine spatial scale in the South Australian abalone (*Haliotis laevigata* and *H. rubra*) fishery

John K. Keesing and Janine L. Baker

Abstract: A description of the South Australian abalone (*Haliotis laevigata* and *H. rubra*) fishery is provided, along with catch history characteristics since the inception of the fishery in 1967. The spatial and temporal patterns in the catch data are complex, and these are described briefly to underscore the value of working with catch data at a small spatial scale (1–10 km of coastline). This is important, given our recent awareness of the limited capacity for dispersal of abalone larvae, which results in numerous, genetically isolated metapopulations. Most abalone fisheries are managed by quota, with divers able to extract abalone throughout all areas of the fishery, potentially depleting particular metapopulations without the decline being evident in a downturn in catch per unit effort (CPUE). Data aggregated over large spatial scales (>10 km) can mask the response of metapopulations to fishing, by averaging variable productivity and resilience to fishing pressure. Examples from the South Australian abalone fishery, where catch and effort data have been collected on a small spatial scale for over 10 years, show that catch data at this small spatial resolution can indicate the status of the fishery in a way not otherwise possible. Catches vary considerably over small spatial scales, and the ability of divers to move freely and independently among all areas of the fishery results in complex spatial and temporal dynamics. Aggregating catches and catch rates over broad spatial scales masks trends in the fishery, as does the aggregation of catch data for divers with largely different fishing patterns. We stress the need for a smaller scale management approach to ensure the commercial sustainability of abalone metapopulations with different dynamics and levels of productivity over space and time.

Résumé : On donne une description de la pêcherie d'ormeaux (*Haliotis laevigata* et *H. rubra*) du sud de l'Australie ainsi que les caractéristiques historiques des prises depuis la création de cette pêcherie en 1967. Les caractéristiques spatiales et temporelles des données sur les prises sont complexes et ces dernières sont décrites brièvement pour souligner l'importance de travailler avec des données sur les prises obtenues à une petite échelle spatiale (1–10 km de côte). Cette question est d'autant plus importante que nous savons maintenant, depuis peu, que la capacité de dispersion des larves d'ormeau est limitée, ce qui entraîne la formation de nombreuses métapopulations génétiquement isolées. La plupart des pêcheries d'ormeaux sont gérées par l'établissement de quotas, les plongeurs pouvant capturer les ormeaux dans tous les recoins de la pêcherie, ce qui pourrait amener l'épuisement possible d'une métapopulation particulière sans que le déclin se manifeste dans une diminution des prises par unité d'effort (PUE). Les données regroupées sur une grande échelle spatiale (> 10 km) peuvent masquer les réactions des métapopulations à la pêche, en établissant une moyenne de la productivité variable et de la tolérance à la pression de pêche. Des exemples tirés de la pêcherie d'ormeaux du sud de l'Australie, où les données sur les prises et sur l'effort sont obtenues sur une petite échelle spatiale depuis plus de 10 ans, montrent que les données sur les prises, à cette faible résolution spatiale, peuvent indiquer l'état de la pêcherie d'une manière qui serait impossible autrement. Les prises varient considérablement sur de petites échelles spatiales et la capacité des plongeurs de se déplacer librement et indépendamment dans tous les secteurs de la pêcherie engendre une dynamique spatiale et temporelle complexe. Le regroupement des prises et des taux de prises sur de grandes échelles spatiales a pour effet de voiler les tendances de la pêcherie, comme c'est le cas lorsqu'on regroupe les données sur les prises pour des plongeurs ayant des habitudes de pêche très différentes. Nous insistons sur la nécessité de recourir à une approche de gestion caractérisée par une échelle spatiale plus petite pour assurer la durabilité commerciale des métapopulations d'ormeaux présentant des dynamiques et des niveaux de productivité différents dans le temps et dans l'espace.
[Traduit par la Rédaction]

J.K. Keesing and J.L. Baker. South Australian Aquatic Sciences Centre, South Australian Research and Development Institute, PO Box 120, Henley Beach, South Australia 5022, Australia.

Correct citation: Keesing J.K., and Baker, J.L. 1998. The benefits of catch and effort data at a fine spatial scale in the South Australian abalone (*Haliotis laevigata* and *H. rubra*) fishery. *In* Proceedings of the North Pacific Symposium on Invertebrate Stock Assessment and Management. *Edited by* G.S. Jamieson and A. Campbell. Can. Spec. Publ. Fish. Aquat. Sci. 125. pp. 179–186.

Introduction

A problem peculiar to abalone fisheries is the disaggregate nature of a stock, which is made up of many metapopulations (Shepherd and Brown 1993). Marine metapopulations may be defined as genetically distinct, reproductively discrete populations of members of one species, or a collection of interacting local populations of the same species which are reproductively connected (Shepherd and Brown 1993). Abalone have limited capacity for widespread larval dispersal (Prince et al. 1988; McShane et al. 1988) and thus conform to the metapopulation

Fig. 1. Abalone management zones within the South Australian abalone fishery and an example of the map and reef code system showing the spatial scale at which catch and effort information is collected in the South Australian abalone fishery.

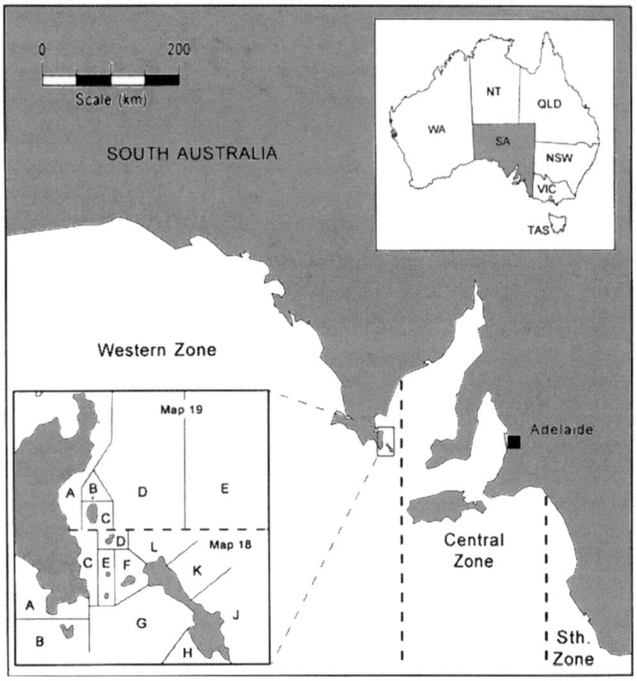

model. Southern Australian greenlip (*Haliotis laevigata* Donovan) and blacklip (*H. rubra* Leach) abalone larvae, for example, are thought to disperse tens to thousands of metres depending upon currents and topographic features (Shepherd et al. 1992; McShane 1995).

In contrast, management measures in abalone fisheries often operate uniformly over a large regional basis (tens to hundreds of kilometres). Most abalone fisheries are managed by a total allowable catch (TAC), which can be fished from anywhere over a large geographical area equally available to all participants. This system potentially allows fishers to serially deplete individual metapopulations, without overfishing being reflected by a downturn in catch rates (Breen 1992). For this reason, catch and effort data are thought to be of limited use in abalone fisheries (Prince 1992). The South Australian abalone fishery suffers this and other typical problems of quota-managed fisheries, such as uniform minimum legal lengths being applied to large geographic areas. This ensures underexploitation of slow-growing metapopulations and overexploitation of fast-growing metapopulations and encourages overexploitation of fishing grounds close to port (Shepherd and Baker 1998). Catch data collected over large spatial scales conceal such problems by averaging the response to fishing of metapopulations with different levels of productivity and mask contraction of the fishery toward the most productive core areas.

In South Australia, the catch and effort database discriminates catch, effort, and size composition of catch over small spatial scales (i.e., 195 separate reef complexes defined along the 1700-km fished length of the coast). The reef complexes are of varying sizes and, most importantly, are not based on a uniform grid, but on practical geographic reef areas delineated according to the distribution of suitable habitat. These reef complexes were identified in consultation with fishers. The reef-based system of data collection has now been operating since 1980.

Examples of analysis from this data set show it is critical to have access to small spatial scale catch and effort data in abalone fisheries. This paper provides the first detailed, published description of the historical pattern of catch and effort in the fishery, and illustrates the complex spatial and temporal scale characteristics of the fishery. Examples shown underscore the importance of having access to catch data collected on a fine spatial scale in a quota-managed abalone fishery. The only previous analysis of the spatial characteristics of catch and effort in the fishery was undertaken in 1984.[1]

Description of the fishery

The South Australian abalone fishery is a limited-entry, quota-managed fishery comprising two species *Haliotis laevigata* Donovan (greenlip) and *H. rubra* Leach (blacklip). The fishery is divided into three zones (Fig. 1), which together cover the entire South Australian coastline. There are 6 licences in each of the Southern and Central zones, and 23 licences in the Western Zone. A total allowable catch (TAC), set annually for each zone, is divided equally between all licence holders in each zone. Licences are transferable and have a high capital value, estimated to be in excess of $100 million. Different minimum sizes apply to each of the species in the different zones. Recreational fishers may take up to five abalone per day, and their impact on the total catch of abalone is unknown, but believed to be low. The amount of illicit catch is not known.

The total catch for the fishery averages about 270 t meat weight (810 t whole weight), with a catch value of A$27 million during 1993/1994 (Fig. 2). The fishery grew rapidly in value in 1985/1986, due to a price rise which followed the introduction of quotas, and the value rose again in 1992/1993.

Historical catches in the commercial fishery for each zone, along with the time course of management regulations, are shown in Fig. 3. Highest catches of both greenlip and blacklip abalone are taken in the Western Zone, where catches of these two species have been similar. The Central Zone is predominantly a greenlip fishery, and the Southern Zone predominantly blacklip (Fig. 3). The fishery in the Western Zone and that for blacklip in the Central Zone show the classic pattern of initial large catches as virgin stocks are fished down, followed by a decline and intervention by government in terms of size limit introduction and licence limitation (the number of licences was reduced from 110 in 1967 to 35 in 1976),[1] and then recovery and relative stability (Fig. 3). The large fluctuation in catch of greenlip abalone in the Central Zone between 1986/1987 and 1988/1989 was a result of the opposition of divers to planned quota introduction in 1988/1989, and also because one of the licences was inactive after 1991. Catches and catch rates have continued to rise gradually in the Southern Zone. Catch rates for all zones of the fishery are shown in Fig. 4.

[1]Lewis, R.K., Shepherd, S.A., Sluczanowski, P.R.W., and Rohan, G. An assessment of the South Australian abalone resource. Unpublished discussion paper prepared by the South Australian Department of Fisheries, March 1984. Available from the librarian at the same address as the authors of the present paper.

Fig. 2. Value of abalone and total catch value of the South Australian abalone fishery since 1973/1974. Financial years begin on 1 July.

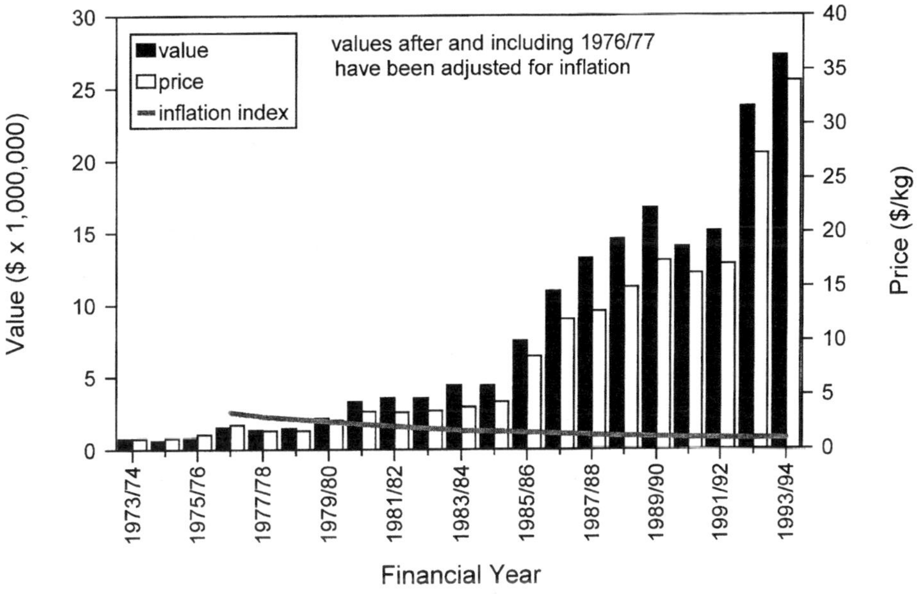

Collection of catch and effort data

Catch and effort information has been collected on the fishery since 1967.[1] The spatial resolution of collected data increased after 1979, from 20 blocks across all zones to a new system of 40 maps with a number of reef codes within each map. The reef codes are based on a system of islands, reefs, and headlands (Fig. 1). There are 110 reef codes on 20 maps in the Western Zone, 43 reef codes on 12 maps in the Central Zone, and 42 reef codes on 7 maps in the Southern Zone.

Catches are reported by fishers every month, on statutory logbook return forms, which detail the daily fishing trips in that month. For each day fished, divers provide details of catch of each species, as well as the combined fishing effort for both species in hours dived, separated into different depth categories. Where processors provide fishers with size gradings (number of abalone meats per unit weight), that information is also recorded. Greater details on the South Australian abalone catch and effort data collection system can be found in Sluczanowski (1979).

Spatial and temporal heterogeneity in the catch

Abalone catches are distributed nonuniformly in both space and time. The catch for each species varies spatially across all map numbers, and temporally among years, as shown in Fig. 5. This pattern may occur for several reasons, including: different distribution of each species; habitat heterogeneity; temporal change and spatial variation in diver behavior, abalone abundance, and catchability. The multitude of factors that result in spatial heterogeneity illustrates the need to examine the fishery on a small spatial scale. The ability of divers to move freely and independently among different metapopulations makes for complex fishery dynamics that need to be understood before conclusions that are based on catch data can be made, even where fishery independent data exist. Spatial variability in catch also occurs within maps, adding to the spatial complexity.

The catch for each of the 11 reef codes within map 18 (see Fig. 1) is shown in Fig. 6. Reef code 18F makes up less than 10% of the map area, but 42% of the catch comes from that code.

Variability in diver behavior and temporal change in individual and aggregate diver efficiency adds another level of complexity in the analysis of catch patterns in abalone fisheries. Variability in catch rate distribution among abalone divers is shown in Fig. 7. There are 20 mapped areas covering the Western Zone fishery, and in 1993 some divers took their quota from as few as 4 areas, while others fished as many as 16 areas. Some of the 20 mapped areas were fished by only 1 diver in that year, while others were fished by as many as 18 of the 22 licence holders operating in 1993. Some divers caught their quota in as few as 42 days, while others took up to 125 days; average catch varied between 63 kg and 196 kg per day.

Uses of small spatial scale catch data

The catch of blacklip abalone in the Western Zone has been relatively stable at about 100 t since quotas were introduced in 1985 (Fig. 8*a*). Since that time, catch rates have varied, but there has been a net positive change of 20%. The composition of small and large abalone in the graded catch of blacklip abalone for the whole Western Zone (1984–1993) is also shown in Fig. 8*b*. Despite stable catches, a slight decline in the proportion of large abalone in the catch has occurred (Fig. 8*b*). Analysis of the data over small spatial scales (between map areas) reveals this pattern to be stronger in some localities, and likely indicates areas of growth overfishing (e.g., from Sheringa Bay to south of Drummond Point, Fig. 8*c*), where the total catch during the two most recent years in the data set (1992 and 1993) was less than half of the total catch recorded during the early to mid-1980's. If Fig. 8*c*, which depicts an area of coastline about 60 km in length, is subdivided down to a single reef (code 11a - Sheringa Bay) along 8 km of coastline (Figs. 8*e* and 8*f*), it is apparent that although catch rates have continued

Fig. 3. Historical characteristics of catch in the three zones of the South Australian abalone fishery together with the time course of management regulation and adjustment. Fishdowns are periods of quota relaxation to allow access to substocks of abalone which never reach the legal minumum length.

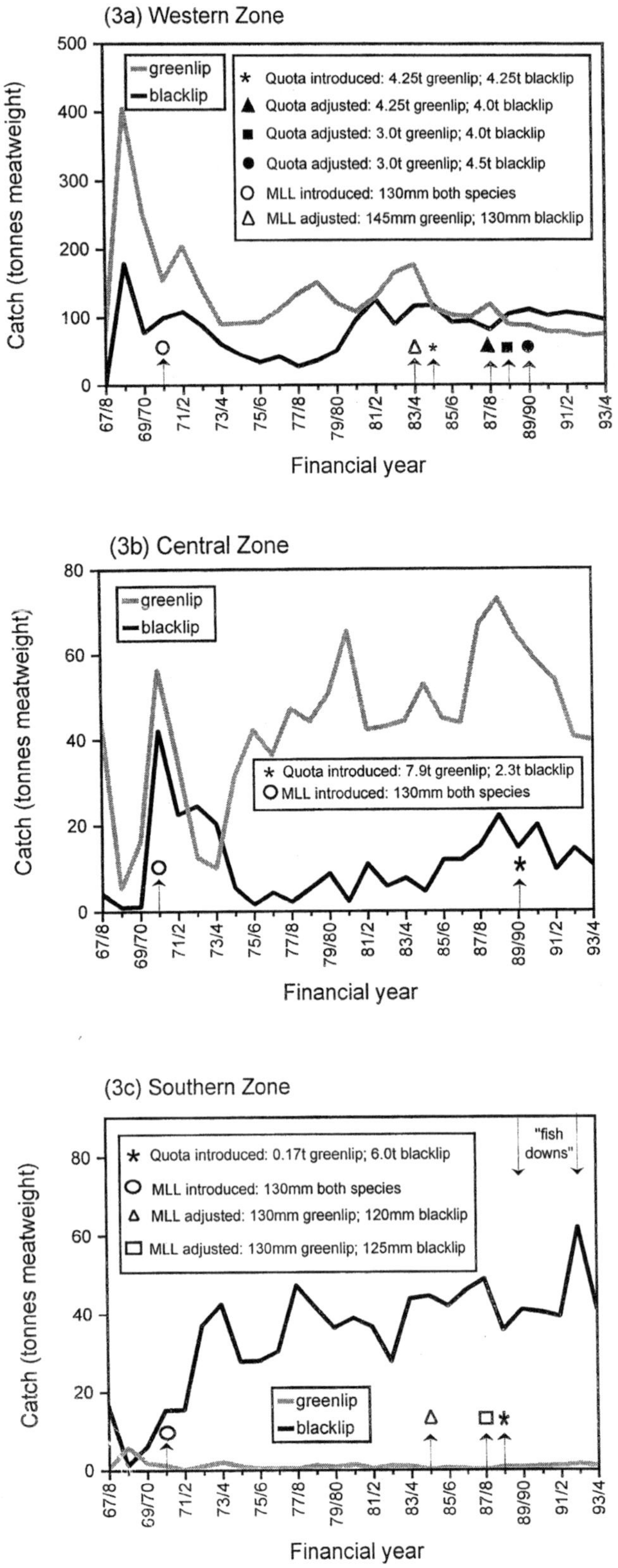

Fig. 4. Historical characteristics of catch rate in the three zones of the South Australian abalone fishery together with the time course of management regulation and adjustment. Fishdowns are periods of quota relaxation to allow access to substocks of abalone which never reach the legal minumum length.

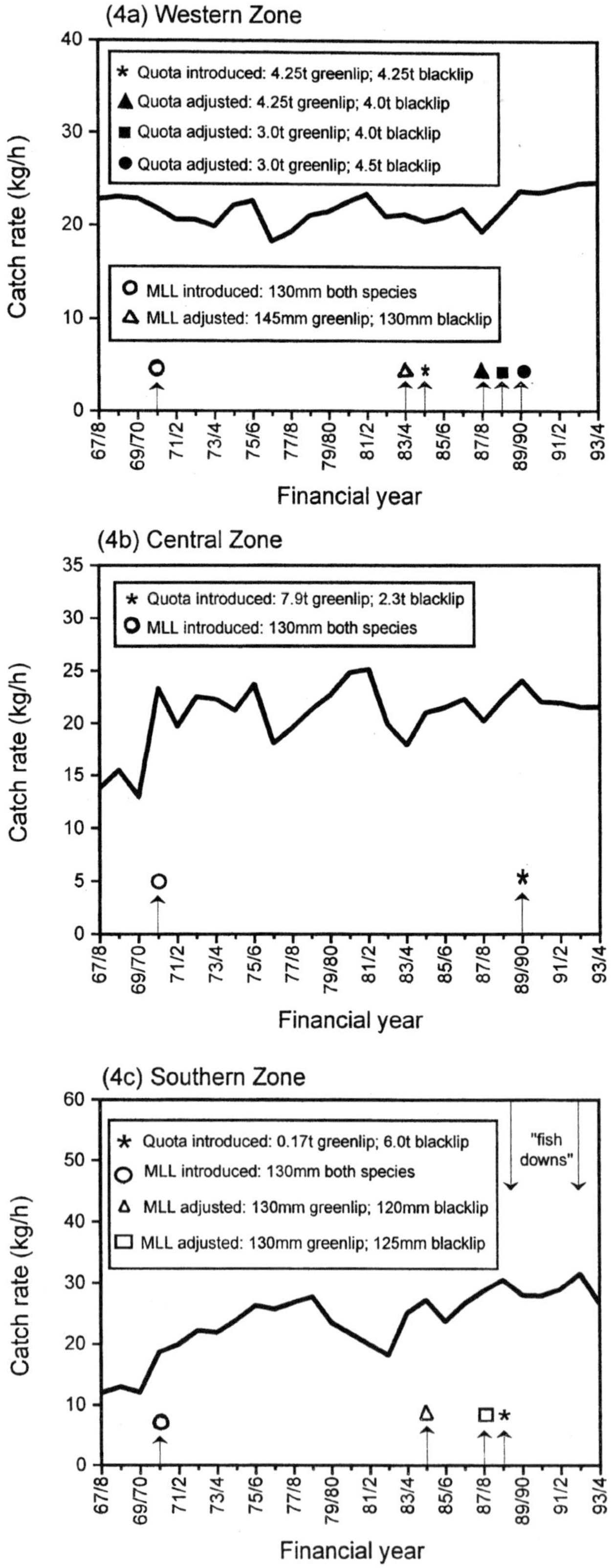

Fig. 5. Temporal and spatial characteristics of catch in the Western Zone of the South Australian abalone fishery: (*a*) greenlip and (*b*) blacklip.

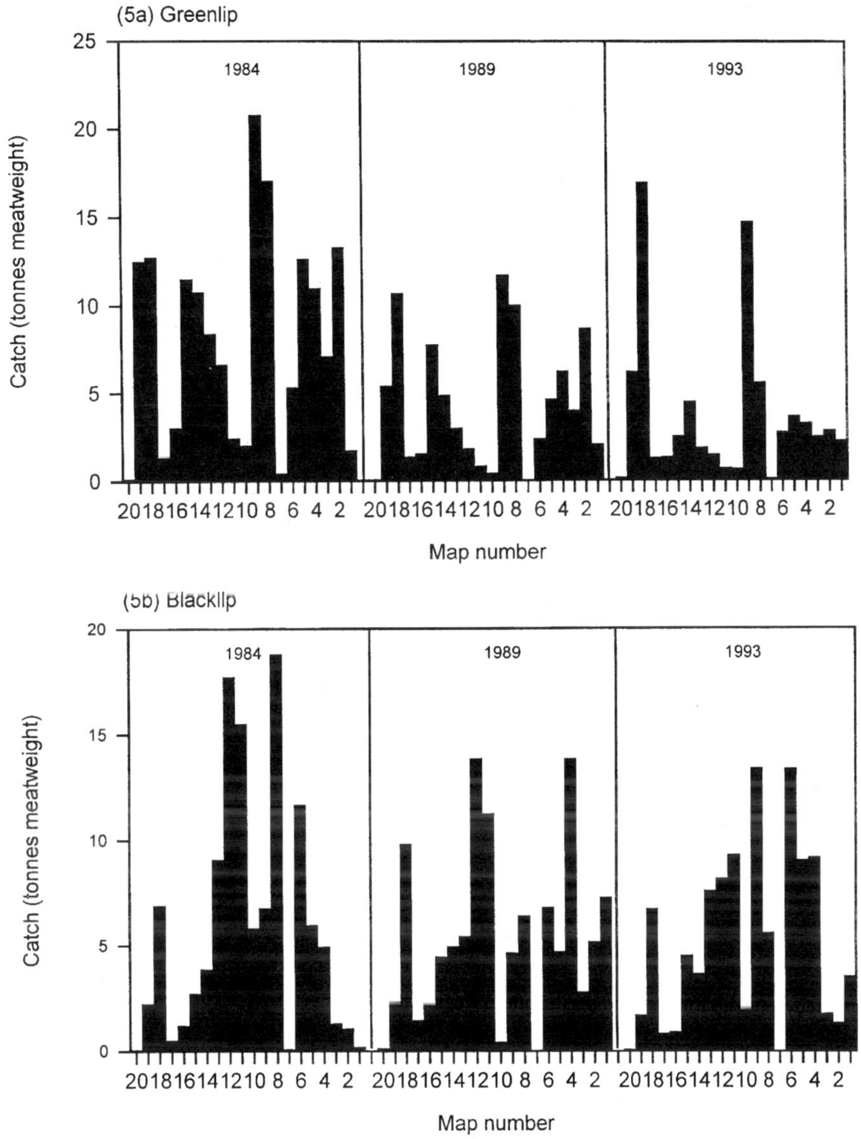

Fig. 6. Catch for each of the 11 codes within Map 18 (see Fig. 1) showing variability possible within map areas. Only data for 1993 shown.

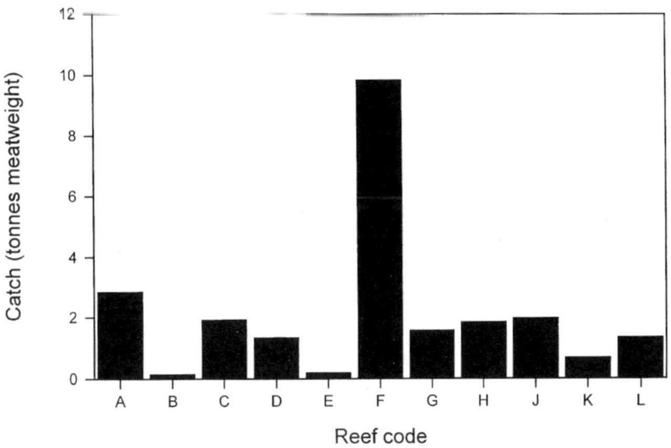

to rise, catches are trending downward and the proportion of large abalone has fallen steadily from 55% to 23% of the catch. At this small scale, these trends suggest overfishing within local stocks. In addition it may indicate that at this scale (<10 km), fishery dependent data may be of some use in interpreting local trends in an abalone fishery.

Data at the scale of map area also highlight other trends in fishers' behavior, such as uneven fishing pressure upon metapopulations according to their location and accessibility.

Some Western Zone fishers, for example, will travel further than others to maximize their catch rates, while others work closer to coastal ports for convenience, and accept lower catch rates from more heavily exploited metapopulations (Fig. 9). A temporal trend of gradually declining catches of the more heavily exploited metapopulations close to the coast is also shown in Fig. 9. If these data were aggregated at a larger spatial scale within the Western Zone, the trends of spatial increases in catch rates away from the coast, and temporal declines in catches close to the coast, would be masked.

Fig. 7. Temporal change in catch rate distribution among divers in the Western Zone of the South Australian abalone fishery.

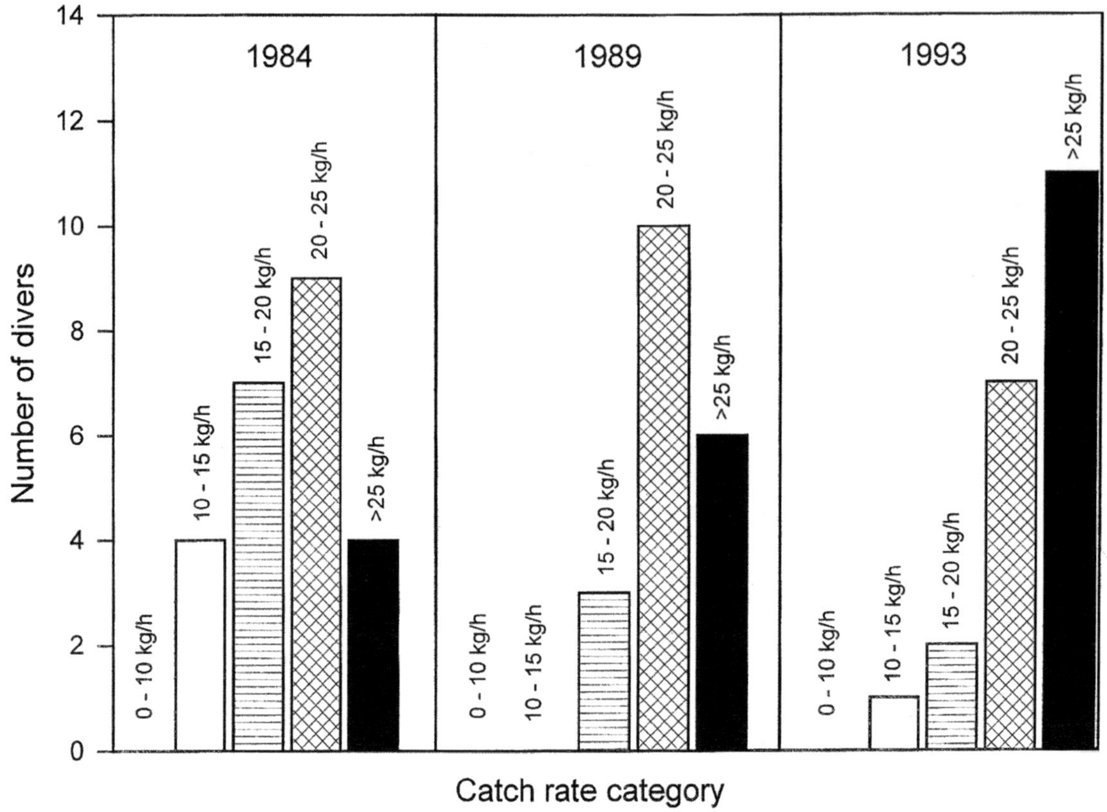

Discussion

Catch and effort data have traditionally been regarded as being of limited utility in abalone fisheries (Breen 1992; Prince 1992). Examining the temporal variability in catches and catch rates over a large number of spatially discrete units, as well as understanding patterns in diver behavior, will allow a more accurate assessment of the value of catch data in measuring the status of metapopulations and their response to fishing. Examples shown here indicate that catch and effort data at small spatial scales, especially when supplemented by other data such as catch size gradings, may be used to interpret local trends in abalone fisheries.

The ability to track temporal change in the dynamics of catch and effort at an appropriate spatial scale is critical in abalone fisheries, because serial depletion, recruitment overfishing, and growth overfishing can occur while traditional indicators such as catch per unit effort (CPUE) remain steady or improve. CPUE is probably more dependent upon idiosyncratic diver behavior (fleet dynamics) than population dynamics. Another factor confounding analysis of catch rates is that effective effort has changed over time, with larger, faster, and more comfortable boats, global positioning systems, and coloured echo sounders. It is likely that now significantly less diving time is spent searching for abalone than in the past. For this reason, data from only the past 10 years are treated in this paper.

The strategy of some divers to fish to maximize catch rates, while others work closer to port and accept lower catch rates, is likely to mask trends in a similar way to the aggregation of catch data over large spatial scales. Examining the interactions between catch, effort, and diver behavior over time, and at a small spatial scale, should allow the underlying dynamics of the fishery to be explained.

The importance of assessing catch and effort data at local scales becomes more important if we consider that growth, fecundity, and recruitment have been shown to vary considerably between metapopulations (Shepherd and Baker 1998). In addition, small abalone metapopulations with variable recruitment have been shown to be particularly vulnerable to overfishing (Shepherd and Baker 1998). Despite this, uniform management measures such as blanket minimum legal sizes and TACs are applied to whole zones of up to hundreds of kilometres of coastline.

There are other clear benefits for fisheries management offered by analysis of small spatial scale data sets, particularly for sedentary species such as abalone. Management options can target the problems affecting specific metapopulations. Utilizing only the global quota tool in such fisheries to protect stocks from overfishing, seems likely to result in a gradual reduction in the TAC, and the concomitant loss in productivity. Alternative, smaller scale management approaches will be important to the long-term sustainability of the fishery, given the variable resilience of various abalone metapopulations to fishing pressure. Sustainable management of abalone in the long term will be aided if total allowable yields from metapopulations are altered over space and time, to reflect the variable dynamics and productivity of various metapopulations. Apart from small-scale manipulation of the TAC according to variable productivity of metapopulations, one potentially useful method of sustaining abalone metapopulations is the use of a

Fig. 8. Catch of blacklip abalone and catch rate for (*a*) the whole Western Zone (800 km coastline), (*c*) Sheringa Bay to Drummond Point (60 km coastline), (*f*) Sheringa Bay (8 km coastline), (*b*, *d*, and *f*) percentage of large (1–2 pieces per pound, ppp) and small (3–5 ppp) in total graded catch for each corresponding area between 1984 and 1993.

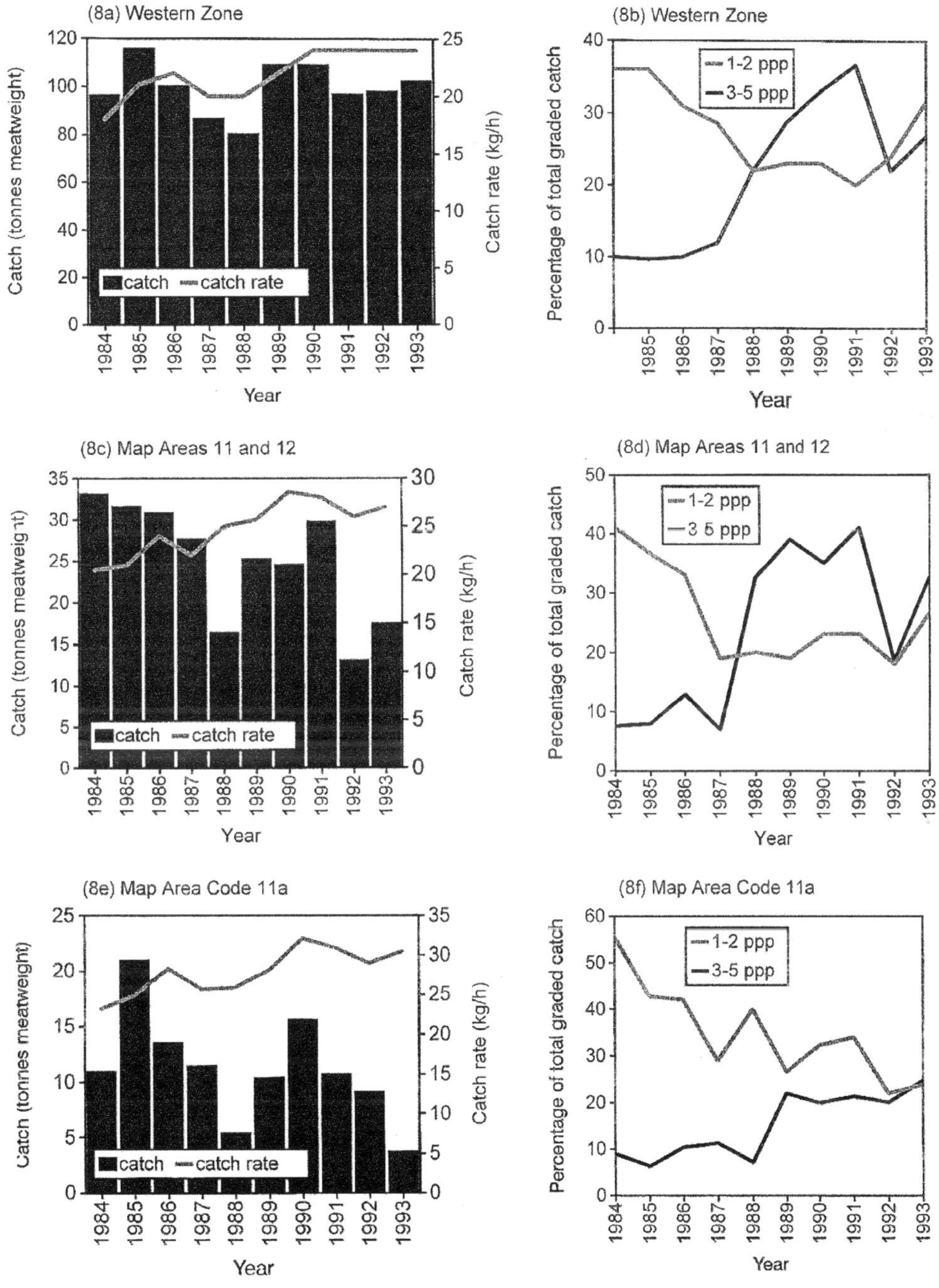

system of temporal closures, or reserves (Sluczanowski 1984, 1986).

Clearly the future for sustainability of abalone fisheries lies in management on an area-by-area basis, with multiple minimum legal sizes and TACs. The need for managers to respond to a meso-scale approach to fisheries management will prove challenging, given the desire for governments to reduce enforcement costs by regulating less, and the resistance of the industry to change. However, if a smaller scale managerial approach is not adopted, serial depletion of metapopulations may continue, and the risk of commercial collapse of the most vulnerable metapopulations will increase (Shepherd and Baker 1998).

Recently Schnute and Richards (1994) challenged fisheries biologists to ensure that the data collected now are of the type that biologists of the future will need.

Decisions about resource allocation for management of abalone fisheries on small spatial subsets of a fishery will be based on the available spatial information on historical catch. For these reasons, catch data collected on small spatial scales, as occurs in South Australia, will be important in managing this change effectively.

Fig. 9. Temporal and spatial variation in effort and catch rates for selected areas in the Western Zone, at different distances from the coast. Sites, from left to right, are approximately 5, 10, 30, 38, 45, and 64 km from port of launching.

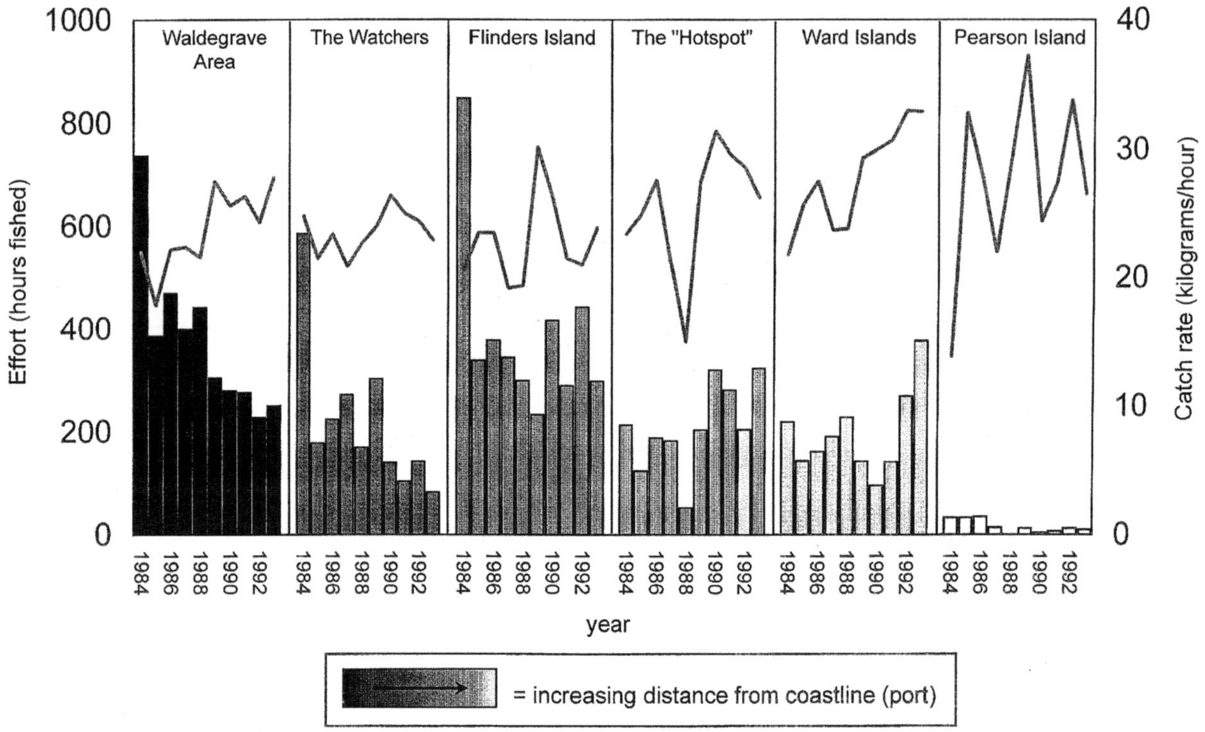

Acknowledgments

Dr. Scoresby Shepherd and the late Dr. Philip Sluczanowski were the architects of the fine spatial scale nature of the South Australian abalone catch and effort database. We are grateful for Dr. Sluczanowski's foresight and contribution to this and other aspects of southern Australian fisheries research and management. Analysis was partly enabled through a grant (94/167) from the Fisheries Research and Development Corporation (FRDC). We thank Gavin Wright for assisting with some of the figures, and S.A. Shepherd for criticizing the manuscript.

References

Breen, P.A. 1992. A review of models used for stock assessment in abalone fisheries. *In* Abalone of the world: biology, fisheries and culture. *Edited by* S.A. Shepherd, M.J. Tegner, and S.A. Guzman del Proo. Blackwell Scientific Publications Ltd., Oxford, U.K. pp. 253–275.

McShane, P.E. 1995. Recruitment variation in abalone: its importance to fisheries management. *In* Progress in abalone fisheries research. *Edited by* S.A. Shepherd, R. Day, and A.J. Butler. Spec. Publ. Mar. Freshwater Res. 46. pp. 555–571.

McShane, P.E., Black, K.P., and Smith, M.G. 1988. Recruitment processes in *Haliotis rubra* (Mollusca: Gastropoda) and regional hydrodynamics in south-east Australia imply localised dispersal of larvae. J. Exp. Mar. Biol. Ecol. **124**: 175–203.

Prince, J.D. 1992. Using a spatial model to explore the dynamics of an exploited stock of the abalone *Haliotis rubra*. *In* Abalone of the world: biology, fisheries and culture. *Edited by* S.A. Shepherd, M.J. Tegner, and S.A. Guzman del Proo. Blackwell Scientific Publications Ltd., Oxford, U.K. pp. 253–275.

Prince, J.D., Sellers, T.L., Ford, W.B., and Talbot, S.R. 1988. Confirmation of a relationship between the localised abundance of breeding stock and recruitment for *Haliotis rubra* Leach (Mollusca: Gastropoda). J. Exp. Mar. Biol. Ecol. **122**: 91–104.

Schnute, J.T., and Richards, L.J. 1994. Stock assessment for the 21st century. Fisheries, **19**(11): 10–15.

Shepherd, S.A., and Baker, J.L. 1998. Biological reference points in an abalone (*Haliotis laevigata*) fishery. *In* Proceedings of the North Pacific Symposium on Invertebrate Stock Assessment and Management. *Edited by* G.S. Jamieson and A. Campbell. Can. Spec. Publ. Fish. Aquat. Sci. 125. pp. 235–245.

Shepherd, S.A., and Brown, L.D. 1993. What is an abalone stock: implications for the role of refugia in conservation. Can. J. Fish. Aquat. Sci. **50**: 2001–2009.

Shepherd, S.A., Lowe, D., and Partington, D. 1992. Studies on South Australian abalone (genus *Haliotis*). XIII. Larval dispersal and recruitment. J. Exp. Mar. Biol. Ecol. **164**: 247–260.

Sluczanowski, P.R.W. 1979. Using new catch and effort return forms to monitor abalone stocks. SAFISH, **3**(2): 10–12.

Sluczanowski, P.R.W. 1984. A management oriented model of an abalone fishery whose substocks are subject to pulse fishing. Can. J. Fish. Aquat. Sci. **41**: 1008–1014.

Sluczanowski, P.R.W. 1986. A disaggregate model for sedentary stocks: the case for South Australian abalone. *In* North Pacific Workshop on Stock Assessment and Management of Invertebrates. *Edited by* G.S. Jamieson and N. Bourne. Can. Spec. Publ. Fish. Aquat. Sci. 92. pp. 393–401.

Concentration profiles and invertebrate fisheries management

Jeremy Prince and Ray Hilborn

Abstract: The spatial distribution of resources and the behavior of a species lead to a spatial pattern of density which has been called the concentration profile of the species. Some species aggregate and may be found almost entirely at high densities, other species space themselves much more uniformly and are usually found at low densities, and some species are found at a very wide range of densities. The interaction between the concentration profile, fishing, and management behavior leads to a number of surprising consequences. Fishermen will naturally tend to fish in the most profitable places, so that the interaction between the concentration profile of the species and the behavior of the fishermen will lead to a complex spatial pattern in abundance, profitability, and catch per unit effort (CPUE). Among the behaviors that emerge from such analysis are (*i*) overall CPUE will frequently decline more rapidly than abundance, (*ii*) CPUE will provide almost no information about abundance that can be used for management except on the smallest of spatial scales, and (*iii*) changes in price or costs of fishing may have significant impacts on the spatial pattern of fishing effort and the associated CPUE. We illustrate these principles in detail with examples from the Tasmanian abalone fishery on a small scale (tens of metres) and a large scale (hundreds of kilometres). Recognition of the concentration profile of the stock suggests that regulatory measures must be carefully tuned to the biology of the species. CPUE will almost certainly be a very poor measure of stock abundance, while fishery-independent surveys will provide the only reliable method of estimating stock status. Complex spatial structure, with differences in growth and mortality will mean that size limits and harvest rates should be finely tuned to the spatial structure of the stock and cannot be set appropriately over large areas. We consider alternative regulatory schemes such as quotas, seasons, size limits, and territorial fishing rights allocations in relation to concentration profiles and suggest that territorial fishing rights offer the greatest potential benefits.

Résumé : La distribution spatiale des ressources et le comportement d'une espèce mènent à une répartition spatiale de la densité que l'on a appelée le profil de concentration d'une espèce. Certaines espèces se rassemblent et se rencontrent presque exclusivement à des densités élevées, certaines se répartissent de manière beaucoup plus uniforme dans l'espace et, enfin, certaines présentent une répartition selon une très grande variété de densités. L'interaction entre le profil de concentration, la pêche et le comportement de gestion mène à des conséquences étonnantes. Les pêcheurs ont naturellement tendance à pratiquer la pêche dans les endroits où cette activité est le plus rentable, de sorte que l'interaction entre le profil de concentration de l'espèce et le comportement des pêcheurs sera à l'origine d'une distribution spatiale complexe en ce qui a trait à l'abondance, à la rentabilité et les prises par unité d'effort (PUE). Parmi les comportements qui ressortent d'une telle analyse figurent les suivants : (*i*) le PUE global déclinera fréquemment plus rapidement que l'abondance, (*ii*) le PUE ne donne presque pas d'information sur l'abondance qui puisse servir à la gestion, sauf à la plus petite des échelles spatiales, et (*iii*) les changements touchant les prix ou les coûts liés à la pêche peuvent avoir des répercussions substantielles sur la distribution spatiale de l'effort de pêche et le PUE correspondant. Nous illustrons ces principes en détail avec des exemples tirés de la pêcherie d'haliotides de Tasmanie à petite échelle (dizaines de mètres) et à grande échelle (centaines de kilomètres). La reconnaissance du profil de concentration du stock laisse supposer que les mesures de réglementation doivent être soigneusement harmonisées avec la biologie de l'espèce. Le PUE constituera presque certainement une mesure médiocre de l'abondance du stock, alors que les relevés indépendants de la pêcherie constitueront la seule méthode fiable pour estimer l'état du stock. La structure spatiale complexe, comportant des différences en ce qui a trait à la croissance et à la mortalité, signifiera que les limites de taille et les taux de récolte devront être ajustés finement à la structure spatiale du stock et qu'ils ne peuvent être appliqués de manière appropriée sur de vastes étendues. Nous prenons en considération d'autres mesures de réglementation comme l'attribution de quotas, les saisons de pêche, les limites de taille et les droits de pêche territoriaux en rapport avec les profils de concentration et estimons que ce sont les droits de pêche territoriaux qui présentent le plus d'avantages potentiels. [Traduit par la Rédaction]

J. Prince. P.O. Box 209, Leederville, West Australia.

R. Hilborn. School of Fisheries, Box 357980, University of Washington, Seattle, WA 98195-7980, U.S.A.

Correct citation: Prince, J., and R. Hilborn. 1998. Concentration profiles and invertebrate fisheries management. *In* Proceedings of the North Pacific Symposium on Invertebrate Stock Assessment and Management. *Edited by* G.S. Jamieson and A. Campbell. Can. Spec. Publ. Fish. Aquat. Sci. 125. pp. 187–196.

Introduction

Few invertebrate resources are spatially homogeneous, individuals are almost always found in higher concentrations in some locations. This distribution of concentrations has many implications for the biology and management of invertebrates and its importance has been noted for many years. Gross and Smyth (1946) suggested that very high density concentrations of oysters may be required for reproductive success. Caddy (1975) and Orensanz (1986) described the spatial pattern of

Fig. 1. Types of concentration profiles.

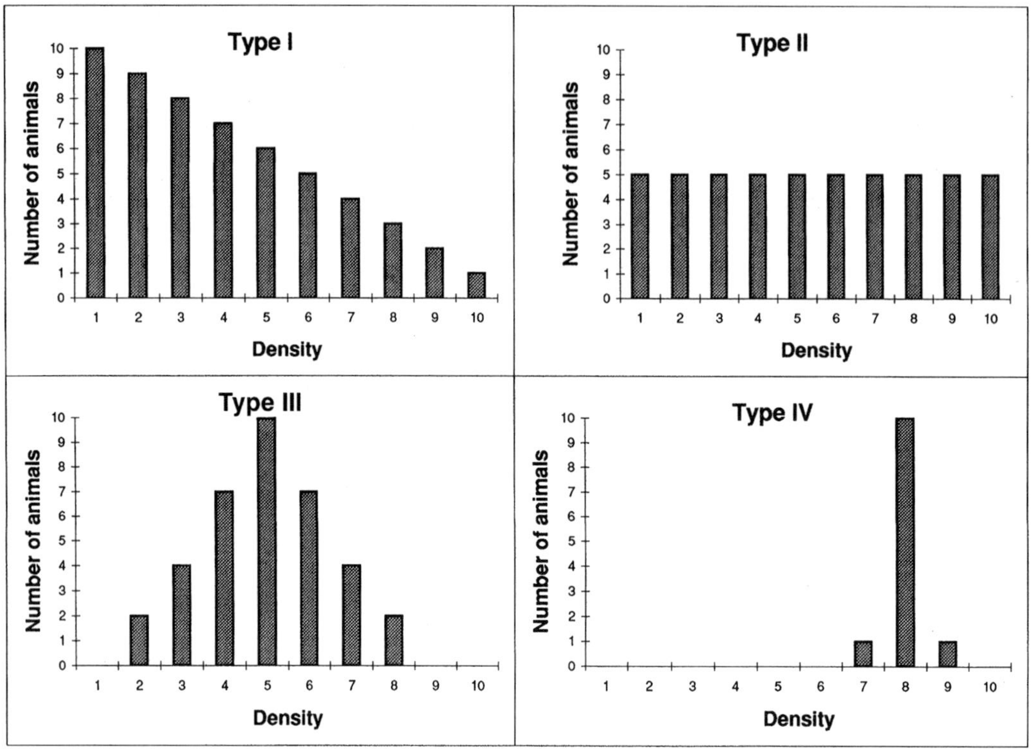

concentration in different scallop populations and discussed biological and management implications of these distributions.

Clark (1982) introduced the term "concentration profile" to the fisheries literature and considered a number of theoretical implications of different concentration profiles on the behavior of fishermen and interpretation of catch-per-unit-effort (CPUE). Clark borrowed the term from mining, where the concentration profile refers to how much of the desired mineral can be found at different concentrations in the ground. For any mineral, the profitability of mining depends primarily on how much recoverable mineral there is per ton of material processed. Below some concentrations it is not economic to process the material. A potential miner is not interested so much in how much mineral is in an area as he is in the amount of ore that exists at profitable concentrations.

This is equally true of fishers. Most fish do not occur at densities that can be profitably harvested, yet the key to an economically viable fishery is the number of fish that can be found at high concentrations. Clark introduced four types of concentration profiles, shown in Fig. 1. Type I profiles have a diminishing number of individuals found at higher densities; type II profiles have a uniform distribution; type III have individuals spread out over a range of densities with most found at an intermediate density; and type IV profiles have almost all individuals found at a single density.

Concentration profiles for anchovy in California and skipjack tuna in the western Pacific were shown in Hilborn and Walters (1992). Both showed type III profiles. The data from Orensanz (1986) are shown as a concentration profile in Fig. 2. It appears to be a type II profile.

Concentration profiles result from the interaction of habitat quality, the behavior of the fish, and the harvesting history. For instance, type IV profiles would be expected to result if the fish were highly aggregating, whereas a type I might result if fish density depended exclusively upon habitat quality and there were many more poor than good areas.

The purpose of this paper is to review what is known about the concentration profiles in invertebrate fisheries and consider the implications of concentration profiles for management. We begin with a discussion of the interaction between concentration profiles and fishermen's behavior and then explore examples of the small-scale and large-scale concentration profiles for abalone in Tasmania.

Fishermen's behavior and concentration profiles

The earliest formal theory of how fishermen will respond to the concentration profile of the fish stock was described by Gordon (1953), who suggested that fishermen would attempt to maximize their profits by optimizing the costs of fishing with the rates of return. If the initial abundance of a fish resource was uniform, fishermen would begin by fishing close to home port, and as local stocks became depleted, they would move farther away, at all times assuring that the expected rate of return in all areas fished was the same. If for some reason the rate of return (income minus costs) were to rise in an area relative to other areas, additional effort would flow into that area, and if the rate of return were to drop relative to other areas, then effort would leave.

Beverton and Holt (1957) provide a similar discussion, and the theory was popularized in the ecological literature by Fretwell and Lucas (1970) and given the name of "ideal free distribution." Hilborn and Ledbetter (1979) showed that

Fig. 2. Concentration profile from Orensanz data.

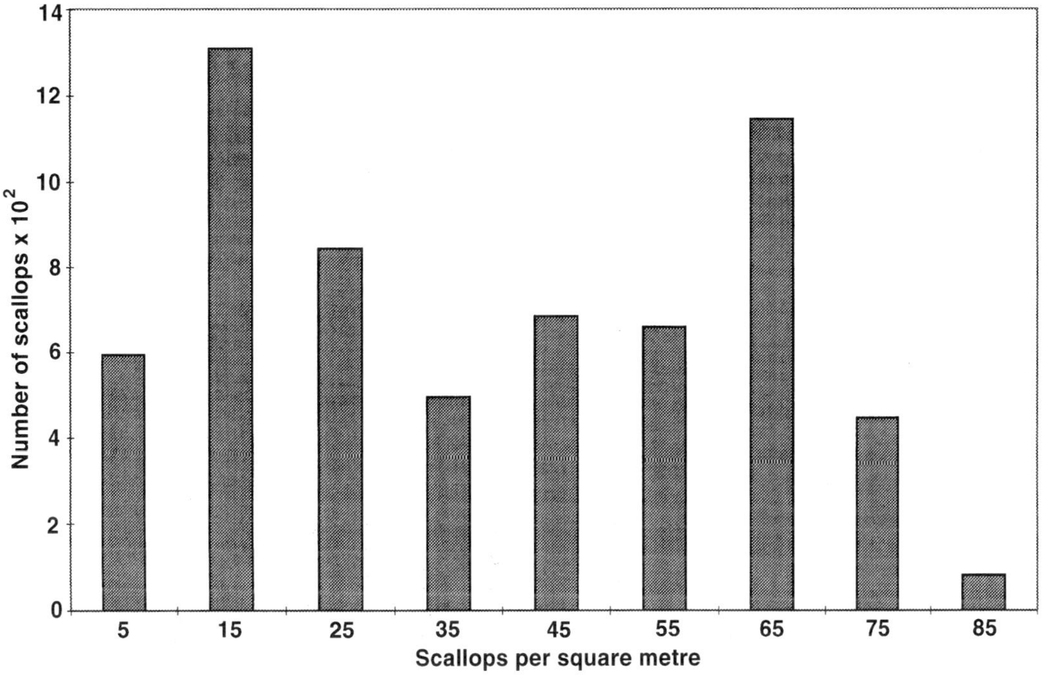

fishermen in the British Columbia purse seine fishery appeared to behave according to these rules, and Clark (1982, 1985) used this theory in his analysis of the implications of concentration profiles. Since then a number of authors have found that fishermen behave as predicted (Abrahams and Healey 1990; Hilborn and Kennedy 1992).

If we assume that fishermen have good information about the location and density of the population or can learn this in the process of fishing, then it is quite natural that fishers will begin fishing at the highest concentrations and gradually work down to lower concentrations until they reach a concentration at which fishing is no longer profitable. The threshold of profitability will depend upon a number of factors associated with the cost of fishing, which may include distance from port, probability of gear loss, depth, etc. In the aggregate, however, we would expect that the natural pattern would be for the highest concentrations to be removed first.

As discussed in Hilborn and Walters (1992) such a behavior would mean that in a type I profile the CPUE would initially decline more rapidly than abundance. The few very good sites would be fished first and, assuming the CPUE depends on the density in the area fished, as fishermen moved to lower density sites, the CPUE would drop. If the concentration profile had a long right tail, the CPUE might initially drop substantially even if only a very small fraction of the total population had been removed. However, with a type IV profile, the CPUE would stay constant while the population was being depleted, a phenomenon that has been observed with a number of strongly schooling fish.

Assuming there are differential costs of fishing in different areas, the CPUE will reflect the costs of fishing at any time in the development of the fishery, so that the profitability in each area will be the same. Areas with high costs will have higher CPUE. Under such circumstances, changes in price will have a significant effect on CPUE. If the price of the product goes up, areas of high CPUE would be relatively more profitable and areas which had not been profitable before would suddenly become profitable, and it has been shown (Hilborn and Walters 1987) that the average CPUE can increase. Similarly, if prices dropped, CPUE would be expected to decrease. Thus aggregate measures of CPUE are subject to changes in the price and cost of fishing in a rather surprising fashion.

Small-scale concentration profiles

The nature of small-scale concentration profiles was demonstrated in an experiment conducted by Prince (1989). A small reef in Tasmania, named George III rock, had been maintained as a reserve for abalone for a number of years. In July of 1987, four volunteer licensed abalone divers spent seven days fishing this reef, recording each dive location and the catch. In the first few days most divers concentrated on the shallow areas of the reef where most abalone were found. As the shallower areas were fished down, the divers moved to deeper areas of the reef where densities had initially been lower. Figure 3 shows the CPUE in abalone per minute at the beginning of the experiment and at the end. Whereas initially the CPUE was much higher in shallow sites, by the end of the 7 days of fishing the CPUE was higher in deeper sites. This is all consistent with the theory that divers prefer to work in shallow water because of health and comfort considerations, and that in a "developed" fishery, the CPUE will be higher at high-cost (in this case deeper) sites.

The reef was divided into 29 roughly equal-sized areas and an analysis of the initial densities in each area revealed a type III or type IV concentration profile. This experiment demonstrated that concentration profiles exist on very small scales; the George III rock was <1 km^2. This distribution of differential densities after the fishery is presumably what the

Fig. 3. Kilograms of abalone captured per minute plotted against depth shown for the first days diving (light bars) and the last days diving (dark bars) from the George III reef.

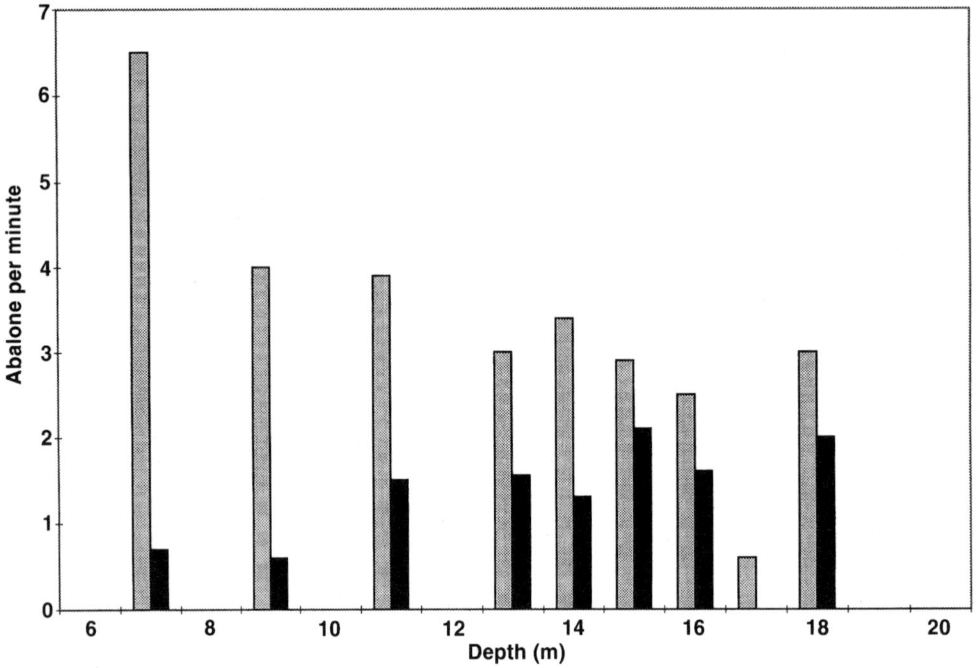

Fig. 4. Total catch (light line) and effort (dark line) for theTasmanian abalonc fishery.

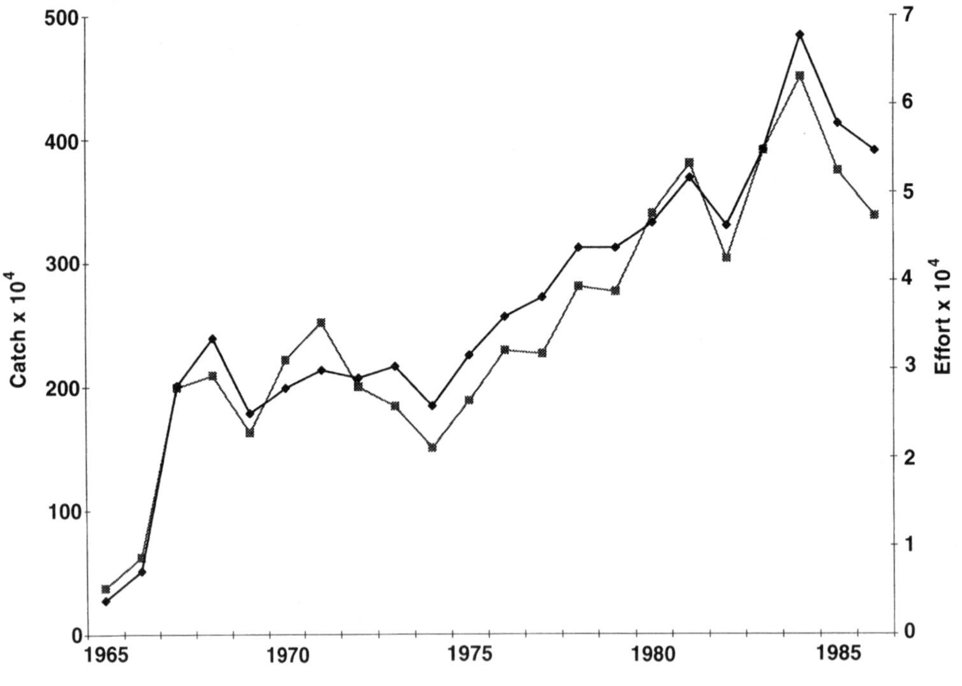

fine scale structure of abalone abundance looks like throughout Tasmania and likely wherever species like abalone occur and are fished. The key observation is that the initial high concentration sites will be fished down and the remaining concentration profile will reflect the cost of fishing, with deep areas having higher concentrations than shallow areas. Areas that have high costs of fishing for other reasons would also be expected to have high concentrations.

Large-scale concentration profiles

We can use this understanding of the determinants of where

Fig. 5. CPUE trends in Tasmanian abalone fishery. The thick line in the middle represents average CPUE, the thin line on top is the CPUE for west coast sites, and the thin line on the bottom is CPUE for east coast sites.

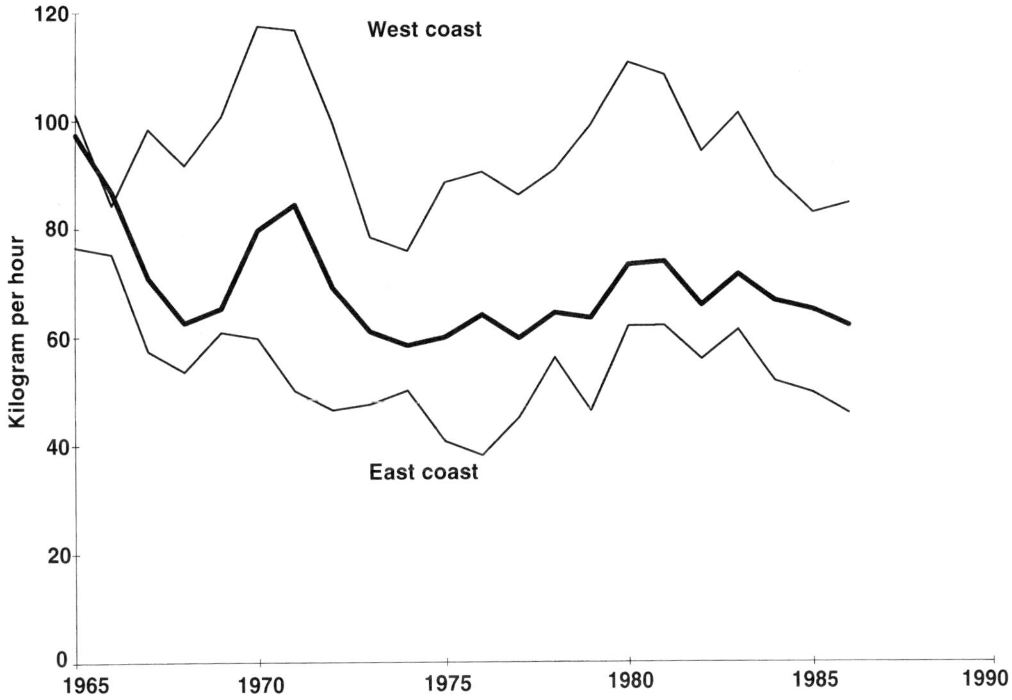

Fig. 6. Map of Tasmania showing statistical areas; circles represent number of individual licensed divers with a home in each community.

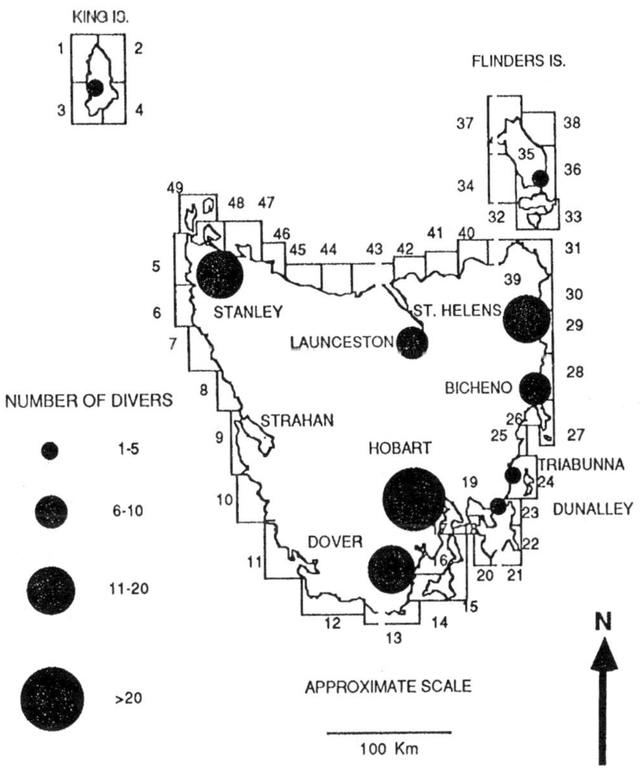

fishers go and what determines catch rates to understand the evolution of the entire Tasmanian abalone fishery, in particular how the small- and large-scale patterns of concentration and costs affect the interpretation of CPUE trends in the fishery.

Historical trends

Prior to 1963, the Tasmania abalone stock was essentially untouched, but by 1963, modern diving equipment and the recognition of Asian markets provided the needed basis for the modern industry. Figure 4 shows the catch and effort trends in the fishery from 1965 to 1987. Total catch rose rapidly from 1963 to 1967, fluctuated between 2500 and 3500 tons between 1967 and 1975, and rose steadily until 1984, when a system of individual transferable quotas was introduced. Effort shows essentially the same trends with somewhat less variation between 1967 and 1975.

Figure 5 shows the trends in catch-per-unit-effort, measured in kilograms per diver-hour under water for the west coast, east coast, and total. The total pattern is rather stable, with a small initial drop in CPUE, a rise in 1970 and 1971, followed by another small decline, and then minor fluctuations. If we were to assume that CPUE is proportional to abundance, our interpretation would be that there had been only a minor decline in abundance since the development of the fishery. The fact that CPUE has remained essentially unchanged since 1974 while effort and catch have tripled, suggests either that the stock is remarkably resilient, or that CPUE is not proportional to abundance. We consider alternative interpretations of the CPUE trend in a later section, but to do this it is necessary to first examine the mechanics of the abalone fishery.

Abalone are taken from rocky bottom areas between 2 and

Fig. 7. Distribution of total catch (vertical bars) and average CPUE (solid line) by statistical area.

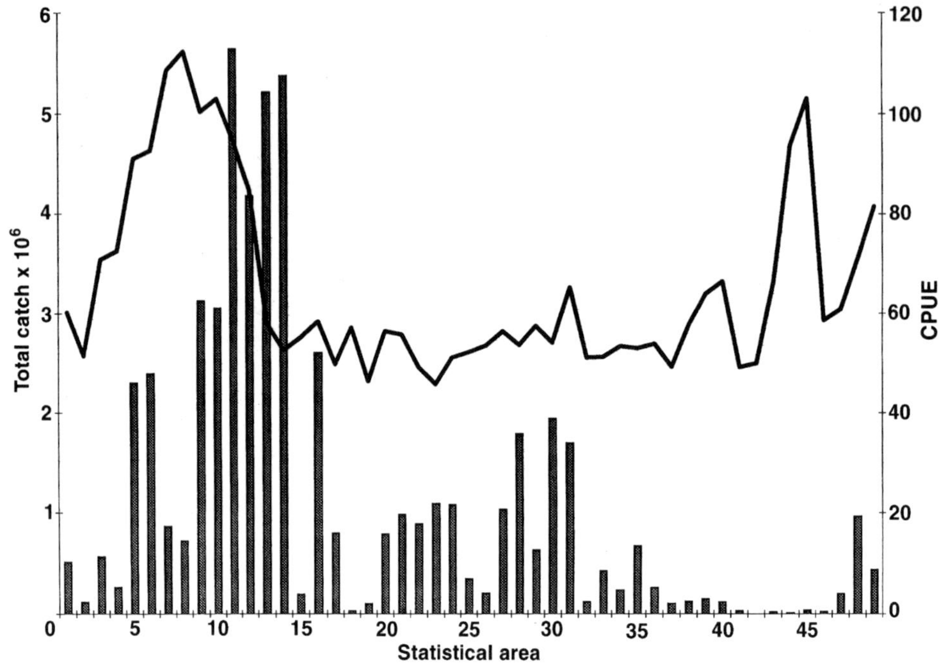

30 m in depth around the entire coast of Tasmania. There are 125 licensed divers who must do their own diving. Divers operate from small, outboard-powered boats, either 4-m dinghies or 7-m runabouts, using hookah gear which pumps air to the diver from the surface. Nearly all divers employ a surface deckhand who manipulates the boat and lifts nets of harvested abalone into the vessel. The divers typically spend 4–5 h per day under water and move along the bottom, harvesting all abalone above the legal size limit. Although diving methods do differ between individuals, divers typically might search 20 $m^2 \cdot min^{-1}$ and encounter 0.1–2.0 abalone$\cdot m^{-2}$.

Most divers prefer to undertake daily fishing trips from their home ports using the larger, twin-hulled, 7-m runabouts. The range on these trips is generally 20–40 km from the point of launching, although distances as great as 80 km may be traveled if the weather is favorable. The best fishing grounds, however, are on the isolated west coast of Tasmania. Divers fish these areas on extended trips (3–7 days), operating from 15 to 30 m motherships, with each diver and his assistant using a 4-m dinghy for daily trips up to 15 km away from the mothership.

Spatial structure

Figure 6 shows a map of Tasmania with the statistical areas used for the collection of abalone data. Also shown are the major home ports of abalone fishermen, with the size of the circle representing the number of divers who live in those communities. Nearly all divers live on the east coast which is well serviced by roads and has many small communities. In contrast, the west coast is very isolated, has only one real access point at Strahan, and is far from the major urban centers of Hobart and Launceston.

Figure 7 shows the distribution of abalone catch (bar graph), with totals from 1965 to 1986 by statistical area. The catch comes predominantly from the west coast and a second area of major productivity on the south and east coasts. The north coast and Flinders and King Islands are unproductive. Figure 7 also shows the average catch rate ($kg \cdot h^{-1}$) from 1965 to 1986 by statistical area. The data lump quite nicely into two groups, the west coast (areas 5–12) and the south and east coasts (areas 13–31). We can ignore areas 1–4 and areas 32–49 simply because the total catch in these areas is trivial. We therefore consider the fishery as consisting of two major areas, the west coast which is far from home, exposed to the prevailing winds and has few safe shelters, and the east coast which is close to home, to the leeward, and has many ports and protected anchorages. The west coast has high catch rates, the east coast has low catch rates.

Alternative hypotheses

Alternative hypotheses to explain the trend in CPUE have been discussed by Harrison (1983) and include increased mechanical efficiency, motivation, and skill. Harrison's primary concern was to attempt to calibrate the data to be used in stock assessment. In this paper we are primarily interested in describing the characteristics of the fishery and how different mechanisms interact to produce observed CPUE trends. We will not attempt to produce a final 'calibrated' CPUE trend and, indeed, we believe this cannot and should not be done.

We begin by discussing alternative hypotheses, the details of the hypotheses, and how the concentration profiles, on the small and large scale, would impact them. In the subsequent section we will look at particular aspects of the spatial structure and the CPUE trends to try to understand the relative importance of the different mechanisms.

Fishermen's movement

The spatial structure of the fishery and the sedentary nature of

Fig. 8. Price of abalone (solid line) and sale price of commercial abalone license.

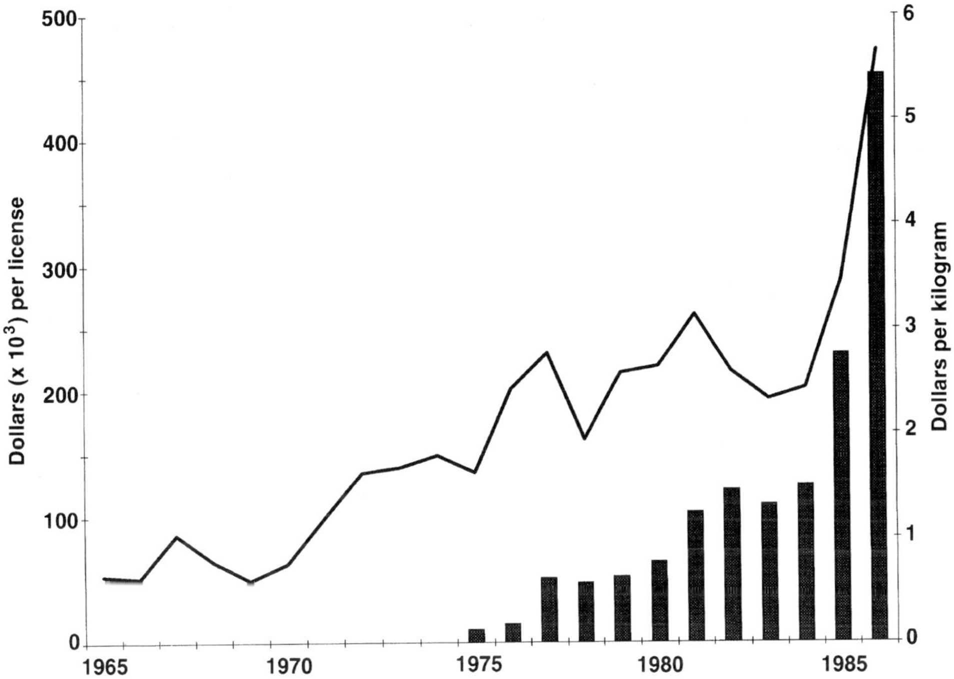

the exploited stocks assure that movement patterns will have an important effect on catch rates. This has been noted previously as an important factor for other abalone fisheries. In the simplest case, we might imagine a pattern of sequential depletion where divers simply clean out reefs that are close to harbors and anchorages and then move farther and farther away. The CPUE could easily stay constant while the total abundance declined drastically. Abalone fisheries would therefore be more like mining operations than fisheries. However, we believe more complex interactions between movement, stock abundance, and motivation of divers continue to occur over the complete range of spatial scales and these will be examined below.

Increased mechanical efficiency

Two major technical changes have taken place since the beginning of the fishery. Before 1969, hookah equipment displaced SCUBA equipment, extending the time available for diving, and reducing the amount of equipment required, and since about 1971, divers employ deckhands who follow them with the vessels and untangle air hoses, allowing unrestricted movement. Moreover, diving suits continue to be more efficient and more comfortable, enabling divers to spend more time underwater. The use of dive computers and motherships have also been noted as technical innovations increasing efficiency.

We expect these increases in mechanical efficiency to affect the fishery in two substantially different ways. The use of hookah equipment and improvements in diving suits increased the number of hours spent diving each day without a major effect on the hourly catch rate, whereas the use of deckhands should have directly affected the catch rates without increasing the total hours worked.

Increasing knowledge and skill

Diver skill can vary within the fishery both individually and collectively. The rudimentary skill of hookah diving is easily acquired, but the real skill of an abalone diver is his ability to quickly locate abalone stocks in unfamiliar areas and to be able to find these stocks again in the future. Another aspect of diver skill involves the knowledge a diver accumulates about stock abundance and his general familiarity with fishing sites. We examined the catch rate of new divers against the number of months they had been in the fishery and found that while their initial catch rates were about 30 $kg{\cdot}h^{-1}$, within five months they were catching 60–70 $kg{\cdot}h^{-1}$ and this did not change with more experience. This rise in catch rate will reflect both increasing skill/experience and may also reflect a willingness of new divers to fish at deeper sites. This we cannot document but it is related to motivation.

Motivation

Once we accept the small-scale and large-scale concentration profiles and the interaction between costs of fishing and CPUE, motivation determines individual diver's choice of where to fish. On the small scale this reflects how frequently they will be willing to dive deep, and on the large scale how frequently they will go to the west coast. The need to earn income can be a primary influence on a diver's motivation. A diver seeking to service debts, establish himself financially, and purchase fishing equipment in the early years of his career will almost certainly be willing to dive deeper or go to the west coast more often than divers who are financially secure.

The price offered for abalone has fluctuated widely at times (Fig. 8) and has also affected the motivation of divers. High prices encourage divers to work hard and take advantage of 'windfall profits,' while low prices may reduce or increase the motivation of divers depending on their financial need. Those with a high need may be forced to work harder, while others may choose not to land catch at the lower price.

Competition is also a source of motivation. Harrison (1983)

reported that established divers without financial needs increased their fishing activity purely to compete with the new, harder-working divers who entered the industry after 1974.

Motivation determines how a diver allocates his effort over a range of fishing sites. Divers will seek to minimize their risk and maximize their comfort both within their general life-style and within their diving practice. To maximize comfort, divers favor short day trips from their home ports rather than extended trips. This gives them more time to enjoy being at home with their families. Many divers prefer to work exclusively from their home ports and avoid extended trips to the west and south coasts. The preference for fishing close to home has been discussed in the literature as theoretically reasonable (Gordon 1954; Beverton and Holt 1957), and has been demonstrated empirically (Hilborn and Ledbetter 1979).

A constant concern is diving-related diseases, particularly dysbaric osteonecrosis, exacerbated by diving time and depth, which causes divers to favor diving in shallow depths (<10 m). This concern and the concerns about the bends are the two major reasons for divers' preference of shallow areas.

Specific events

We can use our understanding of the interaction of concentration profiles and divers' behavior to analyze the historical development of the fishery. In this section we divide this history into a number of periods, characterized primarily by licensing changes and consequent changes on motivation and license price, and discuss what we believe explains the changes that occurred.

1965–1968 Development of the fishery

Until 1969, anyone could enter the fishery for an annual $5.00 license fee. There were few full-time abalone divers since most divers held full-time jobs outside the fishing sector. Figure 5 shows the CPUE trends for the total fishery, as well as west coast and east coast areas separately. The total CPUE declines rather strongly during this period, dropping roughly 35% between 1965 and 1968. This is almost certainly due to declining abalone abundance, moderated by considerable improvement in diving equipment, practices, and knowledge of the diving areas. Harrison (1983) estimated that diving efficiency was twice as high in 1969 as in 1965. Much of this increasing efficiency would have been communal rather than personal since anecdotal information suggests a great turnover of divers during this period. Of particular interest is the marked decline of the fishery on the east coast during the first ten years, without developing a similar pattern on the west coast.

1969–1974 Limited entry, but nontransferable licenses

In 1969, limited entry was introduced. Fishermen had to earn their income primarily from the fishery, forcing them to be full-time divers. These measures, along with the introduction of an annual $100 license fee decreased the number of divers from over 250 to 120 (in 1973, an additional five restricted divers were allowed entry). This resulted in an overall decrease in effort between 1968 and 1969 (Fig. 4). The drop is not proportional to the number of divers leaving the fishery because the number of hours per diver-day increased slightly and the number of days in the water per diver increased by approximately 40%. During this time, licenses could not be transferred; a diver wishing to leave the industry relinquished his diving entitlement to the state fisheries department which reallocated the license to the next person on a waiting list of applicants. Because of this very few divers left the industry during this period. The year 1969 provides a major breakpoint in terms of professionalism and motivation. While the price remained relatively low, the divers involved had the economic need to upgrade their equipment to a professional standard and the desire to establish themselves economically.

Divers' motivation is thought to have dropped considerably during 1972–1974. This is consistent with the fact that the number of hours dived per day declined while the number of days dived remained stable. Moreover, the proportion of effort expended on the west coast declined. This should be expected, since the divers did well financially during 1969–1971. They are likely to have had low levels of debt as they did not have to buy an entitlement to enter the industry. They also had the benefit of fishing relatively virgin stocks on the west coast. As a result, it is likely that they were able to pay for the plant and equipment needed and establish some level of economic security. As prices rose from $0.23 in 1969 to $0.94 in 1974 and the divers began to become aware of the health risks of extensive diving, motivation would have continued to decline. Divers began to lobby the government to allow them to nominate to whom to transfer their entitlement if they left the industry. In other Australian fisheries this had already happened, effectively allowing fishermen to sell their fishing entitlement on the open market. In 1972, the government announced that it would implement this system of license transfer to promote a turnover of divers and ameliorate the effect of diving diseases, but later that year it reversed its decision. The effect was that most divers felt the change would eventually be made and those divers wishing to leave the industry remained in it anticipating a 'windfall' profit when the legislation eventually changed. This can also be expected to have decreased the divers' incentive.

1975–1984 Transferable licenses

In August 1974, the government allowed divers to sell their licenses to people wishing to enter the industry. This period was marked by generally rising levels of effort as divers fished more days of the year and more hours each day. This resulted in rising catches, as would be expected, but it was also associated with a slight overall rise in the CPUE, particularly on the west coast. There is evidence that a range of the factors discussed above influenced these trends.

The initial period of license transferability was marked by an influx of experienced abalone divers, particularly in 1977, from the more northern Australian state of New South Wales (NSW) which resulted in a jump of the price of Tasmanian entitlements from $10 000 to $40 000. The abalone fishery in NSW was being reviewed at this time and the long-term outlook was poor, prompting many NSW divers to buy into Tasmania. These divers had a more professional approach to their diving, having come from the most competitive and heavily exploited abalone fishery in Australia. Their attitude introduced new techniques and a new competitive spirit. Because of the warmer northern waters, NSW divers expected to dive more days per year than their Tasmanian counterparts. This may explain why the total level of effort increased. The rise in CPUE observed on the east coast at this time is almost associated with the widespread introduction of deckhands.

In the period from 1983 to 1984, divers were anticipating the imposition of some restrictive management policy to conserve stocks which greatly increased their incentive as they tried to maximize cash flow before any restrictive legislation was passed. Effort and catch both increased by about 20–25%, effort because of more days and more hours per day dived as well as more fishing on the west coast. This was accompanied by marked and sustained increases in CPUE in 1969–1970 and 1975–1976, but when the level of exploitation was lower in 1983, it resulted in only a slight increase in CPUE, and CPUE declined in 1984.

In 1984, it was announced that individual transferable quotas would be introduced in the following year. Many divers believed that the immediate reduction in the average catch from 36 to 31 tons per diver would substantially reduce the value of entitlements and sold out during 1984. This led to a large number of new and inexperienced divers entering the industry. However, it is difficult to determine what changes this inexperience wrought, since it is confounded with the major structural changes caused by the introduction of individual, transferable quotas (ITQs).

1985–1986 Individual, transferable quotas

In 1985, a system of individual, transferable quotas (28 units per diver, each unit being initially valued at 1.1 tons) was introduced. The level of catch was further reduced in 1986 (each unit being valued at 1.0 ton). The introduction of ITQs directly decreased the level of catch and effort and this, combined with rapidly rising prices ($14.00 per kg in 1986), has meant that there is less incentive for divers to dive in risky or uncomfortable areas. Divers say that the level of competition between them has gone down sharply since the introduction of ITQs, further reducing motivation. This reduction in motivation has been associated with a declining CPUE throughout the state, a fall in the percent of fishing effort allocated to the west coast, and a reduction in the number of hours dived each day and the number of days dived per year. Other factors have undoubtedly also influenced the latest decline in CPUE, including the level of aggregate experience which declined through 1984 and declining stock abundance, both of which have already been discussed.

Summary of Tasmanian abalone fishery

Figure 7 in many ways summarizes the large-scale picture of the Tasmanian abalone fishery. The average CPUE in each area should reflect the abundance at the best fishing areas, and we can see that this is driven by costs of fishing; the west coast has consistently higher CPUE than other areas. It may be obvious, but CPUE is entirely unrelated to total abundance of abalone in a statistical area, high CPUE areas 7 and 8 have no more total removals than most of the east coast areas (20–31). Areas that have almost no catch at all (37–46) still have the same CPUE as most other areas.

We were able to quantitatively discuss changes in one key motivational factor, price, and showed that changes in price are associated with changes in the behavior of divers. We believe that indebtedness is an even more important factor in motivation but have not been able to quantify levels of indebtedness.

The historical approach to fisheries stock assessment was to try to calibrate CPUE sequences from large spatial areas as shown in Fig. 5 and use this as an index of abundance as Harrison (1983) did. With spatially complex fisheries such an approach is very difficult, and we find it hard to imagine a circumstance under which it would prove useful. Since CPUE will be the result of a complex interaction between the spatial structure of the resource, and the fishermen's motivation and cost, it is hard to imagine how all these factors could be accounted for.

Implications of concentration profiles for management

We have shown how concentration profiles make use of any aggregate CPUE data as an index of abundance very difficult. If one could obtain CPUE data on the spatial scale of the resource, that is patch by patch, catch and effort data could form a very effective monitoring tool. One could follow the evolution of effort, catch, and CPUE and probably understand the evolution of the fishery quite well. We know of no fisheries where such a detailed data collection system is currently in place, and there are many obstacles. First it would require very accurate mapping of catch and effort, at least for abalone, at a resolution finer than that offered by current GPS systems. For abalone such a system would have to be used by the diver under water. Even if such a system were possible, it is likely that the CPUE data would simply record when individual patches were fished to commercial extinction; the natural behavior of fishermen would be not to sample many sites with CPUE lower than those obtainable at other sights. Thus one would always be uncertain how many sites were available at CPUE less than the current commercial norm.

The alternative to detailed mapping of CPUE would be to have a very intense stratified survey system, which would record both the number of patches of different densities as well as the total area of the patch. If each of the 120 licensed abalone divers were to dive two days per year in known areas and record the catch for that area, we would have an extremely valuable data base at very little cost to the management agency. In a situation such as this, with ITQs, it would probably be very easy to convince the divers that it was in their own interest to conduct such systematic surveys. Even if only half of them actually participated, the data would be most valuable.

As more fisheries move to ITQ management, concentration profiles will play a major role in affecting the resulting CPUE. When ITQs are imposed on a fishery that has taken place in a very short period as, for instance, in the U.S. and Canadian fisheries for Pacific halibut, we might expect that CPUE would drop as the economic incentives for maximizing catch rate are replaced by economic incentives to maximize profits. However, if the fishery has taken place in a very short period of time, the high CPUE sites were probably subject to intense competition for access, and many participants were likely forced to fish in lower CPUE sites. Under an ITQ fishery, with months instead of days to take the total catch, fishermen may then concentrate on the high CPUE sites and we should not be surprised at all by a rise in CPUE.

A key biological concern is that reproductive success depends on spawning density in many invertebrate species, and the natural behavior of fishermen is to fish the highest density sites and thus destroy the most valuable breeding concentrations. None of the current management tools in use, seasons, size limits, total quota, or ITQ will prevent this rather natural behavior. Two possible methods to prevent the depletion of high-density sites are to set up sanctuaries at these sites, or to set up territorial fishing rights where the territory's owner(s) would have natural incentives to maintain the most valuable high-density concentrations. If reproductive concentrations are seasonal, then temporal closures could present overexploitation of these concentrations.

Concentration profiles pose a number of difficulties for tagging studies. If we are dealing with a large number of distinct populations, it will be natural to tag where high CPUE can be obtained, and indeed it is unlikely we would ever apply many tags in areas of low CPUE. If the population is not evenly mixed, this means we would effectively be estimating the population in the high CPUE sites and ignoring, in our population estimates, those individuals found in low CPUE sites. This may not be critical to the individual study as the tagging study may be aimed at determining attributes of the high CPUE sites, but any attempt to estimate total abundance in a spatially structured fishery is bound to produce an underestimate.

Thus far our discussion of concentration profiles has centered exclusively on the distribution of density. In spatially complex fisheries other factors differ from site to site and these too have an important impact on management. If, for instance, there is a complex spatial pattern of growth rates, as there is for abalone, then the appropriate size limit for one site may be very different from another one. Individuals in low productivity, slow-growth sites may never reach the legal size limit.

When we expand our thinking of concentration profiles beyond pure abundance and begin to think of the invertebrate resources as complex spatial patterns, we see that many of the management approaches we have used are far from optimal.

The presence of concentration profiles poses many challenges to managers. It is difficult both to monitor the stock abundance and find management regulations that will be appropriate for the biology of the stock. One possible solution to both of these problems is territorial fishing rights. If cooperatives, or individuals, are assigned the exclusive right to fish in certain areas, it will be in their interest to set up a survey system that monitors the abundance and to use harvest policies that are appropriate for the spatial variation in growth and mortality. Just as one cannot imagine large-scale regulation of what crops to plant or how much fertilizer to use for farmers facing a range of different soil types, it is impossible to imagine that any large-scale fishing regulations would be appropriate for spatially structured stocks. Detailed top-down monitoring and regulation by small spatial area is certainly another possibility, but this would be very expensive to maintain.

Acknowledgments

We would like to thank Dr. J. Orensanz for numerous ideas and references that assisted the preparation of this paper. Bob Miller and an anonymous reviewer provided helpful comments.

References

Abrahams, M.V., and Healey, M.C. 1990. Variation in the competitive abilities of fishermen and its influence on the spatial distribution of the British Columbia salmon troll fleet. Can. J. Fish. Aquat. Sci. **47**: 1116–1121.

Beverton, R.J.H., and Holt, S.J. 1957. On the dynamics of exploited fish populations. U.K. Min. Agric. Fish., Fish. Invest. Ser. No. 219. 533 p.

Caddy, J.F. 1975. Spatial model for an exploited shellfish population, and its application to the Georges Bank scallop fishery. J. Fish. Res. Board Can. **32**: 1305–1328.

Clark, C.W. 1982. Concentration profiles and the production and management of marine fisheries. *In* Economic theory of natural resources. *Edited by* W. Eichhorn. Physica-Verlag, Wurzburg-Wien, Germany. pp. 97–112.

Clark, C.W. 1985. Bioeconomic modelling and fisheries management. John Wiley & Sons, New York. 291 p.

Fretwell, S.D., and Lucas, H.L. 1970. On territorial behaviour and other factors influencing habitat distribution in birds. Acta Biotheor. **19**: 16–36.

Gordon, H.S. 1953. An economic approach to the optimum utilization of fishery resources. J. Fish. Res. Board Can. **10**: 442–457.

Gross, F., and Smyth, J.C. 1946. The decline of oyster populations. Nature (London), **157**: 540–542.

Harrison, A.J. 1983. The Tasmanian abalone fishery. Tasman. Fish. Res. No. 26.

Hilborn, R., and Kennedy, R.B. 1992. Spatial pattern in catch rates: a test of economic theory. Bull. Math. Biol. **54**: 263–273.

Hilborn, R., and Ledbetter, M. 1979. Analysis of the British Columbia salmon purse seine fleet: dynamics of movement. J. Fish. Res. Board Can. **36**: 384–391.

Hilborn, R., and Walters, C.J. 1987. A general model for simulation of stock and fleet dynamics in spatially heterogeneous fisheries. Can. J. Fish. Aquat. Sci. **44**: 1366–1370.

Hilborn, R., and Walters, C.J. 1992. Quantitative fisheries stock assessment: choice, dynamics and uncertainty. Chapman and Hall, New York. 570 p.

Orensanz, J.M. 1986. Size, environment and density: the regulation of a scallop stock and its management implications. *In* North Pacific Workshop on stock assessment and management of invertebrates. *Edited by* G.S. Jamieson and N. Bourne. Can. Spec. Publ. Fish. Aquat. Sci. 92. pp. 195–227.

Prince, J.D. 1989. The fisheries biology of the Tasmanian stocks of *Haliotis rubra*. Ph.D. thesis. Department of Zoology, University of Tasmania, Tasmania.

Population Dynamics

Intertidal population estimate of razor clams (*Siliqua patula*) at beaches near Masset, Haida Gwaii/Queen Charlotte Islands, and applications to fishery management

R. Russ Jones, Carl Schwarz, and Lynn Lee

Abstract: Intertidal razor clam abundance at beaches near Masset, Haida Gwaii/Queen Charlotte Islands, British Columbia, was estimated using a 3-stage sampling design stratified by beach section. Randomly selected transects in each beach section were sampled at systematically selected beach elevations using a 0.5 m^2 core sampler. Adult and juvenile clams in the sand core sample were captured by fluidizing the sand with pressurized seawater and dipnetting. The razor clam population on 18.8 km of the beach available to the commercial fishery was estimated to be 42 × 10^6 (SE 34%) clams ≥4 mm, 16.3 × 10^6 (SE 16%) clams ≥20 mm, and 5.82 × 10^6 (SE 21%) clams ≥90 mm length. The latter size is the minimum commercial size limit. Schaefer and Fox surplus production models were used to estimate maximum sustainable yield (MSY) as 121 and 115 t, respectively. The MSY estimates were based on an exploitable biomass (≥90 mm size), B_c, of 636 t, 1994 landings of 105 t, and the assumption that fishing mortality at MSY equals the natural mortality ($M = 0.26$). Population and biomass estimates did not include the subtidal population and assumed no movement of clams from subtidal to intertidal areas during the period of the survey. The stock appears relatively robust based on historic landings. The current size limit is unlikely to protect much of the spawning biomass based on age of maturity studies in Alaska and Washington. Fishing quotas are likely unnecessary at this time because B_c may fluctuate considerably from year to year and the number of fishery participants was recently limited.

Résumé : L'abondance du couteau du Pacifique intertidal sur les plages situées près de Masset, Haida Dwaii/îles de la Reine-Charlotte, Colombie-Britannique, a été estimée à l'aide d'un plan d'échantillonnage à trois degrés stratifié par section de plage. Des transects choisis au hasard dans chaque section de plage ont été échantillonnés à des altitudes choisies systématiquement à l'aide d'un échantillonneur à carotte de 0,5 m^2. Les couteaux juvéniles et adultes dans les carottes de sable ont été capturés en fluidifiant la sable avec de l'eau de mer sous pression et en les récoltant à l'épuisette. La population de couteaux du Pacifique sur les 18,8 km de plage ouverts à la pêche commerciale a été estimée à 42 × 10^6 (É-T = 34 %) couteaux ≥4 mm, 16,3 × 10^6 (É-T = 16 %) couteaux ≥20 mm et 5,82 × 10^6 (É-T = 21 %) couteaux ≥90 mm de longueur. Cette dernière taille est la limite minimale de taille commerciale. Les modèles de surproduction de Schaefer et Fox ont été utilisés pour estimer le rendement équilibré maximal (MSY); les résultats obtenus ont été respectivement de 121 et 115 t. Les estimations du MSY ont été fondées sur une biomasse exploitable (taille ≥ 90 mm), B_c, de 636 t, des débarquements en 1994 de 105 t et la supposition que la mortalité par pêche au MSY est égale à la mortalité naturelle ($M = 0,26$). Les estimations de la population et de la biomasse ne comprenaient pas la population infratidale et ne supposaient aucun mouvement de couteaux entres les zones intertidale et infratidale durant la période du relevé. Le stock semble relativement robuste d'après les débarquements historiques. Il est peu probable que la limite de taille actuelle protège une grande partie de la biomasse de reproduction d'après les études sur la maturité réalisées en Alaska et dans l'État de Washington. Les contingents de pêche sont vraisemblablement inutiles à ce moment-ci par ce que B_c peut fluctuer considérablement d'une année à l'autre et que le nombre de participants à la pêche a été limité récemment.
[Traduit par la Rédaction]

R.R. Jones and L. Lee. Haida Fisheries Program, Council of the Haida Nation, P.O. Box 98, Skidegate, Haida Gwaii, BC V0T 1S0, Canada.

C. Schwarz. Department of Mathematics and Statistics, Simon Fraser University, Burnaby, BC V5A 1S6, Canada.

Correct citation: Jones, R.R., Schwarz, C., and Lee, L. 1998. Intertidal population estimate of razor clams (*Siliqua patula*) at beaches near Masset, Haida Gwaii/Queen Charlotte Islands, and applications to fishery management. *In* Proceedings of the North Pacific Symposium on Invertebrate Stock Assessment and Management. *Edited by* G.S. Jamieson and A. Campbell. Can. Spec. Publ. Fish. Aquat. Sci. 125. pp. 199–211.

Introduction

Beaches near Massett in British Columbia have a significant population of razor clams (*Siliqua patula* Dixon) which have supported a major commercial fishery since 1922 and a small but important noncommercial fishery. Commercial landings have varied widely depending on market demand, fishing effort, and fluctuations in populations (Fig. 1). In 1993, commercial fishery landings were low and commercial diggers reported a high proportion of undersize clams in the catch which caused concern among managers about the state of the stock. The purpose of this study was to assess the harvestable

Fig. 1. Haida Gwaii razor clam landings 1923–1994.

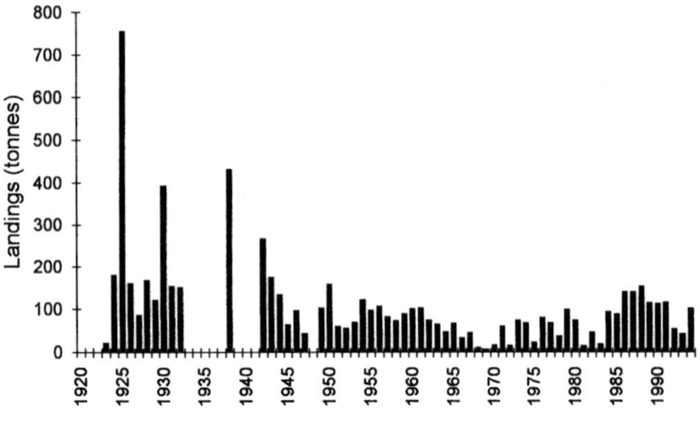

population and estimate sustainable yield in the fishery. A review of current management practices is also presented.

Literature review

Razor clams are found on surf-swept sandy beaches from Pismo Beach, California, to the Aleutian Islands in Alaska (Weymouth and McMillin 1930). There are eight major concentrations sufficient to support fisheries, one in Oregon (Hirschhorn 1962), one in Washington (Tegelberg 1964), two in British Columbia (Bourne 1969), and four in Alaska (Nickerson 1975). The major stocks in British Columbia are located at beaches near Massett in Haida Gwaii with a smaller population at Long Beach on the west coast of Vancouver Island (Bourne 1969).

Razor clams are characterized by a long siphon, a prominent muscular foot, and brittle elongated valves. Razor clams can burrow at rates exceeding 20 $cm \cdot min^{-1}$ and are found up to 25 cm in the sand which makes harvesting a challenge for an inexperienced clam digger. Adults left on the surface of the beach will quickly reburrow. Lateral movement of adults is believed to be small, although juvenile clams have been found to move because of substrate instability (Nickerson 1975). Juvenile clams burrow to a lesser depth and may be washed out and moved because of scouring of the substrate. Although there have been numerous studies of intertidal populations, little is known about subtidal populations in the vicinity of major razor clam beaches (Bourne 1969).

Adult razor clams reach shell lengths up to 160 mm and ages of 18 years. Growth rates vary with latitude with northern populations generally growing slower but reaching a greater age and maximum size (Weymouth and McMillin 1930). Northern populations can be aged by annual winter growth checks, although interpretation depends on ability to identify all the checks (Bourne and Quayle 1970, Nickerson 1975). Survival of intertidal razor clams is likely affected by availability of food, predators, and natural occurrences such as storms or disease. Predators on larger clams include gulls, ducks, crabs, and a few fish species.

Razor clams have separate sexes and are broadcast spawners. Age and size of sexual maturity varies with latitude, but most clams are sexually mature at 2–4 years of age and 97–103 mm in length (Weymouth et al. 1925). Time of spawning varies with location, but generally occurs from April to September, occurring later at northern latitudes. Spawning can be influenced by temperature, upwelling, tidal cycles, currents, food availability, and gonad maturity.

The life history follows a common bivalve pattern of release of gametes into the surrounding water, fertilization, development as a free-swimming pelagic larvae, settlement to the bottom as "spat" and finally development as a sedentary organism. Larval period was estimated to be about 8 wk in Washington (Weymouth et al. 1925). By the time of settlement, larvae can be widely distributed from the point of origin by currents and tides. In laboratory culture at 16.5°C, metamorphosis took place about 20–25 d after fertilization at a shell length of about 300 microns in length. After 3 mo, juveniles reached a mean shell length of 5 mm (Breese and Robinson 1981). Bourne (1979) found modes in shell length of zero-age clams of 2–4 mm in July 1966, 6–8 mm in September, and 12–14 mm the following May. Relatively little is known about environmental factors contributing to survival of razor clams during the vulnerable larval and early juvenile stage of their life history.

Intertidal populations have been estimated using a variety of techniques including mark-recapture, tagging, and more recently, hydraulic core sampling. Bourne (1969) estimated density of clams at North Beach using (*i*) repeated digging of 7 transects on 3 beaches near Massett in 1966, (*ii*) repeated counts of shows in the transects, and (*iii*) tagging and recapture by repeated digging. Repeated digging resulted in recovery of 15–25% of tagged clams. The Washington Department of Fisheries conducts regular assessments of razor clam beaches using (*i*) mark recapture and (*ii*) stratified random digs (Ayres and Simons 1988). Eighteen areas on 4 major beaches are sampled in the spring and a reduced number are sampled in the fall. Stratified random digs provide an indication of density and size composition. Plots are redug a minimum of 2 times until at least 20% of the marks are recovered. Szarzi (1991) estimated razor clam densities at Cook Inlet, Alaska, beaches using hydraulic core sampling and a 2-stage sampling design by beach and distance from a gravel line. Although sampling was carried out on randomly selected transects as part of a 3-stage design, a 2-stage design which ignored the variability between transects was adopted for the final analysis.

Hydraulic sampling was selected for the present study because of the advantages of providing an unbiased density estimate and sample of the population. Digging by hand is selective for larger clams that "show" their siphons. Screening of sand is not an efficient way to sample larger clams.

Management practices in British Columbia, Washington, and Alaska vary with the type and intensity of the fishery. Commercial and recreational razor clam fisheries at North Beach are open year round except for rare paralytic shellfish poisoning (PSP) closures. The commercial fishery is currently managed by a 90 mm size limit. Recreational clam diggers are limited to a daily catch of 75 clams and a possession limit of 150 clams. There is no recreational size limit. Recreational fisheries on Washington beaches are managed by a fixed season (usually 1–2 d) and a daily bag limit of 15 clams with no size limit (Doug Simons, Washington Department of Fisheries, personal communication). There is no size limit in the commercial or recreational fishery. Size limits in commercial fisheries in Washington were eliminated in the 1960's after high wastage of undersize clams was found. Alaska currently has a

Fig. 2. Location of Massett razor clam beach sections.

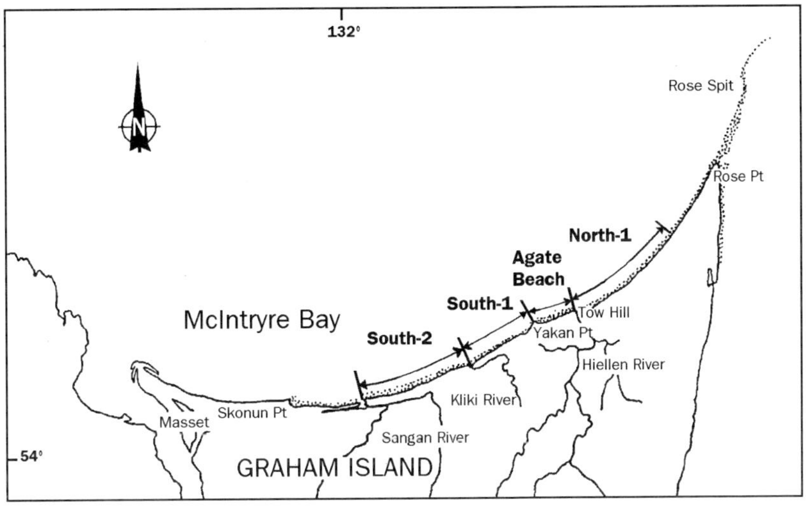

major commercial and recreational fishery near Cook Inlet (Jeff Fox, Alaska Department of Fish and Game, personal communication). There are currently no size limits or quota in the Alaska fishery.

Methods

Data collection

The 3 main beaches near Massett are commonly called North Beach, South Beach, and Agate Beach. The 3 sections of the beach sampled, referred to as North-1, South-1, and South-2, are 7.2 km, 4.6 km, and 6.75 km in length, respectively (Fig. 2). Beach sections were selected based on clam density, beach accessibility, and frequency of use as reported by commercial clam diggers.

North-1 was the main focus of the commercial razor clam fishery in 1994. South-1 is occasionally dug commercially, but clam densities are lower and the distribution is spottier. South-2 is rarely dug. Agate Beach is not dug commercially because there is no truck access. Generally, sampled beaches have gentle slopes from the driftwood line, with 60–75 m of beach exposed per 1 m of elevation. North-1 is a uniform sand beach. South-1 and South-2 are mainly sand beaches interspersed with pockets of gravel. South-2 is more gravelly than South-1. The coarser beach material at Agate and South-2 beaches is said by commercial diggers to make digging and retrieving clams more difficult. Most diggers will not fish North Beach beyond about 7.3 km because vehicles tend to get stuck in the soft substrate.

The intertidal razor clam population was estimated using a 3-stage sampling design which was stratified by beach section (see fig. 2 in Schwarz et al. 1995). Transects were sampled during tides less than 1.0 m in height between March 29 and August 10, 1994. Commercial diggers generally dig only when tides are less than 1.2 m (4.0 ft). Representative samples of the population were collected by hydraulic sampling of sand cores with a 0.5 m^2 sampling tube. In total, 14 transects in 3 beach sections were surveyed and core samples were collected at 99 beach elevations. A total of 533 cores were sampled and 4569 clams of all sizes were captured.

The equipment used was similar to that described by Szarzi (1991). The galvanized steel sampling ring was 0.79 m in diameter (0.5 m^2 in area) and drilled throughout with 3-mm holes to allow easy drainage of water. Four additional oval holes (60 mm long covered with 5-mm wire mesh) were located near the top of the ring to allow further drainage. The sampling ring was forced into the sand to a depth of about 0.5 m. Seawater was pumped using a 4.0 HP Honda pump (Model GX-120) and injected into the sand using a wand (a steel pipe, 1.5 m in length and 38 mm in diameter) attached to the pump by a 51-mm diameter hose about 45 m in length. Upwelling water liquified the sand, dislodging the razor clams. Razor clams were captured by straining the upwelling mixure with a coarse mesh (5 mm) dipnet followed by a fine mesh (<1 mm) dipnet.

Generally the procedure required about 3–5 min per sample with a 5-person crew. One person acted as a tender for the pump intake, 1 person worked the wand that fluidized the sand, 2 others strained the mixture, and 1 person supervised field activities and surveyed the beach elevations, providing assistance wherever necessary. Sampling continued until ~1 min after the last clam was captured. At that time, all clams ≥20 mm in the sample area and the majority of clams <20 mm were thought to be captured. A few clams <20 mm were found to wash through the wire mesh near the top of the ring. A few broken clams were noted, likely due to pressure from the wand as it was inserted. Razor clams from each sample were put into separate labelled bags. The length of individual clams was measured to the nearest millimetre using calipers after field sampling was completed. Wet weight of a systematic random sample of clams was measured using an Ohaus electronic balance. A small amount of breakage from handling clams in sample bags was recorded (115 out of 4569 clams sampled), which occurred mostly for small clams (70 broken clams were <20 mm shell length).

Transects were evenly spaced, 300 m apart on North-1, 200 m on South-1, and 1350 m on South-2. More transects were laid out than were actually sampled, because the number of transects necessary to obtain acceptable confidence limits was not known. Sampled transects were randomly selected and located by driving a fixed distance from an access point as measured with a truck odometer. Just before sampling, each transect was laid out perpendicular to the beach using rebar marked with flagging tape. One transect was sampled per tide. Depending on daylight, sampling generally began 2–3 h before the low tide and continued until 3 h after the low tide.

Beach elevations sampled within a transect were located systematically along the transect, with sampling starting at the surfline at the time of arrival. Distance between sampled elevations ranged from 15 m to 25 m depending on the rate of subsidence (or approach) of the tide. Each beach elevation was flagged with a metal rebar stake during sampling. For each transect, from 4 to 10 beach elevations were sampled depending on the low tide level and time available.

At each beach elevation, the survey crew completed 3–9 core samples. Core samples were situated parallel to the surfline within about 7 m of the transect line with approximately the same number on each side of the transect line. If the tide was falling rapidly, then the survey crew moved to the next lower beach elevation and completed sampling at higher beach elevations on the rising tide.

Elevation was measured using a surveyor's level (Pentax Model AL-240) and rod, using the tidewater level as the reference elevation. At a recorded time near low tide, a profile of the beach and the tidewater elevation relative to the sampled beach elevations was determined using the level and rod. Tidewater elevation above chart datum at the time of sampling was calculated using time and procedures described in Canadian Tide and Current Tables (Department of Fisheries and Oceans 1991). For the purposes of the calculation, Bella Bella was the reference port and Wiah Point was the secondary port. There was a small random error in estimation of tidewater elevation because of ocean swells. Generally, tidewater elevation was estimated to be within ±0.1 m of the true value.

Commercial fishery landings in 1994 occurred from March to September. A systematic random sample of the commercial catch was measured and weighed from March 30 to May 29 (6 samples, $n = 1254$). Clams were measured to the nearest millimetre using calipers. In general, every fifth digger was sampled and every fifth clam was measured to a maximum of 30 clams per digger.

A small noncommercial razor clam fishery (recreational and Haida noncommercial fishers) occurs with effort concentrated during the tourist season from June through August. A random survey of noncommercial clam diggers was conducted from July 30 to September 30 1994 stratified by weekend/weekday and low tide levels (<1 m and 1–2 m). The noncommercial catch was randomly sampled, and individual clams measured on 17 occasions during this period ($n = 1605$).

A random systematic sample of razor clams was collected from 2 transects on North-1 for aging (2 samples, $n = 98$). In addition, a catch sample of larger clams was aged from South-1 ($n = 68$). Clams were aged by counting annual winter growth checks which could be distinguished on the shell after the periostracum was peeled away. Aging was possible by the naked eye aided by a bright light. For most shells, the first 2 annuli were less distinct than subsequent annuli. Annual growth was estimated by measuring the diameter of annuli to the nearest millimetre using a Vernier caliper. Winter growth checks were definitive for all shells examined and no shells were discarded.

Data analysis

The survey design used to collect the samples of clams is a 3-stage sampling design (Cochran 1977) where the first stage is the selection of transects, the second stage is the selection of elevations within a transect, and the third stage is the selection of core samples at a particular elevation. However, the standard methods of analysis for a 3-stage design (Cochran 1977) could not be used because of the different number of cores sampled at each elevation and disproportionate sampling at lower elevations. More core samples were taken at elevations closer to the low tide mark where clam densities were highest. Schwarz et al. (1995) chose a model-based approach where the elevations were first stratified into 3 levels (<1 m, 1–2 m, and >2 m above chart datum) and an ANOVA-based approach was used to estimate the mean density in each level.

Let Y_{ijkl} represent the number of clams in the lth core sample taken at the kth elevation within the jth level of transect i. Preliminary plots showed that the variation in the number of clams in each sample could be closely approximated by a Poisson distribution. The following model was then fit to the square-root transformed response (Krebs 1989, page 448):

$$Y'_{ijkl} = (Y_{ijkl} + 0.5)^{1/2} = \mu + \tau_i + L_j + (\tau L)_{ij} + e_{k(ij)} + \varepsilon_{l(ijk)}$$

where Y_{ijkl} is the observed number of clams in the lth sample taken at the kth elevation within level j on transect i; μ is the overall grand mean; τ_i is a random transect effect and is assumed to be independent of sample size $N(0, \sigma_\tau 2)$; L_j is a fixed level effect and represents the mean transformed density for level j; $(tL)_{ij}$ is a random transect-level interaction effect and is assumed to be independent of $N(0, \sigma_{\tau L}2)$ (This effect will allow density to change among levels differently for each transect.); $e_{k(ij)}$ is a random effect representing the kth elevation, sampled within level j of transect i, and is assumed to be independent of $N(0, \sigma_e 2)$; $\varepsilon_{l(ijk)}$ is a random sample effect for sample l within the kth elevation, sampled within level j of transect i, and is assumed to be independent of $N(0, \sigma^2)$. The effects τ_i, $(\tau L)_{ij}$, $e_{k(ij)}$, and $\varepsilon_{l(ijk)}$ are assumed to be mutually independent.

Because the survey is not balanced, i.e., there are a varying number of elevations sampled in each level, and a varying number of core samples at each elevation, there are no closed-form expressions for the estimates from the above model. We used PROC MIXED (SAS Institute 1992) which is a procedure designed for the analysis of unbalanced mixed-effect models. After the model was fit, the estimates from the fitted model were back transformed to estimate the mean density for each level as described in Schwarz et al. (1995). The overall population estimate was found by multiplying the estimated density for each level by the area of the beach corresponding to the level and adding these estimates over the 3 levels. The SE of the estimate was computed as described by Schwarz et al. (1995).

Results

Age and growth

Information about age and growth of the 1994 population is necessary to interpret clam densities by size fraction and to estimate abundance of year-classes. Table 1 shows age and size measured using winter growth checks and shell length at

Table 1. Razor clam age and growth from shell examination.

	Winter growth checks						Shell length, combined sample	
	South-1 sample		North-1 sample		Combined sample			
Age (years)	Mean length (mm)	*n*	Mean length (mm)	n	Mean length (mm)	n	Mean length (mm)	n
0.5	7.9	68	7.2	98	7.5	166		
1.0							28.7	3
1.5	28.1	68	33.7	95	31.4	163		
2.0							59.5	41
2.5	56.7	68	76.0	54	65.2	122		
3.0							97.3	21
3.5	89.5	68	105.8	33	94.8	101		
4.0							117.9	36
4.5	110.1	57	119.9	8	111.3	65		
5.0							121.8	33
5.5	120.3	28	131.5	4	121.7	32		
6.0							128.3	16
6.5	127.7	15	139.0	1	128.4	16		
7.0							133.1	9
7.5	131.3	6	143.0	1	133.0	7		
8.0							136.6	5
8.5	136.5	2	—	—	136.5	2		
9.0							138.5	2

Fig. 3. Average growth curves by beach section and year-class.

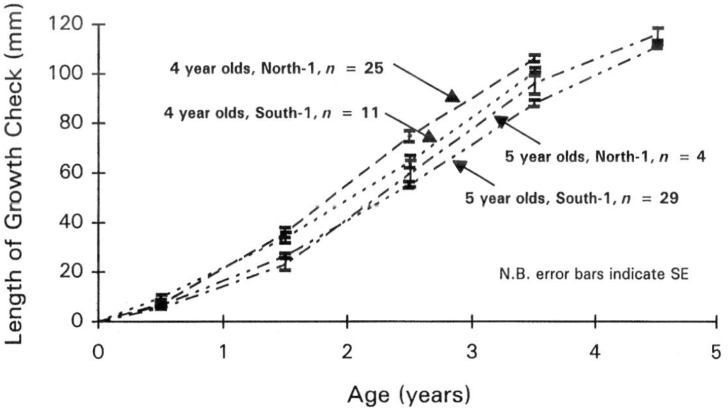

Fig. 4. Length–frequency distribution of age samples.

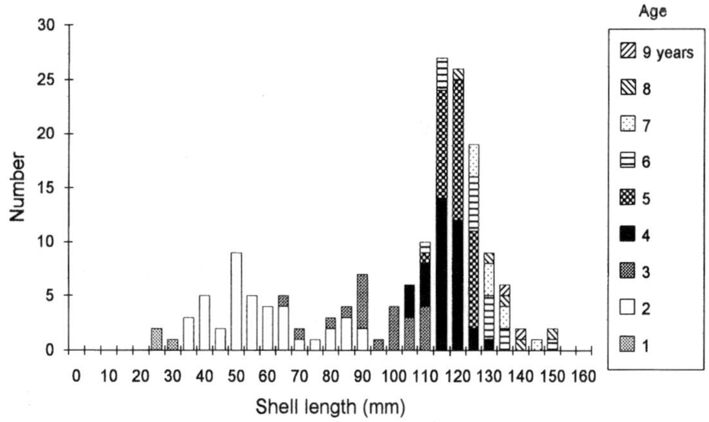

Table 2. Razor clam density data from sample transect.

Distance (m)	Elevation (m)	Clams per sample unit (0.5 m²) 1	2	3	4	5	6	7	Total	Mean	SD
				All clams 4 mm and more in length							
0	0.87	37	47	53	34	47	70	18	306	43.7	16.3
25	1.45	56	56	57	49	—	—	—	218	54.5	3.7
50	1.84	34	39	54	46	—	—	—	173	43.3	8.7
75	2.06	22	21	31	8	17	12	11	122	17.4	7.9
95	2.36	15	12	—	—	—	—	—	27	13.5	2.1
Total									846		
				Clams 20 mm and more in length							
0	0.87	9	5	12	4	6	14	3	53	7.6	4.2
25	1.45	11	11	10	13	—	—	—	45	11.3	1.3
50	1.84	6	8	5	5	—	—	—	24	6.0	1.4
75	2.06	1	4	5	2	4	2	1	19	2.7	1.6
95	2.36	2	4	—	—	—	—	—	6	3.0	1.4
Total									147		
				Clams 90 mm and more in length							
0	0.87	2	3	3	0	2	3	1	14	2.0	1.2
25	1.45	2	2	3	3	—	—	—	10	2.5	0.6
50	1.84	0	1	0	0	—	—	—	1	0.3	0.5
75	2.06	0	0	1	2	1	2	1	7	1.0	0.8
95	2.36	1	0	—	—	—	—	—	1	0.5	0.7
Total									33		

Note: (North-1, T1.3, July 26, 1994)

the time of sampling. Since razor clams spawn in the spring or summer, winter growth checks were assumed to occur at the midpoint in the calendar year (0.5, 1.5, 2.5 years, etc.). Growth checks occur at mean sizes of 8, 31, 65, 95, 111, 122, 128, 133, and 137 mm for clams from 0.5 to 8.5 years of age, respectively. As found by Bourne and Quayle (1970), clams on North Beach appear to grow faster than on South Beach and were 5, 19, 16, and 10 mm larger at 1.5–4.5 years of life, respectively. Growth rates are higher initially on North Beach (0.5–2.5 years) but continue at a higher rate later on South Beach (3.5–4.5 years). The significance of size differences among older clams at North and South Beach is difficult to assess because of the small sample size.

The shell lengths represent the mean size of each year-class at the time of sampling, assumed to be about the end of each year (2.0, 3.0, 4.0 years, etc.). Shell lengths of age-1.0 clams are not representative because age samples did not include clams <20 mm.

Growth showed some variation by beach section and year-class, except for North-1 5 year olds, where comparisons were limited by the sample size (Fig. 3). The data suggests that growth differences may occur for different year-classes as well as beach sections. Differences in growth of year-classes, particularly in the first three years, illustrate the problem with using a razor clam age-length sample from one year to interpret size-frequency data in a different year. However, these differences in mean length are of lesser importance given the wide variation in size at age, particularly for younger clams. Based on the length distribution of age samples (Fig. 4), we classified clams ≥90 mm as age-2.0 and older while clams 20–90 mm were classified as age 1.0–3.0.

Density and population estimates

Table 2 shows an example of the density data collected from core sampling. In all, a total of 14 transects were surveyed with seven on North-1, four on South-1, and three on South-2.

Variation of razor clam mean density by beach level and transect was investigated for three size fractions (≥4 mm, ≥20 mm, and ≥90 mm). Size fractions represent, respectively, all clams captured, overwintered clams age-1.0 and older, and clams larger than the commercial size limit. Since winter growth checks occur at 8 mm (range 3–15 mm) and 31 mm (range 19–53 mm), clams ≥20 mm include some age-1.0 clams and all clams age-2.0 and older. Clams < 20 mm accounted for the majority of the clams sampled in most transects (64%, 41%, and 68% of all clams found on North-1, South-1, and South-2, respectively).

Density displayed a strong variation with elevation (see fig. 2 in Schwarz et al. 1995). Few clams occurred more than 2.5 m above chart datum. There appeared to be a peak density of clams ≥4 mm from about 0.5–1.5 m elevation. In comparison, density of clams ≥90 mm was generally inversely related to elevation and was greatest at low elevations. Variability among transects was high as shown by displacement of density curves upward or downward.

As shown in Table 3*a*, *b*, and *c* (from Schwarz et al. 1995), the test for a transect-level interaction $(\tau L)_{ij}$ was not significant for all beaches and for the three population samples (≥4 mm, ≥20 mm, and ≥90 mm), so this component was dropped from the model allowing a simpler model to be refit. No further simplifications in variance terms were possible. Estimates of variance components (shown as σ_t^2, σ_e^2, and σ^2 for variance by transect, elevation, and sample, respectively) were obtained

Table 3. Population estimate by size fraction for three beach sections.

Beach	Test for interaction between transect and level	Variance component estimates		Test for level effects	Level	Est. area (km^2)	Est. density (0.5 m^2)	Total (millions)	SE total (millions)
				***a.* Clams more than 4 mm in length**					
North-1	χ^2=1.93	σ_τ^2	1.95	F=61.19	<1 m	0.54	17.25	18.63	4.90
	df=1	σ_e^2	0.28	df=2, 217	1–2 m	0.54	12.63	13.64	4.25
	p=0.16	σ^2	0.44	p<0.0001	>2 m	0.36	2.43	1.75	1.31
Totals								34.01	10.22
South-1	χ^2=0.0[a]	σ_τ^2	1.18	F=12.43	<1 m	0.322	5.64	3.64	1.73
	df=1	σ_e^2	0.20	df=2, 146	1–2 m	0.276	3.57	1.97	1.21
	p=0.99	σ^2	0.33	p<0.0001	>2 m[b]	—	0.13	—	—
Totals								5.61	2.89
South-2	χ^2=0.038	σ_τ^2	0.39	F=12.91	<1 m	0.270	2.26	1.22	0.67
	df=1	σ_e^2	0.03	df=2, 71	1–2 m	0.270	1.29	0.69	0.52
	p=0.85	σ^2	0.32	p<0.0001	>2 m[b]	—	0.10	—	—
Totals								1.91	1.16
				***b.* Clams 20 mm or more in length**					
North-1	χ^2=0.00[a]	σ_τ^2	0.10	F=84.15	<1 m	0.54	7.30	7.88	0.84
	df=1	σ_e^2	0.06	df=2, 217	1–2 m	0.54	3.68	3.97	0.65
	p=0.99	σ^2	0.29	p<0.0001	>2 m	0.36	0.73	0.53	0.26
Totals								12.38	1.56
South-1	χ^2=0.156	σ_τ^2	0.13	F=12.15	<1 m	0.322	3.65	2.35	0.56
	df=1	σ_e^2	0.13	df=2, 146	1–2 m	0.276	1.73	0.95	0.37
	p=0.69	σ^2	0.28	p<0.0001	>2 m[b]	—	0.01	—	—
Totals								3.30	0.85
South-2	χ^2=0.00[a]	σ_τ^2	0.02	F=5.18	m	0.270	0.85	0.46	0.14
	df=1	σ_e^2	0.02	df=2, 71	1–2 m	0.270	0.28	0.15	0.10
	p=0.99	σ^2	0.15	p<0.0080	>2 m[b]	—	0.10	—	—
Totals								0.61	0.21
				***c.* Clams 90 mm or more in length**					
North-1	χ^2=0.72	σ_τ^2	0.035	F=42.45	<1 m	0.54	2.75	2.97	0.37
	df=1	σ_e^2	0.055	df=2, 217	1–2 m	0.54	0.85	0.92	0.26
	p=0.40	σ^2	0.170	p<0.0001	>2 m	0.36	0.26	0.19	0.14
Totals								4.07	0.63
South-1	χ^2=0.35	σ_τ^2	0.086	F=12.19	<1 m	0.322	1.81	1.16	0.33
	df=1	σ_e^2	0.076	df=2, 146	1–2 m	0.276	0.44	0.24	0.19
	p=0.55	σ^2	0.148	p<0.0001	>2 m[b]	—	0.01	—	—
Totals								1.41	0.48
South-2	χ^2=0.0[a]	σ_τ^2	0.000	F=5.22	<1 m	0.270	0.56	0.30	0.078
	df=1	σ_e^2	0.016	df=2, 71	1–2 m	0.270	0.07	0.04	0.046
	p=0.99	σ^2	0.073	p<0.0074	>2 m[b]	—	0.02	—	—
Totals								0.34	0.09

[a]Interaction variance component was estimated to be zero. Consequently, there is no change in the log-likelihood when this term is dropped from the model.
[b]Few razor clams are found >2 m above datum on this beach. Consequently, contributions to the overall population are ignored.

from PROC MIXED (SAS Institute 1992). The nonzero variance component for elevations (σ_e^2) reflects the "error" in the simplifying assumption of a constant density within each level. As shown in the density plots (fig. 2 in Schwarz et al. 1995), the relationship between density and elevation does not follow a simple step function. However, the error introduced by this simplifying assumption should be small relative to other sources of variation (e.g., variation among samples and transects).

Population of clams ≥4 mm in length (Table 3a)

This fraction includes clams from 4 to 20 mm length which will not recruit into the fishery for 2–3 years. Variation among transects (σ_t^2) was at least three times as large as other sources

Fig. 5. Length–weight relationship for razor clams.

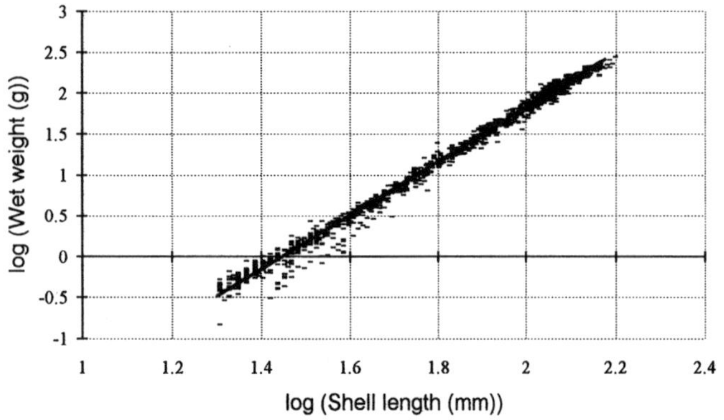

Fig. 6. Length–frequency distribution of razor clams >20 mm shell length.

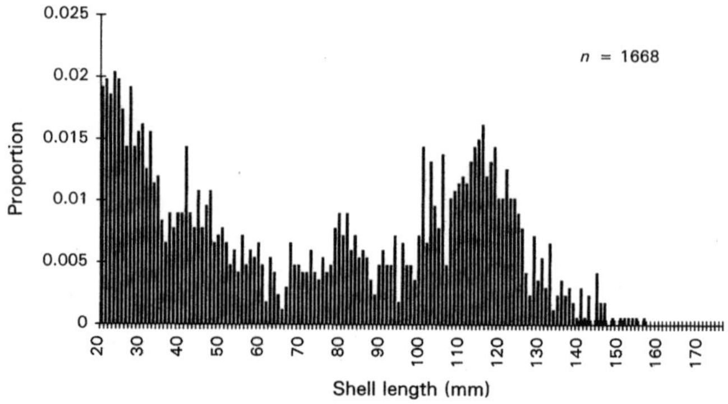

of variation in North and South-1 beaches. For South-2, transect variation was comparable in size to core sample variation (σ^2). The transect variance component reflects large scale changes of density of clams ≥4 mm over the beach. Clams <20 mm are the major size-class in this group. All beaches showed a significant level effect with densities in the <1 m level from 1.5 to 1.9 times as high as in the 1–2 m level. Densities in the >2 m level were much lower. The large transect variance component indicates that overall precision in estimating the population of all clams could be most effectively improved by increasing the number of transects.

Population of clams ≥20 mm in length (Table 3b)

This fraction includes clams available to the fishery (≥90 mm) and clams 1–3 years of age that will recruit to the fishery in 1–2 years. Variation among core samples was larger than variation among transects, and both were significant compared to variation among elevations. This implies that the beach is more homogeneous in density for this subset of the population compared to the entire population.

Population of clams ≥90 mm in length (Table 3c)

For commercial size clams, none of the tests for a nonzero transect variance component σ_t^2 were significant. All tests for a nonzero elevation component were significant but very small compared to sample site variation. These results imply that beaches are much more homogeneous with respect to density of larger clams than smaller clams, and there is less evidence of large scale changes in densities for larger clams. For North-1 clams ≥90 mm, variance due to samples is 3.1 times the variance due to elevations and 4.9 times the variance due to transects. This indicates that overall precision could be most effectively improved by increasing the number of elevation samples within a level. Significant level effects were again found in all three beaches with densities in the <1 m level over 3 times that in the 1–2 m level. Few large clams were found in the >2 m level.

MSY estimate

Neither a time series of catch-at-age nor catch-per-unit-effort data were available for the razor clam fishery at Masset beaches. This precluded the use of age-structured models to estimate maximum sustainable yield (MSY). However, MSY can be estimated using two common surplus production models with an estimate of the mean exploited biomass (B_c), yield in the same year (Y_c) and the fishing mortality at MSY (F_{MSY}) (Garcia et al. 1989):

$$\mathrm{MSY} = F_{MSY} B_c \exp\left(Y_c/(F_{MSY} B_c) - 1\right)$$

Fox (1970) model

$$\mathrm{MSY} = (F_{MSY} B_c)^2/(2F_{MSY} B_c - Y_c)$$

Schaefer (1954) model

The first equation is unstable if B_c approaches zero and at high values of F. The second equation becomes unstable if F approaches $2F_{MSY}$ and Y_c and B_c approach zero estimators. However, this is shown not to be the case for the razor clam fishery.

An estimate for the total biomass, B, and exploitable biomass, B_c, was obtained using a three-step process involving the length–weight relationship, $W(L_i)$, a population size–frequency distribution, $f(L_i)$, and the estimated number of clams within a size range on the beach, N_x. First, a fitted length–weight equation was found for clams ≥20 mm from transect samples ($n = 1406$):

$$\log_{10} W = 3.3058 \log_{10} L - 4.7813$$

where W is the weight in grams and L is the length in millimetres (Fig. 5). This relationship was used to estimate the weight of each 1 mm length class in length–frequency distributions constructed from transect samples, $n = 1668$ (Fig. 6). Only transects sampled early in the season (from March to May 1994) were used in length–weight and length–frequency distributions to reduce the effects of growth on the distribution. Finally, the product of the total number of clams, N_x, the relative frequency in each length class, $f(L_i)$, and the estimate of mean weight in each length class, $W(L_i)$, gives the total biomass estimate of $B = 821$ t for clams ≥20 mm and the exploitable biomass estimate of $B_c = 636$ t for clams ≥90 mm. Clams ≥90 mm make up 77% of the total biomass (Fig. 7).

Biomass could have been calculated more directly if the weight of clam fractions 20–90 mm and ≥90 mm were measured for core samples. However, the appropriate size fractions for analysis had not yet been decided and individual clam length

Fig. 7. Biomass distribution of razor clams >20 mm shell length.

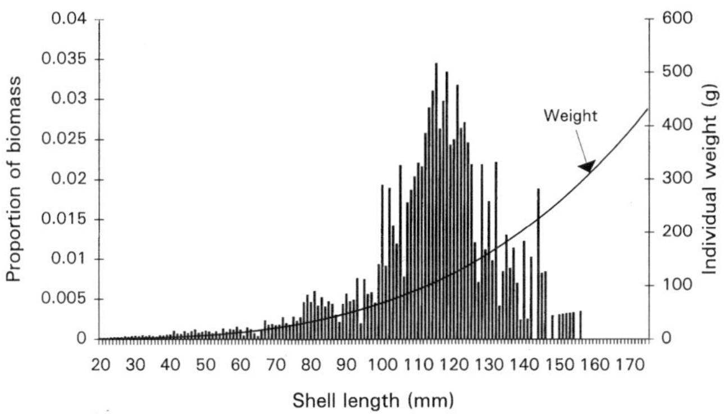

Fig. 8. Length–frequency distribution of commercial catch.

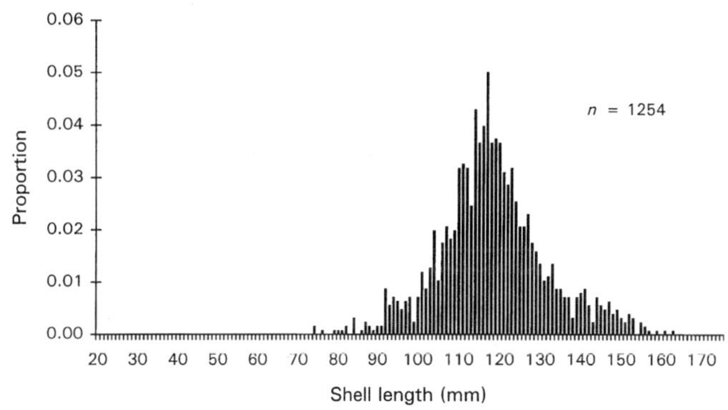

Table 4*a*. Age–length key (number of clams by age and 5 mm length-class).

	Age									
Length	1	2	3	4	5	6	7	8	9	Total
20										0
25	2									2
30	1									1
35		3								3
40		5								5
45		2								2
50		9								9
55		5								5
60		4								4
65		4	1							5
70		1	1							2
75		1								1
80		2	1							3
85		3	1							4
90		2	5							7
95			1							1
100			4							4
105			3	3						6
110			4	4	1	1				10
115				14	10	3				27
120				12	13			1		26
125				2	9	5	3			19
130				1		4	3	1		9
135						2	2	1	1	6
140								1	1	2
145							1			1
150						1		1		2
155										
160										
Total	3	41	21	36	33	16	9	5	2	166
%	1.8%	24.7%	12.7%	21.7%	19.9%	9.6%	5.4%	3.0%	1.2%	100.0%

measurements were still necessary to construct the population size and age structure and estimate natural mortality.

An estimate of Y_c was available from 1994 landings. Commercial landings were 105.4 t from sales slip data (preliminary estimate, Maureen Kostner, DFO Catch and Effort Unit, personal communication). The catch ≥90 mm was estimated to be 0.85×10^6 clams from the landing estimate, the commercial catch size distribution (Fig. 8), and the length–weight relationship, using a calculation similar to that for the biomass estimate. A small proportion (1.5%) of clams in the commercial catch were <90 mm. Noncommercial catch was estimated to be less than 31 000 clams based on mean catch estimates of

Table 4*b*. Estimate of abundance by year-class using age–length key (proportion by 5 mm length-class).

	Age									
Length	1	2	3	4	5	6	7	8	9	Relative number from size–freq. dist.
20	1.00									163
25	1.00	0.00	0.00	0.00	0.00	0.00	0.00	0.00	0.00	135
30	1.00	0.00	0.00	0.00	0.00	0.00	0.00	0.00	0.00	113
35	0.00	1.00	0.00	0.00	0.00	0.00	0.00	0.00	0.00	68
40	0.00	1.00	0.00	0.00	0.00	0.00	0.00	0.00	0.00	85
45	0.00	1.00	0.00	0.00	0.00	0.00	0.00	0.00	0.00	70
50	0.00	1.00	0.00	0.00	0.00	0.00	0.00	0.00	0.00	49
55	0.00	1.00	0.00	0.00	0.00	0.00	0.00	0.00	0.00	50
60	0.00	1.00	0.00	0.00	0.00	0.00	0.00	0.00	0.00	31
65	0.00	0.80	0.20	0.00	0.00	0.00	0.00	0.00	0.00	34
70	0.00	0.50	0.50	0.00	0.00	0.00	0.00	0.00	0.00	37
75	0.00	1.00	0.00	0.00	0.00	0.00	0.00	0.00	0.00	52
80	0.00	0.67	0.33	0.00	0.00	0.00	0.00	0.00	0.00	58
85	0.00	0.75	0.25	0.00	0.00	0.00	0.00	0.00	0.00	37
90	0.00	0.29	0.71	0.00	0.00	0.00	0.00	0.00	0.00	41
95	0.00	0.00	1.00	0.00	0.00	0.00	0.00	0.00	0.00	45
100	0.00	0.00	1.00	0.00	0.00	0.00	0.00	0.00	0.00	86
105	0.00	0.00	0.50	0.50	0.00	0.00	0.00	0.00	0.00	85
110	0.00	0.00	0.40	0.40	0.10	0.10	0.00	0.00	0.00	110
115	0.00	0.00	0.00	0.52	0.37	0.11	0.00	0.00	0.00	110
120	0.00	0.00	0.00	0.46	0.50	0.00	0.00	0.04	0.00	87
125	0.00	0.00	0.00	0.11	0.47	0.26	0.16	0.00	0.00	42
130	0.00	0.00	0.00	0.11	0.00	0.44	0.33	0.11	0.00	31
135	0.00	0.00	0.00	0.00	0.00	0.33	0.33	0.17	0.17	19
140	0.00	0.00	0.00	0.00	0.00	0.00	0.00	0.50	0.50	17
145	0.00	0.00	0.00	0.00	0.00	0.00	1.00	0.00	0.00	7
150	0.00	0.00	0.00	0.00	0.00	0.50	0.00	0.50	0.00	5
155						0.25	0.25	0.25	0.25	1
160										0
Ni 20–160 mm	411.0	528.8	300.7	191.6	115.1	57.1	30.5	21.2	11.9	1668.0
Ni 20–90 mm	411.0	517.1	53.9	0.0	0.0	0.0	0.0	0.0	0.0	982.0

523, 209, 116, and 49 clams/tide, respectively, from sampling on weekend tides <1.0 m, weekday tides <1.0 m, weekend tides between 1 and 2.5 m, and weekday tides between 1 and 2.5 m. Size distribution of the noncommercial catch was found to be similar to the commercial catch. Total noncommercial landings would be less than 3.4 t based on the average weight in the noncommercial catch (110.4 g). Then, Y_c is approximately equal to the commercial landings of 105.4 t. Fishing mortality ($F = Y_c/B_c$) is 16.6% of the exploitable biomass (636 t). Harvest rate is 14.6%, based on 0.85×10^6 clams harvested from an exploitable population of 5.82×10^6 clams.

As a first approximation, F_{MSY} is often assumed to equal the natural mortality (*M*). Razor clam survival was estimated from a catch-at-age curve (Ricker 1975). Abundance of year-classes for the catch-at-age curve was estimated from the population size–frequency distribution and the age–length key (Tables 4*a* and 4*b*) (Gulland and Rosenberg 1992, Kutkuhn 1963). The age–length key was based on a systematic random sample of two North-1 beach transect samples ($n = 36$ and 62 collected April and July 1994, respectively) and one South-1 recreational catch sample ($n = 68$ collected August 1994). Relative abundance by age in each 5 mm length-class, n_i, was then calculated from the population length–frequency distribution for clams ≥90 mm and clams 20–90 mm. Survival of age 3- to 9-year-old clams was estimated to be 59% from $S = (n_4 + n_5 + n_6 + n_7 + n_8 + n_9)/(n_3 + n_4 + n_5 + n_6 + n_7 + n_8 + n_9)$ (Ricker 1975). The rate of change in abundance between year-classes is relatively uniform as shown in the catch-at-age curve (Fig. 9). Age-1 clams are incompletely sampled since many are less than 20 mm. Survival of younger clams (Age-1 and Age-2) may also not be representative of mature clams available to the fishery. As described by Ricker (1975), the method assumes uniform survival rate between age-classes, uniform total mortality rate between age-classes ($Z = F + M$), and that age-classes were recruited in equal numbers to the fishery. Fishing mortality, *F*, was previously estimated to be 15% of the available biomass. Therefore, the estimated natural mortality, *M*, is 26%.

Assuming $F_{MSY} = M$, then MSY is 121 t using the Schaefer

Fig. 9. Abundance–age curve to estimate razor clam mortality.

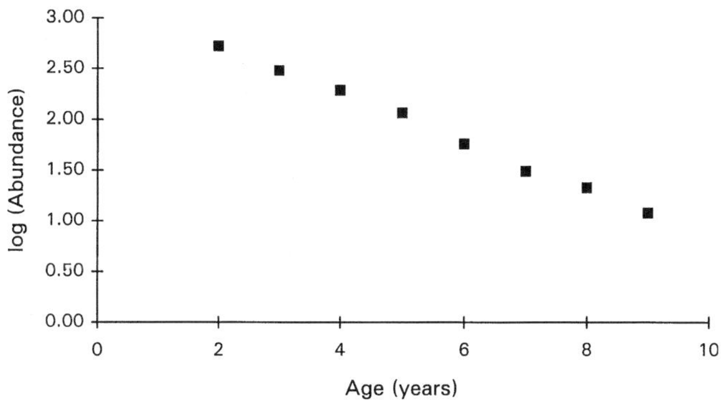

Table 5. Variation of MSY with F_{MSY} based on Schaefer and Fox.

F_{MSY}	$X = F_{MSY}/M$	MSY (tonnes) Schaefer model	MSY (tonnes) Fox model
0.130	0.50	114	109
0.195	0.75	108	107
0.260	1.00	121	115
0.325	1.25	139	127
0.390	1.50	157	139

model and 115 t using the Fox model. The sensitivity of the MSY estimate to F_{MSY} and $X = F_{MSY}/M$ is shown in Table 5.

Based on catch-at-age estimates (Tables 4*a* and 4*b*), clams 20–90 mm in length are made up of 52.7% two year olds and 5.5% three year olds. From the age-length key (also Table 4), it is expected that 82% of two year olds and 100% of three year olds will recruit the following year. Based on the population estimate for clams 20–90 mm (10.5×10^6 clams) and $Z = 0.41$, the estimated 1995 recruitment is 3.0×10^6 clams.

Discussion

Population estimate

The approach, of stratifying density samples by elevation into 2–3 beach levels, accounts for a major factor influencing intertidal variability and allowed a more precise estimate of density. The reference point used to estimate the relative elevations of transects was the tidewater elevation at low tide and chart datum. Measurement accuracy was estimated to be ±0.1 m. Tidewater elevation may be affected by wind set during storms but winds were light on days when sampling occurred. Other surveys have used the high tide line as indicated by beach features as a reference point. Bourne (1969) used distance from the driftwood line at North Beach as a reference point and assumed the slope of the beach to be constant. Szarzi (1991) used a line of gravel deposited on the beach, but measured the profile of some transects. Tidewater elevation is a more constant reference point since the elevation of beach features may vary depending on wind and wave exposure at a particular beach section and may change following storm events.

Density was found to approximate a Poisson distribution about the mean. Density variations likely reflect environmental factors as well as physiological effects. Clams are filter feeders and access to food is directly related to elevation and the duration of time that beds are underwater. Bourne and Quayle (1970) found variations in growth rates in lower, middle, and upper sections of the beach from winter growth checks on shells. Stress may also play a role in reduced growth at higher elevations. Depth provides a refuge from scouring. Small clams bury to a lesser depth than large clams and are more vulnerable to wave action and scouring. Even large clams on North and South Beach are sometimes scoured out of the sand during winter storms and washed up on the beach in windrows leading to a mass mortality. Densities on all beaches below 1.0 m elevation were 1.5–3.0 times greater than that in elevations from 1.0 to 2.0 m. This observation is in keeping with the preference by diggers for tides less than 1.2 m (4.0 ft) above chart datum.

Densities were slightly lower than reported by Bourne (1969) from 1966 surveys using tagging and recovery by repeated digging. He found density in the lowest 60 m of the beach from recovery of tagged clams to average 7.0 clams·m^{-2} on North Beach (4 transects) and 4.0 clams·m^{-2} on South Beach (2 transects). In the 1994 survey, mean density of clams ≥90 mm in the lowest 75 m of the beach (0–1 m elevation) was 5.5 clams·m^{-2} on North-1 (7 transects) and 3.62 clams·m^{-2} on South-1 (4 transects) beach sections.

The 1994 population estimate was 5.8×10^6 (SE 1.2×10^6) clams based on a harvestable beach 19.75 km in length and 4.35 km^2 in area. Bourne (1969) estimated the population of North and South Beaches to be 13.5×10^6 clams (2.3 × 1994 estimate) based on a length of 29.0 km (1.5 × 1994 estimate) and an area of 5.3 km^2 (1.2 × 1994 estimate). Bourne's estimate included 6.3 km at the east end of North Beach.

Several assumptions about harvestable areas were made early in the study, that affect population estimates. Clams were assumed to be accessible down to chart datum (Lowest Normal Tides; 0 m elevation). Twelve tides out of 81 fishable low tides (<1.2 m) were in the range 0–0.3 m above chart datum from March to September 1994 (tide levels at Prince Rupert). Clam diggers use tide levels and times at Prince Rupert because uncorrected Prince Rupert times are closer to actual times than uncorrected Bella Bella times. North Beach presently supports most of the fishing effort and 69% of the exploitable biomass. So, the fishable area defined seems reasonable, but may require revision if fishing patterns change.

MSY estimate

There are several potential sources of error in the estimates of MSY arising through the many steps in estimation of B_c, Y_c and F_{MSY}. The main source of error in B_c is likely to be the population estimate. The size frequency distribution and length–weight relationship should be relatively accurate, providing sampling was representative. Some temporal effects are possible, because the fishery was taking place at the same time as sampling and sampling was extended throughout the season. Similarly, Y_c should be relatively accurate, providing all landings were reported. F_{MSY} was estimated based on the estimate $M = 0.26$ and assumed $F_{MSY} = M$. Age sampling is likely a larger source of variation than the size distribution in estimation of M, but less significant than the assumption about the relationship between F_{MSY} and M.

The Fox and Schaefer models assume equilibrium conditions which rarely occur in real situations. Most invertebrate stocks are subject to highly variable recruitment rates which may cause cyclical fluctuations in population and biomass. Fishing effort can also vary from year to year, although the razor clam fishery appears to be relatively stable in recent years. This is partly a result of its relatively isolated location compared to other British Columbia clam beaches.

Gulland and Rosenberg (1992) recommend an examination of the catch history to see if it may provide an independent approximation of MSY. As shown in Fig. 2, peak landings of 757 t occurred in 1925 and exceeded the 1994 exploitable biomass estimate B_c = 636 t. The 1994 MSY estimate (MSY_{1994}) is 118 t, from the average of results of the two surplus production models. Catches were also more than double MSY_{1994} in 1930, 1938, and 1942 when landings were 394, 433, and 267 t, respectively. In all, catches exceeded MSY_{1994} a total of 17 times from 1923 to 1994. This occurred most recently in 1986, 1987, and 1988 when landings were 142, 142, and 156 t, respectively. Average landings from 1926 to 1932 (a 7-year period) were 177 t. By comparison, average landings from 1985 to 1989 were 129 t and from 1990 to 1994 were 86.6 t. Landings in 1994 were 105.4 t. Since the fishery is not known to have collapsed in the past, either the exploitable biomass (and MSY) fluctuates from year to year, harvest has exceeded MSY in several years, or F_{MSY} is greater than estimated.

In any case, most authors recommend caution in interpreting predictions of F_{MSY} and recommend harvesting levels below MSY for economic reasons, to account for recruitment fluctuations, or for species subject to predation (Caddy and Csirke 1983).

The analysis did not include an estimate of the statistical error in MSY. We are currently investigating utility of a Monte Carlo simulation to assess the sensitivity of the analysis.

Other management considerations

The current size limit of 90 mm may not serve to protect much of the spawning population. Only 22% of the total biomass was found to be <90 mm. Age-of-maturity studies in Alaska indicate that razor clams first mature at 90 mm at Cordova and 85 mm at Swikshak and 50% were mature at 105 mm and 99.1 mm, respectively (Weymouth et al. 1925). In Washington, all clams spawn as two year olds at about 103 mm (Weymouth et al. 1925). In the early days of cannery production, size limits may have served alternate purposes such as keeping small clams off the processing line.

In addition, the fishery has some self-limiting characteristics providing fishing effort is under control. Tides and daylight hours restricted the 1994 fishing season to about 80 tides. The main processer usually does not buy razor clams in July and August because of salmon landings which reduces the fishing season by about 30 tides. Weather and strong northwest winds can further reduce the fishing season. The fishery is low value with landed price of \$1.764/kg (\$0.80/lb). Fishers stop fishing if clams densities are too low. Space on the beach is also limiting, and commercial diggers space themselves out fairly evenly along beach and consider the beach to be already "full."

The rationale for the 1994 survey was concern about the decline in catches and appearance of undersize clams in the 1993 catch. The 1994 catches and survey results provide some reassurance to managers, given the poor fishery in 1993. The Haida Fisheries Program expects to continue surveys in 1995. Biomass estimates are likely not needed on an annual basis because the fishery is not managed by quotas. However, continuing monitoring of stock status is recommended since catch (105 t) is approaching MSY (118 t). Structural factors, price changes, or natural fluctuations in stocks could alter this balance.

Survey effort was relatively modest compared to effort in the fishery. Total field survey effort was approximately 55 person-tides (11 transects times 5 persons). In comparison, daily commercial fishing effort ranged from 30 to 100 diggers per tide.

Data collected in 1994 can be incorporated into an age-structured model in future years. The nonselective sampling method provides an unbiased sample of the population to reconstruct the population size and age structure. In future surveys, clams for aging and maturity studies will be collected directly from transect samples. In 1994, eleven transects allowed a relatively precise estimate of the razor clam population and biomass. Many other fisheries require a variety of assumptions about selectivity of harvest or sampling in order to estimate these stock parameters.

Conclusions

Hydraulic sampling was found to be an effective method for estimating the razor clam population. If transects are randomly selected, the method has the added advantage of providing an unbiased sample of the population. Stratifying the data into two to three beach levels by elevation improved accuracy and resulted in a more efficient use of resources. The population of harvestable clams was 5.8×10^6 (SE 1.2×10^6) clams ≥90 mm and 16.3×10^6 (SE 2.6×10^6) clams ≥20 mm. MSY was estimated to be 121 t based on the Schaefer, and 115 t based on the Fox surplus production model from the calculated exploitable biomass, B_c (636 t), the 1994 yield from reported landings (105.4 t) and the assumption that $F_{MSY} = M$. Natural mortality of exploited razor clams was estimated to be 0.26. MSY will vary from year to year with fluctuations in B_c due to changes in survival, recruitment, and landings. Current fishing mortality is about 0.17 which is about 65% of the estimate of M and F_{MSY}. The current size limit of 90 mm is not thought to protect a significant portion of the spawning biomass based on the 1994 biomass estimate and age at maturity studies in Alaska and Washington (Weymouth et al. 1925). However, given possible fluctuations in B_c, and recent limits on the number of fishery participants, fishing quotas are not thought to be necessary at the present time.

Acknowledgments

Funding for this study was provided by the Aboriginal Fisheries Strategy of the Department of Fisheries and Oceans. Other Haida Fisheries Program staff who participated in the field work included Pat Fairweather, Brad Setso, George Wesley, Irene Bruce, Denise Russ, Cullen Marks, Martin Williams, Dave Monture, and Chris Collison.

References

Ayres, D.L., and Simons, D.D. 1988. The spring and fall 1988 recreational razor clam fisheries with a review of razor clam sampling methodology. Washington Department of Fisheries, Olympia, Wash. Progress Report No. 276.

Bourne, N. 1969. Population studies on the razor clam at Masset, British Columbia. Fish. Res. Board Can. Tech. Rep. No. 118. 24 p.

Bourne, N. 1979. Razor clam, (*Siliqua patula* Dixon) breeding and recruitment at Masset, British Columbia. Proceedings of the National Shellfisheries Association, **69**: 21–29.

Bourne, N., and Quayle, D.B. 1970. Breeding and growth of razor clams in British Columbia. Fish. Res. Board Can. Tech. Rep. No. 232. 42 p.

Breese, W.P., and Robinson, A. 1981. Razor clams, *Siliqua patula* (Dixon): gonadal development, induced spawning and larval rearing. Aquaculture, **22**: 27–33.

Caddy, J.F., and Csirke, J. 1983. Approximations to sustainable yield for exploited and unexploited stocks. Oceanogr. Trop. **18**(1): 3–15.

Cochran, W.G. 1977. Sampling techniques, 3rd edition. John Wiley & Sons, Inc., New York, N.Y. 428 p.

Fox, W.W. 1970. An exponential surplus-yield model for optimizing exploited fish populations. Trans. Am. Fish. Soc. **99**: 80–88.

Department of Fisheries and Oceans. 1991. Canadian tide and current tables. Department of Fisheries and Oceans, Ottawa, Ont.

Garcia, S., Sparre, P., and Csirke, J. 1989. Estimating surplus production and maximum sustainable yield from biomass data when catch and effort time series are not available. Fish. Res. (Amsterdam), **8**: 13–23.

Gulland, J.A., and Rosenberg, A.A. 1992. A review of length-based approaches to assessing fish stocks. FAO Fish. Tech. Paper No. 323. 100 p.

Hirschhorn, G. 1962. Growth and mortality rates of the razor clam *(Siliqua patula)* on Clatsop beaches. Oreg. Fish. Comm. Contrib. No. 27. 55 p.

Krebs, C.J. 1989. Ecological methodology. Harper & Row, New York, N.Y. 654 p.

Kutkuhn, J.H. 1963. Estimating absolute age composition of California salmon landings. Calif. Dep. Fish. Game Fish Bull. No. 20. 47 p.

Nickerson, R.B. 1975. A critical analysis of some razor clam (*Siliqua patula* Dixon) populations in Alaska. Alaska Dep. of Fish and Game, Juneau, Alaska. July 1975. 294 p.

Ricker, W.E. 1975. Computation and interpretation of biological statistics of fish populations. Bull. Fish. Res. Board Can. No. 191. 382 p.

SAS Institute Inc. 1992. SAS/STAT Software: Changes and enhancements. Version 6.07. SAS Technical Report P-229, SAS Institute Inc., Cary, N.C. 620 p.

Schaefer, M. 1954. Some aspects of the dynamics of populations important to the management of the commercial marine fisheries. Bull. Inter-Am. Trop. Tuna Comm. **1**: 27–56.

Schwarz, C.J., Jones, R.R., Lee, L., and Nguyen, M. 1995. Estimating the number of clams in an intertidal zone using a three stage sampling technique. Statistical Consulting Service, Department of Mathematics and Statistics, Simon Fraser University, Burnaby, B.C. Report No. 95-001. 30 p.

Szarzi, N.J. 1991. Distribution and abundance of the Pacific razor clam (*Siliqua patula* Dixon), on the Eastside Cook Inlet beaches, Alaska. M.Sc. thesis. University of Alaska, Fairbanks. 168 p.

Tegelberg, H.C. 1964. Growth and ring formation of Washington razor clams. Wash. Dep. Fish., Fisheries Research Paper 2, No. 3. pp. 69–103.

Weymouth, F.W., and McMillin, H.C. 1930. Relative growth and mortality of the Pacific razor clam (*Siliqua patula* Dixon) and their bearing on the commercial fishery. U.S. Bur. Fish. Bull. No. 46. pp. 543–567.

Weymouth, F.W., McMillin, H.C., and Holmes, H.B. 1925. Growth and age at maturity of the Pacific razor clam, *Siliqua patula* (Dixon). U.S. Dep. Commer., Bur. Fish. Bull. No. 41. pp. 201–236.

An evaluation of surplus production analysis for assessing the fishery for New Zealand red rock lobsters (*Jasus edwardsii*)

Paul A. Breen and Terese H. Kendrick

Abstract: We used simulated data to examine potential biases in surplus production modeling. The study has specific application to the assessment for an important New Zealand fishery with a long time series of catch and effort data. We describe the assessment procedure, discuss important potential problems with the model and data series, then describe an age-structured data simulator with realistic biological inputs. We used the simulator to generate catch and effort time series with various kinds of systematic error. Five indicators were estimated with the assessment model and compared with the "true" values from the simulator. The assessment model performed reasonably well when the data were simulated without systematic error, despite several potentially serious structural problems with the model. The choice of model "shape" influenced the results, but not seriously and not consistently. The various indicators were estimated with different accuracy: the ratios B_{curr}/B_{MSY} and E_{curr}/E_{MSY} were estimated badly compared with virgin biomass, B_0; maximum sustainable yield, MSY; and current surplus production, CSP. Errors in early catch data and systematic error in catch per unit effort, CPUE, or change in catchability, caused the greatest bias in results. Provided that errors in catch and effective effort data are not too severe, and with the caution that appropriate indicators be used, we conclude that surplus-production analysis is a useful tool for this and similar fisheries. The study examined a highly specific situation, and similar studies would be required for different situations.

Résumé : Nous avons utilisé la simulation pour examiner les biais potentiels dans la modélisation de la production excédentaire. Cette étude a des applications spécifiques dans l'évaluation d'une importante pêcherie de Nouvelle-Zélande pour laquelle on possède une longue série chronologique de données sur les prises et l'effort. Nous décrivons la méthode d'évaluation, analysons les problèmes importants que peuvent poser le modèle et la série de données, puis décrivons un simulateur de données structurées selon l'âge, alimenté par des entrées biologiques réalistes. Nous avons employé le simulateur pour produire des séries de données sur les prises et l'effort présentant divers types d'erreur systématique. Cinq indicateurs ont été construits avec le modèle d'évaluation et comparés aux valeurs « vraies » produites par le simulateur. Le modèle d'évaluation a raisonnablement bien fonctionné quand les données étaient simulées sans erreur systématique, même s'il présentait plusieurs problèmes structurels potentiellement graves. Le choix de la « forme » a influé sur le modèle, mais sans gravité ni régularité. Les divers indicateurs ont été estimés avec des degrés variables de précision : les rapports $B_{cour}/B_{PMÉ}$ et $E_{cour}/E_{PMÉ}$ ont été estimés grossièrement par rapport à la biomasse vierge, B_o, à la production maximale équilibrée (PMÉ) et à la production excédentaire courante. Ce sont les erreurs dans les données initiales sur les prises et l'erreur systématique dans l'indice des captures par unité d'effort (CPUE), ou le changement dans la capturabilité, qui ont causé le biais le plus grand dans les résultats. À condition que les erreurs dans les données sur les captures et l'effort réel ne soient pas trop graves, et si l'on prend garde d'employer les indicateurs appropriés, nous concluons que l'analyse de la production excédentaire est un bon outil pour cette pêcherie et pour des pêcheries semblables. Notre étude a porté sur une situation très spécifique, et il faudrait réaliser des études comparables pour des situations différentes. [Traduit par la Rédaction]

Introduction

This paper describes an investigation of potential biases in the stock assessment for the New Zealand spiny lobster *Jasus edwardsii*. The fishery (Booth and Breen 1994) is managed under individual transferable quotas (ITQs) and the management goal is to use a constant harvest strategy to return stocks to the biomass that would produce the maximum sustainable yield (MSY), B_{MSY}. This requires annual assessments to compare total allowable catches with estimated sustainable yields and to compare estimated current biomass with estimated B_{MSY}.

Since 1979 (Saila et al. 1979), surplus-production analysis based on a long time series of catch and effort data has been used as part of the assessment (Booth and Breen 1994). A variety of surplus-production approaches have been used (e.g., Annala and Esterman 1986; Fogarty and Murawski 1986; Breen and Stocker 1994). Surplus-production analysis is a common assessment tool for lobster fisheries (e.g., Marchessault et al. 1976; Morgan 1979; Jensen 1986; Yoshimoto and Clarke 1993).

The simplicity of surplus-production models imparts both advantages and drawbacks (see e.g., Megry and Wespestad 1987 for a full description of assumptions, advantages, and

P.A. Breen and T.H. Kendrick. NIWA, National Institute of Water and Atmospheric Research Ltd., P.O. Box 14901, Wellington, Kilbirnie, New Zealand.

Correct citation: Breen, P.A., and Kendrick, T.H. 1998. An evaluation of surplus production analysis for assessing the fishery for New Zealand red rock lobsters (*Jasus edwardsii*). *In* Proceedings of the North Pacific Symposium on Invertebrate Stock Assessment and Management. *Edited by* G.S. Jamieson and A. Campbell. Can. Spec. Publ. Fish. Aquat. Sci. 125. pp. 213–223.

disadvantages). Many authors have questioned the use of simple surplus-production models in assessment (e.g., Roff 1983; Laloe 1995). Morgan (1979) suggested that "while useful for deriving preliminary estimates, [surplus-production analysis] should not be used as the sole means for monitoring the status of the stock and for making management decisions." Koonce and Shuter (1987) suggested that a detailed population model was superior to a surplus-production model in a wide range of conditions. However, Ludwig and Walters (1985) and Punt (1992) report the opposite under some conditions.

The performance of surplus-production analysis can be affected by many factors. The algebraic form of the model is important (e.g., Yoshimoto and Clarke 1993), but the method of fitting may be more important (Polacheck et al. 1993). These latter authors argue that observation–error time series (Pella and Tomlinson 1969) fitting is the best approach. Assumptions about error are important (Schnute 1989). The exploitation history must have enough "contrast" (Hilborn 1979) and ideally should be more complex than a "one-way trip" (Hilborn and Walters 1992). Violation of the assumption that catch per unit effort (CPUE) linearly reflects abundance (i.e., if catchability changes over time) can affect the results (e.g., Gordoa and Hightower 1991).

Punt (1995) reviews the aspects of population biology that affect the ability of surplus-production analysis to make reasonable estimates. He suggests that low natural mortality rate, low recruitment variability, and high intrinsic rate of increase all favour the method. He also suggests that the best approach is to simulate the biological detail of each specific situation to examine bias, and we do this here.

In this paper we describe the assessment model used in the *Jasus edwardsii* fishery and describe likely problems with the model and fishery data. There are three approaches to examining the performance of an assessment model: (*i*) to apply it to experimental data (e.g., Silliman 1971; Yoshimoto and Clarke 1993); (*ii*) to use it on real fishery data and evaluate performance at "forecasting" catch or CPUE (e.g., Roff 1983); and (*iii*) to apply it to artificially simulated data (e.g., Uhler 1979; Punt 1995). We have chosen the last approach as being the most appropriate (Ludwig and Walters 1985; Polacheck et al. 1993).

We describe an age-structured data simulator, then describe the performance of the assessment model on simulated data with and without systematic errors in the input data. We evaluated the bias in several indicators and included indicators most heavily relied upon in developing management advice from the assessment.

Assessment model

Breen and Kendrick (1995*a*) described a generalized surplus-production model, fit with an observation–error time series to assess the New Zealand rock lobster fishery in 1994 and 1995.

Models such as the Schaefer (Schaefer 1954) and Fox (1975) models implicitly assume that B_{MSY} is a particular percentage of K, 50% and 37%, respectively. With a generalized model, different shapes can be assumed. In practice, shape is difficult to estimate from catch and effort data (Silliman 1971; Andrew and Butterworth 1987) but additional information can be used to constrain the possibilities as we describe below.

The model used was

$$[1] \quad \text{production}_t = rB_t\left((K/B_t)^p - 1\right)$$

where r is a parameter related to productivity (it is the ratio of intrinsic growth rate and p), B_t is recruited biomass at the beginning of year t, and the "shape" parameter p determines the location of B_{MSY} relative to K:

$$[2] \quad B_{\mathrm{MSY}}/K = (1-p)^{1/p}$$

This is a reparameterisation of Pella and Tomlinson's (1969) generalized model.

As p approaches 0, the ratio B_{MSY}/K approaches 37%, the value assumed by the Fox model; at $p = 1.0$, B_{MSY}/K is zero.

Each year's biomass B_t is estimated from the previous year's:

$$[3] \quad B_{t+1} = B_t + [rB_t((K/B_t)^p - 1)] - C_t$$

where C_t is the catch in year t. A simplifying assumption is that

$$[4] \quad B_0 = K$$

This means that we assume the biomass at the beginning of the time series is the same as the mean carrying capacity. This is realistic where the catch time series is available for most of the exploitation history, as is nearly the case for *Jasus edwardsii* (Booth and Breen 1994). This assumption is evaluated by Punt (1990), who concluded that the assumption is reasonable as long as B_0 is not substantially less than K.

CPUE is used as an index of abundance. The CPUE in year t, I^*_t, is assumed to be related to biomass through the catchability coefficient, q:

$$[5] \quad I^*_t = qB_t$$

This equation reflects the assumption that catchability has remained constant. In the work below we examine the effect of that assumption not holding.

The optimum effort, E_{MSY}, is calculated by solving eq. 5:

$$[6] \quad E_{\mathrm{MSY}} = \mathrm{MSY}/(qB_{\mathrm{MSY}})$$

In a New Zealand stock assessment application, the model was fit to a 50-year series of catch and effort data from a substock of *Jasus edwardsii* (Breen and Kendrick 1995*a*). Because the quality of effort data is suspect for the early years, catch data are used from the whole series, but CPUE residuals for the first 18 years do not contribute to the objective function.

In the fitting procedure, an estimate of average recent exploitation rate is used as auxiliary information as described by Hilborn and Walters (1992). The estimate of average recent exploitation rate is obtained from a length-converted catch curve analysis (Sparre et al. 1989) of size data obtained from sampling at sea.

This method will give biased estimates of exploitation rate if the intensity of fishing has changed during the lifespan of animals measured. In the work described below, we examine the effect of using biased estimates.

This auxiliary information is important when analyzing the real data (Breen and Kendrick 1995*a*). Without it, the model estimates high B_0 and low r combinations, resulting in highly unrealistic estimates of the current exploitation.

The model is fit by minimizing the objective function T in

$$[7] \quad T = \sum[\mathrm{wt1}_t(\ln(I^*_{t,\mathrm{pred}}) - \ln(I^*_{t,\mathrm{obs}}))^2] + \mathrm{wt3}\sum[\mathrm{wt2}_t(\ln(\mathrm{ER}_{t,\mathrm{pred}}) - \ln(\mathrm{ER}_{t,\mathrm{obs}}))^2]$$

where $I^*_{t,\mathrm{obs}}$ is the observed CPUE for year t and $I^*_{t,\mathrm{pred}}$ is the

Fig. 1. Annual production (change in biomass plus catch) from the simulation model described in the text, plotted as a function of recruited biomass. The smooth line shows equilibrium production, the lower curve shows production from a model population in decline from a high exploitation rate, the upper curve shows production from a depleted population increasing rapidly under low exploitation.

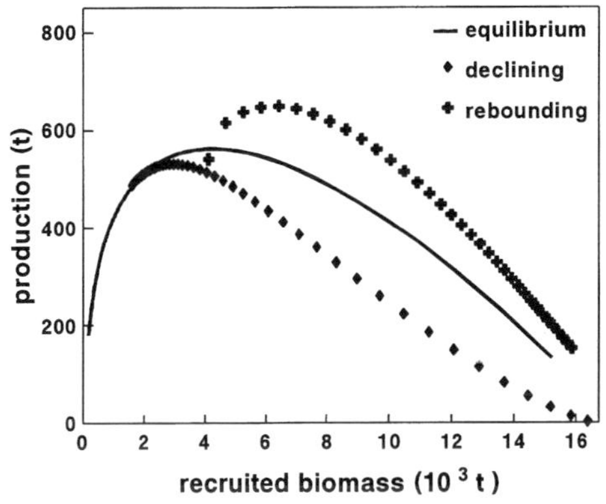

predicted CPUE from eqs. 3 and 5 for year t; similarly $ER_{t,obs}$ and $ER_{t,pred}$ are the observed and predicted values of exploitation rate in year t. The weight $wt1_t$ was used to include or exclude the CPUE estimate from a particular year in the fitting procedure. It was set to 0 for years before 1963 and to 1 for later years, recognizing the suspected poor quality of early CPUE data.

The weight $wt2_t$ was used to include or exclude an exploitation rate estimate from a particular year in the fitting procedure. It was set to 0 for all except the most recent 5 years; it was set to 1 in the years for which an estimate of ER_t is available.

The weight wt3 was used to control the relative contribution of the two components of T. Results are not sensitive to wt3 (Breen and Kendrick 1995a); wt3 was used only to force both contributions to have the same order of magnitude.

Parameters were estimated using least-squares estimation with the minimizing package MicroSimplex (Schnute 1982; Mittertreiner and Schnute 1985). The procedure found values for r, q, and K that minimized T in eq. 6. The value of p was assumed, based on the behavior of an age-structured model as described below (Breen and Kendrick 1995*a*).

Potential problems

There are several potential problems with the procedure just described. We divide these into "structural" problems, which cause the assessment model to differ from reality, and problems with the data.

The model assumes that production is related to biomass at the beginning of the year, without considering the age or size structure of a unit biomass. In reality, the age structure of a population affects productivity. The equilibrium production (annual increase in biomass plus catch) is shown in Fig. 1 as a function of biomass in the age-structured model described in the "Data simulator" section below. Also shown is production versus biomass for populations declining and rebounding at reasonably fast rates, i.e., populations well out of equilibrium.

The model also assumes that CPUE is an index of "biomass." In other words, the assumption is that the productive biomass is the same as the biomass vulnerable to the fishery, reflected by CPUE. This assumption is violated by minimum legal sizes (MLS) in the fishery, because sublegal biomass is not indexed by CPUE. Another violation is caused by a ban on landing egg-bearing females. The ovigerous period is about 6 months while the season is year-round, so there is substantially more mortality on males than females.

A third structural problem may be caused by assuming a production function with the wrong shape. In fitting the New Zealand data, we do not obtain credible results when estimating p as a free parameter (P.A. Breen and T.H. Kendrick, unpublished data), so we assume a shape based on an age-structured model such as described below. The assumed shape may be incorrect. This list of structural problems is not exhaustive, but the three problems listed may be the most serious.

Fishery data have many potential problems. For the New Zealand data we believe the two worst problems may be underestimates of catches in the postwar years when the fishery began and possible systematic changes in the relationship between recruited biomass and CPUE.

Since 1979, fishers have reported catches and effort to the government under fishery rules. Reported catches can be compared with exports and, since 1990, with processor reports. Estimates of the unreported catch and the noncommercial catches can be made (see Breen and Kendrick 1995*a*). By contrast, catches in the early years of the fishery were estimated by inspectors in various ports (Annala and Esterman 1986) and could easily have been significantly underestimated.

The relation between CPUE and recruited biomass is influenced by a multitude of factors (e.g., Hilborn and Walters 1992), including fish and fisher behavior. The relation between abundance and CPUE is unlikely to be linear, but the direction of departure is unknown. For lobsters, the efficiency of fishing has increased dramatically over time as a result of improved technology best documented in Western Australia (Brown et al. 1994). The likely pattern is one in which nominal fishing effort increasingly underestimates effective effort, i.e., one in which catchability increases during the time series.

Finally, there could be errors in the auxiliary information, estimated exploitation rate. These could be caused by a number of problems, especially by biased growth rate estimates and irregular recruitment.

Data simulator

We constructed an age-structured population model to generate catch and abundance data for use in testing the model described above. Lobsters cannot be aged, but lobster population parameters such as growth and mortality are available as annual rates.

The model simulated 50 age-classes of both males and females. Each age had a sex-specific mean individual length and weight, described by von Bertalanffy growth equations (eq. A1) and length–weight relations (eq. A2). Egg production was determined from the number of females in an age-class, the percentage maturity as a function of length based on an ogive (eq. A4), and the individual fecundity of a female of the appropriate length (eq. A5). The essential parameters are given in Table A1.

Table 1. Relations among steepness in the stock-recruitment relation, productivity at optimum biomass (MSY/B_0), "shape" of the production function (B_{MSY} /B_0), and the correct shape parameter *p* in eq. 1.

Steepness	MSY/B_0, %	B_{MSY}/B_0, %	*p*
1.0	3.39	22.4	0.58
0.9	3.25	24.6	0.51
0.8	3.10	26.6	0.45
0.7	2.93	28.7	0.37

Note: The correct shape parameter was determined from the ratio of B_{MSY} to B_0.

Initial recruitment (the number of individuals in year $t = 1$ for each sex) was set arbitrarily at one million. Subsequent recruitment was determined from a Beverton–Holt stock-recruitment function:

$$[8] \quad R_{t+1} = \mathrm{EGGS}_t/(\alpha + (\beta\ \mathrm{EGGS}_t))$$

where R_{t+1} is the expected recruitment to the population at time $t + 1$, EGGS_t is the population egg production in year t, and α and β are parameters. Egg production EGGS_t is the sum of egg production from each age-class, which is the product of numbers (eqs. 13 and 16) and fecundity (eq. A5).

For convenience, the parameter "steepness" was used (Francis 1992). This is the percentage of the virgin level of recruitment that occurs when the population is at 20% of initial biomass. Steepness, *s*, can vary from 0.2 (recruitment directly proportional to biomass) to 1.0 (expected recruitment constant). The parameters are calculated from

$$[9] \quad \alpha = [\mathrm{EGGS}_1/N_1]\ [1 - ((s - 0.2)/(0.8\ s))]$$

$$[10] \quad \beta = (s - 0.2)/(0.8\ s\ N_1)$$

where N_1 and EGGS_1 are the numbers and egg production of the initial population, respectively.

Nothing is known about steepness in *Jasus edwardsii* except that an apparent high resilience to fishing suggests a high steepness. Two values were used as described below.

When randomly varying recruitment was simulated, the coefficient of variation, CVR, was assumed to be 0.4. Random normal deviates (RNDs) were generated, and actual recruitment R'_{t+1} was determined from expected recruitment, R_{t+1}:

$$[11] \quad R'_{t+1} = \exp[(\ln(R_{t+1}) - (0.5\ \ln(\mathrm{CVR}^2 + 1)) + (\mathrm{RND}\sqrt{\ln(\mathrm{CVR}^2 + 1)}))]$$

For individuals smaller than the MLS, numbers at age in the next year for each sex are given by:

$$[12] \quad N_{\mathrm{SEX},k+1,t+1} = N_{\mathrm{SEX},k+1,t}\exp(-M)$$

where M is the instantaneous rate of natural mortality, SEX indexes sex and k indexes age. For individuals smaller than, but within 6 mm of the MLS, eq. 12 was modified to incorporate handling mortality:

$$[13] \quad N_{\mathrm{SEX},k+1,t+1} = N_{\mathrm{SEX},k,t}\exp(-M - hF_t)$$

where F_t is the instantaneous rate of fishing mortality exerted on legal-size males and h is a constant relating sublegal mortality to F_t. It was assumed that all individuals above the MLS are equally vulnerable to fishing except for ovigerous females as explained below.

For males above the MLS, numbers of the next age group in the next year are described by:

$$[14] \quad N_{\mathrm{male},k+1,t+1} = N_{\mathrm{male},k,t}\ \exp(-M - F_t)$$

The catch from age k legal-sized males in year t is given by:

$$[15] \quad C_{\mathrm{male},k,t} = (F_t/(F_t + M))N_{\mathrm{male},k,t}\ W_{\mathrm{male},k}\ (1 - \exp(-M - F_t))$$

where $W_{\mathrm{male},k}$ is the mean weight of age k legal-sized males. For females above the MLS, the model considered the reduced vulnerability of egg-bearing females. Annala and Breen (1989) assumed that the fishing mortality rate for mature females is half the rate for legal-size males and immature females.

The proportion of mature females of age k, AP_k, was calculated from the maturity ogive using the mean length of females at age k. The proportion of immature females is $1 - \mathrm{AP}_k$. The number of females in age group $k + 1$ in year $t + 1$ was calculated by a modification of eq. 14:

$$[16] \quad N_{\mathrm{female},k+1,t+1} = ((1 - \mathrm{AP}_k)N_{\mathrm{female},k,t}\ \exp(-M - F_t)) + (\mathrm{AP}_k\ N_{\mathrm{female},k,t}\ \exp(-M - F_t - 0.5hF_t))$$

The last element of eq. 16 incorporates the handling mortality experienced by berried females. The catch of immature females in age group k in year t is given by:

$$[17] \quad C_{\mathrm{female},k,t} = F_t/(F_t + M)N_{\mathrm{female},k,t}\ W_{\mathrm{female},k}\ (1 - \mathrm{AP}_k)\ (1 - \exp(-M - F_t))$$

and the catch of mature females by:

$$[18] \quad C_{\mathrm{female},k,t} = 0.5F_t/(0.5F_t + M)N_{\mathrm{female},k,t}\ W_{female,k}\ \mathrm{AP}_k\ (1 - \exp(-M - 0.5F_t))$$

The model was driven by effort. An effort time series (50 years) of the general form seen in New Zealand substocks (Booth and Breen 1994) was generated and fishing mortality rate F_t calculated from:

$$[19] \quad F_t = q\ \mathrm{EFFORT}_t$$

Simulator inputs

The data simulator was run with biological estimates from Fiordland, an important fishery area in the southwestern South Island of New Zealand. Specific biological parameters used are given in the Appendix.

Relation between steepness and shape

Relations among the steepness value, the ratios of MSY and B_{MSY} to B_0, and the shape parameter p from eqs. 1 and 2 are shown in Table 1. Values were obtained from the data simulator with the Fiordland data described above. These are deterministic results, CVR = 0. Initial recruitment was arbitrary; B_0 is independent of steepness. The model was run for 50 years, which was sufficiently long to produce equilibrium for the steepness values shown, but not for lower steepness values.

As steepness decreases, MSY decreases as a percentage of B_0, B_{MSY} increases as a percentage of B_0, and the shape parameter p decreases. Results such as these were used to constrain the value of p used in the real stock assessment (Breen and Kendrick 1995*a*).

Fig. 2. Model population biomass trajectories generated from three different q values. From upper to lower they are 1.8×10^{-5}, 3.0×10^{-5}, and 1.0×10^{-4}. The model was run for 50 years with the standard effort data set, steepness of 0.9, and no stochastic recruitment variation. The horizontal line shows B_{MSY}.

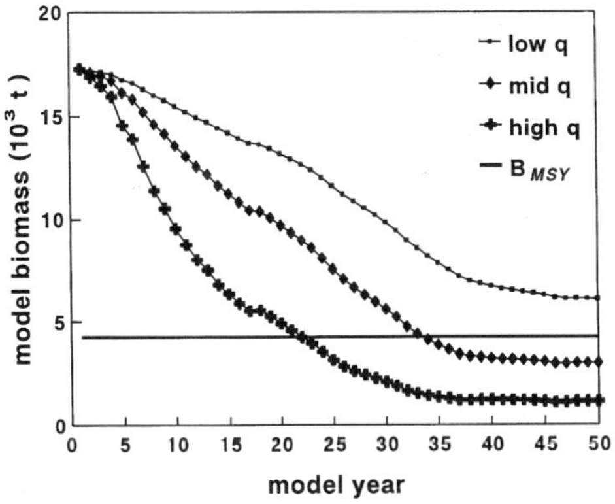

Table 2. Mean and maximum percentage bias for each indicator, combining the results from both steepness values and across all q values, from analyses of data with deterministic recruitment.

	Correct p, %		Incorrect p, %	
Indicator	Mean	Maximum	Mean	Maximum
B_0	–5.5	–15.1	–5.0	–24.6
MSY	–7.2	–17.2	–8.1	–18.1
CSP	–7.8	–14.4	–8.6	–14.4
B_{curr}/B_{MSY}	–6.3	–11.2	–6.5	–22.5
E_{curr}/E_{MSY}	8.4	30.9	10.5	52.8

Note: Percentage bias was calculated as 100 times ("true" value-estimated value)/"true" value.

Indicators

We chose five indicators for comparison between the simulated population and the assessment model output. These were virgin biomass B_0, MSY, current surplus production (CSP), the ratio of current biomass B_{curr} to B_{MSY}, and the ratio of current effort to the equilibrium effort that would produce MSY: E_{curr}/E_{MSY}.

In the assessment model results, B_0 was estimated as K. MSY was estimated by substituting B_{MSY} from eq. 2 into eq. 1, CSP was estimated from eq. 1, B_{curr} was estimated from eq. 5, and E_{curr} is known from the effort time series.

In the data simulator, B_0 was a constant determined by the input parameters described in the Appendix and an arbitrary value for N_1. MSY, B_{MSY}, and E_{MSY} were constant for a given value of steepness.

Performance of the assessment model

We drove the data simulator with a single arbitrary realistic time series of effort. With this, we could generate a variety of population trajectories in the data simulator by varying q. We chose a range of q values to give a range of population trajectories. Three trajectories generated with three values of q and no stochastic recruitment variation are shown in Fig. 2. The range of q values chosen (from 1.8×10^{-5} to 1.0×10^{-4}) led to biomass values after 50 years ranging from 6 to 35% of B_0 and 25–140% of B_{MSY}. This allowed us to examine performance over a range of over- and under-fished situations.

Fig. 3. MSY estimated from data simulated with deterministic recruitment, a steepness of 0.9, and a range of q values. The small circles show estimates made using the correct value of p in the assessment model and large circles show estimates made with the incorrect value. The horizontal line indicates the "true" value. The final model population size is inversely related to the value of q used in the data simulator, so final population size declines from left to right.

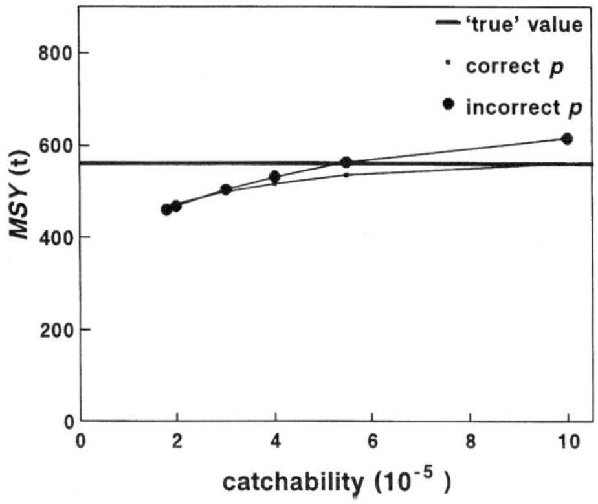

Deterministic data

We generated data sets (catch and effort time series plus exploitation rate for the last five years) from six q values and two values of steepness, with or without stochastic variation in recruitment. Steepness values of 0.9 and 0.7 were chosen arbitrarily. "True" values for the five indicators described above were recorded for each data set.

Then we estimated values for the indicators with the assessment model, using both the correct and incorrect shape parameter, p, for the value of steepness used in the data simulator. The appropriate value of p in the assessment model for data generated with a steepness of 0.9 was 0.51; the value appropriate for steepness of 0.7 was 0.37 (Table 1). For data generated with each steepness value (0.9 and 0.7) we used both p values in the analysis: the correct value and the incorrect value.

The mean and maximum percentage bias for each indicator, combining the results from both steepness values and across all q values, are shown in Table 2. Most results for a particular indicator and data set lay on one side of the "true" value (Breen and Kendrick 1995*b*), so summarizing in this way is realistic. Where results lay on both sides of the "true" value the mean bias may be underestimated.

The model performed reasonably well on these data sets when the correct value of p was used: maximum biases were all <18% except for the indicator E_{curr}/E_{MSY}. The effect of using the incorrect p was generally to increase bias, but this was not consistent. Depending on the indicator examined, the q value and steepness used in generating the data set, sometimes the incorrect p value produced a better estimate than the correct value. This is illustrated in Fig. 3 for MSY estimated from data simulated from steepness = 0.9; the incorrect p value produced more accurate results for data generated from the middle range of q values. Figures such as Fig. 3 are given for all indicators and all situations in Breen and Kendrick (1995*b*).

B_0 and E_{curr}/E_{MSY} were estimated best when B_{curr} was a high

Table 3. Mean bias for each indicator, across all q values and data generated from both steepness values, from analyses of data with stochastic recruitment.

	Correct p, %		Incorrect p, %	
Indicator	Mean	Maximum	Mean	Maximum
B_0	–5.2	–13.7	–3.9	–20.1
MSY	–7.4	–23.4	–8.7	–23.4
B_{curr}/B_{MSY}	–5.7	–13.4	–5.1	–24.3
E_{curr}/E_{MSY}	7.6	35.9	10.6	58.9

Note: The table shows the maximum bias as the highest mean bias for a particular q and steepness value; this allows comparison with results in Table 2.

Fig. 4. B_0 estimated from data simulated with stochastic recruitment, a steepness of 0.9, and a range of q values. The small squares show estimates made using the correct value of p in the assessment model; large circles show estimates made with the incorrect value. The horizontal line indicates the "true" value.

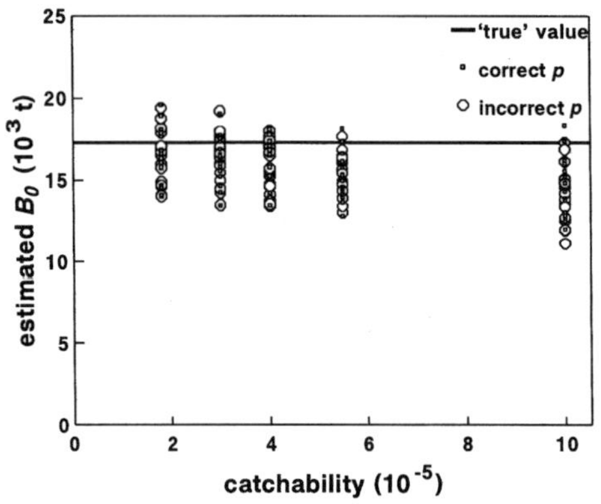

proportion of B_0 (i.e., when a low q was used to generate data). Other indicators were estimated best when B_{curr} was low, i.e., when a high q was used to generate the data and the final population was well below B_{MSY}. This result was predictable: sustainable yields cannot be estimated well until the population has been pushed below B_{MSY} (Hilborn 1979).

Results from this part of the study can be summarized as follows. The model estimated four of the five indicators with reasonable accuracy from data with no stochastic recruitment or other error. The effect of using the incorrect shape parameter was generally increased mean bias and maximum bias, but this was not consistent. B_0 and E_{curr}/E_{MSY} were estimated with the most bias in stressed-fishery data situations; the other three indicators were estimated best in the stressed-fishery situation.

Stochastic data without systematic error

We next generated stochastic data sets by setting CVR to 0.4. For each value of q we generated 19 different data sets. The random number sequences were different for each q value; we made no attempt to use the same sequences for each value of q. The mean bias for each indicator, averaged across data generated from all q values and both steepness values, is shown in Table 3.

These results are generally similar to those from deterministic data in Table 2. Typical results are shown in Figs. 4 and 5, showing the indicators B_0 and B_{curr}/B_{MSY}. Although B_0 was

Fig. 5. The ratio B_{curr}/B_{MSY} estimated from data simulated with stochastic recruitment, a steepness of 0.9, and a range of q values. The small squares show estimates made using the correct value of p in the assessment model and large circles show estimates made with the incorrect value. The straight line indicates the "true" value. The final model population size is inversely related to the value of q used, so final population size declines from left to right.

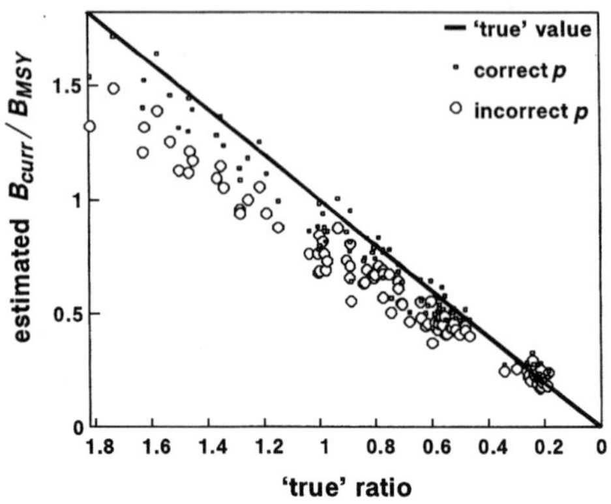

estimated best from data generated with low q, other indicators were estimated best when a high q was used, resulting in low B_{curr}. The effect of using the incorrect p value in the assessment model was again to increase bias slightly, but the effect varied among indicators and between data sets generated with different steepness values (Breen and Kendrick 1995*b*).

In general the difference caused by choice of correct or incorrect p was similar to the background error caused by stochastic recruitment (Figs. 4 and 5). Examination of the effect of choosing the correct or incorrect p was not taken past this stage.

Results from this part of the study can be summarized as follows. When data were generated using stochastic recruitment, estimates for each indicator had roughly the same average bias and average maximum bias as the deterministic results. Choosing the incorrect shape parameter tended to increase bias, but not consistently. Differences in results caused by the choice of shape parameter tended to be lost in the variability caused by stochastic recruitment.

The indicator CSP could not be examined stochastically. "True" CSP in each stochastic run depended on the strength of the model year-class that would recruit in the next year. "True" CSP varied substantially simply because recruitment varied stochastically, so comparing CSP from the assessment and simulation models was not meaningful. In all comparisons reported below, bias in CSP was therefore examined by comparing results from the assessment model using data simulated without stochastic recruitment.

Stochastic data with systematic errors

We generated four kinds of error in the input data: errors in the catch time series, errors in the CPUE caused by increasing catchability, errors in the estimated exploitation rate used as auxiliary information, and all three errors simultaneously. In this part of the study we generated five data sets at each value of q.

We simulated underreported catch data for the early years that gradually changed to perfectly reported data. The simulator reported only 35% of the "true" catch in the first year, rising

Table 4. Mean percentage bias for each indicator estimated from the data sets described in the text.

Experiment	Indicator, % B_0	MSY	CSP	B_{curr}/B_{MSY}, %	E_{curr}/E_{MSY}, %
Deterministic $n = 1$	–5.2	–7.7	–10.5	–6.3	9.4
No systematic errors $n = 19$	–9.9	–6.3		–10.6	12.5
Catch errors $n = 5$	–13.9	1.6	1.9	40.6	–31.0
CPUE errors $n = 5$	–14.4	6.5	10.4	18.7	–23.7
ER errors $n = 5$	0.0	–10.9	–15.5	–9.5	18.2
All errors $n = 5$	–11.5	4.5	15.0	35.8	–33.5

Note: Number of replicates, *n*, is the number of data sets made for each value of *q* in each experiment. Means were taken over *q* values and over the correct and incorrect values of *p*. CSP estimates were made from deterministic recruitment data sets ($n = 1$) and from steepness of 0.9 only.

Table 5. Maximum percentage bias for each indicator estimated from the data sets described in the text and as for Table 4.

Experiment	Indicator, % B_0	MSY	CSP	B_{curr}/B_{MSY}, %	E_{curr}/E_{MSY}, %
Deterministic $n = 1$	–24.6	–18.1	–14.4	–22.5	52.8
No systematic errors $n = 19$	–19.6	–17.7		–21.6	50.3
Catch errors $n = 5$	–46.5	26.5	12.0	65.4	–46.2
CPUE errors $n = 5$	–24.9	35.1	19.1	37.8	–44.8
ER errors $n = 5$	15.8	–31.1	–23.7	–23.4	83.1
All errors $n = 5$	–25.9	18.6	35.3	66.6	–48.6

Note: The maximum bias shown is the maximum of mean bias estimates from data generated from one value of *q*.

linearly to 100% reporting in year 25. This pattern was arbitrary.

To investigate the effects of increasing catchability over time (or with decreasing biomass) we changed the effective value of *q* linearly to 2.5 times the first-year value by the 50th year. We based this on estimates of the annual rate of increase in gear efficiency in the Western Australia rock lobster fishery (Brown et al. 1994). When we analyzed these data, we used CPUE only from the second half of the time series, as we did in the real assessment.

To simulate bias in exploitation rate we modified the "true" ER by 0.85; this value was arbitrary.

The mean bias for each indicator in each experiment (Table 4) was the simple average bias from a number of stochastic data sets (5 or 19 at each value of *q*). For this table we averaged over all *q* values and both the correct and incorrect value of *p* and results are shown only from data generated with steepness = 0.9. Similarly the maximum bias (Table 5) is the maximum of the average biases observed at each value of *q* and *p*. Often the maximum bias occurred at one end of the range of populations simulated. For instance, the worst bias for B_0 when estimated from data with catch errors (Fig. 6A) occurred when the final population was smallest; the worst bias in MSY estimates occurred when the final population was largest (Fig. 6B).

The indicators responded differently to the various errors examined. Except in the lightly fished situation, B_0 was usually underestimated (Fig. 6A, Table 4). This indicator was affected most seriously by systematic errors in the catch time series. When errors in catch data are put aside, the maximum bias when using the correct *p* value was 25% in the whole experiment (Table 5).

Estimated MSY (Fig. 6B) was quite strongly influenced by all the errors examined, especially in the lightly fished situations. In heavily fished situations the estimates showed low bias and were robust to the errors examined. Bias in CSP, estimated from data with deterministic recruitment only, showed a similar pattern (Fig. 6C).

Fig. 6. Mean percentage bias in each of the estimators in the various error situations described in the text: (A) B_0; (B) MSY; (C) CSP; (D) the ratio B_{curr}/B_{MSY}; and (E) the ratio E_{curr}/E_{MSY}. In the "no errors" situation, data were generated with stochastic recruitment variation but with no other error. The other error situations include stochastic recruitment variation plus the errors described in the text. (C) is an exception: for reasons described in the text, there was no stochastic recruitment variation in the data used.

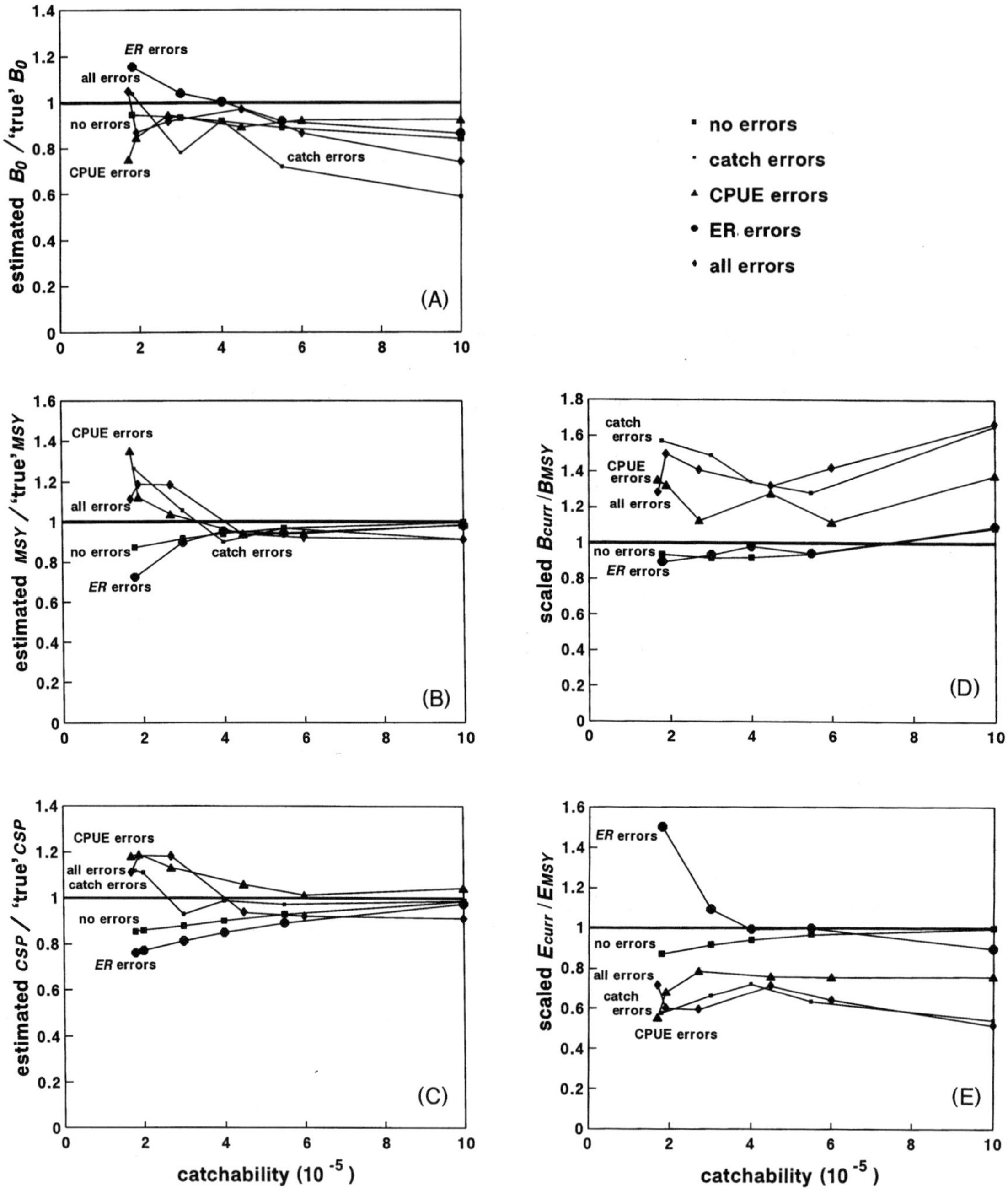

The ratios B_{curr}/B_{MSY} and E_{curr}/E_{MSY} were the indicators most influenced by systematic errors (Figs. 6D and 6E, Tables 4 and 5). Both were biased most strongly by errors in the catch time series. Of the five indicators, E_{curr}/E_{MSY} showed the highest mean and maximum bias and was the most sensitive to the errors examined.

In these experiments, the average bias across all q values was less than 16% for B_0, MSY, and CSP, but was substantially larger for the other two indicators (Table 4). Despite the high bias and sensitivity to error, B_{curr}/B_{MSY} appears to be a useful indicator of a depressed stock, reliably indicating a ratio less than 1.0 when the "true" ratio was less than about 0.8 (Fig. 7). By contrast, the indicator E_{curr}/E_{MSY} performed very poorly when the "true" ratio was high (Fig. 8).

Discussion

The simple assumption of the surplus-production model, eq. 1, is violated badly in our simulated population and by the real

Fig. 7. The ratio B_{curr}/B_{MSY} estimated from the "all errors" situation using stochastic recruitment, a steepness of 0.9, and a range of q values. The small squares show estimates made using the correct value of p in the assessment model and large circles show estimates made with the incorrect value. The straight line indicates the "true" value.

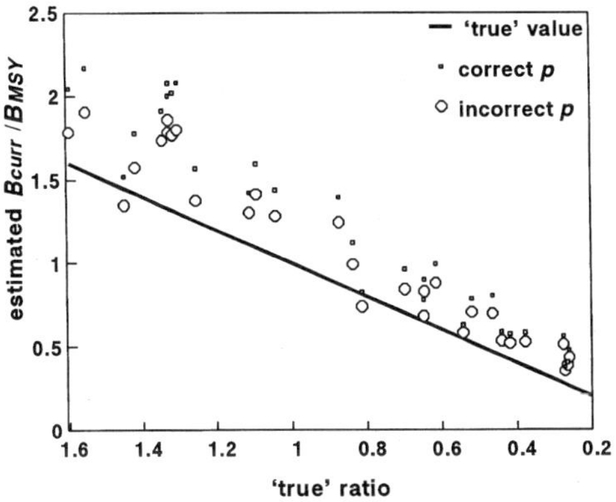

Fig. 8. The ratio E_{curr}/E_{MSY} estimated from the "all errors" situation using stochastic recruitment, a steepness of 0.9, and a range of q values. The small squares show estimates made using the correct value of p in the assessment model; large circles show estimates made with the incorrect value. The straight line indicates the "true" value.

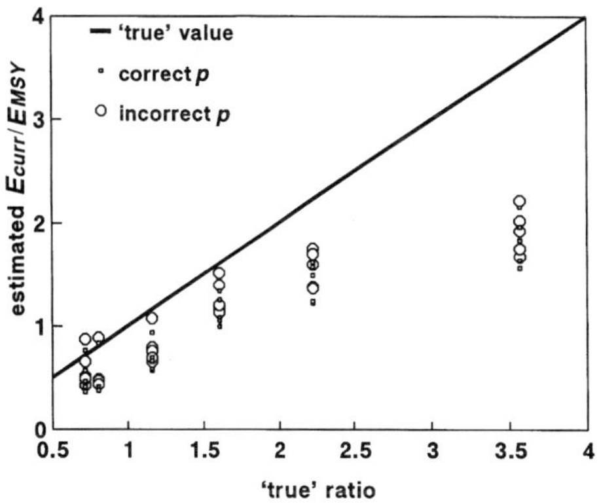

population. The main reasons for this are that productivity in the simulated population depends on the size structure of the population as well as the overall biomass and that time lags between settlement and recruitment are important (Fig. 1). A minimum legal size and protection of egg-bearing females cause significant violation of the assumption in eq. 5.

These problems caused some bias in the indicator estimates we evaluated (Figs. 3–5, Tables 2–5). In the runs with deterministic data or stochastic data with no systematic errors, bias tended to be greatest when populations were lightly fished. This result is well known (e.g., Hilborn 1979; Ludwig and Walters 1989). The B_0 estimates were an exception to this; B_0 was most strongly underestimated when the final population was smallest, probably because eq. 5 is violated most strongly when the population is heavily fished.

Generally, however, the overly simple assumptions did not result in unacceptable bias in estimated indicators. Mean bias was <11% and maximum bias <25% (Tables 4 and 5) if the ratio E_{curr}/E_{MSY} is ignored. The indicators MSY and CSP were estimated reasonably well when the model population had been fished sufficiently hard. These indicators are derived from the parameters of the model. Uhler (1979) found, in testing surplus-production models on simulated data, that derived estimates sometimes had less bias than the model parameter estimates.

From the results based on deterministic data and stochastic data without systematic errors, the simple assessment model appears to be a reasonable one, despite the known structural problems, to apply to the kind of data we have where the population has been fished down slowly over a long time. The structural problems might be more serious in situations where population biomass is reduced more quickly because of the problem illustrated by Fig. 1.

Using the incorrectly shaped model generally increased mean and maximum bias (Tables 2 and 3, Figs. 3–5), but the effect in any specific situation was unpredictable, depending on the indicator and the data. The effect was swamped by the effects of stochastic recruitment (e.g., Fig. 4). The incorrect model performed better than the correct model in some situations (e.g., Fig. 7). In practice, results should be obtained from a range of model shapes. However, the consequences of using the incorrect model shape do not appear to be very serious, at least within the range suggested by independent biological information. This is also suggested by low sensitivity of analyses of real data to the shape parameter (P.A. Breen and T.H. Kendrick, unpublished data).

Potential problems in the data cause much more bias in the estimates than do these problems with the model (Fig. 6, Tables 4 and 5); the different indicators responded differently and the effect depended on how heavily the model population had been fished. Of the three possible errors in data we simulated, errors in estimated ER had the least effect; systematic errors in catch and CPUE had the greatest impact, especially on B_0 and the two ratios B_{curr}/B_{MSY} and E_{curr}/E_{MSY}. The effect of systematic errors in CPUE could be reduced by standardizing effort against known changes in technology (Brown et al. 1994). The sensitivity to errors in early catch data and the impossibility of recovering early data suggest that our assessment should explore a range of possibilities with respect to early catch errors. MSY and CSP estimates were robust to these errors when the population had been heavily fished.

In our real fishery assessments for *Jasus edwardsii*, the most important indicator is B_{curr}/B_{MSY}, used to determine whether the stock is below its deterministic biological optimum. Bias in estimates of this indicator is higher than for MSY and CSP, and the sensitivity to errors is also higher. However, Figs. 5 and 7 suggest this is a still a useful indicator, especially when B_{curr}/B_{MSY} is low. The possible errors we examined caused this to be overestimated; thus an assessment that suggests B_{curr} is less than B_{MSY} is credible. Caution needs to be used when the indicator is near 1.

In fisheries with input controls, such as pot limits used to regulate total effort in Western Australia (Walters et al. 1993), an important indicator might be the ratio E_{curr}/E_{MSY}. Our results show this to be a poor indicator: bias in this was greatest of all the indicators we examined. When the fishery is more heavily fished, the indicator does worse (Fig. 8). This indicator should obviously not be used.

This was a specific case study, as suggested by Punt (1995), rather than a general test of surplus production methods. Some of the features specific to this work are as follows: the form of the model, the narrow range of shape parameter values suggested by the biological estimates, our use of exploitation rate as auxiliary information, a population fished down reasonably slowly over a long time, and good data contrast but in a "one-way trip." In our situation, the method appears to be acceptable despite structural and possible data problems. If the specific features just listed were to be changed, this type of study would need to be repeated.

Acknowledgments

We thank Chris Francis, Mark Maunder, Paul Starr, Marianne Vignaux, and two anonymous reviewers for comments on the manuscript, and Jon Schnute for giving us his package MicroSimplex.

References

Andrew, P.A., and Butterworth, D.S. 1987. Is $F_{0.1}$ an appropriate harvesting strategy for the Cape hakes? *In* The Benguela and comparable frontal systems. *Edited by* A.I.L. Payne, J.A. Gulland, and K.H. Brink. S. Afr. J. Mar. Sci. **5**: 925–935.

Annala, J.H., and Breen, P.A. 1988. Yield- and egg-per-recruit data inputs and model results for the rock lobster *Jasus edwardsii* from 10 areas around New Zealand. Fish. Res. Centre Internal Rep. No. 87. MAF Fisheries, Wellington, N.Z. 46 p.

Annala, J.H., and Breen, P.A. 1989. Yield- and egg-per-recruit analyses for the New Zealand rock lobster, *Jasus edwardsii*. N.Z. J. Mar. Freshwater Res. **23**: 93–105.

Annala, J.H., and Bycroft, B.L. 1987. Fecundity of the New Zealand red rock lobster, *Jasus edwardsii*. N.Z. J. Mar. Freshwater Res. **21**: 591–597.

Annala, J.H., and Esterman, D.B. 1986. Yield estimates for the New Zealand rock lobster fishery. *In* North Pacific Workshop on Stock Assessment and Management of Invertebrates. *Edited by* G.S. Jamieson and N. Bourne. Can. Spec. Publ. Fish. Aquat. Sci. 92. pp. 347–358.

Booth, J.D., and Breen, P.A. 1994. The New Zealand fishery for *Jasus edwardsii* and *J. verreauxi*. *In* Spiny lobster management. *Edited by* B.F. Phillips, J.S. Cobb, and J. Kittaka. Blackwell Scientific, Oxford, U.K. pp. 64–75.

Breen, P.A., and Anderson, O.F. 1993. Fishing intensity on the red rock lobster (*Jasus edwardsii*) North and South Island stock. N.Z. Fish. Assmt. Res. Doc. 93/3. 43 p.

Breen, P.A., and Kendrick, T.H. 1995*a*. The 1994 stock assessment for the red rock lobster (*Jasus edwardsii*) fishery. N.Z. Fish. Assmt. Res. Doc. 95/23. 59 p.

Breen, P.A., and Kendrick, T.H. 1995*b*. How useful is surplus-production analysis for assessing the red rock lobster (*Jasus edwardsii*) fishery? N.Z. Fish. Assmt. Res. Doc. 95/24. 47 p.

Breen, P.A., and Stocker, M. 1994. Evaluating the consequences of constant catch levels on the red rock lobster, *Jasus edwardsii*, population of New Zealand, *In* Proceedings of the International Symposium on Management Strategies for Exploited Fish Populations. *Edited by* G. Kruse, D.M. Eggers, R.J. Marasco, C. Pautzke, and T.J. Quinn. Alaska Sea Grant College Program Report 93-02. pp. 39–59.

Brown, R.S., Caputi, N., and Barker, E. 1994. A preliminary assessment of increases in the fishing power on stock assessment and fishing effort expended in the western rock lobster (*Panulirus cygnus*) fishery. Crustaceana, **68**(2): 227–237.

Fogarty M.J., and Murawski, S.A. 1986. Population dynamics and assessment of exploited invertebrate stocks. *In* North Pacific Workshop on Stock Assessment and Management of Invertebrates. *Edited by* G.S. Jamieson and N. Bourne. Can. Spec. Publ. Fish. Aquat. Sci. 92. pp. 228–244.

Fox, W.W. 1975. Fitting the generalised stock production model by least squares and equilibrium approximation. Fish. Bull. (U.S.) **73**: 23–27.

Francis, R.I.C.C. 1992. Use of risk analysis to assess fishery management strategies: a case study using orange roughy (*Hoplostethus atlanticus*) on the Chatham Rise, New Zealand. Can. J. Fish. Aquat. Sci. **49**: 922–930.

Gordoa, A., and Hightower, J.E. 1991. Changes in catchability in a bottom-trawl fishery for Cape hake (*Merluccius capensis*). Can. J. Fish. Aquat. Sci. **48**(10): 1887–1895.

Hilborn, R. 1979. Comparison of fisheries control systems that utilize catch and effort data. J. Fish. Res. Board Can. **36**: 1477–1489.

Hilborn, R., and Walters, C.J. 1992. Quantitative fisheries stock assessment. Chapman and Hall, New York, N.Y.

Jensen, A.L. 1986. Assessment of the Maine lobster fishery with surplus production models. N. Am. J. Fish. Manage. **6**: 63–68.

Koonce, J.F., and Shuter, B.J. 1987. Influence of various sources of error and community interactions on quota management of fish stocks. *In* International Symposium on Stock Assessment and Yield Prediction. *Edited by* W.J. Christie and G.R. Spangler. Can. J. Fish. Aquat. Sci. **44**, Suppl. 2. pp. 61–67.

Laloe, F. 1995. Should surplus production models be fishery description tools rather than biological models? Aquat. Living Res. **8**(1): 1–16.

Ludwig, D., and Walters, C.J. 1985. Are age-structured models appropriate for catch-effort data? Can. J. Fish. Aquat. Sci. **42**: 1066–1072.

Ludwig, D., and Walters, C.J. 1989. A robust method for parameter estimation from catch and effort data. Can J. Fish. Aquat. Sci. **46**: 137–144.

Marchessault, G.D., Saila, S.B., and Palm, W.J. 1976. Delayed recruitment models and their application to the American lobster (*Homarus americanus*) fishery. J. Fish. Res. Board Can. **33**: 1779–1787.

Megry, B.A., and Wespestad, V.G. 1987. Review of biological assumptions underlying fishery assessment models. *In* Lecture notes on coastal and estuarine studies. 28: Fishery Science & Management. *Edited by* W.S. Wooster. Springer Verlag, Berlin. pp. 31–69.

Mittertreiner, A., and Schnute, J. 1985. Simplex: a manual and software package for easy nonlinear parameter estimation and interpretation in fishery research. Can. Tech. Rep. Fish. Aquat. Sci. 1384. 90 p.

Morgan, G.R. 1979. Assessment of the stocks of the western rock lobster *Panulirus cygnus* using surplus yield models. Aust. J. Mar. Freshwater Res. **30**: 355–363.

Polacheck, T., Hilborn, R., and Punt, A.E. 1993. Fitting surplus production models: Comparing models and measuring uncertainty. Can. J. Fish. Aquat. Sci. **50**(12): 2597–2607.

Pella, J.J., and Tomlinson, P.K. 1969. A generalized stock production model. Inter-Am. Trop. Tuna Comm. Bull. **13**: 421–496.

Punt, A.E. 1990. Is $B_1 = K$ an appropriate assumption when applying an observation error production-model estimator to catch-effort data? S. Afr. J. Mar. Sci. **9**: 249–259.

Punt, A.E. 1992. Selecting management methodologies for marine resources with an illustration for South African hake. S. Afr. J. Mar. Sci. **12**: 943–958.

Punt, A.E. 1995 The performance of a production-model management procedure. Fish. Res. **21**: 349–374.

Roff, D.A. 1983. Analysis of catch/effort data: a comparison of three methods. Can. J. Fish. Aquat. Sci. **40**: 1496–1506.

Saila, S.B., Annala, J.H., McKoy, J.L., and Booth, J.D. 1979. Application of yield models to the New Zealand rock lobster fishery. N.Z. J. Mar. Freshwater Res. **13**: 1–11.

Schaefer, M.B. 1954. Some aspects of the dynamics of populations important to the management of commercial marine fisheries. Inter-Am. Tropical Tuna Comm. Bull. **1**: 27–56.

Schnute, J. 1982. A manual for easy nonlinear parameter estimation in fishery research with interactive microcomputer programs. Can. Tech. Rep. Fish. Aquat. Sci. 1140. 115 p.

Schnute, J. 1989. The influence of statistical error on stock assessment: examples from Schaefer's model. *In* Effects of ocean variability on recruitment and an evaluation of parameters used in stock assessment models. *Edited by* R.J. Beamish and G.A. MacFarlane. Can. Spec. Publ. Fish. Aquat. Sci. 108. pp. 101–109.

Silliman, R.P. 1971. Advantages and limitations of "simple" fishery models in light of laboratory experiments. J. Fish. Res. Board Can. **28**: 1211–1214.

Sparre, P., Ursin, E. and Venema, S.C. 1989. Introduction to tropical fish stock assessment. Part I - Manual. FAO Fish. Tech. Pap. 306/1.

Uhler, R.S. 1979. Least squares regression estimates of the Schaefer production model: some Monte Carlo simulation results. Can. J. Fish. Aquat. Sci. **37**: 1284–1294.

Walters, C.J., Hall, N., Brown, R., and Chubb, C. 1993. Spatial model for the population dynamics and exploitation of the Western Australian rock lobster, *Panulirus cygnus*. Can. J. Fish. Aquat. Sci. **50**: 1650–1662.

Yoshimoto, S.S., and. Clarke, R.P 1993. Comparing dynamic versions of the Schaefer and Fox production models and their application to lobster fisheries. Can. J. Fish. Aquat. Sci. **50**: 181–189.

Appendix. *Biological data used in data simulator*

For growth, the von Bertalanffy equation was used:

[A1] $l_{\mathrm{SEX},k} = L_{\infty \mathrm{SEX}} \, [1 - \exp(-K_{\mathrm{SEX}} \, (k - t_0))]$

where l is the mean length (in mm CL) of animals of sex indexed by SEX and of age k years. Growth constants are shown in Table A1. Constants were estimated from tag–return data summarized in Annala and Breen (1988).

Weight was estimated from length (mm CL), using a power function where $a2$ and $b2$ are the constants published by Saila et al. (1979) and shown in Table A1.

[A2] $W_{\mathrm{SEX},k} = a2_{\mathrm{SEX}}(l_{\mathrm{SEX},k}{}^{b2SEX})$

The model used lengths in both carapace length (CL), for instance in growth and length–weight calculations, and in tail width (TW), for instance in separating legal and sublegal-sized fish. The morphometry is described by Breen and Kendrick (1995*a*):

[A3] $l_{\mathrm{SEX},k}$ (mm TW) = $a3 + b3\ l_{\mathrm{SEX},k}$ (mm CL)

The relationship between the proportion of mature females, AP_m, in each size group MW (in mm tail width) was described by:

[A4] $\mathrm{AP}_m = \exp(a4 + b4\ \mathrm{MW})/(1 + \exp(a4 + b4\ \mathrm{MW}))$

The constants $a1$ and $b1$ were estimated using least-squares estimation with MicroSimplex (Schnute 1982; Mittertreiner and Schnute 1985) from observations of females with and without setose pleopods (a reliable index of maturity) in catch sampling in Fiordland, 1989–1994 (Breen and Kendrick 1995*a*).

Fecundity of individual mature females as a function of length ML (mm CL) was described by Annala and Bycroft (1987):

[A5] $f_{\mathrm{ml}} = a5 + \mathrm{ML}^{\,b5}$

Natural mortality rate M was based on a literature review (Breen and Anderson 1993).

Table A1. Parameter values used in the data simulator.

Equation	Constant	Males	Females
von Bertalanffy	t_0	–0.5	–5
growth	L_∞	225	140
	K	0.10	0.08
Length–weight	$a2$	0.00050	0.00055
	$b2$	3.01	2.99
CL–TW size	$a3$	5.42	–18.27
conversion	$b3$	0.4880	0.7495
Proportion mature	$a4$		–22.614
(AP)	$b4$		0.40118
Fecundity	$a5$		0.0567
	$b5$		3.18
Natural mortality	M	0.1	0.1
Handling mortality	h	0.1	0.1
Minimum legal size	TW (mm)	54	57

An example of how exploitation can increase production and yield in a northern crayfish (*Orconectes virilis*) population

Walter T. Momot

Abstract: The population dynamics of two crayfish (*Orconectes virilis*) populations were followed over a 15-year period. Initially, dynamics of both populations were similar. However, following exploitation of one population, annual production of the two populations unexpectedly diverged. In the unexploited population, maturing males regulated growth and mortality rates and thus restricted the subsequent recruitment of hatchlings and juveniles. In the exploited population, removal of adult males allowed prerecruitment survival rates of crayfish to the fishery to increase. As a result, expansion of the population took place despite increased trapping pressure. Placing restrictive sex and size limitations on crayfish harvest inhibited population growth and caused the population to decrease.

Résumé : La dynamique de deux populations d'écrevisses (*Orconectes virilis*) a été étudiée pendant quinze ans. Au départ, les deux populations présentaient une dynamique similaire. Toutefois, après l'exploitation de l'une des populations, la production annuelle des deux populations a divergé de façon inattendue. Dans la population non exploitée, les mâles en cours de maturation représentaient le facteur qui régularisait les taux de croissance et de mortalité, et limitait ainsi le recrutement subséquent de larves et de juvéniles. Dans la population exploitée, le prélèvement des mâles adultes a permis une hausse du taux de survie des écrevisses avant le recrutement dans la pêche. Par conséquent, la population a augmenté malgré l'intensification de la pression de pêche. L'établissement de limites touchant le sexe et la taille des écrevisses capturées a freiné la croissance démographique et provoqué un déclin de la population. [Traduit par la Rédaction]

Introduction

European crayfish have been exploited for centuries. Habitat loss, pollution, and decimation of native stocks by the crayfish plague fungus, *Aphanomyces astaci*, have forced Europeans in recent years to import crayfish from other areas of the globe. In the northern Great Lakes states and Canada, local crayfish fisheries have begun to supply the European market (Momot 1991). European crayfish fisheries for cool water species employ various restrictive regulations such as size and season limits to ostensibly protect brood stocks and prevent recruitment overfishing. Such limits have also been suggested or have already been adopted for North American cool water species. At this point, we do not know if such regulations are effective or even necessary.

A basic assumption in setting various restrictive regulations is that size of the usable stock is determined by total (fishing plus natural) mortality, recruitment, and growth. Fishing mortality is the only process over which we have direct control. It is assumed that exploitation of a previously unfished stock establishes a new balance between the other three processes: natural mortality, recruitment, and growth. When food is adequate, decrease in abundance of a stock results in greater recruitment, greater growth, and presumably reduced natural mortality. Here, I investigate the fishing mortality level at which both growth and recruitment overfishing (Ricker 1975) of a crayfish population take place, suggest measures to prevent this from occurring, and propose a rational basis for the development of management regulations.

To study these issues I perturbed a crayfish, *Orconectes virilis*, population in Dock Lake, Ontario, Canada, from 1977 to 1990 by periodically increasing harvest levels. The response of various parameters, including density, growth, fecundity, mortality, and annual production of this population to periodic increases in trapping effort was measured. I thought that detectable changes in values of these parameters might serve as early warning indicators of population collapse from either growth or recruitment overfishing. These parameters were also monitored in an unfished reference population (Shallow Lake). I proceeded on the premise that between-lake comparisons should provide insight into processes regulating the population dynamics and production of crayfish. My expectation was that crayfish populations would demonstrate the same responses as a fin fish population to increased fishing. Study results proved otherwise.

W.T. Momot. Department of Biology, Lakehead University, Thunder Bay, ON P7B 5E1, Canada.

Correct citation: Momot, W.T. 1998. An example of how exploitation can increase production and yield in a northern crayfish (*Orconectes virilis*) population. *In* Proceedings of the North Pacific Symposium on Invertebrate Stock Assessment and Management. *Edited by* G.S. Jamieson and A. Campbell. Can. Spec. Publ. Fish. Aquat. Sci. 125. pp. 225–233.

Materials and methods

Description of lakes

Two small senescing, marl lakes, Dock and Shallow, situated within the municipal boundary of Thunder Bay, Ontario, 10.5 km west of Lake Superior (48°28′N, 89°28′W) were chosen for study. Momot (1978) described their chemical and physical characteristics. Dock Lake (1.2 ha) has a mean depth

Table 1. Annual estimates of population numbers of age I and older crayfish present in Dock and Shallow Lakes, Ontario.

	Age-classes							
	Males				Females			
Year	I	II	III	Total	I	II	III	Total
				Dock Lake				
1976	609	268	8	885	627	294	29	950
1977	958	383	27	1368	949	299	14	1262
1978	1260	796	8	2064	1503	724	43	2270
1979	934	469	17	1420	1013	533	39	1585
1980	932	336	33	1301	947	286	9	1242
1981	629	469	8	1106	951	580	30	1561
1982	1650	371	39	2060	1463	737	40	2240
1883	2402	495	88	2985	2763	738	75	3549
1984	2606	693	56	3355	2157	1290	147	3594
1985	3613	970	38	4621	3445	1383	35	4863
1986	1418	1010	173	2601	1452	1965	140	3557
1987	2970	764	358	4092	2854	1063	134	4051
1988	2384	1369	163	3916	1900	1508	131	3539
1989	932	812	260	2004	1022	661	94	1777
1990	1146	280	154	1580	654	312	195	1161
				Shallow Lake				
1976	1137	228	25	1387	594	714	16	1324
1977	940	292	38	1270	516	411	38	965
1978	1188	400	39	1627	1344	357	69	1770
1979	1212	551	27	1790	888	508	60	1456
1980	1348	699	8	2055	1546	374	30	1950
1981	796	981	34	1811	1486	575	18	2079
1982	846	646	46	1538	1030	1324	98	2452
1983	1245	790	368	2403	1904	952	947	3803
1984	1809	989	33	2831	1809	1451	167	3427
1985	1507	1215	128	2850	4225	1222	96	5543

of 2.4 m and a maximum depth of 4.5 m (Momot 1986). Because of summer thermal stratification, a severe hypoliminial oxygen deficit occurs at depths. This event confines crayfish within the littoral zone during the entire growing season, as crayfish avoid hypoliminial deficit water (Fast and Momot 1973). Shallow Lake (1.6 ha) does not stratify and has a mean and a maximum depth of 2.2 and 2.4 m, respectively. Both lakes are in an early stage of eutrophication and exhibit high levels of nutrients, dissolved solids, conductivity, and alkalinity (Momot 1986).

The study ran from May 1976 to September 1988 in Shallow Lake and until September 1990 in Dock Lake. After 1985, beavers gradually inundated the littoral zone (2.5 m) through dam construction in Shallow Lake. This prevented accurate young-of-the-year (YOY) population estimation. In 1988, failure of the beaver dam drained this lake, preventing population assessment of any age-classes and terminating this portion of the study.

Life history events in relation to sampling

The life cycle of *Orconectes virilis* is categorized into three stages: (*i*) juveniles (egg–YOY), from spring hatching to the end of the first summer; (*ii*) maturing juveniles (YOY–age I), from the first fall until females attain sexual maturity and until males recruit to the exploitable stock, which occurs at a size of 24 mm carapace length (CL); and (*iii*) adult (age I–II), after males mature and are capable of molting into the sexually active form I condition. In the genus *Orconectes*, when a male attains sexual maturity, its copulatory stylets become sharply pointed with the central projection having a horny amber appearance. After reproduction, males molt and the exoskeleton, termed form II, is characterized by copulatory stylets which are blunter and less heavily scleroterized. Most males and females live through three growing seasons; the few surviving to a fourth are considered age III.

Population estimates

Population densities (number/lake) in these two closed systems were estimated by mark and recapture annually using the Schumacher–Eschemeyer (S–E) formula (Ricker 1975) (Table 1). In 1979, and then from 1981 until 1990, the number of tagged crayfish occurring in the annual harvest of crayfish was recorded. This allowed direct estimation of annual exploitation rate (Ricker 1975) and, by use of the Petersen (P) method of estimating population abundance, estimation of the population proportion harvested in each year (Table 2). Before 1987, all trapped crayfish were released after measurement but after 1987 measurements were only obtained for males since only males ≥30 mm CL were harvested. Numbers of YOY, yearlings, adult males, and females were estimated in summer from early July to early August during the intermolt period before

Table 2. Comparison of Schumacher–Eschemeyer (S–E) and adjusted S–E with Petersen (P) estimates of population density (number per lake) in Dock Lake, Ontario.

Year	Sex	Adjusted S–E[a]	S–E	P	% Difference[b]
1979	M	1420	1420	1561	+9.9
	F	1585	1585	1965	+23.9
1981	M	1106	940	1118	+1.0
	F	1561	1496	1626	+1.0
1982	M	2060	1954	2164	+5.0
	F	2240	2240	2485	+10.9
1983	M	2985	2763	3174	+6.3
	F	3549	3549	3432	–3.3
1984	M	3355	3355	3883	+15.4
	F	3594	3594	3202	–11.1
1985	M	4621	3170	6205	+34.3
	F	4863	4269	5973	+22.8
1986	M	2601	1377	3083	+18.5
	F	3557	2807	3657	–0.02
1987	M	4092	3261	3040	–34.6
	F	4051	3242	3740	–8.3
1988	M	3916	2504	3701	–5.8
	F	3539	3539	—	—
1989	M	2004	1215	2004	+64.9[c]
	F	1777	1777	—	—
1990	M	1580	925	1580	+70.8[c]
	F	1161	1161	—	—

[a]Schumacher–Eschemeyer (S–E) estimates were adjusted upwards by adding the actual number of age I crayfish captured in the harvest (Table 3) to the number of estimated age II crayfish (Table 1). This gave the total number of age I crayfish estimated to be actually present in those years when the (S–E) estimated age I were less than the actual number harvested.

[b]Difference between Petersen and S–E (adjusted).

[c]Difference between P and unadjusted S–E estimates.

harvest. Harvesting began immediately upon completion of summer population size estimates.

A detailed description of the methods used to measure population parameters, age composition, maturity schedules, and the experimental harvest are given in Momot (1986) and Momot et al. (1990).

Experimental harvest

Our experimental crayfish harvest was a Ricker (1975) type I fishery because all adult population size estimates preceded the harvest. In a type I fishery, the biological year for which growth and mortality rates are estimated starts at the time fishing begins, but all natural mortality occurs after fishing (trapping) ends. Total annual mortality is thus separated into discrete trapping and natural components (Ricker 1975).

Experimental harvesting began in Dock Lake in 1977 with an initial nominal effort of 150 trap days (Table 3). A trap day (TD) is one modified minnow trap fished 24 h. Effort gradually increased over the study (Table 4).

Gear competition occurs when the effective trap radius (the area covered by one sampling unit) exceeds linear trap spacing. With 200 traps, the effective trap radius was 3.16 m and linear trap spacing was 3.05 m, while with 250 traps, effective trap radius was 2.83 m and linear trap spacing was 2.44 m. Thus gear competition in 1984 produced lower harvest rates than those experienced in other years (Table 3) (Morgan 1987). In 1985, effort increased to 6000 TD employing 200 traps per day.

Harvest usually began in late July and was completed by the end of August. In 1988, males ≥30 mm (cephalothorax length) (CL) were harvested (Table 3). During harvests, all females were released. Annual production was measured using Ricker's (1975) instantaneous growth methodology. Environmental variables measured were: mean monthly air temperature, total monthly precipitation, and total degree cooling days below 18°C during the growing season (May through September (Table 4).

Results and discussion

Population estimates

Annual population estimates of age I and older crayfish are presented in Table 1. Confidence limits were usually between 10 and 20% of the estimate (Momot 1967, 1978). In those years when population density calculated by the Petersen method greatly exceeded S–E estimates (Table 2), S–E estimates were adjusted upwards (Table 2). This was necessary in years when the sum of age I crayfish captured in the harvest, together with the number of estimated age II crayfish known to be present, gave a population estimate greater than that recorded by the S–E estimate.

In general, S–E estimates probably underestimated the number of juvenile male crayfish whenever the catch of large

Table 3. Yields in kilograms of crayfish harvested in Dock Lake under various harvest rates.

	Year													
	1977 (50)	1978 (50)	1979 (200)	1980 (200)	1981 (200)	1982 (200)	1983 (200)	1984 (250)	1985 (200)	1986 (200)	1987 (200)	1988 (200)	1989 (200)	1990 (200)
Males	1.584	9.267	5.062	7.873	6.249	17.088	23.319	22.440	32.572	18.469	25.428	29.266	9.958	8.014
Females	0.588	5.150	3.656	4.733	4.603	10.347	13.816	9.959	17.050	10.468	14.033	—	—	—
Total	2.172	14.417	8.718	12.606	10.852	27.435	37.135	32.399	49.622	28.937	39.461	29.266	9.958	8.014
Effort	150	1350	2600	2600	2600	4000	4000	4000	6000	6000	6000	6000	6000	6000

Note: Number in parenthesis (under year) represents the number of traps. Effort is given in trap days (1 trap fished for a 24-h period).

Table 4. The length of growing season as determined by the number of cooling degree days ≤18°C and total precipitation as given in mm from 1976 to 1990 during the months of April through September of each year.

Year	Precipitation (mm)	Cooling degree days <18°
1976	292	697
1977	741	717
1978	371	701
1979	359	772
1980	539	664
1981	519	702
1982	569	789
1983	429	687
1984	429	706
1985	606	756
1986	466	689
1987	405	712
1988	415	575
1989	373	704
1990	466	725
15 year mean	429	739

males in traps was high. Traps select for larger males which if abundant, agonistically exclude smaller males and females. S–E estimates were calculated in years when trapped crayfish were returned to the lake. Petersen (P) estimates were based on recapture of animals which were harvested. Since approximately 50% of large males were removed by day five of the harvest, smaller males and females could then be caught (Morgan 1987). This reduced the ratio of marked to unmarked crayfish, making P estimates usually larger than S–E estimates. Percent difference between P and adjusted S–E estimates of population size ranged from −11.1 to +34.6%.

Climate can also be a factor when estimating population size. 1985, which had the greatest discrepancy between S–E and P estimates, was the coldest and wettest year in the study period (Table 4). This reduced catches of males during sampling to determine S–E estimates. At the time of harvest, water temperatures had increased, as did catches of large males, making P estimates relatively more accurate. When possible, I accepted adjusted S–E estimates (Table 1) as most accurate estimates of population size. This only applied to Dock Lake, since crayfish in Shallow Lake were not harvested. These data were utilized in all subsequent calculations of population size and mortality rates. In spring, only densities of sexually mature females were estimated.

Fig. 1. Relationship between total number of male and female crayfish harvested in Dock Lake, Ontario and total annual effort in trap days. Data for 1988–1990 are not on the figure since only males were harvested.

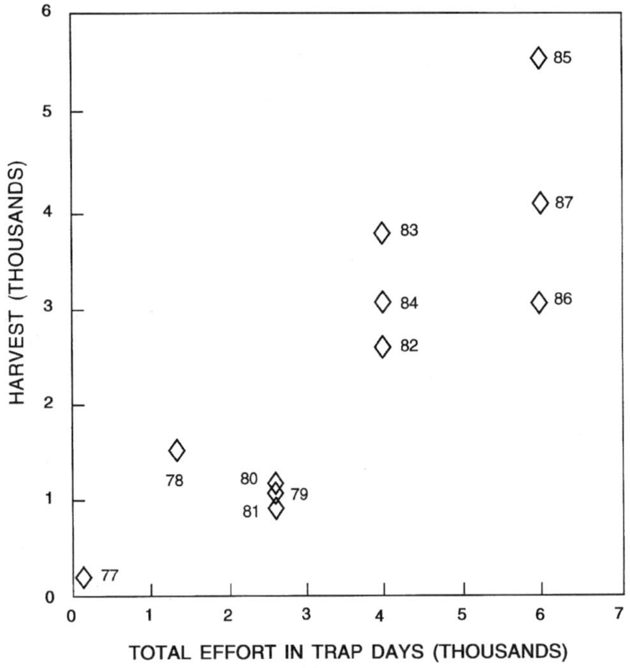

Experimental harvest

Crayfish yield from Dock Lake increased with increased effort (Fig. 1). Catch per unit effort fell until 1981, but then unexpectedly increased (Fig. 2). By 1982, effort removed approximately 60% of the estimated exploitable stock of age I+ crayfish (approximately 70% of the males and 50% of the females) (Momot et al. 1990), however, the stock did not collapse (Fig. 1). Removals peaked in 1985, but remained high up to 1988, when a size and sex limit regulation was imposed. After 1988, the number of males declined (Table 2) but harvest rate remained high at 40–51% of the males over the three year period at a sustained effort of 6000 trap days.

Many freshwater fin fish populations show evidence of overfishing at high exploitation rates (Pearse 1988). Our results prompted us to consider that removing at least 50% of the

Fig. 2. Catch per unit effort (CPUE) of males and females in number per day and total annual effort (trap days) in Dock Lake, Ontario.

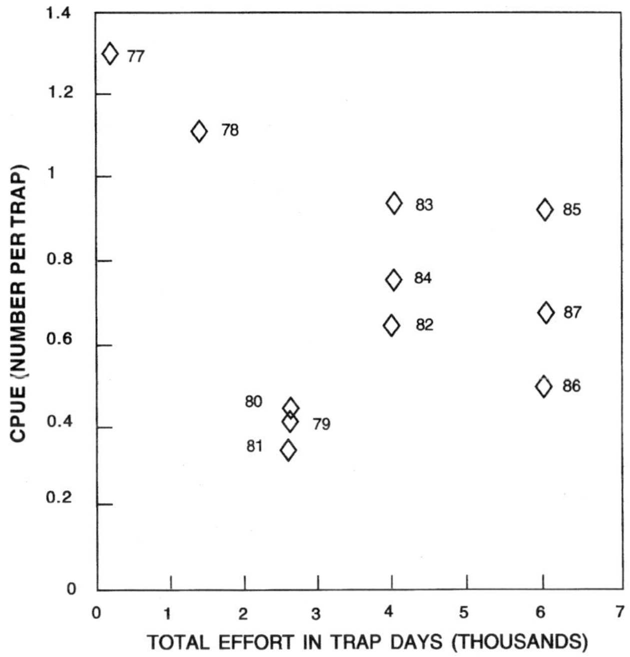

Fig. 3. Relationship between mean size of both male and female young-of-the-year (YOY) in mm carapace length, total precipitation during May–September (A) and cooling degree days <18°C during May–September (B) in Dock Lake, Ontario.

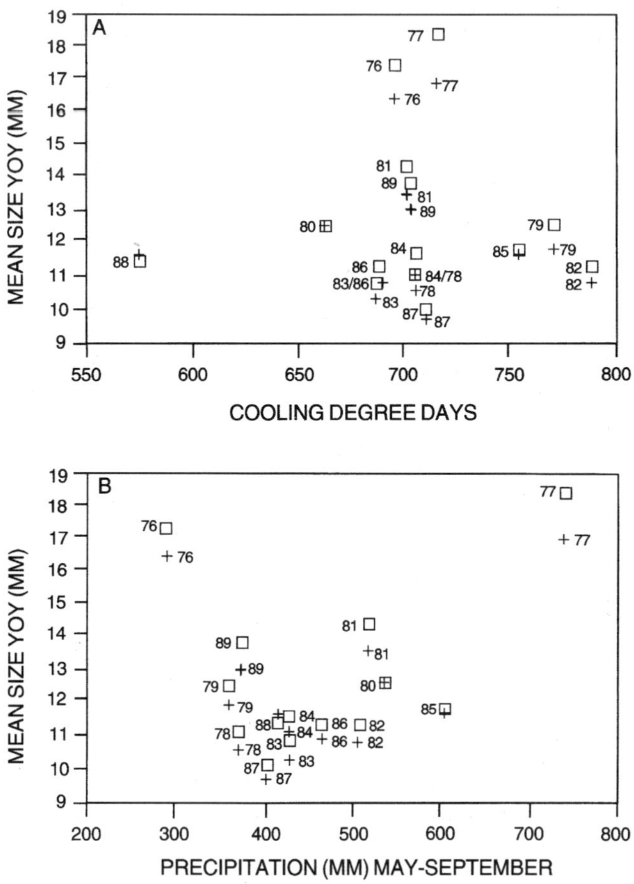

Fig. 4. Annual production of crayfish in kg·ha^{-1}·yr^{-1} in Dock Lake and Shallow Lake. For each data point, e.g., 81, annual production for adults is measured from mid-August 1981 to mid-August 1982, while for YOY it is measured during the growing season of 1981.

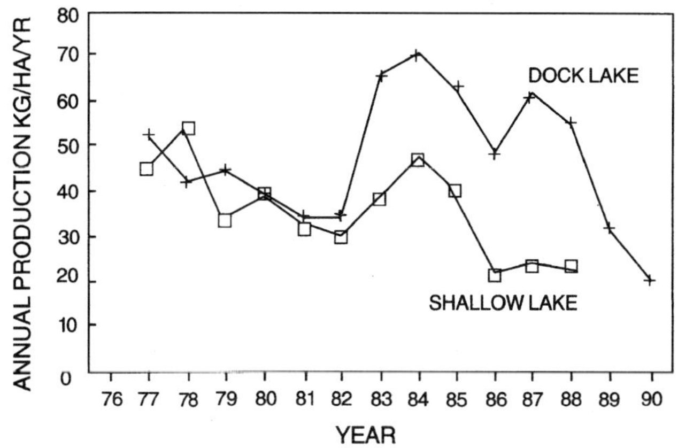

biomass annually after 1982 did not result in recruitment overfishing. This result characterized what could be sustained by this population. Was this unexpected response due to an increase in fishing effort releasing a density-dependent control or perhaps only a fortuitous event that was coincident with some trend in climatic variables?

Environmental variables

Potential relationships between population parameters and environmental variables were evaluated using linear regressions (Zar 1974). Both length of growing season and water temperature varied. These two factors control crayfish metabolic rate and hence feeding activity. Morgan (1987), examining data from 1976 to 1984, found age- and sex-specific growth rates, mortality rates, and population size to be independent of year-to-year variation in solar energy input. We recalculated those data, including additional data from 1985 to 1989, but no consistent patterns were observed between cooling degree days below 18°C and precipitation (May–September) during the growing season, and average size of YOY at the end of the growing season (Fig. 3). Data from Shallow Lake gave similar results.

Comparison of annual production in both populations

Effects of climate variation on crayfish production are apparent when comparison is made of total annual production values for Dock and Shallow Lakes between 1976 and 1985 (Fig. 4). Annual production incorporates both growth and survival of individuals, providing a single measure of a population's response to climate. Parallel, simultaneous fluctuations in annual production were observed in both lakes (Fig. 4). While the reference population is not a strict experimental control, it remains a standard against which experimental results can be compared. At low levels of fishing intensity in Dock Lake (up to 1982), annual production declines in the two lakes were similar (Fig. 4). After 1982, increase in production was greatest in the heavily fished population (Fig. 4). Annual production (Fig. 4) was then compared against cooling degree days <18°C (mean 739 days) and precipitation (mean 429 mm) between May and September (Table 4). Production declined in both lakes in

Fig. 5. Relationship between the number of female and male recruits at age I at year 1 + 1 and the sum of the female breeding population (ages I, II, III) (SUM F) in Dock Lake (A) during the unrestricted phase of the developing fishery (1977–1987) and in unexploited Shallow Lake (B), Ontario during 1977–1987. Each year represents the year in which the brood was recruited.

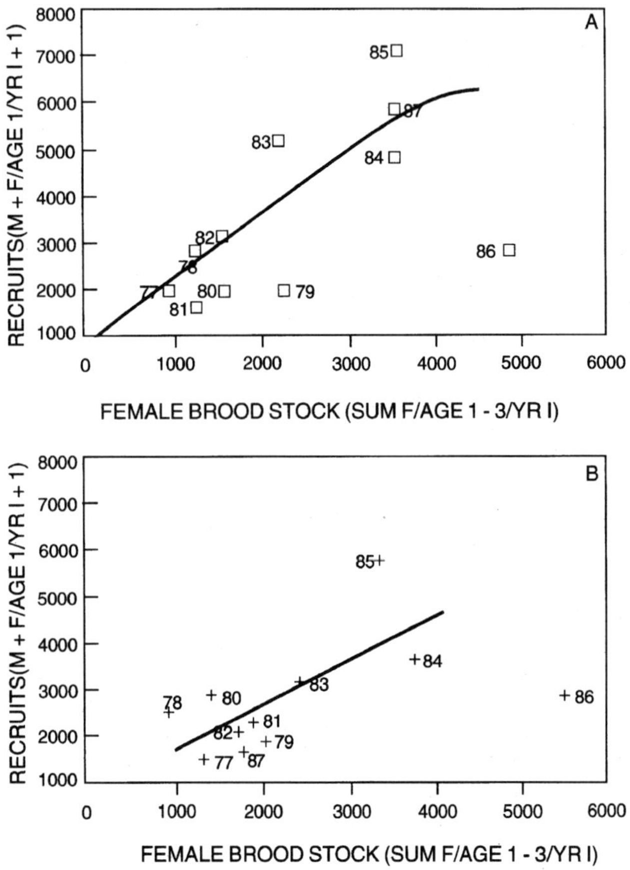

Fig. 6. Comparison of the relationships between the instantaneous rates of mortality (*Z*) of juveniles in Shallow Lake (A) and Dock Lake (C) and maturing juveniles in Shallow Lake (B) and Dock Lake (D) and their instantaneous growth rate (*G*). For Shallow Lake juveniles (6A): $r^2 = 0.85, p < 0.001$ for males ( ); $r^2 = 0.86$, $p < 0.001$ for females (+); for maturing juveniles (6B): $r^2 = 0.34$, $p > 0.05$ for males; $r^2 = 0.86, p < 0.001$ for females. For Dock Lake juveniles (6C): $r^2 = 0.33, p < 0.05$ for males; $r^2 = 0.17, p > 0.05$ for females; for maturing juveniles (6D): $r^2 = 0.15, p > 0.05$ for males; $r^2 = 0.008, p > 0.05$ for females.

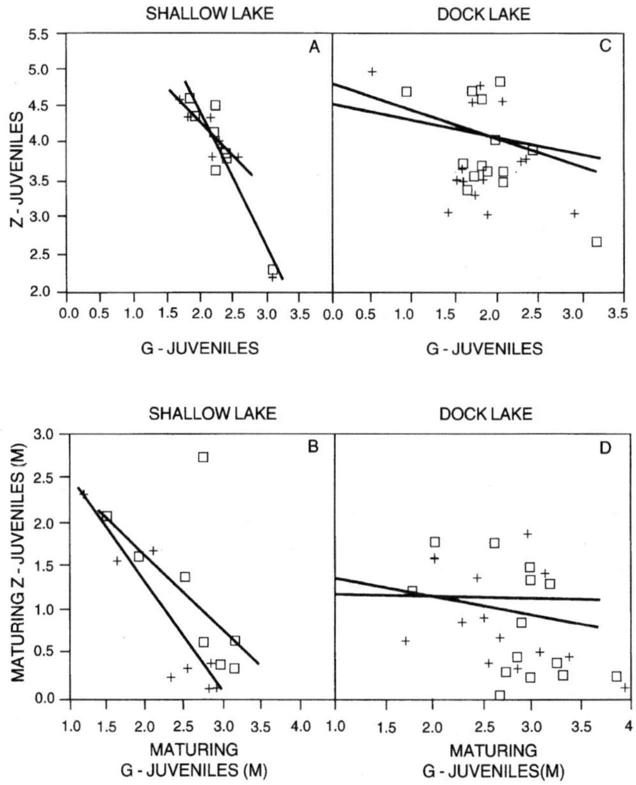

1982 and 1985, both cold, wet years and carried over into 1986 as a result of poor recruitment in 1985 (Fig. 4).

Removal of maturing males (age I to age II) allowed overall population size to increase. Although annual production was measured until 1988 in Shallow Lake, production for 1985–1987 was underestimated due to inadequate sampling of YOY and small juveniles. The average percentage of production contributed by YOY was 25% between 1976 and 1984. Production estimates for Shallow Lake were thus adjusted upwards by 25% for the years 1985–1987. Increased Dock Lake production after 1982 resulted from an increase in recruitment of age I crayfish (Fig. 5) (Table 1). After 1987, we removed only males ≥30 mm CL, allowing abundance of age I and age II males to increase. As expected, annual production and yield dramatically decreased (Fig. 4). As we will explain, this decrease in production and yield was due to a decrease in recruitment.

Stock-recruitment relationship

Plotting recruitment as the number of male and female age I crayfish at year 1 + 1 against the female brood stock estimated as the sum of age I, II, and III females at year 1 during the unrestricted fishing in Dock Lake suggested the familiar linear left-hand side of a Beverton–Holt recruitment curve (Fig. 5A) (Ricker 1975). The low value for 1986 was likely the result of poor recruitment during the cold, wet year of 1985 (Table 4). For Shallow Lake, there was a linear trend (Fig. 5B). ($y = 1364 + 0.6(x)$; $r^2 = 0.38, p < 0.05$), but this may be the result of the 1985 data, when there was a very large number of recruits in both lakes (Figs. 5A and 5B).

Age specific trends in growth and mortality

Increases in annual production and yield were examined by first comparing juvenile growth and mortality rates within both exploited (Dock Lake) and unexploited (Shallow Lake) populations (Table 5). The only significant relationship was for a decrease in male YOY density with increasing density of age II males.

Population regulation

In unexploited Shallow Lake, juvenile mortality and juvenile growth are inversely correlated for both sexes (for males

Table 5. Comparison of regression statistics for the growth and mortality of hatchling and juvenile life stages in Dock Lake (n = 14) and Shallow Lake (n = 8) in relation to increasing density or mortality of age I to age II males.

	Shallow Lake (Unexploited)			Dock Lake (Exploited)		
	a	*b*	r^2	*a*	*b*	r^2
Juveniles (egg to YOY)						
Male YOY density vs. density of age II males	5862	–4.28	0.52*	3079	–0.331	0.008
Female YOY density vs. density of age II males	5728	–3.86	0.30	3474	–0.565	0.01
Male YOY density vs. mortality of age I to age II males	910	–0.24	0.24	3669	–0.995	0.09
Female YOY density vs. mortality of age I to age II males	1557	–0.41	0.17	3607	–0.622	0.02
Maturing juveniles (YOY to age I)						
Male juvenile growth vs. adult male mortality	4.69	–1.42	0.74	0.749	–0.027	0.001
Female juvenile growth vs. adult male mortality	1.79	0.859	0.77	0.704	–0.053	0.006
Male juvenile growth rates vs. adult male mortality	1.86	0.759	0.72	0.105	–0.176	0.10
Female juvenile growth rates vs. adult male mortality	4.62	–1.31	0.62	0.59	–0.054	0.10
Male juvenile growth rates vs. density of age II males	2.78	–0.00076	0.26	2.24	–0.0005	0.10
Female juvenile mortality rates vs. density of age II males	2.88	0.0009	0.30	2.09	–0.0004	0.06
Male juvenile mortality rates vs. density of age II males	2.69	0.0018	0.40	3.36	–0.0008	0.21
Female juvenile mortality vs. density of age II males	2.90	–0.0015	0.28	3.34	–0.0007	0.12
Maturing juveniles						
Male maturing juvenile growth vs. mortality of age II males	3.09	–0.405	0.33	2.467	–0.571	0.14
Female maturing juvenile growth vs. mortality of age II males	2.904	–0.428	0.44	2.43	–0.400	0.06
Male maturing juvenile mortality vs. mortality of age II males	0.839	0.74	0.14	0.825	–0.021	0.0001
Female maturing juvenile mortality vs. mortality of age II males	–0.138	5.74	0.085	0.867	–0.0031	0.0004

*$p < 0.05$

r^2 = 0.85, $p < 0.001$; for females r^2 = 0.86, $p < 0.001$) (Fig. 6A). Maturing juvenile mortality rates for females, but not for males, was significantly negatively correlated with growth ($r^2 = 0.86, p < 0.001$) (Fig. 6B).

In exploited Dock Lake, only juvenile males had a significant negative correlation between growth rate and mortality ($r^2 = 0.33, p < 0.05$) (Fig. 6C). However, data were collected during the (1977–1980) period of relatively low fishing effort. For maturing juveniles there was no significant relationship detected between growth and mortality for either males or females.

In summary, a relationship between growth and mortality occurred with both prerecruitment portions of the life cycle only in the unexploited population. Exploitation diminished these interactions and completely eliminated them with maturing juvenile life stages. In Dock Lake, production of both juvenile stages of crayfish seemed unrelated to the growth–mortality relationships that regulated the Shallow Lake population.

Recruitment

Change in numbers of recruits entering the fishery was affected at two points in the life cycle only in Shallow Lake. In Dock Lake, removing maturing males negated this feedback control, resulting in more crayfish recruits. Increased recruitment at age I consistently produced large year-classes (Table 1) (Fig. 5). This implies a density-dependent relationship between YOY production and subsequent age I recruitment. In Dock Lake, the relationship between fecundity and resulting number of YOY hatched suggested maximum numbers of YOY are produced relative to available nursery habitat. Any increase in female population density, individual female fecundity, or total egg production did not result in production of more hatchlings (Momot 1986; Morgan 1987). In Dock Lake, annual production of hatchlings remained relatively constant. However, juvenile mortality, not the number of crayfish hatched, was critical in determining year-class strength. In population regulation, the selective trap fishery removed proportionately more age I males whenever the population increased. As fishing effort grew (e.g., 2600 TD or greater), yield increased because age I recruitment rates rose.

In unfished Shallow Lake, age II males strongly regulated recruitment of both juveniles and maturing juvenile females. In most years, as the maturing males mortality rate decreased, female growth rate increased and their mortality rate significantly decreased (Figs. 6A and 6B). In Dock Lake age I to age II males were exploited and the mechanism regulating growth and mortality was suppressed (Table 5). This increased survival of maturing juveniles and recruitment of age I males increased (Fig. 5). Fishing continuously removed even large numbers of age I to age II males. At high densities, such males act as population regulators. Their removal increased recruitment, biomass, and density to the carrying capacity of the habitat (Momot 1986).

I suggest that the mechanism underlying such a sensitive recruitment response is the natural agonistic behaviour of male *Orconectes virilis,* enhanced by their dimorphic growth pattern. This allows them to grow larger and dominate females (Bovbjerg 1956). Male distribution overlaps with both berried females and newly hatched YOY just at the time when young juveniles are hatching (Fast and Momot 1973). Males may cannibalize YOY (a readily available protein source for males seeking food prior to undertaking their first annual molt in early spring) or simply suppress normal YOY foraging for food. YOY lacking access to quality food probably died either from predation or from the failure to acquire sufficient energy reserves to complete a molt. Either event could produce high juvenile mortality.

Possible mortality sources

To document potential cannibalism, we collected adult males in 1989 with hand nets and an electroshocker during the period when females released YOY in Dock Lake. From 123 males examined, ranging in size from 25 to 46 mm CL, none contained YOY in their stomachs. Also, no YOY were found in samplings of adult males in 1986 ($n = 37$) or 1987 ($n = 33$).

However, growth mortality interactions shown in Figs. 6A and 6 B suggest that juveniles and maturing juveniles that have low growth mostly die. This situation was not as evident in the exploited population (Figs. 6C and 6D). In the laboratory, starved males placed in the presence of egg-bearing females did not aggressively pursue them or females with attached young. They did not catch active young that voluntarily left a female. In the lab, such males only captured those young that were immobile for a very brief instant during molting. Hanson et al. (1990) and Dye and Jones (1974) also observed no cannibalism in lab studies.

If cannibalism causes an insignificant mortality, suppression of YOY daily activity by adult males may be important. Young constantly disturbed by foraging males may fail to secure sufficient good quality food to have a successful molt. Failure to grow rapidly may increase their susceptibility to insect predators. For example, dragonfly nymphs feed intensively on 4–6 mm CL crayfish and can consume crayfish up to 12 mm CL (Dye and Jones 1974). Fish predation, important to other populations, was not a consideration in either lake here.

Both cover and quality of available food affect survival of the young. Maxwell (1988) showed that age I to age II males can influence growth and mortality of young, suggesting the temporal and spatial juxtaposition of newly hatched young during their first growing season with behaviorally dominant males can reduce early growth of YOY.

Exploitation of males decreases this density-dependent regulation. The vulnerability of adult males to a fishery largely controls recruitment of age I crayfish. Increasing trapping levels to 6000 TD did not impact reproduction since sufficient numbers of males still survived to mate with females. Thus, exploitation at the level conducted stimulated population growth of *Orconectes virilis* in Dock Lake.

Implications for management

In this particular study, exploitation apparently altered and affected the mechanisms that normally regulate an unfished population. Increased yield was not just the result of a fishing-up effect. Number, sex, and size of crayfish harvested, method of harvest, and arrangement of gear deployed were all important factors to consider. This may have implications for crustacean fisheries which harvest only one sex. In this particular case, the fishing appeared to alter inadvertently a mechanism controlling overall population size.

This has important implications for fisheries since theory about population responses to exploitation typically assumes that the same processes controlling population size operate for both unexploited and exploited populations. Relaxation of fishing pressure should result in restoration of a population to its former abundance. This study suggests that, at least for some species, population responses are not predictable by such theory.

This study also enforces the importance of long-term experimental empirical studies to measure impacts of fishing on populations. With crayfish, it is not known if growth or recruitment overfishing are possible and there are no published studies that employ or measure useful indices or parameters which could forecast recruitment or growth overfishing.

Crayfish management regulations are based on the assumption that exploited crayfish populations behave as do fin fish populations. It is assumed that theories (e.g., Beverton–Holt or Ricker models) developed for fin fish will naturally apply to crayfish. This study demonstrates that exploitation affects the relationship between growth, natural mortality, recruitment, fishing mortality, and their roles in regulating a target population. Here exploitation altered the complex relationships between age and sex specific growth and mortality and the role they play in regulating recruitment to an exploitable stock. Accompanying changes in size of the exploited population were not just functions of a straightforward response in growth and survival. An alternative scenario may reduce population density to such a low level that new regulative mechanisms will then describe the recruitment process. Relaxing fishing pressure, as in this case, may not necessarily reestablish the old "population regulating rules."

Acknowledgments

I wish to thank the Natural Science and Engineering Research Council for the support of this project over the years 1975–1990. The Senate Research Committee, Lakehead University and the Department of Fisheries and Oceans, Environment Canada provided supplementary funding at crucial times. Many graduate and undergraduate students contributed their time over the years in the collection of these data. A special thanks to Bea Termaat and P. Lynn Hauta for their assistance in data analysis, Lorraine Scarcello, who typed endless versions of this manuscript, and Kris Nelson for final typing and editing of the manuscript. Finally our thanks to Dr. Evan Powell, Thunder Bay, for the use of his property over all these years.

References

Bovbjerg, RV. 1956. Some factors affecting aggressive behaviour in crayfish. Physio. Zool. **29**: 127–136.

Dye, L., and Jones, P.D. 1974. The influence of density and invertebrate predation on the survival of young-of-the-year, *Orconectes virilis*. Freshwater Crayfish, **2**: 529–538.

Fast, A.W., and Momot, W.T. 1973. The effects of artificial aeration on the depth distribution of the crayfish *Orconectes virilis* (Hagen) in two Michigan lakes. Am. Midl. Natur. **89**: 89–102.

Hanson, J.M., Chambers, P.A., and Prepas, E. 1990. Selective foraging by the crayfish, *Orconectes virilis*, and its impact on macroinvertebrates. Fresh. Bio. **24**: 69–80.

Maxwell, S. 1988. Effects of selected biotic and abiotic factors on growth and survival of young crayfish, *Orconectes virilis* (Hagen), M.Sc. thesis. Lakehead University, Thunder Bay, Ontario. 159 p.

Momot, W.T. 1967. Population dynamics and productivity of the crayfish, *Orconectes virilis*, in a marl lake. Am. Midl. Natur. **78**: 55–81.

Momot, W.T. 1978. Annual production and production/biomass ratios of crayfish, *Orconectes virilis* in two northern Ontario lakes. Trans. Am. Fish. Soc. **107**: 776–784.

Momot, W.T. 1986. Production and exploitation of the crayfish, *Orconectes virilis* in northern climates. Can. Spec. Publ. Fish. Aquat. Sci. 92. pp. 154–167.

Momot, W.T. 1991. Potential for exploitation of freshwater crayfish in coolwater systems: management guidelines and issues. Fisheries, **16**(5): 14–21.

Momot, W.T., Hauta, P.L., and Schaeffer, J.P. 1990. Yield estimates for the virile crayfish, *Orconectes virilis* (Hagen) employing the Schaeffer logistic model. J. Shellfish Res. **9**(2): 373–381.

Morgan, G.E. 1987. Population dynamics of an exploited population of *Orconectes virilis* in northwestern Ontario. M.Sc. thesis. Lakehead University, Thunder Bay, Ontario. 178 p.

Pearse, P.H. 1988. Rising to the challenge. Canadian Wildlife Federation, Ottawa, Ontario. 180 p.

Ricker, W.E. 1975. Computation and interpretation of biological statistics of fish populations. Bull. Fish. Res. Board Can. 191. 382 p.

Zar, J.H. 1974. Biostatistical analysis. Prentice-Hall Inc., New Jersey. 620 p.

Biological reference points in an abalone (*Haliotis laevigata*) fishery

S.A. Shepherd and Janine L. Baker

Abstract: The problem of estimating, for management purposes, a biological reference point at which recruitment to an abalone (*Haliotis laevigata*) stock is reduced below its optimum is considered. We take an empirical approach by examining the performance of six populations with different egg production and recruitment characteristics after 26 years of fishing. Two of the populations became depleted soon after fishing started, three have persisted under exploitation but show signs of decline, and one has been sustained under fishing. Per-recruit population egg production (PREP), the total number of eggs produced by a notional unfished cohort during its lifetime, and %EPR, the percentage of that total number of eggs produced at the known levels of fishing, were calculated for each population. PREP, a measure of population fitness, was found to vary directly with productivity (mean annual catch) and with reef area of the populations. We found no uniform %EPR value that ensured optimal recruitment of all populations, but we propose a %EPR threshold boundary whose value increases with decreasing population productivity and reaches 100% as productivity tends to zero. If egg production is kept above the threshold, recruitment is optimal and exploitation sustainable, but below it the risk of recruitment overfishing is high. A comparison with similar analyses for other abalone stocks shows consistency with the proposed threshold

Résumé : On examine le problème qui consiste à estimer, pour des besoins de la gestion, un point de référence biologique à partir duquel le recrutement dans un stock d'ormeaux (*Haliotis laevigata*) est réduit en deçà de sa valeur optimale. Nous adoptons une approche empirique pour examiner le rendement de six populations présentant des caractéristiques différentes du point de vue de la production d'oeuf et du recrutement après 26 ans de pêche. Deux des populations ont été épuisées peu après que la pêche a débuté, trois ont persisté malgré la pêche, mais ont montré des signes de déclin, et une s'est maintenue de manière durable malgré la pêche. Pour chacune de ces populations, on a calculé la production d'oeuf par population de recrues (POPR), le nombre total d'oeufs produits par une cohorte théorique non exploitée durant sa vie, et le %EPR, c.-à-d. le pourcentage de ce nombre total d'oeufs produits à des niveaux d'exploitation connus. On a constaté que la POPR, qui est une mesure de l'adaptation de la population, variait directement avec la productivité (prises annuelles moyennes) et avec la surface récifale occupée par les populations. Nous n'avons pas trouvé de valeur de %EPR uniforme qui garantissait un recrutement optimal pour toutes les populations, mais nous avons proposé un %EPR seuil dont la valeur augmente avec la diminution de la productivité de la population et qui atteint 100 % lorsque la productivité tend vers 0. Si la production d'oeuf est gardée au-dessus de ce seuil, le recrutement est optimal et l'exploitation est durable, mais en dessous de ce seuil, le risque de surpêche du recrutement est élevé. Une comparaison avec des analyses semblables pour d'autres stocks d'ormeaux montre une cohérence avec le seuil proposé. [Traduit par la Rédaction]

Introduction

Explanation of the decline of many abalone fisheries around the world has given rise to a Hegelian dialectic. Some claim that there is no evidence that overfishing has ever caused the decline of abalone stocks and argue that the high fecundity of a few individuals is able to repopulate an area so that recruitment varies independently of spawning stock except for the trivial case when the stock is zero (Harrison 1986; discussion in Breen 1992; McShane 1992). Others have implicated recruitment overfishing in practically all documented declines (Sluczanowski 1986; Shepherd et al. 1991, 1995; Shepherd and Partington 1995). This dichotomy of opinion has arisen from ignorance of the underlying stock–recruitment relation for abalone (Breen 1986). However, recent studies focusing on the aggregational structure of populations which affects fertilization success of spawned eggs have been more successful in demonstrating a strong stock–recruitment relation for an abalone population and the need to maintain adult densities to ensure adequate recruitment (McShane 1995*a*; Shepherd and Partington 1995). Furthermore, the strong Allee effect at reduced densities (Shepherd and Partington 1995) suggests that a sharp threshold of spawning stock density exists below which the risk of recruitment failure is high.

A problem that besets abalone fisheries is that monitoring more than a few populations for the purposes of stock assessment is rarely possible, so that it is difficult to tell just how close to their thresholds of spawning stock density most populations are being fished. In this paper we ask whether thresholds of egg production in egg-per-recruit (EPR) models can be used as a surrogate for spawning stock density in order to predict the risk of overfishing under given levels of exploitation (Mace and Sissenwine 1993). To answer this question we examine the long-term histories of six exploited populations of

S.A. Shepherd and J.L. Baker. South Australian Research and Development Institute, Aquatic Sciences Centre, PO Box 120, Henley Beach, South Australia 5022, Australia.

Correct citation: Shepherd, S.A., and Baker, J.L. 1998. Biological reference points in an abalone (*Haliotis laevigata*) fishery. *In* Proceedings of the North Pacific Symposium on Invertebrate Stock Assessment and Management. *Edited by* G.S. Jamieson and A. Campbell. Can. Spec. Publ. Fish. Aquat. Sci. 125. pp. 235–245.

Fig. 1. Location of reef areas in South Australia referred to in the text. Inset shows northern and southern areas of the Thorny Passage metapopulation.

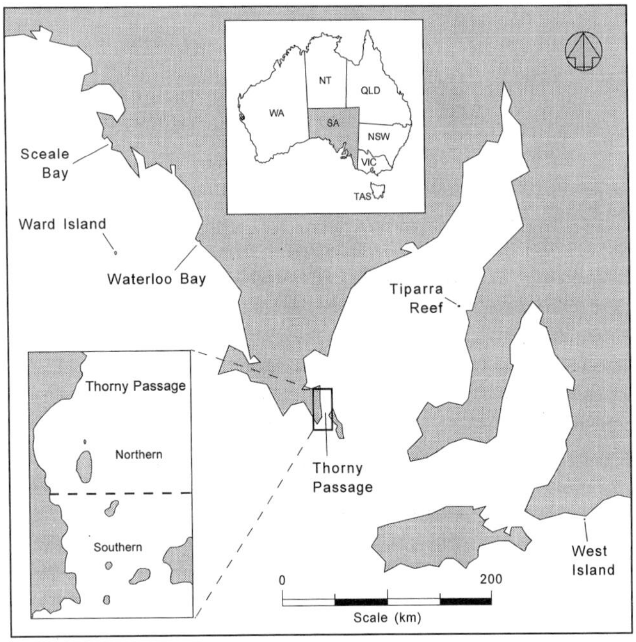

Haliotis laevigata Donovan in South Australia (Fig. 1) because empirical evidence of sustained yield over a long period is the best assurance of population persistence under fishing. We have selected populations which are ecologically, and almost certainly genetically, distinct from neighbouring populations, so that each can be considered a metapopulation, i.e., a group of local populations linked by larval exchange but genetically isolated from all others (Shepherd and Brown 1993). We prefer to use the term metapopulation which has a more precise, but practically indistinguishable, meaning from the now loosely used term stock (Berst and Simon 1981; Hanski and Gilpin 1991). The six populations examined include two which have become depleted, three which show slight evidence of decline, and one which appears stable. Theoretical considerations suggest that the threshold of egg production will vary according to population size, so an analysis of these populations along a gradient of population size or productivity should clarify the location and shape of the egg production threshold curve. The biological reference boundary provided by this curve is more informative than single biological reference points which take no account of the effect of population size: it will be useful for managers to predict the risk of overfishing for other populations, both small and large. Here the term 'overfishing' or 'recruitment overfishing' means any level of fishing that jeopardizes the long-term capacity of a stock to produce an optimum yield (cf. Sissenwine and Shepherd 1987; Thompson 1993). We also assume that an underlying compensatory spawner–recruit relation controls the dynamics of populations of this abalone, as has been shown for the Waterloo Bay population (Shepherd and Partington 1995), and that the populations are in quasi-equilibrium (Caddy 1996).

Historical aspects of the fishery for *H. laevigata* are given by Prince and Shepherd (1992) and Keesing and Baker (1998).

Table 1. Productivity, reef area, recruitment variability (RV), and PREP values for six metapopulations, with the number of survey sites for each metapopulation.

Reef	Period	Catch (t)	SD	PREP	Reef area (km^2)	RV	Survey sites
West Island	1970–1985	0.7	1.8	1.5	0.1	80	6
Waterloo Bay	1977–1994	2.7	3.3	0.6	0.62	69	19
Sceale Bay	1971–1994	19.9	8.4	1.6	2.0	—	—
Ward Island	1972–1996	9.4	3.5	4.2	1.4	37	5
Thorny Passage	1972–1994	47.5	22.1	4.2	9.1	43	18
Tiparra Reef	1974–1994	79.9	34.9	4.4	5.8	48	4

Note: Productivity is estimated as the mean annual catch (total weight) (with standard deviations, SD) for the period given; recruitment variability is the coefficient of variation expressed as a percentage.

The fishery is controlled by a complex of input and output measures of which key controls are minimum legal limits (MLL) and quotas. The fishery was generally underexploited in the 1970's due to stagnation in the industry (Prince and Shepherd 1992), and fully exploited since about 1980. Since then the fishing mortality rate, F, is believed to have been more or less constant, except for a possible increase in 1983 before quota introduction in 1984, and for short-term changes as divers move about the fishing zone (Keesing and Baker 1998).

Materials and methods

Catch and effort data available since 1968 when the fishery began were extracted from institutional statistical records (Keesing and Baker 1998) for the populations investigated. The mean annual catch over a time series during which a population was fully fished was accepted as a reasonable estimate of its productivity (Kesteven 1996).

Annual surveys of selected populations started in 1980 and abalone abundance was measured by the timed swim technique. At each site research divers did 4–6 replicate swims of 10 min each in a given direction and measured all emergent abalone within a 1–m swathe. The mean power of research divers was 19.8 $m^2 \cdot min^{-1}$ (range 15–27 m^2) and their mean searching efficiency was 0.74 (range 0.67–0.78) (Shepherd 1985). An equation relating area searched and swimming time was derived for each diver and estimates of relative density of abalone measured during the swims calculated.

From the length–frequency data produced by the surveys, densities were partitioned between recruits (the first emergent year class, either 2+ or 3+ years according to locality), legal-sized abalone, and total numbers. Survey details for Waterloo Bay are given by Shepherd and Partington (1995). Recruitment variability was measured as the coefficient of variation (the standard deviation of the mean annual recruitment values for a population divided by the mean) and expressed as percentage values (see Pimm 1991 for discussion of measurement of variability).

Table 2. Parameters used in the EPR and YPR analyses.

	West Island	Waterloo Bay <6 yr	Waterloo Bay >6 yr	Thorny Passage	Ward Island	Sceale Bay	Tiparra Reef
Growth parameters							
t_0	1.81	1.11	1.24	1.65	1.80	0.40	1.51
$K{\cdot}yr^{-1}$	0.48	0.59	0.19	0.6	0.41	0.19	0.41
L_∞ (mm)	138	148	191	180	167	186	145
Length–weight relations							
a (× 10^{-5})	1.7	20	20	4.7	10	12	1.26
b	3.41	2.92	2.92	3.16	3.07	3.00	3.51
Fecundity parameters							
c	−0.36	−0.36	−0.36	−1.57	−1.87	−1.13	−1.51
d	0.015	0.004	0.004	0.014	0.008	0.011	0.02
W_0	24	90	90	111	234	103	72
Mortality rates							
M	0.2	0.4	0.2	0.25	0.13	0.25	0.22
F	0.1		0.6	0.7	0.35	0.5	0.5
MLL	120	145	145	145	145	145	130
Catch curve analyses							
N_c	—	—	—	3	2	—	2
N_a	—	—	—	160	185	—	194

Note: The growth parameters, t_0, $K{\cdot}yr^{-1}$, and L_∞, are from the von Bertalanffy growth equation. Weight relations are given by the equation $W = a\,L^b$ (where W = total weight in g, L = shell length in mm, and a and b are constants). Fecundity relations are given by the equation $F = c + dW$ (where F = number of eggs in millions, W = total weight, and c and d are constants) and W_o is the x-intercept. MLL = minimum legal length. In the catch-curve analyses N_c = number of commercial catch samples examined; N_a = mean number of shells aged per catch sample.

Fig. 2. Relative PREP values for unfished and fished metapopulations of *H. laevigata*. Scale is a relative measure of the total number of eggs per female recruit from age zero to maximum age (20 years).

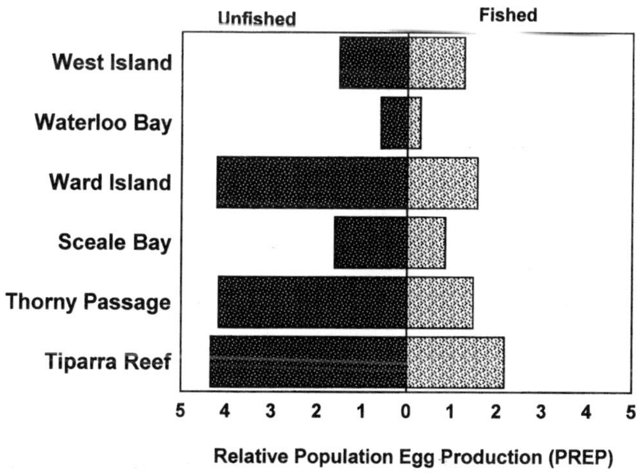

Estimates of the total reef area occupied by populations were obtained, in the case of West Island and Waterloo Bay, from our own detailed surveys, and in other cases from maps supplied by commercial divers familiar with them. Estimates based on such maps are necessarily crude because of the difficulty of estimating areas of patchy reefs scattered among expanses of sandy bottom.

The populations surveyed, the number of survey sites within populations, estimated reef areas, and mean annual catches are given in Table 1.

Per-recruit population egg production, egg-per-recruit and yield-per-recruit

An analytical model was used to examine the effect of size at first capture and exploitation rate on the egg production and yield of each population. Modelling was done with the PRAna software program (Sluczanowski et al. 1992) which is an age-structured dynamic pool model based on Baranov's (1918) classic catch equation. The spawning potential of each age class is summed from age of reproductive maturity to maximum age (20 years) for an unfished stock, to produce the relative population egg production per recruit (PREP); this is a measure of the fitness of the population (see Discussion). The spawning potential of the fished stock, under various levels of fishing mortality, F, is expressed as a percentage of the spawning potential of the unfished stock to produce the percentage egg-per-recruit (%EPR) values. The yield per recruit (YPR) is the sum over all age classes of the catch biomass, and is expressed as a percentage of the maximum biomass obtainable at an optimum age at first capture and optimal F, determined by iteration.

The parameters in Table 2 were used as inputs to the model. The von Bertalanffy growth parameters are from Shepherd and Hearn (1983) and Shepherd et al. (1992*a*). For Waterloo Bay an age–length key was prepared from the extensive mark–recapture data set over 16 years (Shepherd and Hearn 1983 and unpublished data), using the published growth curve to estimate age at marking. From this we derived growth parameters for the migrant fraction of the population (> ca. 6 years) found on reefs of high relief (Shepherd 1986) where the abalone continue to grow to a larger size. Fecundity and length–weight data are from Shepherd (1987) and Shepherd et al. (1992*b*). Natural mortality rates, M, are summarized by Shepherd and Breen (1992) and the reasons for the ranges chosen are discussed

Fig. 3. Plots of YPR (heavy line) and %EPR (light line) versus fishing mortality rate for six metapopulations of *H. laevigata* at two levels of *M.* The two bottom graphs show plots of YPR and %EPR versus minimum legal length for Waterloo Bay and Thorny Passage at fixed levels of *F*.

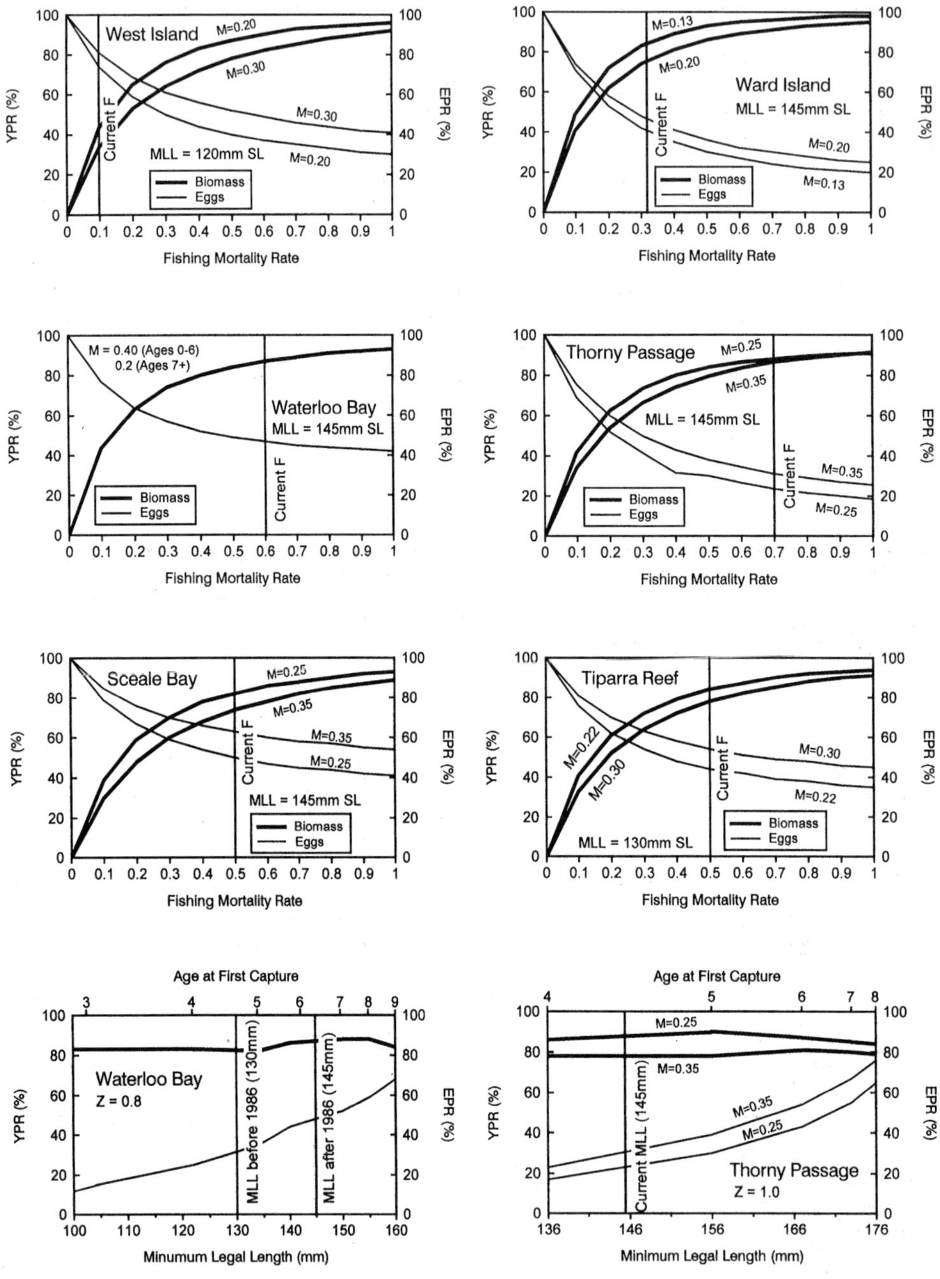

below. Total mortality rates, Z, were obtained by catch-curve analyses (Table 2), using the shell-ageing techniques of Shepherd and Triantafillos (1997) and Triantafillos (1994). For Waterloo Bay we used the age–length key to derive Z from the extensive historical research length–frequency data (Shepherd and Partington 1995). Fishing mortality rates, F, were obtained from the relation $Z = F + M$.

Results

The six populations show large differences between them in productivity, reef area, and PREP values (Table 1) and the ratios of PREP values for fished to unfished conditions also differ between them (Fig. 2). PREP values for unfished conditions were weakly correlated with the mean annual catch ($r = 0.70$; $P < 0.1$) and inversely with recruitment variability ($r = -0.89$; $P < 0.05$). Reef area was weakly correlated with the mean annual catch ($r = 0.75$; $P < 0.1$). YPR and EPR analyses for each population are shown in Fig. 3. The %EPR conserved varies from about 27% to about 83% at YPR values of between 75% and 90%. We now summarize individual population histories.

West Island

This small population has existed in three phases (Fig. 4). The

Fig. 4. West Island, 1967–1995. (*a*) Mean density of recruits (2+ age-class), and adult (>100 mm SL) density. Gaps are missing values. (*b*) Catch (*t*) (total weight) in vertical bars, and catch rate (kg·h^{-1}).

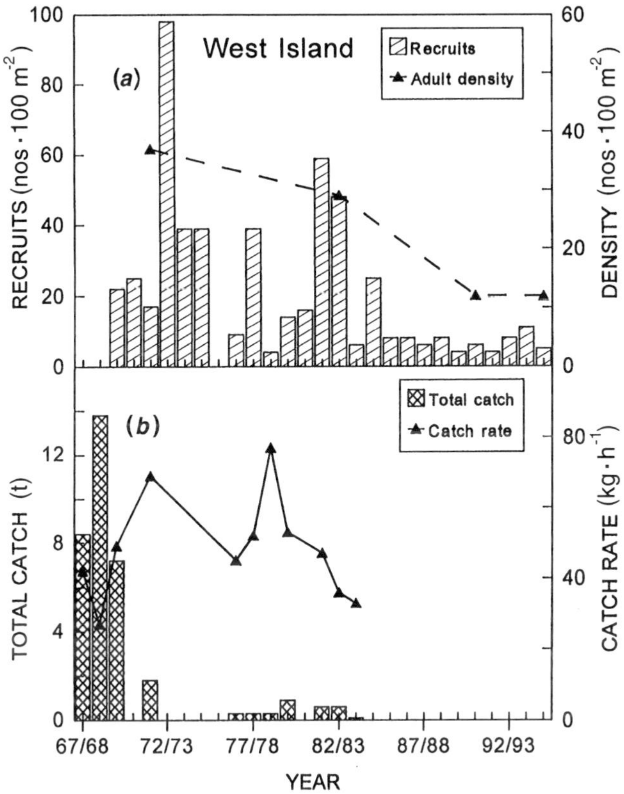

Fig. 5. Waterloo Bay, 1967–1994. (*a*) Total mortality coefficient, *Z*, with standard errors. (*b*) Mean density of recruits (2+ age-class) in vertical bars, and total number of emergent abalone, from 1977 to 1994. Gaps are missing values. (*c*) Catch (*t*) (total weight) in vertical bars and catch rate (kg·h^{-1}).

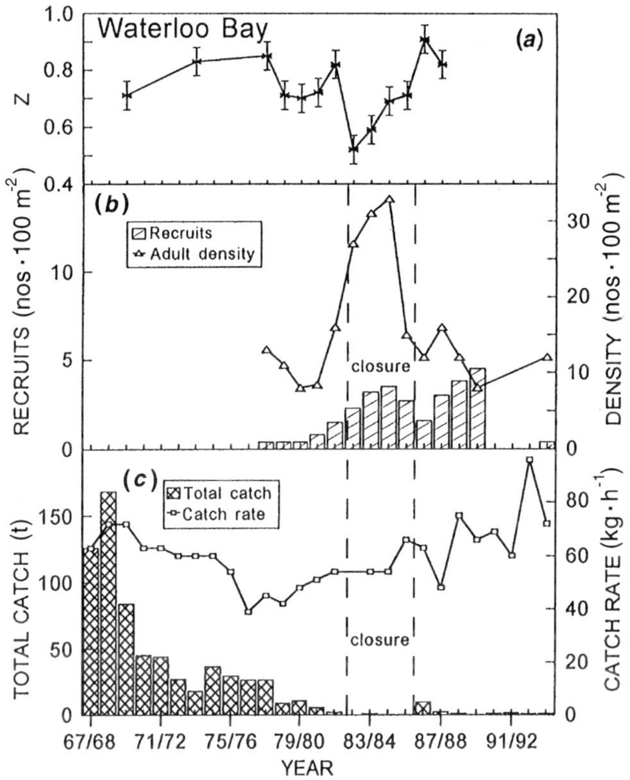

first phase of high density ended in 1970 after three years of fishing. In the next phase from 1970 to 1985 recruitment was irregular and adult *M* rates varied widely from 0.05 to 0.8 according to density (Shepherd 1990). From about 1985 to 1990 the adult population declined to very low densities, apparently because of high mean *M* rates of 0.45 which may have included some light fishing (Shepherd and Brown 1993); after 1990 adult *M* was about 0.1–0.2 (unpublished data).

In the YPR and EPR analyses we assumed $F = 0.5$ for the first phase and 0.1 for the second, $M = 0.3$ for both phases, and MLL = 120 mm shell length (SL) for the second phase, giving %EPR conserved of 51% in phase 1 and 83% in phase 2 (Fig. 3).

Waterloo Bay

The catch history in Waterloo Bay (Fig. 5) shows high initial catches of the virgin stock for three years, followed by slowly declining catches until about 1978 when the population virtually collapsed. The bay was closed to fishing from 1982 to 1986 and the MLL was increased from 130 mm to 145 mm SL. However, the population became depleted again by 1987. Mean adult (>4+ age-classes) and recruit (2+ age-class) densities since 1978 (Fig. 5) are from Shepherd and Partington (1995). Note that catch rates remained high until the closure in 1982 despite the declining abundance of abalone. This was because adult abalone continually migrated from recruitment sites throughout the Bay to reefs of high relief mostly near the bay's entrance, where they were targeted by fishermen.

Our research estimates for *M* were: 0.5 for reefs of low relief (a value which included some emigration) and 0.2 for reefs of high relief where abalone migrated from reefs of low relief (Shepherd and Breen 1992). So in the YPR and EPR analyses we put $M = 0.4$ to age 6 and then 0.2 for older age-classes. We chose a mean value of $Z = 0.8$ (Fig. 5), giving $F = 0.4$ to age 6 and 0.6 thereafter. YPR was about 86%, and %EPR varied from 34% before the closure in 1982 to 48% at the reopening in 1986 (Fig. 3). An increase in the MLL to 160 mm SL would increase %EPR to a safer level of about 68% (Fig. 3).

Sceale Bay

The two abalone species, *H. laevigata* and *H. rubra*, are fished together in this region so fishing effort was divided proportionately between them. The total catch of abalone has declined but that of *H. laevigata* relatively more so as shown by the declining percentage of *H. laevigata* in the overall catch (Fig. 6). CPUE has remained nearly constant. No independent monitoring of the population has been carried out. Assuming $M = 0.25$ and $F = 0.5$ the %EPR conserved is about 53% at a yield of about 85% (Fig. 3).

Ward Island

The catch history data (Fig. 7) suggests the fishery was possibly underexploited during the 1970's and has been stable since

Fig. 6. Sceale Bay, 1968–1994. Catch (*t*) (total weight) of *H. laevigata* in vertical bars, catch rate ($kg{\cdot}h^{-1}$) and percentage of *H. laevigata* in the total catch.

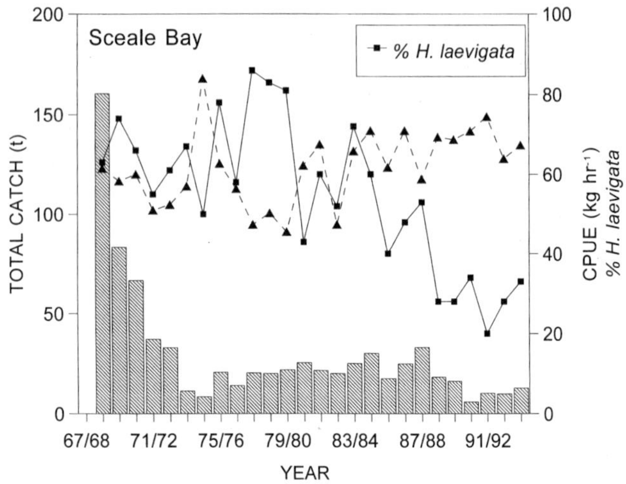

with a catch of about 9 $t{\cdot}yr^{-1}$. The survey data show a significant decline in density of total numbers ($t = 3.2$; $P < 0.01$) and legal-sized abalone ($t = 2.8$; $P < 0.02$). Recruitment has also declined significantly (Fig. 7); mean recruitment in the first six years (mean = $4.4{\cdot}100\ m^{-2}$) was significantly higher than in the next nine years (mean = $2.7{\cdot}100\ m^{-2}$) ($t = 2.6$; $P < 0.05$).

Catch curve analysis of a rarely fished local population on the west side of Ward Island gave $Z = 0.13$ which must be close to the value of M in very exposed conditions, although we would expect M to be up to 0.2 in more sheltered waters. Mean Z values obtained in 1990, 1992, and 1993 from the mainly fished east side of the Island were 0.45 indicating that F is from 0.25 to 0.32. The %EPR conserved under these conditions is about 45% and YPR is about 80% (Fig. 3).

Thorny Passage

This metapopulation comprises numerous local populations, located in southern areas, which largely support the fishery, and northern areas (Fig 1). Long-term catch data (Fig. 8) show likely underexploitation during the first decade after initial removal of the virgin stock and probable full exploitation after about 1980. Since 1981/1982 the catch from northern areas has significantly declined ($t = 6.4$, $P < 0.01$) at a mean rate of about 2 $t{\cdot}yr^{-1}$, whereas the catch from southern areas has shown no significant trend. Similarly the survey data (Fig. 9) suggest a decline of populations in northern areas, but not southern areas, although the data are not yet over a long enough time span to show a clear trend.

For the EPR analysis we used $Z = 1.0$ from catch–curve analysis and assumed an M value of 0.25 with a corresponding F value of 0.75. The %EPR conserved is 27% at the current MLL of 145 mm SL and would increase to 40% with the proposed increase of MLL to 165 mm SL (Fig. 3). The biomass yield is an optimal 90% at an MLL of 155 mm SL and declines only slightly at 165 mm SL (Fig. 3).

Fig. 7. Ward Island, 1968–1996. (*a*) Mean density of recruits (3+ age-class) in vertical bars, legal-sized abalone (>145 mm SL), and total number of emergent abalone from 1980 to 1996. (*b*) Catch (*t*) (total weight) in vertical bars and catch rate ($kg{\cdot}h^{-1}$).

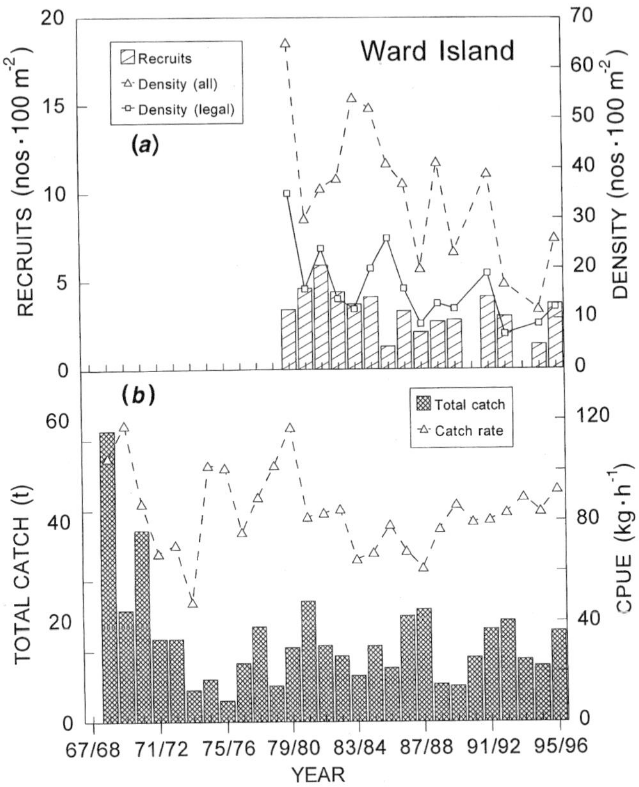

Tiparra Reef

This metapopulation in mid-Spencer Gulf comprises a group of reefs scattered among seagrass beds at 5–12 m depth. The annual catch has fluctuated markedly since 1967 and is currently around 100 t (Fig. 10). Survey data also show fluctuations but no long-term decline of recruitment or density is evident suggesting that the fishery is sustainable at the current levels of fishing. For the EPR analysis we used $M = 0.22$ (Shepherd et al. 1982) obtained at Tiparra Reef and a mean $Z = 0.72$ derived from catch curve analyses giving $F = 0.5$. The %EPR conserved is about 45% at a yield of about 84% (Fig. 3)
.

Discussion

From the individual population case histories we can infer the main factors that have affected their resilience to fishing. Prolonged recruitment failure, well-known as a cause of stock decline (May et al. 1978), appears to have been the proximate cause of the decline of the West Island population. Yet the virtual absence of fishing since 1970 and of recreational fishing since the mid-1980's was not followed by any recovery. This suggests that a new low-density equilibrium in the population may now exist (Connell and Sousa 1983; Myers et al. 1995). The repeated depletion of the Waterloo Bay population by commercial divers was made easier by the strong aggregative behavior of abalone there, so that much of the spawning population could be removed with relatively little fishing effort

Fig. 8. Thorny Passage, 1968–1994. (*a*) Catch (*t*) (total weight) in southern and northern areas. (*b*) Catch (*t*) (total weight) in vertical bars, and catch rate (kg·h^{-1}) for southern and northern areas combined.

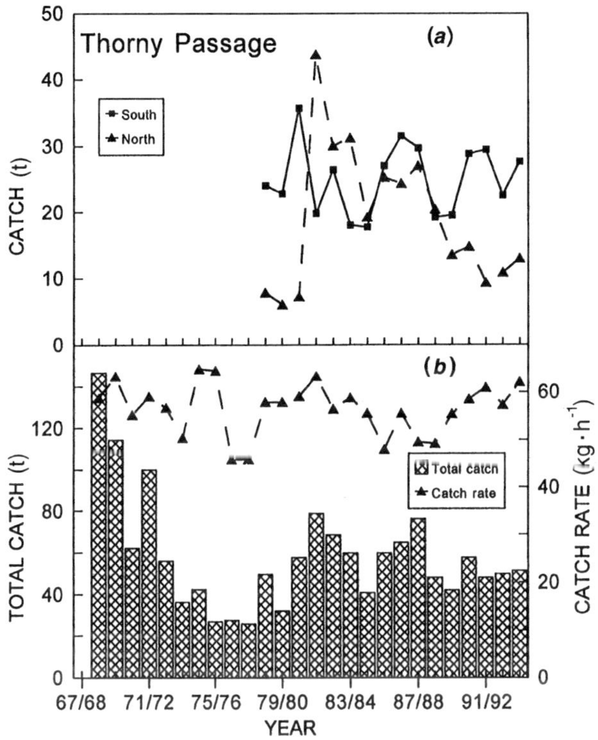

Fig. 9. Thorny Passage, 1986–1995. Mean density of recruits (3+ age-class) in vertical bars, legal-sized abalone (>145 mm SL), and total number of emergent abalone for (*a*) northern areas, and (*b*) southern areas. Gaps are missing values.

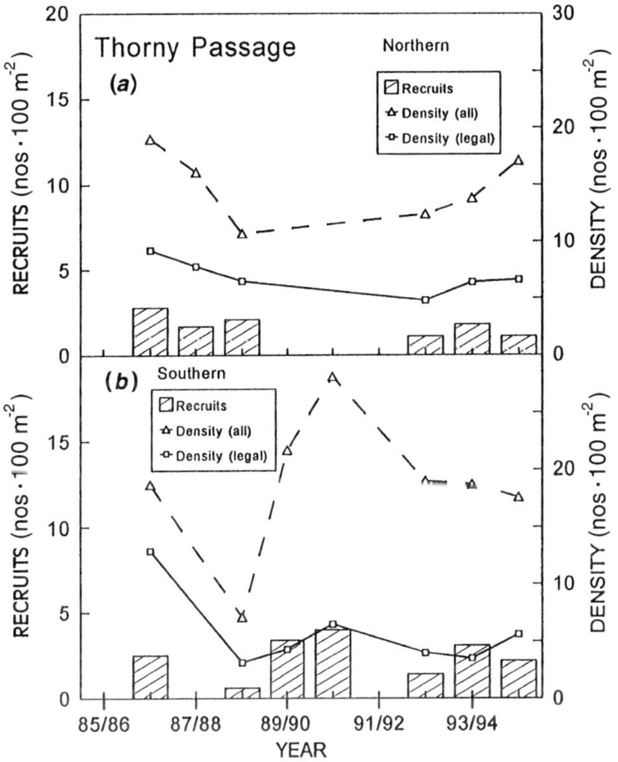

(Shepherd and Partington 1995), leading directly to recruitment failure. The populations of Sceale Bay, Ward Island, and Thorny Passage have been more resilient to fishing, but all show signs of incipient decline. The status of the Sceale Bay population is, however, still ambiguous in the absence of survey data, but the declining proportion of *H. laevigata* in the total abalone catch (Fig. 6) suggests a slow population decline because divers preferentially search for the higher-valued *H. laevigata*. Paradoxically, the survey data for Ward Island show a decline in density and recruitment but the catch has not yet done so. Given the lag time of about four years from the age of measurement of recruitment to age at first capture a decline in the catch may be imminent. In Thorny Passage the declining gradient in recruitment from south to north and the net water current flow northwards led Shepherd and Brown (1993) to postulate that southern populations were larval sources for northern ones. Hence under intense fishing the northern populations would tend to decline first, and this is what appears to be happening. Finally, at Tiparra Reef the catch has been sustained without evidence of long-term decline in recruitment or total abundance.

In all, five out of the six populations have shown varying evidence of decline, and we can conclude that the three showing only slight decline must be close to their threshold %EPR, while the two depleted populations are beyond their threshold. No uniform %EPR threshold appears to exist across these populations. Yet the patterns of decline and the correlations between population productivity, recruitment variability, and PREP values together suggest a model in which the threshold %EPR values vary with some function of population productivity or fitness. A plot of %EPR conserved versus productivity (Fig. 11) suggests a sloping threshold boundary line AB exists below which the risk of recruitment overfishing is high. The curved line AB passes near the points for Sceale Bay, Ward Island, and Thorny Passage which are only slightly overfished, and below the point for Tiparra Reef which is in equilibrium. A minimum viable population (see Shepherd and Brown 1993) has no productivity, by definition, so the line AB must pass through the 100% point at zero productivity. According to this model, populations of low productivity will need to conserve relatively more egg production than those of high productivity. Alternatively, as the productivity of a population declines with overfishing, the threshold level of egg production will gradually increase to 100% at zero productivity. Continued fishing even at a reducing level can cause the collapse of a population, as has occurred with the white abalone, *H. sorenseni*, which is now threatened with extinction (Davis et al. 1996). The precise location of the threshold for metapopulations of *H. laevigata* is necessarily vague with so few points, but as studies extend to the 70 or so others in South Australia the threshold boundary should become clearer.

The model implies that survival from the egg to recruitment to the fishery is relatively poorer in small populations than in large ones. Is there any evidence for this? Our data on recruitment variability (Table 1) show much higher variability in the two smallest populations compared with the three largest ones indicating that in years of poor recruitment larval or juvenile mortality must be very high. Myers and Pepin (1994) found

Fig. 10. Tiparra Reef, 1967–1994. (*a*) Mean density of recruits (3+ age-class) in vertical bars, legal-sized abalone (>130 mm SL), and total number of emergent abalone. (*b*) Catch (*t*) (total weight) in vertical bars, and catch rate (kg·h^{-1}).

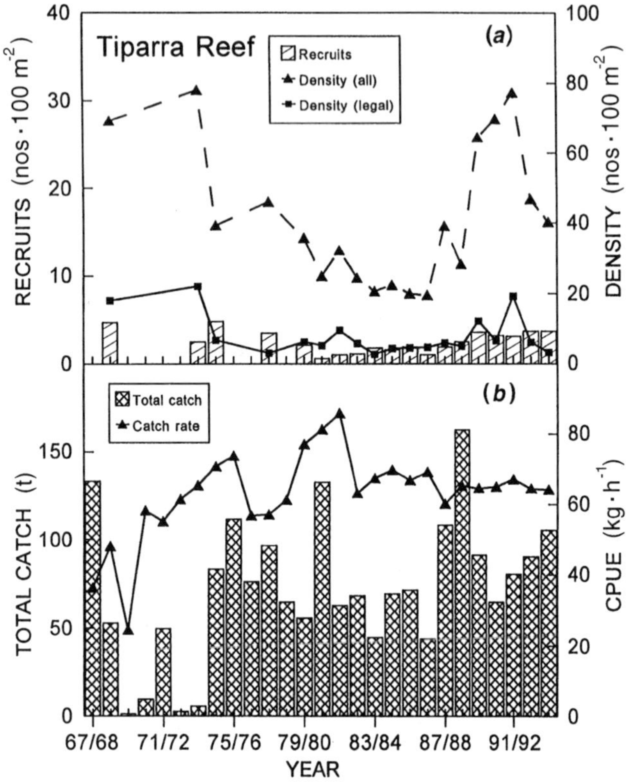

Fig. 11. A plot of %EPR versus productivity (*t*) of six metapopulations of *H. laevigata*. The line AB is the biological reference boundary for recruitment overfishing (see text).

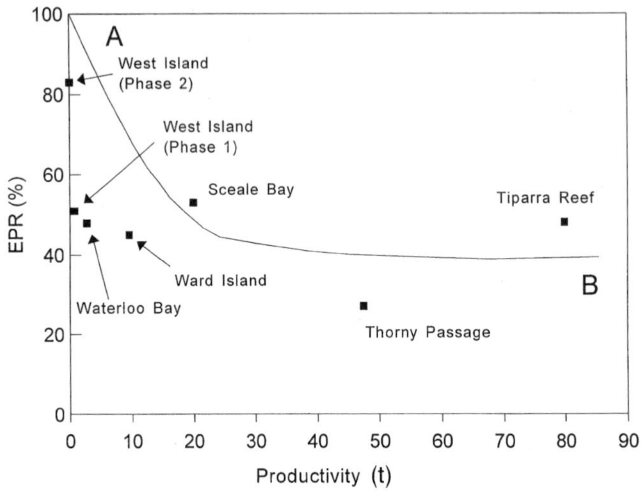

greater recruitment variability in fish populations on isolated offshore banks than in shelf populations, due they believed, to the greater variability of offshore ocean currents. They concluded that populations with greater recruitment variability will be more dependent on occasional strong year-classes to maintain themselves and so will be more susceptible to overfishing during the years of poor recruitment. Similarly, if small abalone populations have greater recruitment variability than large ones then they would need to conserve relatively more egg production to prevent depletion. Both metapopulation theory and island biogeographic theory which have strong conceptual links predict that small isolated populations or "islands" have more variable recruitment and lower persistence probabilities than large ones due to the relatively greater hazards for propagules arriving in the habitat and surviving there (Schoener 1974; Simberloff 1974; review of Taylor 1990).

The question arises why small populations have been able to persist at all even prior to exploitation. While any answer is speculative we note that the three smallest populations all had large accumulated virgin biomasses (Figs. 4–6). The ratio of the mean annual catch in the first three years of fishing when the virgin biomass was removed to the later mean annual catch (Table 1) was 13.3:1 for West Island, 60:1 for Waterloo Bay, and 5.2:1 for Sceale Bay compared with 0.8:1 for Ward Island, 2.3:1 for Thorny Passage, and 1.2:1 for Tiparra Reef (where only two years' fishing was enough to remove the virgin stock). We suggest that the presence of a large, "top-heavy" population size structure is a buffer against prolonged recruitment failure and would enhance the persistence of a population.

What is the significance of the PREP values in these populations? PREP is the equivalent of R_0, the well-known measure of fitness of a population (Stearns 1992), and represents the reproductive response of a population to the totality of its environment; it has presumably evolved toward the maximum possible for each metapopulation (Stearns 1992). Hence values should be low in marginal conditions where the food supply is poor or predation intense, and conversely high in optimal conditions of food supply and habitat. On this hypothesis it is not surprising that PREP values vary between populations (Fig. 2) and tend to increase with productivity and reef area of a metapopulation. In summary we suggest that populations are small when they exist in adverse conditions in terms of food, habitat, or predation. Hence their fitness is low and reproductive output poor. In consequence recruitment is also poor or variable, a factor which may be a sufficient explanation of a small population's vulnerability to overfishing.

EPR models have been criticized because they ignore differences between populations in larval survival (Nash 1992). Indeed, Nash (1992), anticipating the results of this study, predicted that stocks of low productivity would need to conserve a higher level of egg production. Given the inverse relation that seems to exist between recruitment variability and productivity, our model may adequately address this deficiency. Another criticism is that EPR models cannot take into account spatial variation in fecundity, growth, or mortality because they are based on single estimates of these parameters (McShane 1995*b*). This is valid and it means that the investigator needs to be aware of the variability that exists so that the parameter values chosen are representative of the metapopulation. In this study we measured these parameters in the main fishing grounds on the reasoning that the most productive local populations would be more likely to be determinant for the metapopulation as a whole than marginal, less productive local populations. Note that in our model we estimate productivity by the mean catch. This is reasonable in a fully exploited

Table 3. Egg production conserved in various abalone fisheries.

Species	Reference	%EPR	Comments
California *H. cracherodii*	Richards and Davis (1993) Altstatt et al. (1996)	50 (?)	Overfishing caused recruitment failure. Disease now a factor in later decline.
Mexico *H. fulgens*	Shepherd et al. (1991)	17[a]	Fishery became depleted in 1984, and again in 1995–1996 after a slight recovery.
New Zealand *H. iris*	Schiel and Breen (1991)	18[b]	Authors claim current fishing rate is not sustainable.
Canada *H. kamtschatkana*	Breen (1986)	25	Fishery collapsed and now closed. Natural recruitment failure a factor.
Oman *H. mariae*	Shepherd et al. (1995)	29	Fishery is in serious decline.
South Australia *H. rubra*	Sluczanowski (1986)	40–54[c]	Fishery apparently stable.
Tasmania *H. rubra*	Nash (1992)	40–70	Southern stocks (40% EPR) considered vulnerable to overfishing.
Victoria *H. rubra*	Sanders and Beinssen (1996) McShane (1992)	54 50	Fishery stable.
California *H. rufescens* *H. corrugata*	Tegner et al. (1989)	48 51	Both fisheries depleted. Contributing factors are bar cut mortality from recreational fishers, Allee effect, and serial depletion. Fishery closed (in part).

[a]The authors assumed a very high *F*, so underestimated %EPR.
[b]Authors used an equation relating fecundity with length raised to the power of eight, whereas Poore (1973) found a relation with length cubed; this implies that the values given are underestimates.
[c]Sluczanowski (1986) used frequency of fishing visits instead of *F*, so that values are not certain.

population in equilibrium, but if the population is declining even slightly then mean catch will overestimate productivity and our model will underestimate the egg production that should be conserved.

At present a 50% EPR threshold, proposed as a desirable level (Guzmán del Próo et al. 1992, page 62), appears to be an appropriate biological reference point for intermediate-sized metapopulations and a 40% level for large metapopulations of *H. laevigata* in South Australia. Many small metapopulations collapsed in the early years of the fishery (unpublished data) so the appropriate threshold %EPR is not known; our data for Waterloo Bay and West Island suggest it may be as high as 80%. Comparison of our data with other studies on abalone (Table 3) suggests that the thresholds proposed for medium to large metapopulations of *H. laevigata* may be close to those in many other exploited abalone fisheries. Note that globally only fisheries which have conserved levels of 50% or more have so far been sustained in the long term.

Finally, our study provides a mechanism by which serial depletion, postulated as a cause of the collapse of the abalone fisheries of Mexico and California may have occurred (Prince and Guzmán del Próo 1993; Tegner et al. 1989). The fisheries would have comprised many small-to-large metapopulations. Under intense fishing the small metapopulations would have declined first, and then the larger ones more slowly, but inexorably, as fishing effort was increasingly focused on them. The present challenge for biologists and managers is to reverse these trends by focusing management at a fine scale on individual metapopulations as advocated by Keesing and Baker (1998).

Acknowledgments

K.L. Branden, P. Clarkson, A. Dalgetty, B. Davies, C.H. Deane, K. Evans, B. Foureur, D. Graske, L. Gray, J. Keesing, D. Lowe, K. Messner, J.A. Turner, and G. Wright assisted with the diver surveys. Figures were prepared by G. Wright. Drs. J.K. Keesing, G. Morgan, P.E. McShane, G.L. Kesteven, W.J. Nash, and P.A. Breen helpfully criticized earlier drafts of the manuscript. The study was funded in part by the Fishing Industry Research Trust Account.

References

Altstatt, J.M., Ambrose, R.F., Engle, J.M., Haaker, P.L., Lafferty, K.D., and Raimondi, P.T. 1996. Recent declines of black abalone *Haliotis cracherodii* on the mainland coast of central California. Mar. Ecol. Prog. Ser. **142**: 185–192.

Baranov, F.I. 1918. On the question of the biological basis of fisheries. Nauchn. Issled. Ikhtiol. Inst. Izv. **1**: 81–128. [In Russian].

Berst, A.H., and Simon, R.C. (*Editors*). 1981. Proceedings of the Stock Concept Symposium. Can. J. Fish. Aquat. Sci. **38**(12): 1457–1921.

Breen, P.A. 1986. Management of the British Columbia fishery for northern abalone (*Haliotis kamtschatkana*). *In* North Pacific Workshop on Stock Assessment and Management of Invertebrates. *Edited by* G.S. Jamieson and N. Bourne. Can. Spec. Publ. Fish. Aquat. Sci. 92. pp. 300–312.

Breen, P.A. 1992. A review of models used for stock assessment in abalone fisheries. *In* Abalone of the world: biology, fisheries and culture. *Edited by* S.A. Shepherd, M.J. Tegner, and S.A. Guzmán del Próo. Blackwells, Oxford, U.K. pp. 253–275.

Caddy, J.F. 1996. Regime shifts and paradigm changes: is there still a place for equilibrium thinking? Fish. Res. (Amsterdam), **25**: 219–230.

Connell, J.H., and Sousa, W.P. 1983. On the evidence needed to judge ecological stability or persistence. Am. Nat. **121**: 789–824.

Davis, G.E., Haaker, P.L., and Richards, D.V. 1996. Status and trends of white abalone at the California Channel Islands. Trans. Am. Fish. Soc. **125**: 42–48.

Guzmán del Próo, S.A., Tegner, M.J., and Shepherd, S.A. 1992. Abalone of the world: biology, fisheries and culture (supplementary papers). Workshop sessions. Fish. Res. Pap. Dep. Fish. S. Aust. No. 24. pp. 61–62.

Hanski, I., and Gilpin, M. 1991. Metapopulation dynamics: brief history and conceptual domain. Biol. J. Linn. Soc. **42**: 3–16.

Harrison, A.J. 1986. Gastropod fisheries of the Pacific with particular reference to Australian abalone. *In* North Pacific Workshop on Stock Assessment and Management of Invertebrates. *Edited by* G.S. Jamieson and N. Bourne. Can. Spec. Publ. Fish. Aquat. Sci. 92. pp. 14–22.

Keesing, J.K., and Baker, J.L. 1998. The benefits of catch and effort data at a fine spatial scale in the South Australian abalone fishery. *In* Proceedings of the North Pacific Symposium on Invertebrate Stock Assessment and Management. *Edited by* G.S. Jamieson and A. Campbell. Can. Spec. Publ. Fish. Aquat. Sci. 125. pp. 176–186.

Kesteven, G.L. 1996. A fisheries science approach to world fisheries or: three phases of an industrial revolution. Fish. Res. (Amsterdam), **25**: 5–17.

Mace, P.M., and Sissenwine, M.P. 1993. How much spawning per recruit is enough? *In* Risk evaluation and biological reference points for fisheries management. *Edited by* S.J. Smith, J.J. Hunt, and D. Rivard. Can. Spec. Publ. Fish. Aquat. Sci. 120. pp. 101–118.

May, R.M., Beddington, J.R., and Horwood, J.W. 1978. Exploiting natural populations in an uncertain world. Math. Biosc. **42**: 219–252.

McShane, P.E. 1992. Exploitation models and catch statistics of the Victorian fishery for abalone *Haliotis rubra*. Fish. Bull. **90**: 139–146.

McShane, P.E. 1995*a*. Estimating the abundance of abalone: the importance of patch size. *In* Progress in abalone fisheries research. *Edited by* S.A. Shepherd, R.W. Day, and A.J. Butler. Mar. Freshwater Res. **46**: 657–662.

McShane, P.E. 1995*b*. Recruitment variation in abalone: its importance to fisheries management. *In* Progress in abalone fisheries research. *Edited by* S.A. Shepherd, R.W. Day, and A.J. Butler. Mar. Freshwater Res. **46**: 555–570.

Myers, R.A., and Pepin, P. 1994. Recruitment variability and oceanographic stability. Fish. Oceanogr. **3**: 246–255.

Myers, R.A., Barrowman, N.J., Hutchings, J.A., and Rosenberg, A.A. 1995. Population dynamics of exploited fish stocks at low population levels. Science (Washington, D.C.), **269**: 1106–1108.

Nash, W.J. 1992. An evaluation of egg-per-recruit analysis as a means of assessing size limits for blacklip abalone (*Haliotis rubra*) in Tasmania. *In* Abalone of the world: biology, fisheries and culture. *Edited by* S.A. Shepherd, M.J. Tegner, and S.A. Guzmán del Próo. Blackwells, Oxford, U.K. pp. 318–338.

Pimm, S.L. 1991. The balance of nature? Ecological issues in the conservation of species and communities. University of Chicago Press, Chicago, Ill. 434 p.

Poore, G.C.B. 1973. Ecology of New Zealand abalones *Haliotis* species (Mollusca: Gastropoda) 4. Reproduction. N.Z. J. Mar. Freshwater Res. **7**: 67–84.

Prince, J.D., and Guzmán del Próo, S.A. 1993. A stock reduction analysis of the Mexican abalone (Haliotid) fishery. Fish. Res. (Amsterdam), **16**: 25–49.

Prince, J.D., and Shepherd, S.A. 1992. The Australian abalone fisheries and their management. *In* Abalone of the world: biology, fisheries and culture. *Edited by* S.A. Shepherd, M.J. Tegner, and S.A. Guzmán del Próo. Blackwells, Oxford, U.K. pp. 407–426.

Richards, D.V., and Davis, G.E. 1993. Early warnings of modern population collapse in black abalone *Haliotis cracherodii* Leach 1814 at the California Channel islands. J. Shellfish Res. **12**: 189–194.

Sanders, M.J., and Beinssen, K.H.H. 1996. Bio-economic modelling of a fishery under individual transferable quota management: a case study of the fishery for blacklip abalone *Haliotis* rubra in the Western Zone of Victoria (Australia). Fish. Res. (Amsterdam), **27**: 179–201.

Schiel, D.R., and Breen, P.A. 1991. Population structure, ageing and fishing mortality of the New Zealand abalone *Haliotis iris*. Fish. Bull. **89**: 681–691.

Schoener, A. 1974. Experimental zoogeography; colonization of marine mini-islands. Am. Nat. **108**: 715–737.

Shepherd, S.A. 1985. Power and efficiency of a research diver, with a description of a rapid underwater measuring guage: their use in measuring recruitment and density of an abalone population. *In* Diving for Science '85. *Edited by* C.T. Mitchell. American Academy Underwater Science, La Jolla, Calif. pp. 263–272.

Shepherd, S.A. 1986. Movement of the southern Australian abalone *Haliotis laevigata* in relation to crevice abundance. Aust. J. Ecol. **11**: 295–302.

Shepherd, S.A. 1987. Aspects of the biology of the abalone *Haliotis laevigata* and *Haliotis scalaris*. Ph.D. thesis. Deakin University, Victoria, Australia.

Shepherd, S.A. 1990. Studies on southern Australian abalone (genus *Haliotis*). XI. Long-term recruitment and mortality dynamics of an unfished population. Aust. J. Mar. Freshwater Res. **41**: 475–492.

Shepherd, S.A., and Breen, P.A. 1992. Mortality in abalone: its estimation, variability and causes. *In* Abalone of the world: biology, fisheries and culture. *Edited by* S.A. Shepherd, M.J. Tegner, and S.A. Guzmán del Próo. Blackwells, Oxford, U.K. pp. 276–304.

Shepherd, S.A., and Brown, L.D. 1993. What is an abalone stock: implications for the role of refugia in conservation. Can. J. Fish. Aquat. Sci. **50**: 2001–2009.

Shepherd, S.A., and Hearn, W.S. 1983. Studies on southern Australian abalone (genus *Haliotis*) IV. Growth of *H. laevigata* and *H. ruber*. Aust. J. Mar. Freshwater Res. **34**: 461–475.

Shepherd, S.A., and Partington, D. 1995. Studies on southern Australian abalone (genus *Haliotis*) XVI. Recruitment, habitat and stock relations *In* Progress in abalone fisheries research. Mar. Freshwater Res. **46**: 669–680.

Shepherd, S.A., and Triantafillos, L. 1997. Studies on southern Australian abalone (genus *Haliotis*) XVII. A chronology of *H. laevigata*. Molluscan Res. In press.

Shepherd, S.A., Kirkwood, G.P., and Sandland, R.L. 1982. Studies on southern Australian abalone (genus *Haliotis*) III. Mortality of *H. laevigata* and *H. rubra*. Aust. J. Mar. Freshwater Res. **33**: 265–272.

Shepherd, S.A., Guzmán del Próo, S.A., Turrubiates, J., Belmar, J., Baker, J.L., and Sluczanowski, P.R. 1991. Growth, size at sexual maturity and egg-per-recruit analysis of the abalone *Haliotis fulgens* in Baja California. Veliger, **34**: 324–330.

Shepherd, S.A., Clarke, S.M., and Dalgetty, A. 1992*a*. Studies on southern Australian abalone (genus *Haliotis*) XIV. Growth of *H. laevigata* on Eyre Peninsula. J. Malac Soc. Aust. **13**: 99–113.

Shepherd, S.A., Godoy, C., and Clarke, S.M. 1992*b*. Studies on southern Australian abalone (genus *Haliotis*) XV. Fecundity of *H. laevigata*. J. Malacol. Soc. Aust. **13**: 115–121.

Shepherd, S.A., Baker, J.L., and Johnson, D.W. 1995. Yield- and egg-per-recruit analyses of the Omani abalone, *Haliotis mariae*. *In* Progress in abalone fisheries research. *Edited by* S.A. Shepherd, R.W. Day, and A.J. Butler. Mar. Freshwater Res. **46**: 663–668.

Simberloff, D.S. 1974. Equilibrium theory of island biogeography and ecology. Annu. Rev. Ecol. Syst. **5**: 161–182.

Sissenwine, M.P., and Shepherd, J.G. 1987. An alternative perspective on recruitment overfishing and biological reference points. Can. J. Fish. Aquat. Sci. **44**: 913–918.

Sluczanowski, P.R. 1986. A disaggregate model for sedentary stocks: the case of the South Australian abalone. *In* North Pacific Workshop on Stock Assessment and Management of Invertebrates. *Edited by* G.S. Jamieson and N. Bourne. Can. Spec. Publ. Fish. Aquat. Sci. 92. pp. 393–401.

Sluczanowski, P.R., Jones, G.K., Forbes, S., Baker, J., Lewis, R.K., and Wright, G. 1992. Examining the per recruit effects of size limits using PRAna software package. *In* Legal sizes and their use in fisheries management. Australian Society for Fish Biology Workshop, Bureau of Rural Resources Proceedings No. 13. *Edited by* D.A. Hancock. Australian Government Publishing Service, Canberra, Australia. pp. 57–61.

Stearns, S.C. 1992. The evolution of life histories. Oxford University Press, New York, N.Y. 249 p.

Taylor, A.D. 1990. Metapopulations, dispersal, and predator-prey dynamics: an overview. Ecology, **71**: 429–433.

Tegner, M.J., Breen, P.A., and Lennert, C.E. 1989. Population biology of red abalones, *Haliotis rufescens*, in southern California, and management of the red and pink, *H. corrugata*, abalone fisheries. Fish. Bull. (U.S.), **87**: 313–339.

Thompson, G.G. 1993. A proposal for a threshold stock size and maximum fishing mortality rate. *In* Risk evaluation and biological reference points for fisheries management. *Edited by* S.J. Smith, J.J. Hunt, and D. Rivard. Can. Spec. Publ. Fish. Aquat. Sci. 120. pp. 303–320.

Triantafillos, L. 1994. Age determination and population analysis of the greenlip abalone (*Haliotis laevigata*) in South Australia. Graduate diploma thesis. Australian Maritime College, Launceston, N. Tasmania. 40 p.

Stock-recruitment-environment relationships for invertebrate species of Western Australia

N. Caputi, J.W. Penn, L.M. Joll, and C.F. Chubb

Abstract: Research on stock-recruitment relationships (SRR), environmental effects on recruitment, and the effect of fishing on spawning stock has been an integral part of the stock assessment for invertebrate fisheries in Western Australia. Studies on western rock lobsters (*Panulirus cygnus*), king and tiger prawns (*Penaeus latisulcatus* and *P. esculentus*) of Shark Bay and Exmouth Gulf, and scallops (*Amusium balloti*) in Shark Bay, show a variety of patterns. The SRR of tiger prawn stocks in both Exmouth Gulf and Shark Bay show that recruitment overfishing occurred in the early 1980's with the spawning stock declining as a result of high exploitation. The delay to 1990 in implementing management measures to reduce fishing effort in Shark Bay was in contrast to Exmouth Gulf where management measures were introduced as soon as the problem was identified. The subsequent recovery in recruitment, corresponding to the period of reduced fishing and improvement in spawning stock in both fisheries, confirmed recruitment overfishing as the cause of the decline. A large recruitment in the Shark Bay scallop fishery in 1990 resulted in spawning stocks in 1991 and 1992 about 10 times larger than previous levels. However, these stocks did not significantly affect the subsequent recruitment with the main influence being the Leeuwin Current strength. For the rock lobster fishery at coastal locations, environmental factors have a significant effect on the level of puerulus settlement with spawning stock not being significant. However, at the Abrolhos Is. (60 km offshore), a major decline in puerulus settlement during the past 20 years may be due to a decline in egg production throughout the fishery. These studies illustrate that the SRR is an important component of stock assessment of invertebrate fisheries. They emphasize the need for spatial/temporal data on stock and recruitment indices, preferably from research surveys, and highlight the importance of understanding effects of the environment and fishing effort. The adaptive management approach needs greater consideration because of the long-term benefits it provides in improved understanding of the SRR.

Résumé : La recherche sur les relations stock-recrutement (RSR), sur les effets des conditions environnementales sur le recrutement et sur l'effet de la pêche sur le stock de reproduction a fait partie intégrante de l'évaluation des stocks dans les pêcheries d'invertébrés dans l'ouest de l'Australie. Des études sur la langouste *Panulirus cygnus*, les crevettes *Penaeus latisulcatus* et *P. esculentus* de la baie Shark et du golfe d'Exmouth et des pétoncles *Amusium balloti* dans la baie Shark montrent une diversité de situations. La RSR des crevettes *P. esculentus* dans le golfe d'Exmouth et dans la baie Shark montre qu'une surpêche touchant le recrutement a eu lieu au début des années 80, le stock de reproduction déclinant par suite d'une forte exploitation. La situation dans la baie Shark, où l'on a attendu jusqu'à 1990 avant de prendre des mesures de gestion visant à réduire l'effort de pêche, contrastait avec celle du golfe d'Exmouth où des mesures de gestion ont été adoptées dès que le problème a été constaté. Le rétablissement subséquent du recrutement, correspondant à la période de pêches réduite et d'amélioration du stock de reproduction dans les deux pêcheries, a confirmé que la surpêche touchant le recrutement était la cause du déclin. Un recrutement important dans la pêcherie de pétoncles *Amusium balloti* de la baie Shark en 1990 a été à l'origine en 1991 et 1992 d'un stock de reproduction environ dix fois plus important qu'antérieurement. Cependant, ces stocks n'ont pas eu d'effet substantiel sur le recrutement ultérieur, du fait que le facteur principal était la force du courant de Leeuwin. Pour la pêcherie de langoustes *Panulirus cygnus* dans les zones côtières, les facteurs environnementaux ont un effet substantiel sur le degré de fixation des puérulus, alors que l'importance du stock de reproduction n'a pas d'effet significatif. Toutefois, près de l'île Abrolhos (à 60 km de la côte), un déclin important de la fixation des puérulus au cours des 20 dernières années pourrait être attribuable à une diminution de la production d'oeufs dans l'ensemble de la pêcherie. Ces études montrent que la RSR est un élément important de l'évaluation des stocks dans les pêcheries d'invertébrés. Elles soulignent la nécessité de rassembler des données spatiales et temporelles sur les indices des stocks et de recrutement, de préférence à partir de relevés de recherche, et mettent en relief l'importance de comprendre les effets de l'environnement et de l'effort de pêche. L'approche de la gestion adaptative mérite qu'on lui accorde plus d'importance en raison des avantages à long terme qu'elle offre pour mieux comprendre la RSR. [Traduit par la Rédaction]

N. Caputi, J.W. Penn, L.M. Joll, and C.F. Chubb. Fisheries Department of Western Australia, Western Australian Marine Research Laboratories, PO Box 20, North Beach, WA 6020, Australia.

Correct citation: Caputi, N., Penn, J.W., Joll, L.M., and Chubb, C.F. 1998. Stock-recruitment-environment relationships for invertebrate species of Western Australia. *In* Proceedings of the North Pacific Symposium on Invertebrate Stock Assessment and Management. *Edited by* G.S. Jamieson and A. Campbell. Can. Spec. Publ. Fish. Aquat. Sci. 125. pp. 247–255.

Introduction

A number of scientists have drawn attention to the lack of information on the stock-recruitment relationship (SRR) for invertebrate species (Hancock 1973; Gulland 1984; Caddy 1986; Pollock 1991), despite the economic importance of most of these high unit value fisheries. For many fisheries the effects of the environment on the abundance of recruits has been examined but the effect of the spawning stock is often neglected. This neglect leads to a risk of recruitment overfishing.

Fig. 1. The locations of the prawn and scallop fisheries in Shark Bay and Exmouth Gulf and region of the rock lobster fishery in Western Australia. Stars refer to locations of puerulus collectors.

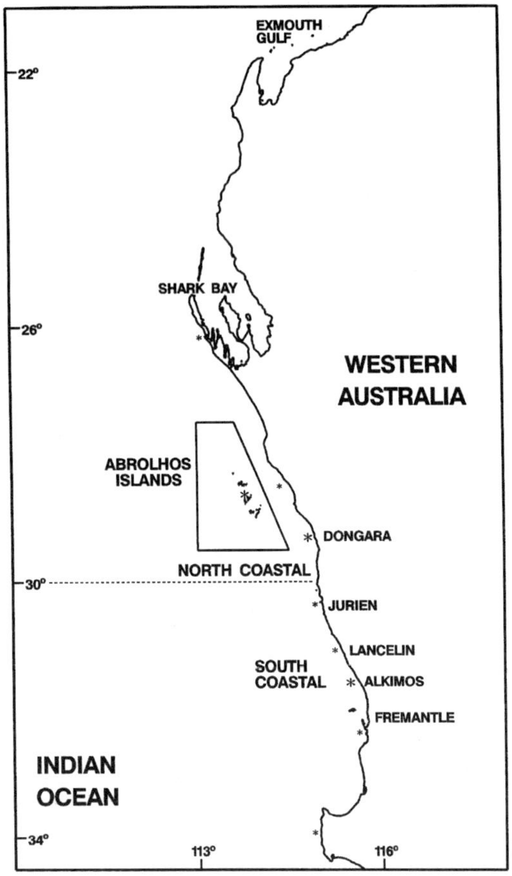

Research on the stock and recruitment relationships (SRR), environmental influences on recruitment, and the effect of fishing on the spawning stock has been an integral part of the stock assessment for a number of invertebrate fisheries in Western Australia over many years. These fisheries on the western rock lobster (*Panulirus cygnus*), king and tiger prawns (*Penaeus latisulcatus* and *P. esculentus*) of Shark Bay and Exmouth Gulf, and scallops (*Amusium balloti*) in Shark Bay, are some of the most valuable in Western Australia. Their combined annual landed value of about A$380 million represents over 90% of the total value of fisheries in Western Australia.

As a result of the desert-dominated climate in Western Australia where river runoff is minimal, the coastal hydrology is comparatively stable except for the dominating influence of the Leeuwin Current. The current consists of warm, low-nutrient waters flowing south along the edge of the continental shelf of the West Australian coast (Lenanton et al. 1991). This strong annual signal against a relatively stable hydrological environment provides an ideal opportunity to examine the relative influence of the breeding stock and the environment on recruitment in a relatively simplified situation. Another factor which has assisted in the study of SRR for these fisheries has been their management as limited-entry fisheries from the 1960's with good catch and effort information available for the entire fleets. Fishery-independent estimates of stock and recruitment indices have also been developed for some fisheries. These have provided the basis to assess the SRR and the recruitment to spawning stock relationship.

The Leeuwin Current has a significant effect on the abundance of a number of invertebrate stocks such as the western rock lobster (Pearce and Phillips 1988; Caputi et al. 1995), western king prawn in Shark Bay (Lenanton et al. 1991), and scallop (Joll and Caputi 1995). The annual variation in the strength of the Leeuwin Current (measured by coastal sea levels) is influenced by ENSO (El Niño/Southern Oscillation) events with the current being weaker during these events (Pearce and Phillips 1988).

The impetus to study SRR in Western Australian invertebrate fisheries is due to a number of factors. For example, the decline in the tiger prawn catches in the early 1980's in both the Exmouth Gulf and Shark Bay fisheries (Fig. 1) resulted in the SRR assessment of these fisheries (Penn and Caputi 1986; Penn et al. 1995) as well as the king prawn fishery which had not experienced any decline in catch.

In the rock lobster fishery, the catch has tended to increase with increasing fishing effort with an average catch of about 10 500 t over the last 10 years. However, there has been increasing concern about an average decrease of 50% in the level of puerulus settlement (first post-larval stage) at the Abrolhos Is. (Fig. 1) from the 1970's to the 1980's and 1990's. At the same time, nominal fishing effort and fishing efficiency of the fishing vessels increased, particularly in deep water (35+ m) where the breeding stock is mainly located and has been declining. This resulted in an assessment of the SRR using the puerulus settlement at three locations as the index of recruitment (Caputi et al. 1995). Caputi et al. (1995) also reviewed the earlier work on SRR of the western rock lobster fishery by Morgan et al. (1982).

The relative effects of an environmental factor (Leeuwin Current) and the spawning stock size were examined when assessing the recruitment variation in the Shark Bay scallop fishery (Joll and Caputi 1995). This study showed that variation in the strength of the Leeuwin Current was the main factor affecting recruitment, with spawning stock not being significant. Since then, an unusually large recruitment from the 1990 spawning resulted in record catches in 1991 and 1992 with the 1992 catch of 20 000 t being over five times greater than the previous highest catch (Joll 1994). This recruitment subsequently resulted in very large spawning stocks in 1991 and 1992 which provided increased contrast in spawning stock to assess its impact on recruitment.

This study reviews and updates the SRR of a number of invertebrate fisheries in Western Australia viz., western rock lobster fishery (Caputi et al. 1995), the king and tiger prawn fisheries of Shark Bay and Exmouth Gulf (Penn and Caputi 1986; Penn et al. 1995), and scallop fishery of Shark Bay (Joll and Caputi 1995). An overview of the above experiences with SRR in invertebrate fisheries is used to indicate problems which need to be considered when assessing SRR.

Methods

The spawning stock indices for the Shark Bay and Exmouth Gulf prawn fisheries are based on standardized catch rates from the location and months of spawning. The catch rates are obtained from a research logbook system which has been completed by nearly all of the fleet since the mid-1960's (Hall and

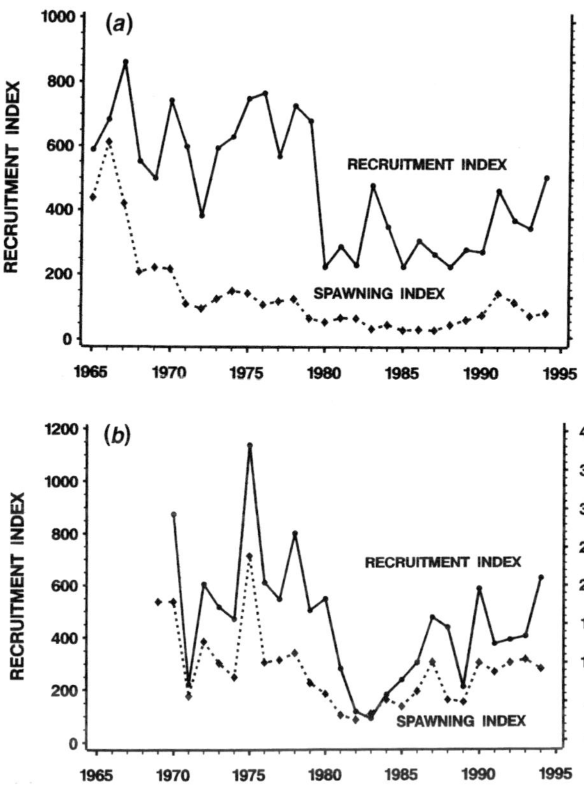

Fig. 2. Spawning and recruitment time series for (*a*) Shark Bay tiger prawns and (*b*) Exmouth Gulf tiger prawns.

Penn 1979). Since 1982 the spawning stock index for the Exmouth Gulf tiger prawn fishery has been based on catch rates from standardized research surveys by charter vessels. The recruitment indices for the prawn fisheries are based on the catches from the months and locations of recruitment and a biomass estimate of prawns present at the beginning of the spawning season. The spawning stock and recruitment indices for the Shark Bay and Exmouth Gulf prawn fisheries are described in detail by Penn and Caputi (1986) and Penn et al. (1995).

The spawning index for the Shark Bay scallop fishery was based on the catch rate from commercial fishermen's logbooks during the spawning period (Joll and Caputi 1995). The recruitment index was based on catch rates of recruits from an annual research survey undertaken in November since 1983. The estimate of the recruits in 1982 was estimated from the catch using a relationship between recruitment index and catch (Joll and Caputi 1995).

The recruitment index for the western rock lobster fishery is the mean number of puerulus caught per collector per year (Phillips 1972). The spawning index is based on the standardized catch rate of breeding size females from an on-board research monitoring of commercial vessels for the coastal fishery and the catch of the Abrolhos Is. fishery since 1972.

Results and discussion

Shark Bay tiger prawn fishery

The tiger prawn catch in the multi-species fishery in Shark Bay declined by about 50% during the 1980's compared to the

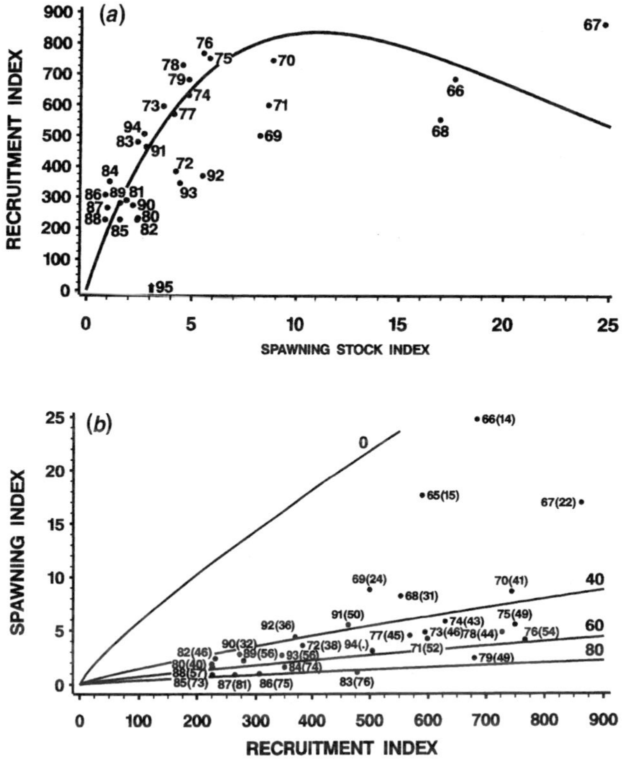

Fig. 3. (*a*) The SRR for the Shark Bay tiger prawn with the year of recruitment indicated. The spawning stock giving rise to the 1995 recruitment is shown. (*b*) The RSR with effort lines of 0, 40, 60, and 80 thousand hours of effective effort on recruits. The year and effective effort in thousands of hours are shown.

previous decade as a result of a decline in the spawning stock affecting the level of recruitment (Penn et al. 1995). Management measures introduced in 1990 and the increased targeting of the high abundance of scallops reduced the fishing effort on tiger prawns and subsequently resulted in a higher level of spawning stock and recruitment (Fig. 2*a*). A Ricker SRR was fitted to the recruitment data from 1966 to 1994 (Fig. 3*a*) updating the results of Penn et al. (1995) which was based on data to 1989. The relationship was fitted using a logarithmic transformation, resulting in the following model:

$$R_t = 207\ S_{t-1}\ \exp(-0.0911\ S_{t-1})$$

where S_{t-1} is the spawning index in year $t-1$ and R_t is the recruitment index of the following year. The fitted log-transformed relationship has a multiple correlation coefficient of 0.64 ($P < 0.01$) and a residual mean square (RMS) of 0.118. The density-dependent parameter estimate (0.0911) was significant at the 0.001 level. The above relationship and subsequent relationships have been adjusted for bias due to the logarithmic transformation using exp(0.5 RMS) (Beauchamp and Olsen 1973). The Durbin–Watson statistic of 1.73 indicates that there is no significant autocorrelation in the residuals. The additional years (1990–1994) provide an improvement in both spawning stock and recruitment with the 1994 recruitment being the highest in 15 years (Fig. 2*a*).

The recruitment to spawning stock relationship (RSR) examines the impact of fishing effort on recruitment survival to

Fig. 4. (*a*) The SRR for the Exmouth Gulf tiger prawn stock. Data points show the year of recruitment and the January and February rainfall (mm). The spawning stock giving rise to the 1995 recruitment is shown. (*b*) The RSR with effort lines of 30, 40, and 60 thousand hours of effective effort. The year and effective effort in thousands of hours are shown.

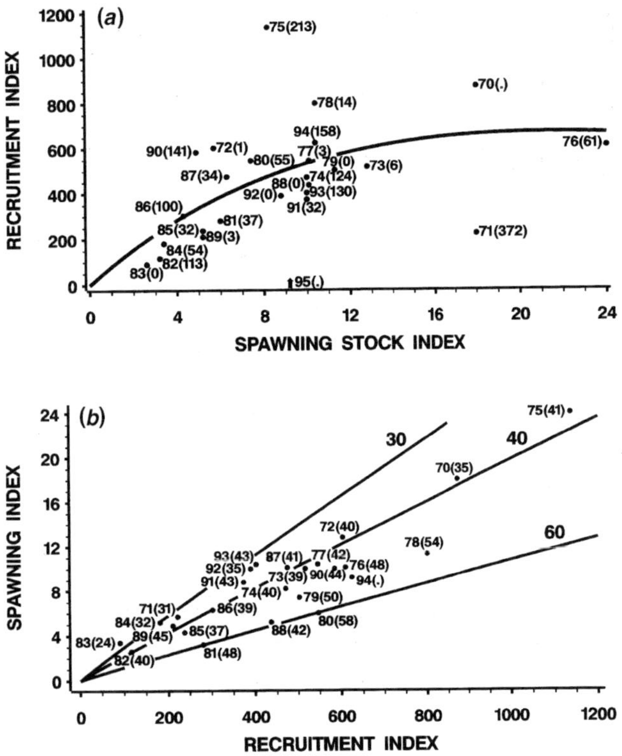

Fig. 5. (*a*) The SRR for the Shark Bay king prawn stock. Data points show the year of recruitment. (*b*) The RSR with effort lines of 0, 25, and 45 thousand hours of effective effort. The year and effective effort in thousands of hours are shown.

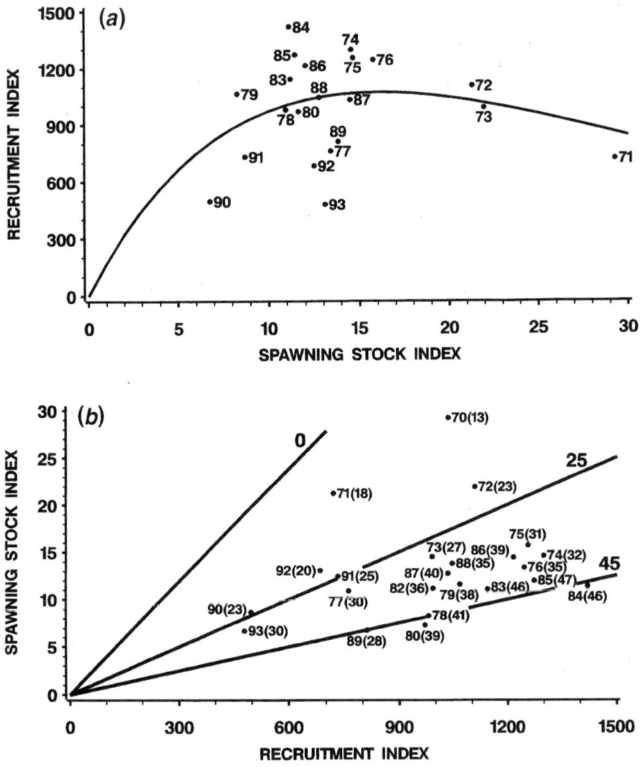

the spawning stock. The form of equation examined allows for a density-dependence between recruitment and spawning and the relative survival due to the level of fishing. The model was fitted using logarithmic transformation resulting in the following (Fig. 3*b*):

$$S_t = 0.131\ R_t 0.823\ \exp(-0.0349\ E_t)$$

where R_t is the recruitment index, S_t is the spawning stock index later in the same year, and E_t is an estimate of the level of effective effort on the year's recruits in thousands of hours trawled. The above relationship has a multiple correlation of 0.96 ($P < 0.001$) and a RMS of 0.069. The parameter estimates associated with effective effort and recruitment are both significant at the 0.001 level. The Durbin–Watson statistic of 1.74 shows that there is no significant autocorrelation in the residuals.

This relationship indicates that the decrease in effective effort from 70 000–80 000 h of trawling to less than 60 000 h in recent years has resulted in improved spawning indices similar to levels obtained in the 1970's.

Exmouth Gulf tiger prawn fishery

The decline in recruitment in the early 1980's was attributed to recruitment overfishing (Penn and Caputi 1986). Management measures were introduced 1982 to reduce fishing effort, and improve the level of spawning stock and the subsequent recruitment (Fig. 2*b*) (Penn et al. 1995). The stock and recruitment time series in recent years (1990–1994) show that the improvement over the stock levels experienced in the early 1980's has been maintained (Fig. 2*b*). The Ricker SRR updating the results of Penn et al. (1995) resulted in the following relationship (Fig. 4*a*):

$$R_t = 85.3\ S_{t-1}\ \exp(-0.0468\ S_{t-1})$$

where S_{t-1} and R_t are the spawning and recruitment indices. The multiple correlation for this relationship is 0.69 with a RMS of 0.20. The density-dependent parameter estimate (0.0468) was significant at the 0.05 level. The relationship shows that the additional years of data (1990–1994), with good spawning and recruitment indices, maintained the strong relationship (Fig. 4*a*).

The additional years of data have not provided any further insight on the impact of the cyclones on the recruitment. Previous results indicated that severe cyclones can have a major positive (1975) or negative (1971) impact on recruitment (Fig. 4*a*) possibly as a result of their timing and intensity (Penn and Caputi 1986). These infrequent summer cyclones result in heavy wave action, increased turbidity, coastal flooding, and heavy rainfall in an otherwise dry region and occur at the time when post-larval and juvenile prawns are either on the shallow seagrass beds or moving into deeper waters.

The complementary relationship between recruitment and surviving spawning stock resulted in the following model (Fig. 4*b*):

Fig. 6. The relationship between recruitment index and the strength of the Leeuwin Current (measured by Fremantle sea level) for the Shark Bay scallop fishery. The 1982 recruitment (*) was estimated from the catch using a relationship between recruitment index and catch.

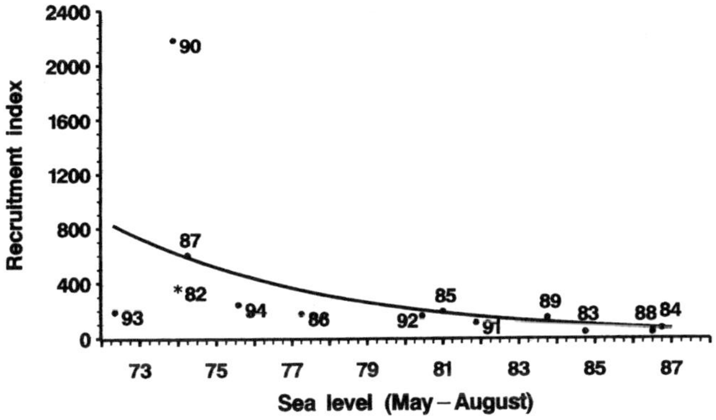

$$S_t = 0.0930\ R_t 0.954\ \exp(-0.0307\ E_t)$$

which has a multiple correlation of 0.94 ($P < 0.001$) and a RMS of 0.034. The parameter estimates associated with recruitment and effective effort are significant at the 0.001 level. The additional data are near the 40 000 h of trawling identified by Penn and Caputi (1986) and Penn et al. (1995) as resulting in spawning stock and recruitment indices on average near the equilibrium points of about 10 and 500, respectively.

Shark Bay king prawn fishery

The relationship between the spawning stock of the Shark Bay king prawn over the May–September period (Penn 1988) and the recruitment the following year is shown in Fig. 5*a*. Fitting of the SRR resulted in the following fitted model:

$$R_t = 173\ S_{t-1}\ \exp(-0.0601\ S_{t-1})$$

where S_{t-1} and R_t are the spawning and recruitment indices. The multiple correlation for this relationship is 0.41 ($P > 0.05$), suggesting that the spawning stock is not significantly related to recruitment. The density-dependent parameter estimate (0.0601) was significant at the 0.001 level.

The RSR resulted in the following model (Fig. 5*b*):

$$S_t = 0.024\ R_t 1.082\ \exp(-0.0359\ E_t)$$

which has a multiple correlation of 0.86 ($P < 0.001$) and a RMS of 0.038. The parameter estimates associated with recruitment and effective effort are both significant at the 0.001 level. The increase in fishing effort in the Shark Bay prawn fishery in the 1970's and 1980's caused a decline in the king prawn spawning stock. However, this decline was not as severe as in the tiger prawn spawning stock, due to differences in catchability (Penn 1984) and the timing of the spawning, and has not caused a consistent decline in recruitment. The lower catch rates in the early 1990's appear to be related to the record catches of scallops (see below) which resulted in prawn trawlers targeting scallops and hence catches and catch rates for king prawns were reduced.

Shark Bay scallop fishery

The relationship between the strength of the Leeuwin Current (measured using Fremantle sea level, FSL_t), and the abundance of recruits (R_t) (Joll and Caputi 1995) was updated using the additional data from 1990–1994 and resulted in the following model (Fig. 6):

$$R_t = 239 \times 10^6\ \exp(-0.174\ FSL_t)$$

This model, fitted using a logarithmic transformation, resulted in a correlation of −0.79 ($P = 0.002$) and a RMS of 0.506.

Based on data to 1989 (Joll and Caputi 1995), an above-average recruitment was expected in the recruitment survey in November 1990 as a result of a weak Leeuwin Current earlier that year. However, the magnitude of the recruitment (Fig. 6) was a surprise. This year-class produced record catches in 1991 and 1992 with the 1992 catch of 20 000 t being over 5 times greater than the previous highest catch (Joll 1994).

The inverse relationship between the sea level and subsequent recruitment indicates that the recruitment will be highest during years when the Leeuwin Current is weak (Fig 6). The variability in the strength of the Leeuwin Current has been linked to ENSO events (Pearce and Phillips 1988). During the study period, 1982–1994, there have been a number of ENSO events, 1982, 1986–1987, and the extended ENSO event in the early 1990's and these have generally corresponded to years of above-average scallop recruitment (Fig. 6).

Two possible mechanisms for the effect of the environment on recruitment are hydrological flushing in years of strong Leeuwin Currents or a temperature effect on spawning or fertilization (Joll and Caputi 1995). Hydrological flushing was considered by Caddy (1989) as an important environmental influence on recruitment success in sedentary molluscs.

The good recruitment in 1990 also resulted in a very large spawning stock in 1991 and 1992 about 10 times larger than previous levels. This increased the contrast in spawning stock which allowed further assessment of its impact on recruitment. The effect of spawning stock, S_t, on recruitment was examined by fitting a Ricker SRR with an environmental variable incorporated and resulted in the following model (Fig. 7):

Fig. 7. The SRR with sea level (SL) as the environmental effect for the Shark Bay scallop fishery.

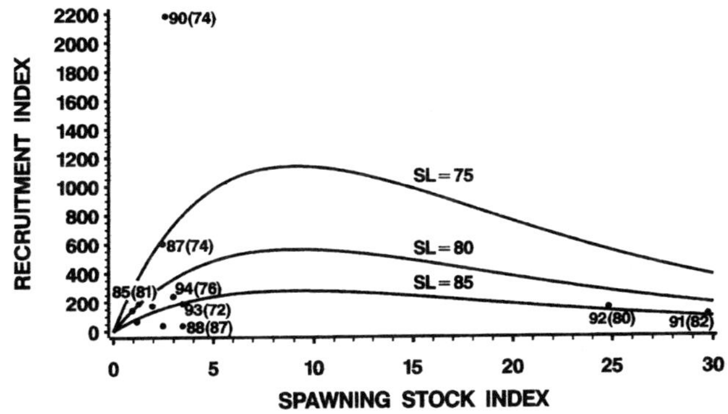

$$R_t = 13.3 \times 10^6\ S_t\ \exp(-0.110\ S_t - 0.141\ FSL_t)$$

with the spawning stock not significantly ($P > 0.05$) affecting the level of recruitment when the strength of the Leeuwin Current was taken into account.

Fig. 8. Multiple regression relationship between the puerulus settlement (mean number per collector) of the western rock lobster fishery at Dongara and the strength of the Leeuwin Current (measured by Fremantle sea level) at two levels for the westerly winds (measured by rainfall at 40 and 90 mm in southern locations during October–November). The year of puerulus settlement is shown with the rainfall (mm) in brackets.

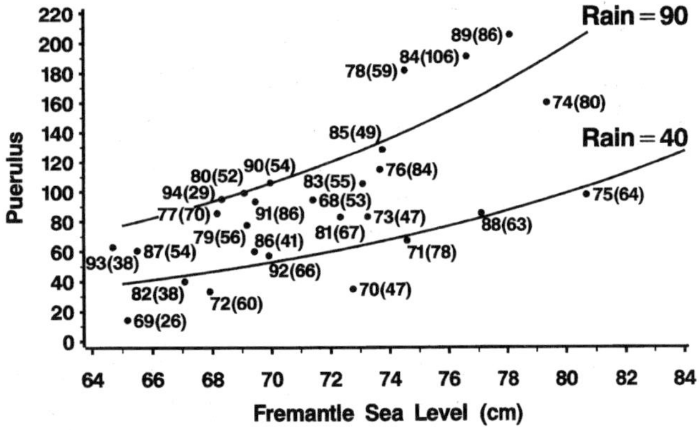

Fig. 9. The stock-recruitment relationship with environmental effects incorporated for (*a*) Dongara and (*b*) Abrolhos Is. of the western rock lobster fishery. The year of puerulus settlement is shown with environmental effects in brackets: (*a*) Fremantle sea level (FSL in cm) and (*b*) rainfall (in mm) in northern locations during October–November. The horizontal arrows indicate years of puerulus settlement when no spawning indices are available.

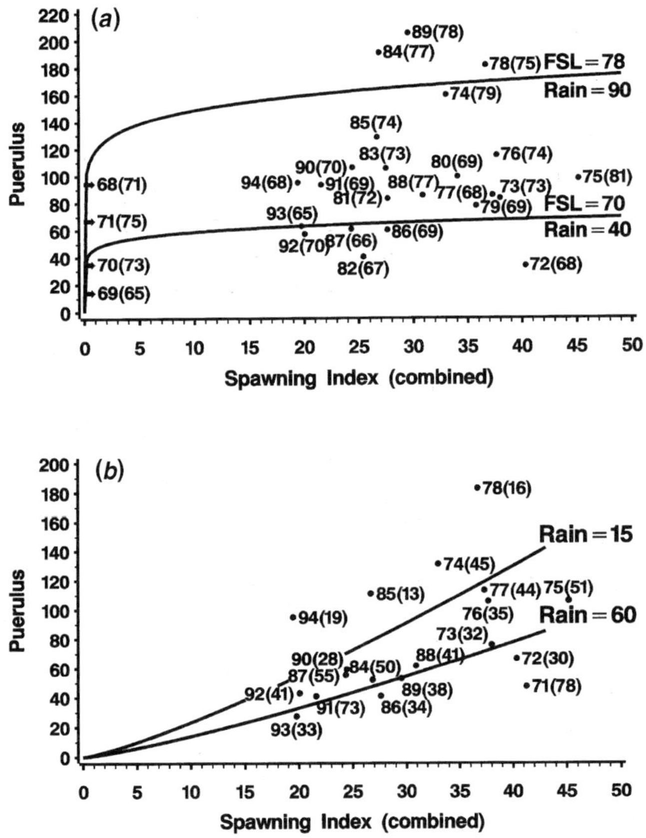

Previous studies on effects of spawning stock on recruitment have suggested that the level of recruitment of scallops may be both positively and negatively related to the spawning stock (Vahl 1982; Orensanz 1986; Young and Martin 1989; Joll and Caputi 1995), depending on the level and timing of fishing. For example, if there is heavy fishing pressure on the recruits with few surviving to spawning, then recruitment overfishing may be possible. If stocks are lightly fished and/or fishing occurs after spawning, then density-dependent effects of the spawning stock may affect the level of recruitment.

Despite an increase in spawning stock by an order of magnitude in 1991 and 1992, there was no detectable effect on recruitment found in this study after taking into account the effect of the Leeuwin Current (Fig. 6). Joll (1994) notes that the average density of scallops was about 2 m^{-2}, with peak densities up to 7 m^{-2}. Thus even at peak density regions the area of the bottom covered by scallops was <10%. At these densities there did not appear to be any food limitations as scallops grew at a similar rate in years of peak recruitment as in years of poor recruitment. Orensanz (1986) showed a negative impact on recruitment of tehuelche scallops due to densities of 10–40 m^{-2} of 2+ year old scallops. He indicated that the carrying capacity was about 100 m^{-2} of 1+ scallops or 40 m^{-2} of 2+ which corresponds to a biomass of about 1500 $g{\cdot}m^{-2}$.

Western rock lobster fishery

Caputi et al. (1995) examined the effect of the environment (Leeuwin Current and westerly winds) and the spawning stock on the level of puerulus settlement. This study reviews this work and examines the impact of two more years of data (1993/1994 and 1994/1995) on the relationships.

The variation in puerulus settlement at Dongara and Alkimos (Fig. 1) was shown to be related to the strength of the Leeuwin Current and the westerly winds (measured by rainfall) (see Fig. 8) with the spawning stock not having a significant effect (Fig. 9*a*) (Caputi et al. 1995). The combination of these two environmental factors provided an improved explanation of the variation in puerulus settlement to the work of Pearce and Phillips (1988) and Caputi and Brown (1993) who considered the individual impacts of the Leeuwin Current and westerly winds, respectively. The additional two years of data have slightly weakened this relationship for Alkimos where the correlation has declined from 0.73 to 0.58.

Caputi et al. (1995) also showed that there was a negative impact of westerly winds on the puerulus settlement at the Abrolhos Is. (60 km off shore, Fig. 1) (Fig. 9*b*). The settlement in 1994/1995, the best for 9 years, was associated with weak westerly wind conditions during the period of peak settlement in October–November. The index of westerly winds was the third lowest and can be compared with the other low values in 1978 and 1985 which corresponded to the highest puerulus settlement in the 1970's and 1980's. Thus the 1994/1995 data has provided additional support for the hypothesis of the negative impact of westerly winds on puerulus settlement at the Abrolhos Is.

This environmental variable does not explain the general decline in the puerulus settlement at the Abrolhos Is. from the

1970's through to the 1990's which may be due to the decline in the spawning stock. Caputi et al. (1995) fitted the SRR with rainfall (RAIN) as an index of the westerly wind and the spawning stock index was an overall index from the coastal and Abrolhos Is. fishery, resulting in the following fitted model:

$$R_t = 6.54\ S_{t-1}0.877\ \exp(-0.131\ \mathrm{RAIN}_t)$$

This results in a multiple correlation of 0.71 with the parameters associated with the spawning stock and rainfall being significant at the 0.05 level. However, this SRR is not conclusive as it consists of a relationship between two time series both of which have declined. Management measures introduced at the start of the 1993/1994 season such as an 18% pot reduction, an increase in minimum size for part of the season, and the release of mature females, were aimed at rebuilding the breeding stock to the level occurring in the late 1970's and early 1980's. The timing of this effort reduction was chosen to coincide with a predicted good recruitment to the fishery to maximize the survival of rock lobsters to the breeding stock and average out catches with the expected below-average catches in 1995/1996 and 1996/1997. This planned increase in spawning stock will provide additional information to assess the impact of spawning on puerulus settlement at the Abrolhos.

If the decline in puerulus settlement at the Abrolhos Is. is due to spawning stock then spawning stocks from other parts of the fishery, e.g., coastal regions, must provide an important contribution to Abrolhos settlement since the spawning stock at the Abrolhos has shown little decline in the last 20 years as breeding commences before they reach legal size (Caputi et al. 1995). For the coastal fishery there has been a significant reduction in spawning stock because of an increase in fishing effort and efficiency (Brown et al. 1995) on rock lobsters which have to survive 1–2 years fishing before reaching maturity. There has also been a large increase in catch in an offshore region north of the Abrolhos Is. since 1987 (Chubb et al. 1994) from 50 t to over 300 t. The impact this may have had on the spawning stock in this area and the overall spawning stock is currently being assessed. The 9–11 month oceanic larval life ensures a certain level of mixing so that it is unlikely that only local spawning stocks contribute to the puerulus settlement at a location. However, whether all areas contribute equally to the puerulus settlement at a given location may also need further investigation. This highlights the need to examine the metapopulation dynamics of the larval dispersion for this fishery as undertaken by Botsford et al. (1998) for Dungeness crabs and Fogarty (1998) for American lobsters.

General discussion

The tiger prawn and rock lobster fisheries illustrate the importance of establishing a time series of basic stock and recruitment data for all major fisheries even if not under threat of overfishing, as increases in fishing effort and efficiency or major environmental shifts may result in unexpected declines in spawning stock or recruitment. Collection of appropriate data to estimate past stock and recruit levels may not be possible after recruitment overfishing has occurred. The above studies demonstrate the benefit of assessing the SRR for stocks which are resilient to recruitment overfishing (e.g., king prawns) and do not show a significant SRR. These studies provide a comparison to stocks which have suffered a decline (e.g., tiger prawns) and enable the differences between the stocks to be assessed. One of the reasons there have been so few published studies on the SRR of invertebrate species may be due to the myth that if a "significant" SRR is not shown then the relationship is not worth reporting.

The multispecies nature of the Shark Bay prawn fishery has enabled a comparison of the SRR between the tiger prawn stock which was susceptible to recruitment overfishing and the king prawn stock which has been resilient at current exploitation levels. This has occurred despite the species habitat overlapping to a large extent and hence both species being subjected to similar levels of fishing effort. Penn and Caputi (1986) and Penn et al. (1995) suggested a number of factors which may cause these tiger prawn fisheries to be more susceptible to overfishing. These factors include the degree of aggregation at spawning, greater vulnerability to trawling (Penn 1984), the level of fishing between recruitment and the start of the spawning period, and the multispecies fishery allowing fishing on low catch rates of tiger prawns. These factors may be used to identify other penaeid stocks which may be potentially at risk from recruitment overfishing.

The general overlap in king and tiger prawn stocks in Shark Bay does not allow the catch of the two species to be optimized separately. The level of fishing effort in the 1980's was too high for the tiger prawn stock but not for king prawns. However, the current reduced level of effort may have improved the recruitment and catches for the tiger prawns but may have also unnecessarily reduced the catch for king prawns. Thus the optimum multispecies catch is less than the sum of the optimum catch of each of the fisheries managed separately. However, the introduction of global positioning systems has enabled the targeting of species more precisely which may allow a greater separation of the effort on each species. This may enable the catches of the two species to be optimized separately and provide a greater overall catch.

The contrast in management approach to the tiger prawn fisheries in Shark Bay and Exmouth Gulf has provided a unique opportunity to test the effect of fishing effort on the spawner-recruit relationship. The management initiatives in Exmouth Gulf, where tiger prawns were the major species caught, commenced as soon as overfishing was suspected as the cause of the recruitment failure in 1982. The reduction in fishing effort allowed the spawning stock and recruitment to build up steadily during the 1980's. However, in Shark Bay, where tiger prawns were not the major species, no significant changes were made to the level of fishing until 1990. This resulted in spawning stock and recruitment remaining relatively low throughout the 1980's (Fig. 2*a*). The introduction of a buy-back scheme in 1990, combined with delays to the start of the fishing season to enable targeting of larger prawns, and increased targeting of the large scallop abundance, allowed the tiger prawn stock to increase. The difference in the timing of reduction in fishing effort between Exmouth Gulf and Shark Bay has provided further evidence that recruitment overfishing was the cause of the decline in recruitment and not some unknown environmental factor. For the rock lobster fishery, the planned increase in the egg production due to a major reduction in fishing effort will soon provide further information on the impact of the coastal spawning stock on the puerulus settlement at the Abrolhos; this information would not have been attainable with a continuation of previous management policies. These "experimental" or adaptive management approaches to fisheries management (Walters and Hilborn 1976; Walters 1986) provide a greater insight to the stock-recruitment dynamics than would otherwise be possible with "traditional" management. These approaches are particularly suitable when there are multiple stocks of species, e.g., tiger prawn fisheries, and management should consider deliberate contrasts in effort to learn about the impact of different levels of spawning stock on recruitment.

One of the factors that has enabled the study of SRR for invertebrate fisheries in Western Australia has been the availability of good catch and effort data from fishermen and catch details from processors over many years. Most of these fisheries have been managed by limited entry since the 1960's and there has been a relatively high level of cooperation between the fishing industry, fisheries management, and research.

Management changes in the prawn fisheries, such as seasonal and area closures, aimed at protecting the recruitment or spawning stock have had a marked effect on the ability to obtain comparable time series of stock/recruitment indices based on data provided by fishermen. Changes in the level of fishing (Caputi 1989) and in fishing efficiency (Brown et al. 1995) may result in catch rates being biased indices of abundance. Errors in the measurement of spawning stock result in the recruitment appearing to be independent of the spawning stock which can lead to overexploitation of stocks (Walters and Ludwig 1981). Standardized surveys using either commercial and/or research vessels are increasingly being used in the prawn, rock lobster, and scallop fisheries to develop unbiased estimates of recruit and spawning stock abundance.

For the rock lobster fishery, the development of an index of recruitment based on the relative abundance of the puerulus stage (first post-larval stage) has been a major factor in understanding the effect of environment and spawning stock on the level of recruitment. If this index was unavailable, the recruitment index would be based on catch rates of a combination of mainly 4- and 5-year-old rock lobsters, i.e., a combination of year-classes recruiting to the fishery. The catch rates would include variability associated with density-dependent effects occurring between the puerulus stage and recruitment to the fishery, environmental effects on catchability, migration between areas, and the effect of fishing efficiency on catch rates. Thus using catch rates would have made it more difficult to detect the environmental and spawning stock effects on recruitment and the spatial and temporal variation of these effects.

As environmental effects play a major role in the recruitment variation, these effects should be taken into account when examining the effect of the spawning stock, otherwise the underlying SRR may be obscured. Omitting an environmental variable from the SRR when the variable is affecting recruitment results in the parameter estimates of the SRR being biased if this variable is correlated to the spawning stock (Caputi 1989). The SRR of the western rock lobster can be used to illustrate this bias. The initial assessment of this SRR showed a strong negative relationship between spawning stock and puerulus settlement at Dongara (Morgan et al. 1982). However, after identifying environmental effects on the level of puerulus settlement (Pearce and Phillips 1988; Caputi and Brown 1993), the spawning stock did not appear to significantly affect the puerulus settlement for the coastal location of Dongara. However, the spawning stock may be positively related to the settlement at the Abrolhos Is. (Fig. 9). For species such as scallops, where the recruitment can vary by an order of magnitude, there is little point in studying the effect of the spawning stock unless some account is taken of the environmental effect on recruitment as this is by far the dominant effect at most levels of spawning stock.

While environmental variables are important in understanding recruitment variations and helping to determine the effect of the spawning stock on recruitment, there is a risk of significant spurious correlations between environmental variables and recruitment (Walters and Collie 1988). This risk is increased if a large number of environmental variables are being tested and the time series are short. These risks highlight the importance of following up the results of these correlations as more data become available to verify the effect of the environment on the recruitment. This study, for example, updates the results of Pearce and Phillips (1988) and Joll and Caputi (1995) supporting the effect of the Leeuwin Current on the rock lobster and scallop recruitment. For the tiger prawn fishery in Exmouth Gulf, while the impact of cyclones can have positive or negative effects on recruitment, the form of this relationship suggested by Penn and Caputi (1986) does not seem appropriate.

If recruitment does decline as a result of environmental conditions then management needs to take this into account. The effect of high fishing pressure on these poor year-classes due to environmental effects can lead to very low spawning stocks which may in turn cause lower recruitment. The decline in puerulus settlement at Dongara in 1992 and 1993 (Fig. 9*a*) assumed to be due to the extended ENSO event experienced in the early 1990's, was a major factor behind the management measures introduced in 1992/1993 to reduce fishing effort.

These studies generally illustrate that studying the SRR of invertebrate species is an important component to the stock assessment of these fisheries and invaluable to the sustainable management of these fisheries. They emphasize the need for timely collection of spatial and temporal data on stock and recruitment indices, preferably from standardized research surveys, and highlight the importance of understanding the environmental effects on recruitment and the effect of fishing effort on the spawning stock. The adaptive management approach needs greater consideration in fisheries management because of the long-term benefits this approach provides in improved understanding of the SRR and hence improved management of fisheries.

Acknowledgments

Tidal data for Fremantle and Carnarvon were supplied by the National Tidal Facility, Flinders University of South Australia, copyright reserved. Alan Pearce (CSIRO Division of Oceanography) provided assistance in accessing these data. The authors thank research scientists of the Western Australian Marine Research Laboratories for critically reading the manuscript and offering many helpful suggestions and Wilf Lehre for figures. They also acknowledge the people with the foresight to set up the databases in the 1960's.

References

Beauchamp, J.J., and Olsen, J.S. 1973. Corrections for bias in regression estimates after logarithmic transformation. Ecology, **54**: 1403–1407.

Botsford, L.W., Moloney, C.L., Largier, J.L., and Hastings, A. 1998. Metapopulation dynamics of meroplanktonic invertebrates: the Dungeness crab (*Cancer magister*) as an example. *In* Proceedings of the North Pacific Symposium on Invertebrate Stock Assessment and Management. *Edited by* G.S. Jamieson and A. Campbell. Can. Spec. Publ. Fish. Aquat. Sci. 125. pp. 295–306.

Brown, R.S., Caputi, N., and Barker, E.H. 1995. A preliminary assessment of the effect of increases in fishing power on stock assessment and fishing effort of the western rock lobster (*Panulirus cygnus* George 1962) fishery in Western Australia. Crustaceana, **68**(2): 227–237.

Caddy, J.F. 1986. Modelling stock-recruitment processes in Crustaceana: some practical and theoretical perspectives. Can. J. Fish. Aquat. Sci. **43**: 2330–2344.

Caddy, J.F. 1989. Recent developments in research and management for wild stocks of bivalves and gastropods. *In* Marine invertebrate fisheries: their assessment and management. *Edited by* J.F. Caddy. John Wiley & Sons, New York, N.Y. pp. 665–700.

Caputi, N. 1989. Aspects of stock-recruitment relationships for crustaceans. Ph.D. thesis, Murdoch University, Western Australia.

Caputi, N., and Brown, R.S. 1993. The effect of environment on puerulus settlement of the western rock lobster (*Panulirus cygnus*) in Western Australia. Fish. Oceanogr. **2**: 1–10.

Caputi, N., Chubb, C.F., and Brown, R.S. 1995. Relationship between spawning stock, environment, recruitment and fishing effort for the western rock lobster, *Panulirus cygnus*, fishery in Western Australia. Crustaceana, **68**(2): 213–226.

Chubb, C.F., Barker, E.H., and Dibden, C.J. 1994. The Big Bank region of the limited entry fishery for the western rock lobster, *Panulirus cygnus*. Fish. Res. Rep. Fish. Dept. West. Aust. **101**: 1–20.

Fogarty, M.J. 1998. Implication of dispersal and larval interchange in American lobster stocks: a metapopulation model. *In* Proceedings of the North Pacific Symposium on Invertebrate Stock Assessment and Management. *Edited by* G.S. Jamieson and A. Campbell. Can. Spec. Publ. Fish. Aquat. Sci. 125. pp. 273–283.

Gulland, J.A. 1984. Introductory guidelines to shrimp management: some further thoughts, *In* Penaeid shrimps, their biology and management. *Edited by* J.A. Gulland and B.J. Rothschild. Fishing News Books, Surrey, England. pp. 290–299.

Hall, N.G., and Penn, J.W. 1979. Preliminary assessment of effective effort in a two species trawl fishery for penaeid prawns in Shark Bay, Western Australia. Rapp. P.-v. Reun. Cons. Int. Explor. Mer. **175**: 147–154.

Hancock, D.A. 1973. The relationship between stock and recruitment in exploited invertebrates. Rapp. P.-v. Reun. Cons. Int. Explor. Mer. **164**: 113–131.

Joll, L.M. 1994. Unusually large recruitment in the Shark Bay (*Amusium balloti*) fishery. Mem. Queensl. Mus. **36**: 261–267.

Joll, L.M., and Caputi, N. 1995. Environmental influences on recruitment in the saucer scallop (*Amusium balloti*) fishery of Shark Bay, Western Australia. ICES Mar. Sci. Symp. **199**: 47–53.

Lenanton, R.C., Joll, L.M., Penn, J.W., and Jones, K. 1991. The influence of the Leeuwin Current on coastal fisheries in Western Australia. *In* The Leeuwin Current: an influence on the coastal climate and marine life of Western Australia. *Edited by* A.F. Pearce and D.I. Walker. J. R. Soc. West. Aust. **74**: 101–114.

Morgan, G.R., Phillips, B.F., and Joll, L.M. 1982. Stock and recruitment relationships in *Panulirus cygnus*, the commercial rock (spiny) lobster of Western Australia. Fish. Bull. **80**(3): 475–486.

Orensanz, J.M. 1986. Size, environment, and density: the regulation of a scallop stock and its management implications. *In* North Pacific Workshop on Stock Assessment and Management of Invertebrates. *Edited by* G.S. Jamieson and N. Bourne. Can. Spec. Publ. Fish. Aquat. Sci. 92: 195–227.

Pearce, A.F., and Phillips, B.F. 1988. ENSO events, the Leeuwin Current, and larval recruitment of the western rock lobster. J. Cons. Int. Explor. Mer. **45**: 13–21.

Penn, J.W. 1984. The behaviour and catchability of some commercially exploited penaeids and their relationship to stock and recruitment. *In* Penaeid shrimps, their biology and management. *Edited by* J.A. Gulland and B.J. Rothschild. Fishing News Books, Surrey, England.

Penn, J.W. 1988. Spawning stock-recruitment relationships and management of the penaeid prawn fishery in Shark Bay, Western Australia. Ph.D. thesis, Murdoch University, Western Australia.

Penn, J.W., and Caputi, N. 1986. Spawning stock-recruitment relationships and environmental influences on the tiger prawn (*Penaeus esculentus*) fishery in Exmouth Gulf, Western Australia. Aust. J. Mar. Freshwater Res. **37**: 491–505.

Penn, J.W., Caputi, N., and Hall, N.G. 1995. Spawner-recruit relationships for the tiger prawn (*Penaeus esculentus*) stocks in Western Australia. ICES Mar. Sci. Symp. **199**: 320–333.

Phillips, B.F. 1972. A semi-quantitative collector of the puerulus larvae of the western rock lobster *Panulirus cygnus* George 1962 (Decapoda, Palinuridae). Crustaceana, **22**: 147–154.

Pollock, D.E. 1991. Population regulation and stock-recruitment relationships in some crayfish and lobster populations. *In* Crustacean Issues 7: crustacean egg production. *Edited by* A. Wenner and A. Kuris. Balkema, Rotterdam. pp. 247–266.

Vahl, O. 1982. Long term variations in recruitment of the Iceland scallop, *Chlamys islandica* (O.F. Muller), from Northern Norway. Neth. J. Sea Res. **16**: 80–87.

Walters, C.J. 1986. Adaptive management of renewable resources. MacMillan Publishing Company, New York, N.Y. 374 p.

Walters, C.J., and Collie, J.S. 1988. Is research on environmental factors useful to fisheries management? Can. J. Fish. Aquat. Sci. **45**: 1848–1854.

Walters, C.J., and Hilborn, R. 1976. Adaptive control of fishing systems. J. Fish. Res. Board Can. **33**: 145–159.

Walters, C.J., and Ludwig, D. 1981. Effects of measurement error on the assessment of stock-recruitment relationships. Can. J. Fish. Aquat. Sci. **38**: 701–710.

Young, P.C., and Martin, R.B. 1989. The scallop fisheries of Australia and their management. Rev. Aquat. Sci. **1**: 615–638.

Testing hypotheses on recruitment, growth, and mortality in exploited bivalves: an experimental perspective

Omar Defeo

Abstract: Experimental manipulation of fishing effort is shown to be a powerful approach in describing the effects of fishing on bivalve populations, and in identifying factors governing the dynamic features of exploited and co-occurring unexploited macrofauna. Here, I reassess results of eight years of monitoring population structure and dynamics along 22 km of exposed beach in Uruguay on two sympatric intertidal bivalves: the exploited yellow clam, *Mesodesma mactroides*, and the unexploited wedge clam, *Donax hanleyanus*. A fishery closure for 32 consecutive months was used to investigate the effects of fishing activities on both populations. Manipulation of the fishery led to important changes in the overall abundance of both harvested and unharvested bivalves, as well as in the processes regulating their population dynamics. Results showed that, (*i*) *M. mactroides* exhibited an overcompensatory stock-recruitment relationship, density-dependent growth, and a natural mortality rate of young-of-the-year directly correlated with fishing effort, the latter suggesting that harvesting activities (viz digging with shovels) might constitute a source of incidental mortality, and (*ii*) abundance of *D. hanleyanus* increased throughout the fishery closure. Spatial variation in wedge clam abundance was inversely correlated with the amount of fishing activity on *M. mactroides*; interspecific, rather than intraspecific, stock-recruitment relationship occurred, i.e., recruitment of *D. hanleyanus* was most correlated with the abundance of juveniles and adults of the yellow clam rather than with the parental abundance of wedge clams. This study shows that exposed sandy-beach bivalve populations may be influenced by more than physical factors, and that further research on these populations should include human influence and potential biotic interactions as potential structuring components.

Résumé : On montre que la manipulation expérimentale de l'effort de pêche est une approche puissante pour décrire l'effet de la pêche sur les populations de bivalves et pour identifier les facteurs qui régissent les caractéristiques dynamiques des macrofaunes exploitées et des macrofaunes cooccurrentes non exploitées. Je réévalue ici les résultats de huit ans de surveillance de la structure et de la dynamique de la population de deux bivalves sympatriques intertidaux, *Mesodesma mactroides*, qui est exploité, et *Donax hanleyanus*, qui n'est pas exploité. Les observations ont eu lieu sur une étendue de 22 km de plage exposée en Uruguay. Une interdiction de la pêche pendant 32 mois consécutifs a été employée pour étudier les effets des activités de pêche sur les deux populations. La manipulation de la pêche a amené des changements importants de l'abondance globale tant des bivalves exploités que des bivalves non exploités, ainsi que dans les processus régulant la dynamique de leur population. Les résultats montrent que, (*i*) *M. mactroides* présentait une relation stock-recrutement de surcompensation, une croissance dépendante de la densité et un taux de mortalité naturelle des jeunes de l'année en corrélation directe avec l'effort de pêche, ce dernier point laisse entendre que les activités de récolte (creusage à la pelle) pourraient constituer un facteur de mortalité accidentelle, et (*ii*) l'abondance de *D. hanleyanus* a augmenté pendant toute la durée de l'interdiction de la pêche. La variation spatiale de l'abondance de *D. hanleyanus* était inversement proportionnelle au degré d'activité de pêche dirigée contre *M. mactroides*; une relation stock-recrutement interspécifique, plutôt qu'intraspécifique, a été observée, c.-à-d. que le recrutement de *D. hanleyanus* était davantage corrélé avec l'abondance des juvéniles et des adultes de *M. mactroides* qu'avec l'abondance parentale de *D. hanleyanus*. Cette étude montre que les populations de bivalves sur les plages sablonneuses exposées peuvent subir l'influence de facteurs autres que physiques et que les recherches futures sur ces populations devraient prendre en compte l'influence humaine et les interactions biotiques potentielles comme éléments de structuration potentiels. [Traduit par la Rédaction]

O. Defeo.[1] Centro de Investigación y de Estudios Avanzados del IPN. A.P. 73 Cordemex, 97310 Mérida, Yucatán, México.

Correct citation: Defeo, O. 1998. Testing hypotheses on recruitment, growth, and mortality in exploited bivalves: an experimental perspective. *In* Proceedings of the North Pacific Symposium on Invertebrate Stock Assessment and Management. *Edited by* G.S. Jamieson and A. Campbell. Can. Spec. Publ. Fish. Aquat. Sci. 125. pp. 257–264.

[1] Address for correspondence: Instituto Nacional de Pesca. Constituyente 1497, C.C. 1612, 11200 Montevideo, Uruguay.

Introduction

The published literature suggests that improved understanding of invertebrate population dynamics comes from experimental manipulation of populations in the field (Jamieson and Caddy 1986; Cobb and Caddy 1989). Experimental management can constitute a useful way to determine patterns of resource response in relation to magnitude of fishing pressure and of linkages and strengths of ecological interactions. This approach is particularly useful in coastal invertebrate fisheries in developing countries, where the overall increase in fishing activity has

Fig. 1. (*a*) Approximated geographic distribution of *M. mactroides* and *D. hanleyanus* in South America, (*b*) study area in the eastern coast of Uruguay, and (*c*) location of the four grounds along the 22 km sampled between the Chuy Stream and the Andreoni Canal. The localities of Barra del Chuy and La Coronilla are also shown.

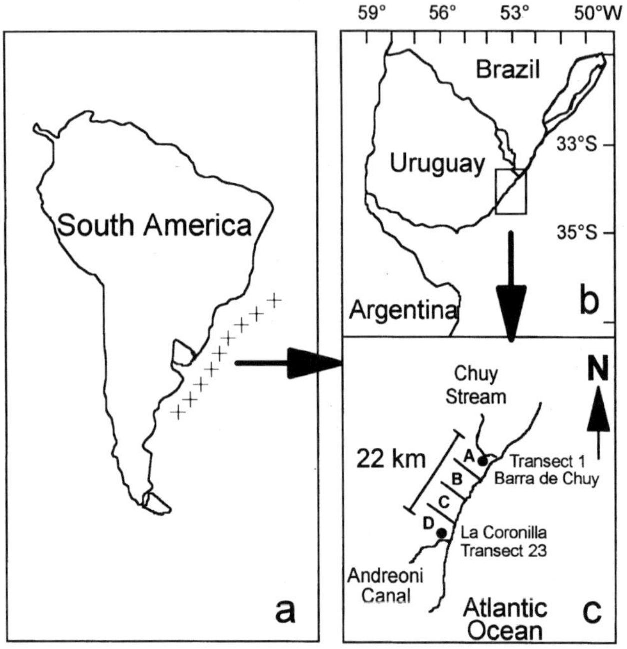

not been accompanied by a corresponding increase in scientific and fishery data (Defeo et al. 1993). The relative absence of demographic and fisheries studies may lead to inadequate fishery management advice.

In the specific case of populations inhabiting exposed sandy beaches, spatio-temporal patterns of distribution, dynamics, and abundance has been assumed to be fully explained by wave climate, sand particle size, and tide range (McLachlan 1983; Jaramillo and McLachlan 1993; McLachlan et al. 1993). However, there has been no quantitative evaluation of the effect of human harvest on the macrofauna of these systems, and information on biological interactions is lacking. The former topic is of particular scientific interest, as varying harvest intensity on species of different trophic levels could produce different community structures, as demonstrated for rocky shores (see e.g., Castilla and Durán 1985; Oliva and Castilla 1986; Durán and Castilla 1989; Bustamante and Castilla 1990; Castilla 1993).

An example of a demographic study of exploited sandy beach invertebrates is that on the yellow clam *Mesodesma mactroides* (Bivalvia: Mesodesmatidae) and the wedge clam *Donax hanleyanus* (Bivalvia: Donacidae) on the Atlantic coast of Uruguay (Defeo 1993; Defeo and de Alava 1995). Both species inhabit 22 km of uninterrupted sand beach from Barra del Chuy to La Coronilla, Uruguay. The yellow clam is artisanally harvested by fishermen with shovels. Although *Donax hanleyanus* is not commercially exploited, it occupies the same intertidal fringe as *M. mactroides*, and so may be also incidentally affected by harvesting activities.

Yellow clam catches have varied from 62 t in 1981 to a peak of 219 t in 1985, after which catch and catch per unit of effort (CPUE) decreased rapidly (Defeo 1989). The fishery was closed for 32 months, from April 1987 to November 1989, along the entire beach (see Defeo 1993).

In the present paper I review published and unpublished data of a long-term study on population structure and abundance of the above bivalves, in the context of fishing activity (catch and effort data) on *M. mactroides*. I evaluate the role of human harvest in determining spatial and temporal variations in abundance of harvested and nonharvested bivalve populations, and quantify possible effects of human harvest on yellow clam population dynamics.

Material and methods

The study was conducted between 1983 and 1990 on 22 km of uninterrupted sand beach along the east coast of Uruguay, between Barra del Chuy (33°40′S, 53°29′W) and La Coronilla (33°50′S, 53°27′W) (Fig. 1). The beach/surf community is biomass dominated by the large yellow clam (Defeo et al. 1992*a*).

Clam populations were sampled monthly from March 1983 through March 1985, and seasonally (at least four times a year) from 1985 to 1990, according to a systematic design developed to quantify the stock of *M. mactroides*. Transects perpendicular to the shoreline were every kilometre from Barra del Chuy (transect 1) to Andreoni Canal (transect 23). Sampling units (SU's) were every 4 m, from the base of the sand dunes to the tide level where two successive SU's without clams were recorded. Sediment at each SU was collected with a cylinder 28.2 cm in diameter and 40 cm deep; some 200 sites were sampled each date. Sediment from each sample was sieved through 0.5 mm mesh, and all clams of both species retained were measured (maximum valve length) and counted. More than 150 000 specimens of both species combined were measured over the study.

Clams of each species were classified into 3 size groups: recruits, juveniles, and adults. Recruits are defined as <10 mm length in the case of the yellow clam (Defeo 1993) and <5 mm for wedge clam (de Alava 1993). Minimum lengths for adults were 43 and 15 mm for yellow and wedge clams, respectively (Masello and Defeo 1986; Penchaszadeh and Olivier 1975). Remaining individuals were classified as "juveniles." Mean density for each size class was calculated by transect, fishing ground (see below), and as an overall average for each sample period. A stock-recruitment relationship (SRR) was determined for each species using mean density estimates of parent stock from December to March and recruits from February to May (Defeo 1993; Defeo and de Alava 1995). In the case of delayed recruitment of yellow clam in 1989, the peak value (August) was used instead. Additional data for 1982 was obtained from Defeo (unpublished). Wedge clam recruitment by fishing ground was not recorded for 1988 and 1989, when only an entire beach estimate was available.

Length–frequency distributions of each of the eight yellow clam cohorts through their first year of life (equivalent to year-classes in this study) from 1983 to 1990 were analyzed. Mean length per transect (L_r) of young-of-the-year was calculated by weighting each mean length (l) per quadrat i pertaining to transect r (l_{ir}) by the respective observed density (ind·m^{-2}: d_{ir}):

$$L_r = \frac{\sum_i l_{ir}\, d_{ir}}{\sum_i d_{ir}}$$

Monthly progressions of mean length were calculated for each cohort. Finite average daily size increments (in mm·day^{-1}) were calculated from differences in mean length between consecutive sampling dates for which the number of clams in both samples was greater than 50. Seven years of data, 1983–1989, were analyzed; the 1990 cohort was excluded because it was not possible to follow it over a complete year cycle.

Natural mortality rate (M) was estimated for the unexploited portion of the yellow clam population <50 mm length (minimum size commercially exploited) when the fishery was open (1983–1987), and for the entire population when the fishery was closed. Estimation of M was done by regressing the natural logarithm of the number of individuals per strip transect (*IST*) against mean estimated age. Weighted mean lengths were converted to age utilizing von Bertalanffy growth model parameters estimated by Defeo (1990) and Defeo et al. (1992*b*). *IST* is expressed as ind·m^{-1} of beach length and is employed in order to avoid biased results due to changing beach profile during rough and calm sea conditions.

Monthly catch data records of the yellow clam are available from 1982. Fishery statistics were divided into four consecutive fishing grounds (see Defeo et al. 1991 for details), with boundaries set at right angles to the coastline: A, transects 1–5; B, transects 6–10; C, transects 11–15; and D, transects 16–22 (Fig. 1). Daily records of fishing effort and catch per fisherman were determined by fishing ground from 1984 to 1987 and from 1990 onwards. Neither clams nor fishing activities were ever observed in transect 23 (Andreoni Canal).

Results

Temporal fluctuations in stock and recruitment, and the SRR

Recruitment of yellow clam *M. mactroides* showed a consistent seasonal pattern (Fig. 2*a*), peaking sharply in summer (especially in February–March) and autumn (mainly in April and May), decreasing during winter, and near to zero during spring. This pattern was observed in seven of the eight years, differing only for 1989, when recruitment occurred only in early winter. Lowest recruitment occurred in 1989, during the harvest closure, and in the autumn of 1990, only four months after fishery reopening. Wedge clam *D. hanleyanus* recruits showed a strong density peak during the summer and autumn (February–May), notably in 1983, 1987, and 1990 (Fig. 2*a*). The four-year period before the closure of the yellow clam fishery had low recruitment, never above 50 ind·m^{-2}, whereas highest recruitment was in the summer of 1990, just after the yellow clam fishery reopened.

Contrasting with recruitment variations, the adult component of the yellow clam population showed two clear density peaks: in 1984 and early 1985, and the highest in 1989, during the third year of fishery closure (Fig. 2*b*). Population recovery following fishing perturbation, i.e., restocking of depleted areas, was observed from spring 1988 on. Adult density in 1989

Fig. 2. Seasonal fluctuations in density of: (*a*) recruits and (*b*) adults of *M. mactroides* (solid line) and *D. hanleyanus* (dashed line) in Uruguay. The closed fishery for *M. mactroides* (April 1987 – November 1989) is indicated.

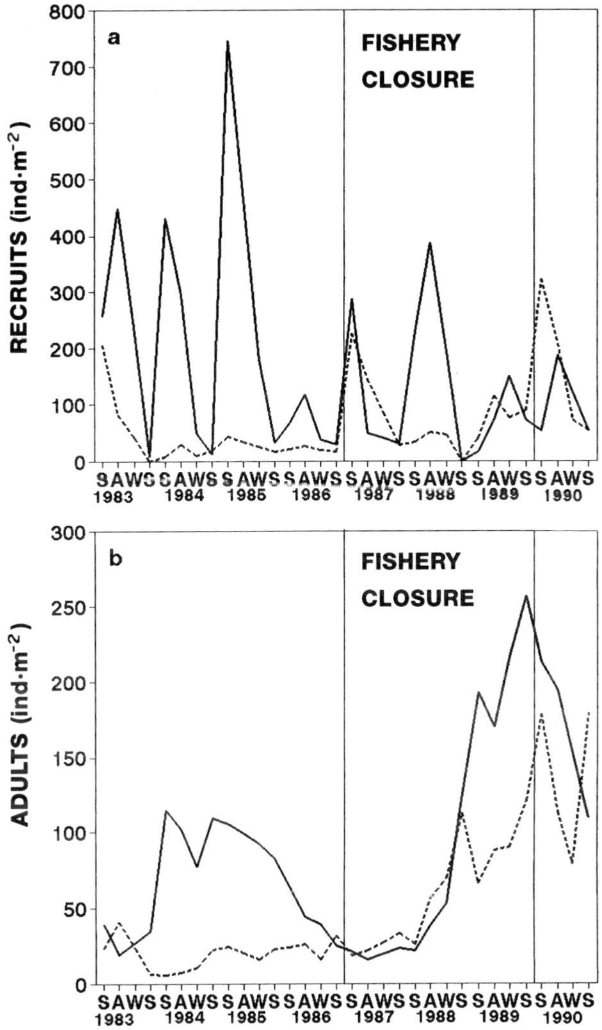

was nearly twice the average level during preclosure years. Average density of adults per sampling period was lowest in 1987, just before the fishery was closed. Seasonal density of wedge clam adults (Fig. 2*b*) was consistently below 50 ind·m^{-2} from the beginning of the study to spring, 1988, when it rose to 112 ind·m^{-2}. From that date, density generally increased.

The yellow clam SRR showed a significant overcompensatory relationship between spawning stock and recruitment density, with an inhibition of recruitment at high adult densities during the fishery closure. Maximum recruitment occurred with moderately low to medium sizes of spawning stock. Variance in recruitment was highest at intermediate levels of spawning stock (1984 and 1985; Fig. 3*a*). The SRR was best described by a Ricker model (Fig. 3*a*) of the form (nonlinear estimate, $r^2 = 0.49$):

$$R_{Mm} = 16.91\, A_{Mm}\, e^{(-0.012 A_{Mm})} + \varepsilon$$

where R is the density of recruits, A is the density of the spawning stock and ε is the error term. Recruitment of wedge clam

Fig. 3. Recruitment curves in (*a*) *M. mactroides*, generated by the minimum and maximum 95% confidence density bounds of recruits (vertical lines) and adults (horizontal lines), and (*b*) *D. hanleyanus*, given by the relationship between recruits of the wedge clam and the density of juveniles plus adults of *M. mactroides*.

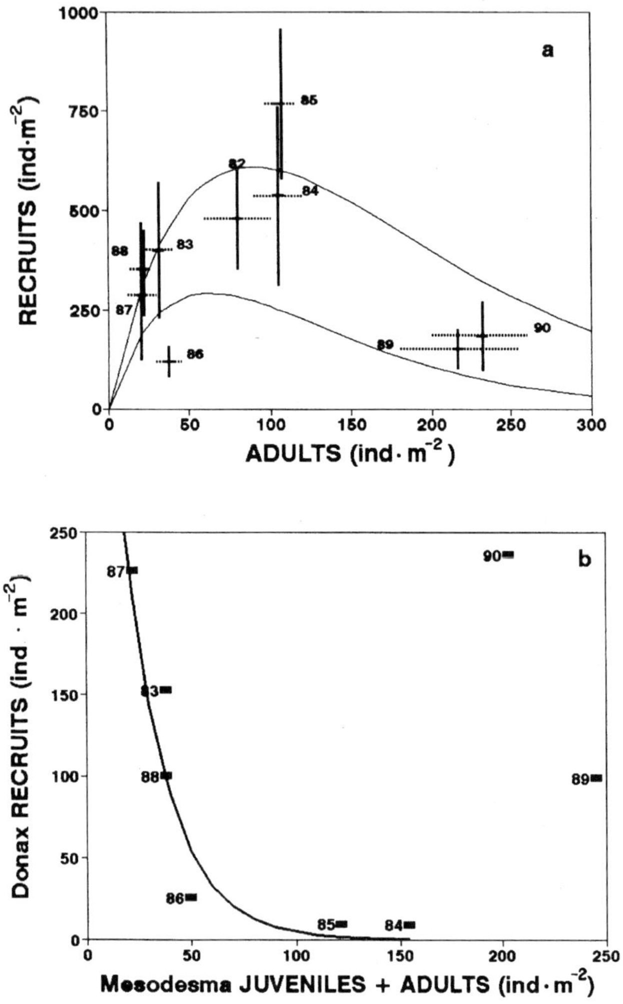

Fig. 4. (*a*) Monthly progression of mean lengths for young-of-the-year *M. mactroides*. Three selected cohorts denote variations in growth rate under different scenarios of adult density: 1985, highest adult densities during the exploited period; 1988, low adult densities during the second year of the fishery closure; and 1989, highest adult densities during the third year of fishery closure; (*b*) variability in estimated annual growth rate with respect to average adult density. Homogeneous groups according to Wilcoxon test results are circled.

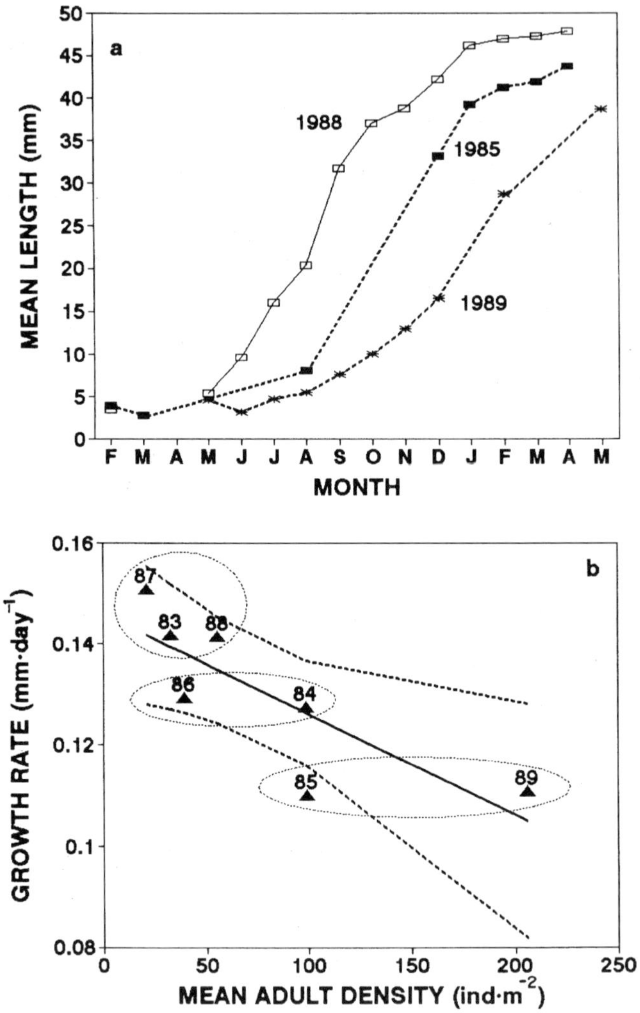

D. hanleyanus showed no evident relationship with the size of the parental stock. However, it could be explained by a monotonically decreasing exponential function of the overall density of yellow clam (size groups pooled) (Fig. 3*b*), suggesting an "interspecific stock-recruitment relationship" (sensu Pauly 1980) of the form ($r^2 = 0.81$; $p < 0.01$):

$$R_{Dh} = 696\ e^{-0.049\,AJ_{Mm}}$$

where R_{Dh} is the density of recruits of wedge clam, and AJ_{Mm} is the density of juveniles and adults of yellow clam. This relationship, however, was evident only from 1983 to 1988. In 1989 and 1990 (excluded from the above fit), highest densities of both species occurred, presumably as a result of the yellow clam fishery closure (de Alava 1993). This, in turn, suggests that temporal variation in wedge clam recruitment could be influenced by the amount of fishing on yellow clam (see "Stock, recruitment, and the fishery," below).

Growth

Growth of the yellow clam was markedly seasonal (Fig. 4*a*). Mean length of ca. 45–50 mm was reached by clams during the first year of life, accounting for nearly 70% of the maximum observed length (76 mm). Most cohorts were recruited between February and May, and mean population length did not vary greatly until September–November, a combined result of slow growth and protracted settlement.

Systematic variation in monthly length increment indicated faster growth in some years than in others. Monthly progression of mean length and length increment per day showed that cohorts 1983, 1987, and 1988 grew fastest, while slowest

Fig. 5. Estimation of M for young-of-the-year *M. mactroides* under both conditions of heavy fishing (1985) and during the second year of the fishery closure (1988).

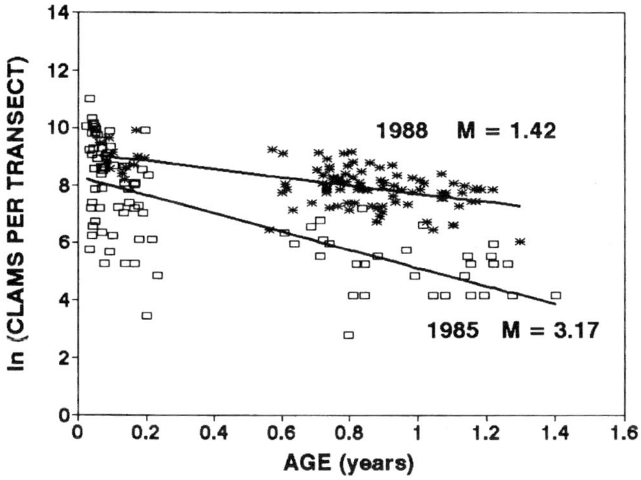

growing cohorts were in 1985, when recruitment was highest, and in 1989 (Fig. 4*a*), at highest adult density. Between-cohort differences in growth can be explained by density-dependence processes, as growth rate of young-of-the-year yellow clams were inversely correlated with average adult density per sample period ($r = -0.81$; $p < 0.027$). Between-cohort differences in growth rate were highly significant (Friedman test X_r^2 = 21.25; $p < 0.003$). A Wilcoxon matched pair test showed that cohorts could be categorized in three groups, (*i*) cohorts 1983, 1987, 1988, with greatest monthly mean lengths, (*ii*) cohorts 1986 and 1984, of intermediate lengths, and (*iii*) cohorts 1985 and 1989, with smallest lengths (Fig. 4*b*).

Natural mortality

Variability in M for yellow clam was high for the eight cohorts studied, with a minimum and a maximum (mean ± SE) of $0.90 \pm 0.50 \cdot yr^{-1}$ and $3.17 \pm 0.34 \cdot yr^{-1}$ for the 1990 and 1985 cohorts, respectively. M was low in 1988, the second year of the fishery closure (Fig. 5), which, in conjunction with fast growth, led to the strong recovery of the resource.

Heterogeneous spatial distribution of fishing effort was a consistent feature in exploitation of yellow clam (Defeo et al. 1991). Highest effort was exerted in the central fishing grounds, B and C (Fig. 6*a*). Manual harvest with shovels causes an incidental damage (broken shells) to the unexploited size classes of yellow clams (i.e., <50 mm length) and physical disturbance of the sediment (Defeo 1993). Indirect fishing mortality of young-of-the-year should be greatest on the central fishing grounds and during years of heavy fishing. Spatial analysis of mortality for young-of-the-year confirmed this hypothesis (Fig. 6*b*). Thus, M values per fishing zone showed a high, positive correlation ($r = 0.74$; $p < 0.002$) with fishing effort (Fig. 6*b*) and with catch ($r = 0.71$; $p < 0.005$) (see Defeo 1993).

Stock, recruitment, and the fishery

Adults and recruits were most abundant in the central zones, but not in zone D (Fig. 7*a*). Spatial dynamics of fishing is

Fig. 6. (*a*) Temporal fluctuations (1984–1990) in fishing effort exerted on yellow clam by fishing ground, and (*b*) bivariate correlation between M for yellow clam young-of-the-year and fishing effort.

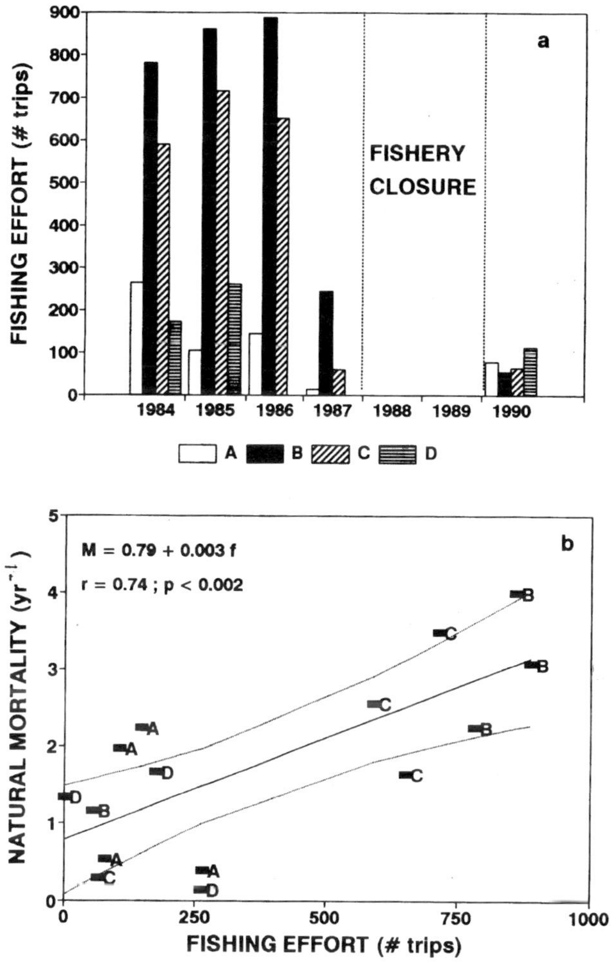

explained by spatial variations in yellow clam abundance (Fig. 7*b*).

With an interspecific SRR for wedge clam only evident from 1983 to 1988, temporal and spatial variation in wedge clam recruitment can be explained by the amount of fishing effort exerted on yellow clam (Fig. 8). A positive correlation between recruitment and catch ($r = 0.70$; $p < 0.05$: Fig. 8*a*) of *M. mactroides* contrasted with the negative correlation ($r = -0.85$; $p < 0.02$) between the annual transformed mean density of wedge clam recruits over the entire beach and yellow clam fishing effort (Fig. 8*b*).

The longshore distribution of wedge clam recruits on grounds A–C can be explained by the yellow clam catch. Fishing effort exerted in ground D was low and a less reliable predictor of wedge clam recruitment abundance (Table 1; see also Defeo and de Alava 1995).

Discussion

Long-term study of sympatric harvested and nonharvested clam populations has shown fishing can influence the structure of invertebrate communities beyond the obvious effects of

Fig. 7. Spatial pattern in: (*a*) mean density ± SE of yellow clam adults and recruits for 1983–1990, and (*b*) mean catch and effort ± SE by fishing ground for 1984–1990.

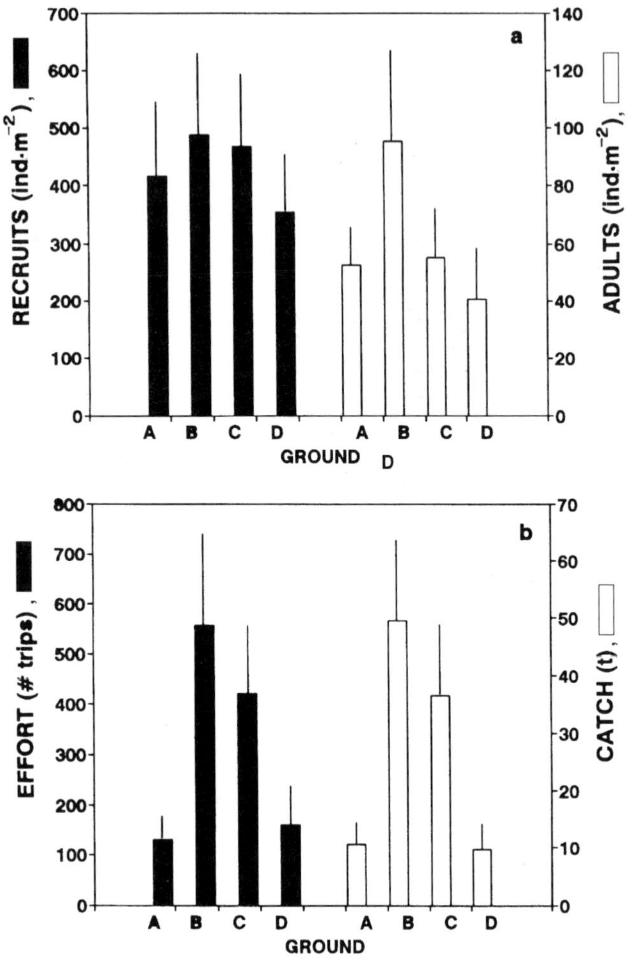

Fig. 8. Correlation (± 95% confidence intervals) between (*a*) yellow clam recruitment and catches, and (*b*) wedge clam recruitment and yellow clam fishing effort.

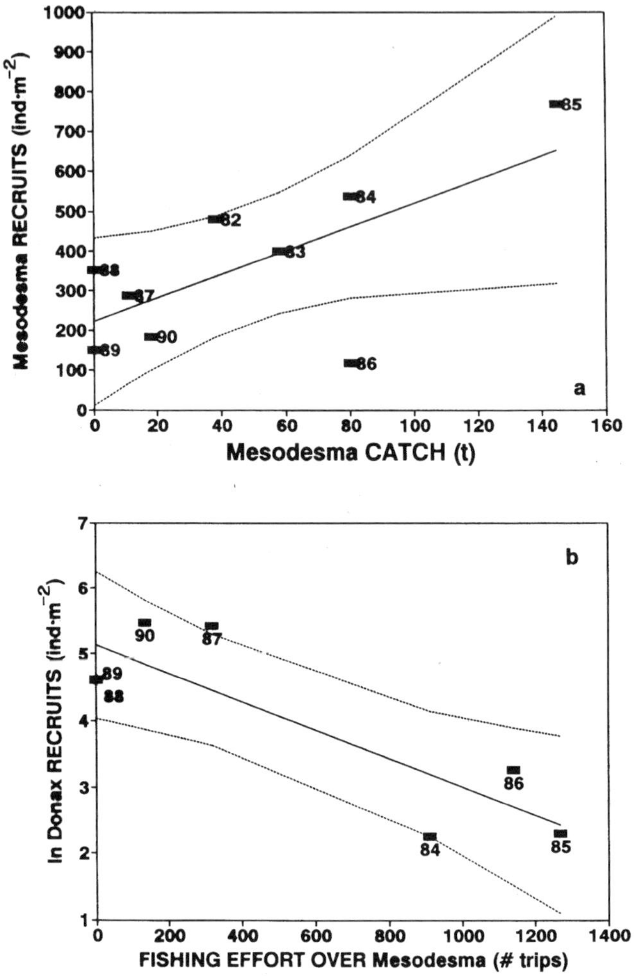

exploitation of the targeted species. The experimental manipulation of fishing effort showed the ecological implications of humans as top predators in the system, and also as a source of physical disturbance associated with harvesting.

Yellow clams display temporal stock-dependent recruitment (sensu Hilborn and Walters 1992), the first time that an overcompensatory relationship has been shown for an exposed sandy beach population (see also Hancock (1973) for an example of the cockle *Cardium (Cerastoderma) edule* in a sheltered environment). The wide range of fishing effort levels created by the yellow clam fishery closure provided a relatively large range of spawning stock sizes, allowing SR parameter estimation with a short time data series. However, the relationship observed for the yellow clam was a "one-way trip" (sensu Hilborn and Walters 1992): highest adult density and poorest recruitment were observed at the end of the study period. A longer time series of data is needed to corroborate observed trends, particularly for the nonharvested wedge clam.

A potential mechanism of overcompensation could be passive filtering of settling larvae by adults, a major process explaining settlement inhibition by established suspension feeders (Woodin 1976; Peterson 1982; Crowe et al. 1987; André and Rosenberg 1991; André et al. 1993). According to analyses of the SRR, the resident adult density (or biomass) that exceeds the limiting carrying capacity appears to be about 100 adults·m^{-2} (Defeo 1993; see Orensanz 1986 for a useful example on carrying capacity features of bivalves). Overcompensation inhibited recruitment, creating a relatively low standing stock a year later. Thus, yellow clam appear to be a quasi-pulse age-class dominated population (Defeo et al. 1992*b*) in which (*i*) older individuals can suppress subsequent cohorts, and (*ii*) the amount of density dependence varies over time (Hughes 1990; Olafsson 1989). Both intraspecific mechanisms are influenced by the amount of fishing which regulates the adult component of the population.

Maximum abundances of recruit and adult yellow clam *M. mactroides* during the exploited period coincided with lowest wedge clam population abundance, thus suggesting a potential ecological interdependence between these two sympatric suspension-feeding bivalves. However, the negative relationship between wedge clam recruits and "adults + juveniles" of yellow clam during the exploited period and the first two years of the fishery closure did not explain the successful wedge clam recruitment in 1989 and 1990, when density of yellow clam was high. Increasing density of the two species during the fishery closure would imply, (*a*) a difference in the

Table 1. Parameter estimates and associated statistics for the linear regression relationship between recruits (ind·m^{-2}) of *D. hanleyanus* (dependent variable) and fishing effort (number of trips) exerted over *M. mactroides* (independent variable) on each fishing ground at settlement (February–May).

	Regression						
Ground	*a*	S.E.	*p*-level	*b*	S.E.	*r*	*p*-level
A	311	64	0.020	–2.34	0.70	–0.89	0.04
B	282	43	0.007	–0.47	0.10	–0.94	0.02
C	130	15	0.003	–0.28	0.05	–0.96	0.01
D	39	34	0.330	–0.09	0.27	–0.19	0.76

Note: S.E., standard error; *r*, correlation coefficient. *p*-levels are for the intercept (*a*) and *r*.

impacts of fishing on the two clams, or (*b*) lack of interaction among them. The former seems the best explanation, as temporal variations in wedge clam recruitment were best explained by the amount of fishing effort exerted on yellow clam. During the exploited period (1983–1987), wedge clam density peaked concurrently with the lowest yellow clam catches and fishing effort, whereas highest densities of wedge clam occurred consistently throughout and immediately after the fishery closure (April 1987 – December 1989).

Newly settled yellow clam cohorts were reduced not only by an overcompensatory mechanism in the SRR but also by density-dependent growth rate. Growth was highest for the young-of-the-year during the first and second year of fishery closure (lowest adult density), and lowest during 1989 (third year into the closed period: highest adult density). Established adults could be depleting suspended foods and depressing individual growth rate, widely documented for bivalves in estuaries, lagoons, and sheltered coastal waters (Peterson 1982; Fréchette and Bourget 1985; Orensanz 1986; Fréchette et al. 1989; Peterson and Black 1987, 1991) but not previously suggested for an exposed sandy beach environment.

Depression of yellow clam growth at high densities also had a density-dependent effect on recruitment produced by the 1985 and 1989 cohorts (see Hancock 1973 for another bivalve example). Fast juvenile growth allows clams to reach approximately 43 mm after one year (Defeo et al. 1992*b*), the size at first maturity (Masello and Defeo 1986). Cohorts with lower growth rate would not reach the same sexual maturity as one-year-olds, reducing population fecundity (assuming that it is a linear function of body size) and recruitment (e.g., 1986 cohort). Relaxation of overcompensation at low adult densities could let clams grow faster and reach sexual maturity prior to the normal spawning season.

Natural mortality for young-of-the-year yellow clam was highest for the abundant 1985 cohort, which had the lowest growth rate and experienced the most fishing activity (see Defeo 1993). The close correlation between *M* for unexploited size-classes and catch or effort per fishing ground suggests that incidental damage (broken shells) and physical stress produced by sediment disturbance during harvesting could be a source of mortality. This anthropogenic disturbance can lead to changes in substrate penetrability (Probert 1984; Peterson and Black 1988; Wynberg and Branch 1992, 1994), which can restrict the movement of burrowing organisms (see Brenchley 1982) and increase *M* (Peterson 1985).

The spatial relationship between the yellow clam resource and fishing activity showed that fishing caused both disturbance and local reduction of clam densities, thereby permitting recolonization of disturbed patches to occur. Highest densities of recruits and adults were always observed concurrent with heavy fishing, suggesting a recurrent pattern of recolonization of suitable areas for settlement (see Caddy 1989; Defeo 1993). The opposite effect was observed in the nonharvested *D. hanleyanus*. Variation in density in its longshore distribution was negatively correlated with the spatial variability of yellow clam fishing (see Defeo and de Alava 1995). Correlation between abundance of both clams and the intensity of fishing activities suggests potential effects of harvesting and disruption of sediment structure affecting long-term patterns in sandy beach populations.

In summary, changes in fishing effort through time has allowed the observation of some population patterns and processes that until the closure were unknown. Spatio-temporal variations in the intensity of harvesting yellow clam affected spatial and temporal patterns of population abundance of both targeted and nontargeted species, and yellow clam growth and mortality. In the absence of harvesting, disturbances produced by the harvesting process and related sediment modifications were attenuated, *M* was diminished, and a strong recovery of both species was promoted. Thus, designation of exposed sandy beach populations as only physically structured may be inappropriate. Further research in sandy beach populations and communities needs to evaluate not only the role of environmental factors as an organizational force, but also the role of human activities and associated disturbance, together with density-dependent processes, as important factors affecting long-term trends.

Acknowledgments

This paper condenses the results of my doctoral dissertation. I am grateful to L. Orensanz, J. Caddy, and G. Jamieson for critical reading of and valuable suggestions to the paper. I thank M.L. Arín, A. Brazeiro, A. de Alava, C. Layerle, A. Masello, and G. Riestra for field and laboratory assistance. Financial support was partially provided by CONICYT (Project No. 1018).

References

André, C., and Rosenberg, R. 1991. Adult-larval interactions in the suspension-feeding bivalves *Cerastoderma edule* and *Mya arenaria*. Mar. Ecol. Prog. Ser. **71**: 227–234.

André, C., Jonsson, P.R., and Lindegarth, M. 1993. Predation on settling bivalve larvae by benthic suspension feeders: the role of hydrodynamics and larval behaviour. Mar. Ecol. Prog. Ser. **97**: 183–192.

Brenchley, G.A. 1982. Mechanisms of spatial competition in marine soft-bottom communities. J. Exp. Mar. Biol. Ecol. **60**: 17–33.

Bustamante, R., and Castilla, J.C. 1990. Impact of human explotation on populations of the intertidal southern bull-kelp *Durvillaea antarctica* (Phaeophyta, Durvilleales) in Central Chile. Biol. Conserv. **52**: 205–220.

Caddy, J.F. 1989. Recent developments in research and management for wild stocks of bivalves and gasteropods. p. 665-700. *In* Marine invertebrate fisheries: their assessment and management. *Edited by* J.F. Caddy. John Wiley & Sons, New York, N.Y.

Castilla, J.C. 1993. Humans: capstone strong actors in the past and present coastal ecological play. *In* Humans as components of ecosystems: the ecology of subtle human effects and populated areas. *Edited by* M.J. McDonnell and S.T.A. Pickett. Springer-Verlag, New York, N.Y. pp. 158–162.

Castilla, J.C., and Durán, R. 1985. Human exclusion from the rocky intertidal zone of Central Chile: the effects on *Concholepas concholepas* (Mollusca: Gastropoda: Muricidae). Oikos, **45**: 391–399.

Cobb, J.S., and Caddy, J.F. 1989. The population biology of decapods. *In* Marine invertebrate fisheries: their assessment and management. *Edited by* J.F. Caddy. John Wiley & Sons, New York. pp. 327–374.

Crowe, W.A., Josefson, A.B., and Svane, I. 1987. Influence of adult density on recruitment into soft sediments: a short-term *in situ* sublittoral experiment. Mar. Ecol. Prog. Ser. **41**: 61–69.

de Alava, A. 1993. Interdependencias ecológicas entre dos bivalvos simpátricos en una playa arenosa de la costa atlántica uruguaya. M.Sc. thesis, CINVESTAV Mérida, Yucatán, México.

Defeo, O. 1989. Development and management of artisanal fishery for yellow clam *Mesodesma mactroides* in Uruguay. Fishbyte, **7**: 21–25.

Defeo, O. 1990. La pesquería de la almeja amarilla *Mesodesma mactroides* (Deshayes, 1854) en playas uruguayas. CINVESTAV Mérida, Yucatán, México.

Defeo, O. 1993. The effect of spatial scales in population dynamics and modelling of sedentary fisheries: the yellow clam *Mesodesma mactroides* of an Uruguayan exposed sandy beach. Ph.D. thesis. CINVESTAV Mérida, Yucatán, México.

Defeo, O., and de Alava, A. 1995. Effects of human activities on long-term trends in sandy beach populations: the wedge clam *Donax hanleyanus* in Uruguay. Mar. Ecol. Prog. Ser. **123**: 73–82.

Defeo, O., Seijo, J.C., Euán, J., and Liceaga, M. 1991. Dinámica espacial del esfuerzo pesquero en una pesquería artesanal de la costa atlántica uruguaya. Invest. Pesq. (Chile), **36**: 17–25.

Defeo, O., Jaramillo, E., and Lyonnet., A. 1992*a*. Community structure and intertidal zonation of the macroinfauna in the Atlantic coast of Uruguay. J. Coastal Res. **8**: 830–839.

Defeo, O., Ortiz, E., and Castilla, J.C. 1992*b*. Growth, mortality and recruitment of the yellow clam *Mesodesma mactroides* in Uruguayan beaches. Mar. Biol. (N.Y.), **114**: 429–437.

Defeo, O., de Alava, A., Valdivieso, V., and Castilla, J.C. 1993. Historical landings and management options for the genus *Mesodesma* in coasts of South America. Biol. Pesq. (Chile), **22**: 41–54.

Durán, L.R., and Castilla, J.C. 1989. Variation and persistence of the middle rocky intertidal community of Central Chile, with and without human harvesting. Mar. Biol. (N.Y.), **103**: 555–562.

Fréchette, M., and Bourget, E. 1985. Food-limited growth of *Mytilus edulis* in relation to the benthic boundary layer. Can. J. Fish. Aquat. Sci. **42**: 1166–1170.

Fréchette, M., Butman, C.A., and Geyer, W.R. 1989. The importance of boundary-layer flows in supplying phytoplankton to the benthic suspension feeder, *Mytilus edulis* L. Limol. Oceanogr. **34**: 19–36.

Hancock, D.A. 1973. The relationship between stock and recruitment in exploited invertebrates. Rapp. P.-V. Reun. Cons. Int. Explor. Mer. **164**: 113–131.

Hilborn, R., and Walters, C.J. 1992. Quantitative fisheries stock assessment. Choice, dynamics and uncertainty. Chapman & Hall, New York, N.Y. 570 p.

Hughes, T.P. 1990. Recruitment limitation, mortality, and population regulation in open systems: a case study. Ecology, **71**: 12–20.

Jamieson, G.S., and Caddy, J.F. 1986. Research advice and its application to management of invertebrate resources: an overview. *In* North Pacific Workshop on Stock Assessment and Management of Invertebrates. *Edited by* G.S. Jamieson and N. Bourne. Can. Spec. Publ. Fish. Aquat. Sci. 92. pp. 117–123.

Jaramillo, E., and McLachlan, A. 1993. Community and population responses of the macroinfauna to physical factors over a range of exposed sandy beaches in South-central Chile. Estuarine Coastal Shelf Sci. **37**: 615–624.

Masello, A., and O. Defeo. 1986. Determinación de la longitud de primera madurez sexual en *Mesodesma mactroides* (Deshayes, 1854). Comun. Soc. Malacol. Urug. **6**(51): 387–392.

McLachlan, A. 1983. Sandy beach ecology — a review. *In* Sandy beaches as ecosystems. *Edited by* A. McLachlan and T. Erasmus. Dr. W. Junk Publishers, The Hague, The Netherlands. pp. 321–380.

McLachlan, A., Jaramillo, E. Donn, T., and Wessels, F. 1993. Sandy beach macrofauna communities and their control by the physical environment: a geographical comparison. J. Coastal Res. **15**: 27–38.

Olafsson, E.B. 1989. Contrasting influences of suspension-feeding and deposit-feeding populations of *Macoma balthica* on infaunal recruitment. Mar. Ecol. Prog. Ser. **55**: 171–179.

Oliva, D., and Castilla, J.C. 1986. The effect of human exclusion on the population structure of key-hole limpets *Fissurella crassa* and *F. limbata* on the coast of Central Chile. P.S.Z.N.I: Mar. Ecol. **7**: 201–217.

Orensanz, J.M. 1986. Size, environment, and density: regulation of a scallop stock and its management implications. *In* North Pacific Workshop on Stock Assessment and Management of Invertebrates. *Edited by* G.S. Jamieson and N. Bourne Can. Spec. Publ. Fish. Aquat. Sci. 92. pp. 195–227.

Pauly, D. 1980. A new methodology for rapidly acquiring basic information on tropical fish stocks: growth, mortality and stock-recruitment relationship. *In* Proceedings of International Workshop on Stock Assessment for Tropical Small-scale Fisheries. *Edited by* S.B. Saila and P.M. Roedel. University of Rhode Island, Kingston, R.I. pp. 154–172.

Penchaszadeh, P., and Olivier, S. 1975. Ecología de una población de "berberecho" (*Donax hanleyanus*) en Villa Gesell, Argentina. Malacologia, **15**: 133–146.

Peterson, C.H. 1982. The importance of predation and intraspecific competition in the population biology of two infaunal suspension-feeding bivalves, *Protothaca staminea* and *Chione undatella.* Ecol. Monogr. **52**: 437–475.

Peterson, C.H. 1985. Patterns of lagoonal bivalve mortality after heavy sedimentation and their paleoecological significance. Paleobiology, **11**: 139–153.

Peterson, C.H., and Black, R. 1987. Resource depletion by active suspension feeders on tidal flats: influence of local density and tidal elevation. Limnol. Oceanogr. **32**: 143–166.

Peterson, C.H., and Black, R. 1988. Density-dependent mortality caused by physical stress interacting with biotic history. Am. Nat. **131**: 257–270.

Peterson, C.H., and Black, R. 1991. Preliminary evidence for progressive sestonic food depletion in incoming tide over a broad tidal sand flat. Estuarine Coastal Shelf Sci. **32**: 405–413.

Probert, P.K. 1984. Disturbance, sediment stability, and trophic structure of soft-bottom communities. J. Mar. Res. **42**: 893–921.

Woodin, S.A. 1976. Adult-larval interactions in dense infaunal assemblages: patterns of abundance. J. Mar. Res. **34**: 25–41.

Wynberg, R.P., and Branch, G.M. 1992. An assessment of bait-collecting for *Callianassa kraussi* Stebbing in Langebaan Lagoon, and of associated avian predation. S. Afr. J. Mar. Sci. **11**: 141–152.

Wynberg, R.P., and Branch, G.M. 1994. Disturbance associated with bait-collection for sandprawns (*Callianassa kraussi*) and mudprawns (*Upogebia africana*): long-term effects on the biota of intertidal sandflats. J. Mar. Res. **52**: 523–558.

Spatial structure of northern shrimp (*Pandalus borealis*) off Labrador and eastern Newfoundland (northwest Atlantic)

G.R. Lilly, D.G. Parsons, and P.J. Veitch

Abstract: Examination of fishing positions of commercial shrimp vessels operating off Labrador and northeastern Newfoundland (northwest Atlantic) revealed that northern shrimp (*Pandalus borealis*) are distributed continuously in deep waters of the continental shelf and upper slope. Ontogenetic changes in the areal distribution of the shrimp were inferred from the distribution of specific length-groups of shrimp found in the stomachs of cod (*Gadus morhua*). The smallest shrimp, assumed to be age-0, were found primarily in deep water on the shelf off northeastern Newfoundland. Older juveniles were found primarily in shallow water. With increasing size, the shrimp moved deeper and onto the commercial fishing grounds. There is evidence of recruitment in the south and subsequent movement to fishing grounds to the north.

Résumé : L'étude des emplacements de pêche des crevettiers commerciaux pêchant au large du Labrador et au nord-est de Terre-Neuve (nord-ouest de l'Atlantique) révèle que la crevette nordique (*Pandalus borealis*) occupe une aire de répartition continue dans les eaux profondes du plateau continental et de la partie supérieure de la pente continentale. Les changements ontogénétiques dans la distribution spatiale de cette espèce sont déduits de la répartition de crevettes trouvées dans l'estomac de morues franches (*Gadus morhua*), par groupes définis en fonction de la longueur. Les crevettes les plus petites, qui entrent dans le groupe d'âge 0, se trouvent principalement dans les eaux profondes du plateau au large de la côte nord-est de Terre-Neuve. Les juvénile plus âgées se rencontrent surtout en eau peu profonde. À mesure que leur taille augmente, les crevettes se déplacent vers les eaux plus profondes et les lieux de pêche commerciale. Selon certaines indications, il y aurait recrutement au sud et déplacement ultérieur vers les lieux de pêche situés au nord.
[Traduit par la Rédaction]

Introduction

Northern shrimp, *Pandalus borealis*, occur in the western Atlantic from Greenland southward to Martha's Vineyard (Squires 1990), but are fished only in relatively discrete areas where densities are high. In the mid-1970's, a commercial fishery began off Labrador in three deep depressions, most notably the Hopedale and Cartwright channels at 56°N and 54.5°N, respectively (Fig. 1). The Hawke Channel at 53°N was fished only sporadically (Parsons and Fréchette 1989). During the period 1987–1991, the fishery expanded both northward to the continental slope off northern Labrador and southward to deep areas of the shelf off northeastern Newfoundland. These new fishing grounds included areas which were previously not known to support shrimp in commercial quantities. In 1992 and 1993, more effort was deployed on the seaward slopes of the continental shelf, particularly off southern Labrador and northeastern Newfoundland.

G.R. Lilly, D.G. Parsons, and P.J. Veitch. Science Branch, Department of Fisheries and Oceans, P.O. Box 5667, St. John's, NF A1C 5X1, Canada.

Correct citation: Lilly, G.R., Parsons, D.G., and Veitch, P.J. 1998. Spatial structure of northern shrimp (*Pandalus borealis*) off Labrador and eastern Newfoundland (northwest Atlantic). *In* Proceedings of the North Pacific Symposium on Invertebrate Stock Assessment and Management. *Edited by* G.S. Jamieson and A. Campbell. Can. Spec. Publ. Fish. Aquat. Sci. 125. pp. 265–271.

Management areas were added to the northern shrimp fishing plan in the late 1980's as new grounds were discovered by the fleet. Although these areas tended to be well separated geographically, no biological basis existed for treating them as distinct management units. In 1994, the total area fished from southern Davis Strait to northeastern Newfoundland was reorganized into fewer assessment–management areas based on new biological data which showed different rates of growth and maturation over broad geographical areas (Anonymous 1994). However, the degree to which the shrimp in each of these areas are isolated from those in other areas remains poorly understood. There are ample opportunities for mixing among areas through larval drift and horizontal migration following settlement. Commercial catch rate data are used extensively to infer stock status within each management area, but it is possible that such data are affected by migrations of shrimp among the commercial grounds and migrations onto the commercial grounds from peripheral areas.

A source of information on shrimp distribution and relative abundance in peripheral areas is the stomach contents of cod (*Gadus morhua*) collected during autumn bottom-trawl surveys designed to assess the status of groundfish. A preliminary examination of cod stomachs collected in 1977–1982 in Northwest Atlantic Fisheries Organization (NAFO) Divisions 2J and 3K revealed that predation by cod on shrimp occurred throughout most of the area, but was most intense in deep regions of the shelf (Lilly 1984). This included the area into which the shrimp fishery expanded after 1986.

The objectives of this paper are to describe an apparent

Fig. 1. Map of the study area, showing major physiographic features and NAFO Divisions. Offshore banks and coastal shelfs mentioned in the text are indicated by their initials: Hamilton Bank (HB), Belle Isle Bank (BIB), Funk Island Bank (FIB), and Fogo Shelf (FS). Depressions on the shelf are indicated by letters: Hopedale Channel (A), Cartwright Channel (B), Hawke Channel (C), St. Anthony Basin (D), and Funk Island Deep (E). The continental shelf is subdivided into the shallow Labrador Shelf, extending from the northern tip of Labrador to southern Hamilton Bank, the shallow Grand Bank in the south, and the deep Northeast Newfoundland Shelf in the middle.

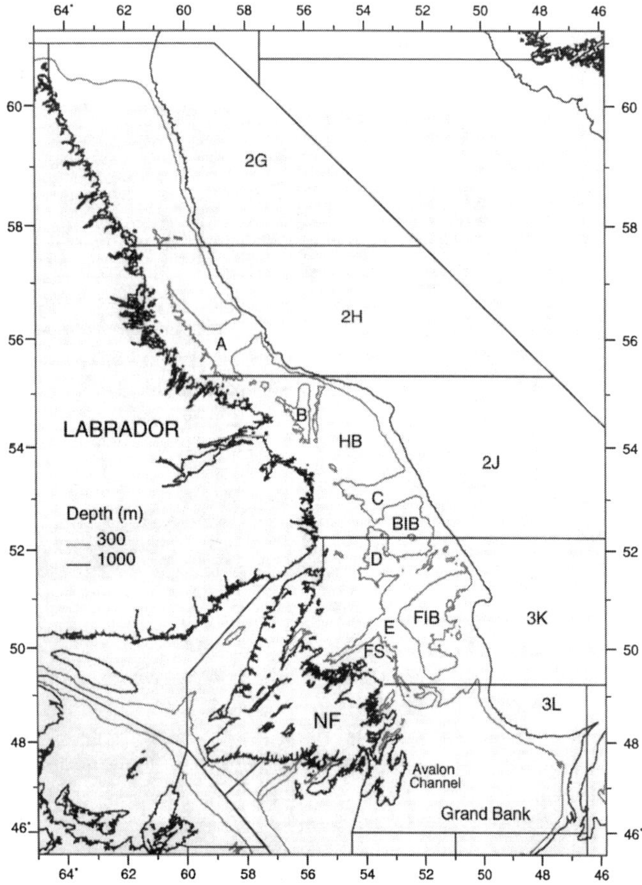

ontogenetic shift in distribution of shrimp as inferred from cod stomach contents and to compare this information to the spatial distribution of fishing effort.

Methods

Study area

The study area in NAFO Subarea 2 and Divisions 3K and 3L encompasses the Labrador Shelf, the Northeast Newfoundland Shelf, the northern Grand Bank, and the adjacent continental slope to a depth of 1000 m (Fig. 1). The continental shelf has several deep areas, most notably the Hopedale and Cartwright channels on the Labrador Shelf and Hawke Channel, St. Anthony Basin, and Funk Island Deep on the Northeast Newfoundland Shelf. The Labrador Current flows southeastward along the shelf and keeps temperatures low (Smith et al. 1937). The cold intermediate layer of the current is deepest and coldest toward the coast and temperatures below 0°C are found to depths of about 200 m in many years (Petrie et al. 1988). Maximum bottom temperatures in the survey area are usually about 4°C.

Stomach collection and examination

Cod were caught during random-stratified bottom-trawl surveys conducted during October–December in Divisions 2J3K (1980–1989) and Division 3L (1981–1989, excluding 1984). Data collected during surveys in more recent years are not included because the cod declined severely in abundance and became increasingly aggregated toward the edge of the shelf (Bishop et al. 1994; Lilly 1994). This resulted in a very uneven distribution of stomach sampling, with relatively high sampling intensity toward the shelf break and very low sampling intensity on the inner part of the shelf. Details regarding ships and gear employed, sampling dates, and changes in survey pattern are provided by Lilly and Davis (1993) and Bishop et al. (1994). Stratum areas and locations are provided by Doubleday (1981) and Bishop (1994). Fishing in all divisions and years was conducted on a 24-h basis.

Stomachs were collected from up to 3 randomly selected cod per 10-cm length-group per station in 1980–1982 and 3 per 9-cm length-group in 1983–1989. Stomachs were not collected from fish that showed signs of regurgitation, such as food in the mouth or a flaccid stomach. Stomachs were individually tagged and preserved in 4% formaldehyde solution in seawater before examination of their contents in the laboratory.

Examination involved separation of food items into taxonomic categories. Decapod crustacea were identified to species when possible. Pandalid shrimp which were not identified to species are not considered in this paper. Items in each taxon were placed briefly on absorbent paper to remove excess liquid and then counted and weighed to the nearest 0.1 g. Carapace length (CL) of shrimp was measured to the nearest mm from the posterior edge of the eye socket to the posterior mid-dorsal margin of the carapace whenever digestive condition permitted. A summary of the number of stomachs examined each year, the percentage of stomachs containing shrimp, and the number of shrimp found is provided by Lilly (1995).

Analysis and presentation of stomach content data

To study the geographical distribution of shrimp, it was assumed that the average quantity of shrimp in the stomachs of cod collected at a specific locality reflected the relative abundance of shrimp at that locality (Fahrig et al. 1993). The average quantity of shrimp of species *s* in the stomachs of the cod in a specified sample was expressed as a mean partial fullness index:

$$PFI_s = \frac{1}{n}\sum_{j=1}^{n}\frac{W_{sj}}{L_j^3} \times 10^4$$

where W_{sj} is the weight (g) of shrimp species *s* in fish *j*, L_j is the length (cm) of fish *j*, and *n* is the number of fish in the sample. This index is based on the assumption that stomach capacity is a power function of length and is analogous to Fulton's condition factor (body weight·length^{-3}).

To study ontogenetic changes in shrimp distribution, the

Fig. 2. Geographic variation in mean partial fullness index (PFI) for *Pandalus borealis* and *P. montagui* occurring in stomachs of cod in 1980–1982 and 1985–1989. Only cod in the length range 18–89 cm were used (Lilly 1995). The data from all stomachs collected in areas of 10′ latitude and 20′ longitude are combined. Areas represented by fewer than 6 stomachs are not plotted.

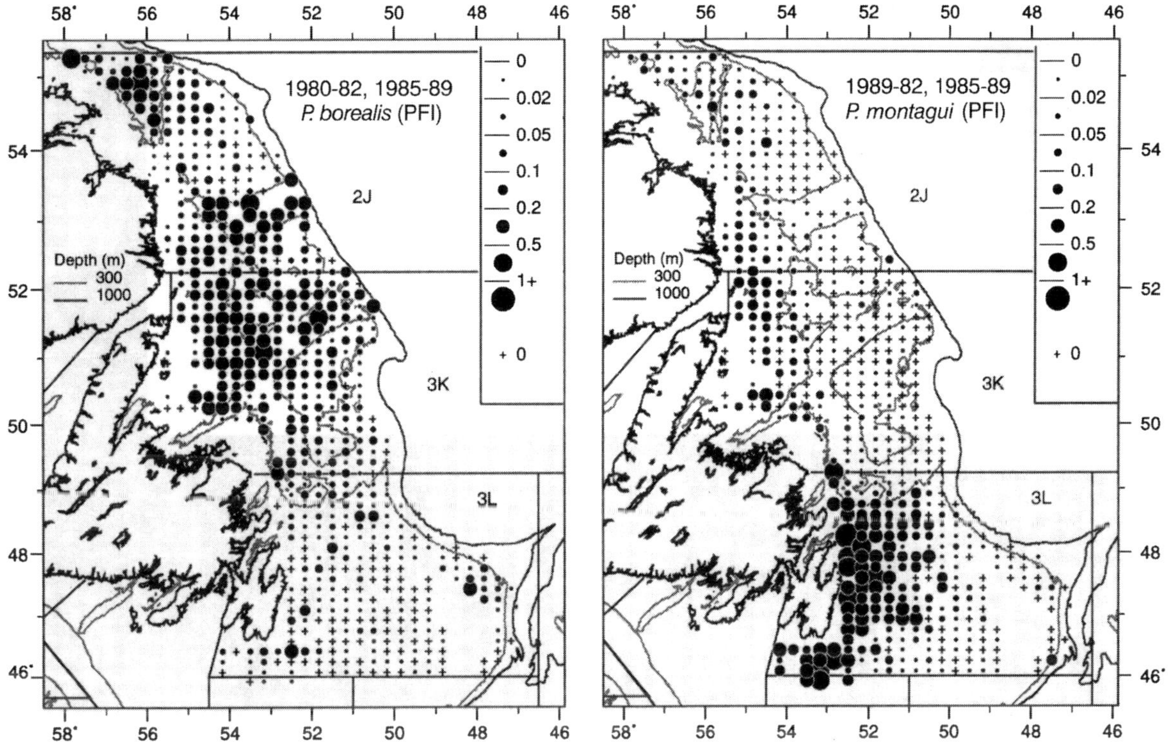

shrimp found in cod stomachs were divided into 4 length-groups: 2–6, 7–11, 12–19, and >19 mm CL. The first two length-groups represent ages 0 and 1, respectively (Parsons et al. 1986). By comparison with lengths-at-age deduced by Parsons et al. (1989), it may be inferred that the third mode represents males of ages 2, 3, and 4. The fourth group would be composed of older males and females. The partial fullness index could not be calculated for each shrimp length-group because the weight of shrimp in cod stomachs was not recorded by shrimp length. However, an index of local shrimp abundance by length-group can be calculated from the number of shrimp at length found in the cod stomachs. (Shrimp which were too digested to be measured were ignored.) A difficulty arises because the number of shrimp of a given length-group found within the cod stomachs varies with cod size in a complex way (Lilly 1995); the maximum size shrimp that a cod can ingest increases with cod length, the number of shrimp of a given size that can be contained within a stomach increases rapidly with cod length because of rapidly increasing stomach capacity, and large cod tend not to feed on small shrimp. These complex patterns were taken into account in part by restricting the size range of the cod used to calculate the mean number of shrimp per stomach to cod of 9–80, 27–98, 36–98, and 54–98 cm, respectively, for the four shrimp length-groups defined above (Lilly 1995).

The geographic distribution of shrimp was determined from visual inspection of expanding symbol plots of spatial variation in (*i*) PFI values and (*ii*) the average number of shrimp found in cod stomachs.

Commercial fishery data

Positions of individual fishing tows by the commercial shrimp fleet from 1977 to 1993 were obtained from logbook records. The logbooks account for most (>75%) of the total effort in most years and are considered an accurate representation of the fishing activity of the fleet.

Commercial shrimp trawls in the northern shrimp fishery have been regulated since the late 1970's by a minimum mesh size of 40 mm in the codend. Animals retained by the gear cover a broad size range from about 10 to 30 mm CL. The fishery is selective for the larger, older animals which, for this protandrous hermaphrodite, are primarily females. Commercial length–frequency distributions were compiled for southern Division 2J and Division 3K to compare with the sizes observed in cod stomachs. The fishery in this area occurs during winter and spring and the length sampling data are representative of the January–May period.

Results

Geographic distribution of *P. borealis* and *P. montagui*

There is concern that partial digestion of the shrimp found in cod stomachs might render them susceptible to misidentification. A test of the reliability of identification is to look for consistency in the patterns of geographic distribution of *P. borealis* and other species with which it may be confused. Although many species of shrimp have been identified in the stomach contents of cod caught in this area, errors in identification of *P. borealis* are most likely to occur with its congener,

Fig. 3. Carapace length distributions of northern shrimp measured from commercial trawl catches and from cod stomach contents in 1988–1989. Most animals from the commercial fishery were caught in January–May, whereas those from cod stomachs were obtained in November–December.

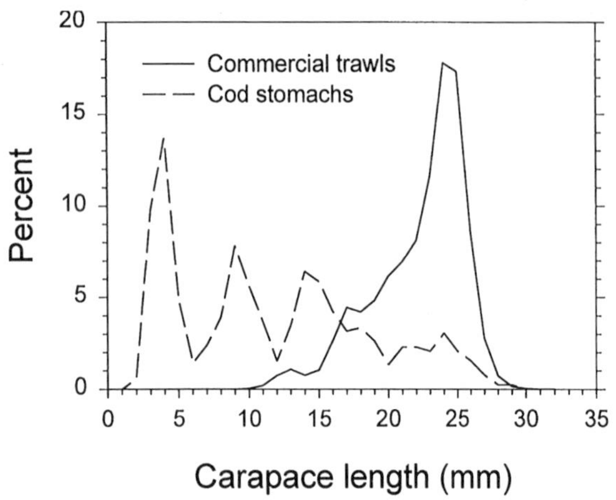

the striped pink shrimp (*P. montagui*). The overall distribution of each of these two species was determined from plots of spatial variation in PFI values (Fig. 2). *Pandalus borealis* were found in largest quantities in stomachs of cod collected in the Cartwright and Hawke channels in Division 2J and in much of the deeper shelf water of Division 3K. In contrast, *P. montagui* were found in stomachs of cod caught over most of the shallower water of Divisions 2J and 3K and in largest quantities in stomachs of cod caught in the western and southern Avalon Channel in Division 3L. These patterns are in broad-scale agreement with those deduced from catches during trawl surveys (Squires 1965, 1990). The spatial consistency in the distribution pattern of each species (Fig. 2) and the relatively small overlap between high PFI_{pb} values and high PFI_{pm} values (Lilly 1995) support the contention that few shrimp were assigned to the wrong species of *Pandalus*.

Length–frequencies

The sizes of shrimp captured in the commercial fishery and the sizes consumed by cod were compared using data collected in southern Division 2J and Division 3K. Analysis was limited to 1988 and 1989, the first two years of relatively large fishing effort in Division 3K and the last two years of cod stomach sampling before the change in cod distribution became severe. Commercial catches, mainly from winter and spring, were composed primarily of larger shrimp with modal length of about 24 mm CL (Fig. 3). Most of these shrimp were females, given that the length at which the male:female ratio = 1 is about 22 mm (D.G. Parsons, unpublished data). Lesser numbers of male shrimp also occurred, but representation of the smaller sizes (<16 mm), corresponding to younger ages, was poor. Modes appeared at approximately 13 and 17 mm CL. In contrast, cod preyed primarily on juvenile shrimp (Fig. 3). The length–frequency of shrimp found in cod stomachs in autumn was dominated by prominent modes at 3–4 mm, 9–10 mm, and 14–15 mm CL.

The large difference between sizes of shrimp caught in the commercial fishery and those consumed by cod was not dependent on the particular combination of years chosen. Annual variation in length–frequencies is small in both the commercial fishery (Anonymous 1994) and cod stomach contents (Lilly 1995).

Distribution of *P. borealis* by size-group

The distribution of shrimp changed with increasing shrimp size (Fig. 4). Age-0 shrimp (2–6 mm CL) were found in highest numbers in cod caught on the flanks of St. Anthony Basin and Funk Island Deep and in the saddles between Belle Isle Bank and Funk Island Bank and between Funk Island Bank and Grand Bank. Very few were found in Division 2J.

Age-1 shrimp (7–11 mm CL) were found in highest numbers in cod caught in shallower water than where the age-0 shrimp were found. Highest numbers occurred to the west of Funk Island Deep and St. Anthony Basin. Moderate numbers were found in Cartwright Channel, west of Hawke Channel, on northern and western Funk Island Bank, and southward from Funk Island Bank to the northern edge of Grand Bank. Densities in Hawke Channel were low.

The 12–19 mm CL group was broadly distributed. Relatively high densities were found in cod caught in the Cartwright Channel, the southern Hawke Channel, and on the slope of the coastal shelf to the west of Hawke Channel, St. Anthony Basin, and Funk Island Deep. There were also relatively large numbers at some sites off the Fogo Shelf and on the northern edge of Grand Bank.

Large shrimp (>19 mm CL) were not as broadly distributed as the 7–11 and 12–19 mm groupings. Highest numbers tended to be found in cod caught in those areas where the commercial fishery has operated, viz. in Cartwright Channel, in northwestern and southern Hawke Channel, on the flanks of St. Anthony Basin and Funk Island Deep, and in the saddle between Belle Isle Bank and Funk Island Bank. The strong similarity between the distribution of large shrimp in cod stomachs and the distribution of commercial fishing effort (Fig. 5) supports the notion that measures of the quantity or number of shrimp in cod stomachs can provide a useful index of shrimp density.

Discussion

Cod as a sampling device for shrimp

The use of a predator such as cod as a sampling device for shrimp has several advantages. The cod stomachs provide a source of information on distribution and length-at-age of shrimp which is independent of that provided by the commercial fishery and research surveys dedicated to shrimp. In addition, the examination of cod stomachs yields individuals of smaller size than are taken by other collection methods and provides information at times and places not covered by shrimp surveys. Finally, the method is relatively inexpensive if the catching of predators is already conducted as part of an ongoing program (e.g., annual groundfish surveys).

There are several potential problems to consider when using predators as sampling tools. (*i*) Digestion processes may render prey susceptible to misidentification. Misidentification is likely to have occurred in this study, but consistency in the patterns of distribution of *P. borealis* and its congener *P. montagui* and the low level of geographic overlap between them support the contention that misidentification was not a serious problem. (*ii*) The predator distribution may not completely

Fig. 4. Geographic variation in mean number of *Pandalus borealis* occurring in stomachs of cod in 1980–1982 and 1985–1989. The shrimp are aggregated into 4 length-groups (2–6, 7–11, 12–19, and >19 mm CL). The cod length ranges used for these shrimp length-groups are 9–80, 27–98, 36–98, and 54–98 cm, respectively (see Lilly 1995). For each of the shrimp groups, the data from stomachs of all cod of the specified length range, collected in areas of 10′ latitude and 20′ longitude, are combined. Areas represented by fewer than 6 stomachs are not plotted.

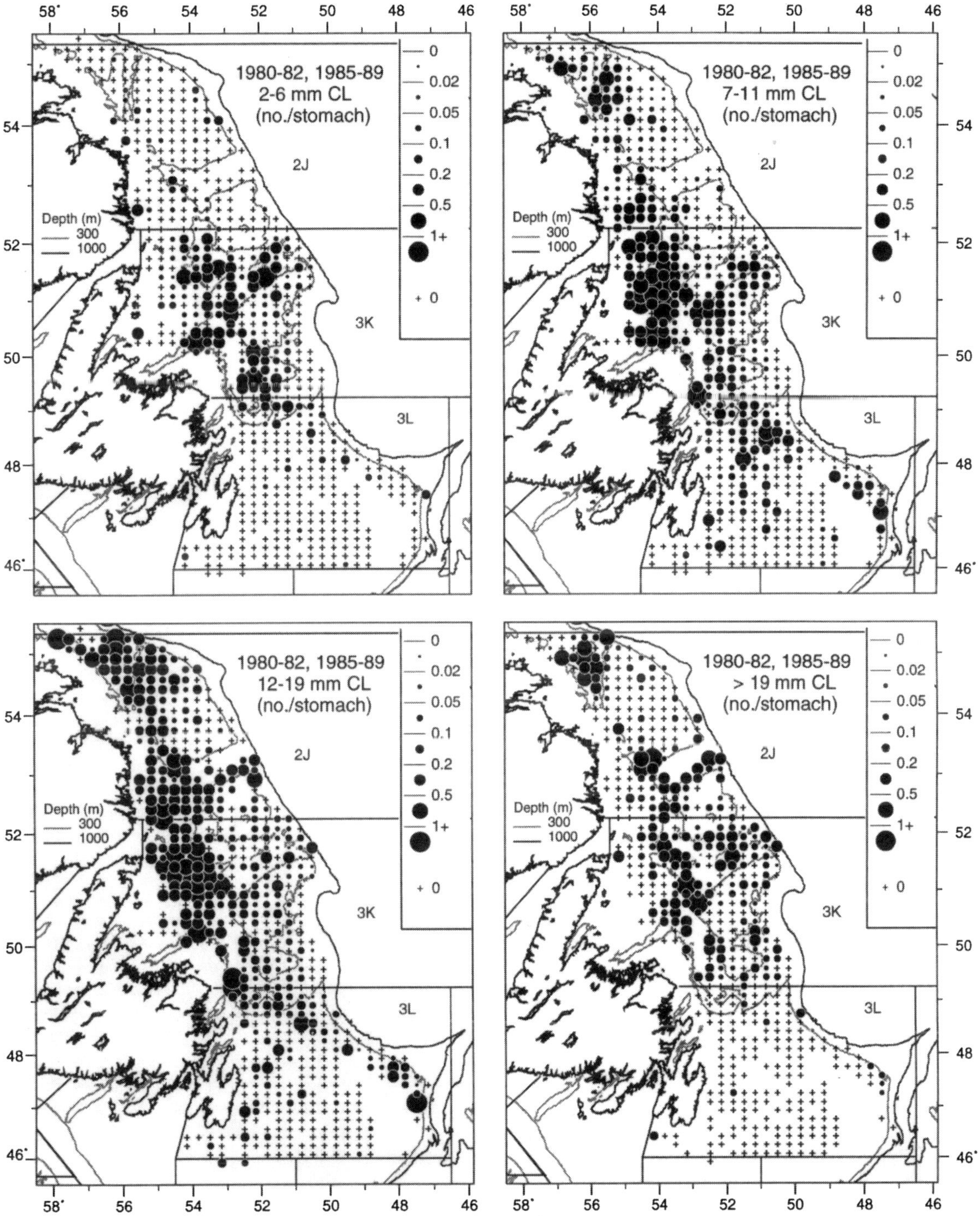

overlap the prey distribution. This was not a problem during the period covered by this study, but the usefulness of cod as a sampling tool for shrimp declined in the 1990's as the cod distribution became reduced (Bishop et al. 1994; Lilly 1994). In addition, the relatively small number of surveys on the Labrador Shelf north of Division 2J and the small catches of cod during those surveys limit the possibilities for exploring the distribution of juvenile shrimp in these more northern areas. (*iii*) The prey may have been ingested at a location other than where the predator was captured. This could clearly be a problem on small spatial scales, but should not be a problem on the spatial scales studied here. We also do not know if the shrimp

Fig. 5. Positions of 16 313 fishing tows conducted by commercial shrimp trawlers from northern Labrador to northeastern Newfoundland (Divisions 2G-3K) in 1977–1993.

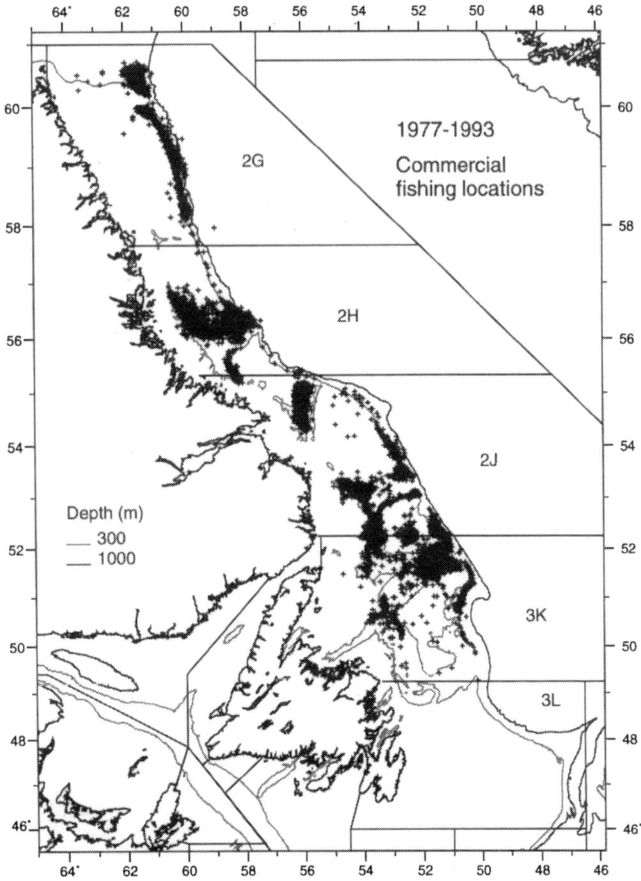

were ingested on the bottom or in the water column. (*iv*) The number of prey of a specific size range present in the stomach of a predator at the time of capture varies with the size of the predator. In the present study, the effect of such variability on estimates of mean number of a specific shrimp length-group per stomach was reduced by excluding cod size-groups which fed very little or not at all on that shrimp length-group. The perceived geographic distribution of a length-group of shrimp may not be greatly dependent on the choice of length range of cod. As a simple test, the distribution of 0-group shrimp was investigated based on their occurrence in 4 cod length-groups: 9–26, 27–44, 45–62, and 63–80 cm (Lilly 1995). Visual inspection of geographic plots derived from these four data sets revealed that they were similar to one another and to the distribution based on all these cod combined. (*v*) The rate of consumption of the prey in question may be affected by the availability of other prey. The possibility that the level of predation by cod on shrimp may vary with the availability of other prey (such as capelin) remains to be tested. This question is of concern if one wishes to examine annual variability in the intensity of predation on shrimp, but is less important if one is looking at long-term (decadal) patterns in distribution.

Distribution of northern shrimp

Cumulative records of fishing locations from commercial shrimp vessels reveal that northern shrimp are distributed continuously at approximately 300–500 m from the northern tip of the Labrador Shelf in Division 2G to southern Funk Island Bank in Division 3K (Fig. 5; Anonymous 1994). It is currently thought that the shrimp throughout this vast area likely constitute a single stock or a complex of stocks (Anonymous 1994). Additional insight into stock structure is provided by the ontogenetic changes in distribution inferred from analysis of the stomach contents of cod caught in Divisions 2J3KL. The smallest shrimp (2–6 mm CL) were found primarily in stomachs of cod caught in deep water on the Northeast Newfoundland Shelf within Division 3K. This result is somewhat surprising because previous studies (Fréchette and Parsons 1983; Shumway et al. 1985) reported that juvenile shrimp occur in shallower water than adult shrimp. Simard et al. (1990) noted transient presence of age-0 shrimp at 120 m in the western Gulf of St. Lawrence, but concluded that these observations were not representative of the actual distribution. They postulated that the age-0 shrimp should occur more commonly at shallow (<100 m) depths. If the absence of the age-0 shrimp from stomachs of cod caught in shallow water in the present study accurately reflects the distribution of that age-group, then they settle at greater depths than previously assumed and move to shallower water some time before their second autumn. The broad distribution of older juveniles in shallow water and their gradual movement with age into deeper water and onto the commercial fishing grounds support the contention that shrimp on the various fishing grounds do not represent separate populations. This observation has been cited as support for combining many of the fishing grounds into larger assessment and management units (Anonymous 1994).

The small numbers of age-0 shrimp in Division 2J and the possibility of movement of age-1 and older shrimp into Division 2J from Division 3K has not previously been recognized. It is likely that at least some of the shrimp caught in northern areas are derived from recruitment in Division 3K. Additional studies are required to identify the nursery grounds and ontogenetic movements of the shrimp found on each of the fishing grounds.

An assessment of the relative importance of fishing and predation by cod as sources of mortality for shrimp is beyond the scope of this paper. However, it appears that predation by cod is broadly distributed over a wide depth range and affects mainly juveniles and young males, whereas the fishery occurs mainly in deep water (300–500 m) where the large shrimp are concentrated and takes mainly older males and females.

Acknowledgments

We thank the many individuals who assisted in the collection of cod stomachs, often under difficult sea conditions, and those who participated as employees of commercial companies in the examination of the contents of the stomachs. B. Moriarity assisted in compiling the data. J. Hoenig, J. Wheeler, P. Livingston, and an anonymous reviewer provided helpful comments on earlier versions of the paper.

References

Anonymous. 1994. Report on the status of the shrimp resource in the Northwest Atlantic in areas off Newfoundland and Labrador, in the Gulf of St. Lawrence, and on the Scotian Shelf. DFO Atl. Fish. Stock Status Report 94/6. 43 p.

Bishop, C.A. 1994. Revisions and additions to stratification schemes used during research vessel surveys in NAFO Subareas 2 and 3. NAFO (Northwest Atl. Fish. Organ.) SCR Doc. 94/43, Serial No. N2413. 23 p.

Bishop, C.A., Anderson, J., Dalley, E., Davis, M.B., Murphy, E.F., Rose, G.A., Stansbury, D.E., Taggart, C., and Winters, G. 1994. An assessment of the cod stock in NAFO Divisions 2J+3KL. NAFO (Northwest Atl. Fish. Organ.) SCR Doc. 94/40, Serial No. N2410. 50 p.

Doubleday, W.G. (*Editor*). 1981. Manual on groundfish surveys in the Northwest Atlantic. NAFO (Northwest Atl. Fish. Organ.) Sci. Counc. Stud. **2**: 7–55.

Fahrig, L., Lilly, G.R., and Miller, D.S. 1993. Predator stomachs as sampling tools for prey distribution: Atlantic cod (*Gadus morhua*) and capelin (*Mallotus villosus*). Can. J. Fish. Aquat. Sci. **50**: 1541–1547.

Fréchette, J., and Parsons, D.G. 1983. Report of shrimp ageing workshop held at Ste. Foy, Quebec, in May and at Dartmouth, Nova Scotia, in November 1981. NAFO (Northwest Atl. Fish. Organ.) Sci. Counc. Stud. **6**: 79–100.

Lilly, G.R. 1984. Predation by Atlantic cod on shrimp and crabs off northeastern Newfoundland in autumn of 1977–82. ICES C.M. 1984/G:53. 25 p.

Lilly, G.R. 1994. Predation by Atlantic cod on capelin on the southern Labrador and Northeast Newfoundland shelves during a period of changing spatial distributions. ICES Mar. Sci. Symp. **198**: 600–611.

Lilly, G.R. 1995. Distribution of northern shrimp (*Pandalus borealis*) in Divisions 2J3KL as inferred from stomach contents of cod (*Gadus morhua*). DFO Atl. Fish. Res. Doc. 95/61. 21 p.

Lilly, G.R., and Davis, D.J. 1993. Changes in the distribution of capelin in Divisions 2J, 3K and 3L in the autumns of recent years, as inferred from bottom-trawl by-catches and cod stomach examinations. NAFO (Northwest Atl. Fish. Organ.) SCR Doc. 93/54, Serial No. N2237. 14 p.

Parsons, D.G., and Fréchette, J. 1989. Fisheries for northern shrimp (*Pandalus borealis*) in the northwest Atlantic from Greenland to the Gulf of Maine. *In* Marine invertebrate fisheries: their assessment and management. *Edited by* J.F. Caddy. John Wiley & Sons, Inc., New York, N.Y. pp. 63–85.

Parsons, D.G., Lilly, G.R., and Chaput, G.J. 1986. Age and growth of northern shrimp *Pandalus borealis* off northeastern Newfoundland and southern Labrador. Trans. Am. Fish. Soc. **115**: 872–881.

Parsons, D.G., Mercer, V.L., and Veitch, P.J. 1989. Comparison of the growth of northern shrimp (*Pandalus borealis*) from four regions of the Northwest Atlantic. J. Northwest Atl. Fish. Sci. **9**: 123–131.

Petrie, B., Akenhead, S., Lazier, J., and Loder, J. 1988. The cold intermediate layer on the Labrador and Northeast Newfoundland Shelves, 1978–86. NAFO (Northwest Atl. Fish. Organ.) Sci. Counc. Stud. **12**: 57–69.

Shumway, S.E., Perkins, H.C., Schick, D.F., and Stickney, A.P. 1985. Synopsis of biological data on the pink shrimp, *Pandalus borealis* Krøyer, 1838. U.S. Dept. of Commerce, NOAA Tech. Rep. NMFS 30. pp. 1–57.

Simard, Y., Brunel, P., and Lacelle, J. 1990. Distribution and growth of pre-recruit cohorts of the northern shrimp (*Pandalus borealis*) in the western Gulf of St. Lawrence as related to hydrographic conditions. Can. J. Fish. Aquat. Sci. **47**: 1526–1533.

Smith, E.H., Soule, F.M., and Mosby, O. 1937. The Marion and General Greene expeditions to Davis Strait and Labrador Sea, 1928–1935. Scientific results, part 2: physical oceanography. U.S. Treasury Dept., Coast Guard Bull. **19**: 1–259.

Squires, H.J. 1965. Decapod crustaceans of Newfoundland, Labrador and the Canadian eastern Arctic. Fish. Res. Board Can. Manuscript Report Series (Biol.) No. 810. 212 p.

Squires, H.J. 1990. Decapod crustacea of the Atlantic coast of Canada. Can. Bull. Fish. Aquat. Sci. 221. 532 p.

Implications of migration and larval interchange in American lobster (*Homarus americanus*) stocks: spatial structure and resilience

Michael J. Fogarty

Abstract: Coastal American lobster (*Homarus americanus*) stocks exhibit strong resilience to exploitation. This resilience may be attributable to strong compensatory mechanisms, the effects of larval subsidies from offshore (or other less intensively exploited) lobster groups, or a combination of both mechanisms. The objective of the present analysis is to explore the implications of linkages between lobster groups for the stability and resilience of exploited stocks. A simple delay–difference model for two subpopulations is developed to represent a hypothetical inshore–offshore system in which the inshore group receives a larval subsidy from offshore. Adult lobsters from inshore are assumed to undertake a directional migration to offshore habitats. For simplicity, it is assumed that the two subpopulations have identical population parameters and generation times. Density-dependent mechanisms are incorporated in the model to represent competition for critical resources such as shelter during the early life history. It is shown that linkage between two subpopulations results in a qualitative change in the behavior of the system even at relatively low levels of interchange. In particular, relatively low levels of larval transport from offshore to inshore regions can allow persistence of inshore stocks at high levels of fishing mortality when offshore populations are subject to low–moderate levels of exploitation. The observed resilience of inshore lobster stocks under exploitation is consistent with a mechanism of this type. Modeling results also suggest that larval source populations should be managed conservatively, particularly under uncertainty concerning the magnitude of transport from the source subpopulation to one or more recipient groups.

Résumé : Les stocks de homards (*Homarus americanus*) des eaux côtières présentent une forte résilience à l'exploitation. Cette résilience peut être attribuée à de puissants mécanismes de compensation, aux effets de l'apport de larves provenant de groupes de homards de haute mer (ou d'autres groupes exploités de façon moins intensive), ou encore à une combinaison de ces deux mécanismes. L'analyse présentée ici porte sur les conséquences des relations entre divers groupes de homards sur la stabilité et la résilience des stocks exploités. Un modèle simple de décalage temporel appliqué à deux sous-populations est en voie d'élaboration, dans le but de représenter un système hypothétique comprenant des sous-populations hauturières et côtières, dans lequel le groupe des eaux côtières recevrait des larves provenant de la haute mer. Le système repose sur une prémisse selon laquelle les homards adultes des eaux côtières entreprendraient une migration en direction de la haute mer. Pour simplifier la méthode, on suppose que les paramètres démographiques et la durée de génération sont identiques dans les deux sous-populations. Le modèle intègre des mécanismes dépendants de la densité afin de refléter la concurrence pour des ressources essentielles comme l'abri au début du cycle biologique. Il est démontré que la relation entre deux sous-populations se manifeste par un changement qualitatif dans le fonctionnement du système, même avec un taux d'échange relativement faible. De façon plus particulière, une intensité assez faible de déplacement de larves hauturières vers la côte peut permettre la persistance du stock côtier malgré des niveaux élevés de mortalité par pêche lorsque les populations de haute mer font l'objet d'une exploitation faible ou moyenne. En ce qui a trait au homard des côtes, la résilience observée dans les stocks exploités est compatible avec ce genre de mécanisme. En outre, les résultats obtenus par modélisation laissent penser que les populations productrices de larves devraient être gérées avec prudence, en particulier si l'on ne connaît pas l'ampleur réelle du déplacement de larves de la sous-population productrice vers un ou plusieurs groupes récepteurs.
[Traduit par la Rédaction]

M.J. Fogarty.[1] National Oceanic and Atmospheric Administration, National Marine Fisheries Service, Northeast Fisheries Science Center, Woods Hole, MA 02543, U.S.A.

Correct citation: Fogarty, M.J. 1998. Implications of migration and larval interchange in American lobster (*Homarus americanus*) stocks: spatial structure and resilience. *In* Proceedings of the North Pacific Symposium on Invertebrate Stock Assessment and Management. *Edited by* G.S. Jamieson and A. Campbell. Can. Spec. Publ. Fish. Aquat. Sci. 125. pp. 273–283.

[1] Present address: Center for Environmental and Estuarine Studies, University of Maryland System, Chesapeake Biological Laboratory, Solomons, MD 20688, U.S.A.

Introduction

The American lobster, *Homarus americanus*, has supported important commercial fisheries in the northeastern United States and the Atlantic provinces of Canada for over a century and a half. Estimated fishing mortality rates in coastal locations are high and concerns have been raised about the viability of the stocks under increasingly intense exploitation. The estimated lifetime reproductive output per female in nearshore waters is extremely low relative to unexploited levels (Northeast Fisheries Science Center 1993, 1996; Fisheries Resource Conservation Council 1995) and the question of whether the inshore stocks are self-sustaining has received increased scrutiny. The potential importance of harvest refugia, defined in space or time, for maintaining the viability of heavily exploited inshore stocks has long been recognized (Anthony and Caddy 1980, page 186). If coastal lobster stocks receive a significant larval subsidy from less intensively exploited source areas, their resilience to exploitation would be substantially enhanced. The possible role of offshore or deep-water lobster stocks in exporting larvae to inshore areas has been accorded particular attention (Anthony and Caddy 1980).

Despite sustained high exploitation rates in the traditional inshore fishing areas, lobster landings have increased markedly during the last three decades throughout the range (Elner and Campbell 1991; Pezzack 1992; Fogarty 1995). Factors underlying the increase in landings are not fully understood although an increase in abundance, escalation of fishing effort, changes in trap size and design, technological advances, and encroachment of the fishery into previously unexploited or lightly exploited areas all appear to have played a role (Fogarty 1995). The geographical extent of the increase in landings points to broad-scale events. Attention has centered on large-scale changes in physical (e.g., temperature and wind fields) and ecological (predator–prey) factors.

Directed exploitation of offshore lobster stocks is comparatively recent. The offshore lobster fishery developed off the northeastern United States with incidental and seasonally dependent directed catches by the groundfish fleet in the late 1950's. By the late 1960's a trap fishery developed which has since dominated the offshore lobster catch (Fogarty et al. 1982; Northeast Fisheries Science Center 1993, 1996; Fogarty 1995). The development of Canadian offshore fisheries was initiated in 1971 under strict regulatory control (Pezzack et al. 1992). Exploitation rates in offshore regions are currently low in comparison to coastal populations in both the United States (Fogarty et al. 1982; Northeast Fisheries Science Center 1993, 1996; Fogarty 1995) and Canada (Pezzack 1992). Landings from the offshore trap sector have remained relatively stable, at much lower levels than the inshore component (not exceeding 20% of the total). In addition to the development of the offshore fishery on the outer continental shelf, there has been a progressive expansion of the traditional inshore grounds further from shore. The development of the offshore fishery and its potential for expansion raises the question of how the persistence of heavily exploited inshore lobster stocks might be affected if the two are linked by larval dispersal and adult migration.

The objective of the present analysis is to explore the implications of linkages among subpopulations for the stability and resilience of exploited lobster stocks. Here, resilience is taken to mean the capacity of a population to withstand a sustained perturbation such as harvesting. A system linked by larval dispersal from offshore regions to inshore sites and return of adults through migration to the source population is considered. A simple age-structured model is employed in a discrete-time formulation for two subpopulations. Here, interest centers on qualitative shifts in the behavior of the system in a comparison of the effects of harvesting in separate populations versus two linked subpopulations.

Background

Motivation for a model linking lobster subpopulations through larval transport and for interchange through directed movements of adults is provided below. The possibility of long-distance dispersal of larvae in wind-driven surface currents has received particular attention (see reviews by Fogarty 1983; Ennis 1995; Cobb 1995). Recent studies have further pointed to the importance of directional orientation and active swimming by postlarvae in reaching coastal locations (Cobb et al. 1989; Katz et al. 1994). Extensive movements of mature lobsters have also been documented (reviewed in Lawton and Lavalli 1995; Fogarty 1995) including migrations between inshore and offshore locations. These observations suggest that exploration of a model of linked subpopulations is appropriate for at least some regions.

Larval transport

The potential importance of larval subsidies to heavily exploited inshore lobster groups has elicited considerable interest and debate (e.g., Rogers et al. 1968; Stasko 1980; Harding et al. 1983; Harding and Trites 1988; Pezzack 1989; Pezzack et al. 1992; Katz et al. 1994). It has been suggested that heavily exploited inshore lobster stocks off Nova Scotia receive substantial larval input from less intensively harvested offshore areas (Stasko 1980; Harding et al. 1983; Harding and Trites 1988) based on joint consideration of larval distribution and circulation patterns. Pezzack (1989) noted, however, that perception of the relative importance of contributions of these areas to larval production depends critically on the geographical definition of the inshore and offshore zones. Pezzack et al. (1992) reviewed evidence for potential linkages between inshore and offshore lobsters in Atlantic Canada and concluded that it is possible that offshore lobsters from the Browns Bank region could serve as a source of larvae to coastal Nova Scotia stocks. The potential contribution from the Georges Bank region to Nova Scotian stocks, however, was deemed negligble.

Cobb et al. (1989) documented the importance of directional swimming behavior in lobster postlarvae and noted that observed orientations and swimming speeds would enhance the probability of a linkage between inshore and offshore lobster groups (see also Cobb et al. 1983). Katz et al. (1994) combined a hydrographic transport model with observations on postlarval behavior to demonstrate that American lobster larvae originating on the edge of the continental shelf off southern New England could reach coastal waters using a combination of directed swimming and drift in prevailing wind-driven surface currents. Examination of the distribution patterns of successive larval stages in synoptic samples along an inshore–offshore transect indicated that later larval stages were more common closer to shore, suggesting that shoreward transport of larvae may in fact be important (Katz et al. 1994).

Movement and migratory patterns

Movement patterns are critical to understanding linkages among subpopulations in two ways. First, seasonal shoalward movements of egg-bearing females prior to hatching of the eggs can reduce the dispersal distances required for larvae to reach inshore habitats. Second, larger individuals may undertake ontogenetic shifts in distribution to deeper water habitats. Substantial evidence exists for seasonal onshore–offshore or deep water – shallow water movements. Less is known about age- or size-specific displacements to different habitats.

Extensive seasonal movements by offshore lobsters in the southern New England region have been documented (Saila

Fig. 1. Graphic representation of stability points for a simple two-stage life history model. The nonlinear curve in both panels represents the relationship between spawning stock size and resulting recruitment. Lines representing the relationship between recruitment and spawning stock size over the life of a cohort are provided for several levels of the instantaneous rate of fishing mortality (*F*). Panel (*a*) represents the case for a closed population; a population collapse is predicted at point labeled "High *F*." Panel (*b*) represents the case where a constant recruitment subsidy is provided to the population from an exogenous source. Note that a complete population collapse cannot occur in this case as long as the exogenous subsidy continues.

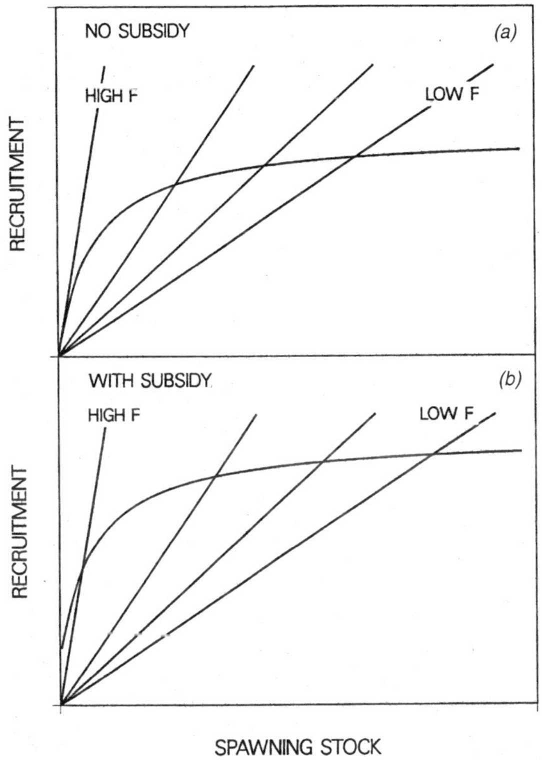

and Flowers 1968; Cooper and Uzmann 1971; Uzmann et al. 1977; Fogarty et al. 1980). Observed shoalward movements of ovigerous females in spring would reduce the dispersal distances required for postlarvae to reach coastal sites in southern New England, increasing the probability of interchange. Cobb (1995) suggested that the onshore–offshore migration pattern may be relict of behaviors evolved as sea level gradually increased following the last glaciation. Long distance movements of mature lobsters have also been demonstrated for lobsters in the Gulf of Maine, Bay of Fundy, and on the Scotian Shelf (Dow 1974; Campbell 1986; Campbell and Stasko 1985, 1986; Campbell et al. 1984; Miller et al. 1989; Pezzack and Duggan 1986). Clear evidence of shoalward movements during late spring and summer has been provided for offshore lobsters in Canadian studies. These movements generally involve displacement from deep-water sites to shallow plateaus such as Browns Bank. Campbell (1986) hypothesized that shoalward movement of ovigerous females to warmer waters enhances the completion of egg development and the transition through the larval stages (see also Cooper and Uzmann 1971).

The extent to which inshore lobsters exhibit distribution shifts to offshore habitats with increasing size or age is not currently known. Ontogenetic shifts in shelter dependence and distribution patterns in different substrate types have been documented (Wahle 1992; Wahle and Steneck 1991, 1992). Smaller lobsters occupy habitat with high structural complexity (Cobb 1971) and associated shelter sites while larger lobsters are less dependent on shelter. Structurally complex habitats are more prevalent in immediate coastal areas while deeper water sites (with the notable exception of submarine canyons) are characterized by lower structural complexity. Reduction in shelter dependence and greater dispersal ranges at larger sizes could lead to ontogenetic shifts in distribution with respect to depth and distance from shore (R. Steneck, Darling Marine Science Center, Walpole, Maine 04573, U.S.A., personal communication).

Determinants of stability and resilience

The response of any population to exploitation depends on the nature and strength of compensatory mechanisms and the importance of immigration and emigration at different life history stages. A population with little or no compensatory capacity will be particularly vulnerable to overexploitation (see Fogarty and Idoine 1986; Pollock 1993 for further discussion). Consider a model for a closed population in which density dependence occurs among members of a cohort during the prerecruit phase. Assume further that population processes following recruitment are density independent. The critical factors affecting the stability and resilience of the population to exploitation can be represented graphically by linking the relationship between egg production (or its proxy, adult population size) and recruitment with a recruitment – egg production relationship (see Fig. 1). The intersection points between the asymptotic egg production – recruitment relationship and the linear recruitment – egg production function are stable equilibrium points. As fishing mortality rates increase, the slope of recruitment – egg production relationship increases. Increasing fishing mortality rates eventually result in a point where no intersection is possible and a stock collapse is predicted (Fig. 1*a*). It is clear that the slope of the recruitment curve at low levels of egg production is critical in determining the levels of exploitation that the stock can sustain without collapse; the steeper the slope at the origin, the higher the resilience of the stock to exploitation.

For the case of a population receiving a constant subsidy of prerecruits from an exogenous source, a different dynamic is found. The effect of the subsidy is to shift the level of the recruitment curve and to result in a nonzero intercept in the egg production – recruitment relationship (Fig. 1*b*). The stable equilibria are again defined by the intersection of the egg production – recruitment and recruitment – egg production relationships. Note, however, that the nonzero intercept in the egg production – recruitment relationship implies that the population can persist under any level of fishing mortality as long as the subsidy continues. These considerations highlight the twin themes to be explored in this paper: the importance of density-dependent processes and the potential role of source populations in permitting persistence of exploited populations.

Model structure

The basic model for a system comprising two local populations in discrete time and space can be written:

Fig. 2. Total numerical yield from an inshore–offshore American lobster complex in which (*a*) the two groups are completely independent and (*b*) the case where an offshore subpopulation provides a larval subsidy to an inshore group (20% of larvae produced offshore are transported inshore and compete with settlers from local production). Adult lobsters (10%) migrate from the inshore to offshore subpopulations. Units are thousands of individuals.

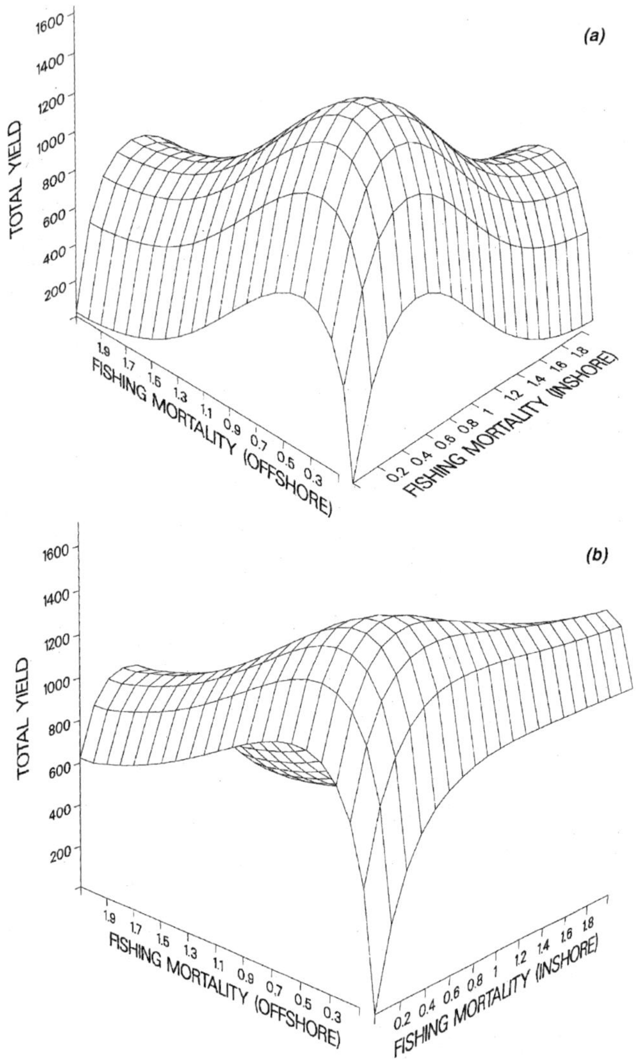

$$N_{1,t} = \delta_1\rho_1 N_{1,t-1} + \gamma_1 f_1(N_{t-r}) + (1-\gamma_2) f_2(N_{t-r}) + \rho_2(1-\delta_2)N_{2,t-1}$$

$$N_{2,t} = \delta_2\rho_2 N_{2,t-1} + \gamma_2 f_3(N_{t-r}) + (1-\gamma_1) f_4(N_{t-r}) + \rho_1(1-\delta_1)N_{1,t-1}$$

where $N_{i,t}$ is the number of adults in the ith subpopulation at time t, δ_i is the proportion of nonmigratory adults, γ_i is the proportion of prerecruits remaining within the natal region, ρ_i is adult survivorship, and $f_i(N_{t-r})$ are functions relating adult population size and recruitment from different sources (note that N_{t-r} is deliberately not indexed by subpopulation but instead can represent a weighted sum of the two groups as described below; r is the reproductive delay). The stock-recruitment relationships consider transport of larvae between subpopulations and potentially include abundance terms for both groups, depending on the sequence of critical life history events and the timing of density-dependent processes. I will consider the special case of $\delta_2 = \gamma_1 = 1$ with $\delta_1 = 0.9$ and $\gamma_2 = 0.8$ to represent an inshore–offshore system with a larval subsidy from the offshore to the inshore group and directional movement of adults from inshore to offshore.

A fraction of the larvae produced by the offshore subpopulation are assumed to be transported inshore where they compete with settlers from the inshore group. For the offshore group, only intragroup competition among settlers is considered. This mechanism leads to an asymptotic relationship between the adult stock and recruitment. This general form is considered to be appropriate for lobster populations based on theoretical and empirical considerations (Caddy 1986; Ennis 1986; Fogarty and Idoine 1986; G.P. Ennis and M.J. Fogarty, in press). It is implicitly assumed that the adult migration from inshore to offshore follows the period of larval release.

The specific model was:

$$N_{1,t} = \delta_1\rho_1 N_{1,t-1} + \gamma_1 \frac{\alpha_1 N_{1,t-r}}{1 + (N_{t-r}/K_1)} + (1-\gamma_2)\frac{\alpha_2 N_{2,t-r}}{1 + (N_{t-r}/K_1)}$$

$$N_{2,t} = \delta_2\rho_2 N_{2,t-1} + \gamma_2 \frac{\alpha_2 N_{2,t-r}}{1 + (N_{2,t-r}/K_2)} + \rho_1(1-\delta_1)N_{1,t-1}$$

where the subscripts 1 and 2 refer to inshore and offshore populations, respectively, α_i is a parameter reflecting fecundity and survivorship during the prerecruit phase, K_i represents a threshold abundance level above which density-dependence dominates, and $N_{t-r} = \gamma_1\alpha_1 N_{1,t-r} + (1-\gamma_2)\alpha_2 N_{2,t-r}$. The numerical values used for the recruitment component of this analysis have been scaled to approximate yields and population sizes from the inshore and offshore southern New England region at observed exploitation rates and mean fecundity and prerecruit survival rates. I have assumed a time delay of five years between the egg stage and recruitment to the adult population. I have further assumed that the juveniles are subjected to one year of harvesting before reaching the adult stage. Knife-edge selection is assumed. Sensitivity analyses show that the qualitative features of the results are not dependent on the length of the reproductive delay. The adult survivorship component is a function of the instantaneous rates of natural (M_i) and fishing mortality (F_i) viz, $\rho_i = \exp(-M_i - F_i)$. Estimates of the instantaneous rate of natural mortality for the late juvenile and adult stages of 0.10 were used in this analysis (Fogarty 1995).

To focus attention on the qualitative differences that emerge in comparisons of linked versus independent group systems, several simplifications are made in the following analyses. It is assumed that only adults are harvested and that the age at maturation (reproductive delay) and the density-independent and density-dependent parameters of the recruitment functions are identical in both groups. These constraints are easily removed to explore the consequences of different life history parameters between groups. No attempt is made to account for sex-specific regulations such as protection of egg-bearing

Fig. 3. Isopleths of total numerical yield for an inshore–offshore American lobster complex corresponding to Fig. 2 for (*a*) the case of discrete populations and (*b*) linked subpopulations as defined in Fig. 2. Units are thousands of individuals.

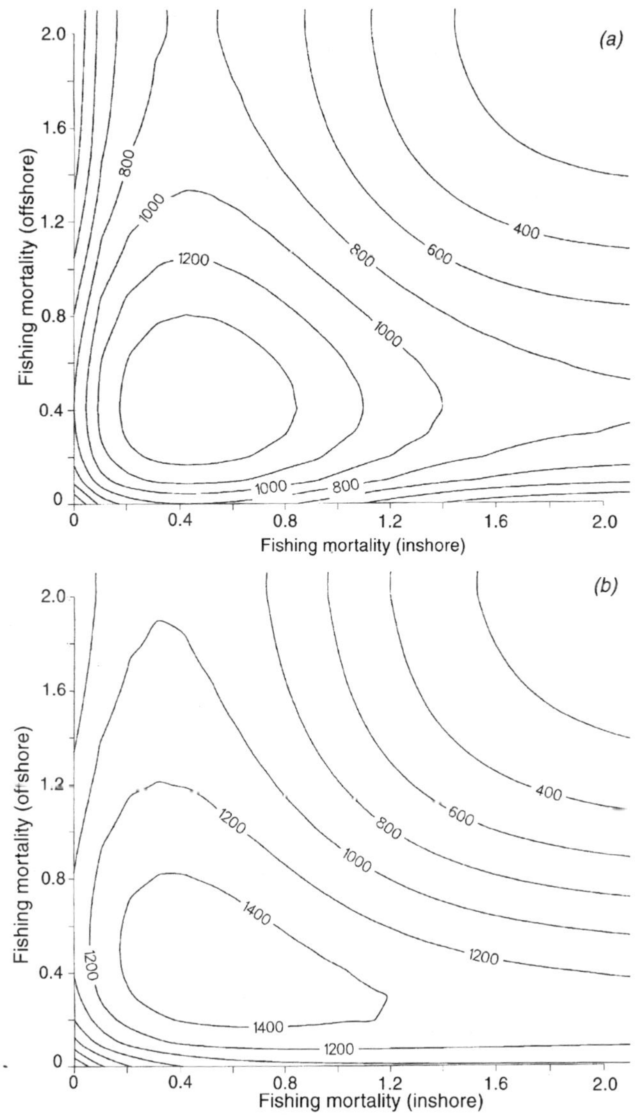

females. Incorporation of these factors would require a more complicated age-structured model that distinguishes between sexes and between ovigerous and nonovigerous females but would not contribute to further understanding of the effects of linkages between subpopulations.

Results

The expected behavior of a system in which two discrete populations are subject to differential levels of exploitation is first considered to provide a frame of reference for subsequent analyses. Increasing levels of fishing mortality in each population eventually result in sharp declines in abundance and yield (Fig. 2*a*). The expected yield as a function of increasing fishing mortality rates is dome-shaped for each group and the combined yield from both groups occurs at intermediate levels of fishing mortality. Exploitation rates in one fishery have no impact on the other. Overexploitation of one component clearly affects the total yield from the complex but cannot interact to affect the other.

The qualitative behavior of the system, however, changes markedly with consideration of linkages between groups through larval dispersal and adult migration. Potentially high levels of fishing mortality can be sustained in one subpopulation if exploitation rates in the other remain low (Fig. 2*b*). For example, the inshore yield changes from convex (independent populations) to asymptotic (linked subpopulations) with increasing inshore fishing mortality for the case where no offshore fishery exists. Similarly, the subsidy provided by migration of adults from inshore to offshore permits higher levels of offshore fishing mortality when inshore exploitation rates are low to moderate than for the case of independent populations. The maximum yield again occurs at low to moderate levels of fishing mortality on both subpopulations (Fig. 2). Yield isopleths for the case of separate versus linked subpopulations demonstrate the loci of maximum yield for both cases and the effects of increasing fishing mortality rates in both fisheries (Figs. 3*a*, 3*b*).

For the case of linked subpopulations and in the absence of an offshore fishery, the highest numerical yield from the inshore fishery occurs at intermediate levels of fishing mortality (Fig. 4*a*). The inshore yield is effectively independent of the coastal fishing mortality rates over a broad range ($F_1 > 0.6$) when no offshore fishery exists. At higher levels of inshore fishing mortality, the development of an offshore fishery results in declining inshore yield as the larval subsidy is reduced due to the effects of the offshore fishery (Fig. 4*a*). If the linkage between the inshore and offshore groups is not recognized, steadily increasing offshore fishing mortality rates could lead to the decline and ultimate collapse of the inshore fishery under these conditions. Conversely, in the absence of an inshore fishery, the yield from the offshore fishery occurs at moderate levels of offshore fishing mortality (Fig. 4*b*). Offshore numerical yield declines with increasing inshore fishing mortality at moderate to high levels of offshore exploitation as the contribution due to migration of adults from inshore to offshore is affected by the fishery. The asymmetry in the impact on the inshore and offshore yield to the development of the fishery in the other component can be attributed to the differences in the linking mechanisms. The magnitude of the contributions from offshore to inshore and the reverse differs in the baseline case considered here. In addition, the larval subsidy provided by the offshore population is subject to density-dependent effects in this formulation while the migratory contribution of adults from inshore to offshore is not.

Comparisons of inshore yield as a function of varying levels of larval subsidy from offshore to inshore reveal a shift in the qualitative behavior between the case of independent populations (no larval subsidy) and even small levels of larval exchange (Fig. 5). For example, with no linkage between inshore and offshore groups, the inshore population and yield decline to extinction with $F_1 > 2.0$. In contrast, with as little as 10% of the offshore larvae transported inshore, the numerical inshore

Fig. 4. Isopleths of numerical yield from (*a*) the inshore subpopulation and (*b*) the offshore population as a function of instantaneous rates of fishing mortality on the inshore and offshore subpopulations. Units are thousands of individuals.

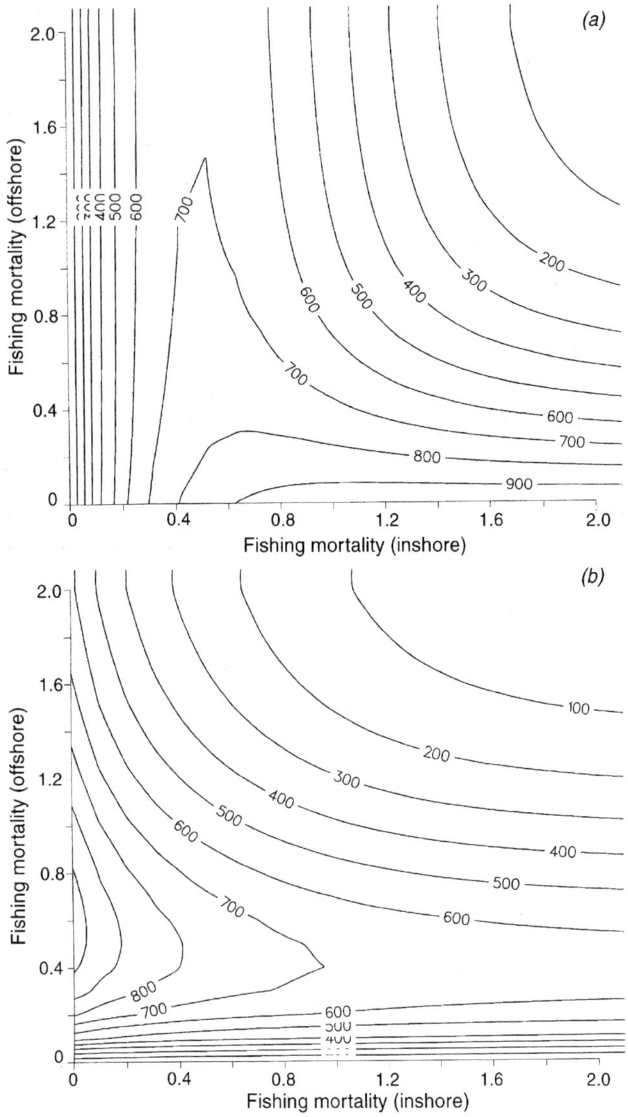

Fig. 5. Inshore numerical yield as a function of the instantaneous rate of fishing mortality in inshore areas at several levels of subsidy from the offshore population ranging from 0 to 50% of the offshore larval production. The instantaneous fishing mortality rate on the offshore population was set at 0.50 and 10% of the inshore adult subpopulation was assumed to migrate offshore. Units are thousands of individuals.

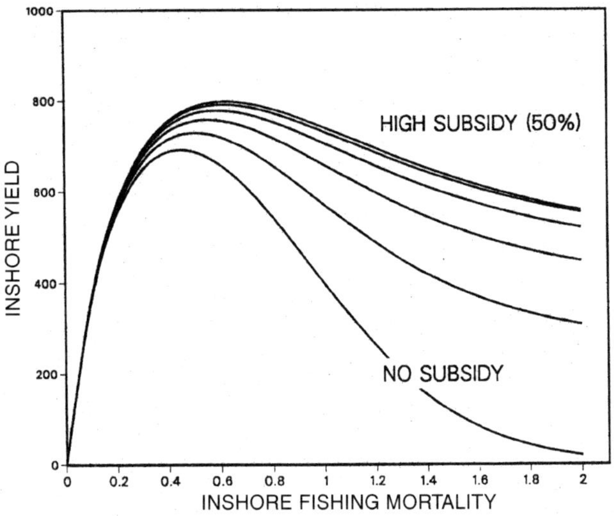

yield at $F_1 = 2.0$ remains relatively high at approximately 50% of the maximum equilibrium yield (attained here at $F_1 \approx 0.5$). Increasing degrees of larval subsidy up to 50% of the offshore larval production results in a less convex relationship between numerical yield and fishing mortality (i.e., reductions in yield at higher levels of fishing mortality become less pronounced at higher levels of larval subsidy; see Fig. 5).

The higher the level of subsidy provided from offshore to inshore subpopulations, the more vulnerable the offshore group is to exploitation. The recruitment function for the offshore region under varying levels of transport to the inshore subpopulation is depicted in Fig. 6. Note that transport (loss) of larvae from the offshore to inshore regions results in a reduction in the slope of the recruitment curve at the origin. In the absence of any return migration of adults from inshore to offshore, in particular, the offshore stock is at increasingly greater risk of overexploitation with increasing levels of larval transport to the inshore group. This suggests that the uncertainty associated with levels of larval subsidy provided from the offshore stock would require greater caution in exploiting this subpopulation.

The analyses above assume that the population sizes of the two groups are equal in the unexploited state. Lobster densities may be substantially higher in inshore areas with favorable habitat than areas further from shore characterized by lower structural complexity. However, the total area encompassed by the offshore areas as defined in this analysis substantially exceeds that in the immediate coastal areas. Accordingly, it is not unreasonable to consider that the total population sizes are similar in each area. To explore the impact of the assumption of equal population size, a sensitivity analysis was carried out in which the density-dependent parameter K_2 for the offshore subpopulation was varied. The equilibrium population size scales monotonically with this parameter. The expected inshore yield, at an inshore fishing mortality of 2.0 and an offshore fishing mortality rate of 0.5, was used as in indicator of the effect of varying the threshold for density dependence for the offshore group by ±50%. Changing the relative population sizes of the inshore and offshore groups did not alter the qualitative behavior of the model. The inshore group continued to exhibit resilience to exploitation when receiving a larval subsidy when the offshore subpopulation was reduced. A 50% decrease in K_2 resulted in a 22.7 decrease in numerical inshore yield at the designated fishing mortality rates; the yield, however, remained substantially above that expected for the case where no larval subsidy was assumed. A 50% increase in K_2 resulted in a 13.8% increase in inshore yield at $F_1 = 2.0$ and $F_2 = 0.5$. The conclusion that resilience is conferred by a larval

Fig. 6. Recruitment to the offshore subpopulation as a function of offshore spawning stock size under varying levels of transport of larvae from offshore to inshore ranging from 0 to 50%. Units are thousands of individuals.

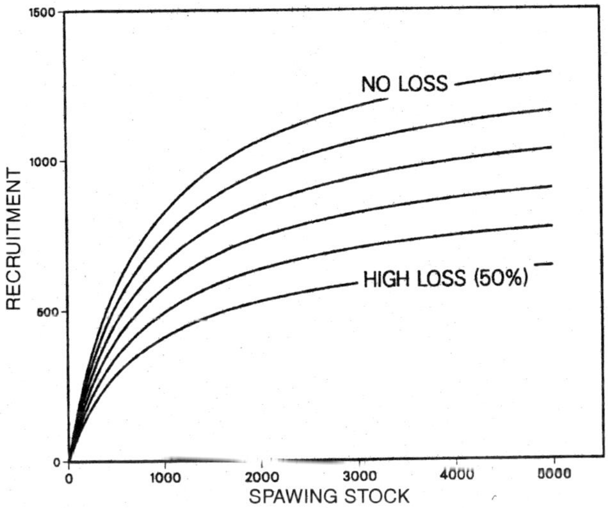

Fig. 7. Total numerical yield from an inshore–offshore system as a function of the instantaneous rate of fishing mortality in inshore and offshore areas under density-dependent migration of adults from inshore to offshore. Units are thousands of individuals.

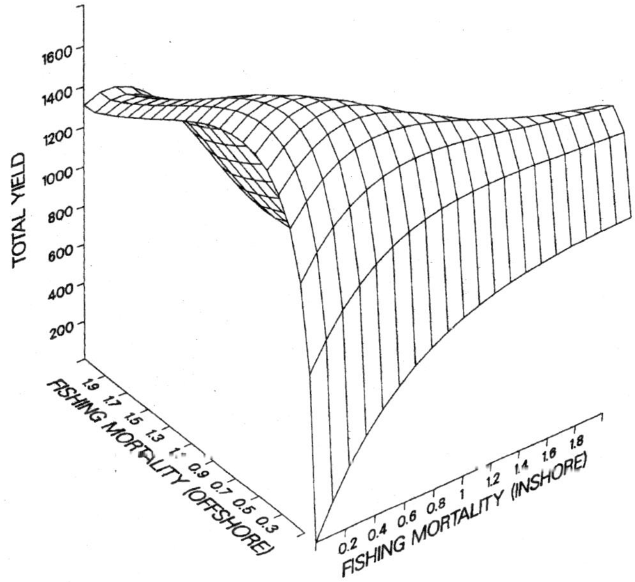

subsidy to the inshore area is robust under varying assumptions concerning the relative sizes of the two groups.

Additional considerations such as density-dependent movement of adults can be readily accommodated. If critical resources such as shelter become limiting, the propensity to undertake directed movements may be increased. To explore the implications of increased probability of movement with increasing abundance, I examined the case where δ_1 was a function of the abundance of inshore adults viz, $\delta_1 = \exp(-(a + bN_1))$, where a and b represent instantaneous movement coefficients. For the purposes of this demonstration, the probability of movement at low population size was scaled to equal the baseline case $\delta_1 = 0.1$. The inshore population remains resilient to exploitation at high fishing mortality rates when offshore fishing intensity remains moderate (Fig. 7). For the particular choice of movement parameters examined here, the relative yield in the offshore region increases as a result of increased export of adults from inshore to offshore; inshore yield correspondingly declines. The overall levels of offshore movement of inshore adults are higher in this analysis than in the baseline case.

Discussion

Inshore lobster populations have long sustained intensive levels of fishing mortality. The historical pattern of the fishery has been characterized by an escalation in exploitation rates in coastal regions and the development of new fisheries in offshore areas within the last several decades. The results of the present analysis suggest that even relatively low levels of larval subsidy from offshore to inshore populations could contribute strongly to the resilience of inshore populations to high levels of exploitation in coastal areas. The development of offshore fisheries could affect this mechanism if exploitation rates are not constrained. Patterns of relatively stable or only slightly decreasing yields under increasingly intense levels of exploitation have been observed in several clawed and spiny lobster populations (e.g., Saila et al. 1979; Fogarty and Murawski 1986; Booth and Breen 1994; Hall and Brown 1994). Refugia of various types are likely to play an important role in the dynamics of these populations. Fishing intensity is generally lower in offshore areas where the capital investment to enter the fishery and operating expenses are higher. Other types of refugia may exist because of topographic features that impede harvesting or other factors.

The linkages through dispersal and migration employed in this analysis imply gene flow between groups. Examination of genetic structure of inshore and offshore lobster groups to date has shown either no differences (Barlow and Ridgway 1971; Odense and Annand 1978; Kornfield and Moran 1990) or differences between groups with known potential isolating mechanisms. Tracey et al. (1975) noted differences between inshore southern New England (near Martha's Vineyard) and offshore lobsters on Georges Bank. Spring and summer circulation patterns on Georges Bank are characterized by an anti-cyclonic gyre driven by strong rotary tidal currents. Retention of larvae released on the bank is expected and little or no interchange is expected with inshore populations. More generally, genetic differences between widely separated groups have been identified (e.g., between lobsters from the Gulf of St. Lawrence and New England; see Tracey et al. 1975; Kornfield and Moran 1990). Evidence for genetic differentiation on a smaller spatial scale within the Gulf of Maine has been reported based on a reexamination of the Tracey et al. (1975) results (Burton and Feldman 1982; Burton 1983; Shaklee 1983). It should be noted that genetic differentiation is still possible even with interchange among groups under certain selective forces (Hedgecock 1986). Morphological and demographic differences between inshore and offshore lobsters have been reported (Saila and Flowers 1969; Cadrin 1995; Russell

1980; Aiken 1980; Aiken and Waddy 1980; Aiken and Waddy 1986). Other discriminators based on parasite infestation in offshore and inshore areas have been reported (Uzmann 1970; Campbell and Brattey 1986). It is not currently known if a genetic basis exists for any of these observed differences. Further work on identifying genetic differences among lobster groups would help in resolving the issue of whether interchange among specific groups is important. Existing evidence is consistent with the hypothesis of interchange among groups on the spatial scales considered in this analysis.

Factors underlying the increase in lobster landings over the last three decades have not been conclusively identified. Temperature is known to affect key aspects of lobster biology (including growth and survival during the early life stages) and behavior (e.g., activity levels affecting movement, feeding, and vulnerability to capture). However, no broad-scale changes in temperature regimes that could explain the increase in landings have been detected (Drinkwater et al. 1996). Large-scale changes in wind fields affecting oceanographic conditions also play a critical role in larval survival (Drinkwater et al. 1991, 1996). It has further been hypothesized that declines in predatory fish populations throughout the northwest Atlantic may be a contributing factor to the increase in lobster abundance (e.g., Pezzack 1992; reviewed in Fogarty 1995) although, again, evidence for this hypothesis is equivocal.

The potential for larval dispersal from offshore (or less intensively exploited) regions to heavily exploited inshore areas could point to possible mechanisms underlying the increase in abundance noted in coastal locations. Specifically, enhanced shoreward transport could result in increased larval supply to inshore sites. Wind-driven transport mechanisms would, in particular, apppear to offer fruitful avenues for further research. Interestingly, while lobster abundance appears to have increased dramatically in the near-shore southern New England area, abundance of offshore lobsters on Georges Bank and off southern New England has generally decreased (Northeast Fisheries Science Center 1996).

Although specific attention has been placed on an inshore–offshore system in this analysis, the results hold for other systems characterized by a spatial partitioning of abundance and fishing intensity. In areas where lobsters are confined to the immediate coastal area, there still may exist refugia operating through other constraints on fishing effort. Included within this classification are areas of low density and marginal catch-per-unit effort in which fishing is not profitable but which nonetheless support viable lobster populations. Refugia defined by temporal partitioning of lobster availability and fishing seasons could also play an important role (Anthony and Caddy 1980).

The resilience afforded by spatial structure observed in this analysis is consistent with previous work on spatially explicit harvesting models (e.g., Beverton and Holt 1957; Caddy 1975, 1993; Fogarty and Murawski 1986; Allen and McGlade 1986; Hilborn and Walters 1987; MacCall 1989; Polacheck 1990; Botsford et al. 1993, 1994; Quinn et al. 1993; Walters et al. 1993; Tyler and Rose (1994); Man et al. 1995; Tuck and Possingham 1995; Ye and Beddington 1996; Holland and Brazee 1996). The importance of source and sink population structures has also received particular attention in other areas of applied ecology, particularly conservation biology (e.g., Pulliam 1988; Howe et al. 1991; Hanski 1991; Hanski and Gilpin 1991) where a system of interconnected subpopulations has been shown to be highly resilient to stress.

The fundamental model structure used in the present analysis could be readily adapted to examine the efficacy of long-term spatial closures or marine reserves. This issue has been approached in the context of metapopulation theory (Man et al. 1995) and in various age-structured population models (e.g., Holland and Brazee 1996). A variant on this theme involving rotating spatial closures has also elicited considerable interest (e.g., Caddy 1993; Botsford et al. 1993). Recent reviews of theory and practice of marine reserve design with implications for harvested systems indicate considerable promise in this approach (Dugan and Davis 1993; Agardy 1994; Bohnsack and Ault 1996; Eichenbaum et al. 1996). Such methods may be easier to implement and enforce than alternative management methods while preserving reproductive potential and serving as a hedge against uncertainties in population and ecosystem dynamics. In the context of the present analysis, setting aside some fraction of the offshore region as a refuge would appear to hold considerable promise if validation of the proposed transport mechanisms is obtained.

Several general themes emerge from a consideration of spatial dynamics in lobster populations. First, a larval subsidy from offshore to inshore populations at even low levels can permit persistence under intense exploitation and could explain, in part, the apparent resilience of coastal stocks to very high fishing mortality rates. The maximum numerical yield from the inshore stocks occurs at moderate levels of fishing mortality. Accordingly, even if a larval subsidy from offshore populations is important, clear benefits to constraining fishing mortality in coastal areas are evident. Excessive exploitation of inshore groups is not the optimal strategy with respect to maximizing numerical yield. Second, contribution of larvae from the offshore to inshore groups results in a reduction in the resilience of offshore stocks to exploitation, particularly for the case of little or no return migration from inshore. Given the uncertainty concerning the magnitude of transport of larvae from offshore to inshore, conservative management of offshore stocks is indicated. This result is consistent with other analyses of metapopulation dynamics which suggest that source populations should be managed more conservatively than sink populations (e.g., Tuck and Possingham 1994).

Fogarty and Idoine (1986) suggested that well-developed compensatory processes play an important role in the resilience of lobster populations to exploitation. It was further noted that refuge areas with low exploitation rates could contribute to strong resilience of inshore lobster groups. Recent advances in modeling advective processes and transport of lobster larvae (e.g., Katz et al. 1994) have reopened the question of how these mechanisms might interact. The present study indicates that even low levels of interchange can be highly stabilizing and reinforces the need to understand the factors affecting the dispersal of the larval and postlarval stages and the issue of ontogenetic shifts in habitat utilization.

Acknowledgments

I am grateful for the helpful comments of two referees and for helpful discussions with my colleagues at the Northeast Fisheries Science Center.

References

Agardy, M.T. 1994. Advances in marine conservation: the role of marine protected areas. Trends Ecol. Evol. **9**: 267–270.

Aiken, D.E. 1980. Molting and growth. *In* The biology and management of lobsters. Vol. 1. *Edited by* J.S. Cobb and B.F. Phillips. Academic Press, New York, N.Y. pp. 91–163.

Aiken, D.E., and Waddy, S.L. 1980. Reproductive biology. *In* The biology and management of lobsters. Vol. 1. *Edited by* J.S. Cobb and B.F. Phillips. Academic Press, New York, N.Y. pp. 215–276.

Aiken. D.E., and Waddy, S.L. 1986. Environmental influences on recruitment of the American lobster, *Homarus americanus*. Can. J. Fish. Aquat. Sci. **43**: 2258–2270.

Allen, P.M., and McGlade, J.M. 1986. Dynamics of discovery and exploitation: The case of the Scotian Shelf groundfish fisheries. Can. J. Fish. Aquat. Sci. **43**:1187–1200.

Anthony, V.C., and Caddy, J.F. (*Editors*). 1980. Proceedings of the Canada-U.S. Workshop on Status of Assessment Science for N.W. Atlantic Lobster (*Homarus americanus*) Stocks. St. Andrews, N.B., Oct. 24–26, 1978. Can. Tech. Rep. Fish. Aquat. Sci. 932.

Barlow, J., and Ridgway, G.J. 1971. Polymorphisms of esterase isozymes in the American lobster (*Homarus americanus*). J. Fish. Res. Board Can. **28**: 15–21.

Beverton, R.J.H., and Holt, S.J. 1957. On the dynamics of exploited fish populations. Fish. Invest. Ser. II Mar. Fish. G.B. Minist. Agric. Fish Food.

Bohnsack, J.A., and Ault, J.S. 1996. Management strategies to conserve marine biodiversity. Oceanogr. Mar. Biol. Annu. Rev. **9**: 93–92.

Booth, J.S., and Breen, P.A. 1994. The New Zealand fishery for *Jasus edwardsii* and *J. verreauxi*. *In* Spiny lobster management. *Edited by* B.F. Phillips, J.S. Cobb, and J. Kittaka. Fishing News Books, Blackwell Sci. Pub., Cambridge, U.K. pp. 64–75.

Botsford, L.W., Quinn, J.F., Wing, S.R., and Brittnacher, J.G. 1993. Rotating spatial harvest of a benthic invertebrate, the red sea urchin, *Strongylocentrotus franisicanus*. *In* Proceedings of the International Symposium on Management Strategies for Exploited Fish Populations. Alaska Sea Grant College Program AK SG-93-02.

Botsford, L.W., Moloney, C.L., Hastings, A., Largier, J.L., Powell, T.M., Higgens, K., and Quinn, J.F. 1994. The influence of spatially and temporally varying oceanographic conditions on meroplanktonic metapopulations. Deep-Sea Res. **41**: 107–145.

Burton, R.S. 1983. Protein polymorphism and genetic differentiation of marine invertebrate populations. Mar. Biol. Lett. **4**: 193–206.

Burton, R.S., and Feldman, M.W. 1982. Population genetics of coastal and estuarine invertebrates: does larval behavior influence population structure? *In* Estuarine comparisons. *Edited by* V.S. Kennedy. Academic Press, New York, N.Y. pp. 537–551.

Caddy, J.F. 1975. Spatial model for an exploited shellfish population and its application to the Georges Bank Scallop fishery. J. Fish. Res. Board Can. **32**: 1305–1328.

Caddy, J.F. 1986. Modelling stock-recruitment processes in Crustacea: some practical and theoretical perspectives. Can. J. Fish. Aquat. Sci. **43**: 2330–2344.

Caddy, J.F. 1993. Background concepts for a rotating harvesting strategy with particular reference to the Mediterranean red coral, *Corallium rubrum*. Mar. Fish. Rev. **55**: 10–18.

Cadrin, S.X. 1995. Discrimination of American lobster (*Homarus americanus*) stocks off southern New England on the basis of secondary sex character allometry. Can. J. Fish. Aquat. Sci. **52**: 2712–2723.

Campbell, A. 1986. Migratory movements of ovigerous lobsters, *Homarus americanus*, tagged off Grand Manan, eastern Canada. Can. J. Fish. Aquat. Sci. **43**: 2197–2205.

Campbell, A., and Brattey, J. 1986. Egg loss from the American lobster, *Homarus americanus* in relation to nermertean, *Pseudocarcinomertes homari*, infestation. Can. J. Fish. Aquat. Sci. **43**: 772–780.

Campbell, A., and Stasko, A.B. 1985. Movements of tagged American lobsters, *Homarus americanus*, off southwestern Nova Scotia. Can. J. Fish. Aquat. Sci. **42**: 229–238.

Campbell, A., and Stasko, A.B. 1986. Movements of lobsters *Homarus americanus* tagged in the Bay of Fundy, Canada. Mar. Biol. (Berlin), **92**: 393–404.

Campbell, A., Graham, D.E., MacNichol, H.J., and Williamson, A.M. 1984. Movement off tagged lobsters released on the continental shelf from Georges Bank to Baccaro Bank, 1971–1973. Can. Tech. Rep. Fish. Aquat. Sci. 1288.

Cobb, J.S. 1971. The shelter-related behavior of the lobster, *Homarus americanus*. Ecology, **52**: 108–115.

Cobb, J.S. 1995. Interface of ecology, behavior, and fisheries. *In* The biology of the lobster, *Homarus americanus*. *Edited by* J. Factor. Academic Press, New York, N.Y. pp. 139–149.

Cobb, J.S., Gulbranson, T., Phillips, B.F., Wang, D., and Syslo, M. 1983. Behavior and distribution of larvae and early juvenile *Homarus americanus*. Can. J. Fish. Aquat. Sci. **40**: 2184–2188.

Cobb, J.S., Wang, D., Campbell, D.B., and Rooney, P. 1989. Speed and direction of swimming by postlarvae of the American lobster. Trans. Am. Fish. Soc. **118**: 82–86.

Cooper, R.A., and Uzmann, J.R. 1971. Migrations and growth of deep-sea lobsters, *Homarus americanus*. Science (Washington, D.C.), **171**: 288–290.

Dow, R.L. 1974. American lobsters tagged by Maine commercial fishermen, 1957–1959. Fish. Bull. (U.S.), **72**: 622–623.

Drinkwater, K.F., Harding, G.C., Vass, W.P., and Gauthier, D. 1991. The relationship of Quebec lobster landings to freshwater runoff and wind storms. Can. Spec. Publ. Fish. Aquat. Sci. 113. pp. 179–187.

Drinkwater, K.F., Harding, G.C., Mann, K.H., and Tanner, N. 1996. Temperature as a possible factor in the increased abundance of American lobter, *Homarus americanus*, during the 1980's and early 1990's. Fish. Oceanogr. **5**: 176–193.

Dugan, J.E., and Davis, G.E. 1993. Introduction to the international symposium on marine harvest refugia. Can. J. Fish. Aquat. Sci. **50**: 1991–1992.

Eichenbaum, W.M., Crosby, M.P., Agardy, M.T., and Laskin, S.A. 1996. The role of marine and coastal protected areas in the conservation and sustainable use of biological diversity. Oceanogr. Mar. Biol. Annu. Rev. **9**: 60–70.

Elner, R.W., and Campbell, A. 1991. Spatial and temporal patterns in recruitment for American lobster, *Homarus americanus*, in the Northwestern Atlantic. Mem. Queensl. Mus. **31**: 349–363.

Ennis, G.P. 1986. Stock definition, recruitment variability, and larval recruitment processes in the American lobster, *Homarus americanus*: a review. Can. J. Fish. Aquat. Sci. **43**: 2072–2084.

Ennis, G.P. 1995. Larval and post-larval ecology. *In* The biology of the lobster, *Homarus americanus*. *Edited by* J. Factor. Academic Press, New York, N.Y. pp. 23–46.

Ennis, G.P., and Fogarty, M.J. 1997. A recruitment overfishing reference point for the American lobster, *Homarus americanus*. Mar. Freshwater Res. In press.

Fisheries Resource Conservation Council. 1995. A conservation framework for Atlantic lobster. Fisheries Resource Conservation Council FRCC95.R.1. Min. Supply Serv. Canada.

Fogarty, M.J. 1983. Distribution and relative abundance of American lobster, *Homarus americanus* larvae: a review. NOAA Tech. Rep. SSRF 775. pp. 3–8.

Fogarty, M.J. 1995. Fisheries, populations, and management. *In* The biology of the American lobster, *Homarus americanus*. *Edited by* J. Factor. Academic Press, New York, N.Y.

Fogarty, M.J., and Idoine, J.S. 1986. Recruitment dynamics in an American lobster (*Homarus americanus*) population. Can. J. Fish. Aquat. Sci. **43**: 2368–2376.

Fogarty, M.J., and Murawski, S.A. 1986. Population dynamics and assessment of exploited invertebrate stocks. Can. Spec. Publ. Fish. Aquat. Sci. 92. pp. 228–244.

Fogarty, M.J., Borden, D.V.D., and Russell, H.J. 1980. Movements of tagged American lobster, *Homarus americanus*, off Rhode Island. Fish. Bull. (U.S.), **78**: 771–780.

Fogarty, M.J., Cooper, R.A., Uzmann, J.R., and Burns, T. 1982. Assessment of the USA offshore American lobster (*Homarus americanus*) fishery. Int. Counc. Explor. Sea. C.M. 1982/K:14 Copenhagen, Denmark.

Hall, N.G., and Brown, R.S. 1994. Modelling for management - the western rock lobster fishery. *In* Spiny lobster management. *Edited by* B.F. Phillips, J.S. Cobb, and J. Kittaka. Fishing News Books, Blackwell Science Inc., Cambridge, U.K. pp. 311–322.

Hanski, I. 1991. Single-species metapopulation dynamics: concepts, models, and observations. Biol. J. Linn. Soc. **42**: 17–38.

Hanski, I., and Gilpin, M. 1991. Metapopulation dynamics: brief history and conceptual domain. Biol. J. Linn. Soc. **42**: 3–16.

Harding, G.C., and Trites, R.W. 1988. Dispersal of *Homarus americanus* larvae in the Gulf of Maine from Browns Bank. Can. J. Fish. Aquat. Sci. **45**: 416–425.

Harding, G.C., Drinkwater, K.F., and Vass, W.P. 1983. Factors influencing the size of American lobster (*Homarus americanus*) stocks along the Atlantic coast of Nova Scotia, Gulf of St. Lawrence, and Gulf of Maine: a new synthesis. Can. J. Fish. Aquat. Sci. **40**: 168–184.

Hedgecock, D. 1986. Is gene flow from pelagic larval dispersal important in the adaption and evolution of marine invertebrates? Bull. Mar. Sci. **39**: 550–564.

Hilborn, R., and Walters, C.J. 1987. A general model for simulation of stock and fleet dynamics in spatially heterogeneous fisheries. Can. J. Fish. Aquat. Sci. **44**: 1366–1369.

Holland, D.S., and Brazee, R.J. 1996. Marine reserves for fisheries management. Mar. Res. Econ. **11**: 157–171.

Howe, R.W., Davis, G.J., and Mosca, V. 1991. The demographic significance of 'sink' populations. Biol. Conserv. **57**: 239–255.

Katz, C.H., Cobb, J.S., and Spaulding, M. 1994. Larval behavior, hydrodynamic transport, and potential offshore recruitment in the American lobster, *Homarus americanus*, Mar. Ecol. Prog. Ser. **103**: 265–273.

Kornfield, I., and Moran, P. 1990. Genetics of population differentiation in lobsters. *In* Life history of the American lobster: Proceedings of a Workshop, November 29–30, 1989, Orono, Maine. *Edited by* I. Kornfield. Lobster Institute, Orono, Maine. pp. 23–24. (Abstract).

Lawton, P., and Lavalli, K.L. 1995. Postlarval, juvenile, adolescent, and adult ecology. *In* The Biology of the lobster, *Homarus americanus*. *Edited by* J. Factor. Academic Press, New York, N.Y. pp. 47–88.

Man, A., Law, R., and Poulin, N.V.C. 1995. Role of marine reserves in recruitment to reef fisheries: a metapopulation model. Biol. Conserv. **71**: 197–204.

MacCall, A. 1989. Dynamic geography of marine fish populations. University of Washington, Sea Grant Press, Seattle, Wash.

McLeese, D.W., and Wilder, D.G. 1958. The activity and catchability of the lobster (*Homarus americanus*) in relation to temperature. J. Fish. Res. Board Can. **15**: 1345–1354.

Miller, R.J., Duggan, R.E., Robinson, D.G., and Zheng, Z. 1989. Growth and movement of *Homarus americanus* on the outer coast of Nova Scotia. Can. Tech. Rep. Fish. Aquat. Sci. 1716.

Northeast Fisheries Science Center. 1993. American lobster. *In* Report of the 16th Northeast Regional Stock Assessment Workshop (16th SAW). Northeast Fisheries Science Center Ref. Doc. 93-18. Woods Hole, Mass. pp. 75–107.

Northeast Fisheries Science Center. 1996. American lobster. *In* Report of the 22th Northeast Regional Stock Assessment Workshop (22nd SAW). Northeast Fisheries Science Center Ref. Doc. 96-13. Woods Hole, Mass. pp. 47–134.

Odense, P.H., and Annand, C. 1978. Isozyme systems of an inshore and offshore lobster population. Int. Counc. Explor. Sea C.M. 1978/K:15.

Pezzack, D.S. 1989. Comments on dispersal of *Homarus americanus* larvae in the Gulf of Maine from Browns Bank. *Edited by* G.C. Harding and R.W. Trites. Can. J. Fish. Aquat. Sci. **46**: 1079–1083.

Pezzack, D.S. 1992. A review of lobster (*Homarus americanus*) landings and trends in the Northwest Atlantic, 1947–1986. J. Northwest Atl. Fish. Sci. **14**: 115–127.

Pezzack, D.S., and Duggan, D.R. 1986. Evidence of migration and homing of lobsters (*Homarus americanus*) on the Scotian Shelf. Can. J. Fish. Aquat. Sci. **43**: 2206–2211.

Pezzack, D.S., Tremblay, M.J., Hudon, C., and Miller, R.J. 1992. The inshore-offshore lobster issue in southwestern Nova Scotia. Can. Manuscr. Rep. Fish. Aquat. Sci. **2165**.

Polacheck, T. 1990. Year around closed areas as a management tool. Nat. Res. Model. **4**: 327–354.

Pollock, D.E. 1993. Recruitment overfishing and resilience in spiny lobster populations. ICES J. Mar. Sci. **50**: 9–14.

Pulliam, H.R. 1988. Sources, sinks, and population regulation. Am. Nat. **132**: 652–661.

Quinn, J.F., Wing, S.R., and Botsford, L.W. 1993. Harvest refugia in marine invertebrate fisheries: models and applications to the red sea urchin *Strongylocentrotus franscicanus*. Am. Zool. **33**: 537–550.

Rogers, B.A., Cobb, J.S., and. Marshall, N.B 1968. Size comparisons of inshore and offshore larvae of the lobster, *Homarus americanus*, off southern New England. Proc. Natl. Shellfish. Assoc. **58**: 78–81.

Russell, H.J. Jr. 1980. The determination of growth rates for American lobsters. Can. Tech. Rep. Fish. Aquat. Sci. 932. pp. 1–8.

Saila, S.B., and Flowers, J.M. 1968. Movements and behavior of berried female lobsters displaced from offshore areas to Narragansett Bay, Rhode Island. J. Cons. Int. Explor. Mer, **31**: 342–351.

Saila, S.B., and Flowers, J.M. 1969. Geographical morphometric variation in the American lobster. Syst. Zool. **18**: 330–338.

Saila, S.B., Annala, J.H., McKoy, J.L., and Booth, J.D. 1979. Application of yield models to the New Zealand rock lobster fishery. N.Z. J. Mar. Freshwater Res. **13**: 1–11.

Shaklee, J.B. 1983. The utilization of isozymes as genetic markers in fisheries management and conservation. Isozymes: Curr. Top. Biol. Med. Res. **11**: 214–247.

Stasko, A.B. 1980. Lobster larval surveys in Canada. Can. Tech. Rep. Fish. Aquat. Sci. 932. pp. 157–165.

Tracey, M.L., Nelson, K., Hedgecock, D., Schlesser, R.A., Pressick, M.L. 1975. Biochemical genetics of lobsters: genetic variation and structure of American lobster (*Homarus americanus*) populations. J. Fish. Res. Board Can. **32**: 2091–2101.

Tuck, G.N., and Possingham, H.P. 1994. Optimal harvesting strategies for a metapopulation. Bull. Math. Biol. **56**: 107–127.

Tyler, J.A., and Rose, K.A. 1994. Individual variability and spatial hetergeneity in fish population models. Rev. Fish. Biol. Fish. **4**: 91–123.

Uzmann, J.R. 1970. Use of parasites in identifying lobster stocks. Proc. Int. Conf. Parasit. **56**: 12–20.

Uzmann, J.R., Cooper, R.A., and Pecci, K.J. 1977. Migration and dispersion of tagged American lobsters, *Homarus americanus*, on the southern New England continental shelf. NOAA Tech. Rep. NMFS SSRF 705.

Wahle, R.A. 1992. Body-size dependent anti-predator mechansisms of the American lobster. Oikos, **65**: 52–60.

Wahle, R.A., and Steneck, R.S. 1991. Recruitment habitats and nursery grounds of the American lobster *Homarus americanus*: a demographic bottleneck? Mar. Ecol. Prog. Ser. **69**: 231–243.

Wahle, R.A., and Steneck, R.S. 1992. Habitat restrictions in early benthic life: experiments on habitat selection and in situ predation with the American lobster. J. Exp. Mar. Biol. Ecol. **157**: 92–114.

Walters, C.J., Hall, N., Brown, R., and Chubb, C. 1993. Spatial model for the population dynamics and exploitation of the western Australian rock lobster, *Panulirus cygnus*. Can. J. Fish. Aquat. Sci. **50**: 1650–1662.

Ye, Y., and Beddington, J. 1996. Modelling interactions between inshore and offshore fisheries: the case of the East China Sea hairtail (*Trichiurus haumela*) fishery. Fish. Res. (Amsterdam), **27**: 153–177.

The impact of coastal circulation on the spatial distribution of invertebrate recruitment, with implications for management

Stephen R. Wing, Louis W. Botsford, and James F. Quinn

Abstract: Population dynamics and abundance of harvested populations depend in part on how physical conditions affect the distribution of recruitment over space. We describe an example of the influence of circulation on patterns of settlement in northern California. We then use a simple model to demonstrate how knowing the spatial distribution of recruitment is valuable to management. During the spring and summer of 1992–1995, we monitored settlement of crabs (primarily *Cancer* spp.) in conjunction with physical variables associated with coastal circulation north and south of Point Reyes in northern California. Temperature, salinity, and wind stress data indicated fluctuations in upwelling strength. During upwelling, warm, low-salinity water collected south of Point Reyes, and during periods of upwelling relaxation this water flowed poleward, north of Point Reyes. Settlement on collectors and samples of planktonic larval distributions indicated that crab larvae are transported northward, alongshore in each relaxation event from a pool of larvae that collects to the south of Point Reyes. This transport mechanism on weekly time scales leads to a predictable pattern of spatial variability in annual settlement: higher settlement south of Point Reyes than to the north. We formulate a simple metapopulation model to illustrate the value of knowing spatial-recruitment patterns in managing to maximize harvest and reduce exposure to risk of overfishing. Model results indicate that the spatial distribution of recruitment influences the effective stock–recruitment relationship of the population. The behavior of this model suggests harvest can be increased and overfishing can be avoided by designing spatially variable harvest rates on the basis of spatial-recruitment mechanisms.

Résumé : La dynamique des populations et l'abondance des populations récoltées dépendent en partie de la façon dont les conditions physiques influent sur la distribution du recrutement dans l'espace. Nous décrivons un exemple de l'effet de la circulation de l'eau sur les caractéristiques de la fixation dans le nord de la Californie. Nous utilisons ensuite un modèle simple pour démontrer dans quelle mesure la connaissance de la répartition spatiale du recrutement est utile pour la gestion. Durant le printemps et l'été de 1992–1995, nous avons surveillé la fixation des crabes (principalement des *Cancer* spp.) en conjonction avec les variables physiques liées à la circulation côtière au nord et au sud de Point Reyes dans le nord de la Californie. Les données sur la température, la salinité et la force d'entraînement du vent ont indiqué des fluctuations dans la force de remontée des eaux. Durant la remontée des eaux, l'eau chaude de faible salinité s'accumulait au sud de Point Reyes et durant les périodes de relaxation de la remontée des eaux, cette eau s'écoulait en direction du pôle, au nord de Point Reyes. La fixation sur des collecteurs et des échantillons visant à établir la distribution du plancton larvaire ont indiqué que les larves de crabe sont transportées en direction nord, le long de la côte à chaque événement de relaxation à partir d'un regroupement de larves qui se forme au sud de Point Reyes. Ce mécanisme de transport, rapporté sur des échelles de temps hebdomadaires, permet de prévoir des modalités de variabilité spatiale dans la fixation annuelle : une fixation plus élevée au sud de Point Reyes qu'au nord. Nous avons élaboré un modèle de type métapopulation simple qui illustre les avantages liés à la connaissance des caractéristiques de la répartition spatiale de recrutement pour gérer la ressource en vue d'une récolte maximale et d'une réduction des risques de surpêche. Les résultats du modèle indiquent que la distribution spatiale du recrutement influe sur la relation stock–recrutement réel de la population. Le comportement de ce modèle indique que la récolte peut être augmentée et que la surpêche peut être évitée par l'établissement de taux de capture variant dans l'espace en se fondant sur les mécanismes du recrutement spatial.
[Traduit par la Rédaction]

S.R. Wing and L.W. Botsford. Wildlife, Fish and Conservation Biology, University of California, Davis, CA 95616-8751, U.S.A.

J.F. Quinn. Division of Environmental Studies, University of California, Davis, CA 95616-8751, U.S.A.

Correct citation: Wing, S.R., Botsford, L.W., and Quinn, J.F. 1998. The impact of coastal circulation on the spatial distribution of invertebrate recruitment, with implications for management. *In* Proceedings of the North Pacific Symposium on Invertebrate Stock Assessment and Management. *Edited by* G.S. Jamieson and A. Campbell. Can. Spec. Publ. Fish. Aquat. Sci. 125. pp. 285–294.

Introduction

Marine invertebrate fisheries are generally composed of species with complex life cycles involving planktonic larvae which are subject to the vagaries of the coastal ocean. Environmental forcing can be strong and variable, masking any clear relationships between spawning stock and recruitment. For many taxa, such as crabs and lobsters, larval stages may last weeks to several months and dispersal distances are potentially vast. In such cases, local populations are rarely closed, rather they rely on external sources for recruitment. As an added complication, many harvested invertebrate species are relatively sedentary as adults, leading to strong spatial heterogeneity of adult density and productivity if recruitment and suitability of habitat vary over space. For these reasons, and the general difficulty in tracking the movements of individual larvae, information on larval sources is generally lacking and managers often have poor estimates for stock–recruitment relationships for invertebrate populations (Hancock 1973; Caputi 1993).

Such obstacles have made it difficult for researchers concerned with population dynamics and managers of invertebrate fisheries to account for spatial aspects of recruitment. In most cases they have had to treat populations as single open subpopulations of adults, or a number of separate stocks with recruitment supplied by an unknown but possibly exogenous source. An alternative to these approaches is to link subpopulations of adults through larval dispersal to comprise a metapopulation (Botsford et al. 1994; Botsford 1995; Fogarty 1998). The dynamic response of metapopulations to harvest depends critically on the transport of larvae between subpopulations and their relative productivity. Unfortunately, the temporally and spatially varying patterns of larval exchange are rarely known.

Management of metapopulations will likely benefit from the increasing number of studies attempting to identify physical influences on larval dispersal and settlement variability. While many studies have provided descriptions of temporal variability in settlement, relatively few have investigated spatial variability in settlement, and fewer still have provided predictable physical mechanisms for spatially varying settlement rates.

For some invertebrate species, in particular those with short larval dispersal distances and those whose life histories are confined to bays, information on larval dispersal between populations has been obtained. For example, investigations into larval dispersal among abalone (*Haliotus* spp.) populations in south Australia suggest that some populations may have short dispersal distances while others may share larvae across relatively large distances (Shepherd and Brown 1993; Shepherd 1998). Investigations of stock–recruitment relationships in Australian prawn fisheries (*Penaeus* spp.) have demonstrated the importance of small source stocks for fisheries management (e.g., Rothlisberg et al. 1985), and that independence of stocks is useful for adaptive management strategies (Caputi 1993).

This kind of data is less common for taxa with larger scale larval dispersal. For example, research into the larval distribution and settlement patterns of lobsters has led to some information on possible transport patterns of American lobsters (*Homarus americanus*) in the northwestern Atlantic (e.g., Harding and Trites 1988; Elner and Campbell 1991; Hudon 1994; Katz et al. 1994), spiny lobsters (*Panulirus argus*) in the Gulf of Mexico (e.g., Herrnkind and Butler 1994), and the Australian rock lobster (*Panulirus cygnus*) in the Indian Ocean (e.g., Caputi et al. 1995*a*, 1995*b*; Pearce and Phillips 1994). However, there remains a great deal of uncertainty in predicting the contribution of any one area to recruitment because of the vast distances over which larvae may travel.

In the California Current, researchers concerned with Dungeness crab (*Cancer magister*) have observed correlations between catch and upwelling (Peterson 1973; Botsford and Wickham 1975), alongshore winds (Johnson et al. 1986), and alongshore geostrophic flow (McConnaughey et al. 1992). Studies involving direct sampling of larvae have proposed mechanisms underlying temporal and, in some cases, spatial variability in recruitment (Jamieson et al. 1989; Hobbs et al. 1992), and one mechanism for the separation of stocks has been proposed (Jamieson and Phillips 1993). In the California Current, recruitment studies of commercially important sea urchins (*Strongylocentrotus spp.*) have suggested that the predictability and density of recruitment is heavily influenced by mesoscale advective features, such as jets associated with headlands (e.g., Ebert and Russell 1988; Ebert et al. 1994; Wing et al. 1995*a*, 1995*b*).

Understanding larval dispersal in the California current system is particularly difficult because larvae can be influenced by two distinctly different current regimes. For example, larvae of the Dungeness crab are released in the winter when the predominant flow is poleward near shore (see Botsford et al. 1989). At the spring transition in March or April upwelling begins and the mean flow shifts to equatorward and offshore (e.g., Strub et al. 1987). Mesoscale variability in the current field during this period includes local centers of upwelling associated with capes and offshore transport in the form of cold filaments extending from capes (e.g., Largier et al. 1993). Investigations into larval dispersal of crabs and sea urchins in this system have suggested that during the upwelling period, areas along the coast with reduced upwelling may retain larvae and possibly act as conduits to the nearshore (e.g., Wing et al. 1995*b*). In the California Current, as elsewhere, researchers are beginning to establish some of the general physical mechanisms that deliver final stages of planktonic larvae to settlement sites (e.g., Williams et al. 1984; Shanks 1986; Farrell et al. 1991; Wing et al. 1995*a*). However, it is rare that the ultimate sources of these successful larvae can be gleaned.

The fact that larval dispersal among subpopulations is usually unknown presents a problem for management of invertebrate populations (e.g., Walters et al. 1993). For discrete adult populations connected by larval dispersal, those subpopulations that receive few propagules may be particularly sensitive to local decline from harvest pressure, yet they could be an important source of propagules for the metapopulation as a whole (e.g., Pulliam 1988). For example, in some invertebrate fisheries, management relies on remote or offshore refuge populations to seed more heavily fished grounds (e.g., American lobster; Harding and Trites 1988; Fogarty 1998). Loss of these potentially sensitive larval sources may result in recruitment overfishing. However, decline of such local populations may not be immediately detectable from fishery catch statistics as effort shifts to the remaining grounds (Hilborn and Walters 1987; Collie and Walters 1991). These considerations are particularly important when a fishery expands over a large area, affecting per capita survival rates in remote areas that have been refuges from harvest and are also sources of recruits to the rest of the population.

Here we describe our recent findings regarding the way in which coastal circulation patterns redistribute *Cancer* spp. and other crab larvae among subpopulations along the northern California coast. This is an example of the way in which circulation in the coastal ocean can influence the spatial structure of recruitment. We then use a simple metapopulation model with two subpopulations to demonstrate the potential value to management of the type of information we are gaining in our studies of temporal and spatial variability in recruitment.

Fig. 1. A general schematic of surface transport during upwelling (shaded arrows) and during relaxation (black arrows). These patterns are a compilation of information from drifter tracks, satellite images, and current meter studies (c.f. Send et al. 1987; Wing et al. 1995*a*). Southeast Farallon Island is indicated by SEFI, Duxbury reef by DBR, and San Francisco by SF. The open circle on the coast at Bodega Head indicates position of the 1992 sampling station, and the four dark circles on the coast indicate positions of 1993–1994 sampling stations.

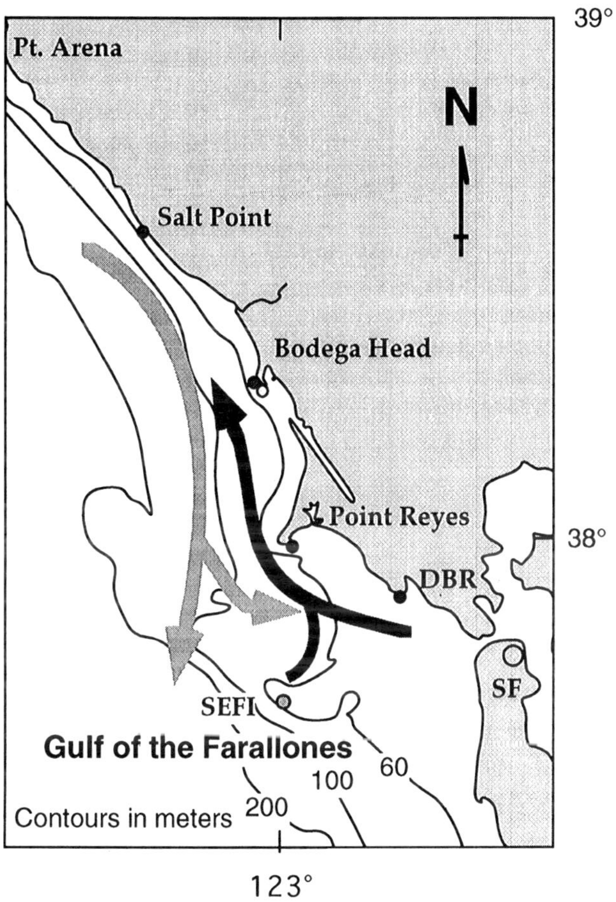

Fig. 2. Total crab settlement and sea surface temperature near Bodega Head during spring and summer of 1992. Note that warming periods (which indicate upwelling relaxation) are associated with increased crab settlement.

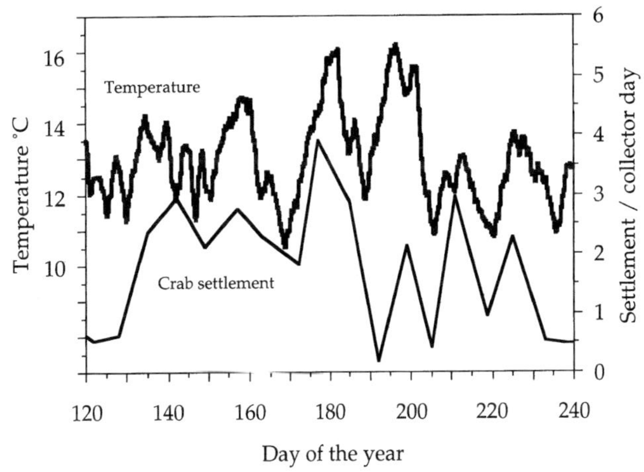

Coastal circulation and settlement variability in northern California

Over the past several years we have studied the way in which coastal circulation patterns influence spatial variability in invertebrate larval settlement in northern California (Wing et al. 1995*a*, 1995*b*). During the spring and summer months of 1992–1995, we monitored settlement of postlarval stages of crabs (primarily *Cancer* spp.) and sea urchins (*Strongylocentrotus* spp.) on scrub brush collectors near the sea bottom from the Gulf of the Farallones (37°50′N) to Point Arena (38°55′N). We concurrently measured physical variables including wind, sea surface temperature, salinity, and sea level height, which provided a description of coastal circulation associated with upwelling and relaxation conditions along the coast (Fig. 1).

Along this part of the coast in the spring, strong equatorward winds produce upwelling with newly upwelled water flowing equatorward and offshore north of Point Reyes, and the alongshore component deflected offshore by Point Reyes (e.g., Largier et al. 1993) (Fig. 1). This circulation pattern and the weakness of upwelling south of Point Reyes results in the accumulation of warm water of low salinity in the lee of Point Reyes in the Gulf of the Farallones. During cessation or slight reversal of upwelling winds, this warm, low-salinity water flows poleward along the coast north of Point Reyes (Send et al. 1987; Largier et al. 1993; Wing et al. 1995*a*, 1995*b*) (Fig. 1). These alongshore currents are partially buoyancy forced and through Coriolis forcing remain coastally trapped as they flow poleward over the inner and midshelf from Point Reyes as far north as Point Arena. Past observations of temperature–current patterns (Send et al. 1987), drifter tracks (Davis 1985) and movement of an oil spill from the Gulf of the Farallones (Breaker and Bratkovitch 1993) are evidence that the Gulf of the Farallones is the source of the warm water that reaches the coast between Point Arena and Point Reyes during upwelling relaxation.

From the observation in 1992 that crab settlement at Bodega Head was associated with episodic increases in temperature (Fig. 2), we proposed that planktonic crab larvae of several species had either accumulated in the Gulf of the Farallones or had moved onshore there, and been transported northward during upwelling relaxation. In 1993 we sampled settlement at four locations along the coast to test this hypothesis (Wing et al. 1995*b*). We found that settlement and larval availability were not homogeneous alongshore, but rather varied in response to intra-annual variability in upwelling and the consequent variability in alongshore flow (Wing et al. 1995*b*). In particular, to the south of Point Reyes settlement events were larger and more frequent than north of Point Reyes, and some settlement occurred during upwelling conditions. Conversely, north of Point Reyes settlement events were smaller and occurred primarily during upwelling relaxations. These results are consistent with the explanation that crab larvae were transported northward, alongshore in each relaxation event from an accumulation in the Gulf (Wing et al. 1995*b*).

During the upwelling season in 1994 and 1995 we sampled alongshore distributions of meroplankton, including crab larvae, during shipboard surveys of the region. We found large concentrations of cancrid crab larvae, including Dungeness crabs, in the northern Gulf associated with warm and stratified

Fig. 3. Annual settlement rates for *Cancer* crabs at four sites in northern California during (*a*) 1993, and (*b*) 1994. Settlement data were collected weekly from April through September.

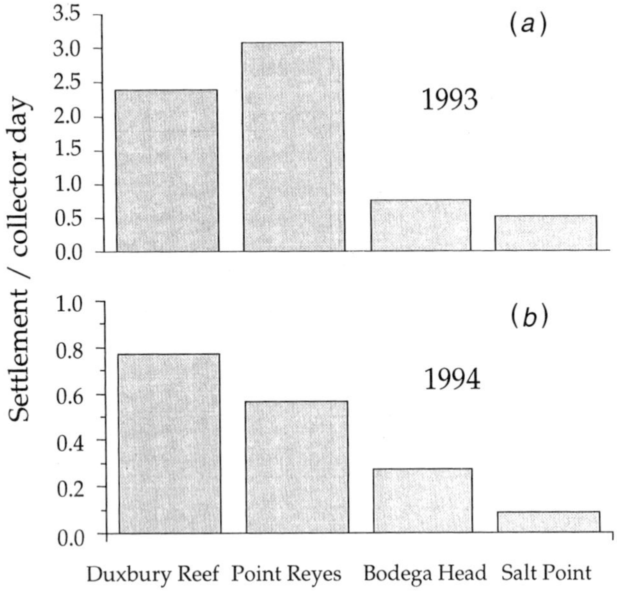

Fig. 4. Diagrammatic representation of a simple metapopulation model with larval dispersal. The multipliers ϕ and δ determine each subpopulation's contribution to larval supply which is then redistributed between subpopulations according to ε.

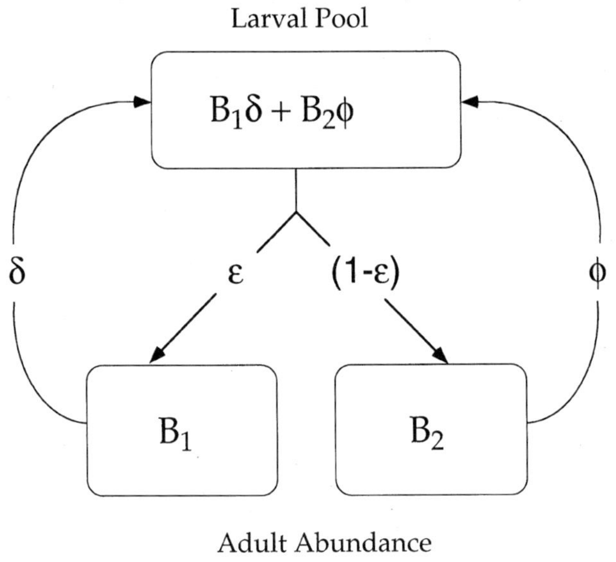

waters (S.R. Wing, L. Botsford, J. Largier, and S. Ralston, unpublished data). This appears to be an "upwelling shadow," similar to the feature observed in northern Monterey Bay (Graham et al. 1992).

This alongshore retention and transport mechanism on weekly time scales during the upwelling period appears to be responsible for a predictable spatial pattern of total annual settlement (Fig. 3). South of Point Reyes there is higher annual settlement because settlement occurs both during upwelling as well as relaxed conditions. Conversely, north of Point Reyes annual settlement is lower because settlement events occur only during occasional relaxation events.

These characteristics were reflected in the settlement of several species of coastal crabs, and more intermittently in the two species of sea urchins (*S. franciscanus* and *S. purpuratus*). The most abundant group, cancrid crabs, includes *Cancer antennarius, C. productus* and *C. magister*; other crabs include *Loxorhynchus* sp., *Pugettia* sp., *Petrolisthes* sp., *Pachycheles* sp., and *Pagarus* sp. (Wing et al. 1995*a*, 1995*b*). The influence of this alongshore flow on settlement of such a large number of species suggests that the mechanism is general and a strong determinant of settlement variability for crabs. To the extent that these settlement rates are reflected in local recruitment rates, they are of potential use in management.

Because this mechanism involves alongshore transport, rather than just temporal variability, it expands the suite of potential physical–biological influences on recruitment of meroplanktonic species along the northern California coast. Earlier studies focused on the effect of the cross-shelf flow that results from upwelling relaxation on temporal variability in recruitment (Roughgarden et al. 1991; Farrell et al. 1991). The alongshore mechanism identified in Wing et al. (1995*b*) has two implications that are of potential use in harvest management: (*i*) alongshore spatial variability in settlement is predictable on the basis of the alongshore advective environment, and (*ii*) the larval sources for populations across the system are shared yet unknown (i.e., local settlement rates probably do not vary independently over time).

While knowing the mechanisms of larval transport that determine the spatial distribution of recruitment contribute to a basic understanding of coastal ecology, it is not immediately obvious that such information will be of value to fisheries management (see e.g., Walters and Collie 1988). We do not know enough about the coastwide settlement pattern for the species involved here to formulate a coastwide management plan. However, we can use a simple model to answer the general questions of how knowing the spatial distribution of recruitment could be used in management, and how much it would improve management.

A two-patch metapopulation model with larval dispersal

Here we present a nonspecific, heuristic model to answer the question of whether knowing transport patterns in the coastal ocean and resultant spatial patterns in settlement can be used in management. For modeling purposes, we divide the coast into two regions with different recruitment as a generalization of the observed annual settlement patterns near Point Reyes; low settlement to the north, and high settlement to the south (Fig. 3). The model consists of coupled difference equations that explicitly describe the relative contribution by each subpopulation to a larval pool, and the redistribution of larvae from the pool to each subpopulation (Fig. 4). Using this form we can mimic a system in which larvae are contributed to a larval pool in various ways and then redistributed along the coastline according to a mechanism such as the one implied by our field observations. We represent the effects of circulation on dispersal by three multipliers, δ, ϕ, and ε. The values of δ and ϕ are the fractions of eggs produced by each population that contribute to a larval pool. The values of ε and $1 - \varepsilon$ represent the effects of circulation on redistribution from the larval pool to each subpopulation (Fig. 4).

In this model we make several simplifying assumptions so that we can focus more clearly on the effects of knowing settlement distribution. In particular, we do not consider the effects of age, size, or interannual environmental variability on survival, growth, or fecundity. Dispersal is by the larval stage only (i.e., there is no movement of adults between subpopulations), and is not subject to random environmental variability. Each of these are important considerations and more knowledge of these effects would provide a more realistic view of specific natural populations. However, to illustrate the general value of knowing spatial recruitment mechanisms, we chose the most rudimentary iteroparous case because it isolates the effects of spatial differences in recruitment on population harvesting. We do not present a complete analysis of harvest policy for this type of population here. Rather we show simply that knowledge of recruitment mechanisms can improve harvest.

The behavior of populations described by this model will depend on the stage in the life cycle at which density-dependence in the stock–recruitment relationship occurs. Here we consider three possibilities: predispersal density-dependence, postdispersal density-dependence, and larval density-dependence. Predispersal density-dependence refers to the situation in which larval production declines with increasing local adult density. An example is density-dependent fecundity where per capita fecundity is reduced at high population levels (for example as proposed for Dungeness crab by McKelvey et al. 1980). In this case we represent recruitment for the two subpopulations R_1 and R_2 as

$$[1]\quad R_1 = [\delta B_1 f_1(B_1) + \phi B_2 f_2(B_2)]\varepsilon$$

$$R_2 = [\delta B_1 f_1(B_1) + \phi B_2 f_2(B_2)](1 - \varepsilon)$$

For each case B_1 and B_2 are adult abundance of each subpopulation and f_1 and f_2 are the stock–recruitment relationships for each subpopulation.

Postdispersal density-dependence refers to a situation where survival of newly settled individuals declines with increasing density of the juveniles or adults present. This incorporates such compensatory effects as cannibalism by adults on young, a process observed in several harvested crustaceans (e.g., Dungeness crab: Gotshall 1977; Stevens et al. 1982; Botsford and Hobbs 1995; snow crab: Jewett and Feder 1982; Waiwood and Elner 1982). In this case we represent recruitment for the two subpopulations R_1 and R_2 as

$$[2]\quad R_1 = [\delta B_1 + \phi B_2]\varepsilon f_1(B_1)$$

$$R_2 = [\delta B_1 + \phi B_2](1 - \varepsilon) f_2(B_2)$$

Larval density-dependence refers to a situation in which survival to recruitment depends on density of the settling cohort. These effects may occur in the larval life stages that reduce per capita survival at high-larval density (e.g., Paulik 1973), or on juveniles independently of adult density (e.g., in Dungeness crab, Fernandez et al. 1993; in spiny lobster: Herrnkind and Butler 1994; and in American lobster: Wahle and Steneck 1991). In this case we can represent recruitment for the two subpopulations R_1 and R_2 as

$$[3]\quad R_1 = [\delta B_1 + \phi B_2] f_1[\delta B_1 + \phi B_2]\varepsilon$$

$$R_2 = [\delta B_1 + \phi B_2] f_2[\delta B_1 + \phi B_2](1 - \varepsilon)$$

The stock–recruitment relationships, f_1 and f_2, could have any form. Here we use the Beverton-Holt stock–recruitment relationship for which $f(B) = 1/(1 + B/K)$ (Beverton and Holt 1957). The results we present here would also hold for an overcompensatory stock–recruitment relationship such as the Ricker form (Ricker 1954) for which $f(B) = e^{-B/K}$.

These three cases are not exhaustive but cover a wide range of possible different forms of a metapopulation. A population could have a combination of several of these density-dependent mechanisms or weak density-dependence. Further, we do not explicitly include depensatory, Allee effects in the stock–recruitment relationships here, although these can be important, especially in predispersal density-dependence. This type of effect is common in free-spawning invertebrates whose fertilization success drops to zero when spawning individuals are spaced far apart, which may occur at low density (e.g., Pennington 1985; Levitan 1991; Levitan and Sewell 1998). This depensatory effect on fecundity can result in catastrophic collapse of harvested populations if they are fished to low density (e.g., Quinn et al. 1993).

To explore the different responses of these models to harvest, we incorporated each of the stock–recruitment relationships into a simple two-subpopulation iteroparous model, using the sum of the fractions surviving at rate s and recruitment to each subpopulation, R_1 and R_2, which are harvested at rates H_1 and H_2, respectively. This model is written as a pair of difference equations

$$[4]\quad B_{1,t+1} = (sB_{1,t} + R_{1,t})(1 - H_1)$$

$$B_{2,t+1} = (sB_{2,t} + R_{2,t})(1 - H_2)$$

Transport between the two subpopulations changes the effective stock–recruitment relationships depending on the values of ε, δ, and ϕ. For example, the equilibrium values of the stock–recruitment relationships for a nonharvested postdispersal density-dependent case with equal proportional contribution of larvae from each subpopulation ($\delta = \phi = 0.5$) and equal redistribution of larvae between subpopulations from the larval pool ($\varepsilon = 0.5$) are identical, provided the values of K in f_1 and f_2 are equal (Fig. 5). However, if larvae are distributed unequally between subpopulations ($\varepsilon = 0.1$) the equilibrium values of the stock–recruitment relationships become locally different, while still dependent on the total contribution to the larval pool (Fig. 5). Equilibrium values are obtained by varying s in eq. 4. As the difference between the effective stock–recruitment relationships widens (i.e., ε becomes small) population 1 becomes a larval source (net exporter of larvae) and population 2 becomes a larval sink (net importer of larvae). Similar changes take place in the models with predispersal, and larval density-dependence as the redistribution, ε, is varied.

The overall stock–recruitment relationship of both subpopulations (i.e., the dependence of $R_1 + R_2$ on $B_1 + B_2$) also depends on the values of ε, δ, and ϕ. From eq. 3 for the model with larval density-dependence, this relationship is independent of ε if the terms of f_1 and f_2 are identical, and depends only on the values of δ and ϕ. From eq. 2 for the model with postdispersal density-dependence, this relationship depends on ε and the values of δ and ϕ. From eq. 1 for the model with predispersal density-dependence, this relationship is independent of ε. For the latter two, which involve adult density-dependence, total

Fig. 5. The stock–recruitment relationships for each subpopulation in the model with $\delta = \phi = 0.5$ and $\varepsilon = 0.5$ (solid line) and $\delta = \phi = 0.5$ and $\varepsilon = 0.1$ (dashed line) for the postsettlement density-dependence case. Equilibrium recruitment to each subpopulation is plotted on the ordinate versus total abundance on the abscissa as the survival s is varied. The Beverton-Holt model is used for both f_1 and f_2.

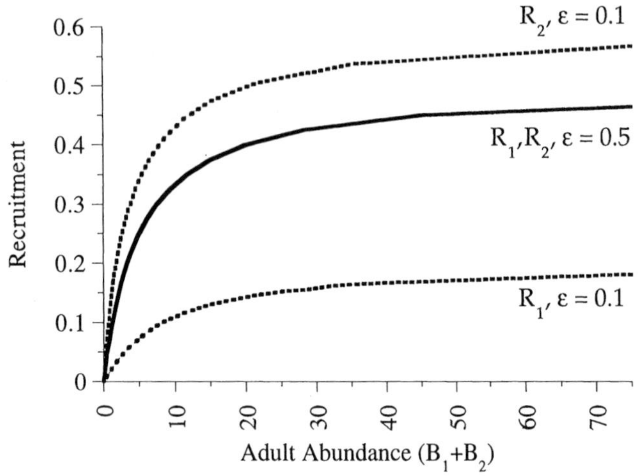

recruitment is reduced as the distribution of larvae between subpopulations becomes unequal (Fig. 6).

To demonstrate the value of knowing the distribution of settlement (i.e., knowing the value of ε), we compared the results of ordinary fishery management policy to those of a management policy based on knowledge of spatial variability in recruitment. For ordinary management policy, the subpopulations were harvested so that an escapement proportional to the abundance of each subpopulation was allowed. For management based on knowledge of the redistribution parameter ε, subpopulations were harvested with harvest rates proportional to their redistribution factor (i.e., ε or $1 - \varepsilon$) and we plotted the results as a function of total escapement so that they could be compared to the cases with ordinary management. For these examples, we assumed that the stock–recruitment relationships were the same for each population.

The case with postdispersal density-dependence demonstrates the possible improvement in harvest. When each subpopulation contributes equally to larval supply (i.e., $\delta = 1.0$, $\phi = 1.0$) and distribution of settlement is equal (i.e., $\varepsilon = 0.5$), as escapement is reduced by increasing harvest rate, equilibrium harvest reaches a peak and then declines to zero at zero escapement (Fig 7*a*). If redistribution of larvae is unequal (i.e., $\varepsilon = 0.1$), equilibrium harvest is less (Fig. 7*a*), as implied by the observed reduction in the total stock–recruitment relationship (Fig. 6*c*). However, if the redistribution factor ε is known and accounted for in the harvest rates, equilibrium harvest is increased by about 20%. The magnitude of this effect depends on the values of δ and ϕ, and is most pronounced in the extreme case when population 1 is a source ($\delta = 1.0$) and population 2 is a pure sink ($\phi = 0.0$) (Fig 7*b*), where the improvement is about 300%.

It is also useful to view these results in terms of the effects of the reduction in uncertainty gained through increased knowledge of recruitment dynamics. The values of ε, δ, and

Fig. 6. The total stock–recruitment relationships for the total of the two subpopulations with $\delta = \phi = 0.5$ and $\varepsilon = 0.5$ (solid line) and $\delta = \phi = 0.5$, $\varepsilon = 0.1$ (dashed line) for (*a*) larval density-dependence, (*b*) presettlement density-dependence, and (*c*) the postsettlement density-dependence case.

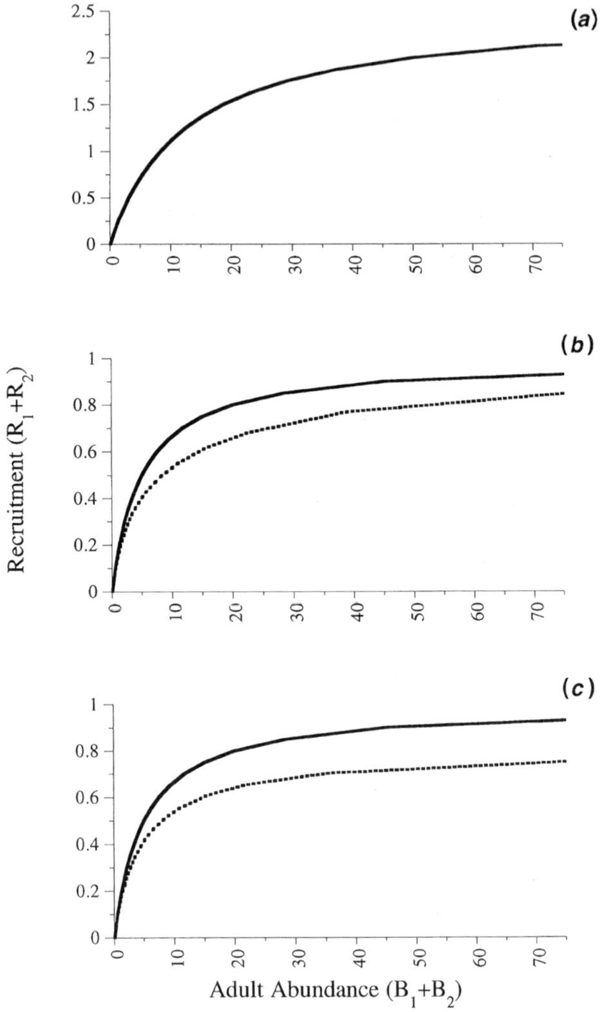

ϕ represent information on larval dispersal among subpopulations that is missing for most harvested species of invertebrates. In northern California, we can estimate that the value of the redistribution parameter ε is less than 0.5, but we do not know the source of the larvae, the values of δ and ϕ. We therefore consider the question: if relative contributions δ and ϕ remain unknown is there any value in knowing the way in which larvae are redistributed (i.e., ε)?

To answer this question, we computed equilibrium states of a fishery for all larval contributions (δ and ϕ) in the range from 0 to 1, for a model with postdispersal density-dependence and redistribution $\varepsilon = 0.1$. When ε is unknown, and hence escapements are proportional to abundance, maximum harvest varies between 0.1 and 1.1 (Fig. 8*a*). When ε is known and accounted for by harvesting in proportion to ε, harvest varies from about 0.3 to 1.3 (Fig. 8*b*). In addition to this increase, the population is much less susceptible to overfishing if ε is known and accounted for. From the lines indicating harvest rate in Fig. 8,

Fig. 7. Equilibrium harvest versus fractional escapement for the postsettlement density-dependent model when (*a*) $\delta = \phi = 1.0$, $\varepsilon = 0.5$ (thin solid line) and $\delta = \phi = 1.0$, $\varepsilon = 0.1$ (dashed line) when per capita escapement is the same for the two subpopulations; and for $\delta = \phi = 1.0$, $\varepsilon = 0.1$ when escapement is scaled with recruitment for each subpopulation (thick solid line). In (*b*) $\delta = 1.0$, $\phi = 0$, $\varepsilon = 0.5$ (thin solid line), $\delta = 1.0$, $\phi = 0$, and $\varepsilon = 0.1$ (dashed line) for subpopulations harvested for equal per capita escapement, and $\delta = 1.0$, $\phi = 0$, and $\varepsilon = 0.1$ when escapement is determined by recruitment (ε) for each subpopulation (thick solid line).

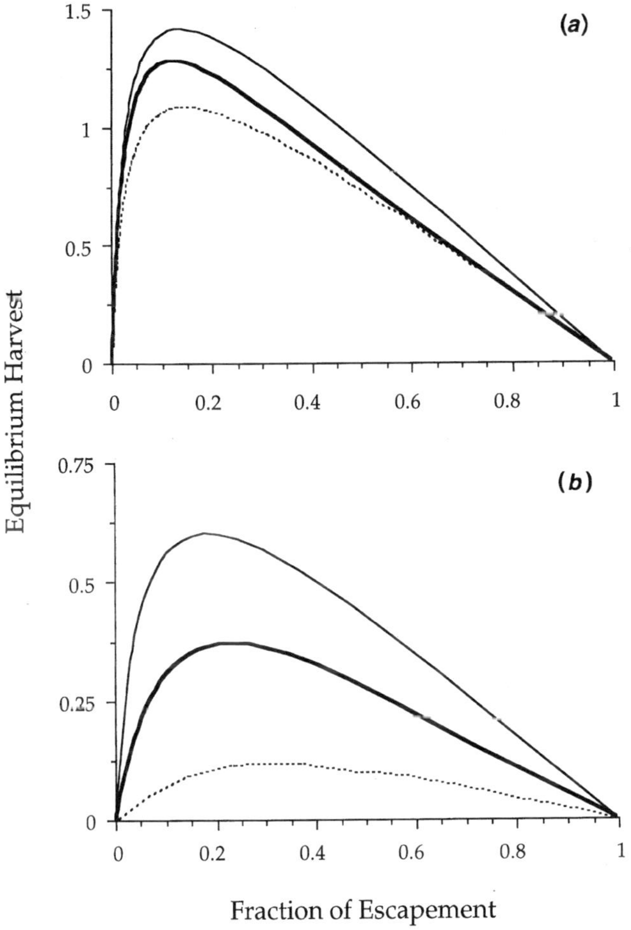

Fig. 8. Equilibrium harvest versus fractional escapement for all states of the postsettlement density-dependent model when (*a*) $\varepsilon = 0.1$ while ϕ and δ can vary between 0 and 1 when subpopulations are fished at a common per capita escapement. Vertical lines indicate equal fishing rates, each line indicates an increase of 0.01 in $(f_1 + f_2)$. In (*b*) each subpopulation is fished so that escapement is proportionate to recruitment to each subpopulation. Note that equilibrium harvest values are positive for a larger range of fishing rates.

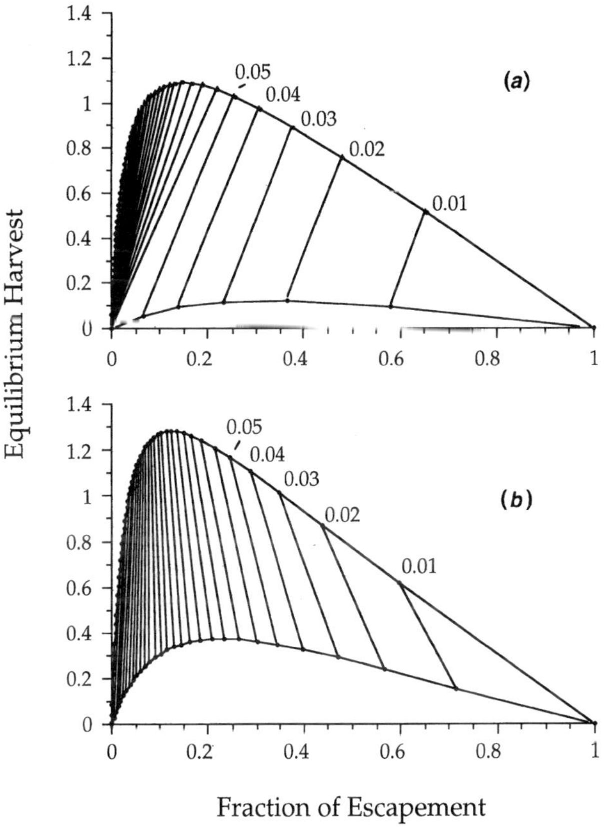

we can see that populations are harvested down to zero harvest at much lower harvest rates when ε is not known and accounted for, especially for very different values of relative contribution, i.e., $\delta = 1$ and $\phi = 0$.

Discussion

The field studies summarized here provide an example of an invertebrate stock for which the physical/biological interactions underlying the spatial distribution of recruitment have been identified. In this case the coastal circulation that results from fluctuations in upwelling winds leads to a specific annual pattern in the spatial distribution. Our current research is oriented toward extending this understanding to the north, across similar embayments in the range of the species of interest. Two of the species are of commercial importance, Dungeness crab and the red sea urchin. Although we do not attempt here to formulate a plan for the spatial management of either of these species, we do broach the question of the value of research on the spatial distribution of recruitment in general.

The model results presented here show that knowledge of the spatial distribution of recruitment across subpopulations that share unknown larval sources can help managers to obtain greater harvest and to avoid collapse of areas with relatively low recruitment. While this model is heuristic and not intended to provide detailed management specifications for any specific population, it does provide insight into the basic behavior of a harvested metapopulation. As such it suggests a general conservative strategy for management of populations with spatially varying recruitment rates. In this context, we use the model as a template for the more complicated spatial relationships that may occur in more complex metapopulations (e.g., Hastings 1991; Gotelli 1991; Roughgarden et al. 1985). We note that information on the location or existence of "source" and "sink" areas in invertebrate fisheries is generally lacking. However, investigators are starting to gain information on spatial variability in settlement and recruitment rates. We explicitly separate larval dispersal into an unknown contribution of each subpopulation to a shared pool and a known redistribution between subpopulations, which is analogous to the information that may be obtainable by managers of invertebrate fisheries. By demonstrating the use of information on spatially varying recruitment, we provide an obtainable management strategy for protecting against recruitment overfishing in spatially structured populations.

When some of the relevant population characteristics not explicitly considered here are included, there may be even worse consequences of not accounting for transport patterns. For example, in systems with depensatory effects on survival or fecundity, stocks may be susceptible to catastrophic collapse at relatively high escapement (e.g., Clark 1974; Quinn et al. 1993). In such cases it may be necessary to preserve some high-density stocks by controlling local fishing effort with complete spatial closure. In this case, management strategies can be implemented to both preserve high-density stocks locally and distribute them along the coastline to maximize the possibility of contributing recruits to neighboring fished areas (Quinn et al. 1993; Man et al. 1995).

The crux of the problem addressed here is illustrated by the difficulty in defining a unit "stock" in invertebrate fisheries management (e.g., Shepherd and Brown 1993; Ennis 1986). Ideally fisheries management is formulated in the context of "unit stocks," each comprised of a dynamic pool of the exploited species. For most harvested invertebrate species that exist as discrete sedentary adult populations connected by larval dispersal, the spatial dimension of a unit stock is as unknown as the complete larval dispersal pattern. Therefore, the problem faced is how to manage a population composed of loosely connected multiple stocks, when the effective stock–recruitment relationships in different areas may be interdependent. Similar considerations may apply to some groundfish fisheries where adults form spatially discrete subpopulations connected by larval dispersal (e.g., Collie and Walters 1991; Hilborn and Walters 1987).

Traditionally, harvest of multiple stocks by a common fishery has been considered for salmon fisheries. In such cases, for multiple stocks with very different stock–recruitment relationships, some subpopulations may be depleted or fished to extinction to reach optimal yield (Ricker 1958; Paulik et al. 1967). An alternative strategy is to maximize yield while reducing chance of extinction of relatively unproductive stocks (Paulik 1973; Hilborn 1985; Kope 1992). In both cases total harvest is reduced when stock–recruitment relationships vary greatly between substocks in a common fishery. Similarly, the sum of equilibrium harvest for metapopulations that have large differences in recruitment distribution will be lower than the sum of harvest of populations that have more evenly distributed recruitment. For example, Tuck and Possingham (1994) examined a form of the predispersal model and conclude that sources will be overharvested and sinks will be underharvested in a common fishery. For most invertebrate populations, the link between physical oceanographic processes and alongshore settlement variability is central to the understanding of local recruitment (i.e., identifying source and sink areas), and in this context valuable to spatial management.

Although we have demonstrated the potential management value of knowledge of larval exchange among subpopulations of a model metapopulation, we have not explored here the optimal management of such stocks. There remains much to do regarding the use of this type of information in management, especially when all relevant factors are included. We are currently pursuing extensions of optimal management of mixed stock fisheries (e.g., Hilborn 1985; Kope 1992; Tuck and Possingham 1994) that incorporate the linkages, and uncertainties in linkages, described here. It is apparent that continued specific accounting of metapopulation structure in management policy, and continuing pursuit of the biological-physical mechanisms underlying larval transport among them can lead to a quantum improvement in the management of harvest of invertebrate (and other) stocks.

Acknowledgments

This article was funded by a grant from the National Sea Grant College Program, National Oceanic and Atmospheric Administration, U.S. Department of Commerce, Washington, DC, under grant no. NA89AA-D-5G138, project no. R/F-150 through the California Sea Grant College, and in part by the California State Resources agency. The authors' views expressed herein do not necessarily reflect those of NOAA or any of its subagencies. The U.S. Government is authorized to reproduce and distribute this article for governmental purposes.

References

Beverton, R.J H., and Holt, S.J. 1957. On the dynamics of exploited fish populations. Fish. Invest. Ser. II, Mar. Fish. G.B. Minist. Agric. Fish. Food, **19**. 533 p.

Botsford, L.W. 1995. Population dynamics of spatially distributed, meroplanktonic, exploited marine invertebrates. ICES Mar. Sci. Symp. **199**: 118–128.

Botsford, L.W., and Hobbs, R.C. 1995. Recent advances in the understanding of cyclic behavior of Dungeness crab (*Cancer magister*) populations. ICES Mar. Sci. Symp. **199**: 157–166.

Botsford, L.W., and Wickham, D.E. 1975. Correlation of upwelling index and Dungeness crab catch. Fish. Bull. **73**: 901–907.

Botsford, L.W., Armstrong, D.A., and Shenker, J.M. 1989. Oceanographic influences on the dynamics of commercially fished populations. *In* Coastal oceanography of the Pacific Northwest. *Edited by* M.R. Landry and B.M. Hickey. Elsevier Science Inc., Amsterdam, The Netherlands. pp. 511–565.

Botsford, L.W., Moloney, C.L., Hastings, A., Largier, J.L., Powell, T.M., Higgins, K., and Quinn, J.F. 1994. The influence of spatially and temporally varying oceanographic conditions on meroplanktonic metapopulations. Deep-Sea Res. II, **41**: 107–145..

Breaker, L.C., and Bratkovich, A. 1993. Coastal-ocean processes and their influence on the oil spilled off San Francisco by the M/V Puerto Rican. Mar. Environ. Res. **36**: 153–184.

Caputi, N. 1993. Aspects of spawner-recruit relationships, with particular reference to crustacean stocks; a review. Aust. J. Mar. Freshwater Res. **44**: 589–607.

Caputi, N., Chubb, C.F., and Brown, R. 1995*a*. Relationships between spawning stock, environment, recruitment and fishing effort for the western rock lobster, *Panulirus cygnus*, fishery in Western Australia. Crustaceana, **68**: 213–226.

Caputi, N., Brown, R., and Chubb, C.F. 1995*b*. Regional prediction of the western rock lobster, *Panulirus cygnus*, commercial catch in Western Australia. Crustaceana, **68**: 245–256.

Clark, C.W. 1974. Possible effects of schooling on the dynamics of exploited fish populations. J. Cons. Int. Explor. Mer. **36**: 7–14.

Collie, J.S., and Walters, C.J. 1991. Adaptive management of spatially replicated groundfish populations. Can. J. Fish. Aquat. Sci. **48**: 1273–1284.

Davis, R.E. 1985. Drifter observations of coastal surface currents during CODE: the statistical and dynamical views. J. Geophys. Res. **90**: 4756–4772.

Ebert, T.A., and Russell, M.P. 1988. Latitudinal variation in size structure of the west coast purple sea urchin: a correlation with headlands. Limnol. Oceanogr. **33**: 286–294.

Ebert, T.A., Schroeter, S.C., Dixon, J.D., and Kalvass, P. 1994. Settlement patterns of red and purple sea urchins (*Strongylocentrotus franciscanus* and *S. purpuratus*) in California, USA. Mar. Ecol. Prog. Ser. **111**: 41–52.

Elner, R.W., and Campbell, A. 1991. Spatial and temporal patterns in recruitment of American lobster, *Homarus americanus*, in the northwestern Atlantic. Mem. Queensl. Mus. **31**: 349–363.

Ennis, G.P. 1986. Stock definition, recruitment variability, and larval recruitment processes in the American lobster, *Homarus americanus*: a review. Can. J. Fish. Aquat. Sci. **43**: 2072–2084.

Farrell, T.M., Bracher, D., and Roughgarden, J. 1991. Cross-shelf transport causes recruitment to intertidal populations in central California. Limnol. Oceanogr. **36**: 279–288.

Fernandez, M., Armstrong, D., and Iribarne, O. 1993. First cohort of young-of-the-year Dungeness crab, *Cancer magister,* reduces abundance of subsequent cohorts in intertidal shell habitat. Can. J. Fish. Aquat. Sci. **50**: 2100–2105.

Fogarty, M.J. 1998. Implications of dispersal and larval interchange in American lobster stocks: a metapopulation model. *In* Proceedings of the North Pacific Symposium on Invertebrate Stock Assessment and Management. *Edited by* G.S. Jamieson and A. Campbell. Can. Spec. Publ. Fish. Aquat. Sci. 125. pp. x–xx.

Gotelli, N.J. 1991. Metapopulation models: the rescue effect, the propagule rain and the core satellite hypothesis. Am. Nat. **138**: 768–776.

Gotshall, D.W. 1977. Stomach contents of northern California Dungeness crabs, *Cancer magister*. Calif. Fish Game, **63**: 43–51.

Graham, W.M., Field, J.G., and Potts, D.C. 1992. Persistent 'upwelling shadows' and their influence on zooplankton distributions. Mar. Biol. (Berlin), **114**: 561–570.

Hancock, D.A. 1973. The relationship between stock and recruitment in exploited invertebrates. Rapp. P-V Reun. Cons. Int. Explor. Mer. **164**: 113–131.

Harding, G.C., and Trites, R.W. 1988. Dispersal of *Homarus americanus* larvae in the Gulf of Maine from Browns Bank. Can. J. Fish. Aquat. Sci. **45**: 416–425.

Hastings, A. 1991. Structured models of metapopulations dynamics. Biol. J. Linn. Soc. **42**: 57–71.

Herrnkind, W.F., and Butler, M.J. 1994. Settlement of spiny lobster, *Panulirus argus* (Latreille, 1804), in Florida: Pattern without predictability? Crustaceana, **67**: 46–64.

Hilborn, R. 1985. Apparent stock recruitment relationships in mixed stock fisheries. Can. J. Fish. Aquat. Sci. **42**: 718–723.

Hilborn, R., and Walters, C.J. 1987. A general model for simulation of stock fleet dynamics in spatially heterogeneous fisheries. Can. J. Fish. Aquat. Sci. **44**: 1366–1369.

Hobbs, R.C., Botsford, L.W., and Thomas, A. 1992. Influence of hydrographic conditions and wind forcing on the distribution and abundance of Dungeness crab, *Cancer magister*, larvae. Can. J. Fish. Aquat. Sci. **49**: 1379–1388.

Hudon, C. 1994. Large-scale analysis of Atlantic Nova Scotia American lobster (*Homarus americanus*) landings with respect to habitat, temperature, and wind conditions. Can. J. Fish. Aquat. Sci. **51**: 1308–1321.

Jamieson, G.S., and Phillips, A. 1993. Megalopal spatial distribution and stock separation in Dungeness crab (*Cancer magister*). Can. J. Fish. Aquat. Sci. **50**: 416–429.

Jamieson, G.S., Phillips, A.C., and Huggett, W.S. 1989. Effects of ocean variability on the abundance of Dungeness crab larvae. Can. Spec. Publ. Fish. Aquat. Sci. 108. pp. 305–325.

Jewett, S.C., and Feder, H.M. 1982. Food of the tanner crab, *Chionoecetes bairdi* near Kodiak Island, Alaska. *In* Proceedings of the International Symposium on the Genus *Chionoecetes*. Lowell Wakefield Fisheries Symposium Series. University of Alaska, Fairbanks, Alaska. Alaska Sea Grant Report No. 82-10. pp. 293–318.

Johnson, D.F., Botsford, L.W., Methot, R.D. Jr., and Wainwright, T.C. 1986. Wind stress and cycles in Dungeness crab (*Cancer magister*) catch off California, Oregon and Washington. Can. J. Fish. Aquat. Sci. **43**: 838–845.

Katz, C.H., Cobb, J.S., and Spaulding, M. 1994. Larval behavior, hydrodynamic transport, and potential offshore recruitment in the American lobster, *Homarus americanus*. Mar. Ecol. Prog. Ser. **103**: 265–273.

Kope, R.G. 1992. Optimal harvest rates for mixed stocks of natural and hatchery fish. Can. J. Fish. Aquat. Sci. **49**: 931–938.

Largier, J.L., Magnell, B.A., and Winant, C.D. 1993. Subtidal circulation over the northern California shelf. J. Geophy. Res. **98**: 18147–18179.

Levitan, D.R. 1991. Influence of body size and population density on fertilization success and reproductive output in a free spawning invertebrate. Biol. Bull. **181**: 261–268.

Levitan, D.R., and Sewell, M.A. 1998. The importance of population size to reproductive success in free-spawning marine organisms. *In* Proceedings of the North Pacific Symposium on Invertebrate Stock Assessment and Management. *Edited by* G.S. Jamieson and A. Campbell. Can. Spec. Publ. Fish. Aquat. Sci. 125. pp. x–xx.

Man, A., Law, R., and Polunin, N.V.C. 1995. Role of marine reserves in recruitment to reef fisheries: a metapopulation model. Biol. Conserv. **71**: 197–204.

McConnaughey, R.A., Armstrong, D.A., Hickey, B.M., and Gunderson, D.R. 1992. Juvenile Dungeness crab (*Cancer magister*) recruitment variability and oceanic transport during the pelagic larval phase. Can. J. Fish. Aquat. Sci. **49**: 2028–2044.

McKelvey, R.D., Hankin, D., Yanosko, K., and Snygg, C. 1980. Stable cycles in multistage recruitment models: an application to the northern California Dungeness crab (*Cancer magister*) fishery. Can. J. Fish. Aquat. Sci. **37**: 2323–2345.

Paulik, G.J. 1973. Studies of the possible form of the stock-recruitment curve. J. Cons. Int. Explor. Mer. **164**: 302–315.

Paulik, G.J., Hourston, A.S., and Larkin, P.A. 1967. Exploitation of multiple stocks by a common fishery. J. Fish. Res. Board Can. **24**: 2527–2537.

Pearce, A.F., and Phillips, B.F. 1994. Oceanic processes, puerulus settlement and recruitment of the Western Rock Lobster *Panulirus cygnus*. *In* The bio-physics of marine larval dispersal. *Edited by* P.W. Sammarco and M.L. Heron. Coastal and Estuarine Studies 45. American Geophysical Union, Washington, D.C. pp. 279–306.

Pennington, J.T. 1985. The ecology of fertilization of echinoid eggs: The consequences of sperm dilution, adult aggregation, and synchronous spawning. Biol. Bull. (Woods Hole), **169**: 417–430.

Peterson, W.T. 1973. Upwelling indices and annual catches of Dungeness crab, *Cancer magister*, along the west coast of the United States. U. S. Natl. Mar. Fish. Serv., Fish. Bull. **71**: 902–910.

Pulliam, H.R. 1988. Sources, sinks and population regulation. Am. Nat. **132**: 652–661.

Quinn, J.F., Wing, S.R., and Botsford, L.W. 1993. Harvest refugia in marine invertebrate fisheries: models and applications to the Red Sea Urchin, *Strongylocentrotus franciscanus*. Am. Zool. **33**: 537–550.

Ricker, W.E. 1954. Stock and recruitment. J. Fish. Res. Board Can. **11**: 559–623.

Ricker, W.E. 1958. Maximum sustainable yields from fluctuating environments and mixed stocks. J. Fish. Res. Board Can. **15**: 991–1006.

Rothlisberg, P.C., Staples, D.J., and Crocos, P.J. 1985. A review of the life history of the banana prawn, *Penaeus merguiensis*, in the Gulf of Carpentaria. *In* Second Australian National Prawn

Seminar. *Edited by* P.C. Rothlisberg, B.J. Hill, and D.J. Staples. NPS2, Cleveland, Australia. pp. 125–136.

Roughgarden, J., Iwasa, Y., and Baxter, C. 1985. Demographic theory for an open marine population with space-limited recruitment. Ecology, **66**: 54–67.

Roughgarden, J., Pennington, J.T., Stoner, D., Alexander, S., and Miller, K. 1991. Collisions of upwelling fronts with the intertidal zone: The cause of recruitment pulses in barnacle populations of central California. Acta Oecol. **12**: 35–51.

Send, U., Beardsley, R.C., and Winant, C.D. 1987. Relaxation from upwelling in the Coastal Ocean Dynamics Experiment. J. Geophys. Res. **92**: 1683–1698.

Shanks, A.L. 1986. Tidal periodicity in the daily settlement of intertidal barnacle larvae and an hypothesized mechanism for the cross-shelf transport of cyprids. Biol. Bull. **170**: 429–440.

Shepherd, S.A. 1998. How good are egg per recruit analyses for managing abalone fisheries? *In* Proceedings of the North Pacific Symposium on Invertebrate Stock Assessment and Management. *Edited by* G.S. Jamieson and A. Campbell. Can. Spec. Publ. Fish. Aquat. Sci. 125. (Unpublished.)

Shepherd, S.A., and Brown, L.D. 1993. What is an abalone stock: implications for the role of refugia in conservation. Can. J. Fish. Aquat. Sci. **50**: 2001–2009.

Stevens, B., Armstrong, D.A., and Cusimano, R. 1982. Feeding habits of the Dungeness crab *Cancer magister* as determined by the index of relative importance. Mar. Biol. (Berlin), **72**: 135–145.

Strub, P.T., Allen, J.S., Huyer, A., Smith, R.L., and Beardsley, R.C. 1987. Seasonal cycles of currents, temperatures, winds, and sea level over the northeast pacific continental shelf: 35°N to 48°N. J. Geophys. Res. **92**: 1507–1526.

Tuck, G.N., and Possingham, H.P. 1994. Optimal harvesting strategies for a metapopulation. Bull. Math. Biol. **56**: 107–127.

Wahle, R.A., and Steneck, R.S. 1991. Recruitment habits and nursery grounds of the American lobster (*Homarus americanus* Milne Edwards): a demographic bottleneck? Mar. Ecol. Prog. Ser. **69**: 231–243.

Waiwood, K.G., and Elner, R.W. 1982. Cod predation of snow crab (*Chionoecetes opilio*) in the Gulf of St. Lawrence. *In* Proceedings of the International Symposium on the Genus *Chionoecetes*. Lowell Wakefield Fisheries Symposium Series. University of Alaska, Alaska Sea Grant Report No. 82-10. pp. 499–520.

Walters, C.J., and Collie, J.S. 1988. Is research in environmental factors useful to fisheries management? Can. J. Fish. Aquat. Sci. **45**: 1848–1854.

Walters, C.J., Hall, N., Brown, R., and Chubb, C. 1993. Spatial model for the population dynamics and exploitation of the western Australian rock lobster, *Panulirus cygnus*. Can. J. Fish. Aquat. Sci. **50**: 1650–1662.

Williams, D. McB., Wolanski, E., and Andrews, J.C. 1984. Transport mechanisms and the potential movement of planktonic larvae in the central region of the Great Barrier Reef. Coral Reefs, **3**: 229–236.

Wing, S.R., Largier, J.L., Botsford, L.W., and Quinn, J.F. 1995*a*. Settlement and transport of benthic invertebrates in an intermittent upwelling region. Limnol. Oceangr. **40**: 316–329.

Wing, S.R., Botsford, L.W., Largier, J.L., and Morgan, L.E. 1995*b*. Spatial variability of upwelling relaxation and crab settlement in the northern California upwelling system. Mar. Ecol. Prog. Ser. **128**: 199–211.

Metapopulation dynamics of meroplanktonic invertebrates: the Dungeness crab (*Cancer magister*) as an example

Louis W. Botsford, Coleen L. Moloney, John L. Largier, and Alan Hastings

Abstract: Many harvested marine invertebrate populations are metapopulations, composed of relatively sedentary subpopulations connected by dispersing larval stages. Stock assessment should take explicit account of this metapopulation structure and its consequences for population dynamics. We describe the implications of recent findings regarding metapopulation dynamics for Dungeness crab (*Cancer magister*) populations along the west coast of the United States. A model of the dependence of larval growth and mortality rates on temperature and salinity shows the potential latitudinal difference in Dungeness crab larval duration and associated differences in overall survival through the larval period. It also shows that for larvae in general, the proposed decline in survival due to decreased development rate, depends critically on the assumption that mortality rate is temperature invariant. Combining the effect of temperature on development rate with the influence of variability in cross-shelf transport using an advection–diffusion approach shows the dependence of settlement time on time of larval release. Because of latitudinal differences in both (*i*) the direction of change in temperature through the larval season and (*ii*) the advective environment, late larval release leads to greater settlement in the north, whereas the reverse is true in the south. A model of the interaction between density-dependent recruitment and larval dispersal among subpopulations shows that, if density-dependent effects on recruitment occur after the dispersing larval phase, regular asynchronous patterns along the coast can result. A similar effect may be present in the Dungeness crab catch record. Simulation of a mechanism recently proposed to explain the cycles in catch, i.e., variability in alongshore flow producing larval loss to the north, does not reproduce the observed cyclic behavior.

Résumé : Bien des populations exploitées d'invertébrés marins sont en fait des métapopulations, composées de sous-populations relativement sédentaires reliées par des stades larvaires qui se dispersent. L'évaluation des stocks devrait explicitement prendre en compte cette structure en métapopulations et ses conséquences pour la dynamique des populations. Nous décrivons les incidences de certaines découvertes récentes sur la dynamique des métapopulations pour des populations de crabe dormeur (*Cancer magister*) le long de la côte ouest des États-Unis. Un modèle de la dépendance des taux de croissance et de mortalité des larves à l'égard de la température et de la salinité montre que la durée du stade larvaire des crabes dormeurs peut varier avec la latitude, de même que certaines caractéristiques associées de la survie globale pendant la période larvaire. Ce modèle montre aussi que, pour les larves en général, le déclin supposé de la survie qui serait dû à la baisse du taux de développement dépend de façon critique de l'hypothèse selon laquelle le taux de mortalité ne varierait pas avec la température. La combinaison de l'effet de la température sur le taux de développement et de l'influence de la variabilité du transport en travers de la plate-forme selon une approche d'advection–diffusion montre que le moment de l'installation sur le fond dépend du moment de la libération des larves. À cause des différences de latitude dans *i*) la direction du changement de la température pendant toute la saison larvaire et *ii*) le régime d'advection, la libération tardive des larves occasionne une plus forte installation sur le fond dans le nord, alors que l'inverse est vrai dans le sud. Un modèle de l'interaction entre le recrutement dépendant de la densité et la dispersion des larves entre les sous-populations montre que, si des effets sur le recrutement dépendants de la densité se produisent après la phase de dispersion des larves, il peut en résulter des schémas asynchrones réguliers le long de la côte. Un effet similaire peut se retrouver dans l'historique des prises de crabe dormeur. La simulation d'un mécanisme récemment proposé pour expliquer les cycles des captures, c'est-à-dire la variabilité de l'écoulement le long du rivage, qui occasionnerait une perte de larves pour le nord, ne reproduit pas le comportement cyclique observé. [Traduit par la Rédaction]

L.W. Botsford. Department of Wildlife, Fish, and Conservation Biology, University of California, Davis, CA 95616-8751, U.S.A.

C.L. Moloney. Marine Science Research Institute, Zoology Department, University of Cape Town, Rondebosch 7700, South Africa.

J.L. Largier. Center for Coastal Studies, Scripps Institution of Oceanography, University of California, San Diego, La Jolla, CA 92093, U.S.A.

A. Hastings. Division of Environmental Studies, University of California, Davis, CA 95616, U.S.A.

Correct citation: Botsford, L.W., Moloney, C.L., Largier, J.L., and Hastings, A. 1998. Metapopulation dynamics of meroplanktonic invertebrates: the Dungeness crab (*Cancer magister*) as an example. *In* Proceedings of the North Pacific Symposium on Invertebrate Stock Assessment and Management. *Edited by* G.S. Jamieson and A. Campbell. Can. Spec. Publ. Fish. Aquat. Sci. 125. pp. 295–306.

Introduction

Many benthic marine invertebrate populations are meroplanktonic metapopulations, groups of subpopulations linked by a dispersing larval stage. Most harvested fish populations are also metapopulations, but invertebrate metapopulations differ in that the stages from postsettlement through adult are relatively sedentary. Despite the ubiquity of this spatial population structure, until recently there have been few explorations of the general dynamic behavior of meroplanktonic metapopulations. This is due in part to the fact that their behavior often depends on the spatio–temporal dynamics of larval dispersal which is often poorly understood. However, even with simple, constant, neighborhood dispersal, the behavior of meroplanktonic metapopulations can be quite different than would be expected on the basis of the dynamics of a population not distributed over space. Here we describe some recent results regarding the general behavior of such populations and evaluate their implications for the Dungeness crab, *Cancer magister*.

The growing number of marine invertebrate populations in which metapopulation aspects have been shown to be important indicates that better understanding of the dynamic implications of metapopulation structure is needed (Botsford 1995). In American lobster populations on the east coast of the U.S., offshore subpopulations provide a critical larval subsidy that apparently sustains intensively harvested inshore stocks (Fogarty 1998). Scallop populations have been viewed as spatially distributed metapopulations (Orensanz 1986; Orensanz et al. 1991) and for some of them there may be enough information on larval dispersal and local population dynamics to evaluate metapopulation dynamics, especially as quantitative descriptions of larval dispersal become available (e.g., Tremblay et al. 1994). For the western rock lobster in Western Australia, predictions of catch from settlement collectors have been improved by separating the metapopulation into three subpopulations (Caputi et al. 1995*b*), but better understanding of outstanding questions (e.g., the decline in settlement on Arbolhos Island (Caputi et al. 1995*a*)) may be achieved by incorporating descriptions of the influence of physical conditions (e.g., Pearce and Phillips 1994) on larval dispersal into models of spatially distributed juveniles and adults (see Walters et al. (1993) for example).

Investigations of the relationships between life-history parameters and the possible forms of population dynamic behavior are a necessary first step in stock assessment and management, the topics of this symposium. Typically, stock assessment involves fitting models to data for their immediate use in management. The goal of such modeling is accurate prediction of future population behavior, enabling well-informed management. Knowledge of the mechanisms that may underlie the dynamics of a population is necessary in the early stages of stock assessment to choose the right model or suite of possible models to be fit. For metapopulations, it is not yet clear which biological mechanisms need to be included in models to allow specific kinds of observed spatio–temporal behavior. For example, in studies of covariability of subpopulations at different locations, synchrony is usually associated with wide-scale environmental forcing or dispersal among populations (e.g., Campbell and Mohn 1983; Cohen et al. 1991; Elner and Campbell 1991; Koslow et al. 1987). However, as we show here, deterministic dispersal in metapopulations can actually lead to a lack of synchrony and regular spatial patterns.

Like the prediction implied in stock assessment, the decision making required in management also depends on knowledge of mechanisms. The theory of optimal harvesting indicates that the decision of whether to remove an individual from a harvested population ultimately depends on comparison of the individual's current market value (minus harvesting costs) with its comprehensive future value, including future growth, reproduction, and effect on density dependence (cf., Botsford 1981). Whether density dependence is present may be indicated by observed temporal and spatial variability. The relationship between density dependence and population variability is one of the primary questions addressed in our investigations of metapopulation dynamics. Metapopulation considerations also introduce the strategy of spatial variability in harvest into the optimal harvest problem (e.g., Botsford et al. 1993; Quinn et al. 1993; Caddy and Seijo 1998; Wing et al. 1998).

The almost synchronous cyclic fluctuations in catch of Dungeness crab between northern California and Washington on the west coast of the U.S. have been the focus of considerable research over the past 20 years (reviewed in Botsford 1986; Botsford et al. 1989; Botsford and Hobbs 1995)(see Fig. 1). That research has been based largely on time series of catches, which for this highly exploited, male-only fishery are a reasonable proxy for recruitment (Methot and Botsford 1982). Predator–prey mechanisms with both salmon and humans as the predators have been discounted as causes of the cycles, but various mechanisms involving environmental forcing and density-dependent recruitment are considered possible. Environmental forcing mechanisms have involved ocean temperature, surface winds, sun spots, and alongshore transport (see reviews by Botsford 1986; Botsford et al. 1989; Botsford and Hobbs 1995). Also, on the basis of higher catches resulting from recruitment during years with lower sea level in Washington, McConnaughey et al. (1994) proposed that observed fluctuations are driven by a mechanism in which years of greater poleward, alongshore flow lead to loss of larvae to the north from this metapopulation. Density-dependent recruitment mechanisms evaluated to date include density-dependent fecundity, an egg-predator worm whose abundance is proportional to crab fecundity, and cannibalism of adults on juveniles (see reviews by Botsford 1986; Botsford et al. 1989; Botsford and Hobbs 1995).

Dungeness crab stocks form a metapopulation consisting of subpopulations of relatively sedentary juveniles and adults on patches of sandy bottom (McConnaughey et al. 1994). Subpopulations are linked by a dispersing planktonic larval phase of about 4 months duration present from December through June. Specific dispersal patterns are unknown. In the few field observations of larval distribution, later zoeal stages are found further offshore (Lough 1976; Reilly 1983) and older members of the last stage, the megalopal stage, are found nearer shore than younger megalopae (Hatfield 1983; Jamieson et al. 1989). A topical question is whether larvae observed up to 400 km offshore (Hobbs et al. 1992) are transported to requisite settlement sites within 15 km of the coast or are simply larval wastage (McConnaughey et al. 1992).

The oceanographic context of Dungeness crab larval dispersal south of the bifurcation of the West Wind Drift off

Fig. 1. Dungeness crab catch along the west coast of the U.S. Catch is a good proxy for recruitment four years earlier (Methot and Botsford 1984) because of the high harvest rate in this size-limited, male-only fishery. [Source: Pacific States Marine Fisheries Commision].

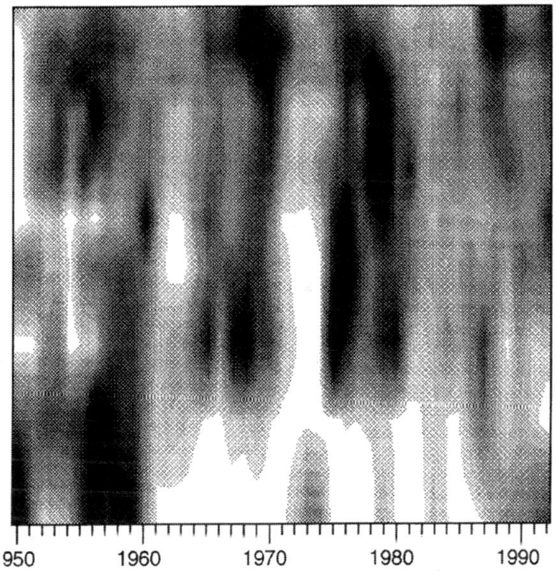

British Columbia is an eastern boundary current, the California Current. Characteristic features of coastal circulation are associated with three seasons along this part of the coast (Largier et al. 1993). The early larval period takes place in the storm season near the beginning of the calendar year when nearshore flow is poleward and onshore, punctuated by frequent winter storms moving from west to east. The late larval period occurs during the upwelling season, which begins when the North Pacific high pressure zone moves poleward near the beginning of April, an event termed the spring transition (Strub et al. 1987). The spring transition marks the beginning of equatorward winds and a shift in surface flows to equatorward and offshore. Differences in coastal morphology, bathymetry, and distance from the North Pacific high pressure system lead to latitudinal differences in coastal circulation during both the early and late part of the larval period. In the winter storm season, onshore flow is stronger off Washington than off California, where it is almost nonexistent. During the upwelling season, offshore flow is strongest in the south off northern California.

Assessment of metapopulation aspects of these Dungeness crab stocks requires investigation of the influence of the physical environment on the larval development and dispersal, then examination of the impact of those results on the way subpopulations are linked to produce metapopulation dynamics. Here we present larval–environment aspects first, then metapopulation dynamics.

Effects of temperature and salinity on larvae

Both biotic and abiotic aspects of the immediate environment of an invertebrate larva can affect growth and development rates in ways that influence recruitment success. To evaluate these, we formulated a partial differential equation model that described the way in which the distribution of larvae over developmental stage changed with time, essentially a stage-structured version of the von Foerster equation (stage here refers to relative development from hatch to metamorphosis, not instar, Moloney et al. 1994). The solution to this model expresses the time course of settlement $S(t)$ as

$$[1] \quad S(t) - R(t - a(t)) \frac{g(1, t)}{g(0, t - a(t))} \exp\left(-\int_0^1 \frac{D(m, t)}{g(m, t)} \, dn\right)$$

where $R(t)$ is the pattern of recruitment to the larval phase (i.e., hatching), $a(t)$ is the age of a larva hatched at time t, $g(m, t)$ is the development rate of a larva that is at stage m at time t, and $D(m, t)$ is larval mortality rate at stage m and time t. Stage of development within the larval period is reflected in the variable m, which varies continuously from 0 (hatch) to 1 (metamorphosis). In this expression, the first term reflects the effect of the pattern of larval release, the second term reflects the effect of changes in development rate, and the third term reflects the effect of mortality.

This expression reflects two important general aspects of the influence of the local environment on larval development and survival. The first is a qualification of the well-known result that small changes in development time can lead to large changes in the fraction surviving larval stage because of the high larval mortality rate (e.g., Underwood and Fairweather 1989). This change in survival affects the integrand in the exponent, $D(m, t)/g(m, t)$. While this term clearly indicates that a change in development rate $g(m, t)$ leads to a change in survival through the larval stage, if mortality rate $D(m, t)$ changes by the same fraction, change in development rate has no effect. Thus, if the change in development rate is due to temperature and mortality is due primarily to predation by poikilotherms, whose consumption rate depends on temperature in a similar

Fig. 2. Ocean temperatures during the Dungeness crab larval period from various shore stations along the coast (column a); settlement resulting from constant release rate, December–February, with no mortality except that due to suboptimal temperatures (column b); and settlement from the same releases with larval mortality rate (0.066 day^{-1}, Hobbs, et al. 1992) included (column c). Shaded bars represent the approximate timing of the spring transition (Strub et al. 1987). Redrawn from Moloney et al. (1994).

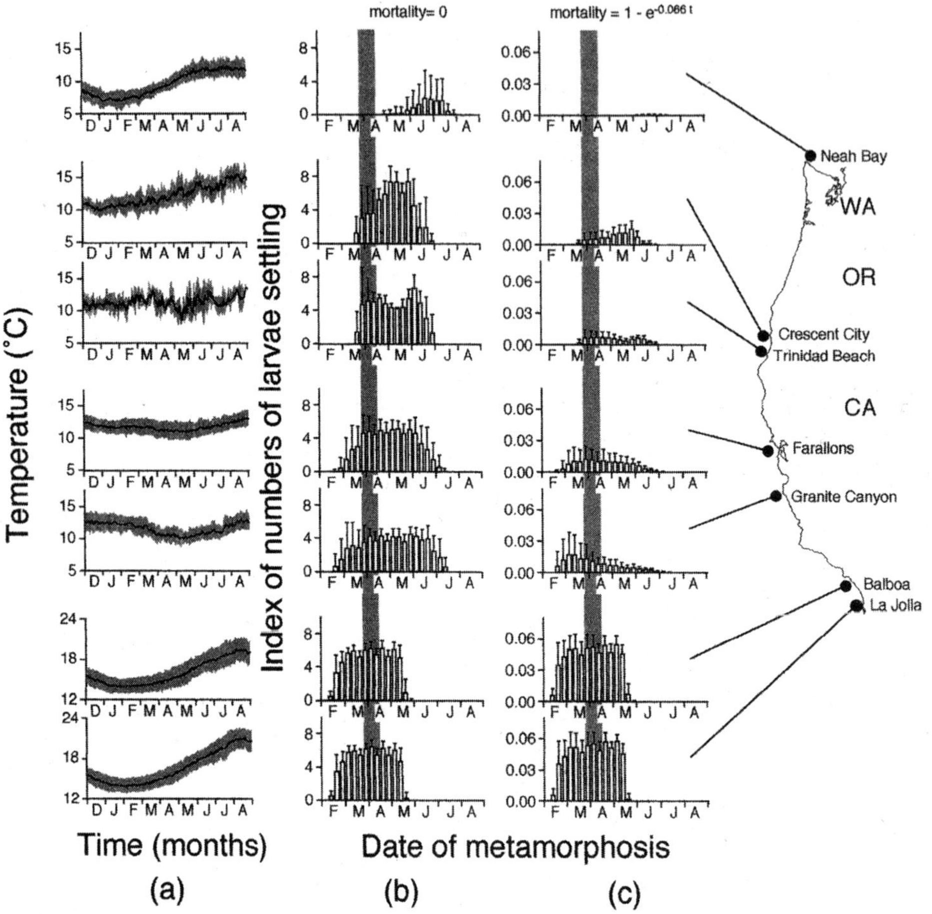

way, a change in temperature will not change overall survival through the larval stage.

The second general implication of this result is that when development rate is increasing (decreasing), the time period over which larvae settle will be less than (greater than) the time period over which they are hatched. This is reflected in the second term of eq. 1 in which rate of settlement is proportional to the ratio of rate of development at the end of the larval period to rate of development at the beginning. The timing of the beginning and end of settlement is reflected in the first term. This result implies, for example, that if development rate is temperature dependent and temperature is increasing through the larval period, timing of larval settlement is less sensitive to timing of larval release (because the settlement period is compressed).

The implications of eq. 1 and these results for Dungeness crab were obtained by formulating a model describing the influence of temperature and salinity on development and mortality from laboratory data, then using available daily temperature and salinity at shore stations to compute settlement at various locations along the coast (see Fig. 2) (Moloney et al. 1994). Laboratory data indicated that development rate depended primarily on temperature, rather than salinity, and that naturally occurring temperature–salinity combinations rarely elevated mortality. The settlement resulting from a 3-month pulse of larvae at low mortality rate (middle column, Fig. 2) demonstrates the compression of the settlement period, where temperature is increasing during the larval period (e.g., in Washington), and the expansion of the settlement period where temperature is declining during the larval period (e.g., in central California where strong upwelling occurs). The difference between settlement patterns at low (column b in Fig. 2) and high mortality rate (column c in Fig. 2) depends critically on the assumption that mortality rates do not vary with temperature.

Cross-shore larval transport

These influences of the local environment on development and survival must be considered in conjunction with the influences of coastal circulation and larval behavior on larval transport. Larvae of many marine invertebrates disperse far from their origin during the larval phase, but must return to specific areas, usually near shore, for successful settlement. There are several modeling methods for evaluating the effects of transport. A direct, focused approach is to compute the transport possible

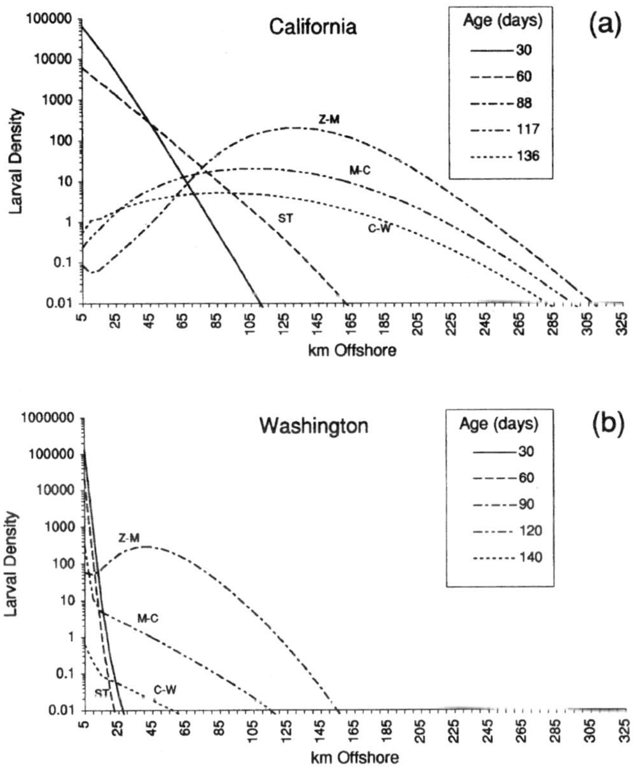

Fig. 3. Cross-shore distribution of Dungeness crab larvae from an advection–diffusion model for larvae released on February 1st off: (*a*) northern California and (*b*) Washington with a spring transition on April 1st. Distributions are given at the times of significant events identified as: ST, the spring transition; Z-M, the transition from zoea to megalopa; M-C megalopae becoming competent; and C-W, the end of the assumed 20-day competency period. Larval ages when each of these occurs, determined from temperature-dependent development rates, are also given.

from specific known currents and larval swimming or vertical migratory behavior (e.g., Rothlisberg et al. 1983; Hobbs et al. 1992; Katz et al. 1994). In a more comprehensive, mechanistic approach, three-dimensional coastal circulation can be computed from first principles, numerical solution of the Navier–Stokes equations with appropriate constraints, forcing, and boundary conditions (e.g., Haidvogel et al.1991; Pares-Sierra and O'Brien 1989; Song and Haidvogel 1995), and model larvae can be inserted in them (Hofmann et al. 1991; Botsford et al. 1994). These models are computation intensive and do not yet adequately describe all of the important mesoscale features. While these models are being developed to the point that they can be used easily by population biologists, some have turned to a simpler formulation, advection-diffusion models (e.g., Hill 1990; Hill and White 1990). Advection–diffusion models calculate the distribution of larvae from specified advective rates (i.e., the velocity of a patch of larvae) and diffusive rates (i.e., how fast the patch spreads out) (see Okubo (1994) for a description of how advection–diffusion models relate to physical oceanographic phenomena). Advective and diffusive rates may be obtained from calculations based on first principles or they may be purely empirical, based on current measurements or other observations.

Larval transport in many benthic, marine invertebrate populations can be viewed in terms of two interrelated aspects, recruitment success and net dispersal. Recruitment success typically requires being near the coast to settle in areas with the appropriate prey base and substratum, as well as minimal predators. Recruitment success therefore typically hinges on cross-shore flows. On the other hand, larval dispersal of benthic, coastal invertebrates relates to alongshore flow. It is the distance moved from the larval source to other suitable settlement habitat. This division of two-dimensional dispersion into its Cartesian elements is, of course, not a statement of independence of the two aspects (e.g., the cross-shelf distribution can be determined by alongshore flows) and exceptions are possible (e.g., the proposal by McConnaughey et al. (1994) links recruitment success to alongshore flow). It is merely a rationale for exploring the impacts of cross-shelf flows first, before combining the effects of flow in both horizontal directions. As a first step we have used an advection–diffusion model describing cross-shore transport of Dungeness crab larvae in the California current. A one-dimensional model is particularly appropriate for the California Current system with its dramatic, seasonal shifts in cross-shelf currents and for the Dungeness crab with its requirement to settle near shore (Carrasco et al. 1985; Jamieson et al. 1989; McConnaughey et al. 1992).

Although our study of the impact of cross-shore flows on larval dispersal (Botsford et al. 1994) involved several general results regarding the relative importance of advection and diffusion, here we focus on questions motivated by the Dungeness crab. These questions involve interactions of the hatching date, the known seasonal shifts from onshore to offshore flow, the temperature-induced variability in development described above, geographical differences in cross-shore flow, and the effect of larval vertical migratory behavior on transport. A description of the empirical basis for parameter values and details of the advection–diffusion model are in Botsford et al. (1994). To compare effects of circulation on settlement in areas of strong upwelling to northern areas with weaker upwelling, we formulated two cases, one for northern California and one for Washington. Advective rates off Washington were 0.02 $m \cdot s^{-1}$ onshore before the spring transition and 0.02 $m \cdot s^{-1}$ offshore after the spring transition and the rates off California were 0.0 $m \cdot s^{-1}$ before the spring transition and 0.05 $m \cdot s^{-1}$ offshore after. Because Dungeness crab larvae migrate vertically during the megalopal stage, when the model larvae reached that stage, these rates were replaced with typical values from computations based on observed winds and vertical migratory behavior by Hobbs et al. (1992), 0.03 $m \cdot s^{-1}$ onshore off Washington and 0.01 $m \cdot s^{-1}$ onshore off California.

The combined effects of these transport mechanisms on the spatial distribution of larvae throughout the larval period can be seen in plots of model results for larvae released on February 1st and a spring transition on April 1st (Fig. 3). For California, the distribution is a result of diffusion only, until the spring transition at 60 days, after which larval distributions extend offshore. With the advent of the megalopal stage there is some onshore movement due to vertically migrating larvae and surface winds (Hobbs et al. 1992). For Washington, onshore

Fig. 4. Settlement of Dungeness crab larvae released on various dates off California and Washington. Variability about each line reflects observed interannual variability in temperature.

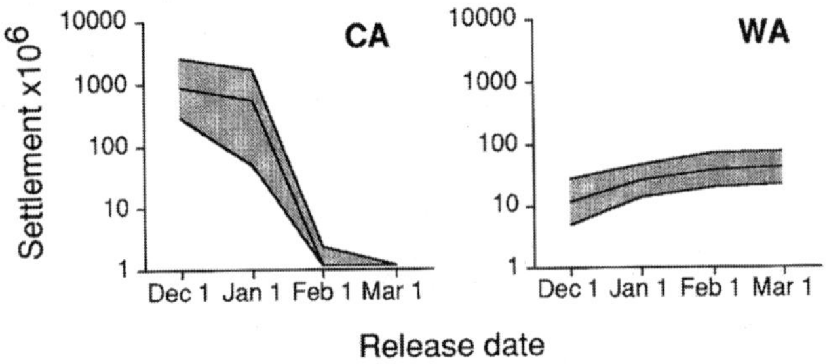

advection during the first two months maintains larvae close to shore and the effects of upwelling after the spring transition are not as great as in California. Onshore movement throughout the larval stage is also greater. Note that for these parameter values the greater net onshore flow off Washington is almost exactly compensated for by the greater total mortality during the longer development periods due to colder temperatures off Washington.

Some aspects of these results are consistent with the four years of larval data available (Hobbs et al. 1992). The fact that the overall spatial cxtent is comparable indicates that the results from this approach may be meaningful, despite ignoring mesoscale oceanographic features that could retain larvae near shore or transport them alongshore. The timing of cross-shelf distributions compares favorably with larval surveys which indicate a nearshore distribution of larvae when sampling occurred during the late megalopal stage (years 1981 and 1982 in Hobbs et al. 1992) and an offshore distribution of larvae when sampled during the early part of the megalopal stage (years 1983 and 1985 in Hobbs et al. 1992). Early sampling in 1985 indicates more megalopae off California than Washington, whereas late sampling in 1981 and 1982 indicates more megalopae off Washington, consistent with the differences in development periods.

Results from this model for Dungeness crab parameters show how the date of egg hatching affects settlement in years of various temperatures and at the two locations (Fig. 4). Date of hatching has a strong effect on settlement in California where upwelling winds are strong and a weaker effect of the opposite sign in Washington where upwelling winds are weaker. In Washington, effects of increasing temperature on development, and hence on survival, are stronger than the effects of seasonal shifts in cross-shore transport. This result, indicating that later hatching leads to greater settlement in Washington but less settlement in California, provides a possible selective force that could underlie the proposed genetic differences between southern and northern crab stocks indicated by later settlement dates to the north (Orensanz and Gallucci 1988; Dinnel et al. 1993; Jamieson and Phillips 1988). Another interesting aspect of the geographical comparison is that resultant levels of settlement at the two locations are similar. The greater settlement off Washington, expected on the basis of lower offshore advection, is not seen because of slower development rate at colder temperatures. These results also indicate that variations in temperature in the ranges observed over the past several decades can lead to variation in settlement of an order of magnitude at both locations.

Thus far, in these two studies we have demonstrated some of the likely potential effects of temperature, salinity, and cross-shore flows on larval settlement, but we are far from a complete understanding of dispersal of Dungeness crab larvae. We are in the process of evaluating the effects of known, broad-scale, alongshore flows in a two-dimensional advection–diffusion model, as well as through field sampling (A. Hastings, L.W. Botsford, J.L. Largier, and J. Pilliod, unpublished data; Wing et al. 1998). Also, little is known of the influence of food on these larvae. These modeling results currently serve as a null hypothesis, regarding spatial and temporal variability in larval distributions and settlement, to be tested against further sampling.

Metapopulation aspects

Even though we do not have a complete understanding of the survival, transport, and redistribution of the larvae of most marine invertebrate species, including Dungeness crab, it is useful to begin to evaluate the kinds of population dynamic behavior that can result from various dispersal patterns. Given the potential problems with increasingly intensive harvest and possible climate change, we need to understand the dynamic behavior and resilience that can occur in harvested invertebrate metapopulations. To date, our approach to this problem has been to examine the effects of linking a number of subpopulation models through a simple form of larval dispersal. Subpopulations have density-dependent recruitment and age structure and the dispersal patterns are Gaussian-shaped such as would result from diffusion.

The behavior of an age-structured subpopulation model with density-dependent recruitment was discussed in the context of Dungeness crab at the last North Pacific Invertebrate Conference (Botsford 1986). Recruitment was described as the product of egg production and recruitment survival, a nonlinear function reflecting the dependence of recruitment on effective population size,

$$[2] \quad R_t = B_t\, \mathrm{f}(C_t)$$

where both egg production, B_t, and effective population size, C_t, were weighted sums over abundance at age. The weighting for each age in the former was fecundity, while the weighting in the latter was the effect of an individual at each age on density dependence in recruitment. The rest of the model reflected the fact that, for ages beyond recruitment, survival did not vary with time,

$$[3] \quad n_{a,t} = R_{t-a}\, s_a$$

where s_a is the survival from recruitment to age a. Behavior of this model involved variability on time scales of roughly twice the mean age as the normalized slope of the recruitment survival function, f, became more negative. If the slope was negative enough, cycles in recruitment and abundance were produced. Cycles were more likely for narrow, peaked age structures than for broad, flat ones. These results and the conditions under which life history characteristics could cause cycles have been used, along with empirical data, to evaluate various proposed causes of the cycles in Dungeness crab (Botsford 1986; McKelvey et al. 1980; Botsford and Hobbs

Fig. 5. Behavior of a metapopulation composed of age-structured subpopulations as the width of the Gaussian-shaped dispersal kernel, σ, is increased. Subpopulations have predispersal density dependence and the parameter values of cyclic Dungeness crab populations.

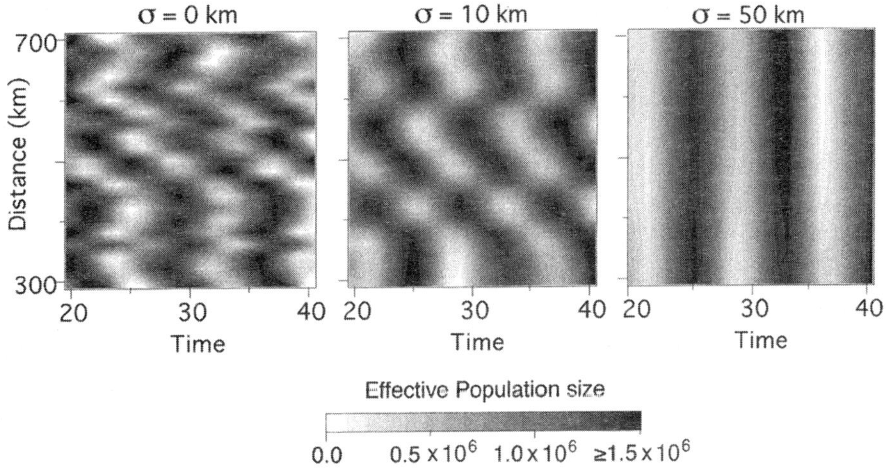

Fig. 6. Behavior of a metapopulation composed of age-structured subpopulations as the width of the Gaussian-shaped dispersal kernel, σ, is increased. Subpopulations have postdispersal density dependence and the parameter values of cyclic Dungeness crab populations.

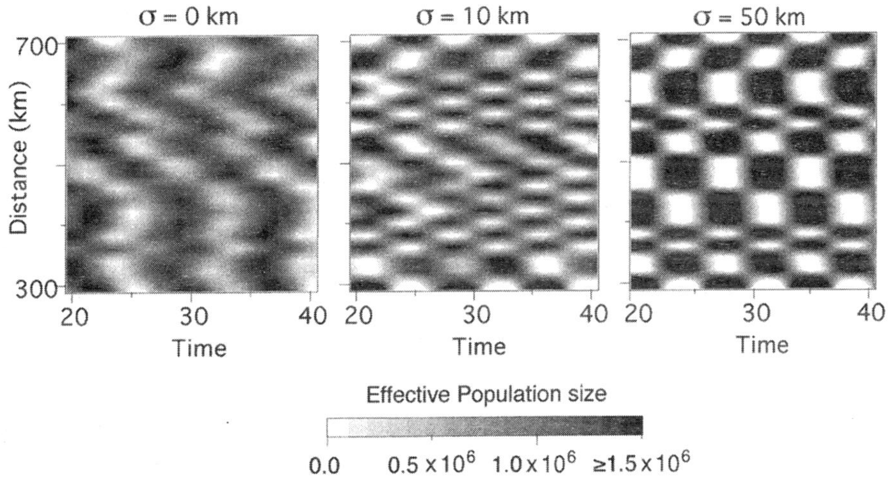

1995). Higgins et al. (1996) have recently shown that, for general age-structured models with density-dependent recruitment represented by a Ricker model (Ricker 1954), as adult survival declines, recruitment may not cycle with periods nearly twice the mean age, but rather can be chaotic (cf., Botsford 1991). However, for the specific age structure of Dungeness crab, the result, indicating limit cycles on time scales of twice the mean age, is robust (i.e., behavior on shorter time scales is not likely) (Botsford et al. 1994).

We have combined a number of subpopulations such as this into a model of a metapopulation along a coastline, linking them together through larval dispersal (Botsford et al. 1994). Recruitment at location x along the coast is

$$[4] \quad R_t(x) = \mathrm{f}\,(C_t(x), x) \int_{x_1}^{x_u} p_t(x, y)\, \mathrm{g}(W_t(y), y)\, B_t(y)\, \mathrm{d}y$$

In this model there are two functions representing density-dependent recruitment survival, one (g) representing density-dependence occurring before dispersal termed predispersal density dependence, and the other (f) representing density-dependent interactions between settling individuals and resident juveniles and adults after dispersal, termed postdispersal density dependence. Both depend on local effective population size (C_t and W_t). For Dungeness crab, for example, the former could represent density-dependent fecundity (cf., McKelvey et al. 1980) and the latter could represent cannibalism (cf., Gotshall 1977; Botsford and Hobbs 1995). This model also includes a dispersal kernel, $p(x, y)$, which represents the fraction of larvae originating at location y that settles at location x. This part of the model would be determined from analyses and observations such as those described in the previous section on the influence of physical conditions on larval survival and transport.

One of the first questions we addressed with this model was whether the fact that the population was distributed over space made any difference in terms of dynamic behavior. Can we predict metapopulation behavior by simply extending the behavior of single subpopulations synchronously along the coast? To answer this question, we connected a number of subpopulations with the parameter values of cyclic Dungeness crab populations along a 1000-km coastline. We tested the effects of dispersal among subpopulations of a metapopulation

Fig. 7. Correlations between Dungeness crab catch at various ports along the west coast of the U.S. Ports are in the order of south to north.

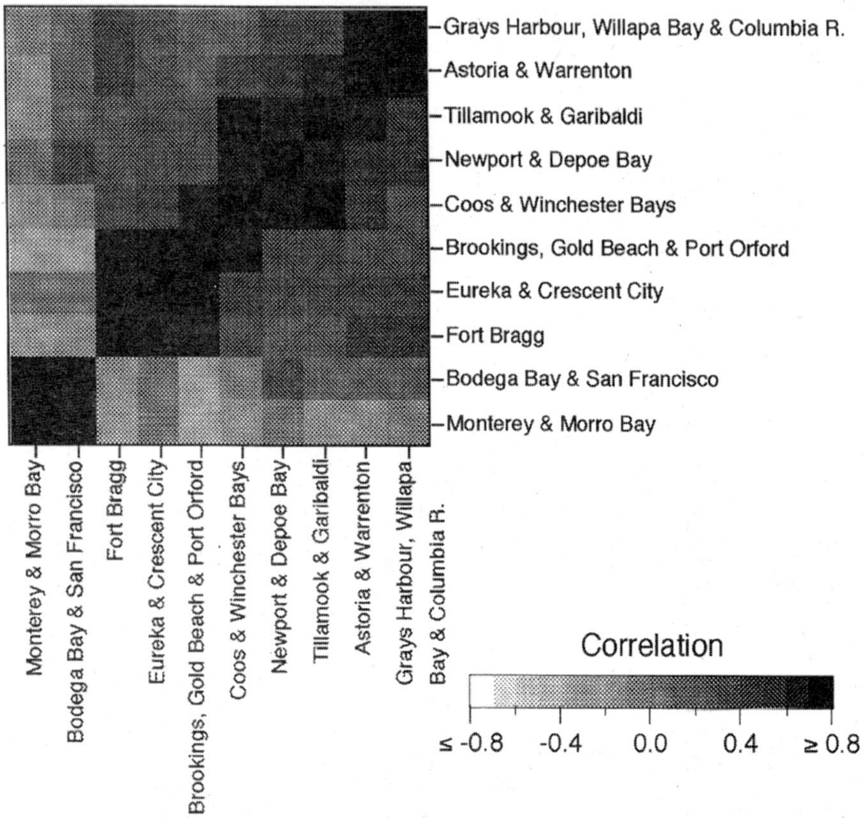

by gradually increasing the distances dispersed by increasing the standard deviation of the Gaussian dispersal kernel, σ. For the case with predispersal density dependence (Fig. 5), when $\sigma = 0$, populations cycled independently. As σ was increased to 10 km, patterns began to form and, for $\sigma = 50$ km, cycles were completely synchronous. For the case with postdispersal density dependence (Fig. 6), as σ was increased from 0 to 10 km, patterns with spatial structure on scales of 20–30 km formed and, when $\sigma = 50$ km, patterns on scales just greater than 100 km formed. Populations remained cyclic in time, but they were out of phase along the coast. Dispersal among subpopulations seems to lead to a qualitative difference in behavior.

This new type of population behavior leads to a paradox with regard to Dungeness crab. When we include two putative characteristics of Dungeness crab populations, postdispersal density dependence in the form of cannibalism and dispersal over moderate distances, we obtain asynchronous behavior with a distinct spatial pattern, contrary to our standing description of synchronous cycles along the coast. There are several possible explanations of this apparent contradiction. First, closer scrutiny of the observed cycles along the coast (Fig. 1) reveals significant alongshore correlation (Fig. 7). Translation of these correlations into actual distances, rather than ports, reveals correlation over distances of several hundred kilometers. These are the spatial scales of mesoscale variability in coastal circulation; hence, this scale may be important (e.g., Wing et al. 1998). A second possible explanation is that the cycles are not due to density-dependent recruitment, but rather to cyclic or near-cyclic variability in recruitment survival. Running this model with larval survival increased by a factor of five every ten years produces synchronous cyclic behavior. This effect, combined with some spatial variability in survival, could produce the observed behavior.

Another possible explanation is the proposal by McConnaughey et al. (1994) that variability in Dungeness crab catch is due to variability in alongshore flow. McConnaughey et al. (1994) computed alongshore flows off Washington from monthly sea-level data from 1947 to 1986 during the Dungeness crab larval period, and noting they were statistically related to catch, concluded that this flow controlled settlement by transporting larvae out of the northern end of the range considered here. Using transport distances from McConnaughey et al. (1994) as the dispersal distance for larvae in this model leads to a population that gradually goes extinct from south to north because the subpopulations cannot sustain themselves with large larval losses (Fig. 8). Adjusting the alongshore transport by setting the mean transport of the 40 years equal to zero, leads to apparent cyclic behavior at each point along the coast, but the north and south are out of phase. Modifying the transport so that a portion of the larvae do not disperse while the rest disperse still leads to populations which do not cycle appropriately. Limiting the transport to the months February–May, rather than the period January–May as used in McConnaughey et al. (1994), leads to populations which do not go extinct as rapidly, but are less cyclic.

Discussion

The modeling results described here represent only an

Fig. 8. A metapopulation model with (*a*) Dungeness crab parameters, Gaussian dispersal with $\sigma = 10$ km, and alongshore transport taken from computations in McConnaughey et al. (1994), and (*b*) using the same alongshore transport data with the mean removed. The arrows indicate the start of the influence of the transport data.

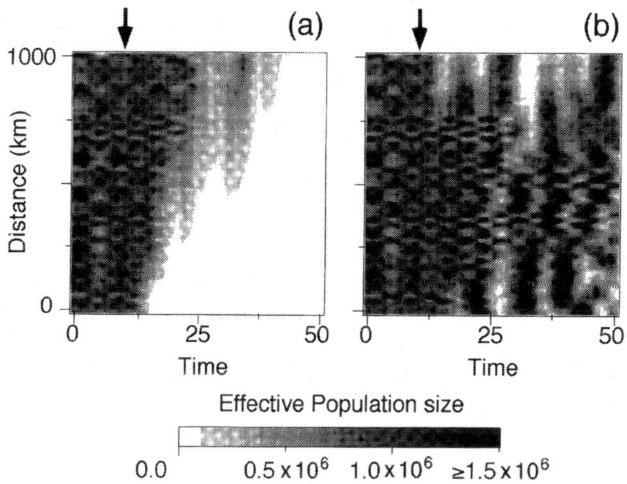

incremental increase in our understanding of meroplanktonic metapopulations in general and that of Dungeness crab in particular. Results regarding the larval phase parallel similar work on other species (cf., Cobb and Wahle 1994; Katz, et al. 1994 for American lobster). For Dungeness crab, our results show that observed levels of interannual variability in temperature and cross-shelf transport can lead to order-of-magnitude changes in settlement. These results have potential implications for past cyclic variability as well as projected effects due to long-term climate change. Two of the proposed effects of climate change, a change in upwelling winds (Bakun 1990) and changes in frequency and intensity of ENSO events (Zebiak and Cane 1991) would alter temperatures and flows in the coastal ocean.

Our results regarding the larval stage also underline the importance of latitudinal differences in development times and cross-shelf flows. These potential differences indicate caution is necessary when drawing conclusions regarding recruitment to the Dungeness crab metapopulation on the basis of cross-shelf flows computed at any one location. For example, the conclusion by McConnaughey et al. (1992) that all Dungeness crab larvae important to settlement remain near the coast throughout the year was based on computations of cross-shelf flows off Washington where flow is onshore during the winter period and upwelling is weak. This conclusion would not necessarily hold off northern California where offshore flows are stronger.

Another implication of latitudinal differences is that they provide a selective mechanism that could underlie the differences in hatch dates that have been proposed to be due to genetic differences. These differences could also be due to differences in ocean temperature, but we have not evaluated that hypothesis here. While Dungeness crab larvae, along the coastline considered here, tend to settle in late spring – early summer (Poole 1966; Lough 1976; Dinnel et al. 1993), larvae in the Puget Sound Basin and points further north tend to settle in the late summer (Orensanz and Gallucci 1988; Dinnel et al. 1993). Jamieson and Phillips (1988) have proposed a mechanism for isolation of these stocks, but genetic differences have not been demonstrated. Geographical differences in the sensitivity of settlement to timing of larval release, depicted in Fig. 4, indicate that earlier release would be selected for in the southern location, whereas later release would be selected for in the north.

The fact that we know little about the effects of prey availability on growth and survival of crab larvae is a substantial gap in our knowledge of crab recruitment processes. However, the type of interaction between development rate and circulation events described here for invertebrates may be more important for invertebrates than finfish. Hjort's (1914) second basic requirement for successful recruitment of larval fish, transport to a favorable location, may be more critical for invertebrates which have much less ability to adjust their location after metamorphosis.

The explicit mathematical description of how larval development and mortality rates interact revealed a relationship not previously emphasized, that the potential dramatic effects of temperature on development rate and therefore survival through the larval stage depend critically on a lack of temperature dependence in the mortality rate. Mortality rate probably does vary with temperature, for Dungeness crab larvae and others. As a consequence, the caveat follows that our results, involving an impact of temperature on development rate and then survival through the larval stage (e.g., Figs. 3 and 4), depend on an assumed constant mortality rate. Similar caveats should accompany other instances in which this mechanism is involved (Underwood and Fairweather 1989).

Results at the metapopulation level indicate that even if we had a complete understanding of the larval-dispersal phase of a species and could predict dispersal patterns we would not have a complete understanding of the behavior of the metapopulation. We need to know how various spatio–temporal patterns arise from specific life-history characteristics and dispersal patterns. It would also be beneficial to know what subpopulation dynamics tell us about metapopulation dynamics.

Increased understanding of similar spatio–temporal patterns in population biology provides some hope that this will be possible. Turing (1952) first showed that spatial patterns could develop in systems with nonlinear dynamics and dispersal. Levin (1974) showed that such spatial patterns arose in spatially distributed predator–prey systems with dispersal and Hastings (1992) showed that age-dependent dispersal coupled with strong age-dependent density dependence could also lead to "Turing-type" instabilities. Essentially these patterns arise because populations that are regulated (kept stable) by interaction between age-classes, lose that regulation when one of the age-classes disperses. Local instabilities grow and spatial patterns result. In the Dungeness crab example described here, for the case with predispersal density dependence, population regulation is maintained within subpopulations and no patterns arise. For the case with postdispersal density dependence, local regulation is lost as dispersal distances increase and spatial patterns emerge.

Although these results help to increase understanding of these mechanisms, other recent results (Hastings and Higgins 1994) indicate that metapopulations present new, poorly understood characteristics. In making predictions, we often assume that population dynamics go to some qualitative type of behavior (e.g., equilibrium, cycles, chaos) and remain there. Calculations with a semelparous version of the model used here indicate that metapopulation behavior can be extremely unpredictable on long time scales (Hastings and Higgins 1994). After exhibiting one type of qualitative behavior for hundreds of years, populations changed suddenly to another type of behavior with no external forcing. Work in progress indicates that this kind of unpredictability persists as long as the postmaturity survival rate of adults remains low, as would be the case in heavily harvested species.

To date, results for metapopulation behavior indicate that we need to expand our view of expected behavior in population analysis and stock assessment where metapopulations are involved (i.e., virtually all harvested benthic invertebrates). In particular, we need to consider possibilities other than just synchrony over large spatial scales and complete lack of synchrony. Metapopulations can exhibit more complex spatio–temporal patterns in which some subpopulations covary, but are out of phase. Discovering these underlying patterns will require ingenuity because they will probably not be as regular as in population models and they may be obscured by observation error.

For Dungeness crab specifically, these results expand rather than contract the suite of possible mechanisms that could underlie population behavior. Even though the Dungeness crab catch record (Fig. 1) does not exhibit the dramatic spatial patterns shown in Fig. 6, the fact that there is some correlation over space implies that dispersal may be responsible for this lack of synchrony. Retention within embayments in the southern half of the range considered here, as suggested by Wing et al. (1998), could determine the spatial scale of dispersal that might underlie these observed patterns.

Better understanding of the Dungeness crab metapopulation and its response to harvesting and climate change will require additional field sampling; therefore, part of the value of our results is in the direction they provide for such sampling. From the latitudinal differences obtained here, it seems clear that field observations of dispersal need to be made at several locations, possibly taking advantage of a comparative approach. The metapopulation results suggest that alongshore dispersal distances are important.

Our next step in modeling is the development of transport models that allow better evaluation of the two-dimensional aspects of larval dispersal. Such models will provide a comprehensive means of evaluating putative aspects of transports in a metapopulation context. For example, the mechanism involving larval wastage due to poleward transport, proposed as a cause of the cycles by McConnaughey et al. (1994), requires that larvae lost to the north not be replaced by advection from the south, a characteristic that would require complex two-dimensional flows to work.

The explicit acknowledgment of spatial distribution inherent in metapopulation approaches is of wide potential importance to invertebrate stocks. The papers presented at this conference indicate much more attention is being paid to spatial variability than 10 years ago at the last North Pacific Invertebrate Conference (however, see Orensanz 1986). We have begun to account for behavioral responses of harvesters to spatial variability in abundance (Prince and Hilborn 1998; Woodby 1998) and to account for variability in biological productivity through spatial management (Caddy and Seijo 1998; Botsford et al. 1993; Walters et al. 1993; Wing et al. 1998). Metapopulation approaches will be required to identify the mechanisms underlying observed spatial behavior in productivity, especially in cases in which sustainability of stocks and their fisheries depend critically on larval links between subpopulations (Fogarty 1998).

Acknowledgments

This material is based on work supported by the National Science foundation under award No. OCE-9016721 in the GLOBEC program funded jointly by NOAA and NSF.

References

Bakun, A. 1990. Global climate change and intensification of coastal ocean upwelling. Science (Washington, D.C.), **247**: 198–201.

Botsford, L.W. 1981. Optimal fishery policy for size-specific, density-dependent population models. J. Math. Biol. **12**: 256–293.

Botsford, L.W. 1986. Population dynamics of the Dungeness crab (*Cancer magister*). *In* North Pacific Workshop on Stock Assessment and Management of Invertebrates. *Edited by* G.S. Jamieson and N. Bourne. Can. Spec. Publ. Fish. Aquat. Sci. 92. pp. 140–153.

Botsford, L.W. 1991. Further analysis of Clark's delayed recruitment model. Bull. Math. Biol. **54**: 275–293.

Botsford, L.W. 1995. Population dynamics of spatially distributed, meroplanktonic, exploited marine invertebrates. ICES Mar. Sci. Symp. **199**: 118–228.

Botsford, L.W., and Hobbs, R.C. 1995. Recent advances in the understanding of cyclic behavior of Dungeness crab (*Cancer magister*) populations. ICES Mar. Sci. Symp. **199**: 157–166.

Botsford, L.W., Armstrong, D.A., and Shenker, J.M. 1989. Oceanographic influences on the dynamics of commercially fished populations. *In* Coastal oceanography of Oregon and Washington. *Edited by* M.R. Landry and B.M. Hickey. Elsevier Oceanogr. Ser. **47**: 511–565.

Botsford, L.W., Quinn, J.F., Wing, S.R., and Brittnacher, J.G. 1993. Rotating spatial harvest of a benthic invertebrate, the red sea urchin, *Strongylocentrotus franciscanus*. *In* Proceedings of the International Symposium on Management Strategies for Exploited Fish Populations. Alaska Sea Grant College Program, No. AK-93-02. pp. 409–428.

Botsford, L.W., Moloney, C.L., Hastings, A., Largier, J.L., Powell, T.M., Higgins, K., and Quinn, J.F. 1994. The influence of spatially and temporally varying oceanographic conditions on meroplanktonic metapopulations. Deep-Sea Res. **41**: 107–145.

Caddy, J.F., and Seijo, J.C. 1998. Application of a spatial model to explore rotating harvest strategies for sedentary species. *In* Proceedings of the North Pacific Symposium on Invertebrate Stock Assessment and Management. *Edited by* G.S. Jamieson and A. Campbell. Can. Spec. Publ. Fish. Aquat. Sci. 125. pp. 359–365.

Campbell, A., and Mohn, R.K. 1983. Definition of American lobster stocks for the Canadian Maritimes by analysis of fishery landing trends. Trans. Am. Fish. Soc. **112**: 744–759.

Caputi, N., Chubb, C.F., and Brown, R. 1995*a*. Relationships between spawning stock, environment, recruitment and fishing effort for the western rock lobster, *Panulirus cygnus*, fishery in Western Australia. Crustaceana, **68**: 213–266.

Caputi, N., Brown, R., and Chubb, C.F. 1995*b*. Regional prediction of the western rock lobster, *Panulirus cygnus*, commercial catch in Western Australia. Crustaceana, **68**: 245–256.

Carrasco, K.R., Armstrong, D.A., Gunderson, D.R., and Rogers, C.

1985. Abundance and growth of Cancer magister young-of-the-year in the nearshore environment. *In* Proceedings of the Symposium on Dungeness Crab Biology and Management. *Edited by* B. Melteff. Lowell Wakefield Fish. Symp. Ser., University of Alaska Sea Grant No. 85-3. pp. 171–184.

Cobb, J.S., and Wahle, R.A. 1994. Early life history and recruitment processes of clawed lobsters. Crustaceana, **67**: 1–25.

Cohen, E.B., Mountain, D.G. and O'Boyle, R. 1991. Local-scale versus large scale factors affecting recruitment. Can. J. Fish. Aquat. Sci. **48**: 1003–1006.

Dinnel, P.A., Armstrong, D.A., and McMillan, R.O. 1993. Evidence for multiple recruitment-cohorts of Puget Sound Dungeness crab, *Cancer magister*. Mar. Biol. (Berlin), **115**: 53–63.

Elner, R.W., and Campbell, A. 1991. Spatial and temporal patterns in recruitment for American lobster, *Homarus americanus*, in the northeastern Atlantic. Mem. Queensl. Mus. **31**: 349–363.

Fogarty, M.J. 1998. Implications of migration and larval interchange in American lobster (*Homarus americanus*) stocks: spatial structure and resilience. *In* Proceedings of the North Pacific Symposium on Invertebrate Stock Assessment and Management. *Edited by* G.S. Jamieson and A. Campbell. Can. Spec. Publ. Fish. Aquat. Sci. 125. pp. 273–283.

Gotshall, D.W. 1977. Stomach contents of northern California Dungeness crabs, *Cancer magister*. Calif. Fish Game, **63**: 43–51.

Haidvogel, D.B., Wilkin, J.L., and Young, R. 1991. A semi-spectral primitive equation ocean circulation model using vertical sigma and orthogonal curvilinear horizontal coordinates. J. Comp. Phys. **94**: 151–185.

Hastings, A. 1992. Age dependent dispersal is not a simple process: density dependence, stability, and chaos. Theor. Popul. Biol. **41**: 388–400.

Hastings, A., and Higgins, K. 1994. Long transients in spatially structured ecological systems. Science (Washington, D.C.), **263**: 1133–1136.

Hatfield, S.E. 1983. Intermolt staging and distribution of Dungeness crab, *Cancer magister*, megalopae. *In* Life history, environment, and mariculture studies of the Dungeness crab, *Cancer magister*, with emphasis on the central California fishery resource. *Edited by* P.W. Wild and R.N. Tasto. Calif. Dep. Fish Game Fish. Bull. **172**: 85–95.

Higgins, K., Hastings, A., and Botsford, L.W. 1996. Density dependence and age structure: nonlinear dynamics and population behavior. Am. Nat. **149**: 247–269.

Hill, A.E. 1990. Pelagic dispersal of Norway lobster (*Nephrops norvegicus*) larvae examined using an advection-diffusion-mortality model. Mar. Ecol. Prog. Ser. **64**: 167–174.

Hill, A.E., and White, R.G. 1990. The dynamics of Norway lobster (*Nephrops norvegicus L.*) populations on isolated mud patches. J. Cons. Int. Explor. Mer, **46**: 167–174.

Hjort, J. 1914. Fluctuations in the great fisheries of northern Europe viewed in light of biological research. Rapports et Proces-verbaux des Reunions. J. Cons. Inter. Explor. Mer, **19**: 3–228.

Hobbs, R.C., Botsford, L.W., and Thomas, A. 1992. Influence of hydrographic conditions and wind forcing on the distribution and abundance of Dungeness crab, *Cancer magister*, larvae. Can. J. Fish. Aquat. Sci. **49**: 1379–1388.

Hofmann, E.E., Hedstrom, K.S., Moisan, J.R., Haidvogel, D.B., and Mackas, D.L. 1991. Use of simulated drifter tracks to investigate general transport patterns and residence times in the coastal transition zone. J. Geophys. Res. **96**: 15041–15052.

Jamieson, G.S., and Phillips, A.C. 1988. Occurrence of Cancer crab (*C. magister* and *C. oregonensis*) megalopae off the west coast of Vancouver Island, British Columbia. Fish. Bull. (U.S.), **86**: 525–542.

Jamieson, G.S., Phillips, A.C., and Huggett, W.S. 1989. Effects of ocean variability on the abundance of Dungeness crab (*Cancer magister*) megalopae. *In* Effects of ocean variability on recruitment and an evaluation of parameters used in stock assessment models. *Edited by* R.J. Beamish and G.A. McFarlane. Can. Spec Publ. Fish. Aquat. Sci. 108. pp. 305–325.

Katz, C.H., Cobb, J.S., and Spaulding, M. 1994. Larval behavior, hydrodynamic transport, and potential offshore recruitment in the American lobster, *Homarus americanus*. Mar. Ecol. Prog. Ser. **103**: 265–273.

Koslow, A.J., Thompson, K.R., and Silvert, W. 1987. Recruitment to northwest Atlantic cod (*Gadus morhua*) and haddock (*Melanogrammus aeglefinus*) stocks: influence of stock size and climate. Can. J. Fish. Aquat. Sci. **44**: 26–39.

Largier, J.L., Magnell, B.A., and Winant, C.D. 1993. Subtidal circulation over the northern California shelf. J. Geophys. Res. **98**(C10): 18147–18179.

Levin, S.A. 1974. Dispersion and population interactions. Am. Nat. **108**: 207–228.

Lough, R.G. 1976. Larval dynamics of the Dungeness crab, *Cancer magister*, off the central Oregon coast. Fish. Bull. (U.S.) **74**: 353–376.

McConnaughey, R.A., Armstrong, D.A., Hickey, B.M., and Gunderson, D.R. 1992. Juvenile Dungeness crab (*Cancer magister*) recruitment variability and oceanic transport during the pelagic larval phase. Can. J. Fish. Aquat. Sci. **49**: 2028–2044.

McConnaughey, R.A., Armstrong, D.A., Hickey, B.M., and Gunderson, D.R. 1994. Interannual variability in coastal Washington Dungeness crab (*Cancer magister*) populations: larval advection and the coastal landing strip. Fish. Oceanogr. **3**: 22–38.

McKelvey, R.D., Hankin, D., Yanosko, K., and Snygg, C. 1980. Stable cycles in multistage recruitment models: an application to the northern California Dungeness crab (*Cancer magister*) fishery. Can. J. Fish. Aquat. Sci. **37**: 2323–2345.

Methot, R.D., and Botsford, L.W. 1982. Estimated preseason abundance in the California Dungeness crab (*Cancer magister*) fisheries. Can. J. Fish. Aquat. Sci. **39**: 1077–1083.

Moloney, C.L., Botsford, L.W., and Largier, J.L. 1994. Development, survival and timing of metmorphosis of planktonic larvae in a variable environment: the Dungeness crab as an example. Mar. Ecol. Prog. Ser. **113**: 61–79.

Okubo, A. 1994. The role of diffusion and related physical processes in dispersal and recruitment of marine populations. *In* The biophysics of marine larval dispersal. American Geophysical Union, Washington, D.C. Coastal Estuarine Stud. **45**: 5–34.

Orensanz, J.M. 1986. Size, environment and density: the regulation of a scallop stock and its management implications. *In* North Pacific Workshop on Stock Assessment and Management of Invertebrates. *Edited by* G.S. Jamieson and N. Bourne. Can. Spec. Publ. Fish. Aquat. Sci. 92. pp. 195–227.

Orensanz, J.M., and Gallucci, V.F. 1988. Comparative study of postlarval life-history schedules in four sympatric species of Cancer (Decapoda: Brachyura: Cancridae). J. Crust. Biol. **8**: 187–220.

Orensanz, J.M., Parma, A.M., and Iribarne, O.O. 1991. Population dynamics and management of natural stocks. *In* Scallops: biology, ecology and aquaculture. *Edited* by S.E. Shumway. Elsevier Science Inc., New York, N.Y. pp. 625–713.

Pares-Sierra, A., and O'Brien, J.J. 1989. The seasonal and interannual variability of the California Current System: a numerical model. J. Geophys. Res. **94**(C3): 3159–3180.

Pearce, A.F., and Phillips, B.F. 1994. Oceanic processes, puerulus settlement and recruitment of the Western Rock Lobster *Panulirus cygnus*. *In* The Bio-Physics of Marine Larval Dispersal. *Edited by* P.W. Sammarco and M.L. Heron. American Geophysical Union, Washington, D.C. Coastal Estuarine Stud. **45**: 279–306.

Poole, R.L. 1966. A description of laboratory-reared zoea of *Cancer magister* Dana, and megalopae taken under natural conditions (Decapoda Brachyura). Crustaceana, **11**: 83–97.

Prince, J., and Hilborn, R. 1998. Concentration profiles and invertebrate fisheries management. *In* Proceedings of the North Pacific

Symposium on Invertebrate Stock Assessment and Management. *Edited by* G.S. Jamieson and A. Campbell. Can. Spec. Publ. Fish. Aquat. Sci. 125. pp. 187–196.

Quinn, J.F., Wing, S.R., and Botsford, L.W. 1993. Harvest refugia in marine invertebrate fisheries: models and applications to the red sea urchin, *Strongylocentrotus franciscanus*. Am. Zool. **33**: 537–550.

Reilly, P.N. 1983. Dynamics of Dungeness crab, *Cancer magister*, larvae off central and northern California. *In* Life history, environment, and mariculture studies of the Dungeness crab, *Cancer magister*, with emphasis on the central California fishery resource. *Edited by* P.W. Wild and R.N. Tasto. Calif. Dep. Fish Game Fish Bull. **172**: 57–83.

Ricker, W.E. 1954. Stock and recruitment. J. Fish. Res. Board Can. **11**: 559–623.

Rothlisberg, P.C., Church, J.A., and Forbes, A.M.G. 1983. Modeling the advection of vertically migrating shrimp larvae. J. Mar. Res. **41**: 511–538.

Song, Y.H., and Haidvogel, D. 1994. A semi-implicit ocean circulation model using a generalized topography-following coordinate system. J. Comp. Phys. **115**: 228–244.

Strub, P.T., Allen, J.S., Huyer, A., and Smith, R.L. 1987. Large-scale structure of the spring transition in the coastal ocean off western North America. J. Geophys. Res. **92**: 1527–1544.

Tremblay, J., Loder, J.W., Werner, F.E., Naimie, C.E., Lynch, D.R., Page, F.H., and Sinclair, M.M. 1994. Drift of sea scallop larvae on Georges Bank: a model study of the roles of mean advetion, larval behavior and larval origin. Deep-Sea Res. **41**: 7–49.

Turing, A. 1952. The chemical basis of morphogenisis. Philos. Trans. R. Soc. Lond. B Biol. Sci. **237**: 37–72.

Underwood, A.J., and Fairweather, P.G. 1989. Supply-side ecology and benthic marine assemblages. TREE, **4**: 16–19.

Walters, C.J., Hall, N., Brown, R., and Chubb, C. 1993. Spatial model for the population dynamics and exploitation of the western Australian rock lobster, *Panulirus cygnus*. Can. J. Fish. Aquat. Sci. **50**: 1650–1662.

Wing, S.R., Botsford, L.W., and Quinn, J.F. 1998. The impact of coastal circulation on the spatial distribution of invertebrate recruitment, with implications for management. *In* Proceedings of the North Pacific Symposium on Invertebrate Stock Assessment and Management. *Edited by* G.S. Jamieson and A. Campbell. Can. Spec. Publ. Fish. Aquat. Sci. 125. pp. 285–294.

Woodby, D. 1998. Adaptive cluster sampling: efficiency, fixed sample sizes, and an application to red sea urchins (*Strongylocentrotus franciscanus*) in southeast Alaska. *In* Proceedings of the North Pacific Symposium on Invertebrate Stock Assessment and Management. *Edited by* G.S. Jamieson and A. Campbell. Can. Spec. Publ. Fish. Aquat. Sci. 125. pp. 15–20.

Zebiak, S.E., and Cane, M.A. 1991. Natural climate variability in a coupled model. *In* Greenhouse-gas-induced climatic change: a critical appraisal of simulations and observations. *Edited by* M.E. Schlesinger. Elsevier Science Inc., New York, N.Y. pp. 457–469.

The Fishery as a Selective Force

Implications of selective harvests in Dungeness crab (*Cancer magister*) fisheries

Glen S. Jamieson, Antan Phillips, and Barry D. Smith

Abstract: Most commercial invertebrate, or shellfish, species are unique in that their hard shells facilitate selective culling of individual animals. Only captured individuals possessing desirable characteristics are retained and animals returned to the sea seem to have little mortality. Robustness to physical handling has resulted in the establishment of fishing techniques and management regimes that often use specific size and/or sex selection criteria. Invertebrates have generally high fecundities and because refuge populations exist for many species, it is often assumed that a high exploitation rate on only sexually mature individuals has little significance for population dynamics. Here, we question this assumption for several types of invertebrate fisheries and investigate possible implications for Dungeness crab (*Cancer magister*) populations where only males above a specific size are harvested. General considerations include possible effects of fishing on reproduction, growth, and availability of individuals in a year-class, but for Dungeness crab we focus on implications for reproduction. We present evidence that newly mature male Dungeness crab in intensively exploited populations have a two-year intermolt period, that few such sublegal-sized crab seem to molt to legal size, and that most mating in the population is done by mature, sublegal-sized crab. We suggest these parameters may be influenced by the degree and nature of fishing and that, if ignored, the long-term consequence may be selection for individuals that never exceed the legal size limit.

Résumé : Presque toutes les espèces commerciales d'invertébrés, ou de crustacés, ont une caractéristique particulière en ce sens que leur carapace dure facilite l'élimination sélective des animaux. Seuls les individus capturés possédant les caractéristiques recherchées sont conservés et ceux qui sont remis à l'eau présentent un faible taux de mortalité. En raison de leur résistance à la manipulation physique, on a utilisé des techniques de pêche et des régimes de gestion fondés sur des critères particuliers de taille et/ou de sélection selon le sexe. En général, chez les invertébrés, le taux de fécondité est élevé et parce qu'il existe des populations refuge chez de nombreuses espèces, on suppose souvent qu'un taux d'exploitation élevé des individus qui ont atteint la maturité sexuelle influe peu sur la dynamique de la population. Dans le cas présent, nous avons des réserves à cet égard pour plusieurs types d'exploitations d'invertébrés et nous examinons les répercussions possibles sur les populations de crabe dormeur (*Cancer magister*) où seuls les mâles de taille supérieure à une taille donnée sont prélevés. Parmi les considérations générales, on mentionne les effets possibles de la pêche sur la reproduction, la croissance et la capturabilité des individus d'une classe annuelle, mais dans le cas des crabes dormeurs, nous nous intéressons aux effets sur la reproduction. Selon certaines indications, les crabes dormeurs mâles qui viennent d'atteindre la maturité sexuelle dans une population fortement exploitée muent à tous les deux ans, peu de ces crabes de taille sub-légale semblent muer et atteindre la taille légale, et presque toute la reproduction de la population repose sur des crabes adultes de taille sub-légale. Il semble que le taux de pêche et sa nature influent sur ces paramètres et que, si l'on n'en tient pas compte, à long terme, on assisterait à une sélection d'individus qui ne dépassent jamais la limite de taille légale. [Traduit par la Rédaction]

Introduction

One of the unique characteristics of invertebrate fisheries, in comparison with many exploited vertebrate species, is that individuals are usually sorted by hand, with very little damage, if any, in the processes of fishing, sorting, or release of nonretained specimens. Exploited invertebrates have no internal air chambers, which might rupture because of rapid pressure change when the animal is brought to the surface, and most have hard, protective exoskeletons. Both size and sex can sometimes be used as effective selection criteria, since rigid shells allow precise measurement, even for small species. Indeed, the management of many invertebrate fisheries is often based almost entirely on minimum legal size limits.

Criteria for determining an optimal minimum size limit for a species are, in many cases, vaguely defined. The main biological objective of most fisheries is conservative exploitation to ensure sustainable harvests and so an optimal minimum size limit ensures reproductive opportunity for all individuals before harvest is permitted. However, for some species, size limit regulations may not be based on biological data, particularly if regulations were established decades ago. Identification and evaluation of possible biological consequences of existing regulations provide an opportunity to understand better a species' population dynamics, which in turn may lead to new understanding of the consequences of inappropriate fisheries management. In this paper, we use Dungeness crab (*Cancer magister*) management in British Columbia, Canada, as an example.

In general, there are three potential consequences of human

G.S. Jamieson and A. Phillips. Department of Fisheries and Oceans, Pacific Biological Station, Nanaimo, BC V9R 5K6, Canada.

B.D. Smith. Canadian Wildlife Service, Pacific Wildlife Research Centre, 5421 Robertson Rd., Delta, BC V4K 3N2, Canada.

Correct citation: Jamieson, G.S., Phillips, A., and Smith, B.D. 1998. Implications of selective harvests in Dungeness crab (*Cancer magister*) fisheries. *In* North Pacific Symposium on Invertebrate Stock Assessment and Management. *Edited by* G.S. Jamieson and A. Campbell. Can. Spec. Publ. Fish. Aquat. Sci. 125. pp. 309–321.

exploitation on the population dynamics of fished species: (*i*) reduced population fecundity owing both to reduced population size and preferential removal of large females. The largest females may produce a disproportionately large number of young most likely to survive, perhaps because of better gamete characteristics (e.g., better quality eggs) or placement of eggs in more optimal habitats. The latter could result from either learned behavior, especially if larger females are the oldest, or from successful competition with smaller individuals; (*ii*) increased mortality of young if parents are harvested while brooding young; and (*iii*) selection for specific stocks or morphs. The former is likely to occur if sympatry in distributions occurs, such as with many salmon stocks and the latter is likely for genetically determined traits, which may include age of maturity or growth rate. The consequences in both cases are that a specific genetic grouping of the population may end up doing much, if not most, of the mating.

Here we will examine these possible consequences in relation to existing invertebrate fisheries and make suggestions as to the implications of selective harvests on population dynamics.

Invertebrate exploitation patterns

Apart from effort restrictions, there are two basic fishery regulation approaches for invertebrates practised worldwide: (*i*) biomass quotas, with no specific age or size selection, and (*ii*) use of minimum size limits, either voluntary or enforced by regulation, with or without quotas.

The former typically occurs with species with no rigid exoskeleton (e.g., sea cucumbers and cephalopods) or of small adult size (e.g., euphausiids and some shrimp). Some gear selection still occurs, but size preference is mostly market or processor driven. Individuals that are too small for the market receive a lower price or, if not cost-effectively processable, are discarded. Harvest procedures include trawls (shrimp and euphausiids), seines (squid), traps (octopus), hooks (squid), and underwater diving (sea cucumbers). Each fishing technique has its own size-selection characteristics. Quotas for a species are typically established on the basis of research surveys and population production parameters.

With size limits, exploitation by sex becomes a consideration.

Both sexes equally exploited

This approach is usually adopted for species with little difference in body size between the sexes (e.g., American (*Homarus americanus*) and spiny lobsters), where sex is difficult to determine through visual observation (e.g., echinoderms, gastropods, and bivalves), and where both sexes have marketable characteristics (e.g., some crabs, such as *Cancer pagurus* in Europe).

Predominantly female exploitation

Females are generally the target with protandric hermaphrodites, i.e., species in which individuals are males at smaller sizes and become females at larger sizes. With pandalid shrimps, size limits are generally not used in trawl fisheries for small offshore species, but with trap fisheries for larger, inshore species, minimum seasonally adjusted population sex ratios can be used to restrict the relative proportion of females harvested (e.g., *Pandalus platyceros* in British Columbia (Jamieson 1986)).

Predominantly male exploitation

Males are primarily targeted in species where mature males are significantly larger than mature females and where females have comparatively less desirable market characteristics, such as reduced meat yield (e.g., many crab species, including Dungeness crab).

Potential reproductive effects of size-selective removals

No obvious effect

Apart from a reduction in number of breeding adults, no obvious effect in adult population dynamics or recruitment rate may occur. This may occur often, as many invertebrates produce tens of thousands of larvae, with few normally surviving to recruitment. Larval mortality may be density-dependent, so even a significant reduction in the number of larvae produced may only marginally affect settlement magnitude. However, no comprehensive recruitment studies with both "before" and "after" fishery commencement monitoring have been conducted to confirm this suggestion. Landing statistics are often the only data available, preventing accurate estimation of prefishery recruitment rate. Where more extensive evaluation has been attempted, studies (e.g., Breen 1986) usually use population age frequencies in the first few years of a new fishery to estimate prefishery recruitment rates. Such an approach provides few comparative data with short-lived species, but there are also problems with long-lived species. The latter may have the greatest number of years of prefishery recruitment data for comparison, but such species also typically have a low average annual recruitment rate. Furthermore, the statistical power of analyses used in identifying cause–effect relationships is generally low.

Alteration of population dynamics

There are four main exploitation patterns that can affect reproductive potential. (*a*) Exploitation below the size of sexual maturity: Populations of some species, such as American lobster, are heavily exploited as immatures. These exploited populations have apparently persisted because of immigration of either larvae or juveniles from refugia or other populations exploited at a size above the size of maturity. (*b*) Removal of mainly larger individuals: For bivalves at least, it has been hypothesized (N. Bourne, DFO, Nanaimo, B.C., personal communication) that even though newly mature animals may be sexually active, their contribution to overall population fecundity may be low. This occurs either because they produce a disproportionately small number of larvae in comparison to older mature animals or because their young have either reduced growth or survival characteristics. Potentially, removal of larger individuals may have an impact on population fecundity much greater than expected on the basis of the proportion of breeding individuals remaining after a fishery. A secondary consequence of removal of larger individuals, particularly if mostly males, is that mating opportunities of smaller individuals may be increased, with potential consequences that will be discussed further later. (*c*) Removal of mainly smaller individuals: Fishery preference for small, mature individuals in a population is typically practised when the monetary return from a harvested individual is determined primarily by some quality aspect and not its absolute size or weight. Sea urchins, for example, are harvested for their roe and larger, older urchins often have roe of poor color and texture in comparison to newly mature urchins. Consequently, fishers target the youngest sexually mature urchins. Such selection may create a "large adult" refuge, although fishers may intentionally destroy larger urchins to prevent them from competing for limited food with smaller urchins if such a refuge is not protected by an upper size limit. Food removal by large urchins may decrease both the growth rate and roe quality of small urchins. Population consequences are dependent on exploitation rate and the proportion of the population which reaches the "large adult" refuge. This proportion may be small if fishing is intense and it may create the false perception of having an abundant, successfully reproducing, large-adult biomass. (*d*) Change in local carrying capacity: For some species selective removal may be more related to the behavior of fished individuals rather than their size. For example, fishing may affect the relative survival of cryptic versus noncryptic individuals, with surviving individuals primarily inhabiting environments with cover. Abalone often forage in open areas, but in areas heavily exploited by humans or a natural predator such as the sea otter (*Enhydra lutris*), surviving individuals are largely confined to crevices and the undersides of rocks (Breen 1986). The converse may occur if habitat preferred by a species is limited. Octopus fishers may locally increase carrying capacity by their traps acting as lairs for octopuses passing by that are searching for shelter. (*e*) Genetic selection: In most wild fisheries, management allows all individuals to have spawning opportunities prior to their potential harvest. The opportunity to select for or against specific traits would thus be limited. However, there are situations in intensively exploited, wild harvest fisheries that have genetic implications: (*i*) Species with limited larval production: The Olympia oyster, *Ostrea conchaphila*, in the Northeast Pacific, incubates its young internally and releases juveniles. Selection above a size limit could be for earlier sexual maturity, a slower growth rate, or smaller maximum size; (*ii*) Species with high larval production but limited larval dispersal: Examples are abalone (*Haliotis* spp.), where many species have minimal larval dispersal because of short larval periods (7–10 days). Fisheries for localized populations, possibly best adapted genetically to local oceanographic conditions, may be commercially extinguished if excessive fishing occurs. Abalone fishing is typically regulated with both a quota and a size limit, but quotas are usually established for relatively large geographic areas, which may encompass a number of abalone concentrations. Fishing in the larger area is typically centered on a few concentrations in which legal-sized individuals, i.e., animals above a minimum size limit, are almost completely removed. Size-limits may not adequately account for this loss of local reproductive capacity, increasing the likelihood that discrete abalone concentrations might disappear; and (*iii*) Species experiencing a regular, almost complete exploitation of a portion (e.g., fastest growing) of recruiting year-classes before that portion of each year-class has opportunity to mate: Examples are some crustaceans, including Dungeness crab, where because of discontinuous growth and seasonal molt timing in relation to fishing and mating events, different year-class components may have unequal spawning opportunities.

A selected case: male-only removal of dungeness crab

Potential effects of male-only removal on crustacean population dynamics are considered using a brachyuran, the Dungeness crab in the Northeast Pacific, as our example. Our purpose is to show that most mating is effected by male crab below the legal size limit in two south coast British Columbia Dungeness crab populations which are heavily fished. Evidence is presented to show that intense fishing has significantly reduced total mating opportunities for legal-sized crab in these populations.

Fishery characteristics

Two fishery patterns were studied, a year-round fishery near Tofino, B.C., and a predominantly summer-only fishery near Vancouver, B.C. For both male-only fisheries, the minimum legal size limit (165 mm carapace width (CW), spine-to-spine) is equivalent to about 155 mm CW, notch-to-notch. The notch-to-notch measurement (used in this paper) is considered the more accurate measure of crab size because lateral spines wear down and can be broken off. Retention of female Dungeness crab of any size was made illegal in British Columbia in 1991, but few, if any, female crab were retained in either the Tofino or Vancouver area fisheries in the decade before 1991.

Both fisheries were intense. In the year-round fishery around Tofino, most legal-sized male crabs were removed by fishing from the population as soon as they became sufficiently hard to be marketed. Fishing mortality (F) in the Tofino and Vancouver area was estimated at 5.1–6.9 (Smith and Jamieson 1989*c*) and 2.3 (G. Jamieson, unpublished data), respectively. In recent years, there was negligible retention of soft-shell crab. In the summer–fall Vancouver fishery, there was substantial recruitment of legal-sized males before the fishery opening. The consequence of a winter–spring closure, which prevented capture of soft-shell crab, has been overcapitalization and a rush among fishers to catch crab as soon as possible after the fishery opens (Fig. 1). The commercial value of Dungeness crab has encouraged an unquantifiable, but thought to be substantial, illegal harvest of prerecruit male crab near Vancouver. A high demand for crab meat, numerous potential landing locations, and the rapid communication among fishers of enforcement officer movements and activities has made effective enforcement difficult.

The waters around Tofino and Vancouver (within Canadian Department of Fisheries and Oceans Statistical Areas (SA) 24 and 29, respectively) have sustained productive crab fisheries (Fig. 2) throughout most of this century (Jamieson 1985) and from 1982 until 1991, accounted for 17 and 23%, respectively, of the total average annual landing (1419 t) of Dungeness crab in British Columbia. Annual Dungeness crab landings were generally consistent over the past decade in SA 24, except in 1987, when crab landings doubled because of a large year-class, hypothesized to have settled in 1983. Tagging targeted this exceptional year-class in 1985 and early 1986 (Fig. 3). Annual landings near Vancouver (SA 29) have also fluctuated

Fig. 1. Weekly landings following the mid-July opening of the Fraser River delta B.C. crab fishery for (A) 1993 and (B) 1994.

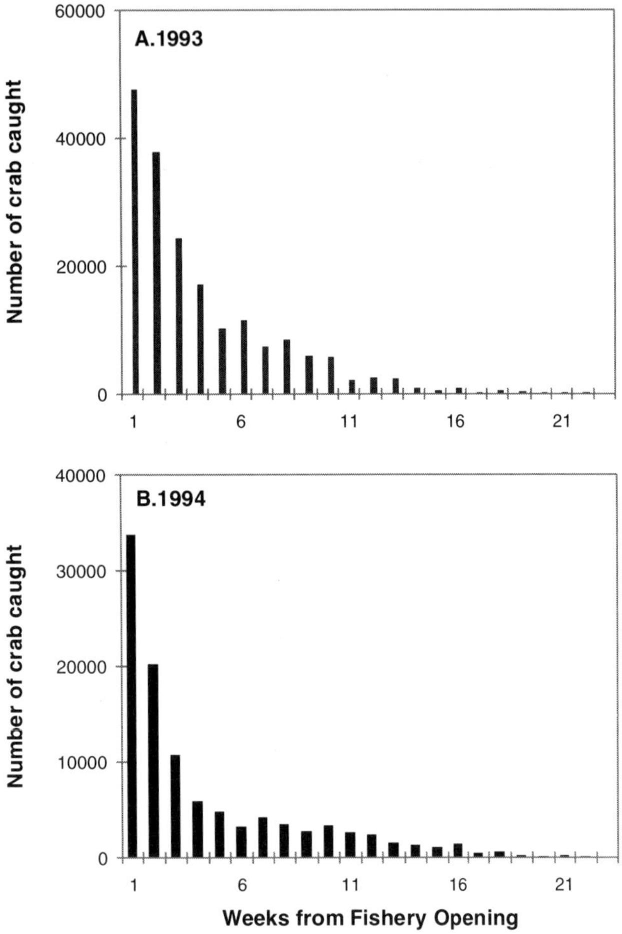

Fig. 2. Annual landings (t) from B.C. Statistical Areas 24 (around Tofino) and 29 (Fraser River delta, Vancouver) from 1980 to 1994.

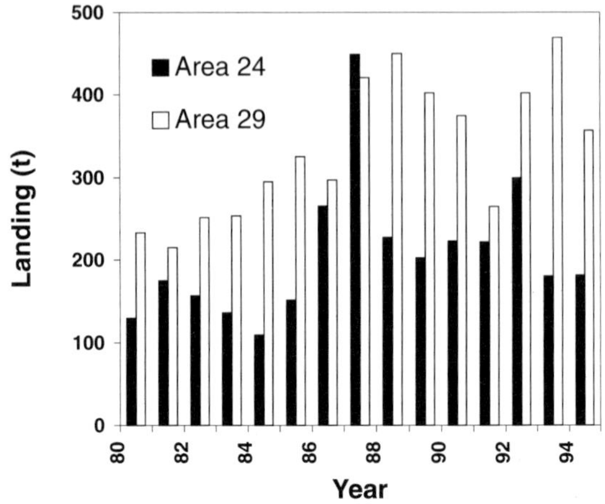

(Fig. 2) but have not shown any exceptional year-classes as in SA 24.

For the Tofino area fishery, the number of legal-sized crab dropped quickly after the major spring molt to legal size (Smith and Jamieson 1989*c*). For the Vancouver fishery, daily landings dropped soon after the summer opening (Fig. 1). Although in both fisheries some crab molted to legal size in all months of the year, few legal-sized male crab remained in either population after the end of summer. An important detrimental consequence in both fisheries is that there were relatively few legal-sized (i.e., large) male crab in the population at the time when mature females were ready to molt, typically July–August around Tofino and August–September around Vancouver.

Methods

Site descriptions and research sampling protocols

Data on Dungeness crab population structure were obtained using a program of regular beam trawl and trap sampling. The former was designed to sample smaller individuals, while the latter focused on larger individuals. The beam trawl used was assumed to be efficient and equally selective for crab, regardless of sex, up to 135–140 mm CW. Few crab >140 mm were captured by trawling. In contrast, our research traps (Type A, escape ports closed, after Smith and Jamieson (1989*b*)), were considered ineffective in retaining male crab <135 mm CW. Increasing male retention in traps with increasing carapace width between 135–170 mm CW occurred, as described in Smith and Jamieson (1989*b*). Trap retention rates of female crab by size are unknown, as our studies to date have not encountered female crab >135 mm CW in sufficient abundance for adequate data analysis.

All beam trawling and trapping were conducted in sheltered coastal waters near Tofino (Smith and Jamieson 1989*a*, 1989*b*, 1989*c*) and Vancouver, B.C. Waters around Tofino were generally 5–15 m depth and well mixed, with an annual seawater temperature range of about 6–12°C. Fishing locations around Vancouver were Sturgeon and Roberts Banks on the delta of the Fraser River, which bisected the study area. Waters around Vancouver were 0–100 m depth, with a pronounced year-round near-surface pycnocline (most pronounced near river outflows) and a summer thermocline at about 10 m depth. Annual temperature ranges above and below the thermocline were about 6–20 and 6–9°C, respectively (Thomson 1981). Substrate at both locations varied from sand in areas of stronger water flow to mud in more sheltered areas.

Research sampling has been conducted around Tofino and Vancouver on a routine basis since 1985 and 1988, respectively. The area around Tofino was sampled monthly from March 1985 until April 1987. Since then, sampling was about every three months. Sampling involved the standardized deployment of 60 research crab traps (Type A of Smith and Jamieson (1989*b*)), 10 double-wired and 50 normal-wired traps, in 6–9 strings of 10 traps each. Soak times were 20–24 h. Six 10-minute beam trawl tows, using gear described by Gunderson et al. (1985), were also conducted at designated stations during each sampling period. The Vancouver sampling occurred biannually since 1991, in June before the fishery opened in July and in late September after most legal-sized crab were caught. Sampling involved the standardized deployment of 60–70 strings of 10 research traps each spread over two weeks within the sampling period. Soak times ranged from 18–30 h. Traps were 10 double-wired and 60 Type A research traps. About 12 10-minute beam trawl tows were also conducted at designated stations during each sampling period. Size frequencies obtained from all research samples were

Fig. 3. Numbers of crab tagged and tagged crab recoveries by month between May 1985 and February 1987 in the Dungeness crab fishery around Tofino, B.C.

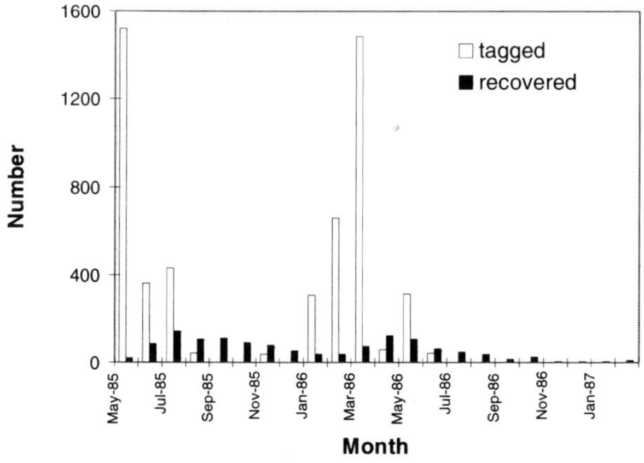

considered representative of commercial catches for crabs >155 mm CW.

For each crab caught, data recorded included sex, size, molt stage, presence of mating marks (see section on mating biology), and date of capture. After 1991, shell age within an instar was also estimated based on shell epiphytes and color, chela abrasion, presence of chitonacious bacterial lesions, and general shell deterioration. Shell molt stage codes 1–5, 6, and 7 represent time periods of <1, 1–2, and 2$^+$ years since the last molt, respectively. All data from each sampling period were combined to create a representative sample for that sampling period and area because there were frequent differences in spatial distribution by sex and size within each study area.

Near Tofino, 4861 prerecruit (<155 mm CW) male crab were tagged with T-bar anchor tags between April 1985 and May 1986 (Smith and Jamieson 1991*a*). Most 1985 tagging occurred during April–May when males were molting, whereas most 1986 tagging occurred before this time period. Fishers were asked to return all tags from tagged crab caught, regardless of their size, along with information on the date of capture and crab size. There were 773 crab recoveries during the 22 months following tagging (Fig. 3). Among these recoveries, 73 crab had molted while at large.

Life history

Mating biology

Dungeness crab show no documented difference in growth rate or behavior between the sexes until females reach sexual maturity at age 2–3 years. Females as small as 70 mm CW have been reported in premating embraces with males (Smith and Jamieson 1991*b*) but most were >80 mm CW. Females only mate immediately after molting and so the smallest size of mated crab is the next larger instar following the instar in which premolt females are first embraced. These females are mainly in instar 2 of Smith and Jamieson (1989*a*) which is equivalent to our instar $x + 1$ in this paper (Table 1).

Premolt female brachyuran crabs typically require a larger, hard-shelled male mating partner (Butler 1960; Snow and Neilsen 1966; Hartnoll 1969; Hankin et al. 1997). We have not observed mating pairs of crabs where the premolt female was larger than the male, although Hankin et al. (1997) reported a

Fig. 4. Annual spring male size frequency distributions (open bars) in the Tofino area, 1987–1994. Solid bars show size frequencies of male crabs with mating marks. Numbers on each plot refer to the last two digits of the year (first two numbers) and the month(s) (e.g., 8701-04 means combined January–April data, 1987). Crab widths are presented in 5 mm groupings, indicated by the largest value in the grouping (i.e., 140 mm grouping = crab 135–140 mm CW).

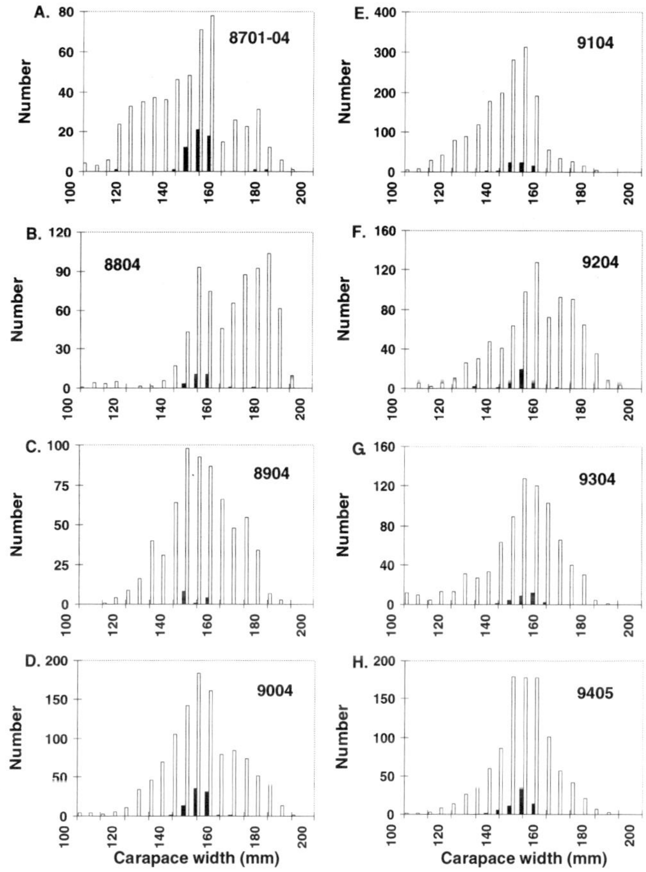

few occurrences. Many male Dungeness crab show exoskeleton abrasions on their chelae, referred to as mating marks (Butler 1960), caused by embracing a female mate. Not all male crab that have mated can be assumed to have recognizable mating marks. Nevertheless, size frequencies of crab with mating marks were assumed to represent the portion of the male population that was capable of mating. For example, male Tofino area crab as small as 120 mm CW had mating marks, but the proportion with marks increased with size. In the population around Tofino, about 10% of male crab between 130–135 mm CW, and up to about 40% above 140 mm CW, showed mating marks (Smith and Jamieson 1991*b*).

There may be some minor geographical variation in the size at which males first mate. Near Tofino (outer coast of Vancouver Island), the first evidence of male mating activity, as provided by mating marks, occurs at about 135 mm CW, whereas in the Strait of Georgia (near Vancouver), the equivalent size is about 130 mm CW (Figs. 4 and 5). This size difference probably reflects different growth patterns under different environmental conditions but may be genetic. Jamieson and Phillips

Table 1. Estimated mean and standard deviation (SD) of carapace widths (CW, mm) for successive instars of Tofino Dungeness crab >80 mm CW (Smith and Jamieson 1989*a*).

	Males			Females		
Instar (*x*)	Instar mean (mm CW)	Instar SD (mm CW)	CV	Instar mean (mm CW)	Instar SD mm CW)	CV
x	80.5	9.3	11.6	81.7	12.5	15.3
$x+1$	104.1	10.5	10.1	102.9	11.0	10.7
$x+2$	129.4	11.7	9.0	121.3	9.7	8.0
$x+3$	156.3	12.9	8.3	137.1	8.5	6.2
$x+4$	185.2	14.2	7.7	150.8	7.6	5.0

Note: CV = coefficient of variation.

Fig. 5. Annual before (A–D) and after (E–H) mating male size frequency distributions (open bars) near Vancouver, 1991–1994. Solid bars show size frequencies of male crabs with mating marks. Numbers on each plot refer to the last two digits of the year (first two numbers) and the month(s) (e.g., 9206 means June data, 1992). Crab widths are presented in 5 mm groupings, indicated by the largest value in the grouping (i.e., 140 mm grouping = crab 135–140 mm CW).

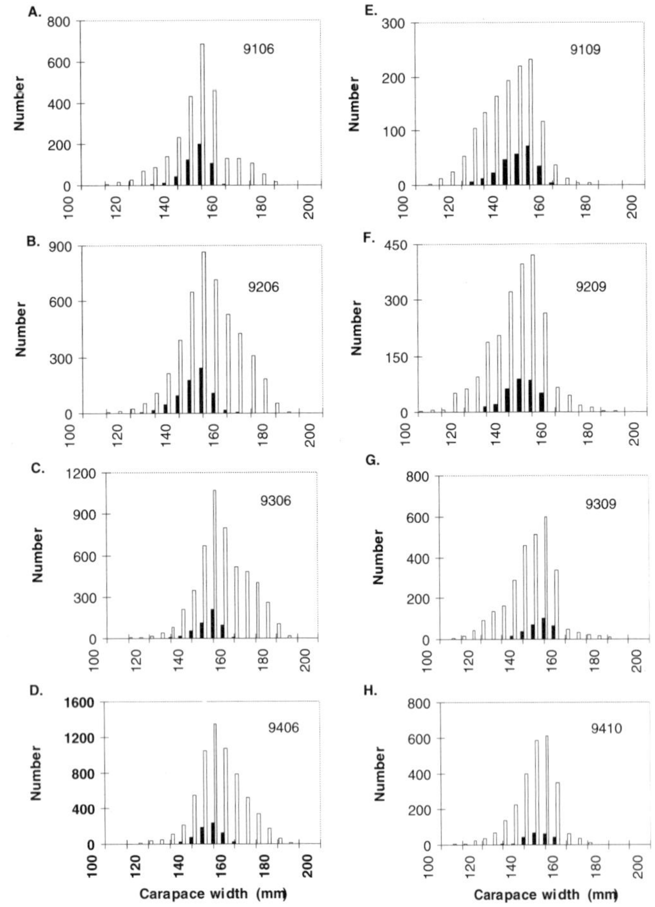

(1993) concluded that Strait of Georgia Dungeness crab were isolated oceanographically from outer coast crab and their larvae had unique diel migratory behavior. However, available data suggests that age-at-sexual maturity (about 3 years) is the same for both crab populations.

At our two study sites, most male crab molting occurred in the spring, before the July–September period when most females molted and consequently mated. Most egg extrusion began in October when females were hard-shelled.

Male growth near legal size

Based on the growth and size frequency analyses of Smith and Jamieson (1989*a*), the mean size of male crab in instar $x+3$ (instar x being the smallest instar size identified by Smith and Jamieson (1989*a*), Table 1) was estimated at 156 mm CW. This value is near the minimum legal size (155 mm CW) suggesting that about half of this instar was above legal size during 1986–1987. Virtually all crab in male instars $x+2$ and $x+4$ were below and above minimal legal size, respectively (Table 1). This size range distribution creates ambiguity with use of the term "prerecruit," which in a fisheries sense generally refers to a size category just below legal size from which individuals recruit into the exploitable size category, generally in the next year. In conflict with this strict definition, there may be two "prerecruit" categories of male Dungeness crab: the larger half of male instar $x+2$ crab, which is expected to molt to legal-sized (instar $x+3$) crab, and the sublegal-sized instar $x+3$ crab which, if they molt, would molt to instar $x+4$. However, there is evidence that most sublegal-sized, instar $x+3$, male crab may never molt to instar $x+4$ (see later) and that many crab in the 145–155 mm CW size range may thus not be true "prerecruits," even though they are just below legal size. Sublegal-sized instar $x+3$ male Tofino area crab (Smith and Jamieson 1991*a*) were reported to have a relatively high annual natural mortality rate ($M \approx 2.5$; Table 2). Fewer than 8% were estimated to survive one year. The high natural mortality rate for these males corresponded with mortality rates estimated for female Dungeness crab (Hankin et al. 1985, 1989) (Table 2).

Total mortality during an instar is affected by the duration of the intermolt period. Evidence from tagging (Fig. 6) and size frequency analysis (see later) suggest a two-year intermolt period for male crab near legal size. However, while natural mortality may be high for sublegal-sized male crab in instar $x+3$, within an instar, natural mortality may not be constant over time (Jamieson 1996) and with a two-year intermolt period, natural mortality, if it is partially the result of a natural senescence, may be higher in the second year than in the first year. Natural mortality rate (M) may thus not be well described by an average annual value over the whole intermolt period,

Table 2. Natural mortality (*M*) estimates by sex and size for Dungeness crab from the published literature.

Study	Value or range of *M*
Hankin et al. (1985, 1989):	
Tagged females > 140 mm CW range	2.0–2.5
Tagged females 125–140 mm CW	~0.7
Smith and Jamieson (1991*a*):	
Tagged prerecruit males	2.5 (95% CL = 2.3–2.8)
Tagged females 135–171 mm CW	1.3

Note: *M* = 1.0, 1.5, and 2.0 represent annual survival rates of 36.6, 22.3, and 13.5%, respectively. CL = confidence limits.

Fig. 6. Time durations (months) between release and recovery of (A) molted and (B) nonmolted tagged, male crab in Tofino between May 1985 and February 1987 (see Fig. 3).

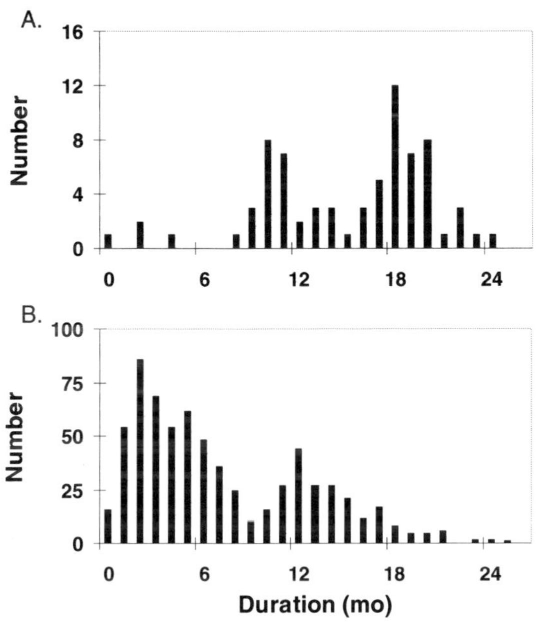

Fig. 7. Time durations (months), by size of crab tagged, between release and recovery of (A) molted and (B) nonmolted male crab near Tofino between May 1985 and February 1987 (see also Fig. 3). The mean ± SD of size and time-at-large are 136.9 ± 9.5 mm CW and 18.8 ± 3.4 months for molted crab and 149.3 ± 5.3 mm CW and 12.0 ± 4.6 months for nonmolted crab, respectively.

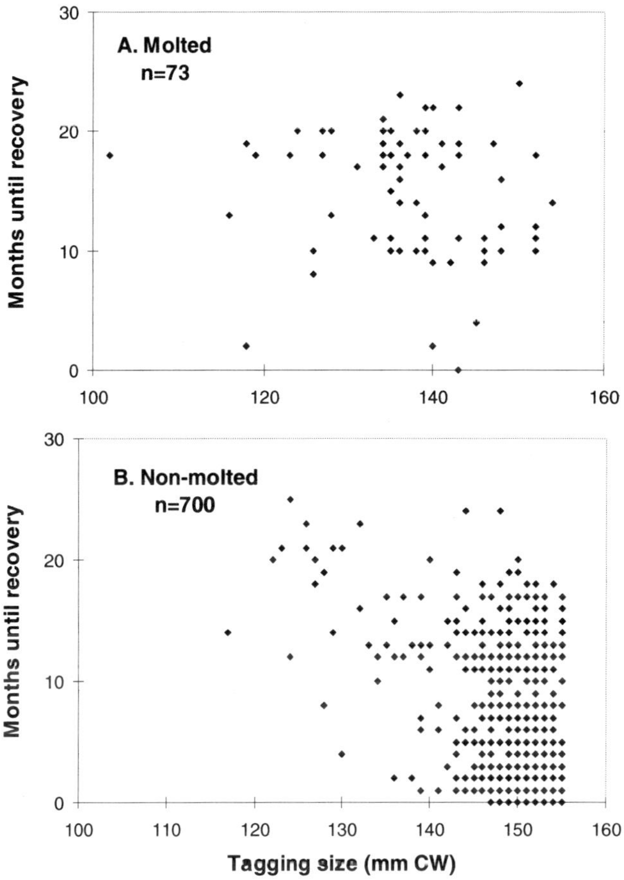

as reported by Smith and Jamieson (1991*a*). Data need to be interpreted accordingly, since given the above, *M* may be different for different cohorts within an instar, depending on their age and molting probabilities. This may also apply to mature female crab, which may also skip an annual molt (Hankin et al. 1989, 1997).

Our tagging data from Tofino suggest an intermoult period of about 2 years for sublegal-sized $x + 3$ instar crab, since recaptures (Fig. 6) before the next molt occurred had largely stopped after this time period. With intensive commercial fishing (Smith and Jamieson 1991*a*), most crab tagged as prerecruits, but recaptured as legal-sized crab ($n = 73$), were assumed to have been captured soon after their shells hardened. Both molted and nonmolted tagged crab were recaptured after time intervals up to 25 months (Fig. 6), indicating that many prerecruit crab had an intermolt period longer than 12 months. We do not know how long tagged crab had been at the size they were tagged before being tagged. This time would have to be added to the time it took the crab to molt in order to determine actual intermolt periods. Later, we use size frequency analysis to argue for an intermolt period of two years for many sublegal-sized male crab near legal size.

There was only weak evidence that size of crab at tagging, regardless of whether the crab subsequently molted, had much effect on the estimate of intermolt period (Fig. 7). The average initial size of crab tagged that were subsequently recovered and had not molted was 149.3 mm CW (SD = 5.3). On the other hand, the average initial size of those recovered that had molted was 136.9 mm CW (SD = 9.5 mm). Interpretation of these data is confounded by the fact that smaller crab, when tagged, tended not to be recovered as soon as larger crab because they are less likely to be caught in traps until they molt to a larger size. Nevertheless there is an indication that smaller sublegal-sized males may have a higher molting rate than larger sublegal-sized males.

Identification of large male instars

A detailed study of the Dungeness crab population around Tofino was completed in the fall of 1987 (Smith 1988) but the population has continued to be monitored every 3–6 months until the present. In most years and samples, crab size frequency was dominated by instar $x + 3$ crab (Fig. 8A) and the size of legal-sized crabs generally ranged from 155 to 175 mm

Fig. 8. Size frequencies (Fall 1987, $n = 1214$; Spring 1988, $n = 1533$; Fall 1988, $n = 840$; Spring 1989, $n = 1234$) and the corresponding predicted fits obtained by simultaneously analyzing molt increment data and all four size frequency data sets (Table 3). The molt increment data were compiled from several sources referenced in Smith and Jamieson (1989*b*) and also include the data in Fig. 9. For each size frequency data set, thick lines indicate the fit to the data by each instar (dashed thick) and to the entire distribution (solid thick). The thin dashed lines project the form of the distributions if the population had not been exploited. Instars 152.5 and 181.1 = instars $x + 3$ and $x + 4$, respectively. Note that the minimum legal size limit of 155 mm CW for instar $x + 3$ is echoed at ~184 mm CW for instar $x + 4$.

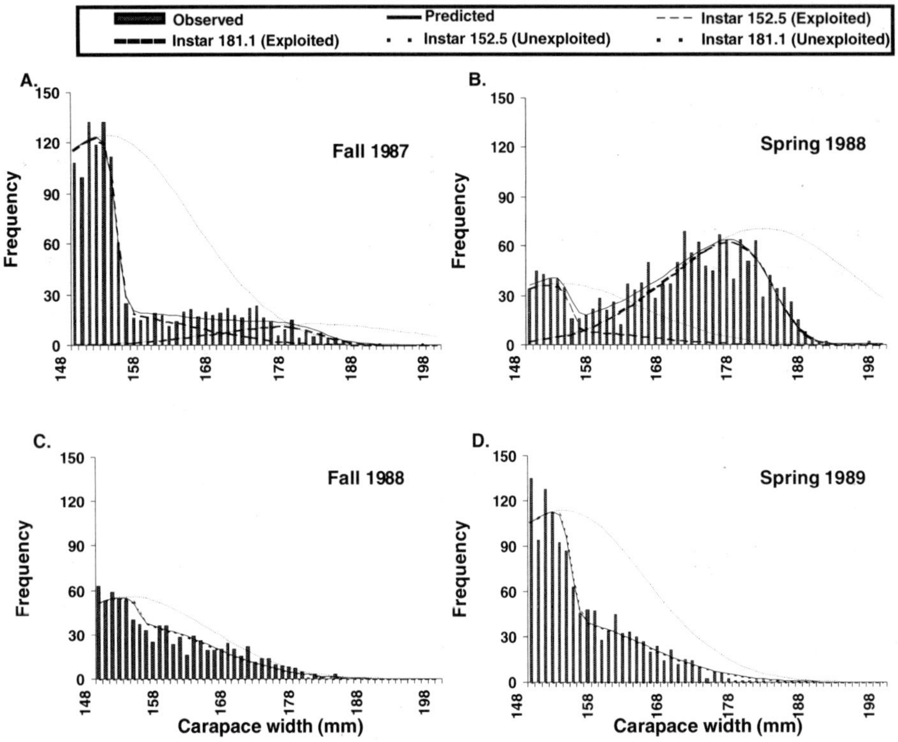

Fig. 9. Male molt increments obtained from tagged crab which molted during the course of the study.

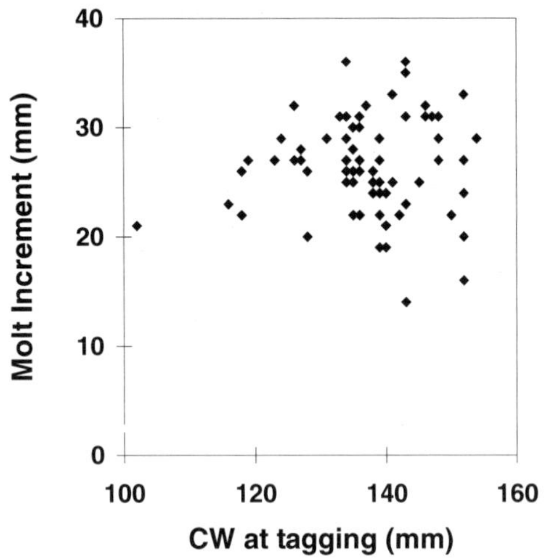

CW. Sublegal-sized $x + 3$ crab, which would mostly be in the size range 145–155 mm CW, would be expected to have an average growth increase at the next molt of 20–30 mm (Fig. 9), which would put them between 165–185 mm as $x + 4$ crab if they molted. Only in the spring of 1988, and for a period lasting less than three months, did trap samples reveal an unusual abundance of particularly large male crab, i.e., >170 mm CW (Fig. 8B). These larger crab were observed two years after tagging was completed and one year after the largest half of the exceptional 1983 year-class had been nearly fully exploited as legal-sized, instar $x + 3$ crab in 1987. These data led us to conclude that these crab arose from the remnant, smaller half of an unusually abundant instar $x + 3$ molting to instar $x + 4$ (Fig. 8B). With the reported high mortality rate for instar $x + 3$ crab (Smith and Jamieson 1991*a*), only a small proportion of sublegal-sized $x + 3$ crab would survive to molt, meaning the relatively small absolute abundance of $x + 4$ instar crab that we detected in the population in all the years sampled except 1988 may be typical for this heavily fished population.

Occasionally, fishers have mentioned to us that decades ago, male Dungeness crab as large as 200 mm CW were common near Tofino and indeed throughout British Columbia coastal waters. Unfortunately, there are no historical population size frequencies to indicate how abundant such large crab were. Analysis of data from a lightly fished population in Holberg Inlet, B.C. (fig. 4 of Smith and Jamieson 1991*b*), where large males were relatively abundant in comparison to the population around Tofino, suggested that in unfished, or lightly fished populations, instar $x + 4$ crab were more abundant.

We used a modified version of the size frequency model of Smith and Jamieson (1989*a*), which facilitated maximum likelihood estimation of instar composition in truncated size frequency distributions to estimate the instar composition of large, trap-caught male crab. Standard size frequency analysis assumes that the frequency count for all cells not explicitly

Table 3. Dungeness crab instar descriptions (mean and SD) using the modified size frequency analysis of Smith and Jamieson (1989*a*).

Date	Sequential instar number	Instar mean (mm CW)	Instar SD (mm CW)	Proportion of sampled distribution	Proportion exploited
Fall 1987	$x+3$	152.5	11.7	0.90	0.85
	$x+4$	181.1	12.5	0.10	0.99
Spring 1988	$x+3$	152.5	11.7	0.33	0.77
	$x+4$	181.1	12.5	0.67	0.99
Fall 1988	$x+3$	152.5	11.7	1.0	0.33
	$x+4$	181.1	12.5	0	—
Spring 1989	$x+3$	152.5	11.7	1.0	0.65
	$x+4$	181.1	12.5	0	—

Note: The "Proportion exploited" represents the proportion of legal-sized crab exploited when those crab were in instar $x + 3$.

Fig. 10. Male size frequency distributions from the Fraser River delta, B.C., crab fishery during the first 4 weeks following the mid-July fishery opening in 1993 and 1994.

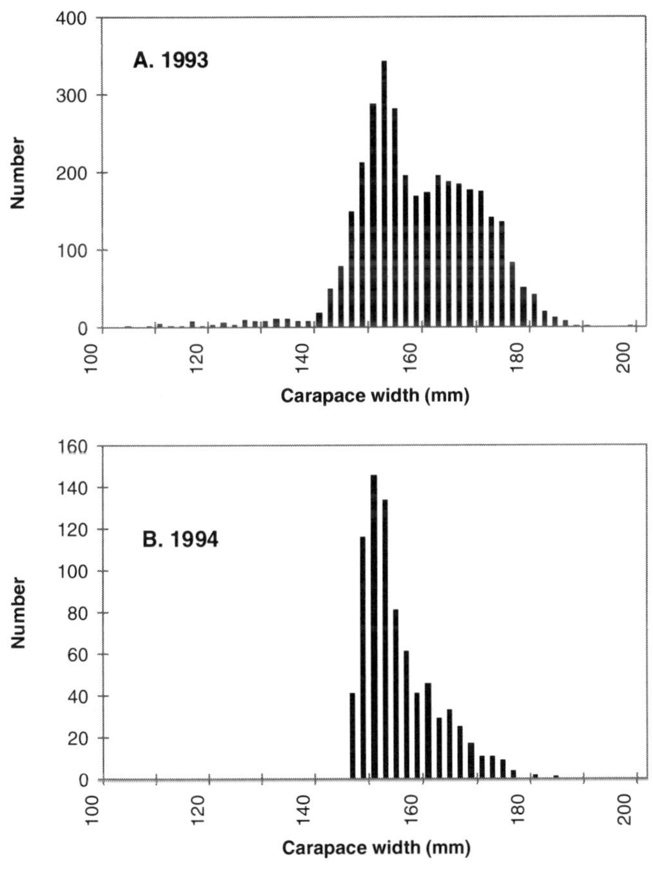

assigned a nonzero count is zero. Our modification allowed us to treat the frequency distributions obtained from crabs collected in traps as truncated at a lower bound of 148 mm. The choice of 148 mm avoided the analysis being severely affected by the differential size selectivity of traps for crab that size and smaller. As in Smith and Jamieson (1989*a*), we assumed a Gaussian distribution for unexploited and exploited instars, except that exploited instars had this shape modified by knife-edged selection at the minimum legal size limit. The exploitation rate of an instar applies to those animals in the instars $x + 3$ and $x + 4$.

Using this size frequency and its associated growth model (Smith and Jamieson 1989*a*), we simultaneously analyzed size frequency distributions obtained from traps near Tofino for four consecutive periods (Fig. 8: Fall 1987, Spring 1988, Fall 1988, and Spring 1989). These data were analyzed in combination with the molt increment-at-size data amalgamated from several sources reported in Smith and Jamieson (1989*a*) and include growth increment data from the Tofino area (Fig. 9). The estimated mean instar sizes (Table 3) were slightly smaller than those reported by Smith and Jamieson (1989*a*) (Table 1). These differences can be attributed to sampling variability and different analytical assumptions but true biological differences can not be ruled out.

Results indicated that the largest crab belonged to instar $x + 4$, which has a mean size near 181 mm CW (Table 3). Individuals belonging to this instar were virtually absent from the Fall 1987, Fall 1988, and Spring 1989 samples, but composed an important proportion of the Spring 1988 sample. Our analysis revealed for the Spring 1988 sample that, firstly, the larger half of instar $x + 4$ was missing (Fig. 8B). The bisected form of the 181 mm CW instar echoes the knife-edged exploitation that took place at the 155 mm CW minimum legal size limit when those crab were in the $x + 3$ instar. In the absence of fishing it is likely that crab as large as 200 mm CW would be present in instar $x + 4$. Secondly, the one-year time lag from the peak abundance observed in the next smaller, $x + 3$ instar (Fig. 8), coupled with our observation of a 1^+-year intermolt period for many tagged crab (Fig. 6), supports a 2-year intermolt period for most of those prerecruit, $x + 3$ instar male crab that molted.

We do not have time series of research trap data when the fishery was open for Fraser River delta crab, but we do have commercial size frequency data from this area for 1993 and 1994 (Fig. 10). These latter data were in agreement with our survey data (Fig. 5) and suggested that while some crab in the 170–180 mm CW range were present, their abundance did not show evidence of peaking at around 180 mm as it did with the April 1988 data from around Tofino. Rather, abundance around 180 mm CW was inversely related to size, which would not be expected if most of the smallest half of instar $x + 3$ had molted to a larger size. There was a size range overlap between the large $x + 3$ and small $x + 4$ crab, but since abundance of small $x + 3$ crab was greatest near the minimum size limit,

Fig. 11. (A) Relative size frequencies of male shell condition indices from a representative Tofino data set, April 1993. (<6 = <1 year old, 6 = 1 year old, 7 = 2 years old). (B) Proportion of crab having shell codes 1–5 with mating marks (solid bars). (C) Proportion of crab having shell code 6 with mating marks (solid bars).

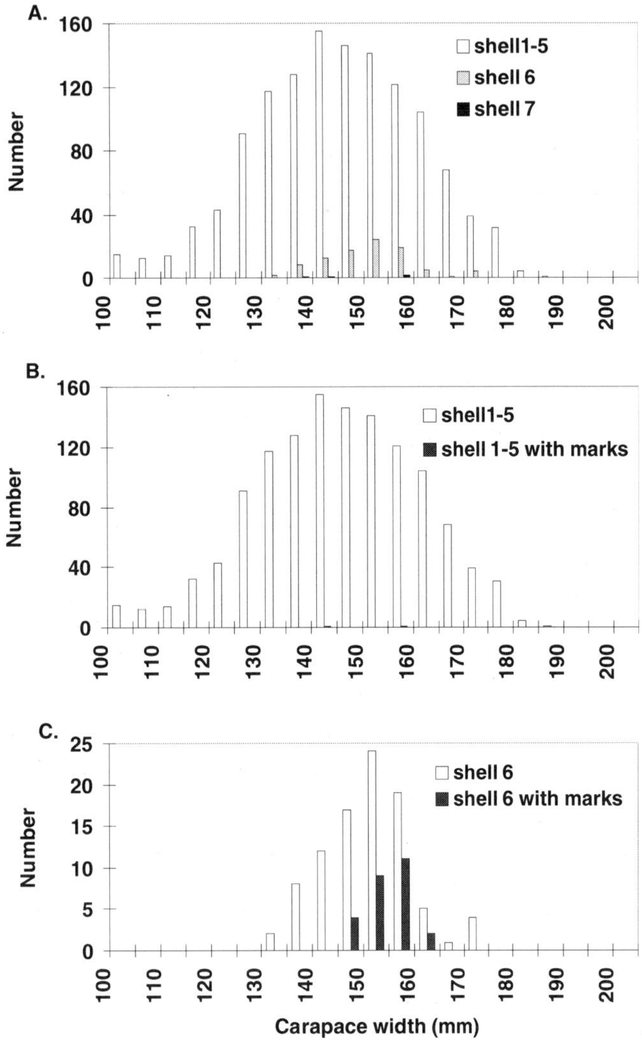

abundance of $x + 4$ crab should also peak near 180 mm CW, assuming F was high. However, we have not defined instars for the Vancouver crab population, so splitting of the $x + 3$ instar by the 155 mm CW size limit was not done.

Male mating activity

Evidence relating to mating activity suggested that the relation between fishery timing and male molting pattern affected crab reproductive dynamics. Comparison of mating mark frequencies on prerecruit and legal-sized male crab caught around Tofino between January and April 1987 and in April 1988–1994 (Fig. 4) indicated that in all years most crab with mating marks were below legal size. An absence of legal-sized males with mating marks was expected, since the fishery removes most legal-sized crab shortly after they achieve legal size. Since mating tends to occur during summer in these crab populations, few legal-sized crab would survive the few months to the time of female molting and even fewer to the following spring. Legal-sized males present in the spring, i.e., before the main male molting period in this year-round fishery, were thought to be primarily early molting males of that year or migrants from elsewhere.

Data on mating mark occurrence immediately after female molting were most abundant for the crab population near Vancouver. There was a longer time interval between female molting (July–August) and our fall quarterly survey (October) in the Tofino area, meaning most legal-sized size male crab in the Tofino area that might have mated would likely have been harvested by the fishery before our sampling occurred there. As expected, legal-sized crab abundance in the Fraser River delta data was relatively low (Fig. 5) and the low frequency of legal-sized males with mating marks suggested that few legal-sized crab had opportunities to mate. Most legal-sized male crab available to mate would be those that had molted to legal size in the few weeks before the female molt and that had not yet been caught.

When the presence of mating marks was related to shell age within an instar, for example with April 1993 data from around Tofino (Fig. 11), only a few crab with a shell <1 year old had mating marks. This was also expected since the male molt was underway in April and females there would not be molting until the summer. In contrast to this, a relatively high proportion of large, sublegal-sized crab with shells about one year old had mating marks. This indicates that these males had mated the previous summer, i.e., within 6 months after their molt to maturity.

Discussion

The ideas presented here may seem of minor consequence, particularly for an exploited species such as Dungeness crab, which has supported significant viable fisheries for decades. However, many aspects of this species' population dynamics, and in particular, responses to high exploitation rates, have never been rigorously studied. Our work suggests that some current Dungeness crab harvest practices may not be optimal for the long term and that population responses, which negatively affect yield, can not be ruled out. We argue that while our data are only from two small, isolated populations, which may be somewhat unique in their population dynamics, there is now sufficient cause to undertake new, more comprehensive investigations of the effects fisheries may be having on this species' productivity. Our conclusions do not necessarily imply that existing Dungeness crab fisheries have to change, although Jamieson (1996) evaluated management scenarios that perhaps should be considered. However, modified management might increase harvestable yield and, at the least, should minimize selection for potentially inherited, less-desired traits, such as smaller molt increments and/or earlier age of maturity.

Researchers and managers should always be challenging their understanding of a species' population dynamics to ensure they are well understood. Often, fisheries and management regulations are established before a species' population dynamics are barely investigated, sometimes with the assumption that a particular species' dynamics may be the same as for another, presumably similar, species. The former was the case with Dungeness crab and minimum legal size and fishery

Fig. 12. Representative annual spring female size frequency distributions (open bars) around (A) Tofino (April 1988) and (B) Vancouver (June 1994). Solid bars show the relative abundance of female crab in the size range (<138 mm CW) which can be mated by sublegal-sized male crab. Crab widths, measured to the nearest mm, are presented in 5 mm groupings, indicated by the largest value in the grouping, i.e., 140 mm grouping = crab 135–140 mm CW.

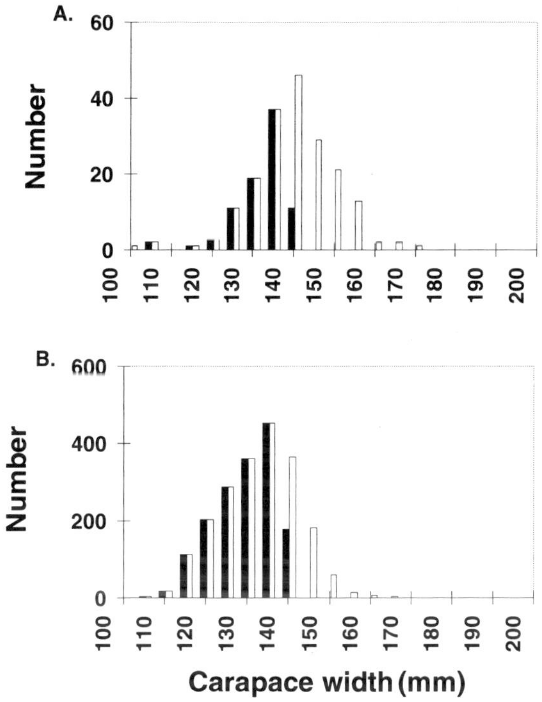

timing regulations largely remain the same as when the fisheries were initially established, at least in Canada.

A consequence of a 2-year intermolt period for male crab near the minimum legal size limit is that size and age categories have less relevance for crab near legal size. Assuming 3-year-old crab molt to instar $x + 3$, this means that prerecruit crab consist of newly molted instar $x + 3$ crab (3 years old) and 1-year-old instar $x + 3$ crab (4 years old). Small, legal-sized crab would be newly molted instar $x + 3$ crab (3 years old), while the largest (most >175 mm CW) legal-sized crab would be instar $x + 4$ crab (5 years old). Many prerecruit crab could thus be 1 year older than many legal-sized crab caught by a fishery, particularly if adjacent year-classes differ greatly in relative abundance.

There is also evidence that female reproduction may be affected by intense exploitation of the male crab cohort responsible for most mating. With the legal-sized crab removed from the population, opportunity for mating the largest females in the population rests with the largest, sublegal-sized male crab (around 154 mm CW) (Hankin et al. 1997). However, data on mating pairs (fig. 3 in Smith and Jamieson 1991*b*; Hankin et al. 1997) suggested this size of male is unlikely, if not unable, to mate with premolt female crab larger than about 135–140 mm CW and so premoult female crab larger than 135–140 mm CW are unlikely to be remated. Since females mature at about 80 mm CW, there would have been several earlier molts (Table 1) where large females probably mated, such that all females in the population are likely carrying sperm (see also Hankin et al. 1997), even if they have not been recently inseminated.

We know that our trap sampling is likely to poorly represent the abundance of crab <135 mm CW, but the proportion of larger female crab (>155 mm CW of those >135 mm CW) in both the intensely exploited Tofino and Vancouver area populations was low, 11.9 (n = 151) and 2.3% (n = 1084), respectively. A larger proportion of comparable female crab (68%; n = 502) was reported for a less exploited B.C. crab population in Holberg Inlet (Smith and Jamieson 1991*b*), which may suggest that intensive male crab exploitation is having some effect on female growth, at least for larger females, and hence possibly reproduction. This hypothesis was argued against by Hankin et al. (1997), but interestingly, they reported intensely exploited crab populations from northern California (1982–1995) had an average of 6.4% of females >155 mm CW (of those >85 mm CW), while Fraser River delta data from 1969–1970 had an average of only 0.6% of females >155 mm CW. While sampling methodology for the Canadian data was not described by Hankin et al. (1997) and crab populations are known to segregate by sex and size over even short distances (G. Jamieson, unpublished data), we also observed that female crab from the population around Vancouver averaged smaller in size than those from the outer coast around Tofino (Fig. 12), suggesting that female Strait of Georgia crab may consistently never get as large as outer coast crab.

We cannot yet provide answers to many of the biological questions our data raise. We do not know why mature, breeding, sublegal-sized male Dungeness crab often did not molt again. Failure to molt was not because of a terminal molt as with majid crabs (Hartnoll 1969); no morphological change with molting to maturity is evident with cancrid crabs. Dungeness crab more than two years old since their last molt generally had badly deteriorated exoskeletons due to chitonacious bacteria, but this may have been a result of not molting, not the cause of failure to molt. However, we suggest later some possible hypotheses, none of which has yet been shown to be biologically feasible but which may be experimentally tested. Elner and Beninger (1995) addressed a similar key question for *Chionoecetes opilio* and offered two, mutually exclusive, hypotheses as to what "triggers" morphologically immature (MI) snow crab to become morphologically mature (MM). However, *Chionoecetes* biology is significantly different from *Cancer*'s, notably the absence of a terminal molt and a shorter life span in *Cancer*.

Although suggested explanatory hypotheses may be somewhat similar, growth options available to the two above crab species are different. Suggested hypotheses for the molting patterns we observed in prerecruit male crab are as follows.

(*i*) Successful mating may inhibit subsequent molting, perhaps by release of a chemical, e.g., a hormone, in the mating male crab, i.e., internal feed-back of inhibitory information. This differs from Elner and Beninger's (1995) hypothesis in that we suggest no subsequent molting will occur, whereas they proposed that an MI male's next molt would be to a terminal state. If the goal of each crab is to successfully pass its genes to the next generation, molting, with its greater risk of mortality, may not offer the maximum reward. Only if larger male crab abound and a prerecruit crab's probability of mating is low, because its probability of losing in aggressive interactions is high (Hankin et al. 1997), does molting to a larger size seem to make evolutionary sense. Arguments against this idea are that no such feed-back loop has been shown to exist in crustaceans and that other ecological factors may be involved, such as increased predation on smaller crab.

(*ii*) The antithesis may also apply — losing in competitive interactions to mate may increase the probability of molting, possibly through internal feed-back of stimulatory information. Again, no such mechanism has been shown to exist in crustaceans. The rationale here is that if a crab is unsuccessful in breeding at a particular size, there is little to be lost in risking mortality associated with a molt. This may be more important for a species like Dungeness crab, where males may only be reproductively active for 2–3 years, as opposed to male snow crab which may mate for 4–5 years after a terminal molt, and 1–2 years before a terminal molt. Elner and Bailey (1986) suggested that recruitment of legal-sized snow crab males is enhanced by removal of old terminal males and Comeau et al. (1991) suggested that the male snow crab molt to maturity may be density dependent and be triggered by low numbers of terminal molt males.

(*iii*) The mating period, which can extend over months, may have a greater demand on stored energy reserves of the male than is currently recognized, resulting in either smaller molt increments or a lower molting probability by mature, sexually active male crab. The former is not supported by our tagging data, although we have little data for prerecruit instar $x + 3$ molt increments (Fig. 9). Hankin et al. (1997) estimated that mean duration of mating activity for individual pairs was 7–9 days and that an average male could conceivably mate with 6–12 females during the mating period for females. The energy requirement of sperm production is small, but male mating behaviour may create energy demands (e.g., greater movement) or inhibit adequate food consumption (no feeding can occur when a female is clasped), both of which in turn may negatively affect potential growth. Females available for mating are likely to be more numerous in comparison to available mature males, primarily because of the removal of legal-sized males by the fishery. A mating period extending over several months may draw down a male's energy reserves below a threshold required for molting. Although advantageous from a breeding perspective, the cost of numerous matings may be a relatively shorter life span. The argument against this hypothesis is that males have the rest of the year to build their energy reserves, so decreased feeding for a few months, even if through optimal summer growing conditions, might not have a detrimental long-term effect.

(*iv*) Finally, in intensively exploited populations, such as those described here, prerecruit crab near the minimum size limit will be frequently caught and released through fishing for legal males. In our studies, we often recaught the same tagged prerecruit male crabs several times over a period as short as a few weeks and fisheries are expected to do the same. This frequent capture and handling may cause physical damage, reduced feeding opportunities, and probably creates stress, which may inhibit subsequent molting. These detrimental indirect fishing effects may be accumulative with the other hypotheses suggested above.

Until this study, the selective effects of fishing large males in modifying the population dynamics of an invertebrate species had only been hypothesized for *Chionoecetes* (Elner and Beninger 1995), where it was suggested that a long-term fishery could promote selection for males to have their terminal molt below the minimum size limit. Our data suggest that Dungeness crab may be similarly affected in that much of a year-class may not molt to legal size and that researchers and managers should consider potential negative effects of fisheries in heavily exploited populations. Long-term consequences of status quo male-only fisheries are difficult to project because of the unknown implications of possible genetic selection for specific attributes. Many crabs (Jamieson et al. 1989; McConnaughey et al. 1995; Sainte-Marie et al. 1996) are also characterized by extensive larval dispersal. Larval dispersal should tend to minimize selection unless most subpopulations have their largest mature males removed by fishing before they have the opportunity to mate.

Since females of most crab species are not exploited, they may provide a gene pool which would minimize the effects of selection in males. The genetic control of any molting aspect of any crustacean has not been studied, making heritability of molting characteristics impossible to predict. However, common sense suggests that all individuals in a population should have a real opportunity to breed at least once, meaning that conservative fisheries management should ensure this potential is met. If fishing occurs before the time of female molting in a given year, seasonal timing of fishing for male crab, if they are exploited in their first year of sexual maturity, should thus be delayed until after most mating occurs, i.e., after the female molt, as suggested by Jamieson (1996).

The potential for fisheries to exert selective effects on species is most likely only a recent phenomena, as only in the last few decades has product transportation become sufficiently cost efficient and rapid that viable fisheries can now often occur in even remote locations. With most populations of exploited invertebrate species now being harvested, the dynamics of many species may be vulnerable to modification by fisheries. For species where both sexes are exploited after individuals have had an opportunity to spawn, these effects may be minimal, if present at all. However, for other species, such as crabs in particular, effects may be more prevalent and impacting and not beneficial over the long term for either the species or the fishers exploiting them.

Acknowledgments

We thank Wayne Harling for technical support in obtaining the field data presented here and the many student employees who assisted us. Dr. Alan Campbell and two anonymous reviewers provided constructive critiques.

References

Breen, P.A. 1986. Management of the British Columbia fishery for northern abalone (*Haliotis kamschatkana*). *In* North Pacific Workshop on Stock Assessment and Management of Invertebrates. *Edited by* G.S. Jamieson and N. Bourne. Can. Spec. Publ. Fish. Aquat. Sci. 92. pp. 300–312.

Butler, T.H. 1960. Maturity and breeding in the Pacific edible crab, *Cancer magister* Dana. J. Fish. Res. Board. Can. **17**: 641–646.

Comeau, M., Conan, G.Y., Robichaud, G., and Jones, A. 1991. Life history patterns and population fluctuations of snow crab (*Chionoecetes opilio*) in the fjord of Bonne Bay on the west coast of Newfoundland, Canada — from 1983 to 1990. Can. Tech. Rep. Fish. Aquat. Sci. 1817. 73 p.

Elner, R.W., and Bailey, R.F.J. 1986. Differential susceptibility of Atlantic snow crab, *Chionoecetes opilio*, stocks to management. *In* North Pacific Workshop on Stock Assessment and Management of Invertebrates. *Edited by* G.S. Jamieson and N. Bourne. Can. Spec. Publ. Fish. Aquat. Sci. 92. pp. 335–346.

Elner, R.W., and Beninger, P.G. 1995. Multiple reproductive strategies in snow crab, *Chionoecetes opilio*: physiological pathways and behavioral plasticity. J. Exp. Mar. Biol. Ecol. **193**: 93–112.

Gunderson, D.R., Armstrong, D.A., and Rogers, C. 1985. Sampling design and methodology for juvenile Dungeness crab surveys. *In* Proceedings of the Symposium on Dungeness Crab Biology and Management. *Edited by* B.R. Melteff. University of Alaska, Fairbanks, Alaska. Sea Grant Rep. 85-3. pp. 135–144.

Hankin, D.G., Butler, T.H., Wild, P.W., and Xue, Q-L. 1997. Does intense fishing on males impair mating success of female Dungeness crab? Can. J. Fish. Aquat. Sci. **54**: 655–669.

Hankin, D.G., Diamond, N., Mohr, M., and Ianelli, J. 1985. Molt increments, annual molting probabilities, fecundity and survival rates of adult female Dungeness crabs in northern California. *In* Proceedings of the Symposium on Dungeness Crab Biology and Management. *Edited by* B.R. Melteff. University of Alaska, Fairbanks, Alaska. Sea Grant Rep. 85-3. pp. 189–206.

Hankin, D.G., Diamond, N., Mohr, M.S., and Ianelli., J. 1989. Growth and reproductive dynamics of adult female Dungeness crabs (*Cancer magister*) in northern California. J. Cons. Inter. Explor. Mer. **46**: 94–108.

Hartnoll, R.G. 1969. Mating in the Brachyura. Crustaceana, **16**: 161–181.

Jamieson, G.S. 1985. The Dungeness crab, *Cancer magister*, fisheries of British Columbia. *In* Proceedings of the Symposium on Dungeness Crab Biology and Management. *Edited by* B.R. Melteff. University of Alaska, Fairbanks, Alaska. Alaska Sea Grant Rep. 85-3. pp. 37–60.

Jamieson, G.S. 1986. A perspective on invertebrate fisheries management — the British Columbia experience. *In* North Pacific Workshop on Stock Assessment and Management of Invertebrates. *Edited by* G.S. Jamieson and N. Bourne. Can. Spec. Publ. Fish. Aquat. Sci. 92. pp. 57–74.

Jamieson, G.S. 1996. Molting patterns in southern British Columbia Dungeness crab and implications for fisheries. *In* High latitude crabs: biology, management and economics. *Edited by* B.R. Melteff. University of Alaska, Fairbanks, Alaska. Sea Grant Rep. 96-2. pp. 397–410.

Jamieson, G.S., and Phillips, A. 1993. Megalopal spatial distribution and stock separation in Dungeness crab (*Cancer magister*). Can. J. Fish. Aquat. Sci. **50**: 416–429.

Jamieson, G.S., Phillips, A., and Hugget, S. 1989. Effects of ocean variability on the abundance of Dungeness crab larvae. *In* Effects of ocean variability on recruitment and an evaluation of parameters used in stock assessment models. *Edited by* R.J. Beamish and G.A. McFarlane. Can. Spec. Publ. Fish. Aquat. Sci. 108. pp. 305–325.

McConnaughey, R.A., Armstrong, D.A., and Hickey, B.M. 1995. Dungeness crab (*Cancer magister*) recruitment variability and Ekman transport of larvae. ICES Mar. Sci. Symp. **199**: 167–174.

Sainte-Marie, B., Sévigny, J-M., Smith, B.D., and Lovrich, G.A. 1996. Recruitment variability in snow crab (*Chionoecetes opilio*): pattern, possible causes, and implications for fishery management. *In* High latitude crabs: biology, management and economics. *Edited by* B.R. Melteff. University of Alaska, Fairbanks, Alaska. Sea Grant Rep. 96-2. pp. 451–478.

Smith, B.D. 1988. An evaluation of the current minimum legal size limit for the Dungeness crab (*Cancer magister* Dana) fishery near Tofino, British Columbia, Ph.D. thesis, University of British Columbia, Vancouver, B.C. 160 p.

Smith, B.D., and Jamieson, G.S. 1989*a*. Growth of male and female Dungeness crabs near Tofino, British Columbia. Trans. Am. Fish. Soc. **118**: 556–563.

Smith, B.D., and Jamieson, G.S. 1989*b*. A model for standardizing Dungeness crab (*Cancer magister*) catch rates among traps which experienced different soak times. Can. J. Fish. Aquat. Sci. **46**: 1600–1608.

Smith, B.D., and Jamieson, G.S. 1989*c*. Exploitation and mortality of male Dungeness crab (*Cancer magister*) near Tofino, British Columbia. Can. J. Fish. Aquat. Sci. **46**: 1609–1614.

Smith, B.D., and Jamieson, G.S. 1991*a*. Movement, spatial distribution and mortality of male and female Dungeness crab *Cancer magister* near Tofino, British Columbia. Fish. Bull. (U.S.), **89**: 137–148.

Smith, B.D., and Jamieson, G.S. 1991*b*. Possible consequences of intensive fishing for males on the mating opportunities of Dungeness crabs. Trans. Am. Fish. Soc. **120**: 650–653.

Snow, C.D., and Neilsen, J.R. 1966. Premating and mating behavior of the Dungeness crab (*Cancer magister* Dana). J. Fish. Res. Board Can. **23**: 1319–1323.

Thomson, R.E. 1981. Oceanography of the British Columbia Coast. Can. Spec. Publ. Fish. Aquat. Sci. 56. 291 p.

Management

Limited entry in the Washington coastal Dungeness crab (*Cancer magister*) fishery: the first step toward rationalizing an overcapitalized and chaotic fishery

P.E. LaRiviere and S.T. Barry

Abstract: A constant size, sex, season, and open-access management approach for the Washington coastal Dungeness crab, *Cancer magister*, commercial fishery prevailed for 68 years until enactment of a limited-entry program in 1995. Washington's coastal Dungeness crab stock abundance reached an all-time high in 1988–1989. The crab fleet responded with dramatic increases in effort. In 1988 and 1989, factions within the industry initiated, but failed to achieve, limited-entry legislation for the fishery. High crab abundance in six of the last nine seasons caused increases in the number of vessels and pots, and when coupled with increases in vessel size and improved shipboard technology, all contributed to an accelerated fishery during the winter months and economic problems for crabbers. By 1991, a coastwide industry committee (Oregon, California, and Washington) agreed to pursue limited entry in all three states. In 1994, Washington Legislature passed a coastal Dungeness crab limited-entry bill and limited entry began on January 1, 1995. By 1996, all three coastal states had crab limited-entry programs that limited the number of licenses but did not further reduce effort by limiting number of pots or trip limits or prohibit night fishing. Ultimately, the ability to control fishing effort effectively off any coastline depends on enactment of statutory measures, either federal or reciprocal between states allowing each coastal American state to regulate all crabbers including nonresident vessels in adjacent offshore (from 4.8 to 322 km) waters. The need for complete regulatory authority is important because (*i*) in Washington, the majority of crab landings originate from outside territorial waters (>4.8 km); (*ii*) it provides the means to reduce overcapitalization of all vessels off the Washington coast, which would decelerate the early season harvest rate and lead to a more even flow of product and dollars throughout the season; and (*iii*) there is a need to properly allocate shellfish resources to treaty Indian tribes.

Résumé : Pendant 68 ans, l'exploitation commerciale du crabe dormeur (*Cancer magister*) des côtes de l'État de Washington s'est effectuée selon une stratégie de gestion fondée sur une taille constante, le sexe, la saison et le libre accès; mais, depuis 1995, on applique un régime d'accès limité. En 1988–1989, ce stock avait atteint le plus haut niveau d'abondance jamais enregistré. La flottille de crabiers avait réagi en augmentant de façon spectaculaire son effort de pêche. En 1988 et 1989, des factions du secteur ont demandé, sans succès, à faire imposer par réglementation l'accès limité à cette pêche. En raison de la grande abondance de crabes pendant six des neuf dernières saisons, le nombre de bateaux et de casiers a augmenté; cette situation, associée à une augmentation de la taille des bateaux et au progrès de la technologie embarquée, a favorisé une accélération de la pêche pendant les mois d'hiver et entraîné des problèmes économiques pour les crabiers. À partir de 1991, un comité de l'industrie à l'échelle de la côte (Oregon, Californie, et Washington) a convenu de demander un accès limité pour cette pêche dans ces trois États. En 1994, l'assemblée législative de l'État de Washington a adopté une loi limitant l'accès à la pêche côtière du crabe dormeur; elle est en vigueur depuis le 1er janvier 1995. Depuis 1996, ces trois États côtiers appliquent des régimes de pêche à accès limité qui restreignent le nombre de permis, mais n'ont pas pris de mesures pour réduire l'effort en limitant le nombre de casiers, en fixant des limites de sortie, ou en interdisant la pêche de nuit. En dernier recours, la capacité de contrôler réellement l'effort de pêche devant les côtes dépend de la promulgation de règlements, soit fédéraux soit d'application réciproque entre les États, permettant à chaque État côtier des États-Unis de réglementer l'activité de tous les crabiers, y compris les bateaux non résidents, dans la zone contiguë à la mer territoriale (de 4,8 km à 322 km). Un pouvoir de réglementation complet est nécessaire parce que *i*) dans l'État de Washington, presque tous les crabes débarqués proviennent de l'extérieur des eaux territoriales (>4,8 km); *ii*) cela donne des moyens pour réduire la surcapitalisation de tous les bateaux devant la côte de Washington, ce qui ralentirait le taux de capture au début de la saison et permettrait un écoulement plus régulier de la ressource et des revenus pendant la saison; et *iii*) il faut allouer correctement la ressource en crustacés aux tribus indiennes visées par un traité. [Traduit par la Rédaction]

P.E. LaRiviere and S.T. Barry. Washington Department of Fish & Wildlife, 48 Devonshire Rd., Montesano, WA 98563, U.S.A.

Correct citation: LaRiviere, P.E., and Barry, S.T. 1998. Limited entry in the Washington coastal Dungeness crab (*Cancer magister*) fishery: the first step toward rationalizing an overcapitalized and chaotic fishery. *In* Proceedings of the North Pacific Symposium on Invertebrate Stock Assessment and Management. *Edited by* G.S. Jamieson and A. Campbell. Can. Spec. Publ. Fish. Aquat. Sci. 125. pp. 325–333.

Introduction

The Washington coastal Dungeness crab (*Cancer magister*) fishery experienced a dramatic rise in crab production during the 1987–1988 and 1988–1989 seasons when an unprecedented increase in capital (number of vessels and pots) entered the fishery. The fishery was open access and a right to harvest was given to anyone who could afford the cost of a commercial license. Collinsworth et al. (1976) documented that the Dungeness crab fishery was overcapitalized in Washington, Oregon, and California as early as the 1970's. The late 1980's influx of vessels and gear further compounded overcapitalization of Washington's coastal fishery, prompting long-time participants who belonged to commercial Dungeness crabber organizations to seek limited-entry regulations for the crab fishery. Limiting the number of licenses wasn't enough. Washington legislation included a reciprocal jurisdiction provision intended to provide regulatory authority in shared (federal) waters. By 1996, Oregon and California had limited-entry programs, but without provisions for reciprocal jurisdiction. That left questions unanswered about both how to achieve regulatory authority for Washington over all vessels fishing territorial waters and how to reduce effort enough to maintain a viable fishery for local fishing communities. In this paper, we examine the history of the commercial Dungeness crab fishery in Washington, crab management both in Washington and coastwide, and the political difficulties encountered by the state and industry in obtaining passage of the Washington coastal Dungeness crab limited-entry program. We also discuss the implications of U.S. west coast limited entry and how to achieve it with Dungeness crab fisheries.

Fishery background

Dungeness crab are highly prized by commercial and sport fishers. They are found in commercial quantities from the Aleutian Islands (Hoops 1973) in Alaska to south of San Francisco in central California. Coastal Dungeness crab live in the intertidal zone out to a depth of 170 m (Phillips 1935). Dungeness crab represent the only commercially important crab within Washington's territorial waters, where there are two distinct fisheries: in Puget Sound and along the outer coast (Fig. 1). Timing of major reproductive and growth (molting) events differs by several months between each area, resulting in different fishing seasons and separate fishery management plans. Coastal production has averaged four times greater than the Puget Sound fishery. Washington's coastal commercial crab grounds extend from the Columbia River to Cape Flattery near Neah Bay and include the estuary of the Columbia River, Willapa Bay, and Grays Harbor.

From their arrival, non-native settlers on Washington's coast commercially exploited Dungeness crab (Cleaver 1949). During the early 20th century, fishing occurred only in nearshore areas, coastal bays, and estuaries. Modernization of transportation, storage, and processing facilities accelerated commercial fishery harvest (Cleaver 1949), and after adopting oblong wooden pots in the mid-1930's, the coastal fishery in 1935 exceeded 450 t. Use of circular steel pots with stainless steel mesh, polypropylene rope, and hydraulic power blocks further revolutionized the commercial fishery in the 1950's. Since 1950, Washington's annual coastal commercial harvest has averaged over 3800 t, ranging from 1160 to 9900 t, (PSMFC 1993; Barry 1985; LaRiviere 1997) (Fig. 2).

Fig. 1. Coastal north Pacific Ocean and Puget Sound waters of the State of Washington, U.S.A.

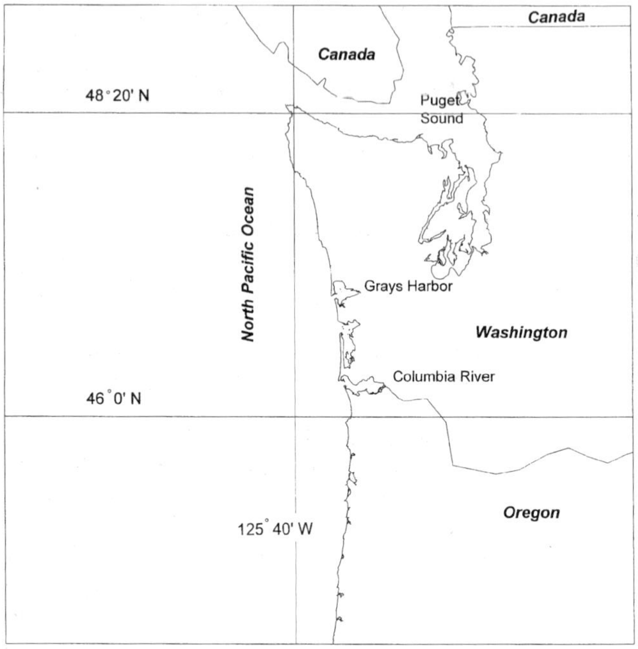

Washington commercial fishers have had few new sustainable fishery opportunities during the late 1980's and 1990's. Other existing fisheries are either limited entry [salmon (*Oncorhynchus* sp.), groundfish trawl (*Sebastes* sp. and some flat fish), ocean pink shrimp (*Pandalus jordani*), red and green sea urchins (*Strongylocentrotus franciscanus* and *S. drobachiensis*, respectively), and sea cucumbers (*Parastichopus californicus*)] or remain open access only under a competitive bidding regime [geoduck clams (*Panopea abrupta*), hardshell clams (*Tapes japonica* and *Protothaca staminea*), and oysters (*Crassostrea gigas* and *Ostrea lurida*)].

Before the 1970's, the Washington coastal crab fleet consisted of about 85 vessels averaging <13 m in length. The fleet increased to between 120 and 140 vessels in the 1970's and further expanded to a high of 331 vessels in 1988–1989 (Fig. 3). Most crabbers participated in other fisheries from the 1950's to the 1970's and crab vessels were also commonly equipped for salmon troll or gill net fishing and albacore tuna (*Thunnus alalunga*) fishing (Methot 1986; Collinsworth et al. 1974). Some crabbers also fished groundfish and ocean pink shrimp. Most multispecies vessels used in salmon, tuna, and Dungeness crab fisheries could not hold more than 4.5 t of product, small in contrast to current capacities of 20–40 t. Some fishers crabbed during the winter but, due to small vessel size, most fished only during the spring in calm weather and stayed relatively close to port. Before 1970, crab landings were usually highest in late spring.

Fishery management background

Dungeness crab management is based primarily on a minimum

Fig. 2. Washington coastal Dungeness crab fishery landings, 1950–1951 to 1995–1996 season.

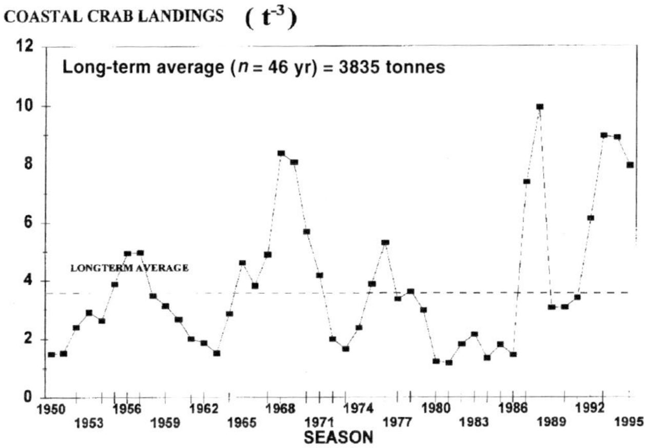

Fig. 3. Washington coastal Dungeness crab fishery, number of boats and pots (t^{-3}) since the 1981–1982 season.

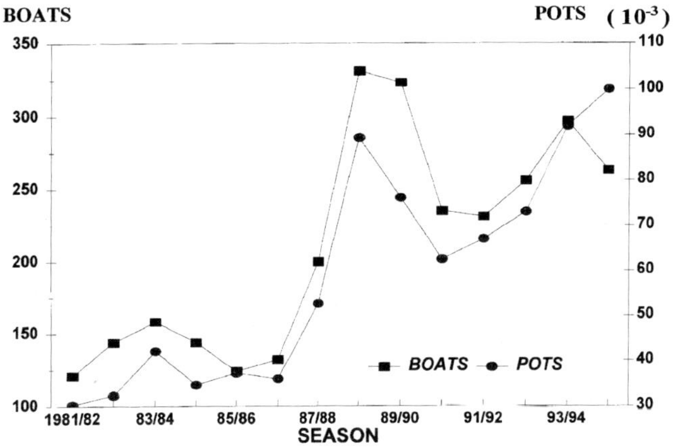

size limit (159 mm in Washington, Oregon, and California, measured at the widest part of the carapace just anterior to the 10th lateral spine), gear regulations, prohibition of female harvest, and a season closure during the male softshell period immediately after the recruitment molt (Hoopes 1973; Miller 1976; Warner 1985; Barry 1985). Minimum size limits are assumed to protect most male crab until after maturity and first breeding (Cleaver 1949; Bulter 1960, 1961). All recruited male crab >158 mm are regarded as harvestable surplus. We estimate, but cannot document due to a lack of stock-assessment data, that >95% of recruited crab are harvested annually. Dungeness crab are sexually dimorphic allowing for easy shipboard sorting of females. The fishery is closed during the male softshell period, generally from about September 15 to November 30, to prevent excessive handling mortality (Kruse et al. 1994) and to maximize product quality.

In 1927, the Washington Legislature enacted crab size and sex regulations and a regulation defining a closure during the summer to protect male softshell crab (State of Washington Ch.298L;27). The closure applied to all state waters but was most likely intended for the intensive Puget Sound crab fishery. In 1943, the state assessed the coastal softshell period (Cleaver 1949) and found it to occur later in the calendar year than in Puget Sound: between mid-September and mid-October. Since 1943, the coastal softshell closure has been in the fall and the coastal crab fishery has been in winter, spring, and summer.

Annual timing of the peak softshell period on the Washington coast varies usually by <4 weeks. Currently, Washington Department Fish & Wildlife (WDFW) uses a systematically designed preseason test fishery completed in late October to estimate peak molting period for the coastal stock (LaRiviere 1996). During the test fishery, biologists measure shell condition and meat recovery rates to determine molt maturity and quality of crab meat. Crab condition determines appropriateness of a December 1 opening. Washington crab usually molt later in the season than crab in Oregon and California. Generally, the commercial Dungeness crab fishery opens December 1 in all three coastal states to minimize major shifts in effort between the states.

Interjurisdictional management activities

Political jurisdiction problems common to many commercial fisheries (Ginter and Rettig 1978; Miller 1976; Erickson 1973) exist in the U.S. Dungeness crab fishery. The resource extends across state boundaries requiring co-ordinated management efforts by all political jurisdictions. By the early 1970's, larger and better equipped vessels created a more competitive atmosphere in the coastwide fleet. Improved vessel mobility meant fishers no longer relied solely on local crab stocks and vessels traversed the western U.S. coast looking for the best harvest opportunity.

Throughout the 1970's, the coastal crab fishery in all three Pacific coastal states grew in number of vessels and fishing capacity, with effort deployed in both state and federal waters. During the drafting of the federal Magnuson Fishery Conservation and Management Act (FCMA) in 1976, federal officials asked regional programs to determine the practicality of co-ordinated fishery management plans (FMP). Simultaneously, representatives of the coastal Washington Dungeness Crab Fishermen's Association (WDCFA) began to lobby state managers to co-ordinate seasons, areas, and gear regulations with other coastal states (Mr. Ernie Summers, Star Route 1, Box 137, Grayland, WA 98547, personal communication). Everyone recognized that the management approach, especially in offshore waters, needed to deal with more than just biological issues (Magnuson 1978; Christy 1973).

Before 1968, Dungeness crab management in all three coastal states consisted mostly of monitoring commercial catches. Managers had little need to concern themselves with issues involving out-of-state vessels, gear conflicts, harvest rates, or product quality. In 1972, federal congressional leaders sought an extension of federal jurisdiction in offshore waters (from 19.3 to 322 km) (Magnuson 1978). In anticipation of extended jurisdiction, both the National Oceanic and Atmospheric Administration (NOAA) and National Marine Fisheries Service (NMFS) initiated regional studies of multijurisdictional fisheries (Collinsworth et al. 1976) to provide a basis for federal FMP. In 1973, the states of California, Oregon, and Washington, together with the NMFS and the Pacific Marine Fisheries Commission (PMFC), now the Pacific States Marine

Fishery Commission (PSMFC), initiated regional Dungeness crab studies. The latter is a nonprofit interstate marine fisheries commission sanctioned by the U.S. Congress to promote better use of shared fishery resources. State and federal fishery agencies entered into co-operative studies known as the State and Federal Fisheries Management Program (SFFMP) for Dungeness crab (Robinson et al. 1977) to begin development of a coastwide FMP using the best management principles available.

The initial phase of the Dungeness crab SFFMP program (Collinsworth et al. 1974) developed preliminary recommendations for seasonal timing and other fishery practices. Phase II collected extensive data on capitalization and economic efficiency. After four years, fishery managers and economists, with participation from industry, summarized biological and economic data in a report to the PMFC (Collinsworth et al. 1976). Meanwhile, in 1976, the U.S. Congress passed the FCMA, which many believed would resolve multistate jurisdiction issues. The SFFMP disbanded, with phase III (Robinson et al. 1977) completing work in progress for a federal FMP required by the FCMA. Coastal Dungeness crab management proposals developed in recent years had their foundation in the 1970's SFFMP.

The Pacific Fisheries Management Council (the current PFMC), created by the FCMA, began development of a federal Dungeness crab FMP in the late 1970's. Existing data, mostly from the SFFMP, was reviewed and a draft federal Dungeness crab FMP was completed in 1979. However, both PFMC priorities, which focused on salmon and offshore jurisdictional disputes between states, and lack of funding prevented completion of a final federal Dungeness crab FMP. Consequently, in 1980, state fishery agency directors entered into a memorandum of understanding (MOU) pledging interstate cooperation in management of coastal Dungeness crab. The MOU called for "mutually supportive actions to maximize the economic and biological exploitation of the Dungeness crab resource coast wide."

Management of state Dungeness crab fisheries continued throughout the 1980's without any formal tri-state process or forum to address fishery problems. Resource agencies continued to overlook, for various reasons, results and recommendations of studies by the SFFMP and the draft federal FMP until a season opening crisis in December 1989. During preseason test fishing, biologists found crab quality poor off the Washington and northern Oregon coast. Washington kept the fishery closed until January 10, 1990, but Oregon officials, to avoid a split season opening along their coast that would cause a massive effort shift within Oregon waters, retained the scheduled December 1 opening. The tri-state MOU proved ineffective in 1989 because it could not reconcile a nonuniform season opening problem caused by poor condition crab, emphasizing the need for a formal tri-state process. As a result, Oregon crabbers landed softshell crab that went to waste because of poor quality.

Motivation for change

Increasing effort

Washington coastal Dungeness crab landings peaked in 1976, followed by 10 years of below average (3836 t, $n = 46$ yr) landings, including seven continuous seasons with production below 2132 t (Fig. 2). In the 1970's and early 1980's, commercial salmon fishing was strong (WDF 1991). Since then, native tribal allocation (Townsend 1990; Benson and Longman 1978; Boldt 1974) and resource problems (PFMC 1992; Radtke 1993) have limited salmon fishing opportunity (WDF 1991). The coastal crab fishery emerged as a logical choice for salmon fishers to try and earn new income. It was open access and the Dungeness crab harvest peaked during the winter period when salmon fishing is at a lull (Radtke 1993), although a mass influx of new vessels into the crab fishery did not occur until after crab production increased in the late 1980's.

Coastal Dungeness crab landings increased in the 1987–1988 season and peaked in 1988–1989 at 9919 t (Fig. 2). As expected (Radtke 1993; Ginter and Rettig 1978; Collinsworth et al. 1976), a huge increase in fishing effort followed. In the 1988–1989 season, crabbers fished an estimated 379 823 pot-months, an all time high and an increase of 115% in just two seasons. Washington crabbers maintained a high effort level of 469 402 and 474 735 pot-months during the 1993–1994 and 1994–1995 seasons, respectively (LaRiviere 1997).

Overcapitalization

Dungeness crab fishery managers considered reducing overcapitalization (Erickson 1973; Collinsworth et al. 1976; Miller 1976; Robinson et al. 1977; PFMC 1979 (draft); PSMFC 1993; WDF 1993) by individual catch quotas, trip limits, pot limits, and catch-area quotas. Some of these measures rely, though, on accurate stock or year-class analyses, which were not available. Collinsworth et al. (1976), Townsend (1990), McHugh (1978), and Christy (1973) all concluded that success of limited entry depended on a combination of vessel and gear reductions.

Regulating an optimal level of effort in the fishery seemed pragmatic to Collinsworth et al. (1976) when 120–170 vessels existed in the Washington crab fleet. The SFFMP recommended a license buyback program as a means to achieve fleet reduction and to help alleviate unemployment and retirement concerns, although at that time, owners could still use their vessels in other fisheries. Since the late 1980's, socioeconomic issues have become more complex. The fleet doubled in size and multiple fishery opportunities disappeared. Consequently, the relative importance of Dungeness crab revenue to crabbers increased. Radtke (1993) reported that, during the calendar year 1991, half the Washington crab fleet relied on Dungeness crab revenue for 86.2% of their total income.

As early as 1978, SFFMP economists determined 60 000 as the optimal number of pots needed to harvest the Dungeness crab resource coastwide (California, Oregon, and Washington), with 26 000 pots for Washington. Economists combined effort estimates with both long-term average landings and expected ex-vessel prices to calculate an optimal number of pots that would generate an optimal revenue per vessel. By 1972, 40 000 to 45 000 pots existed in the Washington fishery (Erickson 1975). Recent estimates, believed to be conservative, show Washington coastal crabbers currently use 100 000 pots (Figs. 3, 4) (LaRiviere 1997). Expansion of the Washington Dungeness crab fleet in the late 1980's and 1990's has brought the fishery to marginal economic status, referred to by Anderson (1986) as static maximum economic yield.

Fig. 4. Washington coastal Dungeness crab fishery, fleet number of pots (10^{-3}) per season, averaged by decade, 1950's to 1990's.

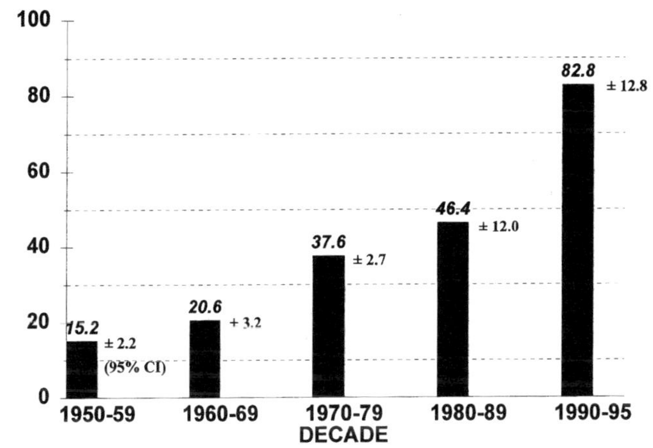

Fig. 5. Washington coastal Dungeness crab fishery, annual landings percentage by season; winter = December, January, and February; spring = March, April, and May; summer = June, July, August, and September; 1980–1981 to 1995–1996.

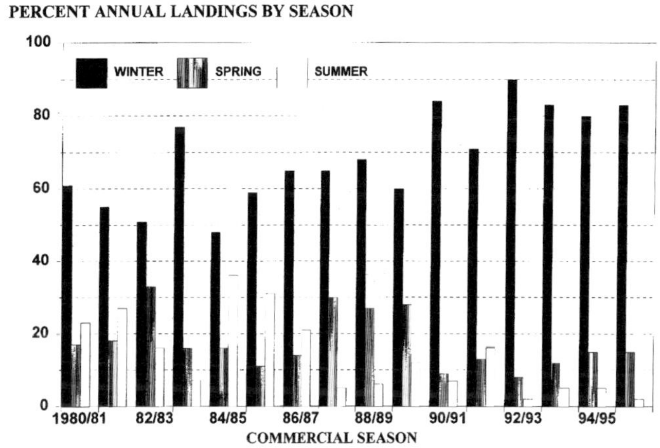

This meant that for many crabbers the cost of operations exceeded income (Radtke 1993).

Accelerated harvest

Effort intensified between 1990–1991 and 1994–1995 to unprecedented levels (Fig. 4). In December 1993, a record 4139 t was landed (LaRiviere 1995). Since 1990–1991, early season (December, January, and February) harvest has exceeded 80% of the total season catch in most years (Fig. 5). The exception in 1991–1992 was because of a 20-day fishery closure as a result of the presence of domoic acid.

The Washington coast experiences more severe winter weather in comparison to Oregon and California. Storms used to keep crab vessels in port for up to 20 days per month. However, this is no longer the case, as high crab abundance has encouraged new large (15.2–30.5 m), high capacity vessels to enter the crab fleet (Table 1). Before and after 1988–1989, the average number of vessels greater than 15.2 m increased from 32 to 73 vessels, respectively. Most larger vessels feature technological advances in shipboard electronics, fishing gear, and weather monitoring equipment, which all facilitate winter fishing (Eddie 1983).

Increasing fishing efficiency not only contributes to accelerated harvest rates (Eddie 1983) but also exacerbates income disparity within the fleet, with those who can afford technological changes reaping greater profits (Waters 1991). In 1988 (January–December), average and median crab revenues for 273 vessels in Washington were $86 900 and $34 530, respectively (Radtke 1993). However, large vessels in Washington did not necessarily profit proportionally because benefits were dissipated through entry of additional large productive vessels (Waters 1991; Anderson 1986; Collinsworth et al. 1976; Christy 1973; Erickson 1973, 1975). Nonetheless, since 1991–1992, large vessel revenue has been progressively higher each season, while the percent of total fleet income of more numerous 9.1–15.2 m vessels has decreased (Table 1).

Industry crusade

Early limited-entry initiatives

Before 1987, coastal Dungeness crabbers expressed little interest in license limitation or other measures restricting effort. During 1987–1988, Washington Department of Fisheries (WDF; now WDFW) staff interviewed fishers and found only 27% wanted some type of effort restriction. Two years later, 48% of the fishers wanted effort restriction and, by 1992, a PSMFC survey (PSMFC 1993) indicated 86% of Washington coastal crab fishers favored limited entry.

In 1987–1988 and 1988–1989, fleet size increased 51 and 65% (Fig. 3), with 44 and 47% of the fleet new entrants, respectively. Dropout rate was only 14 and 13% in 1987–1988 and 1988–1989, respectively. The coastal WDCFA supported limited-entry legislation but this failed because of opposition from Puget Sound crabbers. Puget Sound crabbers, with an existing limited-entry program since 1982, wanted the coastal crab fishery to remain open access for potential future opportunity.

In 1989, the WDCFA sponsored another attempt at limited-entry legislation for the coastal Washington crab fishery, but again, the initiative failed because of opposition by Puget Sound crabbers.

Tri-state Dungeness Crab Committee

In response to widespread concerns expressed by fishers, buyers, retailers, and the public, following 1989–1990 season opening problems, the PSMFC offered to convene a tri-state group of industry representatives and coastal state resource agency advisors to resolve the nonuniform season opening problem, caused by late molting off Washington. The group became known as the Tri-state Dungeness Crab Committee (TDCC).

TDCC meetings in 1990–1991 focused on developing a coast-wide procedure to prevent shifts in early season fishing effort if one state delayed the season opening because of poor crab condition. Limited entry did not become a serious item of

Table 1. Value of Dungeness crab landed in Washington; earnings by vessel length-class by season, coastal Dungeness crab fishery, 1981–1982 to 1995–1996.

Season	Number of vessels	% of fleet	Average ex-vessel $ value per season	% of total fleet ex-vessel value
		<9.1 m		
1981–1982	17	14	6 076	4
1982–1983	26	18	3 908	2
1983–1984	25	16	5 815	2
1984–1985	23	16	4 486	2
1985–1986	14	11	5 419	1
1986–1987	26	20	5 064	3
1987–1988	54	27	7 167	2
1988–1989	87	26	8 151	3
1989–1990	59	18	7 884	4
1990–1991	45	20	6 898	3
1991–1992	39	17	7 367	3
1992–1993	44	17	7 769	2
1993–1994	39	13	17 589	3
1994–1995	26	10	35 728	3
1995–1996	16	7	26 246	2
		9.1–15.2 m		
1981–1982	79	66	20 288	62
1982–1983	87	61	35 540	61
1983–1984	92	59	40 619	50
1984–1985	87	62	27 417	54
1985–1986	80	65	34 754	52
1986–1987	78	60	35 612	64
1987–1988	102	52	91 714	51
1988–1989	158	48	65 908	47
1989–1990	166	52	31 499	54
1990–1991	126	55	42 612	50
1991–1992	151	64	40 625	66
1992–1993	161	61	55 894	65
1993–1994	177	59	73 358	55
1994–1995	153	58	111 381	51
1995–1996	142	60	78 518	52
		>15.2 m		
1981–1982	24	20	35 933	34
1982–1983	30	21	62 974	37
1983–1984	40	25	91 398	48
1984–1985	31	22	63 864	44
1985–1986	29	24	84 145	47
1986–1987	26	20	55 555	33
1987–1988	41	21	208 978	47
1988–1989	86	26	131 608	50
1989–1990	96	30	42 434	42
1990–1991	56	25	91 122	47
1991–1992	46	19	62 436	31
1992–1993	57	22	80 718	33
1993–1994	83	28	119 004	42
1994–1995	85	32	182 377	46
1995–1996	78	33	127 316	46

Note: Source: PacFIN and WDFW data base.

discussion until May 1991 when the TDCC agreed to conduct the first of two coastwide limited-entry surveys. After receiving positive results, the TDCC asked each state to explore license limitation options.

With failure of Washington limited-entry bills in 1988 and 1989, the TDCC, supported by state fish and wildlife agencies, advocated an organized campaign to achieve limited-entry programs in each respective state. The TDCC supported Washington in 1992, when the Dungeness Crab Fishery Study bill was passed, which advised Washington crabbers that future legislation might restrict issuance of licenses to fishers not licensed before September 15, 1991 or to fishers not proving substantial participation in the fishery prior to that date. The bill also directed the WDF to co-operate with PSMFC in a coastwide study and, after comprehensive public review, to report independently on the pros and cons of establishing future limits on the number of Dungeness crab licenses. The PSMFC–TDCC report (PSMFC 1993) satisfied the 1992 bill coastwide study requirement, while the WDF (1993) report to the governor and legislature both provided the pros and cons of Dungeness crab limited entry and detailed the public review process, which was absent in all previous legislative attempts.

After passage of the Washington crab study bill, other states became more resolute in pursuing limited entry for their own Dungeness crab fisheries. In 1992, TDCC members from California supported legislation that resulted in a 3-yearr moratorium on Dungeness crab licenses (Assembly Bill 3189). The California moratorium provided notice to Dungeness crab industries in Oregon and Washington that California, with the largest Dungeness crab fleet on the coast (>600 vessels; PSMFC 1993), seriously wanted to reduce overcapitalization in its fishery. The California bill also required a Dungeness crab limited-entry study analogous to the regional Washington study.

Coincidentally, the Alaskan industry successfully initiated a moratorium for their southeast Dungeness crab fishery from 1992 to 1996. This moratorium had repercussions coastwide, as it represented one less fishing opportunity for California to Washington Dungeness crabbers.

Legislative process for limited entry

1993 — House Bill 1471; Washington coastal Dungeness crab limited entry

The WDCFA, along with the Columbia River's Crab Fishermen's Association (CRCFA), influenced by the California and Alaskan crab moratoriums and escalating effort within the Washington fishery, obtained sponsorship for House Bill (HB) 1471 during the 1993 legislative session. The bill established transferable permits for crab vessels licensed from 1990 through 1993 that landed a minimum of eight times for a minimum total of 2.268 t in two of three seasons from 1988–1989 to 1990–1991. Vessels that were under new construction by September 1991 also qualified if the owner could prove the vessel was being constructed for use in the coastal crab fishery. All permits were limited to one permit transfer per 5-year period to a vessel no more than 3.5 m longer.

Washington State fishery managers expressed concern that the bill was premature because it preceded completion of reports required by the 1992 study bill, which included a comprehensive public process, the tri-state fishery study and WDF recommendations. WDF acknowledged support in concept for a limited-entry program but preferred to complete the public process and provide recommendations later in the year as required by HB 2294. House Bill 1471 did not pass.

Limited-entry opponents realized legislation would be reintroduced in Washington at some future point. They countered by establishing a new crab fisher organization, the Washington Coastal Crab Coalition (WCCC), which began to actively lobby legislators. Meanwhile, crab landings continued to rise, providing new entrants with profits, and more than 50 individuals joined this crabber organization. Similar preemptive actions are common adjuncts to limited-entry legislation and, as Townsend (1990) explains, often result in more eligible license holders because of political pressure.

1994 — Second Engrossed Substitute House Bill 1471; Washington coastal Dungeness crab limited entry

Both proponents and opponents lobbied intensely around another proposed bill in 1994. Legislators looked to placate as many constituents as possible, including crab processors. Two large companies, representing 38% of the Washington coastal crab purchased during the 1993–1994 season, testified against limited entry for the coastal fishery because an open fishery fostered a "highly profitable" attitude. They could buy volume at a low price and sell later at a higher price. In contrast, smaller processors and cash buyers wanted a steady crab supply. Small processors argued that without limiting early season effort (December–February), supply of Dungeness crab would decrease towards the end of the fishery opening (Fig. 5). Second Engrossed Substitute House Bill 1471 (2ESHB 1471), the coastal Dungeness crab limited-entry program, passed and took effect January 1, 1995.

The bill contained provisions capping immediately the number of licenses as of January 1, 1995 (240 licenses). To appease some opponents, mostly new entrants, a fourth season (1991–1992) for eligible landings was allowed. To reduce overcapitalization further, legislators mandated a reduction of the current fleet size. In accord with WDFW recommendations, the bill provided for a two-tier license system (A and B licenses), with B licenses to be forced out of the fishery after five additional seasons. A transition period was also established for the expiration of borderline legitimate qualifiers (B license) to prevent immediate economic loss. The program incorporated an industry-funded license buyback and included the same license transfer provisions from the 1993 HB 1471.

The bill identified a need for future effort restrictions to reduce overcapitalization further. Subsequently, legislators directed WDFW, with the aid of the industry, to prepare a fishery resource plan to achieve even-flow harvesting and long-term stability of the fishery.

Other Dungeness crab limited entry

In 1994, California acted beyond a moratorium scheduled to expire in 1995 and enacted a Dungeness crab limited-entry program. Assembly Bill (AB) 3337 took effect April 1, 1995. The program resembles Washington's but lacks key provisions for reciprocal jurisdiction regulations with Oregon and Washington.

In 1995, Oregon enacted a Dungeness crab limited-entry program in response to the California and Washington programs. Legislators feared displaced vessels from other states would compound economic problems in the existing Oregon crab fleet. Oregon's program, like California's, lacked reciprocal jurisdiction with other coastal states and, like the other states, only limited licenses, not gear or other measures of effort.

Other Dungeness crab limited-entry programs exist on the west coast of North America in Puget Sound, Washington; Alaska (south of Cape Fairweather), and British Columbia (B.C.), Canada. All three programs included secondary effort reduction measures: Puget Sound had a 100-pot limit, Alaska had a variable pot limit with a maximum of 300, and B.C. crabbers must register by submanagement areas.

Rafeedie decision: Indian shellfish rights

In December 1994, a U.S. District court ruling by Judge Edward Rafeedie upheld Indian shellfish rights in Washington, based on treaties signed in the 1850's. Coastal native American tribes chose not to enter the crab limited-entry debate while they sought relief through Judge Rafeedie. The court ruled Indian tribes in Washington have the right to up to 50% of harvestable shellfish in their usual and accustomed (U & A) fishing grounds. The ruling affects the Washington coastal Dungeness crab fishery to the extent that more than half the 241 km Washington coastline could be within tribal U & A fishing grounds (coastal tribal U & A boundaries are in dispute and are being litigated as of 1997). All but four coastal limited-entry permits belong to non-Indian fishers. Before 1995, coastal tribal harvest averaged <110 t per season but managers expect substantial increases in catch as tribal fishing effort increases, furthering compounding overcapitalization of the fishery.

Discussion

Experience shows that transition from open access to limited entry for most commercial fisheries tends to be gradual and contentious (Townsend 1990; Waters 1991). The Washington coastal Dungeness crab limited-entry program will likely be no exception. Legal challenges, license appeals, mandated studies, offshore jurisdictional issues are bound to impede transition.

Beginning with the first limited-entry bill in 1988–1994, proponents relied little on economic data, fishery statistics, and technical information derived from documented, credible public sources to support their position (Miller 1976). Coastal crab representatives from WDCFA and CRCFA insisted that they had enough industry support to persuade legislators to adopt a limited-entry program. They faulted WDF for lack of support during their 1993 limited-entry attempt but failed to recognize the obligation that legislators and the public resource agency have to conduct a comprehensive public review. Studies by PSMFC (1993) and WDF (1993) documented the legitimate needs for limited entry, based on economic and orderly fishery factors, and limited-entry legislation was then successfully passed. Legislators have asked for studies to determine how to further reduce effort to improve program effectiveness (Christy 1973; Collinsworth et al. 1976; McHugh 1978; Townsend 1990). Townsend (1990) noted that successful limited-entry programs are most often enacted essentially in their final form. Washington lawmakers recognized that limited entry without complementary fleet effort reduction counters a main goal of intent of limited entry, which was to improve profitability.

Without tri-state co-ordination, the future of the Washington limited-entry program is tenuous. Jurisdictional reciprocity in offshore waters (federal) was an important goal of the TDCC. Washington enacted the only crab limited-entry program with reciprocal measures that recognized the rights of respective states to regulate nonresident crabbers in adjacent offshore waters. Jurisdictional reciprocity meant complete state jurisdiction over the crab fishery to 322 km offshore. The lack of complete jurisdiction over all vessels fishing in adjacent offshore waters makes unilateral state action impractical. Christy (1973) explained why Washington should be concerned about unilateral state action. With more than 50% of Washington crab catches originating outside 4.8 km, effort and catch by unregulated Oregon and California vessels can escalate while the Washington fleet undergoes a reduction in effort.

Current state laws provide for a simple resolution of within-state territorial matters, but rationalizing a fishery where a major share of the harvest is in federal waters requires both state and federal co-operation. In 1997, instead of jurisdictional reciprocity, jurisdictional stalemate exists. Another attempt at a Dungeness crab federal FMP may be necessary to alleviate jurisdictional disputes and reduce overcapitalization. The current need for federal management arises from even greater economic pressures than those identified in the 1979 draft federal FMP. Industry in the past has viewed federal involvement in management of the fishery as abhorrent, but record levels of effort and economic problems exacerbated by jurisdictional stalemate might change opinions.

In addition, tribal and Washington State managers must determine how to allocate a resource with only minor stock assessment and where more than 50% of the catch is caught outside state territorial waters. Tribal allocation issues may be the catalyst for resolution of jurisdictional issues coastwide. Co-operative, co-ordinated management by the three coastal states, the federal government, and native Americans is essential.

Acknowledgment

We thank Tom Northup and Ed Manary for helpful comments on the paper and for their participation and support of our effort on the tri-state Dungeness Crab Committee and throughout the prolonged legislative process. We thank Drs. Robert Otto and Glen Jamieson and an anonymous reviewer for comments received on earlier drafts. We also thank coastal Dungeness crab industry representatives Mr. Richard Sheldon and Mr. Ernie Summers for their relentless and integral effort on the subject matter of the paper.

References

Anderson, L.G. 1986. The economics of fisheries management. Rev. ed. John Hopkins University Press, Baltimore, Md. 296 p.

Barry, S.T. 1985. Overview of the Washington coastal Dungeness crab fishery. Proceedings of the Symposium on Dungeness Crab Biology and Management. University of Alaska, Alaska Sea Grant Report No. 853. pp. 33–36.

Benson, G., and Longman, R. 1978. The Washington experience with limited entry. *In* Limited entry as a fishery management tool. *Edited by* R.B. Rettig and J.J.C. Ginter. University of Washington Press, Seattle, Wash. pp. 333–352.

Boldt, G. 1974. United States v. State of Washington: 384 Federal Supplement Reporter 312 (W.D. Wash. 1974).

Butler, T.H. 1960. Maturity and breeding of the Pacific edible crab, *Cancer magister* Dana. J. Fish. Res. Board Can. **17**. 641–646.

Butler, T.H. 1961. Growth and age determination of the Pacific edible crab *Cancer magister* Dana. J. Fish. Res. Board Can. **18**: 873–889.

Christy F.T. Jr. 1973. Alternative arrangements for marine fisheries, an overview. Resources for the Future Inc., Washington, D.C. 91 p.

Cleaver, F.C. 1949. Preliminary results of the coastal crab (*Cancer magister*) investigation. Wash. Dept. Fish. Bio. Rep. 49A. pp. 47–82.

Collinsworth, D.W., Silverthorne, S.W., and Stewart, N.E. 1974. Phase I completion report: Proposal for development of a state-federal fisheries management plan for the California, Oregon, and Washington Dungeness crab fishery. Pac. Mar. Fish. Comm., Portland, Oreg. 109 p.

Collinsworth, D.W., Silverthorne, S.W., and Stewart, N.E. 1976. Phase II completion report: proposal for development of a state-federal fisheries management plan for the California, Oregon, and Washington Dungeness crab fishery – effort management in the Dungeness crab fishery. Pac. Mar. Fish. Comm., Portland, Oreg. 85 p.

Eddie, G.C. 1983. Engineering, economics, and fisheries management. Fishing New Books Ltd. Farnham, Surrey, England. 107 p.

Erickson, L.E. 1973. Summary of preliminary work and proposal for further economic study of the Dungeness crab fishery in Washington. Wash. Dept. Fish. Rep. Olympia, Wash. 45 p.

Erickson, L.E. 1975. An economic study of a common property problem: The Dungeness crab fishery of the Washington coast. Ph.D. dissertation. University of Michigan, Ann Arbor, Mich. Diss. Abstr. Inter. **36**(10-A). 209 p.

Ginter, J.J. and Rettig, R.B. 1978. Limited entry revisited. *In* Limited entry as a fishery management tool. *Edited by* R.B. Rettig and J.J.C. Ginter. University of Washington Press, Seattle, Wash. pp. 157–172.

Hoopes, D.T. 1973. Alaska's fishery resources – the Dungeness crab. Fishery Facts 6., U.S. Dept. of Commerce, National Oceanic and Atmospheric Administration. 14 p.

Kruse, G.H., Hicks, D., and Murphy, M.C. 1994. Handling increases mortality of softshell Dungeness crabs returned to the sea. Alaska Fish. Res. Bull. **1**(1): 1–9.

LaRiviere, P.E. 1995. Washington coastal Dungeness crab interjurisdictional fisheries program 1994 project report. Wash. Dept. Fish Wild. Olympia, Wash. 16 p.

LaRiviere, P.E. 1996. Washington coastal Dungeness crab interjurisdictional fisheries program 1995 project report. Wash. Dept. Fish Wild. Olympia, Wash. 16 p.

LaRiviere, P.E. 1997. Washington coastal Dungeness crab interjurisdictional fisheries program 1996 project report. Wash. Dept. Fish Wild. Olympia, Wash. 17 p.

Magnuson, W.G. 1978. Forward. *In* Limited entry as a fishery management tool. *Edited by* R.R. Rettig and J.J.C. Ginter. University of Washington Press, Seattle, Wash. 2 p.

McHugh, J.L. 1978. Limited entry as a conservation measure. *In* Limited entry as a fishery management tool. *Edited by* R.B. Rettig and J.J.C. Ginter. University of Washington Press, Seattle, Wash. pp. 175–187.

Methot, R.D., Jr. 1986. Management of Dungeness crab fisheries. *In* North Pacific Workshop on Stock Assessment and Management of Invertebrates. *Edited by* G.S. Jamieson and N. Bourne. Can. Spec. Publ. Fish. Aquat. Sci. 92. pp. 326–334.

Miller, R.J. 1976. North American crab fisheries: regulations and their rationales. Fish. Bull. **74**(3): 623–633.

Pacific Fishery Management Council (PFMC). 1979. Draft fishery management plan for the Dungeness crab fishery off Washington, Oregon, and California. Pac. Fish. Management Council, Portland, Oreg. 93 p.

Pacific Fishery Management Council (PFMC). 1992. Review of 1991 ocean salmon fisheries. Pac. Fish. Management Council, Portland, Oreg. 83 p.

Pacific States Marine Fisheries Commission (PSMFC). 1993. A review of the California, Oregon, and Washington Dungeness crab fishery. Pac. States Mar. Fish. Comm., Portland, Oreg. 83 p.

Phillips, J.B. 1935. The crab fishery of California. Cal. Fish. Game, **21**(1): 38–60.

Radtke, H.D. 1993. A report to Pacific States Marine Fisheries Commission: capitalization of the Dungeness crab fishery off California, Oregon, and Washington. Pac. States Mar. Fish. Comm. Portland, Oreg. 28 p.

Robinson, J.G., Odemar, M.W., Richards, J.A., and Westley, R. 1977. Phase III completion report: proposal for extension of Dungeness crab state-federal fisheries management plan development for the California, Oregon, and Washington Dungeness crab fishery. Pac. Mar. Fish. Comm., Portland, Oreg. 49 p.

Townsend, R.E. 1990. Entry restrictions in the fishery: A survey of the evidence. Land. Econ. **66**: 359–378.

Warner, R.W. 1985. Overview of the California Dungeness crab, *Cancer magister*, fisheries. Proceedings of the Symposium on Dungeness Crab Biology and Management. University of Alaska, Alaska Sea Grant Report No. 85-3. pp. 11–25.

Waters, J.R. 1991. Restricted access vs. open access methods of management: toward more effective regulation of fishing effort. Mar. Fish. Rev. **53**(3): 1–10.

Washington Department of Fisheries (WDF). 1991. A study of Washington's commercial salmon fisheries. Olympia, Wash. 75 p.

Washington Department of Fisheries (WDF). 1993. Dungeness crab fishery study report to the governor and legislature. Olympia, Wash. 19 p.

A model for assessing season closure management options in the South Australian rock lobster (*Jasus edwardsii*) fishery

Richard McGarvey and James H. Prescott

Abstract: In the northern zone of the South Australian rock lobster (*Jasus edwardsii*) fishery, managed by input controls, effective effort is rising through advances in capture technology while catches are declining. Fishers requested that management strategies be developed to counter this increase and suggested closures of varying time periods in different months of the seven-month season. Two simulation models using monthly time steps were constructed. The first, a biomass-based simulation, yielded estimates of the number of days of closure needed in different months to compensate for specified increases in effective effort. Monthly estimates of catchability were obtained by fitting to monthly effort and catches by weight, assuming Baranov depletion through harvest and natural mortality. Combining seasonal price data, the short-term effect of closures in reducing annual revenue was assessed. A second numbers-based model was employed to evaluate the effect of closure and increased minimum size on egg-per-recruit. A single cohort was simulated over ten years with simplified assumptions for growth. Sensitivity to a range of reasonable values for season-opening biomass and to the inclusion or not of growth and recruitment during the fishing season was analyzed. Model recommendations, of raising minimum size to enhance egg production, combined with three 7–8 day closures of which each vessel would observe two, were voted on and adopted by fishers in 1994–1995. These were sufficient to compensate for a 6% increase in capture power.

Résumé : Dans la zone nord de pêche à la langouste (*Jasus edwardsii*) du sud de l'Australie, gérée par contrôle des participants, l'effort effectif augmente grâce aux progrès de la technologie de capture, alors que les prises sont à la baisse. Les pêcheurs ont demandé que des stratégies de gestion soient développées pour contrer cette tendance et ont proposé que les pêches soient fermées pour diverses périodes, dans différents mois parmi les sept que dure la saison. On a élaboré deux modèles de simulation utilisant des pas de temps mensuels. Le premier, une simulation basée sur la biomasse, donnait des estimations du nombre de jours de fermeture requis chaque mois pour compenser des augmentations données de l'effort effectif. On a obtenu des estimations mensuelles de capturabilité en ajustant l'effort mensuel et les prises en poids, et en postulant une diminution de Baranov par prélèvement et mortalité naturelle. En combinant les données de prix saisonnières, on a pu évaluer les effets à court terme des fermetures sur la réduction du revenu annuel. Un second modèle, basé sur les effectifs, a été utilisé pour évaluer l'effet des fermetures et l'accroissement de la taille minimale sur le nombre d'oeufs par recrue. On a fait une simulation d'une cohorte unique sur dix ans, avec des hypothèses de croissance simplifiées. On a analysé la sensibilité à une gamme de valeurs raisonnables pour la biomasse en début de saison et à l'inclusion ou l'exclusion de la croissance et du recrutement pendant la saison de pêche. Les recommandations issues des résultats du modèle, soit d'augmenter la taille minimale pour accroître la production d'oeufs, en combinaison avec trois fermetures de 7 ou 8 jours dont chaque navire observerait deux, ont été mises au vote et acceptées par les pêcheurs en 1994–1995. Ces mesures ont été suffisantes pour compenser une augmentation de 6 % de la puissance de capture.
[Traduit par la Rédaction]

Introduction

In input-limited fisheries, closures offer an important policy tool in reducing yearly effort (Crutchfield and Zellner 1962; Adkins 1993), in assuring sufficient escapement for recruitment (Shipman and Essig 1984; Basson and Beddington 1993), in enhancing egg- and yield-per-recruit (Die and Watson 1992; Watson et al. 1993), and for controlled experimental investigations of population response to fishing (Leaman and Stanley 1993; Walters et al. 1993). In the northern zone of the South Australian rock lobster (*Jasus edwardsii*) fishery, accelerating investment in capture technology, notably color depth sounders, sonars, larger planing-hulled vessels driven by larger engines, and satellite navigational systems (GPS), is believed to result in a 2–6% annual increase in effective effort. Catches in recent years declined from 1220 t in 1991–1992 to 930 t in 1993–1994. To balance this greater effective effort, a one-week closure was implemented in 1993–1994. In August 1994, fishers requested a way to calculate the number of additional days of closure required in different months to compensate for increased effective effort in the 1994–1995 season, running from November 1 to May 31. Two additional goals were to maximize egg production while minimizing loss of revenue.

Estimating annual increases in effective effort has proven

R. McGarvey and J.H. Prescott. South Australian Research and Development Institute, South Australian Aquatic Sciences Centre, 2 Hamra Ave., West Beach SA 5024, Australia.

Correct citation: McGarvey, R., and Prescott, J.H. 1998. A model for assessing season closure management options in the South Australian rock lobster (*Jasus edwardsii*) fishery. *In* Proceedings of the North Pacific Symposium on Invertebrate Stock Assessment and Management. *Edited by* G.S. Jamieson and A. Campbell. Can. Spec. Publ. Fish. Aquat. Sci. 125. pp. 335–340.

difficult in fisheries worldwide. In the Western Australia rock lobster fishery, considerable work has been devoted to quantifying the increase (Brown et al. 1995), yielding an overall estimate of 1.4% per year from 1971–1972 to 1992–1993, with the impact roughly double in deeper water where technology had allowed the range of effective fishing to expand. The rates of increase accelerated due to the widespread adoption of GPS navigation and color sounders in the late 1980's and early 1990's, allowing the more accurate placement of pots on or near lobster habitat. During the peak seasons of 1988–1989 and 1989–1990, Brown et al. reported that GPS and color sounders produced estimated increases in deeper water of 19% and 14%, respectively. In South Australia, a written poll survey was undertaken in 1991 asking for fishers' personal estimations of the rate of increase in their own fishing power. The questionnaire asked specifically for these estimates over two time periods 1980–1985 and 1985–1990. Seventy-eight fishers responded for the 1980–1985 period and 92 for the more recent 1985–1990 period out of roughly 270 license holders. This yielded an overall estimate of 2.7% per year.

In the northern zone, vessels are at sea for most available days of the fishing season. Temporal closures can therefore be an effective means of reducing total annual effort. Some response by fishers to compensate for imposed reductions in effort was nevertheless expected. Fishery-wide reductions in per-vessel pot allocations of 10% in 1992–1993 resulted in a 7.3% decreasc in total season pot lifts. Assuming a similar compensatory response, a 6% target was chosen to achieve a 5% reduction in overall nominal effort.

To address the designated management goals, we modeled changing biomass through a fishing season. Absolute levels of abundance were not known; a range of initial biomasses were therefore assumed. Monthly catch and effort data were used to estimate monthly catchability. Losses in biomass through capture and natural mortality were then simulated in monthly time steps. To test closure options, simulation fishing effort was reduced in proportion to the numbers of closure days being tested in each month, yielding predictions of the relative declines in annual catch and revenue. A second model was constructed by following a simulated cohort of numbers to evaluate the impacts on egg-per-recruit under (*i*) various choices of closure option and (*ii*) an increase in minimum size of capture by one year's growth. These two related seasonal models are being used in conjunction with a more comprehensive spatial dynamic model acting over yearly time scales (Walters et al. 1998) in assessing management options in both zones of the South Australian rock lobster fishery.

In this paper, first, data and population parameters are summarized. Then the four stages of model construction are outlined in Methods. Reductions in annual catch and increases in lifetime egg production per recruit, predicted by the model under closures and size-increase strategies, are included in Results. The relation of the model to other fisheries models and the recent history of its application in managing the northern zone are summarized in Discussion.

Data and parameters

Daily catch, as rock lobster by weight, and effort, measured as numbers of lobster trap retrievals, were reported monthly by fishers. Fishers generally retrieve each pot once per day, with each vessel carrying approximately 50 pots. Monthly means were obtained by averaging over the five past seasons, from 1988–1989 to 1992–1993, chosen to include years before the exceptional peak harvests of 1990–1991 and 1991–1992.

Parameters employed in the model were as follows: (*i*) The accepted value for instantaneous yearly natural mortality, $M = 0.1$, was assumed to be constant with size. The calculation of lifetime egg production required measures of growth and fecundity. (*ii*) The fecundity relationship of an average mature female at different carapace lengths was taken from field measurements of a range of Tasmanian rock lobsters populations (R.B. Kennedy, Department of Primary Industry and Fisheries, PO Box 192B, Hobart, TAS 7001, Australia, personal communication). (*iii*) A simplified scheme for mature female growth of 8 mm in carapace length per year was assumed. Sensitivity analysis in the Appendix assessed the effect of growth on the outcomes. (*iv*) Fishable biomass in this population had not been estimated. To accommodate this uncertainty, we tested the sensitivity of the management recommendations to a range of four assumed season-opening (November 1) biomasses of 2000, 4000, 6000, and 8000 t. Each biomass scenario yielded a different set of monthly catchabilities and, by consequence, closure period recommendations. Monthly price data were taken from the latest availabler year, 1992–1993.

Methods

The simulation was constructed in four successive stages: (*i*) Data averages of catch, $\overline{C}_m$ (m = 1,..., 7), and effort, $\overline{E}_m$ (m = 1,..., 7), over the five previous years were calculated for each month, m. (*ii*) Coefficients of monthly catchability, q_m (m = 1,..., 7), were estimated by fitting monthly catch and effort averages to the Baranov (continuous-time) catch equation (e.g., Beverton and Holt 1957, eqs. 4 and 6),

$$[1] \quad \overline{C}_m = B_m \frac{q_m \overline{E}_m}{q_m \overline{E}_m + M} [1 - \exp(-(q_m \overline{E}_m + M)\, t_m)] \, .$$

The time interval of each month, t_m, was in units of years. The population variable, B_m, was the model biomass at the beginning of month m. The catch equation (eq. 1), together with eq. 3 (below) for losses due to natural mortality, uniquely defined q_m for each month. Numerical solutions were obtained for each monthly catchability, using Newton forward tangent minimization of $(C_m - \overline{C}_m)^2$ in Excel Solver. Each of the four assumed values of November 1 biomass yielded a different set of monthly estimates for catchability which, multiplied by effort averages, yielded monthly fishing mortalities,

$$[2] \quad F_m = q_m \overline{E}_m \, .$$

(*iii*) The monthly simulation calculated the decrease in fishable stock biomass through the fishing season due to natural mortality, M, and the monthly fishing mortalities, F_m. The four assumed biomasses, from 2000 to 8000 t, were taken as the season-opening biomasses, B_1. Deaths due to natural mortality were calculated by a form analogous to eq. 1:

$$[3] \quad D_m = B_m \frac{M}{F_m + M} [1 - \exp(-(F_m + M)\, t_m)] \, .$$

Biomasses were calculated from losses each preceding month,

$$[4] \quad B_{m+1} = B_m - C_m - D_m$$

Table 1(*a*). Derived estimates for monthly catchabilities, q_m, for the four assumed levels of November 1 biomass.

Biomass	Catchability (10^{-6} pot lifts^{-1})						
(t)	Nov	Dec	Jan	Feb	Mar	Apr	May
2000	9.32	9.94	12.4	15.0	15.2	14.9	13.8
4000	4.56	4.65	5.46	6.17	5.75	5.19	4.47
6000	3.02	3.03	3.50	3.88	3.55	3.14	2.67
8000	2.26	2.25	2.58	2.83	2.57	2.25	1.90

Table 1(*b*). Derived estimates for monthly fishing mortalities, F_m, for the four assumed levels of November 1 biomass.

Biomass	Fishing mortality (yr^{-1})						
(t)	Nov	Dec	Jan	Feb	Mar	Apr	May
2000	1.02	1.10	1.46	1.61	1.73	1.48	0.99
4000	0.50	0.51	0.64	0.66	0.65	0.51	0.32
6000	0.33	0.34	0.41	0.41	0.40	0.31	0.19
8000	0.25	0.25	0.30	0.30	0.29	0.22	0.14

Table 2. Days of closure in given months required to achieve a 6% reduction in annual catch.

Biomass	Days of closure						
(t)	Nov	Dec	Jan	Feb	Mar	Apr	May
2000	23	22	16	15	14	17	26
4000	15	15	12	12	12	16	26
6000	14	14	11	11	12	16	26
8000	13	13	11	11	12	15	26

through the seven months of the fishing season. Closures of given numbers of days were simulated by reducing effort proportionally in the month desired. Reliable parameter estimates for growth and recruitment were not available and omitting their explicit representation from the model could introduce significant error. A sensitivity analysis was therefore undertaken comparing the levels of catch reduction predicted by the model above (eqs. 1–4) with this model when growth and recruitment are explicit. The methods and results of this sensitivity analysis are presented in the Appendix. (*iv*) A second model was developed to estimate the impact of closures and changes in minimum size of capture on lifetime egg production per recruit. This model employed population numbers in an average cohort, rather than biomass, as the fundamental population variable. Eggs-per-recruit was calculated by following the decrease in a hypothetical cohort of one million mature females entering the fishery at 100 mm. Removals from this cohort occur by natural mortality and fishing mortality using the simulation constructed in step (*iii*). Employing the derived levels of monthly fishing mortality during the season and taking into consideration natural mortality over the winter closure, we followed the decrease in numbers of this cohort over a 10-year lifespan. In each November time step (the season of annual spawning of South Australian rock lobster), eggs produced by survivors were added to the lifetime sum.

Several additional features of fishery dynamics were incorporated in the egg-per-recruit model. By regulation, females bearing spawn are returned to the water. We assume that 12% of all captured females were in spawn during November, an

Fig. 1. Model simulation time series for the starting biomass assumption of 6000 t. The line at 1 indicates rescaled means of effort, catchability, and biomass.

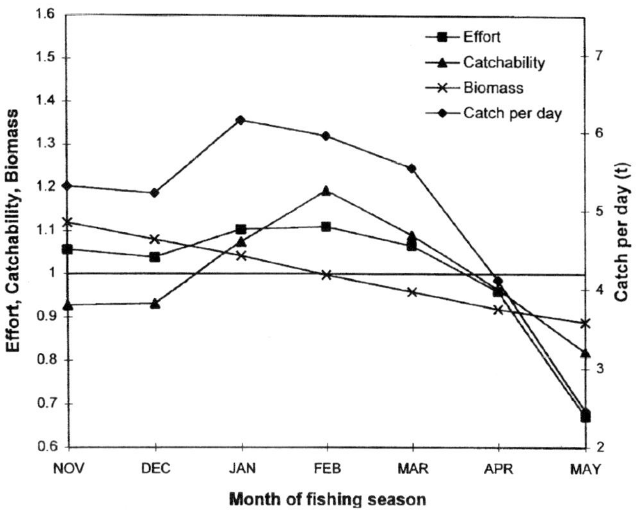

estimate based on catch sampling data. We further assumed that 15% of released spawners died of incidental mortality, principally predation, a rough estimate based on unpublished field observation (S.A. Shepherd, SARDI-Aquatic Sciences, PO Box 120, Henley Beach, SA 5022, Australia, personal communication). Sensitivity analysis showed that varying this incidental mortality factor (to 25%) had little effect on the lifetime egg production estimate.

In addition to closure, we assessed the relative benefit for egg per recruit from raising the minimum size of capture by one year's growth, setting $F = 0$ during the first year of the simulation cohort's 10-year life.

Results

The derived catchability (Table 1*a*) and fishing mortality coefficients (Table 1*b*) for each month and assumed starting biomass exhibited a seasonal trend with higher catch rates in summer, from January through March. Peak catchability occurred in February.

The predicted number of days of closure in each month needed to obtain a 6% reduction in catch, assuming historical average monthly levels of fishing effort and derived catchability through the season, are presented in Table 2. The minimum in January–February reflected a similar trend in catchability (Fig. 1), resulting in fewer days of closure needed to achieve equal effective catch power reductions.

Adding monthly price data to the monthly changes in catch with different closure strategies allowed us to calculate the short-term economic impact on fishermen in loss of income. Prices generally varied through the season inversely with overall catch in Australia and New Zealand. In addition, two- to three-week upsurges in price occurred with holiday demand before Christmas and with the Chinese New Year in late February. Revenue per pot lift by month, determined by price, catch- ability, and assumed biomass (Fig. 2), was derived from the reported 1992–1993 monthly price and catch-per-pot time

Fig. 2. Revenue per pot lift by month from 1992–1993 monthly price and catch-per-pot time series for the northern zone, with price. Assumed initial biomass, as in Fig. 1, is 6000 t.

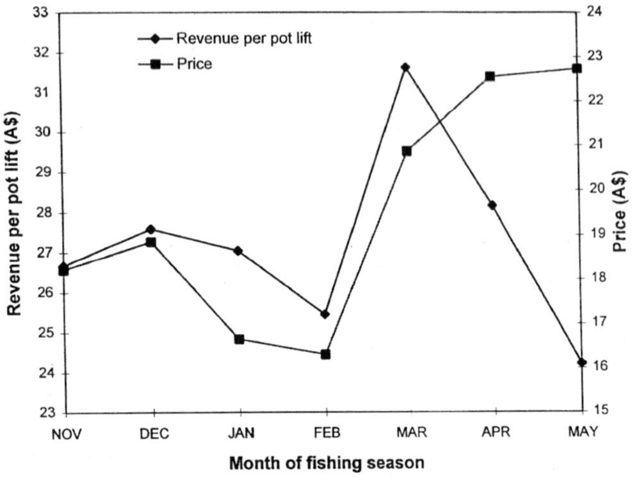

Table 3. Reduction in gross revenue from a 6% reduction in annual catch through closure in month indicated.

Biomass	Reduction in revenue (%)						
(t)	Nov	Dec	Jan	Feb	Mar	Apr	May
2000	5.5	5.9	4.6	4.4	6.7	7.1	7.2
4000	5.7	5.9	5.1	4.9	6.6	7.1	7.2
6000	5.7	5.9	5.1	5.0	6.6	7.1	7.2
8000	5.7	5.9	5.2	5.1	6.6	7.1	7.2

series for the northern zone. The relatively higher levels of price and catchability in March made it an unwise choice of closure month. Biomass, and particularly, catchability, declined in April and May more rapidly than price rose, yielding an overall drop in revenue per pot lift. Minimums in revenue per pot lift were evident in February and May. Declines in income in the first year, under equal reductions of 6% in annual harvest through closure in different months (Table 3), followed price, with a minimum loss in February.

Increases in egg production will also vary with month of closure. Lifetime egg production per female recruit was estimated, comparing closure options of (*i*) November, (*ii*) February, and (*iii*) no closure (Table 4, 100-mm columns). The small advantage of February over November was attributed to the fact that under regulations already in effect, females bearing eggs are released in November during spawning season, but all females are kept in February; therefore, a February closure protected more females overall.

To further address the management goal of enhancing population egg production, a second form of fishing regulation was tested, namely increasing the minimum size of capture by 8 mm. The results for three closure strategies and two minimum sizes of capture (Table 4), indicated that more substantial gains in lifetime eggs per recruit were achieved by raising minimum size by one year's growth than by a 28-day closure.

Table 4. Lifetime egg production (in hundreds of thousands of eggs) of an average female recruit under various combinations of minimum size and closure.

Biomass	No closure (mm)		Nov closure (mm)		Feb closure (mm)	
(t)	100	108	100	108	100	108
2000	3.4	5.5	3.6	5.7	3.7	5.7
4000	7.6	9.7	8	10.1	8	10.2
6000	10.9	12.9	11.3	13.3	11.4	13.3
8000	13.2	15	13.7	15.5	13.7	15.6

Note: The closures were for the numbers of days, specified in Table 2, sufficient to yield a 6% decline in annual catch.

Discussion

These seasonal models appear to be robust. The sensitivity of the model predictions was tested with respect to three inputs: (*i*) levels of biomass, (*ii*) release mortality, and (*iii*) monthly factors of growth and recruitment. The 0.2–2.5% range of errors introduced by omitting explicit terms for growth and recruitment was small by comparison to other inputs, notably uncertainty in the natural mortality coefficient and year-to-year variability in the estimated parameters themselves. Since the recommendations of the model were whole numbers, for instance either 14 or 15 days of closure, a 1–2% error altered the recommendations only when the estimated days of closure fell to either side of the rounding value (of 14.5).

The results were more sensitive to estimated level of season-opening biomass. The estimated numbers of closure days required to achieve a 6% reduction (Table 2) yielded large differences between the 2000 t biomass assumption and higher assumed levels, the former yielding F of 1.0 or higher. One or no days separated the estimates for the two more likely levels of 4000 and 6000 t, with F of approximately 0.5–0.65 and 0.3–0.4. Approximate knowledge of biomass (i.e., of F) is thus necessary for applying this method to achieve high accuracy (in this case, defined as estimating the exact number of closure days needed). In practice, variations in wind and sea conditions often affect the final number of pot retrievals during the season year more strongly than plus or minus one day of fishing.

Watson et al. (1993; Die and Watson 1992) constructed a model of closures for tropical penaeid fisheries, principally to optimize egg- and yield-per-recruit. The dynamics of this model are more complex and, therefore, the parameters more loosely represented those of the prawn fishery studied, as in many detailed models of this form. The fundamental trade-off in task-specific models of this sort is between more detailed dynamic rate equations and more simple fit to data.

The type of model choosen depends on the management goal. Two general objectives were presented in South Australia. First, estimates of the numbers of days of closure in different months needed to give 6% catch reduction were requested. Second, once this was achieved, it was desirable to assess which of those closures was more economically favorable given historical variation in seasonal price and to consider the effect of closures and raising minimum legal size on egg production. Egg production is affected by the release mortality of egg-bearing females and the fecundity of females of different sizes. For questions related to seasonal dynamics acting on monthly rather than yearly time scales or for per-recruit outcomes, long-term reproductive (i.e., stock-recruitment) dynamics can be neglected. In this circumstance, the management advantages of the ability to closely predict the effect of various closure strategies on monthly catch recommended a simple but accurate model. The goal was primarily one of strict accounting of monthly catches rather than a general representation of underlying population dynamics.

A second practical advantage of the model presented is the exact agreement between observed and model catches; q can be found to fit the catches to any desired level of accuracy. This is possible because for each monthly time step, there is effectively one unknown, q, and one equation, namely the catch equation. Apart from numerical error in nonlinear solving, the estimate of q for each month is uniquely determined by the catch and effort totals. The derived monthly q coefficients are thus understood to incorporate changes in temperature and behavior which affect catchability, as well as representing seasonal changes in monthly biomass, such as growth and recruitment. It is for this reason that the outcomes in days of closure needed were insensitive to explicit inclusion of growth. Once the initial biomass was given, only one possible set of catchability coefficients would yield the observed set of monthly catches and efforts. When seasonal growth and recruitment were made explicit, catchabilities were again found to fit the model exactly.

This seasonal model is thus a form of time series model, the data time series uniquely defining the parameters. It differs from ARIMA and other models of Box and Jenkins (1980) in that the equations are nonlinear, assuming the depletion equations of Baranov. This seasonal model also shares three properties in common with virtual population analysis (VPA), where rather than numbers of each cohort depleting over years through its lifetime, the total population biomass depletes over monthly time steps through the fishing season. In both VPA and the seasonal models presented above, (*i*) catch data time series are taken as input, (*ii*) Baranov depletion equations are assumed, and (*iii*) F (or with effort data, q) are obtained as output at each time step. The last point of similarity is the most fundamental, namely that the information provided is sufficient in each time step to solve for F. A simulation of the form presented (using eqs. 1–4) can also be constructed from any (untuned) VPA.

No stock-recruitment relationship has been demonstrated in the South Australian stock. Like the majority of exploited lobster populations worldwide, the variation in yearly recruit numbers exhibits little relationship to prior levels of fishing. Because long-term catches appear to be stable with varying degrees of exploitation, strong density dependence is believed to act at some stage of the life history, possibly at the juvenile stage. Therefore the effect of enhancing egg production on long-term recruitment and thus stock productivity is thought to be less important than for other fish stocks. However, studies on the distribution patterns of the pelagic larvae suggest that, through biological and/or physical mechanisms yet to be identified, *Jasus edwardsii* phyllosomas remain in the general vicinity of the adults (Booth 1992). The population off South Australia would therefore be largely reproductively self-sustaining or at least would contribute signficantly to local puerulus settlement. Fishers therefore have placed strong emphasis on conserving egg production as a principal biological goal for assuring long-term sustainability. Environmental variability, notably changes in the current patterns offshore, can strongly affect the numbers of pelagic larvae reaching the near-shore coastal zone where they settle as puerulus.

In the northern zone where closure strategies are applied, catches vary more strongly with recruitment. This was particularly evident following the large recruitment event that entered the fishable stock in 1991 in both zones. Catches in the northern zone exhibited a sharp rise and subsequent decline of approximately 35%. Recent policies, quota in the southern zone and a 33% reduction in maximum allowable effort in the northern zone, have yielded increases in egg production. This has the potential to enhance yields more in the northern zone because densities are generally lower in the northern zone and annual harvests rely more directly upon years of higher settlement.

These outcomes for different months of closure and different conservation measures (closure versus raising minimum size) were presented to the northern zone fishers in an open forum before the 1994–1995 season. They opted for a mixed strategy, closure plus increase in minimum size. The model was used to make the closures equitable in each month by calculating numbers of days off needed to yield equal catch reduction. Three closure periods of 8 days at the start of the season, 8 days at Christmas, and 7 days at the low-price week prior to the Chinese New Year were adopted. Individual license holders then chose which two of the three closure time periods to remain in port.

Previous analyses have shown that yield-per-recruit for males would be increased by raising minimum size of capture (Prescott and Lewis 1992). Length–frequency samples indicated an initial reduction in catches of 2.4% by returning to the water lobsters between 98.5 and 102 mm. To achieve increases in yield-per-recruit and long-term egg production, fishers opted to increase minimum size to 102 mm.

Together, the combination of closures and increased minimum size equate to a reduction in nominal catch capacity of 6%. These 15- or 16-day closures in 1994–1995 resulted in a reduction in total days fished per license equal to 71% of the closure days removed, slightly more than the 20% compensation anticipated, yielding a reduction in nominal effort of 4.3%.

The model used to allocate closure days in different months was simple to construct and was found to be robust. The results were accurate to less than ±1 day and thus directly were applicable for management. Regarding its use in a lobster fishery, lifestyle and conservation benefits have accrued, allowing fishers the choice of month for taking time off during the season and, in that way, reducing fishing effort

Acknowledgments

We wish to thank the fishers in the northern zone of the South Australian rock lobster fishery, who were responsible for its proposal and implementation, and to Angelo Tsoulos for data. This work was supported by Australian Fisheries Research and Development Corporation grant no. 93/087.

References

Adkins, B.E. 1993. Management of the commercial fishery for spot prawns (*Pandulus platyceros*) in British Columbia. J. Shellfish Res. **12**: 144 (Abstract).

Basson, M., and Beddington, J.R. 1993. Risks and uncertainties in the management of a single-cohort squid fishery: the Falkland Islands

Illex fishery as an example. *In* Risk evaluation and biological reference points for fisheries management. *Edited by* S.J. Smith, J.J. Hunt, and D. Rivard. Can. Spec. Publ. Fish. Aquat. Sci. 120. pp. 253–259.
Beverton, R.J.H., and Holt, S.J. 1957. On the dynamics of exploited fish populations. Fisheries Investment Series 2, Vol. 19. U.K. Ministry of Agriculture and Fisheries, London, U.K.
Booth, J.D. 1992. *Jasus phyllosomas* in southern oceans. The Lobster Newsletter, **5**(2): 8–9.
Box, G.E.P., and Jenkins, G.M. 1970. Time series analysis forecasting and control. Holden-Day, San Francisco, Calif. 553 p.
Brown, R.S., Caputi, N., and Barker, E. 1995. A preliminary assessment of increases in fishing power on stock assessment and fishing effort expended in the western rock lobster (*Panulirus cygnus*) fishery. Crustaceana, **68**(2): 227–237.
Crutchfield, J., and Zellner, A. 1962. Economic aspects of the Pacific halibut fishery. Fishing Industrial Research, Vol.1, No. 1, U.S. Dept. of the Interior, Washington, D.C. 173 p.
Die, D.J., and Watson, R.A.. 1992. A per-recruit simulation model for evaluating spatial closures in an Australian penaeid fishery. Aquat. Living Res. **5**(3): 145–153.
Francis, R.I.C.C. 1988. Maximum likelihood estimation of growth and growth variability from tagging data. N.Z. J. Mar. Freshwater Res. **22**: 42–51.
Leaman, B.M., and Stanley, R.D. 1993. Experimental management programs for two rockfish stocks off British Columbia, Canada. *In* Risk evaluation and biological reference points for fisheries management. *Edited by* S.J. Smith, J.J. Hunt, and D. Rivard. Can. Spec. Publ. Fish. Aquat. Sci. 120. pp. 403–418.
Prescott, J., and Lewis, R. 1992. Summary of South Australian southern zone rock lobster fisheries. South Australian Department of Fisheries, Adelaide, Australia, 43 p.
Shipman, S., and Essig, R.J. 1984. Georgia sounds: an evaluation of shrimp management strategy. J. Shellfish Res. **4**: 100 (Summary).
Walters, C., Goruk, R.D., and Radford, D. 1993. Rivers inlet sockeye salmon: an experiment in adaptive management. N. Am. J. Fish. Manage. **13**: 253–262.
Walters, C., Prescott, J.H., McGarvey, R., and Prince, J. 1998. Management options for the South Australia rock lobster fishery (*Jasus edwardsii*): a case study of co-operative assessment and policy design by fishers and biologists. *In* Proceedings of the North Pacific Symposium on Invertebrate Stock Assessment and Management. *Edited by* G.S. Jamieson and A. Campbell. Can. Spec. Publl. Fish. Aquat. Sci. 125. pp. 377–383.
Watson, R.A., Die, D.J., and Restrepo, V.R. 1993. Closed seasons and tropical penaeid fisheries: a simulation including fleet dynamics and uncertainty. N. Am. J. Fish. Manage. **13**: 326–336.

Appendix

To quantify the error introduced by omitting growth and recruitment from the simulation, a sensitivity analysis was undertaken. Seven closure options were tested, including two of the three which were adopted. For each option, the estimates of catch reduction from the regular model (eqs. 1–4) were compared with those from an extended model version where growth and recruitment were made explicit. For this analysis, an initial biomass of 6000 t was assumed. To incorporate growth and recruitment, a positive term, as a monthly fraction of the surviving biomass, was added to the difference equation describing monthly biomass change. In place of eq. 4, we have, for the growth-explicit model,

$$[5] \quad B_{m+1} = B_m - C_m - D_m + g_m B_m$$

where g_m are monthly coefficients of growth. A formal calculation of these coefficients, which represent the fractional increase of biomass resulting from growth and recruitment in each month of the summer molt, is not required for this sensitivity analysis.

Table 5. Numbers of days needed to yield a 6% decline in annual catch, with days assigned to each month in the indicated proportions of each policy listed.

Closure policy tested	Days to achieve 6%		
	Regular	groR1	groR2
8 days Nov & 8 days Dec	14.2	14.07	14.12
8 days Nov	14.08	13.92	13.90
8 days Nov, 8 days Dec, & 7 days Jan	13.15	12.96	12.91
8 days Dec & 7 days Jan	12.74	12.49	12.43
8 days Mar, 9 days Apr, & 10 days May	16.52	16.39	16.38
9 days Apr & 10 days May	19.78	19.65	19.68
10 days May	25.88	25.65	25.83

Analysis of the growth trends, based on GROTAG maximum likelihood fits of tagging data to von Bertalanffy growth curve (Francis 1988), was undertaken to derive rough estimates of the total fractional increase in biomass that occurs due to growth and recruitment in an average year. The annual increase in the northern zone was estimated as 26 and 18% as a result of growth and recruitment, respectively; thus, the combined annual increase in biomass, as a fraction of those present, was taken as 44%.

Two monthly growth schedules, groR2 and groR1, were chosen for testing. Most females and many males molt in winter and most males and many females molt in summer. Assumed biomass increases were divided evenly among the summer months of December, January, and February with lower growth in November. The groR2 schedule of $g_m = 0.03$, 0.012, 0.012, 0.012, 0, 0, 0 yields a growth during the fishing season, equal to the amount estimated for the entire year, of $\prod_{m=1}^{7} (1 + g_m) = 44\%$. This implies that all of the year's growth occurs in the summer molt; thus, groR2 is taken as the maximum biomass growth case during the seven months of the fishing season. Taking as more probable a schedule which assumes two-thirds of annual growth occurs in summer, we obtained the groR1 schedule of $g_m = 0.02$, 0.0835, 0.0835, 0.0835, 0, 0, 0. For each growth schedule, the set of monthly catchability parameters were reestimated and the simulations run for all 7 closure policy options.

The number of days of closure needed to compensate for a 6% increase in effective effort recommended by the two simulation models that include growth and recruitment, compared to the regular model, were calculated (Table 5). The errors introduced, presented as the fractional difference in predicted catch reduction, ranged from 0.65 to 2.00% for the differences between groR1 and the regular model and from 0.86% to 2.44% between groR2 and the regular model, among the 7 closure strategies tested. These small differences in closure option predictions indicated low sensitivity to the inclusion or not of terms for growth and recruitment.

Small-scale variation in demography and its implications for an alternative size limit in the fishery for blacklip abalone (*Haliotis rubra*) in New South Wales, Australia

D.G. Worthington and N.L. Andrew

Abstract: Enforcement of a minumum width limit has the potential to improve the management of the New South Wales (NSW) abalone (*Haliotis rubra*) fishery. Variation in demography, and growth in length in particular, can occur over very small distances and prevents the enforcement of appropriate length limits for different populations. Estimates of egg production per recruit at sites where abalone grow quickly are below 25% of that from an unfished population. This could be increased by reducing the frequency of fishing visits, or increasing the minimum length. Because of the differences in morphology, a minimum width limit would allow abalone that grow quickly to be removed at larger lengths than slow growers. During an experimental trial of a minimum width limit within the NSW abalone fishery, divers fished at sites which are rarely visited using the present 115 mm length limit as few individuals reach this length. Consequently, a minimum width limit would both increase the length at which fast-growing abalone could be removed, and by allowing the increased exploitation of slow-growing populations, reduce the frequency of fishing at sites where the abalone grow quickly.

Résumé : L'application d'une limite de largeur minimale pourrait améliorer la gestion de la pêcherie d'haliotides (*Haliotis rubra*) en Nouvelle-Galles du Sud (NGS). La variation démographique et, en particulier, la croissance en longueur, peut se manifester sur de très petites distances et compromettre l'application des limites de longueur appropriées pour différentes populations. Les valeurs estimées de la production d'oeufs par recrue dans les sites où les haliotides croissent rapidement sont inférieures à 25 % des valeurs mesurées dans la population non exploitée. On pourrait augmenter cette valeur en réduisant la fréquences des visites de pêche ou en augmentant la longueur minimale. En raison des différences morphologiques, une limite de largeur minimale permettrait la récolte des haliotides à croissance rapide à des longueurs plus grandes que les haliotides à croissance lente. Dans une expérience visant à faire l'essai d'une limite de largeur minimale dans la pêcherie d'haliotides de la NGS, des plongeurs sont allés pêcher dans des sites rarement visités avec la limite de longueur actuelle de 115 mm étant donné que peu d'individus atteignent cette longueur. Par conséquent, une limite de largeur minimale aurait deux effets : elle augmenterait la longueur à laquelle les haliotides à croissance rapide pourraient être récoltées et, en permettant l'exploitation accrue des populations à croissance lente, elle réduirait la fréquence de la pêche dans les sites où les haliotides croissent rapidement.
[Traduit par la Rédaction]

Introduction

Intense spatial variation in demography appears almost characteristic of abalone populations around the world (see reviews in Shepherd et al. 1992). In particular, patterns of growth (Worthington et al. 1995) and mortality (Emmett and Jamieson 1989) have been well studied and appear to vary at spatial scales ranging from metres to hundreds of kilometres. There is also little doubt that the observed variation in fecundity (Nash 1992) and aggregation (Andrew and Underwood 1992) contribute to small-scale variation in reproduction (Clavier 1992) and subsequent recruitment (McShane et al. 1988). Such variation in demography presents difficulties for the management of abalone fisheries, which are often spread over hundreds of kilometres of coastline. In response to these difficulties, there have been several calls to match the scale of management of abalone fisheries to the scale of variation in demography (e.g., Prince and Shepherd 1992).

The enforcement of different size limits has been used in an attempt to improve management of several abalone fisheries (e.g., Guzmán del Próo 1992). For example, minimum sizes may be reduced over large areas with consistently slower growth, or reduced for short periods to allow the collection of abalone from sites where they are stunted (Prince and Shepherd 1992). While these measures attempt to reduce the spatial scale of management, a variety of problems combine to complicate their use. First, there can often be substantial variation in demography within the area where management is modified. Variation in growth, in particular, can occur among abalone separated by only metres (Worthington et al. 1995). Second, intense research effort must be devoted to gain a knowledge of

D.G. Worthington and N.L. Andrew. NSW Fisheries Research Institute, PO Box 21, Cronulla, New South Wales, Australia 2230.

Correct citation: Worthington, D.G., and Andrew, N.L. 1998. Small-scale variation in demography and its implications for an alternative size limit in the fishery for blacklip abalone (*Haliotis rubra*) in New South Wales, Australia. *In* Proceedings of the North Pacific Symposium on Invertebrate Stock Assessment and Management. *Edited by* G.S. Jamieson and A. Campbell. Can. Spec. Publ. Fish. Aquat. Sci. 125. pp. 341–348.

Fig. 1. Total annual commercial catch and the time of introduction of major management changes in the NSW abalone fishery. Data is of limited reliability before 1964.

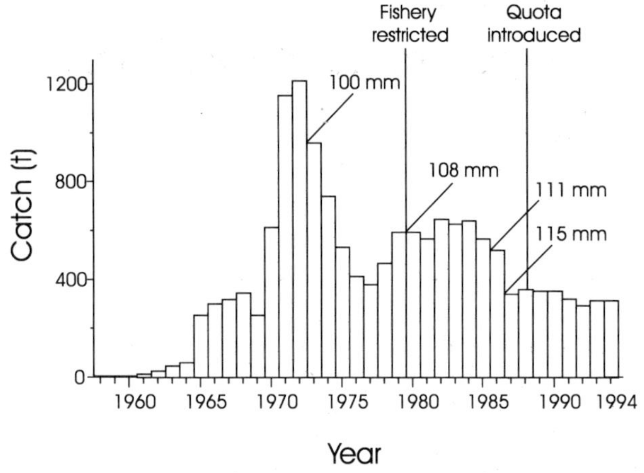

Fig. 2. Frequency of expected average width at 115 mm in length for 52 sites in NSW. Expected average width at 115 mm in length for slow-, medium-, and fast-growth populations used in the per recruit analyses are also shown.

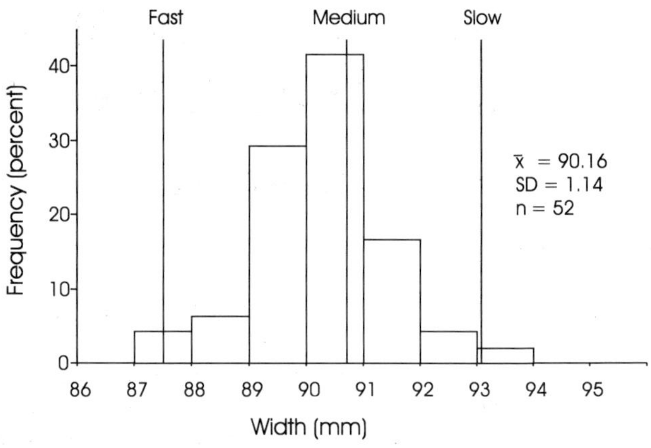

the demography of the abalone within each area that might be managed separately. Finally, the logistics of enforcing different size limits are often difficult, particularly when they need to be enforced over very small spatial scales.

Size limits for abalone have traditionally been based on the maximum diameter (i.e., length) of the shell. Exceptions have occurred in some fisheries (e.g., Sluczanowski 1984; Breen 1986), but each subsequently reverted to a size limit based on length. Length limits for abalone fisheries have historically been set at sizes that optimize the exploitation of populations which grow at average rates. Unfortunately, large variation in rates of growth in length can lead to large variation in the efficiency of exploitation of populations which grow at different rates (see also McShane 1992). Covariation between growth in length and morphology of *Haliotis rubra* in NSW, Australia, suggested several alternatives to the present minimum length (Worthington et al. 1995). In particular, because of the ease with which it can be measured, a size limit based on the width of the shell was suggested. Because abalone that grow slowly are shorter at a given width than those that grow quickly, a minimum width limit should allow slow-growing abalone to be collected at shorter lengths than fast growers (Worthington et al. 1995). In addition, the morphological differences imply that rates of growth in width are less variable than rates of growth in length.

Despite the difference in morphology among fast- and slow-growing abalone, variation in the rate of growth in width does still occur (Worthington et al. 1995). Consequently, the actual size of a width limit needs careful consideration. Similarly, there is a need to assess the practical and theoretical implications of a change to a width limit. Here we review information on spatial variation in demography of *Haliotis rubra*, and then use it to assess the implications in terms of yield and egg production per recruit of size limits based on both length and width. Finally, we detail the results of an experimental trial of a width limit within the NSW abalone fishery.

Description of the fishery

Commercial divers began to collect abalone in NSW during the late 1950's, but it was not until the 1960's that catches started to increase dramatically (Fig. 1). During 1972, catches peaked when almost 300 divers caught over 1200 t of abalone. To prevent the collection of immature abalone, a minimum length of 100 mm was introduced throughout the fishery in early 1973, and catches dropped to 400 t by 1976 (Fig. 1). In an attempt to reduce effort in the fishery, the number of divers was restricted from over 100 to 59 in 1980, and the minimum length was increased to 108 mm. Further increases in the minimum length were made during 1986 and 1987, so that the present length limit throughout the fishery is 115 mm (Fig. 1). This size was chosen as it balanced concerns of yield and egg production for populations with average rates of growth, mortality, and fecundity. An annual 10 t quota per diver was introduced during 1988, and was reduced to 9 t in 1992. Since the quota was introduced catches have been stable around 350 t (Fig. 1). In 1994, there were 37 divers in the fishery.

Divers use trailable boats to gain access to sites separated by up to 800 km, mainly on the south coast of NSW (see map in Worthington et al. 1995). The pattern of effort at a site is determined primarily by its distance from a boat ramp, weather conditions and expected catch rates. Once at a site, most of the abalone over the minimum length will be collected, and then the area will be left to recover (see also Sluczanowski 1984). The rate of recovery is dependent on both the abundance of abalone immediately below the minimum length (i.e., prerecruits) and their growth rates. Populations at sites where abalone are abundant and grow quickly will recover more rapidly and hence be fished more often.

Variation in demography of *Haliotis rubra*

Growth

Variation in growth is perhaps better understood than any other demographic variable. Both age- and length-based estimates of growth exist for *Haliotis rubra*, and it is clear that substantial variation can occur at scales ranging up to hundreds of kilometres (Table 1; Day and Fleming 1992). Despite the variation that can occur over hundreds of km, variation among sites separated by less than 20 km in NSW was almost as great as the entire range previously reported for the species from other

Table 1. Examples of the variation in growth and morphology of *Haliotis rubra* among sites in NSW.

	von Bertalanffy parameters		Expected annual growth at length (mm)				Morphological parameter		
	L_∞ (mm)	k (yr^{-1})	40	65	90	115	a ($\times 10^{-5}$)	b	c
Fast	138	0.58	43.1	32.1	21.1	10.1	7.47	3.13	0.76
Medium	117	0.57	33.5	22.6	11.7	0.9	5.51	3.22	0.79
Slow	115	0.28	18.4	12.3	6.2	0.1	5.61	3.25	0.81

Note: Further details on the estimation of these parameters can be found in Worthington et al. 1995. These estimates are used as standards for calculation of yield and egg production per recruit. The von Bertalanffy parameter, t_0, was assumed to be zero for all analyses. Morphological relationships are of the form *weight* = *a*(*length*)^b and *width* = *c*(*length*).

Table 2. Range of estimates of natural mortality for *Haliotis rubra* >6 months old in NSW and other states of Australia.

State	Author	Range of *M*
NSW	This study	0.11–1.21
Victoria	Day and Leorke 1986	0.81– 0.91
Tasmania	Beinssen and Powell 1979	0.20
Tasmania	Prince et al. 1988	0–1.4
Tasmania	Nash 1992	0.16–0.28
South Australia	Shepherd et al. 1982	0.21–0.36
South Australia	Shepherd and Breen 1992	0.42

states in Australia (Worthington et al. 1995). For example, annual rates of growth of 65 mm abalone at one site were more than double those at another site less than 10 km away (i.e., 28.8 mm vs.12.7 mm). At smaller spatial scales there was also significant variation in growth among individual abalone. For example, within one site in NSW the average annual growth of a 65-mm abalone was 12.7 mm, but half of such animals were expected to grow either slower than 5.2 mm or faster than 20.2 mm (Worthington et al. 1995).

Covariation in growth and morphology of *Haliotis rubra* in NSW allows a rapid assessment of the likely growth rate at a site by measuring the morphology of a sample of individuals (Worthington et al. 1995). One measure of the morphology of individuals at a site is the expected average width of abalone when they reach the minimum length limit of 115 mm. By measuring the length and width of samples of abalone from sites spread along the NSW coast, a distribution of the frequency of different morphologies can be estimated (Fig. 2). Morphology can then be related to growth using estimates of growth derived from tagging (Table 1), to provide an estimate of the frequency of sites with different rates of growth (Fig. 2). Overall, the average expected width of an abalone 115 mm in length is approximately 90 mm.

Mortality

Estimates of the instantaneous natural mortality (*M*) of *Haliotis rubra* span a wide range, although most are between 0.2 and 0.4 (Table 2). The variation in estimates of *M* is at least partly caused by changes in the rate of mortality with the size and age of individuals. Natural mortality is initially high, then declines as individuals grow (Shepherd and Breen 1992). For individuals of approximately the same size or age, variation in natural mortality can also be large, although little is known of the spatial scale over which differences occur (Table 2).

In a series of experiments designed to estimate the variation in natural mortality within NSW, abalone larger than 34 mm in length were marked and released in groups of approximately 50. This was repeated 55 times at sites near Broughton Island (32°5′S, 152°18′E), Sydney (33°52′S, 151°8′E), and Eden (37°6′S, 149°56′E), which span almost 1000 km of the NSW coastline. Abalone at these sites were protected from both commercial and recreational fishing by both reserves and the 115 mm length limit. After periods of up to 21 mo the sites were intensively searched for all tagged abalone. The proportion of abalone recaptured at each site can then be used with estimates of tag loss gained from double tagging experiments (i.e., 1.2% per year) and estimates of emmigration from movement studies (i.e., 3.0% per year) to provide an indication of natural mortality. This can be done by calculating the slope of a regression line relating the proportion of abalone remaining at the end of the experiment (log transformed and adjusted for rates of tag loss and emmigration) to the time since tagging (see methods of Beinssen and Powell 1979).

Natural mortality was significantly higher at Sydney (M = 1.2, SE = 0.33, n = 13) than either Eden (M = 0.48, SE = 0.17, n = 22) or Broughton Island (M = 0.11, SE = 0.16, n = 20). These estimates encompass almost the entire range of previously reported estimates for adult *Haliotis rubra* (Table 2). While there were significant differences in the natural mortality of abalone among these sites separated by almost 1000 km, it was also apparent that mortality varied among consecutive experiments at the same site (see standard error of estimates).

Fecundity

Less is known of the variation in fecundity of abalone than other demographic variables. Fecundity of *Haliotis rubra* in other states of Australia has been shown to be related to length (Prince et al. 1987; McShane et al. 1988) and weight (Sluczanowski 1984) although the relationships are not strong (Table 3). While differences exist among the states, the relative fecundity of abalone of different sizes does not vary greatly (Table 3). In NSW, similar relationships have been found, with little variation in relative fecundity among sites, but massive variation among individuals within a site (Table 3 and Worthington and Andrew 1998).

Yield and egg production per recruit

When sites are intensively fished and then left to recover, the period between visits to a site can be interpreted as an index of the rate of exploitation (Sluczanowski 1984, 1986). Sluczanowski (1984) developed a model similar to that of Beverton and Holt (1957), except the traditional index of exploitation (instantaneous fishing mortality, *F*) is replaced by the period between visits to a site (*P*). The annual yield (in

Table 3. Estimates of the relationship between fecundity (millions of eggs) and length (mm) or weight (g) for *Haliotis rubra* in NSW and other states of Australia.

Relationship	e_1	e_2	r^2	Fecundity at length 100	115	130
			NSW			
Weight	0.004	0.341	0.26	—	—	—
Length	0.021	1.774	0.19	0.35	0.67	0.99
			Victoria (McShane et al. 1988)			
Length	0.029	2.360	0.60	0.54	0.98	1.41
Length	0.040	3.620	0.77	0.38	0.98	1.58
			Tasmania (Prince et al. 1987)			
Length	0.028	2.415	0.36	0.39	0.81	1.23
			South Australia (Sluczanowski 1984)			
Weight	0.015	0.381	—	1.36	2.40	3.89

Note: All relationships are linear and are of the form $fecundity = e_1(size) - e_2$. Estimates of fecundity at length for South Australia were calculated using a weight–length relationship.

biomass) per recruit that can be harvested for a given period between fishing visits (P) and size limit (which corresponds to an age at recruitment, t_r) can then be calculated as

$$\overline{B}(P, t_r) = \int_{t_r}^{\min(t_\lambda, t_r + P)} N_t w_t \, dt/P$$

where N_t is the relative number of abalone, w_t is the weight of abalone of age t, and t_λ is the maximum age. The relative number and weight of individuals can be calculated in a variety of ways, but here

$$N_t = \exp(-M_t)$$

$$w_t = a(l_t)^b$$

where l_t is the length of individuals of age t, calculated using

$$l_t = L_\infty(1 - \exp(-K(t - t_0)))$$

To provide an indication of the potential variation, yield per recuit was calculated using the parameters of the growth curve and weight-length relationship taken from populations chosen to be representative of those at sites where growth was fast, medium and slow (Table 1). As well as differing in rates of growth, abalone at sites where growth is fast reach larger lengths than those at sites where growth is slow (Table 1 and Worthington et al. 1995).

Annual egg production per recruit as a percentage of an unfished population for a given period between fishing visits (P) and size limit (which corresponds to an age at recruitment, t_r) was calculated using

$$\overline{E}(P, t_r) = \int_{t_r}^{\min(t_\lambda, t_r + P)} \int_{t_f}^{t} N_t E_t \, dt \, dt/PA_{max}$$

where E_t is the fecundity of an abalone of age t, A_{max} is the maximum number of eggs per recruit that can be produced by an age-class and t_f is the age at which individuals become mature. The fecundity of an abalone (E_t) was calculated as

$$E_t = e_1 w_t - e_2$$

where the parameters of the fecundity-weight relationship are those shown for NSW in Table 3. In this way, individuals were assumed to mature at a weight of 85.3 g which occurred after 1.5 years at sites where growth is fast, 2.0 years where medium, and 4.2 years where slow (Worthington and Andrew 1998). If individuals that grow at different rates all mature at the same age (see Nash 1992), levels of egg production at the fast-growth sites are likely to be even lower. Both yield and egg production per recruit were calculated by numerical integration at two levels of natural mortality ($M = 0.2$ and 0.4). These levels of M were chosen to span the range of most commonly reported estimates for *Haliotis rubra*.

With the present minimum length limit of 115 mm there will be large differences in the efficiency of exploitation among sites with different rates of growth (Figs. 3 and 4). At sites with medium rates of growth, the present minimum length allows a yield per recruit approximately half the maximum attainable, although the absolute yield is greater for lower natural mortality (Figs. 3*b* and 4*b*). The present minimum length also allows egg production per recruit greater than 50% of an unfished population with similar medium rates of growth (Figs. 3*b* and 4*b*). These levels are relatively insensitive to variation in the frequency of fishing (Figs. 3*b* and 4*b*), and represent an appropriate balance between the sometimes conflicting goals of maximizing yield and protecting egg production.

At sites where abalone grow quickly and to large sizes, yield per recruit will be maximized if the sites are fished more frequently than every 1.5 years (Figs. 3*a* and 4*a*). In practice, divers revisit sites where the abalone grow quickly more frequently than every 6 mo. At this rate of fishing, yield per recruit is very high, but egg production per recruit will be less than 25% of an unfished population (Figs. 3*a* and 4*a*). That is, fast-growing populations are producing high levels of yield per recruit at the expense of egg production. These fast-growth sites could be more efficiently exploited if the frequency of fishing visits were reduced and the minimum length increased (Figs. 3*a* and 4*a*).

At sites where abalone grow slowly they do not reach such large sizes (Table 1). Consequently, at some sites few abalone reach the minimum length of 115 mm. Yield from these sites is very low, and egg production per recruit is similar to an unfished population (Figs. 3*c* and 4*c*). That is, for slow-growing populations the minimum length limit is resulting in very high conservation of egg production at the expense of yield. The only way these sites can be more efficiently exploited is

Fig. 3. Yield (*g*) and egg production (% of an unfished population) per recruit for populations of abalone that grow at (*a*) fast, (*b*) medium, and (*c*) slow rates, with low natural mortality (i.e., $M = 0.2$). Arrows indicate the potential changes if a minimum width limit were introduced (note differences in scale on width axis).

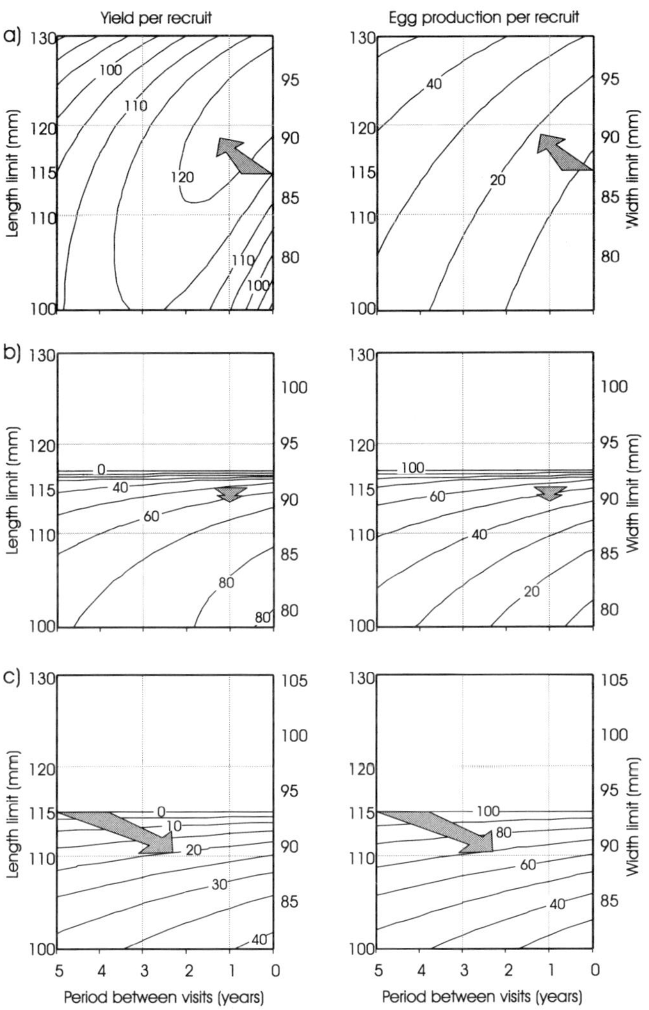

Fig. 4. Yield (*g*) and egg production (% of an unfished population) per recruit for populations of abalone that grow at (*a*) fast, (*b*) medium, and (*c*) slow rates, with high natural mortality (i.e., $M = 0.4$). Note differences in scale on width axis. Arrows indicate the potential changes if a minimum width limit were introduced (note differences in scale on width axis).

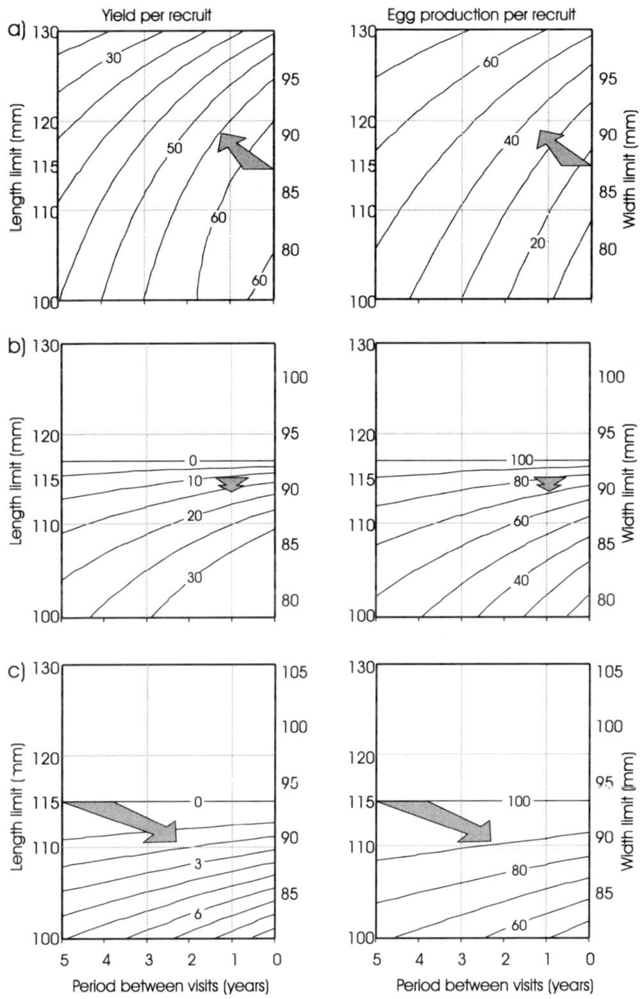

to reduce the minimum length, thus allowing them to be fished more frequently.

As the average expected width of an abalone 115 mm in length is approximately 90 mm, this may be an appropriate size for a width limit. If a width limit of 90 mm were introduced, it would reduce the differences in the efficiency of exploitation among sites. At sites where the abalone grow quickly, the effects of a 90 mm width limit would be similar to increasing the minimum length limit. That is, it would increase levels of egg production per recruit with little change in yield per recruit (Figs. 3*a* and 4*a*). At sites where they grow slowly, the effects of a width limit would be similar to reducing the minimum length. This would allow the collection of many more abalone, and hence increase yield per recruit, for a slight decrease in egg production (Figs. 3*c* and 4*c*). With the increased collection of abalone from sites where they grow slowly, the frequency of fishing at sites where abalone grow quickly would decline, further increasing egg production per recruit with little change to yield. At sites with medium rates of growth, there would be little change in the efficiency of exploitation (Figs. 3*b* and 4*b*).

Trial of a width limit

To investigate the practicality and likely effects of a width limit, 5 divers within the NSW commercial fishery were permitted to collect abalone using an 88 mm minimum width limit. That is, an abalone was considered legally undersize if they passed through an 88 mm ring. Using the width limit, each diver was able to collect a maximum of 150 kg per day for five days between 20 September and 14 October, 1994, with the weight caught considered part of their annual quota. On days

Fig. 5. Length and width distributions at (*a*) a site rarely fished using the length limit, but fished during the experimental trial of the width limit ($n = 102$) and (*b*) a site frequently fished using the length limit ($n = 118$). The percentage of animals over 115 mm in length or over 90 mm in width is also shown.

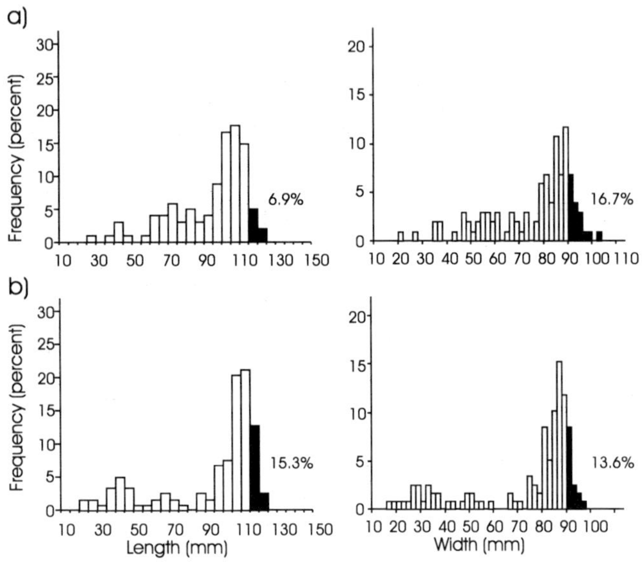

Fig. 6. Relationship between length and width of abalone caught using (*a*) the 88 mm width limit and (*b*) the 115 mm length limit. See text for an explanation of black highlight. The percentage of abalone highlighted in black, and the number of individuals and the number of diver days over which these were sampled is also shown.

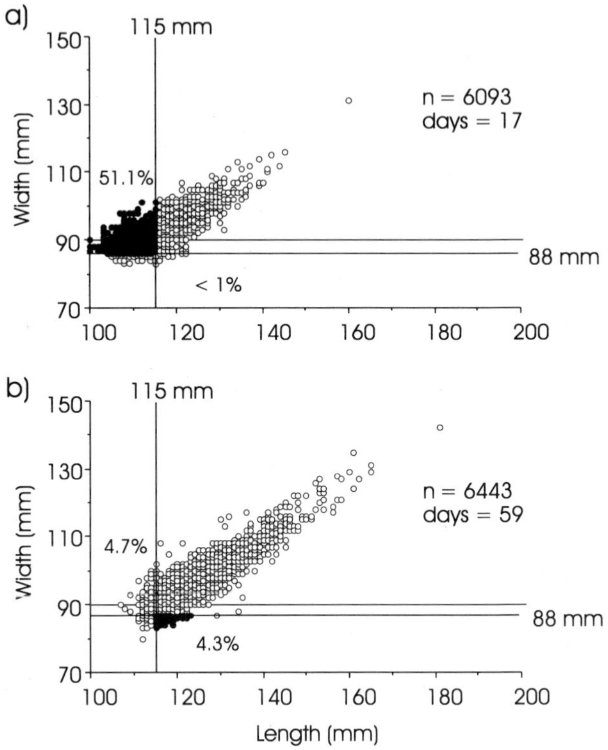

when divers used the width limit they were prohibited from also landing abalone using the 115 mm length limit. Other divers in the fishery continued to collect abalone using the length limit. The length, width, and weight of abalone landed under both width and length limits were sampled to assess changes in the pattern of fishing.

In total, 2068 kg of abalone were landed using the width limit during 17 days of fishing. Divers landing abalone on the width limit had higher average catch rates (30.0 $kg \cdot h^{-1}$, SE = 3.9) than those using the length limit (25.1 $kg \cdot h^{-1}$, SE = 1.5). This is at least partly explained by the divers using the width limit being able to fish at sites not normally visited using the length limit because few abalone reach 115 mm in length. While individuals greater than 115 mm in length were rare at such sites, individuals greater than 88 mm in width were much more common (see Fig. 5*a*). This is in contrast to those sites fished most frequently using the length limit where there is generally a much larger proportion of abalone over 115 mm in length, but a lower proportion of animals over 88 or 90 mm in width (see Fig. 5*b*).

Most of the abalone collected using the 115 mm length limit were also available for collection using the 88 mm width limit (Fig. 6). However, there were two subtle, but important changes, in the frequency with which abalone of different morphologies were caught. First, using the width limit, few abalone greater than 115 mm in length and less than 88 mm in width were landed (Fig. 6*a*). These individuals, that were thin and light for their length (average weight = 252.1 g, SE = 0.16), were caught using the length limit (highlighted black in Fig. 6*b*), and are likely to be abalone that have grown, and will continue to grow, very quickly (see Worthington et al. 1995). Second, during the trial of the width limit, many abalone less than 115 mm in length, but greater than 88 mm in width were landed (highlighted black in Fig. 6*a*). These individuals, that were wide and heavy for their length (average weight = 300.2 g, SE = 0.02), were rarely caught using the length limit, and are likely to be abalone that have grown very slowly, and may not continue to grow much larger (see Worthington et al. 1995).

Discussion

Variation in demography and alternative size limits

Variation in patterns of growth, mortality, and fecundity of *Haliotis rubra* occur at a range of spatial scales. At the largest scale, differences in the demography of abalone between the States have led to the different restrictions used to manage the fisheries (see Prince and Shepherd 1992). Within NSW, there appears to be as much variation in demography as that found between the other States, but there is no variation in management. Unfortunately, a large proportion of the variation in demography of abalone in NSW occurs among populations separated by short distances (i.e., <10 km). Patterns of growth, in particular, can vary among individual abalone separated by only a few metres (Worthington et al. 1995). It is obviously impractical to attempt to enforce different length limits over these small spatial scales.

Despite their almost universal use in abalone fisheries, size limits have been frequently criticized as a technique of management (e.g., Breen 1986). This may appear justified following the collapse of some fisheries with seemingly appropriate size limits (Tegner et al. 1989). In NSW, following the progression of length limit increases during the 1980s, fishing effort was concentrated at sites where the abalone grew quickly to large sizes. Because of this effort, the abalone at such sites were given little opportunity to reproduce, which combined with the limited dispersal of larval abalone (Prince et al. 1987) to make the sites particularly vulnerable to overfishing (see also McShane 1992). In addition to this, the increases in the minimum length resulted in the loss to the fishery of many populations where few individuals reach 115 mm in length. Consequently, the minimum length limit created both large differences in the efficiency of exploitation among sites, and the concentration of effort at those sites that were most vulnerable to overfishing.

Size limits based on the length of the shell have traditionally been used to manage many abalone fisheries. Unfortunately, the large variation in growth in length of *Haliotis rubra* in NSW has combined with the length limit to create inefficiencies in the pattern of exploitation. Because of covariation in the rate of growth in length and shell morphology, size limits based on other measures are more appropriate. In particular, a minimum width limit would allow abalone that grow slowly to be collected at shorter lengths than fast growers (Worthington et al. 1995). This would be equivalent to increasing the length limit for fast-growing populations and decreasing the length limit for slow-growing populations.

Implications of a width limit

If a minimum width limit were enforced it would change the efficiency of exploitation among sites within the NSW abalone fishery. By increasing the effective length limit for fast-growing abalone, egg production per recuit will increase from dangerously low levels. Yield per recuit of fast-growing abalone will be reduced slightly, but if the increased levels of egg production flow through to subsequent recruitment, overall yield may also rise. By decreasing the effective length limit for slow-growing abalone, there will be a large increase in yield per recuit from very low levels. The concomitant decrease in egg production per recruit is small and the level of protection will remain high. A minimum width limit will also open sites to exploitation that are not presently fished using the length limit. With the increased catch of individuals from these sites where abalone grow slowly, the annual quota will further reduce pressure on the most intensely exploited sites where the abalone grow quickly.

The experimental trial of a minimum width limit revealed that it could be practically used within the NSW abalone fishery. This was aided by the width chosen for the trial (i.e., 88 mm) which was chosen to encourage divers to trial the width limit, but is probably too small for ongoing use. A larger width limit (e.g., 90 mm) may be a more appropriate size, as it corresponds to a 115 mm length limit for populations with average rates of growth. During the trial, fishing effort was redistributed from sites where abalone grew quickly to sites where they grew more slowly. Many of the sites that were fished using the width limit are not commonly fished using the length limit as few individuals reach 115 mm in length. There is some evidence that growth of such populations is limited by the availability of food (Shepherd and Hearn 1983), implying that if some individuals could be removed, the growth rate of those remaining may increase.

As a consequence of the increased effort at sites where the abalone grow slowly, more abalone that were wide for their length were caught during the trial of the width limit. If these individuals are disproportionately fecund, a minimum width limit may reduce overall levels of egg production and subsequent recruitment. Little is known of the relative fecundity of populations with fast- and slow-growing abalone, or the exchange of larvae between them. It is likely that, because fast-growing abalone reach much larger sizes than slow growers, the potential egg production of fast-growing populations is much higher (Worthington and Andrew 1998). Alternatively, the increased density of the slow-growing populations that are rarely fished may allow higher rates of fertilization. Regardless of the relative rate of reproduction, patterns of dispersal of larval abalone among fast- and slow-growing sites are particularly important. If dispersal is limited (see Prince et al. 1987), the reduced egg production at sites where abalone grow slowly should not greatly decrease recruitment at sites where abalone grow quickly.

A minimum width limit has the potential to reduce the differences in exploitation among sites in the NSW abalone fishery. This can be achieved without the need to enforce different size limits over small spatial scales. Further trials of a width limit are proposed before any replacement of the present length limit in the NSW abalone fishery. If these trials are successful, a change to a width limit should enable both the exploitation of stocks not presently fished, and the more efficient exploitation of those that are. Considering the similar patterns in growth and morphology of other abalone species (e.g., Breen 1986), alternative size limits may also be able to improve the management of other fisheries.

Acknowledgments

We thank Gary Hamer for letting us present his data on natural mortality, and Nokome Bentley and Penny Brett for helpful discussions. Doug Ferrell and Yong Chen provided helpful comments on earlier versions of the manuscript. Finally, we would like to thank the divers in the NSW abalone fishery who cooperated with us during the experimental trial.

References

Andrew, N.L., and Underwood, A.J. 1992. Associations and abundance of sea urchins and abalone on shallow subtidal reefs in southern New South Wales. Aust. J. Mar. Freshwater Res. **43**: 1547–1559.

Beinssen, K., and Powell, D. 1979. Measurement of natural mortality in a population of blacklip abalone, *Notohaliotis ruber*. Rapp. P.-v. Réun. Cons. Int. Explor. Mer. **175**: 23–26.

Beverton, R.J.H., and Holt, S.J. 1957. On the dynamics of exploited fish populations. Minist. Agric. Fish. Food Lond. Fish. Invest. Ser. 2(19). 533 p.

Breen, P.A. 1986. Management of the British Columbia fishery for northern abalone (*Haliotis kamtschatkana*). *In* North Pacific workshop on stock assessment and management of invertebrates. *Edited by* G.S. Jamieson and N.F. Bourne. Can. Spec. Publ. Fish. Aquat. Sci. 92. pp. 300–312.

Clavier, J. 1992. Fecundity and optimal sperm density for fertilization in the ormer (*Haliotis tuberculata* L.). *In* Abalone of the world: Biology, fisheries and culture. *Edited by* S.A. Shepherd, M.J. Tegner, and S.A. Guzmán del Próo. Fishing News Books, Oxford, UK.

Day, R.W., and Fleming, A.E. 1992. The determinants and measurement of abalone growth. *In* Abalone of the world: Biology, fisheries and culture. *Edited by* S.A. Shepherd, M.J. Tegner, and S.A. Guzmán del Próo. Fishing News Books, Oxford, UK.

Day, R.W. and Leorke, A.E. 1986. Abalone catches — what factors affect them. Aust. Fish. **45**: 32–36.

Emmett, B.S., and G.S. Jamieson. 1989. An experimental transplant of prerecruit abalone in Barkley Sound, British Columbia. Fish. Bull. U.S. **87**: 95–104.

Guzmán del Próo, S.A. 1992. A review of the biology of abalone and its fishery in Mexico. *In* Abalone of the world: Biology, fisheries and culture. *Edited by* S.A. Shepherd, M.J. Tegner, and S.A. Guzmán del Próo. Fishing News Books, Oxford, UK.

McShane, P.E. 1992. Exploitation models and catch statistics of the Victorian fishery for abalone *Haliotis rubra*. Fish. Bull. U.S. **90**: 139–146.

McShane, P.E., Black, K.P., and Smith, M.G. 1988. Recruitment processes in *Haliotis rubra* (Mollusca: Gastropoda) and regional hydrodynamics in southeastern Australia imply localized dispersal of larvae. J. Exp. Mar. Biol. Ecol. **124**: 175–203.

Nash, W.J. 1992. An evaluation of egg per recruit analysis as a means of assessing size limits for blacklip abablone (*Haliotis rubra*) in Tasmania. *In* Abalone of the world: Biology, fisheries and culture. *Edited by* S.A. Shepherd, M.J. Tegner, and S.A. Guzmán del Próo. Fishing News Books, Oxford, UK.

Prince, J.D., and Shepherd, S.A. 1992. Australian abalone fisheries and their management. *In* Abalone of the world: Biology, fisheries and culture. *Edited by* S.A. Shepherd, M.J. Tegner, and S.A. Guzmán del Próo. Fishing News Books, Oxford, UK.

Prince, J.D., Sellers, T.L., Ford, W.B., and Talbot, S.R. 1987. Experimental evidence for limited dispersal of haliotid larvae (genus *Haliotis*; Mollusca: Gastropoda). J. Exp. Mar. Biol. Ecol. **106**: 243–263.

Prince, J.D., Sellers, T.L., Ford, W.B., and Talbot, S.R. 1988. Recruitment, growth, mortality and population structure in a southern Australian population of *Haliotis rubra* (Mollusca: Gastropoda). Mar. Biol. **100**: 75–82.

Shepherd, S.A., and Breen, P.A. 1992. Mortality in abalone: its estimation, variability and causes. *In* Abalone of the world: Biology, fisheries and culture. *Edited by* S.A. Shepherd, M.J. Tegner, and S.A. Guzmán del Próo. Fishing News Books, Oxford, UK.

Shepherd, S.A., and Hearn, W.S. 1983. Studies on southern Australian abalone (Genus *Haliotis*). IV Growth of *H. laevigata* and *H. ruber*. Aust. J. Mar. Freshwater Res. **34**: 461–475.

Shepherd, S.A., Kirkwood, G.P., and Sandland, R.L. 1982. Studies on southern Australian abalone (genus *Haliotis*). III Mortality of two exploited species. Aust. J. Mar. Freshwater Res. **33**: 265–272.

Shepherd, S.A., Tegner, M.J., and Guzmán del Próo, S.A. (*Editors*). 1992. Abalone of the world: Biology, fisheries and culture. Fishing News Books, Oxford, UK.

Sluczanowski, P.R. 1984. A management-orientated model of an abalone fishery whose substocks are subject to pulse fishing. Can. J. Fish. Aquat. Sci. **41**: 1008–1014.

Sluczanowski, P.R. 1986. A disaggregate model for sedentary stocks: The case of South Australian abalone. *In* North Pacific workshop on stock assessment and management of invertebrates. *Edited by* G.S. Jamieson and N.F. Bourne. Can. Spec. Publ. Fish. Aquat. Sci. **92**: 393–401.

Tegner, M.J., P.A. Breen, and Lennert, C.E. 1989. Population biology of red abalones, *Haliotis rufescens*, in southern California and managment of the red and pink, *H. corrugata*, abalone fisheries. Fish. Bull. U.S. **87**: 313–339.

Worthington, D.G., Andrew, N.L., and Hamer, G. 1995. Covariation between growth and morphology suggests alternative size limits for the abalone, *Haliotis rubra*, in NSW, Australia. Fish. Bull. U.S. **93**: 551–561.

Worthington, D.G., and Andrew, N.L. 1998. Does covariation between growth and reproduction compromise the use of an alternative size limit in the fishery for blacklip abalone, *Haliotis rubra*, in NSW, Australia. Fish. Res. In press.

Harvesting and distribution of Pacific geoduck clams, *Panopea abrupta*, in British Columbia

A. Campbell, R.M. Harbo, and C.M. Hand

Abstract: Harvesting of the long-lived Pacific geoduck clam, *Panopea abrupta*, in British Columbia (B.C.) is described in terms of catch and effort trends to indicate behavior of fishers in response to geoduck density, distribution, and management regulations. Recent surveys of commercial beds have indicated that the distribution of geoducks was generally aggregated, with densities higher in sand-shell than in rocky substrates. Geoduck distribution or concentration profiles were influenced by harvesting history in a bed. Depletion experiments showed that catch per unit effort (CPUE) was variable about an asymptote but did not decline dramatically until geoducks were below a threshold density level of about 0.5 geoduck·m^{-2}, below which the low CPUE discouraged fishers to continue fishing. Other factors such as water visibility, variability in geoducks showing their siphons above the substrate, and economic factors influencing effort distribution, also make CPUE a poor index of geoduck abundance. Fishery-independent surveys are recognized as the only reliable method of estimating geoduck biomass on a bed-by-bed basis. The commercial dive fishery for *P. abrupta* began in B.C. during 1976. During the initial exploratory phase of the fishery, effort and landings increased from 44 t to a peak of 5735 t in 1987. Some large geoduck beds were more intensively fished than others due to high quality and accessibility of geoducks. Increased management regulation, such as individual vessel quotas (IVQ's), area closures, and rotation starting in 1989, resulted in a reduced landing of 2226 t in 1994, reduced vessel and diver number, increased effort per diver, spread of effort more evenly throughout B.C., reduced competition between vessels, more constant supply of geoduck throughout the year, and increased product value. Market demand for high quality geoducks and increased value of geoducks have resulted in the potential discarding of low quality geoducks (high grading) in less desirable areas, which would result in higher exploitation than the quotas allow. Although hard to detect, discarding is being discouraged by managers and industry with enforcement and by promoting the sale of all geoducks regardless of quality.

Résumé : Nous décrivons l'exploitation en Colombie-Britannique de la panope du Pacifique (*Panopea abrupta*), espèce longévive, en termes de tendances des captures et de l'effort de pêche, pour définir le comportement des pêcheurs en réaction à la densité des panopes, à leur distribution et à la réglementation qui encadre la gestion de ce coquillage. Certains relevés récents des gisements commerciaux ont révélé une distribution en agrégations, les densités des panopes étant plus forte dans les substrats de sable coquillier que dans les substrats rocheux. La distribution ou les profils de concentration des panopes étaient influencés par l'historique de l'exploitation du gisement. Des expériences de déplétion ont montré que les CPUE (captures par unité d'effort) variaient de façon asymptotique mais ne baissaient pas de façon spectaculaire tant que les panopes n'atteignaient pas un niveau seuil de densité d'environ 0,5 panope·m^{-2}, au-dessous duquel la faiblesse de l'indice CPUE dissuadait les pêcheurs de continuer à pêcher. D'autres facteurs comme la transparence de l'eau, la variabilité de l'apparition du siphon des panopes au-dessus du substrat, et les facteurs économiques influant sur la distribution de l'effort font que les CPUE ne constituent pas un bon indice de l'abondance des panopes. La seule méthode fiable pour estimer la biomasse de panopes gisement par gisement consiste à effectuer des relevés indépendamment de la pêche. La pêche commerciale en plongée de la panope a commencé en Colombie-Britannique en 1976. Pendant la première phase exploratoire de cette pêche, l'effort et les débarquements ont grimpé de 44 t à 5 735 t en 1987. Certains grands gisements, dont les coquillages étaient accessibles et de très bonne qualité, ont été exploités plus intensivement que les autres. Le resserrement de la réglementation, notamment l'imposition des quotas individuels par bateau, les fermetures de certaines zones et la rotation à partir de 1989, ont causé une baisse des débarquements (2 226 t en 1994), une réduction du nombre de bateaux et de plongeurs, un accroissement de l'effort par plongeur, une répartition plus égale de l'effort dans toute la Colombie-Britannique, une baisse de la concurrence entre les bateaux, un approvisionnement plus constant tout au long de l'année et une hausse de la valeur des produits. La demande commerciale de panopes de grande qualité et la hausse de la valeur de ce coquillage peuvent avoir pour effet le rejet des panopes de moindre qualité (écrémage) dans les zones moins recherchées, ce qui risque de causer un niveau d'exploitation supérieur aux quotas alloués. Le rejet est une pratique difficile à détecter, mais à laquelle s'opposent les gestionnaires et l'industrie, tant par le biais de la répression qu'en promouvant la vente de toutes les panopes quelle que soit leur qualité. [Traduit par la Rédaction]

A. Campbell and C.M. Hand. Fisheries and Oceans Canada, Pacific Biological Station, Nanaimo, BC V9R 5K6, Canada.

R.M. Harbo. Fisheries and Oceans Canada, South Coast Division, Nanaimo, B.C. V9T 5K3, Canada.

Correct citation: Campbell, A., Harbo, R.M., and Hand, C.M. 1998. Harvesting and distribution of Pacific geoduck clams, *Panopea abrupta*, in British Columbia. *In* Proceedings of the North Pacific Symposium on Invertebrate Stock Assessment and Management. *Edited by* G.S. Jamieson and A. Campbell. Can. Spec. Publ. Fish. Aquat. Sci. 125. pp. 349–358.

Introduction

The Pacific geoduck clam, *Panopea abrupta* (Conrad, 1849) (hereafter referred to as geoduck), a large hiatellid bivalve, is distributed along coastal areas from southern California to Alaska and west to southern Japan (Bernard 1983; Coan et al. 1998). Geoducks are buried to 1 m deep within mud and sand substrates from the low intertidal to at least 100 m (Jamison et al. 1984). The "geoduck" name is believed to be from the Nisqually (Native North American) phrase for "dig deep" (Quayle 1978; Gordon 1996). Commercial dive fisheries for geoduck are prosecuted in Alaska, British Columbia (B.C.), and Washington State (Goodwin and Pease 1991; Harbo et al. 1995). The geoduck fishery, occurring throughout coastal B.C. (Fig. 1), started during 1976 and grew to a value of C$43.3 million by 1995 (Fig. 2). The geoduck fishery in B.C. has been described in terms of landings, regulations, quota estimation (Cox 1979; Harbo and Peacock 1983; Farlinger and Bates 1985; Farlinger and Thomas 1988; Harbo et al. 1986, 1992, 1993, 1994, 1995, 1997; Jamieson 1986; Hand et al. 1998*a*, 1998*b*), and daily catch and effort data of individual divers and vessels by examination of logbook data (Harbo et al. 1986). A requirement of the geoduck fishery since 1978 has been that fishers have had to maintain daily logbooks providing information on daily catch, effort (number of minutes per dive), and bed location. The objectives of this paper are to describe the distribution and harvesting of geoducks at the diver and bed level, as well as the fishery level in terms of catch and effort trends, and to indicate the behavior of fishers in response to geoduck density, distribution, and management regulations during the first 20 years of the geoduck fishery in B.C.

Fig. 1. British Columbia showing general coastal regions where geoduck are fished and survey areas. Regions: SIW, south inside waters surrounding eastern Vancouver Island; WCVI, west coast Vancouver Island. Survey areas: R, Ritchie Bay where depletion experiment was conducted; C, Comox Bar; G, Goletas Channel; K, Kitasu Bay; and M, Marina Island.

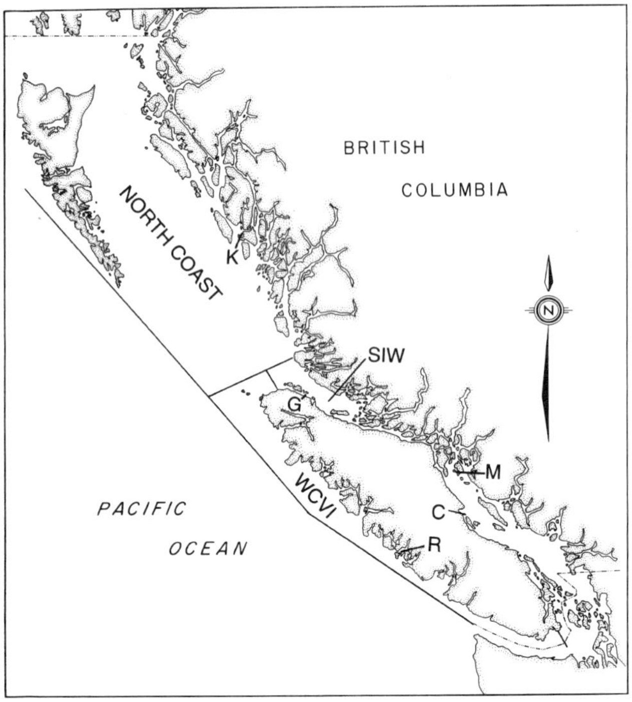

Biology

Only a summary of relevant geoduck biology is reviewed here. Geoducks are long-lived, reaching ages well over 100 years (Shaul and Goodwin 1982; Breen and Shields 1983; Harbo et al. 1983). Adult geoducks have separate sexes, with ripe gonads found in individuals from 4 to 107 years, and spawn annually, usually during summer (Andersen 1971; Goodwin 1976; Sloan and Robinson 1984). Although females have large ovaries containing millions of eggs, they are dribble spawners, usually releasing 1–2 million eggs (maximum observed 20 million) per individual during a spawning event (Goodwin and Pease 1989). Eggs are fertilized and larvae develop in the water column until settling on substrates within 47 days (Goodwin et al. 1979; Goodwin and Pease 1989). Postlarvae attach themselves with byssal threads on substrates and actively crawl and/or use the byssal threads as parachutes to travel with currents (King 1986; Goodwin and Pease 1989). At ~ 2 mm shell length (SL), the post larvae become juvenile clams (Goodwin and Pease 1989) and during the next 2 years, depending on growth rates in shell and neck length, can burrow into the substrate to at least 30–46 cm depth, reaching similar average depths to adults (30–85 cm) by the 3rd or 4th year of age (A. Campbell, unpublished data). On reaching adult size (about 75 mm SL and between 2 and 5 years), geoducks lose the ability to dig actively (Goodwin and Pease 1989). Geoduck juveniles and adults feed by filtering food particles (e.g., phytoplankton) from sea water through their siphons and gills (Goodwin and Pease 1989).

Geoduck growth is variable but most rapid in the first 10 years; thereafter, although growth in shell length is greatly reduced, shell thickness and meat weight continue to increase at a slow rate (Harbo et al. 1983; Goodwin and Shaul 1984; Sloan and Robinson 1984). Burger et al. (1998) found that mean weights (kg) of geoducks sampled from commercial catches varied from one bed to another, although, in general,

Fig. 2. Annual landings and value for geoduck fishery in three regions of British Columbia from sales slip data.

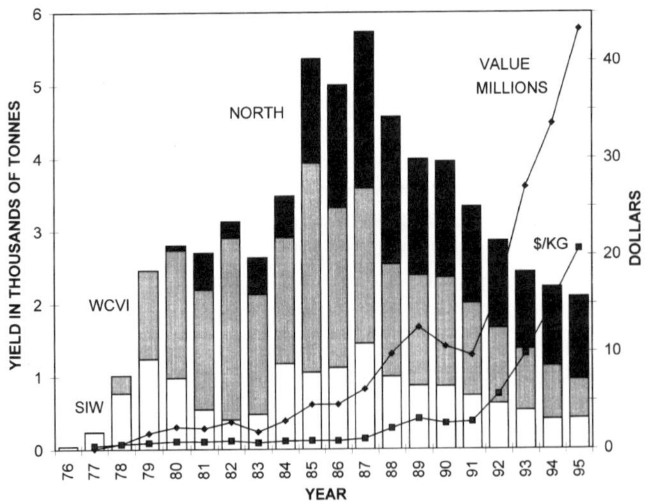

Table 1. Size of geoduck beds with landings in British Columbia based on logbook data and digitized charts up to 1995.

Details	North coast (NORTH)	South inside waters (SIW)	West coast Vancouver Island (WCVI)
Mean size (ha)	11.1	28.2	21.9
Max size (ha)	182.4	526.5	423.2
Number of beds	431	293	286
Percent of total for bed areas (ha)			
1–10	68.7	59.0	62.2
11–50	28.8	26.6	27.3
>50	2.5	14.4	10.3

they were heavier in the north coast of B.C. (NORTH COAST) (1.2 kg) than the west coast of Vancouver Island (WCVI) (1.1 kg) and the south coast of B.C. inside waters (SIW) (1.0 kg) (Fig. 1) (Hand et al. 1998*b*).

Although highly fecund, geoducks are considered to have variable but low recruitment (0.5–5.0%) to adult populations at ≥4 years (Breen and Shields 1983; Harbo et al. 1983; Goodwin and Shaul 1984). Natural mortality rate (*M*) estimates of adult geoducks in B.C. range from 0.01 to 0.04 (Breen and Shields 1983; Harbo et al. 1983; Sloan and Robinson 1984; Noakes 1992; Noakes and Campbell 1992).

Distribution

The distribution of geoducks is influenced by the availability of substrate types that allow burrowing (e.g., mud, sand, and gravel), with the density of geoducks being highest in mud–sand or sand substrates (Goodwin and Pease 1991; Campbell et al. 1996*a*). Geoduck densities are generally low at shallow depths (lower intertidal to 1 m) and increase with depth to about 20 m (Goodwin and Pease 1991; Campbell et al. 1996*a*, 1996*b*). Surveys conducted to date have been depth limited to about 20 m, which is similar to the geoduck fishery at present (Hand et al. 1998*b*). Diving deeper than 20 m restricts the dive time of divers for safety and no-decompression reasons, and consequently, little is known about geoduck distribution and densities deeper than 20 m. Some underwater video work has shown that geoduck densities increased to 24 m depth and decreased below this depth to 110 m (Jamison et al. 1984; Goodwin and Pease 1991). Frequency distributions of density data from surveys indicate that geoducks are generally contagiously distributed (Andersen 1971; Fyfe 1984; Goodwin and Shaul 1984; Goodwin and Pease 1991; Campbell et al. 1996*b*; Campbell et al. 1998*a*).

Data on the spatial distribution of individual geoduck beds and sizes in B.C. come mainly from detailed position information submitted by commercial geoduck fishers with their logbooks. Most beds are <10 ha; however, there are, on average, larger beds in the south coast than in the north coast (Table 1). The contribution to total geoduck biomass (bed area × density × mean weight, after Hand et al. 1998*b*) was greatest for small beds (<25 ha) in each region (Fig. 3).

The distribution of geoducks with different grades of marketable quality can be influenced by habitat type. Visual

Fig. 3. Distribution of total estimated biomass of recruited geoduck in relation to estimated areas (ha) of commercial-sized geoduck beds, by region, in the 3–20 m depth range. Bed sizes are grouped into 25 ha classes. Estimates of bed areas are uncertain (especially for the north coast) and are based on fisher reports in logbooks to 1995. The few geoduck beds in areas permanently closed because of pollution are included.

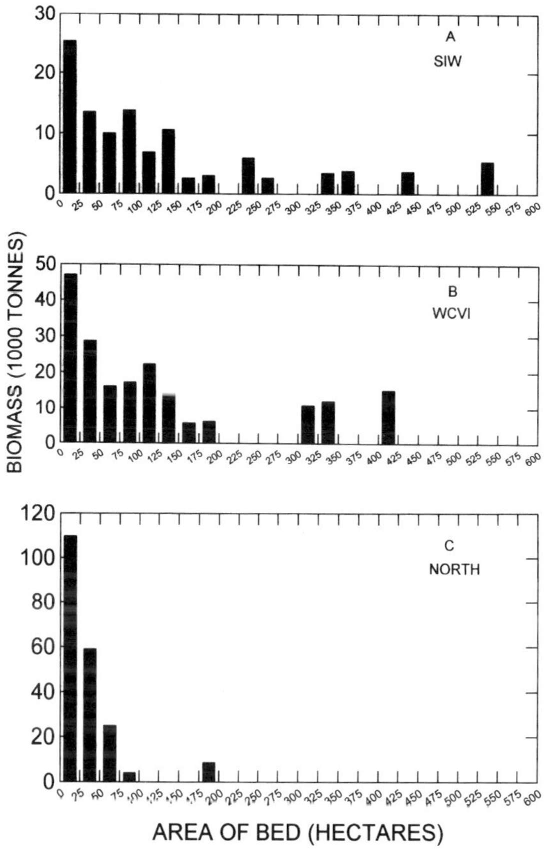

appearance of geoducks is important in determining product quality for the market. The best grade has the whole animal intact and the lightest color (near white) neck meat, whereas the poorest grade has dark brown neck meat and damaged body or shell. Generally, younger geoduck have lighter colour meat. Geoduck quality is related to substrate type, with coarser sand producing higher quality geoducks than mud substrates with a high organic content (Goodwin and Pease 1991).

The apparent distribution of geoducks is also influenced by their behavior. The proportion of geoducks in a bed "showing" their siphons at or above the substrate surface varies under different environmental and physiological conditions. The highest seasonal percentage of shows (>80%) is during the summer (April–September) when feeding and reproduction occurs and the lowest (<35%) is during the winter (November–February) when many geoducks are dormant (Goodwin 1973, 1977; Cox and Charman 1980; Turner and Cox 1981; Fyfe 1984). Shows are highest when local water currents are not excessive and there is no mechanical disturbance of the bottom (Goodwin 1977). A documented proportion of geoduck shows was reduced from about 80 to 65% after a storm in a wave-exposed area but returned to about 80% within 2–3 weeks of the storm event (Campbell et al. 1996*a*).

Harvesting at diver and geoduck bed level

Method of harvest

The basic method of harvesting geoducks has remained the same throughout the fishery and has been described in detail by Goodwin (1973, 1978) and Cox (1979). Briefly, fishing vessels (7–16 m in length) provide support to divers with both a "hookah" surface air supply hose and a high pressure water hose attached to an approximately 1-m-long stainless steel pipe (called a "stinger") that provides a water jet to allow geoduck removal from the substrate. Air and water hoses are ~75 m long, providing a considerable radius for fishing activities while at a single anchor position. After seeing the tip of a geoduck siphon or "show," the diver directs the water jet to liquefy the substrate, allowing the diver to grab the neck or shell and remove the geoduck from the substrate. The diver then places the geoduck into a large mesh bag and looks for another geoduck neck show. The process is repeated, with the diver slowly moving over the substrate until the bag is filled, lifted to the boat for geoduck sorting and storage, and replaced with another empty bag. A diver usually can make 2–3 dives per day, depending on depth, and remains under water for about 150–210 minutes per day, harvesting 1–5 geoducks per minute.

In the first few years of the fishery, a vessel often deployed two divers harvesting concurrently. This dual-diver approach often led to inefficiencies, with one diver "silting" out or reducing water visibility for the other diver. About 35–40 cages of geoduck per day per vessel were expected when using two diver hoses. By the early 1980's, all vessels used only one diver per vessel at a time harvesting ~20 cages of geoduck per day per vessel. A standard plastic cage (45.7 × 30.8 × 27.5 cm inside dimension) can hold between 20 and 27 geoducks (minimum 11, maximum 40), depending on the average geoduck size and how the geoducks are arranged in the cage. As the live product market developed during the 1990's, geoducks were packed between plastic layers to reduce breakage, thus reducing the number of geoducks per cage to ~20.

Usually the dive vessel will remain anchored in one location for as long as the divers can find harvestable quantities of geoducks. Decreased visibility due to suspended sediment in the water, or disturbed geoducks retracting their siphons, both caused by the fishing activity, reduces the catchability of geoducks. Some divers look for groupings of geoducks within a small visual field and retain a mental image of the positions of the geoducks so that the geoducks can be harvested even when fishing activity produces silt to reduce visibility. Geoduck siphons can often be felt by divers by probing holes and depressions with their hands when visual shows or water clarity is poor. Divers may search for geoducks over the same area several times.

Choice of bed

The choice of fishing one bed over another depends on a number of factors, including the following: geoduck density, quality, and seasonal "shows"; substrate type for ease of geoduck removal (sand substrates are preferred over substrates with a lot of rock or shells); shelter from waves; depth; and ease of access. The location of harvest will also be directed by the management strategy and regulations during any particular year (discussed below).

Other factors, such as catch rate, are used to decide when to move a vessel to another location to harvest geoducks. Normally, during the warm months (April–October), a catch threshold of <3 to 5 cages of geoducks per hour, depending on the season (a lower threshold is acceptable in cold periods when "shows" are low), will cause a vessel to move. Interaction between fishers, in terms of sharing information concerning which areas have better densities or quality of geoducks, may also influence fishers to move to another location in the geoduck bed or to another bed in a management area that is open to the fishery. Economic market demands will also influence the timing, location, and quality of geoducks required.

Density versus CPUE: a depletion experiment

Preamble

Catch per unit effort (CPUE) is generally considered to be a poor indicator of abundance of a fished species, particularly those that are contagiously distributed (Hilborn and Walters 1992; Prince and Hilborn 1998). The relationship between CPUE and abundance has not been documented for geoducks in B.C. The objectives of this depletion experiment were to determine (*i*) the relationship between CPUE and geoduck density in a small area of one geoduck bed, and (*ii*) whether any other fishery characteristics (i.e., mean weight, number of cages per hour, percent damage to geoducks) could be related to harvesting of geoducks by divers. Although the experiment was conducted over a 3-year period, we assumed that during the short harvest duration (<30 days) in each year there was no natural mortality or recruitment and that reductions in density were due only to harvest removal of geoducks from the study area. These are reasonable assumptions given the low annual estimates of M and recruitment of geoducks (see Biology section). The survey and removal was conducted on "harvestable-sized" geoducks; the necks of young geoducks (<3 years) are usually difficult to detect in the substrate.

Methods

A 0.25-ha area (50 × 50 m) was delineated with lead line on a geoduck bed characterized by fine sand substrate at a depth range of 10–15 m in Ritchie Bay, Clayoquot Sound, B.C. (125°55.0′W, 49°13.5′N) (Fig. 1). Ritchie Bay has been closed to commercial fishing since 1980, when Fyfe (1984) started his geoduck study. The study area was subdivided into 10-m sections with lead lines to provide five 500-m^2 strips (50 × 10 m). A survey of the geoduck density in the area, using the method described by Campbell et al. (1996*a*, 1996*b*) was conducted in each of 3 years (1 August, 1991; 12 May, 1992; and 19 May, 1993) before harvesting. This survey method involved counting geoduck siphons in 5-m^2 (5 × 1 m) quadrats along both sides of the inner subdivision boundaries and on the inner sides of the study area border.

Because all geoducks may not show their siphons at the substrate surface at any one time, "show" plots were used to determine the percentage of exposed geoducks, i.e., showing their siphons, each day. Three 10-m^2 (each 2 × 5 m) show plots were marked with ropes in the fished area and three control show plots of the same size were placed 20–50 m away from the fished area. Each geoduck showing in the show plots was

Fig. 4. Results of the depletion experiment at Ritchie Bay in a 2440-m^2 area, in terms of the relationship between geoduck de sity and the daily (A) CPUE for each year of harvest (dark dots for 1991, open dots for 1992, and crosses for 1993), (B) mean geoduck weight, (C) percent stinger misses and geoducks damaged, and (D) number of full cages (20–29 geoducks per cage) per hour.

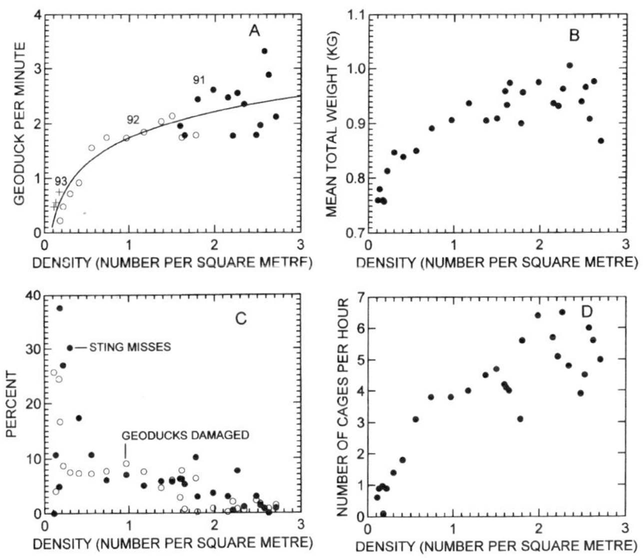

marked with a "flag" (stainless steel wire) in the substrate beside the neck. The proportion of geoducks showing on any given day was determined by dividing the number showing on that day by the cumulative total number of geoducks flagged in each plot after a total of ~30 days. To obtain the mean density of geoducks in the overall bed, the mean observed density was divided by the proportion showing on the day of the survey (Campbell et al. 1998*b*).

Geoducks were harvested from the study area by an experienced fisher using usual commercial harvest methods for 13 days during August/September 1991, 12 days during May/June 1992, and 3 days during May/June 1993. The fisher was instructed to systematically harvest each of the five 500-m^2 strips right up to the rope borders on each day fished. This ensured as much as possible (except for environmental influences such as tidal current and visibility variation) that all geoducks had an equal chance of being harvested. Geoducks within the three 20-m^2 show factor plots, however, were not harvested, making the actual area harvested 2440 m^2. Number of geoducks harvested, dive duration (minutes), and number of times that the diver missed capturing a geoduck with the "stinger" were recorded for each dive. Number of damaged geoducks (broken shell, body or neck damaged), individual shell length (mm) and weight (g), number of geoducks per standard plastic cage (45.7 × 30.8 × 27.5 cm inside dimension), and number of cages filled with geoducks per day were also recorded.

Geoduck density remaining after each harvest day (D_{t+1}) was estimated as:

$$D_{t+1} = D_1 - (\Sigma C_t) / A$$

where A is the fished area (2440 m^2), ΣC_t is the accumulated number of geoducks removed by day t, and D_1 is the geoduck density (number of geoducks·m^{-2}) estimated from the survey just before fishing each year (assumed density before fishing on day 1). The relationship between number of geoducks per minute ($CPUE_t$) and geoduck density (D_t) on day t was determined with the equation $CPUE_t = a + b \log_e D_t$ using the least squares method to estimate the constants a and b.

Results and discussion

Estimated geoduck densities for the whole experimental area before harvest were 2.70, 1.78, and 0.17 geoducks·m^{-2} for 1991, 1992, and 1993, respectively. The mean percent of geoducks showing was >90% during the surveys (before harvest) in all show plots and remained above 84% in the control show plots during all harvest days. No major differences were observed in percent showing between control and harvest show plots except when harvest show plots declined to 72% in 1991, perhaps caused by harvest activity. The slight overlap in estimated densities between the beginning and end of each yearly harvest (Fig. 4A) may be due to variation in density estimates or recruitment of individuals into fishable size. Mean weight generally declined as geoduck density was reduced (Fig. 4B).

There was no proportional relationship between CPUE and geoduck density, with CPUE fluctuating above 1.5 geoducks per minute for densities ≥0.5 geoducks·m^{-2} (Fig. 4A). The relationship between $\log_e$ transformed density and CPUE, expressed by equation $CPUE_t = 1.74 + 0.68 \log_e D_t$, was highly correlated ($r^2 = 0.81$, $n = 28$). This relationship has what Hilborn and Walters (1992) called "hyperstability," where CPUE remains high even as abundance declines until a stock reaches a low threshold density level. They predicted this type of relationship on a small spatial scale when handling time was large and search was highly efficient, with effort concentrating in areas where individuals were most abundant. Only when geoduck densities declined below 0.5 geoducks·m^{-2} did CPUE decline rapidly and the percent of missed stings increase above 10%, suggesting that search efficiency was declining (Figs. 4A, 4C, 4D). At these low geoduck densities, fishers seemed to use the stinger haphazardly while searching the liquefied substrate with their free hand, resulting in more unsuccessful stings and damaging more (>10%) geoducks with the stinger (Fig. 4C). This pattern of results was confirmed by similar depletion experiments conducted at Marina Island and Comox Bar in SIW (Fig. 1) (A. Campbell, unpublished data).

Clearly, CPUE is not a reliable index of abundance for geoducks in B.C. CPUE is affected by many factors including: (*i*) economic motivation influenced by market demand for product quality; (*ii*) a diver's searching and handling ability is affected by experience (at least in the first few weeks of learning harvest techniques and finding geoduck shows), physical fitness, depth, water temperature, and daily motivation, since considerable physical exertion is required by fishing; (*iii*) water visibility caused by plankton, algae, and/or sediment suspension; (*iv*) percent geoducks showing siphons; (*v*) substrate type (sand or loose pebbles allow a faster harvest rate, whereas rocks or layers of shell can slow harvest rate); and (*vi*) geoduck size/weight, and density distribution (clumped or dispersed).

Influence of harvesting on concentration profiles

"Concentration profiles" describe the distribution of a species at any point in time and can be used to interpret the behavior

Fig. 5. Concentration profiles for geoducks obtained from density surveys of the study area in the Ritchie Bay geoduck bed during (A) 1991 before fishing, (B) 1992 after the previous year's harvest, (C) 1993 after 2 years of harvesting, (D) in an unfished bed sand-pebble substrate in west Kitasu Bay, North Coast, 1994, (E) a moderately fished bed with pockets of sand between bedrock and rocks in Goletas Channel, 1994 (Campbell et al. 1998*a*), and (F) in a heavily fished bed with mostly sand substrate at Marina Island, 1992 (Campbell et al. 1996*b*). Density estimates are from 5-m² quadrats along transects. Frequencies for Kitasu, Goletas, and Marina beds were standardized as percentages of total quadrats per bed.

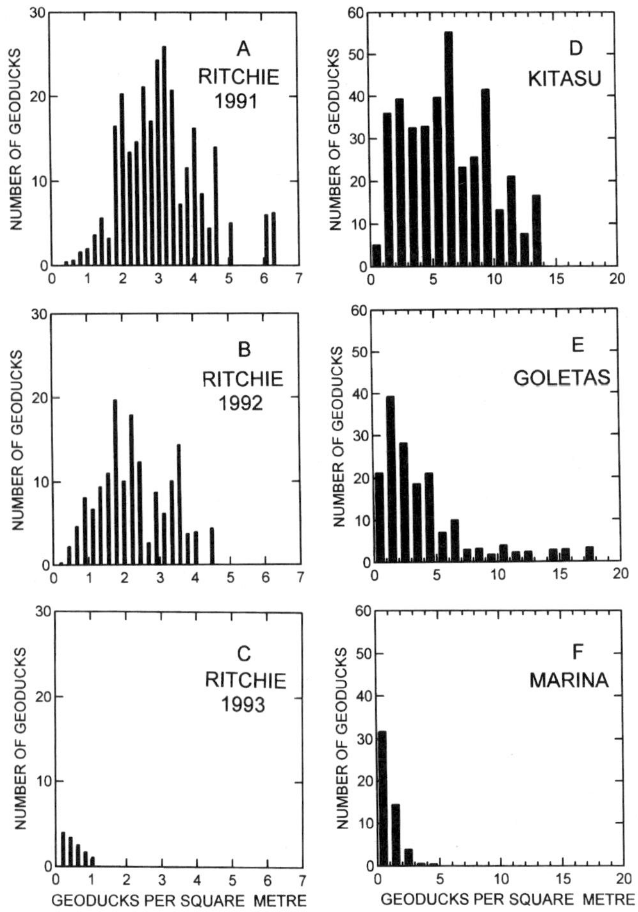

of fishers with changing CPUE (Clark 1982; Hilborn and Walters 1992; Prince and Hilborn 1998). In Type I concentration profiles, there are few locations with high densities of individuals and many locations with low densities; Type II profiles have a uniform density distribution; and Type III profiles have an intermediate density that is most common, with low and high density locations both less abundant. Concentration profiles for geoducks are Type III for small areas within a bed (Figs. 5A, 5B) and for whole beds that are pristine or relatively unfished (e.g., Fig. 5D). However, the profile changes towards Type I when geoduck beds are progressively fished more heavily (e.g., Figs. 5C, 5E, 5F).

With hyperstability, Hilborn and Walters (1992) suggest that if searching is efficient and unit product price is sufficiently high a stock could be susceptible to overfishing. Some beds in B.C., such as those at Marina Island (Figs. 1, 5F), are considered overfished and have been closed to fishing until they recover. A number of management measures have been implemented to reduce the possibility of overfishing geoduck populations in B.C. (see Fishery management section).

Harvesting at fishery level

Stock assessment and management

Geoduck fisheries management in B.C. involves interactions between assessment biologists, managers, fishers, First Nations representatives, economists, and regulation enforcement personnel. Assessment-oriented studies and biological advice for managers are first vetted through peer review by the Pacific Stock Assessment Review Committee (PSARC). Consultation with industry on geoduck management issues, such as regulation or quota changes, is maintained through annual or semiannual meetings of an advisory committee. Proposed changes in annual geoduck fishery regulations are approved at a high regional management level and ultimately by the Minister of the Federal Department of Fisheries and Oceans Canada.

The main management strategy for the geoduck fishery in B.C. is to use a fixed target exploitation rate of 1% of estimated virgin biomass for known geoduck beds to establish annual allowable quotas (Harbo et al. 1995; Hand et al. 1998*b*). For each geoduck bed, virgin biomass is calculated from estimated bed area, mean weight per geoduck, and an assumed mean density. As new surveys for geoduck density are conducted, the mean density for a specified bed area is refined for use in the assessment (Hand et al. 1998*b*). Breen (1982) first recommended the current target harvest rate and cautioned that calculated quotas depend on accurate estimation of biomass with this approach. Improving geoduck density and biomass estimates have been suggested by several authors (e.g., Sloan 1985; Campbell et al. 1998*b*).

Stock status indicators rely heavily on data from catch monitoring by industry-funded independent port observers, validators and on-ground observers, fishers logbooks, and density surveys. Harvest logbook data have been used in quota estimation since 1984 (Harbo et al. 1992, 1993, 1994, 1995; Hand et al. 1998*a*). Bed area and bed-specific catch data are obtained from logbook data and exploited bed locations on charts that are reviewed annually with fishers.

Fishery management

Quota management, license limitation, seasonal closures, and 3-year rotations in areas are used to manage the geoduck fishery in B.C. (Table 2, Fig. 6).

The geoduck fishery began in 1976, with no restrictions imposed on the fishery during 1976–1978. The number of boats fishing geoducks increased dramatically until 1979, when effort constraints were introduced (Table 2, Figs. 6B, 6C). In 1981, license limitation, using a minimum landing criteria, further reduced the fleet size to 55 eligible licenses. In addition to regular quotas, exploratory quotas have been allowed, starting in 1984, to permit exploration for new beds and to provide a year-round fishery (Table 2, Fig. 6C). Individual vessel quotas (IVQ) were introduced in 1989, wherein the coastwide quota was divided equally among license holders (Table 2, Fig. 6). Since then, all landings at designated landing

Table 2. Summary of management actions and other influences on the geoduck fishery in British Columbia, 1976–1995 (after Harbo et al. 1997).

Year	Management actions and other influences on geoduck fishery
1976	Commercial fishery opens in south coast, no quotas.
1979	To reduce increases in effort, annual quotas, management area units, and seasonal closures were introduced and no new licenses granted. A limited fishery was opened in the north coast.
1981	License limitation introduced, based on previous minimum landing criteria of 13.6 t in 1978 or 1979 for license eligibility.
1982	Coastwide opening and quota for both north and south coast. Pearse Commission recommends vessel quota and area licensing system.
1983	Introduction of G license transferable to a vessel of equal or shorter length. Grading of geoducks introduced by processors as about 30% of harvest provided for live geoduck market. Consequently problems of dumping and high grading begin.
1984	Exploratory areas opened in winter to allow year-round fishery.
1985	Seasonal openings and additional exploratory quotas were set to maintain year-round supply of geoducks for live market. Seasonal openings varied by area and influenced by live market demands (especially in December–January) and PSP concerns.
1989	Introduction of individual vessel quota (IVQ), first 3-year area rotation, area licensing (some license quotas stacked, i.e., ≥1 license per boat). License condition includes industry-funded fishing location notification, catch validation, and reporting by independent observer.
1992	Second 3-year area rotation of IVQ. About 76% of harvest provided to live geoduck market.
1995	Third 3-year area rotation of IVQ. About 90% of harvest provided to live geoduck market.

Note: See Fig. 6 for quota, vessel (boat), and license numbers.

Fig. 6. Summary of annual geoduck fishery and management actions in the north and south coasts of British Columbia. (A) number of management units; (B) the number of boats fishing (closed circles) and number of fishing licenses (open circles) (before license limitation, not all licenses issued were used to fish geoducks; after IVQ's were introduced, some boat owners stacked licenses, i.e., more than one license per boat); (C) total geoduck quota for the north and south coast (there was a coastwide quota in 1982; the stippled area indicates additional exploratory quotas); and (D) yield or landings as a percentage of the annual quota (differences between yield and quota were negligible when IVQ's were introduced and landings were monitored by a validation company at designated landing ports). IVQ represents individual vessel (boat) or license quota.

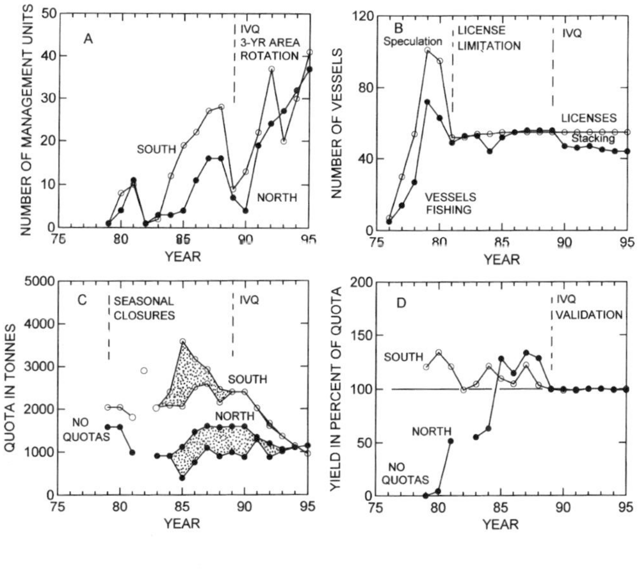

ports have been monitored by an independent contracted validation company, which effectively stopped quota overages (i.e., annual yield > authorized quota) (Fig. 6D). In addition, a 3-year rotational area fishery was initiated when each of three geographic regions (SIW, WCVI, NORTH) was divided into three subunits of similar area for geoduck harvest. Each subunit is rotationally fished once every 3 years at three times the quota that would have been permitted with annual fishing. This regulation, along with seasonal area openings, reduced the number of delivery ports for quota validation and has allowed more detailed monitoring of fished areas. Since 1990, quotas have been reduced due to revised density and bed estimates used in quota calculations, especially in SIW (Fig. 6C). Before 1995, the overall quota was higher in the south coast than the north coast (Figs. 2, 6C). Also, some beds (15.8% in SIW and 10.7% in WCVI of total beds per area), where >50% of the estimated original biomass had been harvested, were removed in 1995 from the potential fishery until geoduck populations recover. As the geoduck fishery developed, the number of management units was increased (Fig. 6A), with fewer beds per management unit, to spread fishing effort over a larger area, to encourage fishers to find new fishing grounds, and to reduce overharvesting in known beds.

Catch and effort

Initially, only the south coast (especially SIW) was fished because of the shorter distance from most fishers' home ports and closeness to processing plants. Harvesting for geoducks in the north coast started in 1980 (Figs. 2, 6). Before 1985, there was insufficient effort or incentive to meet the quota set for the North (Fig. 6D). About 50% of the known geoduck beds in SIW (as of 1995) were being fished annually before 1988 (Fig. 7A). However with introduction of the IVQ and a 3-year rotation, only a third of the bed area for each region was fished each year (Fig. 7A). As fewer new areas were found in the SIW, more new bed areas were being discovered in WCVI and especially in the north coast (Fig. 7B).

Logbooks indicated that the number of divers involved per vessel per year decreased from 14 in 1978 (when 2 divers per boat fished at a time) to around 4–5 in 1992 for SIW (when 1 diver per boat fished at a time) and has fluctuated between 4 and 7 for the other two regions. The mean number of minutes

Fig. 7. Geoduck annual (A) percent bed area fished of known total estimated bed area as of 1995 and (B) percent new bed area fished of the total bed area fished, each year for each region, south inside waters (SIW), west coast Vancouver Island (WCVI), and north coast (NORTH), of British Columbia.

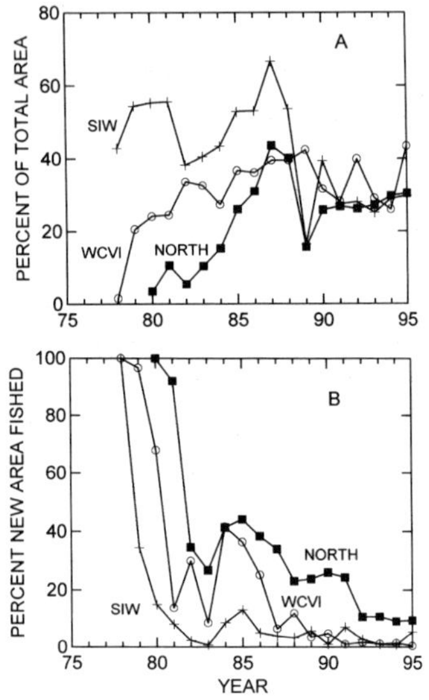

Fig. 8. Annual average trends in catch per unit effort ($kg{\cdot}min^{-1}$) per diver for each region, south inside waters (SIW), west coast Vancouver Island (WCVI), and north coast (NORTH), of British Columbia.

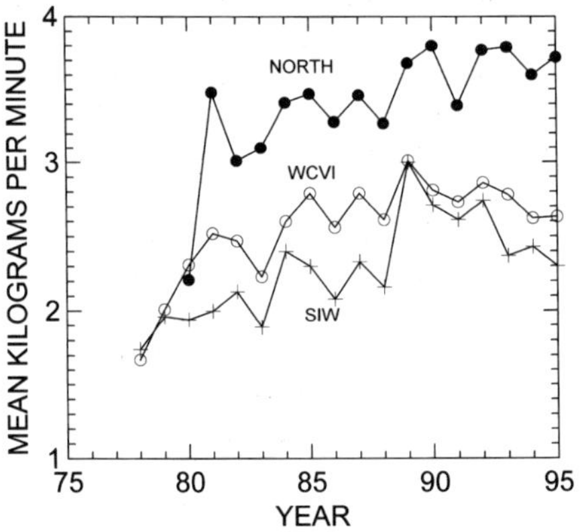

fished per day per diver was higher in the north coast (210) than for either SIW (185) or SIW (165) during 1985–1992. As the fishery progressed, the more experienced divers stayed in the fishery, with the mean number of years in the fishery progressing linearly from almost 0 in 1978 to 5.7 years in 1992. The percent of new divers of total divers entering the fishery per year decreased from 100% in 1976 to about 20% in 1991–1992. In 1995, the number of licenses by region was 30 for the north coast, 11 for SIW, and 14 for WCVI.

Initially, effort was directed to the south coast and subsequently fluctuated at a higher level in WCVI than in SIW, whereas effort started low in the north coast during 1980 but continued to increase until 1990. CPUE increased for all regions until 1981, generally fluctuated around 2.1 $kg{\cdot}min^{-1}$ for SIW and 3.2 $kg{\cdot}min^{-1}$ for the north coast but continued to increase to about 2.8 $kg{\cdot}min^{-1}$ for WCVI by the early 1990's (Fig. 8). The interregional differences in CPUE trends can be explained by heavier mean geoduck weights (Burger et al. 1998), higher geoduck densities (Hand et al. 1998*b*), and greater proportions of visible geoduck siphon shows higher above the substrate (R. Harbo and fishers' personal observations) in the north coast and WCVI compared to SIW. Recent changes in fishing costs, landed value, market demand, and management initiatives have changed the spatial pattern of effort. Fishers have harvested geoducks from the most profitable areas, which would maintain or increase CPUE values. These factors (and others discussed in earlier sections) make CPUE trends poor indicators of overall geoduck abundance.

Market influences

Initially, high landings of geoducks helped develop foreign markets. However, the value of the geoduck fishery and the unit price per kilogram remained low with a high supply of processed product (Fig. 2). There was a gradual increase in value through development of a year-round supply and an increase in live product starting in the mid-1980's (Table 2, Fig. 2). Implementation of an IVQ system in 1989, with reduced quotas, dramatically increased the unit value of the product (Fig. 2). In addition to providing stability to fishers by removing competition between them, IVQ's also allowed more time for fishers to diversify and participate in other dive fisheries (e.g., sea urchin and sea cucumbers). Increased fishery value has also provided resources for the industry to fund catch validation at landing ports, to monitor for paralytic shellfish poisoning and water quality, and to conduct fishery-independent surveys of stock abundance.

However, market demand for high quality geoducks and their increased value have resulted in several detrimental influences on geoduck stocks that have been hard to control. The more desirable beds in each management area, with high quality geoducks, have experienced a higher exploitation rate than beds with low quality product. Discarding, or "high grading," of low quality geoducks continues to occur but is difficult to detect and quantify. Fished geoducks of poor quality that are discarded underwater cannot dig back into the substrate and will die from fishery-related injuries or from predation. Discarding results in undetected overfishing while fishers try to reach the quota with high quality product and may cause underestimates of geoduck mortality in the assessment process. Managers and the fishing industry are trying to reduce discarding through enforcement and by encouraging the sale of all geoducks regardless of quality (S. Heizer, Fisheries and Oceans Canada, Nanaimo, B.C., personal communication). The increase in unit value has also encouraged poaching of geoducks, which is becoming an enforcement concern.

Future of geoduck fishery in B.C.

Although the geoduck fishery in the north coast of B.C. is probably not fully developed, with new geoduck beds still being discovered, a conservative precautionary management approach has been adopted. Management regulations attempt to keep fishing pressure low for this long-lived, slow-growing species. The potential impacts of poaching and high grading of geoducks may be high in some areas and should be regarded as an additional source of mortality in the assessment of stocks. Maintaining refuge areas, where no exploitation of geoducks occurs, assuming that these areas could act as additional sources of recruitment for fished areas throughout B.C., could be considered as an additional precautionary management approach to protect geoduck stocks. Currently, natural unfished refuges already occur in areas of rocky or shell substrate where geoducks are difficult to remove, where geoducks are unacceptable because of poor commercial quality or contaminated by pollution, or in waters deeper than 20 m where no fishing occurs. Initial industry-motivated stock enhancement efforts, through rearing of geoduck larvae in hatcheries and planting juvenile geoduck seed in the wild to rehabilitate heavily fished areas, are showing some promise in B.C. Studies in Washington State, U.S.A., have found juvenile seeding to be uneconomical except in carefully controlled small aquatic farm areas (W. Bradbury, Washington Dept. of Fish and Wildlife, Pt. Whitney Shellfish Lab., 1000 Pt. Whitney Road, Brinnon, WA 98320, personal communication). Enhancement should not be a substitute for good management of this renewable resource, but rather an added tool for maintaining viable geoduck populations in some areas where recruitment has been poor.

Acknowledgments

We thank B. Clapp, J. Clarke, L. Convey, G. Dovey, S. Heizer, K. Marcus (Hobbs), and C. Paltiel for technical assistance, K. Cripps for the Kitasu Bay data, many of the geoduck commercial fishers belonging to the Underwater Harvesters Association for assisting with the field experiments and surveys, J. Austin, K. Montgomery, and S. Renshaw for useful insights in geoduck diver harvesting methods, and W. Bradbury, S. Robinson, and G.S. Jamieson for reviewing earlier drafts of this paper.

References

Andersen, A.M. 1971. Spawning, growth and spatial distribution of the geoduck clam, *Panope generosa* (Gould), in Hood Canal, Washington. Ph.D. thesis, University of Washington Coop. Fish Unit, Seattle, Wash. 133 p.

Bernard, F.R. 1983. Catalogue of the living Bivalvia of the eastern Pacific Ocean: Bering Strait to Cape Horn. Can. Spec. Publ. Fish. Aquat. Sci. 61. 102 p.

Breen, P.A. 1982. Geoducks. *In* Assessment of invertebrate stocks off the west coast of Canada (1981). *Edited by* F.R. Bernard. Can. Tech. Rep. Fish. Aquat. Sci. 1074. pp. 14–16.

Breen, P.A., and Shields. T.L. 1983. Age and size structure in five populations of geoduc clams (*Panope generosa*) in British Columbia. Can. Tech. Rep. Fish. Aquat. Sci. 1169. 62 p.

Burger, L., Rome, E., Campbell, A., Harbo, R., Thuringer, P., Wasilewski, J., and Stewart, D. 1998. Analysis of landed weight information for geoduck clams (*Panopea abrupta*) in British Columbia, 1981–1995. Can. Tech. Rep. Fish. Aquat. Sci. In press.

Campbell, A., Harbo, R., and Heizer, S. 1996*a*. A survey of geoduck population density near Sandy Island, Comox, 1993. Can. Tech. Rep. Fish. Aquat. Sci. 2089. pp. 132–156.

Campbell, A., Harbo, R., and Heizer, S. 1996*b*. A survey of geoduck population density at Marina Island, 1992. Can. Tech. Rep. Fish. Aquat. Sci. 2089. pp. 157–203.

Campbell, A., Clapp, B., Hand, C., Harbo, R., Hobbs, K., Hume, J., and Scharf, G. 1998*a*. A survey of geoduck population density in Goletas Channel, 1994. Can. Tech. Rep. Fish. Aquat. Sci. In press.

Campbell, A., Hand, C.M., Paltiel, C., Rajwani, K.N., and Schwartz, C.J. 1998*b*. Evaluation of some survey methods for geoducks. Can. Tech. Rep. Fish. Aquat. Sci. In press.

Clark, C.W. 1982. Concentration profiles and the production and management of marine fisheries. *In* Economic theory of natural resources. *Edited by* W. Eichhorn. Physica-Verlag, Wurzburg-Wien, Germany. pp. 97–112.

Coan, E.V., Scott, P.H., and Bernard, F.R. 1998. Bivalve seashells of western North America. St. Barbara Mus. Nat. Hist. Contrib. Sci. In press.

Cox, R.K. 1979. The geoduck, *Panope generosa*: some general information on distribution, life history, harvesting, marketing and management in British Columbia. B.C. Mar. Resour. Br. Minist. Environ. Fish. Manage. Rep. 15. 25 p.

Cox, R.K., and Charman, E.M. 1980. A survey of abundance and distribution (1977) of the geoduck clam (*Panope generosa*) in Queen Charlotte, Johnstone and Georgia Straits, British Columbia. B.C. Mar. Res. Br. Fish. Manage. Rep. 16. 122 p.

Farlinger, S., and Bates, K.T. 1985. Review of shellfish fisheries in northern British Columbia to 1984. Can. MS Rep. Fish Aquat. Sci. 1841. 35 p.

Farlinger, S., and Thomas, G.A. 1988. Review of shellfish fisheries in northern British Columbia 1985 and 1986. Can. MS Rep. Fish Aquat. Sci. 1988. 38 p.

Fyfe, D.A. 1984. The effect of conspecific association on growth and dispersion of the geoduck clam, *Panope generosa*. M.Sc. thesis, Simon Fraser University, Burnaby, B.C. 110 p.

Goodwin, C.L. 1973. Subtidal geoducks of Puget Sound, Washington. Wash. Dep. Fish. Tech. Rep. 13. 64 p.

Goodwin, C.L. 1976. Observations of spawning and growth of subtidal geoducks (*Panope generosa,* Gould). Proc. Natl. Shellfish. Assoc. **65**: 49–58.

Goodwin, C.L. 1977. The effects of season on visual and photographic assessment of subtidal geoduck clam (*Panope generosa* Gould) population. Veliger, **20**: 155–158.

Goodwin, C.L. 1978. Some effects of subtidal geoduck (*Panope generosa*) harvest on a small experimental plot in Hood Canal, Washington. Wash. Dep. Fish. Prog. Rep. 66. 21 p.

Goodwin, C.L., and Pease, B. 1989. Species profiles: life histories and environmental requirements of coastal fish and invertebrates (Pacific Northwest) – Pacific geoduck clam. U.S. Wildl. Serv. Biol. Rep. 82(11.120). U.S. Army Corps of Engineers, TR EL-82-4. 15 p.

Goodwin, C.L., and Pease, B.C. 1991. Geoduck, *Panopea abrupta* (Conrad, 1849), size, density, and quality as related to various environmental parameters in Puget Sound, Washington. J. Shellfish Res. **10**: 65–77.

Goodwin, L., and Shaul, W. 1984. Age, recruitment and growth of the geoduck clam (*Panope generosa*, Gould) in Puget Sound, Washington. State Wash. Dep. Fish. Prog. Rep. 215. 30 p.

Goodwin, L., Shaul, W., and Budd, C. 1979. The larval development of the geoduck clam (*Panope generosa*, Gould). Proc. Natl. Shellfish. Assoc. **69**: 73–76.

Gordon, D.G. 1996. Field guide to the geoduck. Sasquatch Books, Seattle, Wash. 48 p.

Hand, C.M., Hobbs, K., Harbo, R., and Thomas, G.A. 1998*a*. Quota options and recommendations for the 1996 geoduck clam fishery. Can. Tech. Rep. Fish. Aquat. Sci. In press.

Hand, C.M., Marcus, K., Heizer, S., and Harbo, R. 1998*b*. Quota options and recommendations for the 1997 and 1998 geoduck clam fisheries. Can. Tech. Rep. Fish. Aquat. Sci. In press.

Harbo, R.M., and Peacock, S.D. 1983. The commercial geoduck clam fishery in British Columbia, 1976 to 1981. Can. MS Rep. Fish. Aquat. Sci. 1712. 40 p.

Harbo, R.M., Adkins, B.E., Breen, P.A., and Hobbs, K.L. 1983. Age and size in market samples of geoduc clams (*Panope generosa*). Can. MS Rep. Fish. Aquat. Sci. 1714. 77 p.

Harbo, R.M., Hand, C., and Adkins, B.E. 1986. The commercial geoduck clam fishery in British Columbia 1981 to 1984. Can. MS Rep. Fish. Aquat. Sci. 1873. 59 p.

Harbo, R.M., Farlinger, S., Hobbs, K., and Thomas, G. 1992. A review of quota management in the geoduck clam fishery in British Columbia, 1976 to 1990 and quota options for the 1991 fishery. Can. MS Rep. Fish. Aquat. Sci. 2178. 135 p.

Harbo, R.M., Thomas, G., and Hobbs, K. 1993. Quotas for the 1992–1993 geoduck clam fisheries. Can. MS Rep. Fish. Aquat. Sci. 2179. 209 p.

Harbo, R.M., Thomas, G., and Hobbs, K. 1994. Quota options and recommendations for the 1994 geoduck clam fishery. Can. MS Rep. Fish. Aquat. Sci. 2228. 115 p.

Harbo, R.M., Thomas, G., and Hobbs, K. 1995. Quota options and recommendations for the 1995 geoduck clam fishery. Can. MS Rep. Fish. Aquat. Sci. 2302. 141 p.

Harbo, R.M., Thomas, G., and Hobbs, K. 1997. Geoduck clam dive fishery 1993/94. Can. MS Rep. Fish. Aquat. Sci. 2369. pp. 11–26.

Hilborn, R., and Walters, C.J. 1992. Quantitative fisheries stock assessment: choice, dynamics and uncertainty. Chapman and Hall, New York, N.Y. 570 p.

Jamieson, G.S. 1986. A perspective on invertebrate fisheries management — the British Columbia experience. *In* North Pacific Workshop Stock Assessment and Management of Invertebrates. *Edited by* G.S. Jamieson and N. Bourne. Can. Spec. Publ. Fish. Aquat. Sci. 92. pp. 57–74.

Jamison, D., Heggen, R., and Lukes, J. 1984. Underwater video in a regional benthos survey. Proc. Pac. Cong. on Mar. Tech., Mar. Tech. Soc., Honolulu, Hawaii.

King, J.J. 1986. Juvenile feeding ontogeny of the geoduck, *Panope abrupta* (Bivalvia: Saxicavacea), and comparative ontogeny and evolution of feeding in bivalves. M.Sc. thesis, University of Victoria, Victoria, B.C. 281 p.

Noakes, D.J. 1992. On growth and mortality of geoduck clams (*Panope abrupta*) (or how fast do all 'ducs go to heaven?). Can. MS Rep. Fish. Aquat. Sci. 2169. pp. 22–34.

Noakes, D.J., and Campbell, A. 1992. Use of geoduck clams to indicate changes in the marine environment of Ladysmith Harbour, British Columbia. Environmetrics, **3**: 81–97.

Prince, J., and Hilborn, R. 1998. Concentration profiles and invertebrate fisheries management. *In* Proceedings of the North Pacific Symposium on Invertebrate Stock Assessment and Management. *Edited by* G.S. Jamieson and A. Campbell. Can. Spec. Publ. Fish. Aquat. Sci. 125. pp. 187–196.

Quayle, D.B. 1978. The intertidal bivalves of British Columbia. B.C. Prov. Mus. Handb. **17**. 104 p.

Shaul, W., and Goodwin, C.L. 1982. Geoduck (*Panope generosa*: Bivalvia) age as determined by internal growth lines in the shell. Can. J. Fish. Aquat. Sci. **39**: 632–636.

Sloan, N.A. 1985. Feasibility of improving geoduck stock assessment: history of the problem, recommended methods and their costs. Can. MS Rep. Fish. Aquat. Sci. 1848. pp. 57–67.

Sloan, N.A., and Robinson, S.M.C. 1984. Age and gonad development in the geoduck clam *Panope abrupta* (Conrad) from southern British Columbia, Canada. J. Shellfish Res. **4**: 131–137.

Turner, K.C., and Cox, R.K. 1981. Seasonal reproductive cycle and show factor variation of the geoduck clam, *Panope generosa,* (Gould) in British Columbia. J. Shellfish Res. **1**: 125.

Application of a spatial model to explore rotating harvest strategies for sedentary species

John F. Caddy and Juan Carlos Seijo

Abstract: Simulations were carried out for sedentary marine resources, with a range of values of natural mortality rate and growth coefficients, distributed contagiously over an array of unit areas. Effects of changing a rotation period of *n* years for different *M*/*K* ratios were investigated. Optimal conditions were searched for according to physical criteria relating to stock biomass density left on the ground, using three arbitrary parameter sets corresponding to low, medium, and high rates of growth and natural mortality. For long-lived species ($M = 0.1$), optimal biomass was achieved for the three growth velocities with a rotation period of 6 years. For fast-growing and long-lived species ($K = 0.6$ and $M = 0.1$), the optimal rotation period was 4 years. Short-lived and fast-growing species had an optimal rotation period of 2 years.

Résumé : Des simulations, effectuées à l'aide de différentes valeurs du taux de mortalité naturelle et du coefficient de croissance, ont porté sur des ressources marines sédentaires réparties de façon contagieuse dans un ensemble d'unités de surface. On a étudié les effets d'une modification de la période de rotation de *n* années pour différents rapports *M*/*K*. On a cherché des conditions optimales en relation avec des critères physiques liés à la densité de la biomasse du stock restant sur le fond, en utilisant trois jeux de paramètres arbitraires correspondant à des taux de croissance et de mortalité naturelle faible, moyen et élevé. Dans le cas des espèces longévives ($M = 0{,}1$), une période de rotation de 6 ans a permis d'obtenir une biomasse optimale pour les trois taux de croissance. Pour les espèces longévives à croissance rapide ($K = 0{,}6$ et $M = 0{,}1$), la période de rotation optimale était de 4 ans. Chez les espèces à courte durée de vie et à croissance rapide, la période de rotation optimale était de 2 ans. [Traduit par la Rédaction]

Introduction

Except for the southern California red sea urchin, *Strongylocentrotus franciscanus*, fishery (Botsford et al. 1993) and large-scale sea ranching of cod, *Codus morua,* in Norway (Dandberg et al. 1993), rotating harvest models have not been described for fisheries. Although location-specific pulse fishing with an indeterminate rotation period is effectively used by fishermen on sedentary resources and pulse fishing is also a common harvest strategy for some distant water exploiting fleets, theoretical evaluations of them are scarce (see, e.g., Sluczanowski 1983). More formal crop rotation schemes are of course widely applied in agriculture and forestry and their possible application to fishery resources seem worth at least some theoretical investigation by marine population modelers.

The rudimentary rotating harvest model described in Caddy (1993) was developed for long-lived species and perhaps has its most obvious justification with long-lived, slow-growing stocks in areas of open access. As explained by Hilborn and Walters (1992), if the annual rate of increase of biomass of a fish population falls below the effective discount rate that is currently applied by the fishing industry, there is a strong incentive for "pulse fishing" of the resource. The resource is then often treated as a commodity to be "mined" down to a low level before the fleet moves on to other resources. Long-lived species with values of natural mortality rate of around 0.1 and von Bertalanffy *K*'s of a similar order tend, therefore, to be the first to disappear from an ecosystem. Rotating harvest schemes that promote a form of "pulse fishing by subarea" are one of a few options to protect such longer-lived species, short of introducing permanent closure zones.

J.F. Caddy. Fisheries Department, F.A.O. Via delle Terme di Caracalla, Rome 00100, Italy.

J.C. Seijo.[1] Centro de Investigación y de Estudios Avanzados del IPN, Unidad Mérida, Apartado Postal 73 Cordemex, Mérida, México.

Correct citation: Caddy, J.F., and Seijo, J.C. 1998. Application of a spatial model to explore rotating harvest strategies for sedentary species. *In* Proceedings of the North Pacific Symposium on Invertebrate Stock Assessment and Management. *Edited by* G.S. Jamieson and A. Campbell. Can. Spec. Publ. Fish. Aquat. Sci. 125. pp. 359–365.

[1] Present address: Centro Marista de Estudios Superiores, Km. 7 Carretera Mérida-Progreso, Mérida, México.

Rationale behind rotating harvest schemes

Estimating the potential yield from harvesting a cohort before or after the age at which, on average, the species provides the maximum yield has predominantly depended on assuming that if ages are mixed in a population, it will be impossible to arrive at harvesting strategies that maximize returns of a particular age-group or groups. This is the basis for the standard yield-per-recruit calculation, which assumes that all age-groups available to the gear must be harvested simultaneously. Strategies could be developed via a system of rotating closures, which would favor the production of standing stocks of those age-groups that are important for markets or as a spawning stock.

A considerable amount of work has already been done on adapting fish stock management approaches to sedentary shellfish management, but in many cases, as for the dynamic pool assumption, basic assumptions are not appropriate. Particular year-classes may dominate a shellfish bed. A degree of mutual exclusion of year-classes may even be apparent (see Hancock 1973) and a "carrying capacity" defined in numbers or biomass is a reasonable starting hypothesis.

The main questions asked are: "Is rotational harvest feasible, and if so, does it provide a higher stock biomass than the best strategy possible with all areas open continuously?" Two additional questions follow necessarily: "What is the rotation period to achieve an optimum or maximum stock biomass and what is the period that yields the maximum net present value?" To answer the first two questions it was necessary to conduct a wide range of simulation experiments with different rotation periods for sets of M and K combinations reflecting life cycles and growth rates of different types of sedentary species.

Methods

To answer the above-mentioned questions the following steps were undertaken: (*i*) selection of a set of common values for relevant bioeconomic parameters such as number of recruits, standard deviation of patch distribution (expressed in %), maximum weight of species, parameter t_0 of growth equation, gear selectivity parameters, area occupied by the stock, area swept per day, number of fishing days per year, average cost per fishing day, and average price of target species, (*ii*) selection and application of an appropriate spatial dynamic model, (*iii*) execution of simulation experiments for sets of M and K combinations with and without rotation harvest strategy, (*iv*) determination of a bioeconomically optimum level of effort, (*v*) usage of that level of effort to undertake simulation experiments of alternative rotation periods, (*vi*) comparison of the theoretical results of biomass, yield, rent, and net present value of the fishery with different rotation periods, and (*vii*) discussion of the main results and identification of additional hypotheses for future theoretical and experimental research on rotating harvest schemes in marine fisheries.

Mathematical model

As pointed out by Caddy (1975), existing fisheries models employ the "unit stock" concept and make no explicit allowance for spatial distribution of biomass and effort per fishing ground. The usefulness of the unit-stock concept rests largely upon the "dynamic pool" assumption. However, this is invalid for sedentary species and difficult to apply when information about spatial distribution by statistical subunits is available, as for some sedentary demersal resources (Caddy 1989). More realistic spatial assumptions might include the following: (*i*) recruitment occurs in patches of random size and location with the constraint that local biomass can not exceed the virgin biomass for each unit area, and (*ii*) the fraction of effort allocated within each unit area of a fishing ground is either determined by available local biomass alone (proportional effort allocation) or in combination with traditional fishing practice.

Predictions of a model based on these assumptions differ significantly from those employing the unit-stock concept. Model behavior is in general as follows: (*i*) under proportional effort allocation, overall yield declines more sharply with increasing effort subsequent to maximum sustainable yield than under dynamic pool assumptions, (*ii*) peak mortality may occur earlier in life than predicted from the gear selection ogive if year-classes are independently distributed and there is no size-limit regulation, and (*iii*) variance in biomass per unit area decreases with age.

Because of the above, the model selected for this theoretical exercise is a new version of YAREA (Caddy 1975; Seijo et al. 1993), termed YAREA-MC, which incorporates the possibility of specifying in advance a rotating harvest scheme. It also allows for bioeconomic and Monte Carlo analyses of the rotating harvest strategy. Some of the main features of YAREA-MC are presented in the following sections.

Specification of maximum biomass and recruitment

To generate a spatial population structure, the user must specify the maximum estimated biomass ($Bmax_{ij}$) in every cell of the gridded area selected to determine species distribution and local maximum carrying capacity. The total number of recruits per year can be constant, time varying, or stochastic. For the purpose of this paper, model YAREA-MC was run under the constant recruitment option.

Spatial distribution of recruitment patches

The number of individuals to be recruited in each patch (NT) was estimated by multiplying the annual number of recruits by a uniformly distributed random number (0, 1). The model then proceeds to randomly select recruitment patch center coordinates and to calculate recruitment density away from patch center

$$[1] \quad R_{ij}(t) = \frac{NT}{\sqrt{2\pi\sigma^2}}\, e^{(-x^2/2\sigma^2)}$$

through the use of a bivariate normal distribution, as represented in eq. 1, where $R_{ij}(t)$ = is the number of recruits located in row i and column j of the species distribution grid area, NT = randomly selected number of recruits in each patch before biomass constraint, x = distance of unit area ij from patch center, and s = standard deviation. It should be pointed out that other distributions such as the negative binomial or gamma functions could be employed to generate patch density distribution.

Distance from patch center (x) is estimated by the following equation:

$$[2] \quad x = \sqrt{(TSG\,(y-i))^2 + (TTG\,(z-j))^2}$$

where TSG = length of cell, TTG = width of cell, $i = 1, 2, 3,..., SG$ and SG = rows of grid, $j = 1, 2, 3,..., TG$ and TG = columns of grid, $y = SG/2$, and $z = TG/2$.

It should be mentioned that spatial analysis of patch distribution is becoming increasingly relevant because many habitats are becoming fragmented in coastal areas due to human activities. Survival of species restricted to patch-type habitats ultimately requires that these species have contacts with individuals from other patches (Allen et al. 1990).

The biomass allocated in each cell ij is estimated by eq. 3 as follows:

$$[3] \quad B_{ij}(t) = \sum_{k=1}^{\text{mage}} N_{ijm}(t) \cdot W_m$$

where $N_{ijm}(t)$ is the number of individuals in cell ij in time t, and mage is the maximum age of the species.

Individual lengths are estimated for different ages by the von Bertalanffy growth function:

$$[4] \quad L_m = L_\infty \cdot (1 - e^{(-K\cdot(m - t_0))})$$

where L_m = length of species at age m, K = curvature parameter of growth function, L_∞ = maximum length of species, and t_0 = initial time parameter of growth equation.

Corresponding weights at different ages (W_m) are obtained by eq. 6 as follows:

$$[5] \quad W_m = W_\infty \cdot (1 - e^{(-K \cdot (m - t_0))})^3$$

where W_∞ is the observed maximum weight of target species.

Spatial catch per unit of effort

Once recruits have been assigned to cells up to the maximum biomass permitted, effort is spatially allocated in proportion to available biomass. A heterogeneous dynamic catch per unit of effort ($\mathrm{CPUE}_{ij}(t)$) in every fishing site ij is estimated in **YAREA** by summing over all ages the product of the age-specific catchability coefficient by the number times weight of individuals of different ages.

Total catch per unit of effort (CPUET(t)) is then estimated by summing the CPUE estimated for each cell of the gridded fishing area. Estimation of $\mathrm{CPUE}_{ij}(t)$ and CPUET(t) are expressed in eqs. 6 and 7, respectively:

$$[6] \quad \mathrm{CPUE}_{ij}(t) = \sum_m (q_m \cdot N_{ijm}(t) \cdot W_m)$$

$$[7] \quad \mathrm{CPUET}(t) = \sum_{ij} \mathrm{CPUE}_{ij}(t)$$

where q_m is the age-specific catchability coefficient per unit area.

Age-specific catchability coefficient per unit area

The definition of fishing mortality by Baranov (1918) from the effective area swept by gear of known selectivity over a uniformly distributed population of stationary organisms can be applied to heterogenous populations if the stock can be subdivided into unit areas within which distribution is effectively uniform (Caddy 1975). Given the above, an age-specific catchability coefficient (q_m) per unit area is estimated by eq. 9 as follows:

$$[8] \quad q_m = -(\ln(1 - (\frac{c \cdot a \cdot S_m}{\mathrm{Area}/[\mathrm{SG} \cdot \mathrm{TG}]})))$$

where c = probability of capture by the gear, AREA = area occupied by the stock, S_m = age-specific gear retention defined by a gear selection ogive, and a = area swept per day.

Age-specific gear retention (S_m) is estimated by the following equation (Sparre et al. 1989; Hilborn and Walters 1992):

$$[9] \quad S_m = \frac{1}{1 + e^{(S_1 - S_2^* L_k)}}$$

where parameters S_1 and S_2 are estimated by eqs. 10 and 11, respectively:

$$[10] \quad S_1 = L_{50\%} \cdot (\frac{\ln 3}{L_{75\%} - L_{50\%}})$$

$$[11] \quad S_2 = \frac{S_1}{L_{50\%}}$$

where L_{50} = length at 50% gear retention and L_{75} = length at 75% gear retention.

Spatial allocation of fishing intensity

As pointed out by Caddy (1975), "fishermen are continually reevaluating density gradients, and if perfect information on stock distribution were available at any one time, effort would tend to be exclusively expended on areas of above average density." It should be mentioned however, that distance to those areas and the corresponding transfer costs may also determine the allocation of fishing intensity. These last considerations are incorporated in **ALLOC** and **CHART** models of software package SPATIAL (Seijo et al. 1993).

In **YAREA**, the unit fishing area (grid ij) receives a constant effort annually by a single gear type. Effort in each unit area is made proportional to local catch per unit of effort at the start of the year. For a specified total effort, fishing intensity in site ij over time, ($f_{ij}(t)$), is calculated by the following equation:

$$[12] \quad f_{ij}(t) = \frac{\mathrm{CPUE}_{ij}(t)}{\mathrm{CPUET}(t)} \cdot \mathrm{FT}$$

where FT is total effort per year.

Note, however, that if fishing effort is concentrated on patches, the vector $\mathrm{FR}_{ijm}(t)$ is a function not only of gear selectivity but also of the spatial allocation of fishing intensity. This dynamic and spatial age-specific exploitation rate constitutes an output of model **YAREA**.

Fishery yield obtained from different sites

To estimate fishery yield from site ij in time t ($\mathrm{YIELD}_{ij}(t)$):

$$[13] \quad \mathrm{YIELD}_{ij}(t) = \sum_m \mathrm{FR}_{ijm}(t) \cdot B_{ijm}(t)$$

where $\mathrm{FR}_{ijm}(t)$ is the site and age-specific exploitation rate.

Spatially distributed population structure

Given a set of initial conditions for the number of individuals of different ages in each site ($N_{ijm}(0)$), survivors at age per unit of area over time are given by Seijo and Defeo (1994):

$$[14] \quad N_{ijm}(t + \mathrm{DT}) = N_{ijm}(t) + \int_t^{t+\mathrm{DT}} (S_{ijm-1}(\tau) \cdot N_{ijm-1}(\tau) - N_{ijm}(\tau))\, d\tau$$

where $S_{ijm}(t)$ is the survival rate of individuals of age m located in site ij. The latter is estimated as one minus the total mortality rate (the rate of exploitation, $\mathrm{FR}_{ijm}(t)$, plus the finite natural mortality rate (MR), sensu Gulland (1983)).

Economic rent

The present value of the economic rent (PVπ) obtained with rotating harvest schemes is estimated through the following equation:

$$[15] \quad \mathrm{PV}\,\pi = \int_0^T ((Y(t) \cdot p - \mathrm{FT} \cdot \mathrm{cu}) \cdot e^{-\delta t})\, dt$$

where $Y(t) = \sum_{ij} \mathrm{Yield}_{ij}(t)$ and p is the average price of species, cu is the unit cost per fishing day and δ is the selected rate of discount.

Model assumptions

The model assumptions are as follows: (i) The model assumes successive annual time periods, in which recruitment is

Table 1. Bio-economic parameters used in the rotating harvest simulation experiment.

Parameter	Description	Unit of measurement
$R = 1 \times 10^9$	Annual recruitment	Individuals·year^{-1}
$W_\infty = 400$	Maximum weight	g
$L_\infty = 300$	Maximum length	mm
$L_{50} = 200$	Length at 50% retention	mm
$L_{75} = 250$	Length at 75% retention	mm
$t_0 = 0$	t_0, v.B. growth equation	Years
SG = 14	Rows of grid area	—
TG = 14	Columns of grid area	—
$c = 0.9$	Probability of capture	%
AREA = 20 000	Area occupied by stock	km^2
$a = 0.5$	Area swept per day	km^2
$p = 3000$	Price of target species	\$·tonne^{-1}
cu = 1000	Unit cost of effort	\$·day^{-1}
$\text{Bmax}_{ij} = 5000$	Maximum local biomass	Tonne·cell^{-1}

instantaneous and occurs at the beginning of each year. (*ii*) The degree of density and extension of recruit patches is defined by implication from the input variance of the bivariate normal distribution used to generate them. (*iii*) There is no location or distance of each unit area implied or calculated with respect to one or more ports of origin. (*iv*) The number of vessels is implied in the model only through the input of total annual fishing effort. (*v*) Gear characteristics of the swept area type are assumed constant in the model. (*vi*) Migration, dispersal, and movement are not considered in the model, as its application is limited to sedentary resources. (*vii*) Natural mortality is assumed constant for all ages. Fishing mortality is made a function of catchability, selectivity at age, and fishing effort.

Bio-economic parameters used in the simulation experiment

Once the model was initialized, parameters used in the application of the above model are given in Table 1.

Design of simulation experiments

With each combination of curvature parameter of growth equation K (0.1, 0.3, and 0.6) and natural mortality coefficient M (0.1, 0.2, and 0.3), the model generated appropriate spatially distributed virgin stocks. Longevity was specified in accordance to the expression $\lambda = t_0 + 3/K$ (Pauly 1991). Then, the generated spatially distributed population structure for each species group (identified with a single K and M combination) was simulated over 24 years, the maximum number of years needed to undertake the largest rotation period (8 years) over the 2 areas in which each stock was divided. Performance variables specified above (biomass, yield, and resource rent) were numerically estimated over space and time for each species group under four alternative fishery management regimes: continuously open areas, and 2, 4, and 6 years of fishery harvest rotation. Level of effort used for each combination of M and K was that which generated the optimal bio-economic level of effort for an open access scenario. The latter, i.e., effort at maximum economic yield, was estimated with the spatial model YAREA-MC.

Results

To illustrate model behavior, the heterogeneous spatial distribution of biomass for a species with $M = 0.3$ and $K = 0.3$ is presented in Fig. 1.

Optimum level of effort and corresponding fishery performance variables for different groups of species characterized by the K and M parameters are given in Table 2.

This optimum effort value was then used in rotating harvest simulations, with results presented in Table 3.

For each combination of parameters K and M, an optimum rotation period was identified in terms of the maximum biomass achieved at the end of the 24-year period. For long-lived invertebrate species ($M = 0.1$) with three growth velocities, optimum biomass was achieved with a rotation period of 6 years. For fast-growing, long-lived species (K of 0.6 and M of 0.1), optimum rotation period was 4 years. Short-lived, fast-growing species had an optimum rotation period of 2 years (Figs. 2 and 3).

Even though there was a substantial improvement in resource biomass with the optimum rotating harvest period identified for each species group (e.g., 29.7 and 38.8% with $K = 0.1$ and $M = 0.1$, and $K = 0.6$ and $M = 0.1$, respectively), yield and net present value were reduced when the rotating harvest period was increased in most K and M parameter combinations. The usual trade-off between biomass and rent remained.

Discussion

Rotating harvest models for even-aged stands have a long history in forestry, dating to as early as 1849 where a simple bio-economic modeling approach, the Faustmann Formula (cited in Clark 1976), was applied. This formulation, postulated by Faustmann, a German forester, assumed a single cohort of trees in his analysis. Although a formulation for stands of mixed ages presents more difficulties and divergences, it is essentially the problem discussed in this paper. Differences with the forestry example relate both to different time scales and, most explicitly, to consideration in this paper of non-uniform spatial distributions for both recruitment and effort.

Comparison of yield-per-recruit calculations using dynamic pool assumptions and a typical trawl selection ogive shows a higher yield per recruit than obtained with a model such as YAREA that assumes aggregation of age-groups into clumps. These concentrations are gradually fished down towards a uniform-age distribution by vessels as fishers seek to maximize their yield by concentrating on higher densities which are inevitably dominated by younger age-groups (Caddy 1975; Seijo et al. 1993).

Another model (Caddy 1993) predicts the most favorable integer rotation time in years by assuming that this is roughly similar to the time elapsed in cohort growth to the year of peak biomass of a single cohort subject to only natural deaths and harvested simultaneously T years after recruitment at mean weight w, as given by:

$$[16] \quad B(T) = N(t) \cdot \bar{w}(T)\, e^{(-M \cdot T)}$$

Corresponding rotation periods were calculated with this simpler model for the same pairs of values of M and K used in this paper in Table 2. Comparison was also made with the time to maximum rate of cohort biomass increase, $\Delta B(t) = B(t + 1) - B(t)$,

Fig. 1. Spatial distribution of biomass (*t*) generated by model YAREA-MC.

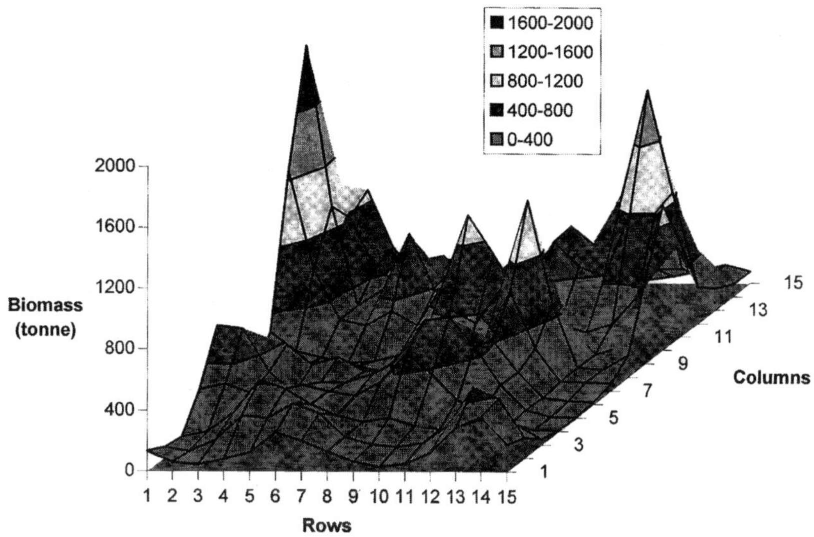

Table 2. Bio-economically optimal effort levels with open areas.

Optimum effort (days·year^{-1})	K	M	Longevity (years)	Biomass (10^{-3} t)	Yield (10^{-3} t)	PVπ (10^{-6} US$)
16 000	0.1	0.1	30	249	34	2324
12 000	0.1	0.2	30	124	11	405
5 000	0.1	0.3	30	73	3	484
18 000	0.3	0.1	10	445	97	4535
20 000	0.3	0.2	10	276	61	2593
26 000	0.3	0.3	10	171	40	1532
28 000	0.6	0.1	5	430	148	5995
30 000	0.6	0.2	5	336	117	4554
36 000	0.6	0.3	5	263	94	3542

Table 3. Fishery performance with biologically optimum rotation periods.

Optimum effort (days·year^{-1})	K	M	Rotation period (years)	Biomass (10^{-3} t)	Yield (10^{-3} t)	PVπ (10^{-6} US$)
16 000	0.1	0.1	6	323	16.7	1966
12 000	0.1	0.2	6	137	7.1	376
5 000	0.1	0.3	6	77	2	384
18 000	0.3	0.1	6	649	57	3223
20 000	0.3	0.2	6	385	31	1876
26 000	0.3	0.3	6	232	22	1054
28 000	0.6	0.1	4	597	77	4035
30 000	0.6	0.2	2	418	76	3388
36 000	0.6	0.3	2	363	49	2355

after which it was presumed with the single cohort model that advantages of longer rotation periods would fall off rapidly (Caddy 1993).

Comparison of these results suggest that YAREA-MC rotation results lie intermediate between those calculated from the simple biomass–recruit model for peak rate of increase and that for peak biomass. This suggests that rotation models which do not to take into account interactions between patchiness and fishing strategy probably overestimate optimal rotation period in a mixed cohort fishery because fishing in a patchy mixed age-group fishery will be directed towards partially recruited younger age-groups. Surprisingly perhaps, short rotation periods of 2–4 years look promising for a wide range of living marine resources and offer practical advantages. Another argument in favor of relatively short rotation periods is that if the total number of zones is a multiple of the rotation period, with more than one zone open in a given year, local communities could each have their own open zone geographically close by. Some problems and advantages of rotating harvest models were discussed for red sea urchin fisheries by Botsford et al. (1993), noting that the State of Washington uses a 3-year rotating harvest system. They suggested a marked increase in

Fig. 2. Biomass over time of a species characterized by parameters $K = 0.1$ and $M = 0.1$ subject to fishing effort of 16 000 days·yr^{-1}, with and without a rotating harvest scheme.

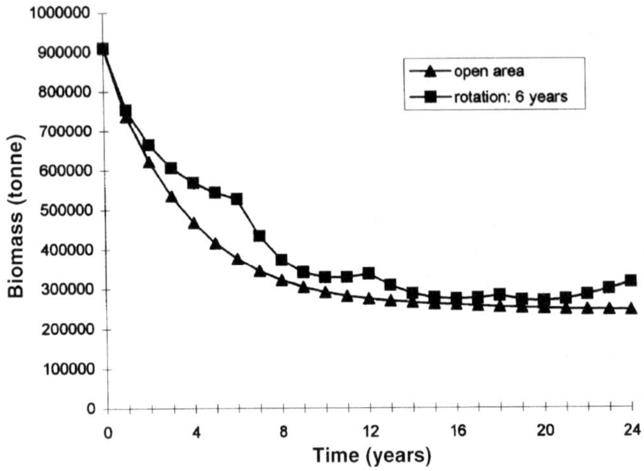

Fig. 3. Biomass over time of a species characterized by parameters $K = 0.6$ and $M = 0.1$ subject to fishing effort of 28 000 days·yr^{-1}, with and without a rotating harvest scheme.

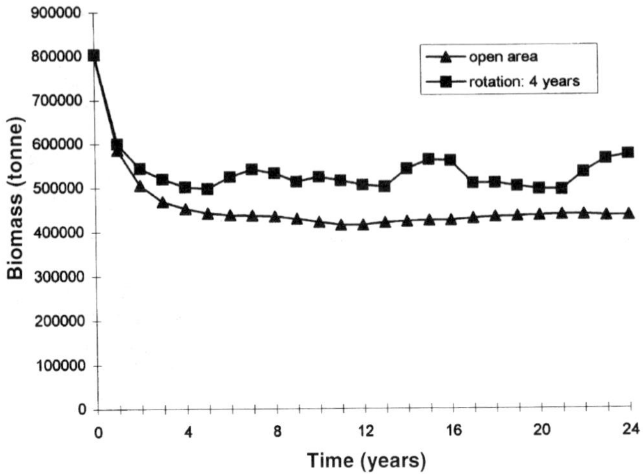

Table 4. Identification of key questions of existing management context for potential rotating harvest schemes.

	Key questions of management context	Potential for rotating harvest management schemes
1	Do de facto exclusive harvesting rights exist?	If not, rotating harvest schemes are difficult to enforce.
2	Is preventing poaching in closed areas/seasons feasible, cost effective, and supported by fishermen?	If not, rotating harvest schemes are unfeasible.
3	Is there a management authority with the authority to allocate fishing rights by area to individual participants?	If not, rotating harvest schemes are unfeasible.
4	Are there a discrete number of population subunits for the resource?	If not, rotating harvest schemes are unfeasible.
5	Can the stock be separated into subunits of comparable size, between which migration is limited?	If not, rotating harvest schemes are unfeasible.
6	Are the number of subunits equal to or greater than a calculated optimum period of harvest rotation?	If not, a suboptimal rotating harvest scheme may still be feasible and desirable.
7	Are there alternative means of employment for local fishermen and/or processors if a local resource area is closed for a number of years?	If not, rotating harvest schemes problematical.
8	Do fishermen have access in each year of the scheme to other stocks?	If not, rotating harvest schemes problematical.
9	Is the method of harvesting selective for the species and sizes most desired?	If not, rotating harvest schemes problematical.

egg production but slight progressive decline in yield with increasing rotation period.

A likely advantage of rotating harvests is that there should be a higher density of mature survivors in unit areas just prior to their harvest than with continuous harvesting and hence greater probability of successful population replacement. To assess fully the potential bio-economic effect of a rotation period would require the input of size-specific prices to the model. Here, however, we were only concerned with maximizing biomass tonnage and the present value of net revenue given uniform unit prices. If older age-classes have a higher unit value, and in a number of species this is evidently the case, then an economically optimal rotating period may exist.

Calculations suggest that dividing grounds into open and closed areas and fishing them in alternate years might be a worthwhile strategy for species with life spans of two years or more but probably not for annual species like some shrimp and squid. Rotation periods of less than one year are unlikely to be practical. In this latter case, closed seasons could include the prerecruitment period. Closed areas protecting animals early in their life history effectively act as size limits and can significantly increase yield (Garcia 1984; Garcia and Demetropoulos 1986).

Multiple fishing zones may also help reduce conflictive problems between adjacent fleets if exclusive allocations were possible. Implementing such an experimental management approach would require (and might promote) community participation, with local involvement likely essential for success.

Co-operation and participation of local fishermen would be required for self-policing in situations where fishery control and surveillance are poor. In Table 4, key questions are identified as being potentially relevant when considering the possibility of establishing a rotating harvest scheme. Questions are institutional (1 and 3), biological and ecological (4, 5, and 6), economic (2, 7, and 8), and technological (9) in nature.

Finally, although this management tool has been applied here to sedentary resources (e.g., bivalves, sea urchins, holothuria, abalone, etc.), its application to other territorially restricted resources (e.g., reef fish which tend to remain territorially resident on and around geographically discrete reef systems) and even motile but not migratory resources also seems relevant. At least in principle, this management tool seems particularly appropriate for coastal environments where relatively narrow shelves and high coastal productivity can be easily subdivided.

Acknowledgments

We wish to express our gratitude to Oscar Novelo who assisted in the design of the interactive version of YAREA-MC. We thank Tiburcio Castro for assisting in the process of running the large number simulation experiments required for this study and Eduardo Perez for his useful comments and suggestions for the improvement of the original version of YAREA. We also wish to express our gratitude to Glen Jamieson, Paul A. Breen, and an anonymous referee for their useful comments to improve the manuscript.

References

Allen, L.J.S., Moulton, M.P., and Rose, F.L. 1990. Persistence in an age-structured population for a patch type environment. Nat. Res. Mod. **4**: 197–214.

Baranov, T.I. 1918. The question of biological basis of fisheries. Proc. Inst. Ichthyol. Invest. **1**: 81–128.

Botsford, L.W., Quinn, J.F., Wing, S.R., and Brittnacher, J.G. 1993. Rotating spatial harvest of a benthic invertebrate, the red sea urchin *Strongylocentrotus franciscanus*. Proc. Int. Symp. on Management Strategies for Exploited Fish Populations, Alaska Sea Grant College Program, AK-SG-93-02. pp. 409–428.

Caddy, J.F. 1975. Spatial models for an exploited shellfish population, and its application to Georges Bank scallop fishery. J. Fish. Res. Board Can. **32**: 1305–1328.

Caddy, J.F. 1989. Recent developments in research and management for wild stocks of bivalves and gastropods. *In* Marine invertebrate fisheries: Their assessment and management. *Edited by* J.F. Caddy. John Wiley, Interscience, New York, N.Y. pp. 665–700.

Caddy, J.F. 1993. Background concepts for a rotating harvest strategy with particular reference to the Mediterranean red coral resource. Mar. Fish. Rev. **55**: 10–18.

Clark, C.W. 1976. Mathematical bio-economics. John Wiley & Sons, Inc., New York, N.Y. 352 p.

Dandberg, P., Oen, R., Danielssen, D.S., and Moksness, E. 1993. Economic consequences of large-scale sea ranching of cod in Norway. Proc. Int. Symp. on Sea Ranching of Cod and Other Marine Species. Arendal, Norway. 61 p.

Garcia, S.M. 1984. Modelisation and exploitation rationelle des stocks de corail precieux: une premiere approache. FAO Informe de Pêche, **306**. 142 p.

Garcia, S.M., and Demetropoulos, A. 1986. Management of Cyprus fisheries. FAO Fish. Tech. Pap. **250**. 43 p.

Gulland, J.A. 1983. Fish stock assessment: A manual of basic methods. John Wiley & Sons, Inc., New York, N.Y. 223 p.

Hancock, D.A. 1973. The relationship between stock and recruitment in exploited invertebrates. Rapp. P.-V. Reun. Cons. Perm. Explor. Mer, **164**: 113–131.

Hilborn, R., and Walters, C.J. 1992. Quantitative fisheries stock assessment: choice, dynamics and uncertainty. Chapman and Hall, New York, N.Y. 570 p.

Pauly, D. 1991. Growth performance in fishes: rigorous description of patterns as a basis for understanding causal mechanisms. Aquabyte, **4**: 3–6.

Seijo, J.C., and Defeo, O. 1994. Dynamics of resource and fishermen behavior in coastal invertebrate fisheries. IFREMER, Proc. VIth IIFET Conf. **2**: 209–222.

Seijo, J.C., Caddy, J.F., and Euan, J. 1993. SPATIAL: space-time dynamics in marine fisheries, a bio-economic software package for sedentary species. FAO Comp. Info. Ser. (Fish.) **6**. 116 p.

Sluczanowski, P.R. 1983. A management oriented model of an abalone fishery whose sub-stocks are subject to pulse fishing. Can. J. Fish. Aquat. Sci. **41**: 1008–1014.

Sparre P., Ursini, E., and Venema, S. 1989. Introduction of tropical fish stock assessment. Part 1. FAO Fish. Tech. Pap. **306/1**. 337 p.

Territorial user's rights and the Australian abalone (*Haliotis* sp.) fishery

Jeremy Prince, Carl Walters, Rebecca Ruiz-Avila, and Philip Sluczanowski

Abstract: On the basis of the Australian experience with the research and management of abalone (*Haliotis* sp.) stocks and our wider experience of fisheries research and management, we argue that the spatial scale of an exploited species should be an important determinant in developing management strategies for any exploited species. The small spatial scale of functional units of abalone stock together with their high level of variability between populations and the "law of the commons" context of the Australian abalone fishery combine to undermine the effectiveness of modern broad-acre management tools such as size limits, closures, limited entry, and individual transferable quotas (ITQs). Despite the sophistication of current management regimes, component units of stock can still be sequentially overexploited because the spatial scale of functional units of stock within the fishery is smaller than the effective scale of management. The "tragedy of the commons" and a "tyranny of scale" renders the existing sophisticated system of management suboptimal for this valuable renewable resource. We suggest that management through territorial user rights would enable individual fishers or small communities of fishers to adjust the scale of management to the small scale appropriate to the species. The difficulty of changing management in this way is discussed.

Résumé : À partir de l'expérience australienne en recherche et en gestion des stocks d'ormeau (*Haliotis* sp.) et de notre expérience plus large de la recherche et de la gestion des pêches, nous posons que l'échelle spatiale d'une espèce exploitée doit être un élément déterminant dans l'élaboration de stratégies de gestion pour une espèce exploitée. La petite échelle spatiale des unités fonctionnelles des stocks d'ormeau et leur niveau élevé de variabilité entre populations ainsi que le contexte de « propriété commune » de la pêche de ce mollusque en Australie se combinent pour miner l'efficacité des outils modernes de gestion à grande échelle comme les limites de taille, les fermetures, l'accès limité et les quotas individuels transférables (QIT). Malgré le raffinement des régimes actuels de gestion, certaines composantes des stocks peuvent se trouver encore surexploitées de façon suivie car l'échelle spatiale des unités fonctionnelles de stock dans la pêcherie est plus petite que l'échelle effective de gestion. La « tragédie de la propriété commune » et une « tyrannie de l'échelle » rendent suboptimal le système actuel complexe de gestion de cette précieuse ressource renouvelable. Nous suggérons d'avoir recours à des droits territoriaux d'exploitation, régime de gestion qui permettrait à des pêcheurs, ou à de petites communautés de pêcheurs, d'ajuster leur intervention à la petite échelle convenant à cette espèce. Nous analysons les difficultés suscitées par un tel changement dans la gestion.
[Traduit par la Rédaction]

Introduction

During the twentieth century fisheries science and management developed rapidly. Before the turn of the century, it was argued that humans would never be able to impact the great fish stocks (Safina 1995). However, by early this century, Russell (1931) was using yield per recruit analysis to theoretically consider the overfishing problem of the North Sea and to argue in support of introducing minimum mesh sizes to the North Sea trawl fishery. Assuming constant recruitment to a uniformly growing stock, yield per recruit analyses estimate the optimal size of capture and by the middle of the century legal minimum size limits were commonly applied to fisheries to maximize yields (Cushing 1968; Hancock 1979). During the second half of the century the technologies developed for war (e.g., radar, acoustics, hydraulics, global position systems) were applied to harvesting marine resources and the impact of fishing on recruitment became evident. Ricker (1954) focused attention on the relationship between the abundance of breeding stock and the level of future recruitment to a fishery.

Hardin (1968) described the "tragedy of the commons" by which "each man is locked into a system that compels him to increase his herd without limit in a world that is limited." He noted that "freedom in a commons brings ruin to all" and described its application to the human use of a wide range of renewable resources, including fisheries. During the 1960's and 1970's managers increasingly moved to control fishing power by limiting entry to fisheries and the amount of fishing gear used by each entrant (Hancock 1979). Limited entry denied the general community the right to catch fish for commercial gain, reserving this right for a limited number of licenced commercial fishers, thus, theoretically at least, solving the tragedy of the commons for fisheries.

J. Prince and R. Ruiz-Avila. Biospherics Pty Ltd, P.O. Box 168, South Fremantle, Western Australia 6162.

C. Walters. Fisheries Centre, University of British Columbia, Vancouver, BC V6T 1Z4, Canada.

P. Sluczanowski. Formerly with South Australian Research and Development Institute, P.O. Box 1625, Adelaide, South Australia 5000.

Correct citation: Prince, J., Walters, C., Ruiz-Avila, R., and Sluczanowski, P. 1998. Territorial user's rights and the australian abalone (*Haliotis* sp.) fishery. *In* Proceedings of the North Pacific Symposium on Invertebrate Stock Assessment and Management. *Edited by* G.S. Jamieson and A. Campbell. Can. Spec. Publ. Fish. Aquat. Sci. 125. pp. 367–375.

Fig. 1. The five southern Australian states with their abalone management zones, and the geographic range of the three commercially exploited abalone species (from Prince and Shepherd 1992).

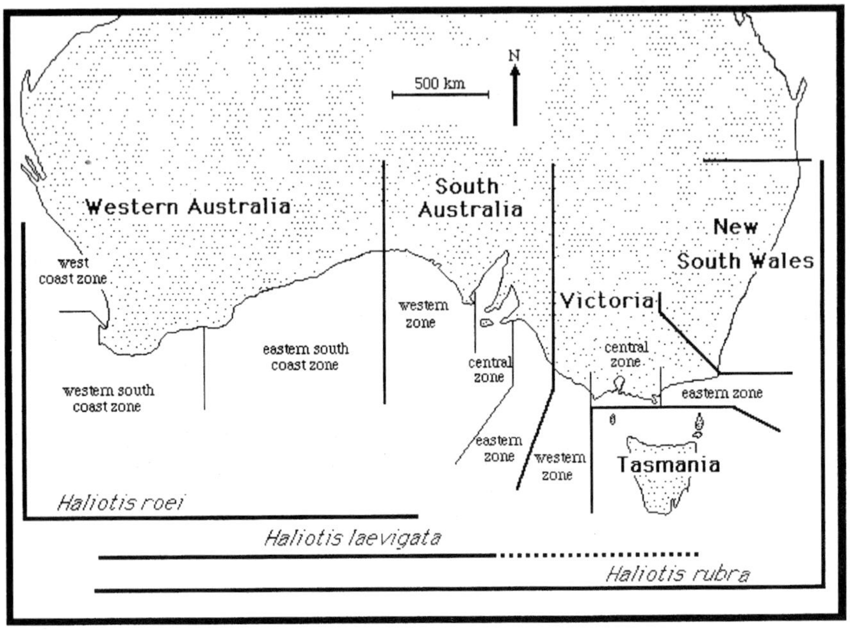

By the 1980's fisheries managers were commonly controlling the size of fish caught, the number of participants, the number and size of vessels, the amount and specifications of fishing gear, and the areas and times fished (Bourne 1986; Harrison 1986). However, a seemingly inevitable upward creep of fishing pressure within limited-entry fisheries (Morgan 1980; Harrison 1986; Robins and Sachse 1994) was widely recognized prompting a move towards directly controlling total landings with catch quotas, often individually transferable quotas (ITQs). In some fisheries ITQs have been applied over existing layers of management so that the size of fish caught, the application of fishing effort, and the quantity of catch can all be controlled by legislation (e.g., Prince and Shepherd 1992).

In this paper, it is argued that for some fished stocks even this sophisticated level of management is inadequate for optimizing the management. When the spatial scale of management and monitoring is larger than the scale of the managed populations, fisheries will remain vulnerable to localized overfishing and ongoing population collapses. This argument is supported by a case study of the Australian experience with the research and management of abalone stocks. However, our wider fisheries experience suggests that the issue is relevant to many other fisheries, particularly sedentary invertebrate species and dive fisheries.

Abalone

Abalone are large marine molluscs that form aggregations on shallow inshore reefs (Shepherd 1986; Shepherd and Godoy 1989) where they are easily accessible to collection by commercial and amateur divers. The Japanese and Chinese consider abalone a delicacy and the market price is set in Asia (Rudd 1994). Worldwide abalone stocks form the lucrative basis of high value – low volume dive or littoral collection fisheries.

World production of abalone peaked in the late 1960's at approximately 27 600 $t{\cdot}yr^{-1}$. However, since that peak, catches from the large Californian and Mexican abalone fisheries collapsed (Tegner 1989; Guzmán del Próo 1992), Japanese catches continued to slowly decline (FAO 1973–1988), and quota reductions occurred in Australia (Prince and Shepherd 1992). By the late 1980's global production declined to around 12 000 $t{\cdot}yr^{-1}$. The real value of abalone increased steadily during the last three decades reflecting declining supply and growing demand in Asia (Rudd 1994). Australia's annual production of about 5000–6000 $t{\cdot}yr^{-1}$ and Japan's of around 3000–4000 $t{\cdot}yr^{-1}$ currently dominate global production figures (FAO 1973–1988).

Australian abalone management

Australia's recorded abalone exports are annually worth approximately A$120 million and by value abalone is about 8% of Australia's fish production (Kailola et al. 1993). Three species of abalone are harvested by commercial divers using surface supplied compressed air diving equipment across the five southern states of Australia (Fig. 1). Each state within the Commonwealth of Australia has jurisdiction for the abalone resource within its waters but despite the different jurisdictions management regimes in each state have evolved along similar lines. For a detailed account of each fishery the reader is referred to Prince and Shepherd (1992).

In Australia, the modern fishery began in the early 1960's when compressed air diving equipment became commercially available. By the mid-1960's most states had imposed minimum size limits around the size of first maturity. During the late 1960's catches increased rapidly and a catch in excess of 8000 t was reported in 1968. Most states moved to limit entry to their abalone fisheries during the late 1960's and early

Fig. 2. Maps illustrating the three spatial scales used to map *H. laevigata* in southwest Western Australia. (*a*) Zone 2 of the Western Australian abalone fishery showing the grid used to record commercial catch data. (*b*) The Augusta area showing the grid used to map the productivity of abalone beds in the area. (*c*) A 20-m diameter survey site showing an indicative distribution of *H. laevigata*.

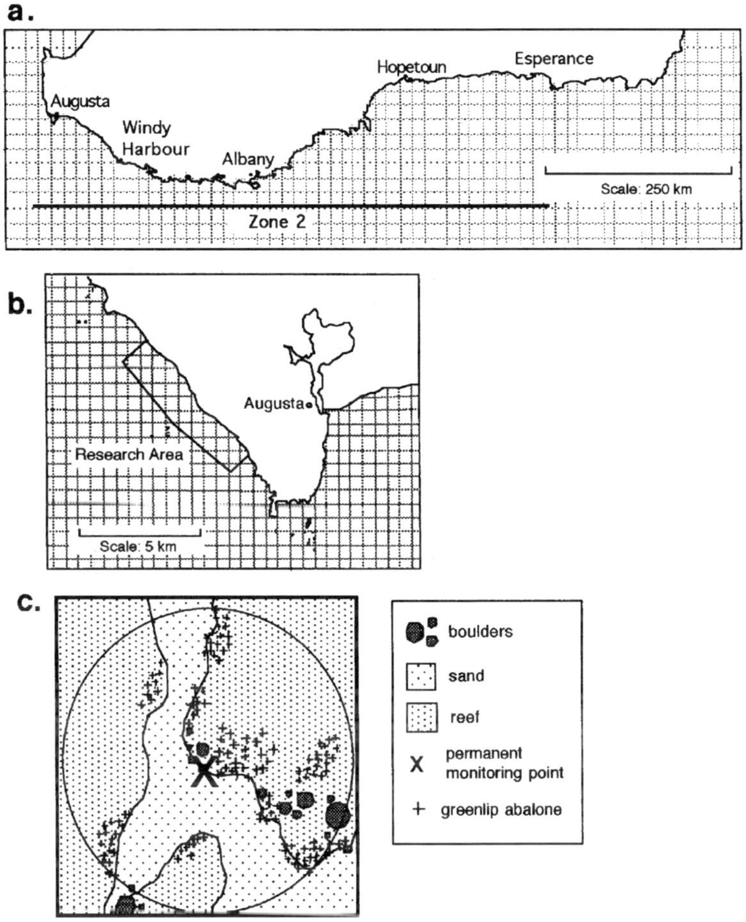

1970's, capping the number of commercial abalone divers at approximately 345 licenced abalone divers across Australia.

At first abalone diving licences were nontransferable; retiring divers relinquished their licences and they were reallocated according to government-maintained waiting lists. However, few divers retired and reported annual catches slowly declined to around 5000 t by 1975. In 1974 the state authorities began to allow the sale of diving licences (Harrison 1986). Abalone divers were allowed to retire from the fishery by nominating a replacement diver and transferring their licence to them. The introduction of licence transferability promoted a further and rapid expansion of landings as new entrants tended to fish harder than the divers they replaced. Between 1975 and 1985 recorded annual catches steadily increased to a peak of approximately 8200 t.

Once again concern at rising catches prompted the authorities to act, introducing individually transferable quota systems during the mid-1980's. After the implementation of ITQs the total allowable commercial catches (TACCs) were generally reduced with industry support. In Tasmania, against the advice of the state fisheries agency, the industry lobbied the government for an overall TACC reduction of 40%. Australia-wide TACCs were around 6000 t by the early 1990's. The TACCs are nominally reviewed annually and most states administer them within 2–3 separate zones each encompassing 100–1000 km of coastline.

Most states also sanction the occasional fishing of "stunted stocks" with otherwise sublegal minimum size limits. This special fishing of stunted stocks is sanctioned on specific days within prescribed areas.

Current understanding of abalone fisheries biology

During the late 1970's and early 1980's as abalone catches rose there was widespread concern about the sustainability of the abalone industry in Australia. The Commonwealth Fishing Industry Research Trust Account responded by supporting parallel research programs into abalone fisheries biology in South Australia, New South Wales, Victoria, and Tasmania. Together these programs of research concluded that a number of factors make abalone stocks difficult to assess and susceptible to overfishing (Prince and Shepherd 1992).

Highly aggregated populations

Abalone are not dispersed through the environments they inhabit; rather they occur in highly concentrated "nuggets" of stock. Three spatial scales used for mapping abalone populations (*Haliotis laevigata*) in southwest Western Australia are

Fig. 3. The spatial distribution of abalone at three scales within Zone 2 of the Western Australian abalone fishery. (*a*) Commercial catch data from the entire Zone 2, scale of aggregation: 25 km. (*b*) Productivity in the Augusta area as reported by commercial divers, scale of aggregation: 0.5 km. (*c*) Distribution of abalone within a single 20-m diameter survey site as measured by scientific divers, scale of aggregation: 0.5 m.

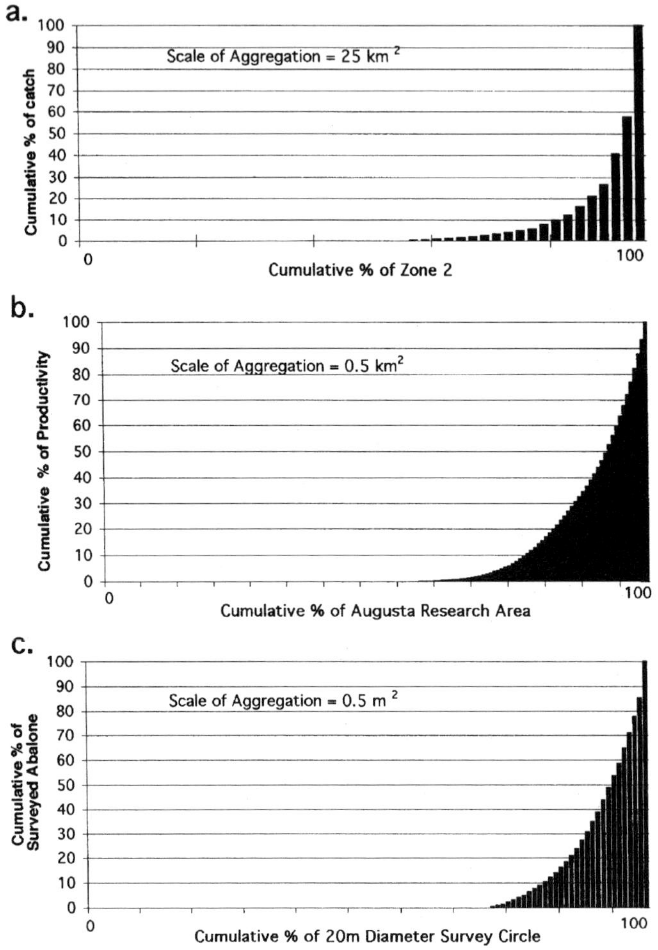

illustrated in Fig. 2: (*i*) the commercial catch statistics aggregated over 10–50 km, (*ii*) the productivity of abalone beds mapped by commercial divers within a research area 8 × 1.5 km, and (*iii*) the location of individual abalone mapped within surveyed circles of 10 m radii.

At all three spatial scales 80% of the abalone are concentrated within 20% of the potential habitat (Fig. 3). In Fig. 4, a similar distribution pattern is shown for *H. rubra* on George III Rock in southeast Tasmania determined using Leslie estimates of 26 cells of equal area (Prince 1989*a*). Approximately 70% of the population was found to be concentrated in 20% of the area. This repeated density profile over a range of spatial scales suggests that abalone distribution patterns are fractal (Sugihara and May 1990).

At the scale of 0- to 10-m, populations of abalone actively aggregate around fixed positions which are favourable for feeding or breeding (Shepherd 1986; Prince 1989*a*). At a scale of 100's to 1000's of metres, these aggregations are clumped within reef complexes to form metapopulations (Shepherd and Brown 1993) that abalone divers call "abalone beds." In turn metapopulations or abalone beds will be common along sections of coastline 50- to 100-km long where rocky substrate and algal growth support them, but nonexistent or limited on adjacent sections of coastline lacking these features.

Visually searching divers learn the locations of the abalone beds and the small component aggregations within them (Prince 1989*a*). In a developed abalone fishery, divers spend little time randomly searching for abalone. Divers check the condition of known abalone beds by looking at the state of its best aggregations before deciding whether or not to dive in that location. One consequence of this is that catches tend to remain proportional to time spent diving, rendering catch rate a poor indicator of stock abundance (Prince 1989*a*).

Restricted movement and dispersal

Larval and adult movements are generally limited to scales of 10's to 100's of metres (Prince et al. 1987, 1988; McShane et al. 1988; Brown 1991; Shepherd and Brown 1993). Some level of interaction is probable between the aggregations within a metapopulation (Shepherd and Brown 1993) but the interaction through adult movement and larval drift may be low or unidirectional. It is unlikely that interactions occur between metapopulations on different reef complexes. Consequently abalone fisheries are made up of many 100's to 1000's

Fig. 4(*a*). A map of George III Rock in southeast Tasmania showing the 29 blocks of equal size for which Leslie estimates of the abalone populations size were derived (from Prince 1989*a*).

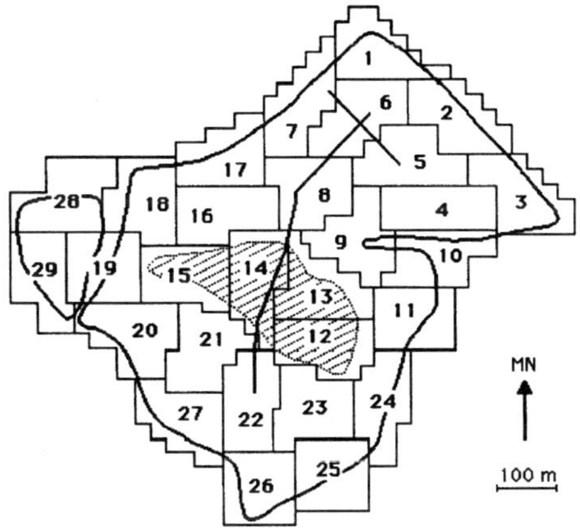

Fig. 4(*b*). Leslie estimates of abalone population size in 29 blocks of equal area across George III Rock in southeast Tasmania (from Prince 1989*a*).

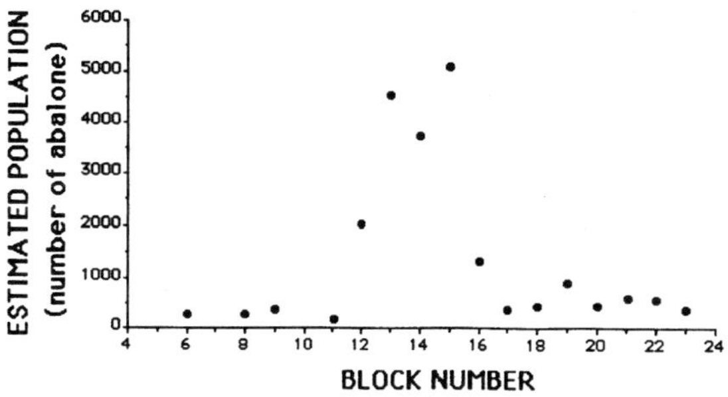

of discrete functional units of stock (Gulland 1969). Serial localized depletions of abalone beds occurring over an extended time frame can apparently lead to the decline of large abalone fisheries as may have occurred in Mexico (Prince and Guzmán del Próo 1993).

Variable patterns of growth and fecundity

Patterns of growth and fecundity vary greatly between and within metapopulations (Shepherd and Laws 1974; Prince 1989*a*; McShane 1991). Maturity is determined principally by age, rather than size (Nash 1992). Abalone of the same species commence breeding at around the same age over broad regions but their size at maturity varies greatly. At maturity cryptic juvenile abalone emerge from under boulders and take up exposed positions and become vulnerable to fishing (Prince 1989*a*). In productive fast growing areas, abalone mature and emerge at larger sizes than in slower growing areas. When legal minimum sizes are applied across an abalone fishery, the breeding stock in slow-growing areas is given a higher level of protection than in fast-growing productive areas (McShane 1991). Moreover fishing pressure tends to concentrate on the faster-growing areas because that is often where legal-sized abalone are most easily found. While protecting breeding stock in many areas a high minimum size limit leads to local depletions in the fastest growing and most productive areas of the fishery (Sluczanowski 1984; Hilborn and Walters 1987).

The uncertain status of Australia's abalone fisheries

Because state fisheries authorities in Australia do not allocate sufficient resources to the monitoring of abalone populations the current status of the Australian abalone fisheries cannot be determined with any quantitative rigour (Prince 1989*a*; Prince and Shepherd 1992). Rather, Australian managers rely qualitatively on regional trends in catch and catch rates to annually set TACCs of abalone (e.g., Keesing and Baker 1998). The stability of TACCs in most states and the fact that quotas are filled each year is used to argue that the introduction of ITQs and the subsequent industry initiated catch reductions of the mid-1980's have stabilized stocks in most states. However, definable threats to the long-term sustainability of the resource exist and their impact remains unmonitored.

The normal legal fishing pattern of licensed commercial harvesters can lead to the serial depletion of individual reefs even if TACCs are set at conservative levels for the fishery. This is because many divers may catch their individual quota allocations from the same abalone bed giving rise to a "tragedy of the commons" situation. Fishing pressure naturally concentrates in certain areas, e.g., abalone beds close to access points and home ports where the costs of fishing are lowest and on shallower reefs where decompression requirements are lowest or where a decompressing diver can kill time by searching for abalone (Prince 1989*a*). Divers also favour diving on fast-growing populations where legal-size abalone are most easily found. Size limits and quotas set over broad zones of the fishery give little protection to these favoured dive areas where fishing pressure focuses.

Illegal fishing pressure also raises concerns over sustainability. In New South Wales, it is estimated that the abalone catch by illegal commercial divers is at least equivalent to that of the legitimate commercial industry and may be double the commercial catch (Prince 1989*b*). The impact of illegal exploitation is particularly damaging to the long-term sustainability of the resource because the amount taken is uncontrolled and legal size limits are usually disregarded. As accessibility to coastal areas increases, amateur fleets grow in size, and diving equipment and modern navigational electronics become increasingly available, it is expected that the illegal catching of abalone for sale will pose an increasing threat to the sustainability of the abalone resources (Prince and Shepherd 1992).

Declining water quality in inshore environments is also an increasing threat to the fishery. Burgeoning coastal development and agricultural inputs increase inshore nutrient loadings and destabilize salinities. This combination of conditions has apparently caused mass mortalities and disease problems off the coast of suburban New South Wales and the mouth of the Blackwood River, Western Australia (D. Leadbetter, Ocean Watch, New South Wales, Australia and T. Adams, Western Australian Fishing Industry Council, Western Australia, Australia, personal communications).

All these impacts can lead to the ongoing loss of abalone populations and an erosion of the abalone resource. In the short term this leads to a loss of potential production. In the long term, if the continuing loss of abalone populations reduces the productive areas beyond the point where current TACCs can be sustained, the fishery will destabilize (Prince and Shepherd 1992). But with no effective system of quantitative stock assessment in place there is no way these processes can be monitored or their aggregate impact assessed.

Reef-by-reef management

The knowledge and techniques required for the optimal, sustainable management of the Australian abalone resource have been developed and are known. Each abalone bed requires its own specifically tailored management plan with total allowable catch, size limits, and monitoring regime (Prince 1989*a*; Shepherd and Brown 1993). However, in Australia there is not a single population of abalone which can be quantitatively shown to be being harvested sustainably. The government agencies charged with this responsibility are incapable of collecting sufficient high quality data or effecting management plans at a small enough scale. The conundrum they face is this: how does a modern, centralized, small government manage this valuable but spatially intricate resource when fiscal reality prevents monitoring, management, and enforcement at an appropriately fine scale? Despite the aggregate value of the resource no centralized, economically rational, small government can monitor, quantitatively assess, or manage the resource optimally. A democratic, liberal society with centralized priority setting will always favour spending scarce tax revenue on hospitals, welfare, and government works, over spending on monitoring and policing the harvest of abalone. The public sector fisheries biologist charged with assessing and managing spatially complex abalone populations with minimal resources faces an impossible task.

As it is currently structured, reef-by-reef management is an impossible dream in the Australian abalone fishery. The tragedy of the commons and a tyranny of scale forces our society to manage this resource suboptimally. The existing management framework is structurally unable to meet the challenge of assessing and managing this spatially intricate renewable resource.

Optimizing Australian abalone management

Hardin (1968) argued that the "tragedy of the commons" does not have a technical solution; rather it is a social issue requiring society to change and voluntarily relinquish existing rights and freedoms.

The Zone 2 abalone divers of south Western Australia (Fig. 2*a*) voluntarily practice "concept fishing" in an attempt to maintain and build production from a depleted high growth area previously overfished using legislated minimum size limits and TACCs (T. Adams, Western Australian Fishing Industry Council, Western Australia, Australia, personal communications). When fishing in the "concept area" (Fig. 2*b*) the Zone 2 divers voluntarily: (*i*) co-ordinate their diving effort to ensure each aggregation of abalone is only harvested once a year and share their daily catches; (*ii*) refrain from harvesting an aggregation of abalone if they see that it has not rebuilt since the previous year; (*iii*) use a self-determined minimum size limit above the legal minimum size limit. The divers only harvest abalone that have finished their rapid growth phase (in terms of both shell length and volume). They judge this by the depth and roundness of abalone shells; and (*iv*) remove no more than 30% of the abalone in an aggregation above their self-determined size limit, selecting the abalone from across the size range available rather than just taking the biggest.

Growing catches and positive reports on stock levels suggest that the Augusta concept plan is rebuilding the abalone beds in their concept area. However, with no legal underpinning the voluntary co-operation needed to foster this experiment in fisheries management easily and often breaks down. There is no guarantee that long-term personal benefit will accrue from the short-term cost of this conscientious stock management. Years of co-operating can be negated by a short period of unco-operative but legal behaviour. The existing legal framework is a blunt instrument in which behavior is often determined by the lowest common denominator and the creed: "if I don't do it, somebody else will."

It has been argued that transferring greater responsibility for management to users of renewable resources fosters economically efficient and sustainable harvesting of renewable resources (Kesteven 1988; Keen 1991; Young 1992). We argue that for abalone stocks and other species with geographically restricted populations this responsibility for management must be linked to defined areas and populations. A form of territorial users rights is necessary in this fishery (TURF) to enable individuals, or small groups of individuals, to manage at a scale appropriate for the functional units of stock.

A TURF system would give abalone divers (or small groups of divers) secure harvesting rights to specific abalone beds and exclude others from harvesting abalone in those areas. Divers could then be expected to look after their own long-term interests by developing the skills needed to manage their own abalone beds. Most experienced abalone divers have observed population trends on the abalone beds they fish regularly and have well-developed ideas on how abalone should be managed. The principles of the Augusta concept plan show how sophisticated these home-grown ideas can be. However, under current arrangements, these ideas are seldom implemented because of the "tragedy of the commons" dilemma. Most abalone divers would willingly tend valuable abalone beds like gardeners tending their gardens if our system of social constraint encouraged rather than discouraged this behavior.

Easily managed nuggets

It is economically feasible to manage abalone populations intensively. Abalone stocks are concentrated into valuable nuggets of stock; a square kilometre of productive reef area such as George III Rock (Figs. 4*a* and 4*b*) may sustainably produce an annual 2–4 t of abalone with a gross value of A$50–150 000 (Prince 1989*a*). Intensive management would involve few costs over and above the existing ones. Most of the infrastructure required is already used by the existing catching sector of the fishery and, if forthcoming, the voluntary behavior of harvesting divers is free. The major cost would be for ongoing population monitoring and quantitative stock assessment if these were required. However, once permanently established with existing techniques, monitoring and stock assessment on an area this size need take no more than 4–6 personnel weeks per annum, worth approximately A$5–10 000. While far beyond the level of resources a centralized agency would devote to a single reef, a private operator could afford this cost over and above minimal costs of normal harvesting.

Under TURF management abalone harvesters would become directly responsible for planning both their harvesting and stock management strategies for defined abalone beds. This would eventually include catch levels and the size of capture. They would also assume responsibility for monitoring and securing their own abalone beds. The government role would shrink to supporting the development of optimal management skills among abalone harvesters and to verifying, on behalf of the community, that predetermined minimum standards of operation are observed.

Under this regime the value of an entitlement to harvest abalone would become linked to the expected production from a defined abalone bed. Economic imperatives would favour the harvester who can optimize long-term harvest rates by optimizing management (Kesteven 1988; Keen 1991; Young 1992). New information and changes in the condition of resources are more efficiently assimilated by individuals than by centralized governments; therefore, we should expect management to become innovative, experimental, and adaptive. Multiple experiences with many different units of stock will present great opportunities to learn through adaptive management (Walters 1986; Walters and Holling 1990).

The issue of incremental resource degradation from diffuse external threats (i.e., pollution and illegal harvesting) is also more likely to be confronted under TURF management. Abalone harvesters would be committed to ensuring long-term productivity from specific areas of reef. At the present time abalone divers give way when a stock declines due to environmental damage, recreational, or illegal fishing pressure. Rather than addressing the threat, harvesters relocate their operation placing greater pressure on the dwindling number of remaining productive beds. Under TURF management harvesters would not be able to relocate without purchasing the rights to new abalone beds. This will create an incentive for harvesters to address problems of eutrophication, siltation, habitat destruction, or illegal use in their areas. With strong financial commitment to the integrity of the natural environment in specific locations, abalone harvesters can be expected to evolve into environmental watchdogs guarding the integrity of the inshore marine environment on behalf of the rest of the community.

Precedence

Territorial user rights fisheries have considerable precedence. In Europe and North America some stocks of intertidal bivalves are managed as private property where it has been found to maximize production and minimize surveillance by managing agencies (Beattie et al. 1982; Bourne 1986).

Japanese prefectures continue to manage their own fisheries on a basis of local corporate ownership of an area of fishing ground (Mottet 1980; H. Kojima, Tokushima Prefectural Fisheries Experimental Station, Hiwasa-cho, Kaifu-gun, Tokushima-ken 779-23, Japan, personal communication). Despite the ongoing slow decline of Japanese abalone catches in recent times, the stability of Japanese catches over hundreds of years of exploitation argues that the Japanese have had considerable success in managing their abalone stocks sustainably.

Most marine resources were managed as territorial rights by the traditional societies of Oceania, the reef tenure providing the motivation for conservation (Ruddle and Johannes 1983). The displacement of traditional TURF-type management and the introduction of a law-of-the-commons framework is now encouraging the use of destructive fishing practices involving poisons and explosives. The power of villages, clans, and chiefs to control their own fishing reefs has been eroded while the governmental authorities which nominally take control, lack sufficient resources to monitor, manage, or enforce (Johannes 1992; De Allessi 1997).

In stark contrast to the general trend in Oceania, a recent innovation in Vanuatu has seen the strengthening of village control over the management of local stocks of the marine gastropod *Trochus niloticus* (R.E. Johannes, 8 Tyndall Court, Bonnet Hill, Tasmania, Australia 7053, personal communication). The results have been spectacular. With the support of some basic biological education, traditional village-based power structures have reasserted control over reef areas determining when they can be fished. Introduced on a trial basis in a few villages, the success of the initiative can be gauged by its rapid spread to other villages and the way the villagers have extended the concept to other species. Similarly a form of TURF management has recently been implemented in the Chilean fishery for the gastropod *Concholepas concholepas* with considerable success (Castilla et al. 1998). As in Vanuatu, early indications in Chile are that small, relatively nontechnical, local communities controlling their own reef areas are capable of sophisticated and creative management decisions.

The apparent effectiveness of TURF in traditional societies with low levels of scientific training and few governmental resources suggests that TURF could also be successful in technologically advanced countries such as Australia.

Challenges to the implementation of TURF for Australian abalone

Collectively we authors have spent considerable time discussing TURF management with a wide cross section of abalone resource stakeholders in Australia. We have spoken publicly at conferences, workshops, and annual meetings of associations and discussed it informally in numerous private conversations. There is widespread support for subdividing the Australian abalone fisheries into smaller zones (100's of km) each with fewer divers in order to reduce competition between divers and encourage divers to "farm manage" the resource (Prince and Shepherd 1992). However, support for TURF management is more mixed. In each fishery there are strong pockets of support for changing to a TURF arrangement but an ill-defined resistance to change is more pervasive.

Concern is expressed about the exclusive nature of the harvesting rights underpinning TURF management. In this context it needs to be noted that territorial user rights need not exclude other compatible uses for TURF areas; they would simply create exclusive harvesting rights for abalone in specific areas. It should also be noted that exclusivity is already a feature of the existing fisheries. Limited entry and ITQ systems are exclusive; a limited amount of commercial divers are allowed to gather abalone for sale. In addition most abalone fisheries already have areas reserved for recreational, indigenous, or preservationist groups and commercial abalone divers are excluded from these areas. This zonation of the existing fisheries provides a starting point from which the broader community can ensure, through political processes, that all stakeholder groups maintain equitable access to the resource.

A change to TURF management would further reduce the existing right of stakeholders to move relatively freely around their fisheries. All stakeholders would be excluded from many of the areas in which they currently operate and restricted to operating in some smaller subset of those areas. Hardin (1968) argues that the tragedy of the commons does not have a technical solution; rather it is a social issue requiring members of a community to voluntarily relinquish existing rights and freedoms.

As might be expected, the willingness of the existing stakeholders to voluntarily relinquish existing rights of free movement seems to relate to how strongly they perceive their own need for change. Most abalone divers we have spoken with acknowledge the long-term gains that would be made by optimizing the management of abalone beds with TURF. However, those from fisheries with good stocks of abalone generally value their freedom of movement and relative lack of responsibility too much to support changing. It is the divers from areas with perceived stock problems and a strong interest in rebuilding stock levels that tend to be most supportive of a change to TURF management. For this latter group, existing rights of free movement are devalued by their concern for the long-term viability of their stocks.

Allocation

The impossibility of converting the existing ITQ allocations into an equitable allocation of areas is another factor commonly cited in conversations as a barrier to changing the existing management arrangements. However one possible strategy for equitably converting allocations has already been devised by Dan and Danielle Pollock of the West Coast Abalone Harvesters of British Columbia.

The method they proposed involves the following steps. (*i*) Grid the available coastline using an appropriately fine scale (1 km^2). (*ii*) Each stakeholder then assigns their own value to each grid square using a scale of 0–10, zero indicating no value and ten indicating maximum value. Individuals define their own personal reasons for assigning value, including stock abundance, fishing history, accessibility, and suitability for diving. (*iii*) The total perceived value of each square is then determined by summing the values assigned to each square by the stakeholders. (*iv*) The total perceived value of the fishery is then determined by summing the perceived value of all the individual 1-km squares. This total perceived value of the fishery can then be compared to the existing TACC and a conversion rate calculated between ITQ units and units of perceived value. (*v*) By consensus the grided 1-km^2 squares may then be amalgamated to form a smaller number of larger areas with approximately equivalent perceived value. These larger units of approximately equal perceived value could be called TURF units. (*vi*) A ballot is then used to allocate TURF units in proportion to the ITQ holdings of each stakeholder. (*vii*) Once this balloted allocation is completed stakeholders can begin trading TURF units in order to rearrange TURF holdings in line with their individual requirements.

There are undoubtedly other ways of allocating areas so that a TURF management strategy can be implemented. However, the system outlined above meets the necessary criteria of being equitable, open, and above manipulation by individuals or groups (Hively 1995).

Concluding comments

Using the specifics of the Australian experience with the research and management of abalone stocks together with our wider fisheries experience, we have argued that the spatial scale of an exploited stock should be an important determinant of management strategies. Where the spatial scale of the functional units of stock within a fishery is smaller than the effective scale of management, a tragedy of the commons situation may arise despite otherwise sophisticated management. Component stocks within the fishery can be overexploited and a serial depletion of stocks may occur. From our experience fished stocks with localized patterns of movement and dispersion, including many tropical reef and invertebrate species, are particularly prone to these localized impacts. We suggest that management through territorial user rights would allow individual or small communities of fishers to adaptively adjust the scale of management to the scale of the stock. There may be other strategies for achieving this same end but our theory is that current trends in modern government mitigate against the current system of zonal size limits, effort limitation, and ITQs optimally managing spatially intricate renewable resources.

References

Beattie, J.H., McMillan, D., and Wiegardt, L. 1982. The Washington state oyster industry: a brief overview. *In* Proceedings of the North American Oyster Workshop. *Edited by* K.K. Chew. World Mariculture Society, Spec. Publ. No. 1. pp. 28–38.

Bourne, N. 1986. Bivalve fisheries: their exploitation and management with particular reference to the Northeast Pacific region. *In* North Pacific Workshop on Stock Assessment and Management of Invertebrates. *Edited by* G.S. Jamieson and N. Bourne. Can. Spec. Publ. Fish. Aquat. Sci. 92. pp. 2–13.

Brown, L.D. 1991. Genetic variation and population structure in the Blacklip abalone *Haliotis rubra*. Aust. J. Mar. Freshwater Res. **42**: 77–90.

Castilla, J.C., Manriquez, P., Rosson, A., Pino, C., Soto, R., Alvarado, J., Oliva, D., Espóz, C., and Defeo, O. 1998. Artisanal "caletas" as units of production and co-managers of benthic invertebrates in Chile. *In* Proceedings of the North Pacific Symposium on Invertebrate Stock Assessment and Management. *Edited by* G.S. Jamieson and A. Campbell. Can. Spec. Publ. Fish. Aquat. Sci. 125. pp. 407–413.

Cushing, D.H. 1968. Fisheries biology. A study in population dynamics. University of Wisconsin Press, Madison, Wisconsin and London, U.K. 200 p.

De Allessi, M. 1997. Holding out for some local heroes. New Sci. **8**(March): 46.

FAO. 1973–1988. Yearbook of fisheries statistics. Catch and landings. UNFAO. pp. 45–60.

Gulland, J.A. 1969. Manual of methods for fish stock assessment. Part 1. Fish population analysis. FAO. Man. Fish. Sci. 4. 154 p.

Guzmán del Próo, S.A. 1992. A review of the biology of abalone and its fishery in Mexico. *In* Abalone of the world: biology, fisheries and culture. *In* Proceedings of the 1st International Symposium on Abalone. *Edited by* S.A. Shepherd, M.J. Tegner, and S.A. Guzmán del Próo. Fishing News Books. Blackwell Science Pty Ltd., Oxford, Victoria, Australia. pp. 341–350.

Hancock, D.A. 1979. Population dynamics and management of shellfish stocks. *In* Population assessment of shellfish stocks. ICES, **175**: 8–19.

Hardin, G. 1968. The tragedy of the commons. Science (Washington, D.C.), **162**: 1243–1248.

Harrison, A.J. 1986. Gastropod fisheries of the Pacific with particular reference to Australian abalone, *In* North Pacific Workshop on Stock Assessment and Management of Invertebrates. *Edited by* G.S. Jamieson and N. Bourne. Can. Spec. Publ. Fish. Aquat. Sci. 92. pp. 14–22.

Hilborn, R., and Walters, C.J. 1987. A general model for simulation of stock and fleet dynamics in spatially heterogeneous fisheries. Can. J. Fish. Aquat. Sci. **44**: 1366–1369.

Hively, W. 1995. Dividing the spoils. Discovery, March 1995. pp. 49–57.

Johannes, R.E. 1992. 6th FFA Technical Subcommittee Workshop Focus: Decentralized nearshore fisheries management in Oceania. 6th Forum Fisheries Agency Committee. April 1992.

Kailola, P.J., Williams, M.J., Stewart, P.C., Reichelt, R.E., McNee, A., and Grieve, C. 1993. Australian fisheries resources. Bureau of Resource Sciences and the Fisheries Research and Development Corporation, Canberra, Australia. 422 p.

Keen, E.A. 1991. Ownership and productivity of marine fishery resources. Fisheries, **16**: 18–22.

Keesing, J.K., and Baker, J.L. 1998. The benefits of catch and effort data at a fine spatial scale in the South Australian abalone (*Haliotis laevigata* and *H. rubra*) fishery. *In* Proceedings of the North Pacific Symposium on Invertebrate Stock Assessment and Management. *Edited by* G.S. Jamieson and A. Campbell. Can. Spec. Publ. Fish. Aquat. Sci. 125. pp. 179–186.

Kesteven, G.L. 1988. The conservation of fishery resources and management of their exploitation: the role of a licensing system. Asian Fish. Sci. **1**: 123–133.

McShane, P.E. 1991. Exploitation models and catch statistics of the Victorian fishery for abalone *Haliotis rubra*. Fish. Bull. **90**: 139–146.

McShane, P.E., Black, K.P., and Smith, M.G. 1988. Recruitment processes in *Haliotis rubra* Leach (Mollusca: Gastropoda) and regional hydrodynamics in southeastern Australia imply localized dispersal of larvae. J. Exp. Mar. Biol. Ecol. **124**: 175–203.

Morgan, G.R. 1980. Increases in fishing effort in a limited entry fishery — the western rock lobster fishery 1963–1976. J. Cons. Int. Explor. Mer. **39**: 82–87.

Mottet, M.G. 1980. Factors leading to the success of Japanese aquaculture with an emphasis on northern Japan. Wash. Dep. Fish., Tech. Rep. 63. 106 p.

Nash, W.J. 1992. An evaluation of egg-per-recruit analysis as a means of assessing size limits for blacklip abalone (*Haliotis rubra*) in Tasmania. *In* Proceedings of the 1st International Symposium on Abalone. Abalone of the world: biology, fisheries and culture. Fishing News Books. Blackwell Science Pty Ltd., Oxford, Victoria, Australia. pp. 318–338.

Prince, J.D. 1989*a*. The fisheries biology of the Tasmanian stocks of *Haliotis rubra*. Ph.D. thesis, University of Tasmania, Hobart, Australia. 174 p.

Prince, J.D. 1989*b*. The amateur and illegal fisheries for abalone in New South Wales. Report prepared for the Abalone Divers Cooperative of New South Wales. 23 p.

Prince, J.D., and Guzmán del Próo, S.A. 1993. A stock reduction analysis of the Mexican abalone (Haliotid) fishery. Fish. Res. **16**: 25–49.

Prince, J.D., and Shepherd, S.A. 1992. Australian abalone fisheries and their management. *In* Proceedings of the 1st International Symposium on Abalone. Abalone of the world: biology, fisheries and culture. *Edited by* S.A. Shepherd, M.J. Tegner, and S.A. Guzmán del Próo. Fishing News Books. Blackwell Science Pty Ltd., Oxford, Victoria, Australia. pp. 407–426.

Prince, J.D., Sellers, T.L., Ford, W.B., and Talbot, S.R. 1987. Experimental evidence for limited dispersal of haliotid larvae (genus *Haliotis*: Mollusca: Gastropoda). J. Exp. Mar. Biol. Ecol. **106**: 243–263.

Prince, J.D., Sellers, T.L., Ford, W.B., and Talbot, S.R. 1988. Conformation of a relationship between the localized abundance of breeding stock and recruitment for *Haliotis rubra* Leach (Mollusca: Gastropoda). J. Exp. Mar. Biol. Ecol. **122**: 91–104.

Ricker, W.E. 1954. Stock and recruitment. J. Fish. Res. Board Can. **11**: 559–623.

Robins, C., and Sachse, M. 1994. A creep called technology: evolution of the fleet. *In* Australia's northern prawn fishery — the first 25 years. *Edited by* P.C. Pownall. NPF25, PO Box 120, Cleveland, Queensland, Australia. pp. 23–34.

Rudd, M. 1994. A review of international abalone trading patterns and pricing. *Presented at* Second International Symposium on Abalone Biology, Fisheries and Culture. February 1994. Hobart, Tasmania.

Ruddle, K., and Johannes, R.E. (*Editors*). 1983. Traditional marine resource management in the Pacific Basin: an anthology. UNESCO/ROSTSEA, Jln. M.H. Thamrin No. 14. Jakarta, Indonesia. 410 p.

Russell, E.S. 1931. Some theoretical considerations on the "overfishing" problem. J. Cons. Int. Explor. Mer. **6**: 3–27.

Safina, C. 1995. The world's imperiled fish. Sci. Am. November 1995. pp. 30–37.

Shepherd, S.A. 1986. Studies on southern Australian abalone (Genus *Haliotis*), VII. Aggregative behaviour of *H. laevigata* in relation to spawning. Mar. Biol. **90**: 231–236.

Shepherd, S.A., and Brown, L.D. 1993. What is an abalone stock: implications for the role of refugia in conservation. Can. J. Fish. Aquat. Sci. **50**: 2001–2009.

Shepherd, S.A., and Godoy, C. 1989. Studies on southern Australian abalone (Genus *Haliotis*), XI. Movement and natural mortality of juveniles. J. Malacol. Soc. Aust. **10**: 87–95.

Shepherd, S.A., and Laws, H.M. 1974. Studies on southern Australian abalone (Genus *Haliotis*), II. Reproduction of five species. Aust. J. Mar. Freshwater Res. **25**: 49–62.

Sluczanowski, P.R. 1984. A management orientated model of an abalone fishery whose substocks are subject to pulse fishing. Can. J. Fish. Aquat. Sci. **41**: 1008–1014.

Sugihara, G., and May, R.M. 1990. Applications of fractals in ecology. TREE, **5**(3): 79–86.

Tegner, M.J. 1989. The California abalone fishery: production, ecological interactions, and prospects for the future. *In* Marine invertebrate fisheries. *Edited by* J.F. Caddy. Wiley Interscience Publications, New York, N.Y. pp. 401–420.

Walters, C.J. 1986. Adaptive management of renewable resources. Biological Resource Management Series. Macmillan Publishing Co., New York, N.Y. 374 p.

Walters, C.J., and Holling, C.S. 1990. Large-scale management experiments and learning by doing. Ecology, **71**(6): 2060–2068.

Young, M.D. 1992. Sustainable investment and resource use: equity, environmental integrity and economic efficiency. *In* Man and the biosphere series. *Edited by* J.N.R. Jeffers. The Parthenon Publishing Co., Canberra, Australia. 176 p.

Management options for the South Australian rock lobster (*Jasus edwardsii*) fishery: a case study of co-operative assessment and policy design by fishers and biologists

Carl Walters, James H. Prescott, Richard McGarvey, and Jeremy Prince

Abstract: A modelling workshop process was used to bring biologists and commercial fishers together to develop a spatial model for population dynamics and harvest regulation of the South Australian rock lobster (*Jasus edwardsii*) fishery. The resulting model provided a credible reconstruction of how the space, time, and size structures of the stock have changed over the history of the fishery, and offers a rich variety of regulatory policy options for exploration of how the stock might have behaved (and might behave in the future) if managed differently. Initial use of the model has been to test options for reducing risk of recruitment overfishing by increasing spawning stock and egg production. A number of regulations ranging from increased size limits to large spatial refuges could accomplish this risk reduction aim. One option is to simply reduce the fishing season length dramatically. The model predicts that short-term yield loss under this strategy would eventually be regained through increased survival and higher catch rates of larger lobsters, and offers the economic advantage of greatly reduced fishing costs. This policy hypothesis can be tested in the field by a management experiment allowing fishers to see for themselves whether an area with a short season does indeed result in catch rates high enough to compensate for fishing time loss.

Résumé : La formule de l'atelier de modélisation a été utilisée pour réunir des biologistes et des pêcheurs commerciaux pour élaborer un modèle spatial dans le but d'examiner la dynamique des populations et les règlements régissant la récolte dans la pêcherie de langouste (*Janus edwardsii*) du sud de l'Australie. Le modèle qui a résulté a fourni une reconstruction crédible de la façon dont le stock a évolué du point de vue du temps, de l'espace et de sa structure de taille au cours de l'histoire de la pêcherie et offre une gamme très riche d'options du point de vue de la politique de réglementation permettant de se faire une idée du comportement passé et futur du stock s'il était géré différemment. Le modèle a été utilisé initialement pour vérifier des options visant à réduire les risques de surpêche au niveau du recrutement en augmentant le stock de géniteurs et la production d'oeufs. Un certain nombre de mesures réglementaires variant de l'augmentation de la limite de taille à la création de grandes réserves pourraient permettre d'atteindre cet objectif de réduction du risque. Une des options consiste simplement à écourter de manière substantielle la saison de pêche. Le modèle prévoit que la perte de rendement à court terme découlant de cette stratégie finirait par être récupérée grâce à une survie accrue et à des taux de capture plus élevés de langoustes de plus grande taille, sans compter que cette stratégie présente l'avantage économique de réduire considérablement les coûts de la pêche. Cette hypothèse de politique peut être vérifiée sur le terrain par une expérience de gestion permettant aux pêcheurs de constater par eux-mêmes si une zone soumise à une courte saison de pêche entraîne effectivement des taux de capture suffisamment élevés pour compenser la réduction du temps de pêche. [Traduit par la Rédaction]

Introduction

Traditional approaches to fishery management have frequently created deep divisions between fishers and regulatory agencies. Fishers take as much as they can, show remarkable inventiveness at finding ways around regulatory measures, and often provide only as much information as absolutely required. Biologists try to make sense of this information for assessment, then regulators often adopt openly paternalistic attitudes when forced to discuss assessments and regulatory options with fishers. No one wins in these situations; assessments are dangerously unreliable; opportunities for better information gathering and co-operative experiments are missed; and regulatory systems are usually ineffective due to both immediate enforcement problems and technological innovations to circumvent them.

There has been an opportunity in the South Australian (SA) lobster fishery to begin moving toward a more co-operative approach to management, building upon the shared goal of both fishers and regulators to ensure a sustainable future for the fishery. The SA lobster fishery is relatively small, with licensing and regulation split into two zones (northern,

C. Walters. Fisheries Centre, University of British Columbia, Vancouver, BC V6T 1Z4, Canada.

J.H. Prescott and R. McGarvey. South Australian Research and Development Institute, South Australian Aquatic Sciences Centre, PO Box 120, Henley Beach, SA 5022, Australia.

J. Prince. Biospherics Pty. Ltd., Perth, Australia.

Correct citation: Walters, C., Prescott, J.H., McGarvey, R., and Prince, J. 1998. Management options for the South Australian rock lobster (*Jasus edwardsii*) fishery: a case study of co-operative assessment and policy design by fishers and biologists. *In* Proceedings of the North Pacific Symposium on Invertebrate Stock Assessment and Management. *Edited by* G.S. Jamieson and A. Campbell. Can. Spec. Publ. Fish. Aquat. Sci. 125. pp. 377–383.

Fig. 1. South Australian rock lobster marine fishing areas (as 1° square blocks) and the two management zones. Southern zone is currently managed through individual transferable quota (ITQ) system, northern zone through traditional effort controls (pot number, season length).

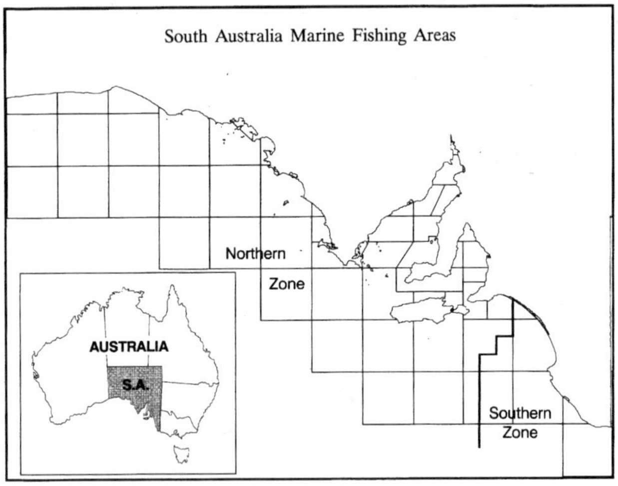

78 vessels; and southern, 187 vessels). Most fishers are relatively prosperous, well-educated, and keenly aware of the biology and population dynamics of the lobster. Many have been in the fishery since its major development in the early 1960's, and have seen very substantial changes in the stock since then. They have also been part of a relatively dynamic management system that has adjusted season length, reduced pot numbers, and in the southern zone actively reduced the number of vessels licensed. The license buy-back was paid for by the remaining fishers with the help of a state government loan, a policy that has since been widely commended by fishers.

Recently southern zone fishers opted to move to a quota management system, albeit one with the same effort regulations (pot limits, seasons, size limits) still in force. As before, there has been much concern about how this new regulatory measure will protect the spawning stock. Most fishers agree that fishing mortality rates are very high; perhaps dangerously high in the southern zone, as evidenced by the scarcity of larger mature lobsters. Northern zone fishers have opted to remain with effort controls to manage their fishery. However, many northern zone fishers fear that the stock may be in danger of being overfished due to improved fishing technology. In 1994, fishers sought a means of compensating for an anticipated increase of 5% in effective effort. A model was developed leading to the adoption of an increase in minimum size and a series of 8–9 day closures (McGarvey and Prescott 1998). Concerned fishers in both zones helped initiate and have backed a research program through financial support and direct participation in a data collection program in place since 1991 to provide data for a robust stock assessment.

One aspect of the research program that was specifically requested by fishers was a spatial model of the fishery with good visual output. Some fishers had seen models of an abalone fishery, AbaSim (Sluczanowski and Prince 1994), and the southern shark fishery, SharkSim (Sluczanowski 1994), and recognized their value for conveying complex information in a way fishers could understand. We intended to produce a model of the lobster fishery with such an interface but decided also to use the model development as a way to provide further motivation for co-operative information gathering. At the same time we tried to capitalize on fishers' knowledge of factors such as distribution of lobster habitat, by involving both fishers and biologists co-operatively in a computer model building process.

The explicit objective of building the model was to provide a device for synthesizing existing data into a useful format for policy analysis. However, the more important objectives were to foster better communication (trying to build a working simulation model requires precise definition of terms and use of information), to demonstrate to fishers exactly how data are used for biological assessment, and hence why much better data (and management experiments) are needed. We saw the model development as a level playing field for all stakeholders, with information and policy analysis suggestions from fishers being potentially just as critical as anything provided by professional biologists. We did not expect that the simulation model produced during this first co-operative effort would be particularly useful for policy analysis, but we did hope it would provide a concrete starting point for further cooperation and development. The ultimate aim would be to develop a policy screening tool that can deal not only with obvious policies such as pot reduction and quotas, but also a wide variety of other regulatory tactics such as size limits, fishing season pattern, etc.

Surprisingly, just a few days spent on model development led to both a very useful policy screening model, and to a possible win-win policy option for both increasing fishers' incomes and providing better protection for the spawning stock. Here we describe the model and policy analysis results obtained to date, and speculate on how co-operative management will develop in the future of the fishery.

We used an Adaptive Environmental Assessment (AEA) workshop process (Holling 1978; Walters 1986) to structure involvement by biologists and fishers in the model development process. AEA workshops proceed from definition of precisely what policy options and performance indicators are to be evaluated, through a series of data analysis and submodel development sessions for developing the actual simulation code, to gaming sessions where workshop participants "test" the model and its predicted policy options and suggest ways to improve it. In this case, the workshop included twenty biologists with a range of experience in lobster fisheries and population dynamics from across Australia and New Zealand, and twenty fishers from various fishing ports and the two South Australian management zones. The model reviewed in the following section thus represents the experience (and consensus) of a remarkably diverse participant/development group.

Spatial model description

The same model accounting structure for spatial population and fishing effort dynamics was chosen as had been developed in a previous analysis of the Western Australian rock lobster fishery (Walters et al. 1993). Here we review only the main features of that structure, which represents population and harvest processes on a spatial grid of cells (1° blocks in this case) laid over the fishing grounds corresponding to statistical reporting blocks for the fishery (Fig. 1). Various policy parameters (license access, refuge closures) and biological parameters (growth patterns, proportions of annual total recruitment, proportions of ocean bottom of suitable habitat for lobsters) are allowed to vary across cells. The cells are linked through three main processes: (*i*) allocation of total fishing effort among cells, (*ii*) larval settlement pattern (allocation of total recruitment over space), and (*iii*) movement of lobsters.

A key initial part of the model development was to have experienced fishers provide rough maps of benthic habitat type within each model spatial cell based on their past fishing success, using the simple classification: suitable for lobsters at all pot setting sites, sparse with small suitable patches requiring careful pot location, and not suitable for lobsters. This classification allowed us to capture some known differences among spatial cells in the effective area for lobsters (and fishing); in particular, much of the northern zone is either unsuitable or sparse habitat, while most of the southern zone cells have very high proportions of suitable habitat. All recruitment and fishing effort calculations for model cells were made relative to the estimated suitable habitat rather than cell size; thus in some cells with relatively little habitat, even a low total fishing effort can generate high simulated fishing mortality rate, while much higher efforts are needed to generate similar fishing rates in a cell with much good habitat.

The lobster subpopulation in each spatial cell is represented in terms of length (rather than age) structure, with the number of lobsters having 82 mm carapace length and larger divided into 8-mm (typical moult increment) categories. In each model cell, growth is represented by size-specific tables of moult frequencies (the proportion of animals moulting in each season/size category is then moved to the next larger size category).

Recruitment to the smallest category (82–90 mm) is calculated from simulated puerulus settlement 3–4 years earlier (using a Beverton-Holt stock-recruit relationship between egg production and total settlement). The Beverton-Holt recruitment relationship was included in the simulation to represent the possibility of recruitment overfishing should the population egg production be reduced sufficiently. The phyllosoma larvae spend a year or more in a pelagic phase in the open ocean, potentially traversing distances of perhaps 1000 km based on typical current speeds in these waters of the Southern Ocean south of the Great Australian Bight. We presume that this precludes development of local subpopulation structure, so that total recruitment for each simulated year is calculated as a grand pool of settling puerulus dependent on antecedent egg production throughout South Australia. But we soon found that no hypothesis or model involving declining recruitment as the fishery developed would predict the observed pattern of catches and relative abundance (as indicated by catch per unit effort, CPUE); this suggests that recruitment has been relatively stable since the early 1970's. The recruitment relationship is left in the model as a functional form with parameter values set so that simulated recruitment is impaired only if egg production is reduced substantially from current (early 1990's) levels. This is not a serious limitation of the model, since there was a very clear consensus among biologists and fishers that they were only interested in exploring "safer" policy options involving regulations to increase egg production and hence move away from that uncertain point on the recruitment relationship where recruitment begins to fail.

For survival, growth, and harvest calculations, each simulated year is divided into 2-week time steps. Two fortnights in the middle of the moulting periods, summer and winter, are designated as the seasonal moulting times. Using a 2-week time step is of course not really necessary for the survival/growth calculation; its value is to allow model users to vary fishing season patterns and season length widely, and to allow more realistic representation of the annual fishery depletion and spatial effort movement process. Discard mortality from undersized lobsters and females bearing spawn which are returned to the water, as well as losses from illegal fishing and a small recreational sector, are also incorporated in the harvest submodel.

Spatial variation in recruitment rate (proportions of total recruits settling in different spatial cells) appears to be critical to the structure of the fishery. We noticed that catches in most spatial cells have been stable for the past decade. This implies that annual recruitment rate per cell (or per unit suitable habitat within each cell) can be estimated from the average catch and estimated yield per recruit (average recruitment in a near-equilibrium situation must be yield divided by yield per recruit). We used fishing effort, natural mortality, and growth estimates to estimate yield per recruit for each cell. The catchability coefficient (fishing mortality rate per unit effort) needed for the fishing mortality part of the yield per recruit calculation was estimated by running the overall simulation model while varying the catchability parameter and historical fishing effort, to find catchability and total fishing mortality that would match changes from early in the fishery to the present in observed length frequencies. The resulting recruitment (yield/yield per recruit) calculation is admittedly crude, but it provides at least a more realistic estimate of spatial variation in recruitment rate than would crude catch or catch-per-effort statistics alone. Catch-per-effort does not in fact vary much over the whole fishing area, indicating that effort is attracted to areas of high lobster density and recruitment quickly enough to cause strong exploitation competition among fishers.

We found a very close relationship between recruitment rate estimated as above and annual fishing effort (averaged for 1989–1993), apparently indicating that effort is strongly responsive to spatial variations in recruitment rate (Fig. 2). Unfortunately we cannot be certain that the strong relationship in Fig. 2 does in fact represent attraction of fishing effort to areas where recruitment is concentrated. The observed pattern could be produced in at least two other ways. First, recruitment could be the same in all cells, but effort could be distributed in some way related to factors like access from port. Then if fishing mortality rate were in fact low in all cells, our recruitment calculation (catch divided by yield per recruit) would be dominated by catch variation due only to effort variation, with some spurious correction in yield per recruit from incorrectly assuming high fishing rates in some areas. The main evidence against this explanation is that length frequencies in areas with high effort indicate that fishing mortality rates are definitely not low in such areas. Second, recruitment could again be the same in all areas, but catchability could vary greatly so as to make the apparent or vulnerable stock look much larger in some areas (and attract more fishing to those areas). We see no way to reject this explanation using data from the fishery; there could indeed be substantial abundances of lobsters that are for some reason "invisible" to the fishery, but it would be plainly unwise

Fig. 2. Estimated relationship between recruitment rate per unit usable habitat and fishing effort for statistical subareas within the South Australian lobster fishery. Recruitment rate is estimated as observed average catch for each statistical subarea divided by estimated yield per recruit for the subarea.

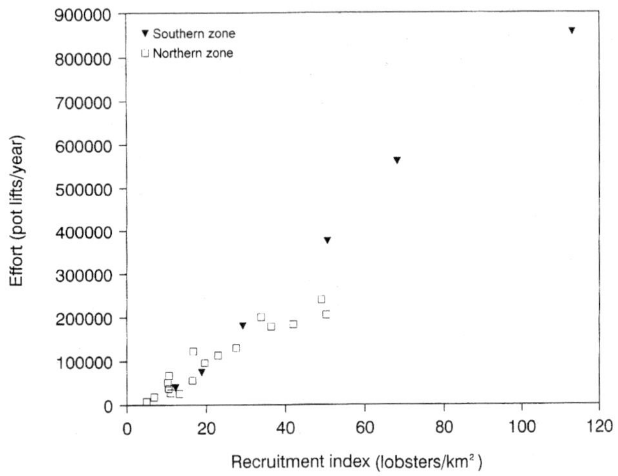

to count on such invisible animals as a source of protection against overfishing (egg and recruitment source).

The model uses the approach of Walters et al. (1993) and Allen and McGlade (1986), simulating the spatial redistribution of fishing effort each biweekly time step according to the desirability for vessels to fish in each cell. This variable, the effort "attractivity" of each cell, is directly proportional to the expected profitability from fishing there and is hypothesized in the model to be a function of spatial patterns in expected fishing success as measured by CPUE. For each simulation fortnight, the model calculates expected attractivity for each cell as a weighted average of the CPUE experienced the previous fortnight that same year and the historical value of the previous year's CPUE for the cell that fortnight of the season, with each predictor weighted equally. Effort is then allocated to each cell according to the attractivity proportion, the attractivity for the cell divided by the sum of expected attractivities over all cells. Over a fishing season of several months, repeating this redistribution calculation results in the pattern shown in Fig. 2; high effort is attracted to areas with high recruitment early in the season where it drives CPUE down. Later, effort spreads out to cells with lower initial recruitment as the more attractive cells are depleted. This effort redistribution submodel is critical for evaluating impacts of a variety of policy options, including spatial refuges that concentrate effort into remaining open areas and reductions in fishing season length that may reduce the tendency for effort to move into less attractive areas later. In addition to abundance, model effort responds to variations in price, both through the season, and as it varies with the supply, taken as the catch in South Australia overall.

Fractions of larger lobsters migrating between the spatial cells will be obtained directly from the results of a large mark-recapture study now being completed. Rates of migration in the model are assumed to be proportional to the density of animals in each cell. The proportion of lobsters leaving a cell is also assumed to decrease with increasing lobster size, so that lobsters with 120+ mm carapace length are not moved in the model. Preliminary tagging data suggest that movements are principally offshore and in distance are generally less than the width of a model cell, so only movement to adjacent cells is simulated. As it turns out, varying the small proportions of lobsters moving between cells does not alter model policy predictions, since increasing simulated movement simply causes simulated fishing effort to move as well. We were concerned about the movement parameters in initial workshop discussions, since we had found in the West Australian model (Walters et al. 1993) that offshore movement rate is critical to assessments of population egg production. Historically, offshore areas in Western Australia receive much less effort than inshore, providing a partial refuge for the spawning population. However, this is not the case in South Australia, where fishers are apparently much more willing to fish offshore than their Western Australian counterparts.

Model user interface and gaming procedures

The model is programmed to provide a series of spreadsheet-like interfaces for changing model parameters and policies, and a complex visual display of reference data and simulation results as each simulation or gaming trial proceeds. When we first presented this interface, biologists were concerned that it would be too complex for fishers to understand. In fact, fishers learned very quickly how to read the display screen that resembles the instrument panel of a modern fishing vessel, where several display blocks each show some relatively simple part of the results. The upper left area of the display shows color-coded maps of overall density changes from year to year, for juvenile and adult (egg producing sizes) lobsters. The yearly simulated size distributions are compared with recent data in two panels in the left center of the screen. The right side of the screen has four panels showing time series plots of long-term change in egg production, effort as pot lifts per year, catch, and CPUE outputs for both zones showing historical data and simulation time series.

Simulation time begins in 1950 (when the fishery started to develop rapidly), so that cumulative effects over time of any biological parameter changes made by the model user are immediately evident in terms of how well the model matches 45-year historical catch/effort and size distribution trends. This protocol helps to avoid the risk of confusing biological parameter changes made in one simulation run with changes made during subsequent runs in a policy evaluation session. The model can be stopped in any simulated year to introduce parameter changes. Normally only policy changes would be made, then the model would be restarted so the effects of change can be evaluated by comparing predicted outcomes to actual experience and to previous model outputs simulating historical policy. This method of comparing policy options is easier to understand (and far more credible) for both biologists and fishers than the usual modelling approach of just simulating alternative futures.

The model was deliberately run with a constant catchability chosen to match recent catch, CPUE estimates, and fishing mortality rate (as evidenced by length frequency) since 1980. Using this parameter value for early simulation years results in predicted catches that are much higher than reported for the

Fig. 3*a*. Predicted catch under four alternative management scenarios in the southern zone (SZ). Scenarios plotted include the baseline model (i.e., the historical management simulation), immediate reduction in fishing season length from seven to two months, gradual reduction in season over eight years, increased minimum size limit from 98 to 144 mm, and a refuge area of 1° square, i.e., one model cell in the zone.

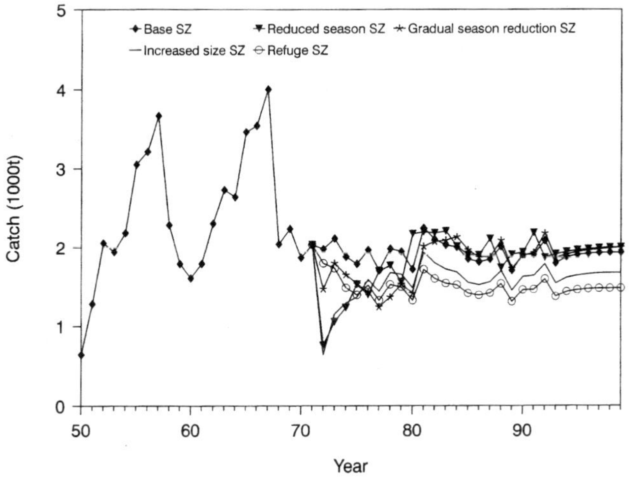

Fig. 3*b*. Predicted catch under four alternative management scenarios in the northern zone (NZ). Scenarios plotted include the baseline model (i.e., the historical management simulation), immediate reduction in fishing season length from seven to two months, gradual reduction in season over eight years, increased minimum size limit from 98 to 144 mm, and a refuge area of 1° square, i.e., one model cell in the zone.

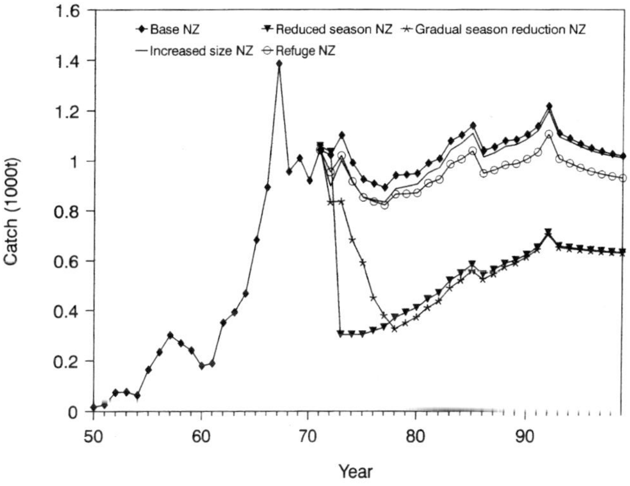

southern zone and lower than reported for the northern zone. We believe that two factors contributed to this result. Conflicting results between the two zones are thought to be the result of the way that total effort was ascribed to each zone prior to their creation in 1968. However, we attribute most of the difference between reported and predicted catch rates in the southern zone to the increase in fishing power that has occurred over time. Basically, the model results with constant catchability indicate that the modern fishing fleet would have achieved roughly three times the catch (and population impact) of its early (1950–1970) counterpart. We could obviously vary the catchability parameter by including an arbitrary time effect to improve the fit to historical data, but in this instance the prediction-data discrepancy was very easily understood by fishers and in fact appeared to make the whole model more credible to them.

Options for reducing risk of recruitment overfishing

Fishers participating in the AEA workshop were given an opportunity to "play" with the model in a session where no biologists were present. They quickly tested a variety of surprisingly intrusive and potentially effective options for reducing risk of recruitment overfishing by increasing average annual egg production. Among these options were: (*i*) very large and immediate reduction in fishing season lengths, (*ii*) very large but more gradually imposed reduction in fishing season lengths, (*iii*) large increases in legal minimum size limits, and (*iv*) establishment of permanent closed areas (egg production "refuges"). Many simpler and more modest options were also examined by fishers, and the history-reference simulation approach described above appeared to help them very quickly grasp why such modest options would not likely have a significant effect.

The most interesting option examined to date is the very simple one of drastically reducing the season length, from seven to two or even one month. Fishers first tried this option as a perturbation test on the southern zone, to see what the model would do with an extreme policy change. They then brought this to the attention of the model development biologists and said that they had found a mistake in the code. The model predicted a dramatic decrease in catch in the years immediately after reducing the season, as expected. However, catch slowly built back to historical levels (Fig. 3*a*) over a period of eight years. We agreed that there must be an error, and had a frantic workshop session trying to find it. We finally realized that there was no mistake, and we were looking at the transient dynamics for a very classical prediction in the yield-per-recruit theory of fishing: starting at zero, equilibrium yield increases rapidly with fishing rate (F), then the relationship becomes very flat or slightly decreasing (provided recruitment overfishing is not a factor) over a wide range of higher F values. This undiminished catch rate is due primarily to increase in yield-per-recruit since the stock-recruitment relationship is effectively flat through the range of model egg production. One qualifier on this prediction of a recovery in catch to near pre-season reduction levels with a large reduction in effort is the assumption of no density dependence in growth or adult mortality, or of recruitment. The history-reference display interface became critical in helping to explain this prediction to fishers; we simply pointed out how the model CPUE and length-frequency structure would recover under reduced fishing time to values near what the fishery produced in early years when total fishing effort was much lower. One uncertainty that fishers raised was whether lobsters can survive to larger sizes today as they did historically when F was lower. To test this critical assumption fishers suggested a planned experimental reduction in fishing mortality in a few spatial areas. They also quickly found a policy involving progressive reduction in season length over about 10 years that would result in substantial increase in population egg production with minor catch

Fig. 4*a*. Effect of alternative southern zone policies shown in Fig. 3*a* on total population egg production: comparison of the baseline output with immediate and gradual fishing season reduction, increased minimum size, and refuge area.

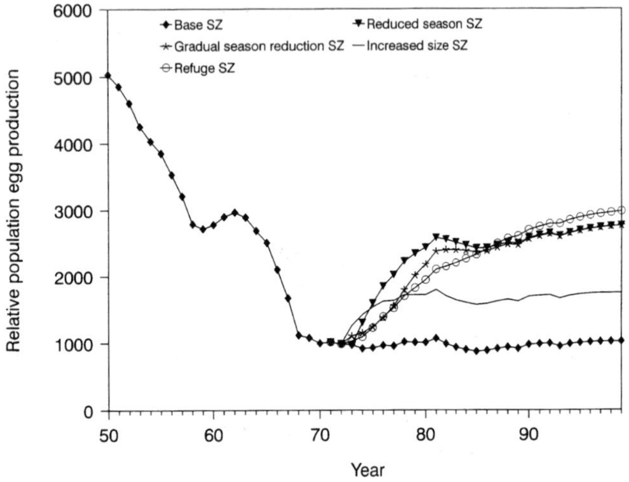

Fig. 4*b*. Effect of alternative northern zone policies shown in Fig. 3*b* on total population egg production: comparison of the baseline output with immediate and gradual fishing season reduction, increased minimum size, and refuge area.

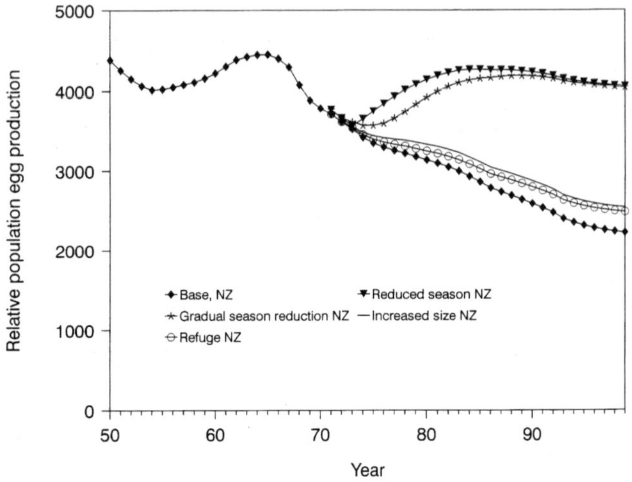

reductions (of the order of 10–20% per year) along the way (Fig. 3*a*, gradual reduction case). Note in Fig. 3*b* that the recovery effect is not predicted to occur in the northern zone, where F is apparently much lower in the first place.

The model predicts large impacts on egg production of closing even a few model grid cells (1° square blocks) to fishing. Figures 4*a* and 4*b* show simulated egg production impacts of closing one southern zone and one northern zone cell, with the northern zone closed cell in a high recruitment area west of Kangaroo Island. In the southern zone, this policy has roughly the same simulated impact as drastically shortening the fishing season. However, this policy disadvantages southern zone fishers by closing a larger percentage of their fishing area, and it intensifies competition within the remaining open areas thereby increasing fishing mortality rate in these areas. Fishers suggested that instead of moving immediately to such refuges, potential refuge areas could instead be subject initially to substantially reduced fishing seasons. This staged approach to refuge area development would then also provide an experimental test of the predictions about season length reduction. The decision to proceed to complete closure of a refuge could then be made on the basis of experimental evidence. Unfortunately, the model and data analysis of Fig. 2 predicts that fishers will respond to any initial increase in CPUE in the experimental areas by shifting effort to them, preventing stock size and CPUE from increasing much. Under regulation this shift could presumably be restrained. Hence it appeared that the policy was failing when, in fact, it would work if these effort shifts could be restrained.

Another policy suggestion from fishers was to substantially shorten the fishing time allowed each license or quota holder, while allowing individual fishers to select their own fishing "seasons." A similar policy was adopted by northern zone fishers in the months following the workshop. Its effect is like a reduced fishing season, but with less gear competition for available pot setting locations. It would also be much like a policy of greatly reducing the total number of licensed fishing pots, since the number of pots fishing in any part of the overall season would be much reduced. The model does not keep track of individual fishing patterns, but we did try simulations of major reductions in number of pots fishing for each 2-week time step. A supplementary spreadsheet model was developed to test the consequences of fisher-chosen season reductions (McGarvey and Prescott 1998) and was used to decide the lengths and months of closures in the 1994/1995 season. This option is very popular with fishers in the northern zone so it may be worth testing the concept further, if fishers are willing to cooperate in establishing and maintaining experimental areas where the annual number of pot lifts is drastically reduced.

With quota management established in the southern zone, it should be possible, in principle, to establish a sequence of quotas over time that would cause the same ramping down in fishing mortality as would shortening fishing seasons. But implementation of this policy has inherent risks when annual assessments of stock size are subject to large errors. Quotas could trigger a decline in stock abundance rather than an increase if stock size were overestimated and quotas initially allocated were too large, because an absolute level of removal is an increasing fraction as the stock declines. However, the southern zone is unusual in that the fishery has maintained the same effort controls, pot number limits, and seven-month season, that were in place prior to quotas being introduced. Without effort controls, to be relatively sure of obtaining the same reduction in risk of severe stock decline as a season reduction policy, quotas would have to be reduced much more, and for much longer, than would catches under season reduction. The dilemma of how to reconcile quota management with need for direct and simple regulation of exploitation rate has not been resolved, and will likely be a matter of much future discussion between biologists and fishers.

Discussion

We entered the modelling exercise thinking that a detailed and realistic spatial model would be needed to link long-term population dynamics considerations with an analysis of specific regulatory options. Indeed, it was perhaps necessary to proceed with the analysis in detail so as to make it credible to biologists and fishers alike. But in the end, the most important policy findings, including reduced season lengths and experimental tests of season length reduction perhaps leading to refuge areas, involve only very simple predictions about the impact of reduced fishing mortality rate on population size structure, CPUE, and fecundity due to changes in yield- and egg-per-recruit. While it seems that we did more analysis than was really necessary, we view the overall findings as a very fortunate outcome. Had the policy exploration and testing uncovered only a few specific policy alternatives whose efficacy depended on particular local and highly uncertain estimates of population parameters, the model would have been a much less useful tool. We would have been left with the tired old complaint that the data are inadequate and more research is needed.

During the model development and testing, we repeatedly found the most critical data to be length-frequency patterns sampled from the fishery. Beyond the obvious use of these data for quantitative assessment of fishing mortality and catchability, we found them necessary in "credibility checks" on policy alternatives. Our only real justification for model predictions that lobsters would on average increase in size under reduced fishing mortality is that, in fact, lobsters were larger when effort was lower. However, it did not take fishers long to find the basic flaw in this argument, and to recognize the strong assumption of stationarity implied by arguing that historical data are good predictors of the effects of future policy change. They noted that the difference in size structure could be due to recent growth rates being lower or unusually high recruitment some years before the early samples were taken. They further noted that we cannot reliably use northern/southern zone size frequency comparisons for mortality assessment, because the higher frequency of large lobsters in the northern area, where effort is lower, could be due simply to large areas in the far western regions and offshore where fishing in the past was rare, leaving substantial numbers of near virgin stock to be harvested now. In the end, we cannot reject these counter arguments and criticisms of the model assumptions by using only available data, or by continued monitoring. The arguments are instead perhaps the best case we can make for the need to immediately establish at least some small, experimental refuge areas to provide reference or baseline information on how population size structure should look under reduced fishing mortality.

Somewhat surprisingly, there was strong support in principle by fishers for setting up management experiments to directly test for such effects as shifts in abundance and size distributions under reduced fishing effort. The same experiments could also reveal if there are significant increases in gear efficiency when fewer pots are in competition. Usually there is a tacit assumption by fishers, and many biologists, that just gathering more data will somehow permit analyses to resolve key uncertainties. But in this case, most stakeholders recognized immediately that the critical uncertainties involve circumstances that no longer occur naturally in the fishery, and must be deliberately created if policy evaluations are to be rigorously tested. Laying the groundwork for a co-operative program to design and conduct real adaptive management experiments is perhaps the most important single achievement of our efforts to date.

We believe that our AEA workshop will serve as a stimulus for other such workshops in fisheries. Participants were quick to appreciate how the data they collected were used to develop an understanding of such a complex and dynamic system. Participants were also able to use the model to test policies almost as soon as they had access to it because of the model's relatively simple graphical interface. Models with such easy-to-use and understand graphical output are relatively rare. This may be because most scientists are able to gain sufficient understanding of model outputs using methods familiar to them such as tables of results and static graphs. While the biologists may understand the results fishers often do not. The power of this type of graphical interface for communicating results was clearly obvious to every biologist in attendance.

Acknowledgments

Motivation and organization for the workshop process was provided by Phillip Sluczanowski; he passed away before the process was actually carried out, and was deeply missed. Key analyses and model development assistance were provided by Norm Hall. Rob Lewis contributed institutional support, statistics, and perspective on the fishery history. We gratefully acknowledge the patience and support of all the fishers and biologists who participated in the AEA workshop process.

References

Allen, P.M., and McGlade, J.M. 1986. Dynamics of discovery and exploitation: the case of the Scotian Shelf groundfish fisheries. Can. J. Fish. Aquat. Sci. **43**: 1187–1200.

Holling, C.S. (*Editor*). 1978. Adaptive environmental assessment and management. Wiley-Interscience, Chichester, England.

McGarvey, R., and Prescott, J.H. 1998. A model for assessing closure season management options in the South Australian rock lobster (*Jasus edwardsii*) fishery. *In* Proceedings of the North Pacific Symposium on Invertebrate Stock Assessment and Management. *Edited by* G.S. Jamieson and A. Campbell. Can. Spec. Publ. Fish. Aquat. Sci. 125. pp. 335–340.

Sluczanowski, P.R.W. 1994. SharkSim: raising industry awareness. Agric. Syst. Inf. Technol. **6**(1): 43–44.

Sluczanowski, P.R.W., and Prince, J.D. 1994. User interface adds value to fisheries model — ABASIM. Agric. Syst. Inf. Technol. **6**(1): 44–46.

Walters, C.J. 1986. Adaptive management of renewable resources. McMillan Pub. Co., New York, N.Y.

Walters, C.J., Hall, N., Brown, R., and Chubb, C. 1993. Spatial model for the population dynamics and exploitation of the Western Australian rock lobster, *Panulirus cygnus*. Can. J. Fish. Aquat. Sci. **50**: 1650–1662.

Co-management of the razor clam *(Siliqua patula)* fishery at Haida Gwaii, British Columbia, Canada

R. Russ Jones and Dolores A. Garza

Abstract: The current political climate of treaty making in Canada and a new federal policy, the Aboriginal Fisheries Strategy (AFS), recently established a policy basis for co-management agreements in British Columbia. An agreement between the Council of the Haida Nation and Canada's Department of Fisheries and Oceans for the razor clam (*Siliqua patula*) fishery in Haida Gwaii (Queen Charlotte Islands) is an example of the type of agreement possible under the AFS. This agreement resolved regulatory issues in the commercial fishery, in which almost all participants were Haida. Income in the fishery is low but employment and benefits are important to the Haida community of Old Masset. This agreement is compared to instances of co-management with indigenous groups in Washington and Alaska. Canada's efforts to improve First Nations access to fishery resources in British Columbia is likely to be slow until treaty negotiations are completed. In the meantime, co-management arrangements are more likely to be established in fisheries where First Nations participation is high, social and political factors heighten the profile of an issue, allocation and jurisdictional issues can be overcome, and conservation may be improved.

Résumé : Récemment, avec les traités qui se négocient au Canada et la nouvelle politique fédérale énoncée dans la Stratégie sur les pêches autochtones, il s'est constitué une assise politique sur laquelle on a pu fonder des ententes de cogestion en Colombie-Britannique. Ainsi, l'accord conclu entre le conseil de la Nation haida et le ministère des Pêches et des Océans du Canada au sujet de la pêche du couteau (*Siliqua patula*) de Haida Gwaii (îles de la Reine-Charlotte) est un exemple du type d'ententes que la Stratégie a rendues possibles. Cet accord a réglé les questions relatives à la réglementation de la pêche commerciale, où presque tous les participants sont des Haidas. Les revenus tirés de cette pêche sont faibles, mais l'emploi et les avantages qu'elle représente sont importants pour les Haidas d'Old Masset. Nous comparons cette entente à des arrangements de cogestion conclus avec des groupes autochtones des États de Washington et d'Alaska. Au Canada, les efforts déployés pour améliorer l'accès des Premières Nations aux ressources halieutiques en Colombie-Britannique risquent de n'avancer que lentement, tant que des traités seront en cours de négociation. En attendant, des arrangements de cogestion seront vraisemblablement conclus pour les pêches où la participation des Premières Nations est importante, où certaines questions attirent plus l'attention à cause des facteurs sociaux et politiques qu'elles mettent en jeu, où les différends en matière d'allocations et de pouvoirs peuvent être résolus et où il est possible d'améliorer la conservation des ressources.
[Traduit par la Rédaction]

Introduction

In August 1994, the Council of the Haida Nation (CHN) and Canada's Department of Fisheries and Oceans (DFO) concluded a co-management agreement for the razor clam fishery in Haida Gwaii (also known as the Queen Charlotte Islands) (Fig. 1). The CHN is a political organization representing approximately 5000 Haidas whose traditional territories include the lands and waters surrounding Haida Gwaii. The agreement followed four years of uncertainty during which DFO unsuccessfully tried to introduce new clam licenses to regulate the mainly Haida commercial razor clam fishery. As a Haida advisor, the primary author was a participant and observer throughout the negotiation process.

An understanding of the policy basis and process for developing co-management agreements is important for managers who work with aboriginal organizations. Management implications of the razor clam agreement are explored, including the relative importance of regulatory, allocation, and conservation issues to the final agreement. The situation in Haida Gwaii is compared with examples of co-management arrangements in Washington and Alaska. Preliminary results are presented, although it is too early to assess the outcome of the agreement. Initiatives such as the razor clam agreement address current issues within the limitations of existing policies. Negotiation of treaties will likely lead to further changes to aboriginal fishing policies in B.C. but workable co-management examples

R.R. Jones. Haida Fisheries Program, Council of the Haida Nation, P.O. Box 98, Skidegate, Haida Gwaii, BC VOT 1SO, Canada.

D.A. Garza.[1] Marine Advisory Program, University of Alaska Fairbanks, 700 Katlian St., Suite D, Sitka, AK 99835, U.S.A.

Correct citation: Jones, R.R., and Garza, D.A. 1998. Co-management of the razor clam *(Siliqua patula)* fishery at Haida Gwaii, British Columbia, Canada. *In* Proceedings of the North Pacific Symposium on Invertebrate Stock Assessment and Management. *Edited by* G.S. Jamieson and A. Campbell. Can. Spec. Publ. Fish. Aquat. Sci. 125. pp. 385–391.

[1] Represents Southeast Alaska on the Alaska Sea Otter Commission.

Fig. 1. Location of North Beach and Haida Gwaii (Queen Charlotte Islands).

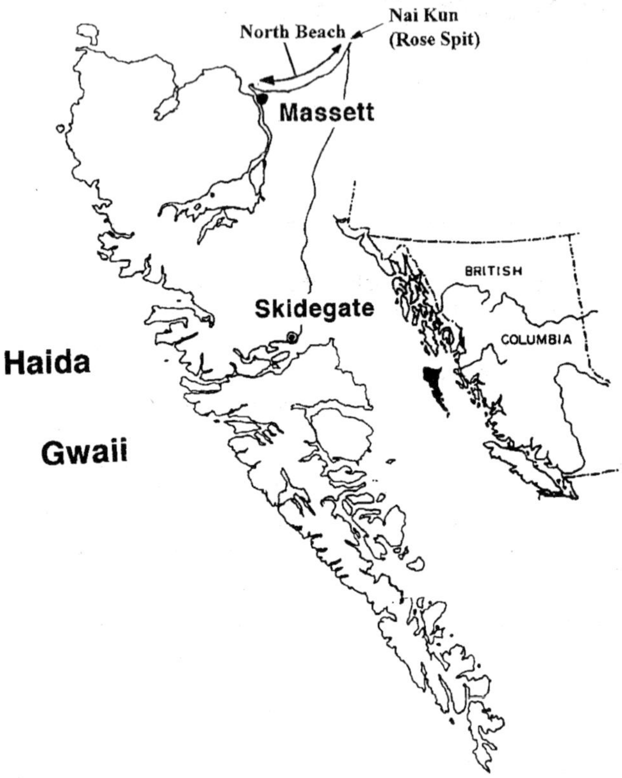

through the Aboriginal Fisheries Strategy (AFS) could have application in future treaties.

Background

Aboriginal policies in British Columbia are evolving in response to legal decisions and a new political climate of treaty making. Implementation of a recent federal AFS has created an opportunity to develop co-management arrangements in advance of treaties that could address a wide range of fishery management objectives.

Fisheries management review

Co-management is a shared decision-making process, formal or informal between two or more groups, usually between a government agency and an indigenous people[2] or other stakeholder, for managing an exploited species or other resource (James Schwarber, Native American Fish & Wildlife Society, Anchorage, Alaska). Co-management arrangements provide another level of consultation and an opportunity to address modern fishery management goals through new processes. Valencia and VanderZwaag (1989) reviewed a wide variety of maritime allocation and management arrangements and conflicts with indigenous people. Each arrangement was unique to the individual groups and their legal and/or cultural settings. Understanding the basis for relationships between indigenous peoples and government is important for fisheries managers as they adapt to and provide for new co-management arrangements and, eventually, treaty obligations in Canada.

Management goals and objectives in Canadian and U.S. fisheries have gone through several distinct periods of change in the past few decades. Following World War II, maximum sustained yield (MSY) was the embodied management objective (Larkin 1977). Economic considerations were incorporated during the 1960's and early 1970's and maximum economic yield (MEY) was an often-used goal. Beginning in the early 1970's policy makers began looking at social, political, and economic considerations.

In 1976, Canada adopted a "best use" policy for commercial fisheries (DOE 1976). Best use was defined as "the sum of net social benefits derived from the fisheries and the industries linked to them" (DOE 1976). The 1995 DFO management plan for intertidal clams identified five goals and objectives which included "the optimal use of shellfish stocks in order to meet social and economic objectives of the people of Canada" (DFO 1994). General goals and objectives such as these support greater involvement by aboriginal peoples in fisheries management.

Canadian legal framework

In the absence of negotiated solutions, court decisions have been a driving force for aboriginal policies in Canada. Two landmark court decisions, the 1973 *Calder* decision and the 1990 *Sparrow* decision, are important to understanding the special relationship of the Crown to aboriginal peoples in Canada.

The *Calder* decision fundamentally changed Canada's policy towards aboriginal title in British Columbia (Canadian Bar Association 1988). In *Calder*, the Nisga'a brought the matter of aboriginal ownership and jurisdiction in their traditional territories to the Supreme Court of Canada. Six of the seven judges agreed that aboriginal title existed. However, these six judges split 3–3 about whether or not aboriginal rights had been extinguished. Following the decision, Canada began a comprehensive claims process to negotiate land claim agreements with First Nations. After 20 years of negotiation, the Nisga'a and Canada adopted an agreement-in-principle in early 1996 that is expected to form the basis of the first modern treaty in British Columbia.

The *Sparrow* decision interpreted the aboriginal right to fish in terms of 1982 amendments to the Canadian Constitution contained in the Canadian Charter of Rights and Freedoms (Supreme Court of Canada 1990). Aboriginal rights are entrenched in Section 35(1) of the Constitution, as follows:

"The existing aboriginal and treaty rights of the aboriginal peoples of Canada are hereby recognized and affirmed."

In *Sparrow*, the court ruled that the aboriginal right to fish for food and for social and ceremonial purposes is constitutionally protected and has priority over commercial and recreational uses. The rights are communal in nature and can evolve with time. The Crown has a trust relationship with aboriginal peoples and government regulations cannot unduly infringe on their rights unless proper justification, such as conservation, can be proven. The decision meant that existing statutes and regulations such as the Fisheries Act could be subject to continued challenge on the basis of interference with the

[2] Throughout this paper the terms indigenous or aboriginal people will be used interchangeably with First Nation, Alaska Native, and Treaty Tribe used in British Columbia, Alaska, and Washington State, respectively.

aboriginal right to fish. The *Sparrow* decision became a driving force for federal policy changes concerning aboriginal fishing.

In Canada, litigation has been necessary to begin to resolve basic issues concerning aboriginal rights and title. Other options for resolving conflicts are negotiation and confrontation (Canadian Bar Association 1988). Negotiation now forms the cornerstone for both a new treaty process in British Columbia and DFO's Aboriginal Fisheries Strategy, an interim policy dealing with aboriginal fisheries.

Emerging policies affecting co-management

The treaty process in British Columbia

In 1992, Canada, British Columbia, and First Nations endorsed a process for negotiating treaties throughout British Columbia (B.C. Claims Task Force 1991). The CHN is one of 43 First Nations currently in the treaty process, which is a six-stage process facilitated by a neutral British Columbia Treaty Commission.

Recent treaties in northern Canada indicate a trend toward co-management of natural resources with aboriginal peoples. The Inuvialuit Final Agreement with the Inuit of the Western Arctic and the James Bay – Northern Quebec Agreement with the Cree and Inuit of Northern Quebec assign specific fishery management responsibilities to aboriginal peoples, involving varying degrees of co-management (Parsons 1993).

The aboriginal fisheries strategy

DFO established the AFS in June 1992, as a policy for addressing issues raised by the *Sparrow* decision (DFO 1992, 1993). AFS is a 7-year, $140 million program which represents a major turning point in the federal approach to aboriginal fisheries policy (Parsons 1993).

A major focus of AFS has been negotiating aboriginal fishery harvest levels and regulating fish sales under community fishery agreements. The process involves reaching formal agreements between DFO and First Nations on resource issues. Program elements include joint management initiatives, a commercial license buyback, and demonstration projects allowing the sale of aboriginal caught salmon. Joint management initiatives have included stock assessment, catch monitoring, salmonid enhancement, enforcement, and economic development. In 1995, DFO concluded 61 agreements with First Nations in British Columbia (DFO 1996).

The objectives of AFS are to: (*a*) improve conservation, management, and enhancement of the resource through negotiating fisheries agreements with aboriginal peoples to provide the framework to manage fisheries in a manner consistent with the *Sparrow* decision; (*b*) improve management of aboriginal fisheries through co-management projects with aboriginal people; (*c*) contribute to economic self-sufficiency of aboriginal communities; (*d*) contribute to the development of self-government and treaties; and (*e*) provide mitigation to current licence holders for transfer of commercial fishing opportunity to aboriginal peoples under AFS (DFO 1996).

New regulations were enacted under the Fisheries Act to enable DFO to issue Aboriginal Communal Fishing Licenses to aboriginal organizations. A communal licence can override other regulations in the Fisheries Act to reflect negotiated agreements between DFO and a First Nation.

A commercial license buyback was initiated to equitably transfer allocations from the commercial fishing sector to First Nations. The program was revised in 1994 so that licences would be communally held but subject to the usual commercial fishery rules and regulations (DFO 1996). In 1993 and 1994, pilot sales agreements allowing commercial sale of salmon were negotiated in three areas where sale of food fish had become a prominent issue: the lower Fraser River, the west coast of Vancouver Island, and the Skeena River. Overall allocation changes from commercial fishers to First Nations as a result of the program have been minimal. DFO decided not to expand this program to other areas because of manageability concerns and opposition by commercial fishers.

The Haida Gwaii razor clam fishery

The razor clam (*Siliqua patula* Dixon,) known as k'aamalh in Haida, is important culturally, socially, and economically to the Haida of the village of Old Masset. Biology and stock size were studied in the 1970's (Bourne 1969; Bourne and Quayle 1970) and a 1994 study indicated that current landings were approximately 90% of MSY (Jones et al. 1998). The stock is the largest in British Columbia and has supported a commercial fishery for more than 70 years. Almost all participants in the commercial fishery are Haida. Small recreational and traditional Haida food fisheries also occur.

The importance of shellfish is noted in Haida oral history and recorded by early ethnographers such as Swanton (1909). Archaeological evidence shows human habitation in Haida Gwaii going back 8 000 – 10 000 years (Fladmark 1989). Shellfish middens at sites such as Blue Jacket Creek near Masset indicate significant utilization of shellfish and a transition to more permanent settlements about 5000 years ago (Fladmark 1989).

Razor and other clams were traditionally preserved by drying (Ellis and Wilson 1981). The black tip of the siphon and the stomach, which are known to accumulate paralytic shellfish poisoning (PSP), were not utilized. Razor clams are still an important Haida food item, collected mainly by hand digging. Old Masset villagers watch for winter storm conditions that can wash scallops and razor clams up on the beach where they can be easily collected.

The commercial razor clam fishery is labor intensive and harvesting methods have changed little over the years despite changes in products and markets. Clams are located by "shows," dimples in the sand produced by a clam's siphon, and dug by hand with a shovel. Haida crews today usually work with the owner of a truck who brings them back and forth from the beach and delivers their clams to a processing plant in Masset on a daily basis.

The commercial fishery started around 1923 when a clam cannery was established at the mouth of the Hiellen River on North Beach. Haidas were the main labor force involved in digging and working on the processing lines. Currently, the market is almost exclusively for a whole frozen product which is sold for crab bait.

The commercial fishery has a relatively low economic value and is passively managed through size limits. The commercial fishery occurs from March to September during daylight low tides. The main conservation measure is a 90-mm shell length (SL) size limit. The fishery is open year-round but has occasional closures due to PSP or domoic acid presence. In 1994, preliminary commercial landings were 103.3 t with a value of C$182 000 and average price of $\$1.76 \cdot kg^{-1}$ (Maureen Kostner, DFO Catch Statistics, Vancouver, February 1996). The average annual value of landings over the past five years (1990–1994) was C$119 000 based on DFO Catch Statistics.

Haida participation in the commercial fishery is high. In 1995, DFO issued five category "Z" clam licences and the CHN issued 244 designation cards. In 1994, there were between 40 and 60 commercial diggers on the beach during favorable low tides (Jones et al. 1998). The fishery has attracted few nonresidents because the work is hard, income is low, and harvesting is spread out through the season.

There are relatively small Haida food and recreational fisheries for razor clams. Recreational fishing is done by island residents and visitors, many staying in campgrounds in Naikoon Provincial Park. In 1994, an average of 3.6 noncommercial digging parties per tide ($3.0 \text{ persons} \cdot \text{party}^{-1}$) were observed on the beach from April to September (Jones et al.1998). The average catch was $81 \text{ clams} \cdot \text{party}^{-1}$ on tides <1.0 m. In 1994, the recreational fishery was regulated by a bag limit of 75 clams, a possession limit of 150 clams, and no size limit. There are no restrictions on the Haida food fishery.

There have been only two studies of the status of the North Beach razor clams. A 1967 population assessment concluded that the stock was relatively lightly exploited (Bourne 1969; Bourne and Quayle 1970). In 1994, the Haida Fisheries Program conducted a population assessment that was funded under AFS. The exploitable biomass was estimated at 636 t with a sustainable yield of about 120 t (Jones et al. 1998). Studies were initiated because of a decline in razor clam landings in 1992 and 1993 and a high proportion of undersized clams reported in the catch. Landings improved in 1994, but population assessments were repeated in 1995 and 1996 as part of the co-management agreement.

Chronology of events

AFS provided a window of opportunity for DFO and the Haida to resolve a dispute over razor clam licencing that began in 1989. A description of the source of the controversy and important events leading up to the razor clam agreement follow.

In 1989, DFO introduced a new regulation requiring commercial diggers throughout British Columbia to obtain an area-specific category "Z" clam license. The measures were introduced in response to increased fishing effort on southern British Columbia beaches where repeated commercial digging and the appearance of undersized clams in the catch led to stock depletion concerns.

From 1989 to the time of the agreement, Haida fishers refused to obtain the DFO "Z" license. The Haida were concerned that they would eventually lose access to the fishery as a result of a new licensing program. While DFO had the option of actively enforcing the new regulations, this did not occur. In 1990, DFO and CHN explored the possibility of a Haida community license, but DFO had no suitable policy framework for such an innovation. Active dialogue from year to year was important in preventing direct confrontation between DFO enforcement staff and Haida fishers.

In 1992, new policies enacted under the AFS allowed for negotiated agreements with First Nations and provided a mechanism for DFO to issue communal licenses to aboriginal organizations. The razor clam fishery was an obvious area for DFO and the CHN to develop an agreement (Sue Farlinger, DFO, personal communication). Under AFS, two agreements were negotiated that related to razor clams: (*i*) a 1992 general agreement that defined relationships, and (*ii*) a 1994 agreement that outlined co-management details.

Framework agreement under AFS

Under AFS, the CHN and DFO signed a Framework Interim Measures Agreement in 1992, which was revised and renewed in 1993 and 1997 (the "Framework Agreement"). The agreement laid out a process listing issues for discussion, provided funding of up to $1 million per year for co-management activities, and set aside fundamental differences expected to be resolved through the treaty process.

The Framework Agreement laid out a three-tiered process for consultation: a negotiating level, a policy development and planning level, and a technical level. At the second level, a Co-operative Management Group (CMG) consists of 3 CHN and 3 DFO representatives responsible for reviewing issues and implementing agreements. Four joint DFO/CHN technical committees were formed, one each for salmon, shellfish, herring, and halibut and groundfish. The Joint Shellfish Technical Committee (JSTC) consists of three DFO and three CHN members. DFO members are from Science Branch, Policy and Planning Branch, and Operations Branch.

The Framework Agreement reflects a balancing of conflicting interests and obligations between the federal government and the Haida Nation. The two parties state a shared interest in conservation and co-operation in fisheries management. Major differences over jurisdiction and ownership are stated as follows:

"The Parties maintain viewpoints that converge regarding protection of the marine and freshwater ecosystems of the Islands and diverge with respect to sovereignty, title, or ownership as follows:

The Haida Nation sees the Islands as Haida territory by virtue of heredity. On this basis the Haida Nation sees that it maintains jurisdiction over, and management responsibility for, the Islands and that citizens of the Haida Nation have priority rights to access the Islands' fishery resources.	The government of Canada sees the marine and anadramous fisheries resources of the Islands and surrounding waters as a resource to be developed for the benefit of all Canadians with management responsibility vested in the Minister of Fisheries and Oceans."

The Framework Agreement provided funding for co-management activities. In 1994, the CHN undertook a razor clam population study under the direction of the JSTC. Also, discussions to develop a razor clam agreement began through the new process.

The Razor Clam Subagreement

The CHN and DFO concluded a Razor Clam Subagreement (RCS) in August 1994 that detailed the co-management process and new management arrangements, including licencing in the commercial fishery. The RCS has a three-year term, beginning in January 1995. The main elements of the RCS are (*i*) New commercial licencing practices are introduced for the Haida commercial razor clam fishery. DFO issues a communal licence to CHN; (*ii*) CHN designates commercial Haida razor clam diggers that fish under the communal licence; (*iii*) DFO continues to issue clam "Z" licenses to non-Haidas and limits entry to non-Haidas beginning in 1995. Licences were only renewed for fishers who held licences and recorded landings in three of the previous five years; (*iv*) CHN and DFO jointly monitor the fishery and enforce regulations; (*v*) CHN and DFO co-operatively manage the entire fishery through the JSTC which develops (*a*) a joint management plan, (*b*) a co-operative program to monitor and assess the fishery, and (*c*) a research plan. The joint management plan is reviewed by CHN and DFO each year; (*vi*) Final decision making is left to each party, but the CMG and JSTC are expected to make consensus recommendations; (*vii*) CHN collects clam samples for PSP testing to address public health and safety concerns.

A stated purpose of the RCS is "to serve as an interim measure to address Haida concerns about the impacts of commercial licensing on Haida participation in the razor clam fishery ...." The RCS is not a treaty or land claim agreement and is without prejudice to aboriginal rights or treaty negotiations. The RCS is important because it defines an interim co-operative arrangement involving co-management of an existing commercial fishery and clearly defines a role for CHN in monitoring and enforcement of the fishery.

The management planning process is based on consensus recommendations by joint committees set up under the Framework Agreement (CMG and JSTC). Potential management issues were addressed in the RCS: a minimum commercial size limit of 90 mm SL was maintained, the management area was defined, only hand digging was allowed, and participant identification and labeling of product was required. Each year of the RCS, the JSTC recommends a joint management plan to the CMG which must recommend a plan to the parties by October 1. The plan is implemented on January 1. This timing allows for review within DFO's usual management planning process for shellfish.

The RCS directs the JSTC to develop a co-operative program to assess sustainable harvest levels for the fishery, monitor and survey the fishery, exchange information about the fishery, and evaluate the RCS.

Main biological issues relate to uncertainty about the size of the stock and concerns about commercial over-harvesting. In 1995 and 1996, the Haida Fisheries Program repeated population surveys begun in 1994 with the aim of assessing sustainable yields in the fishery.

Co-management framework in nearby regions

It is useful to compare the razor clam example with co-management arrangements involving aboriginal groups in other legal and political settings. There are notable differences in treaty rights and aboriginal policies in British Columbia, Washington State, and Alaska that influence the process for reaching agreements in each jurisdiction.

Treaty tribe co-management of salmon in Washington State

In Washington State, Indian tribes and the state government went through a long period of confrontation over fishing rights. The dispute was finally resolved by the Boldt decision in 1974, which held that Indian tribes in Washington State and the government of the United States were bound by earlier treaty obligations to provide for an equal sharing arrangement (Cohen 1986).

Judge George H. Boldt ruled that Indian tribes were entitled to the opportunity to catch enough fish to provide a moderate living, up to a maximum of half the harvestable number of fish in each run. The decision was subsequently affirmed by the United States Supreme Court. At the time of the decision, treaty tribes were catching only 5% of the salmon harvest. Over the next 10-year period, treaty Indian catches gradually increased to approximately 50% of the harvest.

Judge Boldt ruled that the tribes had the authority to regulate their own off-reservation fisheries provided they met certain conditions establishing their management capability. The state and federal governments and tribes subsequently embarked on developing a new management relationship under the direction of the court. Cohen (1989) described the efforts taken by the tribes to further develop their management capabilities. Individual tribes developed tribal rolls and began to license fishermen. They built up their management staff, hired fisheries biologists, and formed the Northwest Indian Fisheries Commission as a co-ordinating body to provide assistance, technical advice, program development, and public information.

Once a court solution was imposed and a process established through the court, the treaty tribes and state agencies were able to develop a productive co-management arrangement. In this allocation dispute, fundamental differences were so great that negotiations failed. This illustrates the limitation of negotiation as a strategy to resolve major allocation issues.

Alaska marine mammal management

In the past few decades, Alaska Natives renewed harvests of whale, walrus, and sea otter stocks that had been depleted by large-scale harvests since the time of contact with Euro-Asians. This situation provided an opportunity for new co-management arrangements involving the federal government, which has fiduciary duties to Alaska Natives, the State of Alaska, and tribal organizations such as marine mammal commissions.

The federal Marine Mammal Protection Act (MMPA) of 1972 places a national moratorium on marine mammal harvest but provides a broad exemption to Alaska Native hunters for "subsistence purposes" or to create "authentic native" handicrafts or clothing. The federal government can regulate native taking of a marine mammal only when a species becomes depleted.

Alaska Native organizations and tribes have been successful at establishing co-managment arrangements with the federal government that limit the need for government regulation. Structures such as the Alaska Eskimo Whaling Commission, the Eskimo Walrus Commission, and the Alaska Sea Otter Commission provide an opportunity for self-regulation in a manner responsive to traditional management practices (Langden 1989).

The federal government has responded to changes in marine mammal status and Alaska Native needs. Alaska Natives have been successful at developing co-operative arrangements for marine mammal management, partly because Alaska Natives are the only harvesters, the resources are mostly in healthy condition, and harvest levels are relatively small.

Discussion

AFS provided a policy mechanism to address the major issues of concern to DFO and CHN in the razor clam fishery. Results are reviewed by fishery management issue: regulation, allocation, and conservation (Hanna and Smith 1992). A discussion of social/economic issues and management implications follows.

Regulatory issues

A main issue addressed by the Razor Clam Subagreement (RCS) was concern about loss of Haida access in the commercial fishery as a result of new licencing measures. Limited entry licencing in the commercial salmon fishery led to declines in aboriginal participation in British Columbia and in Alaska. In British Columbia, Indian participation in the salmon fisheries declined by roughly 60% between 1964 and 1971, despite Canadian government policies to protect Indian participation in commercial fisheries (Pearse 1982). Also, by the 1970's, the fishing and processing industries employed less than half the number of Indians that had been involved two decades earlier (Pearse 1982). Loss of commercial licences by Haida in halibut and salmon fisheries was particularly severe and led to loss of income to and employment of individuals as well as to Haida Gwaii communities.

DFO interests in seeking an agreement with CHN were to achieve a regulated fishery, as well as promoting involvement of the Haida in their razor clam fishery, and to test new approaches as possible building blocks for treaties (Sue Farlinger, DFO, personal communication).

Allocation issues

In Canada, existing laws now require fishery managers to take into account the aboriginal fishing priority for food, social, and ceremonial purposes. Allocation, however, was not a major issue in the RCS. Almost all participants in the commercial fishery were Haida. In general, the RCS did not adversely affect the few non-Haida fishers. In contrast, allocation conflicts over valuable Washington salmon eventually required a court decision before a solution was reached.

The RCS established limited entry for non-Haida fishers. While not a formal allocation, this licencing measure ensured a high level of Haida participation in the fishery and curbed potential increase in non-Haida participation. This addressed managers' concerns about a potential influx of new participants.

Conservation issues

While the RCS promotes conservation, it was not the determining factor in reaching a negotiated agreement. As a result of the RCS, however, the parties now know more about the stock, have a better idea of catch, and can better recognize signs of possible crisis (Sue Farlinger, DFO, personal communication). The RCS supports research on razor clams and involves DFO Science Branch in monitoring and review.

Economic/social issues

From the Haida point of view, there were important economic and social issues at stake. The commercial razor clam fishery has a relatively low overall value but provides income and employment for Haidas early in the year when there are few other fishing opportunities. Unemployment in the village of Old Masset was 68% in 1994 (Clarence Nyce, CEIC Skeena Aboriginal Management Board, Terrace, B.C., 1994). Although average daily earnings of a commercial digger are low (C$40–$200 per day), overall investment is minimal, consisting of a shovel and containers, meaning fishers are the primary beneficiaries.

Razor clam fishing also provides critical employment to help qualify fishers for unemployment insurance benefits. Unemployment insurance for fishermen is an integral part of the economics of many coastal fishing communities (Parsons 1993).

The RCS is consistent with DFO social and economic objectives under AFS including involving participants in management and contributing to economic self-sufficiency of aboriginal communities (DFO 1996).

Management implications

The RCS demonstrates some of the benefits of co-management, highlights the importance of well-defined objectives, and, although not adopted as a long-term policy, may have applications to future treaty arrangements in B.C. fisheries.

The razor clam fishery is relatively unique in B.C. since few other commercial fisheries have as high a level of aboriginal participation. Other AFS agreements have similar goals of providing aboriginal access to commercial fisheries. For example, DFO provided exclusive fishing opportunities through a 1993 AFS agreement to the Heiltsuk Tribal Council in a new commercial clam fishery on the B.C. central coast. The fishery had been closed for a number of years but was reopened subject to PSP sampling by the Tribal Council. Agreements such as the RCS are setting the stage for participation by aboriginal peoples in management. They provide an opportunity for aboriginal fisheries guardians and technicians to develop working relationships with DFO staff in management decision making, fisheries enforcement, and collection of fisheries information.

Negotiated agreements help define the relationship between aboriginal groups and government agencies and the responsibility of each party in a co-management arrangement. To this end, the RCS provides a clear statement of understandings between CHN and DFO and sets criteria for performance and decision making.

Agreements such as the RCS, the Heiltsuk clam agreement, and pilot sales agreements are useful for testing fishing policies or arrangements (DFO 1992). About November 1996, AFS was renewed beyond March 31, 1999, as a link to treaties (Jim Wild, DFO, personal communication). While the Framework Agreement and RCS between the CHN and DFO are interim agreements which do not deal with long-term allocations or management processes, they do provide immediate economic benefits to aboriginal communities and opportunities for self-management of fisheries.

Acknowledgments

A variety of people were key to formulation of provisions of the RCS, including CHN representatives Miles Richardson, Arnie Bellis, and Don Yorke, and DFO representatives Maryantonett Flumian, Marion Lefebvre, and Sue Farlinger. Nancy Morgan was legal counsel for the CHN and provided helpful comments on an earlier draft of this paper.

References

Bourne, N. 1969. Population studies on the razor clam at Masset, British Columbia. Fish. Res. Board Can. Tech. Rep. 118. 24 p.

Bourne, N., and Quayle, D.B. 1970. Breeding and growth of razor clams in British Columbia. Fish. Res. Board Can. Tech. Rep. 232. 42 p.

B.C. Claims Task Force. 1991. The Report of the British Columbia Claims Task Force. June 28, 1991. B.C. Treaty Commission, Vancouver, B.C. 84 p.

Canadian Bar Association. 1988. Report of the Canadian Bar Association Committee on aboriginal rights in Canada: an agenda for action. Canadian Bar Association, Ottawa, Ont. 111 p.

Cohen, F.G. 1986. Treaties on trial. A report prepared for the American Friends Service Committee. University of Washington Press, Seattle, Wash. 229 p.

Cohen, F.G. 1989. Treaty Indian tribes and Washington State: the evolution of tribal involvement in fisheries management in the U.S. Pacific Northwest. *In* Co-operative management of local fisheries, new directions for improved management and community development. *Edited by* E. Pinkerton. University of British Columbia, Vancouver, B.C. pp. 37–48.

Department of Fisheries and Oceans, Canada (DFO). 1992. Notes for remarks by the Hon. John C. Crosbie, Minister of Fisheries and Oceans, to a media conference to announce the Aboriginal Fisheries Strategy, Ottawa – Vancouver, June 29, 1992. Crosbie Announces Aboriginal Strategy for B.C. Fishery. NR-HG-92-55E. Crosbie Announces Agreements with Lower Fraser Aboriginal Groups. NR-HG-56E. Voluntary Licence Retirement Program Announced. NR-HG-92-57E. Aboriginal Fisheries Strategy – Backgrounder I: The Program – Backgrounder II: The Context – Backgrounder III: Native Groups Participate in Management, Enhancement – Backgrounder IV: Sale to be Tested in 1992.

DFO. 1993. Fisheries Act, Aboriginal Communal Fishing Licences Regulations; Regulatory Impact Analysis Statement. Can. Gaz. Pt. II, **127**(13): 2899–2904.

DFO. 1994. Pacific Region 1995 Intertidal Clam Management Plan. Department of Fisheries and Oceans, Vancouver, B.C. 43 p.

DFO. 1996. AFS Discussion Document, slide presentation by DFO Fisheries Management Branch, April 1996, 36 p. Summary of AFS Review Results, Pacific Region First Nations Consultation Meeting, slide presentation by DFO Aboriginal Fisheries Branch, April 25, 1996, Vancouver, B.C. 26 p.

DOE. 1976. Policy for Canada's commercial fisheries. Department of Environment, Ottawa, Ont. 70 p.

Ellis, D.W., and Wilson, S. 1981. The knowledge and usage of marine invertebrates by the Skidegate Haida people of the Queen Charlotte Islands. Queen Charlotte Islands Mus. Soc. Monogr. 1. 42 p.

Fladmark, K.R. 1989. The native culture history of the Queen Charlotte Islands. *In* The outer shores. Based on the Proceedings of the Queen Charlotte Islands First International Symposium. *Edited by* G.G.E. Scudder and N. Gessler. University of British Columbia, Vancouver, B.C. August 1984. pp. 199–221.

Hanna, S.S., and Smith, C.L. 1992. Resolving allocation conflicts in fishery management. Soc. Nat. Resour. **6**: 55–69.

Jones, R.R., Schwarz, C., and Lee, L. 1998. Intertidal population estimate of razor clams (*Siliqua patula*) at North Beach, Haida Gwaii/Queen Charlotte Islands and applications to fishery management. *In* North Pacific Symposium on Invertebrate Stock Assessment and Management, March 6–10, 1995. *Edited by* G.S. Jamieson and A. Campbell. Can. Spec. Publ. Fish. Aquat. Sci. 125. pp. 199–211.

Langden, S.J. 1989. Prospects for comanagement of marine mammals in Alaska. *In* Co-operative management of local fisheries, new directions for improved management and community development. *Edited by* E. Pinkerton. University of British Columbia, Vancouver, B.C. pp. 154–169.

Larkin, P.A. 1977. An epitaph for the concept of maximum sustainable yield. Trans. Am. Fish. Soc. **106**: 1–11.

Parsons, L.S. 1993. Management of marine fisheries in Canada. Can. Bull. Fish. Aquat. Sci. 225. 763 p.

Pearse, P.H. *(Commissioner)*. 1982. Turning the tide: A new policy for Canada's Pacific fisheries. The Commission on Pacific Fisheries Policy, Minister of Supply and Services, Ottawa, Ont. 180 p.

Supreme Court of Canada. 1990. *Ronald Edward Sparrow* v. *Regina et al.* Western Weekly Rep. pp. 411–444.

Swanton, J.R. 1909. Contribution to the ethnology of the Haida. Am. Mus. Nat. Hist. Mem. **8**: 1–300.

Valencia, M.J., and VanderZwaag, D. 1989. Maritime claims and management rights of indigenous peoples: rising tides in the Pacific and northern waters. Ocean Shoreline Manage. **12**: 125–167.

Community-based sea urchin (*Tripneustes gratilla*) grow-out culture as a resource management tool

Marie Antonette Juinio-Meñez, Natalie N.D. Macawaris, and Helen Grace P. Bangi

Abstract: The collapse of the valuable sea urchin (*Tripneustes gratilla*) fishery in Bolinao, Pangasinan, Philippines, has led to the formulation of an alternative approach to sea urchin management which integrates culture technology and local community participation. A conceptual model focusing on family/village-managed reproductive reserves in the form of sea pens or cages where juvenile sea urchins can be grown and selectively harvested (i.e., >70 mm test diameter) is presented. This concept is reviewed with respect to ecological, educational, and economic values based on field and laboratory studies and grow-out culture practice of some fishers in Nalvo, Ilocos Sur. Implications of the need for some limited exclusive use of grow-out areas in a traditionally open-access fishery and the development of new marketing systems that would favor higher prices paid to fishers are discussed.

Résumé : L'effondrement de la précieuse pêcherie d'oursin (*Tripneustes gratilla*) de Bolinao, Pangasinan, Philippines, a entraîné l'élaboration d'une nouvelle approche à la gestion de l'oursin qui intègre la technologie de l'aquaculture et la participation des communautés locales. On présente un modèle conceptuel centré sur la création de réserves de reproduction, par le biais de cages ou d'enclos marins, gérées par la famille ou le village, et dans lesquelles les oursins juvéniles peuvent croître et être récoltés de manière sélective (p.ex. diamètre du test >70 mm). Le concept est examiné des points de vue écologique, éducatif, et économique à partir d'études sur le terrain et en laboratoire et d'exemples de culture de croissance pratiquée par certains pêcheurs de Nalvo, Ilocos Sur. On discute des implications de la nécessité de recourir à une certaine forme d'utilisation exclusive limitée d'aires de croissance dans une pêcherie dont l'accès est traditionnellement libre et de l'élaboration de nouveaux systèmes de commercialisation qui favoriseraient une augmentation des prix payés aux pêcheurs. [Traduit par la Rédaction]

Introduction

The roe, or gonads, of the sea urchin *Tripneustes gratilla* is harvested from adjacent reef areas of Bolinao, Pangasinan, Philippines (Fig. 1: 119°50′E, 16°28′N), for both local consumption and commercial markets. Harvest for export markets was intermittent in the mid-1970's, but became more regular in the early 1980's. Since then, the fishery for this species has become a major source of livelihood for many local coastal families. Based on landed sea urchin catches, the fishery generated millions of pesos per annum from 1988 to 1992 (Talaue-McManus and Kesner 1995). Total net income for sea urchin roe landed at one major landing site, Silaqui Island, peaked in 1989 at PhP 9.6 M (US$348 000) for 5.3 t. About 2.5 M sea urchins are expected to have been collected from the reef area (~32 km^2) near this site in 1989, since 40–50 sea urchins (45–85 mm test diameter, TD) are required to produce 1 kg of sea urchin roe. Value of landed roe decreased to PhP 3.2 M (US$128 000) and PhP 1.9 M (US$76 000) in 1991 and 1992, respectively, due to a drastic decline in the landed catch (Talaue-McManus and Kesner 1995). These data are estimated to be about half the total annual roe harvested from Bolinao.

M.A. Juinio-Meñez, N.N.D. Macawaris, and H.G.P. Bangi. Marine Science Institute, University of the Philippines, Diliman, Quezon City, Philippines 1101.

Correct citation: Juinio-Meñez, M.A., Macawaris, N.N.D., and Bangi, H.G.P. 1998. Community-based sea urchin (*Tripneustes gratilla*) grow-out culture as a resource management tool. *In* Proceedings of the North Pacific Symposium on Invertebrate Stock Assessment and Management. *Edited by* G.S. Jamieson and A. Campbell. Can. Spec. Publ. Fish. Aquat. Sci. 125. pp. 393–399.

Assessment of natural populations from monthly size-frequency data in 1987–1988 suggested that urchin stocks were overexploited (M. Trinidad-Roa and J. Pasamonte, Marine Science Institute, University of the Philippines, Diliman, Quezon City, Philippines, unpublished data). Upon the recommendation of the University of the Philippines Marine Science Institute (UPMSI), a seasonal ban on sea urchin collection from December to February (the assumed period of peak spawning) was then legislated by the local government of Bolinao, starting in 1988. However, this closed season was enforced for only two years. Catch continued to decline and the fishery collapsed during the second half of 1992. To protect this species from further exploitation, commercial harvesting and export of sea urchin roe was banned indefinitely effective January 1993. Field surveys during 1992–1995 showed recruits to be rare from October to January of each year when they used to be abundant in previous years (M. Trinidad-Roa, J. Pasamonte, and M. Juinio-Meñez, Marine Science Institute, University of the Philippines, Diliman, Philippines, unpublished data).

Sadly, the scenario of uncontrolled and nonselective harvesting is not unique (e.g., Wells 1981; Ross 1984; Trinidad-Roa 1988; Juinio et al. 1989) and appears to be largely

Fig. 1. Map of Bolinao showing the two field study sites (filled squares), Lucero and Dewey, for *T. gratilla* growth experiments. Inset: Map of the Philippines showing locations of Bolinao, Pangasinan, and Nalvo, Sta. Maria, Ilocos Sur.

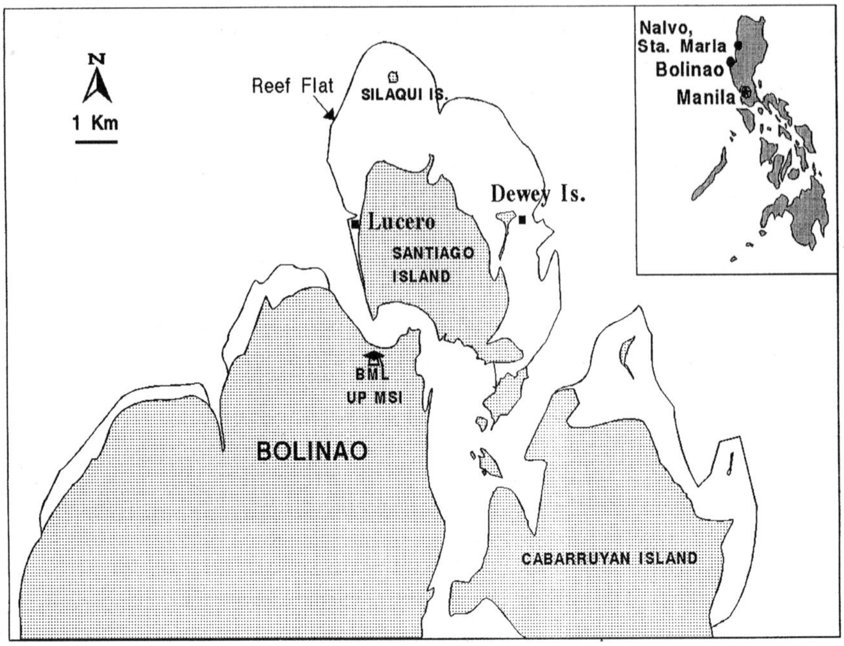

undocumented, with only anecdotal reports from interviews with fishers and traders of depleted populations for many other exploited benthic marine invertebrates in the Philippines. This lack of substantial baseline information on the status of Philippine invertebrate stocks reflects prevalent logistical constraints in monitoring multispecies artisanal fisheries (Lopez 1986). While some conventional fishery regulations (e.g., size limits, closed seasons) exist, these are not well enforced due to various socio-economic constraints (e.g., Garcia 1992). In this situation, there is a need to explore alternative and/or complementary approaches to marine resource management. The objective of this paper is to present the results of field and laboratory studies on the sea urchin *Tripneustes gratilla*, culture practice of fishers, and to propose community-based sea urchin grow-out culture as a management tool. The ecological, educational, and economic factors that would favor the viability of community-based sea urchin culture in Bolinao, Pangasinan, are considered.

Materials and methods

Comparison of the fisheries for *T. gratilla* in Bolinao, Pangasinan, and Nalvo, Sta. Maria, Ilocos Sur (120°10′E, 17°20′N), both on the northwestern coast of Luzon (see inset Fig. 1) was undertaken from site visits and interviews with fishers and buyers at both localities.

Field studies

Monthly field surveys were conducted from March 1992 to October 1994 at selected sites in Bolinao to monitor changes in urchin population density and size structure. Two to three 100-m transect lines were laid at each site by two SCUBA divers and the TD of all sea urchins found within 1 m of each side of the line was measured. Monthly average densities of *T. gratilla* from previous studies in 1987–1989 at one site, Lucero, were incorporated with data gathered here to show the relative change in sea urchin abundance in the Bolinao reef flat before and after collapse of the sea urchin fishery.

Laboratory culture and experimental grow-out studies

After collapse of the sea urchin fishery in Bolinao in 1992, research focused on development of sea urchin culture. Laboratory culture techniques were largely adapted and modified from Shokita et al. (1991). Wild broodstock was collected and spawned artificially with 0.5 M KCl. Larvae were cultured at 25–26°C in 3-litre glass jars, and larval density was 1 individual·ml^{-1}. Early larval stages (2- to 4-arm) were fed daily with 10 000 cells·ml^{-1} of the flagellate *Isochrysis galbana* and from the 6-arm stage to the presettlement stage with 20 000–30 000 cells·ml^{-1} of the diatom *Chaetoceros gracilis*. The benthic diatom *Navicula ramosissima* was provided for newly settled and metamorphosed larvae.

Larval development was monitored every two days until settlement. When juveniles attained a TD of about 10 mm, average growth rates of two to three replicate groups (6–20 individuals per replicate) of juveniles from three spawnings were monitored. Sibling juveniles from each group were either transferred to experimental sea pens (0.5 × 0.5 × 0.5 m polyethylene plastic net) in Dewey and Lucero (Fig. 1) or maintained in laboratory tanks. Larger juveniles were provided unlimited *Sargassum* for food, following the practice of fishers in Nalvo. Average growth (ΔTD) of sea urchins in both sea pens and laboratory tanks were determined weekly from August to December 1994. One-way analysis of variance (ANOVA), followed by Tukey–HSD test, was used to compare average monthly growth rates among sea urchin of the same spawning batch grown in either the laboratory or experimental pens at the two field sites.

Since 1992, meetings with fishers to share information and

Table 1. Comparison of the utilization of sea urchin *Tripneustes gratilla* in Nalvo and Bolinao.

	Nalvo	Bolinao
Fishery	Incidental collection in the wild Small-scale grow-out in pens (1986–1991)	Open access commercial fishery
Market	Local (Occasional export)	Export-oriented (e.g., Manila, Japan)
Product and value Prices (1991–1992) [1 US$ = P 25.00]	(*a*) Whole sea urchin P5.00 per pc (>9 cm) P 3.00 per pc (6–8 cm) P 2.00 per pc (4–5 cm)	(*a*) Brined roe P 65.00–70.00 per kg (*b*) Fresh roe P 70.00 per kg
Regulation	Nonformal agreement (*a*) Standardized prices (*b*) Sequential selling by fishers	Municipal ordinances (*a*) Closed season 1989–1992 (*b*) Moratorium on commercial harvesting 1993–present

discuss natural sea urchin stocks have been conducted regularly. Some fishers provided information on wild harvest areas but the involvement of fishers was greatest in the experimental grow-out studies. Fishers participated in selection of sites for sea pens, committed time and effort in feeding the sea urchins, guarding them from poachers, and assisted in monthly monitoring of growth and survival rates.

Results

Wild harvest and grow-out culture

Sea urchin fisheries in Bolinao and Nalvo are open access. However, there is no large-scale commercial fishery in Nalvo comparable to that in Bolinao. Rather, from 1986 to 1991, a small group of fishers collected juvenile sea urchins and grew them in pens to marketable size.

Utilization of harvested sea urchins also differs between Bolinao and Nalvo. Sea urchins are a regular part of the local diet and are considered a local delicacy in Nalvo while in Bolinao, sea urchins are harvested primarily for export (Table 1). In Nalvo, sea urchins are sold whole at the local market and priced according to size (Table 1). Besides eating the gonads of broiled sea urchin, locals drink the coelomic fluid which they regard as a good source of iodine to prevent goiter. In Bolinao, gonads are shucked from whole urchins and marketed fresh or in brine to local buyers for Manila-based exporters.

Grow-out culture of juvenile sea urchins was undertaken by 11 fisher families in Nalvo from 1986 to 1991. Fishers set up 4 × 4 m, 2-m high sea pens made of woven bamboo strips (US$18 each) at a depth of about 2 m on the reef flat in front of their village. The rocky rubble reef flat was covered by *Sargassum* spp. Pens were stocked with up to 4000 juvenile sea urchins (20–30 mm TD), collected from the wild in October–March. Juveniles were added to pens as they were collected while marketable urchins in the pens were harvested periodically. The grow-out period was a minimum of 3–4 months or until the animals reached a minimum size of 40–50 mm TD. At the onset of the typhoon season in June, pens were dismantled and all marketable sea urchins sold, with smaller ones released onto the reef flat. There was a nonformal agreement among fishers to standardize prices and to take turns selling their cultured sea urchins. Sea urchins were sold only to local buyers since prices offered by roe exporters to Japan were a third of that obtained at the local market.

Between March and May 1987, a total of 10 000 sea urchins were sold by the 11 fisher families, providing about US$100 per family. This additional income was significant considering that the average monthly family income in the area is about US$125. Unfortunately, fishers have been unable to continue their grow-out activities since 1992 due to the lack of natural seedstock.

Population densities

Mean monthly density of *T. gratilla* in Lucero, Bolinao, was highest in 1987 (11.2–214.3 individual·100 m^{-2}) and lowest in 1994 (0.0–0.3 individual·100 m^{-2}) (Fig. 2). No monthly field monitoring of population densities were conducted from June 1989 to February 1992.

Larval development and juvenile growth rates

Average ages of different larval *T. gratilla* developmental stages are summarized in Fig. 3. Total pelagic larval duration (i.e., prior to settlement) ranged from 42 to 52 days. High variability in development rate was observed among larvae from the same spawning batch and among larvae from different batches.

The highest percentage of larvae from a batch which developed into presettlement larvae was 14% with 20–70% of these larvae successfully settling and metamorphosing. From settlement, juveniles took 3–6 months to grow to 10 mm TD. The average growth rate of juveniles from the three spawning batches reared in the sea pens and the laboratory (Figs. 4*a* and 4*b*) was 17 mm·mo^{-1}. At this rate, 10 mm TD juveniles attained the minimum size of sexual maturity (50–60 mm TD as determined by the presence of ripe ovaries and spontaneous spawning in the laboratory tanks and sea pens) in 3–4 months. *T. gratilla* maintained in sea pens at Lucero had an average growth rate of 18 mm·mo^{-1} (±0.81), comparable to the rate of 17.5 mm·mo^{-1} (±1.09) for those maintained in the laboratory tanks. These growth rates were significantly higher (ANOVA, Tukey–HSD test, $P < 0.05$) than growth rates of siblings maintained at Dewey which grew an average of only 14 mm·mo^{-1} (±1.12). The majority of juveniles used in grow-out

Fig. 2. Mean density of *T. gratilla* at Lucero, from August 1987 to May 1989, and from March 1992 to October 1994.

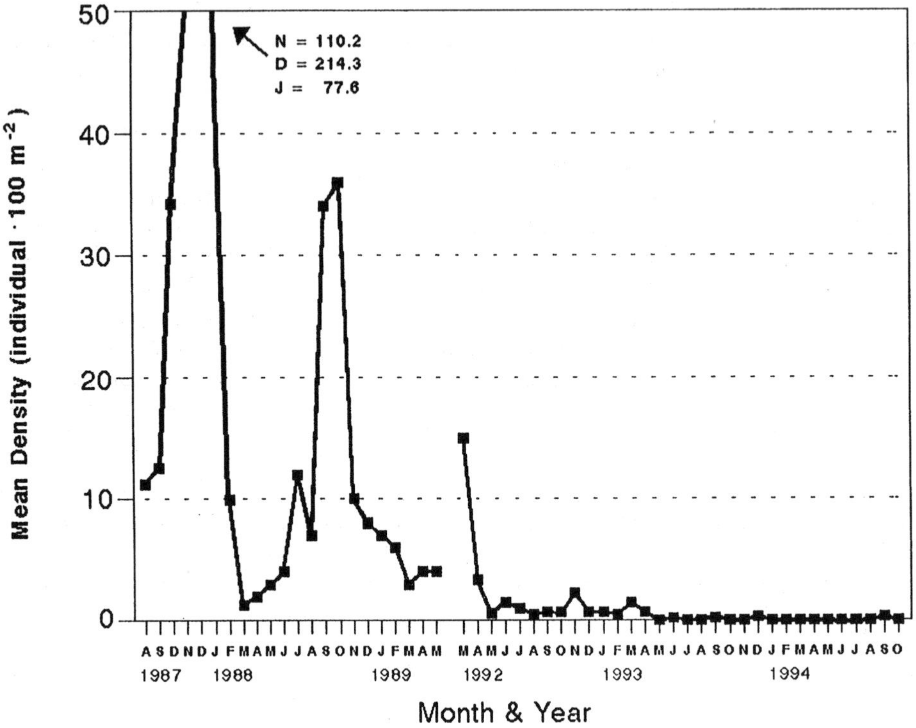

Fig. 3. Average duration of different cultured larval stages of *T. gratilla* at 24–26°C. (R = range of duration where at least 60% of larvae are at that stage).

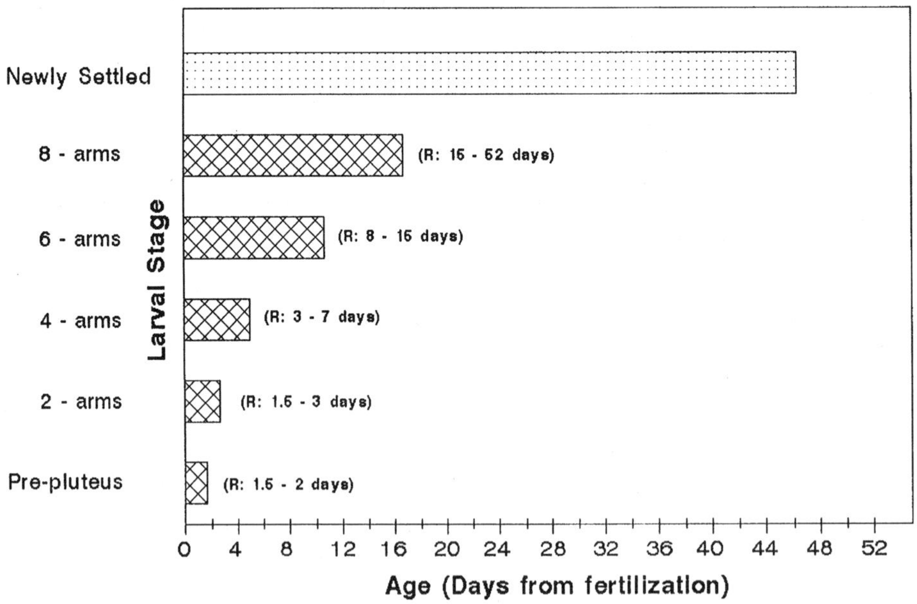

experiments from the three batches attained sexual maturity 7–8 months after artificial fertilization.

Discussion

Conceptual framework

The very low densities of *Tripneustes gratila* observed since 1992 relative to densities in 1987–1988 (Fig. 2) are correlated with the progressive decline in the volume of landed roe from 1989 to 1992 (Talaue-McManus and Kesner 1995). This suggests that either overharvesting or natural cycling of the population effected the collapse of the sea urchin fishery and near depletion of this species in Bolinao, Pangasinan. Regardless of the cause, to enhance both recovery of sea urchin populations and more importantly, management of marine resources in

Fig. 4. Average growth curves of three batches of artificially reared sea urchins grown in sea pens at (*a*) Lucero, and (*b*) Dewey, Bolinao, and in outdoor laboratory tanks in 1994.

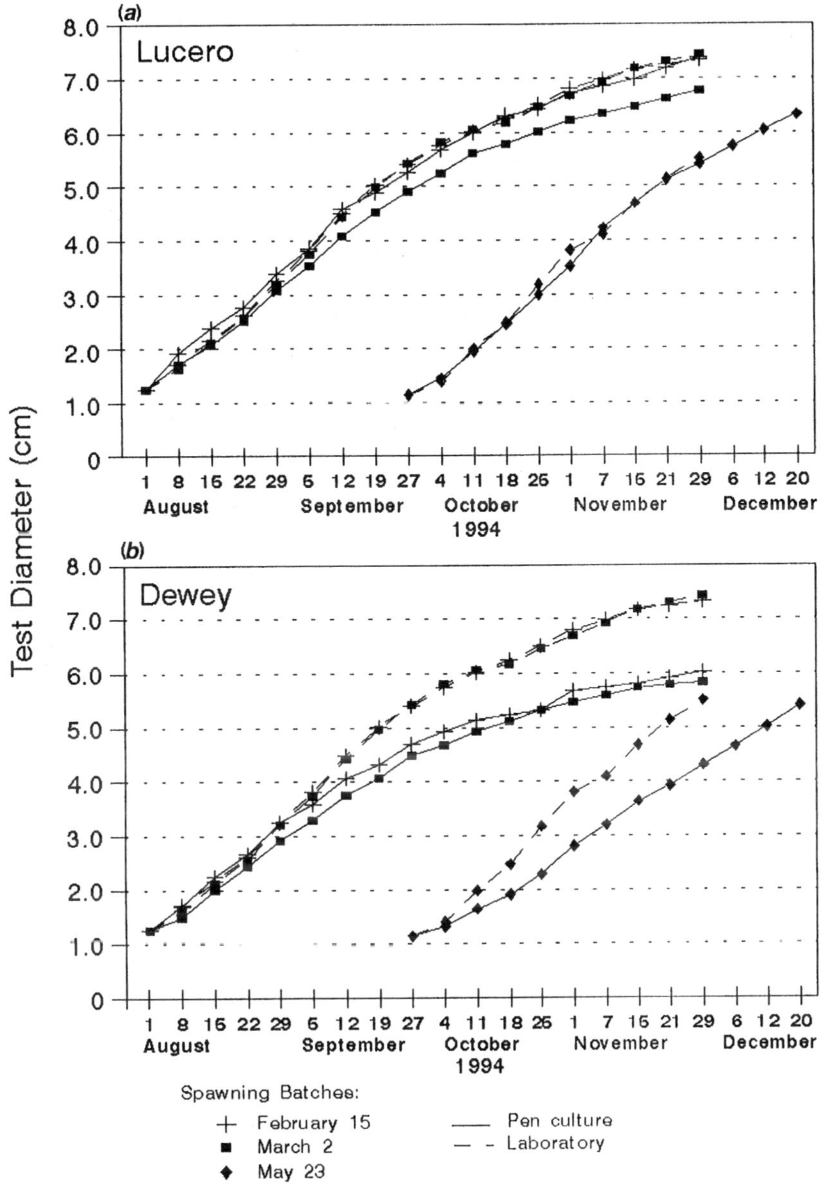

Bolinao in general, a conceptual model (Fig. 5) using community-based sea pen culture as a management strategy was developed. This approach links laboratory (e.g., larval culture) and field research (e.g., growth dynamics) with local community participation to help natural stocks recover, while providing a supplemental source of income or food for fishers. Community participation is deemed as crucial in ensuring the sustainability of any resource management initiative.

Sea urchin grow-out culture provides opportunity to enhance natural stocks if sea urchins are allowed to spawn in the wild. Cultured *T. gratilla* in this study attained sexual maturity at a size of about 60 mm TD, comparable to reports from wild populations in other localities (e.g., Chang-Po and Kun-Hsiung 1981). Growth studies showed that sexual maturity can be reached 7–8 months after fertilization. Populations of *T. gratilla* have been reported to spawn year round although timing of spawning peaks varies among localities (e.g., Tuason and Gomez 1979; Chang-Po and Kun-Hsiung 1981). The aggregation of adults in sea pens or cages increases the probability of egg fertilization by bringing individuals close together (Levitan 1991; Levitan et al. 1992). In localities where natural populations are depleted (e.g., Bolinao), aggregation of *T. gratilla* in pens should result in greater production of larvae and benthic recruits either in the area or in adjacent areas, as has occurred elsewhere (*Strongylocentrotus*: Tegner and Dayton 1977; *Diadema setosum*: Hunte and Younglao 1988). Managed as such, grow-out cultures could function as reproductive reserves. The potential of sea pen/cage culture in promoting natural stock recovery is only achieved if selective harvesting of reproductively mature (>60 mm TD) urchins occurs after they had opportunity to spawn. Holding of mature sea urchins in pens for a longer time is also justified with markets that place a premium value on larger sized urchins, as occurs in Nalvo. In addition, sea urchin grow-out involves low

Fig. 5. Model for community-based management of sea urchins using sea pens.

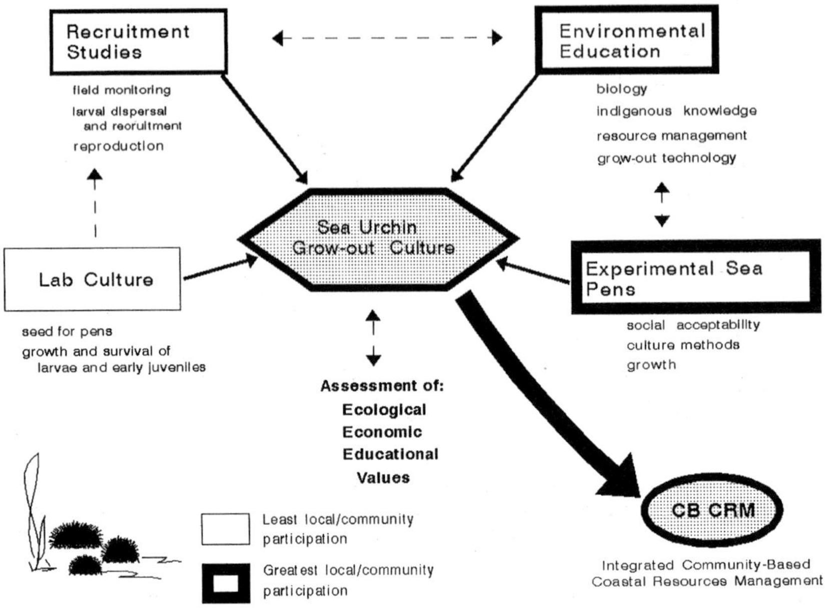

capital outlay and maintenance cost, giving grow-out culture a high potential as a supplemental income and food source for fishers.

Implementation considerations

Considerations in the development of sea urchin grow-out culture as a means to enhance the recovery of natural populations and as a supplemental source of livelihood are: (*i*) availability of seedstock, (*ii*) availability of suitable culture areas on subtidal seagrass or *Sargassum* reef flats, (*iii*) favorable markets, and (*iv*) interest of local fishers. Research on increasing production of juvenile seedstock is currently in progress, along with studies of recruitment dynamics in natural populations. Simulation of larval dispersal patterns indicate that the sea urchin population in Bolinao may be the source of larvae for other sea urchin populations on the northwestern coast of Luzon and conversely, receive larvae from these other populations depending on the monsoon season (Juinio-Meñez and Villanoy 1995). Studies on how the presence of conspecific adults affects settlement and metamorphosis of *T. gratilla* larvae are also being conducted. Seagrass and *Sargassum* beds in Bolinao are considered suitable grow-out sites for urchins because the reef system has until recently supported a substantial wild population. Urchin growth studies at Lucero show high growth rates.

In Bolinao, fishers sold roe by weight to buyers from exporting companies, with gonads later sorted and processed according to size in Manila prior to export. Profit in this system accrued primarily to middlemen, with only about 5% of net export market value going to harvesters (Talaue-McManus and Kesner 1995). A system to ensure greater return to fishers for cultured sea urchins is needed and here fishers can perhaps help themselves. One option could be direct selling by fisher cooperatives to local markets such as specialty restaurants (e.g., Japanese) in Manila, which pay a high price for fresh and large-sized roe.

Wildstock enhancement can best be achieved if the spatial pattern of sea pens is considered. However, with the traditional open access nature of sea urchin harvesting in Bolinao, ground use conflicts may arise if sea pens are set up. The acceptability of limited exclusive use rights by individuals or groups of fishers is crucial and has to be resolved through consultations between the local government and communities. Precedents exist, such as leased concessions for rabbitfish corals and milkfish pens. The Philippine Local Government Code of 1991 gives local governments the mandate to legislate and enforce laws regarding usage of municipal waters. The code provides incentives for development of fisher cooperatives by ensuring preferential rights to fisher groups in the use of coastal waters. This legislation could facilitate the development of an acceptable system for village-based grow-out areas for sea urchins.

At present, scale of grow-out is limited by the availability of cultured seedstock since wild seed is scarce. Nonetheless, even at a pilot scale, sea urchin grow-out culture heightens environmental awareness and improves understanding of the rationale for management by fishing communities whose activities have traditionally been unregulated. In the long term, participation of local communities and governments in such activities can facilitate implementation of other resource management options, such as marine protected areas where reseeding can be undertaken within a program of integrated reef management. The model of village-based sea enclosures as reproductive reserves presented here may be easily adapted and implemented in areas where other natural macrobenthic invertebrate populations are exploited as part of an overall fisheries management scheme to help enhance and sustain natural populations of a number of benthic invertebrate resources.

Acknowledgments

The authors are grateful to our colleagues at the Institute, Mariejo Trinidad-Roa and Liana McManus, and Arsenia Domingo of the University of Northern Philippines for providing valuable baseline information for this study. This study was funded by the International Development Research Centre, Canada, and the Philippine Council for Aquatic and Marine Research and Development through the University of the Philippines Marine Science Institute. Travel support for the first author to present this paper was provided for by the Faculty Development Fund of the University of the Philippines. This is the U.P. Marine Science Institute contribution no. 272.

References

Chang-Po, C., and Kun-Hsiung, C. 1981. Reproductive periodicity of the sea urchin *Tripneustes gratilla* in Taiwan compared with other regions. Int. J. Invertebr. Reprod. **3**: 309–319.

Garcia, M.P. (*Editor*). 1992. Policies and issues on Philippine fisheries and aquatic resources. Philippine Council for Aquatic and Marine Research and Development, Department of Science and Technology Book Series 14. 181 p.

Hunte, W., and Younglao, D. 1988. Recruitment and population recovery of *Diadema antillarum* (Echinodermata; Echinoidea) in Barbados. Mar. Ecol. Prog. Ser. **45**: 109–119.

Juinio, M.A.R., Meñez, L.A., Villanoy, C.L., and Gomez, E.D. 1989. Status of giant clam resources of the Philippines. J. Molluscan Stud. **55**: 431–440.

Juinio-Meñez, M.A.R., and Villanoy, C.L. 1995. Sea urchin recruitment studies for management considerations. *In* Proceedings of the LIPI-JSPS Joint Seminar in Marine Science. Jakarta, November 15–18, 1994. Indonesian Institute of Science (LIPI), Jakarta, Indonesia. pp. 15–20.

Levitan, D.R. 1991. Influence of body size and population density on fertilization success and reproductive output in a free-spawning invertebrate. Biol. Bull. (Woods Hole), **181**: 261–268.

Levitan, D.R., Sewell, M.A., and Chia, F.S. 1992. How distribution and abundance influence the fertilization success in the sea urchin *Strongylocentrotus franciscanus*. Ecology, **73**: 248-254.

Lopez, M.D.G. 1986. An invertebrate resource survey of Lingayen Gulf, Philippines. *In* North Pacific Workshop on Stock Assessment and Management of Invertebrates. *Edited by* G.S. Jamieson and N. Bourne. Can. Spec. Publ. Fish. Aquat. Sci. 92. pp. 402–409.

Ross, M.A. 1984. A quantitative study of the stony coral fishery in Cebu, Philippines. Mar. Ecol. Prog. Ser. **5**: 75–91.

Shokita, S., Kakazu, K., Tomori, A., and Toma, T. (*Editors*). 1991. Aquaculture in tropical areas. Midoro Shobo Co. Ltd., Japan. 360 p. [English edition prepared by M. Yamaguchi.]

Talaue-McManus, L., and Kesner, K.P. 1995. Valuation of a Philippine municipal sea urchin fishery and implications of its collapse. *In* Philippine coastal resources under stress. Selected papers from the Fourth Annual Common Property Conference, Manila, Philippines, June 16–19, 1993. *Edited by* M.A.R. Juinio-Meñez and G. Newkirk. Coastal Resources Research Network, Dalhousie University, Halifax, NS, Canada, and Marine Science Institute, University of the Philippines, Quezon City, Philippines. pp. 229–239.

Tegner, M.J., and Dayton, P.K. 1977. Sea urchin recruitment patterns and implications on commercial fishing. Science (Washington, D.C.), **196**: 324–326.

Trinidad-Roa, M.J. 1988. Beche-de-Mer fishery in the Philippines. Naga, The ICLARM Quarterly, **10**(4): 15–17.

Tuason, A.Y., and Gomez, E.D. 1979. The reproductive biology of *Tripneustes gratilla* Linnaeus (Echinoidea: Echinodermata) with some notes on *Diadema setosum* Leske. *In* Proceedings of the International Symposium on Marine Biogeography and Evolution in the Southern Hemisphere, Auckland, New Zealand, July 17–20, 1978, Vol. 2. New Zealand Department of Scientific and Industrial Research, Wellington, New Zealand. pp. 707–715.

Wells, S.M. 1981. International trade in ornamental corals and shells. Proc. 4th Int. Coral Reef Symp. (Manila), **1**: 323–330.

A case study of the traditional trochus (*Trochus niloticus*) fishery in the Maluku region, Indonesia

Z. Arifin, P. Pradina, and A.H. Purnomo

Abstract: In this paper, we present data on trochus (*Trochus niloticus*, L.) biology and describe the traditional sasi fishery management system. Our calculated von Bertalanffy equation for trochus is $D_t = 123.16[1 - e^{-0.621(t-t_0)}]$, where D = shell diameter (mm) and t is time in years. Spawning occurs every month, but peak spawning is from April to August. We suggest that aspects of the traditional sasi system are still relevant to trochus fishery management and we propose scenarios under which it can be applied. We also suggest that the trochus fishery be open only in the fall after the peak spawning season and that fishing activities in the intertidal zone during fishery closures be minimized.

Résumé : Dans cet article, nous présentons des données sur la biologie du troque (*Trochus niloticus*, L.) et nous décrivons la gestion des pêches suivant la formule traditionnelle sasi. L'équation de von Bertalanffy que nous avons formulée pour le troque se lit comme $D_t = 123.16[1 - e^{-0.621(t-t_0)}]$, où D est le diamètre de la coquille (mm) et t est le temps en années. On peut observer la fraye tous les mois, mais elle est maximale d'avril à août. Nous pensons que, par certains aspects, la formule de gestion traditionnelle sasi demeure pertinente dans le cas du troque et nous présentons des scénarios d'application. Nous proposons par ailleurs de n'ouvrir la pêche du troque qu'en automne, après la saison de fraye maximale, et de limiter autant que possible les activités de pêche dans la zone intertidale lorsque la pêche est fermée. [Traduit par la Rédaction]

Introduction

Trochus (Archaeogastropoda: *Trochus niloticus*, L.), known as lola (in the Maluku region) or siput susu bundar (in Indonesia), is an important marine resource for the Maluku region, Indonesia. It is a traditional food and a leading export item which generates revenue for many fishing communities. Trochus is collected either by skin diving or by walking along the reef at low tide. The meat (foot) of trochus is processed (cooked and dried or salted and smoked) and shells (85% of the animal's live weight) are exported to Japan, Singapore, Germany, and Italy (Usher 1984).

The annual production of trochus shell (1983–1987) is 150–225 tonnes (t) with a wholesale value of about C$10 million (Arifin 1993). The price fishers receive for clean, dry shells ranges from approximately $1.50 to $3.50 kg^{-1} for second and first grade-quality shell, respectively. The final value of the shell is many times the value of the raw product. An average trochus shell costs $0.25 ($2.50 kg^{-1}) but will produce about 30 buttons, worth $0.88 each or $26.50 per shell. In addition, a shell powder by-product of the button industry is used in making paint and nail polish or processed to produce mother-of-pearl chips.

Although trochus are abundant in Maluku, there is growing concern that there is stock depletion in some locations. Decreased production has been reported in many islands of the Maluku region over the last 15 years (Wells 1981; Arifin 1993). To address this problem, it is important to obtain basic information on the biology of trochus, including habitat preference, growth rate, and spawning season, and to evaluate the current fishery management system. The objectives of this study are (*i*) to provide biological information necessary to rationalize the traditional sasi management system, and (*ii*) to evaluate why the sasi system was once very effective for trochus management in the Maluku region.

Trochus management approaches

Many management measures, including minimum size restrictions, closed or restricted areas, and open and closed seasons have been used in trochus management. Goals of management were to reduce fishing mortality and restrict size at first capture, and thus prevent overfishing. In 1987, because of decline in the trochus fishery, the national government (Ministry of Forestry of the Republic of Indonesia, Decree No. 12/KPTS-II/1987) imposed a trochus fishing closure, which is currently still in effect. However, stock rebuilding has been largely unsuccessful as managing institutions either did not have the necessary human resources or the discipline to implement and enforce regulations. The fishing closure also recreated new problems because of the complexity of trochus habitat and the economic importance of this species to traditional fishers. Habitat complexity, i.e., the complex spatial pattern of coral reefs, made monitoring and surveillance difficult to implement. From the point of view of fishers, they simply engaged

Z. Arifin[1] and P. Pradina. Division for Marine Resources R & D, Indonesian Institute of Sciences, Guru-Guru, Poka, Ambon 97233, Indonesia.

A.H. Purnomo. Assessment Institute for Agricultural Technology, Jl Chr. Soplanit Rumahtiga, Ambon 97233, Indonesia.

Correct citation: Arifin, Z., Pradina, P., and Purnomo, A.H. 1998. A case study of the traditional trochus (*Trochus niloticus*) fishery in the Maluku region, Indonesia. *In* Proceedings of the North Pacific Symposium on Invertebrate Stock Assessment and Management. *Edited by* G.S. Jamieson and A. Campbell. Can. Spec. Publ. Fish. Aquat. Sci. 125. pp. 401–406.

[1] Present address: Department of Biological Science, Simon Fraser University, 8888 University Drive, Burnaby, BC V5A 1S6, Canada. e-mail: arifin@sfu.ca

Fig. 1. Trochus fishery locations in the Maluku region.

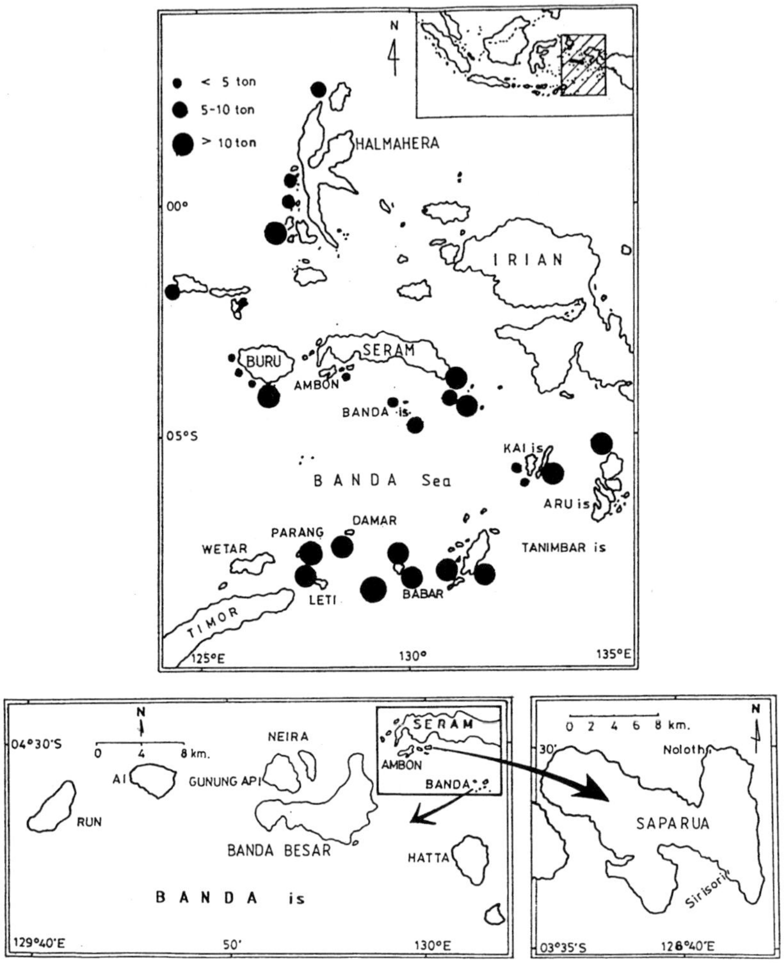

in illegal fishing despite the regulations because other income alternatives were limited.

Fishing closures as modern fishery management regulations, though legally binding, lacked active community support in the Maluku region and in many regions of the country. Problems in enforcement and surveillance remain major obstacles. Violations have become widespread because fishers do not see any benefits for compliance. More importantly, returns from noncompliance are high and exceed the risks involved. In contrast, the traditional fishery sasi system explicitly encouraged active community participation in trochus fishery management. Noncompliance resulted in severe community pressure on individuals for any infringement or violation, with violators sometimes socially ostracized from community activities (Wahyono et al. 1992).

Traditional fishery management is founded on "restrictions," which are referred to as sasi in the local vernacular of the Maluku region (Fig. 1). Sasi parallels closed areas or seasons in modern fishery management, except the local community is directly and actively enjoined by tribal chieftains (*kepala suku*) to observe restrictions or face the consequences. Because communities are relatively small and closely knit, governed by traditions and customs, it is relatively easy for the entire community to monitor any violation of restrictions.

In recent decades, traditional management systems have slowly disintegrated, with many factors contributing to the decline. Firstly, these systems have conflicted with national law and regulations. Zerner (1994) reported breakdown of the original sasi system due to reallocation of rights and decision making from tribal chieftains to the regional government heads (*kepala desa*). Secondly, a growing market economy and changing consumer preference has encouraged regional officials to shorten closed seasons. Finally, the traditional sasi system was increasingly challenged by a lack of biological data on how long sasi should be practiced, when was the best season to harvest trochus, and what minimum size of trochus could be sustainably harvested. The sasi system requires incorporation of such knowledge to make it acceptable today.

Given the local social acceptability of the sasi system, development of a trochus management strategy based on it would seem to benefit both the resource and fishing community. Here, we provide data on some aspects of trochus biology to facilitate re-adoption of the traditional sasi system on a wider scale in the country.

Materials and methods

Trochus habitat and growth rate

The spatial distribution and abundance of trochus were determined at Run, Hatta, Saparua, and the Kai islands (Fig. 1).

Fig. 2. (*a*) Trochus growth rate by shell diameter and (*b*) von Bertalanffy growth curves for trochus. Calculations are based on a 30-day month.

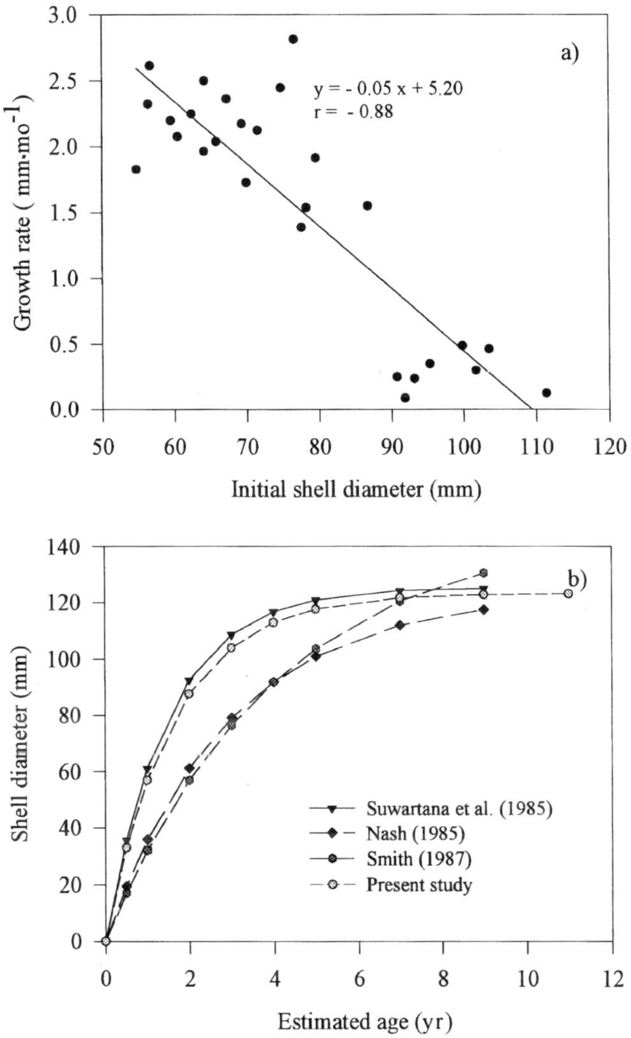

Three 100-m transects parallel to the coast were surveyed in both the intertidal and subtidal zones. Intertidal transects were surveyed during low tides and subtidal transects were done by SCUBA. Trochus encountered within 2.5 m of a rope marking each transect were counted and measured, giving a total area of 500 $m^2 \cdot transect^{-1}$.

Growth studies were carried out at Run Island between December 1992 and July 1993. Growth of trochus was determined by mark and recapture. Trochus were collected by SCUBA and marked by plastic tape attached with plastic epoxy. Shell diameters were measured to the nearest 0.1 mm with vernier calipers in December 1992, April 1993, and July 1993. Length increment data were analyzed by fitting them into the von Bertalanffy growth model,

$$D_t = D_\infty [1 - e^{-k(t-t_0)}]$$

where D_t is shell diameter at time t, t is time, D_∞ is a theoretical maximum shell diameter, k is a constant representing the rate at which the asymptotic shell diameter is approached, and t_0 is a constant representing time when $D_t = 0$.

Spawning season

Because Nash (1985) suggested that trochus spawning was influenced by the lunar cycle on the Australian Great Barrier Reef, monthly samples of 23 trochus (65–102 mm shell diameter) were collected between December 1992 and December 1993 by SCUBA at a depth of 1–5 m every 2–3 days after a new moon. Gonad samples were preserved in Bouin's solution. To determine the reproductive stage, three sectioned conical portions of the visceral mass were studied using Komatsu's procedure (Komatsu 1992). Gonad area was measured with a dissecting microscope and camera lucida on 1-mm grid paper. Gonad index was estimated as follows,

$$\text{Gonad index} = [(\text{Gonad area}) / (\text{Gonad area} + \text{Digestive gland area})] \times 100\%$$

To determine mean oocyte sizes, the mean of two diameter measurements, one perpendicular to the other, was calculated for all oocytes sectioned through the nucleus in a single section from the center of each ovary. Five to fifteen ovaries per month were collected for histological analysis (Luna 1968). Sections 7.0 μm thick were cut from behind the gastric caecum, the center of the ovary, and the tip of ovary. Five consecutive slides from each ovary part were stained in hematoxylin and eosin and measurements from 100 eggs were obtained from each slide.

Results

Habitat

Trochus habitat range extended from the intertidal zone (1.5 m above chart datum) to a subtidal depth of about 7.5 m. They were attached to dead coral boulders beneath soft coral, on dead coral, and in coral crevices. Highest trochus densities were found on reef fronts exposed to strong currents and a constant flow of oceanic water (salinity: 34.0–34.4 ppt; dissolved oxygen: 3.7–4.1 $mL \cdot L^{-1}$). Average trochus density on reefs at Run and Kai islands in December 1992 was 2.5 and 3.0 $trochus \cdot m^{-2}$, respectively.

Size data of trochus occurring in transects at Kai Island were analyzed to determine distribution patterns. Analysis of variance (Zar 1984) of pooled data showed a significant difference ($P < 0.05$) in size frequency distribution between the intertidal reef flat and the subtidal zone. Size range distribution by depth was as follows: intertidal reef flat, 21.9–49.2 mm diameter (mean: SD = 40.0 ± 4.77, *N*: 99); and subtidal zone, 37.3–92.5 mm diameter (55.6 ± 14.29, *N*: 75). Shallow reef flat and intertidal zone substrates inhabited by juvenile trochus typically consisted of cobble and coral rubble that provided both abundant crevices for shelter and algae for food. Juvenile trochus coexisted with juvenile *Tectus pyramis* (Born), while larger trochus shared habitat occupied by other large gastropods such as *Turbo argyrostomus* (Lin.) and *Turbo petholathus* (Lin.).

Growth rate

Of 353 trochus tagged, only 90 trochus were recovered during the study, with recapture rates of 26% (32) from December 1992 to April 1993 and 13% (29) from April to July 1993. Recapture rates were probably affected by the complexity of the habitat, predation, and migration. There was a significant

Fig. 3. (*a*) Mean monthly gonad indices (%) for male and female trochus; (*b*) mean monthly oocyte diameter (m); and (*c*) monthly seawater temperature (°C) measured 1-m below the surface during the study.

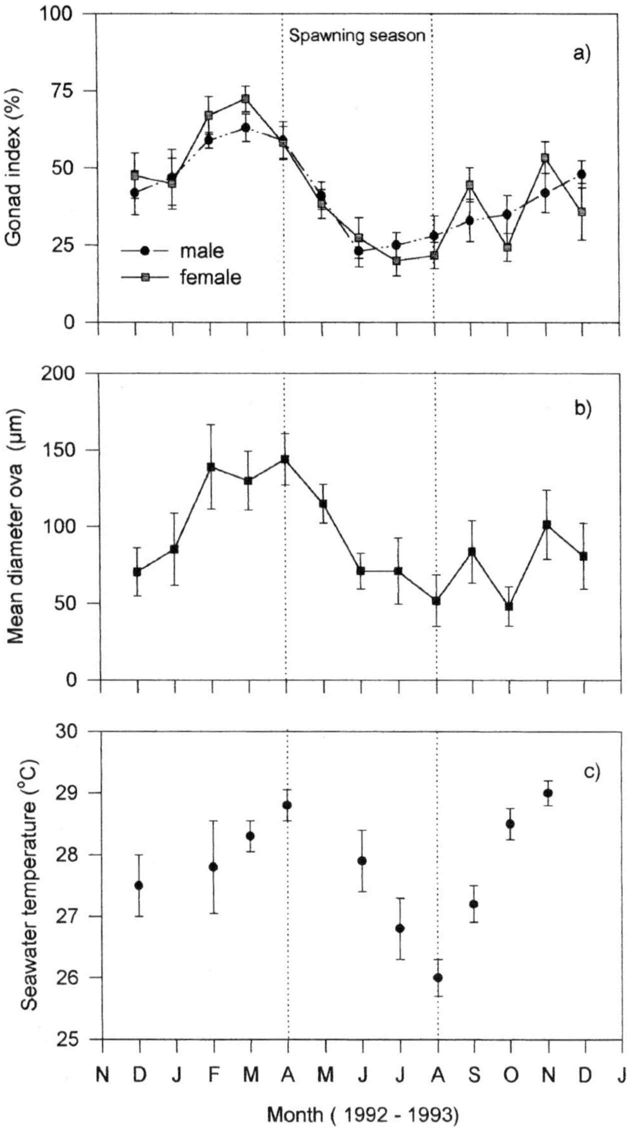

negative correlation ($r = -0.88$, $P < 0.05$) between monthly growth rate and initial shell diameter (Fig. 2*a*). The von Bertallanffy growth model equation is $D_t = 123.16\ [1 - e^{-0.621(t-t_0)}]$, with a prediction of D_∞ at 123 mm at 12 years (Fig. 2*b*). This observation is similar to a previous study by Suwartana et al. (1985) for Tanimbar Island. Compared with other Pacific regions, trochus growth from the Maluku region is comparable with growth rates at Nash's (1985) MC1 site in Australia and at Guam (Fig. 2*b*), although early growth of trochus seems faster in Maluku than in Australia or Guam.

Spawning season

Of 292 trochus sampled, 148 were female. Mean gonad index peaked in March 1993 (Fig. 3*a*). A continued drop in gonad index and average diameter of ova (Fig. 3*b*) from April to August 1993 suggested that the main spawning season was during this time. Spawning appeared to be largely finished by the end of August because mean gonad index started to increase in September. This spawning season is supported by other observations of gametogenic stages (Pradina and Dwiono 1994).

Trochus spawning coincided with the seasonal change from west to east monsoons (June–August), with average monthly seawater temperature decreasing from 28 to 26°C (Fig. 3*c*). The east monsoon is when the Banda Sea is coolest and the highest waves occur, which may be the trigger for mass trochus spawnings on the many reefs.

Table 1. Comparison of trochus von Bertalanffy model parameters in the present study to other published values.

Trochus growth study	*N*	*k*	D_∞ (mm)
Nash (1985)	80	0.344	123.0
Smith (1986)	322	0.246	146.5
Suwartana et al., (1985)	45	0.671	125.3
Present study	32	0.621	123.2

Discussion

Growth rate and spawning season

Growth data from Nash (1985) and Smith (1987) are quite different from that of Suwartana et al. (1985) and from the present study (Fig. 2*b*, Table 1), with the latter indicating a faster growth rate in younger animals. Data from Nash (1985) and Smith (1987) were from unexploited populations while the data of Suwartana (1985) and the present study were from exploited populations, suggesting that interspecific interactions may affect growth rate. The largest trochus collected in the present study was 112 mm diameter, but a shell measuring 123 mm diameter has been found from the Maluku region. Individual trochus measuring 140 mm diameter (Smith 1987) and 142 mm diameter (Hedley 1917) have been reported. Thus, the absence of very large trochus (> 123 mm diameter) in the present study may have led to underestimation of size-at-age and life span estimates.

Seasonal timing of spawning in marine invertebrates ranges widely within and between species, as well as between different localities and years (Pearse 1978). Factors influencing spawning in gastropods include degree of maturation, availability of food, neural and hormonal factors, and intensity of stimuli, e.g., particularly current and temperature (Sloan and Breen 1988). In abalone species, spawning seems to be dependent on temperature: *Haliotis tubercula* spawns around the timing of maximum late summer temperature (Hayashi 1980), *H. rubra* spawns after the summer maxima (Shepherd and Laws 1974), and *H. midae* spawning is associated with increasing temperature. In contrast to the importance of high temperature with abalone, trochus spawning appears to be associated with decreasing temperature in the early rainy season.

Trochus fishery management

The national approach for trochus management has been to impose a year-round fishing closure, which included fishery closures on a number of other coral-associated species (i.e., giant clams, green snails, and black corals). However, this broad

Table 2. Sasi practices in fishing communities in the Central Maluku region.

Fishing practices	Saparua Island	Run Island	Hatta Island
Closed season (yr)	3–4	2	2
Open season (weeks)	1–2	1	1
Minimum size of trochus harvested (mm diameter)	Three fingers width (~60 mm diameter)	~60 mm diameter	~60 mm diameter
Harvest restrictions	Daylight only; other resources can be harvested	During day and night; other resources can be harvested	Daylight only; other resources can be harvested
Participants	Village residents >15 yr old	Village residents only	Village residents, but they may hire people from other islands if they are unable to swim
Profit sharing (%) (fishers:village)	90:10	85:15	94:6

fishery closure has created new problems. The high commercial value of trochus shells has encouraged illegal fishing in many parts of the Maluku region. The closure was enacted in 1987, but since then, about 95.1 t of trochus, mainly in the form of half made and finished buttons valued at 552 million *Rupiah* (C$328 125) have been exported. No official data on trochus shell production exist, presumably because of the fishery closure.

We suggest that the failure of current fisheries management is largely the result of rejection of traditional authority. Many regional government staff do not strictly observe traditional values. For example, fishery staff on Hatta Island broke sasi to collect trochus for a provincial culture project before the open fishery season was declared. This apparent failure of the national strategy for conserving trochus populations has created an opportunity to re-adopt the sasi system.

The traditional sasi fishery management system is still strictly practiced in the Maluku region. Several islands with an average harvest of about 10 t of trochus shells are Kai Island, Tanimbar Island, the Babar Islands, and the Damar and Leti islands (Fig. 1). The length and timing of closed and open seasons, as well as the minimum size of trochus, are considered the primary management factors for maintaining a healthy trochus stock. Sasi and harvest restrictions vary among fishing communities but they commonly agree on the minimum harvestable size of trochus (Table 2).

Trochus shell landings and hypothetical annual harvests (i.e., the trochus shell landing as a function of the duration of closed season) between 1967 and 1993 at three islands in the Central Maluku region are presented in Fig. 4. At Saparua Island, during the period of 1967–1984, closed seasons (sasi) lasted for 3–4 years and the estimated average annual trochus harvest was about 1 t. After 1984, the closed season was shortened to 1–2 years and the estimated average annual trochus harvest declined to about 0.5 t (Fig. 4*a*). Increased indirect fishing mortality because of shorter closed seasons may have caused much of the decline in trochus harvest.

Landings of trochus at Run Island can be divided into pre- and post-El Niño 1982/1983 events (Fig. 4*b*). Pre-El Niño trochus harvest averaged 2 $t \cdot yr^{-1}$ and post-El Niño harvests averaged <1.0 $t \cdot yr^{-1}$. The 1982/1983 El Niño forced fishing communities to abandon their villages due to severe drought, eliminating fishing between 1982 and 1986. Trochus landing data from Hatta Island (Fig. 4*c*) is difficult to interpret, especially because no data was reported from 1970 to 1980. The

Fig. 4. Reported trochus shell landings from three islands in the Central Maluku region, (*a*) Saparua Island, (*b*) Run Island, and (*c*) Hatta Island.

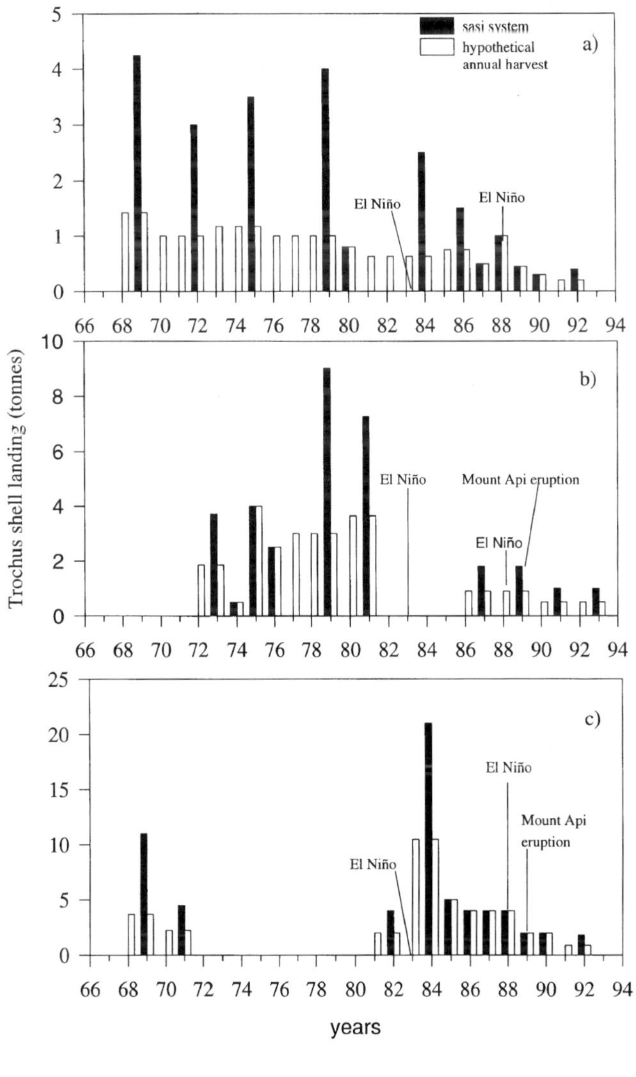

increase in value of the shell landed suggests that catches remained high throughout this period (Usher 1984). In contrast to Run Island, at Hatta Island (Fig. 4*c*), El Niño increased landing with a relatively high harvest of 21.0 t in 1984 compared with the following years.

Another environmental event that influenced the trochus fishery was the eruption of Mount Api-Banda in 1989. The eruption devastated agricultural crops and caused high siltation along the coasts of Run and Hatta islands, resulting in a decline in trochus harvest in 1991. Moreover, increasing market demand for trochus shell and improvement in transportation, coupled with a steady average increase (2.1% annually) in the number of fishers in each fishing community over the last 15 years, has increased human impact on the nearshore environment.

Although accurate information is lacking on the biological impacts of El Niños and the Mount Api eruption on trochus, the interplay of environmental and human-induced factors has caused a decline in trochus landing. Landing data indicates that with the current size limit of 60 mm diameter, a closed season (sasi) of 2 years would only be advisable if the size limit was increased to 80 mm diameter, to ensure a large enough adult population exists to maintain appropriate recruitment. Increasing the minimum size is required to compensate for the reduction in abundance of older year-classes in the breeding population. Nash (1985) found that an increase in size of trochus from 60 to 80 mm diameter resulted in a doubling of average gonad size.

With the current size limit, we suggest that further reduction in the duration of the closed season (i.e., to 1 year) will endanger the stock. However, a one-year sasi better maintains trochus markets. If a 1-year sasi is adopted, a minimum size limit even greater than 80 mm diameter would become necessary to ensure that a minimum abundance of mature trochus is maintained to produce sufficient spawning population.

In addition to the above, it may also be necessary to set an annual community quota with shorter open seasons and fewer divers permitted to participate. This management approach could be combined with a rotating harvest among subareas to permit some sustainable harvesting from the overall population each year (Caddy 1993).

In summary, we recommend development of co-management where the national government and local communities work together for the benefit of the economy and the resource. We suggest that the sasi system can be a useful trochus fishery management strategy if the following scenarios are followed. First, with the current size limit, harvesting is only conservative when the closed season is 3 years. Shorter closed seasons require the size limit to be increased and possibly a rotating harvest strategy to provide more regular landings. Second, with peak spawning between April and August, we recommend that the fishery be open only in the fall each year. Finally, we recommend that coral lifting and fishing for other species in the intertidal zone during the period of the trochus fishery closure be minimized to ensure the maximal survival of juvenile trochus.

Acknowledgments

This research was funded by MAB-UNESCO and EMDI-Canada to ZA and PP. The field assistance of Messrs. P. Makatipu, S. Rumahenga, and D. Hendrik is gratefully acknowledged. We also appreciate the critics and suggestions provided by Dr. G. Jamieson, Dr. M. Yamaguchi, and Dr. G. Edgar.

References

Arifin, Z. 1993. Distribution, habitat and fishery of lola (*Trochus niloticus*) in Maluku waters. Jur. Fak. Perik. Unsrat. **11**(3): 40–48. (In Indonesian.)

Caddy, J.F. 1993. Background concepts for rotating harvesting strategy with particular reference to the Mediterranean red coral, *Corallium rubrum*. Mar. Fish Rev. **55**(1): 10–18.

Hayashi, I. 1980. The reproduction biology of the ormer, *Haliotis tuberculata*. J. Mar. Biol. Assoc. U.K. **60**: 431–437.

Hedley, C. 1917. The economics of trochus niloticus. Aust. Zool. **1**: 69–73.

Komatsu, T. 1992. A study on the reproduction of the green snail (*Turbo marmoratus*) in the Ryukyus, Southern Japan. M.Sc. thesis, University of the Ryukyus, Okinawa, Japan.

Luna, L.G. 1968. Manual of histological staining methods. Harper and Brothers, New York, N.Y. 423 p.

Nash, W. 1985. Aspect of the biology of *Trochus niloticus* and its fishery in the Great Barrier Reef region. A report submitted to Fish. Res. Branch, Queensland Dept. of Primary of Ind. and the Great Barrier Reef Marine Park Authority, Townsville, Queensland, Australia. 210 p.

Pearse, J.S. 1978. Reproductive periodicities of Indo-Pacific invertebrate in the Gulf of Suez. IV. The chiton *Acanthopleura haddoni* Winckworth and *Onithochiton lyelli* (Sowerby), and the abalone *Haliotis pustulata* Reeve. Bull. Mar. Sci. **28**: 92–101.

Pradina, P., and Dwiono, S.A.P. 1994. The characteristic of gonad development stages of *Trochus niloticus* (Mollusca, Gastropoda). Per. Mal. Sek. **8**: 15–21. (In Indonesian.)

Shepherd, S.A., and Laws, H.M. 1974. Studies on Southern Australian abalone. II. Reproduction of five species. Aust. J. Mar. Freshwater Res. **25**: 49–62.

Sloan, N.A., and Breen, P.A. 1988. Northern abalone, *Haliotis kamtschatkana*, in British Columbia: fisheries and synopsis of life history information. Can. Spec. Publ. Fish. Aquat. Sci. 103. 46 p.

Smith, B.D. 1987. Growth rate, abundance and distribution of the introduced top shell (*Trochus niloticus*) on Guam, Mariana islands. Bull. Mar. Sci. **41**(2): 466–474.

Suwartana, A., Dwiono, S.A.P., and Wouthuyzen, S. 1985. Preliminary study on the growth of lola, *Trochus niloticus* (Mollusc: Trochidae). The National Congress of Biology VII, Palembang, Indonesia. 10 p. (In Indonesian.)

Usher, G.F. 1984. Coral reef invertebrate in Indonesia: their exploitation and conservation needs. IUCN/WWF Conservation for Development Programme. Project No. 1688. 100 p.

Wahyono, U., Chong., K.C., Susesno, and Pahlevi, R. 1992. Traditional community-based management practices in Indonesia. FAO/Japan expert consultation on the development of community-based coastal fishery management systems for Asia and the Pacific. FAO Fish. Rep. No. 474. (Suppl. 1). pp. 243–263.

Wells, S.M. 1981. International trade in ornamental shell. IUCN Conservation Monitoring Center. Cambridge, England. 22 p.

Zar, J.H. 1984. Biostatistical analysis. Prentice-Hall, Inc., Englewood Cliffs, New Jersey. 718 p.

Zerner, C. 1994. Tracking sasi: the transformation of central Moluccan reef management institution in Indonesia. *In* Collaborative and community-based management of coral reefs. *Edited by* A. White. Kumarian Press, West Hartford, Conn. pp. 19–32.

Artisanal "Caletas" as units of production and co-managers of benthic invertebrates in Chile

Juan C. Castilla, P. Manríquez, J. Alvarado, A. Rosson, C. Pino, C. Espoz, R. Soto, D. Oliva, and O. Defeo

Abstract: In Chile, small-scale fisheries exist for about 60 benthic invertebrates, with about 15 000 t (US$166 million) landed per year. Mollusc dive fisheries, with more than 10 000 professional registered divers, represent about 70% of total invertebrate landings. The fishery for the gastropod *Concholepas concholepas*, the "loco," was closed between 1989 and 1992. The closure was lifted twice in 1993, for a total of 14 days, under new management regulations institutionalized in the 1991 Chilean Fishery and Aquaculture Law (No. 18,892). The loco fishery landed 8574 t in 1993, with an export value of US$64 million. Existing regulations to control both dive fishery effort and access to fishing grounds are described. Fishery monitoring was conducted during the 1993 loco fishing seasons in 4 Caletas (small-scale fisher associations, each centered around a cove) in central Chile. Annual closures of small coastal MEAs (management and exploitation areas) resulted in larger CPUE (catch per unit effort) and reduced travel time. Larger size loco receiving higher prices resulted. Future co-management options to better achieve sustainable exploitation of Chilean benthic invertebrates are discussed.

Résumé : Au Chili, il existe des pêcheries à petite échelle pour environ 60 invertébrés benthiques et les débarquements s'élèvent à environ 15 000 t (166 millions de dollars US) par année. Les pêcheries de mollusques exploitées par des plongeurs, on dénombre plus de 10 000 plongeurs professionnels enregistrés, comptent pour environ 70 % des débarquements totaux d'invertébrés. La pêcherie du gastropode *Conchelopas conchelopas,* ou « loco », a été fermée entre 1989 et 1992. La fermeture a été levée deux fois en 1993, pour un total de 14 jours, suite à l'adoption de nouveaux règlements de gestion édictés dans la Loi chilienne sur la pêche et l'aquaculture de 1991 (nº 18 892). La pêche au loco a produit des débarquements de 8 574 t en 1993, pour une valeur à l'exportation de 64 millions de dollars US. Le présent article décrit les règlements en vigueur pour contrôler à la fois l'effort de pêche des plongeurs et l'accès à la pêcherie. La surveillance de la pêcherie a été réalisée durant les saisons de pêche du loco de 1993 dans 4 Caletas (petites associations de pêcheurs centrées autour d'une anse) au centre du Chili. Les fermetures annuelles de petites « zones de gestion et d'exploitation » côtières ont entraîné des prises par unité d'effort (PUE) plus élevées et une réduction du temps de déplacement. Résultat : des prises de plus grande taille qui commandent un prix plus élevé. Des options de cogestion futures pour assurer une exploitation plus durable des invertébrés benthiques du Chili sont discutées. [Traduit par la Rédaction]

Introduction

Dive fisheries for benthic invertebrates in Chile are important both economically and socially. Over 60 invertebrates are harvested, current landings are about 150 000 t·yr^{-1}, worth around US$166 million·yr^{-1}, and about 70 000 small-scale (artisanal) fishers participate. Of these fishers, more than 10 000 are registered divers. Molluscs represent about 70% of total shellfish landings, and include the unique high-priced gastropod *Concholepas concholepas* (Bruguiere 1789), locally known as "loco" (Castilla 1982, 1988*a*; Castilla and Jeréz 1986; Geaghan and Castilla 1986, 1988; Castilla et al. 1994; Castilla 1995).

J.C. Castilla, P. Manríquez, J. Alvarado, A. Rosson, C. Pino, C. Espoz, R. Soto, D. Oliva, and O. Defeo. Estación Costera de Investigaciones Marinas, Las Cruces, Facultad de Ciencias Biológicas, Pontificia Universidad Católica de Chile, Casilla 114-D, Santiago, Chile. e-mail: jcastill@genes.bio.puc.cl

Correct citation: Castilla, J.C., Manríquez, P., Alvarado, J., Rosson, A., Pino, C., Espoz, C., Soto, R., Oliva, D., and Defeo, O. 1998. Artisanal "Caletas"as units of production and co-managers of benthic invertebrates in Chile. *In* Proceedings of the North Pacific Symposium on Invertebrate Stock Assessment and Management. *Edited by* G.S. Jamieson and A. Campbell. Can. Spec. Publ. Fish. Aquat. Sci. 125 pp. 407–413.

As a result of rapid development of Asian markets for Chilean shellfish, following aggressive overseas marketing between 1976 and 1982, shellfish export surpassed domestic consumption and some shellfish resources began to show signs of overexploitation. Bustamante and Castilla (1987) recommended immediate conservative management actions to achieve more rational management of these resources.

The Chilean Fishery and Aquaculture Law, Ley de Pesca y Acuicultura, No. 18,892, passed on September 28, 1991, contains several innovative management tools. Aims of management are to rehabilitate shellfish stocks damaged during the export boom, and to ensure rational and sustainable exploitation of small-scale fishery resources (Castilla 1994; Payne and Castilla 1994). With respect to small-scale fishery activity, this law contains two sets of regulations. The first resolves past conflict over access to fishing grounds by both the artisanal and industrial (company-owned) fleets by assigning an exclusive fishing area within 8 km (5 miles) from shore to the artisanal fleet. This fleet comprises about 11 000 vessels of <50 t gross registered weight (Aranda et al. 1989). Approximately 85% of this fleet is composed of small, 7–8 m wooden boats (sail, oar, or outboard powered) with a crew of 2–4 fishers.

The second regulation relates specifically to the management of benthic resources by allowing exclusive fishing rights on defined management and exploitation areas (MEAs) of

Fig. 1. Locations of the 4 Caletas monitored during the 1993 loco fishing seasons. Management and Exploitation Areas are shown in front of their coves. The Las Cruces Marine Coastal Station (ECIM) and its Marine Preserve location (black) is also shown.

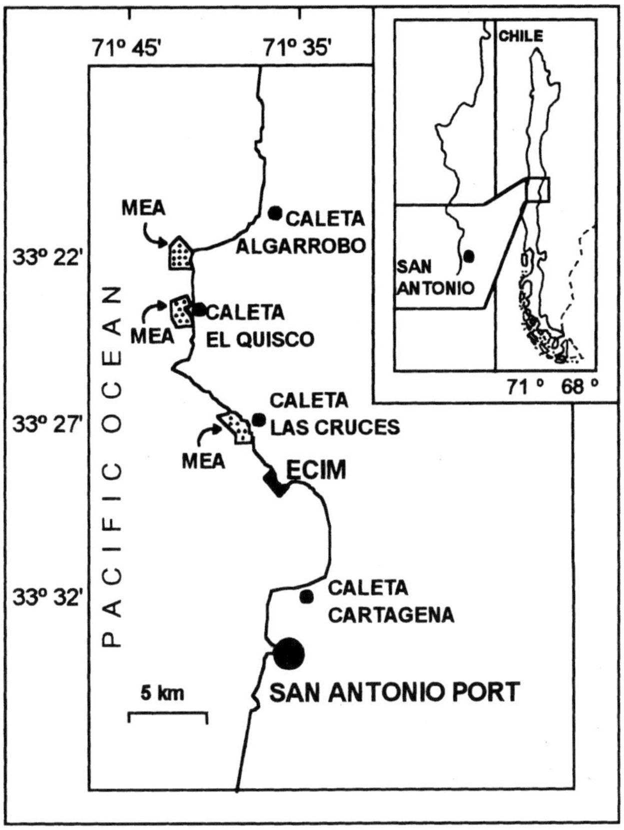

inshore sea bottom to registered organizations or communities of artisanal fishers. In addition, there is a special clause whereby a benthic resource can be declared "fully exploited." In such cases managers can set both fishery and fisher harvest quotas.

This law has been in operation for just over five years, and fishers are still adjusting to changes, particularly regarding establishment of MEAs (Castilla 1994).

Traditional small-scale fishermen's unions, "Sindicatos," are playing a key role in the implementation of MEAs. Sindicatos (Morales 1993; Payne and Castilla 1994; Minn and Castilla 1995) have operated for decades and may be formed by a minimum of 25 members dedicated to artisanal, technical, or professional activities. In the artisanal fishery, Sindicatos operate in small coastal villages known as "Caletas," usually located around coastal coves. In Chile there are about 190 active, recognized Caletas, most of them organized in Sindicatos. According to the Fishery and Aquaculture Law, exclusive fishing rights for benthic resources in an MEA can only be granted to organized groups of small-scale fishers. The system of Sindicatos allowed rapid adoption of MEAs and allows the possibility of co-management of local resources.

Castilla (1994) presented arguments for incorporation of exclusive fishing rights for benthic resources and the MEA concept in the new fishery law. In both cases, the main element catalyzing inclusion was scientific data, mostly from the central coast of Chile, on natural enhancement and recruitment of benthic invertebrates (Castilla 1988*b*) in coastal preserves without human intervention (Castilla and Durán 1985; Durán and Castilla 1989; Durán et al. 1987; Castilla 1990*a*). The precedent-setting, pilot, loco restocking experiment conducted between 1987 and 1990 with the Sindicato of Caleta Quintay in central Chile (Castilla and Jeréz 1986; Geaghan and Castilla 1986), which resulted in Chile's first MEA of 57 ha, was also important. This Sindicato was given exclusive fishing rights for harvest of *C. concholepas* in their MEA in April 1991, prior to the current fishing law being gazetted.

Between 1991 and 1993, several Caletas obtained provisional control over small areas, usually <100 ha of sea bottom, called "Destinacions Marinas" (Payne and Castilla 1994; Minn and Castilla 1995; Pino and Castilla 1995). Before MEAs are formally established, these Destinacions, which require a benthic resource management plan, permit similar benthic resource management.

Social and economic problems inherent in the early loco fishery (Castilla and Jeréz 1986; Geaghan and Castilla 1986, 1987, 1988) also played a key role in rationalizing the benthic resource management tools implemented in the Chilean Fishery Law. Prior to 1991, management options failed to halt declining catches and resulted in unemployment in Caletas. Loss of revenue in turn created problems, including illegal resource harvesting and tax evasion.

Under regulations stated in the 1991 Fishery Law, loco fishing, closed in 1989, was legally reopened twice in 1993, for a total of 14 days. In 1993 the species was declared "fully exploited" to permit experimentation with adaptive fishing options.

Artisanal *C. concholepas* fishery landings are updated and summarized and results from the 14-day, 1993 fishing season are documented and discussed. The effects of the 1991 Chilean Fishery Law on loco harvesting by both MEA-registered and nonMEA-registered Caletas on the central coast of Chile are reviewed and co-management of this resource is evaluated.

Material and methods

Landings and export values (1960–1993) of locos reported in this paper are based on both Annual National Fishery Service Statistics (Anuarios Estadísticos del Servicio Nacional de Pesca, SERNAP) and data from the Custom Bulletin of the Central Bank of Chile (Boletín Aduanero, Banco Central de Chile).

In 1993, the loco fishery was opened twice: 18–22 January (summer) and 26 July – 3 August (winter). A quota of 1500 locos per registered diver per fishing season was permitted. This applied to most Caletas, although in selected ones, the fisher quota was increased to 4000 per season. In the summer fishery, there were about 7000 enrolled divers in the National Small-Scale Fishery Register (Registro Nacional de Pescadores Artesanales de Chile), while in the winter fishery, there were about 10 000 divers. Locos could only be sold in batches of 150, and batch-coupons allowed sales to be tracked both to diver and wholesaler.

Landing points at selected Caletas (Fig. 1) were specified to facilitate fishery monitoring. Field samplers recorded relevant characteristics of the artisanal fleet and fishery at each landing point. As has been the case for the past 30 years, minimum legal size of harvestable loco was 10 cm (maximum diameter of the peristome). The 1993 loco harvesting monitoring program described in this paper was conducted independently by our research team, based at the Estación Costera de Investigaciones Marinas (ECIM), Las Cruces, in four Caletas of central Chile (Fig. 1).

The Caletas

Caleta El Quisco (33°23′ S, 71°42′ W)
Caleta El Quisco is a well-organized Sindicato (Payne and Castilla 1994) having 139 members in 1993, with 24 summer and 42 winter registered divers. The artisanal fleet was composed of 31, 7–8 m wooden boats, each equipped with 25–40 Hp outboard engines and surface hooka air compressors with one or two hose outlets for divers (Castilla and Jeréz 1986). Unilaterally, this Sindicato in 1990 totally banned diving on 57 ha of sea bottom (Fig. 1) next to their cove. The Sindicato took legal possession of this Destinación area on July 9, 1993. Since 1991, our research team has worked jointly with this Sindicato to implement fishery monitoring and assess invertebrate stocks. During the 1993 fishing seasons, El Quisco divers extracted their quotas exclusively from their Destinación.

Three stock assessments in the El Quisco Destinación have been conducted. The first, from February to March 1992, was near the beginning of the annual aggregation of loco for spawning and prior to the January 1993 (summer) fishery. Spawning aggregations, characteristic of muricid snails, occur mainly between March and June (Castilla 1979, 1982). The second survey was from March to May 1993, in the middle of the species spawning aggregation, after the first fishery opening but before the July (winter) fishery. The third survey was during March–May 1994, also at the peak of spawning.

The Destinación was subdivided into 5 areas, each of approximately 10 ha. Number of locos was recorded along 100-m random subtidal transect lines, either perpendicular or horizontal to the coast. Locos 1 m on either side of the line in 1992 (10 transects) and 1.7 m on either side of the line in both 1993 (29 transects) and 1994 (42 transects) were counted. Mean density of locos per transect was estimated using the geometric mean to correct for normality. Confidence intervals were calculated as described by Sokal and Rohlf (1981). Total MEA population was extrapolated for 35 ha of suitable habitat. Population size structure based on randomly collected samples of over 500 individuals per assessment was used to estimate harvestable stock.

Caleta Algarrobo (33°22′S, 71°40′W)
Caleta Algarrobo is a well-organized Sindicato (Payne and Castilla 1994) having 90 members in 1993, with 25 summer and 36 winter registered divers and an artisanal fleet of 28 wooden boats. The Sindicato has forbidden diving in a coastal area of approximately 56 ha since 1991 (Fig. 1), although this has been only partially enforced. In January 1993, they obtained legal possession of this area as a Destinación. In 1993, Algarrobo divers harvested around 90% of locos landed from common diving grounds 32–40 km away from their cove. On the last fishing day of the winter season (August 3), 8300 locos (9.7% of their quota) were harvested from their Destinación.

Caleta Las Cruces (33°33′S, 71°37′W)
Caleta Las Cruces is a poorly organized Sindicato (Payne and Castilla 1994; Minn and Castilla 1995) having 33 members in 1993, with 15 summer and 11 winter registered divers and an artisanal fleet of 8 wooden boats. Sindicato members could not agree to conserve a local coastal area, but they nevertheless obtained, with support from our research group, a Destinación of 84 ha in February 1993 (Fig. 1). In 1993, Caleta Las Cruces divers harvested all their locos from common access diving grounds 8–11 km from their cove.

Caleta Cartagena (33°33′S, 71°36′W)
Caleta Cartagena, also a poorly organized Sindicato (Payne and Castilla 1994), had 27 members with 15 summer and 26 winter divers in 1993 and an artisanal fleet of 25 wooden boats. The Sindicato has taken no steps to conserve a coastal area and has not been granted a Destinación. Cartagena's divers harvested locos in 1993 from common access diving grounds, both relatively close and distant from their cove. They also illegally harvested locos by poaching at night from Destinacións granted to other Caletas.

The 1993 loco fishery

Catches at El Quisco were recorded during both the summer (5 days) and winter (9 days) fishing seasons. Algarrobo and Las Cruces catches were only monitored during the winter. The following data were recorded on a boat-by-day basis: number of divers, travel time (including travel time to diving grounds and underwater searching time), dive time per diver, fishing location, and catch of locos. This permitted calculation of both travel time and CPUE. Individual CPUE was pooled per Caleta. CPUE and travel time were compared using Tukey's multiple comparison procedure.

Maximum diameter of the peristome and weight (including shells) were recorded from 50 randomly selected locos per boat landing. Harvested loco size frequencies by different Caletas were based on daily measurements of 100 (minimum of 2 trips per day) to 1150 locos (23 trips per day). At Caleta Cartagena, only price and catch data were available.

Results

Loco fishery development

Loco catches landed between 1960 and 1993 illustrate the three major phases through which many Chilean invertebrates evolved (Fig. 2A). The first phase (1960–1974) was for domestic consumption only, and landings ranged from 3000 to 6000 t. The second phase (1975–1981) included both export and domestic consumption and was characterized by a lack of fishery management. Landings peaked in 1980 at 24 856 t. Thereafter, landings decreased, in spite of strong Asian market demand, as initial fishery regulations proved ineffective (Fig. 2B). The fishery was then closed for three months per year from 1982 to 1985 to protect reproductive aggregations, and was completely closed between 1989 and 1992. In 1993, the fishery was reopened for a total of 14 days with a landing of 8574 t whole weight (specimens with shell). Exported meat (the foot: frozen, canned, or dried) weight reached 2392 t. Export value increased between 1978 and 1993 from approximately US$5 million to 64 million (Fig. 2A).

Fig. 2. (A) Landings and export value of the fishery for *Concholepas concholepas* between 1960 and 1993. (B) Fishing effort regulatory measures: white bars indicate no regulation; black bars indicate fishing bans.

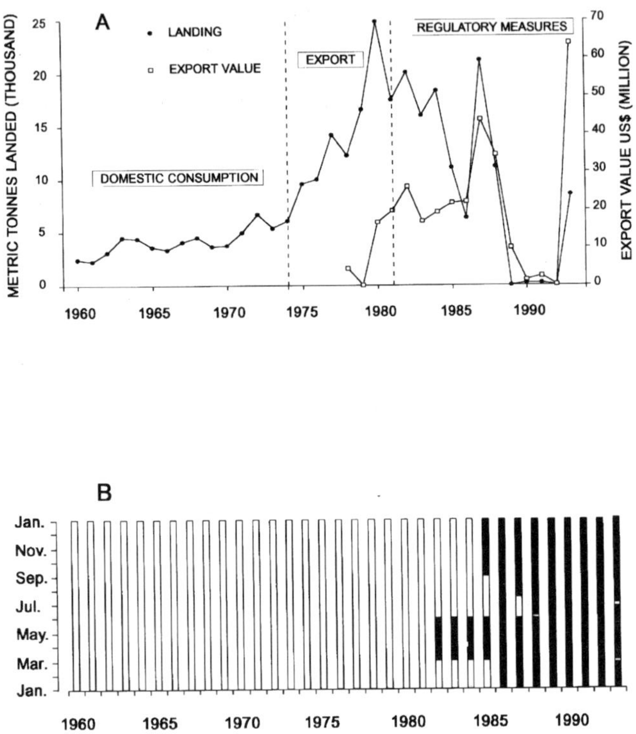

Fig. 3. (A) Locos landed at the 4 monitored Caletas during the 1993 summer and winter fishing seasons. Numbers on top of the bars indicate registered divers. Black bars indicate the number of locos harvested relative to assigned quotas; white bars indicate the quota portion not landed. (B) Landing value of locos. Numbers on top of the bars indicate the mean selling price (US$) per loco. Black bars indicate the total income received by the fishers, and white bars indicate income lost because the quota was not reached.

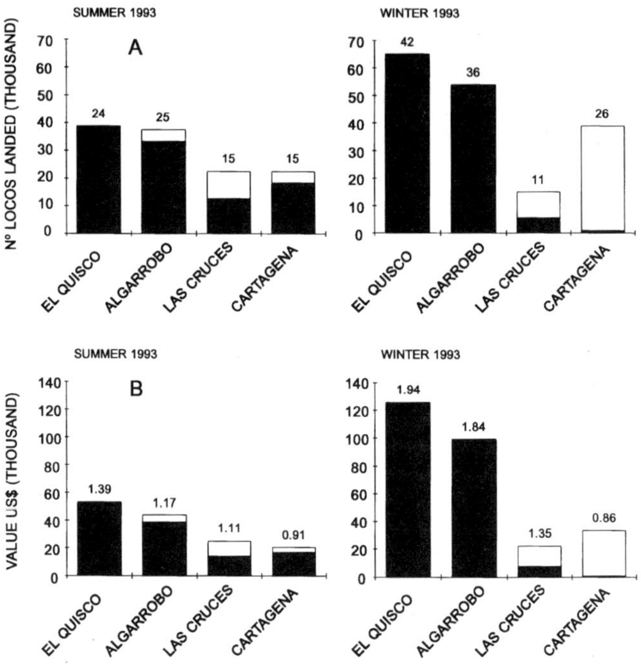

Table 1. Number of loco, *Concholepas concholepas*, in the 35 ha of suitable habitat at El Quisco Destinación.

			Harvestable stock size[b]			
Year	Month	Reproductive stage[a]	Mean	Lower limit	Upper limit	Catches
1992	February–March	1	165438	33331	807174	39347[c]
1993	March–May	2	150660	53085	425017	65446[d]
1994	March–May	2	61965	41933	91424	No catches

Note: Limits represent 95% confidence limits around the mean.
[a]Stage 1 represents initial stage of aggregation and stage 2 represents peak stage of aggregation.
[b]Individuals equal or larger than 10 cm peristome diameter.
[c]Loco harvest during summer 1993.
[d]Loco harvest during winter 1993.

The 1993–1994 loco fishery in Caletas of central Chile

During the 5- and 9-day summer and winter 1993 fisheries, respectively, Caleta El Quisco quotas were 36 000 and 63 000 locos, respectively. Landings were 39 347 and 65 446 locos from their Destinación (Fig. 3A). In the summer fishery, divers harvested 95% of the quota in the first two days, with experienced divers filling their individual quota in 1–2 trips. Individual loco from the summer and winter fishery averaged US$1.39 and 1.94, respectively (Fig. 3B). Caletas fishing on common ground were unable to achieve their assigned quotas (Fig. 3A), and loco selling prices ranged from US$0.91–1.17 (summer) to 0.86–1.84 (winter) (Fig. 3B). El Quisco Destinación's locos commanded higher prices because they were larger than others harvested.

A 1992 stock assessment of the El Quisco Destinación, done one year before the fishery was opened, indicated that the harvestable number of locos (>10 cm) was 165 438 (large confidence limits due to limited surveying, Table 1). In summer 1993, El Quisco divers harvested their quota in 3 days, and also landed their quota during the 1993 winter fishery. CPUE declined during each 1993 fishing season (Fig. 4). El Quisco summer CPUE decreased from 500 to 300 locos·h^{-1}, and winter CPUE dropped from 186 to about 100 locos·h^{-1} (Fig. 4). In 1994, in spite of quota allocations, there were no catches inside

Fig. 4. Daily CPUE (locos/diving hour) at Caletas El Quisco, Algarrobo, and Las Cruces during 1993 fishing seasons. El Quisco MEA summer extraction was preceded by a 30-month closure period (September 1990 – January 1993), and the winter extraction by a 6-month closure (February – July 1993). Bars indicate standard errors.

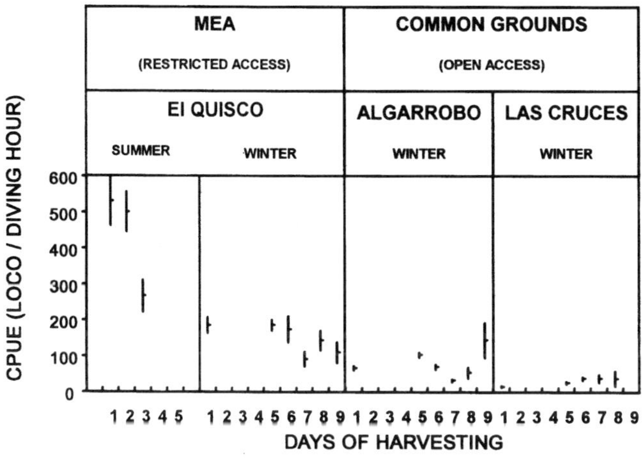

Fig. 5. Mean size of the locos harvested at Caletas El Quisco, Algarrobo, and Las Cruces during 1993 fishing seasons. El Quisco MEA summer extraction was preceded by a 30-month closure period (September 1990 – January 1993), and the winter extraction by a 6-month closure (February – July 1993). Bars indicate standard errors.

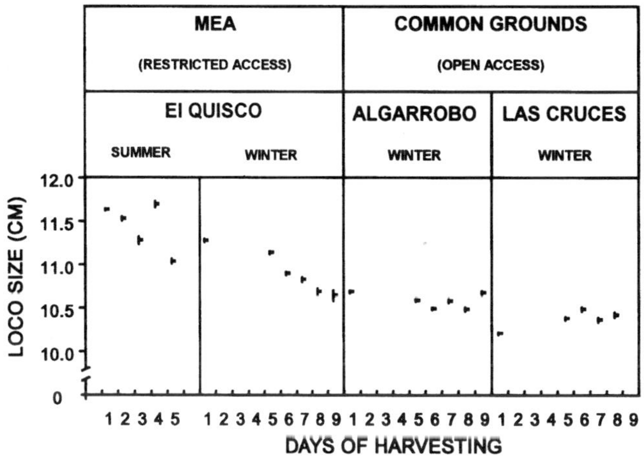

Table 2. Comparison of mean daily CPUE (loco per diving hour) from Caletas El Quisco, Algarrobo, and Las Cruces during 1993 fishing seasons, using Tukey's multiple comparison procedure.

Locality	*N*	Mean	SD	Tukey's multiple comparison grouping
El Quisco (summer)	3	433.02	143.50	a
El Quisco (winter)	6	148.30	40.70	b
Algarrobo (winter)	5	64.89	25.83	bc
Las Cruces (winter)	5	30.93	10.17	c

Note: Means with the same letter are not significantly different ($\alpha = 0.05$).

Table 3. Comparison of the mean daily travel time (h) from Caletas El Quisco, Algarrobo, and Las Cruces during 1993 fishing seasons, using Tukey's multiple comparison procedure.

Locality	*N*	Mean	SD	Tukey's multiple comparison grouping
El Quisco (summer)	3	1.69	0.31	ab
El Quisco (winter)	6	1.51	0.27	a
Algarrobo (winter)	5	2.68	1.06	b
Las Cruces (winter)	5	2.88	0.69	b

Note: Means with the same letter are not significantly different ($\alpha = 0.10$).

the Destinación due to lowering of loco market prices (Table 1).

At Algarrobo, CPUE ranged between 53 and 143 locos·h^{-1}, while at Las Cruces, CPUE ranged between 15 and 38 locos·h^{-1} (Fig. 4). Mean daily CPUE varied greatly both among fished locations and between summer and winter seasons (Table 2). Mean daily travel time in hours varied significantly by fishing locations, with El Quisco having the lowest values (Table 3).

In winter, Algarrobo divers selected diving grounds relatively far away, with a lower average CPUE (Fig. 4). On the last fishing day, Algarrobo divers dove at their Destinación, and achieved a CPUE similar to that obtained by El Quisco divers. Las Cruces divers had the lowest CPUE, and in contrast to Algarrobo divers, spent more time underwater searching for locos than travelling to fishing grounds.

During the summer fishery, El Quisco divers only harvested locos above 11 cm, larger than the minimum legal size (10 cm) and of higher value (Figs. 3 and 5). A declining trend in the size of harvested locos was observed over time (Fig. 5). Algarrobo and Las Cruces divers harvested locos no larger than 10.5–10.7 cm (Fig. 5), and obtained lower prices (Fig. 3B).

Discussion

Organized Caletas that were assigned Destinación (MEAs) and limited fishing on them prior to the 1993 loco fishing season (e.g., Caleta El Quisco) were rewarded with (*i*) an exclusive right to extract loco from their Destinación; (*ii*) highest CPUE during both fishing seasons (Fig. 4); (*iii*) harvest of the largest locos (Fig. 5), and (*iv*) highest market prices (Fig. 3B). Caletas that did not restrict diving activities before 1993 had to search distant common diving grounds, with greater travel time, to land assigned fishing quotas (Table 3) and consequently earned less profit (Fig. 3B). Cartagena could not even fulfill its assigned quotas. Furthermore, divers from these Caletas had CPUE 4–10 times lower than those observed at the El Quisco MEA.

Summer CPUE observed at Caleta El Quisco MEA (around 400 locos·h^{-1}) was comparable with CPUE in the early period of fishery development published by Castilla and Schmiede

(1979) for Caleta Horno in northern Chile (around 550 locos·h^{-1}). CPUE (30–60 locos·h^{-1}) observed on heavily exploited common fishing grounds in 1993 was comparable to that published by Geaghan and Castilla (1986, 1988) for Caleta Quintay in central Chile (80–160 locos·h^{-1}), when loco populations showed evidence of overexploitation.

We present the consequences of alternative exploitation approaches. El Quisco fishers, the best organized, implemented a loco refuge area 2–3 years before the fishery season was opened. Compared with fishers from other Caletas, they did well financially during the 1993 fishing seasons. However, among El Quisco syndicated members this has created an expectation for future loco fisheries that may be unrealistic, as not all factors are under Caleta control (e.g., market and loco abundance). For instance, loco price dropped dramatically in 1994, and as a consequence fishers stopped fishing (Table 1).

Data show that natural restocking of locos can occur, as evidenced by a high CPUE at El Quisco MEA after a 2-year closing (see concept of natural restocking, Castilla 1990*b*). If recruitment is regular, periodic exploitation may be economically attractive (Figs. 3B, 4; Payne and Castilla 1994). Comparison of small-scale, restricted management to open access on "common grounds," i.e., the paradigm of the "tragedy of the commons," has been widely addressed in the literature (Hardin 1968; Berkes 1985, 1987; Ostrom 1990; Castilla 1990*b*; Waters 1991; Bromley 1991; Smith and Berkes 1991; Castilla 1994). Nevertheless, regulations institutionalized in fishery law to deal with this are rare, as many jurisdictions are apprehensive about assigning "ownership" of the sea bottom to interest groups. The Chilean Fishery and Aquaculture Law regarding the management of benthic invertebrates permits testing alternative management approaches which address this dilemma.

Chile is still at an early stage in dealing with "open versus restricted" access issues. Current Chilean experimental management policies for loco can be considered a form of adaptive management (Walters and Hilborn 1978; Walters 1986; Hilborn and Walters 1992). Research priorities are (*i*) to learn about the carrying capacity of coastal MEAs along different stretches of coastline; (*ii*) to understand temporal and spatial variation in recruitment under an open system scenario (Jamieson 1993); (*iii*) to determine genetic spatial variability (Guiñez et al. 1992); and (*iv*) to describe larval dispersal patterns.

It is hoped that "experimental-management" strategies will continue in Chile for benthic invertebrates. Guiding concepts should include (*i*) more effective limited access policies through the establishment of both new MEAs and community-based or co-management scenarios (Alcala and Russ 1990; Castilla 1994); (*ii*) increasing the number of species that can be harvested by Caletas; (*iii*) introduction of harvesting rotation techniques and/or marine protected areas; and (*iv*) creation of a process to establish a network of marine parks and reserves in Chile (Castilla 1996), as has been done in other countries (White 1986; Davis 1989; García-Rubies and Zabala 1990; Roberts and Polunin 1991, 1993; Polunin and Roberts 1993; Bohnsack 1993; De Martini 1993; Carr and Reed 1993).

Acknowledgments

We sincerely acknowledge the continuous help, support, and friendship provided by the fishers of Caleta El Quisco, particularly by the President Señor Francisco Ceballos and the Secretary Señor Miguel Ramírez. Similarly, the artisanal fishers of Caletas Las Cruces and Algarrobo helped with the locos fishery data gathering. Without their help the quality of the information recounted in this paper could not have been achieved, because in small-scale fishery activities much depends on the accurate report of sea-going events along hundreds of fishing trips. Additionally, we sincerely thank our friends Holly Payne, Isabelle Minn, Cristián Pacheco, Claudio Romero, and Sergio Silva for helping with diving and fishery data assembly at the Caletas. The Servicio Nacional de Pesca, SERNAP, provided us with vital information. Financial support was provided to the senior author by FONDECYT/Chile, Project No. 193/0684; The Costal Resource Research Network, Halifax, Canada; Cooperation Program between Italy and the Pontificia Universidad Católica de Chile and the European Community Research Program, Contract N.CI1* - CT93-0338. Our friends Miriam Fernández and Neil Andrew revised the manuscript and suggested important modifications. We also thank an anonymous referee for suggestions and particularly the editor of this volume, Dr. Glen Jamieson. The paper was completed in 1996 thanks to the support of a Pew Fellowship to the senior author. We thank the Pew Foundation and The Center for Marine Conservation, U.S.A.

References

Alcala, A.C., and Russ, G.R. 1990. A direct test of the effect of protective management on abundance and yield of tropical marine resources. J. Cons. Int. Explor. Mer. **46**: 40–47.

Aranda, E., Bustos, E., and Chomali, J. 1989. Estado de situación del sector pesquero artesanal en Chile. Rev. Com. Perm. Pacífico Sur, **18**: 7–33.

Berkes, F. 1985. Fishermen and the "tragedy of the commons." Environ. Conserv. **12**(3): 199–206.

Berkes, F. 1987. The common property resource problem and the fisheries of Barbados and Jamaica. Environ. Manage. **11**(2): 225–235.

Bohnsack, J.A. 1993. Marine reserves: they enhance fisheries, reduce conflicts, and protect resources. Oceanus, Fall 1993: 63–71.

Bromley, D.W. 1991. Testing for commons versus private property: comments. J. Environ. Econ. Manage. **21**: 92–96.

Bustamante, R., and Castilla, J.C. 1987. The shellfishery in Chile: an analysis of 26 years of landings (1960–1985). Biol. Pesq. **16**: 79–97.

Carr, M.H., and Reed, D.C. 1993. Conceptual issues relevant to marine harvest refuges: examples from temperate reef fishes. Can. J. Fish. Aquat. Sci. **50**: 2019–2028.

Castilla, J.C. 1979. *Concholepas concholepas* (Mollusca: Gastropoda: Muricidae): postura de cápsulas en el laboratorio y en la naturaleza. Biol. Pesq. **12**: 91–97.

Castilla, J.C. 1982. Pesquería de moluscos gastrópodos en Chile: *Concholepas concholepas*, un caso de estudio. *In* Segundo seminario taller: bases biológicas para el uso y manejo de recursos naturales renovables: recursos biológicos marinos. *Edited by* J.C. Castilla. Pontificia Universidad Católica de Chile, Santiago. Monogr. Biol. **2**: 199–212.

Castilla, J.C. 1988*a*. Una revisión bibliográfica (1980–1988) sobre *Concholepas concholepas* (Gastropoda, Muricidae): problemas pesqueros y experiencias de repoblación. Biol. Pesq. **17**: 9–19.

Castilla, J.C. 1988*b*. La problemática de la repoblación de mariscos en Chile: diagnóstico, estrategias, y ejemplos. Invest. Pesq. (Chile), **35**: 41–48.

Castilla, J.C. 1990*a*. El erizo chileno *Loxechinus albus*: importancia pesquera, historia de vida, cultivo en laboratorio y repoblación natural. *In* Cultivo de moluscos en América Latina. *Edited by* A. Hernández. Red Regional de Entidades y Centros de Acuicultura de América Latina CIID-CANADA, Bogotá, Colombia. pp. 83–98.

Castilla, J.C. 1990*b*. La técnica de rotación de áreas costeras como una herramienta de manejo de recursos bentónicos y el rol de las reservas marinas y zonas tampones. Abstract X Jornadas de Ciencias del Mar. 28–30 May, 1990. Santiago, Chile. 31 p.

Castilla, J.C. 1994. The Chilean small-scale benthic shellfisheries and the institutionalization of new management practices. Ecol. Int. Bull. **21**: 47–64.

Castilla, J.C. 1995. The sustainability of natural renewable resources as viewed by an ecologist and exemplified by the fishery of the mollusc *Concholepas concholepas* in Chile. *In* Defining and measuring sustainability: the biophysical foundations. *Edited by* M. Munasinghe and W. Shearer. The International Bank for Reconstruction and Development, The World Bank, Washington, D.C. pp. 153–159.

Castilla, J.C. 1996. La futura Red Chilena de Parques y Reservas Marinas y los conceptos de conservación, preservación y manejo en la legislación nacional. Rev. Chil. Hist. Nat. **69**: 253–270.

Castilla, J.C., and Durán, R.L. 1985. Human exclusion from the rocky intertidal zone of central Chile: the effects on *Concholepas concholepas* (Gastropoda). Oikos, **45**: 391–399.

Castilla, J.C., and Jeréz, G. 1986. Artisanal fishery and development of a database for managing the loco, *Concholepas concholepas*, resource in Chile. Can. Spec. Publ. Fish. Aquat. Sci. 92. pp. 133–139.

Castilla, J.C., and Schmiede, P. 1979. Hipótesis de trabajo sobre la existencia de zonas marítimas tampones en relación a recursos marinos bentónicos (mariscos y algas) en la costa de Chile continental. *In* Seminario taller sobre desarrollo e investigación de los recursos marinos de la Octava Región, Chile. *Edited by* V.A. Gallardo. Universidad de Concepción, January 9–13, 1978. Concepción, Chile. pp. 145–167.

Castilla, J.C., Branch, G.M., and Barkai, A. 1994. Exploitation of two critical predators: the gastropod *Concholepas concholepas* and the rocky lobster *Jasus lalandii*. *In* Rocky shores: exploitation in Chile and South Africa. *Edited by* W.R. Siegfried. Springer-Verlag, Berlin Heidelberg. Ecol. Stud. Anal. Synth. **103**: 101–130.

Davis, G.A. 1989. Designated harvest refugia: the next stage of marine fishery management in California. CALCOFI (Calif. Coop. Oceanic Fish. Invest.) Rep. No. 30. pp. 53–58.

De Martini, E.E. 1993. Modeling the potential of fishery reserves for managing Pacific coral reef fishes. Fish. Bull. **91**: 414–427.

Durán, R., and Castilla, J.C. 1989. Variation and persistence of the middle rocky intertidal community of central Chile, with and without human harvesting. Mar. Biol. (Berlin), **10**: 555–562.

Durán, R., Castilla, J.C., and Oliva, D. 1987. Intensity of human predation on rocky shores at Las Cruces in Central Chile. Environ. Conserv. **14**(2): 143–149.

García-Rubies, A., and Zabala, M. 1990. Effects of total fishing prohibition on the rocky fish assemblages of Medes Islands marine reserve (NW Mediterranean). Sci. Mar. **54**(4): 317–328.

Geaghan, J.P., and Castilla, J.C. 1986. Use of catch and effort data for parameter estimates for the "loco" (*Concholepas concholepas*) fishery of central Chile. Can. Spec. Publ. Fish. Aquat. Sci. 92. pp. 168–174.

Geaghan, J.P., and Castilla, J.C. 1987. Population dynamics of the loco (*Concholepas concholepas*) fishery in Central Chile. Invest. Pesq. (Chile), **34**: 21–31.

Geaghan, J.P., and Castilla, J.C. 1988. Evaluación de la actual capacidad de manejo del loco *Concholepas concholepas* (Bruguiere, 1789) (Gastropoda, Muricidae) en Chile. Biol. Pesq. **17**: 57–72.

Guiñez, R., Gómez, M.V., and Castilla, J.C. 1992. Diferenciación genética poblacional en *Concholepas concholepas* (Bruguiere, 1789) (Gastropoda, Muricidae) en su área de distribución Centro-Norte. Biol. Pesq. **21**: 31–41.

Hardin, G. 1968. The tragedy of the commons. Science (Washington, D.C.), **162**: 1243–1248.

Hilborn, R., and Walters, C.J. 1992. Quantitative fisheries stock assessment. Chapman & Hall Inc., New York, N.Y.

Jamieson, G.S. 1993. Marine invertebrate conservation: evaluation of fisheries over-exploitation concerns. Am. Zool. **33**: 551–567.

Minn, I., and Castilla, J.C. 1995. Small-scale artisanal fishing and benthic invertebrate management in caleta Las Cruces, Central Chile. Out of the Shell, **5**: 11–15.

Morales, H.L. 1993. The Chilean and Latin-American experience in horizontal cooperation between artisanal fishermen organizations. European Community Fish. Coop. Bull. **6**(2): 8–9.

Ostrom, E. 1990. Governing the commons. The evolution of Institutions for Collective Action. Cambridge University Press, Cambridge, U.K.

Payne, H.E., and Castilla, J.C. 1994. Socio-biological assessment of common property resource management: small-scale fishery unions in Central Chile. Out of the Shell, **4**(3): 10–14.

Pino, C., and Castilla, J.C. 1995. The key-hole limpets (*Fissurella* spp.) in the Chilean artisanal fishery. Out of the Shell, **5**: 8–10.

Polunin, N.V.C., and Roberts, C.M. 1993. Greater biomass and value of target coral-reef fishes in two small Caribbean marine reserves. Mar. Ecol. Prog. Ser. **100**: 167–176.

Roberts, C.M., and Polunin, N.V.C. 1991. Are marine reserves effective in management of reef fisheries? Rev. Fish. Biol. Fish. **1**: 65–91.

Roberts, C.M., and Polunin, N.V.C. 1993. Marine reserves: simple solutions to managing complex fisheries? Ambio, **22**(6): 363–368.

Smith, A.H., and Berkes, F. 1991. Solutions to the "Tragedy of the Commons": sea urchin management in St. Lucia, West Indies. Environ. Conserv. **18**: 131–136.

Sokal, R.R., and Rohlf, F.J. 1981. Biometry. 2nd ed. W.H. Freeman Press, San Francisco, Calif.

Walters, C.J. 1986. Adaptive management of renewable resources. Macmillan Publishing Co., New York., N.Y.

Walters, C.J., and Hilborn, R. 1978. Ecological optimization and adaptive management. Annu. Rev. Ecol. Syst. **9**: 157–188.

Waters, J.R. 1991. Restricted access vs open access methods of management: towards more effective regulation of fishing effort. Mar. Fish. Rev. **53**(3): 1–10.

White, A.T. 1986. Marine reserves: how effective as management strategies for Philippine, Indonesian, and Malaysian coral reef environments? Ocean Manage. **10**: 137–159.

Regional Perspectives from the North Pacific

Shellfishery biology in Russia: problems and opportunities

Boris G. Ivanov

Abstract: Transition of the Russian economy from planned principles to a free market has resulted in great changes in both Russian fisheries and fishery science. The sudden dramatic increase in fuel cost associated with this transition has excluded the Russian fleet from many distant Soviet fishing grounds and has largely confined it to Russian seas. Government monopoly on exploitation of fish resources has been terminated and many new private companies now participate. Because of their relatively high export prices, fishing pressure on shellfish, in particular, has increased and new methods of harvesting by means of pots and longlines have replaced trawling. At fishery science institutes, the number of research vessels has decreased, organizational structures have been revised, numbers of scientific staff have been reduced, and financial difficulties have become chronic. Institutes are now encouraged to seek additional funds, particularly from Russian–foreign company joint fishing ventures. Scientists now often conduct their work from commercial fishing vessels instead of research vessels, and in invertebrate stock assessments, it has become necessary to use pot catch rates as indices of abundance since research trawl surveys are no longer being conducted. The present state of Russian fishery monitoring has been described as disastrous, but the use of new methodologies in shellfish studies is providing new research opportunities, including realistic spatial mapping of stocks and more relevant data collection from fisheries. In addition, new equipment, particularly computers, has become available in the last 5–7 years. Fishery institutes will survive as long as the Fisheries Committee of the Russian Federation continues to provide support through the allocation of harvestable "research quotas," that are used to generate needed resources for stock monitoring.

Résumé : En Russie, le passage d'une économie planifiée à une économie de marché s'est traduit par des changements importants au niveau des pêches et de l'halieutique. L'augmentation brusque et spectaculaire du coût du carburant, associée à cette transition, a exclus la flottille russe de nombreuses aires de pêche éloignées et a cantonné en général la pêche dans les mers russes. Le gouvernement n'a plus le monopole de l'exploitation des ressources halieutiques et de nombreuses entreprises privées nouvelles participent maintenant à l'exploitation. En raison de leur prix d'exportation assez élevé, la pression de pêche sur les mollusques et crustacés s'est particulièrement accrue, et le chalutage a été remplacé par de nouvelles méthodes de récolte aux casiers et à la palangre. Dans les instituts de recherche halieutique, le nombre de navires de recherche a diminué, les structures organisationnelles ont été revues, le personnel scientifique a été réduit et les difficultés financières sont devenues un phénomène chronique. À l'heure actuelle, on encourage les instituts à chercher d'autres fonds, en particulier auprès de co-entreprises de pêche russo-étrangères. Il est fréquent, à l'heure actuelle, que les scientifiques effectuent leurs travaux à partir de navires de pêche commerciale plutôt qu'à bord de navires de recherche, et dans le cas des évaluations des stocks d'invertébrés, il est maintenant nécessaire d'utiliser les taux de capture des casiers comme indices d'abondance étant donné qu'il n'y a plus de campagne de chalutage. L'état actuel de la surveillance des pêches en Russie est considéré comme catastrophique, mais le recours à de nouvelles méthodes d'étude des mollusques et crustacés offre de nouvelles possibilités de recherche, notamment une cartographie spatiale réaliste des stocks et une collecte de données plus pertinentes sur les pêches. En outre, depuis 5 à 7 ans, on dispose de nouveaux appareils, en particulier d'ordinateurs. Les instituts des pêches survivront aussi longtemps que le comité des pêches de la Fédération de Russie continuera de les soutenir par l'allocation de « quotas de recherches » exploitables, utilisés pour produire les ressources nécessaires à la surveillance des stocks.
[Traduit par la Rédaction]

Introduction

Drastic changes have happened in Russian fisheries since 1988. Transition to a free market economy in Russia has resulted in great increases in the cost of equipment, instruments, materials, and especially oil and fuel. At the same time, budgets to support fisheries research institutes and their scientific staff have been decreased substantially. Availability of research vessels was also greatly reduced, terminating the old trawl system for monitoring abundances of fish and shellfish stocks. Usually these changes have been described as disastrous for Russian fishery science and for the careers of many researchers. Nevertheless, new conditions do provide new opportunities and not all changes should be considered negative (Ivanov 1994*a*, 1994*b*).

This paper reviews Soviet/Russian fishery science from the 1950's and gives a personal perspective on current problems. Special reference is given to invertebrates, which the author has studied for more than 30 years.

B.G. Ivanov. Russian Research Institute of Fisheries and Oceanography (VNIRO), 107140 Moscow, Russia.

Correct citation: Ivanov, B.G. 1998. Shellfishery biology in Russia: problems and opportunities. *In* Proceedings of the North Pacific Symposium on Invertebrate Stock Assessment and Management. *Edited by* G.S. Jamieson and A. Campbell. Can. Spec. Publ. Fish. Aquat. Sci. 125. pp. 417–421.

Heritage of the past

Fisheries

The scientific basis for ongoing macroscale management of fisheries in Russia was, and continues to be, the forecast (prognosis) of allowable catches of fishes and invertebrates by area for the forthcoming year (Elizarov 1993; Petrov 1993; Borisov 1994). A brief description of stock units and recommendations for their harvest is provided in this forecast. All Russian fisheries research institutes participate in development of the fishery forecast. Quotas for fishery enterprises are then assigned, taking into account forecast recommendations. Development of the forecast is considered the ultimate objective of fisheries studies.

In the 1950's and 1960's, Soviet fisheries developed rapidly and entered a "worldwide exploitation period." Soviet fishing vessels could be encountered from Greenland and Spitsbergen to the Antarctic and from tropical west and east African waters to the southeast Pacific. The entire world oceans became the fishing ground for a large Soviet fishing fleet. High seas expansion of Soviet fisheries was planned to meet human food needs caused by inefficient Soviet agricultural practices. The country needed protein and the fishing industry and fishery science were ordered to satisfy this requirement. This goal could not be met without exploitation of the entire world's available fish resources. High seas development of Soviet fisheries was achieved mainly by subsidized oil prices, with oil transported thousands of miles from the Soviet coast to the fishing fleets by an armada of fuel tankers.

High seas oceanic exploitation by Soviet fishers led to a number of stock discoveries and Soviet fisheries, including sardine off west Africa (1957); flatfishes, cod, herring, rockfishes, halibut, and blackcod in the Bering Sea and Gulf of Alaska (late 1950's – early 1960's; at this time, pollock was not yet considered worthy of commercial harvest); cod, hake, redfish, and halibut on the Grand Banks of Newfoundland and the Flemish Cap; marble notothenia, silver fish, tooth-fishes, and icefish in the Antarctic; and horse mackerel in the southeast Pacific. Also, there were new stock discoveries of invertebrates. Squids off Argentina and New Zealand; pandalid shrimps in the Gulf of Alaska, the Bering Sea, off Greenland, Spitsbergen, and in the Barents Sea; penaeid shrimps off west Africa and Mozambique; and deepwater spiny lobsters off Somalia and the Arabian Peninsula. Overall, these discoveries resulted in radical changes in the spatial distribution of Soviet fisheries.

However, dependence on the high seas fishery for protein had its consequences. Fishery plans focused on the capture of large quantities of abundant high seas fish, even if they were coarse and of relatively low price, rather than on capture of smaller quantities of higher quality or more valuable species. Consequently, smaller-scale resources, particularly in coastal, shallow waters, were largely ignored. Development of fleets adapted for inshore fishing was minimal and the coastal infrastructure for distribution, storage, processing, and marketing of landings was poorly developed. For the fishing industry, development of a fishery for Antarctic krill was easier than developing one for native crayfishes. Low-price fuel did not stimulate development of economical methods for capturing species, such as use of pots, traps, and longlines. Usage of such gear was delayed, and consequently, recent adoption of new fishing methods has been difficult. The Majesty Trawl, which required high engine power, dominated among preferred fishing gears.

Fishery research: merits and demerits

Fishery research was of great importance in permitting expansion of Soviet fisheries, but unfortunately, its full potential has not been properly utilized in Russia in recent years. Expansion of Soviet high seas fisheries required adequate scientific information and the structure of research institutes was developed to provide this. Marine laboratories assumed leading roles; for example, the Atlantic Research Institute of Fisheries and Oceanography (AtlantNIRO) at Kaliningrad was entirely oriented to studying distant oceanic areas. Numerous research fleets were created, which included expensive vessels such as tender ships for manned submersibles. More than a hundred research expeditions were conducted every year. No other country has even conducted such extensive high seas fisheries research and monitoring.

A consequence of this extensive high seas program was the training of many marine biologists. Almost all leading fishery scientists in Russia participated in high seas research expeditions. Many, including myself, remember the 1950's and 1960's as the "golden period" of Soviet marine research, a romantic and heroic time, although my experiences are probably colored by youthful perceptions. The R/V "Vitjaz," flagship of the Soviet scientific fleet, beckoned youth to remote lands. In the 1950's, television sets were not common in Russia and quality movies, books, and documentaries by international authorities were not available. Instead, thousands of children in snow-covered Russian cities read popular books by Tarasov (1951) and Bogorov (1954) and dreamed about tropical seas, palms on atoll islands, exotic peoples, and the wonders of coral reefs. High seas fishery exploitation and research provided the only opportunity to make their dreams of travel, at least partially, come true.

Soviet high seas research broadened scientific thinking and gave new understanding of the complexity of marine biological systems. It provided material for comparative ecological and geographical analyses and fostered the belief that Soviet knowledge of high seas resources was more complete, in many cases, than that of scientists from most countries nearer the fishing grounds.

However, it is now clear that Soviet high seas achievements had a negative aspect too. At the human level, there was injustice in salary, with high seas participants paid partially in foreign currency, while participants working in Soviet seas (stormy and more difficult for work) were paid only in rubles. There was also a lack of balance in distribution of research effort. In tropical seas, despite their modest fishery potential, the number of research expeditions was excessively high, at the expense of those in Soviet seas. Ventures which guaranteed visits to duty-free ports such as Singapore, Aden, and Las Palmas were especially attractive. In addition, maintaining high research quality on high seas expeditions was often difficult. Keeping abreast of new techniques, approaches, and publishing in the primary literature was impeded by continuous participation in numerous long (up to 6 months), consecutive cruises. Mountains of manuscript reports could not be transformed into publishable primary articles and loss of samples due to poor preparation for long-term storage, inadequate access to current scientific literature, poor working conditions, and frequent celebrations related to expedition arrivals and departures from ports all had their cost. Censorship also contributed to a reduction in productivity and the usefulness of published Soviet material. Many past publications are impossible to reference, for example, because of the removal by censors of geographic origin of samples. Many past deficiencies in Soviet scientific publications are the result of censor meddling, and not of author's neglect. Limited scientific contacts between Soviet and foreign colleagues also worked against a high scientific standard.

Our research vessels, amazing to foreigners in their number, also amazed foreigners because of their poor equipment. Research efforts were often compromised by equipment failure and data obtained from manned submersibles, for example, were rather modest. This aspect of fishery research is now in a crisis situation, as facilities for gear repair and maintenance are presently of even poorer quality.

New problems and new opportunities

The transition of Russia to a free market economy and the subsequent enormous increase in the cost of ship fuel greatly affected both high seas fisheries and research. Maintenance of both the former research fleet and frequency of expeditions became impossible. The epoch of "extensive" high seas research was finished. Many research vessels were transformed into fishing boats, cargo vessels, or tour boats. Monitoring of even local fish stocks became difficult. This trend is likely to continue and, therefore, the possibility of overfishing at least some populations is relatively high.

Another real problem for fishery research institutes is the current serious shortage of financial support. The State declared that it would stop support and that applied research institutions were to solicit funds directly from contracts with fishing organizations. However, there are few, if any, flourishing fishing enterprises in Russia ready to pay for scientific studies. This sudden cut in funding sent fishery institutes to the edge of destruction. Because scientific staff at institutes was greatly reduced, forcing revisions of the scientific plans and structure of laboratories, 1991 was especially difficult. Many scientists, particularly senior ones, found it very difficult to have their life's science programs terminated. For the first time, institutes had to look for sources of financial support from bodies other than the State.

Reduction of the research fleet resulted in reduction of trawl surveys for stock assessments of crabs and other species. Previously, trawl surveys were the main tool in monitoring programs. This forced consideration of other gear to be used in stock assessments. The problem was partially solved by establishing monitoring procedures on commercial fishing vessels. With crustaceans, monitoring commercial pot catches provided data for use in determining indices of abundance. Changes in types of survey vessel platform and survey gear for sampling are ushering in a new stage of shellfishery biology in Russia. Fishery biology has become even more applied than was previously the case.

Although options to perform trawl survey stock assessments have become more limited, only a few local species were, in fact, studied on an ongoing basis in the past by means of trawl surveys. Multi-year observations on commercial populations were rare, with one of the few species studied this way being red king crab (*Paralithodes camtschaticus*) off western Kamchatka. This was because it was a very important USSR/Russian fishery and because the continental shelf stock was co-operatively exploited, assessed, and managed by both the USSR/Russia and Japan. Before extension of the Russian fisheries zone to 200 miles, the outer shelf was in international waters and co-operative research was required by intergovernmental USSR–Japan agreements. Each country had specific survey obligations, with the USSR/Russia required to conduct two crab trawl surveys off West Kamchatka, in spring and in fall. Now, assessment and management of the crab stock is exclusively the mandate of Russia. Most other invertebrate species were surveyed by trawls only occasionally, meaning the current absence of trawl surveys for these species has had relatively little effect.

The shift in research gear from trawl to pot/trap has had the following advantages: (*i*) As research survey gear and commercial fishing gear are now identical, spatial mapping of commercial stock concentrations (catches, stock density) and interpretation of stock assessment data have become more realistic and more appropriate for the needs of fisheries. Past problems arising from a difference between research and fishery gears is demonstrated with sea snails in the northern Sea of Okhotsk. The sea snail stock was evaluated by trawl and drag surveys, while the fishery used pots. Species catch composition from dragging gear was different from that of pot catches because not every snail species enters pots. Therefore, only a portion of the overall sea snail stock is accessible to pot fishing. Since stock assessment was based on dragging gear, stock size was regularly overestimated, which was a main reason that the actual exploitable stock was overfished. (*ii*) Working on fishing vessels provides opportunities to study the fishery itself, including gear selectivity and size/quality selectivity by fishers, reasons for selections, catch per unit of effort, attractiveness of bait, bycatch species, and fishing strategies of various enterprises. (*iii*) Recognizing that fisheries statistics in the USSR/Russia have always been poor, incomplete, or lacking in detail, observations by independent observers can only help. (*iv*) Commercial monitoring of individual pot catches provides scientists with a greater amount of data than was provided from trawl surveys, where each tow was one unit of effort.

However, this new stage in Russian fishery research raises new problems and learning from the experiences of foreign colleagues dealing with trap data (see review by Miller 1990) should be a priority.

A closer relationship between applied researchers and fishers may be the salvation of Russian fishery biology. In the last 5–7 years, Russian fisheries institutes have received a substantial part of their budget from contracts with fishing organizations for fishery research and stock assessments. This is in contrast to many academic institutes that have not been as successful in finding new funding. Fisheries institutes are surviving and, moreover, have managed to acquire relatively modern personal computers, facsimile and photocopy devices, and other required analytical and communications technology.

It should be emphasized that the acquisition of modern equipment by institutes was largely achieved through joint Russian–foreign fishing ventures, as a part of contract conditions. Institutes are grateful for this modernization, but foreign fishing ventures, as with Russian fishing enterprises, have shown relatively little interest in fishery science per se. They have been most interested in obtaining economic return from high-value species fished during the pot research surveys. Consequently, the major factor determining the continued survival of fishery institutes is the existence of research quotas, i.e., quotas which can only be harvested during the course of research surveys. Such quotas have to be large enough to allow research sampling to provide sufficient indemnity for boat and other expenses and to permit a reasonable profit for collaborating fishing enterprises. Research quotas and the subsidies they provide are a necessary condition now for both the continuation of stock monitoring and the survival of institutes. The Fisheries Committee of the Russian Federation (formerly the Ministry of Fisheries of the USSR) recognizes this and provides institutes with quotas. At present, this Committee is proving to be the main organization in Russia to show an interest in the maintenance of Russian fishery science and the provision of scientific results. The forecast thus continues to be the basis of large scale fisheries management in Russia.

Personnel problems

Work on preparing the forecast is becoming more difficult. The only previous harvester of Russian fishing resources was the State, but now this single body has been replaced by a number of smaller commercial enterprises. Fishing pressure on valuable, high-price species has increased greatly and stocks in many geographical locations are now supporting fisheries. These features make it necessary to monitor many separate stock units and more than a hundred such stock units have been identified in Russian far east seas. In this situation, even the largest institute in the far east, the Pacific Research Institute of Fisheries and Oceanography (TINRO) in Vladivostok, has insufficient staff to provide annual detailed assessments of all shellfish stocks. In response, divisions were created for shellfish study at both the Kamchatka Research Institute of Fisheries and Oceanography (KamchatNIRO) in Petropvalovsk-Kamchatsky and at the Magadan Branch of TINRO in Magadan. These are in addition to the older TINRO Division at the Sakhalin Research Institute of Fisheries and Oceanography (SakhNIRO) in Yuzhno-Sakhalinsk. Nevertheless, a personnel shortage problem remains, which influences the development of various forms of co-operative research, both among Pacific institutes and between the Pacific and European parts of Russia.

A partial solution to this problem is to use contracted observers rather than institute staff on fishing vessels. This is already widely used in the Russian far east, but qualifications of observers are often variable and some data collected has been of poor quality.

Co-operation is also essential for researchers in the Russian North Sea, where in addition to fisheries for the native deepwater prawn (*Pandalus borealis*) and Icelandic scallop (*Chlamys islandica*), a pilot-scale fishery has now begun for Kamchatka (red) king crab (*Paralithodes camtschaticus*), which was introduced into the Barents Sea in the 1960's (Orlov 1994). The Polar Research Institute of Fisheries and Oceanography (PINRO) in Murmansk will hopefully be able to benefit substantially from the far east experience in king crab research. Moreover, far east biologists and fishers have experience in shrimp fishing with pots, particularly for *Pandalus borealis eous*, which may be useful in the development of prawn pot fishing in the inshore area off Murmansk, an area of uneven rocky grounds where traditional trawl fishing for shrimp is impossible.

Wages of scientists working at Russian laboratories remain low, particularly in comparison to the wages of scientists and even technicians and observers when they are on high sea cruises. The difference between land and at-sea earnings also raises new problems. If scientists spend too much time at sea, they have little time for data processing and summarizing observations. Observers urgently need protocol manuals for collecting biological data at sea, adapted to the new methods of data collection. However, preparation of these manuals is not encouraged by relatively low land wages. Because of low institute salaries, many fishery scientists have left marine biology to work for commercial enterprises or even left Russia to look for better working conditions and prosperity abroad. Scientific personnel in many laboratories are not being replaced by younger staff and the average age of staff is becoming higher. Personnel loss may become even more significant if future opportunity to earn higher wages is not improved.

Finally, there are some emerging problems between research institutes and Russian fishing organizations. The latter consider research quotas to be too high and object to the use of foreign boats to exploit quotas. Research institutes prefer to use boats from joint Russian–foreign ventures because this arrangement provides access to modern scientific equipment, operations are more controlled because of contract conditions, and vessels have more advanced navigational equipment. Foreign boats thus provide better platforms for collecting higher quality data on the distributions and abundances of species fished. Solving these contradictions will be a major challenge for the Fisheries Committee.

Conclusions

It is said that Russian fishery biology, particularly related to shellfish stocks, is going through a difficult time. Continuous financial problems and a reduction in the number of research vessels and in research staff seem ongoing. Losses have been great, but there have been some gains. Shellfish stock assessment advice has become more applied and more responsive to the needs of fishers. Laboratories are receiving technologically advanced equipment. Favourable changes have occurred in the general work environment. There is less state control of people and their work, no more political brainwashing, less censorship of publications, more informal contact with foreign colleagues, less difficulty in obtaining approval for foreign travel, progress in computerization, and no compulsory agriculture work. Although scientific study remains difficult and unstable, its quality is becoming more rigorous. Fishery research institutes have survived. Hopefully, if they are not prohibited from receiving additional support from contracts with fishing enterprises and if they continue to be provided with research quotas, their functioning will continue to improve.

Acknowledgments

The author gratefully acknowledges Mr. Yu. B. Ryazantsev (VNIRO) and Dr. G. Jamieson (Department of Fisheries and Oceans, Nanaimo, B.C., Canada) for their kind help in improving the English in this article. My attendance at the Symposium was sponsored by the International Science Foundation (G. Soros Foundation), to which I am very grateful.

References

Bogorov, V.G. 1954. Zhizn' morya. [Life of the sea.] Publishing House "Molodaya gvardiya," Moscow. 301 p. (In Russian.)

Borisov, V.M. 1994. Prognoz po-prezhnemu neobkhodim otrasli. [The industry continues to need forecasts.] Rybnoe Khozyaistvo [Fisheries], **5**: 41–42. (In Russian.)

Elizarov, A.A. 1993. Syr'evaya baza rybnoy promyslennosti: resursy buduschego. [Resources of fishing industry: resources of future.] Rybnoe Khozyaistvo [Fisheries], **6**: 26–28. (In Russian.)

Ivanov, B.G. 1994*a*. Promyslovaya gidrobiologiya Rossii: nasledie, problemy, perspektivy (1). [Commercial hydrobiology of Russia: legacy, problems, perspectives. (Part 1).] Rybnoye Khozyaistvo [Fisheries], **5**: 43–46. (In Russian.)

Ivanov, B.G. 1994*b*. Promyslovaya gidrobiologiya Rossii: nasledie, problemy, perspektivy (2). [Commercial hydrobiology of Russia: legacy, problems, perspectives. (Part 2).] Rybnoye Khozyaistvo [Fisheries], **6**: 30–33. (In Russian.)

Miller, R.J. 1990. Effectiveness of crab and lobster traps. Can. J. Fish. Aquat. Sci. **47**: 1228–1251.

Orlov, Yu.I. 1994. Akklimatizatsiya promyslovykh krabov v Ser-vero-Vostochnoy Atlantike: obosnovanie i pervye rezul'taty. [The acclimatization of commercial crabs in Northeast Atlantic: basis and the first results.] Rybnoe Khozyaistvo. Obzornaya informatsiya. Seriya "Akvakultura': vypusk 1. [Fish Industry. Review Information. Ser. Aquaculture Vol. 1.] *Issued by* VNIERKh, Moscow. 56 p. (In Russian.)

Petrov, O.V. 1993. Nauchno-tekhnicheskoe obespechnie rybokhozyaistvennoy otrasli. [Scientific and technical service for fishing industry.] Rybnoe Khozyaistvo [Fisheries], **6**: 12–15. (In Russian.)

Tarasov, N.I. 1951. More zhivet. [The living sea.] *Published by* Voenno-morskoie izdatel'stvo Voenno-morskogo ministerstva SSSR, Moscow. 276 p. (In Russian.)

Sea urchin (*Strongylocentrotus* spp.) fisheries in Kamchatka: current conditions and problems

A.G. Bazhin

Abstract: Surveys of three species of sea urchin (*Strongylocentrotus droebachiensis*, *S. pallidus*, and *S. polyacanthus*) were conducted along the east coast of Kamchatka and adjacent waters during 1984–1994. *Strongylocentrotus polyacanthus* inhabited shallow exposed open coastal waters with higher salinity and lower temperatures than those for *S. droebachiensis,* which occurred in protected inlets and small bays and in sites near small river mouths with brackish warm shallow waters. *Strongylocentrotus pallidus* occurred mainly in waters deeper than 10 m. Distributions and gonad quality of sea urchins were considered in the implementation of fishery management plans of these species.

Résumé : Entre 1984 et 1994, des relevés de trois espèces d'oursins (*Strongylocentrotus droebachiensis, S. pallidus* et *S. polyacanthus*) ont été effectués sur la côte est du Kamchatka et dans les eaux adjacentes. *Strongylocentrotus polyacanthus* fréquente les eaux côtières peu profondes des zones foraines à salinité élevée et à température plus basse que *S. droebachiensis,* qui habite des bras protégés, de petites baies et des zones situées à proximité de l'embouchure de petites rivières à eaux chaudes saumâtres et peu profondes. *Strongylocentrotus pallidus* vit surtout à une profondeur de plus de 10 m. Les répartitions et la qualité des gonades des oursins ont été prises en compte lors de la mise en oeuvre des plans de gestion des pêches de ces espèces. [Traduit par la Rédaction]

Introduction

There are potential dive fisheries for three sea urchin species (*Strongylocentrotus droebachiensis*, *S. pallidus*, and *S. polyacanthus*) along the east coast of Kamchatka and adjacent waters (Fig. 1). Although previous authors (Djakonov 1949; Baranova 1957; McCaley and Carey 1967; Jensen 1974) reported that there could be 5–8 species occurring in these waters, a recent taxonomic study by Bazhin (1995) indicated there are only three species. Sea urchin roe is a high-value seafood which draws the attention of many fishery institutions. Effective harvest strategies of sea urchin require knowledge of their ecological characteristics as well as their fishery potential. The objectives of this paper are to present the results of a survey of sea urchin abundance in relation to physical factors such as depth, substrate types, and coastal exposure characteristics and to indicate which of these populations would be suitable for commercial harvesting.

Materials and methods

Sea urchins were surveyed by SCUBA diving at sites along the east coast of Kamchatka during May to November, 1984–1994 (Fig. 1). This season coincides with no migration of the animals (personal observation). A preliminary observation of the majority of the sea urchin populations along the coast was conducted in order to define the location of survey stations. The distance between stations ranged from 100 to 500 m, depending on bottom relief and configuration of coast line. Each station consisted of 3–5 samples from different depths. The sampling method consisted of counting the total number of sea urchins of each species within a number of 0.25-m^2 metal frame quadrats thrown randomly on the bottom in each depth interval. The sample size ranged from 3 to 25 quadrats, depending on the density of sea urchins such that no less than 80 specimens were counted per sample. A total of 166 stations were surveyed.

During analysis, the data were grouped into depth intervals of 0–10 m, 10–20 m, and 20–30 m for exposed sites and 0–5 m, 5–15 m, and 15–20 m for protected sites. The data from each depth interval were further grouped into substrate types of rock (e.g., boulders, rock, and large stones), mixed (e.g., pebbles, gravel, shells, sand, mud, and rock in various combinations) and soft (e.g., mud, sand).

Temperature and salinity profiles at 237 hydrological stations were obtained annually during July–September, using portable water analyzers "Horiba" and "Mark-2." The level of surf exposure was defined using Lukin's (1982) classification. Based on these characteristics, the rocky coast of East Kamchatka was classified into four habitat types: (I) exposed open ocean coast, south exposed capes of closed inlets; (II) half-protected inlets, north exposed inlets and capes, shallow sites nearby small river outlets; (III) outer parts of closed inlets, half-protected shallow sites nearby small river outlets; and (IV) inner parts of closed inlets (Fig. 2).

The average relative abundance ($\bar{P}_i$) of sea urchin species i was calculated for each habitat and depth range from the formula

$$[1] \quad \bar{P}_i = P_i / N$$

where N is the total number of samples at each habitat and depth range and P_i is the relative abundance (%) of individuals

A.G. Bazhin. Kamchatka Research Institute of Fisheries and Oceanography, Naberezhnaya 18, Petropavlovsk-Kamchatsky 683002, Russia.

Correct citation: Bazhin, A.G. 1998. Sea urchin (*Strongylocentrotus* spp.) fisheries in Kamchatka: current conditions and problems. *In* Proceedings of the North Pacific Symposium on Invertebrate Stock Assessment and Management. *Edited by* G.S. Jamieson and A. Campbell. Can. Spec. Publ. Fish. Aquat. Sci. 125. pp. 423–427.

Fig. 1. Study area and sea urchin harvesting zones on the southeast coast of Kamchatka.

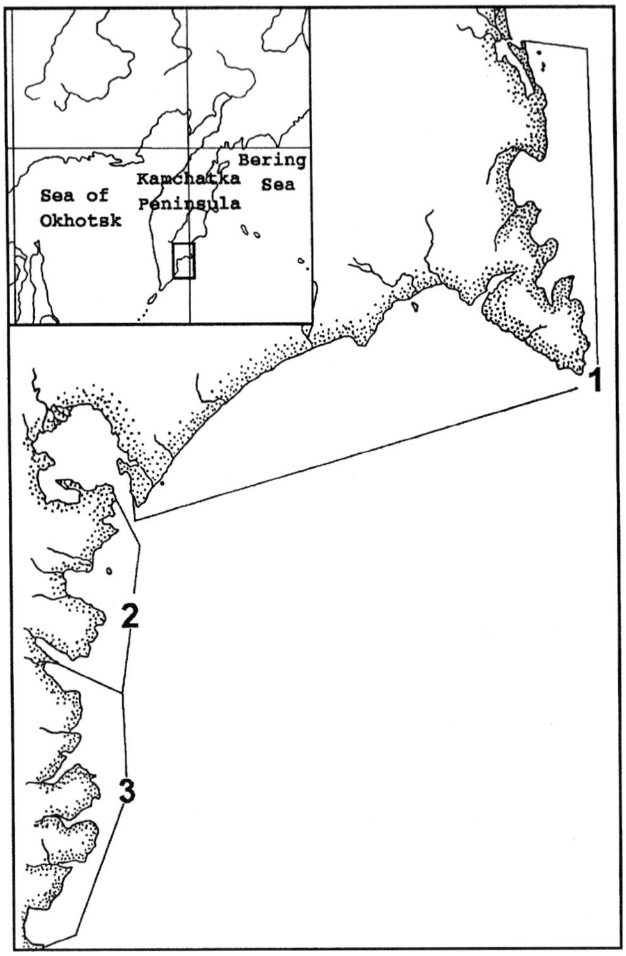

for species i in the sample. This was calculated from the formula

$$[2] \quad P_i = 100\, N_i / T$$

where N_i is the total number of individuals for species i in each sample and T is the total number of urchins of all species in each sample. No less than 30 samples were used for calculating the average relative abundance of urchin species in each depth zone.

The relative occurrence (F_s) of a species in each substrate type s was calculated for each depth range and coastal exposure as

$$[3] \quad F_s = 100\, X_s / Y$$

where X is the number of samples when the species were found in substrate type s and Y is the total number of samples containing a particular sea urchin species in a given depth zone. A range of 40–80 samples were used for the calculation of percentage relative occurrence for each depth zone, except for 15–20 m in protected coast where only 3–5 occurrences of sea urchins were registered.

Sea urchin populations were classified into nine categories in order to simplify the description of their distribution in both scientific papers and reports for fishermen. Samples were classed into three categories, large, medium, and small, according to the percentage of legal individuals in that sample. Samples from each category were then subdivided into three subcategories according to the density of individuals (no.·m^{-2}). The thresholds of the categories were determined by adjusting and manipulating the modes of these characteristics.

Fig. 2. Example of the classification of different habitat types along the southeast coast of Kamchatka. (I) exposed open ocean coast, south exposed capes of closed inlets; (II) half-protected inlets, north exposed inlets and capes, shallow sites nearby small river outlets; (III) outer parts of closed inlets, half-protected shallow sites nearby small river outlets; and (IV) inner parts of closed inlets.

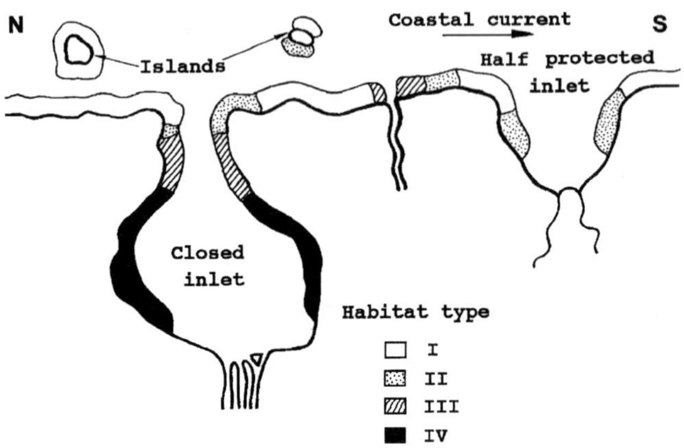

Results and discussion

Habitat type

There was a gradient in the hydrological parameters between habitats I–IV (Table 1). This gradient can be used to show important differences in the distribution of sea urchin species in the upper subtidal zone (Fig. 3).

Strongylocentrotus polyacanthus predominated over the other species in habitat I and II type sites, characterized by high wave exposure and high salinity, from 0 to 20 m depths. It was less abundant in habitat III sites and was not present in habitat IV sites, which were protected and had low salinity and relative warmer water.

The opposite trend was observed for *S. droebachiensis* distribution, which predominated in the closed protected inlets (Fig. 3). The relative number of *S. droebachiensis* decreased in the more exposed and deep water sites and almost completely disappeared in habitat I type sites.

The relative abundance of *S. pallidus* increased with increasing depth in all but habitat IV sites. *Strongylocentrotus pallidus* predominated in habitats I and III at depths >20 m.

Clearly, *S. polyacanthus* inhabited shallow exposed coastal waters with higher salinity and lower temperature compared to *S. droebachiensis,* which occurred in protected inlets and small bays and in sites near river mouths with brackish warm waters. *Strongylocentrotus pallidus* occurred mainly in deeper water in all but habitat IV.

Table 1. Hydrological parameters of habitats I–IV in shallow waters of East Kamchatka.

	Habitat			
Hydrological parameters	I	II	III	IV
Salinity (‰) 0.5–10 m	32–33	29–32	27–32	22–31
Temperature (°C) at 0.5 m				
Maximum	12.6	no data	no data	15.2
Average	10.4 ± 0.4			13.1 ± 0.3
Temperature (°C) at 5 m				
Maximum	12.4	no data	no data	13.7
Average	9.9 ± 0.3			11.9 ± 0.2
Temperature (°C) at 10 m				
Maximum	12.0	no data	no data	13.6
Average	9.1 ± 0.2			10.5 ± 0.3
Depth of thermocline (m)	25–40	25–35	15–20	10–15
Level of exposure[a]	5–6	4–6	2–4	1–2

[a]Relative scale with 1 the lowest exposure and 6 the highest exposure.

Fig. 3. Sea urchin distributions in four habitat types and depth zones found in the upper subtidal zone of the southeast coast of Kamchatka. X-axis: sea urchin species 1, *S. polyacanthus*; 2, *S. droebachiensis*; and 3, *S. pallidus*. Y-axis: relative abundance (%) of each species by depth zone. Z-axis: depth zones.

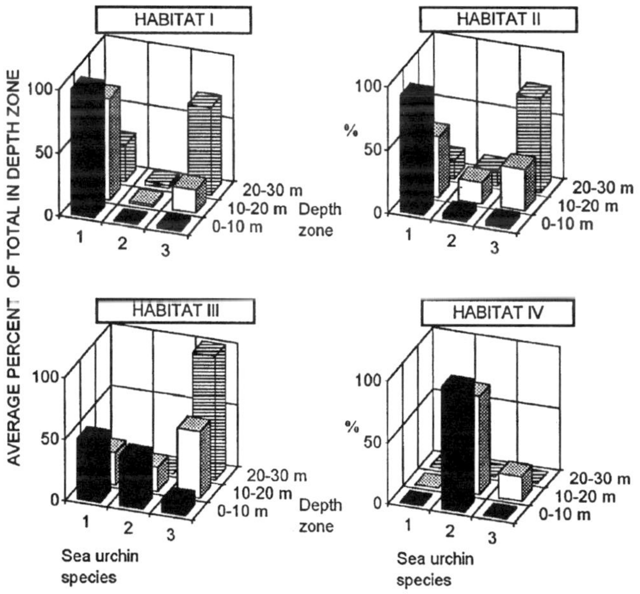

Fig. 4. Relative distribution of occurrence among substrate types of *S. polyacanthus*, *S. droebachiensis*, and *S. pallidus* by depth range and coastal exposure in the waters of the southeast coast of Kamchatka. X-axis: depth zones. Y-axis: relative occurrence (%) of individuals in each substrate type. Z-axis: substrate type.

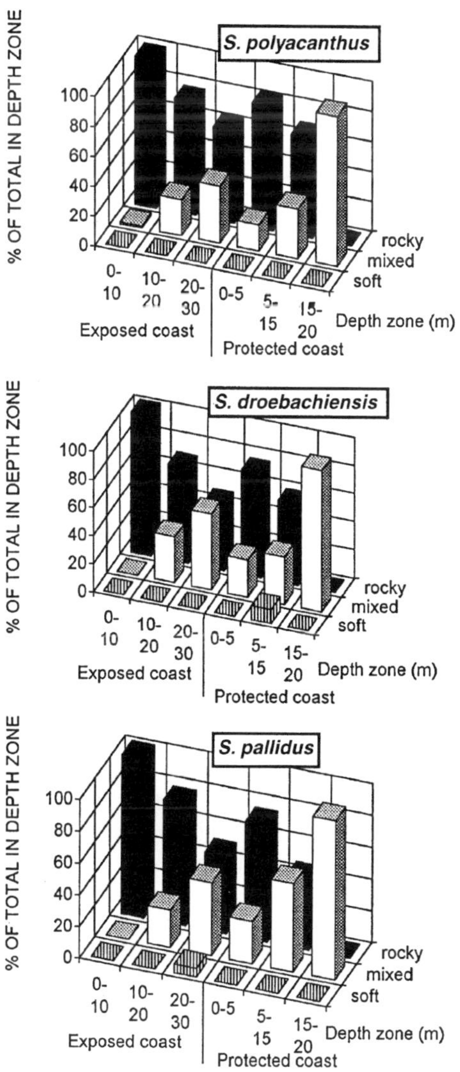

Substrate types

All three sea urchin species occurred exclusively on rocky substrates in shallow water of the exposed ocean zone (Fig. 4). There was a wider range of substrates inhabited by sea urchins at depths of 10–30 m on the exposed coast. Sea urchins were not found to occur on pure soft substrate in the exposed zone at any depth.

A different pattern of substrate affinity was observed on well-protected coasts (Fig. 4). All three species of sea urchin were found on both rocky and mixed substrate types at all but the deepest depth range. The rocky ground affinity weakened with increasing depth for all species and all sea urchin species were found on mixed substrates by 15–20 m. There was an

Table 2. Classification of sea urchin population types in shallow waters of Kamchatka.

Type of population	% legal individuals	Density of individuals (no.·m^{-2})	Density of legal individuals (no.·m^{-2})	Biomass (kg·m^{-2})
LD	60–100	>30	>20	>3.0
LM	60–100	10–30	8–20	1.0–3.0
LS	60–100	1–10	<8	<1.0
MD	15–60	>50	>15	>2.5
MM	15–60	25–50	5–15	0.9–2.5
MS	15–60	<25	<10	<1.7
SD	0–15	>40	0–15	0.4–2.5
SM	0–15	20–40	0–5	<1.0
SS	0–15	<20	0	<0.4

Note: LD, large dense; LM, large moderate; LS, large sparse; MD, middle dense; MM, middle moderate; MS, middle sparse; SD, small dense; SM, small moderate; SS, small sparse.

absence of rocky ground at these depths because of the intensive sedimentation process occurring in well-protected coasts. The distribution of sea urchins by depth may be directly related to the available habitat. At the shallowest depths on both exposed and protected coasts, there is good representation of all three substrate types. With increasing depth, the incidence of mixed and especially soft substrates increases. However, considering that all three substrate types are present in all but one depth range, it can be concluded that the distribution of substrate types has no strong influence on the data presented here.

Fishery management implications

In spite of a large stock of *S. pallidus* (estimated at about 3000 t), this species is not considered for a potential fishery in Kamchatka waters because of the small size (30–50 mm TD, test diameter) of individuals in high-density populations and low density of individuals in large size (70–90 mm TD) populations. There is also a lack of a deep-water harvesting method. Since both *S. droebachiensis* and *S. polyacanthus* are shallow water species and can reach 80–100 mm TD, they have more potential for a fishery using traditional SCUBA diving harvesting techniques.

Both *S. droebachiensis* and *S. polyacanthus* spawn typically in late summer and fall, but their spawning periods generally do not overlap (Arkhipova and Yakovlev 1994; A. Bazhin, unpublished data). *Strongylocentrotus droebachiensis* spawns from the end of August through September and *S. polyacanthus* spawns from the end of September through October. The spawning periods can vary between years and localities, possibly as a result of environmental conditions. For instance, a delay of about one month was observed for both species in 1994, which coincided with a delay in the annual decline of the nearshore water salinity in summer and extremely high levels of winter precipitation (A. Bazhin, unpublished data).

As result of the differences between species in sea urchin habitat and spawning periods, fishers could allocate their fishing effort effectively by switching from one species to another depending on weather and roe condition. Fishers could harvest *S. droebachiensis* in protected localities during storms and *S. polyacanthus* in open coastal areas during calm periods.

For the past three years, the fishery has been attempting to efficiently harvest these species. Progress has been slow, however, with the highest annual total catch being only 80 t. Several reasons may account for this. First, there is lack of experience of the companies in marketing and the fishers in diving under difficult harvesting conditions during the severe weather off the Kamchatka coast. Second, the quota system used to regulate the fishery was inappropriate for the biological and fisheries characteristics of these species. Third, the large individuals (75–90 mm TD), which were the primary target of the fishery, tended to have small dark coloured gonads of poor commercial quality. The incidence of poor quality roe varied from 3 to 60% in different populations. Poor roe quality could be a result of senility in old, slow growing individuals. The quantity of senile sea urchins in a population seems to depend on their survival rate in a given habitat. Estes et al. (1978) suggested that large *S. polyacanthus* may be as old as 20 years. However, numerous attempts to estimate the ages of the animals of various sizes of both species using the growth-line method were unsuccessful.

Ecologists from the Kamchatka Research Institute of Fishery and Oceanography have been monitoring natural populations of *S. polyacanthus* and *S. droebachiensis* for 10 years. After several unsuccessful attempts to use the quota system for the control of the sea urchin fishery, another approach was developed. The region was divided into three large zones (Fig. 1). To better describe the distribution of different sea urchin populations within the zones, a special classification was designed (Table 2). The diverse populations were categorized into 9 population types, based on several criteria. The criteria were: percentages of legal individuals in population (legal size 50 mm TD); total density of individuals; density of legal individuals; and total biomass. A special passport or identification card with an instruction manual for each fishery zone was designed. The passport included the estimated total and harvestable (i.e., legal > 50 mm TD with high quality roe) stocks of sea urchins by population type.

Additional information in the instruction manual for each population type included a description of bottom characteristics (slope and topography) and quantity of macroalgae. For instance, LD population type, i.e., dense population of large individuals, typically occurs on moderate slopes with rocky bottoms or on rocky platforms with a large quantity of brown algae such as *Laminaria* or *Alaria*. SD population type, i.e., dense population of small individuals, is the well-known sea urchin barren grounds that typically occur on flat rocky platforms where only single seaweeds *Agarum* or *Thalassiophillum* are present.

Only four of nine population types (LD, LM, MD, and MM) could be considered as potentially harvestable due to quantity considerations (Table 2). This proposed system of classification simplifies the procedure for both stock assessment and allocation of fishery pressure on sea urchins stocks. The status of each of the four harvestable population types changes after harvest to one position lower in the classification as a result of a decrease in sea urchin abundance.

Each fishery zone is supposed to be leased by a fishery company, which is obliged to fund scientific research, provide harvest information, and adjust their harvest to ensure a sustainable yield. The fishery company is encouraged to conduct cultivation measures to eliminate the large senile individuals, thereby allowing smaller sea urchins into better foraging areas. The expected result is a faster growth rate and better gonad quality. We expect that this will result both in an enhanced reproductive potential for a given population and an enhanced stock of commercially harvestable sea urchins.

Acknowledgments

I thank Alan Campbell and Claudia Hand for reviewing and suggesting modifications to an earlier draft of the paper.

References

Arkhipova, E.A., and Yakovlev, S.N. 1994. Annual gonad cycles of the sea urchins *Strongylocentrotus polyacanthus* and *S. droebachiensis* in Avacha inlet (eastern Kamchatka). Biol. Morya (Vladivostok), **20**: 402–404. (In Russian.)

Baranova, Z.I. 1957. Echinoderms of the Bering Sea. Invest. Far Eastern Seas USSR. Moscow, Leningrad: USSR Academy of Science Publisher, Vol. 4. pp. 149–266. (In Russian.)

Bazhin, A.G. 1995. Species composition, conditions of habitats and distribution of sea urchins genus *Strongylocentrotus* of Russia seas. Unpublished Ph.D. thesis. Institute of Marine Biology, Vladivostok, Russia. pp. 1–126. (In Russian.)

Djakonov, A.M. 1949. Definitions of Echinoderms from far eastern seas. Bull. Pac. Inst. Fish. Oceanogr. **30**: 1–130. (In Russian.)

Estes, J.A., Smith, N.S., and Palmisano, J.F. 1978. Sea otter predation and community organization in the western Aleutian Islands, Alaska. Ecology, **59**: 822–833.

Jensen, M. 1974. The Strongylocentrotidae (Echinoidea), a morphologic and systematic study. Sarsia, **57**: 113–148.

Lukin, V.I. 1982. Bionomical types of upper subtidal zone of far eastern seas of USSR. *In* Biology of shelf zones of world ocean. Proceedings of Second Marine Biology All-Union Conference. Vladivostok, USSR Far Eastern Scientific Center, Vladivostok, Russia. Vol. 1. pp. 44–45.

McCaley, J.E. and Carey, A.G. 1967. Echinoidea of Oregon. J. Fish. Res. Board Can. **24**: 1385–1401.

New directions in invertebrate fisheries management in British Columbia, Canada

Rick Harbo

Abstract: There has been an increase and shift in value and landings in invertebrate fisheries of British Columbia, Canada, from the traditional net and trap fisheries to dive fisheries. Before the 1980's, the traditional fisheries for crustaceans underwent few management changes and maintained relatively stable annual landings over time. The major invertebrate fisheries since the 1980's are the dive fisheries for geoduck clams, sea urchins, and some minor species. These fisheries have moved to individual license quotas with more emphasis on marketing and product quality than the volume of landings. Although landings and the number of licenses decreased, landed values have increased from C$21.5 million in 1985 to C$125 million in 1995. The objectives of fisheries management to conserve, manage, and enhance for the greatest benefit to all user groups are being challenged by aboriginal groups. Consequently, a new process of consultation and decision making is being developed. New initiatives in funding from industry and co-operative ventures with aboriginal groups for stock assessment have been initiated in the 1990's. Some small areas have been set aside for research or experimental fishing. Marine-protected area strategies are also under consideration. The most effective means of controlling effort in many of the commercial fisheries has been in co-operation with industry to set license quotas or gear limits. Industry has contributed to the costs of having on-grounds observers, designated landing ports, validation of landings at the dock, and monitoring of landings at processing plants.

Résumé : En Colombie-Britannique (Canada), la place plus importante de la pêche en plongée par rapport à la pêche traditionnelle au filet et à la trappe s'est traduite par une augmentation de la valeur et du volume des débarquements de fruits de mer. Avant les années 80, la gestion des pêches traditionnelles aux crustacés n'a subi que quelques changements, et le volume des débarquements est demeuré relativement stable d'année en année. Depuis les années 80, les principales pêches aux fruits de mer sont les pêches en plongée des panopes, des oursins et d'autres espèces de moindre importance. Ces pêches sont maintenant régies par des quotas individuels liés aux permis, et accordent une plus grande importance à la commercialisation et à la qualité du produit qu'au volume des débarquements. Bien que le volume des débarquements et le nombre de permis délivrés aient diminué, la valeur des débarquements est passée de 21,5 M$Can à 125 M$Can entre 1985 et 1995. Les objectifs de la gestion des pêches (conservation, gestion et valorisation des stocks, pour le plus grand bénéfice de tous les utilisateurs) sont remis en question par des groupes autochtones. Il a donc été nécessaire de revoir les processus de consultation et de prise de décision. De nouvelles initiatives liées au financement par l'industrie et des projets de coopération avec les groupes autochtones ont été mis en oeuvre au cours des années 90. On a créé de petits secteurs réservés à la recherche et à la pêche expérimentale. On envisage également d'appliquer des stratégies en vue de protéger certaines régions marines présentant un intérêt particulier. Les mécanismes les plus efficaces pour contrôler l'effort dans de nombreuses pêches commerciales ont consisté à collaborer avec l'industrie en vue d'établir les quotas liés aux permis ou les limites applicables aux engins. L'industrie a assumé une partie des coûts liés à l'affectation d'observateurs sur les lieux de pêche et a participé à l'assignation de ports de débarquement, à la validation des débarquements au quai et à la surveillance des débarquements aux usines de transformation. [Traduit par la Rédaction]

Introduction

New directions in fisheries management have been essential in response to conservation concerns in world fisheries (Caddy and Mahon 1995; FAO 1993*a*, 1993*b*; Garcia 1994; Smith et al. 1993). This paper examines the application of several new fishery management programs to the rapidly growing invertebrate fisheries of British Columbia (B.C.), Canada. Since the 1980's there has been an increase in consultation with all users leading to co-operative management and research, individual quota allocations, cost recovery programs, and marine-protected areas. Aboriginal fishing strategies (AFS) are also under development to meet the important food, social, and ceremonial needs of First Nations. There is a new emphasis on stock assessment that requires clear definition of biological, management, and industry objectives to deal with problems that range from the fishery closure of abalone since 1990 to the proposed development of new experimental fisheries.

Jamieson (1986) put forward a perspective on invertebrate fisheries management in B.C. that underestimated the rapid changes and importance of invertebrate fisheries that have occurred to date. The case histories for geoduck clams and northern abalone fisheries are updated and other fishery issues presented for intertidal clams, green sea urchins, red sea urchins, and sea cucumbers.

There are over 35 species harvested from the major groups

R. Harbo. Fisheries and Oceans Canada (DFO), Fisheries Management, South Coast Division, 3225 Stephenson Point Road, Nanaimo, BC V9T 1K3, Canada.

Correct citation: Harbo, R. 1998. New directions in invertebrate fisheries management in British Columbia, Canada. *In* Proceedings of the North Pacific Symposium on Invertebrate Stock Assessment and Management. *Edited by* G.S. Jamieson and A. Campbell. Can. Spec. Publ. Fish. Aquat. Sci. 125. pp. 429–437.

Table 1. List of common and scientific names of commercially exploited species of invertebrates in British Columbia, including the commercial license catagories, harvest gear types, and number of licenses eligible to harvest in 1995.

Phylum – Class common name	Scientific name	Gear type	# of eligible licenses	Comments
(*a*) Phylum Mollusca				
Class Gastropoda				
Abalone (northern, pinto)	*Haliotis kamschatkana*	Dive	(26)	Fishery closed Dec. 1990
Class Bivalvia				
Geoduck (king clam)	*Panopea abrupta*	Dive	55	
Fat horse clam (gaper)	*Tresus capax*	Dive	55	
Pacific horse clam (gaper)	*Tresus nuttallii*	Intertidal	55	
Manila clam	*Venerupis philippinarum*	Intertidal	2452	Not limited entry[a]
Littleneck (native) clam	*Protothaca staminea*	Intertidal	2452	Not limited entry[a]
Butter clam	*Saxidomus gigantea*	Intertidal	2452	Not limited entry[a]
Razor clam	*Siliqua patula*	Intertidal		Not limited entry[a]
Blue (bay) mussel	*Mytilus edulis complex*	Intertidal		Experimental, closed
California (sea) mussel	*Mytilus californianus*	Intertidal		Experimental, closed
Pink (smooth, swimming) scallop	*Chlamys rubida*	Dive/net	72	Not limited entry[a]
Spiny (pink, swimming) scallop	*Chlamys hastata*	Dive/net	72	Not limited entry[a]
Deepwater littleneck	*Compsomyax subdiaphana*	Net		Experimental, closed
Class Cephalopoda				
Pacific octopus	*Octopus dofleini*	Dive/trap	137	Not limited entry[a]
Opal squid	*Loligo opalescens*	Net	81	Not limited entry[a]
(*b*) Phylum Arthropoda				
Class Crustacea				
Subclass Malacostraca				
Euphausiids (krill)	*Euphausia pacifica*	Net	19	
Prawn (spot shrimp)	*Pandalus platyceros*	Trap	260	
Smooth pink shrimp	*Pandalus jordani*	Trap/net	W-260, S-249[b]	
Northern (spiny) pink shrimp	*Pandalus borealis eous*	Trap/net	W-260, S-249[b]	
Sidestripe shrimp	*Pandalopsis dispar*	Trap/net	W-260, S-249[b]	
Coonstripe shrimp	*Pandalus danae*	Trap/net	W-260, S-249[b]	
Humpback shrimp	*Pandalus hypsinotus*	Trap/net	W-260, S-249[b]	
Dungeness crab	*Cancer magister*	Trap	226	
Red rock crab	*Cancer productus*	Trap	226	
Red (Alaska) king crab	*Paralithodes camtschatica*	Trap	226	
Golden (brown) king crab	*Lithodes aequispina*	Trap	226	
Tanner crab	*Chionoecetes bairdi*	Trap		Experimental offshore
Subclass Cirripedia				
Gooseneck barnacles	*Policipes polymerus*	Intertidal	64	Not limited entry[a]
(*c*) Phylum Echinodermata				
Class Echinoidea				
Red sea urchin	*Strongylocentrotus franciscanus*	Dive	110	
Green sea urchin	*Strongylocentrotus droebachiensis*	Dive	49	
Purple sea urchin	*Strongylocentrotus purpuratus*	Dive		Experimental, 1989/1990, west coast Vancouver Is.
Class Holothuriodea				
California sea cucumber	*Parastichopus californicus*	Dive	85	

[a]The number of licenses issued annually is "unlimited," i.e., any number of eligible fishers may purchase a license.
[b]W = Prawn and shrimp by trap license; S = shrimp by trawl license.

of molluscs, crustaceans, and echinoderms. Fifteen commercial fishing plans were developed for 1995, including 6 dive fisheries, 4 trap fisheries, 3 net fisheries, and 2 intertidal fisheries (Table 1). There is currently a moratorium on new fisheries for invertebrates, which has been in effect since 1992. Oysters (*Crassostrea gigas*), Manila clams, (*Venerupis philippinarum*), and Japanese scallops (*Mizuhopecten caurinus*) are farmed and fall under the jurisdiction of the provincial government of British Columbia.

For the purposes of this paper, only the wild fisheries will be considered. Invertebrate landings in 1995 accounted for 11% of the total landings of all fisheries and 30% of the total landed value in B.C. (Figs. 1, 2). Although the total invertebrate landings peaked in 1993 at 31 234 tonnes (t), the total

Fig. 1. B.C. commercial fisheries 1995: (A) landings, thousands of tonnes and (B) landed value, millions of Canadian dollars.

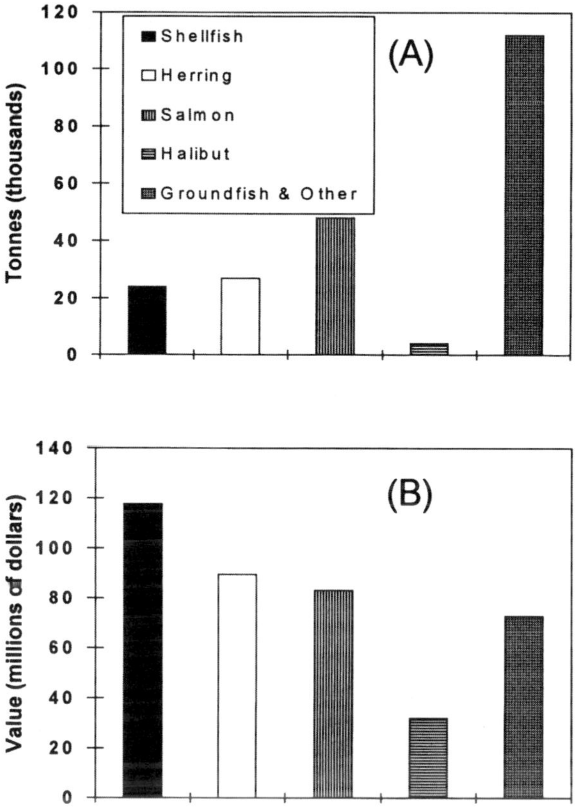

Fig. 2. B.C. shellfish fisheries 1995: (A) landings by species and (B) by gear type, thousands of tonnes.

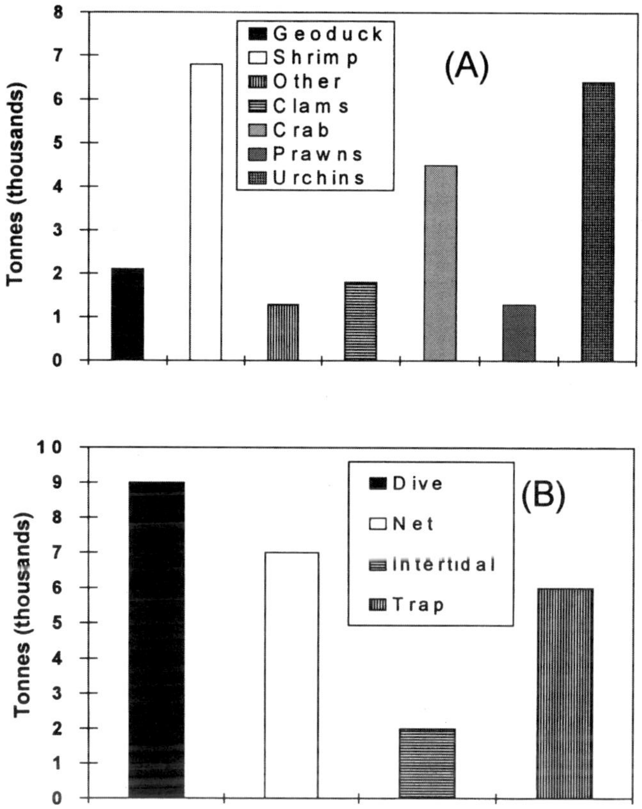

landed value has increased annually to greater than C$125 million in 1995 (Tables 2, 3). The dive fisheries have increased in landings and values surpassing the traditional trap and net fisheries. In 1995, dive fisheries accounted for 38% of the invertebrate landings and 48% of the landed value (Figs. 3, 4).

The trap fishery for Dungeness crabs, *Cancer magister*, has maintained its importance being the single species with the highest landings, 6002 t in 1994, and second in total landed value to geoduck clams (Table 3).

Development of annual fishing plans

The annual cycle of the development of fishing plans begins with requests from fishery managers and clients for stock assessments or with requests from clients to make changes to the current fishing plan. The requests and the management objectives need to be clearly stated. The assessments are often a multistage process and results such as total allowable catches (TAC's) may take several years to prepare. Consequently, interim precautionary strategies are needed for existing and new fisheries that may be initiated.

The Pacific Stock Assessment Review Committee (PSARC) has two invertebrate subcommittee meetings annually to present and review fishery updates and working papers detailing stock assessments. Once this biological advice is approved, the regional Shellfish Working Group, representing the many sectors within DFO, meet and draft a fishing plan. The working group includes representatives from stock assessment, fishery managers, enforcement, licensing, fish inspection, habitat protection, program planning and economics, and fishery statistics. Draft fishing plans are reviewed by users and consultation takes place at numerous advisory committee meetings. Following the input and advice of clients, the Shellfish Working Group finalizes the fishing plan for approval of senior managers and the Minister. For biological and management reasons for some species, the dates of the annual cycle and license year have been shifted from the calendar year.

Aboriginal Fisheries Strategy (AFS)

The Department of Fisheries and Oceans (DFO) recognizes the importance of fish, including invertebrates, to the aboriginal peoples of British Columbia. The 1992 *Sparrow* decision of the Supreme Court of Canada gave aboriginal communities, who have traditionally relied on natural resources, a right to "fish for food, social, and ceremonial purposes" that is second only to conservation in priority. The court also recognized a duty on the part of the federal government to consult with native communities before imposing any restrictions on the exercise of such rights.

Through the AFS, the department seeks to consult and negotiate, with aboriginal organizations, on allocations for food, social, and ceremonial purposes. This has resulted in area or seasonal closures to commercial fisheries in some areas and other management actions. Effective consultation will be a significant and demanding responsibility for fishery managers.

AFS intertidal clam initiatives

Several initiatives are under way as of 1995 in the commercial intertidal clam fisheries. A pilot intertidal clam fishery was

Table 2. Landings (tonnes) of invertebrates in British Columbia, 1985–1995.

	1985	1986	1987	1988	1989	1990	1991	1992	1993	1994[a]	1995a
Intertidal clams –											
commercial fishery											
Razor	90	142	142	155	117	114	117	55	44	105	140
Butter	251	158	68	134	92	109	42	132	104	174	101
Manila[b]	1913	1893	3607	3909	2764	1456	982	914	1059	1376	1292
Nat. Ln.	191	284	373	290	433	465	201	116	131	94	140
Mixed	477	371	87	27	159	339	137	124	133	87	3
Subtotal: commercial fishery	2922	2848	4277	4515	3565	2483	1479	1341	1469	1836	1676
Farmed clams	4	7	25	30	31	39	169	300	300	500	800
Total intertidal clams	2926	2855	4302	4545	3596	2522	1648	1641	1769	2336	2476
Geoduck	5370	5006	5734	4567	3985	3956	3333	2864	2455	2235	2061
Horse clam	6	96	355	325	115	124	110	2	23	62	1
Shrimp	678	768	2644	2561	2299	1940	3265	2683	3283	3192	6780
Prawn	514	550	620	720	820	761	961	1168	1215	1309	1300
Crab[c]	1165	1321	1631	1508	1522	2168	1887	3355	6306	6002	4594
Abalone	42	52	49	49	49	50	NA	NA	NA	NA	NA
Octopus	32	53	129	209	217	198	131	117	145	72	74
Sea urchin	1815	2067	2223								
Red				2116	2658	3158	6945	12018	6388	5829	6255
Green				444	609	475	607	1042	714	332	87
Sea cucumber[d]	346	786	1722	1922	1144	870	1340	1242	812	556	588
Scallop	53	68	66	67	75	69	82	91	90	104	93
Plankton	131	166	130	247	360	530	450	380	53	333	579
Squid	111	79	86	88	70	72	116	93	13	175	78
Mussels	trace	2	2	3	4	1	trace	0	0	0	0
Gooseneck barnacles	trace	2	32	49	30	37	40	38	30	19	7
Oysters, farmed	3420	2864	3482	3702	3721	4547	4482	4500	4000	4900	5300
Total tonnes	16609	16735	23207	23122	21274	21478	25397	31234	27296	27456	30273

[a]Preliminary landings for 1994 and 1995.
[b]The sum of commercial fishery landings, depurated and aboriginal licensed harvest, and not including production from clam tenures.
[c]Crab landings include tanner, king, Dungeness crabs, and others (1989–1995).
[d]Landings are round weight.

Table 3. Landed value of invertebrates in thousands of Canadian dollars in British Columbia, 1985–1995.

	1985	1986	1987	1988	1989	1990	1991	1992	1993	1994[a]	1995a
Intertidal clams –											
commercial fishery											
Razor	95	127	126	137	124	130	129	82	67	186	247
Butter	138	75	40	63	44	53	34	81	60	103	71
Manila[b]	2278	2762	6003	7175	6003	3761	2574	2253	2761	3776	4158
Nat. Ln.	202	327	474	359	588	710	327	193	212	147	241
Mixed	575	510	132	36	196	625	238	252	271	198	7
Subtotal: commercial fishery	3288	3801	6775	7770	6955	5279	3302	2861	3371	4410	4724
Farmed clams	4	14	43	59	96	140	556	1000	1200	1900	3800
Total intertidal clams	3292	3815	59	7829	7051	5419	3858	3861	4571	6310	8524
Geoduck	4605	4294	6184	9762	12967	10582	9659	16237	26994	33426	42518
Horse clam	6	63	309	300	144	274	119	2	46	111	2
Shrimp	1180	1240	4609	2802	2985	2637	4430	2831	3494	4772	13796
Prawn	3379	3734	4326	5724	7083	7006	7728	8380	10121	12133	17752
Crab[c]	4719	5661	6452	5945	6088	9311	8688	11203	18761	25686	23562
Abalone	442	734	973	1076	1170	1347	NA	NA	NA	NA	NA
Octopus	82	136	381	651	707	657	415	350	447	231	269
Sea urchin	763	1011	1276								
Red				1241	1631	1953	4187	8660	5271	8038	11269
Green				584	1020	948	1795	4424	3777	2122	648
Sea cucumber	94	236	768	961	998	1168	1029	1363	982	1035	947
Scallop	95	212	244	285	316	317	387	420	423	490	465
Plankton	89	113	102	192	223	415	390	318	41	259	369
Squid	184	127	132	113	94	81	148	135	17	199	97
Mussels	0	trace	trace	trace	trace	1	trace	0	0	0	0
Gooseneck barnacles	1	4	211	479	343	413	418	448	320	181	71
Oysters, farmed	2613	2515	2548	2725	2938	3613	3465	3600	4700	4500	5500
Total value	21544	23895	28574	40669	45758	46142	46716	62232	79965	99493	125789

[a]Preliminary values for 1994 and 1995.

[b]The sum of commercial fishery values, estimated depurated and aboriginal licensed harvest value, and not including production from clam tenures.

[c]Landings include tanner, king, Dungeness crabs and others (1989–1995).

Fig. 3. B.C. shellfish fisheries 1995: (A) landed value by species and (B) by gear type, millions of Canadian dollars.

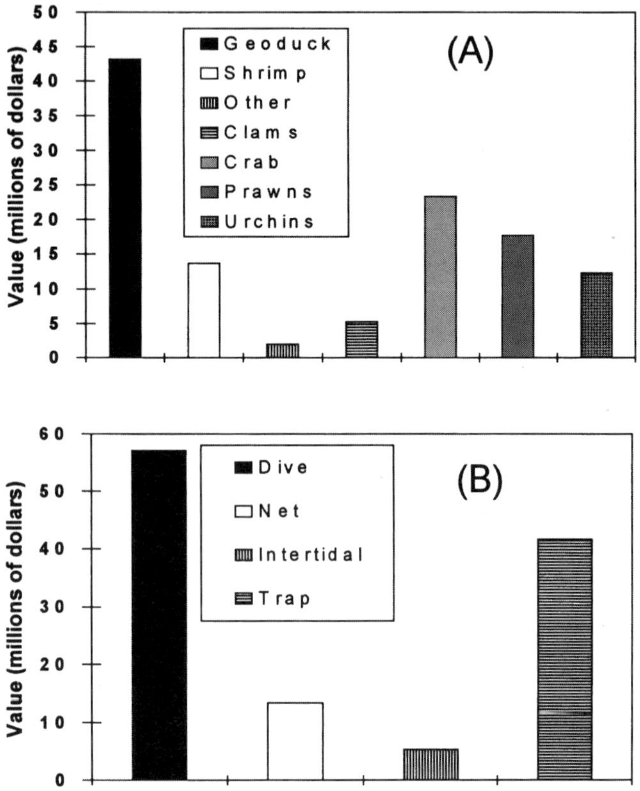

Fig. 4. Mangement of geoduck quotas prior to and after implementation of individual quotas and dockside validation in 1989.

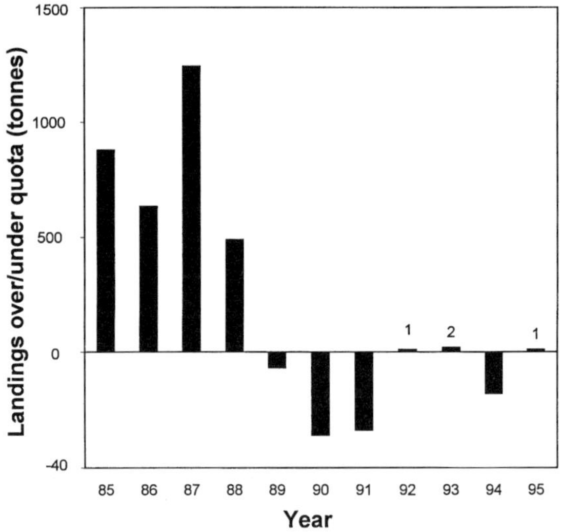

started in 1992 with the Heiltsuk First Nations in the central coast of B.C. The fishery operates under an aboriginal communal fishing license for clam harvesting, with an annual quota of 250 000 lb (113.4 t) each of Manila, *Venerupis philippinarum*, littleneck, *Protothaca staminea*, and butter clams, *Saxidomus gigantea*.

The Haida Nation signed an agreement with DFO, 1995–1997, for co-operative management of the razor clam, *Siliqua patula*, fishery in the Queen Charlotte Islands.

The Klahoose Band has signed an agreement (1994) to harvest Manila and littleneck clams in Squirrel Cove, Cortes Island. Pilot harvests have also been initiated at Kulleet Bay with the Chemainus Tribal Council (1994) and at Kuper Island in the Strait of Georgia with the Mid-Island Tribal Council (1995).

Management plans require the bands to fulfill certain management responsibilities including water quality and paralytic shellfish poison (PSP) sampling in the north coast, harvest monitoring, an enforcement plan, and stock assessment.

Discussions have been on-going with other organizations such as the Nuu-chah-nulth Tribal Council (west coast of Vancouver Island) and the Aweena K'ola Management Project (northern Vancouver Island) to develop community management boards for intertidal clam fisheries.

Co-operative research committees: industry, First Nations, and government

Co-operative research committees were set for red sea urchins in 1993, geoducks in 1994, and more recently, green sea urchins in 1995. Co-operative surveys have been undertaken under the scientific direction of DFO, which provides the sampling strategies. The First Nations have also funded consultants to provide advice on sampling designs and statistical approaches. Field surveys have been carried out under the supervision of a biologist provided by First Nations. The commercial fishers provide a vessel, an experienced skipper, and divers. Survey teams are comprised of commercial divers and divers from the band in the area of the survey. The costs of the surveys are shared by industry and the First Nations.

Voluntary commercial license eligibility retirement program

This retirement program was initiated by the federal DFO to ensure that commercial operators are not affected when fishing opportunities are transferred to aboriginal communities through negotiations under the AFS. License retirement will be limited to 5% per fishery over 6 years. Licenses will be issued to aboriginal communities and will operate as communally held licenses, subject to the usual commercial fishery rules and regulations. There have been initial retirements of licenses in the red sea urchin, green sea urchin, and sea cucumber fisheries, and the program is being expanded to other invertebrate fisheries.

Cost recovery programs

During 1995, there were few management programs with direct cost recovery to the Government of Canada for invertebrate fisheries. Several industry associations, however, have funded new services or programs through management fees to their association. The associations contract most of the programs, although there have been some direct agreements with DFO and the federal Department of the Environment (DOE) for personnel support.

Individual quota programs

Abalone fishery

The first program with individual quotas (IQ's) was for the

northern abalone, *Haliotis kamschatkana*, fishery (Sloan and Breen 1988; Farlinger and Campbell 1992). Jamieson (1986) presented the abalone fishery as an example of a significant fishery taking place when little or no stock assessment information was available. Despite annual reductions in quotas since 1986, surveys in 1989 and 1990 (Farlinger et al. 1991; Thomas et al. 1992) suggested continuing declines in both legal and prerecruit abundance indices.Consequently, the fishery was closed in December 1990.

Since 1985, commercial abalone fishers participated in the annual surveys and the joint surveys continued until the fishery closed in 1990. Subsequent surveys of index sites have not found increases in abundances of sublegal or legal-sized abalone (Winther et al. 1995; Thomas and Campbell 1996).

Geoduck clam fishery

The geoduck, *Panopea abrupta*, fishery began in B.C. in 1976. Jamieson (1986) provided a case history to 1984. Regulation of the geoduck fishery was initiated at a time when much of the biology of geoducks was unknown. Quota options were based on a conservative annual harvest of approximately 1% of the estimated original biomass (Harbo et al. 1992, 1993, 1994).

The geoduck license holders agreed in 1989 to pay for a contract for an independent company to monitor landings in the geoduck fishery under an individual vessel quota system. This program has been successful in allowing the industry to manage the supply of clams to new markets for live geoduck clams on a year-round basis and enhancing the value of the fishery. Implementation of dockside monitoring and validation significantly improved the ability to monitor area quotas and close areas when quotas were reached. Before IQ's in 1989, there were conservation concerns arising from substantial landings in excess of the quotas. As a consequence of having IQ's, not all quotas were taken (Fig. 4) because of marketing considerations. Fishers took a voluntary risk to save quota to the end of the year when prices were traditionally the highest. However, weather conditions or closures because of paralytic shellfish poison (PSP) sometimes restricted the ability of fishers to meet their allocation.

In 1995, the 55 geoduck license holders paid C$21 000 per license to their Underwater Harvesters Association, totaling C$1.155 million. This payment represented C$0.55/kg of geoduck, based on the 1995 IQ of 84 030 lb (37.7 t). The association fee is ~5.6% of the 1993 average landed value of $9.81/kg, but only 3.6% of the increased 1994 value of C$15.14/kg.

The actual cost of the quota management program was estimated to be about only 1% of the landed value and the association fee covered many other activities including water quality monitoring carried out by DOE, PSP sampling in the north coast, and the development of a geoduck hatchery and enhancement techniques. Approximately C$200 000 were allocated to co-operative research programs with DFO and First Nations to survey geoduck beds and other research programs (<1%; 0.6% of the landed value).

Red sea urchin voluntary vessel quotas

A review for the red sea urchin, *Strongylocentrotus franciscanus*, fishery was presented by Harbo and Sloan (1987), Harbo (1990*a*), Campbell and Harbo (1991), and Harbo and Thomas (1992).

In 1992, a plant monitoring service was provided by a contractor paid by individual license holders. Payment was tied directly to the quality of the red sea urchins rather than just volume.

Area licensing was introduced in 1994, with a voluntary section of one of two areas, north and south coasts. During the 1994 season, the fishers in the north coast met and prepared an agreement that divided the north coast quota equally among the 77 licenses and initiated a voluntary dockside validation program. The program was later expanded to the south coast in the summer and fall of 1994, where there were 33 licenses.

By guaranteeing the fishers a fixed portion of the overall harvest, the plan allowed industry to control volume, enhance quality, and be more competitive on the international market. The plan allowed individual fishers freedom to fish in an area and time according to their schedule and when the market requires the product. The value of red urchins increased almost immediately, nearly doubling to C$1.26/kg upon implementation of the IQ program by the fishers.

Again in 1995, the license holders voluntarily agreed to pay a fee of C$0.077/kg to their association, approximately 6.1% of the landed value of red sea urchins in 1994 when the voluntary IQ's came into effect. In 1995, an independent charter vessel was funded by industry to monitor fishing activity and ensure vessels were fishing within the boundaries of the open subareas and to detail harvest areas on charts. The charter provided daily information on fishing activity, packers, and their destination.

Green sea urchin fishery equal individual quotas for the south coast

A fishery review for the green sea urchin, *Strongylocentrotus droebachiensis*, was presented in Campbell and Harbo (1991) and Harbo (1990*b*, 1992). The licenses for this fishery were limited in 1991 and there were 49 licenses eligible in 1995. In the fall of 1994, the fishers developed an agreement to divide the remaining quota equally among the 49 licenses. A contractor printed and issued a validation book to each license holder/operator. A cost of C$0.15/kg was charged to the association for validation of the loads at the dock at the first point of landing. This represented 2.6% of the average landed price in 1993, C$5.98/kg. An additional charge of C$0.29/kg was proposed to be contributed to a research fund, but few license holders contributed. The total charge represented 8.4% of the 1993 landed value.

Sea cucumber proposed voluntary quota, 1994

There have been limited assessments on the sea cucumber, *Parastichopus californianus*, resource (Sloan 1986; Sloan and Harbo 1987; Harbo 1990*c*; Heizer and Thomas 1992). The fishery has been managed by precautionary arbitrary quotas and license limitation in 1991. There are 84 licences eligible to harvest sea cucumbers, including five licences issued to native bands.

License holders and operators proposed a voluntary equal individual license quota for the 1994 fishery. They proposed to pay C$0.018/kg (6.4%) based on a 5000 lb (2268 kg) split weight quota and the 1993 average price of $2.78/kg split weight.

Cost recovery recommendations
Industry representatives have advised that an "up front" management fee collected by DFO at the time of license issuance is the most manageable system. Industry associations have found it difficult to collect fees from license holders on a voluntary basis and to collect from processors or cash buyers during the season.

Logbook cost recovery program

Often, logbook and sales slip data are the only information available to evaluate the stock status, since there have not been sufficient funds for fishery independent surveys and studies for the many invertebrate fisheries.

Mandatory logbooks have been an important tool in documenting harvest locations, catch, and effort. A condition of fishing licenses requires the vessel master to complete and submit harvest logs in both hard copy and electronic formats that are approved by DFO. Since 1991, a cost recovery program for some of the costs associated with logbooks has been supported by industry through the purchase of logbooks from a contractor or through an agreement between a fishing association and a contractor. In 1995, cost recovery was expanded and the contractors services included the costs of printing and distributing logbooks, coding of logsheets, data entry, and keypunching.

Experimental fisheries

There are constant demands to initiate new invertebrate fisheries either using a new gear type or exploiting a new species in B.C. However, because of the limited resources to assess and manage the existing fisheries, a moratorium on experimental fisheries has been in effect since 1992. Experience has shown that funding is required for fishery-independent data for stock assessment. Fishery data do not provide the biological data, in particular, growth, age, and mortality, required for a comprehensive stock assessment.

A process of developing a limited experimental fishery with cost recovery and partnerships is being developed. Limited experimental and developmental licenses are currently being issued for fisheries and aquaculture activities such as: the harvest of contaminated clams for depuration, a deep water clam fishery, and subtidal geoduck enhancement. Provincial agencies license new aquaculture activities such as the subtidal culture of scallops and geoduck clams and the ponding of sea urchins prior to marketing. Activities for the culture of abalone and sea cucumbers have also been initiated.

Marine protected areas (MPA)

In response to requests from communities and recreational scuba diving interests, there have been a number of small areas closed to harvest of all invertebrate species by commercial divers. There are, however, only two small areas, considered as pilot marine-protected areas, that are closed to the harvest of all species by all gear types, including commercial and sport hook and line fishing. These areas are in Howe Sound, fronting Whytecliff Park, West Vancouver, and Porteau Cove Provincial Park. A few small areas have also been designated for research and experimental harvest. For example, there are area closures to the harvest of geoducks (e.g., Gabriola Island; Ritchie Bay, Meares Island) and red sea urchins (Discovery Passage, Campbell River; Moser Point, Father Charles Channel in Clayoquot Sound; and an area near Johnstone Strait).

Conclusions and recommendations

Increased consultation with fishery clients, aboriginal peoples, and the public has improved the management and assessment of invertebrate stocks. Individual quota fisheries have guaranteed the opportunity to land a fixed portion of the overall harvest and allowed industry as a whole to be more flexible in the harvest and be more competitive on the international market. Fisheries with IQ's allow harversters more freedom to fish in an area and time according to their schedule and market demands.

Improved manageability and conservation
Individual vessel quota programs have assisted in the management and conservation of stocks by improving manageability and control of quotas. However, there has also been increased illegal fishing as a result of the high prices received in these fisheries. This is a chronic problem that threatens abalone stocks even though all abalone fisheries in B.C. are closed.

Fishing less product for conservation and a higher value
More conservative quotas have been set for many fisheries without impacting the industry. They have benefited significantly by fishing and landing less product but being more aware of quality and marketing strategies. A higher quality product receives a much higher unit price.

Flexible fishing plans allow fishers to plan their harvest and marketing activities. They can fish limited amounts and plan the timing of their fishing seasons. In-season changes are made in consultation with fisher advisory groups and DFO.

Increased stakeholder participation
The users of the resource, aboriginal communities, and the public are taking a more active role in stock assessment and the management of fishery resources. There have been several voluntary programs by industry to contribute to these programs. Additional cost recovery measures may be necessary as governments attempt to reduce spending.

Fishery dependent data cannot be used alone in stock assessment and fishery management
Industry must be convinced about the need for fishery independent studies to balance fishery data. Initially, co-operative programs have been very successful. However, the management of an increasing number of projects is a challenge to DFO stock-assessment staff and fishery managers.

Additional cost recovery and co-operative programs
As governments reduce their spending there will be an increased need for user pay and cost recovery in management and assessment programs. These programs can benefit industry if they create programs to improve fish quality and examine handling, transportation, and marketing of the catch.

Logbook programs were maintained by a user-pay program initiated in 1991. Co-operative surveys using industry divers and vessels began in 1993, surveying red sea urchins and geoducks.

Marine-protected areas

There is a need to create marine-protected areas for research and potential brood-stock areas. There has been a limited number of small areas closed to single species but no closures that protect communities. MPA's should be considered for existing fisheries and be part of new experimental fisheries.

Acknowledgments

I wish to thank Kerry Hobbs, Mike Kattilakoski, and Steve Heizer for their assistance in preparing summaries and figures of commercial landings. Information on industry programs came from Jamie Austin, Mike Featherstone, Eric Gant, and Phil Montgomery. The paper benefited from the reviews of Doug Woodby and Alan Campbell.

References

Caddy, J.F., and Mahon, R. 1995. Reference points for fisheries management. FAO Fish. Tech. Pap. 347.

Campbell, A., and Harbo, R. 1991. The sea urchin fisheries in British Columbia, Canada. *In* Biology of Echinodermata. *Edited by* T. Yanagisawa, I. Yasumasu, C. Oguru, N. Suzuki, and T. Motokawa. Balkema, Rotterdam. pp. 191–199.

FAO (Food and Agriculture Organization of the United Nations). 1993*a*. Reference points for fisheries management: their potential application to straddling and highly migratory resources. FAO Fish. Circ. 864.

FAO (Food and Agriculture Organization of the United Nations). 1993*b*. The precautionary approach to fisheries with reference to straddling fish stocks and highly migratory fish stocks. FAO Fish. Circ. 871.

Farlinger, S., and Campbell, A. 1992. Fisheries management and biology of northern abalone, *Haliotis kamschatkana*, in the northeast Pacific. *In* Proceedings of the 1st International Symposium on Abalone. *Edited by* S. Shepherd, M.J. Tegner, and S.A. Guzman Del Proo. pp. 395–406.

Farlinger, S., Thomas, G.A., Winther, I., and Carolsfeld, W. 1991. Abalone resurvey in the Estevan Group and Aristazabel Island, June 1989. Can. MS Rep. Fish. Aquat. Sci. 2104. 39 p.

Garcia, S.H. 1994. Precautionary principle: its implications in capture fisheries management. Ocean Coast. Manage. **22**: 99–125.

Harbo, R. 1990*a*. Green urchin fishery. *In* Shellfish stock assessments for the west coast of Canada in 1990. *Edited by* G. Thomas. Can. MS Rep. Fish. Aquat. Sci. 2099. pp. 261–266.

Harbo, R. 1990*b*. Red urchin fishery. *In* Shellfish stock assessments for the west coast of Canada in 1990. *Edited by* G. Thomas. Can. MS Rep. Fish. Aquat. Sci. 2099. pp. 251–260.

Harbo, R. 1990*c*. Sea cucumber fishery. *In* Shellfish stock assessments for the west coast of Canada in 1990. *Edited by* G. Thomas. Can. MS Rep. Fish. Aquat. Sci. 2099. pp. 267–277.

Harbo, R. 1992. Green sea urchins. *In* Shellfish stock assessments for the west coast of Canada in 1991. *Edited by* G. Thomas. Can. MS Rep. Fish. Aquat. Sci. 2169. pp. 266–270.

Harbo, R., Farlinger, S., Hobbs, K., and Thomas, G. 1992. A review of quota management in the geoduck clam fishery in British Columbia, 1976 to 1990, and quota options for the 1991 fishery. Can. MS Rep. Fish. Aquat. Sci. 2178. 135 p.

Harbo, R., and Sloan, N. 1987. Red sea urchin. *In* Status of invertebrate fisheries off the Pacific coast of Canada (1985/85). *Edited by* R.M. Harbo and G.S. Jamieson. Can. Tech. Rep. Fish. Aquat. Sci. 1576. pp. 95–100.

Harbo, R., Thomas, G., and Hobbs, K. 1993. Quotas for the 1992–1993 geoduck clam fisheries. Can. MS Rep. Fish. Aquat. Sci. 2179. 201 p.

Harbo, R., Thomas, G., and Hobbs, K. 1994. Quota options and recommendations for the 1994 geoduck clam fishery. Can. MS Rep. Fish. Aquat. Sci. 2228. 115 p.

Harbo, R., and Thomas, G. 1992*a*. Red sea urchins. *In* Shellfish stock assessments for the west coast of Canada in 1991. *Edited by* G. Thomas. Can. MS Rep. Fish. Aquat. Sci. 2169. pp. 271–279.

Heizer, S., and Thomas, G. 1992*b*. Sea cucumbers. *In* Shellfish stock assessments for the west coast of Canada in 1991. *Edited by* G. Thomas. Can. MS Rep. Fish. Aquat. Sci. 2169. pp. 280–285.

Jamieson, G.S. 1986. A perspective on invertebrate fisheries management — the British Columbia experience. *In* North Pacific Workshop on Stock Assessment and Management of Invertebrates. *Edited by* G.S. Jamieson and N. Bourne. Can. Spec. Publ. Fish. Aquat. Sci. 92. pp. 57–74.

Sloan, N.A. 1986. World jellyfish and tunicate fisheries and the northeast Pacific echinoderm fishery. *In* North Pacific Workshop on Stock Assessment and Management of Invertebrates. *Edited by* G.S. Jamieson and N. Bourne. Can. Spec. Publ. Fish. Aquat. Sci. 92. pp. 23–33.

Sloan, N.A., and Breen, P.A. 1988. Northern abalone (*Haliotis kamtschatkana*) in British Columbia: fisheries and synopsis of life history information. Can. Spec. Publ. Fish. Aquat. Sci. 103. 46 p.

Sloan, N.A., and Harbo, R. 1987. Sea cucumber. *In* Status of invertebrate fisheries off the Pacific coast of Canada (1985/85). *Edited by* R.M. Harbo and G.S. Jamieson. Can. Tech. Rep. Fish. Aquat. Sci. 1576. pp. 101–106.

Smith, S.J., Hunt, J.J., and Rivard, D. 1993. Risk evaluation and biological reference points for fisheries management. Can. Spec. Publ. Fish. Aquat. Sci. 120. 442 p.

Thomas, G., and Campbell, A. 1996. Abalone resurvey in Aristazabal Island, the Estevan group and Banks Island, June 1993. Can. Tech. Rep. Fish. Aquat. Sci. 2089. pp. 97–109.

Thomas, G., Farlinger, S., and Carolsfeld, W. 1992. Abalone resurvey in the southeast Queen Charlotte Islands in 1990. Can. MS Rep. Fish. Aquat. Sci. 2166. 93 p.

Winther, I., Campbell, A., Thomas, G.A., Adkins, B.E., and Clapp, B.G. 1995. Abalone resurvey in the southeast Queen Charlotte Islands, 1994. Can. MS Rep. Fish. Aquat. Sci. 2273. 46 p.

Overview

The assessment and management of spatially structured stocks: an overview of the North Pacific Symposium on Invertebrate Stock Assessment and Management

J.M. (Lobo) Orensanz and Glen S. Jamieson

Abstract: We present an overview of themes addressed at the North Pacific Symposium on Stock Assessment and Management of Invertebrates (March 1995) contained in Jamieson and Campbell (1998) and compare them to those from an essentially analogous event held 11 years earlier (Jamieson and Bourne 1986). Our review includes quotations from the written transcripts of session reviews by scientists, managers, and industry representatives. The diversity of resources, scales, dynamics, and harvest styles involved in the assessment and management of invertebrate resources are highlighted and categorized. Emphasis is placed on specific problems presented by stocks of benthic organisms with meroplanktonic larvae, which were the focus and/or source of motivation for most of the presentations. These highly spatially structured resources, together with their habitats, constitute a template that has molded a diversity of unique management systems, often based on traditional or informal social institutions. Our review is focused around two main themes: the development of spatially explicit approaches to assessment and management and the participation of fishers and other stakeholders in the management process. The emerging framework provides management alternatives that may prove robust in the face of irreducible uncertainty, at least for resources emphasized here.

Résumé : Nous présentons une vue d'ensemble des thèmes couverts à l'occasion du Symposium international du Pacifique Nord sur l'évaluation et la gestion des stocks d'invertébrés (mars 1995) et présentés dans Jamieson et Campbell (1998), et nous les comparons à ceux d'un événement analogue qui s'était tenu onze ans plus tôt (Jamieson et Bourne, 1986). Notre étude rassemble des citations extraites des synthèses des sessions rédigées par des scientifiques, des gestionnaires et des représentants de l'industrie. Nous faisons ressortir la diversité des ressources, des échelles, des dynamiques et des modes d'exploitation qui entrent dans l'évaluation et la gestion des ressources d'invertébrés, et nous les classons en catégories. Nous mettons l'accent sur les problèmes spécifiques posés par les stocks d'organismes benthiques à larves méroplanctoniques, qui ont suscité ou orienté la plupart des communications. Ces ressources à structure spatiale très contraignante, ainsi que leurs habitats, constituent une grille qui a modelé une série de systèmes de gestion particuliers et diversifiés, souvent fondés sur des institutions sociales traditionnelles ou informelles. Notre étude est axée sur deux grands thèmes : l'élaboration d'approches spatialement ciblées de l'évaluation et de la gestion et la participation des pêcheurs et des autres intervenants au processus de gestion. Le cadre qui en émane offre diverses options de gestion qui peuvent se révéler robustes dans un contexte d'incertitude irréductible, au moins pour les ressources dont il est question ici.
[Traduit par la Rédaction]

Introduction

Commercial invertebrates, including shellfish,[1] are the subject of a distinct subset of fishery science and fisheries management. Specific questions are addressed at international meetings (e.g., Thomas 1979; Jamieson and Bourne 1986; Bannister and Conan 1995; Jamieson and Campbell 1998), in specialized journals (e.g., Journal of Shellfish Research), and have been consolidated in a well-known book (Caddy 1989*a*). The International Council for the Exploration of the Sea (ICES) has a special Shellfish Committee (informal documents published in the "K" section of ICES's CM document series) and shellfish or invertebrate divisions are a common element to most fisheries agencies. Yet, we begin this overview with the claim that "shellfish" is a misleading concept in the sense that nothing is unique to the assessment and management of commercial invertebrate stocks.

Papers from the preceding Workshop (Jamieson and

J.M. Orensanz. School of Fisheries, University of Washington, Seattle, WA 98195, U.S.A.

G.S. Jamieson. Department of Fisheries and Oceans, Science Branch, Pacific Biological Station, Nanaimo, BC V9R 5K6, Canada.

Correct citation: Orensanz, J.M., and Jamieson, G.S. 1998. The assessment and management of spatially structured stocks: an overview of the North Pacific Symposium on Invertebrate Stock Assessment and Management. *In* Proceedings of the North Pacific Symposium on Invertebrate Stock Assessment and Management. *Edited by* G.S. Jamieson and A. Campbell. Can. Spec. Publ. Fish Aquat. Sci. 125. pp. 441–459.

[1] Shellfish generally means edible marine invertebrates (including shell-less animals like octopus, sea cucumbers, or jellyfish), while commercial invertebrates also include forms marketed for uses other than food (e.g., ornamental shells, precious corals, etc.). Papers contained in this Proceedings focus on shellfish resources.

Bourne 1986), held in 1984, suggested that some assessments of shellfish stocks were inadequate and that available assessment paraphernalia were not always appropriate:

> "The stock assessment tools that have been developed for finfish stocks, such as virtual population analysis, production models, and stock recruitment analysis are frequently not appropriate for analysis of invertebrate populations" (Hilborn 1986).
>
> "Age-structured models are impractical in [abalone stocks] because monitoring age structures in each of a large number of independent abalone substocks appears to be beyond the finances of most management agencies. Surplus production models rely on the axiom that catch-per-unit-effort (CPUE) is proportional to the population biomass. This assumption is unlikely to be justified [...]. The Beverton–Holt model has the disadvantage that it requires a measure of fishing mortality and a function relating it to fishing effort. [...]. Both of these may be impossible to obtain" (Sluczanowski 1986).
>
> "The spatial structuredness of bivalve stocks presents difficulties which are seldom acknowledged in approaches patterned after the "finfish experience," but, from the viewpoint of researchers and managers, it also offers advantages which are, unfortunately, seldom exploited. Scattered in the shellfish literature are the elements for new conceptual models of shellfish population structure, closer to applied plant ecology than to finfish stock dynamics" (Orensanz 1986).

In retrospect, we suggest that problems identified in the application of conventional "finfish" fishery science are not so much specific to "shellfish" but rather to stocks with strong and persistent spatial structure,[2] whether the animals are finned, shelled, or otherwise. The significance of life-history traits that determine temporal scales relevant to population dynamics is well acknowledged (e.g., survival rates and longevity, multi- vs. semelparity, frequency of reproductive events). By contrast, the spatial dimension of such processes has received comparatively little attention in population theory and in its application to fisheries. Models were developed for aggregated populations in which spatial structure was ignored for the sake of formal simplicity. Most important in defining spatial scales of population processes are both the degree of motility of reproductive adults and the existence of a dispersive phase. The vast majority of papers presented at this Symposium (Jamieson and Campbell 1998, henceforth referred to as this Proceedings) dealt with, or were inspired by, metapopulations of benthic organisms with meroplanktonic larvae. Such stocks, characterized by strong and persistent spatial structure, have been often assessed within the umbrella of conventional stock-assessment models, built on two basic tenets (Caddy 1975): (*i*) The stock is considered a self-contained population unit, closed to immigration and emigration and, most significantly in our context, to recruitment through the settlement of larvae originating in other regions. This is the so-called unit stock concept. (*ii*) The fishing process is homogeneous within the region occupied by the stock. Under this, the dynamic pool assumption, local effects of fishing are spread through the entire stock by virtue of the movements of individuals and mixing.

Both tenets are obviously inappropriate for populations of benthic organisms with pelagic larvae, which are widely open to recruitment by larvae originating in other regions. Sedentary organisms either do not move or the scale of their movements is small relative to that required to fulfill the dynamic pool assumption. Many problems specific to spatially structured benthic stocks were surprisingly clear to early shellfish biologists (e.g., Belding 1910; Gross and Smyth 1946; Fairbridge 1953) but later were ignored. The reason may be found in the influence of two stronger traditions: classical fin fishery stock assessment, which often lent inappropriate models, and shellfish aquaculture, which contributed its own focus and emphasis. Assessing and modeling stocks with strong and persistent spatial structure have too often been regarded as marginalia in the broader fisheries arena. However, such stocks can be important economically; for example, much of the cash harvest in the NE Pacific (SE Alaska to Baja California) originates from spatially structured benthic stocks with meroplanktonic larvae: sea urchins, sea cucumber, abalone, geoduck (*Panopea abrupta*), spiny lobster, Dungeness crab (*Cancer magister*), king crabs, rockfish (*Sebastes* sp.), and others. Problems encountered in the assessment of spatially structured stocks were emphasized by Caddy (1975) and have been discussed since by Conan (1984), Caddy (1989*b*), Prince (1989), and Orensanz et al. (1991).

While assessment and modeling problems created by spatial structure have been long recognized, the fact that the complex spatial structure of many coastal resources led to development of unique management systems has been less considered. Classical fishery science was developed with offshore, open-access industrial fisheries in mind. However, the immense majority of individual fisheries worldwide, many of which are for shellfish, are small-scale and inshore. Some of the world's most productive fisheries are offshore, industrial fisheries, but the societal significance of small-scale fisheries is overwhelmingly higher, as reflected, at least, by the number of fishers (FAO 1995; McGoodwin 1990). Small-scale fishery systems remain poorly understood regarding connections between both the structure and dynamics of harvested resources and the fishing process, which is influenced by the social and cultural matrix of fishing communities that benefit from them. Here, where traditional fishery science does not apply, lies much of the challenge in ensuring the future sustainability of world shellfisheries.

In writing this overview we had three advantages. First, the collection of refereed articles included in this Proceedings, produced by an eclectic group of authorities, gave us a unique window to peer into the field. Second, this being a reenactment of a similar event that took place 11 years earlier (which actually involved many of the same actors!), we were able to gage shifts and trends in focus and approach. Third, we had written transcripts of the six session summaries (Jamieson and Campbell 1998, Appendix 1), each consisting of overviews by three speakers: a scientist, a manager, and a representative from industry. The latter were rich in insight, broad in scope, and full of spontaneous perception and opinion.

We begin by typifying major idiosyncrasies and sources of heterogeneity in shellfisheries: categories of resources, hierarchy of scales, and diversity of harvest styles. Throughout, we focus on spatially structured resources, implications of heterogeneity as opposed to assumptions of homogeneity, and on

[2] We use the expression "strong and persistent" intentionally, to specify a particular type of spatial structure. Stocks of pelagic fish may have strong spatial structure, but the spatial pattern is transient, as opposed to the more static (= persistent in time) patterns typical of benthic stocks.

small-scale inshore fisheries rather than offshore industrial fisheries.

Sources of diversity

Categories of resources

There is considerable confusion in the literature regarding what is unique to the assessment and management of invertebrate or shellfish resources, particularly as compared with finfish resources. There have been many attempts at identifying commonalties among invertebrate populations and their fisheries. Consider, for example, the following statement:

> "The biology of harvested invertebrates is often quite different from that of the majority of fin-fishes, since invertebrates frequently possess very high reproductive output, large variation in year-class strength, rapid growth, and complex spatial distribution" (Hilborn 1986).

However, populations of harvested invertebrates vary greatly with respect to life history attributes: reproductive output ranges from very high (e.g., most bivalves) to low (e.g., octopus, many gastropod species, crayfish) and although some shellfish grow fast and live only a few years (e.g., many squid, octopus, and penaeid shrimps), others are among the slowest-growing and longest-lived animals (e.g., ocean quahogs (*Arctica islandica*) and geoducks). This range is comparable with that found among finfish species. Further, while many shellfish species show high variability in year-class strength (e.g., many scallop stocks), this is not higher than that found among pelagic fishes and many shellfish stocks show only modest fluctuations. Finally, the spatial distribution of sedentary benthic shellfish is not particularly "complex." Rather, being essentially 2-dimensional and relatively static, it is far less complex than the dynamic 3-dimensional patterns displayed by pelagic schooling finfishes.

The only commonality that unites harvested invertebrates as opposed to finfish is that their members share the plesiomorphy of lacking a vertebral column, a feature largely irrelevant to stock assessment, conservation, and management. Whether finned, shelled, or otherwise and whether they are harvested for food or other purposes, harvested stocks fall into four major types: (*i*) Sessile organisms of modular architecture which regenerate somatic material lost to a harvest. This is the case of some algal resources (e.g., *Macrocystis*) and corals. (*ii*) Sedentary benthic organisms, including forms ranging from reduced motility (e.g., scallops, abalone and limpets, sea urchins, sea cucumbers, some clams, etc.) to sessile (e.g., oysters, barnacles, ascidians, most clams, some algae, and coelenterates, etc.), endowed with a limited behavioral repertoire. Most invertebrates falling into Type 2 are broadcast spawners with external fertilization and pelagic larval development. Adults, when not sessile, have little capability of directed movement towards stimuli. (*iii*) Motile benthic or demersal organisms, typically residing for protracted periods in specific or structurally complex habitats (e.g., rocky or coral reefs) and often displaying social behavior (e.g., territoriality and resource defense mating systems). Examples include crabs and some shrimps, lobsters, octopus, and many fish species. (*iv*) Highly motile demersal or pelagic species, not associated with specific refuge areas, most often exhibiting some form of scramble polygynic mating system and seasonal or ontogenetic migrations. This includes the majority of the finfish stocks harvested by industrial fisheries (tunas and mackerels, sardines and anchovies, gadoids, etc.) and, among shellfish, squid, euphausiids, and some shrimps.

Table 1. A summary of the species groups and resource types that received specific attention during the 1984 Workshop (Jamieson and Bourne 1986) and the 1995 Symposium (Jamieson and Campbell 1998).

Resource type	Group	1984	1995
Type I	—	0	0
Type II	Sea urchins	0	4
	Abalone	2	6
	Other gastropods	2	2
	Bivalves (general)	1	0
	Scallops	2	3
	Intertidal clams	0	3
	Geoducks	0	1
Subtotal		7	19
Type III	Rock/spiny lobsters	1	6
	Lobsters	3	2
	Crayfish	1	1
	Spot prawn	1	0
	King crabs	4	2
	Snow crabs	2	3
	Blue crab	1	0
	Dungeness crab[a]	2	3
Subtotal		15	17
Type IV	Penaeid shrimp	0	1
	Northern shrimp	0	1
	Squat lobster	1	0
	Cephalopods (squid)	2	0
Subtotal		3	2
General		4	4
Regional overviews		5	3

[a] The position of Dungeness and blue crab in this scheme is somewhat intermediate, as mobility and use of structured habitats change through ontogeny.

Resources and types of resources receiving specific attention during the 1984 Workshop and the 1995 Symposium are summarized in Table 1.

Hierarchy of scales

Analysis of the spatial dimension of population processes requires the identification of meaningful scales. Here we adopt the following scales, modified from Orensanz et al. (1991):

Megascale, which corresponds to the zoogeographic range of the Artenkreis and Rassenkreis, respectively, collections of closely related species or genetically distinct stocks (Endler 1977; Sinclair 1987).

Macroscale (large scale), corresponding to mega or metapopulations. Examples are sedentary, benthic, spatially disjunct subpopulations linked by a dispersing pelagic larval stage (Orensanz 1986; Roughgarden et al. 1985; Botsford 1995).

Mesoscale (intermediate scale), corresponding to populations within a metapopulation. This is typically the scale of fishing grounds and beds, within which homogeneity assumptions with regards to the fishing process and/or the dynamics of the harvested stocks are made in the classical theory.

Microscale (small scale), corresponding to the neighborhoods of individuals. This is the pertinent scale for interactions taking place in small, dense "patches." The "neighborhood" (Orensanz 1986; Addicott et al. 1987), "ambit" (Lloyd 1967), or "area of influence" (Hall 1983) of an individual is the spatial volume within which a certain interaction between the individual and its environment is effective. Various neighborhoods can be defined, each relative to a specific process: trophic (relative to resource utilization); reproductive (ambit of interaction with mates); sensory (sphere of perception of environmental conditions, including the presence of conspecific and predators); and so on.

Nanoscale, characterized by processes taking place at low Reynolds numbers, such as particle collection in the feeding structures of suspensivores, encounters between eggs and spermatozoids of broadcast spawners, or the settlement of pelagic larvae in the benthic boundary layer.

Among papers contained in this Proceedings, there are examples of the macroscale, mesoscale, and microscale. Macroscale studies focused on metapopulation structure and dynamics (Wing et al. 1998; Botsford et al. 1998; Fogarty 1998; Lilly et al. 1998). Studies dealing with assessment of abundance (direct or indirect) and related parameters generally focused at the mesoscale level (e.g., Smith and Robert 1998; Woodby 1998; McShane 1998; Campbell et al. 1998; Chen et al. 1998; Warren 1998; Moriyasu et al. 1998; Collie and Kruse 1998; Jones et al. 1998; Breen and Kendrick 1998; Lai and Bradbury 1998; Zheng et al. 1998). Presentations focused at the microscale level addressed aggregation patterns and implications for depensatory dynamics (Orensanz et al. 1998*b*; Levitan and Sewell 1998; B. Dew, NMFS, Kodiak, Alaska, unpublished data) and details of the fishing process in the neighborhood of stationary gear (Addison and Bannister 1998). In the case of abalone, where metapopulations can be observed over comparatively small geographic regions, studies addressing the dynamics of the fishing process (Keesing and Baker 1998; Prince and Hilborn 1998) zoomed over multiple scales.

Scale-specific dynamics

The dynamics of proximate significance for stock assessment and management pertain to the macroscale, mesoscale, and microscale.

Source-sink (SS) dynamics (large scale, metapopulations)

The pattern of connectance between components of a metapopulation is the scaffolding underlying SS dynamics, ranging between the ideal model of a well-mixed larval pool (maximal connectance) at one extreme to a collection of closed self-sustaining populations (minimal connectance) at the other. In reality, patterns of connectance are never symmetrical or random. Maximum asymmetry is found in nonreproductive pseudo-populations which depend exclusively on imported recruits and do not contribute larvae to other populations, well exemplified by pink shrimp (*Pandalus borealis*) populations from cold fjords (Parsons and Frechétte 1989) and Pacific oysters (*Crassostrea gigas*) throughout most of British Columbia (Jamieson and Francis 1986; Quayle 1988).

Contraction–expansion (CE) dynamics (individual populations or subpopulations)

The mesoscale is the domain of traditional fisheries models, in which emphasis is on change in population size over time. In recent years, there has been increasing interest in the CE of geographic range that accompanies change in aggregate abundance. This is of central importance to the resource types, which are of most interest here. Range CE's can result from the movements of individuals (e.g., in pelagic fishes) or from larval dispersal and settlement (benthic invertebrates with meroplanktonic larvae). In the former, change can be rapid, e.g., a pelagic stock contracting towards a core region following intensive fishing. In the latter, change will be slow and perhaps less predictable. A popular model of CE dynamics was McCall's (1990) extension of the ideal free distribution (Fretwell and Lucas 1970). A power functional relation between area occupied (A) and abundance (N), i.e., $A \propto N^{\beta}$, where β is a constant, is more or less implicit in many models, particularly in the "proportional density" and "constant density" ($\beta = 0.5$ and 1, respectively) verbal models of Hilborn and Walters (1992). Caddy (1989*b*) presented a graphical model that belongs in this general family, in which the area occupied generally gets larger as population size increases.

Dynamics of density-dependent (DD) small-scale processes (interaction involving neighborhoods)

The DD mechanisms that control recruitment to benthic stocks, whether compensatory or depensatory, always have small operational spatial scales. Recruitment effects, however, may occur at varying spatial scales depending on whether DD effects occur before or after dispersal (Botsford and Hobbs 1995). Consider, for example, the case of sea urchin stocks targeted by commercial divers in coastal areas of the world, whose dynamics were the subject of several presentations at the Symposium (Jamieson and Campbell 1998). At small scales, fertilization rate of these broadcast spawners depends on concentration (density experienced by individuals in their immediate neighborhoods). When concentration is low, fertilization may not occur. This is actually the case that led to the discovery of the "Allee effect," the first depensatory mechanism to be thoroughly considered by population dynamicists. As crowding increases, condition, i.e., the energy reserve of individuals, deteriorates and per capita reproductive output declines (a compensatory effect). Depensation and overcompensation can, in fact, balance each other (Levitan 1991). Typical postdispersal overcompensatory DD mechanisms are various forms of inhibition of settlement or recruitment by high densities of local residents, ranging from adult–larval interactions in bivalves (e.g., André and Rosenberg 1991) to cannibalism of juveniles by conspecific adults in crabs and other crustaceans (e.g., Fernández 1998). Examples of postdispersal depensation include gregarious settlement of bivalve larvae in the neighborhood of adult conspecifics and higher survival of juvenile sea urchins under the "spine canopy" of adults (Tegner and Dayton 1977; Botsford et al. 1993). Postdispersal DD effects can occur at the microscale or mesoscale, depending on the mobility of the benthic stage.

Diversity of harvest styles

Shellfish resources are diverse and so too are the fisheries that target them. The Proceedings (Jamieson and Campbell 1998) consider a wide range of fisheries, described as follows.

Offshore capital-intense fisheries, exemplified by Alaskan crab fisheries (Collie and Kruse 1998; Zheng et al. 1998; Otto 1998) and the scallop (Smith and Robert 1998), snow crab (Chen et al. 1998; Moriyasu et al. 1998), and northern shrimp (Lilly et al. 1998) fisheries of eastern Canada.

Inshore capital-intensive fisheries. Typical examples are commercial dive fisheries for sea urchins (Woodby 1998; Lai and Bradbury 1998) and geoduck (Campbell et al. 1998), the Dungeness crab fishery from the NE Pacific fisheries (Jamieson et al. 1998*b*), and abalone from Australia and New Zealand (McShane 1998; Keesing and Baker 1998; Shepherd and Baker 1998; Prince and Hilborn 1998; Worthington and Andrew 1998).

Inshore low-investment fisheries. These include commercial, subsistence, and recreational fisheries and are illustrated by the razor clam fishery in Haida Gwaii, British Columbia (Jones et al. 1998*)*, the yellow clam fishery in Uruguay (Defeo 1998), and sea urchin harvest in the Philippines (Juinio-Meñez et al. 1998).

Capture methods used in these fisheries include dredging (scallops), trawling (northern shrimp), intertidal digging (some clams), diving (abalone, geoduck, sea urchin), and traps (crabs, lobsters, rock lobsters). Each type of capture gear presents specific problems to analysis of the fishing process, as discussed later.

Stock dynamics

The relation between stock and recruitment

During the Symposium, Carl Walters (University of British Columbia (UBC), Canada) stated:

> "The main reason that all of us are in this room is concern about recruitment overfishing and sustainability of fisheries. Yet, very few of the papers really directly look at what it takes to sustain fisheries in terms of spawning stock sizes."

Indeed, only 6 of the 53 papers presented discussed this issue. It was directly addressed by Shepherd and Baker (1998) for Australian abalone stocks and by Caputi et al. (1998) for several other Western Australian shellfish stocks (lobsters, shrimp, scallops). Defeo (1998) investigated stock-recruitment relations (intra- and inter-specific) in two clam species from an exposed sandy beach in Uruguay. Lai and Bradbury (1998) evaluated reference points for harvested sea urchin stocks in Puget Sound. Stock-recruitment relations were implicit in models of aggregate biomass dynamics explored by Breen and Kendrick (1998) against an operating model inspired by the rock lobster stocks of New Zealand. Jones et al. (1998) used a stock-production model to estimate the maximum sustainable yield of a razor clam population from Haida Gwaii.

However, there was consistent attention to the emerging issue of what it takes to sustain fisheries in terms of spawning stock structure, specifically in terms of the locations of individuals in space. The spatial distribution of abundance is at least as important as total abundance and certainly more important at low abundance levels, when depensatory mechanisms are likely to be effective. For example, the reproductive contribution of a thousand abalone dispersed over an extensive shelf is not likely to be the same as that of a thousand abalone concentrated in a small reef where they are within effective fertilization distance of each other during the spawning season. Two aspects received attention at the Symposium: (*i*) at the microscale, location of individuals relative to each other, with emphasis on the characterization of DD processes, particularly depensation; and (*ii*) at the macroscale, location of aggregations of individuals relative to hydrographic scenarios for larval advection-retention, with emphasis on the SS dynamics of metapopulations.

Bob Elner (Canadian Wildlife Service (CWS), Canada) summarized this focus adjustment in stock dynamics in his session review:

> "We have metapopulation considerations, we have microscale considerations, we have a big focus now on DD processes, which five or ten years ago we very much preferred to ignore entirely."

Decoupling of spatial scales relevant to life history stages involved in the recruitment process implies that, in principle, one should not expect to find evidence of the aggregate stock-recruitment relations proposed for Type-4 unit stocks (e.g., as described by the Beverton–Holt and Ricker stock-recruitment models), which are at the core of traditional fishery theory. Based on empirical experience, such relations have often been found inadequate for invertebrates:

> "Invertebrates often show no relationship between spawning stock and recruitment, which may be due to high reproductive output or the presence of invulnerable individuals in unfished or unfishable areas" (Hilborn 1986).

> "With most, if not all, [shell]fisheries [from British Columbia], stock:recruitment relationships are uncertain [...] and precise estimation of biomass is often impractical" (Jamieson 1986).

We suggest that, a priori, one should not expect to find a stock-recruitment relation while examining a segment of a metapopulation that is open to recruitment. So stated, this notion seems plain common sense. Yet, assumption of compact stock-recruitment relationships is common among fishery scientists and managers, partly because of adherence to a consensual paradigm, but also because such relationships offer a conveniently simple, albeit distorted, representation of the renewal process. One naive proposition to get around this problem would be to expand the definition of the stock under scrutiny in order to absorb the entire metapopulation and treat the later as internally homogeneous. However, this introduces new difficulties stemming from the fact that internal heterogeneity is likely to erase signatures of the processes of interest in the aggregate dynamics of a metapopulation. The best evidence of an aggregate stock-recruitment relation presented at this Symposium was with penaeid shrimp stocks of Western Australia, a Type-4 resource (Caputi et al. 1998).

Metapopulations: SS dynamics

Metapopulations illustrating extremes of connectance were exemplified at the Symposium with long-lived pelagic larvae from rock lobster stocks of Western Australia (Caputi et al. 1998: maximal connectance) and short-lived larvae from abalone metapopulations from South Australia (Shepherd and Baker 1998 and Keesing and Baker 1998: minimal connectance). Caputi et al. (1998) reported that rock lobster recruitment is unrelated to local stock size in segments of the shelf but was well correlated with environmental conditions in the Leeuwin Current system.

The dynamics of benthic metapopulations with

meroplanktonic larvae have been little explored and have been generally ignored in fisheries textbook presentations. Hilborn and Walters (1992) is an exception, but they addressed only the simple scenario of minimal connectance, inspired by the Tasmanian abalone (*Haliotis rubra*). Cases of moderate to high connectance are more widespread and present a formidable challenge to modeling. Investigation of the dynamics of metapopulations of Dungeness crab (*Cancer magister*) from the West Coast of North America, one of the most thoroughly studied among commercial benthic invertebrates, is producing intriguing results (e.g., Hastings and Higgins 1994). The dynamics of metapopulations with long-lived larvae were addressed in the Symposium with cancrid crabs (Wing et al. 1998; Botsford et al. 1998), homarid lobsters (Fogarty 1998), and shrimp (Lilly et al. 1998). While metapopulations are appealing as a conceptual framework, relevance of the latter in management decision making is less clear, as stated by Doug Woodby (Alaska Department of Fish & Game (ADFG), U.S.A.):

> "It was really important how Lilly et al. (1998) tried to answer questions. This gets into a paradigm that managers need to be brought into: [...] is there a stock structure that they need to be aware of? How do you identify that? And how is management going to be effective?"

In spite of dynamic complexity, the spatial arrangement of benthic metapopulations is often remarkably persistent: concentrations (e.g., location of a fishing bed) tend to occur at locations where they occurred in the past (Sinclair 1987). At the mesoscale, defined by the distribution patterns of benthic stages, this is the result of habitat selection by, or differential survival of, larvae, juveniles, or adults in complex benthic seascapes. Seasonal and onthogenic migrants (e.g., many crabs and lobsters) track gradients, such as aggregated depth-temperature-sediment gradients, which influence onshore-offshore seasonal displacements. These can confer substantial structure to populations of mobile organisms. At the macroscale, pattern may reflect recurrent oceanographic features (currents, eddies, estuarine, and shelf fronts, etc.) significant for larval transport-retention. Where the requisite benthic habitat to which larvae settle is restricted (e.g., island or narrow coastal shelves, estuaries), larvae often have a suite of behaviors which provide for "directional" dispersal or minimize it altogether (e.g., Sulkin 1984; Jamieson and Phillips 1993; Crawford and Jamieson 1996). All such mechanisms contribute to the persistent spatial structure of metapopulations.

While total abundance is hard to estimate (see discussion later) and the dynamics of populations are difficult to model, persistency of structure calls for coarse, albeit robust, approaches to assessment and management based on qualitative macroscopic diagnostics. Identification of SS configurations and exploration of its implications (e.g., through simulation) should generally take precedence over analysis of aggregated stock-recruitment relations, both from scientific and management-oriented perspectives. This was recognized by Doug Woodby (ADFG, U.S.A.):

> "Mike Fogarty [...] not only talked about the existence of metapopulations [...] He actually showed us how we can get into a decision framework of harvest rates of metapopulation [components], i.e., offshore and inshore lobsters [...] In his simulation, a decision to leave the offshore lobsters unharvested could provide a cushion from extinction for inshore lobsters."

Small-scale dynamics and intra-population gradients

Compared with mobile animals, analysis of growth, mortality, and production of sedentary organisms presents both advantages and difficulties. Natural mortality and growth rates are comparatively easy to estimate at specified locations; this is particularly true for intertidal or shallow subtidal sites amenable to manipulative experimentation or repeated visits. On the other hand, parameters vary among locations and along environmental gradients. Variability can be significant over even small distances, as shown by Worthington and Andrew (1998) for abalone from New South Wales, Australia. Geographic variation was addressed during the Symposium in several presentations. Shepherd and Baker (1998) compared egg production per recruit among exploited populations of abalone from South Australia and found it was related to reef area. Otto (1998) investigated variation in size-at-maturity in snow crab at the scale of the eastern Bering Sea and discussed its relation to various factors and its implications for management. Worthington and Andrew (1998) discussed implications of spatial variation in the demographic schedules of abalone for size limit regulation. While over dramatically different geographical ranges, snow crab and abalone variability present similar problems for both analysis and management.

The most significant aspect of intra-population variation in growth, survival, fecundity, and settlement relates to DD effects. The magnitude of DD effect depends on how density is experienced by individuals of the population (the concentration profile), discussed by Orensanz et al. (1998*b*). In populations of sedentary organisms with aggregated patterns of spatial distribution (the rule for benthic resources), DD processes may be dynamically important even if they only have a significant effect within small beds or patches. A notable case of aggregation in adult red king crab (*Paralithodes camtschaticus*) was documented by B. Dew (NMFS, Kodiak, Alaska, unpublished) at the Symposium.

DD mechanisms that regulate shellfish metapopulations were not widely appreciated by fishery scientists until relatively recently. DD effects on fertilization rate in broadcast spawners (reviewed by Levitan and Sewell 1998), for example, attracted little attention before the publication of experimental results on sea urchins by Pennington (1985). Until a decade ago, male-only crab fisheries were assumed to be beyond the risk of recruitment overfishing, but the spectacular collapse of some managed fisheries (e.g., the red king crab fisheries from Kodiak and the Eastern Bering Sea during the early 1980's) prompted new questions on limits to effective male polygyny and female sperm retention (Paul 1984; Elner and Beninger 1992; Paul and Paul 1992).[3]

A suggested future direction for yield-per-recruit analysis in broadcast spawners is the incorporation of DD effects in per-capita fertilization rate, with concentration profiles observed or predicted under various harvesting scenarios (Prince and Hilborn 1998; Orensanz et al. 1998*b*). Such results would bear directly on the issue of depensatory dynamics and recruitment overfishing in abalone, urchins, and other species with external fertilization. At a time when the white abalone (*Haliotis sorenseni*) of California might become the first

[3] Sperm limitation in male-only crab fisheries was the subject of a special workshop at the 1997 Scientific Symposium of ICES (Baltimore, October 1997).

marine invertebrate species ever to undergo biological extinction (Davis et al. 1996) as a result primarily of fishing, synthesis of these three venues of research may be timely.

The fishing process

The fishing process is the sequence of actions through which a shellfish resource is located, harvested, and depleted by a fishing force, composed of discrete fishing units. The nature of fishing units varies greatly in fisheries we deal with, ranging from individual clam diggers to offshore trawlers. The fishing force is usually referred to as the fleet in the fisheries literature. In our context, however, a fishing force not composed of boats or vessels is far from exceptional. The fishing force has a dynamics of its own: fishing units are "recruited" and vanish as they become obsolete or transfer to other fisheries. These changes reflect the responses of fishers to the varying relative abundances of harvested stocks, to economic and social conditions, and to management. The field of fleet dynamics has received little attention outside a small body of primary literature (see Hilborn and Walters 1992).

Somewhat related is the subject of effort allocation by fishers. Individual decisions that determine the spatial pattern of effort allocation are responses to the spatial pattern of abundance, distance from port, traditional or legal territorial boundaries, accessibility, exposure, depth, etc., as summarized by Julian Addison (Ministry of Agriculture, Food and Fisheries (MAFF), U.K.):

> "in our crab fisheries there is a relatively easy life to be made with lower catch rates closer to shore in the inshore fisheries, where there are shorter steaming times, less chance of losing your gear, and rather safer conditions. You have to trade that off against fishing further offshore. [...] There you have higher catch rates, but the cost of reaching the grounds is greater, there is more likelihood of you losing your gear by it being towed away, and it's more dangerous to fish out there."

Given the importance of spatial controls in the management of coastal stocks of sedentary organisms, the mapping and modeling of effort allocation have received surprisingly little attention, perhaps due to traditional emphasis on aggregation in both data collection and modeling. Caddy (1975) pioneered the field with the introduction of proportional allocation as a reference model (related to the ideal free distribution; see Gillis et al. (1993) for a fisheries application), later modified by Walters et al. (1993) with the addition of an attractiveness index. Concentration profiles were discussed at the Symposium as a powerful instrument in the investigation of effort allocation (Prince and Hilborn 1998; Campbell et al. 1998) and implications of the latter for depensatory dynamics (Levitan and Sewell 1998), as best synthesized in a summary session by Mike Featherstone (Pacific Urchin Harvesters Association (PUHA), Canada):

> "In discussions during this session we saw the interactions between fishing activity and population density and how this can affect recruitment. Fishermen choose where they conduct their operations based on the ease of access, returns from effort, and other factors. Hopefully by studying these effects we can develop fishing strategies which ensure sustainability."

The fishing process was discussed at the Symposium mostly in relation to commercial diving fisheries for abalone, whose study has already provided much innovative thinking on this subject. McShane (1998), Keesing and Baker (1998), and Prince and Hilborn (1998) showed that because of interaction between the spatial pattern of a stock and fishers' behaviour, catch-per-unit-effort (CPUE) is not a reliable index of abundance (except perhaps at a very small spatial scale). Keesing and Baker (1998), in particular, discussed very clearly how serial depletion can lead to overfishing not being detected in the CPUE trend.

The relation between abundance and CPUE can depart from proportionality because of more subtle reasons, as pointed out by Jamie Austin (Underwater Harvesters Association (UHA), Canada):

> "Market conditions greatly affect CPUE. In some fisheries, commercial diving being a good example, there is a hands-on approach to harvesting, and market price often makes fishers work slower and more carefully to make sure that a better quality product is delivered to the market. Average catch/day of geoduck divers is less now than five years ago, because fishers want to maintain their market price. This reduces their CPUE, but it does not mean the industry is in decline."

While dive fisheries are best understood, perhaps because the fishing process is directly observable, other types of fisheries and capture gear present a plethora of equally interesting and important problems. In the case of dredging and trawling, the area effectively fished (swept) during each elemental fishery operation (a haul in this case) is reasonably easy to measure. Assessing efficiency of the gear is more difficult. Somehow there is often the wrong perception that uncorrected trawl–dredge data measures absolute abundance. Results of trawl–dredge surveys have even been taken as 100% efficient in assessments used in major management recommendations (e.g., crab total allowable catches (TAC's) based on the eastern Bering Sea trawl survey (Otto 1986)). Opposite to dredges and trawls, hand gatherers are easy to observe, but the effective area fished during a single dive, or during a day spent digging on a beach, is almost impossible to define because vision-guided fishers can recognize pattern and allocate effort accordingly, even at the smallest spatial scales. This can produce nonlinearities in the relation between average CPUE and estimated abundance (e.g., as estimated from a survey sample) that may be wrongly attributed to a functional response (e.g., Beinssen 1979) or other effects. The most challenging problems of interpretation and modeling may be posed by traps, as they are affected by complex behavioral interactions between target and nontarget individuals in and around the trap. Details of trapability were the subject of an experimental and modeling study by Addison and Bannister (1998).

Assessment of abundance

In traditional stock assessment, emphasis is on the estimation of total abundance and its standard error. In the cases on which we have focused, where spatial structure is of primary significance for management, mapping an index of abundance, or even a correlate of abundance, is often of more significance than the estimation of total abundance. In the following, we consider both aspects.

Mapping

The scale desired for mapping depends on the scale of the process of interest. Investigation of fertilization rate and depensatory mechanisms in sedentary broadcast fertilizers, for example, requires mapping of the location of individuals at small spatial scales. At an intermediate scale of fishing beds or grounds, different methods can be used to interpolate abundance between sampling stations. Among these are geostatistical methods (e.g., kriging), which can be used to model the variance–covariance structure of the spatial distribution. While the merit of these methods for the estimation of abundance is subject to debate (Warren 1998), their usefulness in spatial interpolation and mapping are illustrated by Warren (1998) and Moriyasu et al. (1998).

Formal methods in the sense of spatial statistics are of no practical value in the mapping of entire metapopulations spread along complex coastscapes, including estuarine systems, coral or rocky reefs, mangrove swamps, and so on. In such cases, estimation of total abundance is often a hopeless proposition. Management of such systems has a strong spatial component, with options often determined or constrained by factors other than abundance itself. Under such conditions, mapping of features relevant to management is the first, and often the most significant, aspect of any assessment. Information that is of interest to mapping includes: (*i*) spatial correlates of resource distribution (bottom topography, sediment types, and benthic communities (e.g., mangroves, seagrass or kelp beds, coral reefs)); (*ii*) fishing effort allocation and its determinants, primarily CPUE; (*iii*) various constraints affecting fishing operations, such as accessibility, exposure, pollution, and closures; and (*iv*) user rights (property and other forms of tenure).

Mapping of many of these features may be expediently conducted by means of remote sensing (e.g., Long et al. 1993) or underwater acoustic techniques, as illustrated during the Symposium by the application of RoxAnn (Greenstreet et al. 1997; Burns et al. 1995) to mapping bottom textures in geoduck fishing grounds. The end result is an overlay of maps; the appropriate tool for presenting these is a Geographic Information System, which can then be used in spatially explicit decision-making models (e.g., Williams 1996). Management options, often being determined by opportunities and constraints, are immediately suggested by inspection of the maps themselves.[4]

Assessment of total abundance

All methods used in the assessment of abundance fall into one of two broad families: direct sampling of the population and modeling the process of depletion by the fishery.

Direct methods

The most typical form of sampling consists of the collection of a quantitative sample in the course of a survey. Sampling units can consist of trawl or dredge hauls, visual counts in quadrats or along stripes, line transects, etc. Among other advantages, these data are free of problems associated with the use of commercial CPUE as an index of abundance, discussed earlier. Results are also appropriate to geo-referenced data analysis, particularly in the case of systematic designs. Difficulties encountered in the surveying of benthic invertebrates are, however, multiple. Hydroacoustic techniques, which provide an expedient way to survey large areas for many pelagic or demersal species (fish or invertebrate), are of little use here. The efficiency and selectivity of direct sampling gear is strongly affected by variation in the substrate and many substrates (e.g., reefs) are nontrawlable or nondredgeable, problems not encountered in midwater trawl surveys. Samples consisting of "quadrats" are popular in benthic ecology, but of little use in the assessment of abundance over large regions. Use of visual surveys (including underwater TV and photography) is largely restricted to epibenthic organisms and in the case of divers by depth. Because collected individuals of many benthic species can not redistribute themselves after being released, mark–recapture methods are of limited value, although they are useful in the assessment of movements, migrations, and growth rate.

Interesting developments in the surveying of benthic populations in recent years have involved model-based approaches, in contrast to the design-based emphasis that dominated the field until the mid-1980's (Thompson 1992). Among model-based techniques, geostatistical analyses, such as kriging, are becoming extensively used. Much of the appeal of geostatistics resides in the fact that beyond providing an estimate of mean abundance and its standard error, application of these techniques can produce a smoothed map.

With regards to design-based approaches, two important questions are addressed in this Proceedings. The first was use of ancillary information in design-based estimation. Smith and Robert (1998), using a scallop survey example for demonstration, illustrate the incorporation of both survey design (a stratified random design) and ancillary information (sediment type) into estimations of mean and total abundance, with the aim of increasing precision. The possibility of mapping surface sediment type over extensive areas of sea bed, demonstrated at the Symposium with RoxAnn, adds to the promise of this approach. The second issue addressed was the problem of highly aggregated spatial distributions, highlighted by Jamie Austin (UHA, Canada):

> "Surveys do not always give a true indication of density in certain areas. In certain cases, placing of transect lines may miss areas that are most productive, thereby affecting and reducing density estimates. Some resources (e.g., geoducks) tend to be dispersed along a bed, with greater concentration in certain spots than others. To miss these areas and only count sometimes less productive areas, we feel, is not an accurate way to estimate stock abundance."

Two problematic cases of aggregation were presented at the Symposium: multiscale aggregation (individuals within patches, patches within stocks) in abalone stocks (McShane 1998) and "podding" in adult red king crabs (B. Dew, NMFS, Kodiak, Alaska, unpublished). Woodby (1998) applied, probably for the first time in shellfish research, ideas on adaptive designs and estimators developed in recent years (Thompson and Seber 1996).

Indirect methods

An alternative to, or complement of, surveys is to utilize commercial catch and effort data to model the processes of recruitment and depletion and estimate abundance and related parameters. Variations on this theme range from single-season depletion analyses (recruitment, natural mortality, and age–sex structure of the harvested population), multiyear studies in which catch can somehow be partitioned among "recruits" and

[4] These issues were recently addressed at the workshop on Spatial Data and Remote Sensing Invertebrate Fisheries Habitat, Research and Management, Fort Walton Beach, Florida, April 1997, in conjunction with the 89th Annual Meeting of the National Shellfisheries Association.

"postrecruits" (e.g., Collie and Kruse (1998) for red king crab), and cases in which information on age or size structure of the population are available. Variations in catch-at-size models in which growth is described by a transfer matrix are presented for Puget Sound sea urchins (*Stronglocentrotus franciscanus*) (Lai and Bradbury 1998) and eastern Bering Sea Tanner crabs (*Chionoecetes opilio* and *C. bardi*) (Zheng et al. 1998).

The natural domain of application of these models is in assessment of industrial fisheries targeting mobile organisms for which reasonably long time series of data on catch, effort, and composition of the catch are available. There are serious difficulties in their application to inshore fisheries and spatially structured stocks that we are mostly concerned with in this overview: (*i*) In the case of sedentary organisms the dynamic pool assumption can not be invoked to redistribute density after each fishing event and spatial allocation of effort by fishers is never random. Abundance tends to drop faster than CPUE as the stock is depleted ("hyperdepletion"; see Hilborn and Walters 1992), rendering even within-season data useless. (*ii*) Trends in size–age composition of the catch are likely to be more influenced by spatial switching in effort than by trends in time, unless there is both a high degree of regional coherence in recruitment and no spatial gradients in growth and mortality rates. Needless to say, these conditions are never met in stocks of sedentary organisms. (*iii*) In the case of fisheries focused on here, time series of catch, effort, and composition of the catch are rarely available, but even if they are, they may be dangerously misleading, for reasons outlined earlier.

Besides limitations derived from the effect of nonrandom allocation of fishing effort during the depletion process, the use of CPUE as an index of abundance presents other difficulties that are common to all types of fisheries. These include saturation of traps and dredges, the functional response of hand gatherers to shellfish abundance, the effect of measurement errors in analysis of DD effects on catchability, and the mismatch of frames encountered when survey and fishery-dependent data are compared. While the shortcomings of CPUE in stock assessment were well illustrated in presentations and thoroughly discussed in the review sessions, Julian Addison (MAFF, U.K.) pointed out that

> "[these problems are] something that those of us working in fisheries assessment and management have known about for a long time, but it is one of those things that we tend to forget about when we are providing advice to managers, administrators, and politicians. [...] Despite all of these problems, we still collect CPUE data and use it. Why? Well, I guess because it is so easy to monitor landings and fishing effort. Also, it is relatively cheap."

Management

In discussing the design of fishery management systems, Charles (1995) made an important distinction between

> "a strategic institutional structure and an operational set of management measures. The former may take the form of traditional central government management, "co-management" power sharing, or property rights-based options, such as individual fisher quotas, community-based management and territorial use rights for fishers (TURF's). On the other hand, operational management tools could include gear limitations, licensing rules and total allowable catch levels."

We used this distinction to organize issues brought up at the Symposium. Later we emphasize first spatially explicit operating measures and then co-management among strategic institutional structures as primary elements in the management of highly structured coastal resources.

Spatially explicit operating measures

Given the sedentary life-history characteristics of invertebrates and the nature of their fishing process, management measures that explicitly acknowledge spatial structure are most suitable. These include the following: (*i*) reproductive refugia and marine protected areas (MPA's); (*ii*) territorial property/use rights (leases, traditional tenure systems, TURF's, etc.); (*iii*) rotation of the harvest, often combined with pulse fishing and/or thinning; (*iv*) experimental management with spatial controls, contrasting treatments, and replication; and (*v*) localized enhancement, including habitat manipulation, seeding, and control of predators.

These measures, which we discuss below, can be combined with other tactics or operational measures, such as closed seasons, minimum legal size, direct controls on fishing effort, and so on.

Reproductive refugia and MPA's

Carl Walters (UBC, Canada) stated in a summary session that

> "If we look at fisheries that have been successful over the long term, the reason for their success is not to be found in assessment, learning and management models, but in the existence of a spatial accident, something about the spatial structure of population dynamics interacting with regulatory systems or about the behavior of the species and fishers that creates a large scale refuge for a substantial segment of the spawning population."

When such "natural accidents" do not exist, the first priority of any sensible management plan should be to create their equivalence: reproductive refugia that, through regulation and enforcement, are off limits to fishery operations. Reproductive refugia have received renewed attention in recent years, primarily in situations that combine complex spatial structure, little available information, and difficulties of enforcement (Roberts and Polunin 1991, 1993; Rowley 1994; Bohnsack 1996). More recently, they have been proposed as a major component in the precautionary management of fisheries in general (Clark 1996; Lauck 1996).[5]

Some criteria have been prescribed for the design of refugia. Caddy (1989*b*), for example, suggested that

> "For a contagiously distributed, sedentary species, an alternative or supplementary hypothesis for the maintenance of adequate spawning levels is that the critical density of mature animals located in an area where hydrographic features allow a high probability of larval return should be maintained. This would support the maintenance of refugia (closed areas) in areas of high density where multiple age groups provide evidence for past successful recruitments."

There are three important aspects to be taken into account in the design of a refuge system. First, the position of refuge

[5] Reproductive refugia, MPA's, and other forms of spatial closures were extensively debated at several meetings during 1997, including the Design and Monitoring of Marine Reserves Workshop (Vancouver, B.C.) (Pitcher 1997), the Science and Management of Protected Areas Conference (Calgary, Alberta), and the Annual Meeting of the Society for Conservation Biology (Victoria, B.C.), which focused on marine conservation issues.

areas relative to current systems (Roberts 1997) and potential mechanisms for larval advection, and second, the concentration profile of the spawning stock. A third important question, the size and number of refugia (e.g., few large vs. many small), relates to connectedness within the metapopulation. Some of these issues have been more formally explored by Quinn et al. (1993).

While the creation of refugia is urgently needed for the conservation of many shellfish stocks, information relevant to their design is often scarce. Even in cases where fisheries agencies have gathered time series of fisheries-dependent data, these often consist of coarsely aggregated catch, effort, and catch-at-size or catch-at-age data, which may not provide required rationale. Macroscopic diagnostics of metapopulation structure, including the preliminary identification of sources and sinks, may provide the best basis for the initial design of a system of refugia. However, even an imperfect system in place is preferable to no system at all. The initial creation of a network of reproductive refugia and MPA's should be framed in an adaptive management protocol, with a plan for learning about the system from its responses (Parma et al. 1998).

Territorial property–user rights

This includes leases, traditional tenure systems, and TURF's. A significant contribution of social scientists to fishery science and management over recent years has been unraveling the complex, varied sea tenure systems along the world's coasts (Cordell 1989). This new understanding is important for several reasons: (*i*) management strategies developed for open-access industrial fisheries, the main client of recent fishery science, are inconsistent with traditional or spontaneous tenure systems; (*ii*) territorial user rights are better suited to the spatially explicit controls that are natural in the case of coastal stocks of sedentary organisms; (*iii*) territorial user rights can create greater incentives for investment in conservation and enhancement by users; and (*iv*) territorial user rights motivate users to experiment, either formally or informally, in a cautious manner.

Different types of sea tenure were discussed at the Symposium, including the traditional sasi management system for the *Trochus* fishery in Indonesia (Arifin et al. 1998) and the artisanal caleta (Spanish for cove) system for local fishery cooperatives along the coast of Chile (Castilla et al. 1998). The latter is now recognized by current Chilean fishery legislation. Study of these systems can provide fresh ideas for coastal fisheries in developed countries, where strategies conceived under the unit stock paradigm and an open-access mentality are often questioned as perhaps being inadequate. Prince et al. (1998) discuss the difficulty of managing highly structured abalone fisheries with "modern broad-acre management tools" (size limits, closures, limited entry, individual transferable quotas (ITQ's)) within a "law of the commons" framework. As an alternative, they propose the introduction of TURF's to facilitate management at an appropriate spatial scale.

Rotation, often combined with pulse fishing and/or thinning

Gear used to harvest benthic resources is often nonselective, as best illustrated by trawls and dredges. Even when gear is selective, individuals of sublegal or nonmarketable size may be killed or damaged by harvesting operations or their exposure to various sources of natural mortality (e.g., predation by crabs, fishes, or birds) is increased. Under such circumstances, spatial rotation strategies are an option, provided that the harvested region can be partitioned into subregions or plots. Rotation strategies are based on two rationales (Botsford et al. 1993): (*i*) a pulse fishing rationale (each plot or subregion can be realistically harvested every once in a while). Virtually the entire economically harvestable stock is removed (harvested or otherwise exposed to mortality) on each occasion. The stock recovers between successive harvests; and (*ii*) a metapopulation rationale: recruits (settling larvae) originate in the regional reproductive pool from the subregions that are not exploited that year.

Management controls and practices associated with rotational strategies include time duration between harvests, spatial pattern in partitioning the harvested region into a number of subregions or plots, and possible enhancement following a harvest to reduce recovery time and increase yield. Motivation to adopt rotational harvests include increased sustainability and continuity of the fishery, reduced enforcement and harvest costs, simpler logistics, and optimality criteria (Clark 1976).

Spatial rotation has been empirically practiced for centuries as a common-sense strategy for the sustainable exploitation of sedentary resources. Precious red coral (*Corallium rubrum*) has been harvested for centuries in the Mediterranean: a 9-year rotation period was apparently practiced by Arab fishermen as early as the 10th century (Grigg 1989). Sasi traditional management in Indonesia (Arifin et al. 1998) shows that rotational strategies have been used for centuries by Pacific indigenous peoples. An elder of the Lummi Indian tribe (Washington State, U.S.A.) described an annual rotation in terms that captures its essence:

> "Chief Chow-it-hoot marked the clam beds that were to be dug each season. They dug each bed in turn, until all had been dug. They repeated this method each season. When they gathered clams from [each] bed again, they had grown to a usable size" (compiled by Beck 1955).

Spontaneous rotational tactics are employed by fishermen in many unregulated shellfisheries (e.g., Lasta et al. 1986), although they are poorly documented. In spite of its recurrent empirical discovery as a workable strategy, rotation has attracted little attention from shellfish resource managers in developed countries. Bed rotation, however, is currently practiced in the British Columbia geoduck (Campbell et al. 1998) and red urchin fisheries (Campbell and Harbo 1991). Caddy and Seijo (1998) presented simulation results on the optimal rotational harvest of a contagiously distributed stock of sedentary organisms, under different combinations of life-history parameters and fishing mortality.

Experimental management

With sedentary species, the area occupied by a stock can be easily partitioned, facilitating the requirements of replication, control, and contrast between treatments in a manner that is virtually impossible in the case of most stocks of mobile organisms. Rotational strategies (described earlier) are a natural candidate for empirical experimentation, since recovery rate of harvested plots depends not only on the local dynamics of the target population but also on subtle and unpredictable ecological responses of the benthic ecosystem to harvesting operations. Quayle (1952) pioneered experimental management in an intertidal bed of butter clams in British Columbia.

Fig. 1. A drawing from Scoresby Shepherd (SARDI, Australia) showing a 3-dimensional chart relating fishers' income (X-axis: low to high), fishery management style (Y-axis: paternalistic/centralist to co-operative/co-management/local), and legal control of coastal waters (Z-axis: centralized to local). He suggested that management becomes more advanced as it moves from the lower left front to the upper right back. See text for further discussion.

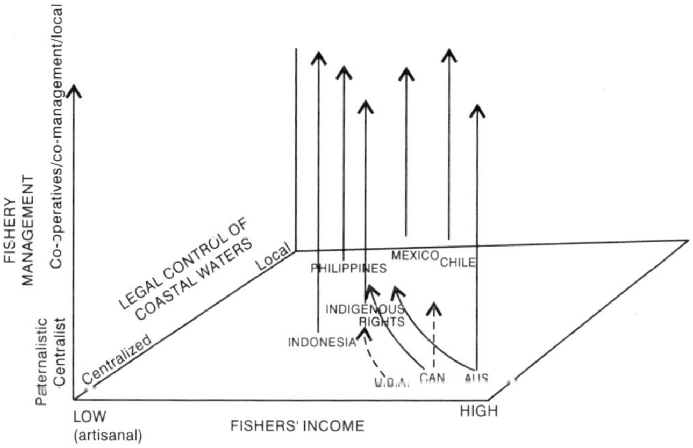

While local population dynamics are amenable to experimentation (e.g., DD effects on growth, rotation periods), uncertainty about sources of settling larvae makes it difficult to interpret results in the context of SS dynamics. The size and network design of refugia, for example, is likely to be guided by macroscopic descriptions of hydrographic scenarios or by constraints or conflicts in the use of the coastal zone. Species with short-lived pelagic larvae (e.g., abalone) or direct development (e.g., whelks and some octopus) offer specific opportunities for management experimentation. In other fisheries, there is opportunity to learn from management interventions contrasted over time. The simplest case is that of systems harvested on a strictly experimental basis, such as the lakes in which Momot (1986, 1998) manipulated crayfish harvests. Large study areas require the active participation of fishers and actions that maintain their support may be a key part of the experiment. The experimental closures described by Defeo (1998) in a beach clam fishery in Uruguay constitutes a well-documented and informative study case. Another example is the ongoing experimental perturbation of red urchin populations in British Columbia (A. Campbell and G.S. Jamieson, unpublished).

Enhancement

Human interventions designed to enhance population productivity at intermediate to small scales are feasible, at least in principle, particularly when motivation is facilitated by territorial user rights. Interventions may include habitat improvement or restoration, creation of artificial reefs, control of predators, thinning and seeding, and actions which are part of a continuum of managerial practices leading ultimately to aquaculture. Juinio-Meñez (1998) describes a program under way in Bolinao (Pangasinan, Philippines) to restore a collapsed sea urchin fishery through community-managed sea-pens or cages.

Enhancement programs have potential as a tool in the management of sedentary resources (particularly bivalves) but the benefits or failures of many efforts have been undocumented. Enhancement projects are relatively easily sold to politicians and to the general public because of visible results and quick cost recovery: there are more clams in plots where predators are excluded and fish are regularly caught around artificial reefs. Many of those beneficial effects, however, are ambiguous, e.g., it remains controversial whether reefs enhance productivity or concentrate fish (Pickering and Whitmarsh 1997) and cost-benefit balances remain untested in many well- publicized projects (e.g., McCay 1989).

Strategic institutional structures: co-management

Lively debate during the Symposium was elicited by a comparative synthesis of shellfishery management structures given by Scoresby Shepherd (South Australian Research and Development Institute (SARDI), Australia). His overview was a conceptual categorization (Fig. 1) which summarized the spectrum of co-management arrangements from centralized government control to strictly community-based management, similar to that used by social fishery scientists (Sen and Nielsen 1997). This continuum of management systems was related to fisher's income, high in developed countries (e.g., U.S.A., Canada, Australia) and lower in lesser developed countries (LDC's) (e.g., Indonesia, Philippines, Mexico, Chile). Three specific aspects (discussed further later) were singled out: participation by fishers in the management process, the exact meaning of community-based management, and the special fishing rights of aboriginal peoples in western societies.

Participation of fishers in the management process

A subject frequently addressed and intensely discussed during session reviews was the interaction between managers, scientists, and fishers. Lorne Clayton (Sea Cucumber Harvesters Association (SCHA), Canada) summed up traditional perceptions among members of these three groups:

> "It was not too many years ago, that if you asked a fisheries manager to define fishermen one answer might be: 'those people whose main purpose is to take all the fish until there is nothing left, and then they become loggers'. If you asked a fisherman to define what fisheries managers were one answer might be: 'those who prevent fishers from making a living harvesting fish'. If you asked either what a scientist is, they might collectively agree that a scientist is the one who makes all the mistakes if anything goes wrong with the fishery."

Current concerns about communication of scientific knowledge, its use in decision making, the role of the industry in the management process, and the nature of institutional frameworks that are best suited to accommodate interactions relating to sustainable resource utilization, all reflect a recent upsurge of interest in the human component of fishing, catalized to a large extent by the collapse of cod fisheries in eastern North America. Participation of fishers in the management process (McCay 1996) was unanimously perceived as desirable by speakers. There were basically three reasons invoked:

Input of knowledge. Fishers have substantive empirical knowledge about important aspects of the system, provide much of the information used in stock assessment, and their participation is often crucial to interpret that information (an outstanding case being trends in CPUE):

"Fishermen are ultimately the people that have the greatest potential to provide the information that drives the whole process, whether they are contributing it directly through log books, or indirectly through interviews" (Howard McElderry, Archipelago Marine Research Ltd., Canada).

"I think most managers have had the experience of presenting to an industry group some assessment based on their log or catch data, only to be told that you actually don't know what you're talking about. Then they provide a perfectly good explanation, based on the fisher's experience, of why a trend is the way it is." (Marilyn Joyce, Department of Fisheries and Oceans (DFO), Canada).

"In virtually every research or development project in which we are involved, industry plays a pivotal role in providing biological information on spawning, behavior, distribution, or hands expertise in diving, vessel, or gear operations" Lorne Clayton (SCHA, Canada).

Costs of assessment and management. Sharing the costs of stock assessment and related research with industry is considered increasingly desirable by governments in the climate of financial restraint experienced by most management agencies today and considering that the industry is the primary beneficiary of resources that, ultimately, are perceived to be shared by society at large. Industry, however, is unlikely to share willingly any costs if it is left out of the decision-making process:

"[Considering] the cash flow problems our government is having, [...] it is not inconceivable to me that some sort of cost sharing agreement could be worked out, whereby those with the largest stake in any fishery, namely the fishers, support the cost of ongoing stock assessment research" (Tim Webster, Gulf Crab Fishery Association (GCFA), Canada).

"An ever more increasing level of research funding is coming from industry itself [...] Successful fishers will be providing much of the future dollars for research, at least in Canada" (Lorne Clayton, SCHA, Canada).

"With the individual vessel quota (IVQ) system in place in B.C., fishermen now feel much more responsibility for the stewardship of the resource and are funding enforcement, enhancement, research, and mapping" (Mike Featherstone, PUHA, Canada).

Fishers' trust. Effective implementation of management plans, no matter how good they are, requires support and compliance and these in turn require understanding and trust. Credibility of the assessment and trust of ensuing management recommendations are always likely to be improved by participation of fishers in the process, as is the quality of the assessment itself:

"What is required for implementation, clearly, is that science be credible to the industry in order to elicit their support and compliance" (Marilyn Joyce, DFO, Canada).

"Fishermen are invariably opposed to management decisions based on research they can't necessarily trust. However, if they become part of the experiment and thereby see the proof gathered before their very eyes under real working conditions, there remains little arm twisting to be done" (Tim Webster, GCFA, Canada).

"I'd like to see the gaps [between management, industry and scientists] filled in more and more. Bring [the industry] on board, making them feel comfortable, as I again refer to the credibility of our work" (Paul LaRiviere, Washington Department of Fish and Wildlife, U.S.A.).

Integration of scientists, managers, and fishers, while desirable, is also challenging. Important questions remain open to debate. For example, what level of participation should fishers have in setting research priorities, even when they share the costs of that research? Jamie Austin (UHA, Canada) was clear in this respect:

"Cost recovery through industry and declining revenues to fisheries departments means that industry is going to expect to have a greater say in what information is obtained. We think we should prioritize projects and conduct surveys that will have the greatest direct effect on industry."

Still open to debate is also the issue of how should fishers' short-term, resource-specific demands be weighed against longer-term, broader scope societal expectations regarding ecosystem sustainability and conservation.

Community-based management

Scoresby Shepherd (SAFDI, Australia), in his session review, went beyond simply organizing the spectrum of management systems around the co-management theme (Fig. 1). He categorized centralized management as "primitive" and various forms of substantive co-management as "advanced" and described a perceived evolution towards various forms of community-based management in shellfisheries from developed western countries, sustained by commentary from other participants. He suggested there are lessons to be learned by resource managers in developed countries from experiences in the management systems of LDC's. However, as Scoresby himself stressed:

"Arifin [et al.]'s (1998) description of the sasi system in Indonesia carries an important warning to us. That is that local control, although it is undoubtedly the way to go, is no guarantee of sustainability."

The issue of the level at which managerial decisions are made is a thorny one. Management co-ordination must relate to the scale of dynamics of a managed stock. For example, highly migratory shrimp cannot be effectively managed by local communities in isolation from each other. The distinction between scales of management decision and implementation is sometimes less clear. The point is more easily made in the case of sessile organisms with pelagic larvae. A regionally co-ordinated plan to manage an oyster stock could delegate management authority to local communities in the form of TURF's or leases, within which fishers can harvest or enhance as they deem appropriate, while at the same time creating reproductive refugia as an insurance policy. Here, there are two tiers of management implementation: larger scale (management authority) zoning (e.g., designation of areas as TURF's and refugia) and smaller scale (local community) harvest strategies (intra-TURF management).

Different scales for managerial decision and management implementation are also part of other spatially explicit strategies, e.g., the control of the harvest rate through spatial "windows" (Walters and Maguire 1996). The ultimate issue is perhaps not so much whether management should be centralized or community-based but rather whether and how it can be organized so that decisions are taken and measures implemented at appropriate levels within a hierarchical management system, where communities would have substantial control over a designated area. Experiences from developing countries, as stated by Scoresby Shepherd, offer a unique opportunity to learn about the prospects and shortcomings of these little understood management systems. Along those lines, this

Proceedings contains first-hand insights by Juinio-Meñez (1998, Philippines), Arifin et al. (1998, Indonesia) and Castilla et al. (1998, Chile).

Fishing rights of aboriginal peoples

This is a subject of high importance in non-European western societies in which a wealthier sector of European ancestry co-exists with ethnically and/or culturally distinct aboriginal peoples. Jones and Garza (1998) discussed an interesting case: co-management of clam resources in Haida Gwaii (Queen Charlotte Islands, B.C.). First Nation access to shellfish and other resources in Canada is presently facilitated under the political climate created by the new federal Aboriginal Fisheries Strategy (AFS) policy. Although perhaps unknown to many Canadians, Canada's initiative has been influential elsewhere, as described in the session summary by Scoresby Shepherd (SARDI, Australia):

> "The Sparrow's case [in Canada] a few years ago and then the clause in the 1982 Constitution, which awarded native fishing rights, was actually put before the Australian Parliament in 1993. It led to inclusion in the [Australian] Native Title Act of an almost exact replica of the Canadian section, which now preserves Native fishing rights in Australia."

Next door to British Columbia, in Washington State (U.S.A.), a recent court ruling known as the Rafeedie shellfish decision (December 20, 1994) upheld the treaty rights of several Native American tribal nations to collect shellfish on public and private lands in the Puget Sound Basin (Johnson 1996). Implementation of this decision has led to a complete overhaul of the management of shellfisheries in the region. Co-management between federal and/or state/provincial fisheries authorities and aboriginal peoples is likely to remain a major factor shaping shellfish management policy, at least in Canada, the U.S.A., Australia, New Zealand, and South Africa.

Conservation and environmental issues

Fishers are not the sole stakeholders of harvested marine ecosystems. The general public is increasingly interested in fostering other specific agendas, particularly those centered around conservation and sustainability (Jamieson 1993). Nongovernment organizations involved with marine conservation have sprouted up everywhere to express those concerns and development agencies and financial institutions have become sensitive and even proactive in the conservation side of fishing. Major fishery collapses have brought the issue of overfishing to the attention of a public that not too long ago was sensitive only to the fate of charismatic marine mammals and birds. Special issues addressing overfishing crises, a subject rarely heard of by the public ten years ago, were recently published by popular magazines targeting general audiences (e.g., Fairlie 1995; Parfit 1995).

There are two broad families of conservation/environmental issues that involve shellfisheries: those created by other uses of the marine ecosystems that affect shellfish resources and those created by shellfish harvesting. The first, collectively the major subject of concern to managers and fishers of coastal resources, includes various forms of pollution, habitat degradation, and urban development, as vividly expressed in comments by Tim Webster (GCFA, Canada):

> "To tell the truth, stock assessment is not something that the average crabber lies awake at night fretting about. [One of the two] issues that gets crabber's attention and really make them toss and turn is pollution. The relentless destruction of crab habitat [in British Columbia] is alarming to say the least [...]. There are numerous sources of pollution, many of which are hard to detect and even harder to control. In other words, the average crabber sees little point in assessing a stock already mired in politics and pollution."[6]

Fishers targeting shellfish in other coastal regions of the world would also likely rate this concern equally high. Juinio-Meñez et al. (1998) presented a dramatic story, the result of rapid growth in the human population along the coast of the Philippines.

Introduction of exotic species, either accidental or intentional, is increasingly frequent in coastal waters. Some exotics have become the basis of entire industries, as exemplified in the NE Pacific (Washington and British Columbia) by the Pacific oyster (west Pacific), the Manila clam (*Venerupis philippinarum*; west Pacific), the softshell clam (*Mya arenaria*; west Atlantic), and the blue mussel (*Mytilus galloprovincialis*; east Atlantic), which together contribute the bulk of current intertidal and cultured bivalve harvest in the region. Other less desirable exotics, though, include predators, competitors, and diseases. The green crab (*Carcinus maenas*; east Atlantic), introduced to San Francisco Bay, California, and now spreading northwards, is expected to invade Washington and British Columbia (Jamieson et al. 1998*a*). The potential to impact coastal ecosystems is still speculative, but fears of a negative effect on bivalve fisheries and aquaculture seem warranted. Whether beneficial or not to local economies, introduction of exotics is changing irreversibly the biodiversity of the world's coastal ecosystems. This, in itself, is a subject of concern in the conservation agenda.

There are environmental problems caused by the physical effects of fishing. The most visible are the ecological side effects of benthic fishing gear: digging, hydraulic devices, dredges, and bottom trawls. These all can have a substantial impact on the whole benthic ecosystem, including disruption of the sedimentary matrix, siltation, destruction or elimination of certain types of habitat, and incidental mortality of nontarget organisms. Less visible, albeit not less significant, are indirect effects of selective species removal by a fishery. In the case of pelagic fisheries, ecological side effects of fishing are usually a matter of speculation, but in the case of benthic fisheries, effects are often unequivocal and dramatic. Examples are cases in which major grazers (e.g., sea urchins) or top predators (e.g., sea otters) are removed by harvest (Paine 1994).

Conservation and environmental issues such as habitat loss or degradation, pollution, direct ecological impacts of the fishing gear, introduction of exotic species, cascading ecological effects of the removal of target species, and so on, are hardly ever heard in relation to the assessment and management of industrial pelagic–demersal fisheries. By contrast, they are central to the management of small-scale coastal fisheries worldwide. Conservation issues may have been the most significantly underrepresented topic at this Symposium. We hope we have somewhat compensated for that imbalance here and suggest that in 10 years time, this issue may well be a dominant one.

---

[6] The other expressed subject of concern was "the ramifications for commercial crab fishing created by the Canada's Aboriginal Fisheries Strategy," discussed earlier.

Epilogue: 1984 vs. 1995 and beyond

Considering the latitude of the issues addressed in previous sections, it was not surprising to find that much of the change in focus between 1984 and 1995 took place in two main directions: (*i*) the extent to which models of populations and the fishing process (whether conceptual or formal) are spatially explicit; and (*ii*) participation of fishers in the management process. Later we address both themes through a comparison of the Proceedings of the 1984 and 1995 Symposia.

Although the significance of large-scale processes was implicit in studies of larval dispersal, explicit reference to SS dynamics was mostly incidental in the 1984 Workshop (Jamieson 1986, p. 67, on abalone; Bannister 1986, pp. 190–191, on scallops and mussels; Elner and Bailey 1986, pp. 342–343, on snow crab). The word metapopulation itself was absent from the Proceedings (Jamieson and Bourne 1986), although Orensanz (1986) used equivalently the now forgotten term megapopulation. By contrast, the metapopulation was an ubiquitous conceptual model in the 1995 Symposium, receiving substantial specific attention (Keesing and Baker 1998; Wing et al. 1998; Botsford et al. 1998; Fogarty 1998).

Methods using survey-sample or commercial catch data to assess abundance directly received little attention in 1984 (with the exception of Robert and Jamieson 1986); survey designs were briefly mentioned in the "Methods" section of several articles but there was no attention to methodological innovation. In 1995, by contrast, there was a set of strong contributions. Interestingly, all of them, whether stressing design-based (Smith and Robert 1998; Woodby 1998; McShane 1998) or model-based (Warren 1998; Moriyasu et al. 1998) methods, dealt with estimates of total abundance and its spatial distribution.

Interest in small-scale DD processes centered around compensation in 1984, primarily with relation to the growth overfishing problem (Orensanz 1986; Fogarty and Murawski 1986). In 1995, attention had shifted dramatically towards depensation, recruitment overfishing being the primary motivation for concern. DD effects on fertilization rate in broadcast spawners has become the subject of ingenuous field experiments and theoretical treatment (Levitan and Sewell 1998; Orensanz et al. 1998*b*).

The spatial dynamics of effort allocation is a problem that is now recognized by scientists, fishers, and managers (Campbell et al. 1998). Interaction between the spatial structure of benthic resources, fisher behaviors in response, and the interaction between DD fishing mortality and depensation were highlighted in 1995 by fresh insights from abalone fisheries (Keesing and Baker 1998; Prince and Hilborn 1998).

Areas in which spatially explicit approaches really took central stage in 1995 were those related to management. In 1984, for example, a review of strategies available to the management of invertebrate resources (Hilborn 1986) made no reference to spatially explicit tactics, although rotational strategies were thoroughly discussed by Sluczanowski (1986). By contrast, in 1995 there was consideration of reproductive reserves (e.g., Castilla et al. 1998), spatially variable harvest rates (Wing et al. 1998; Walters et al. 1998), rotation (Caddy and Seijo 1998; Campbell et al. 1998), spatially contrasted experimental management (Caputi et al. 1998; Momot 1998; Shepherd and Baker 1998), localized enhancement (Juinio Meñez et al. 1998), and territorial tenure systems (Castilla et al. 1998; Prince and Hilborn 1998; Prince et al. 1998). This blooming interest in spatially explicit management has had a tremendous feedback on the focus of research and assessment, prompting new emphasis in topics outlined earlier. This shifting focus is well in line with Carl Walters' (UBC, Canada) take-home message during a final session:

> "A lot of our research should shift towards understanding more space-time structures that lead to sustainable fisheries in spite of the foibles and follies and messes that we are certain to continue making with our assessment methods and our science."

The second theme, participation of fishers in the management process, was also virtually absent during discussions of the 1984 Workshop. A few articles (e.g., Otto 1986) made reference to input from industry to management, but by and large, fishers appear in the Workshop Proceedings as a ghostly presence, a collective subject of analysis and enforcement and never as active participants. A recap article summarizing the sessions of five working groups (Jamieson and Caddy 1986) pondered both biological and economical considerations and the interaction between scientists and managers, but there was no reference to an active role of fishers or industry in the management process. In the intervening years, there has been a dramatic change in fisher participation and in interest in participation, fostered by the spread of ITQ's and related forms of quota allocation, the unveiling by fishery anthropologists of complex de facto management systems (e.g., Cordell 1989), the renewed strength of claims to fishing rights by indigenous peoples in developed western countries, and the inability of traditional systems to handle specific situations, which in the end has led to prominent fishery collapses (Orensanz et al. 1998*a*; Walters and Maguire 1996). Alternative forms of participation of fishers in the management process were intensely debated at the 1995 Symposium, but most significantly, they were debated against a background of almost complete consensus about the need of industry participation.

Frameworks and mechanisms for interaction between scientists, fishers, and managers are still in a fluid state. Communication of methods and results to managers are often made difficult by the intricacy of technicalities, a problem that was as pressing in 1984 (Jamieson 1986) as it still was in 1995:

> "Documentation of biological advice alone still tends to give fisheries management an expectation of precision beyond that which can reasonably be expected even after lengthy study. Extensive development of models and analytical procedures in recent years has further enhanced this perception and has made it even more difficult for managers and fishermen, who often have little sophisticated analytical experience, to understand fully advice being provided" (Jamieson 1986).

> "Given the variety of [stock assessment] methods, these methods can be confusing for managers" (Doug Woodby, ADFG, U.S.A.).

Walters et al. (1998) discussed the results of modeling workshops where biologists and commercial fishers were successfully brought together to develop a spatial model of population dynamics and harvest regulation of the South Australian rock lobster. Such workshops are a promising hands-on, expedient form of communication, providing unexpected insights to all participants. Rick Harbo (DFO, Canada) made an interesting suggestion as to an adequate institutional framework to conduct such exercises in:

> "Something we don't do as much as we could is to have a

facilitator come in between interest groups (coastal communities or groups of fishers) and government and to have one there as an independent body. That is a role perhaps the universities should take a stronger, more active role in."

In our view, this is also an appropriate prescription for action in the developing world, one that could be beneficial for all local universities and the societies in which they are embedded (Pascual et al. 1997).

Analysts of the fishery as a phenomenon in toto have called recurrently for a stronger role of social scientists (economists, anthropologists, sociologists, political scientists) in providing insight and advice during the decision-making process, an arena perceived as dominated by fishery biologists (subscribers to the FISHFOLK discussion list can attest to this subject's recurrence). In shellfisheries managed by a centralized authority with mandates of optimality and stability, consideration of a social component collapses to adjustments of effort estimates at the stock assessment stage. It was not until fishers were recognized as real actors in the management process that social scientists were invited as a meaningful source of insight. At that point, understanding and modeling the dynamics of fisher's behavior became strategically significant. Both the 1984 Workshop and the 1995 Symposium were, however, dominated by biologists, with participation of only a few biologically minded, true statisticians. However, in 1995, there were three notable exceptions: Tony Charles (St. Mary's University, Nova Scotia), Gordon Monro, and Tim Lauck (both UBC, B.C.), but unfortunately no articles were produced for inclusion in this Proceedings.

The essence of Tim's talk has been elaborated in Lauck (1996). Tim brought up the issue of uncertainty, barely addressed at both the 1984 Workshop (but see Hilborn 1986) and the 1995 Symposium. This issue was recognized by Jim Prescott (SARDI, Australia):

"[...] nobody talked about the uncertainty of their results. Uncertainty is, of course, of great interest to not only the managers, but industry and the public. Industry is frequently presented with a single figure — if you do this you will get that — but we all know that is not likely to happen in reality. Then they tend to lose confidence in what you are giving them [...] It is good to explicitly address any uncertainty that there may be in the findings."

Scientists gauge and try to communicate uncertainty to managers, managers make decisions under uncertainty, and fishers, inevitably, make risky investments in the face of uncertainty (i.e., gamble). Inability by managers, and their bosses, politicians, to often appreciate fully the difficulties many scientists have in dealing with uncertainty is likely to remain a serious problem. Mr. John Crosbie, former Canadian fisheries minister, in his evaluation of the role scientists played in giving advice to ministers, stated:

"It was thought that scientists could give you precise and certain knowledge of what the state of the fish stocks was and it turned out that the scientists could not give you that. While their advice was always slavishly followed, it wasn't realized, certainly not by me, that it wasn't really possible to advise you to the tonne what a safe [catch quota] was," (the Canadian "Globe and Mail," Sept. 5, 1997, p. A6).

Most important, as Tim Lauck emphasized, there is always some irreducible level of uncertainty, which cannot be attenuated by any increase of scientific knowledge. This was paraphrased by Jeremy Prince (Biospherics Pty., Ltd., Australia):

"We are faced with irreducible uncertainty and even if we had all the research dollars we wanted, we are still going to be presented with situations we can't resolve [...] The arguments are never going to end no matter how many research dollars are thrown at them."

Once this is acknowledged, an appropriate approach should seek management philosophy and strategies that can cope with this type of uncertainty. Munro and Lauck (1995) gave as an analogy the operation of a securities analyst who does not chart the ups and downs of the market yet still manages to get a good long-term yield on his money. They stated mixed management strategies can be important to the development of a robust overall plan. Lauck (1996) developed this important rationale further based on the principles of hedging and portfolio balance, suggesting that marine reserves can provide a hedge against potential collapse.

This concept was extended by Jeremy Prince (Biospherics Pty., Ltd., Australia):

"With TURF's, at least in fisheries where there is a localized stock [e.g., abalone], we can actually look at having a range of management alternatives. This is like putting eggs in a whole range of different baskets. [...] If we have eggs in a range of baskets we are in a situation where we can learn from failures and successes."

Lauck provided a framework within which economists, ecologists, and fishery managers may find common ground. Marine reserves, territorial user rights, rotation, and other spatially explicit tactics have the potential to become major components of robust strategies for the sustainable management of spatially structured stocks in the face of irreducible uncertainty. These are all steps in the direction leading from a gambler's philosophy towards that of an investor's, consistent with current emphasis on the precautionary approach to fisheries management (Gordon and Munro 1996).

Acknowledgments

We are most grateful to the participants of the Symposium for sharing their viewpoints and experiences with us, but most particularly to the scientists, managers, and members of the fishing industry who made presentations during the Summary Sessions (Jamieson and Campbell 1998, Appendix 1). The transcripts of these presentations were a most valuable source of insight, which we hope to have conveyed at least partially to our readers. LO acknowledges support from a NMFS grant, made available through Washington Sea Grant, while writing this overview. Drs. Ana Parma and Alan Campbell thoroughly critiqued the MS. Ana also helped by keeping LO on leash.

References

Addicott, J.F., Aho, J.M., Antolin, M.F., Padilla, D.K., Richardson, J.S., and Soluk, D.A. 1987. Ecological neighborhoods: scaling environmental patterns. Oikos, **49**: 340–346.

Addison, J.T., and Bannister, R.C.A. 1998. Quantifying potential impacts of behavioral factors on crustacean stock monitoring and assessment: modeling and experimental approaches. *In* Proceedings of the North Pacific Symposium on Invertebrate Stock Assessment and Management. *Edited by* G.S. Jamieson and A. Campbell. Can. Spec. Publ. Fish. Aquat. Sci. 125. pp. 167–177.

André, C., and Rosenberg, R. 1991. Adult-larval interactions in the suspension-feeding bivalves *Cerastoderma edule* and *Mya arenaria*. Mar. Ecol. Prog. Ser. **71**: 227–234.

Arifin, Z., Pradina, P., and Purnomo, A.H. 1998. A case study of the traditional trochus (*Trochus niloticus*) fishery in the Maluku Region, Indonesia. *In* Proceedings of the North Pacific Symposium on Invertebrate Stock Assessment and Management. *Edited by* G.S. Jamieson and A. Campbell. Can. Spec. Publ. Fish. Aquat. Sci. 125. pp. 401–406.

Bannister, R.C.A. 1986. Assessment and population dynamics of commercially exploited shellfish in England and Wales. *In* North Pacific Workshop on Stock Assessment and Management of Invertebrates. *Edited by* G.S. Jamieson and N. Bourne. Can. Spec. Publ. Fish. Aquat. Sci. 92. pp. 182–194.

Bannister, R.C.A., and Conan, G.Y. (*Editors*). 1995. Shellfish life histories and shellfishery models. ICES Mar. Sci. Symp. **199**.

Beck, E.F. 1955. Lummi Indian how stories. The Caxton Printers, Ltd., Caldwell, Idaho.

Beinssen, K. 1979. Fishing power of divers in the abalone fishery of Victoria, Australia. Rapp. P.-v. Reun. Cons. Int. Explor. Mer, **175**: 20–22.

Belding, D.L. 1910. A report upon the scallop fishery of Massachusetts; including the habits, life history of *Pecten irradians*, its rate of growth and other factors of economic value. The Commonwealth of Massachusetts. 150 p.

Bohnsack, J.A. 1996. Marine reserves, zoning, and the future of fishery management. Fisheries, **21**: 14–16.

Botsford, L.W. 1995. Population dynamics of spatially distributed, meroplanktonic, exploited marine invertebrates. ICES Mar. Sci. Symp. **199**: 118–128.

Botsford, L.W., and Hobbs, R.C. 1995. Recent advances in the understanding of cyclic behavior of Dungeness crab (*Cancer magister*) populations. ICES Mar. Sci. Symp. **199**: 157–166.

Botsford, L.W., Quinn, J.F., Wing, S.R., and Brittnacher, J.G. 1993. Rotating spatial harvest of a benthic invertebrate, the red sea urchin, *Stronglyocentrotus franciscanus*. *In* Proceedings of the International Symposium Management Strategies for Exploited Fish Populations. *Edited by* G. Kruse, D.M. Eggers, R.J. Marasco, C. Pautzke, and T.J. Quinn II. Alaska Sea Grant College Program, Fairbanks, Alaska. Rep. AK-SG-93-02. pp. 409–428.

Botsford, L.W., Moloney, C.L., Largier, J.L., and Hastings, A. 1998. Metapopulation dynamics of meroplanktonic invertebrates: the Dungeness crab (*Cancer magister*) as an example. *In* Proceedings of the North Pacific Symposium on Invertebrate Stock Assessment and Management. *Edited by* G.S. Jamieson and A. Campbell. Can. Spec. Publ. Fish. Aquat. Sci. 125. pp. 295–306.

Breen, P.A., and Kendrick, T.H. 1998. An evaluation of surplus production analysis for assessing the fishery for New Zealand red rock lobsters (*Jasus edwardsii*). *In* Proceedings of the North Pacific Symposium on Invertebrate Stock Assessment and Management. *Edited by* G.S. Jamieson and A. Campbell. Can. Spec. Publ. Fish. Aquat. Sci. 125. pp. 213–223.

Burns, D., Egan, T., Seward, E., and Naidu, K.S. 1995. Identification of scallop beds using an acoustic ground discrimination technique (abstract only). Book of abstracts, 10th International Pectinid Workshop, Cork, Ireland.

Caddy, J.F. 1975. Spatial model for an exploited shellfish population, and its application to the Georges Bank scallop fishery. J. Fish. Res. Board Can. **32**: 1305–1328.

Caddy, J.F. 1989*a*. Marine invertebrate fisheries: their assessment and management. John Wiley & Sons, Inc., New York, N.Y. 752 p.

Caddy, J.F. 1989*b*. Recent developments in research and management for wild stocks of bivalves and gastropods. *In* Marine invertebrate fisheries: their assessment and management. *Edited by* J.F. Caddy. John Wiley & Sons, Inc., New York, N.Y. pp. 665–700.

Caddy, J.F., and Seijo, J.C. 1998. Application of a spatial model to explore rotating harvest strategies for sedentary species. *In* Proceedings of the North Pacific Symposium on Invertebrate Stock Assessment and Management. *Edited by* G.S. Jamieson and A. Campbell. Can. Spec. Publ. Fish. Aquat. Sci. 125. pp. 359–365.

Campbell, A., and Harbo, R.M. 1991. The sea urchin fisheries in British Columbia, Canada. *In* Biology of Echinodermata. *Edited by* T. Yanagisawa, I. Yasumasu, C. Oguro, N. Suzuki, and T. Motokawa. A.A. Balkema, Rotterdam, The Netherlands. pp. 191–199.

Campbell, A., Harbo, R.M., and Hand, C.M. 1998. Harvesting and distribution of Pacific geoduck clams, *Panopea abrupta*, in British Columbia. *In* Proceedings of the North Pacific Symposium on Invertebrate Stock Assessment and Management. *Edited by* G.S. Jamieson and A. Campbell. Can. Spec. Publ. Fish. Aquat. Sci. 125. pp. 349–358.

Caputi, N., Penn, J.W., Joll, L.M., and Chubb, C.F. 1998. Stock-recruitment-environment relationships for invertebrate species of Western Australia. *In* Proceedings of the North Pacific Symposium on Invertebrate Stock Assessment and Management. *Edited by* G.S Jamieson and A. Campbell. Can. Spec. Publ. Fish. Aquat. Sci. 125. pp. 247–255.

Castilla, J.C., Manríquez, P., Alvarado, J., Rosson, A., Pino, C., Espoz, C., Soto, R., Oliva, D., and Defeo, O. 1998. Artisanal "Caletas" as units of production and co-managers of benthic invertebrates in Chile. *In* Proceedings of the North Pacific Symposium on Invertebrate Stock Assessment and Management. *Edited by* G.S. Jamieson and A. Campbell. Can. Spec. Publ. Fish. Aquat. Sci. 125. pp. 407–413.

Charles, A.T. 1995. Bio-socio-economic modelling of management options in invertebrate fisheries. *In* North Pacific Symposium on Invertebrate Stock Assessment and Management, Program and Abstracts. *Edited by* G. Jamieson. March 6–10, 1995. Nanaimo, B.C. pp. 27. (Abstract only.)

Chen, C.-L., Hoenig, J.M., Dawe, E.G., Brownie, C., and Pollock, K.H. 1998. New developments in change-in-ratio and index-removal methods, with application to snow crab (*Chionoecetes opilio*). *In* Proceedings of the North Pacific Symposium on Invertebrate Stock Assessment and Management. *Edited by* G.S. Jamieson and A. Campbell. Can. Spec. Publ. Fish. Aquat. Sci. 125. pp. 49–61.

Clark, C. 1976. Mathematical bioeconomics: the optimal management of renewable resources. John Wiley & Sons, Inc., New York, N.Y. 352 p.

Clark, C. 1996. Marine reserves and the precautionary management of fisheries. Ecol. Appl. **6**: 369–370.

Collie, J.S., and Kruse, G.H. 1998. Estimating king crab (*Paralithodes camtschaticus)* abundance from commercial catch and research survey data. *In* Proceedings of the North Pacific Symposium on Invertebrate Stock Assessment and Management. *Edited by* G.S. Jamieson and A. Campbell. Can. Spec. Publ. Fish. Aquat. Sci. 125. pp. 73–83.

Conan, G.Y. 1984. Do assumptions commonly used for modelling populations of finfish apply to shellfish species? International Council for the Exploration of the Sea, Charlottenlund, Denmark. ICES CM 1984/K:49. 20 p.

Cordell, J. (*Editor*). 1989. A sea of small boats. Cultural Survival Report 26. Cultural Survival Inc., Cambridge, Mass. 418 p.

Crawford, W.R., and Jamieson, G.S. 1996. Modelling advection of Dungeness crab (*Cancer magister*) in Dixon Entrance and northern Hecate Strait, British Columbia. *In* High latitude crabs: biology, management and economics. *Edited by* B.R. Melteff. University of Alaska, Fairbanks, Alaska. Sea Grant Rep. 96-2. pp. 489–506.

Davis, G.E., Haaker, P.L., and Richards, D.V. 1996. Status and trends of white abalone at the California Channel Islands. Trans. Am. Fish. Soc. **125**: 42–48.

Defeo, O. 1998. Testing hypotheses on recruitment, growth, and mortality in exploited bivalves: an experimental perspective. *In* Proceedings of the North Pacific Symposium on Invertebrate Stock Assessment and Management. *Edited by* G.S. Jamieson and A. Campbell. Can. Spec. Publ. Fish. Aquat. Sci. 125. pp. 257–264.

Elner, R.W., and Bailey, R.F.J. 1986. Differential susceptibility of Atlantic snow crab, *Chionoecetes opilio*, stocks to management. *In* North Pacific Workshop on Stock Assessment and Management of Invertebrates. *Edited by* G.S. Jamieson and N. Bourne. Can. Spec. Publ. Fish. Aquat. Sci. 92. pp. 335–346.

Elner, R.W., and Beninger, P.G. 1992. The reproductive system and fertilization in snow crab, *Chionoecetes opilio*. Am. Zool. **32**: 524–533.

Endler, J.A. 1977. Geographic variation, speciation and clines. Monogr. Popul. Biol. 10. Princeton University Press, Princeton, N.J. 246 p.

Fairbridge, W.S. 1953. A population study of the Tasmanian "commercial" scallop *Notovola meridionalis* (Tate) (Lamellibranchia, Pectinidae). Aust. J. Mar. Freshwater Res. **4**: 1–40.

Fairlie, S. (*Guest editor*). 1995. Overfishing: its causes and consequences. A special double issue. Ecologist, **25**. 128 p.

FAO. 1995. The state of world fisheries and aquaculture. United Nations, Geneva, Switzerland.

Fernández, M. 1998. Cannibalism in Dungeness crab, *Cancer magister* Dana: the effects of predator-prey size ratio and habitat complexity. Mar. Ecol. Prog. Ser. In press.

Fogarty, M.J. 1998. Implications of migration and larval interchange in American lobster (*Homarus americanus*) stocks: spatial structure and resilience. *In* Proceedings of the North Pacific Symposium on Invertebrate Stock Assessment and Management. *Edited by* G.S. Jamieson and A. Campbell. Can. Spec. Publ. Fish. Aquat. Sci. 125. pp. 273–283.

Fogarty, M.J., and Murawski, S.A. 1986. Population dynamics and assessment of exploited invertebrate stocks. *In* North Pacific Workshop on Stock Assessment and Management of Invertebrates. *Edited by* G.S. Jamieson and N. Bourne. Can. Spec. Publ. Fish. Aquat. Sci. 92. pp. 228–244.

Fretwell, S.D., and Lucas, H.L. 1970. On territorial behavior and other factors influencing habitat distribution in birds. I. Theoretical development. Acta Biotheor. **19**: 16–36.

Gillis, D.M., Peterman, R.M., and Tyler, A.V. 1993. Movement dynamics in a fishery: application of the ideal free distribution to spatial allocation of effort. Can. J. Fish. Aquat. Sci. **50**: 323–333.

Gordon, D.V., and Munro, G.R. (*Editors*). 1996. Fisheries and uncertainty — a precautionary approach to resource management. University of Calgary Press, Calgary, Alta.

Greenstreet, S.P.R., Tuck, I.D., Grewar, G.N., Armstrong, E., Reid, D.G., and Wright, P.J. 1997. An assessment of the acoustic survey technique, RoxAnn, as a means of mapping seabed habitat. ICES J. Mar. Sci. **54**: 939–959.

Grigg, R.W. 1989. Precious coral fisheries of the Pacific and Mediterranean. Marine invertebrate fisheries: their assessment and management. *Edited by* J.F. Caddy. John Wiley & Sons, Inc., New York. N.Y. pp. 637–645.

Gross, F., and Smyth, J.C. 1946. The decline of oyster populations. Nature (London), **157**: 540–542.

Hall, M.A. 1983. A spatial approach to the population dynamics of the Manila clam (*Tapes philippinarum*). Ph.D. dissertation, University of Washington, Seattle, Wash.

Hastings, A., and Higgins, K. 1994. Persistence of transients in spatially structured ecological models. Science (Washington, D.C.), **263**: 1133–1136.

Hilborn, R. 1986. A comparison of alternative harvest tactics for invertebrate fisheries. *In* North Pacific Workshop on Stock Assessment and Management of Invertebrates. *Edited by* G.S. Jamieson and N. Bourne. Can. Spec. Publ. Fish. Aquat. Sci. 92. pp. 313–317.

Hilborn, R., and Walters, C.J. 1992. Quantitative fisheries stock assessment. Chapman & Hall, Inc., New York, N.Y. 570 p.

Jamieson, G.S. 1986. A perspective on invertebrate fisheries management — the British Columbia experience. *In* North Pacific Workshop on Stock Assessment and Management of Invertebrates. *Edited by* G.S. Jamieson and N. Bourne. Can. Spec. Publ. Fish. Aquat. Sci. 92. pp. 57–74.

Jamieson, G.S. 1993. Marine invertebrate conservation: evaluation of fisheries over-exploitation concerns. Am. Zool. **33**: 551–567.

Jamieson, G.S., and Bourne, N. (*Editors*). 1986. North Pacific Workshop on Stock Assessment and Management of Invertebrates. Can. Spec. Publ. Fish. Aquat. Sci. 92. 430 p.

Jamieson, G.S., and Caddy, J.F. 1986. Research advice and its application to management of invertebrate resources: an overview. *In* North Pacific Workshop on Stock Assessment and Management of Invertebrates. *Edited by* G.S Jamieson and N. Bourne. Can. Spec. Publ. Fish. Aquat. Sci. 92. pp. 416–424.

Jamieson, G.S., and Campbell, A. (*Editors*). 1998. Proceedings of the North Pacific Symposium on Invertebrate Stock Assessment and Management. Can. Spec. Publ. Fish. Aquat. Sci. 125. 462 p.

Jamieson, G.S., and Francis, K. 1986. Invertebrate and marine plant resources of British Columbia. Can. Spec. Publ. Fish. Aquat. Sci. 91. 89 p.

Jamieson, G.S., and Phillips, A. 1993. Megalopal spatial distribution and stock separation in Dungeness crab. Can. J. Fish. Aquat. Sci. **50**: 416–429.

Jamieson, G.S., Grosholz, E.D., Armstrong, D.A., and Elner, R.A. 1998*a*. Potential ecological implications from the introduction of the European green crab, *Carcinus maenas*, to British Columbia, Canada, and Washington, USA. J. Nat. Hist. In press.

Jamieson, G.S., Phillips, A., and Smith, B.D. 1998*b*. Implications of selective harvests in Dungeness crab (*Cancer magister*) fisheries. *In* Proceedings of the North Pacific Symposium on Invertebrate Stock Assessment and Management. *Edited by* G.S. Jamieson and A. Campbell. Can. Spec. Publ. Fish. Aquat. Sci. 125. pp. 309–321.

Johnson, R. 1996. Rights to shellfish in Western Washington: roots of the tribal claims — a panel of discussion. School of Marine Affairs, University of Washington (Seattle). New Directions in Marine Affairs, Report No. 3. 4 p.

Jones, R.R., and Garza, D.A. 1998. Co-management of the razor clam (*Siliqua patula*) fishery in Haida Gwaii, British Columbia, Canada. *In* Proceedings of the North Pacific Symposium on Invertebrate Stock Assessment and Management. *Edited by* G.S. Jamieson and A. Campbell. Can. Spec. Publ. Fish. Aquat. Sci. 125. pp. 385–391.

Jones, R.R., Schwartz, C., and Lee, L. 1998. Intertidal population estimate of razor clams (*Siliqua patula*) at beaches near Masset, Haida Gwaii/Queen Charlotte Islands, and applications to fishery management. *In* Proceedings of the North Pacific Symposium on Invertebrate Stock Assessment and Management. *Edited by* G.S. Jamieson and A. Campbell. Can. Spec. Publ. Fish. Aquat. Sci. 125. pp. 199–211.

Juinio-Meñez, M.A., Macawaris, N.N.D., and Bangi, H.G.P. 1998. Community-based sea urchin (*Tripneustes gratilla*) grow-out culture as a resource management tool. *In* Proceedings of the North Pacific Symposium on Invertebrate Stock Assessment and Management. *Edited by* G.S. Jamieson and A. Campbell. Can. Spec. Publ. Fish. Aquat. Sci. 125. pp. 393–399.

Keesing, J.K., and Baker, J.L. 1998. The benefits of catch and effort data at a fine spatial scale in the South Australian abalone (*Haliotis laevigata* and *H. rubra*) fishery. *In* Proceedings of the North Pacific Symposium on Invertebrate Stock Assessment and Management. *Edited by* G.S. Jamieson and A. Campbell. Can. Spec. Publ. Fish. Aquat. Sci. 125. pp. 179–186.

Lai, H.-L., and Bradbury, A. 1998. A modified catch-at-size analysis model for a red sea urchin (*Strongylocentrotus franciscanus*) population. *In* Proceedings of the North Pacific Symposium on

Invertebrate Stock Assessment and Management. *Edited by* G.S. Jamieson and A. Campbell. Can. Spec. Publ. Fish. Aquat. Sci. 125. pp. 85–96.

Lasta, M., Parma, A.M., Pascual, M.S., and Zampatti, E.A. 1986. Consideraciones sobre la explotación del mejillón (*Mytilus edulis platensis*) en la costa bonaerense. Resultados de la Campaña de Prospección del "Banco Faro Querandí" (junio 1981). Rev. Invest. Des. Pesq. **6**: 45–55.

Lauck, T. 1996. Uncertainty in fisheries management. *In* Fisheries and uncertainty — a precautionary approach to resource management. *Edited by* D.V. Gordon and G.R. Munro. University of Calgary Press, Calgary, Alta. pp. 91–105.

Levitan, D. 1991. Influence of body size and population density on fertilization success and reproductive output in a free-spawning invertebrate. Biol. Bull. **181**: 261–268.

Levitan, D.R., and Sewell, M.A. 1998. Fertilization success in free-spawning marine invertebrates: review of the evidence and fisheries implications. *In* Proceedings of the North Pacific Symposium on Invertebrate Stock Assessment and Management. *Edited by* G.S. Jamieson and A. Campbell. Can. Spec. Publ. Fish. Aquat. Sci. 125. pp. 159–164.

Lilly, G.R., Parsons, D.G., and Veitch, P.J. 1998. Spatial structure of northern shrimp (*Pandalus borealis*) off Labrador and eastern Newfoundland (northwest Atlantic). *In* Proceedings of the North Pacific Symposium on Invertebrate Stock Assessment and Management. *Edited by* G.S. Jamieson and A. Campbell. Can. Spec. Publ. Fish. Aquat. Sci. 125. pp. 265–271.

Lloyd, M. 1967. 'Mean crowding.' J. Anim. Ecol. **36**: 1–30.

Long, B.G., Poiner, I.R., and Harris, A.N.M. 1993. Method of estimating the standing stock of *Trochus niloticus* incorporating Landsat satellite data, with application to the trochus resources of the Bourke Isles, Torres Strait, Australia. Mar. Biol. (Berlin), **115**: 587–593.

McCall, A.D. 1990. Dynamic geography of marine fish populations. Washington Sea Grant Program. University of Washington Press, Seattle and London. 153 p.

McCay, B.J. 1989. Co-management of a clam revitalization project: the New Jersey "Spawner Sanctuary" Program. *In* Co-operative management of local fisheries. *Edited by* E. Pinkerton. University of British Columbia Press, Vancouver, B.C. pp. 103–124.

McCay, B.J. 1996. Participation of fishers in fisheries management. *In* Fisheries resource utitilization and policy. *Edited by* R.M. Meyer, C. Zang, M.L. Windsor, B.J. McCay, L.J. Hushak, R.M. Muth, and R.J. Wolotira. Proceedings of the World Fisheries Congress, Theme 2. Science Publ. Inc., Enfield, England. pp. 61–75.

McGoodwin, J.R. 1990. Crisis in the world's fisheries — people, problems, and policies. Stanford University Press, Stanford, Calif. 235 p.

McShane, P.E. 1998. Assessing stocks of abalone (*Haliotis* spp.): methods and constraints. *In* Proceedings of the North Pacific Symposium on Invertebrate Stock Assessment and Management. *Edited by* G.S. Jamieson and A. Campbell. Can. Spec. Publ. Fish. Aquat. Sci. 125. pp. 41–48.

Momot, W.T. 1986. Production and exploitation of the crayfish, *Orconectes virilis*, in northern climates. *In* North Pacific Workshop on Stock Assessment and Management of Invertebrates. *Edited by* G.S. Jamieson and N. Bourne. Can. Spec. Publ. Fish. Aquat. Sci. 92. pp. 154–167.

Momot, W.T. 1998. An example of how exploitation can increase production and yield in a northern crayfish (*Orconectes virilis*) population. *In* Proceedings of the North Pacific Symposium on Invertebrate Stock Assessment and Management. *Edited by* G.S. Jamieson and A. Campbell. Can. Spec. Publ. Fish. Aquat. Sci. 125. pp. 225–233.

Moriyasu, M., Wade, E., Sinclair, A., and Chiasson, Y. 1998. Snow crab, *Chionoecetes opilio*, stock assessment in the southwestern Gulf of St. Lawrence by bottom trawl survey. *In* Proceedings of the North Pacific Symposium on Invertebrate Stock Assessment and Management. *Edited by* G.S. Jamieson and A. Campbell. Can. Spec. Publ. Fish. Aquat. Sci. 125. pp. 29–40.

Munro, G., and Lauck, T. 1995. Acknowledging uncertainty: robust strategies for the management of invertebrate population and harvest. *In* North Pacific Symposium on Invertebrate Stock Assessment and Management, Program and Abstracts. *Edited by* G.S. Jamieson. March 6–10, 1995, Nanaimo, B.C. p. 60 (Abstract only.)

Orensanz, J.M. 1986. Size, environment, and density: the regulation of a scallop stock and its management implications. *In* North Pacific Workshop on Stock Assessment and Management of Invertebrates. *Edited by* G.S. Jamieson and N. Bourne. Can. Spec. Publ. Fish. Aquat. Sci. 92. pp. 195–227.

Orensanz, J.M., Parma, A.M., and Iribarne, O.O. 1991. Population dynamics and management of natural stocks. *In* Scallops: biology, ecology and aquaculture. *Edited by* S.E. Shumway. Dev. Aquat. Fish. Sci. **21**: 625–713.

Orensanz, J.M., Armstrong, J., Armstrong, D., and Hilborn, R. 1998*a*. Decline of crustacean fisheries in the Greater Gulf of Alaska — an historical perspective. Rev. Fish Biol. Fish. In press.

Orensanz, J.M., Parma, A.M., and Hall, M.A. 1998*b*. The analysis of concentration and crowding in shellfish research. *In* Proceedings of the North Pacific Symposium on Invertebrate Stock Assessment and Management. *Edited by* G.S. Jamieson and A. Campbell. Can. Spec. Publ. Fish. Aquat. Sci. 125. pp. 143–157.

Otto, R.S. 1986. Management and assessment of eastern Bering Sea king crab stocks. *In* North Pacific Workshop on Stock Assessment and Management of Invertebrates. *Edited by* G.S. Jamieson and N. Bourne. Can. Spec. Publ. Fish. Aquat. Sci. 92. pp. 83–106.

Otto, R.S. 1998. Assessment of the eastern Bering Sea snow crab, *Chionoecetes opilio*, stock under the terminal molting hypothesis. *In* Proceedings of the North Pacific Symposium on Invertebrate Stock Assessment and Management. *Edited by* G.S. Jamieson and A. Campbell. Can. Spec. Publ. Fish. Aquat. Sci. 125. pp. 109–124.

Paine, R.T. 1994. Marine rocky shores and community ecology: an experimentalist's perspective. Ecology Institute, Oldendorf, Germany.

Parfit, M. 1995. Diminishing returns. Natl. Geogr. Mag. **188**: 2–37.

Parma, A.M., and NCEAS Working Group on Population Management. 1998. What can adaptive management do for our fish, forests, food and biodiversity? J. Intregative Biol. In press.

Parsons, D.G., and Frechètte, J. 1989. Fisheries for northern shrimp (*Pandalus borealis*) in the Northwest Atlantic from Greenland to the Gulf of Maine. *In* Marine invertebrate fisheries: their assessment and management. *Edited by* J.F. Caddy. John Wiley & Sons, Inc., New York, N.Y. pp. 63–86.

Pascual, M., Orensanz, J.M., Parma, A.M., and Saba, S. 1997. The Patagonian challenge: melding conservation with development. *In* Conservation biology for the coming decade. *Edited by* P. Fiedler and P. Kareiva. Chapman and Hall, Inc., New York, N.Y. pp. 384–409.

Paul, A.J. 1984. Mating frequency and viability of stored sperm in the Tanner crab *Chionoecetes bairdi* (Decapoda, Majidae). J. Crust. Biol. **4**: 375–381.

Paul, A.J., and Paul, J.M. 1992. Second clutch viability of *Chionoecetes bairdi* Rathbun (Decapoda, Majidae) inseminated only at maturity molt. J. Crustac. Biol. **12**: 438–441.

Pennington, J.T. 1985. The ecology of fertilization of echinoid eggs: the consequences of sperm dilution, adult aggregation and synchronous spawning. Biol. Bull. **169**: 417–430.

Pickering, H., and Whitmarsh, D. 1997. Artificial reefs and fisheries exploitation: a review of the 'attraction versus production' debate, the influence of design, and its significance for policy. Fish. Res. (Amsterdam), **31**: 39–59.

Pitcher, T.J. (*Editor*). 1997. The design and monitoring of marine reserves. Univ. Brit. Col. Fish. Cen. Res. Vancouver, B.C. Rep. 5(1). 47 p.

Prince, J.D. 1989. The fisheries biology of the Tasmanian stocks of *Haliotis rubra*. Ph.D. dissertation, University of Tasmania, Hobart, Australia. 174 p.

Prince, J., and Hilborn, R. 1998. Concentration profiles and invertebrate fisheries management. *In* Proceedings of the North Pacific Symposium on Invertebrate Stock Assessment and Management. *Edited by* G.S. Jamieson and A. Campbell. Can. Spec. Publ. Fish. Aquat. Sci. 125. pp. 187–196.

Prince, J., Walters, C., Ruiz-Avila, R., and Sluczanowski, P. 1998. Territorial user's rights in the Australian abalone (*Haliotis* sp.) fishery. *In* Proceedings of the North Pacific Symposium on Invertebrate Stock Assessment and Management. *Edited by* G.S. Jamieson and A. Campbell. Can. Spec. Publ. Fish. Aquat. Sci. 125. pp. 367–375.

Quayle, D.B. 1952. The relation of digging frequency to productivity of the butter clam, *Saxidomus giganteus* Deshayes. J. Fish. Res. Board Can. **8**: 369–373.

Quayle, D.B. 1988. Pacific oyster culture in British Columbia. Can. Bull. Fish. Aquat. Sci. 218. 241 p.

Quinn, J.F., Wing, S.R., and Botsford, L.W. 1993. Harvest refugia in marine invertebrate fisheries: models and applications to the red sea urchin, *Strongylocentrotus franciscanus*. Am. Zool. **33**: 537–550.

Robert, G., and Jamieson, G.S. 1986. Commercial fishery data isopleths and their use in offshore sea scallop (*Placopecten magellanicus*) stock evaluations. *In* North Pacific Workshop on Stock Assessment and Management of Invertebrates. *Edited by* G.S. Jamieson and N. Bourne. Can. Spec. Publ. Fish. Aquat. Sci. 92. pp. 76–82.

Roberts, C.M. 1997. Connectivity and management of Caribbean coral reefs. Science (Washington, D.C.), **278**: 1454–1457.

Roberts, C.M., and Polunin, N.V.C. 1991. Are marine reserves effective in management of reef fisheries? Rev. Fish Biol. Fish. **1**: 65–91.

Roberts, C.M., and Polunin, N.V.C. 1993. Marine reserves: simple solutions to managing complex fisheries? Ambio, **22**: 363–368.

Roughgarden, J., Iwasa, Y., and Baxter, C. 1985. Demographic theory for an open marine population with space-limited recruitment. Ecology, **66**: 54–67.

Rowley, R.J. 1994. Marine reserves in fisheries management. Aquat. Conserv. Mar. Freshwater Ecosyst. **4**: 233–254.

Sen, S., and Nielsen, J.R. 1997. Fisheries co-management: a comparative analysis. Mar. Policy **20**: 405–418.

Shepherd, S.A., and Baker, J.L. 1998. Biological reference points in an abalone (*Haliotis laevigata*) fishery. *In* Proceedings of the North Pacific Symposium on Invertebrate Stock Assessment and Management. *Edited by* G.S. Jamieson and A. Campbell. Can. Spec. Publ. Fish. Aquat. Sci. 125. pp. 235–245.

Sinclair, M. 1987. Marine populations. An essay on population regulation and speciation. University of Washington Press, Seattle, Wash. 252 p.

Sluczanowski. P. 1986. A disaggregate model for sedentary stocks: the case of South Australia abalone. *In* North Pacific Workshop on Stock Assessment and Management of Invertebrates. *Edited by* G.S. Jamieson and N. Bourne. Can. Spec. Publ. Fish. Aquat. Sci. 92. pp. 393–401.

Smith, S.J., and Robert, G. 1998. Getting more out of your survey designs: an application to George Bank scallops (*Placopecten magellanicus*). *In* Proceedings of the North Pacific Symposium on Invertebrate Stock Assessment and Management. *Edited by* G.S. Jamieson and A. Campbell. Can. Spec. Publ. Fish. Aquat. Sci. 125. pp. 3–13.

Sulkin, S.B. 1984. Behavioral basis of depth regulation in the larvae of brachyuran crabs. Mar. Ecol. Prog. Ser. **15**: 181–205.

Tegner, M., and Dayton, P.K. 1977. Sea urchin recruitment patterns and implications for commercial fishing. Science (Washington, D.C.), **196**: 324–326.

Thomas, H.J. (*Editor*). 1979. Population assessments of shellfish stocks. ICES, Rapports et Procès-Verbaux des Réunions 175.

Thompson, S.K. 1992. Sampling. John Wiley & Sons, Inc., New York, N.Y. 343 p.

Thompson, S.K., and Seber, G.A.F. 1996. Adaptive sampling. John Wiley & Sons, Inc. New York, N.Y. 265 p.

Walters, C., and Maguire, J.J. 1996. Lessons for stock assessment from the northern cod collapse. Rev. Fish Biol. Fish. **6**: 125–137.

Walters, C.J., Hall, N., Brown, R., and Chubb, C. 1993. Spatial model for the population dynamics and exploitation of the Western Australian rock lobster, *Panulirus cygnus*. Can. J. Fish. Aquat. Sci. **50**: 1650–1662.

Walters, C., Prescott, J.H., McGarvey, R., and Prince, J. 1998. Management options for the South Australia rock lobster (*Jasus edwardsii*) fishery: a case study of co-operative assessment and policy design by fishers and biologists. *In* Proceedings of the North Pacific Symposium on Invertebrate Stock Assessment and Management. *Edited by* G.S. Jamieson and A. Campbell. Can. Spec. Publ. Fish. Aquat. Sci. 125. pp. 377–383.

Warren, W.G. 1998. Spatial analysis of marine populations: factors to be considered. *In* Proceedings of the North Pacific Symposium on Invertebrate Stock Assessment and Management. *Edited by* G.S. Jamieson and A. Campbell. Can. Spec. Publ. Fish. Aquat. Sci. 125. pp. 21–28.

Williams, I. 1996. Integrated Fraser salmon model. Facet Decision Systems, Inc., Vancouver, B.C. 4 p.

Wing, S.R., Botsford, L.W., and Quinn, J.F. 1998. The impact of coastal circulation on the spatial distribution of invertebrate recruitment, with implications for management. *In* Proceedings of the North Pacific Symposium on Invertebrate Stock Assessment and Management. *Edited by* G.S. Jamieson and A. Campbell. Can. Spec. Publ. Fish. Aquat. Sci. 125. pp. 285–294.

Woodby, D. 1998. Adaptive cluster sampling: efficiency, fixed sample sizes, and an application to red sea urchins (*Strongylocentrotus franciscanus*) in southeast Alaska. *In* Proceedings of the North Pacific Symposium on Invertebrate Stock Assessment and Management. *Edited by* G.S. Jamieson and A. Campbell. Can. Spec. Publ. Fish. Aquat. Sci. 125. pp. 15–20.

Worthington, D.G., and Andrew, N.L. 1998. Small-scale variation in demography and its implications for alternative size limit in the fishery for blacklip abalone (*Haliotis rubra*) in New South Wales, Australia. *In* Proceedings of the North Pacific Symposium on Invertebrate Stock Assessment and Management. *Edited by* G.S. Jamieson and A. Campbell. Can. Spec. Publ. Fish. Aquat. Sci. 125. pp. 341–348.

Zheng, J., Kruse, G.H., and Murphy, M.C. 1998. A length-based approach to estimate population abundance of Tanner crab, *Chionoecetes bairdi*, in Bristol Bay, Alaska. *In* Proceedings of the North Pacific Symposium on Invertebrate Stock Assessment and Management. *Edited by* G.S. Jamieson and A. Campbell. Can. Spec. Publ. Fish. Aquat. Sci. 125. pp. 97–105.

Appendix 1: Program

Day 1, Monday, March 6, 1995

Introductions: Dr. Glen Jamieson

Opening addresses: Dr. John C. Davis and Dr. Don Noakes

Session One: Estimation of abundance within specific areas

Chair: Dr. A. Campbell

S.J. Smith: Getting more out of your survey information: models and survey designs

Y.J. Chiasson, E. Wade, and M. Moriyasu: The use of geostatistics for snow crab, *Chionoecetes opilio*, stock assessment: a case study

J.S. Collie and G. H. Kruse: Estimating king crab abundance from commercial catch and research survey data

D. Woodby: Adaptive sampling of aggregated invertebrate populations

R.R. Jones, C. Schwartz, and L. Lee: Intertidal population estimate of razor clams (*Siliqua patula*) at North Beach, Haida Gwaii/Queen Charlotte Islands and applications to fishery management

W.G. Warren: Spatial analysis of marine populations: factors to be considered (Presented by J.M. Hoenig.)

W.J. Nash, J.C. Sanderson, A.J. Cawthorn, and S.R. Talbot: Estimation of abundance, catchability, survival, and movement of blacklip abalone (*Haliotis rubra*) by mark–recapture in three spatially stratified study sites

S.D. Frusher, R.B. Kennedy, and I.D. Gibson: Preliminary estimates of exploitation rates in the Tasmanian rock lobster (*Jasus edwardsii*) fishery using the change-in-ratio and index-removel techniques with tag–recapture data

Session summary

Science Reviewer: N. Andrew, NSW Fisheries Research Institute, Australia

Manager Reviewer: P. LaRiviere, State of Washington Department of Fish and Wildlife

Industry Reviewer: J. Austin, Underwater Harvesters Association, Canada

Session Two: Spatial distribution of abundance and implications

Chair: Dr. L. Botsford

R. Hilborn and J. Prince: Concentration profiles and invertebrate fisheries management

C.B. Dew and P.A. Cummiskey: In-situ observations of the podding behavior of adult red king crab, *Paralithodes camtschaticus*

D.R. Levitan and M.A. Sewell: The importance of population size to reproductive success in free-spawning marine organisms

Day 2, Tuesday, March 7, 1995 (Session Two continued)

S.A. Shepherd: How good are egg per recruit analyses for managing abalone fisheries?

A. Campbell and R. Harbo: Harvesting and distributiion of geoduck clams in British Columbia

S.R. Wing, L.W. Botsford, J.L. Largier, and J.F. Quinn: Spatial variability in settlement of benthic invertebrates and their management implications in a northern California upwelling system

Session summary

Science Reviewer: J. Addison, Ministry of Agriculture, Fisheries & Food, U.K.

Manager Reviewer: M. Joyce, Department of Fisheries and Oceans, Canada

Industry Reviewer: M. Featherstone, Pacific Underwater Harvesters Association, Canada

Session Three: Dynamics of harvested stocks and ecosystems

Chair: Dr. M. Fogarty

J.M. Orensanz, A.M. Parma, and M.H. Hall: Applications of models for spatial point processes to the analysis of benthic populations

L.W. Botsford: Metapopulation dynamics of meroplanktonic invertebrates

W.T. Momot: The role of exploitation in altering the process regulating crayfish populations

R.S. Otto: Assessment and management consequences of geographic and annual variations in the size at maturity of male snow crab, *Chionoecetes opilio*, in the eastern Bering Sea

B.G. Stevens: Environmental modulation of mating behavior in *Chionoecetes bairdi*: an hypothesis

O. Defeo: Ecological effects of fishing on bivalve populations in an exposed sandy beach (Presented by J.M. Orensanz.)

J.T. Addison and R.C.A. Bannister: Behavioral interactions around baited traps and the modeling of harvesting strategies in crustacean fisheries

G.S. Jamieson: Implications of selective harvests in invertebrate fisheries

J.K. Keesing and S.A. Shepherd: The benefits of collecting fine spatial scale catch and effort data in quota managed fisheries: an example from the South Australian abalone fishery

Session summary

Science Reviewer: R.W. Elner, Canadian Wildlife Service

Manager Reviewer: J. Prescott, South Australian Research and Development Institute

Industry Reviewer: D. Saxby, Specialty Marine Products Ltd., Canada

Day 3, Wednesday, March 8, 1995

Session Four: Population modeling and parameter estimation

Chair: Dr. B.D. Smith

N. Caputi: Stock-recruitment-environment relationships for invertebrate species of Western Australia

J.M. Hoenig: Advances in change-in-ratio and index- removal estimation of population size and related parameters

D.G. Parsons and G.R. Lilly: Spatial structure of northern shrimp (*Pandalus borealis)* in the Northwest Atlantic

M.J. Fogarty: Implications of dispersal and larval interchange in American lobster stocks: a metapopulation model

P.E. McShane: Assessing stocks of abalone (*Haliotis* spp.): methods and constraint

Day 4, Thursday, March 9, 1995 (Session Four continued)

Chair: Dr. J.M. Hoenig

H-L. Lai and A. Bradbury: Exploitation dynamics and management strategies for the red sea urchin (*Strongylocentrotus franciscanus*) in Washington State, U.S.A.

P. Breen and T.H. Kendrick: Simulated data analysed with surplus-production analysis

J. Zheng, G.H. Kruse, and M.C. Murphy: A length-based approach to estimate population abundance of Tanner crabs, *Chionoecetes bairdi*, in the eastern Bering Sea

B.D. Smith, L.W. Botsford, S. Wing, and J.F. Quinn: Estimation of growth and mortality parameters from size frequency distributions lacking age patterns: an application to the red sea urchin (*Strongylocentrotus franciscanus*)

C.M. Raubenheimer: Simulation models for South African abalone, *Haliotis midae*

Session summary

Science Reviewer: C. Walters, University of British Columbia, Canada

Manager Reviewer: D. Woodby, Alaska Department of Fish and Game, U.S.A.

Industry Reviewer: T. Webster, Gulf Crab Fishery Association, Canada

Session Five: Invertebrate fisheries management

Chair: Dr. B.G. Ivanov

J.F. Caddy and J.C. Seijo: Rotating harvest models for management of sedentary resources: their relationship to growth and natural mortality rate and their physical and economic optima

J.C. Castilla, P. Manriquez, A. Rosson, C. Pino, R. Soto, J. Alvarado, D. Oliva, C. Espoz, and O. Defeo: The Chilean artisanal "caletas": units of production and the basis for community-based management schemes

N.L. Andrew and D.G. Worthington: Reconciling scales of management with demography: an example from the New South Wales abalone fishery

R.R. Jones: A co-management case study: the razor clam (*Siliqua patula*) fishery in Haida Gwaii, British Columbia, Canada

R. McGarvey and J. Prescott: A model for assessing closure season management options in South Australian rock obster fishery: catch and egg production

R. Harbo: New directions in invertebrate fisheries management: the British Columbia experience

Z. Arifin and P. Purwati: Sustainable use of trochus (gastropod) resource: a case study of the trochus fishery in Maluku islands, Indonesia

N. Carrick: The significance of spatial distribution and dispersal to the management of two mono-specific penaeid shrimp fisheries in South Australia

Day 5, Friday, March 10, 1995 (Session Five continued)

Chair: Dr. J. Castillia

J. Prescott: Co-operative research and management in South Australia's lobster fishery: promises and pitfalls

A.G. Bazhin: The sea urchin fishery in Kamchatka: current conditions and problems

J. Prince, C. Walters, R. Ruiz-Avila, and P. Sluczanowski. Territorial users' rights in the Australian abalone fishery

L.R.S. Castro: Management options for the commercial dive fisheries for sea cucumbers in Baja California, Mexico

M.A. Juinio-Meñez: Community-based sea urchin pen culture as a resource management strategy

Session summary

Science Reviewer: S. Shepherd, South Australian Research and Development Institute

Manager Reviewer: G. Thomas, Department of Fisheries and Oceans, Canada

Industry Reviewer: A. Thomson, Alaska Crab Coalition, U.S.A.

Session Six: Decision making in invertebrate fisheries management

Chair: Dr. G.H. Kruse

C. Walters, J.H. Prescott, R. McGarvey, and J. Prince: Management options for the South Australian rock lobster (*Jasus edwardsii*) fishery: a case study of co-operative assessment and policy design by fishers and biologists

G. Munro and T. Lauck: Acknowledging uncertainty: robust strategies for the management of invertebrate population and harvest

A.T. Charles: Bio-socio-economic modeling of management options in invertebrate fisheries

P.E. LaRiviere and S.T. Barry: Limited entry in the Washington coastal Dungeness crab fishery: the first step toward rationalizing an overcapitalized and sometimes chaotic fishery

B.G. Ivanov: Shellfishery biology in Russia: problems and perspectives

R.J. Conser: A modified DeLury modeling framework designed to support fishery management decisions under uncertainty

V.S. Levin: Resources of bottom fishery invertebrates in Kamchatka: results and perspectives of exploitation

Session summary

Science Reviewer: J. Prince, Biospherics Pty. Ltd., Australia

Manager Reviewer: R. Harbo, Department of Fisheries and Oceans, Canada

Industry Reviewer: H. McElderry, Archipelago Marine Research, Canada

Concluding remarks: Dr. Glen Jamieson